AI-Powered Innovation in Materials Science

The Role of Language Models in Discovery and Design

Xue Jiang
Yanjing Su

WILEY VCH

Authors

Prof. Xue Jiang
Affiliation: University of Science and Technology Beijing, Beijing, 10083, China

Prof. Yanjing Su
Affiliation: University of Science and Technology Beijing, Beijing, 10083, China

Cover Design: Wiley
Cover Images: © Krot_Studio/Shutterstock, © VectorMine/Shutterstock

Typesetting: Lumina Datamatics, Inc.
Printing: CPI Group (UK) Ltd, Croydon CR0 4YY

Library of Congress Card No.: applied for

British Library Cataloguing-in-Publication Data A catalogue record for this book is available from the British Library.

Bibliographic information published by the Deutsche Nationalbibliothek
The Deutsche Nationalbibliothek lists this publication in the Deutsche Nationalbibliografie; detailed bibliographic data are available on the Internet at <http://dnb.d-nb.de>

Print ISBN: 9783527356355
ePDF ISBN: 9783527855834
ePUB ISBN: 9783527855827
oBook ISBN: 9783527855841

C9783527356355_180326

Contents

Preface

The rapid advancements in artificial intelligence (AI) have ushered in a new era of innovation across various fields, with materials science emerging as a quintessential domain for its application. As the demand for novel materials intensifies, driven by evolving technologies and industries, the need for a more efficient, data-driven approach to material discovery has never been greater. Traditional methods, which often rely on trial-and-error and expert intuition, are reaching their limits in the face of increasingly complex material systems. The integration of AI offers unprecedented opportunities to transform the materials science landscape, enabling researchers to accelerate the design, discovery, and optimization of new materials.

In recent years, AI for Materials has emerged as a promising interdisciplinary field, leveraging cutting-edge techniques such as machine learning, natural language processing (NLP), large language models (LLMs), and autonomous research systems to reshape the material discovery process. The field has rapidly developed, driven by advances in AI algorithms, data infrastructure, and computational tools that allow us to tackle previously insurmountable challenges in material design. By enabling AI to not only predict material properties but also suggest new synthesis routes, optimize manufacturing processes, and automate experimental workflows, AI is proving to be a catalyst for faster, more efficient material innovation.

This book aims to explore the transformative role of AI—particularly LLMs—in materials science, highlighting how these technologies are reshaping the way researchers approach material discovery, design, and application. The chapters of this book delve into the intersection of AI, NLP, and materials science, addressing the fundamentals of AI in materials science, the development of LLMs, and how LLMs can enhance knowledge extraction from literature, support predictive modeling, and power autonomous research agents. We will also share case studies, materials research applications, and insights into the future directions of this exciting field.

I would like to express my sincere gratitude to my co-authors—Professor Yanjing Su and the talented students Weiren Wang, Shaohan Tian, Zhihua Jing, Hao Wang, Xiaowen Yang, Jian Zhou, and Zhiqiang Zhu—for their invaluable contributions to this book. I am also deeply grateful to Professor Turab Lookman for his enduring collaboration and unwavering support throughout this endeavor. Their expertise, dedication, and collaborative spirit have been essential in bringing this project to life.

This book is intended for researchers, engineers, and students working in materials science, AI, and related interdisciplinary fields. It will also be of interest to those exploring the applications of AI in scientific discovery, data-driven research, and computational modeling. By combining the strengths of both materials science and AI, we hope to inspire a new wave of innovation that will help solve the world's most pressing challenges, from sustainable energy to advanced manufacturing.

Together, we embark on a journey toward a future where AI-driven materials innovation leads to breakthroughs that shape the technologies and industries of tomorrow.

Xue Jiang
Yanjing Su
March, 2026
Beijing

1

The Revolution of AI for Materials

1.1 Introduction

The advancement of technology and the economy depend fundamentally on the creation of new materials, which serve as the foundation for transformative innovation in diverse sectors. To speed up this process, significant public and private resources are now being channeled into artificial intelligence (AI) and big data technologies. These converging efforts are forming a distinct interdisciplinary field often termed AI for Materials (AI4Mater), which synergizes computational modeling, data-driven science, and hands-on experimentation. AI acts as a unifying platform, merging the stages of design, manufacturing, and practical use, while simultaneously refining both digital simulations and physical lab processes [1]. Consequently, the entire pipeline for materials—from initial ideation to final application—is undergoing a fundamental restructuring. Machine learning integration is proving particularly valuable in tackling persistent issues in multi-scale computational models, and AI-enabled automation is significantly raising the speed and success rate of experimental research. With the help of more advanced data systems, these AI-driven improvements are set to change traditional research and development, modernize industrial processes, and make operations more efficient throughout materials science.

This chapter explains the goals of AI4Mater and looks at recent progress in areas such as materials database, machine learning, advanced computing, automated experiments, and intelligent manufacturing. It provides a plan for global cooperation, focusing on common data standards, shared and compatible databases, and open access to computing tools and AI models in materials research.

1.2 What Is AI4Mater?

1.2.1 Definition

AI4Mater is defined as a comprehensive ecosystem designed to operationalize AI throughout the entire materials science landscape. The framework is architecturally composed of three interdependent pillars: the materials data infrastructure, technical

AI methodologies, and domain-specific applications, as visualized in Figure 1.1. The foundation of this hierarchy is the data infrastructure, which supports the system through rigorous protocols, ontologies, collaborative digital platforms, and advanced processing tools. Resting upon this bedrock are the specific AI4Mater techniques, which synthesize machine learning, high-throughput computing, autonomous experimentation, and intelligent manufacturing. These techniques act as the functional engine, enabling diverse applications ranging from initial material discovery and R&D to final manufacturing and performance validation. Collectively, these components drive scalability, maximize efficiency, and foster innovation across the materials value chain.

1.2.2 History

Over the past 10 years or so, AI has changed how materials science is done. It has moved from scattered tests of single algorithms to a clear, data-focused way of doing research, as shown in Figure 1.2. This change can be described in three main stages:

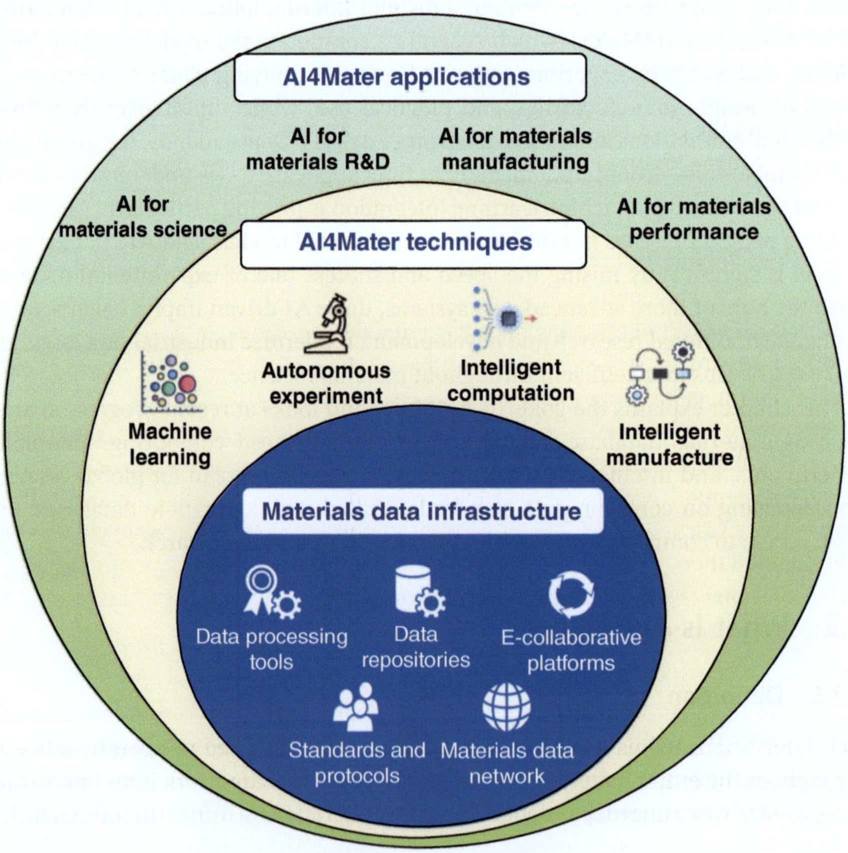

Figure 1.1 The components of AI4Mater.

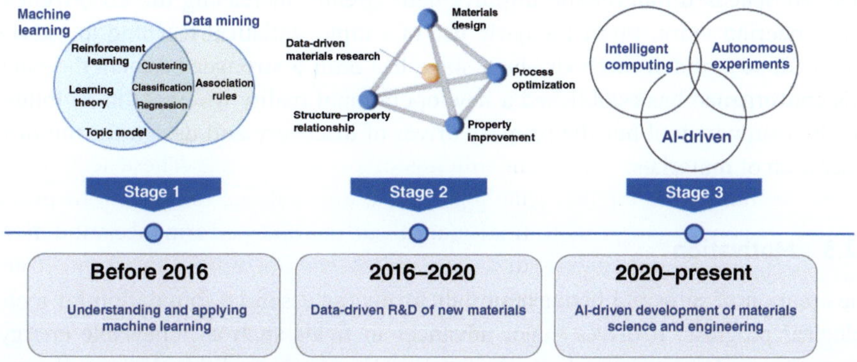

Figure 1.2 The history of AI4Mater.

Phase 1 (before 2016): This early stage was marked by the materials science community first learning and adapting to machine learning methods. During this phase, the primary objective was to construct a computational baseline. Researchers began deploying a diverse algorithmic toolkit, ranging from regression, classification, and clustering to dimensionality reduction and early reinforcement learning—to interpret materials data. The emphasis was placed on validating the utility of these tools, establishing core computational protocols, and laying the groundwork for data-informed analysis.

Phase 2 (2016–2020): The second period was defined by the ascendancy of informatics-driven design and the optimization of processing parameters. During this period, there was a significant increase in the use of large, complex datasets to understand the connections between materials structure and properties. This greatly shortened the time needed to develop new materials. As the velocity of both experimental and computational data generation increased, the focus shifted toward the effective management, integration, and exploitation of this information, cementing data governance as a critical pillar of modern materials research.

Phase 3 (2020–present): The integration of advanced computation, data-driven analysis, generative AI, and automated experimentation is reshaping how materials science research is conducted. This era transcends mere efficiency; it is driven by data-led theoretical discovery and the autonomous generation of tailored material solutions. As these technologies mature, they promise to further disrupt the field, facilitating rapid, targeted innovation in response to specific industrial needs.

The historical progression of AI in this domain illustrates a consistent trend toward the unification of disparate technological elements into a singular, synergistic ecosystem. By weaving AI into the fabric of computation, experimentation, and manufacturing, the field has achieved unprecedented levels of automation and intelligence. AI's capacity to navigate complex, multidimensional data allows for the seamless integration of modeling and validation through iterative, data-driven feedback loops. By linking design, experiment, and production into a continuous cycle, AI speeds up automatic improvement, greatly increasing the effectiveness of engineering work. From the early use of simple statistical learning to today's advanced, self-guided methods, each stage has built a stronger base for the next. This convergence has established a new operational reality where AI is no longer merely a support tool but the primary driver of discovery and design for the next generation of materials.

1.2.3 Motivation

The creation of new and better materials forms the essential foundation for technological progress. It drives major advances in fields such as renewable energy, aerospace, and next-generation electronics. However, traditional research and development methods, which depend heavily on trial and error and basic computer simulations, are now often seen as too slow and limiting. These approaches are often labor-intensive, capital-heavy, and constrained by current theoretical boundaries, thereby stalling the critical innovations required to tackle global imperatives like resource scarcity and energy sustainability. AI4Mater changes the current approach by offering a new way forward. It speeds up discovery, reduces risks in development, creates fresh scientific insights, and smoothly connects theoretical design with real-world production.

AI greatly speeds up research by predicting material properties before they are made, which cuts down on costly and repetitive lab tests. Data-driven analysis quickly identifies the most promising material combinations, while AI-assisted simulations and automated labs make the process of creating and testing materials faster and more efficient. Together, these abilities allow us to explore new areas of materials science and make discoveries faster than traditional methods ever could.

Across the development pipeline, AI reduces uncertainty by predicting performance, manufacturability, and durability. Early triage filters out weak candidates before major investment, and AI-optimized processes raise yield and improve sustainability.

AI does more than just speed up work—it also creates new knowledge. By combining experimental data, multi-scale simulations, and theory, AI uncovers latent patterns and produces clear design guidelines. Understandable models make complex relationships between structure, process, and properties clearer, sparking new ideas and deepening our fundamental knowledge.

AI-powered methods speed up the development of novel, high-performing materials—those that are stronger, tougher, more conductive, or have specially designed surfaces. By shortening the journey from idea to real-world use, AI enables major leaps forward in fields like energy storage, lightweight structures, and self-repairing materials, pushing the boundaries of what is possible.

AI connects design with manufacturing to support complete life cycle management. From choosing raw materials and improving processes to using in-line sensors, digital twins (DTs), and evaluating recyclability, AI provides flexible, smart control at every stage. This connected approach supports sustainability and helps turn lab discoveries into real-world benefits.

1.3 Foundations and Frontiers

In the United States, the launch of the Materials Genome Initiative (MGI) in 2011 marked a turning point for the field. MGI aimed to cut the cost and development time of new materials by half. This was to be achieved by combining fast, automated experiments, computer simulations, and modern digital data systems. For over a decade, the MGI has served as the backbone of the American materials innovation ecosystem, catalyzing cooperation between federal agencies, academia, and the private sector. The initiative's mandate was reinforced through strategic updates in 2014 and 2021, which deepened national investment in AI-guided discovery and intelligent R&D. The critical nature of these technologies was further validated by the National Academy of Sciences in its 2019 Decadal Survey, which positioned data science and AI as indispensable for resolving systemic inefficiencies in materials research. Most recently, the President's Council of Advisors on Science and Technology underscored this strategic imperative in their 2024 report, Supercharging Research. This report treats "AI for Designing Advanced Semiconductors" and "A Phase Change for Materials Discovery" as pivotal frontiers where AI is set to drive scientific breakthroughs.

Europe has similarly established a robust foothold in data-centric materials research. The Accelerated Metallurgy (AccMet) project, launched in 2011 by the European Space Agency (ESA) with 31 partner organizations, led the way in using high-speed alloy design and simulation. This approach successfully shortened research and development timelines and lowered costs. From 2012 to 2022, the Metallurgy Europe Initiative mapped 17 key material requirements and 50 interconnected research areas across sectors including clean energy, sustainable transport, and health care. Between 2014 and 2020, the European Commission's Horizon 2020 framework institutionalized cross-border collaboration and the uptake of AI methodologies across EU member states, amplifying these efforts. Germany's Industry 4.0 strategy, launched in 2013, established the digital foundations for intelligent manufacturing in parallel with Europe-wide initiatives. It used Internet of Things (IoT) systems to connect data-driven design with improved production processes.

From a global perspective, the International Science Council (ISC) has played a coordinating role in advancing the use of AI across scientific disciplines. Its 2025 report, Preparing National Research Ecosystems for AI, offers strategic guidance for embedding AI within national research infrastructures. The report stresses fostering interdisciplinary collaboration, anticipating ethical and regulatory issues, and building resilient, inclusive research ecosystems. Through sharing effective methods and comparing real-world examples, the ISC promotes fair and sustainable AI-based research systems. These frameworks are designed to address worldwide issues while making sure AI is used responsibly.

China's national key R&D program, Key Technologies and Supporting Platforms for Materials Genome Engineering, was launched in 2015. In 2020, follow-on initiatives focused on expanding adoption and translating Materials Genome Engineering from basic research into broader industrial practice. In 2024, the Construction

Plan for the National Materials Big Data Center was launched, building on earlier advances. The program seeks to build a powerful platform for AI research. This foundation is now accelerating breakthroughs in basic science, critical software and equipment, and materials discovery. The increasing combination of AI with well-organized data is paving the way for new, intelligent approaches to materials research and development, enabling ongoing innovation.

At the same time, Beijing and Shanghai have concurrently implemented "AI + Materials Action Plans" to strengthen links between AI and materials research. By setting up AI-powered materials innovation centers, these plans focus on high-speed computational models, smart materials databases, and AI-enhanced experimental platforms.

1.4 Previous Works

1.4.1 Materials Data Infrastructure

As big data and AI continue to grow quickly, materials databases are evolving from basic storage spaces into active, data-powered systems. Older database designs that use fixed formats and rigid layouts often cannot handle the growing variety and mixed forms of materials data. Template-based database systems have emerged to address this challenge, allowing custom data formats and flexible integration. These systems make it easier to share and connect data, let structures be defined after data is gathered, and help encourage open teamwork and new ideas in materials field.

1.4.1.1 Materials Database

Big data and AI have expanded the variety and format of materials information, reshaping database requirements. Before 2010, most databases mainly stored data and allowed users to search it, helping researchers find and reuse information. Platforms like Granta Design, Springer Materials, and MatNavi encouraged data sharing across the community. The US MGI sped up a data-focused approach to materials innovation. Consequently, databases have evolved into comprehensive data centers, offering not only raw data but also computer simulations and analytical tools. This progression has ultimately transformed them into active platforms that directly facilitate the discovery and research of novel materials.

Creating a materials database is essentially a process of abstraction, designed to translate materials science concepts into a form that computers can use. This process usually involves two main steps. First, materials scientists and database engineers work together to turn real-world knowledge into an information model. Scientists outline the key ideas—such as materials, properties, and how they relate—and engineers then structure these concepts into a logical model with clear rules. Second, engineers convert this logical model into a physical one, using database systems like MySQL, Oracle, PostgreSQL, or MongoDB to store and manage the data.

The growing need to integrate diverse data sources and support AI-driven workflows has exposed several limitations of the traditional, schema-first approach.

Predefining detailed schemas raises the design and construction barrier because it requires frequent, close coordination between domain scientists and engineers. Any change to logical or physical storage tends to require engineering effort, which reduces scalability. Finally, application developers must understand the database design before they can access and integrate data, which complicates data exchange.

To meet these needs, the field has shifted toward using template-based database systems. These systems allow users to create their own data descriptions and build reusable data types. They offer adaptable templates that can be tailored for different engineering uses, research teams, organizations, and individual researchers. They also allow for material data structures to be defined after data is collected and stored in a consistent way. Notable examples are the MGEDATA and Materials Data Curation System (MDCS).

MDCS [2], created by NIST, offers reusable basic types such as numbers with high precision, units, and types for uncertainty. Combining these basic types to form more complex data types, the users can create custom templates for uploading their information. MGEDATA was developed by researchers at the University of Science and Technology Beijing [3, 4]. It is a flexible container system that allows users to define their own storage structures. It supports 10 flexible data types, such as string, table, container, array, and generator (see Figure 1.3). The system utilizes a MongoDB document database with BSON format, enabling the flexible description and storage of materials data from diverse sources. Figure 1.3 shows the visual interface of this flexible container system.

1.4.1.2 E-collaborative Platforms

E-collaborative platforms are online systems that bring together open-source tools to improve teamwork. They give users remote access to both experimental and computational resources. They enable end-to-end tracking of collaborative workflows and attribution of progress across geographically and organizationally dispersed teams. Designed for cross-disciplinary efforts, open-source and open-access ecosystems facilitate seamless exchange of information, expertise, and assets among diverse stakeholders.

Purdue University's Network for Computational Nanotechnology (NCN) fostered collaborative research through the creation of the nanoHUB platform [5]. It offers on-demand analysis, tools for building applications, and a wide and expanding set

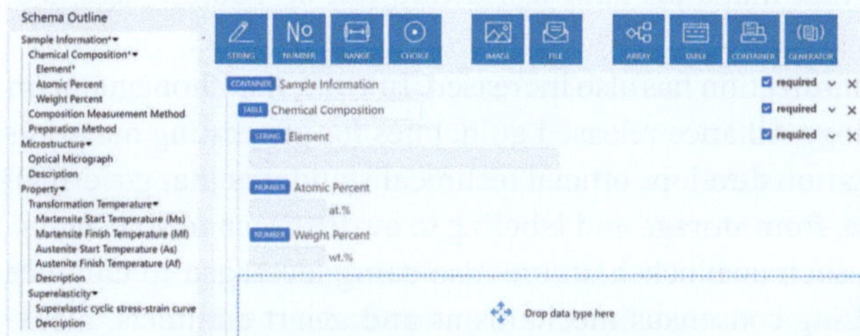

Figure 1.3 MGEDATA dynamic container system GUI.

of simulation software. Running in the cloud, it provides a complete scientific computing environment that supports research, teaching, and teamwork in nanotechnology and materials science.

Following 2011, European initiatives, including AccMet and Metallurgy Europe, catalyzed coordinated development of computational materials data efforts, culminating in the creation of the NOMAD database. Switzerland's MARVEL center introduced the Materials Cloud platform, which brings together datasets and open-source workflows to encourage modeling tools sharing. Materials Cloud assists with publishing and finding data, registering resources, browser-based development, integrating models, and gathering user feedback. This helps build a modern environment for computational materials research.

China launched the National New Materials Database Platform in 2021 to combine distributed development with centralized data sharing. The platform creates a digital workspace that combines data collection, publishing, sharing, merging, searching, analysis, and online applications. It provides cloud-based, on-demand analysis, integrated tools, application building, and quick support for materials research and development. Visual, customizable workflows allow for low-code machine learning and data analysis, while a web-based Jupyter Notebook environment with containerized tools makes it easier to integrate materials design software.

1.4.1.3 Standards and Protocols

Since 2000, international efforts have moved forward with creating standard formats for materials data. MatML, an XML schema created by NIST in 2000 [6], provides a uniform structure for documents containing materials information. This allows the data to be read and processed across different programming languages. NIST first gathered a working group to outline the project and released draft Annotated MatML DTD documents for feedback. The MatML Coordination Committee, formed in 2003, brought together representatives from academia and industry. Today, the ongoing development of the MatML standard is managed by OASIS, which runs a dedicated working group for this purpose.

To tackle the challenge of making different databases work together, the MARVEL center partnered with 31 international databases, like NOMAD, Materials Project, and AFLOW, to create the OPTIMADE alliance. This group is focused on building a common application programming interface (API) to standardize how data is accessed and shared. OPTIMADE plays a key role in supporting the FAIR principles, helping to ensure that materials data worldwide can be Found, Accessed, Worked with, and Reused [7].

In China, work on standardization has also increased. In 2019, the Zhongguancun Materials Testing Technology Alliance released guidelines for generating materials genome data. The organization develops official technical standards that govern all stages of the data life cycle, from storage and labeling to evaluation and sharing.

Emerging technologies such as blockchain are also being leveraged to enhance data governance. By utilizing consensus mechanisms and smart contracts, blockchain creates an immutable, transparent ledger of data activities. This technology provides a verifiable life cycle record that ensures traceability and supports the dynamic assessment of data quality, thereby fostering a trustworthy data ecosystem.

1.4.2 Machine Learning in Materials

Conventional experiments and simulations struggle with the growing complexity of materials research, whereas AI offers a transformative approach: detecting hidden patterns, modeling complex structure–property relationships, and accelerating exploration of vast design spaces. Here, we examine how AI approaches—from machine learning and deep learning to generative AI, optimization, knowledge fusion, and large language models—are reshaping materials research. These tools are creating new ways to work across every stage of developing a material.

1.4.2.1 Classical Machine Learning for Materials Discovery

Progress across many industries is fundamentally driven by the discovery and optimization of new materials [8–10]. Early works in materials machine learning frequently employed regression models for relatively straightforward targets. Neural networks can predict both ceramic–matrix composite mechanics and polymer glass-transition temperatures [11, 12]. Northwestern University used ensemble learning to evaluate formation energies. Using this method, roughly 4,500 stable compounds were uncovered from 1.6 million candidates, achieving a computational speed nearly a million times faster than standard DFT [13]. Additionally, support vector machines (SVMs) can assist in planning chemical reactions. By learning from unsuccessful experiments, SVMs greatly reduced the amount of lab work needed while also making predictions more accurate [14].

Shanghai University's researchers applied a random forest method integrated with symbolic regression to predict the elastic modulus and glass-forming ability of amorphous alloys, achieving results with strong general applicability [15]. A machine learning framework at Northeastern University improved predictions of RAFM steels' tensile properties, exceeding the accuracy and generality of conventional physical-metallurgy models [16]. Gradient-boosting classifiers were used at Southeast University to study 19,841 perovskite compositions. They then applied a follow-up regression model to identify 151 promising materials for ferroelectric–photovoltaic applications, which have suitable bandgaps [17].

To facilitate deployment in real research settings, machine learning tools are increasingly delivered as integrated software. University of Chicago introduced a tool, named Matminer, to streamline dataset retrieval, feature generation, and results visualization [18]. Shanghai University introduced the MLMD platform to make AI tools available to materials researchers without coding [19].

Despite these advances, important challenges continue to exist. Models often have difficulty applying what they learn to different types of materials and managing very large sets of potential candidates. Moving forward will likely require closer integration of machine learning with high-speed experimental and computational systems, as well as adaptive models that keep learning from new data and improve their predictions in real time.

1.4.2.2 Optimization Design for Materials

Discovering new materials often requires exploring huge ranges of possibilities and solving problems with multiple, sometimes competing, goals. Heuristic algorithms and active learning have been shown to be very useful here. They are good at searching broadly and making the most of limited experimental or computing resources.

Genetic algorithms (GAs) are well suited to problems where exhaustive enumeration is infeasible. By emulating evolutionary selection, GAs identify high-performing solutions in high-dimensional spaces. For example, a team at the University of Ottawa used GAs to adjust the metal–organic frameworks (MOFs) functional groups, leading to a significant boost in CO_2 capture. By examining 141 experimentally confirmed MOFs, they found modified versions that increased CO_2 adsorption by over 400%. This highlights how GAs can be effective for tackling complex optimization challenges in materials design [20].

In situations where data is limited, active learning—especially Bayesian optimization (BO)—has become a key approach. By building surrogate models, BO efficiently highlights the most promising candidates, lowering the required experiments or simulations. At Kyoto University, support vector regression (SVR) was integrated with BO and DFT to select materials exhibiting optimal thermal conductivity [21]. At Xi'an Jiaotong University, adaptive BO methods were developed to balance exploration of new candidates with exploitation of top-performing solutions (see Figure 1.4a, b). Their approach integrates model predictions, uncertainty estimates, and repeated experimental checks to keep the search productive even with little data [22–26].

Using BO, researchers at the University of Science and Technology Beijing refined the composition and processing of novel Cu–Ni–Co–Si alloys, reducing reliance on costly elements. With only 32 experimental runs, the approach identified ideal processing parameters, yielding an alloy with more than 50% less cobalt than ASTM C70350 yet preserving equivalent mechanical and electrical performances [27]. At Tsinghua University, BO was used to optimize CO_2 adsorption and identify candidate photosensitizers from extensive molecular collections [28, 29]. Researchers have also expanded BO to a wider array of design problems [30].

While GA and BO deliver substantial gains, they can require many iterations for highly complex landscapes and may struggle in extremely high-dimensional, multicomponent spaces. Merging these techniques with deep learning is expected to streamline discovery and strengthen the robustness of predictions. Overall, optimization methods continue to be essential tools for thoroughly exploring large sets of design options and directing focused experimental work.

1.4.2.3 Applications of Deep Learning in Materials Research

Conventional machine learning approaches often falter when confronted with the high dimensionality and strong nonlinearity characteristics of modern materials datasets. By contrast, deep learning, rooted in multilayer neural architectures inspired by biological networks, can automatically learn hierarchical, high-level representations from large corpora [31, 32]. This capability opens new avenues for tackling computationally intensive problems in materials science.

A prominent case is the deep-learning DFT Hamiltonian (DeepH). It deduces the elements of a quantum mechanical Hamiltonian directly from DFT data (see Figure 1.4c), allowing for accurate and efficient electronic-structure predictions for a wide variety of materials [33]. DeepH achieves accuracy close to that of first-principles

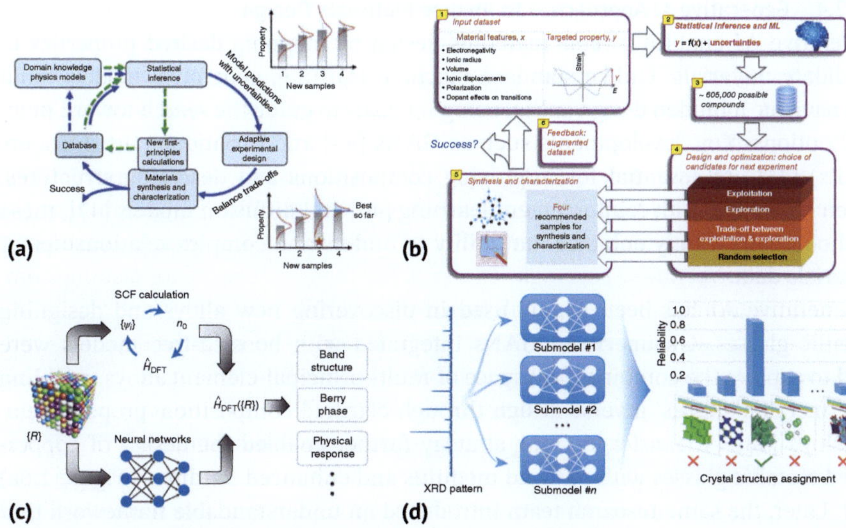

Figure 1.4 Panels (a) and (b) demonstrate BO, while panels (c) and (d) showcase deep learning in DFT electronic structure calculations and crystal structure identification, respectively. *Source*: Panel (a): [22] / Springer Nature / CC BY 4.0; Panel (b): [24] / John Wiley & Sons; Panel (c): [33] / Springer Nature / CC BY 4.0; Panel (d): [40] / with permission of American Chemical Society.

calculations, but at a much lower computational cost. It has also been adapted to work with magnetic superstructures and hybrid-functional setups [34–36].

Deep learning has also improved crystal-structure analysis. Researchers combined graph-theoretical methods with structural chemistry to create a fast and accurate workflow. This method tracks how crystal structures change during both the synthesis process and in real-world use [37, 38]. Graph-based AI approaches have been used to investigate lithium fast-ion conductors and analyze X-ray diffraction patterns. This enables new ways of identifying and exploring crystal structures, as shown in Figure 1.4d [39, 40].

For designing materials based on their properties, Shanghai Jiaotong University created a deep transfer-learning framework to predict adsorption energies, which are important measures for surface reactions. This approach greatly reduces computing costs in applications such as heavy-metal ion adsorption and lithium–sulfur cathodes [41, 42]. At the same time, graph neural networks (GNNs) have been used to model adsorption in 2D materials, speeding the discovery of catalysts for the hydrogen evolution reaction [43]. These examples illustrate how deep learning enhances both predictive fidelity and the pace of discovery.

Despite these advances, important challenges remain. Large, high-quality datasets are still scarce for many emerging material classes, and the opacity of many architectures limits physical interpretability. Addressing these limitations will require data-efficient and interpretable models that embed domain knowledge within the learning process. Nevertheless, deep learning has shown a strong ability to handle the scale, complexity, and nonlinear patterns found in materials data. As the techniques continue to evolve, they offer strong potential to accelerate materials discovery.

1.4.2.4 Generative AI Approaches to Inverse Materials Design

Generative AI enables inverse materials design by mapping desired properties to candidate materials. Unlike standard machine learning, generative models build and navigate a hidden design space, using set goals to guide the search toward practical options. Key developments such as GANs [44] and variational autoencoders (VAEs) [45] are essential for improving compositions and designing structures. When combined with reinforcement learning [46] and diffusion models [47], these methods significantly enhance our ability to understand complex relationships of materials data.

Generative AI has been widely used in discovering new alloys and designing metallic glasses. Unsupervised GANs integrated with boosted-tree models were used to explore the combinatorial space of multi-principal-element alloys, enabling bulk metallic glasses' inverse design through clarified composition–property relationships [48]. A transfer-learning strategy further enabled the design of copper-based metallic glasses with reduced modulus and enhanced hardness (Figure 1.5a) [49]. Later, the same research team introduced an understandable framework that links conditional variational autoencoders (CVAEs) with neural networks, generating complex alloys ranging from systems with four to six principal elements [50]. Xi'an Jiaotong University's researchers merged RL with CVAEs to effectively search through wide ranges of compositions. By using RL to guide repeated optimization steps, along with lab testing, they created a TiNi-based alloy with an exceptionally high transformation enthalpy. This result surpassed what conventional design methods had achieved [51]. An inverse-design framework was further developed to map hand-drawn heat-flow targets to TiNi compositions and processing conditions, as illustrated in Figure 1.5b [52]. This approach directly incorporates human design goals into a machine-guided discovery process.

Generative methods have also proven useful for designing functional materials. For instance, researchers merged GAN and VAE techniques to generate potential superconductor candidates that fulfill designated high-temperature requirements [48]. Later works introduced diffusion models with a doping-adaptive three-channel representation, yielding 200 candidate high- T_c superconductors and informing subsequent experimental efforts [53, 54].

Metamaterials present additional complexity because performance depends critically on intricate topological architectures. Generative AI has also been effective in this area. By integrating active learning, VAEs, and 3D printing, researchers enabled the design of orthopedic implants with customized microstructures [55]. VAE-based generative models were used to map desired thermal properties to topological unit cells for thermal metamaterial design [56]. These developments show how generative models can connect structural design with functional property goals, opening new paths in metamaterial engineering.

1.4.2.5 Materials Knowledge Fusion

As data-driven practices become central to materials research, tightly coupling domain expertise with AI is increasingly crucial—particularly when experimental data are limited. Machine learning methods are excellent at exploring large design possibilities, but they often fall short in explaining the complex behaviors they predict.

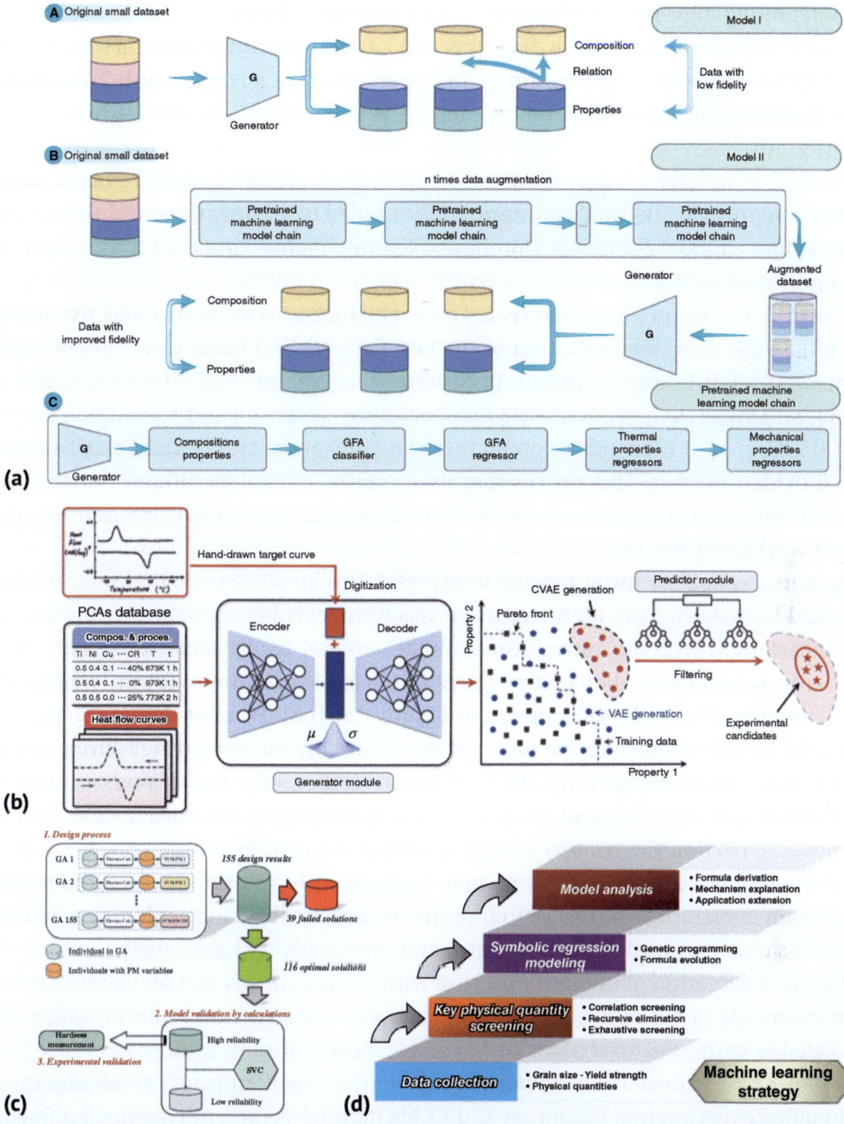

Figure 1.5 Panel (a) illustrates the use of GANs to design copper-based metallic glasses. *Source*: [49] / CityUHK Scholars. Panel (b) shows the application of CVAE to design shape memory alloys. *Source*: [52] / John Wiley & Sons. Panel (c) demonstrates a physical metallurgy-guided machine learning approach. *Source*: [66] / with permission of Elsevier. Panel (d) presents a method that integrates physical mechanism knowledge. *Source*: [67] / with permission of Elsevier.

To address this limitation, hybrid approaches incorporate expert knowledge, physical laws, and theoretical principles directly into the AI workflow. This enhances prediction accuracy and reinforces the robustness of the results [13, 57, 58].

AI excels at uncovering latent patterns within large datasets that could easily evade conventional analytical methods [13, 57, 58]. Performance can be enhanced by integrating physics-based features and specialized descriptors. For instance, descriptors derived from phase diagrams and Landau–Devonshire theory have been employed to steer the exploration of high-performance ferroelectrics. This shows how fundamental scientific principles can directly make AI predictions more effective [59, 60].

Because materials science often involves a large number of possible descriptors, feature-engineering methods are commonly applied to identify the most useful variables before a model is trained. Combining feature engineering with GAs, researchers optimized models and descriptors for accurate prediction of phase stability in HEAs [61]. The same team also created a novelty metric to measure the active learning efficiency. It enabled selection of the best features and faster discovery of high-temperature HEAs with outstanding compressive strength [62]. The researchers at Shanghai University proposed sure independence screening and sparsifying operator (SISSO) [63]. It successfully predicts the stability of perovskites and clarifies how metal oxides interact. This proves that using mathematical techniques to focus on key variables can strengthen the role of materials science knowledge in computational workflows [64, 65].

A more explicit pathway toward interpretability involves encoding quantitative physical laws within the architecture of machine learning models. Researchers at Northeastern University introduced a model guided by physical metallurgy that explicitly accounts for alloy composition and crystal structure to improve property predictions substantially as depicted in Figure 1.5c [66]. Beyond improving predictions, AI has also helped refine basic physical laws. By incorporating factors identified through machine learning, the Hall–Petch relationship was refined, enhancing both prediction accuracy and insight into the governing mechanisms, as shown in Figure 1.5d [67]. In the context of high-dimensional compositional searches where traditional optimization algorithms may stagnate in local minima, the embedding of domain constraints into AI search protocols aids in directing exploration toward solutions that are physically meaningful and practically realizable [59].

The amalgamation of domain expertise with AI has already yielded demonstrable improvements in predictive power and discovery velocity. By combining scientific knowledge with advanced algorithms, researchers can now explore complex possibilities and discover novel materials efficiently. The continued development of automated experimental platforms and LLMs highlights the growing need to incorporate materials knowledge into AI-driven workflows [68, 69].

1.4.2.6 Materials Large Language Models

LLMs, propelled by progress in NLP, are transforming the use of materials science publications, becoming central to AI-driven discovery efforts. LLMs have been successfully applied to pull materials data from research papers, focusing on details like properties, chemical compositions, and how materials are made. As early as 2011, efforts such as ChemDataExtractor began recognizing chemical entities in the literature [70]. An automated NLP system was proposed to gather information on alloy compositions, properties, and preparation methods [71, 72]. ChatExtract used conversational LLM and well-designed prompts to locate, pull out, and check data from tables with very little setup. It can also ask follow-up questions for confirmation [73].

Beyond just data extraction, LLMs can convert insights from scientific papers into compact numerical representations. For instance, researchers used unsupervised methods to turn materials science text into dense word vectors. This revealed connections between structure and properties that helped in discovering novel thermoelectric materials [74]. Also, the researchers used the skip-gram algorithm to create 200-dimensional vectors representing high-entropy alloys based on 6.4 million materials science abstracts. This method identified the BCC Senkov and FCC Cantor systems as promising HEAs well before they gained broader attention in the field [75].

LLMs also accelerate materials research by enabling increasingly autonomous end-to-end workflows, from experimental planning to execution. Integrated AI agents with capabilities for memory retention, logical reasoning, task decomposition, and summarization can orchestrate complex pipelines. For instance, a GPT-4-driven system from Carnegie Mellon University autonomously designs, plans, and conducts complex organic synthesis experiments, using tools to browse documentation, control laboratory robotics, and coordinate other LLMs [76]. ChatMOF uses an LLM to retrieve data, predict material property, and generate MOF structure, achieving 96.9%, 95.7%, and 87.5% accuracies, respectively, with GPT-4 [77].

A key challenge for using LLMs in materials design is their ability to reliably perform quantitative reasoning—for instance, accurately predicting material properties based on descriptions of their composition and how they are processed. Important questions remain: how accurately can LLMs predict material properties, and to what extent can they accelerate design and optimization? To tackle this, researchers at Georgia Tech trained a model called polyBERT (based on DeBERTa) using 100 million hypothetical polymers. polyBERT converts polymer notation (PSMILES strings) into numerical fingerprints. These fingerprints are then used to predict multiple polymer properties at once [78]. SteelBERT (Figure 1.6) transforms scientific

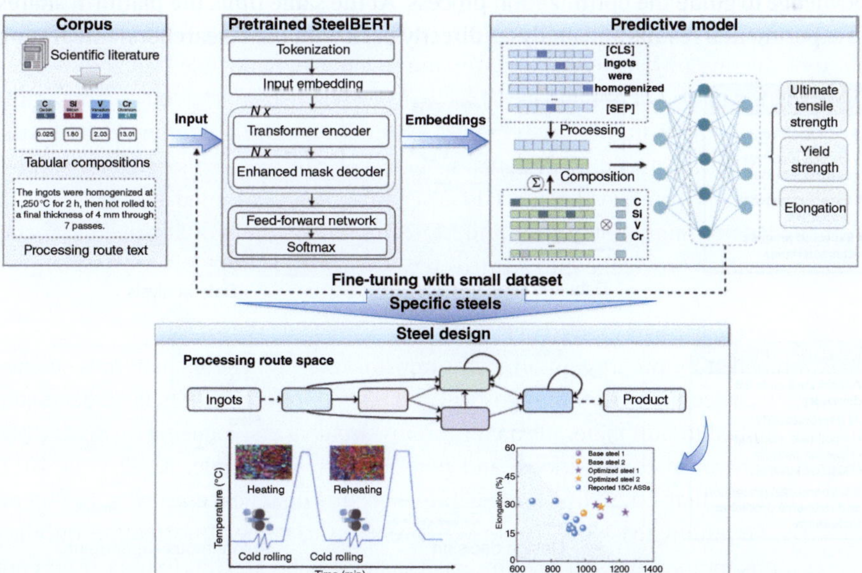

Figure 1.6 Large language model-driven steel design. *Source*: [79] / with permission of Elsevier.

text on steels into numerical property predictions. Trained on 55,000 full-text articles and 4.2 million abstracts, the model leverages a tailored attention mechanism to produce steel-specific representations from textual descriptions [79]. When predicting three key mechanical properties, the system obtained an R^2 value around 0.80. Following fine-tuning on 64 experiments, the model successfully predicted the properties of a novel 15Cr composition. The result demonstrates the potential for LLM-based pipelines to handle quantitative prediction tasks from natural language descriptions and to guide the design of new alloys and processing routes.

LLMs are transforming materials science by automating data collection, revealing hidden patterns, guiding experiments, and predicting properties. Continued development of generative models, LLMs, and expert-knowledge approaches will improve prediction accuracy and efficiency and provide greater insight into material mechanisms. Looking ahead, closely linking AI methods, scientific knowledge, and instant feedback from experiments points toward faster, more intelligent, and more sustainable materials development.

1.4.3 Autonomous Experiments

By integrating AI, robotics, and optimization with lab equipment, autonomous experimentation enables automated, intelligent control of experiments. Its core is a tightly coupled loop: AI analyzes experimental data and, guided by predefined goals and constraints, proposes experimental plans; robotic systems carry out the operations and gather new data; and the results are reanalyzed to refine the next plan. Figure 1.7 illustrates this workflow. The system continuously refines material compositions and processing conditions in the search for materials with superior performance. By interacting with the system, human experts can provide their specialized knowledge to guide the optimization process. At the same time, the platform shares the experimental results and analyses directly back with the researchers. Integrating

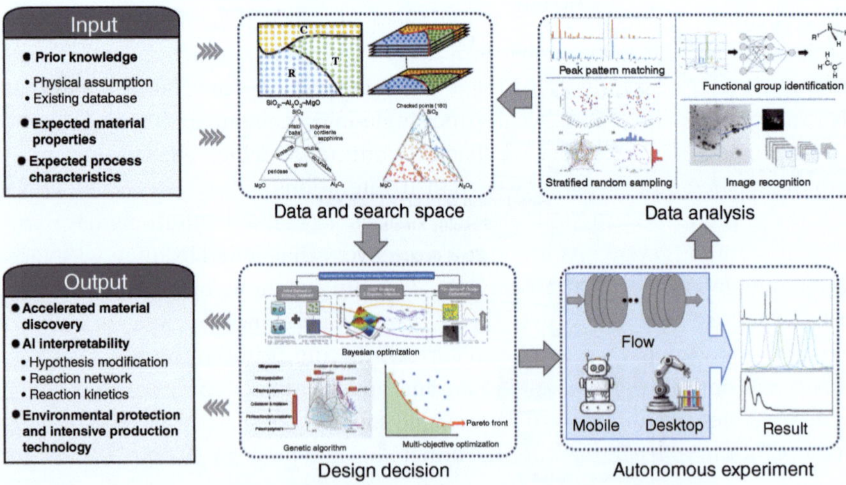

Figure 1.7 The process of autonomous experiments. *Source*: [80] / Elsevier / CC BY-NC-ND 4.0; [81] / Springer Nature / CC BY 4.0.

materials computation, experiments, and data streams raises the platform's intelligence, and combining information technology, networking, and big data enables modular scheduling and resource management. These features improve the efficiency, effectiveness, and use of both experimental and computational resources.

1.4.3.1 Autonomous Laboratory

The first step toward fully autonomous materials experimentation came in 2010 with ARES, developed jointly by the Air Force Institute of Technology, United States, and the Air Force Research Laboratory. This initiative sought to fundamentally restructure the scientific method by integrating robotic instrumentation with AI to execute experimental design, operation, and analysis without human intervention. In 2014, ARES combined a six-axis robot, a multichannel CVD reactor, and real-time analysis tools such as Raman spectroscopy and electron microscopy [82]. By using a computational strategy that combined random forests with BO, the system was able to run about 100 experiments per day with a prediction accuracy of 92%. This capacity allowed it to navigate a 10-dimensional parameter space to successfully pinpoint the specific temperature and hydrocarbon pressure settings governing the wall count of carbon nanotubes.

While the initial iteration relied on offline model training, ARES effectively validated the concept of closed-loop autonomous discovery. The research team used logistic regression to simplify the data and Latin hypercube sampling to identify the most important factors for wall-layer growth across 237 experiments. This method was about 300 times more effective than traditional approaches. It used a strategy that first identified the most important parameters before applying more detailed refinement techniques. In 2016, the system was deployed to investigate carbon nanotube growth rates [83]. Following 84 preliminary trials to establish priors, a random forest model directed a GA to search the reaction space. Running more than 600 automated experiments, the system produced carbon nanotubes at specific growth rates and revealed a new synthesis route along with a unique kinetic trap.

The versatility of the ARES architecture was further demonstrated in 2021 with the release of an autonomous platform dedicated to additive manufacturing [84]. Combining syringe extrusion printing and cloud-based ML, the platform automatically tuned parameters to produce desired single-layer structures. Automated image analysis provided the necessary feedback for online BO which allowed the system to achieve desired outcomes in fewer than 100 iterations.

Following the precedent set by ARES, various academic institutions have engineered autonomous systems for materials discovery that generally align with table, flow, or mobile architectural archetypes. A leading example of a table-based system is Ada, created at the University of British Columbia [85]. Ada functions as a modular robotic platform designed for self-driving thin-film research wherein model-based optimization drives the autonomous tuning of composition and processing variables to enhance optical and electronic properties. By automating solution preparation, spin coating, thermal annealing, and property testing, Ada ensures high reproducibility and creates structured datasets by controlling subtle variables like dispensing nozzle height and inter-step timing.

Later improvements to Ada's hardware and software allowed it to optimize multiple material properties at the same time [86]. A seven-step automated process used a four-axis robot for deposition and analysis and a six-axis robot for sample transfer, allowing multiple thin-film variants to be prepared and tested autonomously. Experimental inputs and outcomes were processed by the qEHVI multi-objective optimizer to determine subsequent experimental steps. The upgraded platform successfully managed a larger variable set and autonomously executed between 40 and 60 experiments to minimize human error and significantly improve data quality and stability.

In the domain of mobile architectures, the University of Liverpool developed a mobile robot known as the Mobile Chemist for photocatalytic experimentation [87]. This unit navigates the laboratory autonomously using laser scanning together with touch feedback to achieve exact positioning across a large working area. The robot executes dexterous operations across multiple stations with a precision comparable to human researchers and maintains an operational uptime of 21.5 hours per day. This system considers ten 10 experimental dimensions simultaneously and delivers a speed increase of approximately 1,000 times compared to manual methods. When paired with BO, the Mobile Chemist autonomously conducted 688 experiments over eight days to identify a catalyst formulation that increased hydrogen production rates sixfold. Its design facilitates operation in dark or photosensitive environments and complies with safety standards for human–robot collaboration while its modular nature allows for the easy integration of new experimental capabilities.

A platform for discovering molecular nanocluster photocatalysts creates extensive molecular combinations and uses automated high-throughput flow synthesis to rapidly identify hydrogen evolution catalysts [88]. By combining small-scale screening with production-ready methods, the system successfully identified the optimal MTPA-CA–CNP147 nanocluster, which shows outstanding hydrogen evolution performance. Subsequent theoretical analysis correlated moderate donor–acceptor binding energies with heightened catalytic activity and thereby established a viable protocol for the integrated discovery and mass preparation of catalysts.

The AI Chemist system establishes a data-driven, fully automated platform for chemical research [89]. This framework combines data mining from scientific literature with automated experimental planning and product optimization. Its goal is to find catalysts suitable for oxygen evolution reactions under simulated Martian constraints. The system uses mobile robots, a central computing unit, cloud servers, and 14 specialized workstations. It operates a two-stage workflow, starting from raw material analysis and moving through to the refinement of catalysts. Following the screening of millions of candidate formulations the system rapidly identified catalysts capable of sustained oxygen generation in Martian environments. This fully automated pipeline negates the necessity for human intervention while enhancing reproducibility and throughput by mitigating human error. The modular nature of the system facilitates adaptation for other extraterrestrial material investigations and synthetic chemistry objectives which supports prospective extraterrestrial resource utilization endeavors.

Artificial Chemist, a flow-based system that uses intelligent process control for synthesizing quantum dots, integrates precursor preparation and flow reaction modules with real-time analysis features [90]. A high-precision module mixes multiple reagents, which then flow into a reactor for continuous synthesis. A short-path flow cell enables rapid exploration of quantum dot compositions by measuring photoluminescence at high concentrations without dilution. This system continuously identifies optimal synthesis pathways and saves the key data for scaling up production with minimal starting materials.

In 2024, RoboChem was introduced as a robotic platform to autonomously optimize and enhance photocatalytic reaction processes [91]. Using a graphical interface, users define parameters, while Python-controlled instruments carry out experiments and relay data in real time to a BO framework. This closed-loop methodology reduces the reliance on manual intervention or expert oversight and concomitantly enhances operational safety. The platform efficiently traverses extensive parameter spaces to ascertain optimal conditions for various substrates and augments both reaction yields and space–time yields.

The convergence of autonomous platforms with three-dimensional printing represents a critical trajectory for the field as demonstrated by the Bayesian autonomous experimental researcher (BEAR) developed to navigate the design space of additive manufacturing [92]. BEAR integrates five FDM printers with a six-axis robotic arm and includes a mass balance and universal testing machine to evaluate material toughness. Bespoke orchestration software synchronizes these instruments to execute high-throughput, closed-loop experiments that autonomously optimize metamaterial architectures and processing parameters. The platform rapidly quantifies uncertainty regarding mechanical properties such as toughness and delineates structure–performance relationships which results in a nearly 60-fold reduction in the experimental volume required to identify high-performance designs compared to a grid search. When coupled with topology optimization and analogous design methodologies, the BEAR system elucidates novel avenues for mechanical materials discovery and the identification of previously inaccessible properties.

1.4.3.2 Software Platform

The ARES team proposed in 2018 that autonomous materials discovery could accelerate in a Moore's Law-like manner. By traversing multidimensional parameter spaces with a speed exceeding human capability, these systems effectively manage the high-dimensional intricacies inherent to contemporary materials science [84]. This concept was put into practice in 2021 with the launch of ARES OS, an open-source operating system built to fully manage ARES hardware [84]. The architecture of ARES OS is predicated on three primary modules encompassing interfaces alongside data analysis and design planning. The interface module guarantees fluid interoperability with peripheral hardware and software to ensure precise command execution and data transmission. Concurrently, the data analysis component utilizes sophisticated algorithms to interpret experimental outputs and detect latent patterns or anomalies to drive optimization. The design planning module synthesizes experimental objectives with available resources and analytical feedback to formulate strategic protocols such as the optimization of reaction parameters for carbon nanotube generation. IBM's RXN platform employs a Transformer model with an attention mechanism to automatically learn reaction rules from literature, enabling the accurate prediction and optimization of both forward and reverse synthesis pathways for target compounds [93].

1.4.3.3 Autonomous Experiments Driven by LLMs

The arrival of LLMs has further accelerated autonomous experimentation, exemplified by the 2023 A-Lab, which leveraged computational modeling, machine learning, and active learning to produce 41 new inorganic compounds in 17 days [94]. This project selected various oxides and phosphates using phase stability predictions, then refined natural language model-proposed syntheses through thermodynamic active learning—a process where analyzing failures yielded practical insights, ultimately demonstrating the high success rate and effectiveness of merging computational science with robotic automation.

Powered by advanced LLMs such as GPT-4 [76], Coscientist autonomously handles the full research cycle, from information gathering and experiment design to code generation and remote laboratory control. Coscientist's software collects publicly available compound data from the Internet and academic sources. After training, it designs a synthesis route and determines the necessary experimental steps, then remotely directs a pipetting robot to carry them out. By using chemical data in SMILES format, the system automatically writes code to enable communication between its modules. Coscientist successfully combines lab automation with LLMs, overcoming traditional experimental barriers and opening new possibilities for autonomous intelligence in chemistry research.

1.4.4 Intelligent Computation

AI-driven materials computation integrates high-throughput computing, multiscale modeling, and Integrated Computational Materials Engineering (ICME), enabling systems to autonomously guide exploration, method selection, resource allocation, and screening. These autonomous processes operate across the entire material life cycle—from design and preparation to processing and final use—forming a closed feedback loop of virtual tests that generate optimal designs for experimental and engineering applications. Computational intelligence helps overcome the cognitive biases common in traditional human analysis by independently exploring variables outside conventional search areas. Such refinement of technical pathways and design concepts boosts computational efficiency and improves the reliability of materials screening.

1.4.4.1 Machine Learning Interatomic Potentials

Molecular simulations have traditionally been built on empirical potentials informed by experiments or quantum mechanics. However, these methods become significantly less efficient as system complexity increases. Machine-learned interatomic potentials (MLIPs) provide an effective remedy to these computational limitations [95, 96]. These potentials utilize algorithmic learning to automatically derive energy functions from reference data which facilitates the rapid and precise prediction of particle interactions. Relative to standard quantum mechanical techniques, MLIPs provide substantial gains in computational speed and possess the adaptability to model diverse complex materials by capturing intricate interactions that elude classical descriptions.

Pioneering efforts in this field predominantly utilized artificial neural networks to predict potential energy based on all atomic coordinates within specific systems [97]. While effective in isolated contexts, the reliance of these early models on their training systems severely restricted their broader applicability and accuracy. To surmount these hurdles, deep-learning algorithms have ascended to a dominant position within MLIP architecture. Deep neural networks are capable of decoding complex atomic interactions via nonlinear mappings across multiple neuronal layers and demonstrate superior performance when applied to high-dimensional systems and massive datasets. Furthermore, GNNs enhance MLIPs by learning interatomic relationships directly from graph-structured data.

Deep learning has catalyzed the creation of diverse MLIP frameworks such as the DeePMD-kit developed by DP Technology [98]. By learning the underlying many-body interaction patterns, deep neural networks allow the platform to predict material potential energies with higher accuracy than classical approaches. The integration of molecular modeling, HPC, and machine learning enabled molecular dynamics (MD) simulations of up to 100 million atoms [99–101]. Another important software for large-scale atomic simulations is GPUMD [102]. Integrating GPU acceleration with neural network potentials, GPUMD efficiently models atomic interactions and maintains high accuracy across diverse materials.

1.4.4.2 Efficient Computation and High-throughput Screening

Conventional computational approaches struggle with high-dimensional data, but AI combined with DFT, MD, and high-throughput computation (HTC) is revolutionizing materials screening and analysis.

Machine learning has dramatically sped up DFT, enabling property predictions orders of magnitude faster than conventional approaches [34, 35, 103]. This computational expediency arises from the capacity of these models to assimilate DFT-generated data which subsequently facilitates rapid experimental iteration and process automation. Machine learning algorithms trained on extensive datasets also enhance predictive reliability by mitigating errors and approaching the inherent noise floor of the data which is a critical attribute for compound ranking. By embedding physical principles and refining feature sets, these models effectively navigate data imperfections and improve transferability by internalizing electronic structure information. Deep learning integrated with DFT allows a neural network to represent the Hamiltonian efficiently, bypassing costly self-consistent field calculations while preserving accuracy for twisted van der Waals systems [33]. Furthermore, AI augments the scalability of DFT for macroscopic architectures by generalizing knowledge from localized atomic environments to facilitate the rigorous examination of complex material structures.

Conventional MD simulations often become prohibitively resource intensive as system complexity escalates and result in voluminous datasets that are difficult to interpret. By uncovering hidden patterns in massive simulation logs, AI overcomes limitations of conventional analysis methods [104, 105]. By training models to detect these concealed patterns, automated data processing dramatically increases research productivity. By applying a fully connected neural network to 180,000 membrane designs, researchers quickly identified optimal candidates based on structure–conductivity correlations [106]. Such predictive capabilities offer precise initialization points for subsequent simulations and obviate the burdensome assumptions inherent in manual trial and error. Additionally, AI is pivotal in optimizing simulation parameters such as time steps and force fields which historically required extensive empirical adjustment. By learning from historical archives, machine learning models can forecast optimal settings based on specific simulation objectives and system characteristics to enhance both the efficiency and fidelity of MD research.

HTC inherently involves massive datasets and complex procedures which frequently leads to the curse of dimensionality when managing high-dimensional variables. Machine learning provides robust support in this arena by employing data-driven methodologies to discern valuable patterns and optimize workflows for materials screening [107]. These algorithms analyze historical records to identify critical variables and recognize previously neglected factors to refine the scientific rationale underpinning material design. When screening materials, models trained on existing datasets can prioritize candidates, filtering promising options from a large pool and offering reliable evaluations early in the design process. DeepMind's GNoME model, trained on first-principles data, generated 2.2 million crystal structures, with an estimated 380,000 likely stable. This breakthrough expanded the catalog of known stable crystals more than 10-fold and set a new standard for discovery [108].

ALKEMIE (Artificial Learning and Knowledge Enhanced Materials Informatics Engineering), a key development in Materials Genome Engineering, uses its AMDIV framework—automation, modularity, database integration, and visualization—to enable data-driven research. ALKEMIE stores all structured and multimodal data in the Matter Cloud, which offers high-level APIs for easy querying and sharing. The repository currently holds over 600,000 crystal structures, 10,000 phonon spectra, 100,000 DFT band structures, 1,000-GW-accurate band structures, and 200,000 machine learning potential datasets [109]. In a parallel vein, researchers from the Chinese Academy of Sciences engineered the MatCloud platform [110], which utilizes a hybrid quantum mechanics–machine learning strategy to forecast material properties. The ProME platform, as shown in Figure 1.8, tackles materials challenges in extreme environments with a suite of validated proprietary

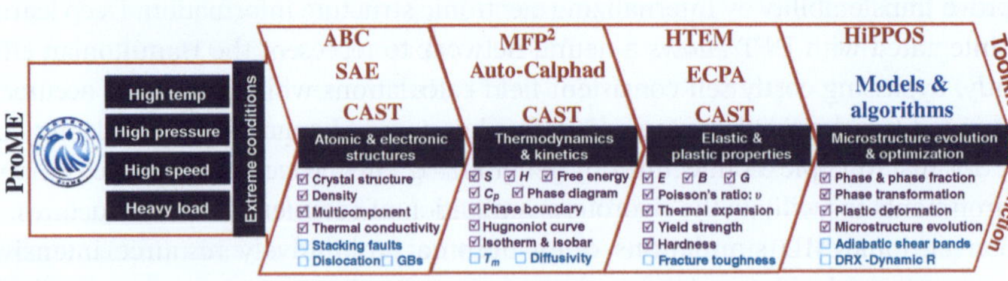

Figure 1.8 Toolkits and their functions within the ProME platform. *Source*: [111] / John Wiley & Sons / CC BY 4.0.

tools—including those for creating ordered and disordered crystal lattices and modeling their properties—centered around a self-adaptive core that manages workflows and corrects errors during HTCs [111].

1.4.4.3 AI-enhanced ICME

ICME combines experiments, computational modeling, and engineering design. It uses multi-scale simulations to link atomic and macroscopic processes, optimizing production paths and predicting performance before manufacturing. Key tools include material databases, property-prediction models, microstructural models, and cost analysis [112]. The increasing complexity of materials design presents challenges for traditional ICME in handling high-dimensional, nonlinear problems. AI integration offers a solution by analyzing large datasets to uncover hidden patterns, optimize parameters, and complement physical models, thereby enhancing accuracy, efficiency, and progress in the field.

AI coupled with ICME now leverages extensive datasets to clarify microstructure–property links, enabling more precise materials predictions [113]. Where explicit physical models are incomplete or unavailable, AI provides distinctive advantages. AI-enhanced ICME in multi-scale modeling combines cross-scale data integration with automated parameter tuning to improve predictive accuracy and robustness [114, 115]. This capability is especially important for high-performance and application-specific materials.

The combined use of AI and ICME has delivered concrete progress. Researchers established composition to processing to structure to property relationships for cemented carbides using a multi-loop machine learning framework, enabling intelligent design [116]. Data-driven ICME was also used to design lightweight and high-entropy alloys [117]. Integrating phase-field modeling with deep VAEs enables microstructure design and accurate thermal conductivity predictions [118]. These examples show the strength of combining machine learning, knowledge-based modeling, and established ICME systems.

AI-assisted design and manufacturing are now core, data-intensive elements of the ICME era [112, 117]. As shown in Figure 1.9, the smart materials paradigm is shifting from knowledge- and data-driven ICME toward an AI^+ era—moving from “designing materials” to “designing with materials” [111, 112, 117]. This progression advances from Embedding AI through Copilot AI toward emerging AI Agent systems [111].

While the MGI focuses on experiments, computation, and databases, Human–Cyber–Physics and Materials Genome Engineering frameworks center on human–machine interaction [119]. Their intersection points toward future advanced-AI applications. Machine learning can speed up fundamental science by revealing key data relationships, strengthening Human–Cyber–Physics interpretation and leading to new principles [120]. Industry 4.0, driven by data and AI, extends digital integration through cyber–physical systems, now giving rise to Materials 4.0 [121]. Figure 1.9 compares top-down engineering and bottom-up computational design [122].

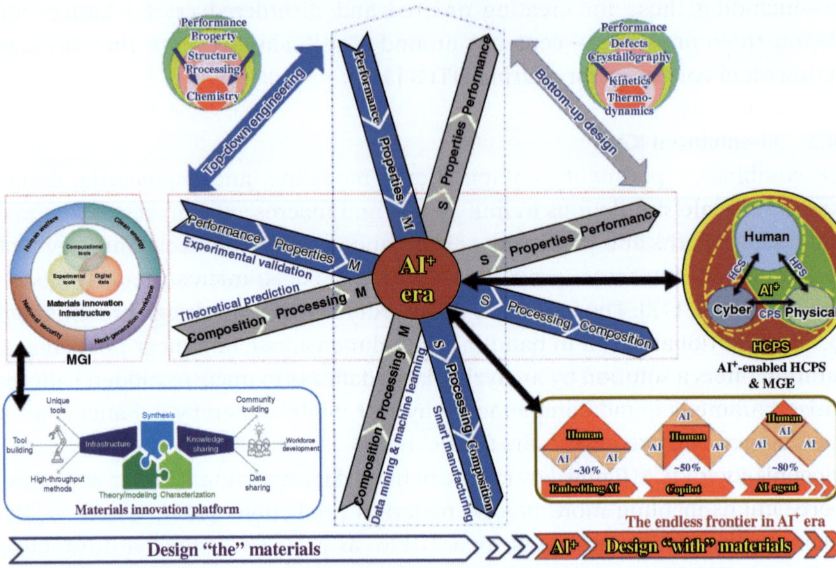

Figure 1.9 The field is moving beyond data-driven ICME, entering an AI+ era of intelligent materials design and manufacturing [112, 117]. Inset of the Materials Innovation Platform adapted from [129].

DTs enable real-time, information-physical integration and are key enablers for smart manufacturing in aerospace, rail, and engine sectors [111, 117, 123]. Process-to-structure and structure-to-property models underscore how materials knowledge turns data into manufacturing intelligence [122, 123]. With advances in LLMs and next-generation AI [124–126], human–cyber interaction will evolve significantly, continuing the shift from material design paradigms toward designing with materials in the AI^+ era [127, 128].

1.4.5 Intelligent Manufacture

1.4.5.1 AI-assisted Additive Manufacturing

Subtractive manufacturing can be inefficient and costly for complex components. Additive manufacturing, or 3D printing, circumvents these limitations by fabricating intricate parts directly from digital designs, streamlining production and reducing material and time costs. Selective laser melting produces precise parts like the C919's winglet bar and NASA's fuel nozzle with minimal scrap [130], while also enabling novel materials such as NASA's GRX-810 alloy [131] and China's adaptable mechanical metamaterials [132, 133].

Advances in materials, technology, and AI have rapidly expanded the potential of 3D printing, enabling its use with metal [134], polymer [135], ceramic [136], composite [137], and biological materials [138] to transform digital manufacturing. AI transforms every step of the process—from choosing material compositions [139] and designing structures [133] to optimizing the printing process [140]. It analyzes large datasets to propose new material formulas, designs stronger and lighter shapes, and adjusts printing settings in real time to improve precision and reduce errors [141]. These innovations make 3D printing more adaptable, sustainable, and precise, advancing smart manufacturing.

1.4.5.2 AI-integrated DT

DT is an effective tool to bridge the physical and digital worlds in smart manufacturing, using continuous data synchronization to strengthen decision-making, enable predictive adjustments, and support autonomous optimization. For instance, researchers developed a Hall–Petch model guided by decision trees that, using key parameters like grain boundary energy and valence electron distance, allowed metallic properties to be optimized in real time without empirical fitting [67]. DT methods have refined laser powder bed fusion for Inconel 718 and accelerated homogenization, and when combined with CNNs, they improve creep fracture life predictions in titanium alloys [142, 143]. In smart manufacturing, researchers developed a deep-learning-integrated DT for real-time defect control in hot-rolled coils [144] and a separate DT model using real-time vibration and temperature data to predict CNC machine tool failures [145].

1.4.5.3 AI Agent-driven Material Manufacturing

LLMs' rapid progress is advancing automation toward smart manufacturing through the rise of intelligent agents [146]. An agent is an intelligent system that perceives its environment, makes decisions, and executes tasks. It learns, adapts, and collaborates, continuously interacting with its surroundings and adjusting through feedback to form a closed-loop process [147].

Agents enhance smart manufacturing at every level. Locally, they improve process optimization, quality control, and equipment management. They replace slow, error-prone manual checks with fast, consistent, and data-driven automated inspections. By integrating AI with vision and sensor technologies, manufacturing systems can detect defects, anticipate equipment wear, and dynamically adjust processes for customized outputs [148, 149].

Though still emerging in smart manufacturing, agents are already showing strong benefits in materials research [146]. For instance, the Material Intelligence Scientist System, developed by the Shanghai Institute of Ceramics, integrates autonomous learning, data extraction, and lab interaction to efficiently optimize materials. The platform autonomously generated a lithium-rich cathode with 282 mAh/g capacity and 100% retention after 20 cycles, surpassing previously reported performance.

Figure 1.10 shows an LLM-driven workflow for optimizing microstructure and classifying defects in welded titanium alloys [111, 150]. It integrates automated coding, image processing, and machine learning to analyze high-throughput welding images, improving defect detection efficiency and reliability. This demonstrates the potential of AI agents to advance smart manufacturing through enhanced process control and innovation.

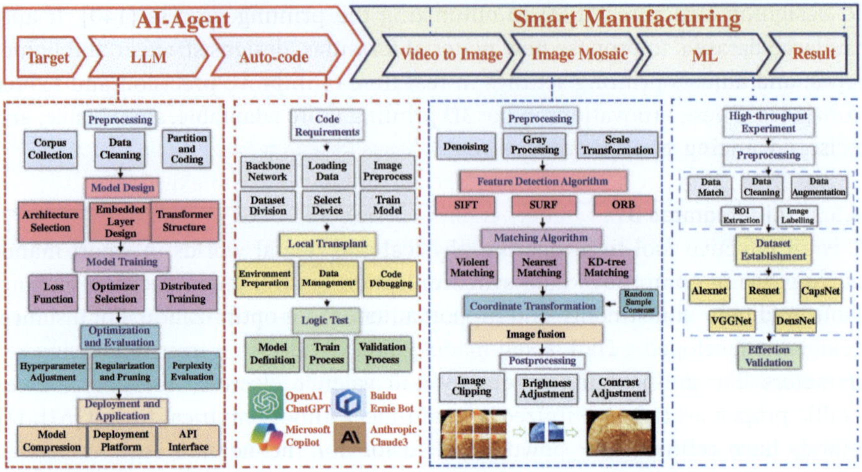

Figure 1.10 An LLM-driven workflow for optimizing microstructure and classifying defects in welded titanium alloys. *Source*: [150] / OAE Publishing Inc. / CC BY 4.0.

References

1 Xie J, Su Y, Xue D, Jiang X, Fu H, Huang H. Machine learning for materials research and development. Acta Metallurgica Sinica. 2021;57(11):1343–61.

2 GitHub. Materials Data Curation System [Internet]. Available from: https://github.com/usnistgov/MDCS

3 Liu S, Su Y, Yin H, Zhang D, He J, Huang H, et al. An infrastructure with user-centered presentation data model for integrated management of materials data and services. npj Computational Materials. 2021;7(1):88.

4 Gong H, He J, Zhang X, Duan L, Tian Z, Zhao W, et al. A repository for the publication and sharing of heterogeneous materials data. Scientific Data. 2022;9(1):787.

5 Klimeck G, McLennan M, Brophy SP, Adams III GB, Lundstrom MS. nanoHUB.org: Advancing education and research in nanotechnology. Computing in Science & Engineering. 2008;10(5):17–23.

6 Kaufman JG, Begley EF. MatML: A data interchange markup language: This article updates and expands upon previously published information on MatML, a new extensible markup language designed specifically for the exchange of materials information. In particular, it describes the strengths of MatML from the perspective of materials specialists dealing with a wide variety of types of materials information. Advanced Materials & Processes. 2003;161(11):35–7.

7 Andersen CW, Armiento R, Blokhin E, Conduit GJ, Dwaraknath S, Evans ML, et al. OPTIMADE, an API for exchanging materials data. Scientific Data. 2021;8(1):217.

8 National Science and Technology Council. Materials Genome Initiative for Global Competitiveness. Whitepaper. Office of Science and Technology Policy (OSTP), Washington, DC; 2011.

9 Agrawal A, Choudhary A. Perspective: Materials informatics and big data: Realization of the "fourth paradigm" of science in materials science. APL Materials. 2016;4(5):053208.
10 Morgan D, Jacobs R. Opportunities and challenges for machine learning in materials science. Annual Review of Materials Research. 2020;50(1):71–103.
11 Rao HS, Mukherjee A. Artificial neural networks for predicting the macromechanical behaviour of ceramic-matrix composites. Computational Materials Science. 1996;5(4):307–22.
12 Yu XL, Yi B, Wang XY. Prediction of the glass transition temperatures for polymers with artificial neural network. Journal of Theoretical and Computational Chemistry. 2008;07(05):953–63.
13 Meredig B, Agrawal A, Kirklin S, Saal JE, Doak JW, Thompson A, et al. Combinatorial screening for new materials in unconstrained composition space with machine learning. Physical Review B. 2014;89(9):094104.
14 Raccuglia P, Elbert KC, Adler PDF, Falk C, Wenny MB, Mollo A, et al. Machine-learning-assisted materials discovery using failed experiments. Nature. 2016;533(7601):73–6.
15 Xiong J, Shi SQ, Zhang TY. A machine-learning approach to predicting and understanding the properties of amorphous metallic alloys. Materials & Design. 2020;187:108378.
16 Wang C, Shen C, Cui Q, Zhang C, Xu W. Tensile property prediction by feature engineering guided machine learning in reduced activation ferritic/martensitic steels. Journal of Nuclear Materials. 2020;529:151823.
17 Lu S, Zhou Q, Ma L, Guo Y, Wang J. Rapid discovery of ferroelectric photovoltaic perovskites and material descriptors via machine learning. Small Methods. 2019;3(11):1900360.
18 Ward L, Dunn A, Faghaninia A, Zimmermann NER, Bajaj S, Wang Q, et al. Matminer: An open source toolkit for materials data mining. Computational Materials Science. 2018;152:60–9.
19 Ma J, Cao B, Dong S, Tian Y, Wang M, Xiong J, et al. MLMD. A programming-free AI platform to predict and design materials. npj Computational Materials. 2024;10(1):59.
20 Collins SP, Daff TD, Piotrkowski SS, Woo TK. Materials design by evolutionary optimization of functional groups in metal-organic frameworks. Science Advances. 2016;2(11):e1600954.
21 Seko A, Maekawa T, Tsuda K, Tanaka I. Machine learning with systematic density-functional theory calculations: Application to melting temperatures of single-and binary-component solids. Physical Review B. 2014;89(5):054303.
22 Xue D, Balachandran PV, Hogden J, Theiler J, Xue D, Lookman T. Accelerated search for materials with targeted properties by adaptive design. Nature Communications. 2016;7:1–9.
23 Xue D, Balachandran PV, Yuan R, Hu T, Qian X, Dougherty ER, et al. Accelerated search for $BaTiO_3$-based piezoelectrics with vertical morphotropic phase boundary using Bayesian learning. Proceedings of the National Academy of Sciences of the United States of America. 2016;113(47):13301–6.

24 Yuan R, Liu Z, Balachandran PV, Xue D, Zhou Y, Ding X, et al. Accelerated discovery of large electrostrains in $BaTiO_3$-based piezoelectrics using active learning. Advanced Materials. 2018;30(7):1702884.
25 Tian Y, Yuan R, Xue D, Zhou Y, Ding X, Sun J, et al. Role of uncertainty estimation in accelerating materials development via active learning. Journal of Applied Physics. 2020;128(1):014103.
26 Tian Y, Yuan R, Xue D, Zhou Y, Wang Y, Ding X, et al. Determining multi-component phase diagrams with desired characteristics using active learning. Advanced Science. 2021;8(1):2003165.
27 Zhang H, Fu H, Zhu S, Yong W, Xie J. Machine learning assisted composition effective design for precipitation strengthened copper alloys. Acta Materialia. 2021;215:117118.
28 Yuan X, Suvarna M, Lim JY, Pérez-Ramírez J, Wang X, Ok YS. Active learning-based guided synthesis of engineered biochar for CO_2 capture. Environmental Science & Technology. 2024;58(15):6628–36.
29 Xu S, Li J, Cai P, Liu X, Liu B, Wang X. Self-improving photosensitizer discovery system via Bayesian search with first-principle simulations. Journal of the American Chemical Society. 2021;143(47):19769–77.
30 Xiong J, He JC, Leng XS, Zhang TY. Gaussian process regressions on hot deformation behaviors of FGH98 nickel-based powder superalloy. Journal of Materials Science & Technology. 2023;146:177–85.
31 LeCun Y, Bengio Y, Hinton G. Deep learning. Nature. 2015;521(7553):436–44.
32 Choudhary K, DeCost B, Chen C, Jain A, Tavazza F, Cohn R, et al. Recent advances and applications of deep learning methods in materials science. npj Computational Materials. 2022;8(1):59.
33 Li H, Wang Z, Zou N, Ye M, Xu R, Gong X, et al. Deep-learning density functional theory Hamiltonian for efficient ab initio electronic-structure calculation. Nature Computational Science. 2022;2(6):367–77.
34 Gong X, Louie SG, Duan W, Xu Y. Generalizing deep learning electronic structure calculation to the plane-wave basis. Nature Computational Science. 2024;4:752–60.
35 Wang Y, Li Y, Tang Z, Li H, Yuan Z, Tao H, et al. Universal materials model of deep-learning density functional theory Hamiltonian. Science Bulletin. 2024;69:2514–21.
36 Tang Z, Li H, Lin P, Gong X, Jin G, He L, et al. A deep equivariant neural network approach for efficient hybrid density functional calculations. Nature Communications. 2024;15(1):8815.
37 Weng M, Wang Z, Qian G, Ye Y, Chen Z, Chen X, et al. Identify crystal structures by a new paradigm based on graph theory for building materials big data. Science China Chemistry. 2019;62:982–6.
38 Li S, Chen Z, Wang Z, Weng M, Li J, Zhang M, et al. Graph-based discovery and analysis of atomic-scale one-dimensional materials. National Science Review. 2022;9(6):nwac028.
39 Zhang W, Weng M, Zhang M, Chen Z, Wang B, Li S, et al. Rapid mining of fast ion conductors via subgraph isomorphism matching. Journal of the American Chemical Society. 2024;146(27):18535–43.

40 Chen L, Wang B, Zhang W, Zheng S, Chen Z, Zhang M, et al. Crystal structure assignment for unknown compounds from X-ray diffraction patterns with deep learning. Journal of the American Chemical Society. 2024;146(12):8098–109.

41 Zhang H, Wang Z, Ren J, Liu J, Li J. Ultra-fast and accurate binding energy prediction of shuttle effect-suppressive sulfur hosts for lithium-sulfur batteries using machine learning. Energy Storage Materials. 2021;35:88–98.

42 Wang Z, Zhang H, Ren J, Lin X, Han T, Liu J, et al. Predicting adsorption ability of adsorbents at arbitrary sites for pollutants using deep transfer learning. npj Computational Materials. 2021;7(1):19.

43 Wu S, Wang Z, Zhang H, Cai J, Li J. Deep learning accelerates the discovery of two-dimensional catalysts for hydrogen evolution reaction. Energy & Environmental Materials. 2023;6(1):e12259.

44 Goodfellow I, Pouget-Abadie J, Mirza M, Xu B, Warde-Farley D, Ozair S, et al. Generative adversarial networks. Communications of the ACM. 2020;63(11):139–44.

45 Kingma DP. Auto-encoding variational Bayes. arXiv preprint arXiv:1312.6114; 2013.

46 Sutton RS, Barto A. Reinforcement Learning: An Introduction. Cambridge, MA: The MIT Press; 1998.

47 Ho J, Jain A, Abbeel P. Denoising diffusion probabilistic models. In: Larochelle H, Ranzato M, Hadsell R, Balcan MF, Lin H, editors. Advances in Neural Information Processing Systems. New York: Curran Associates, Inc.; 2020. pp. 6840–51.

48 Zhou Z, Shang Y, Liu X, Yang Y. A generative deep learning framework for inverse design of compositionally complex bulk metallic glasses. npj Computational Materials. 2023;9(1):15.

49 Shang Y, Zhou Z, Han R, Wang H, Liu X, Yang Y. Customized design of amorphous solids by generative deep learning. The Innovation Materials. 2024;2(2):100071.

50 Chen ZQ, Shang YH, Liu XD, Yang Y. Accelerated discovery of eutectic compositionally complex alloys by generative machine learning. npj Computational Materials. 2024;10(1):204.

51 Xian Y, Dang P, Tian Y, Jiang X, Zhou Y, Ding X, et al. Compositional design of multicomponent alloys using reinforcement learning. Acta Materialia. 2024;274:120017.

52 Dang P, Hu J, Xian Y, Li C, Zhou Y, Ding X, et al. Thermal battery: Ultrahigh heat-storage capacity based on generative learning-designed phase-change alloys. Advanced Materials. 2024;37:2412198.

53 Zhong C, Zhang J, Wang Y, Long Y, Zhu P, Liu J, et al. High-performance diffusion model for inverse design of high T_c superconductors with effective doping and accurate stoichiometry. InfoMat. 2024;6(5):e12519.

54 Zhong C, Wang Y, Long Y, Liu J, Hu K, Zhang J, et al. Enhancing superconductor critical temperature prediction: A novel machine learning approach integrating dopant recognition. ACS Applied Materials & Interfaces. 2024;16(44):60472–81.

55 Peng B, Wei Y, Qin Y, Dai J, Li Y, Liu A, et al. Machine learning-enabled constrained multi-objective design of architected materials. Nature Communications. 2023;14(1):6630.

56 Wang Y, Sha W, Xiao M, Qiu CW, Gao L. Deep-learning-enabled intelligent design of thermal metamaterials. Advanced Materials. 2023;35(33):2302387.

57 Childs CM, Washburn NR. Embedding domain knowledge for machine learning of complex material systems. MRS Communications. 2019;9(3):806–20.

58 Kirklin S, Saal JE, Meredig B, Thompson A, Doak JW, Aykol M, et al. The Open Quantum Materials Database (OQMD): assessing the accuracy of DFT formation energies. npj Computational Materials. 2015;1(1):1–15.

59 Yuan R, Tian Y, Xue D, Xue D, Zhou Y, Ding X, et al. Accelerated search for $BaTiO_3$-based ceramics with large energy storage at low fields using machine learning and experimental design. Advanced Science. 2019;6(21):1901395.

60 Yuan R, Xue D, Xue D, Li J, Ding X, Sun J, et al. Knowledge-based descriptor for the compositional dependence of the phase transition in $BaTiO_3$-based ferroelectrics. ACS Applied Materials & Interfaces. 2020;12(40):44970–80.

61 Zhang Y, Wen C, Wang C, Antonov S, Xue D, Bai Y, et al. Phase prediction in high entropy alloys with a rational selection of materials descriptors and machine learning models. Acta Materialia. 2020;185:528–39.

62 Zhang Y, Wen C, Dang P, Lookman T, Xue D, Su Y. Toward ultra-high strength high entropy alloys via feature engineering. Journal of Materials Science & Technology. 2024;200:243–52.

63 Ouyang R, Curtarolo S, Ahmetcik E, Scheffler M, Ghiringhelli LM. SISSO: A compressed-sensing method for identifying the best low-dimensional descriptor in an immensity of offered candidates. Physical Review Materials. 2018;2(8):083802.

64 Bartel CJ, Sutton C, Goldsmith BR, Ouyang R, Musgrave CB, Ghiringhelli LM, et al. New tolerance factor to predict the stability of perovskite oxides and halides. Science Advances. 2019;5(2):eaav0693.

65 Wang T, Hu J, Ouyang R, Wang Y, Huang Y, Hu S, et al. Nature of metal-support interaction for metal catalysts on oxide supports. Science. 2024;386(6724):915–20.

66 Shen C, Wang C, Wei X, Li Y, van der Zwaag S, Xu W. Physical metallurgy-guided machine learning and artificial intelligent design of ultrahigh-strength stainless steel. Acta Materialia. 2019;179:201–14.

67 Jiang L, Fu H, Zhang H, Xie J. Physical mechanism interpretation of polycrystalline metals' yield strength via a data-driven method: A novel Hall–Petch relationship. Acta Materialia. 2022;231:117868.

68 Liu S, Wen T, Pattamatta ASLS, Srolovitz DJ. A prompt-engineered large language model, deep learning workflow for materials classification. Materials Today. 2024;80:240–49.

69 Hippalgaonkar K, Li Q, Wang X, Fisher III JW, Kirkpatrick J, Buonassisi T. Knowledge-integrated machine learning for materials: Lessons from gameplaying and robotics. Nature Reviews Materials. 2023;8(4):241–60.

70 Swain MC, Cole JM. ChemDataExtractor: A toolkit for automated extraction of chemical information from the scientific literature. Journal of Chemical Information and Modeling. 2016;56(10):1894–904.

71 Wang W, Jiang X, Tian S, Liu P, Dang D, Su Y, et al. Automated pipeline for superalloy data by text mining. npj Computational Materials. 2022;8(1):9.

72 Wang W, Jiang X, Tian S, Liu P, Lookman T, Su Y, et al. Alloy synthesis and processing by semi-supervised text mining. npj Computational Materials. 2023;9(1):183.

73 Polak MP, Morgan D. Extracting accurate materials data from research papers with conversational language models and prompt engineering. Nature Communications. 2024;15(1):1569.

74 Tshitoyan V, Dagdelen J, Weston L, Dunn A, Rong Z, Kononova O, et al. Unsupervised word embeddings capture latent knowledge from materials science literature. Nature. 2019;571(7763):95–8.

75 Pei Z, Yin J, Liaw PK, Raabe D. Toward the design of ultrahigh-entropy alloys via mining six million texts. Nature Communications. 2023;14(1):54.

76 Boiko DA, MacKnight R, Kline B, Gomes G. Autonomous chemical research with large language models. Nature. 2023;624(7992):570–8.

77 Kang Y, Kim J. ChatMOF: An artificial intelligence system for predicting and generating metal-organic frameworks using large language models. Nature Communications. 2024;15(1):4705.

78 Kuenneth C, Ramprasad R. polyBERT: A chemical language model to enable fully machine-driven ultrafast polymer informatics. Nature Communications. 2023;14(1):4099.

79 Tian S, Jiang X, Wang W, Jing Z, Zhang C, Zhang C, et al. Steel design based on a large language model. Available from: http://dx.doi.org/10.2139/ssrn.4870777

80 Jiang X, Xue D, Bai Y, Wang WY, Liu J, Yang M, et al. AI4Materials: Transforming the landscape of materials science and engineering. Review of Materials Research. 2025;1(1):100010. https://doi.org/10.1016/j.revmat.2025.100010

81 Zhang Y, Apley DW, Chen W. Bayesian optimization for materials design with mixed quantitative and qualitative variables. Scientific Reports. 2020;10(1):4924.

82 Nikolaev P, Hooper D, Perea-Lopez N, Terrones M, Maruyama B. Discovery of wall-selective carbon nanotube growth conditions via automated experimentation. ACS Nano. 2014;8(10):10214–22.

83 Nikolaev P, Hooper D, Webber F, Rao R, Decker K, Krein M, et al. Autonomy in materials research: A case study in carbon nanotube growth. npj Computational Materials. 2016;2(1):1–6.

84 Tabor DP, Roch LM, Saikin SK, Kreisbeck C, Sheberla D, Montoya JH, et al. Accelerating the discovery of materials for clean energy in the era of smart automation. Nature Reviews Materials. 2018;3(5):5–20.

85 MacLeod BP, Parlane FGL, Morrissey TD, Häse F, Roch LM, Dettelbach KE, et al. Self-driving laboratory for accelerated discovery of thin-film materials. Science Advances. 2020;6(20):eaaz8867.

86 MacLeod BP, Parlane FGL, Rupnow CC, Dettelbach KE, Elliott MS, Morrissey TD, et al. A self-driving laboratory advances the Pareto front for material properties. Nature Communications. 2022;13(1):995.

87 Burger B, Maffettone PM, Gusev VV, Aitchison CM, Bai Y, Wang X, et al. A mobile robotic chemist. Nature. 2020;583(7815):237–41.

88 Zhang W, Yu M, Liu T, Cong M, Liu X, Yang H, et al. Accelerated discovery of molecular nanojunction photocatalysts for hydrogen evolution by using automated screening and flow synthesis. Nature Synthesis. 2024;3(5):595–605.

89 Zhu Q, Huang Y, Zhou D, Zhao L, Guo L, Yang R, et al. Automated synthesis of oxygen-producing catalysts from Martian meteorites by a robotic AI chemist. Nature Synthesis. 2024;3(3):319–28.

90 Epps RW, Bowen MS, Volk AA, Abdel-Latif K, Han S, Reyes KG, et al. Artificial chemist: An autonomous quantum dot synthesis bot. Advanced Materials. 2020;32(30):2001626.

91 Slattery A, Wen Z, Tenblad P, Sanjosé-Orduna J, Pintossi D, den Hartog T, et al. Automated self-optimization, intensification, and scale-up of photocatalysis in flow. Science. 2024;383(6681):eadj1817.

92 Gongora AE, Xu B, Perry W, Okoye C, Riley P, Reyes KG, et al. A Bayesian experimental autonomous researcher for mechanical design. Science Advances. 2020;6(15):eaaz1708.

93 Schwaller P, Hoover B, Reymond JL, Strobelt H, Laino T. Extraction of organic chemistry grammar from unsupervised learning of chemical reactions. Science Advances. 2021;7(15):eabe4166.

94 Szymanski NJ, Rendy B, Fei Y, Kumar RE, He T, Milsted D, et al. An autonomous laboratory for the accelerated synthesis of novel materials. Nature. 2023;624(7990):86–91.

95 Mortazavi B, Zhuang X, Rabczuk T, Shapeev AV. Atomistic modeling of the mechanical properties: The rise of machine learning interatomic potentials. Materials Horizons. 2023;10(6):1956–68.

96 Butler KT, Davies DW, Cartwright H, Isayev O, Walsh A. Machine learning for molecular and materials science. Nature. 2018;559(7715):547–55.

97 Behler J, Parrinello M. Generalized neural-network representation of high-dimensional potential-energy surfaces. Physical Review Letters. 2007;98(14):146401.

98 Wang H, Zhang L, Han J, Weinan E. DeePMD-kit: A deep learning package for many-body potential energy representation and molecular dynamics. Computer Physics Communications. 2018;228:178–84.

99 Jia W, Wang H, Chen M, Lu D, Lin L, Car R, et al. Pushing the limit of molecular dynamics with *ab initio* accuracy to 100 million atoms with machine learning. In: Proceedings of the International Conference for High Performance Computing, Networking, Storage and Analysis [Internet]. IEEE Press; 2020. pp. 1–14. Available from: https://dl.acm.org/doi/abs/10.5555/3433701.3433707

100 Zhang L. 2020 ACM Gordon Bell Prize awarded to team for machine learning method that achieves record molecular dynamics simulation. News Release [Internet]. Association for Computer Machinery; 2020. Available from: https://awards.acm.org/award_winners/zhang_7608842

101 Guo Z, Lu D, Yan Y, Hu S, Liu R, Tan G, et al. Extending the limit of molecular dynamics with ab initio accuracy to 10 billion atoms. In: Proceedings of the 27th ACM SIGPLAN Symposium on Principles and Practice of Parallel Programming [Internet]. Association for Computing Machinery; 2022. pp. 205–18.

102 Fan Z, Wang Y, Ying P, Song K, Wang J, Wang Y, et al. GPUMD: A package for constructing accurate machine-learned potentials and performing highly efficient atomistic simulations. The Journal of Chemical Physics. 2022;157(11):114801.

103 Li Y, Tang Z, Chen Z, Sun M, Zhao B, Li H, et al. Neural-network density functional theory based on variational energy minimization. Physical Review Letters. 2024;133(7):076401.

104 Kiirikki AM, Antila HS, Bort LS, Buslaev P, Favela-Rosales F, Ferreira TM, et al. Overlay databank unlocks data-driven analyses of biomolecules for all. Nature Communications. 2024;15(1):1136.

105 Yao N, Chen X, Fu ZH, Zhang Q. Applying classical, *ab initio*, and machine-learning molecular dynamics simulations to the liquid electrolyte for rechargeable batteries. Chemical Reviews. 2022;122(12):10970–1021.

106 Zhang Q, Yuan Y, Zhang J, Fang P, Pan J, Zhang H, et al. Machine learning-aided design of highly conductive anion exchange membranes for fuel cells and water electrolyzers. Advanced Materials. 2024;36(36):2404981.

107 Xu D, Zhang Q, Huo X, Wang Y, Yang M. Advances in data-assisted high-throughput computations for material design. Materials Genome Engineering Advances. 2023;1(1):e11.

108 Merchant A, Batzner S, Schoenholz SS, Aykol M, Cheon G, Cubuk ED. Scaling deep learning for materials discovery. Nature. 2023;624(7990):80–5.

109 Wang G, Peng L, Li K, Zhu L, Zhou J, Miao N, et al. ALKEMIE: An intelligent computational platform for accelerating materials discovery and design. Computational Materials Science. 2021;186:110064.

110 Yang X, Wang Z, Zhao X, Song J, Zhang M, Liu H. MatCloud: A high-throughput computational infrastructure for integrated management of materials simulation, data and resources. Computational Materials Science. 2018;146:319–33.

111 Wang WY, Zhang S, Li G, Lu J, Ren Y, Wang X, et al. Artificial intelligence enabled smart design and manufacturing of advanced materials: The endless Frontier in AI^+ era. Materials Genome Engineering Advances. 2024;2(3):e56.

112 Wang WY, Li J, Liu W, Liu ZK. Integrated computational materials engineering for advanced materials: A brief review. Computational Materials Science. 2019;158:42–8.

113 Hashemi SM, Parvizi S, Baghbanijavid H, Tan ATL, Nematollahi M, Ramazani A, et al. Computational modelling of process–structure–property–performance relationships in metal additive manufacturing: a review. International Materials Reviews. 2022;67(1):1–46.

114 Wang X, Xiong W. Uncertainty quantification and composition optimization for alloy additive manufacturing through a CALPHAD-based ICME framework. npj Computational Materials. 2020;6(1):188.

115 Qi J, Fan X, Hoyos DI, Widom M, Liaw PK, Poon J. Integrated design of aluminum-enriched high-entropy refractory B2 alloys with synergy of high strength and ductility. Science Advances. 2024;10(49):eadq0083.

116 Du Y, Schmid-Fetzer R, Wang J, Liu S, Wang J, Jin Z. Computational Design of Engineering Materials: Fundamentals and Case Studies. Cambridge, UK: Cambridge University Press; 2023.

117 Wang WY, Tang B, Lin D, Zou C, Zhang Y, Shang SL, et al. A brief review of data-driven ICME for intelligently discovering advanced structural metal materials: Insight into atomic and electronic building blocks. Journal of Materials Research. 2020;35(8):872–89.

118 Attari V, Khatamsaz D, Allaire D, Arroyave R. Towards inverse microstructure-centered materials design using generative phase-field modeling and deep variational autoencoders. Acta Materialia. 2023;259:119204.

119 Wang WY, Li P, Lin D, Tang B, Wang J, Guan Q, et al. DID code: A bridge connecting the materials genome engineering database with inheritable integrated intelligent manufacturing. Engineering. 2020;6(6):612–20.

120 Umehara M, Stein HS, Guevarra D, Newhouse PF, Boyd DA, Gregoire JM. Analyzing machine learning models to accelerate generation of fundamental materials insights. npj Computational Materials. 2019;5(1):34.

121 Craig PL. Modeling Software for Materials 4.0. 2021.

122 Wang WY, Yin J, Chai Z, Chen X, Zhao W, Lu J, et al. Big data-assisted digital twins for the smart design and manufacturing of advanced materials: From atoms to products. Journal of Materials Informatics. 2022;2(1):1–27.

123 Yi Y, Yan Y, Liu X, Ni Z, Feng J, Liu J. Digital twin-based smart assembly process design and application framework for complex products and its case study. Journal of Manufacturing Systems. 2021;58:94–107.

124 Xu Y, Liu X, Cao X, Huang C, Liu E, Qian S, et al. Artificial intelligence: A powerful paradigm for scientific research. The Innovation. 2021;2(4):100179.

125 Nicola N. Can artificial intelligence create the next wonder material. Nature. 2016;533:22–5.

126 Oganov AR, Pickard CJ, Zhu Q, Needs RJ. Structure prediction drives materials discovery. Nature Reviews Materials. 2019;4(5):331–48.

127 Zhang X, Wang L, Helwig J, Luo Y, Fu C, Xie Y, et al. Artificial intelligence for science in quantum, atomistic, and continuum systems. arXiv preprint arXiv:2307.08423; 2023.

128 Wang H, Fu T, Du Y, Gao W, Huang K, Liu Z, et al. Scientific discovery in the age of artificial intelligence. Nature. 2023;620(7972):47–60.

129 National Science and Technology Council. Materials Genome Initiative Strategic Plan. A Report by the Subcommittee on the Materials Genome Initiative Committee on Technology; 2021.

130 Gao X, Lin X, Yan Q, Wang Z, Yu X, Zhou Y, et al. Effect of Cu content on microstructure and mechanical properties of in-situ β phases reinforced Ti/Zr-based bulk metallic glass matrix composite by selective laser melting (SLM). Journal of Materials Science & Technology. 2021;67:174–85.

131 Smith TM, Kantzos CA, Zarkevich NA, Harder BJ, Heczko M, Gradl PR, et al. A 3D printable alloy designed for extreme environments. Nature. 2023;617(7961):513–8.

132 Khalid MY, Arif ZU, Tariq A, Hossain M, Umer R, Bodaghi M. 3D printing of active mechanical metamaterials: A critical review. Materials & Design. 2024;246:113305.

133 Yu S, Chai H, Xiong Y, Kang M, Geng C, Liu Y, et al. Studying complex evolution of hyperelastic materials under external field stimuli using artificial neural networks with spatiotemporal features in a small-scale dataset. Advanced Materials. 2022;34(26):2200908.

134 Liu Z, Ding M, Xie J. Advancements in digital manufacturing for metal 3D printing. Acta Metallurgica Sinica. 2023;60(5):569–84.

135 Liao E, Zhang Y, Su Y, Zhang C, Geng C, Li C, et al. Active mixing of two-component viscoelastic silicone ink at molecular level for spatiotemporally controlled 3D/4D printing of cellular silicones. Polymer. 2024;302:127053.

136 Zhang F, Li Z, Xu M, Wang S, Li N, Yang J. A review of 3D printed porous ceramics. Journal of the European Ceramic Society. 2022;42(8):3351–73.

137 Ma T, Zhang Y, Ruan K, Guo H, He M, Shi X, et al. Advances in 3D printing for polymer composites: A review. InfoMat. 2024;6(6):e12568.

138 Ma Z, Wang J, Qin L, Chortos A. 3D printing in biofabrication: From surface textures to biological engineering. Chemical Engineering Journal. 2024;500:156477.

139 Zhu Z, Ng DWH, Park HS, McAlpine MC. 3D-printed multifunctional materials enabled by artificial-intelligence-assisted fabrication technologies. Nature Reviews Materials. 2021;6(1):27–47.

140 Ren Z, Gao L, Clark SJ, Fezzaa K, Shevchenko P, Choi A, et al. Machine learning-aided real-time detection of keyhole pore generation in laser powder bed fusion. Science. 2023;379(6627):89–94.

141 Lee JA, Park J, Sagong MJ, Ahn SY, Cho JW, Lee S, et al. Active learning framework to optimize process parameters for additive-manufactured Ti-6Al-4V with high strength and ductility. Nature Communications. 2025;16(1):931.

142 Chen M, Du Q, Shi R, Fu H, Liu Z, Xie J. Phase field simulation of microstructure evolution and process optimization during homogenization of additively manufactured Inconel 718 alloy. Frontiers in Materials. 2022;9:1043249.

143 Zong B, Li J, Zhou C, Wang P, Tang B, Yuan R. Enhancing creep rupture life prediction of high-temperature titanium alloys using convolutional neural networks. Materials Genome Engineering Advances. 2024;2(4):e68.

144 Xiang F, Zhou S, Zuo Y, Tao F. Digital twin driven end-face defect control method for hot-rolled coil with cloud-edge collaboration. IEEE Transactions on Industrial Informatics. 2022;19(2):1674–82.

145 Luo W, Hu T, Zhu W, Tao F. Digital twin modeling method for CNC machine tool. In: Proceedings of the 2018 IEEE 15th International Conference on Networking, Sensing and Control [Internet]. IEEE; 2018. pp. 1–4.

146 Yu S, Ran N, Liu J. Large-language models: The game-changers for materials science research. Artificial Intelligence Chemistry. 2024;2(2):100076.

147 Song T, Luo M, Zhang X, Chen L, Huang Y, Cao J, et al. A multiagent-driven robotic AI chemist enabling autonomous chemical research on demand. Journal of the American Chemical Society. 2024;147:12534–45.

148 Prince MH, Chan H, Vriza A, Zhou T, Sastry VK, Luo Y, et al. Opportunities for retrieval and tool augmented large language models in scientific facilities. npj Computational Materials. 2024;10(1):251.

149 Ruan Y, Lu C, Xu N, He Y, Chen Y, Zhang J, et al. An automatic end-to-end chemical synthesis development platform powered by large language models. Nature Communications. 2024;15(1):10160.

150 Zhang S, Wang WY, Wang X, Li G, Ren Y, Gao X, et al. Large language models enabled intelligent microstructure optimization and defects classification of welded titanium alloys. Journal of Materials Informatics. 2025;4:34.

2

Fundamentals of Language Models and NLP

2.1 Introduction

Natural language processing (NLP), a central branch of artificial intelligence (AI), aims to endow machines with the ability to understand, generate, and interact using human language. It serves as a critical bridge between human cognition and machine intelligence. The evolution of NLP—from early rule-based systems to today's large language models (LLMs)—parallels AI's broader shifts: first from hand-coded rules to data-driven learning, and now toward artificial general intelligence (AGI).

Since its emergence in the late 1940s alongside early experiments in machine translation, NLP has undergone several major technological transformations. In the mid- to late 20th century, rule-based approaches laid the theoretical foundation for formal language processing but struggled with ambiguity and scalability. From the 1990s to the early 2000s, statistical methods, leveraging probabilistic models and large corpora, enabled the commercialization of applications such as machine translation and search engines. The rise of deep learning after 2010 brought about breakthroughs in word embeddings, attention mechanisms, and the Transformer architecture, propelling NLP into the era of pretrained LLMs. These advances dramatically expanded the field's capabilities—from basic text comprehension to complex generation and multimodal interaction.

Today, NLP technologies are deeply embedded in applications such as information retrieval, intelligent question answering, content generation, and cross-modal interaction, with significant potential in sectors including healthcare, education, and finance. However, these advances also present new challenges: the heavy computational and data requirements of LLMs, the problem of hallucinated content, data bias, and emerging ethical concerns—all of which require coordinated efforts between academia and industry.

This chapter provides a comprehensive review of the development trajectory of NLP, tracing the evolution from statistical language models to the Transformer revolution and the cutting edge of large pretrained models. It offers a detailed analysis of key architectures—including language models, encoder-decoder frameworks, and attention mechanisms—emphasizing their design principles and historical development. Furthermore, it explores optimization strategies for training LLMs, such as pretraining methods, efficient fine-tuning techniques, and retrieval-augmented generation (RAG). Prominent model families (e.g., Word2Vec, Bidirectional Encoder Representations from Transformers [BERT], generative pretrained transformer [GPT], and Text-to-Text Transfer Transformer [T5]) and practical toolkits (e.g., Hugging Face and Natural Language Toolkit [NLTK]) are introduced, along with their typical use cases.

This chapter offers a comprehensive overview of the NLP technological ecosystem, serving as a systematic reference for researchers and practitioners—spanning foundational theories to cutting-edge applications—while inspiring new paradigms for materials research empowered by NLP.

2.2 Historical Evolution of NLP

2.2.1 Statistical Language Models

The rule-based and symbolic phase of NLP began in the mid-20th century. In the late 1940s and early 1950s, NLP emerged in close connection with machine translation. In 1949, inspired by wartime code-breaking efforts, Warren Weaver proposed the concept of machine translation [1]. This vision materialized in 1954 through the Georgetown–IBM experiment, which demonstrated simple Russian-to-English translation based on dictionary lookups and word-order transformation rules—marking the first major milestone in NLP [2]. Around the same time, linguist Noam Chomsky introduced the theory of generative grammar in 1957, advocating the use of formal rules to describe syntactic structures, thereby laying a theoretical foundation for rule-based systems [3].

During the 1960s, rule-based systems began exploring practical applications. In 1966, Joseph Weizenbaum created ELIZA, a pioneering chatbot that simulated psychotherapy sessions through keyword matching and predefined response patterns. Despite its simplicity, ELIZA revealed the potential of rule-based methods in surface-level interaction [4]. The 1970s are often considered the golden era of rule-based NLP. In 1970, Terry Winograd introduced SHRDLU, a system that integrated syntactic analysis with logical reasoning to allow users to manipulate objects in a constrained "blocks world" using natural language commands [5]. In 1972, the LUNAR system extended the applicability of such approaches by enabling scientists to query lunar geological data from the Apollo missions in English—representing one of the earliest question-answering systems [6].

In the 1980s, the focus of rule-based systems shifted toward knowledge representation and semantic networks. Launched in 1984, the CYC project aimed to construct an extensive ontology of commonsense knowledge through manual encoding, but its progress was slow due to the overwhelming complexity [7]. In contrast, WordNet, introduced in 1985, emerged as a lexical semantic network that systematically organized word relationships (such as Figure 2.1) and became a foundational resource for many subsequent NLP tasks [8].

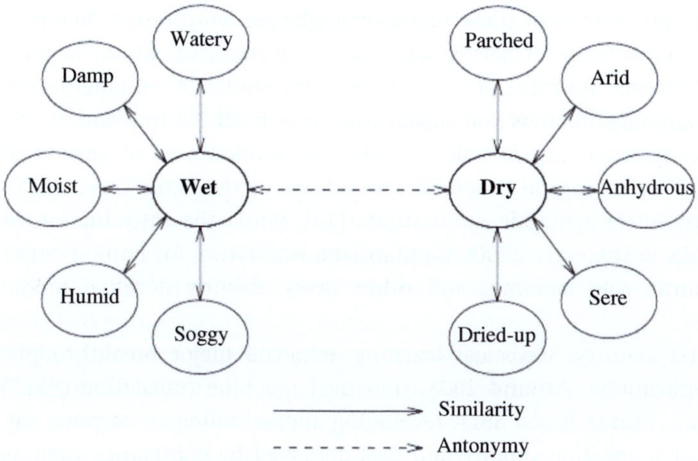

Figure 2.1 Bipolar adjective structure. *Source*: [9] / Oxford University Press.

Despite these advancements, the limitations of rule-based methods became increasingly apparent: static rules were inadequate for handling linguistic ambiguity, the systems were costly to maintain, difficult to scale, and heavily dependent on human-curated knowledge, making them ill-suited for the complexities of real-world language. By the late 1980s, with increased computational power and the growing availability of corpora, statistical machine learning approaches—such as hidden Markov models (HMMs)—began to rise, gradually shifting NLP toward a data-driven paradigm. Nevertheless, the legacy of symbolic methods continues to influence modern NLP technologies. From syntactic parsers to the construction of knowledge graphs, the formal logic and structured reasoning developed during the rule-based era still provide essential theoretical underpinnings and problem-framing methodologies for the field today.

2.2.2 Machine Learning Models

The statistical learning era of NLP began in the 1990s and lasted until around 2012. From the late 1980s to the early 1990s, the accumulation of digital text data and advances in computational power catalyzed a shift in NLP from rule-based systems to data-driven statistical models. This transformation was grounded in information theory and probabilistic modeling. HMMs, introduced in the late 1980s, were employed for tasks such as part-of-speech tagging and speech recognition, modeling sequential data through state transitions and observation probabilities [10]. Concurrently, Bayesian networks and Maximum Entropy (MaxEnt) models became central to text classification and named entity recognition in the early 1990s, emphasizing automatic feature weighting learned directly from data [11].

By the mid-1990s, significant progress was made in text representation techniques. The bag of words (BoW) model and term frequency-inverse document frequency (TF-IDF) weighting were widely adopted to convert textual data into numerical vectors [12]. Although these methods neglected word order, they provided foundational tools for information retrieval and text classification. N-gram models, such as bigrams, introduced local word-order statistics, enhancing the expressiveness of language models and supporting early machine translation and speech recognition systems [13]. Simultaneously, the development of annotated corpora accelerated technological progress. The release of the Penn Treebank in 1992 provided large-scale syntactic annotations [14], while the introduction of CoNLL shared tasks in the early 2000s standardized evaluation for named entity recognition, semantic role labeling, and other tasks, driving iterative model improvements [15].

Entering the 21st century, statistical learning achieved major breakthroughs in commercial applications. Around 2003, statistical machine translation (SMT) became mainstream. Phrase-based SMT, leveraging aligned bilingual corpora, significantly improved translation quality and was deployed by companies such as Google for multilingual services [16]. In the field of search engines, the combination of the PageRank algorithm (1998) with TF-IDF enhanced search result ranking through link analysis and keyword statistics (such as Figure 2.2), laying the groundwork for modern web search technologies [17]. Additionally, support vector machines (SVMs) and MaxEnt models were widely applied in spam detection and sentiment analysis, promoting the practical adoption of text classification [18].

However, the limitations of statistical methods gradually became evident. These models heavily relied on manually engineered features—such as part-of-speech tags and syntactic rules—and struggled to capture deep semantic structures. N-gram models were restricted to local context and failed to model long-range dependencies. In the early 2000s, as computational resources grew and neural

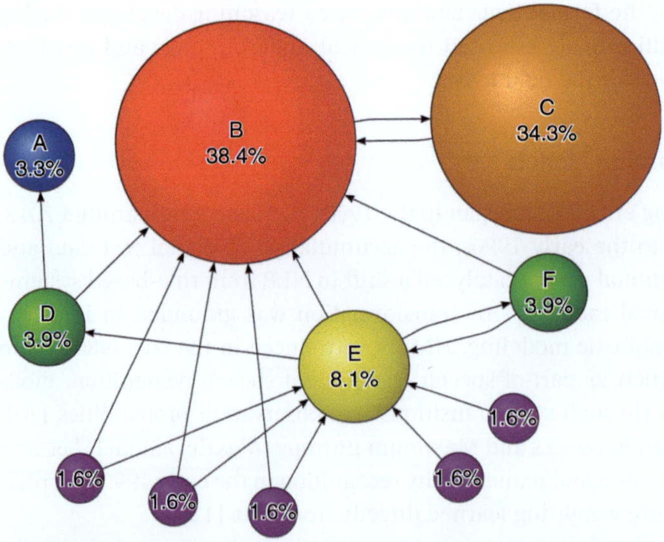

Figure 2.2 A simple illustration of the PageRank algorithm. The percentage shows the perceived importance, and the arrows represent hyperlinks. *Source*: [19] / 345Kai / Wikimedia Commons / Public domain.

networks experienced a resurgence, statistical learning began to transition toward deep learning. The introduction of neural probabilistic language models in 2003, and the breakthrough of Word2Vec in 2013, enabled distributed semantic representations of words, laying the groundwork for the rise of recurrent neural networks (RNNs) and Transformer-based architectures. By around 2012, the statistical learning era had formally given way to a new paradigm centered on end-to-end deep learning.

The statistical learning era marked a shift from rule-based to data-driven NLP. Through probabilistic modeling and large-scale corpora, it not only enabled the commercialization of technologies such as machine translation and search engines, but also established foundational principles in feature learning and distributed representation—providing the algorithmic and engineering groundwork for the deep learning revolution that followed.

2.2.3 Deep Learning Models

The deep learning takeoff in NLP began in the early 21st century and has since fundamentally transformed the field by significantly advancing capabilities in semantic understanding and text generation. In 2001, Yoshua Bengio et al. introduced the neural probabilistic language model, which used a feedforward neural network to predict word sequences while also proposing the concept of word embeddings [13]. This laid the foundation for distributed semantic representations. In 2013, Mikolov et al. introduced Word2Vec, a breakthrough technique for efficiently training word vectors using the Skip-Gram and Continuous Bag of Words (CBOW) architectures [16]. These vectors encoded semantic similarity in geometric space and became a core technique for text representation. That same year, RNNs and long short-term memory (LSTM) networks gained popularity for sequential modeling tasks (such as Figure 2.3), addressing the challenge of long-term dependencies in machine translation and text generation [21, 22].

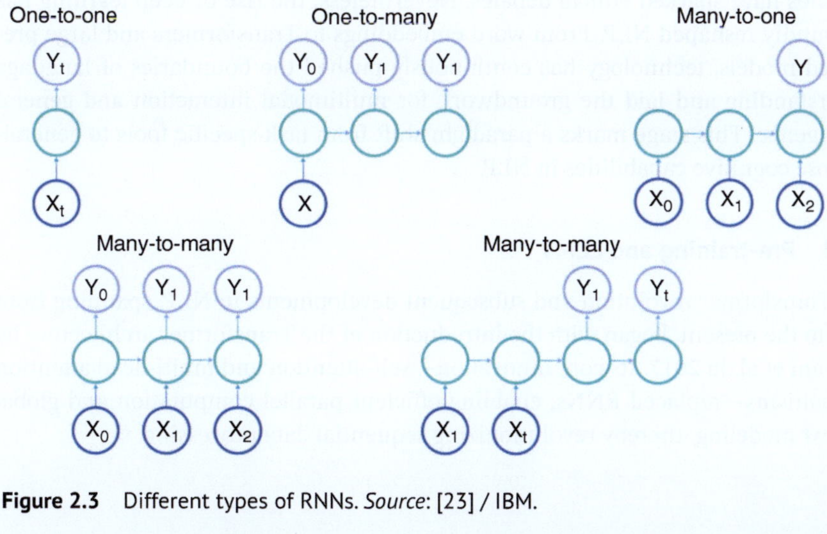

Figure 2.3 Different types of RNNs. *Source*: [23] / IBM.

In 2014, Google proposed the sequence-to-sequence (Seq2Seq) model, which combined an encoder-decoder framework with LSTMs to enable end-to-end machine translation [24]. This architecture soon became a general framework for generative tasks, such as dialogue systems. In 2015, the introduction of the attention mechanism marked a major breakthrough by dynamically weighting input tokens, thus improving the handling of long-range dependencies and significantly enhancing translation quality [25]. In 2017, Google released the Transformer architecture, replacing recurrent structures with multi-head self-attention mechanisms. This allowed for efficient parallel computation while capturing global context, establishing a new foundational model architecture [26]. Since 2017, deep learning in NLP has advanced rapidly with the Transformer enabling efficient parallel training and global context capture, replacing sequential RNN computations. In 2018, pre-training + fine-tuning became the mainstream paradigm, with large-scale unsupervised pre-training providing strong semantic representations for downstream tasks, demonstrating deep learning's scalability [27]. Around 2019–2020, scaling laws were discovered, showing that larger models with more data and compute consistently improved performance, further pushing deep learning boundaries in NLP [28]. Techniques like attention optimization, mixed-precision training, and distributed parallelism were developed to train models with hundreds of billions of parameters efficiently [29–31]. Post-2021, multimodal deep learning integrating text, vision, and speech emerged, reflecting deep learning's generalization capability across modalities [32, 33]. Recently, instruction tuning and reinforcement learning from human feedback (RLHF) have enhanced alignment with human intentions [34]. Overall, deep learning has evolved from small-scale sequential training to large-scale, multimodal, instruction-following systems, fundamentally transforming NLP's capabilities.

This period has also brought challenges and critical reflections. The reliance of large models on vast computational resources and data has raised concerns about cost and energy consumption. Long-context reasoning and commonsense integration remain imperfect, while issues such as data bias, content misuse, and privacy breaches have sparked ethical debates. Nevertheless, the rise of deep learning has profoundly reshaped NLP. From word embeddings to Transformers and large pre-trained models, technology has continuously pushed the boundaries of language understanding and laid the groundwork for multimodal interaction and general intelligence. This stage marks a paradigm shift from task-specific tools to general-purpose cognitive capabilities in NLP.

2.2.4 Pre-training and LLMs

The Transformer revolution and subsequent developments in NLP, spanning from 2017 to the present, began with the introduction of the Transformer architecture by Vaswani et al. in 2017. Its core innovation—self-attention and multi-head attention mechanisms—replaced RNNs, enabling efficient parallel computation and global context modeling, thereby revolutionizing sequential data processing.

In 2018, the era of pretrained language models (PLMs) began in earnest. Google introduced BERT, which leveraged masked language modeling (MLM) and bidirectional context encoding to set new performance benchmarks in tasks such as question answering and text classification [27]. In the same year, OpenAI released GPT-1, initiating the autoregressive generative model paradigm [28].

In 2019, GPT-2 demonstrated zero-shot generation capabilities through large-scale training and multitask learning (MTL), showcasing the potential of generative language models for a wide range of NLP tasks [29].

By 2020, GPT-3, with 175 billion parameters, enabled few-shot learning across diverse tasks, including code generation, poetry composition, and conversational simulation, thereby heralding the era of LLMs while also revealing limitations such as the tendency to generate "hallucinated" content [30]. Concurrently, Google's T5 and Microsoft's Turing-NLG, each with hundreds of billions of parameters, pushed semantic understanding and generation to near-human levels [31].

In 2021, advances extended into multimodal domains with OpenAI's release of DALL·E (for text-to-image generation) and CLIP (for joint vision-language representation), expanding LLM applications beyond text [32, 33].

In 2022, OpenAI released ChatGPT, a landmark conversational AI system enhanced through RLHF, significantly improving dialogue safety and coherence. In 2023, OpenAI launched GPT-4, which further integrated multimodal capabilities by enabling joint processing of text and images, thus extending NLP's reach into cross-domain applications such as healthcare and education [34, 35]. That same year, the open-source ecosystem began to flourish with Meta's release of LLaMA, lowering the technical entry barrier for LLM development, while the continued application of multimodal models such as DALL·E and CLIP facilitated exploration of joint vision-language representations [13].

By 2024, models such as GPT-4V and GPT-4o expanded to include audio interaction, enabling more natural and fluid cross-modal tasks and further advancing multimodal capabilities within the NLP domain.

Since 2024, research focus has shifted toward improving reasoning capabilities while reducing computational costs. OpenAI's o1 model introduced the chain-of-thought (CoT) reasoning technique to emulate complex human reasoning processes, enhancing logical inference capabilities in LLMs [36]. In 2025, China's DeepSeek-R1 and DeepSeek-V3 adopted a Mixture-of-Experts (MoE) architecture to achieve high-quality logical reasoning and long-context understanding at a lower cost, as shown in Figure 2.4, while embracing open-source principles to promote equitable access to AI technologies [37].

Despite these rapid advancements, persistent challenges remain, including hallucinations, dependency on massive computational resources, and ethical concerns such as data bias and privacy risks. Looking ahead, the field is steadily advancing toward AGI, focusing on the integration of multimodal perception, deep reasoning, and ethical governance. This trajectory aims to transform NLP from a language understanding tool into a broader framework for cognitive intelligence and socially embedded AI applications.

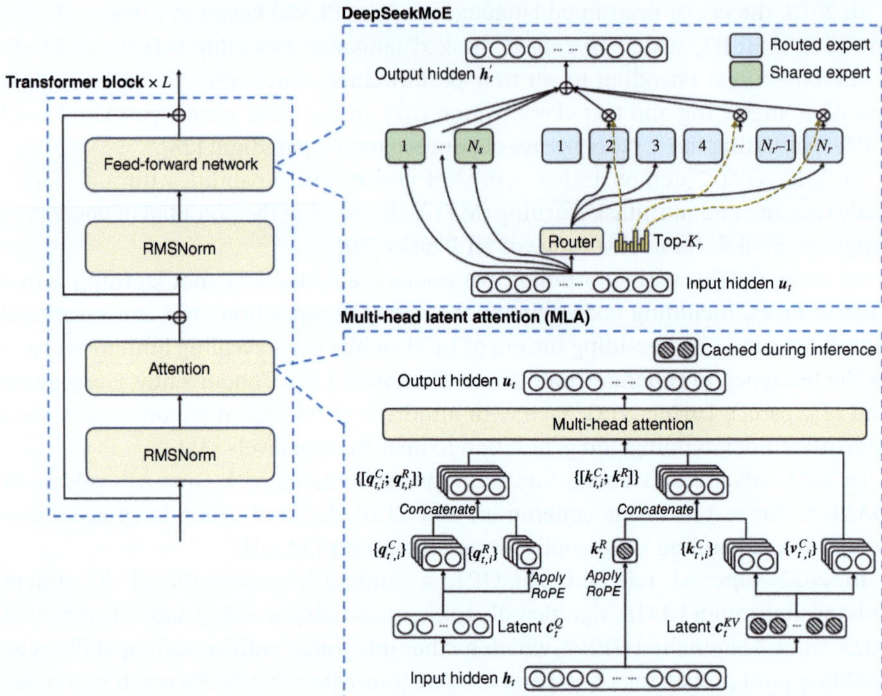

Figure 2.4 Illustration of the basic architecture of DeepSeek-V3. *Source*: [35] / arXiV.

2.3 Core Architectures in Modern NLP

2.3.1 Language Models

Language models lie at the heart of NLP, aiming to estimate the probability distribution over sequences of words. Traditionally, statistical language models such as n-grams approximated this distribution based on fixed-length context windows, but suffered from data sparsity and limited contextual scope [13]. The emergence of neural language models marked a paradigm shift, enabling the modeling of long-range dependencies and richer semantic representations. Early neural approaches, including feedforward and recurrent models, improved fluency and generalization but remained constrained by sequential computation and memory limitations.

The introduction of PLMs—particularly those built upon the Transformer architecture—revolutionized the field. These models are trained on large corpora to capture broad linguistic patterns and are then fine-tuned for downstream tasks. One key innovation is MLM, as used in BERT, where a portion of input tokens is randomly masked and predicted using bidirectional context. Another is causal (autoregressive) language modeling, as used in GPT, where models learn to predict the next token given a unidirectional context, facilitating coherent text generation. The evolution of language models is shown in Figure 2.5.

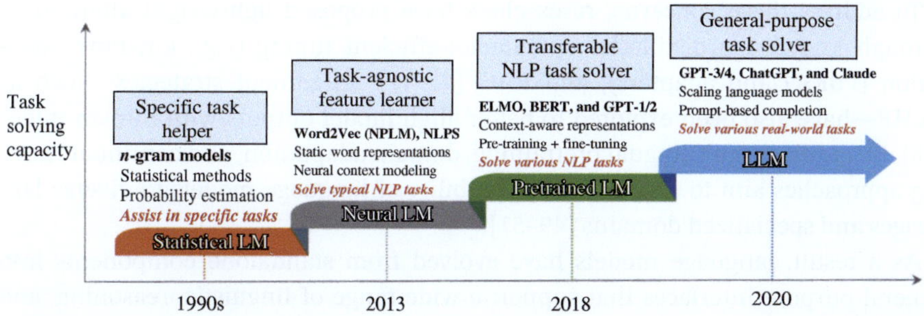

Figure 2.5 An evolution process of the four generations of language models from the perspective of task-solving capacity. *Source*: [41] / arXiV.

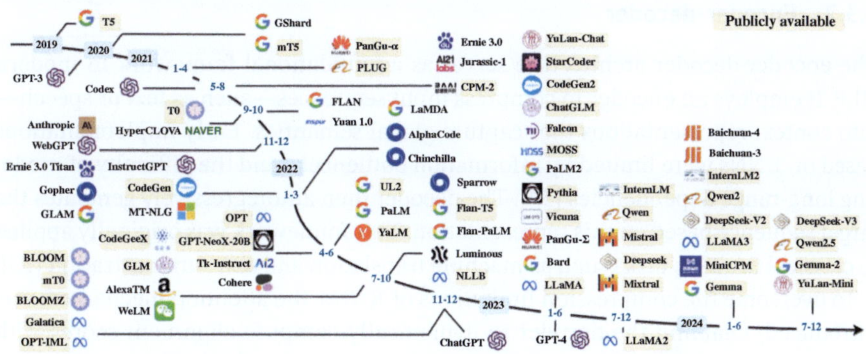

Figure 2.6 A timeline of existing LLMs (having a size larger than 10 billion) in recent years. The timeline was established mainly according to the release date (e.g., the submission date to arXiv) of the technical paper for a model. If there was no corresponding paper, the date of a model as the earliest time of its public release or announcement is shown. The LLMs in yellow color are with publicly available model checkpoint. Due to the space limit of the figure, only the LLMs with publicly reported evaluation results are included. *Source*: [41] / Cornell University / CC0 1.0.

Language models can be broadly categorized into three types (as shown in Figure 2.6): encoder-based models (e.g., BERT and RoBERTa) optimized for understanding tasks such as classification and question answering; decoder-based models (e.g., GPT-2 and GPT-3) suited for generation tasks; and encoder-decoder hybrids (e.g., T5 and BART) that unify encoding and generation in a single framework [36, 37]. These models have achieved state-of-the-art results across a wide range of NLP benchmarks, including the General Language Understanding Evaluation (GLUE) benchmark, the Stanford Question Answering Dataset (SQuAD), and SuperGLUE [38–40].

The scaling of model parameters and training data has become a central trend, with LLMs demonstrating emergent abilities such as in-context learning, few-shot generalization, and zero-shot reasoning [42–44]. However, this rapid growth introduces new challenges: increased computational cost, energy consumption, and environmental impact, as well as risks related to factual inaccuracy, bias propagation, and privacy leakage.

To address these concerns, researchers have proposed lightweight alternatives through knowledge distillation, parameter-efficient tuning (e.g., low-rank adaptation [LoRA] and adapters), and RAG [45–47]. Alignment strategies—such as RLHF—have also been explored to better align model outputs with human intent [48]. In addition, multilingual pretraining, domain adaptation, and continual learning approaches aim to extend the applicability of language models to diverse languages and specialized domains [49–51].

As a result, language models have evolved from standalone components into general-purpose interfaces that support a wide range of linguistic, reasoning, and multimodal tasks. Their centrality in NLP continues to grow, driving progress toward more versatile, scalable, and responsible AI systems.

2.3.2 Encoder-decoder

The encoder-decoder architecture serves as a foundational framework in modern NLP. It employs an encoder to compress input sequences—such as text or speech—into context representations that capture global semantics. Early implementations based on RNNs were limited by information bottlenecks and the difficulty of modeling long-range dependencies [52]. The decoder then autoregressively generates the target sequence based on this representation. This framework was originally applied to classical Seq2Seq tasks such as machine translation and text summarization [53].

To overcome the compression limitations of RNNs, the attention mechanism was introduced, enabling the decoder to dynamically compute alignment scores with different parts of the input sequence during generation [25]. This significantly improved the ability to model long sequences. The Transformer architecture, proposed in 2017, entirely replaced RNNs with a self-attention mechanism and multi-head attention to capture long-range dependencies in parallel. It incorporates positional encoding to retain sequence order and leverages residual connections and layer normalization to accelerate training. The Transformer has since become a new paradigm for efficient context modeling.

This architecture has led to the emergence of three major variants (as shown in Figure 2.7): (1) encoder-decoder models such as BART, which improves generation tasks through bidirectional encoding and denoising pretraining, and T5, which unifies multiple NLP tasks under a "text-to-text" framework; (2) encoder-only models like BERT, which pretrain semantic representations through MLM; and (3) decoder-only models such as the GPT series, which perform zero-shot learning via autoregressive generation.

Further developments have expanded the capabilities of these models. Non-autoregressive decoding (NAT) accelerates inference by generating outputs in parallel [54]. Sparse attention mechanisms reduce computational overhead, making it feasible to process long-form texts [55]. Multimodal integration—such as Vision Transformers (ViT) paired with decoders for image captioning—extends the framework beyond text-only data [56].

Nonetheless, challenges remain, including exposure bias between training and inference phases, as well as issues such as repetition and logical inconsistencies in long-text generation. To address these, researchers have proposed methods

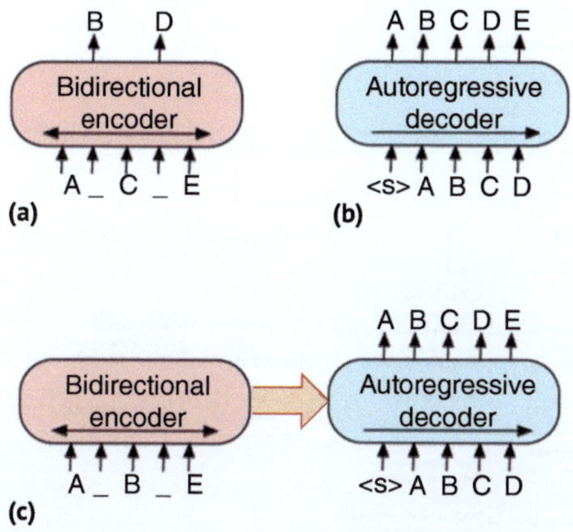

Figure 2.7 A schematic comparison of BART with BERT and GPT. (a) BERT: random tokens are replaced with masks, and the document is encoded bidirectionally. Missing tokens are predicted independently, so BERT cannot easily be used for generation. (b) GPT: tokens are predicted auto-regressively, meaning GPT can be used for generation. However, words can only condition on leftward context, so it cannot learn bidirectional interactions. (c) BART: inputs to the encoder need not be aligned with decoder outputs, allowing arbitrary noise transformations. Here, a document has been corrupted by replacing spans of text with mask symbols. The corrupted document (left) is encoded with a bidirectional model, and then the likelihood of the original document (right) is calculated with an autoregressive decoder. For fine-tuning, an uncorrupted document is input to both the encoder and decoder, and we use representations from the final hidden state of the decoder. *Source*: [37] / arXiv.

like scheduled sampling, coverage mechanisms, and nucleus sampling [57–59]. Additional efforts include model distillation and MoE systems to improve computational efficiency [60–62]. Advances in controllable generation (e.g., prompt engineering), multilingual and cross-modal alignment, and ethical safeguards—such as reducing bias and harmful outputs—continue to evolve, driving NLP toward more general, reliable, and efficient AI systems.

The attention mechanism, a foundational breakthrough in modern NLP, originated with additive attention [25]. This approach introduced learnable nonlinear transformations, such as $\text{Score}(Q, K_i) = v^\top \tanh\left(W_q Q + W_k K_i + b\right)$, to dynamically align encoder and decoder hidden states, thereby mitigating the long-term dependency issues in RNNs—albeit at a relatively high computational cost. Subsequently, Luong et al. [63] proposed multiplicative attention, which simplified the computation to dot products $\left(Q^\top K_i\right)$ or scaled dot products $\left(\frac{Q^\top K_i}{\sqrt{d_k}}\right)$, improving efficiency by removing the nonlinear layer and reducing the risk of gradient explosion [63]. Self-attention further advanced the paradigm by mapping the input sequence into three sets of vectors—queries, keys, and values—and computing dynamic attention weights via $\text{Attention}(Q,\ K,\ V) = \text{softmax}\left(\frac{QK^\top}{\sqrt{d_k}}\right)V$. This enabled the direct modeling of global dependencies without the sequential constraints of RNNs, allowing for parallel computation. Figure 2.8 shows the taxonomy of attention mechanisms.

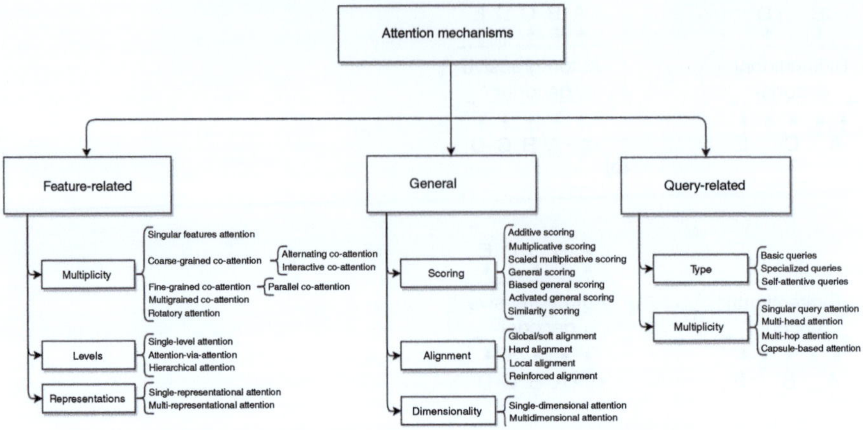

Figure 2.8 A taxonomy of attention mechanisms. *Source*: [64] / IEEE.

Building on this, multi-head attention splits Q, K, and V into multiple heads, each undergoing independent linear projections $\left(QW_i^Q, KW_i^K, VW_i^V\right)$ to capture diverse features across different subspaces, such as local syntactic patterns or long-range coreference. The outputs of all heads are concatenated and linearly transformed via $(\text{Concat}(\text{head}_1, \ldots, \text{head}_h)\, W^O)$, forming the core component of the Transformer architecture that underpins models like BERT (encoder-only), GPT (decoder-only), and T5 (encoder-decoder) [26].

Subsequent variants have extended the boundaries of attention mechanisms. Sparse attention reduces computational complexity from $O(n^2)$ to $O(n)$ using local windows (e.g., Longformer), hashing (e.g., Reformer), or global tokens (e.g., BigBird), enabling document-level processing [29, 65, 66]. Masked attention is used in decoders to enforce causality during autoregressive generation. Cross-attention facilitates multimodal integration, such as in vision-language models like ViT with decoders.

Despite effectively addressing the information bottlenecks of traditional models, attention mechanisms face several challenges. These include the quadratic growth in computation with sequence length (e.g., millions of operations for 1,000-token inputs), questionable interpretability of attention weights regarding true semantic relevance, and issues such as repetition or incoherence in long-form generation.

To address these issues, researchers have proposed efficient attention mechanisms—such as the kernelized approximation in Performer and low-rank projections in Linformer—along with dynamic sparsification based on input relevance and cross-modal extensions like CLIP's vision-text alignment [67–69]. Reinforcement learning (RL) and nucleus sampling have also been adopted to improve generation quality [70, 71].

Looking ahead, attention mechanisms are expected to evolve toward lower computational costs, greater interpretability, and enhanced cross-modal generalization, becoming a central driver of more efficient, controllable, and versatile language models.

2.3.3 Transformers

The Transformer architecture represents a pivotal advancement in NLP, replacing traditional recurrent structures with a fully attention-based framework. Unlike RNN-based encoders and decoders that process tokens sequentially and struggle with long-range dependencies, the Transformer employs self-attention mechanisms to model token interactions in parallel, allowing for efficient and scalable context learning. Each token attends to all others in a sequence, capturing both local and global dependencies through multi-head attention, which enables the model to focus on different subspaces of the input simultaneously.

Introduced in the seminal 2017 paper "*Attention Is All You Need*," the Transformer architecture discards recurrence and convolutions, relying instead on stacked layers of self-attention and feedforward networks. Positional encodings are added to input embeddings to preserve sequence order, compensating for the model's lack of inherent sequential structure. Layer normalization and residual connections facilitate deeper model training by stabilizing gradients and promoting convergence. This design dramatically improves training efficiency and performance on large-scale data.

The Transformer has since become the foundation of most modern NLP systems, enabling the emergence of versatile architectures tailored to specific tasks. Encoder-only models like BERT excel in understanding tasks through MLM, while decoder-only models such as GPT leverage autoregressive objectives for fluent text generation. Hybrid encoder-decoder models like T5 and BART combine bidirectional context understanding with powerful generative capabilities, offering a unified framework for diverse NLP applications.

Extensions of the Transformer include innovations such as sparse attention and efficient Transformers (e.g., Longformer and Performer) to reduce the quadratic complexity of standard attention and scale to longer inputs (such as Figure 2.9).

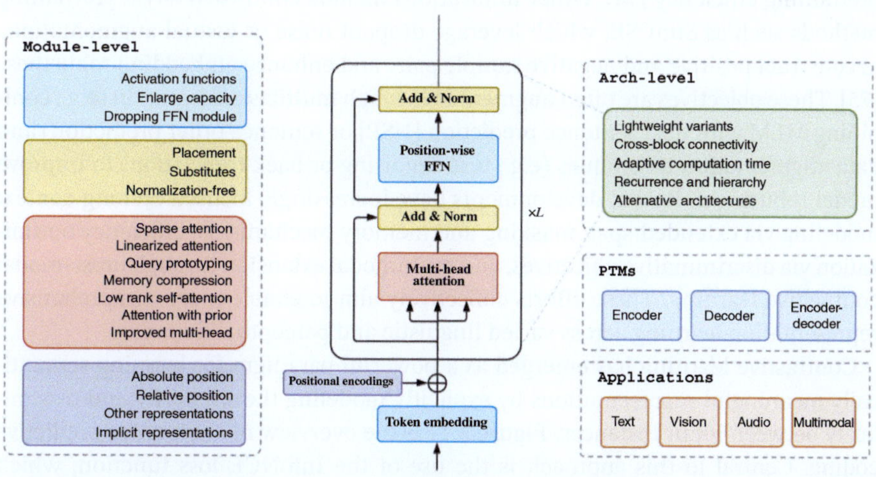

Figure 2.9 Categorization of Transformer variants. *Source*: [72] / arXiv.

Variants like ViT have also demonstrated the architecture's adaptability beyond text, enabling cross-modal tasks through integration with image, audio, or tabular data. Non-autoregressive models speed up decoding, while MoE techniques reduce training costs by dynamically routing computations.

Despite its success, the Transformer architecture still faces limitations. Exposure bias, hallucinations, and resource-intensive training remain active research areas. Solutions include advanced decoding strategies like top-k and nucleus sampling, improved pretraining objectives, and RLHF. Ongoing research into multilingual modeling, interpretability, and responsible AI practices continues to shape the future trajectory of Transformer-based NLP systems, pushing toward more general, ethical, and robust language understanding.

2.4 Training and Optimization Methods for LLMs

2.4.1 Pretraining Strategy

Self-supervised objectives represent a foundational component of LLM pretraining. By automatically constructing supervisory signals from unlabeled corpora, these objectives enable the acquisition of general-purpose semantic representations, as shown in Figure 2.10. Among the most influential strategies is autoregressive modeling, which predicts the next token in a sequence based solely on prior context, as exemplified by the GPT family [73]. This approach prioritizes generative fluency but is inherently constrained by its unidirectional structure. In contrast, MLM, utilized in models such as BERT, masks a random subset of tokens and trains the model to recover them using bidirectional context, thus facilitating richer semantic understanding. An alternative formulation is replaced token detection, pioneered by ELECTRA, which employs a generator-discriminator framework wherein the model learns to distinguish real from substituted tokens, thereby achieving higher pretraining efficiency [74]. Other innovations include contrastive-style pretraining methods such as SimCSE, which leverage dropout noise or textual augmentations to construct positive and negative sample pairs and enhance embedding robustness [75]. These objectives are often augmented through multitask integration (e.g., combining MLM with next sentence prediction [NSP] or sentence order prediction) and data augmentation techniques (e.g., text denoising or back-translation) to improve model robustness. Recent developments have increasingly focused on long-context modeling via extended-span masking and memory mechanisms, efficiency optimization via discriminative objectives, and multimodal extension through cross-modal contrastive learning. These efforts collectively aim to enable more comprehensive representation learning across varied linguistic and perceptual domains.

Contrastive learning has emerged as a powerful paradigm for learning semantically meaningful representations by explicitly modeling the similarity and dissimilarity between input instances. Figure 2.11 is the overview of contrastive predictive coding. Central to this approach is the use of the InfoNCE loss function, which encourages the model to pull semantically similar (positive) pairs closer in the embedding space while pushing dissimilar (negative) pairs apart [77]. Positive

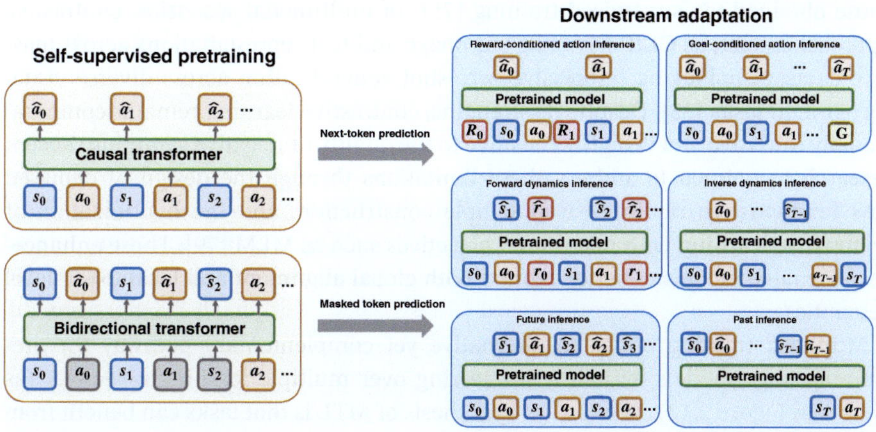

Figure 2.10 Pretrain-then-Adapt pipeline for decision foundation model. Left: self-supervised pretraining involves two basic pretraining objectives based on the Transformer architecture: next token prediction and masked token prediction. However, using different RL components. Right: downstream inference tasks can be roughly divided into three categories: action inference, dynamics inference, and trajectory inference. *Source*: [76] / arXiv.

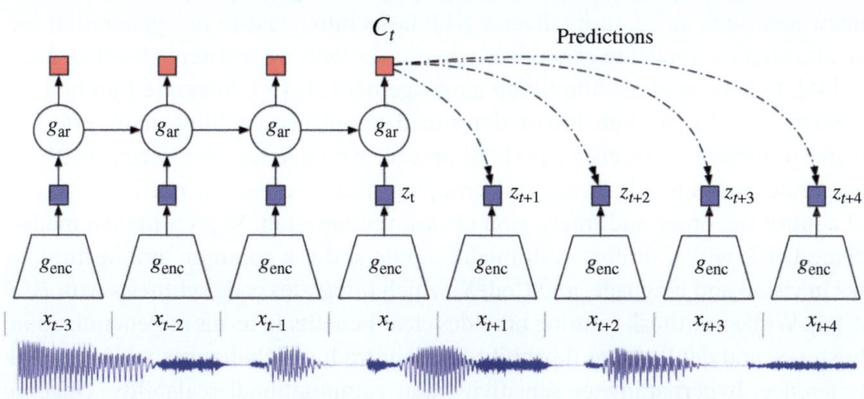

Figure 2.11 Overview of contrastive predictive coding. Although this figure shows audio as input, the same setup for images, text, and RL is used. *Source*: [77] / arXiv.

samples are often generated through in-batch augmentations such as dropout, paraphrasing, or back-translation, whereas negative samples are commonly drawn from other instances within the batch or curated via adversarial sampling strategies to increase training difficulty. Architectural choices such as dual-encoder frameworks or shared-parameter encoders facilitate efficient representation extraction, while temperature scaling is employed to modulate the sharpness of the similarity distribution. Furthermore, techniques like MoCo incorporate momentum encoders and dynamic memory banks to store large pools of negative samples across training steps, thereby stabilizing training and enhancing contrastive discrimination [78]. In the domain of NLP, SimCSE has demonstrated the effectiveness of unsupervised contrastive learning in generating sentence embeddings that outperform those obtained via supervised training [75]. In multimodal scenarios, contrastive approaches such as CLIP have aligned image and text representations across massive datasets, achieving impressive zero-shot generalization across diverse visual recognition tasks [33]. Despite its strengths, contrastive learning remains computationally intensive and is highly sensitive to the quality of negative sampling. Recent research has aimed to address these limitations through the design of debiased loss functions, curriculum-based sample construction, and the hybridization of contrastive learning with token-level objectives such as MLM [79]. These enhancements enable models to better capture both global alignment and local token-level semantics.

Multitask training offers an alternative yet complementary pathway for pretraining large models by jointly optimizing over multiple learning objectives (as shown in Figure 2.12). The central hypothesis of MTL is that tasks can benefit from shared inductive biases and representations, thereby improving generalization and sample efficiency. In practice, this is often implemented using a shared backbone architecture—frequently based on Transformer encoders—alongside task-specific output heads. Dynamic task weighting mechanisms, such as GradNorm or uncertainty-based reweighting, are employed to balance the learning rates across tasks and mitigate optimization conflicts [80]. Task design plays a pivotal role: some frameworks, such as T5, unify diverse NLP tasks into a text-to-text generation format, allowing the model to generalize across tasks with a consistent objective function [36]. Others, such as fine-tuned language net (FLAN), integrate hundreds of downstream tasks through instruction-tuned templates, enabling more effective transfer to unseen instructions [81]. To prevent negative transfer, recent methods incorporate curriculum learning, task grouping, and selective task routing strategies that control the order and interaction of task presentation. Representative models include UniT, which unifies multimodal inputs under a common architecture for tasks in vision and language, and CodeX, which integrates code synthesis with NLP [82, 83]. While multitask training provides clear benefits in terms of generalization, robustness, and deployment flexibility, it also introduces challenges related to task interference, hyperparameter sensitivity, and computational scalability. Ongoing

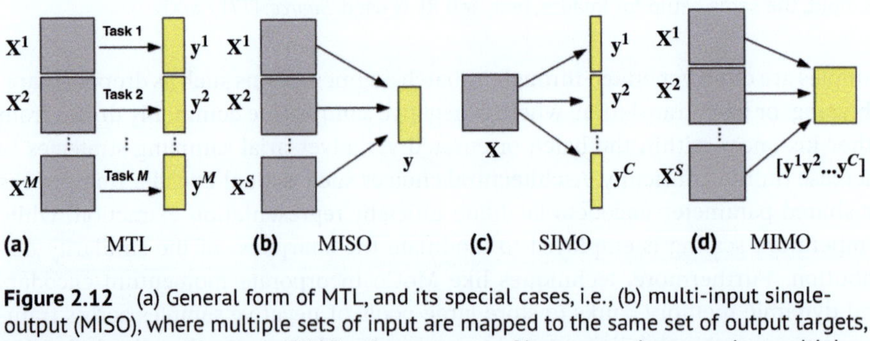

Figure 2.12 (a) General form of MTL, and its special cases, i.e., (b) multi-input single-output (MISO), where multiple sets of input are mapped to the same set of output targets, (c) single-input multi-output (SIMO), where one set of input maps is mapped to multiple sets of output targets, and (d) multi-input multi-output (MIMO), where multiple sets of input are mapped to the same multiple sets of output targets. (X: input data; y: target output). *Source*: [85] / Springer Nature.

research aims to decouple shared and task-specific representations, develop routing mechanisms that dynamically select relevant subtasks, and combine MTL with contrastive and instruction-based fine-tuning to enhance transfer learning capabilities [84]. These strategies collectively contribute to the advancement of more general and adaptable AI systems.

2.4.2 Fine-tuning Strategy

Prompt-based fine-tuning represents a paradigm shift in the way PLMs are adapted to downstream tasks. Unlike traditional fine-tuning, which updates the full set of model parameters, prompt-based approaches maintain the pretrained model's weights and instead reformulate downstream tasks into a format resembling the original pretraining objective—typically MLM or causal language modeling (CLM). This strategy seeks to exploit the knowledge already encoded in large-scale language models by crafting task-specific prompts that guide the model to generate the desired output.

Prompt engineering can be broadly categorized into manual prompting, automatic prompting, and prompt tuning. In manual prompting, domain experts design input templates based on intuition and task requirements, such as Figure 2.13. For instance, in a sentiment classification task, a sentence such as "The strength was high!" might be transformed into "The strength was high! It was [MASK].," where the model is expected to predict "great" or a similar positive token. While effective, manual prompts are often brittle and suboptimal.

To address this limitation, automatic prompt search methods such as AutoPrompt and PromptGen have been proposed [87, 88]. These approaches leverage algorithms (e.g., gradient-based search or RL) to identify optimal prompts without relying heavily on human intuition. However, these discrete prompt search techniques often suffer from instability and limited search space, especially when operating over token sequences.

A more scalable alternative is prompt tuning, a form of continuous prompt optimization. Introduced by Lester et al. [89], prompt tuning freezes the pretrained model and learns a small number of soft prompt embeddings, which are prepended to

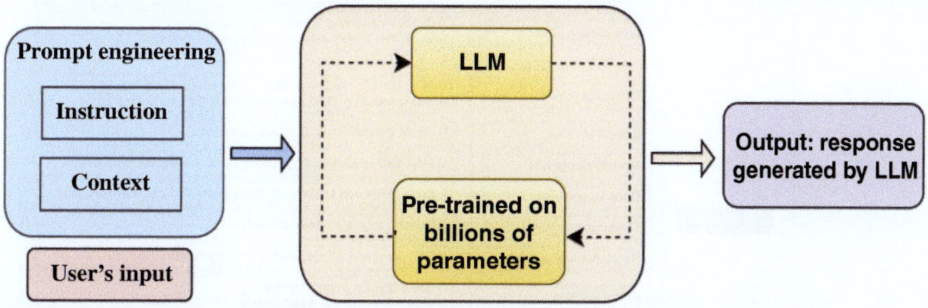

Figure 2.13 Visual breakdown of prompt engineering components: LLMs trained on extensive data, instruction, and context as pivotal elements shaping the prompt, and a user input interface. *Source*: [86] / arXiv / CC BY 4.0.

the input sequence and optimized via gradient descent [89]. These embeddings are task-specific, low-dimensional representations that effectively condition the model's outputs without modifying the original model weights. Despite their compactness, prompt tuning has demonstrated competitive performance across a range of NLP tasks, especially when applied to large-scale language models such as GPT-3 or T5.

Prompt-based fine-tuning offers several advantages. First, it minimizes computational and storage overhead, as only a small number of parameters (typically in the order of thousands) need to be trained. Second, it reduces the risk of catastrophic forgetting since the underlying model remains unchanged. Third, it enhances task generalization by maintaining alignment with the pretraining objective. However, its limitations include potential prompt instability, difficulty in prompt selection, and reduced effectiveness in low-resource or domain-specific scenarios where the pretrained knowledge may not sufficiently align with the downstream task.

Overall, prompt-based fine-tuning serves as a lightweight and efficient mechanism for task adaptation, making it particularly attractive in environments where computational resources are constrained or where multiple tasks must be supported concurrently with minimal model duplication.

LoRA is a parameter-efficient fine-tuning (PEFT) method designed to overcome the limitations of full-model fine-tuning in large-scale neural networks, as shown in Figure 2.14. Proposed by Hu et al. [46], LoRA injects a small number of trainable parameters into the model in the form of low-rank matrices, which are added to the weight update paths during training [46]. The core idea is that the weight updates

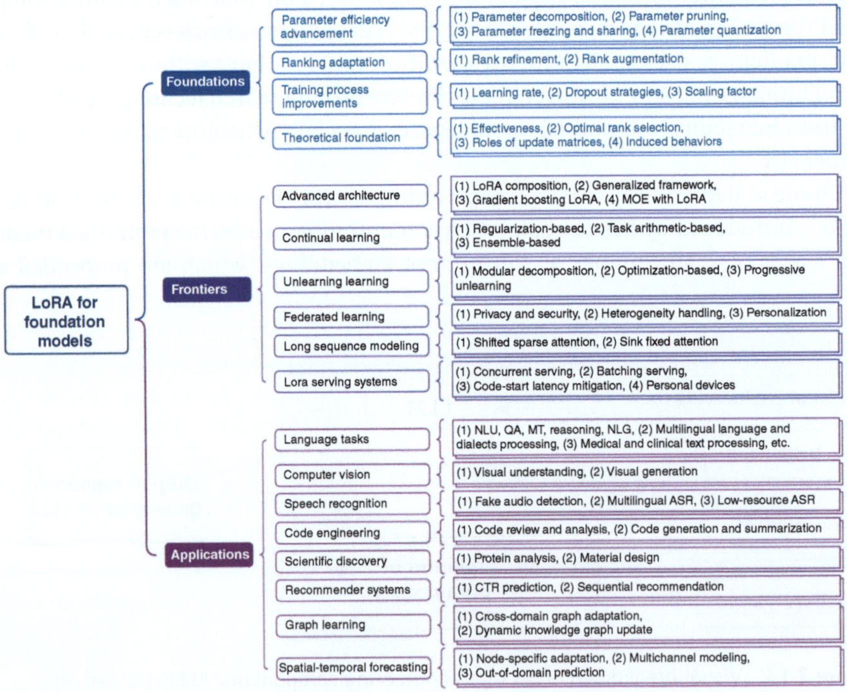

Figure 2.14 Structure of LoRA for foundation models. *Source*: [91] / arXiv.

required to adapt a pretrained model to a downstream task often lie in a low-rank subspace. By constraining the update to this subspace, LoRA enables efficient and effective fine-tuning while preserving the original model's weights.

LoRA has been particularly effective in Transformer-based architectures, where it is commonly applied to the query, key, and value projection matrices in self-attention layers. Because Transformers contain many such projection matrices, and each matrix is large, reducing the trainable parameter count by applying LoRA yields substantial efficiency gains.

One of the primary advantages of LoRA is modularity. Since LoRA modules are injected as external low-rank adapters, they can be selectively applied, removed, or combined without affecting the base model. This makes LoRA ideal for MTL or deployment scenarios where a single backbone model serves multiple downstream applications. For example, task-specific LoRA modules can be loaded on demand without duplicating the full model, saving both memory and inference time. Additionally, LoRA supports rapid experimentation and domain adaptation. Because LoRA modules are lightweight and quick to train, practitioners can efficiently explore new tasks, domains, or languages without retraining or overwriting the base model. This is especially valuable in low-resource or rapidly changing environments, such as dialogue systems, scientific research, or real-time analytics.

From a theoretical perspective, LoRA's low-rank constraint acts as a regularizer, potentially reducing overfitting and improving generalization. Empirical studies have shown that LoRA can match or outperform full fine-tuning in many tasks, including text classification, summarization, and machine translation, despite using significantly fewer trainable parameters. Moreover, LoRA exhibits excellent compatibility with other PEFT techniques [90]. It can be integrated with prompt tuning, adapter layers, or prefix tuning to further enhance flexibility and performance. The combination of LoRA and prompt tuning, for example, has been explored in multimodal and instruction-following models where both textual prompts and adapter weights are fine-tuned to accommodate complex tasks [91].

However, LoRA also presents certain limitations. The selection of the rank (r) and the placement of LoRA modules require empirical tuning. Too low a rank may underfit the task, while too high a rank negates parameter efficiency. Additionally, LoRA may be less effective for tasks that demand substantial structural reconfiguration of the model or in domains far removed from the original pretraining corpus. Despite these challenges, LoRA has gained widespread adoption due to its balance of efficiency, scalability, and performance. It has become a cornerstone of efficient fine-tuning strategies in LLMs, especially in settings where deployment cost and flexibility are key constraints.

2.4.3 RAG

RAG is a paradigm that enhances the capabilities of LLMs by incorporating a retrieval mechanism into the generation process, thereby allowing models to dynamically access and integrate external knowledge at inference time [92]. Unlike traditional LLMs, which rely solely on parametric memory encoded within their weights and are inherently limited by their pretraining corpus, RAG systems combine both parametric and nonparametric knowledge sources, significantly improving factual accuracy, domain specificity, and temporal relevance, as shown in Figure 2.15.

The architecture of a RAG model typically includes two core components: a retriever and a generator. The retriever—often based on dense representation learning such as Dense Passage Retrieval (DPR)—identifies a set of relevant documents or passages from an external corpus given a user query [93]. These documents are then passed to a generator, usually a Seq2Seq model like BART or T5, which conditions its output not only on the input query but also on the retrieved documents. This conditioning can be achieved through simple concatenation of the input with retrieved content or through more sophisticated cross-attention mechanisms. The core idea is that the model learns to generate responses by marginalizing over possible retrieved documents, estimating a weighted sum of outputs based on document relevance scores. This fusion of retrieval and generation brings several key benefits. First, it drastically reduces hallucination rates by grounding model outputs in external factual data. Second, it improves the model's ability to handle long-tail or out-of-distribution queries, particularly those involving rapidly changing information or niche domains. Third, it enables real-time updating of model knowledge without retraining, simply by updating the retrieval corpus.

Despite these advantages, RAG also introduces new challenges, such as the sensitivity of performance to retriever quality, latency due to retrieval operations, and difficulties in jointly optimizing retrieval and generation components. Recent advancements have explored end-to-end training of retriever-generator pipelines, the integration of rerankers, and the use of hybrid retrievers (combining sparse and dense signals) to further improve relevance and robustness [94]. RAG has been successfully applied across various NLP tasks, including open-domain question

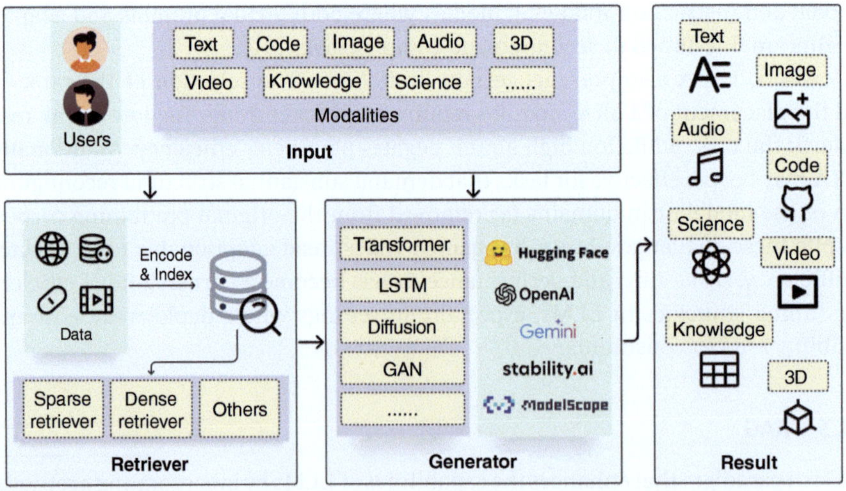

Figure 2.15 A generic RAG architecture. The user queries, spanning different modalities, serve as input to both the retriever and the generator. The retriever extracts relevant information from data sources. The generator interacts with the retrieval results and ultimately produces outcomes of various modalities. *Source*: [96] / Cornell University / CC0 1.0.

answering, customer support bots, biomedical literature synthesis, and legal document summarization, and it serves as a foundational strategy for knowledge-intensive AI applications that require both language fluency and up-to-date factual grounding [95].

2.4.4 Agent

The concept of an AI agent represents an evolution in the functionality of LLMs, moving beyond static completion or question-answering systems toward autonomous, goal-driven systems capable of planning, reasoning, tool use, and multistep decision-making. In this context, an agent is not merely a model executing isolated prompts, but an orchestrated system that perceives its environment, interprets objectives, selects and sequences actions, and iteratively interacts with external tools or data sources to accomplish complex tasks. Figure 2.16 shows the conceptual framework of an LLM-based agent.

The agentic paradigm relies on several key capabilities: (1) task decomposition, where high-level objectives (e.g., "generate a research summary from a scientific database") are broken down into manageable subtasks; (2) tool invocation, in which the agent interfaces with APIs, search engines, code interpreters, databases, or other external functions; and (3) stateful memory, allowing the agent to maintain context and learning across multi-turn or long-horizon workflows. Architecturally, language model agents are typically implemented by coupling an LLM with a controller or policy loop that monitors the system's state, determines the next action, and invokes necessary tools. Prominent frameworks like OpenAI's function calling, LangChain,

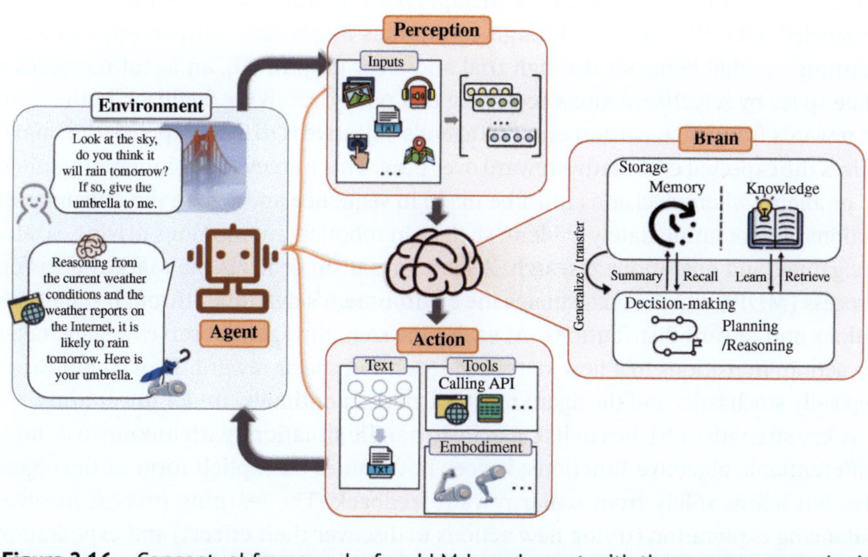

Figure 2.16 Conceptual framework of an LLM-based agent with three components: brain, perception, and action. Serving as the controller, the brain module undertakes basic tasks like memorizing, thinking, and decision-making. The perception module perceives and processes multimodal information from the external environment, and the action module carries out the execution using tools and influences the surroundings. *Source*: [97] / arXiv.

and AutoGPT exemplify this architecture by enabling LLMs to reason about when and how to call predefined functions or APIs, handle error recovery, and iterate over complex queries. For instance, in a data analysis agent, the model might first retrieve relevant datasets, invoke a code interpreter to preprocess the data, generate visualizations via a plotting library, and summarize findings in natural language—all without direct user supervision. The agentic loop is governed by structured reasoning mechanisms such as ReAct (reasoning and acting), CoT, or scratchpad memory.

One of the most transformative features of agents is emergent autonomy: they can operate semi-independently, adaptively respond to changes in input or environment, and optimize toward a specified objective function. In research and enterprise contexts, agents are being deployed for automated report generation, literature review, personalized tutoring, complex workflow automation, and scientific discovery. However, building robust agents remains a nontrivial challenge. It requires reliable grounding, controllable memory mechanisms, robust error handling, and transparent reasoning chains to avoid compounding errors across long trajectories. Additionally, safety, ethical considerations, and interpretability become more critical as the complexity and autonomy of agents increase. Nonetheless, agentic LLM systems represent a significant step toward general-purpose, interactive, and adaptive AI, extending the capabilities of foundation models beyond single-turn completions to full-spectrum cognitive automation.

2.4.5 Reinforcement Learning

RL is a foundational paradigm in machine learning that addresses the challenge of sequential decision-making in dynamic environments. Unlike supervised learning, which learns from labeled data, or unsupervised learning, which seeks patterns in unlabeled data, RL operates through continuous interaction with an environment, learning optimal behavior through trial and error [98]. In RL, an agent navigates a state space by selecting actions according to a policy, receiving feedback in the form of rewards from the environment. The agent's objective is to learn a policy that maximizes the expected cumulative reward over time. This framework is inherently suited to problems where decisions must be made in sequence and where the outcomes of actions are not immediately evident, such as in robotics, autonomous driving, strategic games, and operations research. A core formalism of RL is the Markov decision process (MDP), which encapsulates the environment's dynamics through state transitions and reward distributions. At each time step, the agent observes a state, takes an action, transitions to a new state, and receives a scalar reward. The dynamics are typically stochastic, and the agent must learn to act optimally under uncertainty.

A key strength of RL lies in its capacity to handle situations with unknown or non-differentiable objective functions; it does not require an explicit form of the objective but learns solely from scalar reward feedback. The learning process involves balancing exploration (trying new actions to discover their effects) and exploitation (choosing known optimal actions), a trade-off that is central to RL. Modern RL often employs function approximation via deep neural networks to represent policies and value functions, leading to the subfield of deep RL. For example, in a maze navigation task, the agent learns to move through a grid by receiving positive rewards for reaching the goal and negative rewards for hitting walls or taking unnecessary steps. The optimal policy minimizes the number of steps while avoiding obstacles, which the agent discovers through repeated interactions with the maze environment. This paradigm extends to more complex settings where observations are high-dimensional (such as images), requiring the use of encoders to extract latent state representations.

Additionally, world models may be incorporated to allow agents to simulate and plan within their learned representations of the environment [99]. Conceptually inspired by behavioral psychology and neuroscience, RL mirrors biological learning processes, making it a compelling tool for building autonomous systems. Its general formulation—an agent optimizing behavior based on reward signals through interaction with an environment—makes it broadly applicable and a cornerstone of intelligent decision-making systems. Figure 2.17 shows the framework of LLM-enhanced RL in classical agent-environment interactions.

2.4.6 Corpus Building

Corpus building is a foundational process in computational linguistics, serving as the backbone for empirical language research, model training, and performance evaluation across a variety of downstream applications. A corpus (plural: corpora) refers to a structured, often annotated collection of natural language data, typically consisting of large volumes of text or speech samples. These corpora are meticulously curated to represent specific linguistic, domain-specific, or task-relevant properties, allowing researchers and practitioners to extract statistical patterns, develop language models, and evaluate linguistic hypotheses in a controlled and replicable manner. The corpus building process is far from trivial; it encompasses the selection of source data, acquisition, preprocessing, normalization, annotation, and continuous refinement stages. Each stage demands careful planning and methodological rigor to ensure the quality, representativeness, and usability of the final corpus. Data selection is usually governed by the intended use case of the corpus—whether it aims to support general language modeling, sentiment analysis, information extraction, domain adaptation, or multilingual translation. Source materials may come from a diverse array of channels, including newspapers,

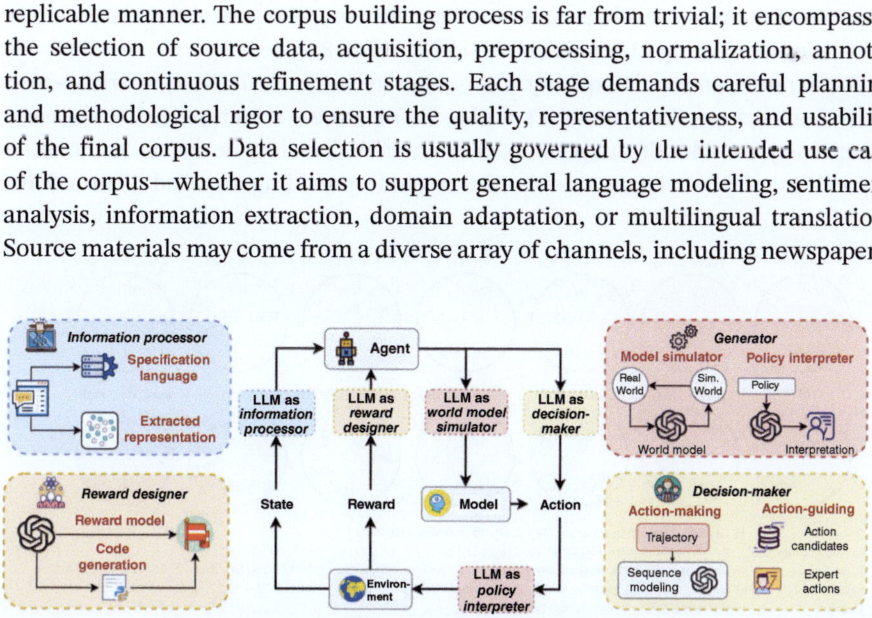

Figure 2.17 Framework of LLM-enhanced RL in classical agent-environment interactions, where LLM plays different roles in enhancing RL. *Source*: [100] / IEEE.

scientific literature, social media, spoken transcripts, government documents, or user-generated content. A critical consideration at this stage is coverage: a high-quality corpus must account for lexical variety, syntactic complexity, genre diversity, speaker or author demographics, and temporal variation to avoid skewed learning outcomes and biases in downstream systems.

Once sources are identified, data acquisition typically involves automated crawling, API-based collection, OCR (optical character recognition) for printed text, or transcription in the case of audio corpora. Legal and ethical concerns must be addressed early in this process—ensuring that licensing agreements, copyright laws, data privacy regulations (e.g., GDPR), and consent protocols are strictly adhered to. Following data collection, preprocessing routines are applied to clean and normalize the raw data. These routines may involve tokenization, sentence segmentation, lowercasing, spelling correction, punctuation removal, and the elimination of boilerplate content. In multilingual or cross-domain corpora, language identification, domain classification, and character encoding normalization are particularly crucial to ensure consistency and interoperability. Depending on the linguistic or computational goals, annotation is then added, transforming a plain text corpus into a richly labeled resource. Annotation types span a broad spectrum, from low-level linguistic layers—such as part-of-speech tags, syntactic parses, lemmatization, and named entity tags—to high-level discourse features like sentiment labels, coreference chains, topic boundaries, and pragmatic functions. Annotation can be manual, semi-automated, or fully automated, with varying implications for cost, scalability, and reliability. Manual annotation by trained linguists or crowdworkers offers high-quality labels but is resource-intensive and time-consuming. Conversely, automated annotation via pretrained models offers speed and scalability, though often at the expense of domain-specific accuracy and consistency. The pretraining data sources and ratios for existing LLMs are shown in Figure 2.18.

To guarantee annotation quality, inter-annotator agreement (IAA) metrics—such as Cohen's kappa, Krippendorff's alpha, or F1-score—are employed to measure consistency and identify ambiguous or contentious labeling decisions. Annotation guidelines play a pivotal role in this regard, providing detailed instructions,

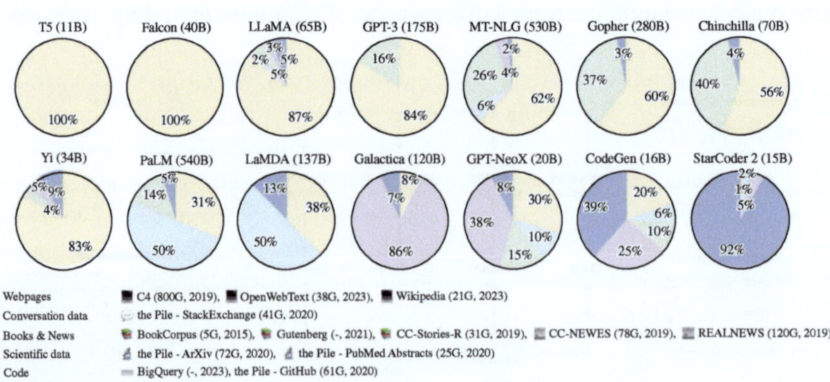

Figure 2.18 Ratios of various data sources in the pretraining data for existing LLMs. *Source*: [41] / arXiv.

examples, and decision rules to ensure replicability and minimize subjective variance. In recent years, there has been a growing emphasis on developing corpora that are not only large-scale but also balanced and fair. Balanced corpora ensure that underrepresented linguistic forms, demographic groups, and discourse functions are adequately covered, reducing systemic biases in language models. Fairness in corpus construction also involves careful auditing for gender, racial, political, and cultural biases, which—if unchecked—can propagate harmful stereotypes through downstream applications such as chatbots, summarization systems, or search engines. As a response, many corpus-building initiatives now include "bias tags," data statements, or documentation protocols (e.g., datasheets for datasets) to increase transparency and accountability in corpus design.

Moreover, with the rise of domain-specific NLP, domain adaptation has become a major concern in corpus building. Standard general-purpose corpora such as Wikipedia or newswire text may not reflect the terminological, stylistic, or pragmatic features of specialized domains like legal texts, biomedical literature, or social media discourse. Consequently, domain-specific corpora—such as PubMed abstracts for biomedicine, arXiv papers for scientific writing, or Reddit threads for informal online dialogue—are constructed to address these gaps. These corpora often require specialized preprocessing and annotation pipelines, for example, using UMLS or MeSH ontologies in medical texts or discourse act classification in online forums [101]. In multilingual settings, parallel and comparable corpora are key to supporting tasks like machine translation, cross-lingual transfer learning, and multilingual semantic analysis. Parallel corpora consist of sentence-aligned translations across languages (e.g., Europarl and OpenSubtitles), while comparable corpora contain topic-aligned but independently authored texts (e.g., news articles in different languages about the same event) [102]. Constructing such corpora introduces additional challenges such as alignment accuracy, translation consistency, and linguistic equivalence, often requiring bilingual annotators or alignment algorithms.

Technological advances have also reshaped corpus construction methodologies. With the advent of large-scale pretraining and self-supervised learning, massive corpora on the scale of billions of tokens (e.g., Colossal Clean Crawled Corpus [C4], Common Crawl, and The Pile) have become standard practice [103–105]. These corpora fuel Transformer-based architectures like BERT, GPT, and T5, pushing the boundaries of language understanding and generation. However, these large-scale corpora are often noisy and heterogeneous, requiring sophisticated filtering techniques such as toxicity detection, deduplication, quality scoring, and topic balancing. Furthermore, streaming-based or continual corpus construction methods are gaining attention, allowing models to stay updated with emerging vocabulary, topical trends, and linguistic shifts over time. This is particularly relevant for real-time applications like search engines or conversational agents. Evaluation of corpora is a critical yet often overlooked step. Corpora are evaluated along dimensions such as linguistic coverage, annotation accuracy, domain fidelity, usability, and scalability. Metadata documentation, sampling methodology, license type, and intended use cases must be clearly reported to facilitate reproducibility and ethical reuse. The community has also embraced open-source corpus sharing platforms like LDC,

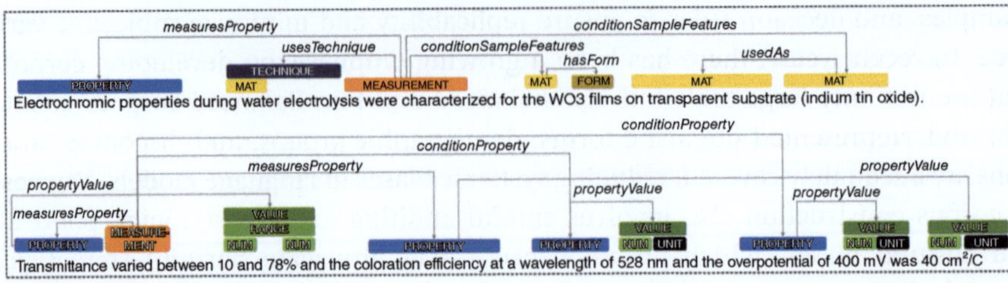

Figure 2.19 Multilayer materials science corpus: named entity, relation, and semantic role annotations. *Source*: [110] / arXiv.

ELRA, Hugging Face Datasets, and Kaggle to foster collaboration, benchmarking, and standardization [106–109].

The future of corpus building points toward interactive, adaptive, and ethically aware paradigms. Interactive corpus building involves integrating human feedback loops during annotation and selection to improve quality dynamically. Adaptive corpus construction allows for context-aware or task-specific data sampling, optimizing performance with minimal supervision. Ethical corpus construction mandates transparency, accountability, and inclusivity, with attention to marginalized dialects, languages, and communities historically underrepresented in NLP research. In sum, corpus building is not merely a preparatory step for language modeling; it is a scientifically grounded, ethically sensitive, and technically sophisticated process that profoundly influences the outcomes of NLP systems. As models grow more capable, the role of curated, well-annotated, and responsibly sourced corpora becomes even more crucial, shaping the very limits of what AI-driven language systems can understand, generate, and decide. The corpus is, in essence, the epistemic infrastructure of NLP—an invisible yet indispensable architecture upon which the reliability, fairness, and effectiveness of all downstream tasks depend. Figure 2.19 shows a multilayer materials science corpus.

2.5 Major Language Model Families

2.5.1 Word2Vector

Word2Vec, introduced by Mikolov et al. at Google in 2013, represents one of the most influential developments in the evolution of language representation learning [111]. Prior to Word2Vec, traditional NLP systems relied on sparse, high-dimensional representations such as one-hot encoding or TF-IDF vectors, which failed to capture the semantic relationships or contextual similarities between words. Word2Vec radically shifted this paradigm by introducing dense, low-dimensional vector representations—also called word embeddings—that encode semantic information based on the context in which words appear. The model operates on the distributional hypothesis: words that appear in similar contexts tend to have similar meanings. This idea is formalized through the unsupervised learning of word vectors from large corpora using shallow neural networks.

Word2Vec comprises two primary architectures: CBOWs and Skip-Gram. Figures 2.20 and 2.21 show the structure of CBOW and Skip-Gram. Both models are trained to predict word-context relationships but differ in directionality. The CBOW model predicts a target word given its surrounding context words, essentially averaging the context embeddings to infer the central word. Conversely, the Skip-Gram model does the opposite: it takes a single target word and learns to predict surrounding context words within a fixed-size window. While CBOW is generally faster and more suitable for large datasets, Skip-Gram tends to perform better for infrequent or rare words because it explicitly models word-context pairs.

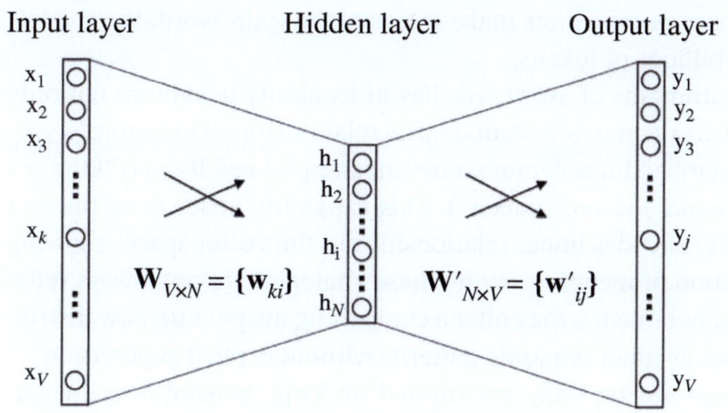

Figure 2.20 CBOW model. *Source*: [112] / arXiv.

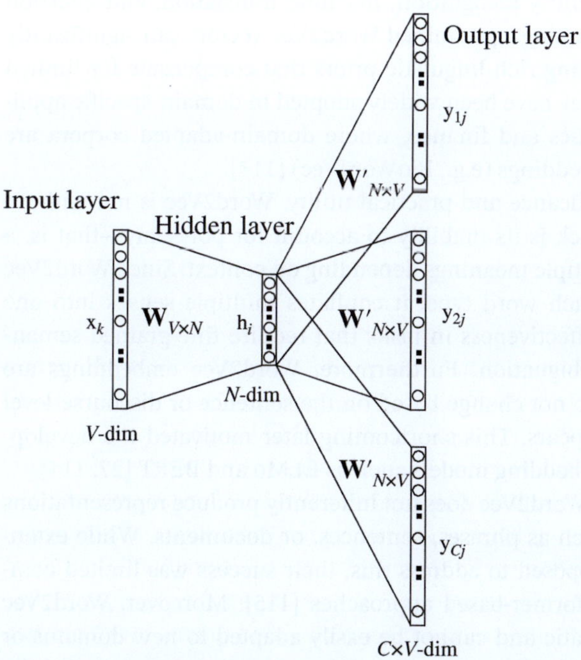

Figure 2.21 The Skip-Gram model. *Source*: [112] / arXiv.

Both models use a simple feedforward architecture with a single hidden layer and are trained using variants of stochastic gradient descent.

A core challenge in training Word2Vec lies in the computational cost of calculating softmax over the entire vocabulary during the output prediction phase. To mitigate this, the original authors proposed two alternative strategies: hierarchical softmax and negative sampling. Hierarchical softmax replaces the flat softmax with a binary tree representation of the vocabulary, enabling logarithmic time complexity in output computation. Negative sampling, on the other hand, simplifies training by sampling a small number of "negative" words that do not co-occur with the target word and only updating their embeddings, thus reducing the complexity of each training step. These optimizations make it feasible to train Word2Vec models on corpora containing billions of tokens.

One of the defining strengths of Word2Vec lies in its ability to capture not only semantic similarity but also syntactic and analogical relationships. For example, vector arithmetic on word embeddings demonstrates meaningful results: vec("king") – vec("man") + vec("woman") ≈ vec("queen"). This capability arises from the fact that Word2Vec implicitly encodes linear relationships in the vector space, allowing for algebraic manipulation of meaning. While these analogies are not always reliable across all linguistic phenomena, they offer a compelling insight into how distributed representations can abstract semantic patterns without explicit supervision.

Word2Vec embeddings are typically pre-trained on large general-domain corpora such as Google News, Wikipedia, or Common Crawl. These embeddings can then be used as fixed feature extractors for a variety of downstream tasks, including sentiment analysis, named entity recognition, machine translation, and question answering. In low-resource settings, pretrained Word2Vec vectors can significantly boost performance by providing rich linguistic priors that compensate for limited labeled data. Additionally, they have been widely adopted in domain-specific applications, such as bioinformatics and finance, where domain-adapted corpora are used to train specialized embeddings (e.g., BioWordVec) [113].

Despite its historical significance and practical utility, Word2Vec is not without limitations. One key drawback is its inability to account for polysemy—that is, a single word form having multiple meanings depending on context. Since Word2Vec assigns a single vector to each word type, it conflates multiple senses into one embedding. This limits its effectiveness in tasks that require fine-grained semantic understanding or disambiguation. Furthermore, Word2Vec embeddings are context-independent: they do not change based on the sentence or discourse-level context in which a word appears. This shortcoming later motivated the development of context-sensitive embedding models such as ELMo and BERT [27, 114].

Another limitation is that Word2Vec does not inherently produce representations for larger linguistic units such as phrases, sentences, or documents. While extensions like Doc2Vec were proposed to address this, their success was limited compared to more recent Transformer-based approaches [115]. Moreover, Word2Vec embeddings are relatively static and cannot be easily adapted to new domains or vocabularies without retraining from scratch, a process that is computationally expensive and prone to overfitting if data is limited. These issues underscore the need for more flexible, context-aware, and scalable approaches to language representation learning.

Nevertheless, Word2Vec's influence on the field of NLP is undeniable. It catalyzed a major shift toward distributed, data-driven representations and demonstrated the power of unsupervised learning at scale. Its introduction paved the way for a series of subsequent advancements, including GloVe, FastText, and ultimately Transformer-based models like ELMo, BERT, and GPT [20, 116]. Many of the design principles that underpin Word2Vec—such as learning from co-occurrence statistics, the use of negative sampling, and embedding space interpretability—continue to inform modern representation learning techniques.

In pedagogical contexts, Word2Vec remains a highly instructive model for understanding the principles of neural language modeling, offering an intuitive and computationally efficient framework for exploring vector semantics. It is often the first model introduced in NLP courses to explain the transition from symbolic to distributed representations. Its implementation simplicity and interpretability make it an excellent teaching tool and a baseline for academic research.

In conclusion, Word2Vec marks a pivotal moment in the history of NLP, bridging the gap between symbolic linguistics and neural computation. While it has been superseded by more powerful and expressive models, its conceptual clarity, efficiency, and practical effectiveness have secured its legacy as a cornerstone in the development of modern language understanding systems. The continued relevance of Word2Vec in many research and industrial settings is a testament to its elegant design and foundational role in the broader NLP landscape.

2.5.2 BERT

BERT was introduced by Devlin et al. in 2018 and marked a significant milestone in the development of language models [27]. Unlike earlier models that processed language in a unidirectional or context-limited manner, BERT introduced a deeply bidirectional architecture that enables each token to attend to both left and right context simultaneously. This design allowed BERT to capture richer syntactic and semantic representations, leading to remarkable improvements on a wide range of NLP benchmarks. At its core, BERT is built entirely upon the Transformer encoder architecture. The model uses stacked layers of multi-head self-attention and feedforward networks to process sequences in parallel, which provides superior efficiency and performance over recurrent models. Each input token is represented by the sum of its word embedding, positional embedding, and segment embedding, and the resulting vectors are propagated through multiple Transformer layers, enabling the final output to encode deep contextual dependencies from the entire sentence. Figure 2.22 shows pre-training and fine-tuning procedures for BERT.

One of BERT's most distinctive innovations lies in its pretraining objectives. Instead of the traditional left-to-right language modeling task, BERT employs an MLM strategy in which 15% of the input tokens are randomly selected and replaced with a [MASK] token, a random token, or left unchanged. The model is trained to predict these masked tokens based on the surrounding context, allowing it to learn representations that consider both preceding and succeeding words. This bidirectional context modeling was unprecedented at the time and significantly enriched the model's ability to understand polysemy and sentence structure. In addition to MLM, BERT introduced another pretraining objective known as NSP, where the

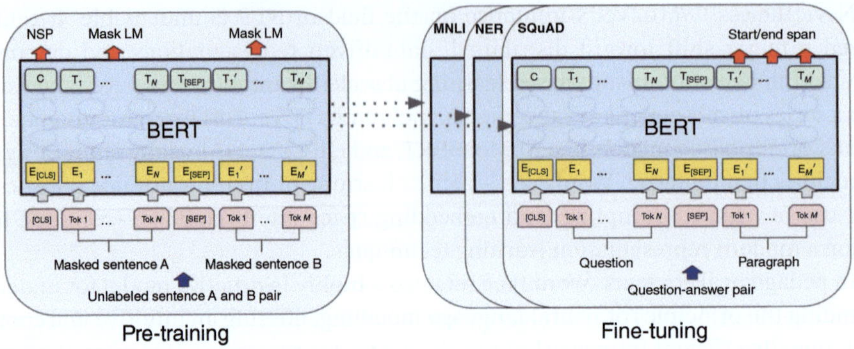

Figure 2.22 Overall pre-training and fine-tuning procedures for BERT. *Source*: [27] / arXiv.

model is trained to classify whether two input segments are contiguous sentences in the original text. This objective was intended to help the model grasp inter-sentence coherence, which is valuable for tasks such as natural language inference and question answering.

The pre-training of BERT was conducted on large-scale corpora, including the BookCorpus (800 million words) and English Wikipedia (2.5 billion words). BERT-Base, the original version, contains 12 layers, 768 hidden units, and 12 attention heads, totaling 110 million parameters. A larger variant, BERT-Large, scales this up to 24 layers and 340 million parameters. This level of scale, combined with the richness of bidirectional context and task-agnostic training, enabled BERT to achieve state-of-the-art performance on multiple NLP benchmarks with minimal task-specific engineering. Unlike earlier approaches that required building separate architectures for each downstream task, BERT introduced a unified fine-tuning paradigm. By adding a lightweight classification layer on top of the pretrained model and fine-tuning the entire network using supervised task data, BERT could be adapted to diverse tasks such as sentiment classification, named entity recognition, and machine reading comprehension.

The impact of BERT was immediate and transformative. It achieved new records on the GLUE, SQuAD, and other prominent NLP challenges. More importantly, it shifted the community toward a pretrain-then-fine-tune paradigm, where large, general-purpose models are first trained using massive unlabeled corpora and then adapted to specific applications. BERT's success also underscored the power of transfer learning in NLP, showing that a single architecture and set of pretrained weights could serve as the foundation for a wide range of downstream tasks. However, BERT's approach was not without limitations. One of the most frequently cited issues is the inefficiency of the MLM objective. Since only 15% of the tokens contribute to the loss in each training step, a significant portion of computational resources is underutilized. Moreover, the artificial introduction of the [MASK] token creates a discrepancy between pretraining and inference, as the model never sees [MASK] tokens during fine-tuning or deployment.

Another challenge is BERT's fixed input length, typically limited to 512 tokens, which restricts its ability to handle long documents. Given that self-attention scales quadratically with sequence length, extending BERT to longer inputs requires architectural modifications or approximations. Furthermore, BERT relies on a fixed

WordPiece vocabulary, which, while mitigating the out-of-vocabulary issue, can still fragment rare or domain-specific terms into subword sequences that may not preserve semantic coherence. BERT is also computationally expensive at inference time, particularly in real-time or resource-constrained settings. As a result, numerous lighter and more efficient variants have been proposed. These include DistilBERT, a smaller model trained using knowledge distillation; ALBERT, which uses parameter sharing and factorized embeddings to reduce model size; and TinyBERT and MobileBERT, designed for edge devices and mobile deployment [117–120].

Beyond efficiency improvements, BERT has inspired a rich ecosystem of derivative models and methodological refinements. RoBERTa, developed by Facebook AI, improved upon BERT by removing the NSP objective, training on a larger dataset, and employing dynamic masking during pretraining [121]. SpanBERT modified the pretraining task to predict contiguous spans of text rather than single masked tokens, better capturing span-level relationships useful for coreference and relation extraction [122]. Multilingual versions such as mBERT and XLM-R extended BERT's architecture to support over 100 languages, enabling cross-lingual transfer learning and zero-shot classification [123, 124]. These developments demonstrate BERT's adaptability and extensibility across different languages and domains.

BERT has also been widely adopted in specialized domains where language differs significantly from general text (such as Figure 2.23). In biomedical applications, BioBERT and ClinicalBERT were pretrained on scientific literature and electronic health records, respectively, improving performance on tasks like medical NER and relation extraction [125, 126]. Similar domain-adapted models have emerged in the fields of law, finance, and materials science, where pretraining on technical corpora enhances the model's ability to handle domain-specific jargon and syntax [127]. Additionally, BERT has become a focal point for research on interpretability and linguistic probing. Numerous studies have examined how information is distributed across BERT's layers and attention heads, revealing insights into the model's understanding of syntax, semantics, and discourse. These analyses have advanced our understanding of how deep language models encode linguistic structure and have guided the design of more interpretable and efficient architectures.

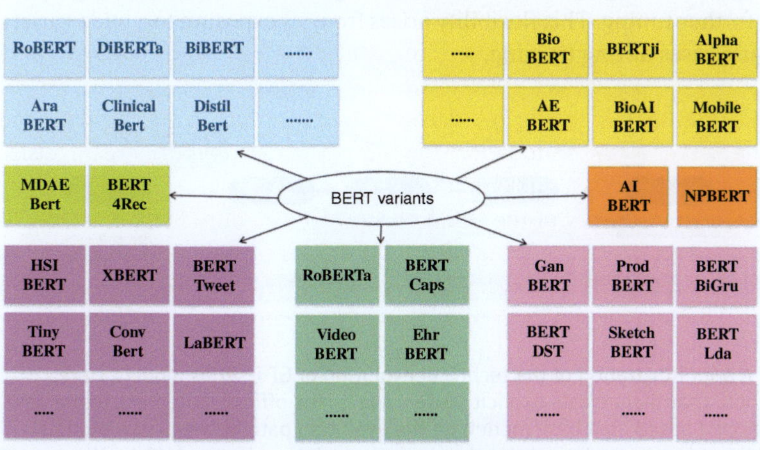

Figure 2.23 Various types of BERT. *Source*: [128] / IEEE.

In summary, BERT represents a foundational advance in the development of language models. By introducing bidirectional pretraining with the Transformer encoder and a unified fine-tuning approach, BERT fundamentally altered the trajectory of NLP research and applications. Its architectural elegance, combined with empirical success and extensibility, has made it a cornerstone of modern NLP. Although subsequent models have built upon and surpassed BERT in terms of scale and performance, its conceptual contributions remain central. BERT's legacy continues in the design of newer architectures, in the refinement of pretraining objectives, and in the growing ecosystem of pretrained models that drive language understanding across disciplines.

2.5.3 GPT

The GPT series, as shown in Figure 2.24, developed by OpenAI, marks a major milestone in the evolution of large-scale language modeling. As the name implies, GPT is designed to generate humanlike language through a process of extensive pretraining on vast textual corpora. It follows a unidirectional, autoregressive modeling approach, meaning it learns to generate text by predicting each word one at a time, based only on the words that precede it. This directional constraint stands in contrast to bidirectional models like BERT, but it is precisely what enables GPT to excel at coherent and contextually rich text generation.

GPT builds upon the Transformer architecture proposed in 2017, specifically utilizing only the decoder portion of the model. This design is sufficient for language generation tasks, as it focuses solely on producing the next word in a sequence. During training, the model is exposed to millions or even billions of sentences and learns statistical patterns in how words and phrases are used. As the training progresses, the model becomes increasingly proficient at capturing grammar, syntax, style, and even some aspects of reasoning and factual knowledge.

Unlike earlier models trained for a specific task, GPT is trained with a single, generic objective: to continue a piece of text. This simple goal turns out to be remarkably effective. After pretraining, the model can be used in a wide range of tasks—writing, answering questions, summarizing documents, translating text—without any further tuning. This flexibility arises from its exposure to a wide variety of language use cases during training.

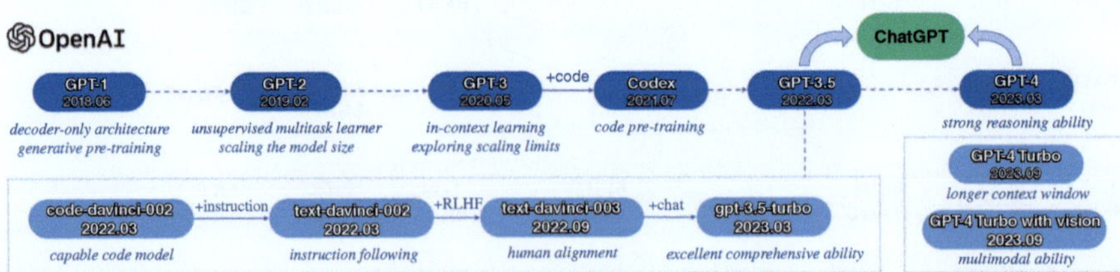

Figure 2.24 A brief illustration of the technical evolution of GPT-series models. Here, solid lines denote that there exists explicit evidence (e.g., the official statement that a new model is developed based on a base model) on the evolution path between two models, while dashed lines denote a relatively weaker evolution relation. *Source*: [41] / arXiv.

One of the key innovations introduced with GPT models is the concept of using prompting as a mechanism for task control. Instead of needing to retrain or adapt the model for every new task, users simply craft prompts—natural language instructions or examples—that guide the model to behave in the desired way. For example, a prompt like "Translate the following English sentence into French: 'How are you?'" is often enough for GPT to return an accurate translation, even though it was not explicitly trained on translation data.

The effectiveness of GPT scales with the size of both the training data and the model itself. GPT-2, released in 2019, already demonstrated impressive generation capabilities with 1.5 billion parameters. Its successor, GPT-3, pushed the scale further to 175 billion parameters and was trained on a mixture of books, websites, and other text sources amounting to hundreds of billions of words. With such scale, GPT-3 was able to perform tasks it had never been directly trained on, simply by being exposed to examples in the prompt. This behavior is often referred to as few-shot or zero-shot learning and has become one of the most discussed emergent properties of LLMs.

One particularly influential development in the GPT line is the use of human feedback to improve model alignment. In models such as GPT-3.5 and GPT-4, developers incorporated a technique called RLHF [129]. After the base model is pretrained, human annotators rate model outputs according to criteria such as usefulness, safety, and accuracy. These preferences are then used to guide the model toward producing more appropriate and desirable responses; the method is shown in Figure 2.25. This process has made newer GPT models more suitable for conversational applications and professional use cases.

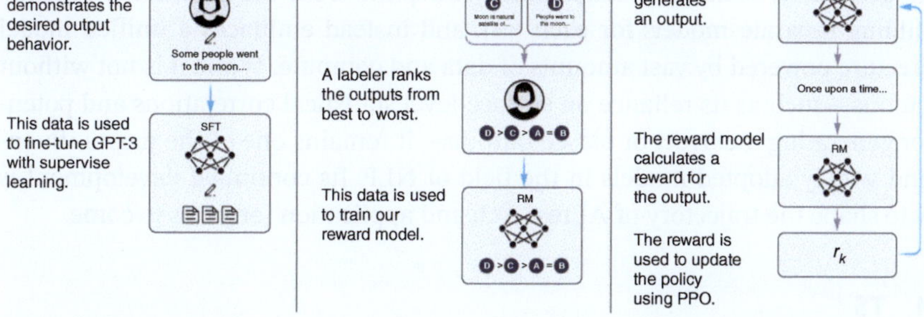

Figure 2.25 A diagram illustrating the three steps of the RLHF method: (1) supervised fine-tuning (SFT), (2) reward model (RM) training, and (3) RL via proximal policy optimization (PPO) on this reward model. Blue arrows indicate that this data is used to train one of our models. In Step 2, boxes A–D are samples from the models that get ranked by labelers. *Source*: [129] / arXiv.

GPT models are widely used in both research and industry. In academic settings, they assist with automatic summarization of literature, generation of research proposals, and even scientific writing. In software development, they are integrated into code completion tools that help programmers by predicting the next lines of code or suggesting bug fixes. For everyday users, GPT is the engine behind many AI-powered writing assistants, customer service bots, and virtual agents.

The versatility of GPT has also extended into domain-specific settings. Although the base model is trained on general-purpose data, organizations often adapt GPT through fine-tuning or prompt engineering for specialized tasks, such as legal document analysis, clinical text summarization, or financial report generation. In this sense, GPT acts as a foundational model upon which more tailored applications can be built.

At the same time, the widespread deployment of GPT models has raised important questions around ethical use, data privacy, and model transparency. Because GPT is trained on publicly available web content, it may inadvertently reproduce societal biases, misinformation, or sensitive information. OpenAI and the broader research community have responded by introducing filters, usage policies, and model auditing practices. Nonetheless, the challenges of safety and fairness remain active areas of investigation.

Another ongoing area of research is the interpretability of GPT models. While attention maps and visualization techniques provide some insight into how the model processes language, the decision-making processes inside such large neural networks remain largely opaque. Researchers continue to explore methods to better understand what information is stored inside the model's parameters and how it reasons about language.

Recent developments have also extended GPT beyond text-only capabilities. Multimodal versions, such as GPT-4 with vision, are able to process both textual and visual inputs. This opens the door to applications such as image captioning, document understanding, and multimodal reasoning. As the line between modalities blurs, GPT and its descendants are increasingly positioned as general-purpose AI systems capable of handling a wide variety of human tasks.

In conclusion, GPT exemplifies a new generation of language models that prioritize generality, scale, and adaptability. It departs from the traditional pipeline of building separate models for each task and instead embraces a unified model architecture powered by vast amounts of data and compute. While it is not without limitations—such as its reliance on surface-level statistical correlations and potential for generating incorrect or biased outputs—it remains one of the most influential and widely adopted models in the field of NLP. Its continued development is likely to shape the trajectory of AI research and application for years to come.

2.5.4 T5

T5 represents an influential shift in how NLP tasks are conceptualized and implemented. Introduced by Google Research in 2020, T5 proposes a unifying framework in which all language problems are cast as text-to-text tasks [36]. This means that both the input and the output of the model are treated as plain text strings,

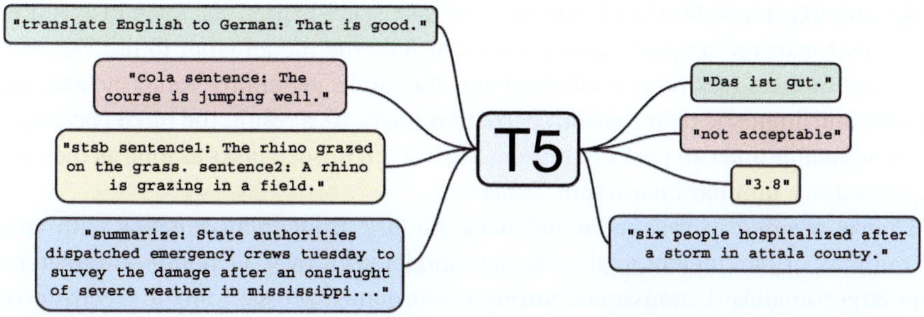

Figure 2.26 A diagram of the text-to-text framework. *Source*: [36] / arXiv.

regardless of the nature of the task—whether it be translation, summarization, classification, or question answering.

The motivation behind T5 is to simplify the NLP pipeline by using a single model architecture and objective function across all tasks. Traditionally, different NLP tasks required customized models with specific output formats—for example, a classifier for sentiment analysis, a sequence labeler for named entity recognition, and a decoder for summarization. T5 replaces this fragmented approach with a single, versatile model that can handle all of these problems through appropriate textual reformulations (such as Figure 2.26). For instance, instead of building a classification model to determine whether a movie review is positive or negative, T5 is prompted with a textual instruction such as "classify the sentiment of this review" followed by the review text. The model then outputs either "positive" or "negative" as plain text.

T5 is based on the standard Transformer encoder-decoder architecture. The encoder processes the input sequence and creates contextual representations, while the decoder generates the output text step-by-step. This is in contrast to models like BERT or GPT, which use only the encoder or decoder, respectively. By leveraging both components, T5 is capable of performing conditional generation tasks in a more flexible and powerful way.

One of the notable contributions of the T5 framework is the release of the C4, a large-scale dataset derived from the Common Crawl web scrape. The authors applied extensive preprocessing and filtering techniques to remove boilerplate text, low-quality content, and non-English segments, producing a high-quality text corpus used for pretraining the model. This dataset, along with the design philosophy of T5, helped set a new benchmark for MTL in NLP.

T5 employs a two-stage training strategy. In the first stage, known as pretraining, the model is trained on a large corpus using a self-supervised objective. Instead of predicting missing words or the next token, as in BERT or GPT, T5 is trained to fill in spans of text that have been masked out. This span corruption objective encourages the model to learn a deeper understanding of syntax and semantics by reconstructing meaningful portions of text. In the second stage, the model is fine-tuned on a mixture of downstream tasks that have been reformulated into the text-to-text format. These include summarization (e.g., "summarize: ..."), question answering (e.g., "question: ... context: ..."), and translation (e.g., "translate English to French: ...").

An important insight of T5 is the value of task-specific prefixes. Each task is preceded by a natural language instruction that tells the model what to do. This not only standardizes task inputs but also helps the model generalize better by learning to associate linguistic patterns with particular goals. As a result, the model becomes more adaptable and can perform reasonably well on unseen tasks as long as they are expressed in a familiar instruction format.

T5 was released in various model sizes, ranging from small versions with tens of millions of parameters to the full-size model with over 11 billion parameters. The larger models demonstrated strong performance across a broad spectrum of NLP benchmarks, including GLUE, SuperGLUE, CNN/DailyMail summarization, and SQuAD question answering. The unified training approach and scale of the T5 model contributed to its high generalization ability and robustness.

One of the key contributions of T5 is not merely its performance, but its role in shaping how NLP researchers think about model design. By enforcing a text-in, text-out paradigm, T5 simplifies the model interface and lowers the barrier to applying transfer learning across diverse tasks. In this regard, T5 can be viewed as a precursor to later instruction-tuned and prompt-based models, including those in the FLAN family and large-scale instruction-following models like InstructGPT [129, 130].

Moreover, T5 has proven to be highly extensible. Its framework allows for easy adaptation to new domains or languages by fine-tuning on customized text-to-text tasks. Researchers in biomedical NLP, legal document processing, and multilingual translation have all adapted T5 to their specific needs with relatively minor modifications. The consistent input-output format facilitates this process, as no task-specific architectural changes are required.

T5 also provides a natural bridge to the emerging field of prompt engineering. Since the model was trained to interpret textual instructions as part of its input, it is well-suited to applications where users craft natural language prompts to elicit specific behaviors. This has made T5 popular in low-code or no-code AI applications, where non-specialists can interact with the model through intuitive instructions.

Despite its versatility, T5 is not without limitations. Like other large-scale models, it is computationally intensive, requiring significant resources for both training and inference. Moreover, while its unified text-to-text format simplifies many aspects of model deployment, it may introduce inefficiencies when used for tasks with more structured outputs, such as parsing or classification, which might otherwise benefit from more compact representations.

Another challenge lies in the quality and nature of the prompts. Because the model's performance is often sensitive to how the task is phrased, poorly worded prompts can lead to suboptimal results. While this is partially mitigated through fine-tuning on diverse task instructions, the reliance on good prompt design remains a practical consideration in real-world applications. In terms of ethical and safety concerns, T5 shares many of the issues associated with LLMs trained on public web data. These include the risk of propagating harmful stereotypes, generating factually incorrect content, or leaking sensitive information. Responsible deployment of T5 requires careful evaluation, filtering mechanisms, and usage controls.

In conclusion, T5 represents a significant advancement in the pursuit of general-purpose natural language understanding and generation. By framing all tasks within a single, consistent text-to-text paradigm, it simplifies the application of deep learning to language processing and demonstrates the power of large-scale MTL. Its design philosophy continues to influence subsequent model families and serves as a foundational step in the evolution toward more general and controllable language intelligence.

2.6 Practical Tools and Libraries for NLP

2.6.1 Hugging Face

Hugging Face has emerged as one of the most influential organizations in the NLP community, particularly for its contributions in democratizing access to powerful language models and creating a rich ecosystem of tools for both research and production. Founded in 2016, Hugging Face initially started as a chatbot startup but quickly evolved into a central hub for open-source machine learning. Its flagship project, the Transformers library, has become a de facto standard for implementing, training, and deploying Transformer-based models. The official website of Hugging Face is shown in Figure 2.27.

At the core of the Hugging Face ecosystem is the Transformers library, which offers a unified API for a vast collection of pretrained models, including BERT, GPT, T5, RoBERTa, DistilBERT, and many others. These models are available in both PyTorch and TensorFlow backends, making the library highly versatile and accessible to a broad user base. Researchers and developers can load pretrained models with just a few lines of code, perform inference on text classification, summarization, translation, and question-answering tasks, and fine-tune models on custom datasets with minimal overhead.

One of the key innovations of Hugging Face is the Model Hub, a public repository that hosts thousands of pretrained models contributed by the community and

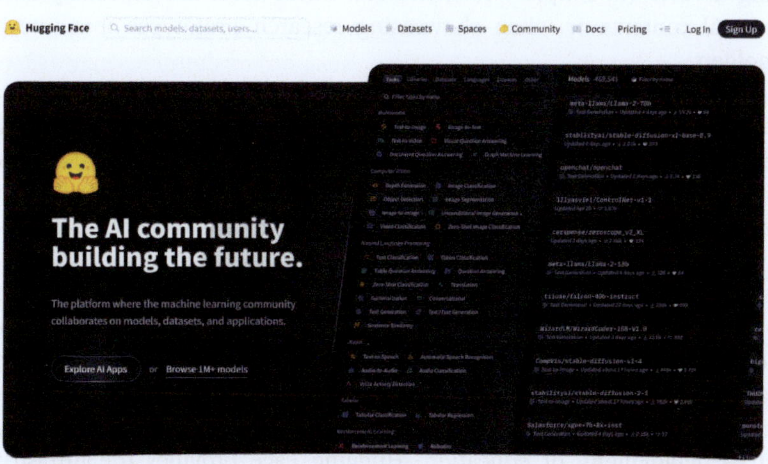

Figure 2.27 The official website of Hugging Face (Hugging Face, Inc. https://huggingface.co/. Last accessed on 25 Aug 2025).

leading research groups. This platform significantly lowers the barrier to entry for NLP work by eliminating the need for expensive training from scratch. Users can easily search, download, and deploy models fine-tuned for specific domains or languages, making it especially valuable in industrial and low-resource settings.

In addition to model distribution, Hugging Face provides tools to support the entire NLP workflow. The datasets library is a lightweight and efficient toolkit for downloading, preprocessing, and streaming large-scale datasets. It integrates smoothly with the Transformers library and supports popular benchmarks such as GLUE, SQuAD, and Common Crawl corpora. The library emphasizes memory efficiency and scalability, allowing users to manipulate gigabyte- or even terabyte-scale data with a simple interface.

For model evaluation and benchmarking, Hugging Face includes the Evaluate library, which standardizes performance metrics and provides a reproducible framework for model comparison. This tool is particularly helpful when experimenting with new architectures or fine-tuning procedures, as it ensures consistent measurement across different runs and tasks.

Another notable component of the Hugging Face platform is the Trainer API, which abstracts the training and evaluation loop for Transformer models. This high-level API supports key functionalities such as mixed-precision training, multi-GPU distributed computing, checkpointing, early stopping, and metric logging. By handling these complexities internally, the Trainer interface allows researchers to focus on high-level model design and analysis, accelerating the development process.

Beyond code libraries, Hugging Face has established a strong emphasis on community and transparency. Its platform encourages sharing through model cards and dataset documentation, which provide structured metadata, intended use cases, ethical considerations, and performance statistics. This practice promotes responsible AI development and facilitates reproducibility across different research groups.

The Hugging Face ecosystem has also expanded to include tools such as Gradio and Spaces, which enable interactive demonstrations and deployment of machine learning models via simple web interfaces. These tools are especially useful in educational contexts or when showcasing research prototypes to nontechnical stakeholders.

Moreover, Hugging Face plays a pivotal role in the growing trend of open science. By partnering with academic institutions, AI research labs, and industry, it actively contributes to open benchmarks, large-scale model releases, and collaborative training initiatives. Projects like BLOOM (a multilingual LLM) and BigScience illustrate Hugging Face's commitment to transparency, inclusiveness, and scalability in language model development.

In sum, Hugging Face is not merely a library but an integrated ecosystem that empowers users at all levels of expertise to experiment with, fine-tune, and deploy state-of-the-art NLP models. Its modular architecture, community-driven model hub, and seamless integration with widely used deep learning frameworks make it a cornerstone of contemporary NLP practice. Whether for research prototyping, industrial deployment, or education, Hugging Face provides the infrastructure and tools to make advanced language technology more accessible, efficient, and collaborative.

2.6.2 PyTorch

PyTorch is one of the most widely adopted deep learning frameworks in NLP research and industry. Originally developed by Facebook's AI Research Lab (FAIR) and released in 2016, PyTorch quickly gained popularity for its intuitive syntax, dynamic computation graph, and strong support for GPU acceleration [131]. It is now the foundation of many NLP systems and a critical backend for several high-level libraries, including Hugging Face Transformers, AllenNLP, and Fairseq [132–134].

A major appeal of PyTorch lies in its eager execution model, which allows developers to build and modify computational graphs on the fly. This dynamic behavior makes PyTorch especially suitable for natural language tasks, where input lengths, structures, and operations often vary. Researchers can easily write and debug custom models without needing to predefine fixed computation graphs, which was a limitation of earlier static-graph frameworks like TensorFlow 1.x. As a result, PyTorch has become the default framework for rapid experimentation and prototyping in NLP.

From an architectural standpoint, PyTorch offers modular building blocks that support both low-level tensor operations and high-level abstractions. Its torch.nn module includes a comprehensive suite of layers, activation functions, and loss criteria that facilitate the construction of deep neural networks. Examples are shown in Figure 2.28. For NLP tasks, common components such as embedding layers, RNNs, LSTMs, gated recurrent units (GRUs), attention mechanisms, and Transformer encoders are readily available and easily extendable.

Training models in PyTorch is streamlined through the use of DataLoader and Dataset classes, which offer customizable data pipelines for mini-batch processing, shuffling, and augmentation. This design is especially beneficial for handling large-scale text datasets, where efficiency and flexibility are essential. Furthermore, PyTorch supports automatic differentiation via its autograd engine, which simplifies the computation of gradients and backpropagation during training.

One of PyTorch's key strengths is its extensive ecosystem of tools and libraries, many of which are tailored for NLP. For instance, the torchtext package provides utilities for text preprocessing, vocabulary construction, tokenization, and dataset loading, although it has become less central with the rise of external tokenizers like those from

```
class LinearLayer(Module):
    def __init__(self, in_sz, out_sz):
        super().__init__()
        t1 = torch.randn(in_sz, out_sz)
        self.w = nn.Parameter(t1)
        t2 = torch.randn(out_sz)
        self.b = nn.Parameter(t2)

    def forward(self, activations):
        t = torch.mm(activations, self.w)
        return t + self.b
```

```
class FullBasicModel(nn.Module):
    def __init__(self):
        super().__init__()
        self.conv = nn.Conv2d(1, 128, 3)
        self.fc = LinearLayer(128, 10)

    def forward(self, x):
        t1 = self.conv(x)
        t2 = nn.functional.relu(t1)
        t3 = self.fc(t1)
        return nn.functional.softmax(t3)
```

Figure 2.28 A custom layer used as a building block for a simple but complete neural network. *Source*: [131] / arXiv.

Hugging Face. Libraries such as AllenNLP build on top of PyTorch to offer reusable components, prebuilt models, and standardized workflows for a range of NLP tasks, including semantic parsing, coreference resolution, and textual entailment [133].

In addition to research, PyTorch is widely used in production settings, thanks to the introduction of TorchScript, a subset of Python that enables the conversion of PyTorch models into static graphs for deployment. This feature allows trained models to run efficiently outside of the Python runtime, such as in C++ environments or mobile applications. TorchServe, an official model serving library, further facilitates deployment by providing tools for model packaging, versioning, RESTful API integration, and scalable inference.

The framework is also highly interoperable with modern hardware accelerators. It supports multi-GPU training through DataParallel and DistributedDataParallel modules and integrates seamlessly with NVIDIA's CUDA and cuDNN libraries. This compatibility ensures that even large-scale Transformer-based models can be trained efficiently on modern GPU clusters. Furthermore, PyTorch has increasingly added support for emerging hardware such as Apple Silicon, TPUs (via PyTorch/XLA), and custom AI chips.

Community and industry support for PyTorch has grown significantly in recent years. Major conferences such as ACL, EMNLP, and NeurIPS feature a high proportion of papers using PyTorch in their experimental implementations. Many leading AI companies, including Meta, OpenAI, Microsoft, and Tesla, rely on PyTorch for both research and production. Its open-source development is active and community-driven, with frequent updates, extensive documentation, and a robust ecosystem of tutorials, pretrained models, and educational resources.

In the context of NLP, PyTorch serves as the computational backbone for most modern Transformer models. Libraries like Hugging Face Transformers abstract away the boilerplate, but the underlying computations, optimization, and tensor operations are handled directly by PyTorch. This makes it crucial for NLP practitioners to understand PyTorch's core mechanics, especially when customizing architectures or scaling models across different hardware environments.

In summary, PyTorch is not only a versatile deep learning framework but also an enabler of innovation in NLP. Its dynamic graph design, comprehensive API, and wide hardware support make it ideal for both rapid prototyping and industrial deployment. As language models continue to grow in scale and complexity, PyTorch remains a central tool in building, training, and deploying cutting-edge NLP systems.

2.6.3 NLTK

The NLTK is one of the earliest and most comprehensive Python libraries for NLP. Developed primarily for educational and research purposes, NLTK has served as an essential introduction to NLP for students, linguists, and data scientists since its initial release in 2001 [135]. It provides a modular and accessible interface to a wide range of linguistic data structures, corpora, and processing tools, making it particularly well-suited for learning and experimenting with fundamental NLP techniques.

At its core, NLTK offers a rich set of text processing functionalities, including tokenization, stemming, lemmatization, part-of-speech tagging, chunking, parsing, and named entity recognition. These tools are designed to operate on raw textual data and follow linguistically informed rules and statistical models. For instance, users can tokenize English text using word or sentence tokenizers that account for punctuation and common abbreviations. Similarly, stemming algorithms like Porter and Snowball, as well as lemmatizers based on WordNet, allow for the normalization of words to their base forms [8, 136, 137].

One of NLTK's distinguishing features is its integration with a wide array of linguistic corpora and lexical resources. The toolkit includes access to dozens of text corpora and linguistic datasets, such as the Brown Corpus, Gutenberg Corpus, Reuters news articles, and the Penn Treebank. These corpora support tasks like frequency analysis, concordance search, and part-of-speech distribution studies. In addition, NLTK provides a Python interface to the WordNet lexical database, enabling semantic queries such as synonym and antonym retrieval, hyponym identification, and semantic similarity computation.

The design of NLTK emphasizes transparency and educational clarity, often favoring interpretability and simplicity over computational efficiency. For example, its syntactic parsers—both rule-based and statistical—allow users to observe the parsing process in detail, examine parse trees, and even define their own grammar rules. This level of control and visibility makes NLTK particularly suitable for academic instruction, where understanding the underlying principles of language processing takes precedence over high-throughput performance.

In practice, NLTK is commonly used for prototyping and pedagogical demonstrations rather than building production-level NLP systems. Its rule-based nature and limited scalability make it less suitable for modern deep learning applications, especially those involving large-scale datasets or Transformer-based models. However, its clarity and ease of use continue to make it a valuable tool for exploring the foundations of linguistic structure, probabilistic models, and text classification techniques.

NLTK also includes predefined pipelines for classical machine learning tasks in NLP, such as document classification, sentiment analysis, and topic modeling. These pipelines leverage traditional algorithms like Naive Bayes, Decision Trees, and MaxEnt models, and can be trained using annotated text corpora included in the toolkit. While these methods may not match the performance of neural architectures, they are instrumental for introducing concepts such as feature extraction, training-test splits, and model evaluation metrics like accuracy and F1-score.

In recent years, NLTK's dominance in the educational domain has been complemented (and sometimes supplanted) by libraries like spaCy and Hugging Face Transformers, which offer more efficient pipelines and support for pretrained neural models. Nevertheless, NLTK remains valuable for understanding the rule-based and statistical roots of NLP, and it is often used alongside other libraries when transparency, modularity, or corpus-based linguistic analysis is required.

Furthermore, NLTK serves as a historical bridge between traditional NLP and the modern deep learning era. Many of the concepts that underpin today's Transformer models—such as tokenization, word embeddings, and syntactic structures—can be first explored in NLTK before transitioning to more advanced libraries. Its documentation, tutorials, and accompanying textbook "*Natural Language Processing with Python*" (commonly known as the NLTK book) provide a comprehensive educational foundation that remains relevant for NLP learners.

In summary, while NLTK may no longer be the default choice for high-performance NLP development, it continues to play a crucial role in education, rapid prototyping, and exploratory linguistic analysis. Its accessible interface, rich linguistic resources, and pedagogical design make it an ideal starting point for understanding the building blocks of NLP before advancing to modern neural approaches.

References

1 Weaver W. Translation. In: Proceedings of the Conference on Mechanical Translation [Internet]. Massachusetts Institute of Technology; 1952 [cited 2025 May 18]. Available from: https://aclanthology.org/1952.earlymt-1.1/

2 Hutchins WJ. The Georgetown-IBM experiment demonstrated in January 1954. In: Frederking RE, Taylor KB, editors. Machine Translation: From Real Users to Research. Berlin, Heidelberg: Springer; 2004. pp. 102–14.

3 Chomsky N. Syntactic Structures. Walter de Gruyter; 2002. p. 137.

4 Weizenbaum J. ELIZA—A computer program for the study of natural language communication between man and machine. Communications of the ACM. 1966;9(1):36–45.

5 Winograd T. Procedures as a representation for data in a computer program for understanding natural language [Internet] [Thesis]. Massachusetts Institute of Technology; 1970 [cited 2025 May 19]. Available from: https://dspace.mit.edu/handle/1721.1/15546

6 Woods WA. Progress in natural language understanding: An application to lunar geology. In: Proceedings of the June 4–8, 1973, National Computer Conference and Exposition [Internet]. New York: Association for Computing Machinery; 1973 [cited 2025 May 18]. pp. 441–50. (AFIPS '73). Available from: https://dl.acm.org/doi/10.1145/1499586.1499695

7 Lenat DB, Guha RV. Building Large Knowledge-based Systems: Representation and Inference in the Cyc Project. 1st ed. Addison-Wesley Longman Publishing Co., Inc.; 1989. p. 372.

8 Miller GA. WordNet: A lexical database for English. Communications of the ACM. 1995;38(11):39–41.

9 Miller GA, Beckwith R, Fellbaum C, Gross D, Miller K. Introduction to WordNet: An on-line lexical database. International Journal of Lexicography. 1990;3(4):235–44.

10 Rabiner LR. A tutorial on hidden Markov models and selected applications in speech recognition. Proceedings of the IEEE. 1989;77(2):257–86.

11 Berger AL, Della Pietra SA, Della Pietra VJ. A maximum entropy approach to natural language processing. Computational Linguistics. 1996;22(1):39–71.

12 Sparck Jones K. A statistical interpretation of term specificity and its application in retrieval. Journal of Documentation. 1972;28(1):11–21.

13 Bengio Y, Ducharme R, Vincent P, Janvin C. A neural probabilistic language model. Journal of Machine Learning Research. 2003;3(null):1137–55.

14 Taylor A, Marcus M, Santorini B. The Penn Treebank: An overview. In: Abeillé A, editor. Treebanks: Building and Using Parsed Corpora [Internet]. Dordrecht: Springer; 2003 [cited 2025 May 18]. pp. 5–22. Available from: https://doi.org/10.1007/978-94-010-0201-1_1

15 CoNLL 2025 | CoNLL [Internet]. [cited 2025 May 18]. Available from: https://www.conll.org/

16 Mikolov T, Sutskever I, Chen K, Corrado GS, Dean J. Distributed representations of words and phrases and their compositionality. Advances in Neural Information Processing Systems. 2013;26:3111–9.

17 Brin S, Page L. The anatomy of a large-scale hypertextual Web search engine. Computer Networks and ISDN Systems. 1998;30(1–7):107–17.

18 Wang Q, Guan Y, Wang X. SVM-based spam filter with active and online learning. In: Proceedings of the Fifteenth Text Retrieval Conference. Gaithersburg, USA: NIST, TREC; 2006.

19 PageRank. In: Wikipedia [Internet]. 2025 [cited 2025 May 19]. Available from: https://en.wikipedia.org/w/index.php?title=PageRank&oldid=1288076239#Relevant_patents

20 Pennington J, Socher R, Manning C. GloVe: Global vectors for word representation. In: Moschitti A, Pang B, Daelemans W, editors. Proceedings of the 2014 Conference on Empirical Methods in Natural Language Processing (EMNLP) [Internet]. Doha, Qatar: Association for Computational Linguistics; 2014 [cited 2025 June 8]. pp. 1532–43. Available from: https://aclanthology.org/D14-1162/

21 Mikolov T, Karafiát M, Burget L, Černocký J, Khudanpur S. Recurrent neural network based language model. In: Interspeech; 2010 [cited 2025 May 19]. pp. 1045–8. Available from: https://www.isca-archive.org/interspeech_2010/mikolov10_interspeech.html#

22 Sundermeyer M, Schlüter R, Ney H. LSTM neural networks for language modeling. In: Interspeech 2012 [Internet]. ISCA; 2012 [cited 2025 May 19]. pp. 194–7. Available from: https://www.isca-archive.org/interspeech_2012/sundermeyer12_interspeech.html

23 What is a Recurrent Neural Network (RNN)? | IBM [Internet]. 2021 [cited 2025 May 19]. Available from: https://www.ibm.com/think/topics/recurrent-neural-networks

24 Song K, Tan X, Qin T, Lu J, Liu TY. MASS: Masked sequence to sequence pre-training for language generation. arXiv preprint arXiv:1905.02450. 2019.

25 Bahdanau D, Cho K, Bengio Y. Neural machine translation by jointly learning to align and translate [Internet]. arXiv; 2016 [cited 2025 May 19]. Available from: http://arxiv.org/abs/1409.0473

26 Vaswani A, Shazeer N, Parmar N, Uszkoreit J, Jones L, Gomez AN, et al. Attention is all you need. Advances in Neural Information Processing Systems. 2017;30:6000–10.

27 Devlin J, Chang MW, Lee K, Toutanova K. Bert: Pre-training of deep bidirectional transformers for language understanding. arXiv preprint arXiv:1810.04805. 2018.

28 Kaplan J, McCandlish S, Henighan T, Brown TB, Chess B, Child R, et al. Scaling laws for neural language models [Internet]. arXiv; 2020 [cited 2025 July 23]. Available from: http://arxiv.org/abs/2001.08361

29 Beltagy I, Peters ME, Cohan A. Longformer: The long-document transformer [Internet]. arXiv; 2020 [cited 2025 May 25]. Available from: http://arxiv.org/abs/2004.05150

30 Rajbhandari S, Rasley J, Ruwase O, He Y. ZeRO: Memory Optimizations Toward Training Trillion Parameter Models [Internet]. arXiv; 2020 [cited 2025 July 23]. Available from: http://arxiv.org/abs/1910.02054

31 Rasley J, Rajbhandari S, Ruwase O, He Y. DeepSpeed: System optimizations enable training deep learning models with over 100 billion parameters. In: Proceedings of the 26th ACM SIGKDD International Conference on Knowledge Discovery & Data Mining [Internet]. Virtual Event. ACM; 2020 [cited 2025 July 23]. pp. 3505–6. Available from: https://dl.acm.org/doi/10.1145/3394486.3406703

32 Wang W, Bao H, Dong L, Bjorck J, Peng Z, Liu Q, et al. Image as a Foreign Language: BEiT Pretraining for all vision and Vision-Language tasks [Internet]. arXiv; 2022 [cited 2025 July 23]. Available from: http://arxiv.org/abs/2208.10442

33 Radford A, Kim JW, Hallacy C, Ramesh A, Goh G, Agarwal S, et al. Learning transferable visual models from natural language supervision [Internet]. arXiv; 2021 [cited 2025 June 2]. Available from: http://arxiv.org/abs/2103.00020

34 Stiennon N, Ouyang L, Wu J, Ziegler DM, Lowe R, Voss C, et al. Learning to summarize from human feedback [Internet]. arXiv; 2022 [cited 2025 July 23]. Available from: http://arxiv.org/abs/2009.01325

35 DeepSeek-AI, Liu A, Feng B, Xue B, Wang B, Wu B, et al. DeepSeek-V3 technical report [Internet]. arXiv; 2025 [cited 2025 July 22]. Available from: http://arxiv.org/abs/2412.19437

36 Raffel C, Shazeer N, Roberts A, Lee K, Narang S, Matena M, et al. Exploring the limits of transfer learning with a unified Text-to-Text transformer [Internet]. arXiv; 2023 [cited 2025 May 19]. Available from: http://arxiv.org/abs/1910.10683

37 Lewis M, Liu Y, Goyal N, Ghazvininejad M, Mohamed A, Levy O, et al. BART: Denoising Sequence-to-Sequence Pre-training for natural language generation, translation, and comprehension. In: Jurafsky D, Chai J, Schluter N, Tetreault J, editors. Proceedings of the 58th Annual Meeting of the Association for Computational Linguistics [Internet]. Association for Computational Linguistics; 2020 [cited 2025 May 25]. pp. 7871–80. Available from: https://aclanthology.org/2020.acl-main.703/

38 Sarlin PE, DeTone D, Malisiewicz T, Rabinovich A. SuperGlue: Learning feature matching with graph neural networks [Internet]. arXiv; 2020 [cited 2025 May 25]. Available from: http://arxiv.org/abs/1911.11763

39 Wang A, Singh A, Michael J, Hill F, Levy O, Bowman SR. GLUE: A multi-task benchmark and analysis platform for natural language understanding [Internet]. arXiv; 2019 [cited 2025 May 25]. Available from: http://arxiv.org/abs/1804.07461

40 Rajpurkar P, Zhang J, Lopyrev K, Liang P. SQuAD: 100,000+ questions for machine comprehension of text [Internet]. arXiv; 2016 [cited 2025 May 25]. Available from: http://arxiv.org/abs/1606.05250

41 Zhao WX, Zhou K, Li J, Tang T, Wang X, Hou Y, et al. A survey of large language models [Internet]. arXiv; 2025 [cited 2025 May 25]. Available from: http://arxiv.org/abs/2303.18223

42 Wang W, Zheng VW, Yu H, Miao C. A survey of zero-shot learning: Settings, methods, and applications. ACM Transactions on Intelligent Systems and Technology. 2019;10(2):1–37.

43 Finn C, Abbeel P, Levine S. Model-agnostic meta-learning for fast adaptation of deep networks. In: International Conference on Machine Learning. Sydney, Australia: PMLR; 2017. pp. 1126–35.

44 Hendel R, Geva M, Globerson A. In-context learning creates task vectors [Internet]. arXiv; 2023 [cited 2025 May 25]. Available from: http://arxiv.org/abs/2310.15916

45 Gao Y, Xiong Y, Gao X, Jia K, Pan J, Bi Y, et al. Retrieval-augmented generation for large language models: A survey [Internet]. arXiv; 2024 [cited 2025 May 25]. Available from: http://arxiv.org/abs/2312.10997

46 Hu EJ, Shen Y, Wallis P, Allen-Zhu Z, Li Y, Wang S, et al. LoRA: Low-rank adaptation of large language models [Internet]. arXiv; 2021 [cited 2025 May 25]. Available from: http://arxiv.org/abs/2106.09685

47 Houlsby N, Giurgiu A, Jastrzebski S, Morrone B, Laroussilhe Q de, Gesmundo A, et al. Parameter-efficient transfer learning for NLP [Internet]. arXiv; 2019 [cited 2025 May 25]. Available from: http://arxiv.org/abs/1902.00751

48 Christiano PF, Leike J, Brown TB, Martic M, Legg S, Amodei D. Deep reinforcement learning from human preferences. In: Proceedings of the 31st International Conference on Neural Information Processing Systems [Internet]. [cited 2025 May 25]. Available from: https://dl.acm.org/doi/10.5555/3294996.3295184

49 Jiang J, Shu Y, Wang J, Long M. Transferability in deep learning: A survey [Internet]. arXiv; 2022 [cited 2025 May 25]. Available from: http://arxiv.org/abs/2201.05867

50 Wang C, Lai H, Nissim M, Bos J. Pre-trained language-meaning models for multilingual parsing and generation [Internet]. arXiv; 2023 [cited 2025 May 25]. Available from: http://arxiv.org/abs/2306.00124

51 Wang L, Zhang X, Su H, Zhu J. A comprehensive survey of continual learning: Theory, method and application [Internet]. arXiv; 2024 [cited 2025 May 25]. Available from: http://arxiv.org/abs/2302.00487

52 Cho K, Van Merrienboer B, Gulcehre C, Bahdanau D, Bougares F, Schwenk H, et al. Learning phrase representations using RNN encoder-decoder for statistical machine translation [Internet]. arXiv; 2014 [cited 2025 May 25]. Available from: http://arxiv.org/abs/1406.1078

53 Sutskever I, Vinyals O, Le QV. Sequence to sequence learning with neural networks [Internet]. arXiv; 2014 [cited 2025 May 25]. Available from: http://arxiv.org/abs/1409.3215

54 Gu J, Bradbury J, Xiong C, Li VOK, Socher R. Non-autoregressive neural machine translation [Internet]. arXiv; 2017 [cited 2025 May 25]. Available from: http://arxiv.org/abs/1711.02281

55 Child R, Gray S, Radford A, Sutskever I. Generating long sequences with sparse transformers [Internet]. arXiv; 2019 [cited 2025 May 25]. Available from: http://arxiv.org/abs/1904.10509

56 Dosovitskiy A, Beyer L, Kolesnikov A, Weissenborn D, Zhai X, Unterthiner T, et al. An image is worth 16x16 words: Transformers for image recognition at scale [Internet]. arXiv; 2021 [cited 2025 May 25]. Available from: http://arxiv.org/abs/2010.11929

57 Holtzman A, Buys J, Du L, Forbes M, Choi Y. The curious case of neural text degeneration [Internet]. arXiv; 2020 [cited 2025 May 25]. Available from: http://arxiv.org/abs/1904.09751

58 Mihaylova T, Martins AFT. Scheduled sampling for transformers [Internet]. arXiv; 2019 [cited 2025 May 25]. Available from: http://arxiv.org/abs/1906.07651

59 See A, Liu PJ, Manning CD. Get to the point: Summarization with pointer-generator networks [Internet]. arXiv; 2017 [cited 2025 May 25]. Available from: http://arxiv.org/abs/1704.04368

60 Hinton G, Vinyals O, Dean J. Distilling the knowledge in a neural network [Internet]. arXiv; 2015 [cited 2025 May 25]. Available from: http://arxiv.org/abs/1503.02531

61 Jacobs RA, Jordan MI, Nowlan SJ, Hinton GE. Adaptive mixtures of local experts. Neural Computation. 1991;3(1):79–87.

62 Lepikhin D, Lee H, Xu Y, Chen D, Firat O, Huang Y, et al. GShard: Scaling giant models with conditional computation and automatic sharding [Internet]. arXiv; 2020 [cited 2025 May 25]. Available from: http://arxiv.org/abs/2006.16668

63 Luong MT, Pham H, Manning CD. Effective approaches to attention-based neural machine translation [Internet]. arXiv; 2015 [cited 2025 May 25]. Available from: http://arxiv.org/abs/1508.04025

64 Brauwers G, Frasincar F. A general survey on attention mechanisms in deep learning. IEEE Transactions on Knowledge and Data Engineering. 2023;35(4):3279–98.

65 Kitaev N, Kaiser Ł, Levskaya A. Reformer: The efficient transformer [Internet]. arXiv; 2020 [cited 2025 May 25]. Available from: http://arxiv.org/abs/2001.04451

66 Zaheer M, Guruganesh G, Dubey A, Ainslie J, Alberti C, Ontanon S, et al. Big bird: Transformers for longer sequences [Internet]. arXiv; 2021 [cited 2025 May 25]. Available from: http://arxiv.org/abs/2007.14062

67 Ramesh A, Dhariwal P, Nichol A, Chu C, Chen M. Hierarchical text-conditional image generation with CLIP latents [Internet]. arXiv; 2022 [cited 2025 May 19]. Available from: http://arxiv.org/abs/2204.06125

68 Choromanski K, Likhosherstov V, Dohan D, Song X, Gane A, Sarlos T, et al. Rethinking attention with performers [Internet]. arXiv; 2022 [cited 2025 May 25]. Available from: http://arxiv.org/abs/2009.14794

69 Wang S, Li BZ, Khabsa M, Fang H, Ma H. Linformer: Self-attention with linear complexity [Internet]. arXiv; 2020 [cited 2025 May 25]. Available from: http://arxiv.org/abs/2006.04768

70 Bai L, Shao S, Zhou Z, Qi Z, Xu Z, Xiong H, et al. Zigzag diffusion sampling: Diffusion models can self-improve via self-reflection [Internet]. arXiv; 2024 [cited 2025 May 25]. Available from: http://arxiv.org/abs/2412.10891

71 Wu T, Zou J, Liang J, Zhang L, Ma K. VisualQuality-R1: Reasoning-induced image quality assessment via reinforcement learning to rank [Internet]. arXiv; 2025 [cited 2025 May 25]. Available from: http://arxiv.org/abs/2505.14460

72 Lin T, Wang Y, Liu X, Qiu X. A survey of transformers [Internet]. arXiv; 2021 [cited 2025 May 25]. Available from: http://arxiv.org/abs/2106.04554

73 Achiam J, Adler S, Agarwal S, Ahmad L, Akkaya I, Aleman FL, et al. GPT-4 Technical Report. arXiv preprint arXiv:2303.08774. 2023.

74 Clark K, Luong MT, Le QV, Manning CD. ELECTRA: Pre-training text encoders as discriminators rather than generators [Internet]. arXiv; 2020 [cited 2025 June 2]. Available from: http://arxiv.org/abs/2003.10555

75 Gao T, Yao X, Chen D. SimCSE: Simple contrastive learning of sentence embeddings [Internet]. arXiv; 2022 [cited 2025 June 2]. Available from: http://arxiv.org/abs/2104.08821

76 Liu X, Jiao J, Zhang J. Self-supervised pretraining for decision foundation model: Formulation, pipeline and challenges [Internet]. arXiv; 2024 [cited 2025 June 3]. Available from: http://arxiv.org/abs/2401.00031

77 Van den Oord A, Li Y, Vinyals O. Representation learning with contrastive predictive coding [Internet]. arXiv; 2019 [cited 2025 June 3]. Available from: http://arxiv.org/abs/1807.03748

78 He K, Fan H, Wu Y, Xie S, Girshick R. Momentum contrast for unsupervised visual representation learning [Internet]. arXiv; 2020 [cited 2025 June 2]. Available from: http://arxiv.org/abs/1911.05722

79 Verwimp E, Yang K, Parisot S, Lanqing H, McDonagh S, Pérez-Pellitero E, et al. CLAD: A realistic continual learning benchmark for autonomous driving [Internet]. arXiv; 2022 [cited 2025 June 2]. Available from: http://arxiv.org/abs/2210.03482

80 Chen Z, Badrinarayanan V, Lee CY, Rabinovich A. GradNorm: Gradient normalization for adaptive loss balancing in deep multitask networks [Internet]. arXiv; 2018 [cited 2025 June 2]. Available from: http://arxiv.org/abs/1711.02257

81 Wei J, Bosma M, Zhao VY, Guu K, Yu AW, Lester B, et al. Finetuned language models are zero-shot learners [Internet]. arXiv; 2022 [cited 2025 June 3]. Available from: http://arxiv.org/abs/2109.01652

82 Hu R, Singh A. UniT: Multimodal multitask learning with a unified transformer [Internet]. arXiv; 2021 [cited 2025 June 3]. Available from: http://arxiv.org/abs/2102.10772

83 Chen M, Tworek J, Jun H, Yuan Q, Pinto HP de O, Kaplan J, et al. Evaluating large language models trained on code [Internet]. arXiv; 2021 [cited 2025 June 3]. Available from: http://arxiv.org/abs/2107.03374

84 He J, Li K, Zang Y, Fu H, Fu Q, Xing J, et al. Not all tasks are equally difficult: Multi-task deep reinforcement learning with dynamic depth routing [Internet]. arXiv; 2024 [cited 2025 June 3]. Available from: http://arxiv.org/abs/2312.14472

85 Thung KH, Wee CY. A brief review on multi-task learning. Multimedia Tools and Applications. 2018;77(22):29705–25.

86 Sahoo P, Singh AK, Saha S, Jain V, Mondal S, Chadha A. A systematic survey of prompt engineering in large language models: Techniques and applications [Internet]. arXiv; 2025 [cited 2025 June 3]. Available from: http://arxiv.org/abs/2402.07927

87 Shin T, Razeghi Y, Logan IV RL, Wallace E, Singh S. AutoPrompt: Eliciting knowledge from language models with automatically generated prompts [Internet]. arXiv; 2020 [cited 2025 June 3]. Available from: http://arxiv.org/abs/2010.15980

88 Zhang Y, Fei H, Li D, Li P. PromptGen: Automatically generate prompts using generative models. In: Carpuat M, de Marneffe MC, Meza Ruiz IV, editors. Findings of the Association for Computational Linguistics: NAACL 2022 [Internet]. Seattle: Association for Computational Linguistics; 2022 [cited 2025 June 3]. pp. 30–7. Available from: https://aclanthology.org/2022.findings-naacl.3/

89 Lester B, Al-Rfou R, Constant N. The power of scale for parameter-efficient prompt tuning [Internet]. arXiv; 2021 [cited 2025 June 3]. Available from: http://arxiv.org/abs/2104.08691

90 Han Z, Gao C, Liu J, Zhang J, Zhang SQ. Parameter-efficient fine-tuning for large models: A comprehensive survey. arXiv; 2024 [cited 2025 June 3]. Available from: https://arxiv.org/abs/2403.14608v7

91 Yang M, Chen J, Zhang Y, Liu J, Zhang J, Ma Q, et al. Low-rank adaptation for foundation models: A comprehensive review [Internet]. arXiv; 2024 [cited 2025 June 3]. Available from: http://arxiv.org/abs/2501.00365

92 Lewis P, Perez E, Piktus A, Petroni F, Karpukhin V, Goyal N, et al. Retrieval-augmented generation for knowledge-intensive NLP tasks. arXiv; 2020 [cited 2025 June 8]. Available from: https://arxiv.org/abs/2005.11401v4

93 Karpukhin V, Oğuz B, Min S, Lewis P, Wu L, Edunov S, et al. Dense passage retrieval for open-domain question answering. arXiv; 2020 [cited 2025 June 8]. Available from: https://arxiv.org/abs/2004.04906v3

94 Sawarkar K, Mangal A, Solanki SR. Blended RAG: Improving RAG (Retriever-Augmented Generation) accuracy with semantic search and hybrid query-based retrievers. arXiv; 2024 [cited 2025 June 8]. Available from: https://arxiv.org/abs/2404.07220v2

95 Izacard G, Lewis P, Lomeli M, Hosseini L, Petroni F, Schick T, et al. Atlas: Few-shot learning with retrieval augmented language models. arXiv; 2022 [cited 2025 June 8]. Available from: https://arxiv.org/abs/2208.03299v3

96 Zhao P, Zhang H, Yu Q, Wang Z, Geng Y, Fu F, et al. Retrieval-augmented generation for AI-generated content: A survey [Internet]. arXiv; 2024 [cited 2025 July 15]. Available from: http://arxiv.org/abs/2402.19473

97 Xi Z, Chen W, Guo X, He W, Ding Y, Hong B, et al. The rise and potential of large language model based agents: A survey [Internet]. arXiv; 2023 [cited 2025 June 8]. Available from: http://arxiv.org/abs/2309.07864

98 Kaufmann T, Weng P, Bengs V, Hüllermeier E. A survey of reinforcement learning from human feedback [Internet]. arXiv; 2024 [cited 2025 June 8]. Available from: http://arxiv.org/abs/2312.14925

99 Ha D, Schmidhuber J. Recurrent world models facilitate policy evolution [Internet]. arXiv; 2018 [cited 2025 June 8]. Available from: http://arxiv.org/abs/1809.01999

100 Cao Y, Zhao H, Cheng Y, Shu T, Chen Y, Liu G, et al. Survey on large language model-enhanced reinforcement learning: Concept, taxonomy, and methods. IEEE Transactions on Neural Networks and Learning Systems. 2025;36(6):9737–57.

101 Bodenreider O. The Unified Medical Language System (UMLS): Integrating biomedical terminology. Nucleic Acids Research. 2004; 32(Database issue):D267–70.

102 Koehn P. Europarl: A parallel corpus for statistical machine translation. In: Proceedings of Machine Translation Summit X: Papers [Internet]. Phuket, Thailand; 2005 [cited 2025 June 8]. pp. 79–86. Available from: https://aclanthology.org/2005.mtsummit-papers.11/

103 Dodge J, Sap M, Marasović A, Agnew W, Ilharco G, Groeneveld D, et al. Documenting large Webtext Corpora: A case study on the colossal clean crawled corpus [Internet]. arXiv; 2021 [cited 2025 June 8]. Available from: http://arxiv.org/abs/2104.08758

104 Gao L, Biderman S, Black S, Golding L, Hoppe T, Foster C, et al. The pile: An 800GB dataset of diverse text for language modeling. arXiv; 2020 [cited 2025 June 8]. Available from: https://arxiv.org/abs/2101.00027v1

105 Common Crawl - Open Repository of Web Crawl Data [Internet] [cited 2025 June 8]. Available from: https://commoncrawl.org

106 Home | Linguistic Data Consortium [Internet] [cited 2025 June 8]. Available from: https://www.ldc.upenn.edu/

107 Language Resources and Evaluation Journal | ELRA [Internet] [cited 2025 June 8]. Available from: https://www.elra.info/dissemination/language-resources-and-evaluation-journal/

108 Hugging Face – The AI community building the future [Internet]. 2025 [cited 2025 June 8]. Available from: https://huggingface.co/

109 Find Open Datasets and Machine Learning Projects | Kaggle [Internet] [cited 2025 June 8]. Available from: https://www.kaggle.com/datasets

110 Schrader TP, Finco M, Grünewald S, Hildebrand F, Friedrich A. MuLMS: A multi-layer annotated text corpus for information extraction in the materials science domain [Internet]. arXiv; 2023 [cited 2025 June 8]. Available from: http://arxiv.org/abs/2310.15569

111 Mikolov T, Chen K, Corrado G, Dean J. Efficient estimation of word representations in vector space. arXiv preprint arXiv:1301.3781. 2013.

112 Rong X. word2vec parameter learning explained [Internet]. arXiv; 2016 [cited 2025 June 8]. Available from: http://arxiv.org/abs/1411.2738

113 Zhang Y, Chen Q, Yang Z, Lin H, Lu Z. BioWordVec, improving biomedical word embeddings with subword information and MeSH. Scientific Data. 2019;6(1):52.

114 Peters ME, Neumann M, Iyyer M, Gardner M, Clark C, Lee K, et al. Deep contextualized word representations. arXiv; 2018 [cited 2025 June 8]. Available from: https://arxiv.org/abs/1802.05365v2

115 Le QV, Mikolov T. Distributed representations of sentences and documents [Internet]. arXiv; 2014 [cited 2025 June 8]. Available from: http://arxiv.org/abs/1405.4053

116 Joulin A, Grave E, Bojanowski P, Mikolov T. Bag of tricks for efficient text classification [Internet]. arXiv; 2016 [cited 2025 June 8]. Available from: http://arxiv.org/abs/1607.01759

117 Jiao X, Yin Y, Shang L, Jiang X, Chen X, Li L, et al. TinyBERT: Distilling BERT for natural language understanding. arXiv; 2019 [cited 2025 June 8]. Available from: https://arxiv.org/abs/1909.10351v5

118 Sanh V, Debut L, Chaumond J, Wolf T. DistilBERT, a distilled version of BERT: smaller, faster, cheaper and lighter [Internet]. arXiv; 2020 [cited 2025 June 8]. Available from: http://arxiv.org/abs/1910.01108

119 Lan Z, Chen M, Goodman S, Gimpel K, Sharma P, Soricut R. ALBERT: A lite BERT for self-supervised learning of language representations. arXiv; 2019 [cited 2025 June 8]. Available from: https://arxiv.org/abs/1909.11942v6

120 Sun Z, Yu H, Song X, Liu R, Yang Y, Zhou D. MobileBERT: A compact task-agnostic BERT for resource-limited devices. arXiv; 2020 [cited 2025 June 8]. Available from: https://arxiv.org/abs/2004.02984v2

121 Liu Y, Ott M, Goyal N, Du J, Joshi M, Chen D, et al. RoBERTa: A robustly optimized BERT pretraining approach [Internet]. arXiv; 2019 [cited 2025 June 8]. Available from: http://arxiv.org/abs/1907.11692

122 Joshi M, Chen D, Liu Y, Weld DS, Zettlemoyer L, Levy O. SpanBERT: Improving pre-training by representing and predicting spans [Internet]. arXiv; 2020 [cited 2025 June 8]. Available from: http://arxiv.org/abs/1907.10529

123 Pires T, Schlinger E, Garrette D. How multilingual is Multilingual BERT? [Internet]. arXiv; 2019 [cited 2025 June 8]. Available from: http://arxiv.org/abs/1906.01502

124 Conneau A, Khandelwal K, Goyal N, Chaudhary V, Wenzek G, Guzmán F, et al. Unsupervised cross-lingual representation learning at scale. arXiv; 2019 [cited 2025 June 8]. Available from: https://arxiv.org/abs/1911.02116v2

125 Lee J, Yoon W, Kim S, Kim D, Kim S, So CH, et al. BioBERT: A pre-trained biomedical language representation model for biomedical text mining. arXiv; 2019 [cited 2025 June 8]. Available from: https://arxiv.org/abs/1901.08746v4

126 Huang K, Altosaar J, Ranganath R. ClinicalBERT: Modeling clinical notes and predicting hospital readmission. arXiv; 2019 [cited 2025 June 8]. Available from: https://arxiv.org/abs/1904.05342v3

127 Tian S, Jiang X, Wang W, Jing Z, Zhang C, Zhang C, et al. Steel design based on a large language model. Acta Materialia. 2025;285:120663.

128 Aftan S, Shah H. A survey on BERT and its applications. In: 2023 20th Learning and Technology Conference (L&T) [Internet]. 2023 [cited 2025 June 8]. pp. 161–6. Available from: https://ieeexplore.ieee.org/document/10092289/

129 Ouyang L, Wu J, Jiang X, Almeida D, Wainwright CL, Mishkin P, et al. Training language models to follow instructions with human feedback [Internet]. arXiv; 2022 [cited 2025 June 15]. Available from: http://arxiv.org/abs/2203.02155

130 Chung HW, Hou L, Longpre S, Zoph B, Tay Y, Fedus W, et al. Scaling instruction-finetuned language models [Internet]. arXiv; 2022 [cited 2025 June 15]. Available from: http://arxiv.org/abs/2210.11416

131 Paszke A, Gross S, Massa F, Lerer A, Bradbury J, Chanan G, et al. PyTorch: An imperative style, high-performance deep learning library [Internet]. arXiv; 2019 [cited 2025 June 15]. Available from: http://arxiv.org/abs/1912.01703

132 Ott M, Edunov S, Baevski A, Fan A, Gross S, Ng N, et al. fairseq: A fast, extensible toolkit for sequence modeling. arXiv; 2019 [cited 2025 June 15]. Available from: https://arxiv.org/abs/1904.01038v1

133 Gardner M, Grus J, Neumann M, Tafjord O, Dasigi P, Liu N, et al. AllenNLP: A deep semantic natural language processing platform [Internet]. arXiv; 2018 [cited 2025 June 15]. Available from: http://arxiv.org/abs/1803.07640

134 Wolf T, Debut L, Sanh V, Chaumond J, Delangue C, Moi A, et al. HuggingFace's transformers: State-of-the-art natural language processing [Internet]. arXiv; 2020 [cited 2025 June 15]. Available from: http://arxiv.org/abs/1910.03771

135 Loper E, Bird S. NLTK: The Natural Language Toolkit [Internet]. arXiv; 2002 [cited 2025 June 15]. Available from: http://arxiv.org/abs/cs/0205028

136 Porter MF. An algorithm for suffix stripping. In: Jones KS, Willett P, editors. Readings in Information Retrieval. San Francisco, CA: Morgan Kaufmann Publishers Inc.; 1997. pp. 313–6.

137 Agichtein E, Gravano L. Snowball: Extracting relations from large plain-text collections. In: Proceedings of the 5th ACM International Conference on Digital Libraries; 2000 [cited 2025 June 15]. Available from: https://doi.org/10.7916/D8V413FM

3

Reinforcement Learning in Materials

3.1 Introduction

Reinforcement learning (RL) is similar to human learning methods: it learns through continuous trial and error and attempts, and uses the rewards brought by doing something as the basis for guiding behavioral improvement. The fundamental structure of RL is composed of two parts: the agent and the environment, they constantly learn and adapt to each other through interaction, as shown in Figure 3.1. The agent is always in an environment and perceives and makes decisions based on its current state. When an agent is in a certain state, it will select an action based on the current strategy and execute it, thereby entering a new state and triggering a state transition in the environment: that is, the environment feeds back a new state based on this action and simultaneously provides an immediate reward signal to evaluate the quality of the behavior. This process is a continuous closed-loop interaction. The agent gradually adjusts its strategy through continuous experimentation and learning, with the aim of obtaining as many cumulative rewards as possible in the long term. The core objective of RL is precisely to enable agents to learn to formulate optimal decision-making strategies in complex and potentially uncertain environments, thereby maximizing the expected total return (also known as return or long-term reward).

This chapter focuses on the fundamental concepts of RL and its applications in several cutting-edge fields. First, it introduces the basic structure and working mechanisms of RL, as well as its similarity to human learning methods, helping readers build a general understanding of the approach. Then, the development history of RL is reviewed, and classical RL algorithms are categorized and analyzed, including value-based methods, policy optimization methods, and model-based approaches. On this basis, the chapter presents recent research progress on the application of RL in materials science, including representative examples in material design and process optimization. Finally, it further explores the application of RL in large language models (LLMs), with a focus on how algorithms play a key role in training such models. Through this chapter, readers will gain a systematic

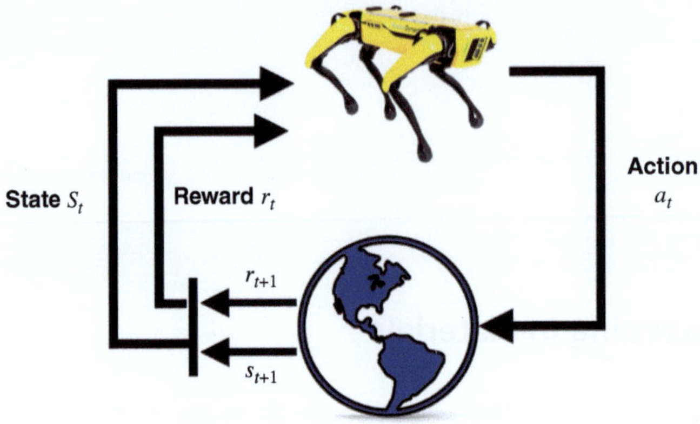

Figure 3.1 Basic schematic diagram of RL. *Source*: [1] / Cornell University / CC BY-SA 4.0.

understanding of the fundamental theories, technological developments, and real-world applications of RL.

3.1.1 The Basic Concepts of RL

In RL, there is a set of core concepts that together constitute the basic framework of the entire learning process. These concepts define how agents perceive the environment, take actions, obtain feedback, and gradually improve their decision-making strategies. These key concepts mainly include state, action, reward, strategy, etc. Furthermore, during the learning process, one must also confront key issues such as the trade-off between exploration and utilization, the estimation of long-term returns, and methods for strategy optimization. The following is a detailed introduction to several key concepts in RL.

The agent is the core of the entire learning system and the main body of learning and decision-making. In RL, the agent is responsible for observing the state of the environment, choosing actions, and continuously optimizing its behavioral strategies based on the obtained feedback, thereby obtaining more rewards in the long term. The environment is the system in which the agent is located, the external background or task scenario of the agent's behavior. When the agent takes an action, the environment will accept this action and return a reward signal and a new state to the agent. The environment can be a simulation system, a game, the real world, or even other agents. The goal of RL is to train such an agent to achieve long-term behavioral optimization through trial and error and experience accumulation in a specific environment.

The state is the representation of the agent's current environment, the basis for the agent's decision-making, and describes the relationship between the agent and the environment, that is, "where the agent is and what situation it is facing." For example, for the grid world in Figure 3.2, each grid is a state, so there are a total of 16 different states: $a_1, a_2, ..., a_{16}$. A state usually contains all the key information that an agent needs in order to select a reasonable action at the current moment. Therefore, a good state design should describe the environmental information of the current agent as completely as possible, while

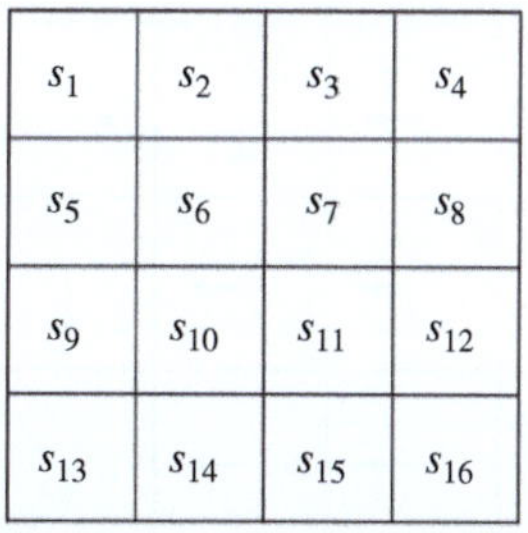

Figure 3.2 The grid-word example.

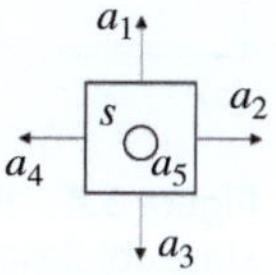

Figure 3.3 The grid-word example's action space.

remaining compact and processable, so that the learning algorithm can extract useful information from it efficiently. The composition of the state generally varies depending on the specific task, mainly including the agent's own attributes, such as position, speed, and direction, as well as the current characteristics of the environment, such as surrounding obstacles and the position of the target object. The state space refers to the collection of all possible states, which defines all the situations that an agent may encounter in the environment. For example, for the grid world in Figure 3.2, the state space is $S = s_{i_{i=1}}^{16}$. The state space can be discrete, such as in grid mazes or board games, where the number of states is limited. The state space can also be continuous, such as the position or speed of a robot. The variables of the state can take an infinite number of values. The state space can also be high-dimensional, such as image data. In RL modeling, it is very important to formulate a reasonable state space because the size of the state space has a direct impact on the design and training efficiency of the learning algorithm. The larger the state space, the more difficult the learning.

An action is a behavioral choice made by an agent in a certain state. When an agent is in a certain state, it selects an action, causing a state change and obtaining a reward. The action space is the set of all the actions that the agent can choose in each state. For example, in the grid world, each state has five possible actions: up, down, left, right, and staying in place, as shown in Figure 3.3. The action space of a certain state is $A(s_i) = a_{i_{i=1}}^{5}$. Different task scenarios generally have different action spaces. In game control, the action space of an agent in a certain state might be {movement, jumping, shooting}, while in autonomous driving, the state space of a certain state might be {acceleration, braking, steering}, etc. The action space generally has the following three different types: a discrete action space with a limited number of actions that can be enumerated. For example, in a discrete maze problem, the action space might be {up, down, left, right}; an action is a continuous action space of one or more continuous variables, such as the speed and corner of a vehicle, etc. It might also be in a mixed action space that contains both discrete and continuous action components simultaneously.

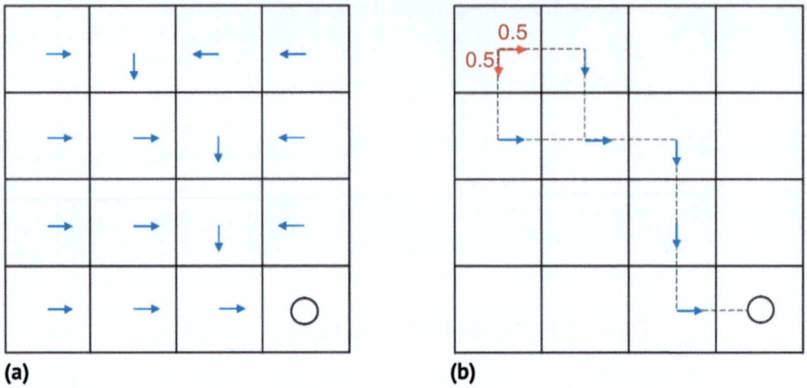

Figure 3.4 Policy types in RL. (a) Deterministic policy: outputs a fixed action for each state. (b) Stochastic policy: outputs a probability distribution over actions.

Policy is the behavioral guideline for what actions the agent should take in each state. In other words, policy is the "behavioral rules" or "decision-making methods" of the agent. Policy is usually represented by a function, denoted as $\pi(a|s)$, which indicates the probability of taking action in state. Depending on the specific form, policy can be divided into two categories: deterministic policy and stochastic policy. The deterministic policy means that the agent directly takes the action with the greatest possibility and always selects action a in state s: $a = \pi(s)$. For example, when the agent is in state s_1, it can only take action a_2, as shown in Figure 3.4a.

The stochastic policy is a function that can be represented as:

$$\pi(a|s) = P(a_t = a, s_t = s) \tag{3.1}$$

When a state is input, the probabilities of all possible actions of the agent are output. Sampling based on this probability distribution is necessary to obtain the action a that the agent will take. For example, when the agent is at s_1, there is a 0.5 probability of taking action a_2 to the right and a 0.5 probability of taking action a_3 to the down, as shown in Figure 3.4b.

In RL, a reward is a feedback signal from the environment for an agent to take an action, which is used to measure the quality of the action in the current state.

It can be expressed as:

$$R_t = r(s_t, a_t) \tag{3.2}$$

Among them, s_t is the state of the agent at time t, and a_t is the action taken at time t. After an agent takes a certain action in a certain state, it will receive a reward signal from the environment. In RL, the positive or negative value of the reward signal directly reflects the superiority or inferiority orientation of the agent's behavior. When an action is executed and the environment returns a positive reward, this usually means that the action is "recognized" or "encouraged" in the current state, and it helps to achieve the task goal. Therefore, positive rewards will drive the agent to be more inclined to repeat the behavior in future decisions. On the contrary, negative rewards represent a punishment mechanism, indicating that the current action is inefficient, unreasonable, or even harmful. It will, to a certain extent, inhibit the recurrence of this action in a similar state. Through the comparison of positive and negative rewards, the agent gradually learns to identify which behaviors are worth adopting and which ones should be avoided in the process of continuous interaction with the environment. For example, in a path planning task, the reward can be designed as follows: the agent may receive a positive reward of +1 for each step taken closer to the goal; a negative reward of −10 for hitting a wall, backtracking, or entering a dangerous area; and a substantial reward of +100 upon successfully reaching the target point. The core objective of the reward mechanism design is to guide the agent to gradually construct a policy evaluation system oriented towards long-term benefits through a continuous trial-and-error process. In other words, the reward function should not only reflect the immediate advantages and disadvantages of an action at a certain moment, but also prompt the agent to understand that choosing which action in a specific state can obtain the maximum cumulative return in the future.

Trajectory is the basic unit that describes the interaction process between an agent and the environment. It is an ordered data sequence, usually recording a series of actions taken by the agent from the initial state under the guidance of a specific strategy, and the resulting state transitions and reward situations. We can formally represent a trajectory as:

$$\tau = (s_0, a_0, r_0, s_1, a_1, r_1, \ldots, s_T) \tag{3.3}$$

Among them, s_t represents the position of the agent at time t, a_t indicates the action taken by the agent in state s_t, r_t represents the reward obtained by the agent after taking action a in state s_t, and s_{t+1} represents the next state entered after performing action a_t.

The trajectory records the entire process of an agent from its beginning to its end, which can be either finite or infinite. In discrete tasks, the trajectory is finite and ends when a certain termination state is reached or a certain termination condition is met. In some continuous tasks or environments without termination conditions, the trajectory can extend infinitely. The trajectory itself is a collection of data, and what we care about is, "Is this trajectory good?" How much return can it bring? To answer this question, RL introduces the concept of reward.

The return refers to the total reward that the agent can accumulate and obtain in the subsequent interactions starting from a certain time step. It is defined as follows:

$$G_t = \sum_{k=0}^{\infty} \gamma^k r_{t+k} \tag{3.4}$$

Among them, G_t is the cumulative return starting from the time step t; $\gamma \in [0,1]$ is the discount factor, which is used to control the attenuation degree of future rewards; r_{t+k} is the immediate reward for step k in the future. The significance of the return lies in the fact that it provides an evaluation criterion for the performance of the strategy. The agent optimizes the strategy by maximizing the expected return, that is, it attempts to take a series of actions to make the future rewards starting from the current state as many as possible. In practical operations, returns are often associated with value functions, dominance functions, etc., and are used to construct the objective function of RL. An episode refers to the entire process from the beginning of a task to its completion or termination. It usually corresponds to a complete trajectory and is an encapsulation of a complete interaction between the environment and the agent. Each episode starts from the initial state of the environment and conducts multistep interactions under the guidance of the agent's execution strategy until the task termination state (such as victory, failure, collision, etc.) is reached, or the maximum number of steps limit is triggered. Formally, an episode is a trajectory, but semantically, an episode places more emphasis on the concept of "task units." Each round contains a complete "attempt"; it is the basic unit of strategy learning and evaluation; it is usually used to evaluate the overall performance of an agent. For example, "average return per round" is a commonly used evaluation metric. Many operations of RL algorithms are carried out based on the granularity of an episode. For example, in actual training, a large number of episodes are usually sampled from the environment to build an experience replay pool, train strategy networks, or estimate environment models, etc. That is to say, an episode is the natural boundary between data generation and the learning process in RL tasks.

3.1.2 The Development History of RL

RL as a vital branch of artificial intelligence, has evolved through decades of interdisciplinary convergence and continuous technological innovation. From its early inspirations in psychology to algorithmic breakthroughs, and from theoretical exploration to practical deployment, RL has developed into a well-established framework encompassing a variety of algorithmic approaches. It has been widely applied in real-world domains such as autonomous driving, recommendation systems, and robotic control, yielding remarkable results. As a result, RL is increasingly becoming one of the central directions in AI research.

3.1.2.1 Origins of Early RL Ideas (1950s–1980s)

The conceptual roots of RL can be traced back to the early 20th-century psychological studies on animal learning behavior. Its foundational theories initially emerged from experimental discoveries related to trial-and-error learning and reward mechanisms. In his classical conditioning experiments, Ivan Pavlov [2] demonstrated the associative relationship between stimulus and response through the salivation response of dogs, providing a prototype model for the perception and response to reinforcement signals.

Building on this, the theory of operant conditioning was developed by B.F. Skinner [3] in the 1930s and 1940s. He emphasized the critical role of reward and punishment in shaping behavior, advancing a theoretical framework in which environmental feedback drives behavioral learning. This became the core of the behaviorist learning school. These psychological studies were the first to systematically demonstrate the importance of environmental feedback in learning behavior, laying a cognitive and experimental foundation for the design of reward functions in modern RL systems.

Meanwhile, theories from engineering and mathematics, cybernetics [4], optimal control [5], and Markov decision process (MDP) [6] began to explore how to achieve optimal decision-making in uncertain environments from a mathematical perspective. These research directions helped establish the theoretical underpinnings of RL. In 1957, Richard Bellman introduced the method of dynamic programming (DP) in his seminal book *Dynamic Programming* [7]. He also formulated what later became the cornerstone of RL—the Bellman equation. This equation describes, in a recursive manner, how the value function evolves across state-action decision sequences, thereby providing theoretical support for computing optimal policies in learning processes.

3.1.2.2 Theoretical Emergence and Framework Construction of RL (1980s–1990s)

From the late 1980s to the 1990s, RL gradually evolved from a collection of heuristic methods into a research paradigm characterized by mathematical rigor and theoretical depth. During this period, the foundational framework and core algorithms of RL were established. Three landmark breakthroughs temporal difference (TD) [8], Q-learning [9], and policy gradient methods [10] laid a solid foundation for the future development of deep RL (DRL).

In 1988, Richard S. Sutton [8] proposed the TD learning method in his paper "Learning to Predict by the Methods of Temporal Differences." This approach combined the recursive update mechanism for value functions from DP with the empirical estimation philosophy of Monte Carlo (MC) methods. As a result, TD learning enabled efficient learning through interaction with the environment without requiring a complete model. TD methods are particularly well-suited to online learning scenarios and have since become core tools in value function approximation across many RL algorithms.

In 1989, Christopher J. C. H. Watkins [9] introduced the now-classic Q-learning algorithm, and later elaborated on its theoretical foundations in a 1992 co-authored paper with Peter Dayan. Q-learning was the first RL algorithm to successfully implement fully off-policy learning, enabling the learning process to converge to the *optimal action-value function* even when the behavior policy differs from the target policy. This property significantly improved the practicality and flexibility of the algorithm, making Q-learning one of the most widely used methods in RL and forming the basis for later models such as deep Q-networks (DQNs).

In 1992, Ronald J. Williams [10] proposed the REINFORCE algorithm, the first method to explicitly estimate policy gradients via sampling. REINFORCE pioneered a path that bypassed value function estimation entirely, optimizing directly over the policy distribution. This opened up a crucial new direction in RL research, namely policy optimization methods. REINFORCE not only improved RL's capability in handling high-dimensional and continuous action spaces, but also gave rise to the development of actor-critic (AC) architectures, which have become a central branch of modern RL.

3.1.2.3 The Scalable Expansion of RL (1990s–2010s)

As RL continued to expand into more diverse application scenarios, the nature of the problems it tackled gradually shifted from small-scale, discrete state spaces to high-dimensional, continuous, and even hybrid state-action spaces. Traditional tabular methods, which rely on explicit enumeration of state-action pairs to represent policies or value functions, became increasingly inadequate due to the computational cost and poor generalization in large-scale problems. Therefore, how to effectively represent and approximate policies or value functions emerged as a critical challenge for the further development of RL.

Against this backdrop, *function approximation* techniques began to gain prominence. In 1995, IBM researcher Gerald Tesauro [11] developed the landmark TD-Gammon system, which combined TD learning with multilayer feedforward neural networks to learn backgammon strategies. The system learned entirely through self-play interactions without relying on handcrafted rules and was eventually able to achieve a performance level comparable to or even exceeding that of human experts. This work was the first to empirically demonstrate the potential of combining neural networks with RL, even though "deep learning" had not yet risen to prominence. Its underlying architecture already hinted at the future paradigm of DRL: using deep nonlinear models to approximate value or policy functions and updating them through experience-driven interactions.

At the same time, researchers began to realize that function approximation alone was insufficient to guarantee the efficiency and stability of policy updates. Since policy iteration is essentially a non-convex and unstable optimization process, overly aggressive or misaligned updates could easily lead to performance degradation or training instability. As a result, increasing attention was paid to the stability and efficiency of policy updates. Enhancing expressive power while designing more robust policy optimization methods became a central research focus.

In 2001, Sham Kakade proposed the natural policy gradient method [12], which introduced Riemannian geometry into the policy space. By weighting updates with the Fisher information matrix, the algorithm directed parameter updates toward the direction of "maximum improvement with minimal change." This approach significantly improved the convergence efficiency of policy optimization and enhanced its robustness in high-dimensional parameter spaces. It became the theoretical foundation for a series of subsequent policy optimization algorithms, including Trust Region Policy Optimization (TRPO) [13] and Proximal Policy Optimization (PPO) [14].

In summary, from the mid-1990s to the early 2010s, RL evolved from the "tabular era" into the "function approximation era." During this period, pioneering practices emerged around using neural networks for strategy learning, along with the development of theoretically sound and empirically effective policy optimization techniques. Although the field had not yet fully transitioned into the deep learning paradigm, the foundational ideas and technical pathways established during this stage laid a solid groundwork for the subsequent rise of DRL.

3.1.2.4 The Rise of DRL (2013–)

In 2013, the DeepMind team introduced the groundbreaking DQN [15, 16], which, for the first time, combined convolutional neural networks with Q-learning and applied this integration to vision-based tasks in the Atari gaming environment. The system demonstrated superhuman performance across a variety of games, marking the formal beginning of the DRL era. Two pivotal mechanisms in DQN experience replay and the target network effectively mitigated instability caused by correlated data and nonstationary training dynamics, laying a foundational framework for subsequent DRL algorithms.

As DRL gained traction, RL rapidly expanded from classical control tasks to more complex and diverse domains. In 2015, John Schulman et al. proposed TRPO [13], which introduced explicit constraints on policy updates based on the KL divergence. Building on the idea of natural gradients, TRPO addressed the instability caused by large policy shifts during training, significantly enhancing optimization performance in high-dimensional and continuous action spaces. Its streamlined successor, PPO [14], introduced in 2017, quickly became one of the most widely used policy optimization algorithms, marking the maturation of policy gradient methods.

During the same period, RL achieved breakthrough results in strategic games. The emergence of AlphaGo [17] in 2016 not only defeated world champions but also brought RL into the public spotlight. Its successors, AlphaGo Zero [18] and AlphaZero [19], fully discarded human gameplay data and learned entirely through self-play, demonstrating a profound integration of RL, tree search, and predictive modeling. In 2020, MuZero [20] pushed these ideas even further: without requiring prior knowledge of environment dynamics, it learned an implicit model purely from interaction data and used it for planning—representing a new apex of *model-free* RL.

Meanwhile, algorithmic innovation continued at a pace. From asynchronous methods like A3C [21] to widely adopted algorithms such as PPO [14], and later to more efficient approaches like Soft Actor-Critic(SAC) [22] and Twin Delayed Deep Deterministic Policy Gradient (TD3) [23], RL steadily expanded its applicability to continuous action spaces and high-complexity control tasks. These advancements empowered applications in robotics, physical simulation, and real-world deployment. Beyond improving training stability and policy quality, such algorithms helped transition RL from the laboratory into real-world scenarios.

This phase marked a period of exponential progression in RL, characterized by:

- Deep neural networks becoming the primary tool for function approximation;
- Policy optimization methods maturing and becoming highly engineered;
- The fusion of model learning and planning significantly boosting performance;
- Application domains expanding across gaming, robotics, autonomous driving, and beyond.

In this phase of transformative growth, RL achieved not only a qualitative leap in theory and algorithms but also a broad expansion of its capability boundaries. Evolving from early systems reliant on handcrafted rules or expert data, RL matured into a paradigm capable of autonomous perception, decision-making, and planning. These advances showcased unprecedented levels of intelligence across a range of real-world problems, laying a solid technical foundation for its future widespread adoption.

3.1.2.5 Future Outlook: Towards Artificial General Intelligence (AGI) (2020s–Beyond)

Entering the 2020s, RL has continued to evolve in theory, algorithms, and applications, presenting a more diverse and systematized development trend. This phase is characterized not only by the steady improvement in algorithmic performance but also by the deep integration and synergistic evolution of RL with other branches of artificial intelligence. RL is gradually transitioning from a narrow task solver to an intelligent system equipped with broad generalization and abstract reasoning capabilities, laying a technical foundation for the advancement toward AGI.

As research progresses, the boundaries of RL continue to expand, with several key emerging trends:

- Multi-agent RL (MARL) [24]: This focuses on the interactions among multiple agents, including cooperation, competition, and equilibrium strategies, with applications in robotic collaboration, autonomous vehicle fleets, and more.
- Offline RL [25]: Addressing the challenge of learning effective policies solely from historical data in situations where online interaction is costly or impractical. This has high value in domains such as industrial automation and healthcare, where exploration risk must be minimized.
- Model-based RL (MBRL): By learning a model of the environment's dynamics, MBRL aims to improve sample efficiency. A representative achievement in this area is *MuZero* [20], which demonstrates powerful predictive capabilities even without access to an explicit environment model.
- RL from human feedback (RLHF): RLHF [26] is one of the most transformative directions in recent RL research. It centers on using human preferences as reward signals to guide agent behavior, addressing the limitations of designing reward functions in traditional RL. Organizations such as OpenAI have employed RLHF to train LLMs like InstructGPT [27] and ChatGPT [28], significantly enhancing alignment and safety in language generation. The rise of RLHF marks a shift in RL from being a purely optimization-based method to becoming a more adaptable and ethically grounded human-AI co-learning paradigm.

In summary, as RL gradually moves beyond closed environments and into the open world, its mission of evolving from a task-specific expert to a general-purpose intelligent agent becomes increasingly apparent. As a core mechanism of autonomous decision-making in intelligent systems, RL is poised to play an indispensable foundational role in the realization of AGI.

3.2 Key Algorithm

RL is a fundamental approach for enabling agents to learn autonomous decision-making. Its algorithmic framework can be systematically constructed around the Bellman equation, which serves as the core theoretical foundation. Based on how policies are represented and optimized, RL algorithms are generally categorized into three major classes: value-based methods, policy-based methods, and hybrid methods that combine value and policy learning (AC). This section will systematically review the key concepts, representative algorithms, and developmental trajectories of these three categories, aiming to provide a comprehensive understanding of the RL algorithmic landscape.

3.2.1 Bellman Equation

3.2.1.1 Markov Decision Process

In RL, an agent continuously learns to make better decisions through interactions with its environment. Specifically, the agent receives the current state information from the environment and selects an action based on that state. The environment then responds to the agent's action by transitioning to a new state and feeding that state back to the agent. This cycle repeats, forming a closed-loop interaction between the agent and the environment. This decision-making and feedback process can be effectively modeled by an MDP [6]. As such, the MDP serves as the fundamental framework for RL.

In a stochastic process, the Markov property means that the future state of the process depends only on the current state and is independent of the past history. This is one of the core modeling assumptions in RL. It can be expressed in probabilistic terms as:

$$P(s_{t+1} | s_t, s_{t-1}, \ldots, s_0) = P(s_{t+1}, s_t) \tag{3.5}$$

If a process satisfies the Markov property, then future transitions are independent of the past and only depend on the present. The Markov property is the basis of all Markov processes (MPs).

An MP can be defined as a tuple:

$$\mathcal{M} = (S, P) \tag{3.6}$$

Among them, S is the finite state space, and $P(s'|s)$ is the transition from state s to state s' and satisfies:

$$\sum_{s'} P(s'|s) = 1 \ \forall s \in S \tag{3.7}$$

When we sample a sequence of states $(s_0, s_1, s_2, \ldots)$ from a process that satisfies the Markov property along the time axis, the resulting trajectory forms a *Markov chain*. A Markov chain is a typical discrete-time stochastic process and serves as the foundation for the sampling phase of many algorithms in RL.

The process of generating such a Markov chain is commonly referred to as Markov sampling, which involves continuously interacting with the environment to produce sequences of states (or state-action pairs) within an MDP. Markov sampling not only reflects the structural characteristics of a Markov chain but also provides essential data for key tasks such as policy evaluation and value function estimation.

To apply an MP to decision-making problems, a reward function needs to be introduced. This extends the MP into a Markov Reward Process (MRP).

An MRP can be defined as a quadruple:

$$\mathcal{M} = (S, P, R, \gamma) \tag{3.8}$$

$R(s)$ is the expected immediate reward, representing the expected reward received when the agent is in state s, that is:

$$R(s) = \mathbb{E}[r_t | s_t = s] \tag{3.9}$$

γ is the discount factor, a constant located within the interval [0,1], which is used to control the importance of future rewards in the current decision. It defines the degree to which an agent attaches importance to future rewards when calculating long-term returns. Specifically, the cumulative return is defined as:

$$G_t = r_t + \gamma r_{t+1} + \gamma^2 r_{t+2} + \ldots = \sum_{k=0}^{\infty} \gamma^k r_{t+k} \tag{3.10}$$

When γ approaches 0, the agent pays more attention to the current immediate rewards, and the influence of future rewards weakens rapidly. When γ approaches 1, the intelligent body will pay more attention to long-term returns and consider more rewards in the future. The discount factor not only reflects the "short-sightedness" or "foresight" of the behavior, but also plays a mathematical convergent role, making the indefinite reward sequence have a finite value.

To enable an agent to make autonomous decisions, an action space must be introduced, leading to the formulation of an MDP, which serves as the core modeling framework in RL.

The MDP can be defined as a five-tuple:

$$\mathcal{M} = (S, A, P, R, \gamma) \tag{3.11}$$

Overall, the key elements of MDP include:

State, the set of states S; action, the set of actions $A(s)$ is associated for state $s \in S$; state transition probability $P(s'|s,a)$, at state s, taking action a, the probability to transit to state s'; $R(s,a)$, at state s, taking action a, the reward received from the environment; γ, the discount factor; policy $\pi(a,s)$, describe the probability of choosing action a in state s; satisfy the Markov property, the memoryless property.

3.2.1.2 Bellman Equation

In RL, the goal of an agent is to maximize its long-term cumulative reward through continuous interaction with the environment. In other words, the agent needs to select actions at each time step to obtain immediate rewards and influence future states and rewards. Specifically, for a given policy π, the sum of future rewards obtained by the agent after time step t is referred to as the return, denoted as G_t. This return is usually defined as the weighted sum of future rewards, as shown in Eq. 3.10.

What we are concerned about is: if we start from state s and follow policy π, what is the total return that can be expected to be obtained? The expectation of G_t is defined as *the state-value function* or simply state value:

$$v_\pi(s) = \mathbb{E}[G_t | S_t = s] \tag{3.12}$$

The state-value function describes the expected total return that can be obtained starting from the current time step t when acting in accordance with the policy π in state s.

This state-value function $v_\pi(s)$ is a function of state s and simultaneously depends on the adopted policy π. That is to say, for different policies π, the corresponding state-value functions are different. The policy π determines what actions the agent chooses in each state, thereby influencing the sequence of states it will experience and the rewards it will obtain in the future. Therefore, different policies lead to different long-term returns in the end.

While state value is important, but it is very difficult to calculate this expectation directly. Since the future rewards depend on many factors such as the strategy, the probability of state transition of the environment, and the reward function, we need a recursive way to calculate this function. This is the core meaning of the "Bellman equation."

We note that the future total return G_t can actually be decomposed into two parts:

$$\begin{aligned} G_t &= R_{t+1} + \gamma R_{t+2} + \gamma^2 R_{t+3} + \ldots \\ &= R_{t+1} + \gamma(R_{t+2} + \gamma R_{t+3} + \ldots) \\ &= R_{t+1} + \gamma G_{t+1} \end{aligned} \tag{3.13}$$

This equation reveals a very crucial fact: The future reward G_t is not simply composed of the simple accumulation of all future rewards, but can be decomposed into two parts, including "the immediate reward" R_{t+1} at the current moment $t + 1$ and "the future reward" G_{t+1} calculated from the next moment $t + 1$. In other words, the current future return not only includes the reward obtained at present, but also the discounted sum of all subsequent rewards calculated recursively in the same way, starting from the next moment. This recursive definition form reflects the time dependence of the return.

Substitute this recursive relation into the definition of the state-value function and obtain:

$$\begin{aligned} v_\pi(s) &= \mathbb{E}[G_t | S_t = s] \\ &= \mathbb{E}[R_{t+1} + \gamma G_{t+1} | S_t = s] \\ &= \mathbb{E}[R_{t+1} | S_t = s] + \gamma \mathbb{E}[G_{t+1} | S_t = s] \end{aligned} \tag{3.14}$$

First, calculate the first term $\mathbb{E}[R_{t+1} | S_t = s]$, which is the mean of immediate rewards:

$$\begin{aligned} \mathbb{E}[R_{t+1} | S_t = s] &= \sum_a \pi(a|s) \mathbb{E}[R_{t+1} | S_t = s, A_t = a] \\ &= \sum_a \pi(a|s) \sum_r p(r|s,a) r \end{aligned} \tag{3.15}$$

Second, calculate the second term $\mathbb{E}[G_{t+1} | S_t = s]$, which is the mean of future rewards:

$$\begin{aligned} \mathbb{E}[G_{t+1} | S_t = s] &= \sum_{s'} \mathbb{E}[G_{t+1} | S_t = s, S_{t+1} = s'] p(s'|s) \\ &= \sum_{s'} \mathbb{E}[G_{t+1} | S_{t+1} = s'] p(s'|s) \\ &= \sum_{s'} v_\pi(s') p(s'|s) \\ &= \sum_{s'} v_\pi(s') \sum_a p(s'|s,a) \pi(a|s) \end{aligned} \tag{3.16}$$

Therefore, can derive the Bellman equation:

$$\begin{aligned} v_\pi(s) &= \mathbb{E}[R_{t+1} | S_t = s] + \gamma \mathbb{E}[G_{t+1} | S_t = s] \\ & \sum_a \pi(a|s) \sum_r p(r|s,a) r + \gamma \sum_a \pi(a|s) \sum_{s'} p(s'|s,a) v_\pi(s') \\ & \sum_a \pi(a|s) \left[\sum_r p(r|s,a) r + \gamma \sum_{s'} p(s'|s,a) v_\pi(s') \right], \forall s \epsilon S \end{aligned} \tag{3.17}$$

For the state set $S = \{s_1, s_2, \ldots, s_n\}$, the Bellman equation can be further expressed in the form of a matrix vector:

$$v_\pi = r_\pi + \gamma P_\pi v_\pi \tag{3.18}$$

Among them, $v_\pi = [v_\pi(s_1), \ldots, v_\pi(s_n)]^T \in \mathbb{R}^n$, $r_\pi = [r_\pi(s_1), \ldots, r_\pi(s_n)]^T \in \mathbb{R}^n, P_\pi \in \mathbb{R}^{n \times n}$.

The Bellman equation is one of the most fundamental mathematical tools in RL. It reveals a recursive relationship: the value of a state is equal to the expected immediate reward obtained by following a given policy π from that state, plus the expected value of the subsequent state (discounted over time). This relationship provides a way to iteratively evaluate the performance of a policy and approximate the state-value function $v_\pi(s)$. Rather than being a closed-form solution, the Bellman equation defines a system of functional equations over all states. This property makes it the theoretical foundation for DP algorithms such as policy evaluation, value iteration, and policy iteration. Moreover, the Bellman equation naturally leads to the definition of the action-value function and the conditions for optimality. It is important to emphasize that the Bellman equation is based on the fundamental assumptions of an MDP, where state transitions and reward distributions depend solely on the current state and action, independent of the past.

After understanding the definition and matrix form of the Bellman equation, a natural question is: how to solve the state-value function $v_\pi(s)$, that is, how to find the expected return of each state under the given environmental model and policy π. Under the premise of a finite state space, there are two main solution methods: the direct solution method and the iterative policy method.

The first is the direct solution, which can be obtained by rearranging the matrix form of the Bellman equation to derive an explicit analytical expression:

$$v_\pi = (I - \gamma P_\pi)^{-1} r_\pi \tag{3.19}$$

This approach is, in theory, the most direct and precise method of solving the problem, but it has a major drawback: high computational complexity.

For most real-world problems, we more commonly use an iterative solution method, also known as Policy Evaluation, which repeatedly updates the state value using the following update rule:

$$v_{k+1} = r_\pi + \gamma P_\pi v_k \tag{3.20}$$

The iteration continues until $\|v_{k+1} - v_k\| < \varepsilon$, where ε is a small threshold that determines the desired level of convergence accuracy.

In the previous chapters, the state-value function $v_\pi(s)$ was defined, representing the expected total return when starting from state S and following policy π. However, in practical problems, interest often extends beyond the value of a state to include the value of taking a specific action in that state. To address this, the action-value function, also known as the Q-function, is introduced and denoted as:

$$q_\pi(s,a) = \mathbb{E}_\pi[G_t | S_t = s, A_t = a] \tag{3.21}$$

It indicates: if the state s takes action a, and follows the policy from the next moment π, the expected total return that can be obtained. It is a function of the state-action pair (s,a), depending on policy π.

Similar to the state-value function, the action-value function also satisfies a recursive relationship, known as the Bellman Expectation Equation for action value:

$$q_\pi(s,a) = \sum_r p(r|s,a)r + \gamma \sum_{s'} p(s'|s,a)v_\pi(s') \tag{3.22}$$

There is a close connection between Eq. 3.17 of the state-value function and Eq. 3.22 of the action-value function. Under the premise that the policy is known, the value of the state can be obtained by the weighted average of the behavioral values of all possible actions in that state. The weights are given by the policy, and its mathematical expression is:

$$v_\pi(s) = \sum_a \pi(a|s)q_\pi(s,a) \tag{3.23}$$

3.2.1.3 Bellman Optimality Equation

In the previous two sections, the state-value function $v_\pi(s)$ and the action-value function $q_\pi(s,a)$ under a given policy π were introduced, along with their corresponding Bellman Expectation Equations. These equations fundamentally describe how an agent estimates future cumulative returns while following a fixed policy. However, the central objective in RL goes beyond evaluating a predetermined policy. The goal is to identify an optimal policy, under which the agent can achieve the maximum possible long-term cumulative return from any given state. This leads to the focus of the current section: the Bellman Optimality Equation.

First of all, define the optimal policy. The optimal policy refers to the policy of choosing the action that can maximize the expected long-term return in each state. In other words, the optimal policy is the one that can maximize the state-value function or behavior value function. Here, the optimal policy is represented by π^*.

For the state-value function, there is:

$$v_{\pi^*} = \max_\pi v_\pi(s) \tag{3.24}$$

For the action-value function, there is:

$$\pi^*(a|s) = arg\ \max_a\ q^*(s,a) \tag{3.25}$$

Based on the above optimal policy, the optimal state-value function is:

$$\begin{aligned} v(s) &= \max_\pi \sum_a \pi(a|s)\Big(\sum_r p(r|s,a)r + \gamma \sum_{s'} p(s'|s,a)v(s')\Big) \\ &= \max_\pi \sum_a \pi(a|s)a(s,a), \forall s \epsilon S \end{aligned} \tag{3.26}$$

The matrix-vector form is:

$$v = \max_\pi (r_\pi + \gamma P_\pi v) \tag{3.27}$$

Starting from state s, the agent can maximize the expected return by choosing an optimal action a.

Similarly, there also exists an optimal action-value function:

$$q^*(s,a) = \sum_{s',r} p(s',r|s,a)\left[r + \gamma \max_{a'} q^*(s',a')\right] \tag{3.28}$$

This means that by taking action a in state s, the environment transitions to the next state s' and yields an immediate reward r. From that point onward, the agent selects the action a' that yields the maximum return. The expected return of this entire process corresponds to the optimal action value of the current state-action pair.

In order to move from theory to practice, three common types of methods in RL will be introduced successively in the following three parts: value-based algorithms, policy-based algorithms, and the AC algorithms that take both into account.

3.2.2 Value-based Algorithm

Value-based methods are among the most classic and fundamental classes of algorithms in RL. The core idea is to learn a state-value function $V(s)$ or an action-value function $Q(s,a)$, and thereby indirectly derive the optimal policy. These methods do not optimize the policy directly; instead, they guide decision-making by estimating the expected return from the environment.

This chapter begins with the theoretical foundation, that is, DP, and gradually transitions to MC methods and TD learning. It then introduces a series of classic action-value-based RL algorithms, including SARSA, Q-learning, and DQN.

This development trajectory reflects a progression from model-based to model-free methods, from non-incremental to incremental learning, and from tabular representations to function approximation. Through this evolution, value-based methods become increasingly capable of handling complex state spaces in real-world problems, laying a solid foundation for more advanced policy optimization techniques.

3.2.2.1 Dynamic Programming

DP [7] is a fundamental class of methods in RL for solving optimal policies. Its core idea lies in leveraging a known model of the environment, specifically, the state transition probabilities and reward functions defined in an MDP, to iteratively compute the state- or the action-value functions, thereby indirectly deriving the optimal policy.

A key prerequisite of DP is that the environment's dynamics are fully known. In other words, DP is a model-based approach. For any given state-action pair (s,a), the system must provide precise knowledge of the transition probability $P(s' \mid s,a)$ and the immediate reward $R(s,a)$. This "known model" assumption enables DP methods to directly apply Bellman equations for numerical reasoning and policy optimization.

The essential idea of DP is to repeatedly apply the Bellman equations to update the value functions, thereby gradually approximating the optimal value function or optimal policy. Two classic methods in this category are value iteration and policy iteration, which will be introduced in detail below.

As mentioned in Section 3.2.1.2, the Bellman equation can be solved through an iterative method, and the solution approach for the Bellman optimal equation is similar:

$$v_{k+1} = f(v_k) = \max_{\pi}(r_\pi + \gamma P_\pi v_k), k = 1,2,3... \tag{3.29}$$

Value iteration is an iterative method that directly updates based on the Bellman optimal equation. Its core lies in approximating the optimal state-value function through the maximization operation at each step. This process starts from any initial function v_0, and its update formula can be divided into two steps.

Step 1: policy update. Start from v_0 which is given, this step is to solve:

$$\pi_{k+1} = arg \max_{\pi}(r_\pi + \gamma P_\pi v_k) \tag{3.30}$$

Step 2: value update.

$$v_{k+1} = r_{\pi_{k+1}} + \gamma P_{\pi_{k+1}} v_k \tag{3.31}$$

Policy iteration is a method that alternates policy evaluation with policy improvement, which can gradually enhance the performance of the policy in each round of iteration. This process starts from any initial policy π_0, and its update formula can also be divided into two steps.

Step 1: policy evaluation. Start from a random initial policy π_0, this step is to calculate the state value of π_k:

$$v_{\pi_k} = r_{\pi_k} + \gamma P_{\pi_k} v_{\pi_k} \tag{3.32}$$

Due to the inconvenience of a direct solution, this step requires many iterations to approach v_{π_k} repeatedly.

Step 2: policy improvement:

$$\pi_{k+1} = arg \max_{\pi}\left(r_\pi + \gamma P_\pi v_{\pi_k}\right) \tag{3.33}$$

Take the new policy as the input for the next round, continue to evaluate and improve the policy until the policy no longer changes, and then converge to the optimal policy.

As a theoretical cornerstone of RL, DP offers strong mathematical rigor. Under the assumption of a known model, its iterative process exhibits good convergence properties and can guarantee that the resulting policy is optimal.

However, DP also faces several practical limitations. First, it is highly dependent on the availability of accurate model information. In most real-world environments, the state transition probabilities and reward functions are either unknown or too complex to enumerate exhaustively. Second, DP methods suffer from the "curse of dimensionality" when applied to high-dimensional state spaces, where the computational and memory costs grow rapidly. In addition, traditional DP struggles to handle continuous state spaces or high-dimensional perceptual inputs (such as images), which severely limits its scalability to large and complex problems.

3.2.2.2 MC Learning

To address the challenge that DP relies heavily on accurate model information and is therefore difficult to apply in practice, the development of RL has gradually shifted from model-based DP methods to model-free, experience-driven approximation approaches, starting with MC methods.

MC methods [29] are a class of RL approaches that do not require prior knowledge of the environment model. Instead, they estimate state- or action-value functions based solely on interaction trajectories with the environment. Unlike DP, MC methods do not assume any prior information about the transition probabilities $P(s'|s,a)$ or the reward function $R(s,a)$, making them more broadly applicable in real-world problems.

According to Eq. 3.33, it can be known that the reason why the DP algorithm needs to rely on the transition probabilities is that an environmental model is required at the policy improvement step:

$$\begin{aligned}\pi_{k+1}(s) &= arg\max_{\pi}\sum_{a}\pi(a|s)\left[\sum_{r}p(r|s,a)r+\gamma\sum_{s'}p(s'|s,a)v_{\pi_k}(s')\right]\\ &= arg\max_{\pi}\sum_{a}\pi(a|s)q_{\pi_k}(s,a)\end{aligned} \tag{3.34}$$

The key is $q_{\pi_k}(s,a)$. If its value can be known, there is no longer a need to know the model.

The basic idea of the MC method is: by sampling the complete trajectory (episode) from the environment, the mean of multiple return samples is used to estimate the state value function or action value function. As shown in the following equation:

$$\begin{aligned}q_{\pi_k}(s,a) &= \mathbb{E}[G_t|S_t=s,A_t=a]\\ &\approx \frac{1}{N}\sum_{i=1}^{N}g^{(i)}(s,a)\end{aligned} \tag{3.35}$$

Among them, $g(s,a)$ is the discounted return.

In terms of the sampling mechanism, DP methods rely on a known environment model, that is, complete knowledge of the state transition probabilities and reward function. Based on this, they compute value functions by systematically sweeping through all possible states and performing updates. In contrast, MC methods are entirely based on experience obtained through interaction with the environment, requiring no prior knowledge of the environment's dynamics. They estimate state- or action-value functions by sampling multiple complete episodes and using the observed returns from each interaction.

Apart from the difference in sampling mechanisms, MC methods are largely consistent with DP methods in terms of value function representation, policy improvement strategies, and the use of Bellman equations. Therefore, aspects of MC algorithms that are identical to DP will not be reiterated.

As a fundamental approach to policy evaluation in RL, MC methods exhibit several distinctive characteristics. First, MC methods do not rely on a model of the environment; they require no prior knowledge of state transition probabilities or the reward function, making them particularly suitable for environments where the model is unknown or highly complex. Second, from a theoretical perspective, MC methods can converge to the true value function under certain conditions. Additionally, MC methods are versatile in that they can be applied to both on-policy and off-policy learning settings.

However, MC methods also have certain limitations. One of the most notable drawbacks is that they must wait until the end of a complete episode before performing an update, which reduces the timeliness of learning. In environments with sparse rewards, the convergence of MC methods tends to be slow, limiting their efficiency. Moreover, during the early stages of learning, the high variance in sampled returns can lead to large fluctuations in value estimates, resulting in instability.

3.2.2.3 TD Learning

To address the issue of delayed updates in the learning process of MC methods, the TD learning [8] approach was introduced. Unlike MC methods, TD learning does not require waiting until the end of a full episode to perform updates. Instead, it can adjust the value function estimates immediately after each interaction with the environment. This step-by-step update mechanism makes TD methods significantly more efficient than MC methods in practice.

More importantly, TD learning combines the strengths of both DP and MC methods: it inherits the bootstrapping nature of DP by using current value estimates for updates, while also embracing the sampling-based nature of MC by not requiring a model of the environment. As a result, TD learning enables a more efficient and flexible RL process without relying on complete trajectories or full knowledge of the environment's dynamics.

The emergence of TD methods marks a pivotal shift in RL algorithms from non-incremental approaches to incremental ones, enabling more responsive and scalable learning in complex, real-world settings.

The core idea of TD learning is to update the value estimate of the current state based on the difference between its current estimate and the estimate of the next state, which is known as the TD error. Taking the prediction of the state-value function as an example, the basic update rule of the TD method can be expressed as follows:

$$v_{t+1}(s_t) = v_t(s_t) - \alpha_t(s_t)\left[v_t(s_t) - \left[r_{t+1} + \gamma\, v_t(s_{t+1})\right]\right] \tag{3.36}$$

Where $t = 0,1,2,\ldots,v_t(s_t)$ is the estimated state value of $v_\pi(s_t)$, $\alpha_t(s_t)$ is the learning rate of s_t at time t.

Among them, the TD target is:

$$\overline{v_t} = r_{t+1} + \gamma\, v_t(s_{t+1}) \tag{3.37}$$

The TD error is:

$$\delta_t = v_t(s_t) - \left[r_{t+1} + \gamma\, v_t(s_{t+1})\right] = v_t(s_t) - \overline{v_t} \tag{3.38}$$

TD algorithms can be applied not only to the learning of state-value functions, but also directly to the learning of action-value functions. In this case, the algorithm estimates the value of each state-action pair to guide policy updates, thereby providing a more direct and fine-grained means of control in RL.

SARSA [30] is a classic on-policy TD control algorithm. Its name derives from the five elements involved in each update step: $(s_t,a_t,r_{t+1},s_{t+1},a_{t+1})$. These represent the current state, the action taken, the immediate reward received, the next state, and the next action chosen, respectively.

The core idea of SARSA is to update the action-value function immediately after each interaction with the environment using the new experience. The update rule is given by:

$$q_{t+1}(s_t,a_t) = q_t(s_t,a_t) - \alpha_t(s_t)\left[q_t(s_t,a_t) - \left[r_{t+1} + \gamma q_t(s_{t+1},a_{t+1})\right]\right] \tag{3.39}$$

As an on-policy method, SARSA follows the current behavior policy π, which is typically implemented using an ϵ-greedy policy: with probability $1-\epsilon$, the best-known action is chosen; with probability ϵ, a random action is selected. This ensures a balance between the exploitation of known information and the exploration of new possibilities.

Q-learning [9] is one of the most representative and widely used TD control algorithms in RL. Unlike SARSA, Q-learning is an off-policy method, meaning that it updates the action-value function using the assumption that the agent always selects the optimal action, regardless of the action actually taken during learning. This allows the algorithm to directly move toward the optimal policy during the learning process.

Q-learning aims to solve the BOE:

$$q(s,a) = \mathbb{E}\left[R_{t+1} + \gamma \max_a q(S_{t+1},a) \middle| S_t = s, A_t = a\right], \forall s,a \tag{3.40}$$

The core update rule of Q-learning is as follows:

$$q_{t+1}(s_t,a_t) = q_t(s_t,a_t) - \alpha_t(s_t)\left[q_t(s_t,a_t) - \left[r_{t+1} + \gamma \max_{a\in A} q_t(s_{t+1},a_t)\right]\right] \tag{3.41}$$

As this equation shows, Q-learning updates the value of the current action based on the maximum estimated value of the next state, rather than the value of the action that was actually taken.

TD methods not only address the update latency problem inherent in MC methods but also inherit the iterative refinement mechanism of current value estimates found in DP. By enabling updates after each interaction step, TD learning realizes an incremental learning process, significantly improving sample efficiency and flexibility. TD methods include algorithms such as SARSA (on-policy) and Q-learning (off-policy), which provide reliable strategies for control updates in both on-policy and off-policy scenarios. In particular, Q-learning's off-policy nature allows it to converge toward the optimal policy, theoretically offering stronger performance guarantees.

However, traditional TD learning methods typically rely on a tabular representation to store and update the value function, which limits their applicability to environments with small state and action spaces. When faced with real-world applications involving high-dimensional, continuous, or even perceptual-level inputs such as images or speech, tabular methods become inadequate.

3.2.2.4 Value Function Approximation and DQN

As application scenarios become increasingly complex, tabular methods such as SARSA and Q-learning reveal a critical limitation: their inability to handle high-dimensional state spaces. This is because tabular approaches require storing and updating a separate value for every possible state-action pair, leading to exponential growth in storage and computational costs.

To address this challenge, the field of RL introduced the concept of Value Function Approximation, which aims to use parameterized functions to approximate the value function. This approach enables generalization across complex state spaces and allows for more efficient representation and computation.

In value function approximation methods, RL no longer maintains a separate numerical value for each individual state or state-action pair. Instead, it employs a parameterized function to approximate the value function. This approach significantly enhances the generalization ability over complex state spaces and alleviates the burden of storage and computation.

Specifically, the state- and action-value functions can be expressed in the following parameterized forms:

$$\hat{v}(s, w) \tag{3.42}$$

$$\hat{q}(s,a, w) \tag{3.43}$$

Here, w is a vector of parameters, which determines the shape and output of the function. These parameters are adjusted through learning, based on experience, to make the function $\hat{v}$ or $\hat{q}$ better approximate the true value.

DQN [15, 16], proposed by DeepMind in 2015, is an RL algorithm that integrates Q-learning with deep convolutional neural networks. It was the first to achieve superhuman performance in complex game environments such as Atari 2600 using only raw pixel inputs without the need for any handcrafted features. This groundbreaking achievement marked the official arrival of the DRL era.

The core idea of DQN is to use a deep neural network, denoted as $\hat{q}(s,a, w)$, to approximate the action-value function, where w represents the network's weight parameters. During each training iteration, DQN updates the network by minimizing the following loss function:

$$J(w) = \mathbb{E}_{(s,a,r,s')\sim\mathcal{D}}\left[\left(r + \gamma \max_{a' \in A(s')} \hat{q}(s',a',w^-) - \hat{q}(s,a,w)\right)^2\right] \tag{3.44}$$

Here, $(s,a,r,s') \sim \mathcal{D}$ denotes a transition tuple sampled from the experience replay buffer; the term $\hat{q}(s,a,w)$ represents the current Q-network's estimate of the action value, parameterized by w, while $\hat{q}(s',a',w^-)$ refers to the value estimated by the target network for the next state-action pair, with parameters w^-that are periodically copied from the main network and kept fixed for a number of steps to improve training stability. In this context, γ is the discount factor and r is the immediate reward. The loss function updates the network parameters by minimizing the mean squared error (MSE) between the target Q-value and the current estimate.

To further improve training stability and convergence speed, DQN introduces two key mechanisms: first, experience replay, the agent interacts with the environment, and each experience tuple (s,a,r,s') is stored in a finite replay buffer. During training, DQN does not use the most recent sample, but instead randomly samples minibatches from this buffer. This approach breaks the temporal correlation between samples, reduces variance, improves sample efficiency, and significantly enhances training stability and generalization. Second, fixed target network, when computing the TD target y, DQN uses a target network whose parameters w^- are updated only periodically from the main network. This delayed update mechanism provides a more stable and smooth learning signal, which helps prevent the frequent fluctuations in the target value that can destabilize training.

In addition, DQN typically leverages modern gradient optimization algorithms such as Adam to update the network parameters efficiently. By combining the representational power of deep learning with the optimization capabilities of RL, DQN has become a foundational algorithm in the field of DRL and laid the groundwork for numerous improved variants, such as double DQN [31], dueling DQN [32], and rainbow DQN [33].

3.2.3 Policy-based Algorithm

In the previous section, we introduced value-based algorithms such as DP, MC methods, and TD learning, as well as core RL methods including Q-learning, SARSA, and DQN. These approaches share a common feature: they derive policies indirectly by estimating a form of value function, such as the state-value function $V(s)$ or the action-value function $Q(s,a)$, with policy improvement typically achieved by greedily selecting the action that appears optimal under the current value estimate. This framework has proven highly effective in tasks with discrete action spaces, such as classical control problems and strategic game environments. However, as RL has been extended to more complex and realistic domains, value-based methods have shown notable limitations. Specifically, they struggle in scenarios where the action space is continuous (e.g., controlling the precise angles of robotic joints), where optimal behavior requires inherently stochastic policies (e.g., due to exploration ambiguity or partial observability), or where policies must possess complex internal structure (e.g., distributed decision-making or memory-dependent behavior), making it impractical to compute or apply value functions directly. To overcome these challenges, the RL community has developed an alternative family of methods known as policy-based algorithms, or policy gradient methods, which learn and optimize policies directly rather than relying on value function approximation.

Unlike value-based methods, policy optimization approaches directly model and optimize the policy itself without relying on a value function. The policy is defined as a parameterized function $\pi(a|s,\theta)$, representing the probability of taking action a given state s, where θ denotes the learnable parameters. The objective of policy optimization is to maximize a form of expected long-term return.

Specifically, the objective function is given by:

$$J(\theta) = \mathbb{E}_{\tau\sim\pi_\theta}\left[\sum_{t=0}^{T} \gamma^t r_t\right] \tag{3.45}$$

where $\tau = (s_0, a_0, r_0, \ldots, s_T)$ denotes a trajectory generated by the policy π_θ.

To maximize the policy objective function, policy gradient methods use gradient ascent, which involves computing the gradient of the objective with respect to the policy parameters $\nabla_\theta J(\theta)$ and updating the parameters to improve the policy. The parameter update rule is:

$$\theta \leftarrow \theta + \alpha \nabla_\theta J(\theta) \tag{3.46}$$

A key question is how to compute $\nabla_\theta J(\theta)$. The Policy Gradient Theorem provides the expression for the gradient, stating that:

$$\nabla_\theta J(\theta) = \mathbb{E}_{\pi_\theta}\left[\nabla_\theta \log \pi_\theta(a_t|s_t) \cdot Q^\pi(s,a)\right] \tag{3.47}$$

In practice, since the expectation is hard to compute exactly, it is estimated through sampling. The most basic algorithm is *REINFORCE* [10], which uses MC sampling to estimate the gradient by generating complete trajectories from executing the policy and computing the cumulative return G_t starting from time step t. The policy gradient in the REINFORCE algorithm is:

$$\nabla_\theta J(\theta) = \mathbb{E}_{\pi_\theta}[\nabla_\theta \log \pi_\theta(a_t|s_t) \cdot G_t] \tag{3.48}$$

where is $G_t = \sum_{k=t}^{T} \gamma^{k-t} r_k$ the actual return from the time step t.

Policy-based methods directly parameterize the policy, avoiding reliance on value functions, which allows them to naturally handle continuous action spaces and stochastic policies. REINFORCE, as the most basic policy gradient algorithm, estimates the gradient using MC sampling. This algorithm is intuitive and easy to implement but suffers from high variance, often resulting in slow convergence and unstable training.

3.2.4 AC Methods

Pure policy gradient methods, such as the REINFORCE algorithm, often suffer from unstable training and slow convergence. To overcome these challenges, the field of RL has developed a class of hybrid algorithms that combine the advantages of value-based and policy-based methods, known as AC methods.

3.2.4.1 AC Framework and Basic Algorithm

Around the year 2000, the foundational AC method [34] was proposed. The AC method combines a policy function (the actor) with a value function (the critic) to perform the complementary tasks of policy optimization and policy evaluation. The actor is responsible for selecting actions given the current state, typically using a parameterized policy function $\pi_\theta(a|\theta)$, which represents the probability of taking action a in state s, while the critic estimates the state-value function $V^\pi(s)$ or the action-value function $Q^\pi(s,a)$ under the current policy to assess the quality of the actor's decisions. In this setup, the critic's value estimates serve as a learning signal that guides the actor's policy updates, enabling the agent to gradually improve its behavior toward maximizing long-term rewards.

A typical update step of the AC method is as follows:

First, sampling. Sample action a_t from state s_t according to the current policy $\pi_\theta(a|\theta)$, perform action a_t to obtain the reward r and the next state s_{t+1}.

Second, a critic update. The estimation error of the value function is calculated using the TD method:

$$\delta_t = r_t + \gamma V_w(s_{t+1}) - V_w(s_t) \tag{3.49}$$

Use δ_t as the TD error and minimize δ_t^2 to update the critic's parameter w.

Third, actor update. Use the TD error to approximate the dominant function and update the policy parameters:

$$\theta \leftarrow \theta + \alpha \cdot \nabla_\theta \log \pi_\theta(a_t | s_t) \cdot \delta_t \tag{3.50}$$

The AC method improves the directionality and sample efficiency of policy gradients through value estimation by the critic, while retaining the flexibility of policy optimization. This makes it particularly well-suited for tasks with high-dimensional and continuous action spaces.

However, it also has limitations; for example, inaccurate value estimates from the critic can lead to biased policy updates, and the simultaneous training of the actor and critic can result in instability. As a result, more advanced algorithms such as A2C, A3C, DDPG, PPO, and SAC have been developed to enhance stability and performance.

3.2.4.2 Trust Region Policy Optimization

In the process of policy optimization, policy gradient methods iteratively improve the policy by maximizing the expected cumulative return. The previously introduced policy-based methods include policy gradient algorithms and AC algorithms. Although these methods rely on standard policy gradient updates that are simple and intuitive, they suffer from a key issue: the magnitude of the update is difficult to control. A slight misstep can cause the policy to degrade or even collapse. This instability arises because if the policy update step is too large, it may lead to a significant discrepancy between the old and new policies, thereby harming training stability.

To recap, policy-based methods involve parameterizing the agent's policy and defining an objective function that reflects the policy's performance. The goal is to maximize this function via gradient ascent to achieve optimal behavior. Specifically, suppose θ represents the parameter of policy π_θ, define $J(\theta) = \mathbb{E}_{s_0}\left[V^{\pi_\theta}(s_0)\right] = \mathbb{E}_{\pi_\theta}\left[\sum_{t=0}^{\infty} \gamma^t r(s_t,a_t)\right]$, the goal of the policy-based method is to find $\theta^* = arg \max_\theta J(\theta)$, and the policy gradient algorithm mainly iteratively updates the policy parameter θ along the direction of $\nabla_\theta J(\theta)$. The core idea of policy gradient methods is to iteratively update the policy parameters along the gradient direction to improve expected returns. However, a major drawback arises when the policy network is a deep model; large steps along the gradient may lead to a drastically worse policy, which in turn disrupts the training process.

To address this problem, researchers introduced the concept of trust region optimization, leading to the development of more stable policy optimization methods. One of the most representative algorithms in this category is TRPO [13]. TRPO's core idea is to define a "trust region" within which policy updates are constrained to ensure performance improvement or at least prevent degradation. Proposed in 2015, TRPO provides a theoretical guarantee of monotonic improvement in policy performance and has demonstrated superior empirical results compared to standard policy gradient methods.

Before introducing TRPO, let's first introduce the concept of the advantage function. In RL, the advantage function measures the superiority or inferiority of a certain action relative to the average performance of the current policy. Its definition is as follows:

$$A^\pi(s,a) = Q^\pi(s,a) - V^\pi(s) \tag{3.51}$$

Compared with directly using state value functions or action value functions, the advantage function can more effectively distinguish between good and bad actions, and thus plays an important role in strategy optimization.

However, in practical applications, the estimation of the advantage function is prone to being affected by high variance. To address this issue, the Generalized Advantage Estimation (GAE) was proposed. GAE introduces a tunable parameter λ between the accuracy and variance, and combines multi-step TD errors to effectively balance bias and variance. Its form is as follows:

$$\widehat{A}_t^{\mathrm{GAE}(\gamma,\lambda)} = \sum_{l=0}^{\infty} (\gamma\lambda)^l \delta_{t+l} \tag{3.52}$$

where $\delta_t = r_t + \gamma V(s_{t+1}) - V(s_t)$ is the TD error. γ is the discount factor that determines the importance of future rewards, while λ is the hyperparameter used in GAE, controlling the trade-off between bias and variance. When $\lambda = 0$, GAE reduces to a TD method, which has higher bias but lower variance. When $\lambda = 1$, it becomes equivalent to the MC method, which has lower bias but higher variance.

In practical RL training, policy optimization is typically not a one-time process but an iterative one. In each update, the policy parameters are adjusted based on the current policy's performance to improve expected future returns. However, this process presents a key challenge: data is collected under the old policy (i.e., with the current parameters), yet the goal is to evaluate and optimize the new policy. To address this mismatch between the data distribution and the target policy, RL commonly adopts a technique from probabilistic inference known as importance sampling (IS).

Suppose we need to calculate the expectation of the function $f(x)$ under the distribution $p(x)$:

$$\mathbb{E}_{x \sim p}[f(x)] = \int p(x) f(x) dx \tag{3.53}$$

However, data can only be sampled from another distribution $q(x)$, such as off-policy. The data generated by the old policy is used to update the new policy, and the bias is corrected through the importance weight $q(x)/p(x)$:

$$\mathbb{E}_{x \sim p}[f(x)] = \mathbb{E}_{x \sim q}\left[\frac{p(x)}{q(x)} f(x)\right] \tag{3.54}$$

However, IS itself has a major drawback: when the difference between the old and new policies becomes too large, the importance weights can fluctuate dramatically, leading to high variance in the objective function estimates and even unstable updates. This issue is especially severe in high-dimensional state or action spaces.

To address this problem, TRPO was introduced. TRPO adopts the idea of a "trust region," explicitly constraining how much the new policy is allowed to deviate from the old one during each update. This is typically enforced by limiting the KL divergence between the two policies to remain below a predefined threshold. By doing so, even when using IS to estimate expectations, TRPO ensures that the policy improves performance steadily with each iteration, avoiding catastrophic drops caused by overly aggressive updates.

The TRPO objective function integrates the advantage function and IS while incorporating constrained optimization principles. This represents a significant advancement in the stability and convergence of policy gradient methods.

In the process of policy optimization, our objective is to maximize the expected cumulative return of the policy π_θ:

$$J(\theta) = \mathbb{E}_{\tau \sim \pi_\theta}\left[\sum_{t=0}^{T} \gamma^t r_t\right] \tag{3.55}$$

According to the policy gradient theorem, the gradient can be expressed as:

$$\nabla_\theta J(\theta) = \mathbb{E}_{\pi_\theta}\left[\nabla_\theta \log \pi_\theta(a|s) Q^{\pi_\theta}(s,a)\right] \tag{3.56}$$

To reduce the variance, the Q function is usually replaced by the advantage function:

$$\nabla_\theta J(\theta) = \mathbb{E}_{\pi_\theta}\left[\nabla_\theta \log \pi_\theta(a|s) A^{\pi_\theta}(s,a)\right] \tag{3.57}$$

In actual training, we cannot resample a large amount of data for each update. Therefore, we hope to use the data sampled by the old strategy to estimate the performance difference of the new strategy. At this point, IS is introduced:

$$\nabla_\theta J(\theta) = \mathbb{E}_{s,a \sim \pi_{\theta_{\text{old}}}}\left[\frac{\pi_\theta(a|s)}{\pi_{\theta_{\text{old}}}(a|s)} \nabla_\theta \log \pi_\theta(a|s) A^{\pi_{\theta_{\text{old}}}}(s,a)\right] \tag{3.58}$$

TRPO adopted the following surrogate objective as the optimization target:

$$\begin{aligned} L^{\text{TRPO}}(\theta) &= \mathbb{E}_{s,a \sim \pi_{\theta_{\text{old}}}}\left[\frac{\pi_\theta(a|s)}{\pi_{\theta_{\text{old}}}(a|s)} A^{\pi_{\theta_{\text{old}}}}(s|a)\right] \\ &= \mathbb{E}_{s,a \sim \pi_{\theta_{\text{old}}}}\left[r(\theta) A^{\pi_{\theta_{\text{old}}}}(s|a)\right] \end{aligned} \tag{3.59}$$

During the strategy update, the range of the strategy update is constrained by the KL divergence (Kullback–Leibler divergence) to form a "trust region," which ultimately constitutes the optimization problem of TRPO:

$$\begin{aligned} &\max_\theta \mathbb{E}_{s,a \sim \pi_{\theta_{\text{old}}}}\left[\frac{\pi_\theta(a|s)}{\pi_{\theta_{\text{old}}}(a|s)} A^{\pi_{\theta_{\text{old}}}}(s|a)\right] \\ &s.t.\ \mathbb{E}_{s \sim \pi_{\theta_{\text{old}}}}\left[D_{\text{KL}}(\pi_{\theta_{\text{old}}}(\cdot|s) \| \pi_\theta(\cdot|s))\right] \le \delta \end{aligned} \tag{3.60}$$

3.2.4.3 Proximal Policy Optimization

Although TRPO introduces a theoretically grounded trust region optimization framework, it faces several practical challenges in terms of implementation and computation. First, it relies on second-order information; specifically, it requires estimating and using the Fisher information matrix to constrain the step size of policy parameter updates, which incurs significant computational cost. Second, its optimization objective involves solving a constrained optimization problem with a KL divergence limit at each update step, making the implementation relatively complex. Furthermore, in order to strictly satisfy the KL constraint, TRPO often adopts conservative update steps, which can slow down policy convergence and reduce overall training efficiency.

Therefore, OpenAI proposed a simpler and more effective alternative solution PPO [14].

PPO continues the idea of "limiting policy updates," but replaces the KL divergence constraint with a more easily implementable clipping function. Its objective function is:

$$L^{\mathrm{CLIP}} = \mathbb{E}_t\left[\min(r_t(\theta)A_t, clip(r_t(\theta), 1-\epsilon, 1+\epsilon)A_t)\right] \tag{3.61}$$

where $r_t(\theta) = \dfrac{\pi_\theta(a_t|s_t)}{\pi_{\mathrm{old}}(a_t|s_t)}$, the clip operation constrains the ratio r_t within the range $[1-\epsilon, 1+\epsilon]$.

Although PPO is categorized as an on-policy method, it exhibits certain characteristics of "near off-policy" behavior. Specifically, PPO uses the current policy π_t to interact with the environment and collect data. Instead of discarding this data immediately after sampling, it reuses the same data over multiple training epochs to update the next policy π_{t+1}. To maintain "local consistency" between the old and new policies, PPO introduces a clipping mechanism (clip ratio) that limits the extent of policy updates. This approach improves sample efficiency while enhancing training stability.

3.2.4.4 Deep Deterministic Policy Gradient

The previous chapters introduced policy-gradient-based algorithms such as REINFORCE, AC, and two improved algorithms, TRPO and PPO. These algorithms share a common characteristic: they are all on-policy methods, which means their sample efficiency is relatively low. Let's recall the DQN algorithm, which directly estimates the optimal Q-function and thus can perform off-policy learning. However, DQN can only handle environments with discrete action spaces because it needs to select the action with the highest Q-value from all possible actions. When the action space is continuous or infinite, discretizing it is possible but crude and limits fine control. Is there a way to apply a similar idea to handle continuous action spaces while still using an off-policy learning algorithm? To address this, the deep deterministic policy gradient (DDPG) [35] algorithm was proposed. It constructs a deterministic policy and uses gradient ascent to maximize the value function.

DDPG mainly consists of two neural networks: an actor network $\mu(s|\theta^\mu)$ and a critic network $Q(s,a|\theta^\mu)$. The actor network outputs a deterministic action $a = \mu(s)$, while the critic network estimates the state-action value function, representing the expected cumulative reward of taking action a in state s, formally expressed as:

$$Q^\mu(s,a) = \mathbb{E}_{r_t, s_{t+1} \sim E}\left[r(s,a) + \gamma Q^\mu(s_{t+1}, \mu(s_{t+1}))\right] \tag{3.62}$$

During training, the critic network updates its parameters θ^Q by minimizing the Bellman error, with the loss function defined as:

$$L(\theta^Q) = \mathbb{E}_{(s,a,r,s')}\left[\left(Q(s,a|\theta^Q) - y\right)^2\right] \tag{3.63}$$

where the target value y is:

$$y = r + \gamma Q\left(s', \mu(s'|\theta^\mu)|\theta^{Q'}\right) \tag{3.64}$$

Here, $\theta^{Q'}$ and $\theta^{\mu'}$ are the parameters of the target networks, which are updated softly toward the main network parameters as follows:

$$\theta^{Q'} \leftarrow \tau\theta^{Q} + (1-\tau)\theta^{Q'}, \theta^{\mu'} \leftarrow \tau\theta^{\mu} + (1-\tau)\theta^{\mu'} \tag{3.65}$$

where $\tau \ll 1$ is the soft update coefficient, ensuring smooth updates and training stability.

The actor network is updated by maximizing the Q-value estimated by the critic for the current policy, that is:

$$\nabla_{\theta^{\mu}} J(\pi_{\theta}) = \mathbb{E}_{s\sim D}\left[\nabla_{a} Q\left(s,a|\theta^{Q}\right)|_{a=\mu(s)} \nabla_{\theta^{\mu}} \mu(s|\theta^{\mu}\right] \tag{3.66}$$

The gradient is computed via the chain rule to optimize the actor parameters θ^{μ}.

To improve training stability, DDPG uses an experience replay buffer to store tuples (s,a,r,s') collected from agent-environment interaction, and samples mini-batches randomly during training. The introduction of target networks prevents oscillations and divergence during learning.

By employing a deterministic policy, DDPG avoids the high variance issues caused by policy sampling, making it particularly suitable for continuous control tasks. It has been widely applied in robotics, autonomous driving, financial trading, and other domains. In material design, DDPG can continuously adjust design parameters to optimize the performance of complex material structures, demonstrating the strong potential of DRL in autonomous design of high-dimensional, complex systems.

Despite its advantages in performance and convergence speed, DDPG faces challenges such as hyperparameter sensitivity and noise susceptibility.

3.2.4.5 Soft Actor-critic

In DRL tasks involving continuous action spaces, the DDPG algorithm improves sampling efficiency and training stability by adopting a deterministic policy. However, DDPG still suffers from several limitations, such as sensitivity to hyperparameters, training instability, and limited exploration capability. To address these issues, the SAC algorithm [22] was proposed. SAC is an off-policy method based on the principle of maximum entropy RL, which aims to enhance both the performance and robustness of the learned policy.

The objective of standard RL is to maximize the expected cumulative reward. In contrast, maximum entropy RL augments the reward with an entropy term, resulting in the following objective:

$$J(\pi) = \mathbb{E}_{\pi}\left[\sum_{t=0}^{\infty} \gamma^{t}\left(r(s_t,a_t) + \alpha\mathcal{H}(\pi(\cdot|s_t))\right)\right] \tag{3.67}$$

Where $\mathcal{H}(\pi(\cdot|s_t)) = -\mathbb{E}_{a_t\sim\pi}\left[\log\pi(a_t|s_t)\right]$ is the entropy of the policy at state s_t, and α is the temperature coefficient that balances the trade-off between reward and entropy. The introduction of entropy encourages stochasticity in the policy, thereby enhancing exploration and reducing the risk of early convergence to suboptimal solutions.

SAC adopts an AC architecture and consists of the following components: policy network (actor): outputs a stochastic action distribution, usually modeled as a Gaussian; two Q-networks (critic): uses double Q-learning to reduce overestimation of Q-values; replay buffer: enables off-policy learning by reusing past experiences.

Based on the Bellman equation, the Q-function is updated by minimizing the following loss:

$$L_Q = \mathbb{E}_{(s,a,r,s')\sim D}\left[\left(Q(s,a) - \left(r + \gamma \mathbb{E}_{a'\sim\pi}\left[\widetilde{Q}(s',a') - \alpha \log\pi(a'|s')\right]\right)\right)^2\right] \quad (3.68)$$

where $\widetilde{Q}(s',a') = \min(Q_1(s',a'), Q_2(s',a')$ is the minimum of the two Q-networks to reduce overestimation bias.

The policy is optimized by minimizing the following objective:

$$L_\pi = \mathbb{E}_{s\sim D}\left[\mathbb{E}_{a\sim\pi}\left[\alpha \log\pi(a|s) - Q(s,a)\right]\right] \quad (3.69)$$

This loss function reflects the maximum entropy principle by encouraging the policy to both maximize entropy (exploration) and expected Q-value (performance).

SAC introduces an automatic entropy temperature tuning mechanism. The loss function for adjusting α is:

$$L(\alpha) = \mathbb{E}_{a\sim\pi}\left[-\alpha \log\pi(a|s) - \alpha \mathcal{H}_{\text{target}}\right] \quad (3.70)$$

where $\mathcal{H}_{\text{target}}$ is a predefined target entropy level that controls the desired degree of randomness in the policy.

The SAC algorithm integrates the advantages of maximum entropy RL, double Q-networks, and off-policy training. Its ability to automatically balance exploration and exploitation, coupled with high sample efficiency and strong robustness, makes it one of the most powerful and practical RL algorithms to date. SAC has demonstrated excellent performance not only in benchmark environments but also in complex, real-world applications such as digital material (DM) design and robotic control.

3.3 Typical Applications of RL

With the development of RL theory and algorithms, its application scope has been continuously expanding from traditional control and game-playing tasks to more challenging domains such as scientific research and the training of artificial intelligence systems. In recent years, RL has demonstrated strong exploratory capabilities in materials science and has been widely applied in tasks such as material design, performance optimization, and experimental process control. At the same time, RL has also played a critical role in the development of large-scale pretrained language models. Algorithms such as PPO have been successfully applied in the post-pretraining stage through RLHF, optimizing the behavior policies of language models so that the generated content better aligns with human preferences and safety requirements. The following two subsections will respectively introduce typical applications of RL in materials science and LLMs, showcasing its cross-domain innovation potential and practical value.

3.3.1 Applications of RL in Materials Science

In recent years, as the complexity of problems in materials science continues to increase, traditional methods such as trial-and-error experiments based on empirical knowledge or optimization approaches relying on static models have begun to show limitations in terms of exploration efficiency, design accuracy, and multi-objective trade-offs. RL as an interactive and adaptive decision-making framework, is naturally suited to sequential optimization and sparse reward scenarios, and has demonstrated significant potential in materials design. In particular, DRL, which integrates deep learning techniques, enables algorithms to automatically learn from and iteratively optimize within high-dimensional, nonlinear design spaces. The following three representative studies exemplify the innovative application of RL in alloy composition design, tuning the properties of two-dimensional (2D) materials, and inverse design of complex structured materials, highlighting its versatile value in the field of materials science.

3.3.1.1 RL-based Design of Multicomponent Alloy Compositions

The design of multicomponent alloys, especially high-entropy alloy systems with shape memory effects and complex phase transformation behaviors faces a central challenge: how to efficiently identify optimal or near-optimal compositions from a combinatorial space that grows exponentially. Alloy composition design is inherently a high-dimensional, non-convex, and highly nonlinear combinatorial optimization problem. As the number of selectable elements and their compositional degrees of freedom increase, the design space can expand to an astronomical scale, reaching up to 10^{18} possible combinations far beyond the capabilities of human intuition or traditional algorithms.

On the other hand, unlike chemical processes, high-throughput synthesis and characterization techniques for alloy systems are still underdeveloped. The cost of acquiring high-quality experimental data remains high, and available datasets are often small and sparse. This severely limits the applicability of traditional data-driven optimization methods in such contexts. As a result, the so-called "combinatorial explosion" renders conventional search techniques, such as grid search or heuristic approaches like Bayesian optimization, progressively ineffective in high-dimensional design spaces. To overcome this, there is an urgent need for more sample-efficient and intelligent optimization strategies capable of navigating vast design spaces under data-scarce conditions.

Against this backdrop, RL has emerged as a powerful tool to address the "black box" nature of multicomponent alloy design problems, thanks to its intrinsic ability to autonomously explore and continuously optimize in unknown environments. Xian et al. [36] introduced RL into the optimization of high-entropy alloy compositions. They proposed a closed-loop alloy design framework based on a "decision-feedback-optimization" paradigm, formulating the alloy composition optimization task as an MDP.

In this framework, the RL agent iteratively selects elements and their concentrations, receives reward signals from a continuously updated surrogate model, and gradually learns an optimal strategy. To address the challenge of sparse experimental data, the authors incorporated a surrogate model-based reward estimation mechanism and adopted uncertainty-aware acquisition functions such as expected improvement as reward signals. Remarkably, with only 112 initial data points, the RL framework successfully navigated a candidate space of over 200 million combinations and identified a TiNiHfZr alloy composition ($Ti_{27.2}Ni_{47}Hf_{13.8}Zr_{12}$) with a phase transformation enthalpy of −37.1 J/g (−39.0 J/g after calibration)—the highest reported within the known TiNi-based alloy family.

Specifically, the entire alloy design workflow consists of an inner loop and an outer loop, as illustrated in Figure 3.5. The inner loop is primarily responsible for training the RL agent. In this stage, the agent repeatedly explores various alloy composition combinations in a virtual environment and progressively optimizes its decision-making policy based on the rewards it receives. The reward signal here is the target material property phase transformation enthalpy (ΔH) which is not obtained through direct experimental measurement, but predicted by a Gaussian process (GP) surrogate model.

This surrogate model is trained on existing experimental data and uses selected elemental features (such as electronic structure, atomic radius, melting point, etc.) as input. These features are combined using composition-weighted averaging, reflecting the molar ratios of the elements in the alloy, to predict the ΔH of a given composition. The RL agent leverages this model to efficiently evaluate the performance of candidate compositions within the design space and refine its strategy toward identifying higher-performing alloys.

The RL algorithm employed is based on the classical DQN approach. In each episode, the agent incrementally selects alloy components in a sequential manner. When the policy converges (i.e., after approximately 5,000 training episodes), the agent no longer discovers significantly better compositions with improved ΔH the system assumes that the current policy has acquired a strong capability for composition selection. At this point, the trained Q-value network is used to identify promising high-performance candidate compositions for real-world validation in the outer loop.

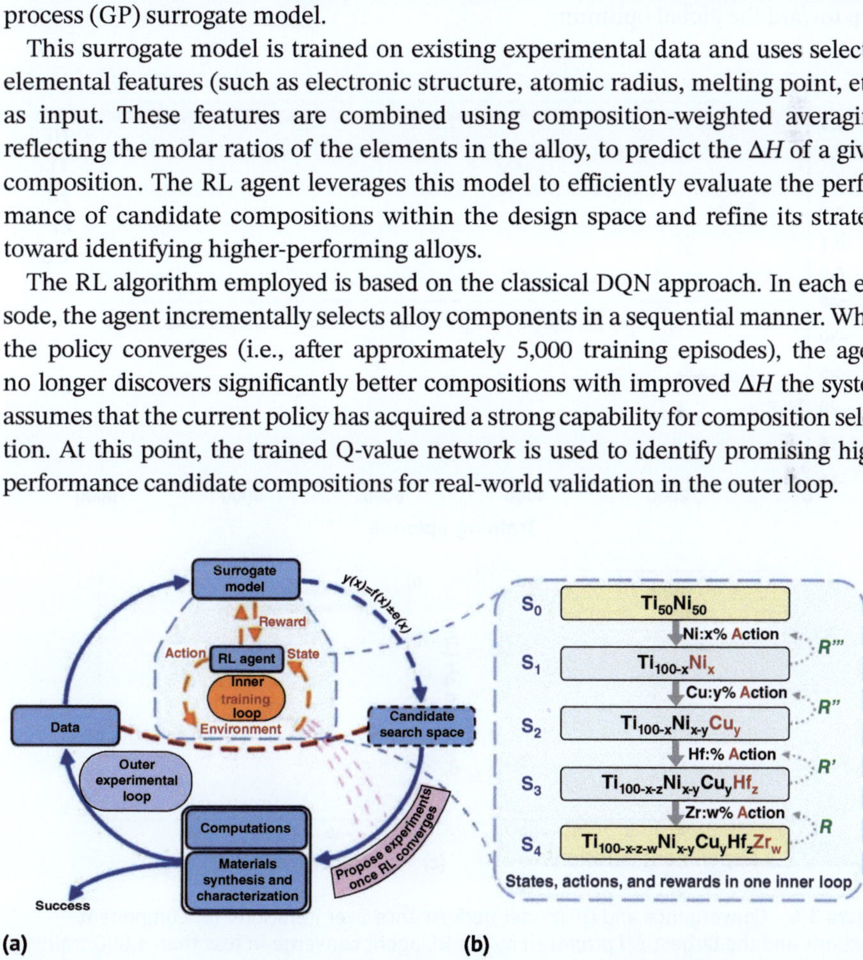

Figure 3.5 Schematic of the proposed alloy compositional design with RL. (a) RL training and reward feedback loop via surrogate model. (b) Example of a single composition design episode starting from $Ti_{50}Ni_{50}$. *Source*: [36] / with permission of Elsevier.

The outer loop corresponds to the actual experimental workflow. Based on the compositions recommended by the inner loop, researchers synthesize the alloys in the laboratory and perform performance testing to obtain new, real-world data. These experimental results are then fed back to the GP surrogate model to update its parameters and improve its prediction accuracy. This updated model is subsequently used in the next round of inner loop training, thereby forming a closed-loop system of decision-making, feedback, and continuous optimization.

The convergence trajectory of the RL agent is shown in Figure 3.6. As the iterations progress, the accuracy of the surrogate model continuously improves.

Through this closed-loop mechanism of simulation-based evaluation, experimental validation, and model update, the RL agent is able to learn effectively from a very limited amount of initial data, while also progressively expanding its knowledge boundaries as more experimental data is accumulated. This allows the agent to break free from the constraints of the initial data distribution and move step by step toward the global optimum.

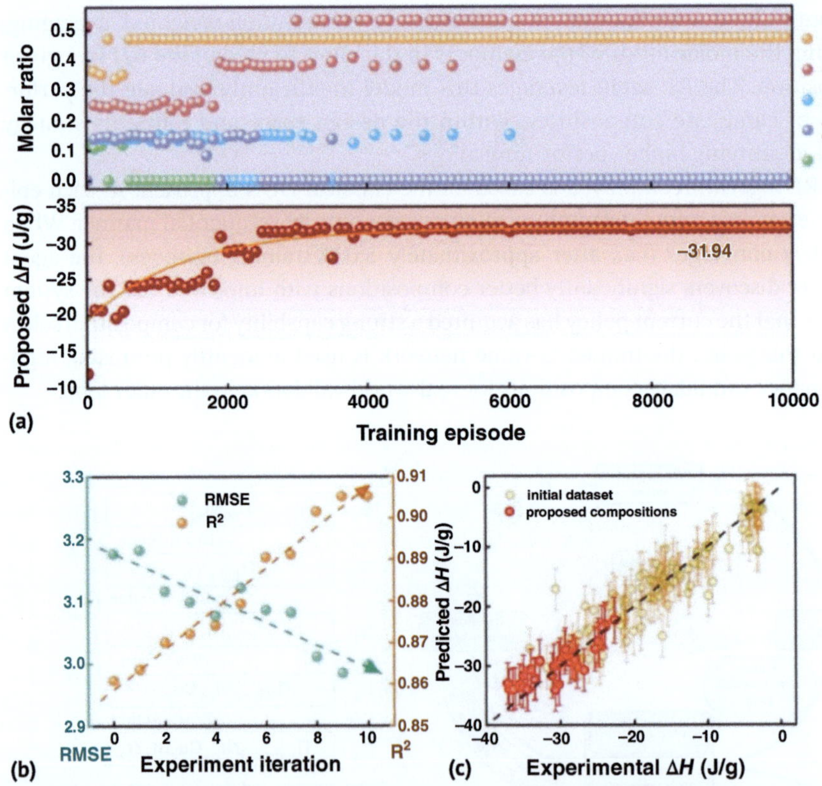

Figure 3.6 Convergence and GP model performance over iterations. (a) Component fractions and the largest Δ*H* proposed by the RL agent converge in less than 4,000 training episodes before synthesis and characterization. (b) Performance of the GP regressor as the iterations proceed. (c) Diagonal plot for the GP showing in red the proposed compositions and yellow representing the initial dataset. *Source*: [36] / with permission of Elsevier.

The framework strikes a careful balance between exploration and conservatism, ensuring design efficiency while minimizing the number of ineffective experiments. As such, it represents one of the most representative paradigms in current intelligent materials design.

This study demonstrates the significant efficiency advantage of RL in the compositional design of multicomponent alloys, as shown in Figure 3.7. Traditional methods, such as Bayesian global optimization (BGO), scale linearly with the number of candidates ($O(N)$), which becomes computationally infeasible when the design space reaches sizes on the order of 10^{18}. In contrast, RL leverages policy learning to limit the number of surrogate model calls to a fixed five per training episode, achieving convergence after approximately 4,000 episodes. This reduces the total number of surrogate interactions from $O(N)$ to approximately $O(5,255 \times N^{0.06})$, effectively overcoming the curse of dimensionality in combinatorial optimization. Even under a randomized action sequence, the complexity only increases to $O(715 \times N^{0.176})$, which remains significantly lower than that of conventional approaches. Such a dramatic reduction in search cost highlights RL as a highly promising intelligent optimization tool for identifying optimal alloy compositions in high-dimensional spaces, especially when experimental data is scarce.

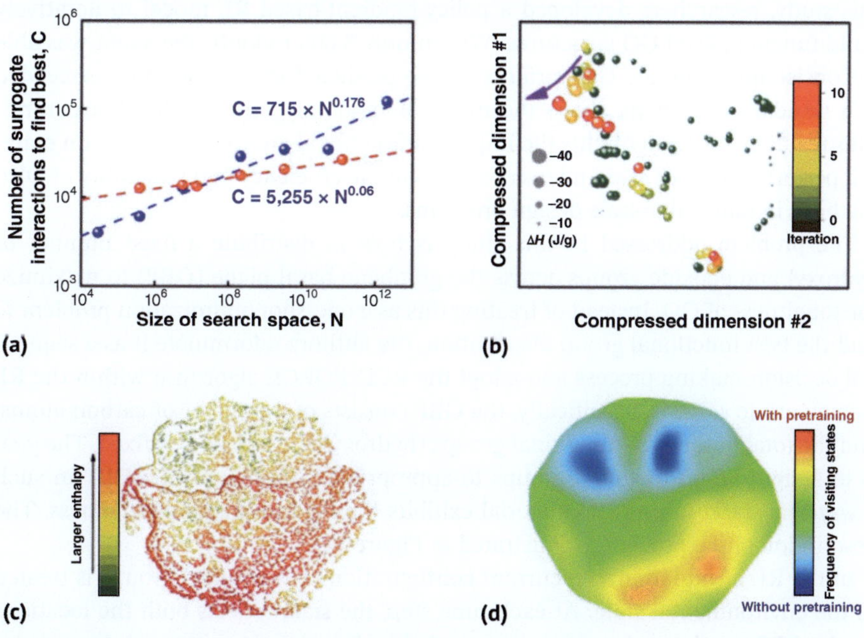

Figure 3.7 Surrogate interaction efficiency and compositional space exploration in RL-based alloy design. (a) The required number of interactions with the surrogate to train the RL agent with increasing pool size N for the two cases where (i) the actions are in a predefined order of component substitution (red) and (ii) where the agent decides on the order of substitution (blue), (b) T-SNE showing all experimentally obtained compositions containing Ti, Ni, Cu, Hf, and Zr. For the pretraining case, (c) T-SNE shows all visited states (compositions) where the color represents the corresponding predicted enthalpy, and (d) frequency of compositional states visited. Shown is the difference between pretraining and no pretraining. *Source*: [36] / with permission of Elsevier.

3.3.1.2 Accelerated Discovery of Thermoelectric Materials via RL and High-throughput Screening

The design of graphene oxide (GO) materials with tailored mechanical properties presents a unique combinatorial challenge: how to optimally arrange diverse functional groups on a 2D carbon lattice to enhance properties such as toughness. Despite GO's growing prominence in electronic, energy, and structural applications, the strategic placement of oxygen-containing groups remains largely unexplored due to the enormous size of the configuration space. Even under simple conditions, the number of possible group distributions can reach the order of 10^{11}, rendering brute-force enumeration or random sampling infeasible.

Moreover, the relationship between functional group configuration and macroscopic mechanical behavior is highly nonlinear and difficult to model analytically. Traditional design methods, such as empirical rules or statistical learning, are insufficient in capturing such spatially dependent interactions, especially when data are sparse. In this context, DRL offers a promising alternative. By framing the functionalization process as a sequential decision-making task, RL agents can learn to navigate the vast design space efficiently, using mechanical toughness as a reward signal.

Zheng et al. [37] applied DRL to the molecular-scale design of GO materials. In this study, researchers developed a policy-gradient-based RL model to iteratively build functionalized GO structures. Within only 5,000 rollouts, the agent was able to consistently generate GO variants whose predicted toughness values exceeded two standard deviations above the mean of randomly functionalized structures. This result not only highlights the impact of functional group distribution on material performance, but also underscores the sample efficiency and scalability of DRL methods in molecular-scale design problems.

The problem addressed in this study is how to distribute a fixed number of hydroxyl and epoxide groups across the graphene basal plane (GBP) to maximize the toughness of GO. Instead of treating this as a one-shot optimization problem to find the best functional group distribution, the authors reformulate it as a sequential decision-making process and adopt the REINFORCE algorithm within the RL framework to solve it. Specifically, the GBP consists of a number of carbon atoms, and the total number of functional groups (hydroxyl and epoxide) is fixed. The goal is to assign these functional groups to appropriate positions on the GBP in such a way that the resulting GO material exhibits the highest possible toughness. The description of the problem is illustrated in Figure 3.8.

In the RL formulation, the current configuration of functional groups is treated as the environmental state. At each time step, the state reflects both the locations on the GBP that have already been occupied by functional groups and the remaining available positions. The state space is a discrete set encompassing all possible configurations of functional group assignments, based on the placement of individual carbon atoms or pairs of neighboring carbon atoms, which is distinct from a continuous Euclidean space. At each time step, the RL agent selects an action, which corresponds to assigning a functional group to a specific site on the GBP. This action is determined by a policy function, parameterized by a neural network. The agent selects actions probabilistically according to this policy π_θ, where θ denotes

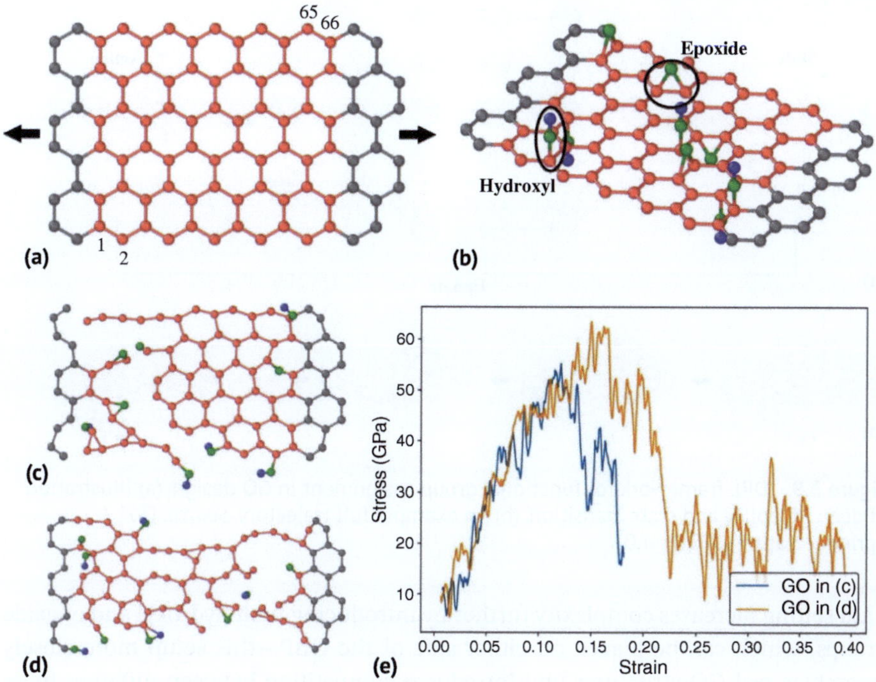

Figure 3.8 Mechanical behavior of GO under tensile loading and the effect of functional groups. (a) Schematic of GO-based polymer (GBP) structure, showing red atoms as host sites for functional groups and gray atoms as non-functionalized atoms subjected to tensile loading. (b) Structural representations of hydroxyl and epoxide functional groups; green and blue spheres represent oxygen and hydrogen atoms, respectively. (c) Tensile fracture pattern of a low-toughness GO sample. (d) Tensile fracture pattern of a high-toughness GO sample. (e) Stress–strain curves of the GO samples are shown in (c) and (d). *Source:* [37] / Springer Nature / CC BY 4.0.

the policy parameters, enabling the agent to explore the vast design space effectively. After an action is taken, the state transitions to a new configuration reflecting the updated distribution of functional groups, and the agent proceeds to the next decision step. This process continues until all functional groups have been assigned, forming a full trajectory, which is a sequence of states and actions. The state transition and trajectory generation process are illustrated in Figure 3.9. To evaluate the superiority or inferiority of the allocation scheme, the model only calculates the toughness value corresponding to the final state through molecular dynamics simulation after completing the allocation of all functional groups as the reward signal. The reward is standardized to ensure the stability and robustness of the training.

To optimize the mechanical performance of GO using RL, the researchers designed a set of progressively complex experimental settings to systematically challenge the DRL model and comprehensively evaluate its performance under different levels of complexity. The experiments were divided into four difficulty levels: easy, medium, hard, and extra hard. In the easy setting, only hydroxyl groups are assigned to one side of the GBP; in the medium setting, hydroxyl groups can be assigned to both sides of the GBP, effectively doubling the size of the state and action space. The

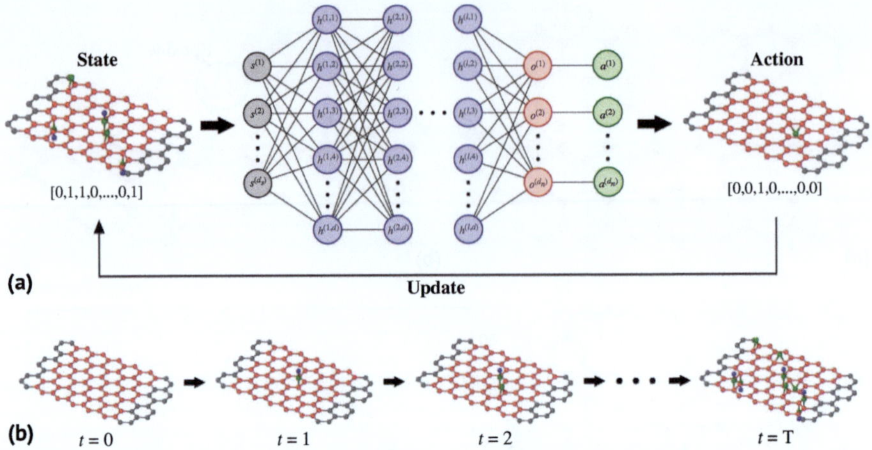

Figure 3.9 DRL framework for functional group assignment in GO design. (a) Illustration of deep RL policy and state transition. (b) An example full trajectory. *Source*: [37] / Springer Nature / CC BY 4.0.

hard setting increases complexity further by introducing both hydroxyl and epoxide groups, which can be placed on either side of the GBP—this setup more closely resembles real GO structures and introduces competition between different types of functional groups. The extra hard setting builds upon the hard configuration by increasing the size of the GBP, expanding the number of host carbon atoms from 66 to 120, to test the scalability of the model to larger design spaces.

In addition, each difficulty level includes two oxidation levels (i.e., oxygen-to-carbon ratios): a low oxidation level of approximately 15%, and a high oxidation level that is double that. This is intended to test the adaptability of the RL strategy to varying functional group densities. In total, eight distinct experimental settings were used, and each was evaluated across four different random seeds to ensure robustness and generalizability of the results. The convergence results under the four difficulty levels are shown in Figure 3.10. Notably, the convergence curves exhibit noticeable fluctuations, which reflect an inherent limitation of the REINFORCE algorithm used in this study.

This study's design enables the RL agent to interact with the environment repeatedly and progressively learn to make informed decisions at each step, ultimately constructing GO structures that exhibit superior toughness. Compared to traditional combinatorial optimization methods, this policy-gradient-based DRL approach effectively handles the high dimensionality and nonlinear complexity of the design space, avoiding the limitations of manually crafted heuristic strategies and demonstrating remarkable sample efficiency and scalability. Overall, the study successfully transforms the complex problem of GO functional group spatial distribution into a data-driven sequential optimization task. By integrating this with high-fidelity molecular dynamics simulation feedback, it achieves intelligent, data-driven optimization in the field of materials design and significantly advances the development of high-performance GO materials.

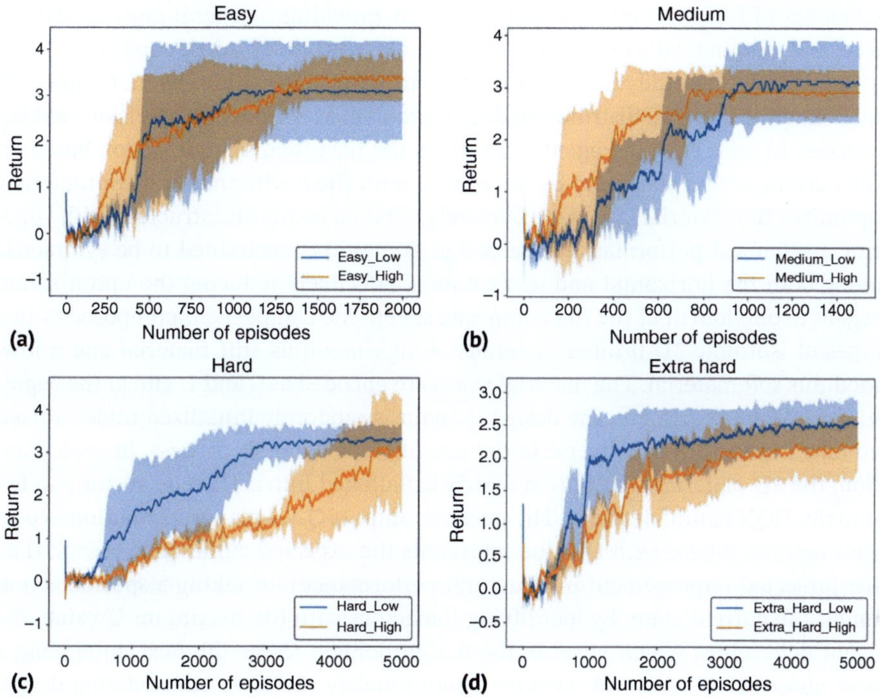

Figure 3.10 DRL performances on various experiments. (a) Easy experiment. (b) Medium experiment. (c) Hard experiment. (d) Extra hard experiment. *Source*: [37] / Springer Nature / CC BY 4.0.

3.3.1.3 DMs Design via DRL

The design of DMs, especially the optimization of microstructural layouts in composite structures, represents a highly complex and challenging problem in materials science. DMs offer an elegant mathematical framework for representing structures with complex geometries and heterogeneous material distributions. However, the enormous scale of their design space makes solving inverse problems particularly difficult. In the absence of prior knowledge from human designers, automatically generating performance-driven designs within such large state–action spaces remains a cutting-edge research frontier.

Sui et al. [38] introduced RL into the optimization of high-entropy alloy compositions. They formulated the materials design process as a self-updating sequential decision-making task guided by reward signals derived from mechanical performance metrics. By constructing a collaborative DQN architecture, their system enables multiple agents to work together across design spaces of varying sizes, performing localized modifications to material distributions while satisfying constraints such as material volume fractions. This RL system efficiently explores optimal design paths in high-dimensional state-action spaces and achieves approximately a 15.9% improvement in final design performance compared to traditional genetic algorithms, demonstrating significantly enhanced optimization efficiency and solution quality. The approach highlights the adaptive capability and decision-making efficiency of DRL in complex materials design, providing a general-purpose, expert-free, and automatically evolving tool applicable to arbitrary initial conditions.

Specifically, this study centers around the use of a DQN within the RL framework, modeling the material distribution design problem as a sequential decision-making process. In each step, the agent determines the next modification action based on the current design state, and by interacting with the environment, it continuously optimizes the material layout to ultimately obtain a composite structure with superior mechanical performance. The design domain is constrained to be symmetric along both the horizontal and vertical axes, effectively reducing the optimization region to one-fourth of the full composite sheet. The composite is composed of two types of isotropic 3D-printed materials: a high-modulus stiff material and a low-modulus soft material. The material states are encoded as 0 and 1, and at the beginning of each experiment, the design domain is randomly initialized under a fixed volume ratio to provide diverse initial conditions for the RL process. In each iteration, the 2D material distribution matrix is flattened into a 1D state vector and fed into the DQN neural network. The network outputs Q-values corresponding to possible actions, where each Q-value represents the expected cumulative reward (i.e., the projected improvement in structural performance) for taking a specific action under the current state. By identifying the action with the maximum Q-value, the agent determines which voxel in the design domain to modify next, producing a new observation state and forming a continuously iterative, self-updating design loop. After each action is executed, the current design state is mirrored to generate a complete symmetric structure, which is then evaluated under two loading conditions using finite element analysis (FEA). The structural performance is quantified by the average equivalent modulus, which serves as the reward signal fed back to the RL agent, guiding adjustments to the subsequent decision-making strategy. The overall RL workflow is illustrated in Figure 3.11.

In the results section of this study, the authors systematically compared the performance of the collaborative DQN with that of the traditional genetic algorithm across design domains of varying sizes. The convergence trajectory of the RL model is illustrated in Figure 3.12. Specifically, three different sizes of design domains (3×3, 5×5, and 7×7) were established, with the material volume fractions held constant in each

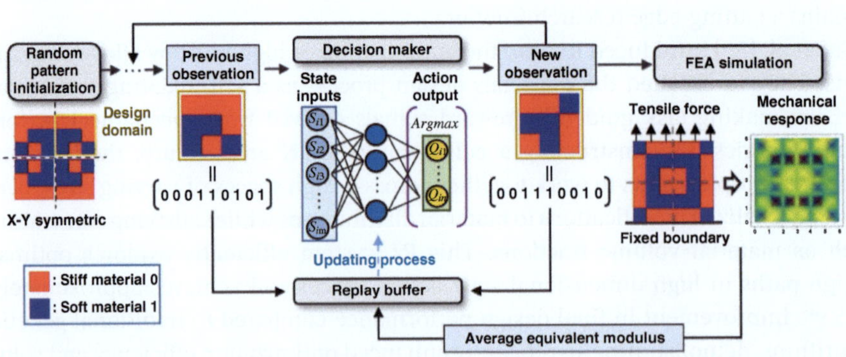

Figure 3.11 Schematic of the collaborative DRL system pipeline. *Source*: [38] / with permission of American Chemical Society.

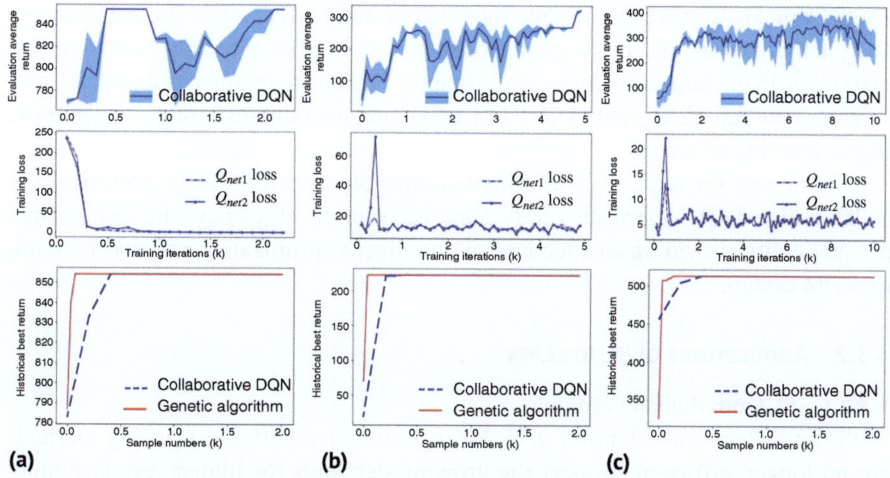

Figure 3.12 Evaluation average return, training loss, and historical best return according to the stiff-soft material ratio, with a shaded area indicating the standard deviation. (a) 3 × 3 domain, 7 stiff and 2 soft units. (b) 5 × 5 domain, 10 stiff and 15 soft units. (c) 7 × 7 domain, 20 stiff and 29 soft units. *Source*: [38] / with permission of American Chemical Society.

case. The comparison focused on the number of samples required to achieve the optimal structural design, the final design performance, and the learning efficiency of both methods. For the collaborative DQN, the evaluation-stage average return (i.e., the expected equivalent modulus of the structure) and the training loss were recorded to characterize the stability and convergence of the learning process.

Under the smallest design domain (3 × 3), the total number of possible combinations is only 126. As a result, both the collaborative DQN and the genetic algorithm can quickly identify the global optimum. For example, under the constraint of "7 stiff units + 2 soft units," the optimal structure achieves an average equivalent modulus of 853.9 MPa. The genetic algorithm finds this optimal solution using only 70 samples, while the collaborative DQN requires 400 samples, making it relatively less efficient. This indicates that in small-scale combinatorial problems, DRL does not show a significant advantage.

However, as the design space increases in size, the number of possible combinations grows exponentially, and the complexity of the design task rises dramatically. In the 5 × 5 design domain (with 10 stiff and 15 soft units), the genetic algorithm reaches its historical best average modulus of 306.0 MPa after 9,540 samples, while the collaborative DQN achieves a higher modulus of 323.5 MPa with only 6,400 samples. This demonstrates not only better final performance but also higher sample efficiency. Although the genetic algorithm converges faster in the early stage, the collaborative DQN begins to outperform it after about 1,400 samples, showing progressively stronger learning capabilities. In most cases, the final design quality achieved by the collaborative DQN is equal to or better than that obtained by traditional methods.

When the design domain expands to 7 × 7, the task becomes even more challenging. Under the constraint of "20 stiff units + 29 soft units," the genetic algorithm requires

23,300 samples to reach an average modulus of 306.5 MPa, while the collaborative DQN achieves a significantly higher modulus of 355.1 MPa with only 7,000 samples, marking a 15.9% improvement in performance. Furthermore, the collaborative DQN surpasses the genetic algorithm after just 1,800 samples, demonstrating a significantly higher learning efficiency.

In summary, for large-scale and high-complexity material design problems, the collaborative DQN approach offers superior sample efficiency and better final design quality, providing an effective and intelligent optimization pathway for complex DM design.

3.3.2 Applications of RL in LLMs

3.3.2.1 RL From Human Feedback

With the widespread adoption of LLMs, traditional supervised learning methods are no longer sufficient to meet the growing demands for higher response quality, safety, and value alignment. Especially in complex tasks such as open-ended question answering, content generation, and conversational systems, models are expected not only to produce linguistically coherent sentences but also to align with human preferences in terms of semantic understanding and moral values. Against this backdrop, RL driven by human feedback, known as RLHF [26], has emerged as a key enabling technology in the development of LLMs.

Language models such as GPT-3 [39], GPT-4 [28], Claude [40], and Gemini, their training typically follows a two-stage process. The first stage, pretraining, involves unsupervised learning on large-scale internet corpora, allowing the model to acquire linguistic patterns and world knowledge. While this provides powerful generative capabilities, the outputs often lack goal alignment and can include factual inaccuracies, unsafe content, or responses misaligned with user intent. The second stage, supervised fine-tuning (SFT), uses high-quality question–answer datasets to improve alignment to a certain extent. However, it still faces critical limitations such as limited data coverage, inability to dynamically adapt to diverse user preferences, and inconsistent response quality. To address these challenges, researchers proposed RLHF, which leverages human evaluations to construct reward signals and applies RL to optimize the model's policy. RLHF thus serves as a crucial bridge between the generative power of language models and the human values and expectations they are intended to meet.

RLHF typically involves three main steps: (1) SFT, which serves as the starting point of the RLHF process. Researchers collect high-quality question–answer samples written by human annotators (i.e., prompt and ideal response pairs) to fine-tune the pretrained model under supervised learning, resulting in an initial policy model π_0. This stage improves the relevance and readability of the model's outputs; (2) reward model training, where multiple responses to the same prompt are generated from π_0, and human evaluators rank or score these outputs to produce preference data pairs (e.g., response A is preferred over response B). This data is used to train a reward model that takes a prompt–response pair as input and outputs a scalar reward value that approximates human preferences. The training process typically follows a pairwise ranking approach, such as the Bradley–Terry model or RankNet; (3) policy optimization via RL, where the trained reward model is used to provide reward signals for optimizing the policy model $\pi(\theta)$. After the model generates a response to a prompt, the reward model scores it, and this reward guides the RL process to continually update the model's policy, aligning it more closely with human preferences and values.

To effectively apply RL methods such as RLHF to the training and fine-tuning of LLMs, it is essential to establish a clear mapping between traditional RL terminology and their counterparts in the LLM context. This alignment allows researchers and practitioners to frame language generation as a sequential decision-making process, facilitating the application of RL algorithms to optimize model behavior. The key term correspondences are summarized as follows:

- **Environment:** In classical RL, the environment provides observations and rewards in response to the agent's actions. In the context of LLMs, the environment refers to the task system, including the input prompt, the token generation process, and the reward model (e.g., a learned preference model or human evaluator).
- **Agent:** The agent corresponds to the LLM itself (e.g., GPT), which learns to generate sequences of tokens in order to maximize the cumulative reward.
- **State:** A state is defined as the current textual context, including the initial prompt and the sequence of tokens generated so far. It represents the decision basis at each generation step.
- **Action:** An action is the next token to be generated by the model. The action space is the entire vocabulary, and each step involves sampling or selecting a token from this space.
- **Policy $\pi(a|s)$:** The policy is a conditional probability distribution over the vocabulary, determining how likely the model is to generate each token given the current context (state). In LLMs, the policy is represented by the model's output logits and softmax probabilities.
- **Reward:** After generating a token or completing a sequence, a reward signal is provided based on the quality of the output. This can come from a reward model trained on human preferences, task-specific metrics, or even rule-based evaluators.
- **Return:** The return is the total cumulative reward for an entire sequence, used to evaluate the overall quality of generation under a given policy.
- **Action-value Function $Q(s,a)$:** This function estimates the expected return of choosing a particular token (action) in a given context (state), under policy π. It helps to identify which tokens are more beneficial in long-term reward terms.
- **State-value Function $V(s)$:** This function measures the expected return from a given state under the current policy, providing a baseline to assess the relative quality of different actions.
- **Advantage Function $A(s,a)$:** Defined as $A(s,a) = Q(s,a) - V(s)$, it measures how much better or worse a particular action is compared to the average expected value of the state. Advantage functions are widely used in policy gradient algorithms like PPO to stabilize training.

This terminology alignment is foundational for adapting RL algorithms to LLM training. The reward model replaces a traditional reward function and provides a scalar feedback signal to fine-tune the LLM policy. By treating token generation as a series of actions in a long-horizon MDP, RL methods can optimize LLMs not just for next-token prediction accuracy but for alignment with human preferences, safety constraints, or task-specific performance.

3.3.2.2 Reinforcement Learning in DeepSeek-R1

In order to better understand the application of RL in large-scale prediction models, this paper takes the RL training process of DeepSeek-R1 as an example to explore its practical application in the training of real large-scale models.

The RL training of DeepSeek-R1 [41, 42] adopts the GRPO algorithm proposed by their team, which is an improvement based on the PPO algorithm. GRPO differs significantly from traditional PPO in its policy update mechanism, as illustrated in Figure 3.13. PPO is an AC-based method that requires a value function or advantage function to evaluate the quality of the policy in order to compute the policy gradient. In this process, the quality of the critic network's training directly impacts the effectiveness of policy learning. In contrast, GRPO eliminates the need to construct or train a critic. It does not rely on accurate estimates of absolute advantage values, but instead guides policy updates through relative ranking within groups of candidate samples. Specifically, GRPO divides multiple generated samples into groups and ranks them based on rewards within each group. The higher-ranking samples are then used to compute the policy gradient relative to others in the same group. Since GRPO only considers ranking rather than absolute values, it avoids the high variance introduced by critic networks and greatly simplifies the implementation and training process. As a result, GRPO demonstrates greater stability and scalability in RL scenarios involving LLMs.

In traditional LLM training paradigms, the process typically consists of two phases: SFT to establish foundational language capabilities, followed by RLHF to align the model with human preferences. DeepSeek-R1-Zero, however, breaks away from this convention by omitting the SFT stage entirely and applying RL directly to the base pretrained model. This novel approach is termed "self-discovery via pure RL." Specifically, DeepSeek-R1-Zero starts with the pretrained DeepSeek-V3-Base as the initial policy network and performs optimization without any human-labeled data. The model generates multiple candidate responses to prompts—particularly

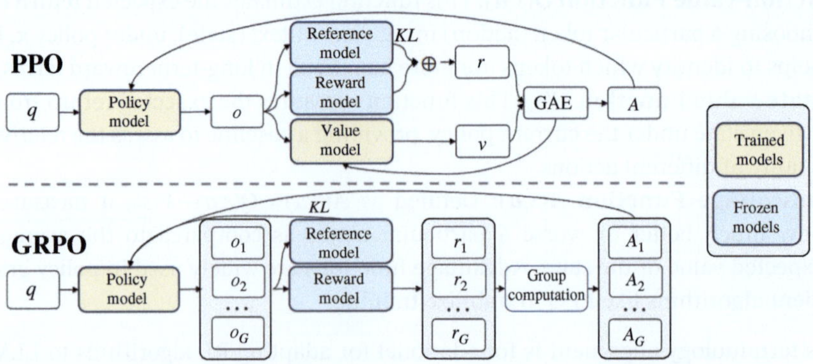

Figure 3.13 Architectural comparison between GRPO and PPO. *Source*: [42] / arXiv.

in reasoning tasks using different chain-of-thought (CoT) paths—and assigns rewards to each candidate based on correctness, semantic structure, and adherence to specific output formats. Notably, these reward signals are derived from simple rules and programmatic scoring, without relying on a separate reward model. Policy updates are performed using GRPO, an algorithm that eliminates the need for a critic network by relying on relative ranking among candidate samples within a group. This ranking-based update avoids the high variance and complexity associated with critic training, leading to a more stable and scalable RL process. Through this setup, DeepSeek-R1-Zero spontaneously develops advanced reasoning behaviors such as self-verification, reflective generation, and multistep reasoning, none of which are explicitly trained, but emerge naturally from the RL process.

Building on this foundation, DeepSeek-R1 introduces a multistage training framework that integrates RL with supervised learning to address limitations in linguistic coherence and stylistic consistency found in pure RL models. The training is conducted in four stages: first, a cold-start SFT phase is applied using a small set of high-quality reasoning samples to establish a baseline for output structure and language fluency. Second, RL is applied to strengthen the model's reasoning capabilities, starting from a more stable and expressive base than R1-Zero. In the third stage, a data augmentation and re-SFT phase is introduced, where large amounts of high-quality CoT samples generated from the RL model are filtered and added to the training corpus to enhance multitask and stylistic generalization. Finally, an alignment reinforcement phase incorporates additional reward dimensions, such as linguistic consistency and user neutrality, to further improve dialogue quality and alignment with human expectations. This staged training framework effectively combines the stability of SFT with the emergent capabilities of RL, allowing DeepSeek-R1 to outperform GPT-3.5 and OpenAI's o1 model on multiple reasoning benchmarks, thereby demonstrating the effectiveness and superiority of its training strategy; the performance evaluation results are shown in Figure 3.14.

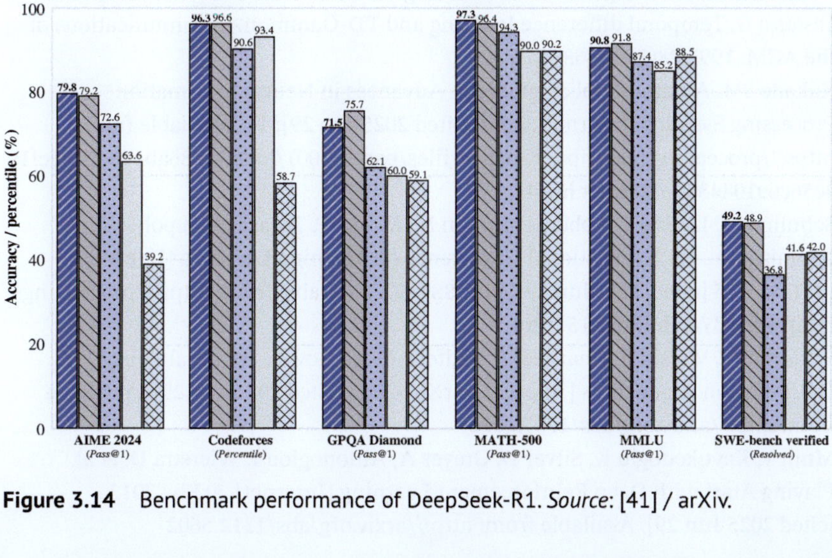

Figure 3.14 Benchmark performance of DeepSeek-R1. *Source:* [41] / arXiv.

This section presents a systematic review and analysis of the application of RL in LLMs. From the practical implementation of RLHF in training large models to specific examples such as the DeepSeek-R1 series, it is evident that RL is gradually becoming a key bridge between the capabilities of pretrained models and the demands of real-world tasks. It not only enhances the depth of reasoning and controllability of model behavior, but also provides strong technical support for further generalization and alignment with human values. As algorithms continue to evolve and training paradigms are continuously optimized, RL will play an increasingly central role in building more intelligent and safer large-scale language model systems.

References

1 Zhang A, Lipton ZC, Li M, Smola AJ. Dive into Deep Learning [Internet]. arXiv; 2023 [cited 2025 Jun 29]. Available from: http://arxiv.org/abs/2106.11342

2 Pavlov PI. Conditioned reflexes: An investigation of the physiological activity of the cerebral cortex. Annals of Neurosciences. 2010;17(3):136.

3 Skinner BF. The Behavior of Organisms: An Experimental Analysis. BF Skinner Foundation; 2019.

4 Wiener N. Cybernetics or Control and Communication in the Animal and the Machine. MIT Press; 2019.

5 Vinter R. Optimal Control [Internet]. Boston: Birkhäuser Boston; 2010 [cited 2025 Jun 29]. Available from: https://link.springer.com/10.1007/978-0-8176-8086-2

6 Puterman ML. Markov Decision Processes: Discrete Stochastic Dynamic Programming. John Wiley & Sons; 2014.

7 Bellman R. Dynamic programming. Science. 1966;153(3731):34–7.

8 Sutton RS. Learning to predict by the methods of temporal differences. Machine Learning. 1988;3(1):9–44.

9 Watkins CJCH, Dayan P. Q-learning. Machine Learning. 1992;8(3–4):279–92.

10 Williams RJ. Simple statistical gradient-following algorithms for connectionist reinforcement learning. Machine Learning. 1992;8(3–4):229–56.

11 Tesauro G. Temporal difference learning and TD-Gammon. Communications of the ACM. 1995;38(3):58–68.

12 Kakade SM. A natural policy gradient. Advances in Neural Information Processing Systems [Internet]. 2001 [cited 2025 Jun 29]; 14. Available from: https://proceedings.neurips.cc/paper_files/paper/2001/hash/4b86abe48d358ecf194c56c69108433e-Abstract.html

13 Schulman J, Levine S, Abbeel P, Jordan M, Moritz P. Trust region policy optimization. In: International Conference on Machine Learning [Internet]. PMLR; 2015 [cited 2025 Jun 29]. pp. 1889–97. Available from: https://proceedings.mlr.press/v37/schulman15.html

14 Schulman J, Wolski F, Dhariwal P, Radford A, Klimov O. Proximal Policy Optimization Algorithms [Internet]. arXiv; 2017 [cited 2025 Jun 29]. Available from: http://arxiv.org/abs/1707.06347

15 Mnih V, Kavukcuoglu K, Silver D, Graves A, Antonoglou I, Wierstra D, et al. Playing Atari with Deep Reinforcement Learning [Internet]. arXiv; 2013 [cited 2025 Jun 29]. Available from: http://arxiv.org/abs/1312.5602

16 Mnih V, Kavukcuoglu K, Silver D, Rusu AA, Veness J, Bellemare MG, et al. Human-level control through deep reinforcement learning. Nature. 2015;518(7540):529–33.

17 Silver D, Huang A, Maddison CJ, Guez A, Sifre L, Van Den Driessche G, et al. Mastering the game of Go with deep neural networks and tree search. Nature. 2016;529(7587):484–9.

18 Silver D, Schrittwieser J, Simonyan K, Antonoglou I, Huang A, Guez A, et al. Mastering the game of Go without human knowledge. Nature. 2017;550(7676):354–9.

19 Silver D, Hubert T, Schrittwieser J, Antonoglou I, Lai M, Guez A, et al. A general reinforcement learning algorithm that masters chess, shogi, and Go through self-play. Science. 2018;362(6419):1140–4.

20 Schrittwieser J, Antonoglou I, Hubert T, Simonyan K, Sifre L, Schmitt S, et al. Mastering Atari, Go, Chess and Shogi by planning with a learned model. Nature. 2020;588(7839):604–9.

21 Mnih V, Badia AP, Mirza M, Graves A, Lillicrap T, Harley T, et al. Asynchronous methods for deep reinforcement learning. In: International Conference on Machine Learning [Internet]. PmLR; 2016 [cited 2025 Jun 29]. pp. 1928–37. Available from: https://proceedings.mlr.press/v48/mniha16.html?ref=.

22 Haarnoja T, Zhou A, Abbeel P, Levine S. Soft actor-critic: Off-policy maximum entropy deep reinforcement learning with a stochastic actor. In: International Conference on Machine Learning [Internet]. PMLR; 2018 [cited 2025 Jun 29]. pp. 1861–70. Available from: https://proceedings.mlr.press/v80/haarnoja18b

23 Fujimoto S, Hoof H, Meger D. Addressing function approximation error in actor-critic methods. In: International Conference on Machine Learning [Internet]. PMLR; 2018 [cited 2025 Jun 29]. pp. 1587–96. Available from: https://proceedings.mlr.press/v80/fujimoto18a.html

24 Lowe R, Wu YI, Tamar A, Harb J, Pieter Abbeel O, Mordatch I. Multi-agent actor-critic for mixed cooperative-competitive environments. Advances in Neural Information Processing Systems [Internet]. 2017 [cited 2025 Jun 29]; 30. Available from: https://proceedings.neurips.cc/paper/2017/hash/68a9750337a418a86fe06c1991a1d64c-Abstract.html

25 Levine S, Kumar A, Tucker G, Fu J. Offline Reinforcement Learning: Tutorial, Review, and Perspectives on Open Problems [Internet]. arXiv; 2020 [cited 2025 Jun 29]. Available from: http://arxiv.org/abs/2005.01643

26 Christiano PF, Leike J, Brown T, Martic M, Legg S, Amodei D. Deep reinforcement learning from human preferences. Advances in Neural Information Processing Systems [Internet]. 2017 [cited 2025 Jun 29]; 30. Available from: https://proceedings.neurips.cc/paper/7017-deep-reinforcement-learning-from-human-preferences

27 Ouyang L, Wu J, Jiang X, Almeida D, Wainwright C, Mishkin P, et al. Training language models to follow instructions with human feedback. Advances in Neural Information Processing Systems. 2022;35:27730–44.

28 OpenAI, Achiam J, Adler S, Agarwal S, Ahmad L, Akkaya I, et al. GPT-4 Technical Report [Internet]. arXiv; 2024 [cited 2025 Jun 29]. Available from: http://arxiv.org/abs/2303.08774

29 Sutton RS, Barto AG. Reinforcement Learning: An Introduction. Vol. 1. Cambridge: MIT Press; 1998.

30 Rummery GA, Niranjan M. On-line Q-learning Using Connectionist Systems. Vol. 37. Cambridge: Department of Engineering, University of Cambridge; 1994.

31 Van Hasselt H, Guez A, Silver D. Deep reinforcement learning with double q-learning. In: Proceedings of the AAAI Conference on Artificial Intelligence [Internet]. 2016 [cited 2025 Jun 29]. Available from: https://ojs.aaai.org/index.php/AAAI/article/view/10295

32 Wang Z, Schaul T, Hessel M, Hasselt H, Lanctot M, Freitas N. Dueling network architectures for deep reinforcement learning. In: International Conference on Machine Learning [Internet]. PMLR; 2016 [cited 2025 Jun 29]. pp. 1995–2003. Available from: http://proceedings.mlr.press/v48/wangf16.html

33 Hessel M, Modayil J, Van Hasselt H, Schaul T, Ostrovski G, Dabney W, et al. Rainbow: Combining improvements in deep reinforcement learning. In: Proceedings of the AAAI Conference on Artificial Intelligence [Internet]. 2018 [cited 2025 Jun 29]. Available from: http://ojs.aaai.org/index.php/AAAI/article/view/11796

34 Konda V, Tsitsiklis J. Actor-critic algorithms. Advances in Neural Information Processing Systems [Internet]. 1999 [cited 2025 Jun 29]; 12. Available from: https://proceedings.neurips.cc/paper/1999/hash/6449f44a102fde848669bdd9eb6b76fa-Abstract.html

35 Lillicrap TP, Hunt JJ, Pritzel A, Heess N, Erez T, Tassa Y, et al. Continuous control with deep reinforcement learning [Internet]. arXiv; 2019 [cited 2025 Jun 29]. Available from: http://arxiv.org/abs/1509.02971

36 Xian Y, Dang P, Tian Y, Jiang X, Zhou Y, Ding X, et al. Compositional design of multicomponent alloys using reinforcement learning. Acta Materialia. 2024;274:120017.

37 Zheng B, Zheng Z, Gu GX. Designing mechanically tough graphene oxide materials using deep reinforcement learning. npj Computational Materials. 2022;8(1):225.

38 Sui F, Guo R, Zhang Z, Gu GX, Lin L. Deep reinforcement learning for digital materials design. ACS Materials Letters. 2021;3(10):1433–9.

39 Brown T, Mann B, Ryder N, Subbiah M, Kaplan JD, Dhariwal P, et al. Language models are few-shot learners. Advances in Neural Information Processing Systems. 2020;33:1877–901.

40 Bai Y, Kadavath S, Kundu S, Askell A, Kernion J, Jones A, et al. Constitutional AI: Harmlessness from AI feedback. arXiv preprint arXiv.2212.08073 [Internet]. 2022 [cited 2025 Jun 29]; Available from: https://ai-plans.com/file_storage/4f32fa39-3a01-46c7-878e-c92b7aa7165f_2212.08073v1.pdf

41 DeepSeek-AI, Guo D, Yang D, Zhang H, Song J, Zhang R, et al. DeepSeek-R1: Incentivizing Reasoning Capability in LLMs via Reinforcement Learning [Internet]. arXiv; 2025 [cited 2025 Jun 29]. Available from: http://arxiv.org/abs/2501.12948

42 Shao Z, Wang P, Zhu Q, Xu R, Song J, Bi X, et al. DeepSeekMath: Pushing the Limits of Mathematical Reasoning in Open Language Models [Internet]. arXiv; 2024 [cited 2025 Jun 29]. Available from: http://arxiv.org/abs/2402.03300

4

Materials Word Embedding Models

4.1 Introduction

The overwhelming majority of scientific knowledge is published as text, which is difficult to analyze by either traditional statistical analysis or modern machine learning methods. By contrast, the main source of machine-interpretable data for the materials research community has come from structured property databases [1, 2], which encompass only a small fraction of the knowledge present in the research literature. Beyond property values, publications contain valuable knowledge regarding the connections and relationships between data items as interpreted by the authors.

There is a growing trend toward the development of material-specific pretrained language models to provide high-quality word embeddings for chemical substances, elements, and other materials information, such as Word2Vec [3] and BERT (Bidirectional Encoder Representations from Transformers) [4]. These embeddings can help capture latent syntactic and semantic similarities among words in literature, facilitating candidate materials screening and serving as input vectors for property prediction. Initially, word embeddings for materials science have been constructed with Word2Vec using information about the co-occurrences of words in the scientific literature. During training, target words are represented as vectors with ones at their corresponding vocabulary indices and zeros everywhere else (one-hot encoding). These one-hot encoded vectors are used as inputs for a neural network with a single linear hidden layer, which is trained to predict all words mentioned within a certain distance (context words) from the given target word. For similar materials such as "iron" and "steel," when trained on a suitable body of text, such methods should produce a vector representing the word "iron" that is closer by cosine distance to the vector for "steel" than to the vector for "organic." Words with similar meanings often appear in similar contexts, and the analogies are expressed by finding the nearest word to the result of subtraction and addition operations between the embeddings.

Assignment of high-dimensional vectors (embeddings) to words in a text corpus in a way that preserves their syntactic and semantic relationships is one of the most fundamental techniques in natural language processing (NLP). Word embeddings are usually constructed using machine learning algorithms such as GloVe [5] or

Word2Vec [6, 7], which use information about the co-occurrences of words in a text corpus. For example, when trained on a suitable body of text, such methods should produce a vector representing the word "iron" that is closer by cosine distance to the vector for "steel" than to the vector for "organic."

4.2 Unsupervised Word Embeddings Capture Latent Knowledge

In recent years, the confluence of NLP and materials science has opened new frontiers for knowledge discovery in domains long constrained by data sparsity and unstructured information [1]. While structured databases—such as the Materials Project or ICSD—offer invaluable repositories of experimentally or computationally derived properties, they represent only a fraction of the latent insights embedded in the vast and ever-growing corpus of scientific literature [2]. The overwhelming majority of domain knowledge, including contextual relationships, inferred mechanisms, and heuristics used by experts, remains locked within unstructured textual sources. Extracting and operationalizing this knowledge at scale requires moving beyond rule-based or supervised methods, which depend on hand-labeled data, toward unsupervised language models capable of autonomously internalizing complex semantics.

The landmark study by Tshitoyan et al. [8] marks a transformative contribution to this effort. By training unsupervised word embeddings on more than 3.3 million abstracts in materials science, the authors demonstrated that vector representations derived solely from word co-occurrence statistics can encode chemically meaningful relationships, analogies, and predictive insights—despite the absence of explicit labels, property annotations, or domain-specific heuristics. Using the Word2Vec skip-gram model, the authors constructed a 200-dimensional vector space in which terms related to materials, structures, properties, and applications occupy positions reflective of their functional and scientific proximity.

The following sections elaborate on three core thematic aspects of this study: (1) the foundations and rationale for applying unsupervised word embeddings in materials science, (2) the encoding of scientific knowledge through semantic relationships in vector space, and (3) the emergence of chemically intuitive element embeddings that mirror the periodic table and predict elemental properties.

4.2.1 Foundations of Unsupervised Word Embeddings for Materials Science

The central premise of Tshitoyan et al.'s approach is that the collective semantic structure of materials science literature can be learned by an unsupervised language model trained to capture word co-occurrence patterns. The study draws on the skip-gram architecture of Word2Vec, wherein a shallow neural network learns to predict context words surrounding a target word within a specified window, optimizing embeddings that reflect contextual similarity [6, 7]. As illustrated in Figure 4.1, this process results in the transformation of one-hot encoded word vectors into dense, low-dimensional embeddings that encode distributional semantics.

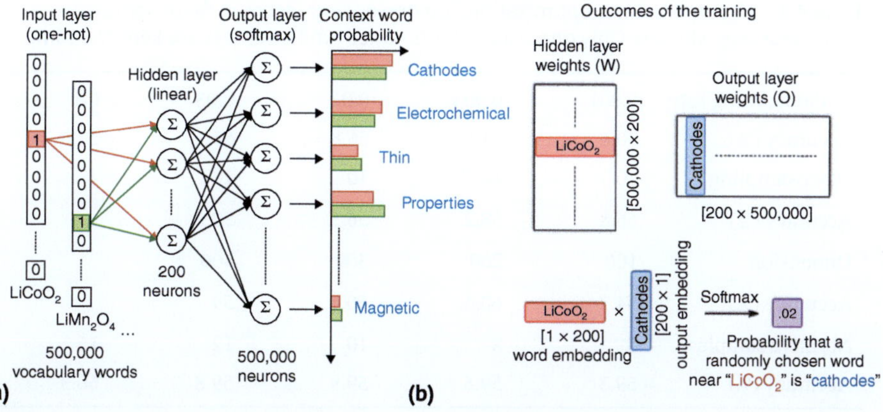

Figure 4.1 Word2Vec skip-gram architecture and embedding representation. (a) Schematic of the skip-gram model used to learn word embeddings. A shallow neural network with a single linear hidden layer is trained to predict surrounding context words for a given target word. In the case of battery cathode materials such as $LiCoO_2$ and $LiMn_2O_4$, which frequently appear in similar textual contexts, the model assigns similar weights to the hidden layer, resulting in closely aligned word embeddings. The output layer uses the softmax function to normalize probabilities across the vocabulary. (b) The outcome of training yields two key matrices: W, the hidden layer weights (word embeddings), and O, the output layer weights (output embeddings). Each row of W represents a dense vector for a word in the vocabulary, while each column of O corresponds to the likelihood of context words appearing near the target word. The dot product between these embeddings determines the probability of word co-occurrence in the corpus. *Source*: Adapted from [8].

The corpus used in the study comprises 3.3 million abstracts drawn from over 1,000 journals spanning materials science, physics, and chemistry, filtered to retain only those abstracts relevant to inorganic materials. The authors employed extensive preprocessing steps, including tokenization of chemical formulas, normalization of units and abbreviations, and selective filtering of irrelevant documents using a classifier trained to identify domain relevance. A particularly notable feature of the training process is the treatment of chemical compositions as individual tokens. For instance, "$LiCoO_2$" and "$Fe_{50}Ni_{50}$" are recognized as compositional units rather than arbitrary word strings. This semantic preservation enables the model to infer relationships between compounds and their constituent elements, forming the basis for chemically meaningful embeddings.

Unlike supervised NLP methods that require task-specific annotations, this unsupervised approach makes no assumptions about chemical structure or properties. The model's knowledge is distilled purely from the frequency and proximity of words in text, aligning with the hypothesis that expert insights are embedded implicitly in the literature. As such, the Word2Vec model functions not merely as a linguistic tool but as a mechanism for scientific knowledge extraction, uncovering latent patterns that align with domain expertise.

The resulting embeddings are validated across multiple dimensions, including analogy resolution, compositional similarity, and performance in downstream prediction tasks. Importantly, the choice of corpus quality over corpus size emerges as a key insight. The authors show that embeddings trained on focused materials

Table 4.1 Hyperparameter optimization. Top one analogy score in % for various hyperparameter choices. Only one parameter is varied while the rest are kept the same.

Initial learning rate	0.001	0.003	0.01	0.03	0.1
Accuracy (%)	50.6	54.6	56.8	55.1	52.6
Downsampling	10^{-3}	10^{-4}	10^{-5}	10^{-6}	
Accuracy (%)	56.8	58.2	56.5	50.6	
Dimension	100	200	300	400	
Accuracy (%)	54.7	60.4	60.5	59	
Negative samples	5	8	10	12	15
Accuracy (%)	59.3	59.5	59.8	59.8	60.3

Source: [8] / with permission of Springer Nature.

Table 4.2 Importance of the text corpus.

Text corpus	Materials	Grammar	All	Corpus size
Wikipedia	2.6	72.8	51	2.81B words
Wikipedia elements	2.7	72.1	41.4	1.08B words
Wikipedia materials	2.2	72.8	41.3	781M words
All abstracts	43.3	58.3	51	643M words
Relevant abstracts	48.9	54.9	52	290M words
Pretrained model from Kim et al. [9]	10.4	47.1	30.8	640k papers

Source: Adapted from [8].

science abstracts significantly outperform those trained on more expansive but heterogeneous corpora such as Wikipedia (Tables 4.1 and 4.2).

4.2.2 Encoding Scientific Knowledge Through Semantic Relationships

A major highlight of the study is the demonstration that word embeddings trained in an unsupervised manner are capable of encoding scientific relationships akin to those uncovered through years of expert research. The Word2Vec embeddings support analogical reasoning, compositional arithmetic, and semantic vector operations that reveal domain-specific knowledge internalized during training.

Analogical reasoning tasks, popularized by [6, 7], are extended in Tshitoyan et al.'s work to the materials science domain. For instance, the model correctly captures analogies such as "NiFe" is to "ferromagnetic" as "IrMn" is to "antiferromagnetic." This type of analogy is computed through vector arithmetic: embedding ("ferromagnetic") − embedding ("NiFe") + embedding ("IrMn") ≈ embedding ("antiferromagnetic") and demonstrates that the learned vector space reflects scientific distinctions between magnetic ordering in materials.

Analogies extend beyond physical properties to encompass structural relationships. Figure 1b of the original paper visualizes how embeddings of elements (Zr, Cr, and Ni) and their respective oxides and crystal structures maintain consistent offsets in vector space. For example, the difference between "Zr" and "ZrO_2" mirrors that between "Cr" and "Cr_2O_3," and similarly between "Ni" and "NiO"—indicating that the operation "oxide of" is consistently encoded. This compositional reasoning holds true for crystal structures as well, with transformations such as "Zr - HCP ≈ Ni - FCC" suggesting that embeddings capture crystallographic knowledge typically acquired through formal training.

Beyond analogies, embeddings cluster semantically related materials and properties in high-dimensional space. Figure 4.2a shows a two-dimensional t-SNE (t-distributed Stochastic Neighbor Embedding) projection of embeddings for over

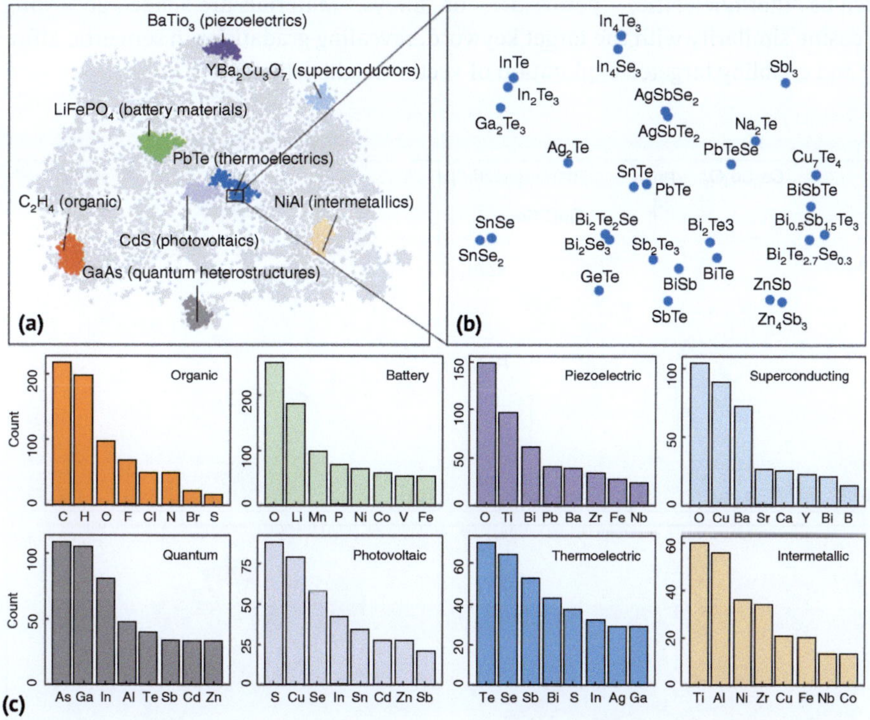

Figure 4.2 Semantic mapping of material embeddings. (a) A two-dimensional t-SNE projection of 12,340 word embeddings, each representing a distinct material stoichiometry mentioned at least 10 times in the corpus. The spatial arrangement reflects context-based semantic similarity, where materials that appear in similar textual environments are positioned closer together. Clustering is performed using the DBSCAN algorithm, which identifies high-density regions without supervision. Within each cluster, the most "connected" material—determined using the PageRank algorithm applied to cosine similarity-weighted embedding graphs—is labeled as the representative. (b) A zoomed-in view of the region around PbTe, a prototypical thermoelectric material, revealing its proximity to related compounds. (c) Histogram of the eight most frequently occurring elements in each cluster from panel (a), counted once per material regardless of stoichiometric proportion, highlighting the elemental composition trends within each functional group. *Source*: Adapted from [8].

12,000 material formulas. The resulting "materials map" reveals coherent clusters corresponding to functional classes such as thermoelectrics, photovoltaics, and battery materials. These clusters are not generated through supervised classification but emerge organically from co-occurrence patterns, suggesting that the embeddings internalize domain-specific taxonomies.

A zoom-in on the cluster containing PbTe (Figure 4.2b) reveals surrounding compounds such as SnSe and $CuGaTe_2$—known thermoelectric materials—further corroborating the semantic validity of the vector space. Clustering is performed using DBSCAN (density-based spatial clustering of applications with noise), and the most central materials are identified via PageRank, providing interpretable summaries of dense regions in the embedding space (Figure 4.2c).

The interpretability of these relationships is further enhanced by dynamic projections (Figure 4.3), which show how embeddings respond to specific keywords such as "thermoelectric," "perovskite," or "alloys." Materials are shaded according to cosine similarity with the target keyword, revealing gradations in semantic affinity and enabling targeted exploration of vector neighborhoods.

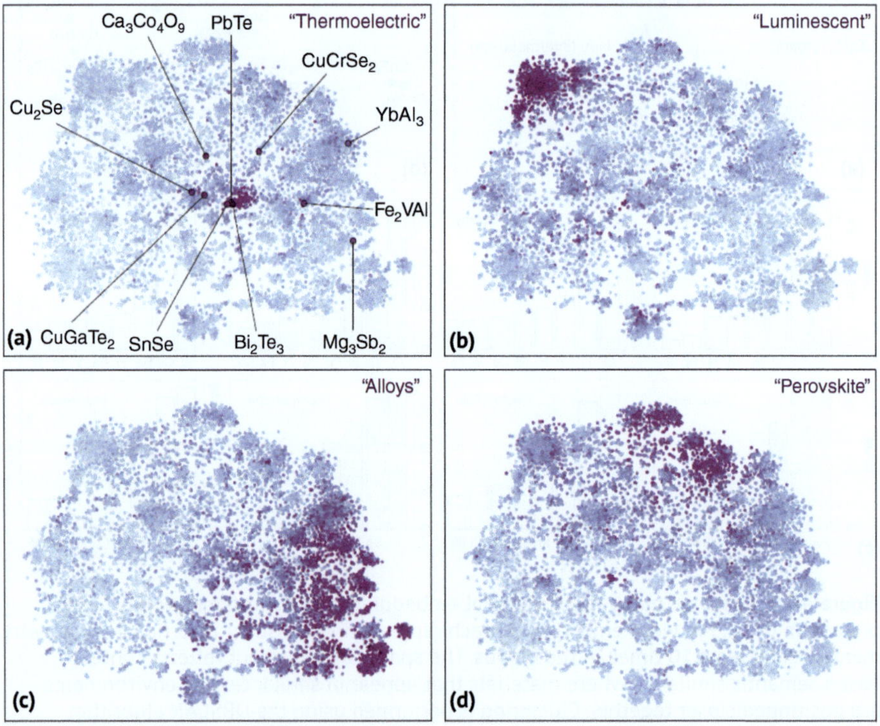

Figure 4.3 Dynamic material maps. Material maps highlighted according to various keywords. Darker colors correspond to more similarity. *Source*: Adapted from [8].

4.2.3 Element Embedding Spaces and Periodic Table Correlations

One of the most compelling outcomes of Tshitoyan et al.'s work is the emergence of chemically meaningful element embeddings. Despite the absence of any direct input about the periodic table, valence, or electron configurations, the model organizes elemental embeddings in a way that reflects established chemical hierarchies (Figure 4.4).

Figures 4.4a visualizes two-dimensional projection of elemental embeddings using t-SNE. Clusters corresponding to alkali metals, transition metals, chalcogens, and noble gases are distinctly visible. Elements that frequently co-occur in similar materials (e.g., Ga and In, or La and Ce) are found in close proximity, while chemically inert or less frequently discussed elements (e.g., noble gases) form separate clusters. This emergent organization mirrors the periodic table not through explicit rules, but through context-based usage in scientific literature.

Notably, certain deviations from periodic positioning arise due to polysemy—where element symbols overlap with common English words. For instance, "Be"

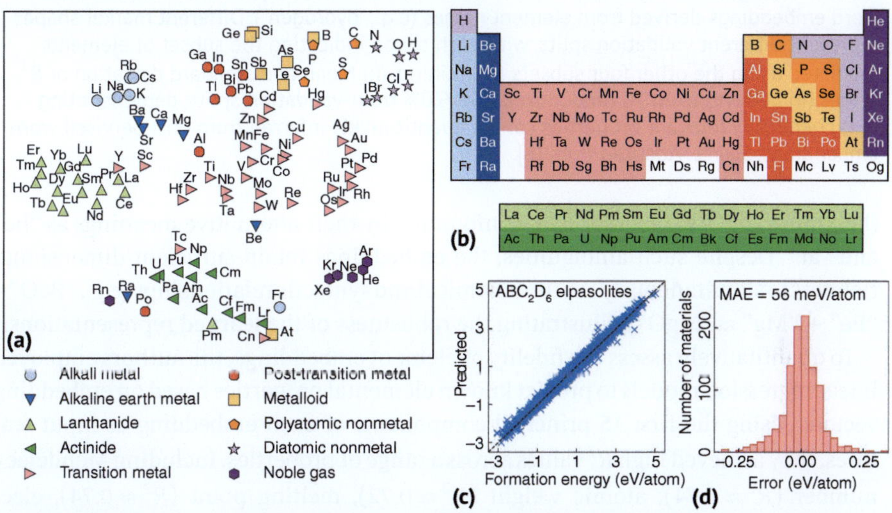

Figure 4.4 Element embeddings and formation energy predictions using word vectors. (a) Two-dimensional t-SNE projection of word embeddings for 100 chemical elements shows clustering by chemical similarity, reflecting a topology reminiscent of the periodic table. (b) Periodic table annotated with element classifications corresponding to those observed in the embedding space in panel (a). (c) Predicted versus actual DFT-calculated formation energies for ~10,000 ABC_2D_6 elpasolite compounds using a neural network model with word embeddings as input features. (d) Distribution of prediction errors on a 10% test set, showing a mean absolute error of 0.056 eV per atom—significantly lower than previously reported baselines using hand-crafted or structure-derived features. *Source*: [8] / with permission of Springer Nature.

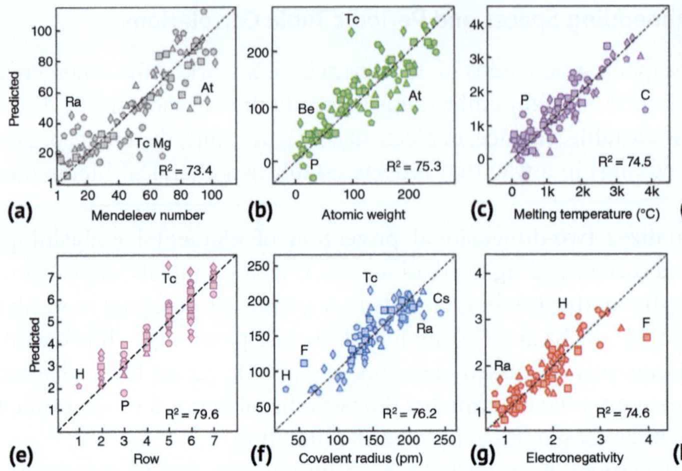

Property	R^2
Mendeleev number	74 ± 9
Atomic weight	72 ± 10
Melting T	74 ± 10
Covalent radius	82 ± 7
Electronegativity	76 ± 12
Row	81 ± 7
Column	69 ± 16

Figure 4.5 Prediction of elemental properties using word embeddings. (a–g) Fivefold cross-validated predictions of seven elemental properties—Mendeleev number, atomic weight, melting point, covalent radius, electronegativity, and periodic table row and column—using linear regression models trained on the first 15 principal components of word embeddings derived from element names (e.g., "hydrogen"). Different marker shapes represent different validation splits, with each shape indicating the subset of elements predicted using the other four subsets for training. (h) Mean and standard deviation of R^2 scores (in percent) from 20 randomized 80%/20% train-validation splits, demonstrating that chemically relevant properties can be quantitatively inferred from unsupervised word embeddings. *Source*: Adapted from [8].

(beryllium) or "At" (astatine) are confounded by their alternative meanings as "be" and "at." Despite such ambiguities, the embeddings retain sufficient dimensional richness to simultaneously capture chemical and syntactic relationships (e.g., "BeO" − "Be" + "Mg" ≈ "MgO"), illustrating the robustness of the learned representations.

To quantitatively assess the fidelity of element embeddings, the authors employed linear regression models to predict known elemental properties based on embedding vectors. Using the first 15 principal components of each embedding as input features, they achieved high R^2 values across a range of properties, including Mendeleev number ($R^2 \approx 0.74$), atomic weight ($R^2 \approx 0.72$), melting point ($R^2 \approx 0.74$), electronegativity ($R^2 \approx 0.76$), period and group ($R^2 \approx 0.69$–0.81) (see Figure 4.5 for full cross-validated results).

These findings indicate that the embeddings internalize chemically significant directions in vector space, which can be harnessed for predictive modeling. The embedding vectors thus function not merely as linguistic artifacts but as chemically expressive feature representations that rival hand-engineered descriptors used in conventional materials informatics.

4.2.4 Discovering and Predicting Materials with Word Embeddings

Among the most impactful findings of Tshitoyan et al. is the use of word embeddings not merely as representational tools, but as active predictors of functional materials—specifically those with desirable applications such as thermoelectricity.

Crucially, these predictions are made in an entirely unsupervised fashion, derived from the semantic proximities between material names and application keywords in the learned vector space.

To illustrate this, the authors focused on the keyword "thermoelectric" and computed cosine similarities between its embedding and those of thousands of material compositions extracted from the corpus. A high similarity score between a material and the target application term is interpreted as an implicit signal that the material may possess relevant functional properties—even if such an association has not been explicitly documented in the literature.

The validation process began by comparing these semantic predictions with a dataset of over 48,000 computed thermoelectric power factors from prior density functional theory (DFT) calculations [10–12]. The authors isolated 7,663 materials from this dataset that were present in the corpus yet had never co-occurred in any abstract with thermoelectric-related terms (e.g., ZT, Seebeck, and thermopower). These formed the candidate set for prediction. Cosine similarity between each candidate material's embedding and the word "thermoelectric" was used to rank these materials (Figure 4.6).

As shown in Figure 4.6b of the main paper, the top 10 predictions from this ranking displayed significantly higher average computed power factors (mean = 40.8 μW/K^2/cm) than both the general candidate pool (11.5 μW/K^2/cm) and even the set of known thermoelectrics (17.0 μW/K^2/cm). Remarkably, the best predictions approached the 99th percentile of all known materials in terms of computed thermoelectric performance. These findings suggest that word embeddings encode latent, predictive signals derived from historical usage patterns in the literature.

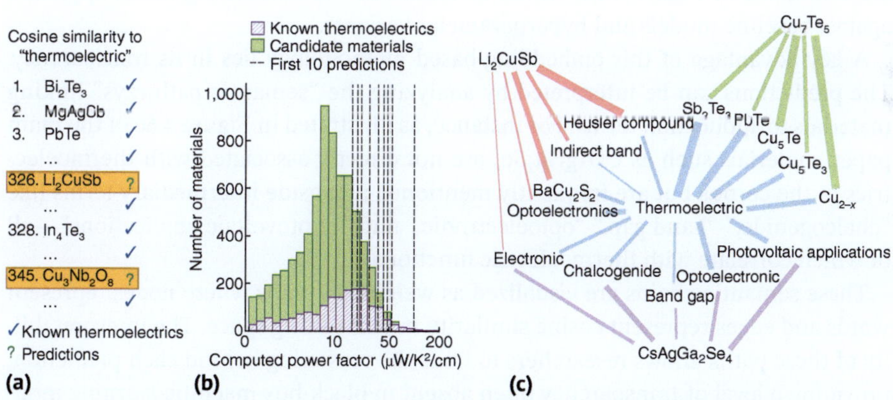

Figure 4.6 Thermoelectric materials are ranked by cosine similarity between their embeddings and that of the word "thermoelectric." (a) High-ranking materials not co-mentioned with thermoelectric-related terms are considered novel predictions for future study. (b) Power factor distributions from DFT are shown for 1,820 known thermoelectrics (purple) and 7,663 unstudied candidates (green). Dashed lines mark power factors of the top 10 predicted candidates with available data. (c) A graph links predicted thermoelectric materials to the word "thermoelectric" through shared context words, with edge widths reflecting cosine similarity. Context analysis shows predictions stem from structural similarity, co-mentions, application overlap, and key property phrases. *Source*: [8] / with permission of Springer Nature.

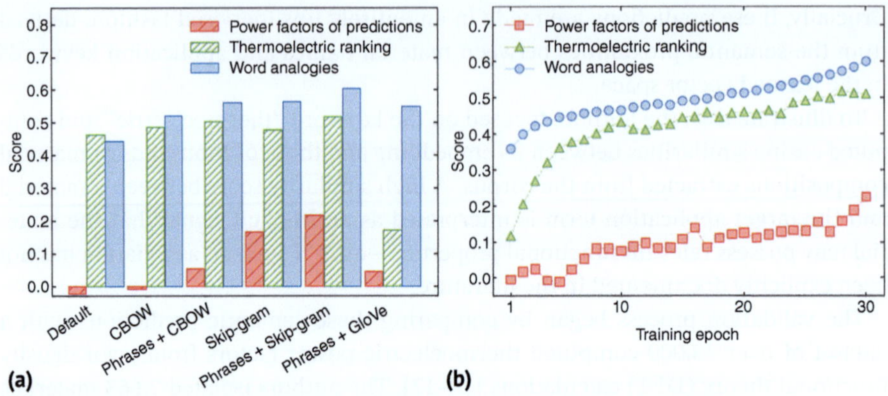

Figure 4.7 Evaluation of word embedding models for analogy and prediction tasks. (a) Comparison of performance metrics for different embedding algorithms and configurations, including analogy accuracy (blue), thermoelectric ranking correlation (green), and normalized power factor score (red). (b) Evolution of the three performance metrics shown in panel (a) across 30 training epochs for the optimized "phrases + skip-gram" model, with a linearly decreasing learning rate from 10^{-2} to 10^{-4}. *Source*: Adapted from [8].

In addition to DFT-based validation, the authors compared their rankings to experimental datasets. Among 83 materials common to both the corpus and an experimental thermoelectric dataset [13], Spearman's rank correlations between cosine similarity scores and experimentally measured performance were 59% for power factor and 52% for maximum zT—substantially outperforming the DFT predictions, which achieved only a 31% correlation (Figure 4.7) [14]. These results are visualized in Figure 4.7a, which benchmarks the embedding-based approach against baseline models and hyperparameter variants.

A key advantage of this embedding-based methodology lies in its transparency. The predictions can be interpreted by analyzing the "semantic pathways" linking materials to application terms. For instance, as illustrated in Figure 4.6c of the main paper, materials such as $CsAgGa_2Se_4$ are not directly associated with thermoelectrics in the corpus but are frequently mentioned alongside intermediary terms like "chalcogenide," "band gap," "optoelectronic," and "photovoltaic applications"—all of which correlate with thermoelectric functionality.

These semantic chains are visualized as weighted graphs where nodes represent words and edges represent cosine similarity in embedding space. The interpretability of these paths allows researchers to trace the reasoning behind each prediction, providing a level of transparency often absent in black-box machine learning models. It also highlights the model's capacity to go beyond compositional similarity and identify novel candidates from adjacent scientific domains.

4.2.5 Historical Validation and Temporal Trends in Discovery Prediction

To further validate the predictive capacity of embeddings, Tshitoyan et al. conducted a historical analysis by simulating how the model would have performed at various points in time. This retrospective approach involved constructing 18 different corpora, each containing only abstracts published before a given cutoff year (e.g., 2001, 2005, and 2010), and training separate word embeddings for each. For each historical model, the authors calculated cosine similarities between the word "thermoelectric" and all candidate materials present up to that year, generating a top 50 list of predicted thermoelectrics. These predictions were then tracked in the subsequent years to determine if they were later studied and reported as thermoelectric materials.

Figure 4.8a of the main paper presents the cumulative percentage of predictions that eventually appeared in thermoelectric-related abstracts over a five-year horizon. The results show a substantial improvement over random baselines: materials in the top 50 predictions were eight times more likely to be studied as thermoelectrics compared to a random unstudied material, and three times more likely than a random material with a nonzero DFT bandgap.

This finding provides compelling evidence that latent information in past literature contains signals indicative of future scientific developments—a form of "collective foresight" embedded in text. Furthermore, the accuracy of predictions improves over time as the corpus grows, evidenced by the steeper slopes for models trained on more recent data.

Figure 4.8b of the main paper delves into the evolution of individual high-ranking predictions. For example, $CuGaTe_2$—one of the top predictions made using pre-2009 abstracts—was independently reported as a high-performance thermoelectric in 2012 [15]. Similarly, ReS_2 and $CdIn_2Te_4$ were proposed as promising thermoelectric materials roughly 8–9 years after being flagged by the model. These materials are particularly noteworthy, as they were not part of mainstream thermoelectric classes at the time, underscoring the novelty of the insights.

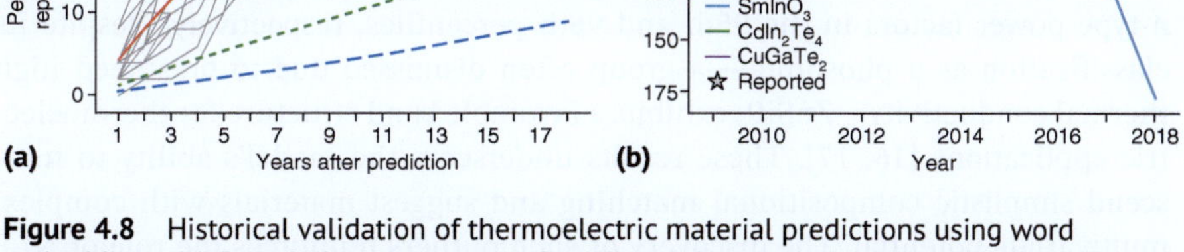

Figure 4.8 Historical validation of thermoelectric material predictions using word embeddings. (a) Cumulative percentage of predicted thermoelectric materials (based on word embeddings trained on abstracts prior to each year) that were later reported as thermoelectrics, with average performance (red) compared to baselines using all materials (blue) and those with nonzero DFT bandgaps (green). (b) Evolution of prediction ranks for the top five thermoelectric candidates identified using the 2009 corpus, with markers indicating the year each was first reported as a thermoelectric in the literature. *Source*: [8] / with permission of Springer Nature.

Some predictions, such as HgZnTe and $SmInO_3$, have yet to be confirmed, partly due to practical issues like toxicity or cost. Nonetheless, the case studies illustrate the model's ability to surface plausible candidates well before they attract experimental attention.

The historical validation framework offers a rigorous, temporally grounded method for assessing model utility and provides a compelling paradigm for future evaluations of AI-generated scientific hypotheses.

4.2.6 Unconventional Discoveries and Knowledge Beyond Composition

Beyond its predictive capabilities, the Word2Vec model demonstrates the potential to uncover scientifically meaningful relationships that defy straightforward compositional or structural similarity. This emergent capability is especially evident in the model's ability to identify unconventional materials that do not belong to established functional classes, yet exhibit promising characteristics.

Figures 4.2 and 4.3 introduce a two-dimensional "materials map" generated by projecting the embeddings of over 12,000 commonly mentioned materials using t-SNE. Clusters are identified using DBSCAN, and central members of each cluster are ranked using PageRank. These maps reveal not only known groupings (e.g., lithium-ion battery cathodes, thermoelectric chalcogenides, and photovoltaic semiconductors) but also regions populated by lesser-known materials that may exhibit dual-use properties or entirely new applications.

Color-coded overlays (Figure 4.3a–d) allow the user to explore how different application keywords influence similarity scores across the map. For instance, the embedding for "thermoelectric" highlights a broad swath of materials, including some that are traditionally classified as optoelectronic or photovoltaic—demonstrating the semantic fluidity of scientific concepts across subfields.

A number of top-ranked thermoelectric candidates identified by the model, such as KAg_2SbS_4 and BiOCl, represent unconventional chemistries rarely explored for such applications. These materials do not closely resemble known thermoelectrics but score highly in both semantic similarity and DFT-computed power factors. For instance, BiOCl (a layered oxychloride) was the top prediction from the 2010 corpus and ranked in the 93rd percentile of computed thermoelectric power factors. Similarly, $ZnSiP_2$, predicted in 2005, displayed p-type and n-type power factors in the 95th and 96th percentiles, respectively. Despite its classification as a phosphide—a group often dismissed due to presumed high thermal conductivity—$ZnSiP_2$ exhibits a favorable band structure for thermoelectric applications [16, 17]. These results underscore the model's ability to transcend simplistic compositional matching and suggest materials with complex, multivariate potential. The discovery of such outliers reinforces the role of NLP in hypothesis generation, enabling the surfacing of ideas that may otherwise remain buried in the literature.

While thermoelectric materials served as the primary case study, the authors demonstrated that the same methodology can be readily extended to other application domains. Using the same embeddings, they substituted the target word

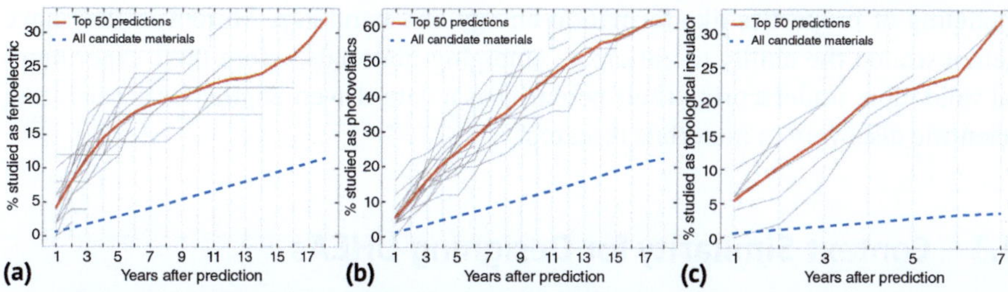

Target word / phrase	Indicator words of a potentially existing study
Ferroelectric	Ferroelectric, antiferroelectric, ferroelectrics, ferroelectricity, ferro-electricity, relaxor, paraelectric, para-electric, multiferroics, multiferroic, anti-ferroelectric, paraelectricity, ferrolectric, para-electric, ferro-electric, piezoelectric, PZT, pyroelectric, piezo-electric, pyro-electric, magnetoelectric, magnetoelectricity
Photovoltaics	Solar, photovoltaic, PV, photodevices, photoelectronics, optoelectronic, optoelectronics, nano-optoelectronics, nano-optoelectronic, opto-electronic, opto-electronics, photodiodes, photodiode, photodetectors, photodetector, photosensor, photosensors, photosensing, LED, LEDs
Topological insulator	Topological, topologically

(d)

Figure 4.9 Historical validations of functional material predictions. (a) Ferroelectric material predictions using word embeddings from historical datasets (2001–2018). (b) Photovoltaic material predictions using word embeddings from historical datasets (2001–2018). (c) Topological insulator predictions using word embeddings from historical datasets (2011–2018; limited by vocabulary emergence). (d) Target words and phrases used to rank materials (by cosine similarity) and their corresponding study indicators. *Source*: [8] / with permission of Springer Nature.

with alternatives such as "photovoltaic," "topological insulator," and "ferroelectric," and observed comparable trends in prediction accuracy and discovery lead time (Figure 4.9). This generalizability points toward a unified framework for application-focused material exploration using unsupervised language models. By simply altering the semantic target, researchers can direct the model toward new domains, fostering cross-disciplinary insight.

4.2.7 Conclusion

Unsupervised word embedding models trained on materials science literature offer a powerful means of extracting latent scientific knowledge directly from unstructured text. By learning vector representations of words based solely on their contextual co-occurrence, these models capture chemically meaningful relationships among elements, compounds, structures, and functionalities without relying on labeled data or predefined ontologies. The resulting embedding spaces reveal patterns that align with scientific intuition—such as periodic trends among elements, analogical reasoning between materials and their properties, and compositional arithmetic reflecting compound formation. These embeddings enable a range of downstream applications, including similarity-based discovery of functional materials, clustering of chemically related compounds, visualization of materials spaces, and predictive modeling of properties like formation energy and band gap. Moreover, they have demonstrated the ability to anticipate emerging materials before their experimental validation, underscoring their potential as a data-driven engine for accelerating scientific discovery in materials research.

4.3 Context Similarity for Designing UHEAs

One notable and pioneering application of these techniques is in the design of ultrahigh-entropy alloys (UHEAs). High-entropy alloys (HEAs)—multicomponent solid solutions with four or more principal elements—present a virtually infinite design space [18, 19]. Identifying promising compositions is computationally and conceptually challenging. Pei et al. [20] addressed this challenge by introducing a novel methodology based on word embeddings to identify "context similarity" among chemical elements, thereby capturing latent materials science knowledge embedded in the literature. By mining over six million materials science abstracts, they developed a word vector model capable of suggesting alloy compositions likely to form stable solid solutions—even those not yet reported.

This section explores the role of word embeddings in the design of UHEAs, emphasizing how such models capture latent chemical and physical relationships among elements, facilitate materials clustering and similarity analysis, and enable novel discovery. It provides a comprehensive summary and conceptual synthesis of the approach proposed by Pei et al., situating their contribution within the broader theme of LLMs in materials science.

4.3.1 From Linguistics to Materials Science

At the heart of the method proposed by Pei et al. lies the use of a skip-gram word embedding model—a neural network architecture originally designed to capture semantic similarity between words based on their co-occurrence in text corpora [6, 7]. When repurposed for materials science, the skip-gram model is trained on abstracts from 6.4 million scientific papers, allowing it to learn vector representations for words—including chemical elements—based on the frequency and proximity of their appearances within the literature.

Each word in the corpus is represented as a high-dimensional vector, and the proximity between two vectors (measured via cosine similarity) reflects the contextual similarity of the words they represent. Importantly, these vectors do not encode only the physical or chemical similarity of elements as defined in the periodic table, but also how researchers have used and discussed these elements in the context of alloy design. This approach thus offers a data-driven and community-informed notion of similarity, rooted in decades of scientific practice.

To ensure accurate learning, particular care was taken in standardizing the representation of alloys in the corpus. For example, the widely studied Cantor alloy

(CoCrFeMnNi) may be cited in numerous permutations, such as CrCoFeMnNi or NiMnFeCrCo. By alphabetizing chemical symbols and removing naming inconsistencies, the model avoids misrepresenting the same alloy as multiple distinct entities [21].

4.3.2 Element Embeddings and Chemical Intuition

A powerful feature of word embeddings is the ability to visualize and interpret the learned relationships among elements. In Figure 4.10 of Pei et al.'s study, a PCA was applied to reduce the dimensionality of the learned embeddings, enabling a two-dimensional representation of chemical elements in latent space. This "context-similar periodic table" reveals clusters of elements that are commonly used together in the literature, offering a radically different organization than the traditional Mendeleevian periodic table.

For instance, aluminum (Al) and manganese (Mn)—distant neighbors in the periodic table—are positioned closely in the latent space because they frequently co-occur in alloy design contexts, particularly in steels and magnesium-based alloys. This observation underscores the utility of context similarity: it reflects not intrinsic properties per se, but how human researchers associate elements based on their functional roles in materials.

Figure 4.11 further illustrates the model's semantic insight. The most contextually similar elements to Fe include Mn, Co, Ni, and Cr, all of which are known constituents of robust solid solutions and HEAs. Similarly, Ni's top neighbors include Co, Cu, Sn, and Mo—indicating rich interconnectivity in experimental literature. These visualizations demonstrate that embedding-based models can surface chemically meaningful patterns embedded in textual data, even when they depart from conventional groupings.

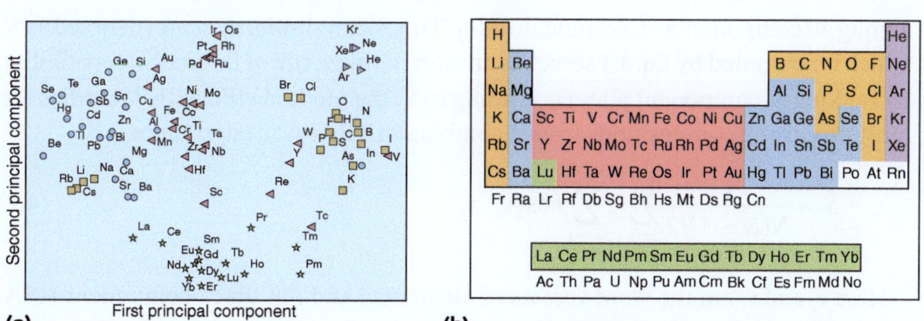

Figure 4.10 The context-similar elements and their applications to design HEAs. (a) Chemical elements are mapped in latent space using PCA on their word vectors. The first two principal components define the horizontal and vertical axes, with element colors matching the periodic table scheme in (b). *Source*: [20] / Springer Nature / CC BY 4.0.

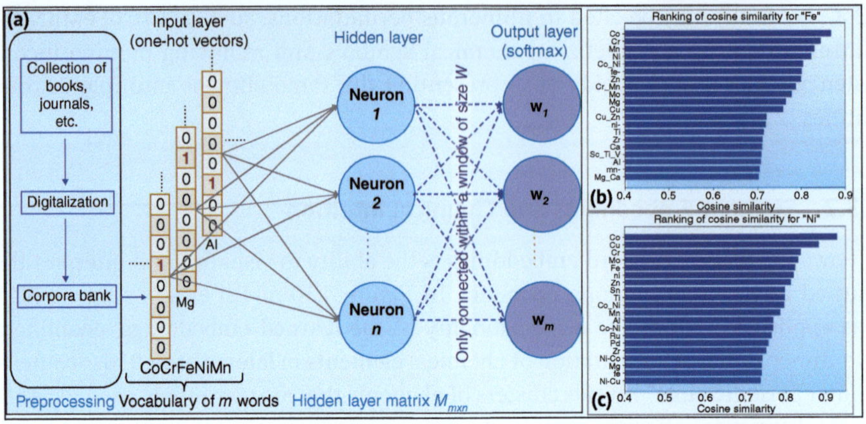

Figure 4.11 Schematic for the word-embedding model (skip-gram). (a) The model is a neural network with a single hidden layer trained on tokenized text corpora from an online database. Words are represented as one-hot vectors, connected through the hidden layer to predict contextual word probabilities using the skip-gram algorithm. (b, c) After training, the hidden layer captures semantic relationships; examples show words similar to "Fe" and "Ni" based on their vector representations. *Source*: [20] / Springer Nature / CC BY 4.0.

4.3.3 From Similarity Scores to Alloy Discovery

Pei et al. propose two principal strategies for leveraging context similarity in alloy design. The first approach is a "seeded" search, where a designer begins with one preferred element—say, Fe—and retrieves the top-k most similar elements in the embedding space. These top-k candidates are interpreted as likely co-formers of solid solutions. For example, given Fe as the seed, the model retrieves Mn, Co, Ni, and Cr as top companions, all of which have indeed been employed in successful HEAs.

The second strategy involves calculating the average pairwise cosine similarity among all components of a candidate alloy. This mean similarity score (denoted as $\overline{S}$, which is computed by Eq. 4.1 serves as an alloy-level metric of internal compatibility. By ranking N-component alloys according to $\overline{S}$, the method efficiently identifies the most promising compositions from a combinatorially vast design space.

$$\overline{S} = \frac{1}{N(N-1)/2} \sum_{i=0}^{N-1} \sum_{j=i+1}^{N-1} \cos\left(v_i,\ v_j\right) \tag{4.1}$$

Here v_i and v_j are the word vectors of elements i and j in the N-component HEA. They rank the alloy candidates according to $\overline{S}$ and pick the top candidates as the most promising ones.

This approach yielded striking results. Notably, it retrospectively ranked the Cantor and Senkov alloys—benchmark FCC and BCC HEAs, respectively—among the top candidates well before their actual discovery. The Senkov alloy was consistently ranked within the top three candidates across multiple temporal slices of the literature, including years prior to its experimental synthesis in 2011 [22, 23].

4.3.4 Correlation with Thermodynamic Predictors

To validate that the linguistic similarity captured by embedding corresponds to physical feasibility, Pei et al. compared their similarity metric $\overline{S}$ with a thermodynamics-based descriptor γ, which quantifies the relative stability of multicomponent solid solutions over their binary subsystems [24]. A high γ value (greater than 1) indicates a strong propensity for solid-solution formation. As shown in Figure 4.12 of the original paper, a strong linear correlation was observed between $\overline{S}$ and γ, providing physical validation for the context similarity approach. This finding suggests that linguistic context—informed by collective scientific usage—encodes thermodynamically relevant information, despite being derived solely from textual data.

Furthermore, when combined with the lattice distortion parameter δ, which accounts for atomic size mismatch, this hybrid framework enables effective screening for single-phase HEAs [25]. Figure 4.12 illustrates that only a narrow subset of high-$\overline{S}$, high-γ, low-δ alloys are likely to form desirable solid solutions.

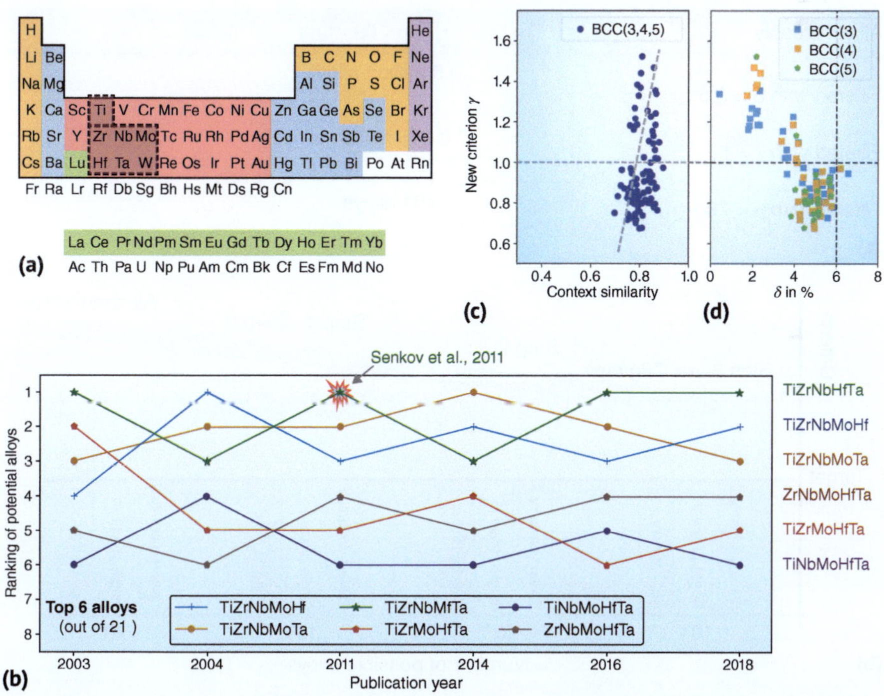

Figure 4.12 The context-similar elements and their applications for BCC HEAs. (a) Using transition elements (Ti, Zr, Nb, Mo, Hf, Ta, and W) as an example, the tendency to form solid solutions aligns with the defined context similarity. For clarity, only three-, four-, and five-component equiatomic alloys are considered. (b) Five-component alloys are ranked by context similarity across publication years, with top examples including the Senkov alloy (TiZrNbHfTa), TiZrNbMoHf, and TiZrMoHfTa. Other notable alloys are ZrNbMoHfTa, TiNbMoHfTa, and TiZrNbMoTa. (c) Context similarity shows a strong correlation with the thermodynamics-based γ parameter. (d) Combining γ with lattice distortion δ improves predictions of solid solution formation. *Source*: [20] / Springer Nature / CC BY 4.0.

4.3.5 Designing Lightweight HEAs

The true power of embedding-based models lies in their scalability. Pei et al. employed their framework to explore the vast space of six- and seven-component equiatomic alloys composed of 30 transition elements—amounting to over 2.6 million candidates. By applying a multistage filtering process based on context similarity ($\overline{S} > 0.6$), solid-solution likelihood ($\gamma > 1$), and density constraints ($\rho < 7.8$ g/cm^3), they reduced the design space to just 494 promising compositions suitable for experimental synthesis.

This workflow, illustrated in Figure 4.13 of the original paper, demonstrates how embedding-based models integrate seamlessly with physical and computational constraints to produce experimentally actionable predictions. Alloys such as TiCrFeCoNiMo and TiVCrMnFeZn emerged as top candidates, balancing context-derived compatibility with thermodynamic and mechanical considerations.

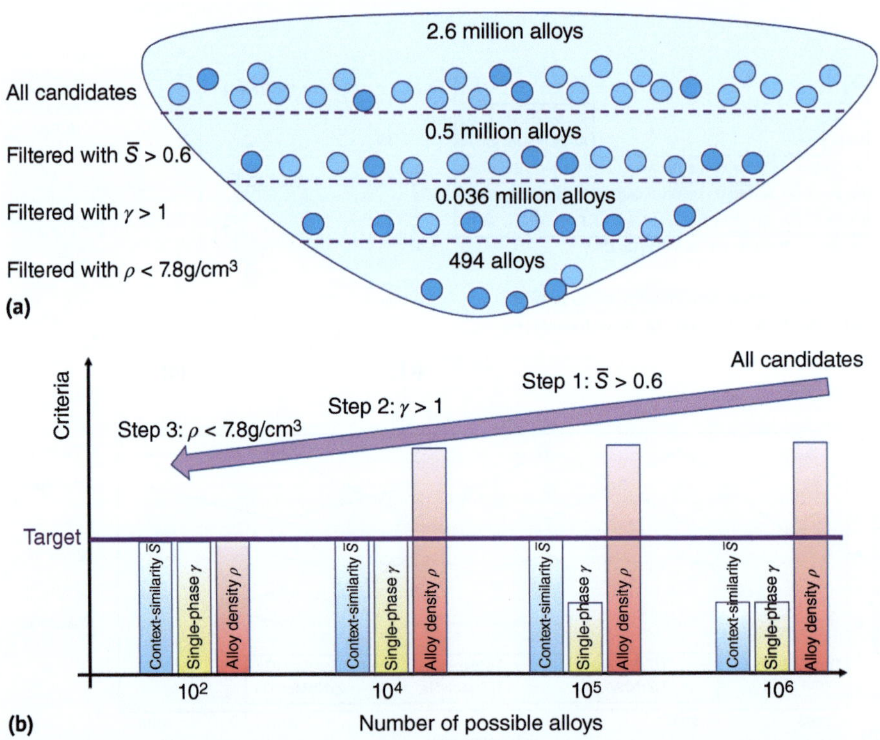

Figure 4.13 An exemplary design of lightweight HEAs with more than five components. (a) An example study on six- and seven-component equiatomic alloys among 30 transition metals follows a three-step filtering process. Alloys are selected based on high average context similarity, $\gamma > 130$ for solid solution likelihood, and density $\rho < 7.8$ g/cm^3. (b) This filtering reduces the number of candidate alloys from 10^6 to 10^2. The resulting shortlist highlights promising materials for synthesis and testing. *Source*: [20] / Springer Nature / CC BY 4.0.

4.3.6 Integration with ICME and KGs

The embedding-based context similarity model serves as a front end to more sophisticated materials informatics pipelines, including Integrated Computational Materials Engineering (ICME) frameworks. Once candidate alloys are identified, ICME tools such as phase diagram computation, ab initio simulation, and structure-property modeling can be used to further refine selections based on target application requirements (e.g., yield strength, creep resistance, and thermal stability) [26, 27].

In parallel, Pei et al. developed a knowledge graph (KG) tailored to HEAs, standardizing alloy names and linking them to publications via DOIs [28, 29]. This "alloyKG" enables rapid retrieval of known compositions, facilitating redundancy checks and literature audits. By combining embeddings with KGs, researchers can avoid redesigning previously studied alloys—a crucial capability in the age of information overload.

4.4 Conclusion

The application of word embedding models to the design of UHEAs underscores the broader potential of language-based representations in materials science. By translating unstructured textual knowledge into continuous vector spaces, these models enable the quantitative analysis of semantic and scientific relationships that were previously accessible only through expert intuition or laborious manual review. The concept of "context similarity," as operationalized by Pei et al., illustrates how co-occurrence patterns in literature can serve as proxies for chemical compatibility, guiding the search through an immense compositional space. More broadly, this work exemplifies how word embeddings can bridge the gap between textual knowledge and materials discovery, turning language into a functional tool for hypothesis generation, materials design, and knowledge-driven exploration. As such, it reinforces the growing role of unsupervised language models not merely as tools for information retrieval, but as engines for scientific reasoning in data-rich yet label-scarce domains like materials science.

References

1 Hill J, Mulholland G, Persson K, Seshadri R, Wolverton C, Meredig B. Materials science with large-scale data and informatics: Unlocking new opportunities. MRS Bulletin [Internet]. 2016;41:399–409. https://doi.org/10.1557/mrs.2016.93
2 Butler KT, Davies DW, Cartwright H, Isayev O, Walsh A. Machine learning for molecular and materials science. Nature. 2018;559(7715):547–55.
3 Thavareesan S, Mahesan S. Sentiment lexicon expansion using Word2vec and fastText for sentiment prediction in Tamil texts. In: 2020 Moratuwa Engineering Research Conference (MERCon). IEEE; 2020. pp. 272–276.

4 Devlin J. BERT: Pre-training of deep bidirectional transformers for language understanding. 2018. Available from: https://arxiv.org/abs/1810.04805
5 Pennington J, Socher R, Manning CD. Glove: Global vectors for word representation. In: Proceedings of the 2014 Conference on Empirical Methods in Natural Language Processing (EMNLP). Doha, Qatar: Association for Computational Linguistics; 2014. pp. 1532–43.
6 Mikolov T, Chen K, Corrado G, Dean J. Efficient estimation of word representations in vector space. 2013. Available from: https://arxiv.org/abs/1301.3781
7 Mikolov T, Sutskever I, Chen K, Corrado GS, Dean J. Distributed representations of words and phrases and their compositionality. In: Burges CJC, Bottou L, Ghahramani Z, Weinberger KQ, editors. Advances in Neural Information Processing Systems. Lake Tahoe, Nevada: United States; 2013. pp. 3111–9.
8 Tshitoyan V, Dagdelen J, Weston L, Dunn A, Rong Z, Kononova O, et al. Unsupervised word embeddings capture latent knowledge from materials science literature. Nature. 2019;571(7763):95–8.
9 Kim E, Huang K, Tomala A, Matthews S, Strubell E, Saunders A, et al. Machine-learned and codified synthesis parameters of oxide materials. Scientific Data [Internet]. 2017;4:170127. https://doi.org/10.1038/sdata.2017.127
10 Ricci F, Chen W, Aydemir U, Snyder GJ, Rignanese GM, Jain A, et al. An ab initio electronic transport database for inorganic materials. Scientific Data [Internet]. 2017;4:170085. https://doi.org/10.1038/sdata.2017.85
11 Hohenberg P, Kohn W. Inhomogeneous electron gas. Physical Review. 1964;136:B864–71.
12 Kohn W, Sham LJ. Self-consistent equations including exchange and correlation effects. Physical Review. 1965;140:A1133–8.
13 Gaultois MW, Sparks TD, Borg CKH, Seshadri R, Bonificio WD, Clarke DR. Data-driven review of thermoelectric materials: Performance and resource considerations. Chemistry of Materials. 2013;25(15):2911–20.
14 Spearman C. The proof and measurement of association between two things. The American Journal of Psychology [Internet]. 1904; 15. https://doi.org/10.2307/1412159
15 Plirdpring T, Kurosaki K, Kosuga A, Day T, Firdosy S, Ravi V. et al. Chalcopyrite $CuGaTe_2$: A high-efficiency bulk thermoelectric material. Advanced Materials [Internet]. 2012;24:3622–26. https://doi.org/10.1002/adma.201200732
16 Pandey C, Sharma R, Sharma Y. Thermoelectric properties of defect chalcopyrites. AIP Conference Proceedings [Internet]. 2017; 1832:110009. https://doi.org/10.1063/1.4980633
17 Zhao LD, Lo SH, Zhang Y, Sun H, Tan G, Uher C, et al. Ultralow thermal conductivity and high thermoelectric figure of merit in SnSe crystals. Nature [Internet]. 2014; 508. https://doi.org/10.1038/nature13184
18 Yeh JW, Chen SK, Lin SJ, Gan JY, Chin TS, Shun TT, et al. Nanostructured high-entropy alloys with multiple principal elements: novel alloy design concepts and outcomes. Advanced Engineering Materials. 2004;6:299–303.

19 Cantor B, Chang I, Knight P, Vincent A. Microstructural development in equiatomic multicomponent alloys. Materials Science and Engineering: A. 2004;375:213–8.

20 Pei Z, Yin J, Liaw PK, Raabe D. Toward the design of ultrahigh-entropy alloys via mining six million texts. Nature Communications. 2023;14(1):54.

21 Weston L, Tshitoyan V, Dagdelen J, Kononova O, Trewartha A, Persson KA, et al. Named entity recognition and normalization applied to large-scale information extraction from the materials science literature. Journal of Chemical Information and Modeling. 2019;59(9):3692–702.

22 Senkov O, Wilks G, Miracle D, Chuang C, Law P. Refractory high-entropy alloys. Intermetallics. 2010;18:1758–65.

23 Senkov O, Scott J, Senkova S, Miracle D, Woodward C. Microstructure and room temperature properties of a high-entropy TaNbHfZrTi alloy. Journal of Alloys and Compounds. 2011;509:6043–8.

24 Pei Z, Yin J, Hawk JA, Altman DE, Gao MC. Machine-learning informed prediction of high-entropy solid solution formation: Beyond the Hume-Rothery rules. npj Computational Materials. 2020;6:1–8.

25 Toda-Caraballo I, Rivera-Diaz-del Castillo PE. Modelling solid solution hardening in high entropy alloys. Acta Materialia. 2015;85:14–23.

26 Council NR, others. Integrated Computational Materials Engineering: A Transformational Discipline for Improved Competitiveness and National Security. The National Academies Press; 2008.

27 Sundman B, Lukas H, Fries S. Computational Thermodynamics: The Calphad Method. New York: Cambridge University Press; 2007.

28 Nie Z, Liu Y, Yang L, Li S, Pan F. Construction and application of materials knowledge graph based on author disambiguation: Revisiting the evolution of $LiFePO_4$. Advanced Energy Materials. 2021;11:2003580.

29 Pei Z, Yin J, Liaw PK, Raabe, D. Toward the design of ultrahigh-entropy alloys via mining six million texts. Nature Communications. 2023;14(1):54.

5
Materials Transformer-based Models

5.1 Introduction

Beginning with early word embedding approaches such as Word2Vec and GloVe, the field has rapidly advanced toward Transformer-based architectures like BERT and its numerous variants. While general-purpose models provide powerful contextual representations, they often fail to capture the specialized terminology and semantics of scientific literature. To address this limitation, domain-adapted models such as BatteryBERT and MatSciBERT have been introduced, demonstrating significant improvements in tasks like document classification, named entity recognition (NER), relation extraction, and scientific question answering (QA). Beyond these applications, LLMs are increasingly leveraged for automated database construction, knowledge graph generation, and large-scale literature mining, thereby transforming how materials knowledge is discovered and organized [1–6]. This chapter provides an overview of encoder-based, decoder-based, and encoder-decoder models tailored for materials research, highlighting representative studies, practical applications, and future directions for accelerating data-driven scientific discovery.

5.2 Encoder-based Models for Materials

Word2Vec captures the semantic meanings of words, allowing to find words that have similar meanings. The model is shallow with two-layer neural networks and is relatively efficient to train. However, the embeddings produced by Word2Vec are static and do not account for the context in which a word is used. Also, it cannot handle out-of-vocabulary words at inference time.

To overcome these shortcomings, Transformer-based language models, which rely on self-attention mechanisms to weigh the importance of different parts of the input data, have the capability to capture long-range dependencies in the text. The attention mechanism was first introduced in 2017 as an extension to an encoder-decoder model [7]. An encoder-decoder model organizes two recurrent neural networks. The first network encodes a source sequence, while the second decodes the source sequence into the target sequence. With the attention mechanism, a model focuses on parts of a source sentence (a word or phrase) where the most relevant information is concentrated. BERT was introduced by Google in 2018 and produces contextualized word embeddings, meaning that the embedding for a word is dependent on the surrounding words in the sentence. This is different from traditional word embedding methods Word2Vec or GloVe, where each word has a single, static embedding. It has also inspired numerous follow-up studies and variations, such as RoBERTa, ALBERT, and DistilBERT, which aim to improve upon BERT's performance and efficiency. The original BERT model utilized a 30-million-token vocabulary in its dictionary and was pretrained on the BooksCorpus (800 million words) and English Wikipedia (2,500 million words). However, the original pretrained BERT lacks materials domain knowledge and therefore does not give reliable embeddings for materials information. To enhance BERT's domain adaptation capabilities within the scientific field, Beltagy et al. [8] introduced SciBERT, a pretrained language model based on BERT. SciBERT was trained on a large multi-domain corpus of scientific publications comprising 1.14 million papers, with 18% from the computer science domain and 82% from the broader biomedical domain using a single TPU v3 with eight cores, resulting in a total corpus size of 3.17 billion tokens. SciBERT significantly outperformed BERT-base and achieves new state-of-the-art results on several downstream scientific NLP tasks.

5.2.1 BatteryBERT

While systems like ChemDataExtractor can parse chemical names and extract certain material–property relationships, they typically depend on extensive hand-written rules or supervised templates and are prone to brittleness when faced with linguistic variation. They also lack the capacity to generalize beyond their rule set or to contextualize terms based on domain-specific semantics. In this context, encoder-based language models—especially those built on the Transformer architecture—have emerged as a promising alternative. These models, such as BERT [9], use self-attention mechanisms to encode bidirectional contextual representations of tokens in a sentence. The key innovation lies in their ability to capture complex syntactic and semantic relationships across diverse linguistic contexts, enabling zero- or few-shot generalization to new downstream tasks.

While models such as SciBERT [8], BioBERT [10], and ClinicalBERT [11] have demonstrated impressive results in their respective domains, their adaptation to materials science has been sparse. The pioneering contribution of Huang and Cole lies in introducing a set of Transformer-based encoder models—collectively referred to as BatteryBERT—tailored to the domain of battery materials research. By pretraining and fine-tuning BERT variants on large corpora of battery-related literature, the authors demonstrate that such models can significantly improve the performance of NLP tasks central to scientific data mining: document classification, NER, and extractive QA.

5.2.1.1 Construction of Domain-specific Language Models for Batteries

To realize a language model attuned to the unique terminology and semantics of battery science, the authors constructed six Transformer-based models using the

BERT architecture: BatteryBERT, BatteryOnlyBERT, and BatterySciBERT, each in both cased and uncased versions. These models differ primarily in their pretraining corpora and tokenizer vocabularies, enabling an investigation into how domain specificity and corpus composition affect model performance on battery-related NLP tasks.

BatteryBERT was initialized with weights from the original BERT-base model, pretrained on general English corpora including Wikipedia and BooksCorpus [12], and then further pretrained on a domain-specific corpus of battery papers drawn from publishers such as the Royal Society of Chemistry (RSC), Elsevier, and Springer. This two-phase training leverages the general linguistic knowledge of BERT while infusing domain-specific semantics.

BatterySciBERT follows a similar strategy but begins from the weights of SciBERT, which had been pretrained on biomedical and computer science texts using the SciVocab tokenizer. Further pretraining was conducted on the battery corpus, making BatterySciBERT a hybrid model that bridges related scientific domains.

In contrast, BatteryOnlyBERT was trained entirely from scratch, both in terms of weights and vocabulary. It uses only battery research papers for both its tokenizer (BatteryVocab) and model weights [13]. This model serves as a testbed for pure domain adaptation without any influence from general or cross-domain knowledge.

Figure 5.1 offers a comparative word cloud visualization of the vocabularies from the BERT, SciBERT, and battery-specific corpora. The battery corpus (Figure 5.1a) exhibits high-frequency terms such as "electrode," "energy," and "battery," reflecting its focused domain semantics. In contrast, SciBERT (Figure 5.1b) contains terms like "patient," while BERT (Figure 5.1c) features general English words like "said," underscoring the need for domain-adapted vocabularies in materials research.

The size and content of the training corpora also differ substantially. Table 5.1 shows that BatteryOnlyBERT was trained on 1.87 billion tokens, all from battery papers, whereas BatteryBERT and BatterySciBERT were trained on 3.3B + 1.87B and 3.17B + 1.87B tokens, respectively, incorporating both general and domain-specific data.

Vocabulary divergence is quantitatively illustrated in Figure 5.2, where the overlap between BatteryVocab and the original BERT vocabulary is shown to be only

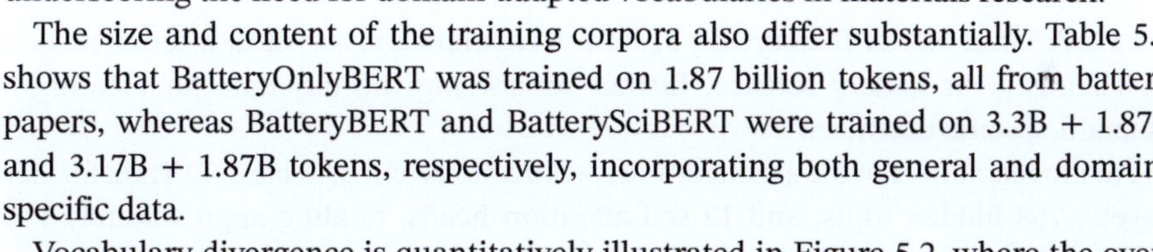

(a) (b) (c)

Figure 5.1 Word clouds illustrating the most commonly occurring terms in the vocabularies derived from (a) the battery-specific corpus, (b) the SciBERT training corpus, and (c) the original BERT corpus. *Source*: Adapted from [14].

Table 5.1 Token counts employed during the training of BatteryBERT, BatterySciBERT, and BatteryOnlyBERT models.

Model	Number of tokens
BatteryBERT	3.3B + 1.87B
BatterySciBERT	3.17B + 1.87B
BatteryOnlyBERT	1.87B

Source: Adapted from [14].

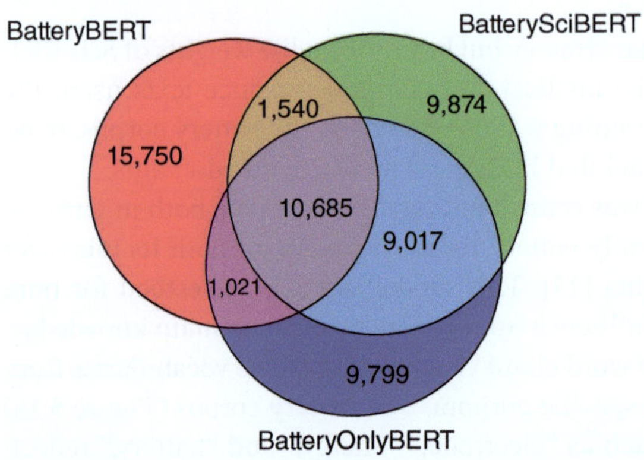

Figure 5.2 Vocabulary comparison among the BatteryBERT, BatterySciBERT, and BatteryOnlyBERT models. *Source*: Adapted from [14].

40.4%. Around 60% of BatteryOnlyBERT's vocabulary consists of new terms related to chemistry or battery science, further reinforcing the rationale for creating a domain-specific tokenizer.

All models adhere to the architectural specifications of BERT-base: 12 Transformer layers, 768 hidden units, and 12 self-attention heads, totaling approximately 110 million parameters (Table 5.1). Training was conducted over 5–7 days using high-performance GPUs, and pretraining involved masked language modeling (MLM) as the primary learning objective, with next-sentence prediction (NSP) excluded due to its limited benefit [15].

This careful engineering of multiple variants allows for systematic comparison across models with differing degrees of domain specificity, and for evaluating trade-offs between general language understanding and scientific domain adaptation.

5.2.1.2 Document Classification and QA

Once pretrained, the BatteryBERT models were fine-tuned on two practical downstream tasks central to battery research: (1) document classification to identify whether a scientific abstract pertains to battery research; and (2) extractive QA to identify specific battery device components (anode, cathode, and electrolyte) from text.

The first task involves binary classification of abstracts as either battery-related or non-battery. The need arises from the fact that simple keyword-based retrieval—such as scraping papers containing the term "battery"—yields many false positives (FPs). For example, papers in biomedical domains that discuss batteries as part of an instrument are irrelevant to battery materials research. To train and evaluate this task, the authors created a labeled dataset of 46,663 papers across 1,058 journals. Journals such as *Journal of Power Sources* were labeled as "battery," while journals such as *Human Genetics* were labeled as "non-battery." The dataset was split into 70% training, 20% validation, and 10% testing (Table 5.2). Performance was benchmarked against a traditional logistic regression (LR) classifier using TF–IDF features. As shown in Table 5.3, BatteryOnlyBERT-cased achieved the highest classification accuracy (97.34%), significantly outperforming the LR baseline (90.95%) and original BERT (96.83%). These results clearly demonstrate the superiority of encoder-based models—especially those pretrained on domain-specific corpora—in scientific classification tasks.

The second task involved fine-tuning the models to answer specific questions such as "What is the anode?" given a contextual passage. The models were trained on the SQuAD v1.1 dataset [16] and then evaluated using a manually labeled battery-specific QA set consisting of 272 records. Each record includes a context, a question (e.g., "What is the cathode?"), and an answer extracted from the text. The fine-tuning process involved adding a span-classification head on top of the model to predict the start and end tokens of the answer. Confidence scores were derived using the product of softmax probabilities for the predicted span.

Table 5.2 Training, validation, and test datasets of the document classification.

	Battery papers	**Non-battery papers**	**Total**
Training	20,629	12,034	32,663
Validation	5,895	3,438	9,333
Test	2,948	1,719	4,667
Total	29,472	17,191	46,663

Source: Adapted from [14].

Table 5.3 Performance accuracy of binary classification models on the test set comprising battery and non-battery abstracts.

Model	**LR (%)**	**BatteryBERT (%)**	**BatterySciBERT (%)**	**BatteryOnlyBERT (%)**	**Original BERT (%)**
Accuracy (cased)	90.95	96.85	97.19	97.34	96.83
Accuracy (uncased)	90.12	96.94	97.47	97.08	96.29

Source: Adapted from [14].

Evaluation metrics included EM-single (Exact Match), the proportion of answers exactly matching the ground truth, and F_1-single, a token-level harmonic mean of Precision and Recall, which is more forgiving of partial overlaps. Table 5.4 presents the results. BatteryBERT-cased achieved the highest F_1-single score (89.16%), outperforming both SciBERT variants and the original BERT (88.58%). Figure 5.3 provides annotated examples of QA pairs used during evaluation, illustrating how the models extract structured data from textual descriptions of materials. Further, Table 5.4 shows performance on the domain-specific QA dataset using Precision-dataset and Recall-dataset metrics. BatteryBERT again outperformed other models with a Precision of 70.74% and Recall of 84.19% for the cased version.

To optimize Precision–Recall trade-offs in practical applications, a confidence score threshold of 0.4 was adopted (Figure 5.4), balancing high confidence with sufficient Recall for extracting reliable data. Together, these downstream tasks illustrate the transformative role of encoder-based models in enabling scalable, accurate, and domain-adapted knowledge extraction from scientific literature—tasks that were previously limited by rule-based heuristics or constrained supervised learning.

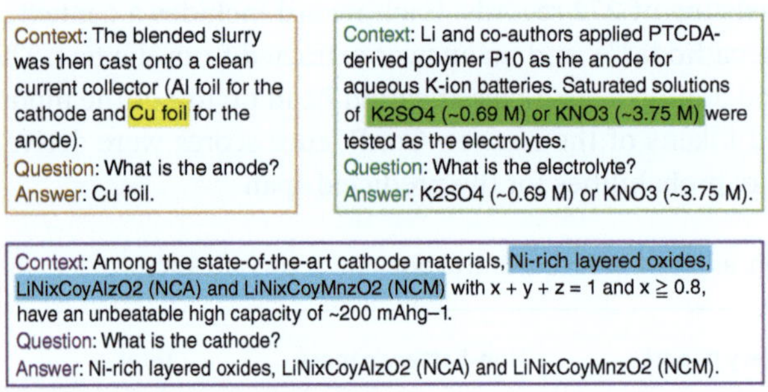

Figure 5.3 Sample entries from the QA evaluation dataset.
Source: Adapted from [14].

Table 5.4 EM-single and F_1-single scores for each model evaluated on the SQuAD validation dataset.

Model	EM-single score (%)	F_1-single score (%)
BERT-base-cased	81.30	88.58
BatteryBERT-cased	**81.54**	**89.16**
BatterySciBERT-cased	79.66	87.43
BatteryOnlyBERT-cased	79.61	87.30
BERT-base-uncased	80.93	88.20
BatteryBERT-uncased	**81.08**	**88.41**
BatterySciBERT-uncased	79.81	87.66
BatteryOnlyBERT-uncased	79.53	87.22

Source: Adapted from [14].

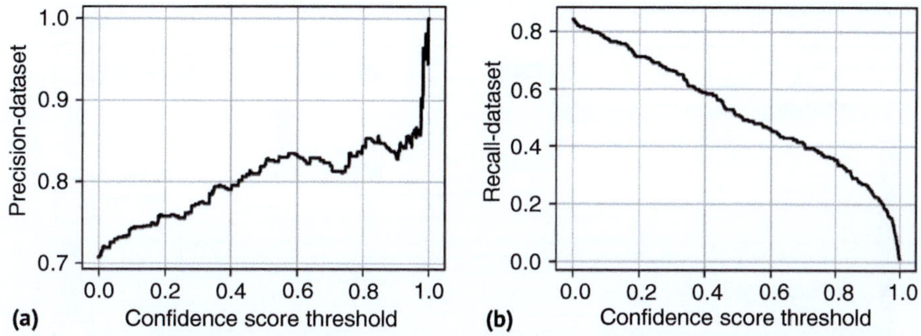

Figure 5.4 (a) Precision-dataset and (b) Recall-dataset versus the confidence score threshold for the BatteryBERT-cased model. *Source*: Adapted from [14].

5.2.1.3 Battery Database Enhancement via Transformer Models

One of the most impactful applications of BatteryBERT lies in its use for enhancing structured databases of battery materials through large-scale automated information extraction. By coupling the fine-tuned language models with rule-guided post-processing, the authors developed a comprehensive pipeline to augment an existing battery materials database with newly extracted relational data, particularly focused on device components (anodes, cathodes, and electrolytes) and material–property associations. The Transformer-based enhancement pipeline begins with a corpus of full-text articles retrieved from major scientific publishers including RSC, Elsevier, and Springer. Pretrained BatteryBERT models are fine-tuned for two purposes: document classification and extractive QA. The document classifier first removes papers irrelevant to battery research. Then, the QA-enabled model identifies device materials by posing structured questions such as "What is the anode?" to contextually relevant passages in the paper.

This process is illustrated in Figure 5.5, which contrasts the traditional ChemDataExtractor-based pipeline with the Transformer-based enhancement pipeline [17]. In the latter, contextual QA is guided by programmatic triggers. For instance, when a term like "anode" is detected in a paragraph, the fine-tuned BatteryBERT model automatically attempts to extract the associated material. A confidence threshold mechanism plays a critical role in refining the precision of the extracted data. The authors empirically determined an optimal threshold value (0.4), balancing Precision (correctness of extracted data) with Recall (completeness). This is shown in Figure 5.4, where Precision increases and Recall decreases with higher confidence thresholds. At the selected threshold, the Precision-dataset score reaches 79.01%, with a Recall-dataset score of 58.82%, which was deemed a robust trade-off for downstream applications.

To further ensure quality, the extracted text is filtered using chemical named entity recognition (CNER) tools from ChemDataExtractor to confirm that the extracted terms are valid material names. In addition, a name normalization step ensures consistency across different spellings and nomenclatures.

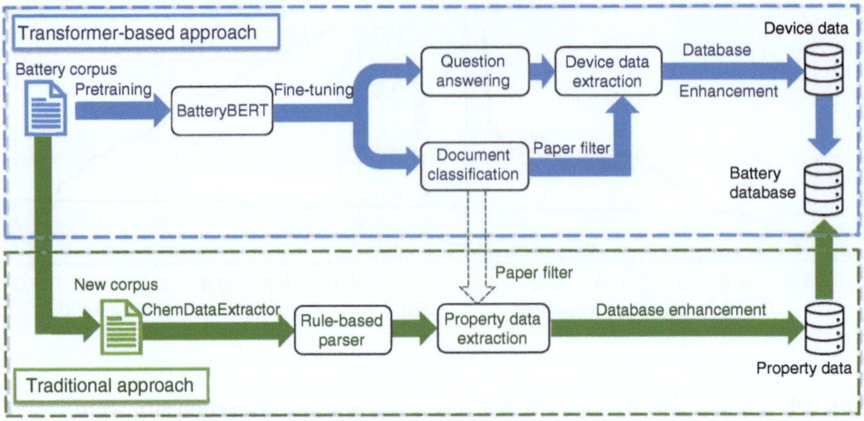

Figure 5.5 Database enhancement pipeline: Transformer-based approach and traditional approach. *Source*: Adapted from [14].

Table 5.5 Summary of data records in the device component database.

Data	Description	Data type
Type	Device component types	String
Name	Chemical compound names	String
Extracted_name	Normalized chemical name	List of dictionaries
Score	Confidence score	Float
Context	Context of original data	String
DOI	Source article DOI	String
Journal	Published journal	String
Date	Published date	String
Title	Source article title	String

Source: Adapted from [14].

The resulting battery device component database includes over 308,000 valid data records, classified into anode, cathode, and electrolyte categories (Table 5.5). Among these, 14,052 are unique entries, indicating high redundancy in reported materials—a finding that is itself scientifically informative. Figure 5.6 displays the distribution of device components and the most common representative materials in each category. For example, lithium (Li) dominates as the most common anode, while $LiCoO_2$ (lithium cobalt oxide) and KOH emerge as the most frequent cathode and electrolyte, respectively.

Furthermore, the integration of newly classified documents and extracted property data led to substantial updates to the previously existing battery-material database. A comparison is presented in Table 5.6, showing that while the overall number

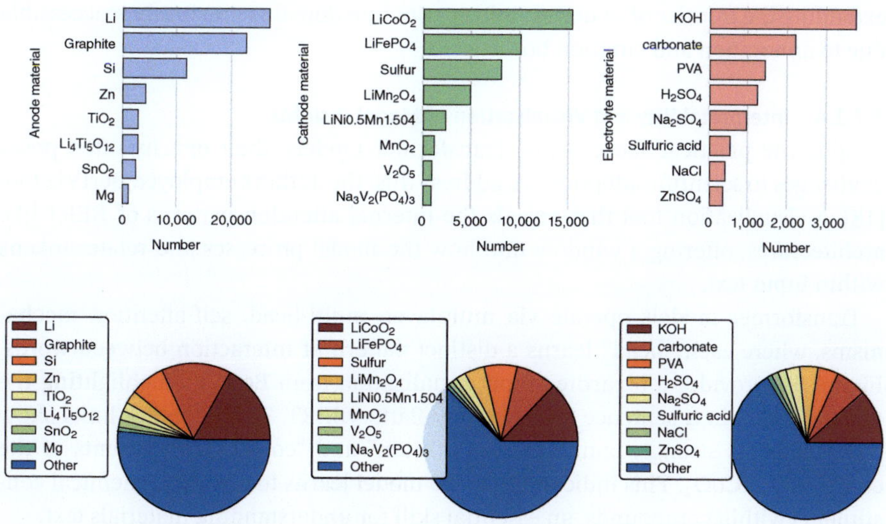

(a) Anode material **(b)** Cathode material **(c)** Electrolyte material

Figure 5.6 Distribution of device component types along with their eight most frequently occurring materials. *Source*: Adapted from [14].

Table 5.6 Comparison of databases: with and without paper filtering for RSC and Elsevier sources, and with filtering including Springer data.

Property	**Number of data records (without paper filters, without Springer data)**	**Number of data records (with paper filters, with Springer data)**	**Number of data records (with paper filters, without Springer data)**
Capacity	144,359	139,118	129,232
Conductivity	7,168	12,697	11,992
Coulombic efficiency	11,033	12,692	11,821
Energy	15,543	14,771	13,542
Voltage	115,240	52,067	46,197
Total	292,313	231,345	

Source: Adapted from [14].

of data records dropped from 292,313 to 231,345 (due to the filtering of irrelevant papers), the quality and domain relevance of the database significantly improved. Notably, entries for conductivity and Coulombic efficiency increased, while less-specific data like voltage declined, reflecting a reduction in domain ambiguity.

These enhancements demonstrate the power of Transformer-based NLP models to convert unstructured text into structured scientific data at scale, thereby extending the frontier of materials informatics into domains previously inaccessible due to annotation and curation bottlenecks.

5.2.1.4 Interpretability and Visualization of Model Insights

Despite the practical successes of Transformer models, their opacity often poses challenges to scientific adoption. To address this, the authors employed BertViz tool [18], a visualization tool that reveals the internal attention patterns of BERT-like architectures, offering a window into how the model processes and relates tokens within input text.

Transformer models operate via multilayer, multi-head, self-attention mechanisms, where each "head" learns a distinct pattern of interaction between words. Figure 5.7 provides a bipartite graph visualization from BertViz, highlighting the token "oxide" in a sentence processed by BatteryBERT. The attention links show that "oxide" is strongly connected to "lithium" and "cobalt," components of the compound $LiCoO_2$. This indicates that the model learns to associate chemical constituents within compounds, an essential skill for understanding materials text.

The visualization also reveals meaningful syntactic structures. For instance, the token "[SEP]," used to separate sentence segments in BERT input, receives strong attention, reflecting the model's learned awareness of contextual boundaries.

An even more detailed overview is given in Figure 5.8, which shows a heatmap of attention weights across 12 layers and 12 heads in the BatteryOnlyBERT-uncased

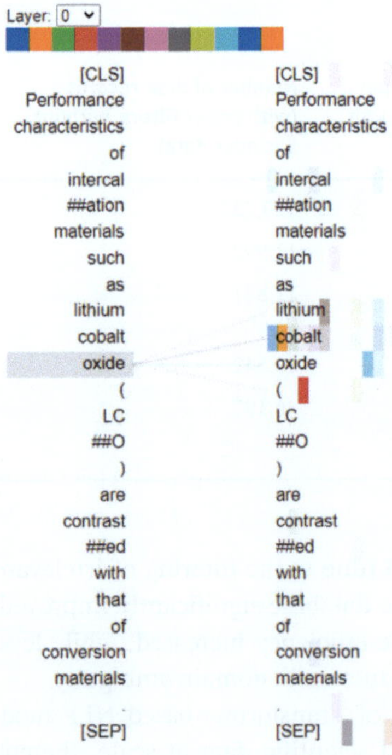

Figure 5.7 Visualization of Layer 0 in the BatterySciBERT-cased model architecture. *Source:* Adapted from [14].

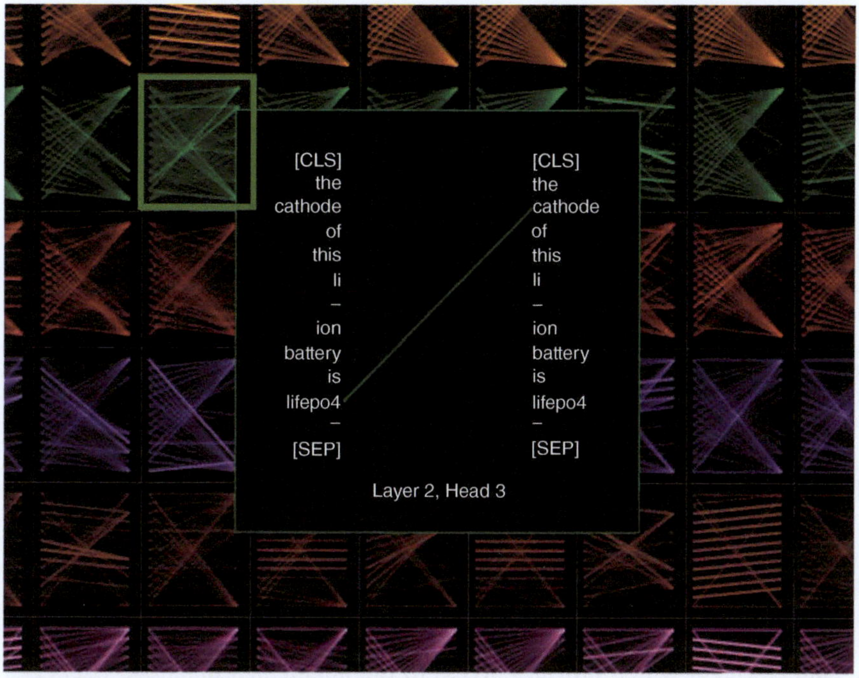

Figure 5.8 Visualization of Layer 2, Head 3 of the fine-tuned QA BatteryOnlyBERT-uncased model architecture. (https://huggingface.co/batterydata/batteryonlybert-uncased-squad-v1) [14].

model, fine-tuned on the QA task. In the example sentence "The cathode of this Li-ion battery is $LiFePO_4$," attention from the token "$LiFePO_4$" points back to "cathode," indicating that the model has learnt a contextual semantic relationship between the material and its functional role.

These visualizations are significant for two reasons. First, they provide interpretability, helping researchers validate that the model is making decisions based on chemically meaningful relationships. Second, they suggest that Transformer-based models internalize domain-specific associations, offering a foundation for developing explainable AI tools in materials science.

The authors note that model performance can be further improved on shorter passages, where attention is less diffused and contextual boundaries are more clearly defined. This insight paves the way for future research on chunking strategies and improved context selection in scientific text mining.

5.2.1.5 Web-based Applications and Future Directions

To maximize accessibility and utility, the authors deployed BatteryBERT in a public-facing web application, available at https://www.materialsforbatteries.org. This platform demonstrates the model's capabilities in real time, enabling users to interact with pretrained models without needing technical knowledge of NLP or model fine-tuning.

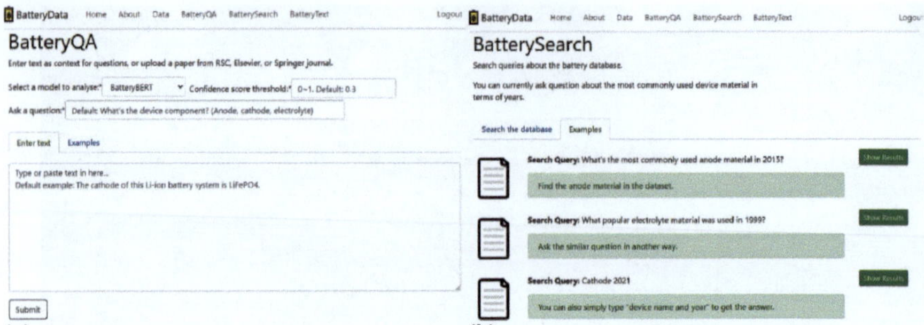

(a) (b)

Figure 5.9 Snapshot of the BatteryData web app with (a) BatteryQA and (b) BatterySearch sections. *Source*: Battery Data. https://www.materialsforbatteries.org/ (last accessed on 26 August 2025); [14].

The web interface comprises two main components:

1. BatteryQA, where users can input raw text or upload a paper, select a BatteryBERT variant, adjust the confidence threshold, and pose standard questions like "What is the cathode?" The system returns the predicted device component along with contextual highlights and confidence scores.
2. BatterySearch, which provides access to the augmented database. Users can query relational data such as "What is the most common electrolyte in 2019?" using the structured data entries extracted by the model. Snapshots of the interface are shown in Figure 5.9. These tools not only support expert users in mining scientific information but also democratize access to cutting-edge NLP tools for nonspecialists in AI.

In terms of broader impact, the BatteryBERT project sets a precedent for similar efforts in other subfields of materials science. The combination of pretraining on domain-specific corpora, fine-tuning on task-relevant datasets, and visualization of attention patterns offers a generalizable framework for building domain-adapted encoder-based models across various scientific disciplines. Moreover, all models and data have been made publicly available via the HuggingFace model hub (https://huggingface.co/batterydata) [19], promoting transparency and reproducibility. This commitment to open science is especially important in interdisciplinary research, where collaboration between materials scientists and machine learning practitioners depends on shared tools and standards.

Looking ahead, the authors propose expanding the capabilities of the BatteryBERT system to include more complex tasks such as multi-hop QA, event extraction, and reasoning over experimental protocols. Further enhancements to the database, particularly in integrating time-series data or incorporating ontologies for materials taxonomy, are also envisioned.

5.2.1.6 Conclusion

The development and deployment of BatteryBERT exemplify a broader methodological shift in materials informatics: the transition from rule-based and manually engineered text mining approaches to scalable, encoder-based language models capable of autonomously extracting latent scientific knowledge from unstructured literature. As demonstrated in this work, domain-specific encoder architectures—particularly those built upon the Transformer paradigm—offer a powerful framework for modeling scientific language in context, enabling both coarse-grained tasks such as document classification and fine-grained tasks like extractive QA. Crucially, the construction of such models requires deliberate choices in corpus design, tokenizer vocabulary, and pretraining strategy, all of which directly influence downstream performance and generalizability. The BatteryBERT framework underscores the importance of aligning model architecture with domain semantics through targeted pretraining and domain-aware fine-tuning. Furthermore, it highlights the need for interpretability tools and deployment pathways—such as visualization interfaces and web applications—to translate model outputs into actionable insights for materials scientists. Together, these practices point toward a systematic blueprint for applying encoder-based models in materials science: one that integrates domain-specific data curation, model customization, performance evaluation, and end-user accessibility into a unified pipeline for scientific discovery.

5.2.2 MatSciBERT

NLP, and more specifically, Transformer-based encoder models like BERT [9], have revolutionized the extraction of information from unstructured text. However, general-purpose models such as BERT or even cross-domain models like SciBERT [8], which is trained on scientific and biomedical literature, struggle to interpret the nuanced vocabulary, specialized symbols, and domain-specific syntax of materials science. This limitation has spurred the development of domain-adapted, encoder-based models such as MatSciBERT, a Transformer architecture fine-tuned specifically for the materials domain. MatSciBERT leverages the encoder capacity of the BERT architecture while grounding it in a corpus derived from ~150,000 full-text materials science papers, enabling accurate extraction, classification, and semantic understanding of materials-specific knowledge.

The following subsections detail the motivations behind developing MatSciBERT, the construction and pretraining of its domain-specific corpus, and its performance across key downstream tasks. Together, these subsections illustrate how encoder-based models, when tailored to domain-specific language, can significantly enhance computational tools for materials discovery.

5.2.2.1 The Need for Domain-specific Language Models in Materials Science

Transformer-based language models such as BERT and its scientific variant SciBERT have achieved remarkable success in general NLP and scientific text mining. Nevertheless, when applied directly to the materials science domain, these models exhibit reduced efficacy due to a lack of exposure to domain-specific terminologies, formulations, and naming conventions. For instance, general-purpose tokenizers often fragment materials-specific terms such as "yttria-stabilized zirconia" into subword tokens like "yt," "##tri," "##a," "-," "stabilized," "zircon," and "##ia," thereby diluting semantic cohesion. A domain-aware tokenizer might instead preserve tokens like "yttria" and "zirconia" to better reflect their chemical identity.

Moreover, the vocabulary and ontology of materials science differ significantly from other scientific fields. Materials literature includes dense formulations of properties (e.g., "open-circuit voltage"), complex composite names (e.g., "$LiNi_{0.5}Mn_{1.5}O_4$"), and specialized experimental terminology (e.g., "annealing temperature," "degradation rate," or "crystallographic orientation"). These are rarely present in biomedical or computer science corpora that dominate pretrained models like SciBERT. Consequently, encoder models not tailored to the materials domain often fail to capture contextual cues and chemical semantics crucial for accurate information extraction [20].

The sheer volume of unstructured textual data in the materials literature compounds the problem. Manually extracting relevant insights from this data is labor-intensive and prone to inconsistency. Recognizing this gap, Gupta et al. developed MatSciBERT [6], a domain-specific encoder model designed to address the limitations of general models and to automate text mining and information extraction in materials science. As illustrated in Figure 5.10, their methodology involves building a dedicated corpus, adapting the pretrained SciBERT model through further domain-specific training, and evaluating its performance across multiple materials-relevant NLP tasks.

5.2.2.2 Corpus Construction and Pretraining of MatSciBERT

A central innovation of MatSciBERT lies in its domain-aware training corpus, the materials science corpus (MSC), which comprises approximately 150,000 research articles selectively drawn from over one million documents available in the Elsevier ScienceDirect database. The curated corpus spans four core sub-domains: inorganic glasses and ceramics, metallic glasses, alloys, and cement and concrete. Importantly, the corpus includes both full-text articles and abstracts, allowing for broad coverage of both experimental and computational studies.

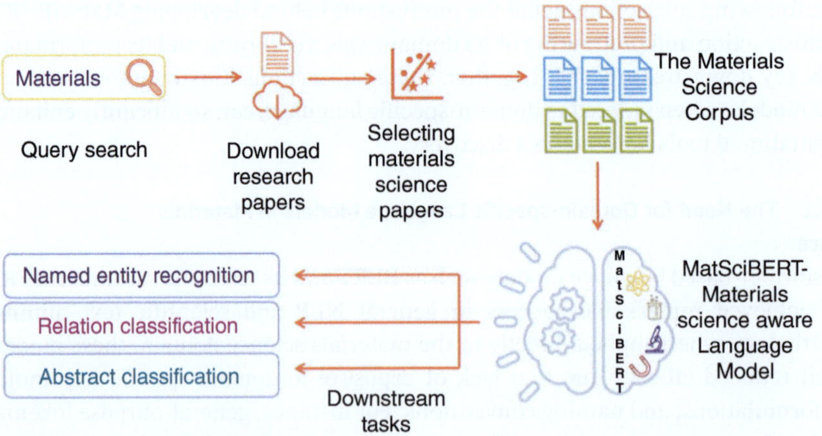

Figure 5.10 Methodology for training MatSciBERT. *Source*: [6] / Springer Nature.

Table 5.7 Representation of the word counts of important strings relevant to the several subfields in materials science. It should be noted that the corpus encompasses manuscripts from several important subfields such as thermoelectrics, nanomaterials, polymers, and biomaterials.

String	Training corpus	Validation corpus
Thermoelectric	5,104	1,902
Nanomaterial	5,114	873
Polymer	173,738	31,234
Biomaterial	5,259	887

Source: Adapted from [6].

According to Table 5.7, the MSC comprises approximately 285 million words. While this represents only 9% of the total word count in SciBERT's original training corpus (3.17B words), the domain-specificity of MSC enables more effective representation of materials-related text. Table 5.7 further details the frequency of technical terms relevant to materials science, confirming the corpus's domain representativeness.

To reduce manual effort in corpus construction, the authors applied a classifier trained on labeled abstracts to identify relevant articles among the initial one million downloads. Classifiers fine-tuned from SciBERT were used to tag papers belonging to specific categories, such as "glass science" or "magnesium alloy composites," ensuring a high-quality input dataset for model training.

For pretraining MatSciBERT, the authors adopted the domain-adaptive pretraining strategy introduced by Gururangan et al. [6], which involves continuing the training of an existing language model on a domain-specific corpus. The authors compared vocabulary overlap between their corpus and existing models: MSC's vocabulary had 53.64% overlap with SciBERT's uncased vocabulary and only 38.90% with BERT's, supporting the decision to initialize MatSciBERT using SciBERT's weights.

Architecturally, MatSciBERT uses several techniques from RoBERTa to improve training efficiency and downstream performance:

1. Dynamic whole word masking, which masks entire words rather than subword tokens to preserve contextual integrity.
2. Removal of NSP loss, a training simplification found to improve generalization.
3. Full-length sequence training, where input sequences are not truncated prematurely.
4. Large batch sizes, improving model convergence and generalization.

Pretraining was conducted over 360 hours on dual NVIDIA V100 32-GB GPUs, achieving a final perplexity of 2.998, comparable to or better than benchmarks set by BERT (3.99) and RoBERTa. As shown in Figure 5.11, the training trajectory indicated consistent improvement in MLM loss and accuracy, demonstrating the effectiveness of MatSciBERT's domain adaptation.

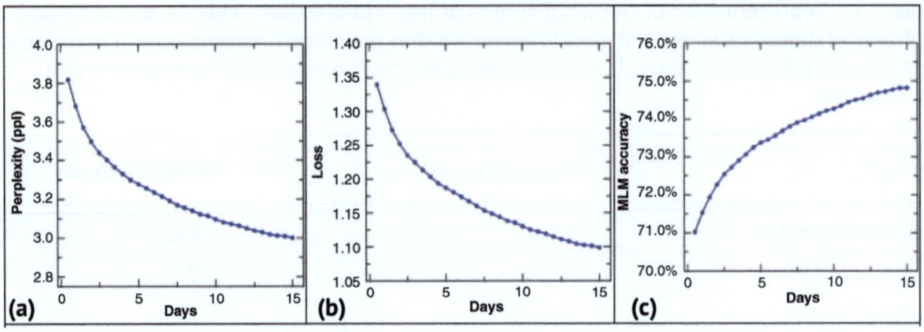

Figure 5.11 Visualization of MatSciBERT's MLM performance on the validation set over training time, showing trends in (a) perplexity, (b) loss, and (c) accuracy as a function of training days. *Source*: Adapted from [6].

5.2.2.3 Evaluation on Downstream Tasks: Benchmarking MatSciBERT

The effectiveness of MatSciBERT was validated through its performance on three key NLP tasks central to materials science: NER, Relation Classification, and Paper Abstract Classification. These tasks were chosen to represent real-world applications such as materials database construction, literature classification, and semantic information retrieval.

MatSciBERT was evaluated on three NER datasets: Matscholar [21], containing seven materials-relevant entity types (e.g., MAT [Inorganic Material], PRO [Material Property], and CMT [Characterization Method]); SOFC and SOFC-Slot [22], which include fine-grained entities relevant to solid oxide fuel cells, such as "anode material," "degradation rate," and "open-circuit voltage."

Table 5.8 presents Macro-F_1 scores for all models tested. MatSciBERT consistently outperformed both SciBERT and BERT across all architectures (Linear, CRF, and BiLSTM-CRF). For instance, on the SOFC-Slot dataset using LM-CRF, MatSciBERT achieved a Macro-F_1 score of 65.35%, exceeding SciBERT's 59.07% and BERT's 58.26%. Its performance on the Matscholar dataset was similarly impressive, with a Macro-F_1 score of 86.38%, surpassing the previous state-of-the-art of 85.10% [21].

Figure 5.12 illustrates entity-level F_1-scores on the SOFC-Slot validation set. MatSciBERT demonstrated superior performance in all materials-related categories, including "electrolyte material" and "support material," underscoring its domain awareness and enhanced capacity to extract nuanced entities from complex sentences.

Relation Classification task is evaluated using the Materials Synthesis Procedures dataset [23], involving classifying semantic relationships between pairs of entities in synthesis protocols. MatSciBERT significantly outperformed baseline models (MaxPool, MaxAtt) and general-purpose Transformers. Table 5.9 shows a Macro-F_1 score of 89.02% for MatSciBERT, compared to 87.22% for SciBERT and 85.40% for BERT. The improved accuracy demonstrates MatSciBERT's ability to comprehend and disambiguate fine-grained relationships critical to synthesis planning.

In the Paper Abstract Classification task, the model was asked to label abstracts as pertaining to "glass" or "non-glass" materials. Table 5.10 reports that MatSciBERT

Table 5.8 Average Macro-F_1 performance on the test dataset for Matscholar, computed across three different random seeds [6].

Architecture	LM=MatSciBERT (%)	LM=SciBERT (%)	LM=BERT (%)	SOTA (%)
LM-Linear	85.46±0.13 (87.83±1.21)	83.80±0.32 (86.05±0.55)	82.10±0.81 (82.79±0.20)	85.10 (85.41)
LM-CRF	86.38±0.49 (88.66±0.88)	85.04±0.77 (88.07±0.96)	84.07±0.19 (84.61±0.81)	
LM-BiLSTM-CRF	86.09±0.46 (89.15±0.57)	85.66±0.24 (87.66±0.29)	83.39±0.20 (84.07±0.29)	

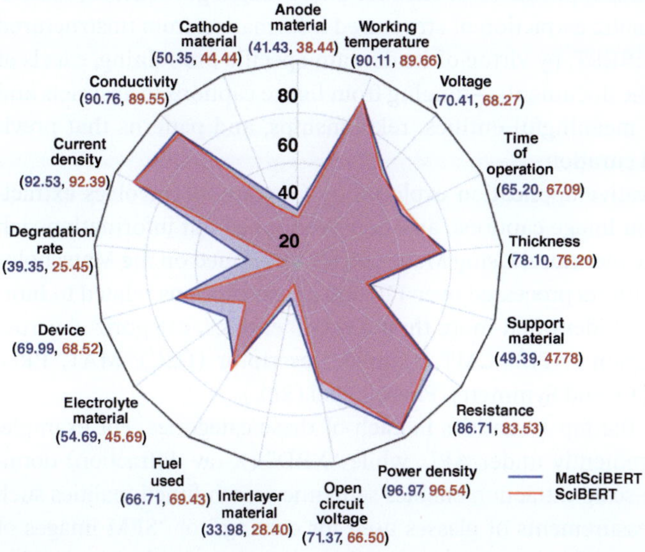

Figure 5.12 Comparison of MatSciBERT and SciBERT on validation sets of SOFC-Slot dataset. *Source*: Adapted from [6].

Table 5.9 Test set results for Materials Synthesis Procedures dataset averaged over three seeds [6].

	MatSciBERT (%)	SciBERT (%)	BERT (%)	MaxPool (%)	MaxAtt (%)
Macro-F_1	89.02±0.27 (88.31±0.14)	87.22±0.58 (87.21±0.17)	85.40±1.45 (85.95±0.78)	81.19±1.54 (80.93±0.71)	80.39±0.85 (81.53±2.23)
Micro-F_1	91.94±0.20 (91.50±0.20)	91.04±0.32 (91.03±0.08)	90.16±0.69 (90.44±0.54)	86.81±1.55 (86.68±0.84)	87.16±0.60 (87.62±1.34)

Table 5.10 Test set results for glass vs. non-glass dataset averaged over three seeds.

	MatSciBERT (%)	SciBERT (%)	BERT (%)	MaxPool (%)	MaxAtt (%)
Accuracy	96.22±0.16 (95.33±0.27)	93.44±0.57 (94.00±0.00)	93.89±0.68 (93.33±0.98)	91.44±0.31 (92.22±0.56)	91.44±0.68 (91.22±0.16)

achieved 96.22% accuracy, outperforming SciBERT (93.44%) and BERT (93.89%). This performance difference reinforces the model's superior contextual understanding of domain-specific cues and terminology.

5.2.2.4 Information Extraction Applications in Materials Research

One of the most impactful applications of encoder-based language models in materials science is the automatic extraction of structured information from unstructured textual sources. MatSciBERT, by virtue of its domain-specific pretraining, excels at parsing diverse scientific documents—ranging from figure captions to abstracts and full texts—to uncover meaningful entities, relationships, and patterns that previously required manual curation.

A particularly innovative application explored by Gupta et al. involves extracting named entities from image captions, an often overlooked but information-rich component of scientific literature. Using MatSciBERT fine-tuned on the Matscholar NER dataset [6], the authors processed over 110,000 image captions related to inorganic glasses. The model identified more than 429,000 entities, categorized across seven classes: Application (APL), CMT, Sample Descriptor (DSC), MAT, PRO, Synthesis Method (SMT), and Symmetry/Phase Label (SPL).

Figure 5.13 presents the top-10 entities in each of these categories. For example, "coating" appears prominently under APL, while "XRD" (X-ray diffraction) dominates CMT. Such large-scale extraction enables semantic search functionalities such as querying: "XRD measurements of glasses used for coating," or "SEM images of doped bioglasses with Ag." This represents a significant leap over previous manual or rule-based methods, facilitating rapid discovery of relevant visual data in literature.

In Figure 5.14, the authors compare MatSciBERT's automated tagging against human-annotated single-label captions. Remarkably, MatSciBERT extracts multiple meaningful entities from the same caption, capturing a richer set of information than prior manual methods. This capacity allows researchers to retrieve figures based on multiple intersecting criteria, increasing the utility of image-based scientific communication.

To further structure the extracted data, the authors construct Materials Caption Graphs (MCGs). Each graph connects entities extracted from a single caption, and these are then merged into a large graph structure linking thousands of nodes. Two specific examples of these subgraphs are shown in Figure 5.15.

In Figure 5.15a, the authors demonstrate how entities not directly linked in any single caption (e.g., "T_g" and "anneal") can be shown to share common neighbors—such as "glass," "amorphous," and "XRD." This suggests that the property T_g (glass transition temperature) is frequently studied in relation to amorphous and doped

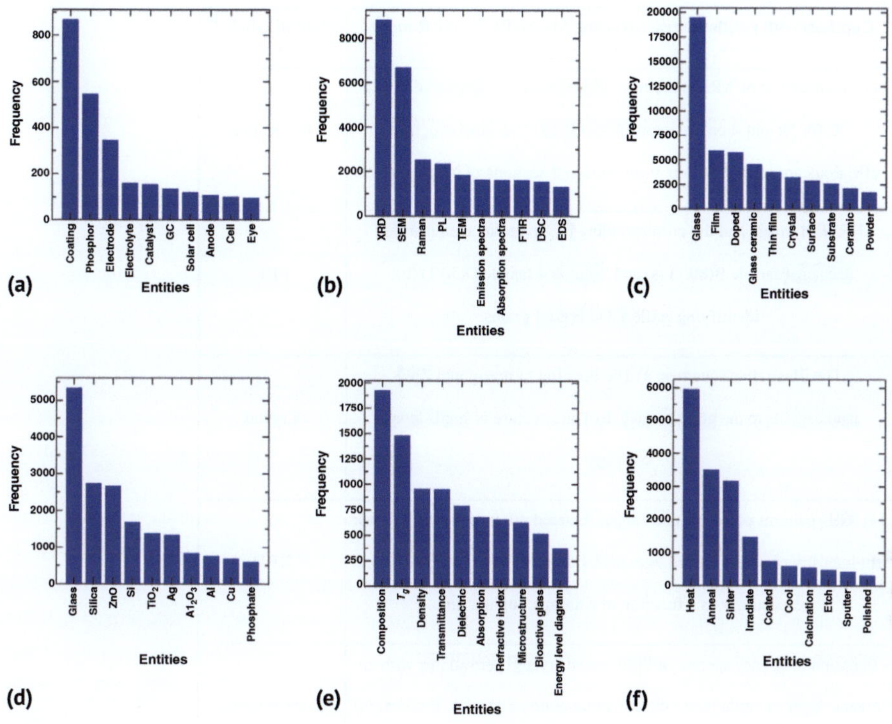

Figure 5.13 Top-10 entities for various categories: (a) APL, (b) CMT, (c) DSC, (d) MAT, (e) PRO, and (f) SMT. *Source*: Adapted from [6].

glasses and characterized using XRD, with annealing likely serving as the synthesis route. Such inferences align with established domain knowledge, thus validating the model's semantic understanding.

In Figure 5.15b, the model explores entity neighborhoods around "XRD," showing its association with different materials (e.g., ZnO, YPSZ), synthesis methods (e.g., annealing, aging), and applications (e.g., phosphors). These graphs can guide hypothesis generation or experimental planning by illustrating connections within the literature that may not be explicitly stated.

5.2.2.5 Topic Modeling and Knowledge Graph Construction

In addition to information extraction, encoder-based models like MatSciBERT enable deeper semantic structuring of materials literature via contextualized topic modeling and automated KG generation.

Traditional topic modeling methods, such as latent Dirichlet allocation (LDA), rely on word frequency and co-occurrence statistics. These methods are effective in broad contexts but falter in scientific domains where word sense disambiguation and contextual subtleties are essential. For example, the word "glass" can refer to a material, a transition temperature, or even a container, depending on the context. To address this, Gupta et al. use the [CLS] token embeddings from MatSciBERT to generate context-aware vector representations of abstracts. These embeddings are projected

Captions with entities extracted using MatSciBERT NER model	Manual labels [10]
The comparison of XRD patterns of glass ceramic heat treated at 725 °C for 5h and rhombohedral Ba4Gd3F17. The superstructure reflections are marked with o. Inset: enlarged sections of XRD patterns.	Reflection
HRTEM image and the corresponding FFT pattern taken from as-deposited sample B(80/1) (a) and annealed sample D(30/1) (b); identifying rutile TiO2 crystal grains.	FFT
The illustrative schemes: a) The bonding of hexagonal ZnO nanocrystals to the glass surface. b) The structure of multi-layers coatings.	Crystal
(a) XRD patterns of the glass-ceramics sintered different holding times; (b) intensity of μ- and α-cordierite peaks count at (101) and (110) plane respectively as a function of sintering holding time.	XRD
Photoluminescence spectra of PbBr-based layered perovskites with an organic layer of naphthalene-linked ammonium molecules. Profiles: (a) 1; (b) 2; (c) 3; (d) 4; (e) 5.	Luminescence
Labels: Application (APL), Characterization method (CMT), Descriptor (DSC), Material (MAT), Property (PRO), Synthesis method (SMT), Symmetry/phase label (SPL)	

Figure 5.14 A comparison between NER tags generated by MatSciBERT and those assigned manually. *Source*: [6] / Springer Nature.

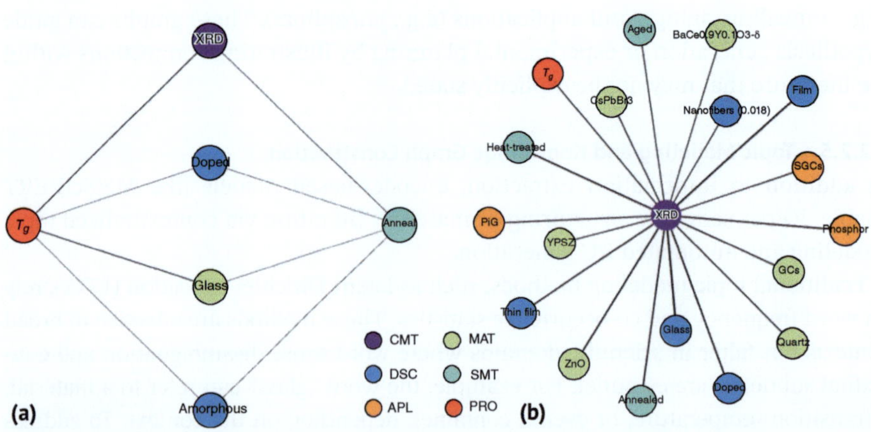

Figure 5.15 MCGs: (a) establishing a relationship between two previously unrelated components; and (b) investigating elements associated with the characterization technique "XRD." *Source*: [6] / Springer Nature.

Table 5.11 Top-10 topics obtained based on LDA along with the top-10 keywords associated with each topic [6].

Topic no.	Keywords
0	["cells," "protein," "cell," "activity," "binding," "structure," "proteins," "dna," "human," "enzyme"]
1	["water," "properties," "concentration," "ph," "content," "showed," "results," "temperature," "polymer," "different"]
2	["glass," "composite," "fiber," "fibers," "properties," "mechanical," "materials," "material," "strength," "results"]
3	["based," "use," "new," "paper," "structure," "methods," "study," "research," "data," "systems"]
4	["temperature," "phase," "glass," "transition," "magnetic," "fe," "structure," "properties," "dielectric," "samples"]
5	["soil," "study," "patients," "treatment," "species," "group," "results," "fluoride," "significantly," "rare"]
6	["particles," "size," "particle," "nanoparticles," "ag," "powder," "surface," "nm," "silver," "diameter"]
7	["mantle," "ma," "pb," "high," "sr," "rocks," "ree," "ratios," "samples," "composition"]
8	["model," "results," "data," "experimental," "flow," "method," "structure," "parameters," "time," "transition"]
9	["la," "ce," "et," "da," "ra," "des," "en," "ta," "sa," "une"]

into two-dimensional space using the uniform manifold approximation and projection (UMAP) algorithm, followed by k-means clustering. The resulting clusters are analyzed for dominant keywords, offering a more coherent and interpretable topic structure.

The comparison between traditional LDA and MatSciBERT-based topic modeling is summarized in Tables 5.11 and 5.12. While the LDA-derived topics contained noisy or generic keywords (e.g., French terms or ambiguous tokens), the MatSciBERT clusters yielded semantically rich and recognizable themes. Expert reviewers could interpret these clusters as meaningful research topics, such as "oxide thin-film synthesis," "luminescence in glasses," and "materials for energy storage." This example underscores the capacity of encoder-based models to enhance unsupervised knowledge discovery in scientific literature, where terminological overlap and nuanced semantics often obscure insights.

As discussed in the previous subsection, MCGs represent a novel form of domain-specific KGs derived directly from figure captions. Unlike traditional curated ontologies, these graphs are generated automatically using NER-labeled entities and co-occurrence logic. Each caption yields a fully connected subgraph of its entities, and these are combined to create a holistic view of entity relationships across the literature.

The ability to generate such graphs at scale offers multiple benefits: semantic exploration: researchers can explore related methods, materials, and properties in a graph format. Experimental design: by identifying frequently co-occurring synthesis and characterization pairs, one can infer standard protocols or innovative

Table 5.12 Top-10 topics obtained based on MatSciBERT along with the top-10 keywords associated with each topic.

Topic no.	Keywords
0	["materials," "used," "radiation," "laser," "model," "process," "paper," "thermal," "use," "based," "chapter," "solar," "experimental," "applications," "energy"]
1	["dielectric," "ceramics," "conductivity," "sintering," "doping," "phase," "ceramic," "sintered," "constant," "doped," "grain," "solid," "frequency," "electrical," "ferroelectric"]
2	["tio2," "photocatalytic," "light," "visible," "zno," "nanoparticles," "activity," "efficiency," "polymer," "uvÃ¢," "synthesized," "vis," "doped," "film," "crystal"]
3	["laser," "silica," "surface," "ion," "radiation," "measured," "time," "energy," "induced," "diffusion," "irradiation," "index," "observed," "used," "glasses"]
4	["complexes," "ii," "complex," "ligand," "compounds," "fluorescence," "synthesized," "nmr," "ligands," "1h," "compound," "crystal," "iii," "ir," "h2o"]
5	["luminescence," "crystals," "doped," "ions," "conductivity," "glasses," "excitation," "band," "observed," "absorption," "centers," "emission," "energy," "crystal," "laser"]
6	["zno," "films," "deposited," "film," "gap," "substrates," "substrate," "electrical," "band," "deposition," "xrd," "annealing," "doping," "resistivity," "optical"]
7	["adsorption," "ph," "fluorescence," "protein," "fluoride," "water," "phosphate," "ii," "binding," "removal," "detection," "acid," "concentration," "solution," "ions"]
8	["films," "film," "deposition," "deposited," "substrates," "si," "substrate," "annealing," "optical," "zno," "band," "silicon," "thickness," "gap," "sputtering"]
9	["waste," "melt," "melts," "water," "silicate," "alteration," "composition," "data," "compositions," "mantle," "dissolution," "ash," "elements," "rich," "co2"]

alternatives. Literature analytics: graph centrality measures could identify influential materials, techniques, or properties within a given domain.

Together, these capabilities represent a shift toward automated, scalable knowledge engineering in materials science—a development largely enabled by domain-specific encoder models like MatSciBERT.

5.2.2.6 Conclusion

The development of encoder-based models such as MatSciBERT exemplifies a critical trajectory in the application of LLMs to materials science: the necessity of domain adaptation to capture the linguistic, structural, and semantic intricacies of scientific discourse. Unlike generic Transformer models, domain-specific encoders are trained on corpora that reflect the terminological density and contextual subtleties of materials research, enabling them to internalize and operationalize domain knowledge at scale. This section has demonstrated that the success of encoder-based approaches in materials science hinges on three core pillars: the construction of high-fidelity domain corpora, the strategic adaptation of existing pretrained architectures through continued pretraining, and the rigorous evaluation of model capabilities across representative downstream tasks. MatSciBERT illustrates how encoder models can move beyond passive text representation to active agents of scientific knowledge extraction—fueling applications such as automated NER, relational reasoning, topic discovery, and materials KG construction. As the field progresses, the design of encoder-based models will increasingly depend not only on algorithmic advancements, but also on the co-evolution of annotated datasets, shared benchmarks, and interdisciplinary collaboration between materials scientists and NLP practitioners. Encoder architectures thus offer a powerful and extensible foundation for building intelligent systems capable of navigating and synthesizing the vast, unstructured frontier of materials science literature.

5.3 Decoder-based Models for Materials

Traditional NLP offers simplicity and tailored solutions but struggles with flexibility and complex tasks. Recently, GPT models have used the Transformer decoder architecture consisting of self-attention and feed-forward layers to generate text. Conversational LLMs such as GPT-4 have demonstrated their remarkable capability for efficiently extracting data from extensive collections of research papers. The application of LLMs to analyze information, extract pertinent details, and create responses, is becoming increasingly popular in several research fields in a standalone mode. This harnesses LLMs by use of prompt engineering and fine-tuning and their integration with other scientific tools.

5.3.1 Chemistry Assistant: A Decoder-based Model for MOF Synthesis Text Mining and Prediction

The advent of LLMs based on decoder-only architectures, such as OpenAI's GPT series [24, 25], has transformed how unstructured scientific knowledge can be processed and utilized in materials science. Traditionally, the extraction of synthesis conditions, chemical properties, and reaction protocols from scholarly literature required expert-driven, rule-based systems or supervised learning approaches, often restricted by narrow scopes and high maintenance burdens [26, 27]. Decoder-based models, with their ability to follow instructions, generate coherent text, and handle domain-specific reasoning without retraining, have emerged as a powerful alternative.

In this context, Zheng et al. [28] introduced the ChatGPT chemistry assistant (CCA), a system built upon GPT-3.5 and GPT-4, to automate the mining of MOF synthesis conditions from heterogeneous scientific literature and to facilitate downstream tasks such as machine learning modeling and QA [29]. The work represents a paradigmatic shift in materials informatics, illustrating how decoder-based models can reduce reliance on coding and domain-specific parsing rules while maintaining high accuracy and interpretability. The following subsections explore three key pillars of this contribution: the ChemPrompt Engineering framework, the multi-process workflow, and a rigorous performance evaluation of the system.

5.3.1.1 ChemPrompt Engineering: Tailoring Prompts for Accurate Chemistry Text Mining

At the core of this study is the concept of ChemPrompt Engineering, a principled approach for instructing decoder-based LLMs to perform high-fidelity text mining in the chemistry domain using natural language alone. Recognizing that decoder-only models are sensitive to prompt phrasing, Zheng et al. established three design principles that underpin successful instruction of ChatGPT for scientific information extraction: minimizing hallucinations, providing detailed instructions, and requesting structured output [30, 31].

Principle 1: Minimizing hallucinations

One major challenge with LLMs is their tendency to generate plausible-sounding but incorrect information—a phenomenon known as hallucination [32]. For example, when asked to specify synthesis parameters for a nonexistent MOF (e.g., MOF-99999), ChatGPT may fabricate conditions unless explicitly told not to. To mitigate this, the authors proposed adding clarifying statements such as "If uncertain, respond with 'I do not know'." Empirical results shown in Table 5.13 confirm that such prompt modifications significantly reduce hallucination rates for both known and unknown compounds.

Principle 2: Providing detailed instructions

The second principle involves explicitly defining the scope and expected behavior of ChatGPT. In particular, the prompts specify which types of reactions to ignore (e.g., organic linker synthesis or post-synthetic modifications), how to report missing values (e.g., use "N/A"), and how to interpret multiple measurement units (e.g., separating them by commas). This is critical for avoiding the inclusion of irrelevant or confusing content and ensuring domain fidelity.

Principle 3: Requesting structured output

Finally, ChemPrompt Engineering emphasizes the use of standardized tabular formats to ensure that the outputs are immediately machine-readable. This

Table 5.13 Improvements in ChatGPT response accuracy utilizing a basic prompt engineering strategy.

Initial query	Guided prompt	ChatGPT response
Which metal is used in the synthesis of MOF-5?	If you are uncertain, please reply with "I do not know."	Zinc (correct)
Which metal is used in the synthesis of MOF-519?		I do not know (correct)
What is the linker used in the synthesis of MOF-99999?		I do not know (correct)
What is the linker used in the synthesis of MOF-419?		I do not know (correct)
What is the linker used in the synthesis of ZIF-8?		2-Methylimidazole (correct)

Source: [28] / with permission of American Chemical Society.

ChatGPT Processing Demo

Prompt: Answer the question as truthfully as possible using the provided context. Please summarize the following details in a table: compound name or chemical formula (if the name is not provided), metal source, metal amount, organic linker(s), linker amount, modulator, modulator amount or volume, solvent(s), solvent volume(s), reaction temperature, and reaction time. If any information is not provided or you are unsure, use "N/A".

Please focus on extracting experimental conditions from only the MOF synthesis and ignore information related to organic linker synthesis, MOF postsynthetic modification or metalation, high throughput (HT) experiment details, or catalysis reactions.

If multiple conditions are provided for the same compound, use multiple rows to represent them. If multiple units or components are provided for the same factor (e.g. g and mol for the weight, multiple linker or metals, multiple temperature and reaction time, mixed solvents, etc.), include them in the same cell and separate by comma.

The table should have 11 columns, all in lowercase:I compound name I metal source I metal amount I linker I linker amount I modulator I modulator amount or volume I solvent I solvent volume I reaction temperature I reaction time I

Input:

Synthesis of MOF-808. In a 100 mL media bottle were dissolved 1,3,5-benzenetricarboxylic acid (210 mg) and ZrOCell2·8H2O (970 mg) in a solution containing DMF (30 mL) and formic acid (30 mL). The bottle was sealed and heated in a 100 °C isothermal oven for a day. White powder was collected by centrifugation (8000 rpm, 3 min), washed with DMF three times (60 mL × 3) over a 24 h period, and then washed with acetone three times (60 mL × 3) over a 24 h period.

Output:

compound name	metal source	metal amount	linker	linker amount	modulator	modulator amount or volume	solvent	solvent volume	reaction temperature	reaction time
MOF-808	ZrOCl2-8H2O	970 mg	1,3,5-benzenetricarboxylic acid	210 mg	formic acid	30 mL	DMF	30 mL	100 °C	24 h

Figure 5.16 A schematic depiction of prompt elements intended to assist ChatGPT in extracting synthesis conditions, formulated in accordance with the three key principles of chemistry-specific prompt engineering. *Source*: [28] / with permission of American Chemical Society.

includes instructing the model to return data in fixed columns, such as "metal source," "linker," "solvent," "modulator," "reaction time," and "temperature." Figure 5.16 provides an example of the tabular structure generated from a real synthesis paragraph.

Together, these principles were applied iteratively in what the authors describe as interactive prompt refinement—a conversational method wherein ChatGPT is asked to improve its own prompts based on human feedback. This meta-programming capability is uniquely enabled by decoder-based models and contributes to rapid iteration cycles without code changes.

5.3.1.2 Multi-process Workflow for MOF Synthesis Text Mining

To extract synthesis data from over 200 MOF-related papers with diverse narrative formats and section structures, Zheng et al. designed a three-tiered workflow composed entirely of ChatGPT-assisted processes (Figure 5.17). Each process builds on the previous one, balancing automation, accuracy, and speed.

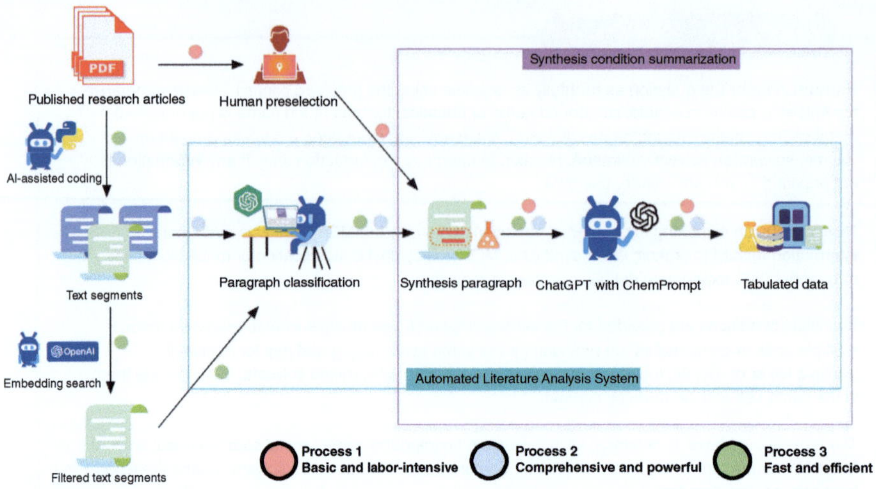

Figure 5.17 The CCA workflow. *Source*: [28] / with permission of American Chemical Society.

Process 1: Summarization from manually selected sections
The first process involves manually selecting paragraphs from the experimental sections of research articles and prompting ChatGPT to convert these into structured synthesis data. While requiring human preselection, this process ensures maximum accuracy by feeding only relevant content to the model. As described in the main text and shown in Figure 5.16, the model outputs an 11-column table capturing key synthesis parameters such as metal quantity, linker type, solvent volume, reaction time, and more.

Process 2: Paragraph classification via few-shot learning
Process 2 introduces paragraph classification to automate the identification of synthesis-related sections from full papers. ChatGPT is prompted with a few-shot classification prompt (Figure 5.18) and asked to return "Yes" or "No" for each paragraph. Only "Yes"-labeled paragraphs are passed to Process 1 for summarization. This mirrors the reading logic of a human researcher scanning for experimental content and avoids redundant manual labor.

Process 3: Embedding-based filtering for efficiency
To reduce the computational load of classifying every paragraph, Process 3 uses text embeddings generated via OpenAI's text-embedding-ada-002 model [33, 34]. Each paragraph is converted into a 1,536-dimensional vector, and semantic similarity to a synthesis-relevant anchor prompt is computed via cosine similarity. Low-scoring paragraphs are filtered out, and the remainder are passed to Process 2. Figure 5.19 visually demonstrates distinct clusters of synthesis, gas sorption, characterization, and reference sections in a 2D projection of 18,248 embedding vectors.

5.3.1.3 Performance Evaluation of the CCA

To assess the validity and robustness of the workflow, the authors conducted a full-scale performance audit across all three processes. Rather than relying on sampling,

ChatGPT Processing Demo

Prompt: Determine whether the provided context includes a comprehensive MOF synthesis with explicit reactant quantities or solvent volumes, and answer with either Yes or No.

Context: In a 4-mL scintillation vial, the linker H2PZVDC (91.0 mg, 0.5 mmol, 1 equiv.) was dissolved in N,N-dimethylformamide (DMF) (0.6 mL) upon sonication. An aqueous solution of ALC13·6H2O (2.4 mL 0.2 M) was added dropwise, and the resulting mixture was heated in a 120 °C oven for 24 hours.
Question: Does the section contain a comprehensive MOF synthesis with explicit reactant quantities or solvent volumes?
Answer: Yes.

Context: A 0.150 M solution of imidazole in DMF and a 0.075M solution of Zn(NO3)2·4H2O in DMF were used as stock solutions, and heated in an 85 °C isothermal oven for 3 days.
Question: Does the section contain a comprehensive MOF synthesis with explicit reactant quantities or solvent volumes?
Answer: Yes.

Context: Solvothermal reactions of Co(NO3)·6H2O, Hatz, and L1/L2 in a 2:2:1 molar ratio in DMF solvent at 180 °C for 24 h yielded two crystalline products, 1 and 2, respectively.
Question: Does the section contain a comprehensive MOF synthesis with explicit reactant quantities or solvent volumes?
Answer: No.

Context: A 22.9% weight loss was observed from 115 to 350 °C, which corresponds to the loss of one DEF molecule per formula unit (calcd: 23.5%).
Question: Does the section contain a comprehensive MOF synthesis with explicit reactant quantities or solvent volumes?
Answer: No.

Input # 1:
In a 100 mL media bottle were dissolved 1,3,5-benzenetricarboxylic acid (210 mg) and ZrOC12·8H2O (970 mg) in a solution containing DMF (30 mL) and formic acid (30 mL).

Output # 1:
Yes.

Input # 2:
Single crystal X-ray analyses were performed at room temperature on a Siemens SMART platform diffractometer outfitted with an APEX II area detector and monochromatized Mo Kα radiation.

Output # 2:
No.

Figure 5.18 An illustration demonstrating the application of few-shot prompting to direct ChatGPT in identifying synthesis-related paragraphs. The figure includes two sample inputs along with their corresponding classification outputs for clarity. *Source*: [28] / with permission of American Chemical Society.

they manually verified the extraction of 11 parameters for approximately 800 MOF synthesis entries—totaling over 26,000 data points.

Each extracted value was labeled as: true positive (TP): correct parameter identified; FP: incorrect or hallucinated value; and false negative (FN): valid parameter omitted. The team computed Precision, Recall, and F_1-score for each parameter across all three processes. As summarized in Figure 5.20c,d, all processes achieved Precision > 95%, Recall > 90%, and F_1 > 92% on average. Errors primarily arose in solvent volumes and modulator identification due to ambiguity in mixed solvent systems and stock solution use.

Reaction time and temperature were most consistently extracted due to standardized unit patterns. In terms of efficiency, Figure 5.20a,b shows that: Process 1

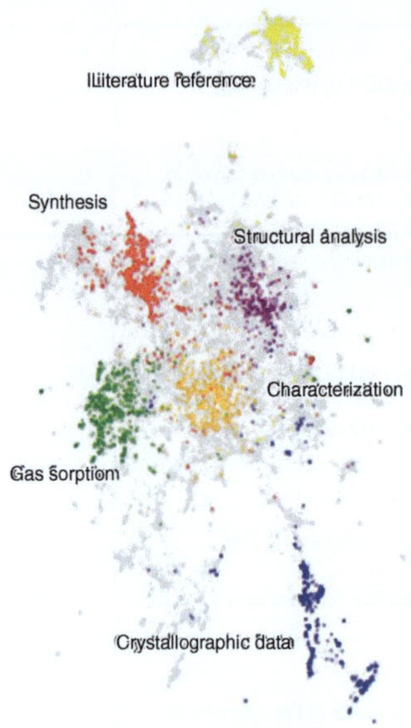

Figure 5.19 Two-dimensional visualization of 18,248 text segments embedding. *Source*: [28] / with permission of American Chemical Society.

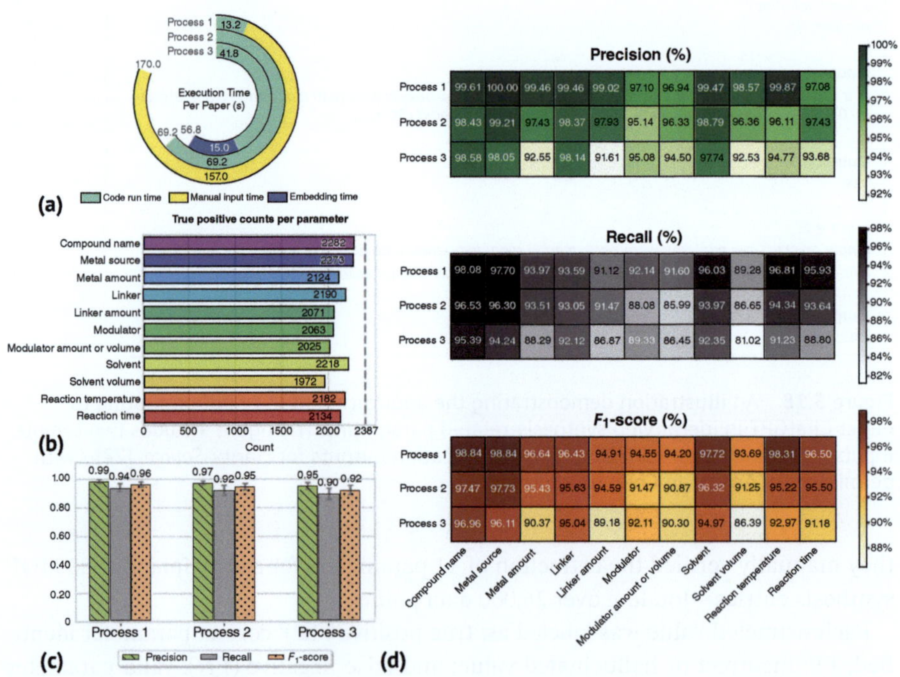

Figure 5.20 Multifaceted performance analysis of ChatGPT-based text mining processes. (a) Average time per paper for each process, comparing their relative execution efficiency. (b) True positive counts for 11 synthesis parameters aggregated over 2387 conditions, showing strong consistency across processes and effective parameter extraction. (c) Overall average Precision, Recall, and F_1-scores for each process, with gray error bars indicating standard deviations. (d) Heat map of parameter-wise Precision, Recall, and F_1-score across the three processes, revealing detailed extraction performance for each synthesis parameter. *Source*: [28] / with permission of American Chemical Society.

required human input and averaged ~2.5 minutes per paper for section selection, with summarization itself taking ~13 seconds; Process 2, being fully automated, improved throughput significantly; and Process 3 delivered the fastest performance by pre-filtering documents, reducing classification time by ~30%.

One interesting phenomenon emerged from the analysis of ambiguous synthesis conditions. In cases where phrases like "using similar procedures" replaced full protocols for derivative MOFs, ChatGPT conservatively reported "N/A" rather than duplicating prior conditions—reflecting a cautious, non-hallucinatory design.

5.3.1.4 Data-driven Prediction of Crystallization Outcomes

The predictive utility of decoder-based language models is further exemplified in the authors' extension of the mined MOF synthesis dataset to machine learning-based crystallization prediction. Leveraging over 800 distinct MOF entries and more than 26,000 synthesis parameters extracted via ChatGPT, the authors constructed a binary classification model to distinguish synthesis conditions yielding single-crystal (SC) products from those resulting in polycrystalline (P) outcomes.

The initial step involved extracting qualitative morphological descriptors of synthesized MOFs (e.g., "needle-like crystals," "red powder") from literature. Using a modified ChatGPT prompt (Figure 5.21), the assistant was instructed to classify

ChatGPT Data Processing

Prompt: Please classify the provided descriptions of materials based on their crystalline structure into either single-crystal (SC) or polycrystalline (P). The descriptions should be parsed and classified as per the following rules: If the description includes words like "powder", "solid", "microcrystalline", "crystalline product", or "N/A", classify it as "P" (polycrystalline). If the description includes words like "single-crystal" or "crystal", classify it as "SC" (single-crystal).
Please present the results in a two-column table. The first column should contain the original description, and the second column should display the classification result "SC" or "P". Ensure that each row in the table represents one description and its corresponding classification.

Input:

```
"""
red solid
Green block crystals
colorless rodlike crystals
pink precipitate
yellow crystalline product
purple hexagonal platelike crystals
large yellow crystalline product
Colorless rod-shaped crystals
colorless rhombus-shaped block crystals
Off-white powdered product
Tetragonal colorless crystals
Sky-blue powdered product
Yellow needle crystals
fine powdered MOF
dark yellow cubic crystals
white microcrystalline product
Needlelike colorless crystals
"""
```

Output:

Original Description	Classification
red solid	P
Green block crystals	SC
colorless rodlike crystals	SC
pink precipitate	P
yellow crystalline product	P
purple hexagonal platelike crystals	SC
large yellow crystalline product	P
Colorless rod-shaped crystals	SC
colorless rhombus-shaped block crystals	SC
Off-white powdered product	P
Tetragonal colorless crystals	SC
Sky-blue powdered product	P
Yellow needle crystals	SC
fine powdered MOF	P
dark yellow cubic crystals	SC
white microcrystalline product	P
Needlelike colorless crystals	SC

Figure 5.21 Utilization of GPT-4 in classifying synthesized MOFs by crystalline morphology. *Source*: [28] / with permission of American Chemical Society.

each synthesis outcome into SC or P. This illustrates the model's capacity for contextual understanding and soft classification, enabled entirely through prompt-based instruction—an affordance unique to decoder-based architectures.

To build a predictive model, the authors designed six categories of materials-specific descriptors representing the chemical and physical characteristics of MOF synthesis. These included: (1) metal, linker, solvent, and modulator identities; (2) stoichiometric ratios (e.g., linker-to-metal, solvent-to-metal); (3) reaction conditions (time, temperature); and (4) crystallization method (e.g., solvothermal, vapor diffusion). These structured features were derived from ChatGPT outputs and standardized using both in-model prompts and ChatGPT-generated Python scripts.

Using this curated dataset, the team trained a random forest (RF) classifier to predict crystallization outcomes [35]. After applying recursive feature elimination (RFE) with fivefold cross-validation, the model retained 33 features and achieved accuracy of ~87% on the held-out set, F_1-score of 92%, and robustness to class imbalance and noisy data (Figure 5.22a,b).

The top predictive features, including "modulator-to-metal ratio," "solvent-to-metal ratio," and "reaction time," aligned well with domain knowledge of crystal growth kinetics and thermodynamics (Figure 5.22c).

Notably, this approach does not require retraining ChatGPT or using decoder-based models directly for prediction [36]. Instead, ChatGPT is employed to convert unstructured text into structured features, demonstrating the utility of decoder models as data generation and knowledge structuring tools in downstream ML applications. Figure 5.22 presents six illustrative MOF cases (e.g., MOF-520, MOF-74, and ZIF-8) whose predicted outcomes matched experimental reports, confirming the practical value of the hybrid ChatGPT–ML pipeline.

5.3.1.5 Building an MOF-focused Chemistry Chatbot

A distinctive innovation in Zheng et al.'s work is the construction of an MOF synthesis chatbot, powered by embeddings and grounded responses derived from the mined dataset. This data-to-dialog transformation marks a significant advance in applying decoder-based models as interactive chemistry assistants rather than one-off text processors.

Each MOF entry, composed of synthesis parameters and bibliographic metadata (authors, publication year, and DOI), is embedded using OpenAI's text-embedding-ada-002 [37]. Upon receiving a user query, the system compares the query's embedding against the precomputed dataset, identifies the most relevant synthesis record, and sends it as context into a ChatGPT prompt.

This architecture ensures that the response is grounded in real literature, rather than relying on the model's internal prior. Figure 5.23 schematically outlines this pipeline.

To maintain factual accuracy, the chatbot is instructed via prompt design to refrain from answering queries beyond the dataset scope. If no relevant data is found, the chatbot asks users to rephrase or narrow their question. This strategy (emphasized in Figure 5.24) substantially reduces hallucination risks and enables safe, reliable, dialog-based exploration.

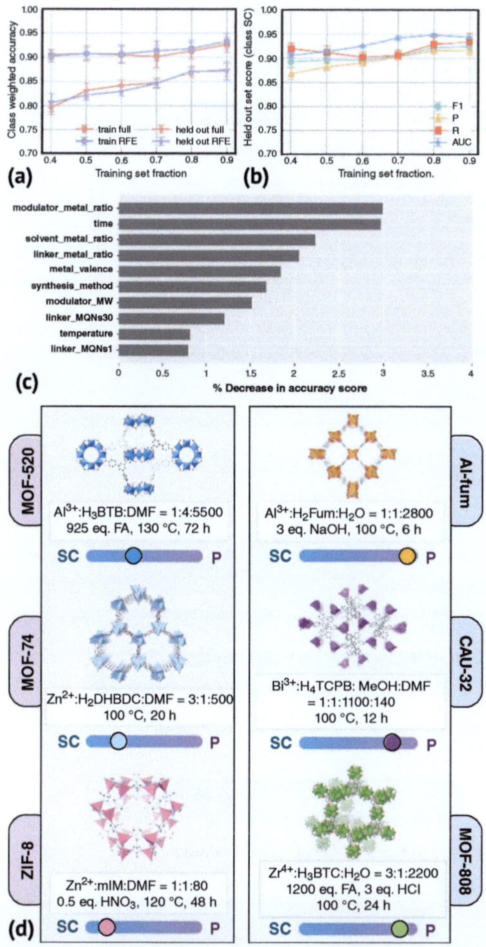

Figure 5.22 Performance of the classification models in predicting the crystalline state of MOFs from synthesis. (a) Learning curves with 1σ error bars show improved model accuracy as the training set size increases. (b) F_1-score, Precision, Recall, and AUC metrics demonstrate strong classification performance across training ratios. (c) The top-10 descriptors, ranked by accuracy contribution, highlight key factors influencing MOF crystallization. (d) Model predictions for six MOFs accurately match experimental outcomes, distinguishing SC from P states. *Source*: [28] / with permission of American Chemical Society.

Another notable feature is the preservation of prior context across conversational turns. When a user asks follow-up questions, ChatGPT retains the synthesis data and metadata from the previous round, enabling multi-turn, grounded dialog. This stands in contrast to the default ChatGPT web interface, which lacks domain-specific memory tethering to external structured knowledge.

The authors suggest that this MOF chatbot serves not only expert chemists but also students and nonspecialists, providing precise, literature-backed synthesis details with citations (e.g., DOIs). By converting a static dataset into an interactive knowledge interface, the model supports guided exploration, hypothesis generation, and independent learning—hallmarks of LLM-based tools in scientific research.

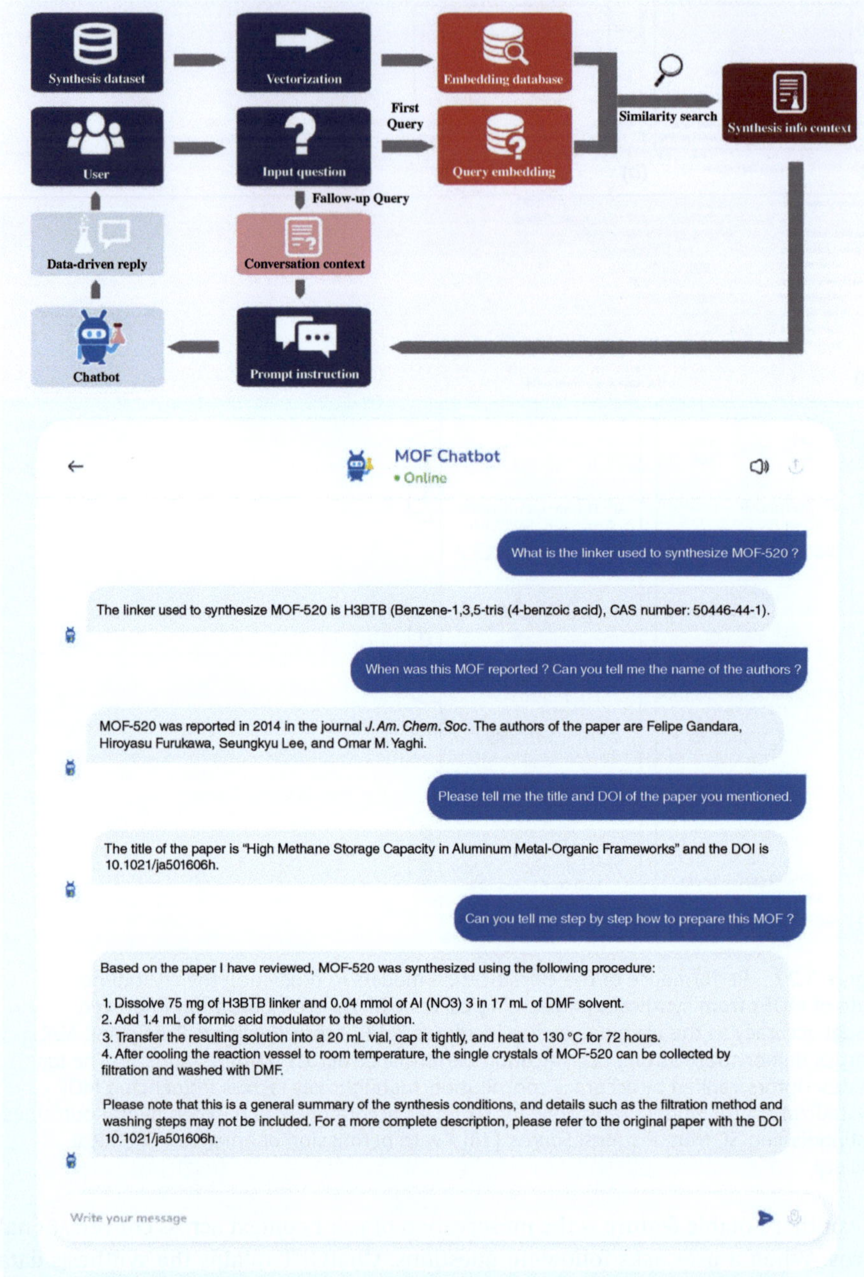

Figure 5.23 Integrated workflow of the MOF chatbot. *Source*: [28] / with permission of American Chemical Society.

5.3.1.6 Conclusion

The application of decoder-based language models, as demonstrated by the CCA, marks a shift in how materials scientists interact with unstructured literature. Unlike static rule-based systems, models like GPT-3.5 and GPT-4 enable instruction-driven, adaptive workflows—capable of classifying, summarizing, and transforming

ChatGPT API

```python
def chatbot(question, past_user_messages=None, initial_context=None):
    if past_user_messages is None:
        past_user_messages = []
    past_user_messages.append(question) # Store Synthesis and Paper Information Cards
    df_with_emb = pd.read_csv("xxx.csv")# Get Information Cards and Embeddings

    if initial_context is None:
        # Find the context based on the first question
        first_question = past_user_messages[0]
        question_return = openai.Embedding.create(model="text-embedding-ada-002",
input=first_question)
        question_emb = question_return['data'][0]['embedding']

        df_with_emb_sim = add_similarity(df_with_emb, question_emb)
        num_paper = 3
        top_n_synthesis_str = top_similar_entries(df_with_emb_sim, num_paper)
        initial_context = top_n_synthesis_str

    message_history = [
        {
            "role": "system",
            "content": "You are a chemistry assistant that specifically handles questions
related to MOF synthesis conditions based on the papers you have reviewed. Answer the
question using the provided context. If the question is not relevant to the context or the
MOF is not mentioned in the context, respond with 'Based on the information available from
the MOF paper I have read so far, I cannot provide a reliable answer to this question. Please
revise your question.'\n\nContext:\n" + initial_context
        },
    ]

    for user_question in past_user_messages:
        message_history.append({"role": "user", "content": user_question})

    response = openai.ChatCompletion.create(
        model='gpt-3.5-turbo',
        messages=message_history
    )
    answer = response.choices[0].message["content"]
    return answer, initial_context, past_user_messages

def add_similarity(df, given_embedding):
    def calculate_similarity(embedding):
        return cosine_similarity([embedding], [given_embedding])[0][0]
    df['similarity'] = df['embedding'].apply(calculate_similarity)
    return df
```

Figure 5.24 Python code snippet demonstrating the utilization of the ChatGPT API for establishing a literature-grounded chatbot. *Source*: [28] / with permission of American Chemical Society.

scientific text into structured data through natural language alone. Central to this paradigm is prompt engineering, which guides model behavior without retraining. This flexibility allows decoder-based models to address varied tasks—from synthesis extraction to dialog-based knowledge retrieval—making them scalable, low-code tools for accelerating materials discovery and bridging the gap between human expertise and machine reasoning.

5.3.2 NatureLM: Unlocking the Language of Nature for Scientific Discovery

In recent years, decoder-only LLMs have fundamentally reshaped how NLP systems interpret and generate complex information. Their autoregressive architecture, exemplified by models such as GPT-3, LLaMA, and Mixtral, lends itself naturally to tasks requiring generative synthesis, context-aware reasoning, and multimodal integration. While early successes focused on human languages, emerging efforts have begun extending decoder-based models into scientific domains—particularly in areas where data can be represented as tokenized sequences. One such groundbreaking effort is the Nature Language Model (NatureLM) introduced by Xia et al. [38], which reimagines scientific sequences across domains—small molecules, proteins, DNA, RNA, and materials—as components of a unified "language of nature."

NatureLM is a generative science foundation model built on decoder-based Transformer architectures. Unlike traditional models that operate in siloed domains, NatureLM is pretrained on a corpus spanning 143 billion tokens of domain-specific and textual data, enabling unified modeling across disparate fields. Of particular relevance to materials science, NatureLM demonstrates state-of-the-art performance on material generation, property prediction, and crystal structure modeling tasks—while also enabling new paradigms such as instruction-based materials design and cross-domain interaction (e.g., protein-to-material inference). This subsection delves into three key aspects of NatureLM's development and relevance to materials science: (1) its foundational motivation and architecture; (2) its multi-domain pretraining and representation schema; and (3) its specific applications in material generation and design.

5.3.2.1 A Unified Science Foundation Model: Motivation and Architecture

Traditional efforts to apply machine learning in the sciences have typically focused on domain-specific models fine-tuned for narrow tasks. For instance, protein folding, retrosynthetic analysis, and materials property prediction each historically relied on separate model architectures and datasets. While effective within their own domains, these models fail to capitalize on the latent interconnectivity of scientific knowledge. Recognizing that many scientific entities—from SMILES-encoded molecules to materials compositions—can be expressed as sequences, Xia et al. proposed NatureLM as a unified, sequence-centric science foundation model (see Figure 5.25).

At its core, NatureLM adopts a Transformer-based decoder architecture, built upon and extending LLaMA 3 (8B) and Mixtral (8 × 7B) models. The decision to use decoder-only models is crucial: such architectures enable flexible autoregressive generation, ideal for forward synthesis tasks, iterative optimization, and instruction-guided modeling. Three versions of the model were developed:

1. NatureLM (1B): a compact model with 16 Transformer layers and reduced hidden dimensions.
2. NatureLM (8B): based on LLaMA 3, with 32 layers and a vocabulary size of 130,239.
3. NatureLM (8 × 7B): a mixture-of-experts (MoE) model totaling 46.7 billion parameters, composed of eight 7B expert models, with sparse activation for efficiency.

To support scientific reasoning, NatureLM augments standard language model vocabularies with specialized domain tokens (e.g., ⟨mol⟩, ⟨material⟩, and ⟨protein⟩).

These tokens demarcate sequence boundaries for domain-specific entities and are critical for teaching the model to differentiate between, for example, a chemical structure and a DNA sequence. A two-stage pretraining strategy ensures that these tokens are effectively learnt: in the first stage, only the new token embeddings are trained while the rest of the model is frozen, reducing instability. In the second stage, full fine-tuning across the entire network integrates these scientific tokens into the model's broader generative capabilities.

Figure 5.25 illustrates NatureLM's architecture and training flow, highlighting the integration of heterogeneous data sources (text, SMILES, FASTA, and materials data) into a unified pretraining pipeline. Table 5.14 provides architectural

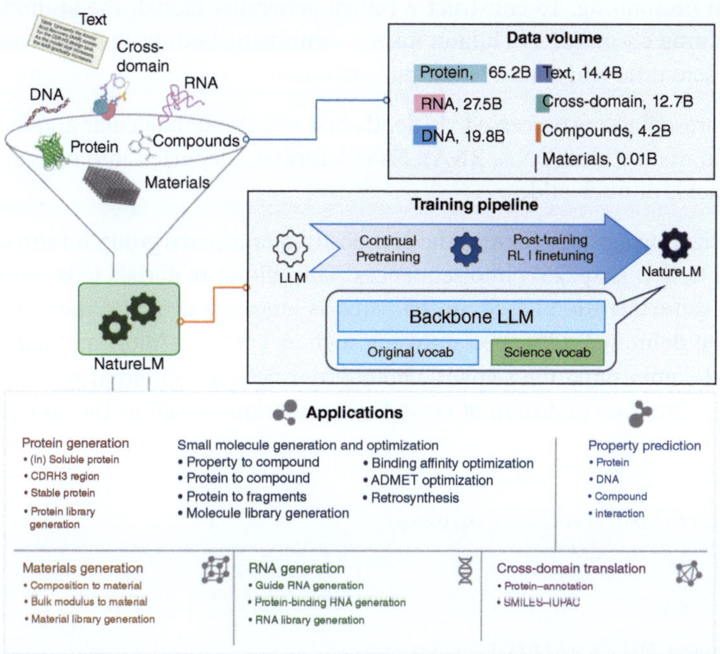

Figure 5.25 Schematic overview of the NatureLM model. *Source*: [38] / arXiv.

Table 5.14 Model parameters of different sizes of NatureLM.

Model parameter	**1B**	**8B**	**8 × 7B**
Hidden dimensions	2,048	4,096	4,096
FFN dimensions	5,504	14,336	14,336
Attention heads	32	32	32
KV heads	8	8	32
Number of layers	16	32	32
Vocabulary size	130,239	130,239	38,078

Source: [38] / arXiv.

specifications across the three model sizes, revealing how scaling up the parameter count improves the model's ability to capture complex dependencies across scientific domains.

The motivation for such a model is clear: scientific tasks often require contextual reasoning across textual descriptions, symbolic notations, and structured data. A generalist model like NatureLM, capable of synthesizing this information in an instruction-following generative setting, offers a transformative step forward in materials informatics and computational design.

5.3.2.2 Multi-domain Pretraining and Cross-domain Representation

One of the most distinguishing aspects of NatureLM is its systematic approach to multi-domain pretraining. To construct a robust generalist model, the authors curated a pretraining corpus of 143 billion tokens comprising both pure sequences and interleaved scientific-text data. This corpus includes:

- SMILES representations processed via depth-first search on molecular graphs.
- Proteins and nucleotides (DNA, RNA): FASTA format, tokenized into individual amino acids or nucleotides.

A novel linearization strategy flattens the composition and space group information (e.g., "Li Li O O O O sg225") into sequences. This allows materials to be processed with the same attention-based mechanisms as language or molecular data. All sequences are delimited with special tokens such as ⟨mol⟩...⟨/mol⟩ or ⟨material⟩...⟨/material⟩, preserving the semantic boundaries between different types of data (see Figure 5.26). The inclusion of crystal structure data—such as the atomic

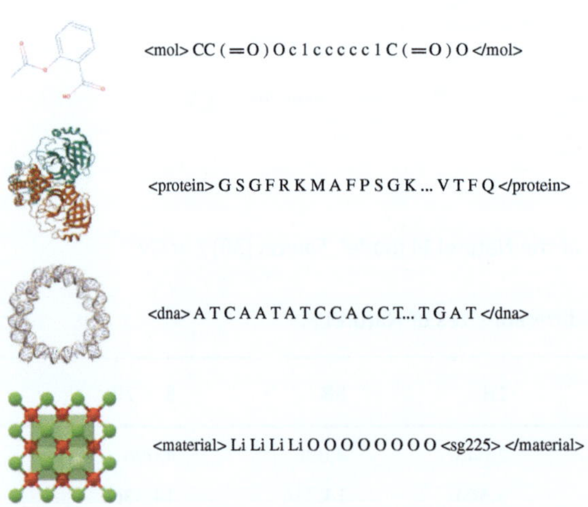

Figure 5.26 Representative examples of data used across domains. The small molecule shown is Aspirin (PubChem CID: 2244), rendered using RDKit [39]. The protein structure is sourced from the Protein Data Bank (PDB ID: 7CAM) [40]. DNA is represented by chains I and J from PDB entry 1KX5 [41], visualized via UCSF Chimera [42]. The materials example is taken from the Materials Project database (ID: mp-1960) [43]. *Source*: [38] / arXiv.

composition and space group (e.g., for mp-1960 from the Materials Project)—allows NatureLM to reason about materials-specific patterns in generative tasks.

The pretraining corpus was not limited to isolated single-domain entries. A major innovation of NatureLM is the construction of cross-domain sequences—synthetic documents that integrate multiple scientific representations. These were created using three key methods:

1. Interleaved sequences: Entities in biomedical literature were detected using BERN2 and replaced with their respective sequences (e.g., SMILES or FASTA), enabling training on contexts where text and scientific data co-occur.
2. Parallel alignment: Descriptive metadata from PubChem, UniProt, and the Materials Project were paired with structured data, aligning text with entities like bandgap or energy-above-hull.
3. Central dogma linking: DNA sequences were translated into their corresponding proteins, capturing transcription-translation mappings fundamental to molecular biology.

Figure 5.27 presents the token distribution of the pretraining data. While protein and RNA sequences dominate the corpus, materials-related data—though comprising only 0.01% of tokens—play a pivotal role in domain-specific fine-tuning (e.g., Matbench). Table 5.15 further quantifies the cross-domain samples, showing that over 12.7 billion tokens involve explicit cross-domain interactions.

This comprehensive treatment of domain and cross-domain data enables NatureLM to not only perform unimodal scientific tasks but also engage in complex cross-domain generation, such as inferring material properties from textual prompts or designing materials that resemble protein-like structural motifs.

5.3.2.3 Materials Science Applications of NatureLM

While NatureLM supports tasks across biology and chemistry, its applications to materials science are both technically innovative and strategically significant. The model is capable of performing a range of generative and predictive tasks relevant to

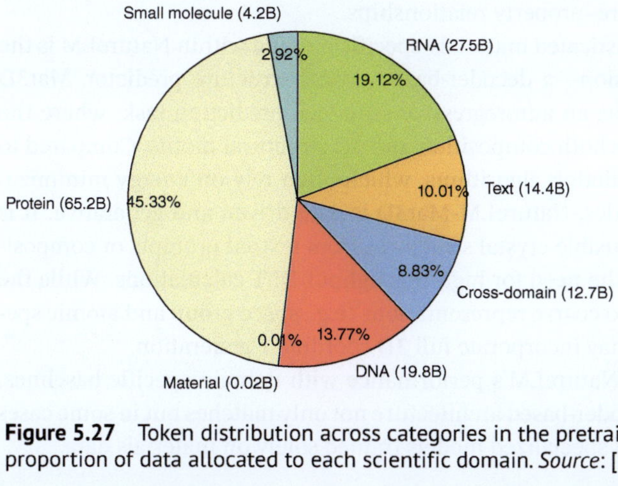

Figure 5.27 Token distribution across categories in the pretraining dataset, indicating the proportion of data allocated to each scientific domain. *Source*: [38] / arXiv.

Table 5.15 Statistics of cross-domain data.

	Samples (by million)	Tokens (by billion)	Samples (%)	Tokens (%)
Interleaved sequence	4.3	4	10.2	31.3
Text-SMILES	33	3	78.8	24
Text-protein	1.9	1.4	4.6	10.8
Text-material	1.7	0.2	4	1.6
DNA-protein	1	4.1	2.4	32.3
Total	41.9	12.7	100	100

Source: [38] / arXiv.

materials discovery, including unconditional materials generation, property-guided synthesis, and crystal structure prediction.

In this task, the model is prompted with the ⟨material⟩ token and asked to generate chemically plausible compositions and crystal structure descriptors. Validity is assessed by checking if the generated output can be parsed into known chemical compositions and valid space group representations. Although this is a syntactic task, it evaluates the model's ability to internalize domain-specific grammar—a prerequisite for more complex material synthesis.

Beyond unconditional generation, NatureLM can also perform conditioned materials generation. In the composition-to-material task, given a chemical formula, the model predicts additional structural descriptors. In the bulk modulus-to-material task, the model is provided with a target mechanical property (e.g., bulk modulus value) and asked to generate a material likely to satisfy this constraint. These tasks are analogous to inverse design: generating candidates that meet predefined target properties. The performance on property-conditioned generation is visualized in Figure 5.28, where the generated materials cluster around the target property values, indicating strong controllability. The authors also report high Spearman correlations between input properties and generated outputs, validating the model's understanding of structure–property relationships.

Perhaps the most sophisticated materials-specific module within NatureLM is the NatureLM-Mat3D extension—a decoder-based crystal structure predictor. Mat3D formulates the problem as an autoregressive sequence prediction task, where the output sequence includes both composition and 3D structural motifs. Compared to traditional structure prediction algorithms, which often rely on energy minimization or geometric heuristics, NatureLM-Mat3D is data-driven and generative. It is capable of predicting plausible crystal structures from textual prompts or compositional inputs, bypassing the need for high-throughput DFT calculations. While the initial version is limited to coarse representations (e.g., space group and atomic species), future extensions may incorporate full 3D coordinate generation.

The authors compare NatureLM's performance with domain-specific baselines, highlighting that the decoder-based architecture not only matches but in some cases surpasses the accuracy of specialized models trained solely on materials data.

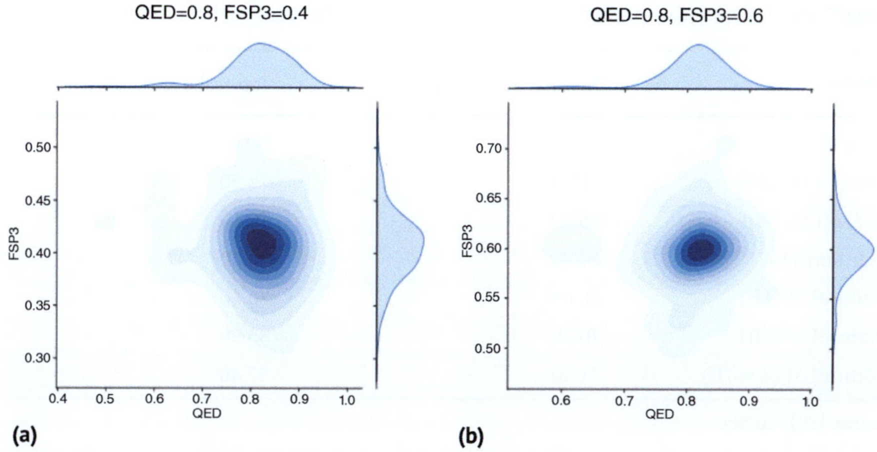

Figure 5.28 Heatmap of molecule generation based on QED and fraction of sp^3 (FSP3) properties. Each generated compound's QED and FSP3 values are calculated using RDKit and visualized in the heatmap. *Source*: [38] / arXiv.

5.3.2.4 Generalist Performance in Scientific Discovery Tasks

A defining contribution of NatureLM lies in its generalist capability across a wide spectrum of scientific tasks. By design, the model treats materials, molecules, nucleotides, proteins, and text as instances of a unified sequence modality. This flexible representation enables transfer learning across scientific domains and allows NatureLM to perform complex tasks ranging from synthesis planning to biological sequence design—often rivaling or surpassing domain-specific models.

In the context of materials science, this generalist competence is particularly beneficial when integrated into broader pipelines for synthesis, optimization, and interpretation of materials properties. While the paper by Xia et al. primarily benchmarks this generalist capacity on tasks involving small molecules and proteins, several insights and architectural choices are directly transferable to materials applications.

Retrosynthesis prediction—an essential task in chemistry and materials precursor planning—is an instructive example of NatureLM's decoder-based capabilities. The model is prompted with a target molecule's SMILES string and generates the reactant sequences, relying on its autoregressive architecture to unfold complex chemical transformation patterns.

On the USPTO-50K benchmark [44], NatureLM achieves top-one accuracy of 71.9% using the 8 × 7B model, surpassing both domain-specific methods such as LocalRetro [45] and even powerful general-purpose models like GPT-4 (22.4%). This performance showcases the advantage of pretraining on structured scientific sequences. Although retrosynthesis pertains to chemistry, the underlying mechanics—learning inverse design mappings from product to precursor—are closely aligned with materials discovery processes such as phase diagram inference or precursor pathway planning. Table 5.16 shows NatureLM's superior performance in top-one and top-three retrosynthesis accuracy, validating its ability to emulate expert reasoning. Figure 5.29 provides a real-world case study where NatureLM correctly

Table 5.16 Retrosynthesis prediction results on USPTO-50K dataset.

Model	Top-one accuracy (%)	Top-three accuracy (%)
GPT-4	22.40	N/A
LocalRetro [45]	51.50	76.50
R-SMILES [46]	56.00	79.10
EditRetro [47]	60.80	80.60
NatureLM (1B)	68.60	86.80
NatureLM (8B)	70.20	85.90
NatureLM (8 × 7B)	71.90	87.40

Source: [38] / arXiv.

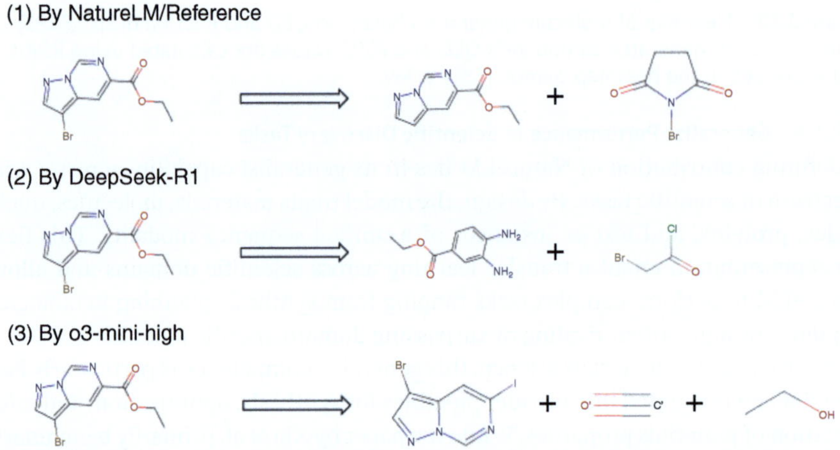

Figure 5.29 Case study on retrosynthesis prediction. It evaluated the performance of NatureLM, DeepSeek-R1, and o3-mini-high for retrosynthesis prediction using a reaction from US Patent ID US12018024B2 (not included in training set), granted to Novartis on 25 June 2024. NatureLM successfully proposed the ground-truth reactants from the patent, while the outputs from the other two methods required further refinement to achieve the same level of accuracy. *Source*: [38] / arXiv.

proposes a synthesis route for a molecule extracted from a June 2024 US patent (US12018024B2), outperforming competitive LLMs.

Cross-domain generation is another distinctive strength of NatureLM, made possible through its unified token vocabulary and instruction tuning. For example, the model can generate a guide RNA sequence from a DNA template, a molecule optimized for protein binding, a protein from a descriptive prompt, and a material candidate that satisfies a mechanical property constraint (e.g., bulk modulus).

These tasks simulate real-world scientific workflows where researchers begin with an objective (e.g., thermal resistance, electronic bandgap) and work backward to propose a candidate system. NatureLM's decoder-based setup is ideal for such

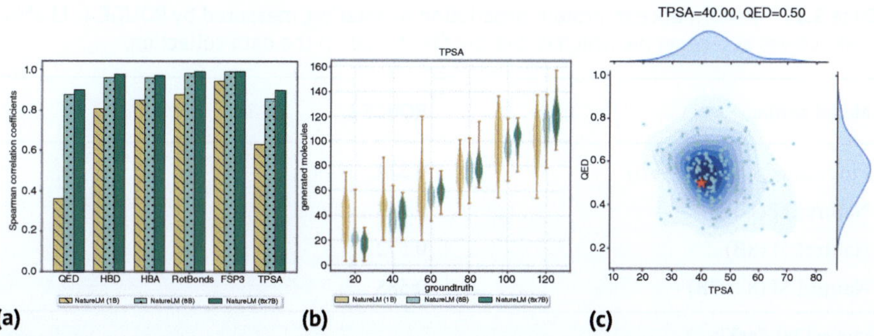

Figure 5.30 Evaluation of property-to-molecule generation. (a) Bar plot of the Spearman correlation coefficients between the input property values and generated molecules' property values. (b) Violin plot showing the input TPSA values and generated molecules' TPSA values. (c) The joint distribution of the generated molecules' TPSA and QED given the input "TPSA=40, QED=0.5." *Source*: [38] / arXiv.

workflows, as it allows conditioning on arbitrary combinations of structured tokens and natural language prompts.

Figures 5.28 and 5.30 demonstrate how NatureLM handles multi-property generation tasks, such as simultaneously targeting QED and TPSA values for molecules. This same approach is extensible to materials, where multiple properties (e.g., formation energy, stability, and elasticity) must be jointly optimized.

Beyond generation, NatureLM also excels at explanation. It can produce textual annotations for scientific sequences, thereby bridging structured data and human-readable insights. For instance, when given a protein sequence, it generates a natural language description that accurately captures its function, orthologs, or localization. Such functionality has implications for materials science as well—for example, generating textual summaries of phase stability, elemental trends, or synthesis feasibility from a given material composition. The decoder architecture enables this modality shift from structured input to textual output.

Table 5.17 and Figure 5.31 show that NatureLM outperforms fine-tuned LLaMA 3 models in annotation tasks, with the 8 × 7B model achieving a ROUGE-L score of 0.585. This suggests that decoder-based scientific LLMs can serve as interpretable agents in AI-accelerated discovery workflows.

5.3.2.5 Model Scaling, Instruction Tuning, and Evaluation

Model scaling plays a pivotal role in NatureLM's performance. Xia et al. systematically evaluated three variants of NatureLM—1B, 8B, and 8 × 7B parameters—across 22 categories of tasks, including molecular optimization, RNA design, and material generation. The results consistently show that larger models outperform smaller ones, especially in tasks involving fine-grained scientific reasoning.

Figure 5.32 visualizes model performance rankings across domains. The 8 × 7B MoE model achieves top performance in 19 out of 22 categories, while the 8B dense model ranks first in three. This indicates that decoder-based models benefit significantly from parameter scaling when dealing with structured scientific data.

Table 5.17 Performance of protein description generation, measured by ROUGE-L. LLaMa 3 (8B) serves as a baseline, which is directly fine-tuned on the data collection.

Model setting	ROUGE-L
Fine-tuned LLaMa 3 (8B)	0.324
NatureLM (1B)	0.548
NatureLM (8B)	0.572
NatureLM (8 × 7B)	0.585

Source: [38] / arXiv.

Gene ID	834896	68201	161973
Reference	Member of putative ligand-gated ion channel subunit family.	Orthologous to human ccdc34 (coiled-coil domain containing 34).	Orthologous to human fbxo 15 (F-box protein 15).
LLaMa 3 (8B)	Contain amino acid sequence motifs characteristic of ATP-binding cassette (ABC) transporters.	Orthologous to human FAM107A (family with sequence similarity 107 member A).	Is expressed in germline cell.
NatureLM (1B)	Orthologous to human SLC25A37 (solute carrier family 25 member 37).	Orthologous to human CCDC74A (coiled-coil domain containing 74A) and CCDC74B (coiled-coil domain containing 74B).	Orthologous to human C2orf7 (chromosome 2 open reading frame 76).
NatureLM (8B)	Encodes a plasma membrane protein involved in the positive regulation of organ size development. Overexpression results in organ size enlargement.	Orthologous to human CCDC34 (coiled-coil domain containing 34).	Orthologous to human FBXW12 (F-box and WD repeat domain containing 12).
NatureLM (8×7B)	Member of putative ligand-gated ion channel subunit family.	Orthologous to human CCDC34 (coiled-coil domain containing 34).	Orthologous to human FBXO15 (F-box protein 15).

Figure 5.31 Generated descriptions for different proteins. *Source*: [38] / arXiv.

NatureLM employs supervised fine-tuning on over 5.1 million instruction–response pairs spanning six domains (small molecules, proteins, DNA, RNA, materials, and general text). The responses are structured using a consistent prompt format and trained such that loss is computed only over the response tokens—allowing stable optimization even for highly varied instruction types. Figure 5.33 shows the distribution of post-training sequences. Materials-related instructions account for 5.8% of the fine-tuning dataset, enabling downstream applications in property-driven generation, bulk modulus prediction, and conditional design.

Instruction tuning is critical for enabling models to follow arbitrary commands—for example, "Generate a stable perovskite with a bulk modulus over 80 GPa." The decoder architecture is ideally suited for such instruction–response tasks, as it can generate coherent sequences based on learned representations of both input prompts and output formats.

Given the high computational cost of large decoder-based models, the NatureLM team also implements inference acceleration using PagedAttention and Selective Batching, deployed via the vLLM framework. This results in significant

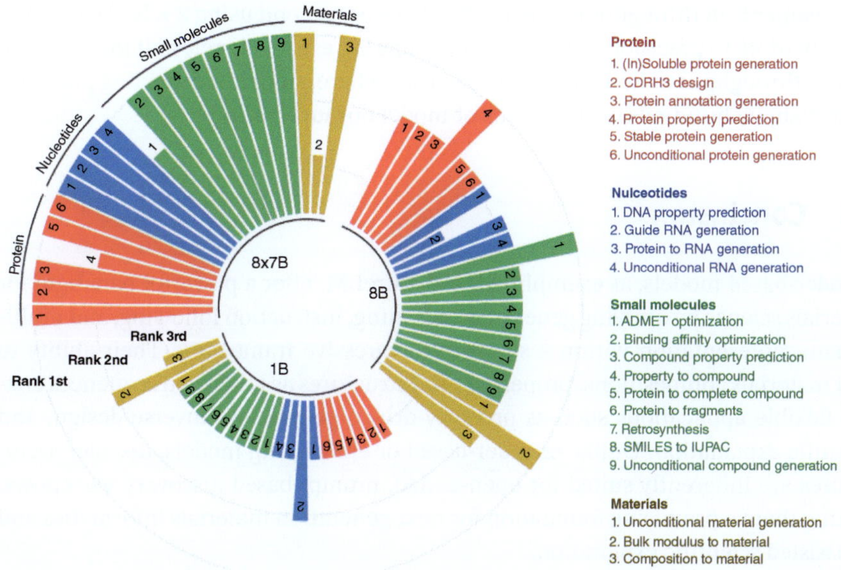

Figure 5.32 The scaling effect in NatureLM is obvious. The chart depicts the overall ranking of models with varying sizes, where a better rank is represented by the "outsider" bar. The 8 × 7B model achieves top performance in 19 tasks, while the 8B model excels in three tasks. Eighteen categories exhibited performance improvements with increasing model size (i.e., 8 × 7B demonstrated the best performance, followed by 8B, and then 1B), highlighting the potential of large foundation models for scientific applications. *Source*: [38] / arXiv.

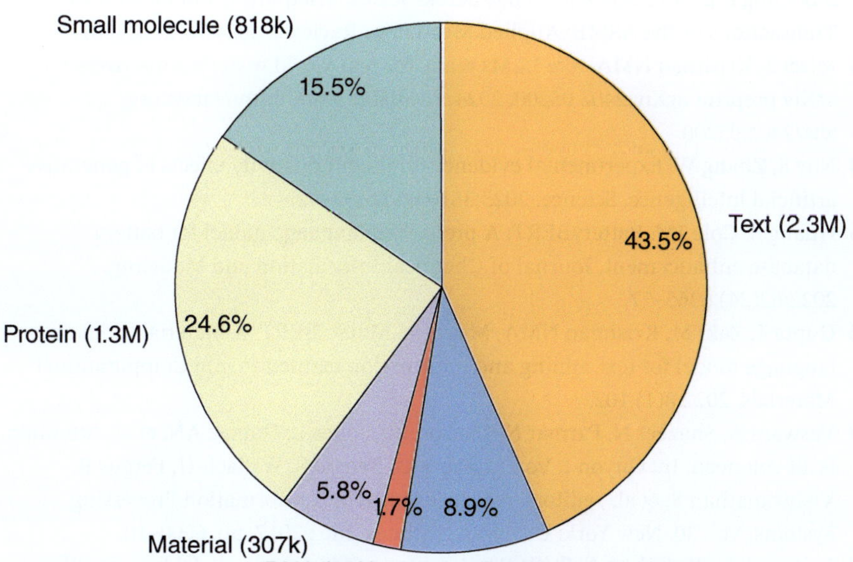

Figure 5.33 Statistics of post-training data, measured by the number of sequences. *Source*: [38] / arXiv.

improvements in throughput—up to 525 tokens per second using 2 × NVIDIA A100 GPUs with BF16 precision. These optimizations are especially relevant for real-time or high-throughput materials generation tasks, where candidate screening pipelines must evaluate hundreds or thousands of model outputs per minute.

5.4 Conclusion

Decoder-based models, as exemplified by NatureLM, offer a powerful paradigm for materials science by unifying generative reasoning, instruction following, and multi-domain representation within a single autoregressive framework. Their ability to treat materials compositions, properties, and structures as tokenized sequences enables flexible applications such as property-driven generation, inverse design, and scientific explanation. Unlike encoder-based or embedding models, decoder architectures are inherently suited for open-ended, prompt-based discovery workflows, making them a promising foundation for next-generation materials informatics and AI-assisted scientific exploration.

References

1 OpenAI. GPT-4 technical report. arXiv preprint arXiv:2303.08774; 2023. Available from: https://arxiv.org/abs/2303.08774
2 Buehler MJ. MechGPT, a language-based strategy for mechanics and materials modeling that connects knowledge across scales, disciplines, and modalities. Transactions of the ASME: Applied Mechanics Reviews. 2024;76(2):021001.
3 Miret S, Krishnan NMA. Are LLMs ready for real-world materials discovery? arXiv preprint arXiv:2402.05200; 2024. Available from: http://arxiv.org/abs/2402.05200
4 Noy S, Zhang W. Experimental evidence on the productivity effects of generative artificial intelligence. Science. 2023;381(6654):187–92.
5 Huang S, Cole JM. BatteryBERT: A pretrained language model for battery database enhancement. Journal of Chemical Information and Modeling. 2022;62(24):6365–77.
6 Gupta T, Zaki M, Krishnan NMA, Mausam. MatSciBERT: A materials domain language model for text mining and information extraction. npj Computational Materials. 2022;8(1):102.
7 Vaswani A, Shazeer N, Parmar N, Uszkoreit J, Jones L, Gomez AN, et al. Attention is all you need. In: Guyon I, Von Luxburg U, Bengio S, Wallach H, Fergus R, Vishwanathan S, et al. , editors. Advances in Neural Information Processing Systems, Vol. 30. New York: Curran Associates, Inc.; 2017. pp. 6000–10.
8 Beltagy I, Lo K, Cohan A. SciBERT: A pretrained language model for scientific text. arXiv preprint arXiv:1903.10676; 2019.
9 Devlin J. BERT: Pre-training of deep bidirectional transformers for language understanding. arXiv preprint arXiv:1810.04805; 2018.

10 Lee J, Yoon W, Kim S, Kim D, Kim S, So CH, et al. BioBERT: A pre-trained biomedical language representation model for biomedical text mining. Bioinformatics. 2020;36(4):1234–40.

11 Huang K, Altosaar J, Ranganath R. ClinicalBERT: Modeling clinical notes and predicting hospital readmission. arXiv preprint arXiv:1904.05342; 2019.

12 Pechenick EA, Danforth CM, Dodds PS. Characterizing the Google Books corpus: Strong limits to inferences of socio-cultural and linguistic evolution. PLoS ONE. 2015;10(10):e0137041.

13 Wu Y, Schuster M, Chen Z, Le QV, Norouzi M, Macherey W, et al. Google's Neural Machine Translation system: Bridging the gap between human and machine translation. arXiv preprint arXiv:1609.08144; 2016.

14 Huang S, Cole JM. BatteryBERT: a pretrained language model for battery database enhancement. Journal of chemical information and modeling. 2022;62(24):6365–77.

15 Liu Y, Ott M, Goyal N, Du J, Joshi M, Chen D, et al. RoBERTa: A robustly optimized BERT pretraining approach. arXiv preprint arXiv:1907.11692; 2019.

16 Rajpurkar P, Zhang J, Lopyrev K, Liang P. SQuAD: 100,000+ questions for machine comprehension of text. arXiv preprint arXiv:1606.05250; 2016.

17 Swain MC, Cole JM. ChemDataExtractor: A toolkit for automated extraction of chemical information from the scientific literature. Journal of Chemical Information and Modeling. 2016;56(10):1894–904.

18 Vig J. A multiscale visualization of attention in the transformer model. arXiv preprint arXiv:1906.05714; 2019.

19 Jain SM. Hugging face. In: Introduction to Transformers for NLP: With the Hugging Face Library and Models to Solve Problems. Berkeley, CA: Apress; 2022. pp. 51–67.

20 de Pablo JJ, Jackson NE, Webb MA, Chen LQ, Moore JE, Morgan D, et al. New frontiers for the materials genome initiative. npj Computational Materials. 2019;5(1):41.

21 Weston L, Tshitoyan V, Dagdelen J, Kononova O, Trewartha A, Persson KA, et al. Named entity recognition and normalization applied to large-scale information extraction from the materials science literature. Journal of Chemical Information and Modeling. 2019;59(9):3692–702.

22 Friedrich A, Adel H, Tomazic F, Hingerl J, Benteau R, Maruscyk A, et al. The SOFC-Exp corpus and neural approaches to information extraction in the materials science domain. arXiv preprint arXiv:2006.03039; 2020.

23 Mysore S, Jensen Z, Kim E, Huang K, Chang HS, Strubell E, et al. The Materials Science Procedural Text corpus: Annotating materials synthesis procedures with shallow semantic structures. arXiv preprint arXiv:1905.06939; 2019.

24 Brown TB. Language models are few-shot learners. arXiv preprint arXiv:2005.14165; 2020.

25 Radford A, Wu J, Child R, Luan D, Amodei D, Sutskever I, et al. Language models are unsupervised multitask learners. OpenAI Blog; 2019. Available from: https://cdn.openai.com/better-language-models/language_models_are_unsupervised_multitask_learners.pdf

26 Jensen Z, Kim E, Kwon S, Gani TZH, Román-Leshkov Y, Moliner M, et al. A machine learning approach to zeolite synthesis enabled by automatic literature data extraction. ACS Central Science. 2019;5(5):892–9.

27 Park S, Kim B, Choi S, Boyd PG, Smit B, Kim J. Text mining metal–organic framework papers. Journal of Chemical Information and Modeling. 2018;58(2):244–51.

28 Zheng Z, Zhang O, Borgs C, Chayes JT, Yaghi OM. ChatGPT Chemistry Assistant for text mining and the prediction of MOF synthesis. Journal of the American Chemical Society. 2023;145(32):18048–62.

29 Yaghi OM, O'Keeffe M, Ockwig NW, Chae HK, Eddaoudi M, Kim J. Reticular synthesis and the design of new materials. Nature. 2003;423(6941):705–14.

30 Huang L, Yu W, Ma W, Zhong W, Feng Z, Wang H, et al. A survey on hallucination in large language models: Principles, taxonomy, challenges, and open questions. ACM Transactions on Information Systems. 2025;43(2):1–55.

31 Jablonka KM, Schwaller P, Ortega-Guerrero A, Smit B. Is GPT all you need for low-data discovery in chemistry? 2023. Available from: https://doi.org/10.26434/chemrxiv-2023-fw8n4-v2

32 Bubeck S, Chadrasekaran V, Eldan R, Gehrke J, Horvitz E, Kamar E, et al. Sparks of artificial general intelligence: Early experiments with GPT-4. arXiv preprint arXiv:2303.12712; 2023.

33 Mikolov T. Efficient estimation of word representations in vector space. arXiv preprint arXiv:1301.3781; 2013.

34 Kusner M, Sun Y, Kolkin N, Weinberger K. From word embeddings to document distances. Proceedings of Machine Learning Research. 2015;37:957–66.

35 Breiman L. Random forests. Machine Learning. 2001;45:5–32.

36 Park H, Kang Y, Choe W, Kim J. Mining insights on metal–organic framework synthesis from scientific literature texts. Journal of Chemical Information and Modeling. 2022;62(5):1190–8.

37 Le Q, Mikolov T. Distributed representations of sentences and documents. Proceedings of Machine Learning Research. 2014;32(2):1188–96.

38 Xia Y, Jin P, Xie S, He L, Cao C, Luo R, et al. Nature Language Model: Deciphering the language of nature for scientific discovery. arXiv preprint arXiv:2502.07527; 2025. Available from: http://arxiv.org/abs/2502.07527

39 Landrum G. RDKit Documentation. Release 2013.09.1. San Franscisco, CA: RDKit.

40 Wang YC, Yang WH, Yang CS, Hou MH, Tsai CL, Chou YZ, et al. Structural basis of SARS-CoV-2 main protease inhibition by a broad-spectrum anti-coronaviral drug. American Journal of Cancer Research. 2020;10(8):2535–45.

41 Davey CA, Sargent DF, Luger K, Maeder AW, Richmond TJ. Solvent mediated interactions in the structure of the nucleosome core particle at 1.9 Å resolution. Journal of Molecular Biology. 2002;319(5):1097–113.

42 Pettersen EF, Goddard TD, Huang CC, Couch GS, Greenblatt DM, Meng EC, et al. UCSF Chimera—a visualization system for exploratory research and analysis. Journal of Computational Chemistry. 2004;25(13):1605–12.

43 Jain A, Ong SP, Hautier G, Chen W, Richards WD, Dacek S, et al. The materials project: A materials genome approach to accelerating materials innovation. APL Materials. 2013;1:011002. Available from: https://doi.org/10.1063/1.4812323

44 Schneider N, Stiefl N, Landrum GA. What's what: The (nearly) definitive guide to reaction role assignment. Journal of Chemical Information and Modeling. 2016;56(12):2336–46.

45 Chen S, Jung Y. Deep retrosynthetic reaction prediction using local reactivity and global attention. JACS Au. 2021;1(10):1612–20.

46 Zhong Z, Song J, Feng Z, Liu T, Jia L, Yao S, et al. Root-aligned SMILES: A tight representation for chemical reaction prediction. Chemical Science. 2022;13(31):9023–34.

47 Han Y, Xu X, Hsieh CY, Ding K, Xu H, Xu R, et al. Retrosynthesis prediction with an iterative string editing model. Nature Communications. 2024;15(1):6404.

6

Materials Data Extraction from Literature by NLP and Large Language Models

6.1 Introduction

Artificial intelligence (AI) and machine learning (ML) are fundamentally reshaping research paradigms in materials research. Data-driven methodologies are gradually replacing traditional trial-and-error approaches, significantly accelerating the pace of new materials development [1–8]. Within this paradigm, the availability of rich, high-quality data is essential to the successful application of AI and ML techniques. At present, scientific data are primarily obtained through two main channels: (1) data generated experimentally or computationally by researchers in the laboratory; and (2) data manually extracted from peer-reviewed academic publications.

Peer-reviewed publications collectively constitute a treasure trove of rigorously validated scientific data. However, with the ever-increasing volume of scholarly articles written in natural language, manual data extraction by domain experts has become both time-consuming and inefficient. This manual approach is increasingly unable to keep pace with the rate of new publications [9]. For individual researchers, the task of maintaining and continuously updating structured databases from the literature has become an overwhelming burden. Consequently, there is an urgent need to develop fast and accurate methods for extracting structured data in scientific literature.

Natural language processing (NLP) techniques have emerged as a key approach in chemistry and materials science, enabling the automated retrieval of information pertaining to both organic and inorganic materials [10–14]. Among these techniques, named entity recognition (NER) and relation extraction play a pivotal role in accurately retrieving information from texts. Recently, large language models (LLMs) have been introduced into materials science, providing a new paradigm for extracting information by leveraging prompt engineering and fine-tuning, in contrast to traditional rule-based or statistical NLP methods. Leveraging their exceptional semantic understanding and generalization capabilities, LLMs have demonstrated impressive performance in parsing complex scientific language, reducing the reliance on large annotated corpora, and enabling rapid adaptation across diverse materials domains. As a result, LLM-based approaches are becoming a dominant strategy for materials information extraction. This chapter offers an in-depth analysis of how NLP and LLMs can be used to transform scientific literature information in the field of materials research into structured data. We begin by reviewing the traditional NLP pipeline, outlining its key components such as corpus acquisition, text preprocessing, and named entity and relation extraction. We then compare this conventional approach with emerging LLM-based methods, emphasizing their workflows, strengths, and current limitations. To illustrate these techniques in practice, we present case studies of representative tools—including ChemDataExtractor and MaterialsBERT—as well as specialized pipelines developed for extracting data on chemical compounds, polymers, and alloys. Collectively, these examples highlight the transformative potential of integrating classical NLP methods with modern LLMs to advance data-driven research in materials science.

6.2 The Pipeline for Automatically Extracting Materials Data Using NLP

The rapid development of the Internet and the World Wide Web in the 1990s dramatically accelerated the sharing and dissemination of information, leading to a historic transformation in global communication. This revolution catalyzed an exponential increase in the number of scientific publications. As illustrated in Figure 6.1, an analysis of publication trends in the Web of Science database reveals that, since the early 21st century, the number of materials science publications across various subfields has grown at an exponential rate.

Textual information in scientific literature typically appears in unstructured and highly diverse formats, which significantly complicates manual data extraction. This process is not only time-consuming and labor-intensive but also prone to errors. To address this challenge, NLP techniques have been applied in domains

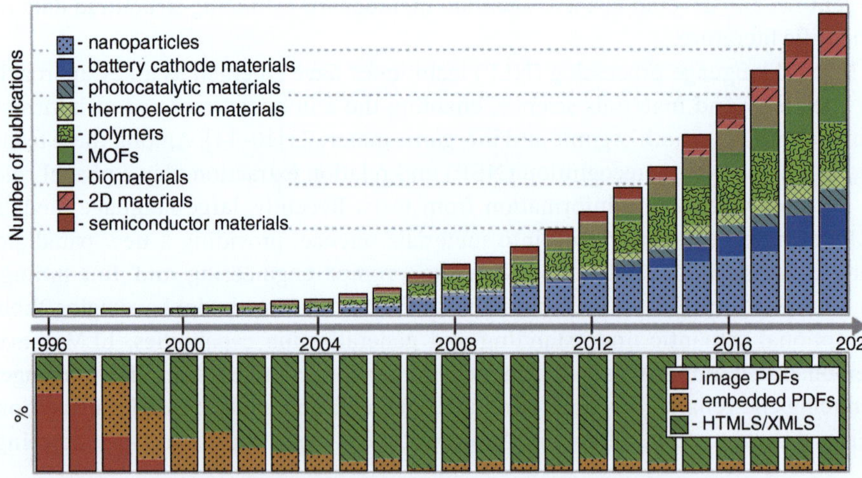

Figure 6.1 Publication volumes across different scientific fields from 1996 to 2020. *Source*: [9] / with permission of Elsevier.

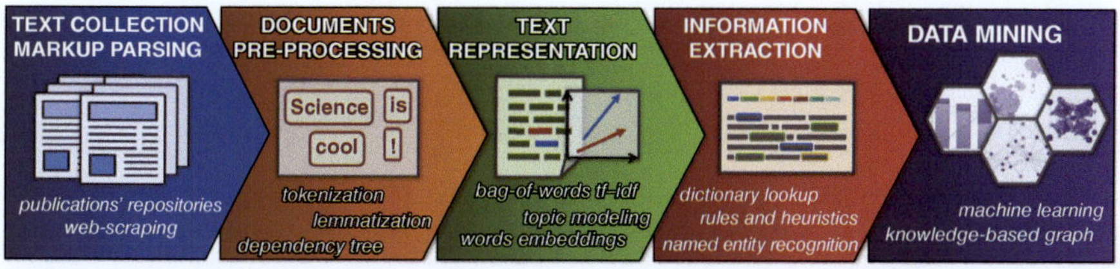

Figure 6.2 Standard workflow of information extraction from scientific publications using NLP techniques. *Source*: [9] / with permission of Elsevier.

such as biomedicine and chemistry. These methods aim to mine valuable information and knowledge from large-scale scientific texts, supporting applications such as disease treatment, drug discovery, and compound synthesis. Unlike standard texts found in general domains such as news or mass media, scientific literature is written in a domain-specific and highly technical language. It contains a large volume of specialized terminology and distinctive expressions, which adds another layer of complexity to automated text processing.

In computational linguistics, a text corpus refers to a large and systematically organized collection of documents. To date, NLP techniques have been successfully applied to text data extraction and knowledge discovery across a wide range of scientific disciplines, contributing significantly to the development of materials databases and innovative materials. As illustrated in Figure 6.2, existing studies indicate that the application of NLP to scientific literature typically follows a standardized workflow. This process generally includes the retrieval and acquisition of relevant documents, text preprocessing, text representation modeling, information extraction, and data mining.

6.2.1 Overview of NLP and Its Differences from LLMs in Data Extraction

NLP encompasses tasks, including information retrieval, NER, relation extraction, text classification, topic modeling, semantic textual similarity, machine translation, and question answering (QA). Leveraging these capabilities, researchers have constructed automated pipelines capable of automatically extracting materials data from journal articles, encompassing chemical compositions, material properties, and synthesis methods for organic and inorganic compounds, as well as alloys. The extraction of materials-related information generally proceeds along two pathways, conventional NLP and strategies leveraging LLMs. The two pathways share common stages that include corpus acquisition, preprocessing, information extraction, and interdependency resolution. Conventional workflows employ dedicated steps, including text classification, table parsing, NER, and relationship extraction between entities. In contrast, information extraction using LLMs primarily relies on prompt engineering and fine-tuning when needed, allowing information to be retrieved through conversational or instruction-driven interactions, as shown in Figure 6.3.

The initial step involves the automated collection of a relevant corpus of articles. Many journals and publishers provide developer APIs that expose content in HTML,

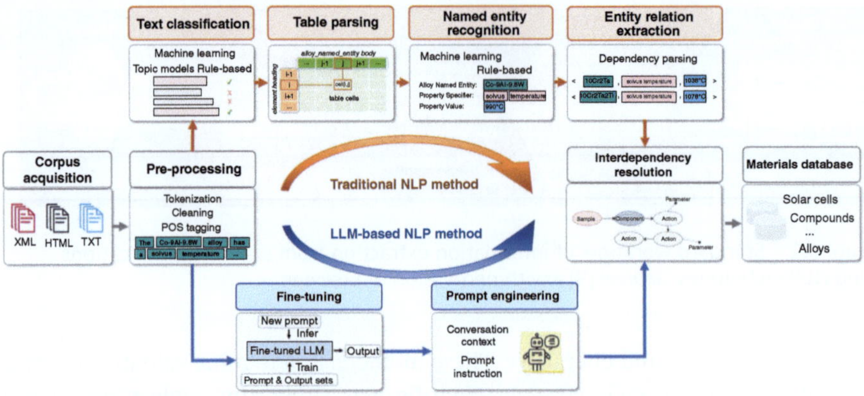

Figure 6.3 The workflow of NLP pipeline for extracting materials data using traditional NLP approaches and LLM-based methods. *Source*: [15] / Springer Nature.

plain text, or XML, making them considerably easier to process programmatically than PDF files [16]. Retrieved content typically includes the full text and metadata such as journal, title, abstract, and author information. Once the corpus has been compiled, it undergoes standard preprocessing procedures such as corpus cleaning, tokenization, and part-of-speech tagging. Corpus cleaning involves removing HTML and XML markup as well as extraneous metadata from programmatically retrieved articles. Tokenization divides the text into sentences, phrases, words, or subword units to facilitate individual processing [11]. These steps yield a normalized, refined corpus suitable for subsequent information extraction.

In conventional NLP, preprocessing is followed by targeted text classification to locate sentences or paragraphs that contain the desired compositions, properties, or synthesis procedures. This can be implemented with a binary classifier trained on positive samples that represent relevant paragraphs and negative samples that represent all others, or by topic modeling or rule-based methods. Tables are particularly useful for extracting materials information due to their ability to condense large volumes of data, with table parsing transforming both captions and content into a structured format. NER and relation extraction form the foundation of materials data extraction, identifying chemical compounds, alloy compositions, properties, synthesis processes, and relevant parameters, while simultaneously establishing semantic connections—such as linking a compound to its measured properties for a specific material. Subsequently, interdependency resolution brings together compositional information, processing routes, and dispersed property fragments that correspond to the same material instance.

In LLM-based NLP, preprocessing is followed by prompt engineering. Carefully constructed prompts steer models to produce outputs that satisfy extraction requirements during inference, all without altering the model's parameters. Through iterative refinement of these prompts, the model's responses can often be brought into alignment with task objectives, reducing or even removing the need for additional training [17–19]. If prompt engineering alone is insufficient, fine-tuning on a materials-specific corpus can be used. Training the model further on domain-specific examples strengthens its ability to address the unique requirements of materials information extraction in that context [20].

6.2.2 Traditional NLP Pipeline

Techniques for NER and relation extraction span a spectrum from dictionary- and rule-based approaches to semi-supervised and ML-driven methods [10, 11, 21–23], as shown in Figure 6.4. The selection of method depends chiefly on the size of the annotated corpus. Directly supervised models usually require large labeled datasets and often assess similarity by comparing representations in an embedding space [24]. With an adequately sized labeled corpus, a bidirectional LSTM network combined with a conditional random field layer can be trained to carry out both NER and entity relation extraction. The bidirectional LSTM processes text in forward and backward directions using LSTM cells that capture long-term dependencies and encode richer sentence-level context. Typically, a BiLSTM takes as input a layer of word embeddings, with each embedding transforming a plain-text token into a dense, fixed-length vector of real numbers. The BiLSTM produces label scores (or probabilities) for each token in the sequence. These scores are then passed to a CRF layer, which models correlations between neighboring labels and jointly decodes the most likely label sequence for the input sentence. When annotated corpora are limited, semi-supervised ML strategies [10, 25–27] and rule-based methods [22, 28, 29] tend to work better.

Extensive effort has focused on extracting material compositions and properties using conventional NLP pipelines, aided by advances in chemical information extraction. As early as 2011, tools such as OSCAR [29], ChemicalTagger [30],

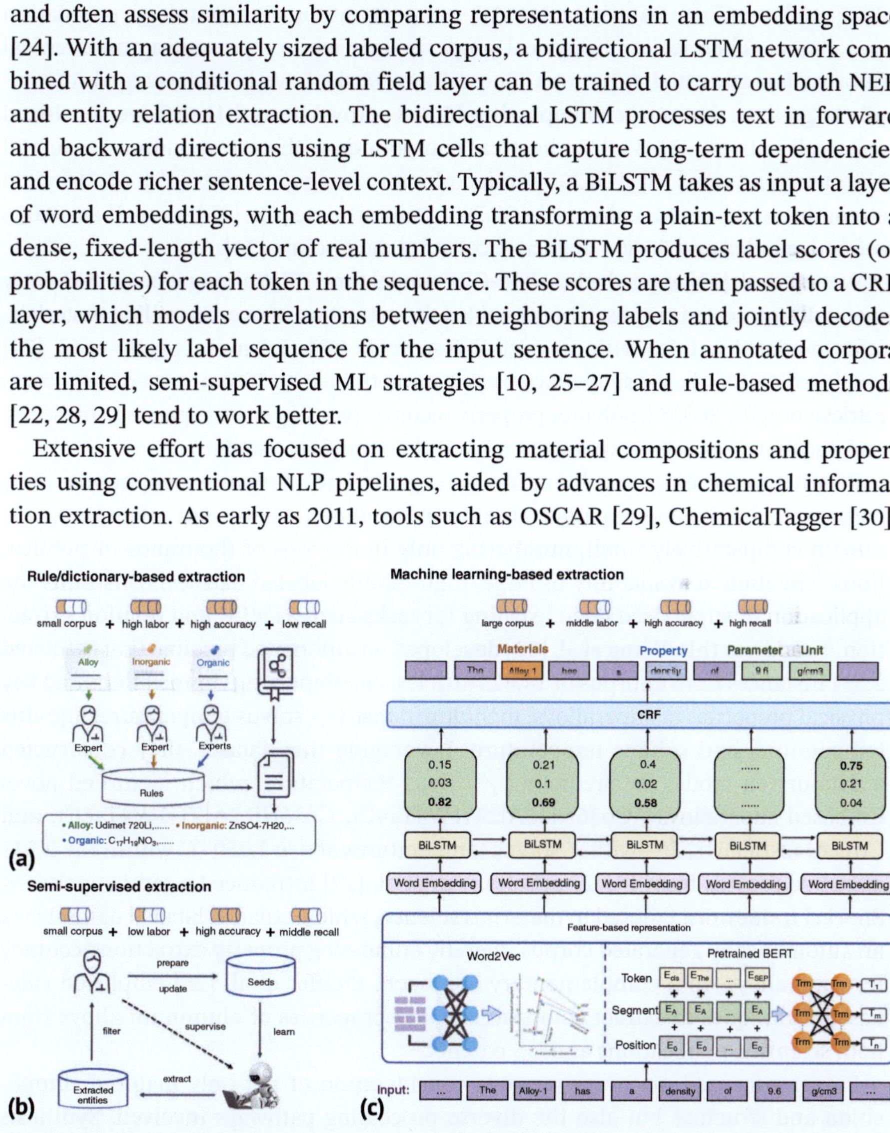

Figure 6.4 Three primary strategies are employed for materials NER: (a) methods based on rules or dictionaries; (b) semi-supervised approaches; and (c) extraction techniques driven by ML. *Source*: [15] / Springer Nature / CC BY-NC-ND 4.0.

and ChemDataExtractor [11] were introduced to identify chemical entities in the literature and to automate the capture of entities, properties, and measurements for population of structured databases. In 2016, Swain and Cole [11] introduced ChemDataExtractor, a hybrid platform that integrates rule- and dictionary-based methods with ML to provide a chemistry-aware workflow for tokenization, part-of-speech tagging, NER, and phrase parsing.

In 2018, Court and Cole [10] enhanced ChemDataExtractor by incorporating a semi-supervised probabilistic component for extracting quaternary relationships, primarily leveraging a modified version of the Snowball algorithm. The 2021 update, ChemDataExtractor 2.0, supports automated ontology population and extracts chemical and physical property data while representing hierarchical information as nested structures. These frameworks have been leveraged to compile large-scale datasets, encompassing 15,818 perovskite and 41,680 dye-sensitized solar cell entries [31], 39,822 chemical compounds with corresponding Curie and Néel temperatures [10], 100,236 semiconductor bandgap records [32], 64,269 yield strength measurements alongside 30,285 grain size records [33], as well as 49,076 refractive index and 60,804 dielectric constant data points [34].

Weston et al. [23] applied a BiLSTM-CRF model for NER, training it on an 800-entry manually annotated corpus, which enabled the extraction of over 80 million materials-science-related entities with an overall *F*-score of 87%. Likewise, Shetty et al. [24] employed a BERT-based single-layer architecture trained on 750 annotated abstracts to retrieve roughly 300,000 polymer property records spanning supercapacitors, fuel cells, and polymer solar cells, revealing notable and nonobvious trends.

Relative to the extensive literature on organic and inorganic compounds, alloy corpora encompassing superalloys, aluminum alloys, steels, and related systems remain comparatively small, numbering only in the tens of thousands of publications. The limited availability of large, high-quality labeled datasets constrains the application of supervised deep learning for tasks such as NER and relation extraction. To address this, Wang et al. [22] developed an automated pipeline that extracted 2,531 instances from a corpus of 14,425 articles, encompassing compositions and key physical properties of superalloys, including density, γ' solvus temperature, liquidus temperature, and solidus temperature. Leveraging this dataset, they constructed a data-driven model for predicting γ' solvus temperature, which identified novel Co-based superalloys—Co36Ni12Al2Ti1W4Ta4Cr, Co36Ni12Al2Ti1W4Ta6Cr, and Co12Al4.5Ta35Ni2Ti—with γ' solvus temperatures above 1,250 °C within the 2,531 extracted records from 14,425 articles. Yan et al. [27] introduced a semi-supervised Snorkel framework tailored to materials science, which expands labeled datasets via an automatically generated corpus, thereby enhancing property extraction accuracy for superalloys. In a complementary approach, Pfeiffer et al. [28] employed rule-based techniques to extract compositions and properties of aluminum alloys from both scientific publications and US patents.

Property-focused discovery requires consideration of not only material composition and structure but also the diverse processing pathways involved. Synthesis and processing are described as continuous sequences of actions that vary in type, expression, and associated conditions, and are often interleaved with discussions of phenomena and intermediates, which complicates the isolation of discrete actions and parameters.

Studies have focused on extracting inorganic synthesis procedures, as illustrated in Figure 6.5. Kim et al. [35, 36] applied a combination of neural network and parsing-based techniques to collect synthesis parameters from over 640,000 articles spanning 30 oxide systems. Wilary and Cole [37] introduced ReactionDataExtractor, a software tool designed to automatically capture information from multistep reaction schemes. Huo et al. [25] employed a semi-supervised latent Dirichlet allocation framework to categorize inorganic synthesis narratives and reconstructed likely procedural flowcharts using a Markov chain representation of step sequences. Other approaches integrated BiLSTM-CRF models, dependency parsing, and rule-based methods to extract comprehensive inorganic synthesis recipes, including precursors, targets, operations, conditions, and reactions [35, 38–40], as well as procedures specific to solution-processed inorganic materials and gold nanoparticles [41, 42]. A knowledge base comprising 29,900 text-mined solid-state synthesis recipes was subsequently leveraged to predict precursor selections and recommend synthesis routes

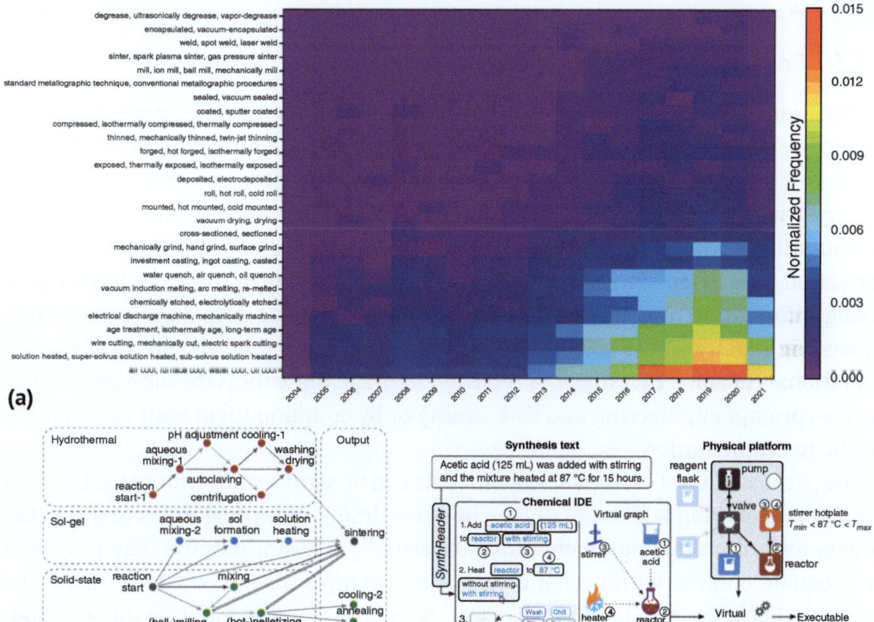

Figure 6.5 Instances of extracting materials' synthesis and processing information. (a) A heatmap illustrating the frequency of extracted actions related to superalloy synthesis and processing, as reported in the literature from 2004 to 2021. *Source*: [26] / Springer Nature / CC BY 4.0. (b) A flowchart generated by an ML model, illustrating the transitions between experimental steps for various types of material synthesis. *Source*: [25] / Springer Nature / CC BY 4.0. (c) A system workflow illustrating the conversion of synthetic procedures, described in natural language, into a hardware-independent chemical code that is internally stored as XML but can be represented and visually edited as natural language. *Source*: [15] / Springer Nature.

for novel targets, achieving success rates of at least 82% [39, 40]. Expanding the utility of these extracted datasets, Vaucher et al. [43] defined a standardized set of synthesis actions with associated properties and implemented a Transformer-based sequence-to-sequence model to convert textual procedures into structured action sequences. Additionally, Kim et al. [12, 13] developed an unsupervised conditional variational autoencoder to link literature information with context-aware insights for synthesis planning.

In alloy corpora, action annotations are applied at token and chunk levels depending on phrase length, whereas chemical synthesis corpora typically rely on token-level entities. Action descriptions shift with their position in an alloy processing workflow, so an aging treatment may appear as primary aging or secondary aging, which complicates consistent manual labeling. Wang et al. [26] introduced a semi-supervised approach for recommending token-level actions, complemented by a multitier bootstrapping strategy targeting chunk-level actions, to retrieve parameters from sequences of synthesis and processing steps within a superalloy corpus. Applying this framework, they automatically identified 9,853 superalloy synthesis and processing actions along with their corresponding compositions across 16,604 scientific articles.

6.2.3 Recent Developments Using LLMs

Conventional NLP methodologies often function well for specific applications but frequently encounter limitations regarding adaptability and the management of intricate tasks. To address these deficits conversational LLMs such as GPT-4 have emerged which employ the Transformer-based decoder architecture comprising self-attention and feed-forward layers to facilitate text generation and efficient data extraction from extensive research corpuses. The utilization of these models for analyzing information and extracting salient details alongside response generation is expanding rapidly across scientific disciplines largely driven by their autonomous operational capacity. Investigators typically leverage these tools through techniques such as prompt engineering and fine-tuning or by coupling them with external scientific instrumentation.

The performance of conversational models can be significantly enhanced through prompt engineering which entails the iterative design of input queries to direct the system toward producing precise and contextually relevant outputs. The provision of adequate context within well-constructed prompts also serves to mitigate the risk of hallucination in model responses. Numerous studies validate this methodology. As illustrated in Figure 6.6a, Zheng et al. [17] introduced a framework that directed ChatGPT to extract 26,257 distinct synthesis parameters for 800 metal–organic frameworks, achieving Precision, Recall, and F_1-scores within the 90–99% range. Similarly, da Silva et al. [44] exploited these models to automatically parse synthesis details for reticular materials by crafting comprehensive prompts that amalgamated task instructions with illustrative examples. This approach achieved high performance using a very limited number of examples, attaining an F_1-score of 0.98 for paragraph classification and an accuracy of 0.96 in information extraction.

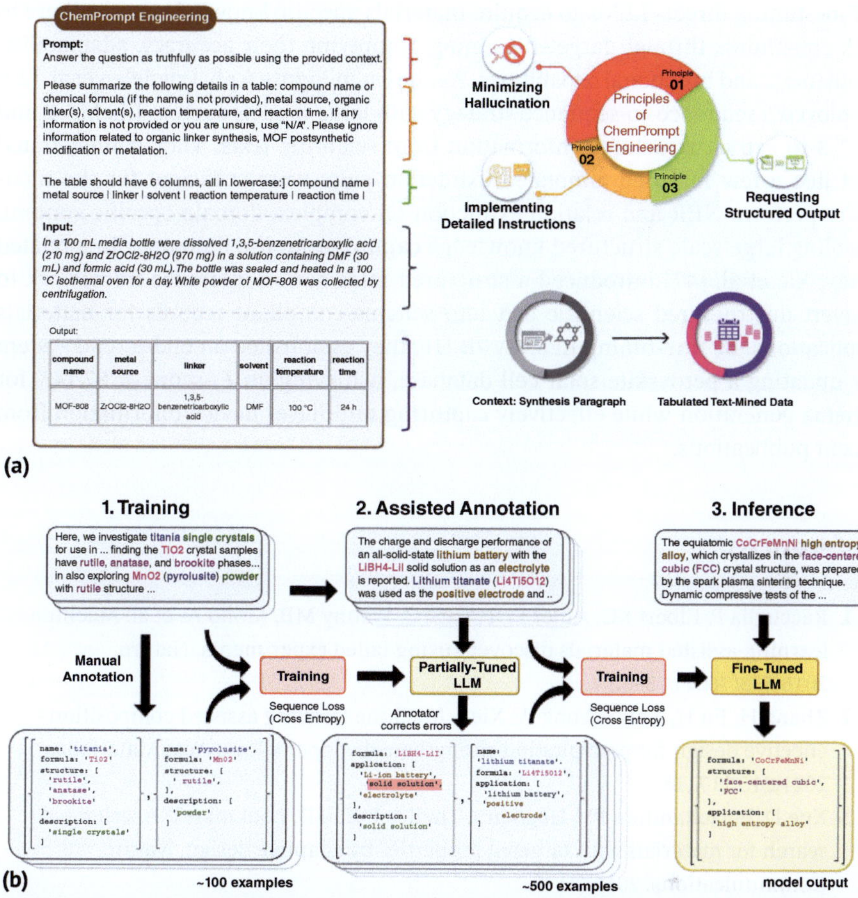

Figure 6.6 LLMs-based information extraction. (a) ChemPrompt Engineering leverages three fundamental principles to systematically identify and condense synthesis conditions from specific sections of literature, subsequently compiling the extracted information into a structured table. *Source*: [17] / with permission of American Chemical Society. (b) Document-level joint NER and relation extraction for materials achieved by partially training LLMs with a human in the loop annotation workflow. *Source*: [20] / Springer Nature / CC BY 4.0.

In another application, Thway et al. [45] evaluated a GPT-3.5 framework for parsing synthesis recipes of ternary chalcogenides to expedite thermoelectric materials development. By employing a gold standard dataset curated by domain experts, they achieved an extraction accuracy of 73% across 61 of the 168 targeted research articles. Building on this, Polak and Morgan [46] developed ChatExtract, a framework that utilizes ChatGPT in a zero-shot configuration combined with carefully designed prompts to recover material–value–unit triplets. This method facilitated high-precision extraction with minimal initial effort and limited domain knowledge, yielding 87.7% Recall and 90.8% Precision for bulk modulus, and 83.6% Recall and 91.6% Precision for the critical cooling rates of metallic glasses.

Fine-tuning directs LLMs to acquire materials-specific knowledge and adhere to task constraints through targeted training, improving their accuracy, adaptability, robustness, and functional capabilities. As shown in Figure 6.6b, Dagdelen et al. [20] employed a sequence-to-sequence strategy with large models such as LLaMA 2 and GPT-3 to extract structured information from scientific texts. They demonstrated that just a few hundred annotated extraction pairs were sufficient for the models to perform NER and relation extraction on complex, domain-specific content, enabling large-scale structured knowledge capture from the literature. In a related study, Xie et al. [47] introduced a structured information inference framework to convert unstructured scientific text into schema-compliant records for materials applications. By fine-tuning LLaMA 7B HF, they established an end-to-end system for updating a perovskite solar cell database, achieving an F_1-score of 87.14% for schema generation while effectively capturing multilayer device information from recent publications.

References

1 Raccuglia P, Elbert KC, Adler PDF, Falk C, Wenny MB, Mollo A, et al. Machine-learning-assisted materials discovery using failed experiments. Nature. 2016;533(7601):73–6.
2 Zhang H, Fu H, Zhu S, Yong W, Xie J. Machine learning assisted composition effective design for precipitation strengthened copper alloys. Acta Materialia. 2021;215:117118.
3 Xue D, Balachandran PV, Hogden J, Theiler J, Xue D, Lookman T. Accelerated search for materials with targeted properties by adaptive design. Nature Communications. 2016;7:1–9.
4 Wen C, Zhang Y, Wang C, Xue D, Bai Y, Antonov S, et al. Machine learning assisted design of high entropy alloys with desired property. Acta Materialia. 2019;170:109–17.
5 Segler MHS, Preuss M, Waller MP. Planning chemical syntheses with deep neural networks and symbolic AI. Nature. 2018;555(7698):604–10.
6 Wen C, Wang C, Zhang Y, Antonov S, Xue D, Lookman T, et al. Modeling solid solution strengthening in high entropy alloys using machine learning. Acta Materialia. 2021;212:116917.
7 Zhang Y, Wen C, Wang C, Antonov S, Xue D, Bai Y, et al. Phase prediction in high entropy alloys with a rational selection of materials descriptors and machine learning models. Acta Materialia. 2020;185:528–39.
8 Jiang X, Jia B, Zhang G, Zhang C, Wang X, Zhang R, et al. A strategy combining machine learning and multiscale calculation to predict tensile strength for pearlitic steel wires with industrial data. Scripta Materialia. 2020;186:272–7.
9 Kononova O, He T, Huo H, Trewartha A, Olivetti EA, Ceder G. Opportunities and challenges of text mining in materials research. iScience. 2021;24(3):102155.
10 Court CJ, Cole JM. Auto-generated materials database of Curie and Néel temperatures via semi-supervised relationship extraction. Scientific Data. 2018;5(1):180111.

11 Swain MC, Cole JM. ChemDataExtractor: A toolkit for automated extraction of chemical information from the scientific literature. Journal of Chemical Information and Modeling. 2016;56(10):1894–904.

12 Kim E, Huang K, Jegelka S, Olivetti E. Virtual screening of inorganic materials synthesis parameters with deep learning. npj Computational Materials. 2017;3(1):53.

13 Kim E, Jensen Z, van Grootel A, Huang K, Staib M, Mysore S, et al. Inorganic materials synthesis planning with literature-trained neural networks. Journal of Chemical Information and Modeling. 2020;60(3):1194–201.

14 Krallinger M, Rabal O, Lourenço A, Oyarzabal J, Valencia A. Information retrieval and text mining technologies for chemistry. Chemical Reviews. 2017;117(12):7673–761.

15 Jiang X, Wang W, Tian S, Wang H, Lookman T, Su Y. Applications of natural language processing and large language models in materials discovery. npj Computational Materials. 2025;11(1):79.

16 Lammey R. CrossRef's text and data mining services. Learned Publishing. 2014;27(4):245–50.

17 Zheng Z, Zhang O, Borgs C, Chayes JT, Yaghi OM. ChatGPT Chemistry Assistant for text mining and the prediction of MOF synthesis. Journal of the American Chemical Society. 2023;145(32):18048–62. https://doi.org/10.1021/jacs.3c05819

18 White J, Fu Q, Hays S, Sandborn M, Olea C, Gilbert H, et al. A prompt pattern catalog to enhance prompt engineering with ChatGPT. arXiv preprint arXiv:2302.11382; 2023.

19 Liu Y, Deng G, Xu Z, Li Y, Zheng Y, Zhang Y, et al. Jailbreaking ChatGPT via prompt engineering: An empirical study. arXiv preprint arXiv:2305.13860; 2023.

20 Dagdelen J, Dunn A, Lee S, Walker N, Rosen AS, Ceder G, et al. Structured information extraction from scientific text with large language models. Nature Communications. 2024;15(1):1418.

21 Huang S, Cole JM. BatteryDataExtractor: Battery-aware text-mining software embedded with BERT models. Chemical Science. 2022;13(39):11487–95.

22 Wang W, Jiang X, Tian S, Liu P, Dang D, Su Y, et al. Automated pipeline for superalloy data by text mining. npj Computational Materials. 2022;8(1):9.

23 Weston L, Tshitoyan V, Dagdelen J, Kononova O, Trewartha A, Persson KA, et al. Named entity recognition and normalization applied to large-scale information extraction from the materials science literature. Journal of Chemical Information and Modeling. 2019;59(9):3692–702.

24 Shetty P, Rajan AC, Kuenneth C, Gupta S, Panchumarti LP, Holm L, et al. A general-purpose material property data extraction pipeline from large polymer corpora using natural language processing. npj Computational Materials. 2023;9(1):52.

25 Huo H, Rong Z, Kononova O, Sun W, Botari T, He T, et al. Semi-supervised machine-learning classification of materials synthesis procedures. npj Computational Materials. 2019;5(1):62.

26 Wang W, Jiang X, Tian S, Liu P, Lookman T, Su Y, et al. Alloy synthesis and processing by semi-supervised text mining. npj Computational Materials. 2023;9(1):183.

27 Yan R, Jiang X, Wang W, Dang D, Su Y. Materials information extraction via automatically generated corpus. Scientific Data. 2022;9(1):401.

28 Pfeiffer OP, Liu H, Montanelli L, Latypov MI, Sen FG, Hegadekatte V, et al. Aluminum alloy compositions and properties extracted from a corpus of scientific manuscripts and US patents. Scientific Data. 2022;9(1):128.

29 Jessop DM, Adams SE, Willighagen EL, Hawizy L, Murray-Rust P. OSCAR4: A flexible architecture for chemical text-mining. Journal of Cheminformatics. 2011;3(1):41.

30 Hawizy L, Jessop DM, Adams N, Murray-Rust P. ChemicalTagger: A tool for semantic text-mining in chemistry. Journal of Cheminformatics. 2011;3:1–13.

31 Beard EJ, Cole JM. Perovskite- and dye-sensitized solar-cell device databases auto-generated using ChemDataExtractor. Scientific Data. 2022;9(1):329.

32 Dong Q, Cole JM. Auto-generated database of semiconductor band gaps using ChemDataExtractor. Scientific Data. 2022;9(1):193.

33 Kumar P, Kabra S, Cole JM. Auto-generating databases of yield strength and grain size using ChemDataExtractor. Scientific Data. 2022;9(1):292.

34 Zhao J, Cole JM. A database of refractive indices and dielectric constants auto-generated using ChemDataExtractor. Scientific Data. 2022;9(1):192.

35 Kim E, Huang K, Saunders A, McCallum A, Ceder G, Olivetti E. Materials synthesis insights from scientific literature via text extraction and machine learning. Chemistry of Materials. 2017;29(21):9436–44.

36 Kim E, Huang K, Tomala A, Matthews S, Strubell E, Saunders A, et al. Machine-learned and codified synthesis parameters of oxide materials. Scientific Data. 2017;4(1):170127.

37 Wilary DM, Cole JM. ReactionDataExtractor: A tool for automated extraction of information from chemical reaction schemes. Journal of Chemical Information and Modeling. 2021;61(10):4962–74.

38 Kononova O, Huo H, He T, Rong Z, Botari T, Sun W, et al. Text-mined dataset of inorganic materials synthesis recipes. Scientific Data. 2019;6(1):203.

39 He T, Sun W, Huo H, Kononova O, Rong Z, Tshitoyan V, et al. Similarity of precursors in solid-state synthesis as text-mined from scientific literature. Chemistry of Materials. 2020;32(18):7861–73.

40 He T, Huo H, Bartel CJ, Wang Z, Cruse K, Ceder G. Precursor recommendation for inorganic synthesis by machine learning materials similarity from scientific literature. Science Advances. 2023;9(23):eadg8180.

41 Wang Z, Kononova O, Cruse K, He T, Huo H, Fei Y, et al. Dataset of solution-based inorganic materials synthesis procedures extracted from the scientific literature. Scientific Data. 2022;9(1):231.

42 Cruse K, Trewartha A, Lee S, Wang Z, Huo H, He T, et al. Text-mined dataset of gold nanoparticle synthesis procedures, morphologies, and size entities. Scientific Data. 2022;9(1):234.

43 Vaucher AC, Zipoli F, Geluykens J, Nair VH, Schwaller P, Laino T. Automated extraction of chemical synthesis actions from experimental procedures. Nature Communications. 2020;11(1):3601.

44 da Silva VT, Rademaker A, Lionti K, Giro R, Lima G, Fiorini S, et al. Automated, LLM enabled extraction of synthesis details for reticular materials from scientific literature. arXiv preprint arXiv:2411.03484; 2024.

45 Thway M, Low AK, Khetan S, Dai H, Recatala-Gomez J, Chen AP, et al. Harnessing GPT-3.5 for text parsing in solid-state synthesis—case study of ternary chalcogenides. Digital Discovery. 2024;3(2):328–36.

46 Polak MP, Morgan D. Extracting accurate materials data from research papers with conversational language models and prompt engineering. Nature Communications. 2024;15(1):1569.

47 Xie T, Wan Y, Zhou Y, Huang W, Liu Y, Linghu Q, et al. Creation of a structured solar cell material dataset and performance prediction using large language models. Patterns. 2024;5(5):100955.

7

Case Studies of Chemical Information Extraction

7.1 Introduction

This chapter focuses on the automated extraction of chemical information from scientific literature, a critical step toward building structured, machine-readable datasets for materials informatics. Traditional manual curation of chemical data is slow, error-prone, and unable to keep pace with the rapid growth of publications [1–8]. To address this challenge, domain-specific natural language processing (NLP) tools have been developed, combining rule-based grammars, named entity recognition (NER), and transformer-based models [9–10]. Among them, ChemDataExtractor stands out as a fully automated toolkit capable of parsing unstructured text, tables, and figures into structured chemical records, while more recent pipelines such as MaterialsBERT demonstrate the potential of domain-adapted language models for large-scale property extraction. Together, these methods illustrate how automated pipelines can accelerate the discovery, organization, and analysis of chemical and polymer property data, providing a foundation for data-driven research in materials science.

7.2 ChemDataExtractor

ChemDataExtractor, introduced by Swain and Cole, addresses this gap by offering a fully automated, end-to-end toolkit for extracting structured chemical information from scientific documents [11]. Unlike supervised deep-learning systems that often require large manually labeled datasets [12–14], ChemDataExtractor adopts a rule-driven, decoder-based architecture that aligns with the classical definition of decoder models: systems that transform unstructured inputs into semantically structured outputs [15]. In this context, the toolkit acts as a domain-specific decoder, interpreting chemical narratives, captions, and tables to generate machine-readable records.

Crucially, ChemDataExtractor demonstrates how carefully engineered rule-based grammars, domain-aware tokenization, and modular NER pipelines can match or

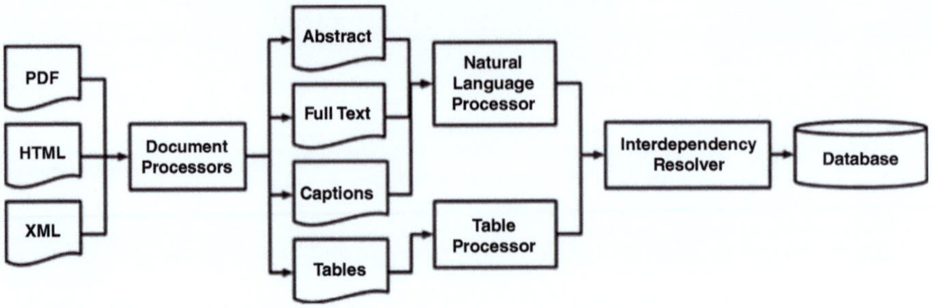

Figure 7.1 Overview of the ChemDataExtractor pipeline, illustrating the conversion of diverse document formats into a unified linear model of elements—such as paragraphs and tables—that are independently processed and merged to generate structured chemical records. *Source*: [11] / with permission of American Chemical Society.

even exceed the performance of purely neural systems in structured information retrieval. This chapter explores the foundational motivations for such a toolkit, its architectural design, and the NLP techniques that underlie its success in materials science applications.

The architecture of ChemDataExtractor is a modular and extensible system built around a pipeline that transforms heterogeneous document formats—PDF, HTML, and XML—into structured chemical records. As illustrated in Figure 7.1, the pipeline consists of three core phases: (1) document processing; (2) NLP; and (3) data post-processing and integration.

The first stage involves document ingestion and structural unification. The toolkit parses HTML and XML documents using existing markup to extract titles, headings, paragraphs, captions, and tables. For PDF documents—which lack semantic markup—the system utilizes PDFMiner-based layout analysis to reconstruct document structure from the spatial positions of text and images. Critically, all document formats are transformed into a linear stream of elements that preserves their logical reading order. This normalization ensures that subsequent NLP modules operate on a uniform representation regardless of the original format, facilitating seamless processing across diverse publication sources.

Tables are retained in their native structure using nested lists, with each cell treated as a semantically independent domain. This design enables the downstream table parser to infer row-wise chemical entities and column-wise properties. Captions and figure labels are likewise isolated as standalone elements to allow dedicated parsing grammars to extract contextual data. By converting semi-structured documents into a consistent representation, ChemDataExtractor decouples its NLP pipeline from document formatting idiosyncrasies. This is particularly valuable in the materials domain, where journal formatting practices vary widely, and information may be distributed across figures, captions, and main text. The standardized document model thus serves as a foundational decoder structure, enabling uniform rule application across disparate inputs.

At the heart of ChemDataExtractor's information extraction capabilities lies a finely tuned NLP pipeline tailored for the chemistry domain. The pipeline comprises several decoder-style components, including tokenization, POS tagging, chemical

NER, and phrase parsing. Each component is designed to translate unstructured text into increasingly structured representations, culminating in the generation of machine-readable chemical records.

Tokenization begins with sentence splitting using the unsupervised Punkt algorithm, trained on a corpus of over 3,500 chemistry articles from the American Chemical Society, Royal Society of Chemistry, and Springer. The domain-specific training enables ChemDataExtractor to accurately handle chemical abbreviations and punctuation ambiguities common in scientific texts, such as "Fig." or "et al."

Following sentence segmentation, a custom word tokenizer is employed to split text into tokens. This tokenizer incorporates domain-aware rules to prevent fragmentation of chemical names and formulas. For instance, it distinguishes between hyphenated chemical names and simple punctuation, and ensures consistent treatment of units (e.g., "25 °C") regardless of spacing.

Normalization routines subsequently standardize token representation by resolving variant spellings, converting nonstandard Unicode symbols, and removing control characters. This facilitates consistent downstream processing while maintaining traceability to original text locations.

The POS tagger in ChemDataExtractor employs a CRF model augmented with Brown clustering—a form of unsupervised word representation based on distributional similarity [16]. Clustering was performed on over 20 million words extracted from chemistry articles, yielding 1,500 clusters with hierarchical binary identifiers.

As shown in Figure 7.2, these clusters group semantically related words, such as chemical quantities, unit symbols, or spectral descriptors. The CRF-based tagger integrates cluster features with standard lexical cues, such as prefixes, suffixes, capitalization patterns, and punctuation. Evaluation against biomedical and news corpora demonstrates that this hybrid model achieves high generalizability and robustness across chemistry-specific content (Table 7.1).

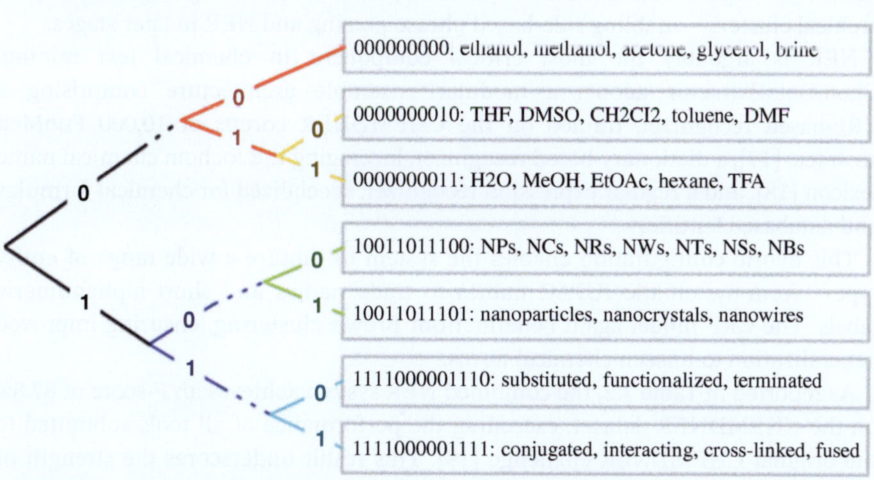

Figure 7.2 Representative frequent terms from seven sample word clusters, each identified by a binary path indicating its position in the hierarchical structure; broader cluster groupings correspond to shorter binary prefixes. *Source*: [11] / with permission of American Chemical Society.

Table 7.1 POS tagging accuracy of various training configurations assessed using the WSJ and GENIA benchmark corpora.

Training system	WSJ (%)	GENIA (%)
WSJ	97.19	83.50
WSJ + clusters	97.23	84.15
GENIA	78.88	98.53
GENIA + clusters	81.19	98.62
WSJ + GENIA	97.02	98.26
WSJ + GENIA + clusters	97.08	98.34

Source: [11] / with permission of American Chemical Society.

Table 7.2 Performance metrics (Precision, Recall, and *F*-score) for CRF, dictionary-based, and regular expression chemical NER systems evaluated individually and as a combined approach.

System	Precision (%)	Recall (%)	*F*-score (%)
CRF	90.50	80.00	84.90
Dictionary	88.60	70.20	78.30
Regular expression	89.40	11.00	19.60
Combined system	89.10	86.60	87.80

Source: [11] / with permission of American Chemical Society.

This approach exemplifies a decoder-style design in which structured syntactic labels (POS tags) are assigned to raw text using domain knowledge encoded in hierarchical clusters—enabling rule-based phrase parsing and NER in later stages.

NER is arguably the most critical component in chemical text mining. ChemDataExtractor adopts a modular ensemble architecture comprising a CRF-based recognizer, trained on the CHEMDNER corpus of 10,000 PubMed abstracts [17], a dictionary-based recognizer, leveraging the Jochem chemical name lexicon [18], and a regular expression recognizer, specialized for chemical formulas and database identifiers.

This hybrid configuration enables the system to capture a wide range of entity types—from systematic IUPAC names to trade names and short alphanumeric labels. The CRF model again benefits from Brown clustering, ensuring improved generalization to unseen chemical terms.

As reported in Table 7.2, the combined NER system achieves an *F*-score of 87.8% on the CHEMDNER dataset, exceeding the performance of all tools submitted to the original CHEMDNER challenge [19]. This result underscores the strength of domain-specific, decoder-based pipelines in high-precision entity recognition.

Importantly, the modularity of the NER pipeline allows users to extend or modify recognizers as new corpora or chemical ontologies become available. Such flexibility is particularly valuable for emerging materials subfields, such as battery electrolytes or metal–organic frameworks, which may contain novel terminology not captured by existing models.

In addition to sentence-level extraction, ChemDataExtractor's ability to accurately interpret and extract information from tabulated data and chemically descriptive phrases constitutes a defining strength. This capability is particularly vital for materials science literature, where property values, spectroscopic features, and experimental conditions are often summarized in tables or brief captions rather than full sentences. Here, decoder-based models manifest as rule-based grammars that parse complex yet formulaic inputs into structured records.

Whereas general-purpose NLP tools struggle with the precision and variability of chemical language, ChemDataExtractor uses a library of specialized grammars tailored to common patterns in chemical documentation. Unlike the monolithic grammars used by systems such as ChemicalTagger [20], which attempt to parse all variations within a single framework, ChemDataExtractor adopts a modular approach, deploying multiple focused grammars optimized for specific tasks.

Each grammar consists of nested rules that match sequences of POS and named entity tags. As illustrated in Figure 7.3a, these rules are implemented in Python using composable objects such as T (tag matcher), W (exact word matcher), and R (regular expression matcher). A sample rule matches phrases of the form "ChemicalName (Label)," associating an alphanumeric label with a full chemical name. Figure 7.3b visually represents the parse tree generated by this grammar.

This rule-driven architecture exemplifies a decoder approach, wherein a hierarchical syntax tree is constructed from linear input text based on deterministic rules. The result is a structured representation that explicitly captures chemical identities and their syntactic roles in the sentence—crucial for downstream tasks such as property linking and disambiguation.

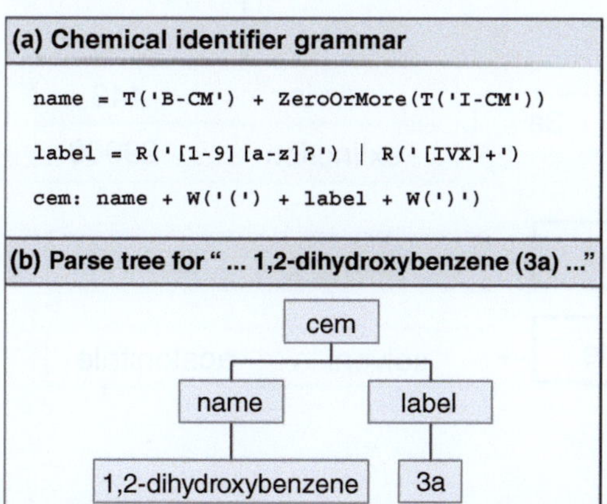

Figure 7.3 Illustrative rule-based grammar designed to identify chemical entities accompanied by alphanumeric identifiers. (a) Example grammar rules defining how chemical names and labels are identified and combined based on named entity recognition tags and pattern matching. (b) Tree-structured representation generated by applying the grammar to an example sentence, illustrating the hierarchical parsing result. *Source*: [11] / with permission of American Chemical Society.

Tables in chemistry literature often encapsulate high-density, semi-structured information. ChemDataExtractor's table parser is designed to exploit this structure by interpreting rows as records of chemical entities and columns as attributes such as melting point, absorption wavelengths, or solvent environment.

The parsing process begins with column heading analysis, wherein rule-based grammars identify the property type and units associated with each column. Next, each individual table cell is parsed using grammars tailored to the identified property type. This two-tiered parsing ensures both vertical and horizontal coherence in the resulting data records.

Figure 7.4 exemplifies this approach using a simplified UV–vis absorption table. After classifying the columns (e.g., "λ_max" as wavelength), the parser processes each row to construct structured chemical records. Notably, the system also extracts contextual data such as solvent or temperature from headers or captions when present, applying it uniformly to relevant entries.

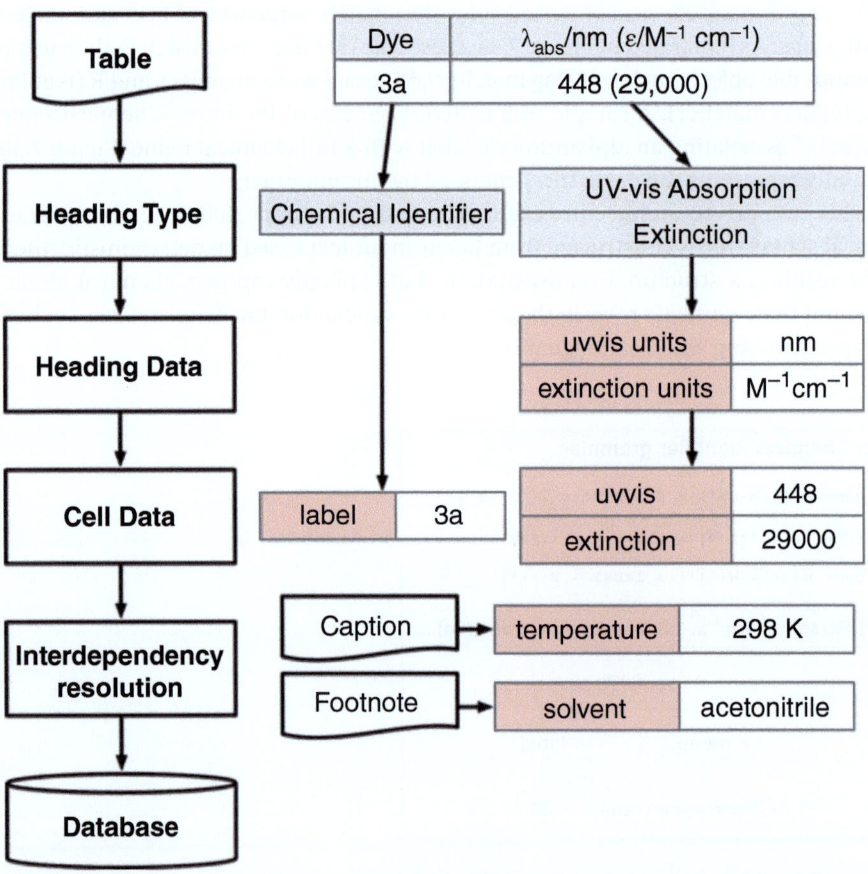

Figure 7.4 Summary of the table parsing workflow (left) demonstrated using a simplified example table (right); column headers are first analyzed to determine attribute types, followed by row-wise parsing to generate compound records, with contextual information from captions and surrounding text integrated during the final resolution stage. *Source*: [11] / with permission of American Chemical Society.

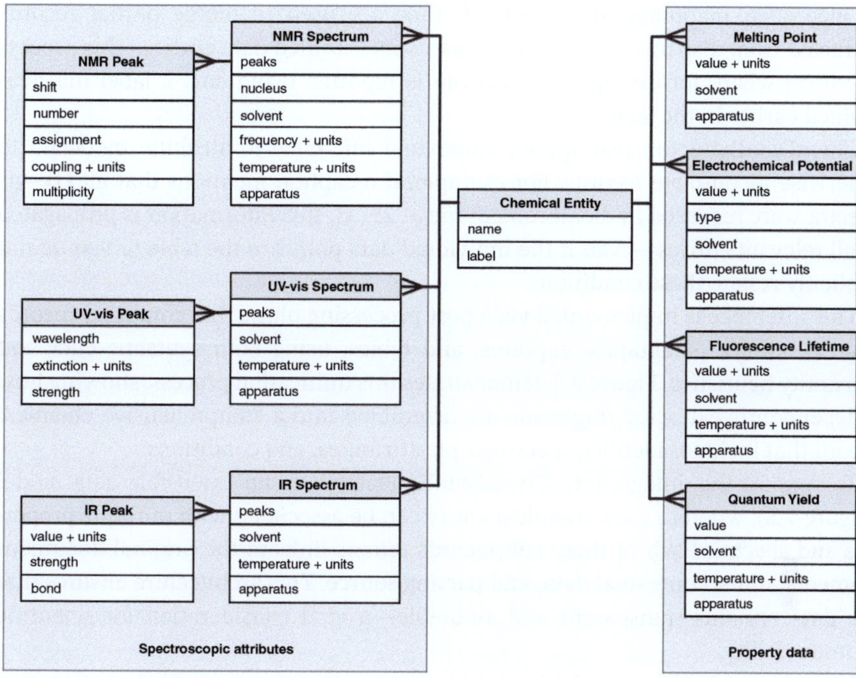

Figure 7.5 ChemDataExtractor's data model linking chemical entities with their experimental properties and spectroscopic attributes, designed for extensibility through user-defined parsers. *Source*: [11] / with permission of American Chemical Society.

The result of this parsing process is a set of semantically enriched records that align with the final data model depicted in Figure 7.5. Each chemical entity is linked to multiple properties and spectra, with metadata such as units, measurement context, and source location preserved.

This structured representation is essential for populating downstream ML pipelines, enabling robust cross-document analyses, statistical property estimation, and materials design exploration—all while retaining provenance for reproducibility and verification.

Scientific documents often distribute information across multiple modalities—e.g., a figure caption may mention the measurement apparatus, a paragraph may define a compound label, and a table may report the actual property values. ChemDataExtractor is distinguished by its capacity to resolve such interdependencies at the document level, enabling the unification of fragmentary data into coherent chemical records.

A central challenge in chemical text mining is that the same chemical entity may be referred to by different names or symbols across a document. To address this, ChemDataExtractor incorporates a decoding routine based on the Schwartz–Hearst algorithm [21], which detects definitional patterns such as "Tetrakis(triphenylphosphine)palladium(0) ($Pd(PPh_3)_4$)." This allows the system to map alphanumeric labels (e.g., "1a," "$Pd(PPh_3)_4$") back to their full chemical names.

Once such mappings are identified, they are used to merge partial records extracted from disparate document elements into unified data entries. This process is critical when, for example, a spectrum is reported using only a label that was defined earlier in the text.

ChemDataExtractor also applies contextual inference to attribute metadata to otherwise incomplete records. For example, if a caption mentions that all UV–vis spectra were recorded in dichloromethane at 25 °C, this information is propagated to all relevant entries—even if the individual data points in the table or text do not explicitly restate these conditions.

This inference is implemented via a post-processing phase that combines records from headings, paragraphs, captions, and tables, using both syntactic cues and proximity heuristics. Figure 7.6 demonstrates this unification process, showing how independently extracted fragments are assembled into a comprehensive chemical record that includes identity, spectroscopic attributes, and conditions.

To support this integration, ChemDataExtractor uses an extensible data model (Figure 7.5), wherein each chemical entity can be associated with multiple properties and spectra. Each of these sub-records retains links to the original document elements, units, contextual data, and parsing source. This architecture ensures that the data remains transparent and auditable—a vital consideration for scientific reproducibility.

Users can extend the model by defining new property or spectrum types using Python-based schemas and parsers. This modularity makes the toolkit well suited for evolving materials domains, allowing rapid adaptation to novel classes of compounds or measurement techniques.

ChemDataExtractor has undergone rigorous evaluation to validate its performance across multiple types of documents and extraction tasks. These assessments

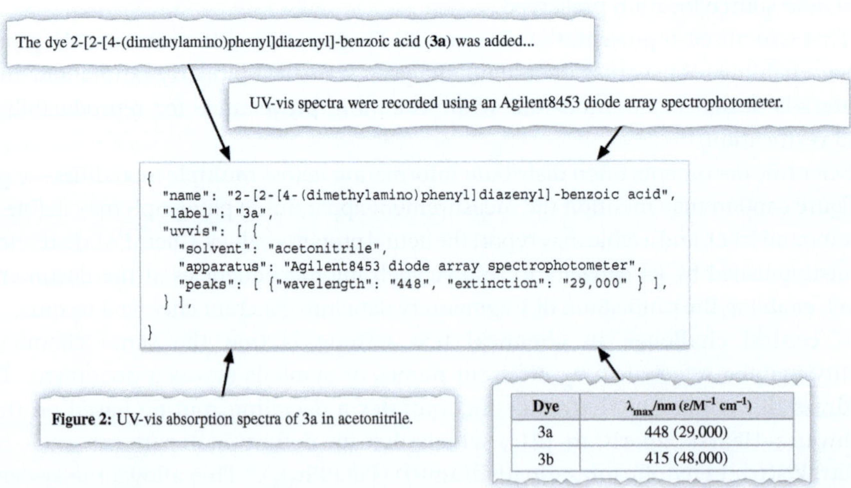

Figure 7.6 Illustration of how textual, caption, and tabular information is integrated to construct a unified chemical data record. *Source*: [11] / with permission of American Chemical Society.

underscore the toolkit's viability as a practical solution for building structured materials databases.

The toolkit was tested on a gold-standard set of 50 open-access chemistry articles from journals published by the ACS, RSC, and Springer. Extraction accuracy was evaluated in terms of Precision, Recall, and *F*-score across chemical identifiers, spectroscopic attributes, and chemical properties.

Table 7.3 presents the key metrics: chemical identifiers: *F*-score of 93.4%; spectroscopic attributes: *F*-score of 86.8%; and chemical properties: *F*-score of 91.5%.

Further breakdowns (Table 7.4) reveal that individual components—such as spectrum types and numerical values—achieve near-perfect Precision, particularly in structured formats like tables. Apparatus and temperature values, which rely more heavily on cross-domain inference, show slightly lower Recall but retain high Precision, illustrating the importance of document-level integration.

To assess performance in a less-structured environment, ChemDataExtractor was applied to a sample of 2,000 US patents and benchmarked against the LeadMine-ChemicalTagger pipeline developed by Tetko et al. [22]. The task focused on extracting melting point values.

ChemDataExtractor identified 18,180 unique melting point records, compared to 13,198 extracted by the LeadMine pipeline. Of these, 8,978 records overlapped

Table 7.3 Evaluation of extraction performance across different information types using Precision, Recall, and *F*-score metrics.

	Precision (%)	**Recall (%)**	***F*-score (%)**
Chemical identifier records	94.10	92.70	93.40
Spectrum records	88.30	85.40	86.80
Chemical property records	93.50	89.60	91.50

Source: [11] / with permission of American Chemical Society.

Table 7.4 Performance evaluation of chemical property extraction based on Precision, Recall, and *F*-score.

	Precision (%)	**Recall (%)**	***F*-score (%)**
Property value	100	95.90	97.90
Property units	100	94.80	97.40
Chemical subject	93.50	89.60	91.50
Solvent	100	94.40	97.10
Temperature	100	88.90	94.10
Apparatus	100	87.50	93.30
Full property records	93.50	89.60	91.50

Source: [11] / with permission of American Chemical Society.

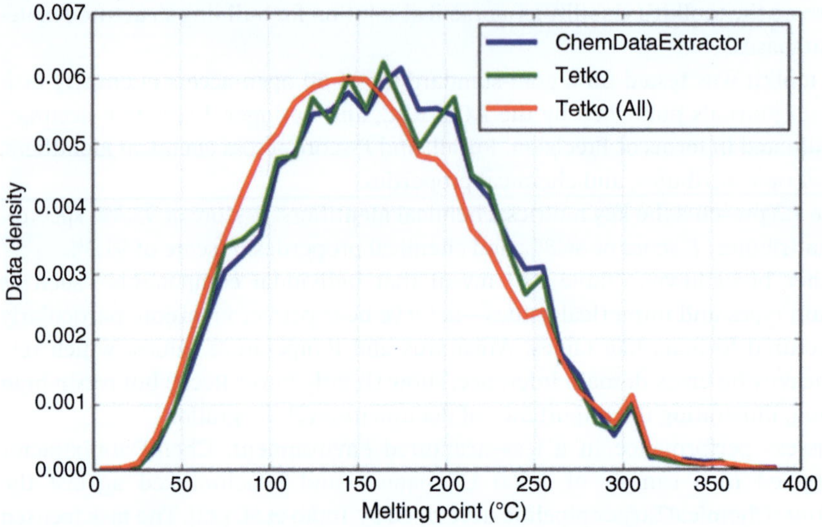

Figure 7.7 Comparison of melting point temperature distributions extracted by ChemDataExtractor and Tetko et al. from 2,000 sample patents, alongside the full Tetko dataset, with prior duplicate removal applied to the latter. *Source*: [11] / with permission of American Chemical Society.

Table 7.5 Summary of melting point counts and corresponding averages for temperature (T), molecular weight (MW), and non-hydrogen atom count (NA) from data extracted by ChemDataExtractor and Tetko et al. using a 2,000-patent evaluation sample.

Dataset	Count	Average T (°C)	Average MW	Average NA
ChemDataExtractor	18,180	166.3	385.5	26.9
Tetko	13,198	164.1	385.1	27
Tetko (all)	241,958	159	357	25

Source: [11] / with permission of American Chemical Society.

exactly between the two systems, while the remainder reflected either additional extractions or differences in treatment of duplicates.

Distributional comparisons (Figure 7.7) showed strong agreement in the range and frequency of extracted values, confirming the system's reliability across document styles. Table 7.5 further showed parity in molecular weight and atom counts of extracted compounds, supporting the comparability of the datasets.

These results demonstrate that decoder-based systems like ChemDataExtractor, when enhanced with domain-specific grammars and robust document modeling, can perform on par with or exceed hybrid neural-rule pipelines, particularly in high-precision tasks.

Individual NLP modules were also benchmarked separately. The POS tagger trained on combined biomedical and news corpora achieved 98.34% accuracy on the GENIA corpus, demonstrating robust generalization. The chemical NER module achieved an *F*-score of 87.8% on the CHEMDNER corpus—outperforming all official submissions to the CHEMDNER challenge at the time of publication.

The development and deployment of ChemDataExtractor exemplify the capabilities of domain-specific, decoder-style architectures for automated extraction of chemical and property information from scientific literature. Through a modular pipeline that integrates rule-based grammars, domain-adapted tokenization, hybrid NER, and robust table parsing, ChemDataExtractor transforms unstructured or semi-structured inputs into structured, machine-readable chemical records. Crucially, its ability to resolve document-level interdependencies and unify fragmentary data across captions, tables, and narrative text represents a significant advancement in information integration—particularly in the materials domain where contextual information is often dispersed. This case study underscores the importance of combining deterministic rules with statistical models tailored to chemistry-specific language, and highlights how decoder-based NLP pipelines can effectively meet the challenges of accuracy, flexibility, and extensibility in large-scale materials data extraction efforts.

7.3 A General-purpose Extraction Pipeline for Polymer Properties

The rapid proliferation of scientific literature in materials science presents a formidable barrier to extracting, aggregating, and utilizing material property data at scale. This challenge is particularly acute in polymer science, where domain-specific linguistic complexity, inconsistencies in nomenclature, and structural representation difficulties obstruct automated knowledge extraction. The work by Shetty et al. addresses this challenge through the development of a general-purpose NLP pipeline for extracting polymer property data from a massive corpus of scientific abstracts. This pipeline exemplifies the application of decoder-based methodologies—specifically, transformer-based NER models—in enabling scalable, high-fidelity extraction of chemical and physical property information.

This section presents an in-depth account of three foundational components of the study: (1) the motivation and challenges unique to polymer property extraction; (2) the design and fine-tuning of the MaterialsBERT language model; and (3) the construction of the PolymerAbstracts corpus and its associated ontology. These components form the methodological and conceptual core that drives the pipeline's success.

The extraction of property data from scientific literature is a cornerstone of modern materials informatics. However, polymers pose unique obstacles that distinguish them from inorganic materials and small molecules. First, the nomenclature of polymers is inconsistent and often ambiguous. The same material may appear in literature under multiple variant names, such as "polyethylene," "poly(ethylene)," and "polyethylene," depending on the author's stylistic preference. These aliases cannot be reliably mapped to a canonical structure using rule-based or trivial lookup methods.

Second, unlike small organic molecules or inorganic compounds, polymers typically do not have standardized structure encodings such as SMILES (Simplified

Molecular Input Line Entry System) strings readily available from text. This absence of machine-readable representations significantly hampers the integration of textual data into structure-aware property prediction pipelines.

Third, the sheer volume of polymer-related publications—compounded by a 6% annual growth in materials science literature—renders manual curation infeasible. Despite the efforts of human experts, databases such as PoLyInfo, while rich and reliable, are inherently constrained by their dependence on human annotation, leading to issues of scale and latency.

The solution proposed by Shetty et al. is rooted in adapting techniques from the broader NLP literature—namely, pretrained language models and task-specific fine-tuning—to create a scalable pipeline that begins with raw text and culminates in structured, machine-readable records of material–property relationships. This approach contrasts with traditional rule-based or semiautomated systems, which often target a narrow subset of materials or properties.

Importantly, polymers exacerbate several NLP challenges such as co-referencing (e.g., linking a full material name to its acronym within an abstract), relation extraction across sentence boundaries, and context-sensitive disambiguation of units and values. For instance, associating a tensile strength value with the correct material requires understanding not only proximity but also domain-specific heuristics. Addressing these challenges requires a combination of supervised learning, domain-specific ontologies, and heuristic post-processing, which the authors integrate into a unified framework.

1. **Model initialization and fine-tuning**
   MaterialsBERT was initialized using PubMedBERT, a BERT-based model pretrained on 14 million biomedical abstracts and full-text articles. PubMedBERT was selected as the base model because of its domain-specific vocabulary, which exhibits significant lexical overlap with materials science, especially polymers, due to shared terminologies in biomedical and polymer chemistry contexts [23].

   The authors fine-tuned PubMedBERT using a corpus of 2.4 million materials science abstracts. The rationale for focusing exclusively on abstracts—rather than full texts—was twofold: abstracts are readily available and free from the access restrictions often imposed on full texts, and previous research has demonstrated that models trained on abstracts alone outperform those trained on full-text corpora when the downstream tasks are similarly abstract-based.

   This fine-tuning process, conducted over 90 hours on four RTX6000 GPUs, allowed the model to learn contextual embeddings tailored to the lexical and syntactic characteristics of materials science literature. These embeddings then served as input representations for a shallow linear classifier used to predict entity types during NER.
2. **Performance evaluation**
   The resulting model, MaterialsBERT, achieved state-of-the-art performance on a custom polymer dataset (PolymerAbstracts) and outperformed other baseline BERT-based models across several external benchmarks. Table 7.6 in the original paper compares MaterialsBERT with PubMedBERT, MatBERT, ChemBERT, BioBERT, and the general-purpose BERT-base on

Table 7.6 Comparison of different BERT-based encoders evaluated on the PolymerAbstracts test set.

Model	Precision (%)	Recall (%)	F_1-score (%)
MaterialsBERT (ours)	62.5	70.6	66.4
PubMedBERT	61.4	70.7	65.8
MatBERT	60.9	70.1	65.2
BioBERT	59.2	66.3	62.6
ChemBERT	52.2	62.6	57.0
BERT-base	52.1	61.0	56.2

Source: [24] / Springer Nature / CC BY 4.0.

PolymerAbstracts. MaterialsBERT attained the highest F_1-score of 66.4%, outperforming PubMedBERT (65.8%) and MatBERT (65.2%), while significantly surpassing general-purpose models such as ChemBERT (57%) and BERT-base (56.2%).

In further benchmarking on four publicly available materials science datasets—including CHEMDNER and ChemRxnExtractor—MaterialsBERT achieved the highest F_1-score on three out of five tasks (Table 7.7), confirming its generalizability across subdomains of materials science.

These results underscore the benefit of domain-specific language modeling, affirming that fine-tuning a transformer model on a materials-focused corpus yields superior performance compared to relying on general or biomedical models alone.

3. **Construction of the PolymerAbstracts corpus and ontology**

For supervised fine-tuning of the NER model, the authors created a high-quality annotated dataset named PolymerAbstracts. This corpus comprises 750 polymer-relevant scientific abstracts, manually annotated by domain experts using a custom-designed ontology.

The ontology used for annotation defines eight entity types tailored to capturing material–property relationships in polymer literature:

- POLYMER: specific named polymers (e.g., polystyrene)
- POLYMER_CLASS: generic classes of polymers (e.g., polyimide)
- PROPERTY_NAME: names of properties (e.g., tensile strength)
- PROPERTY_VALUE: numerical values associated with properties (e.g., 120 MPa)
- MONOMER: repeat units of polymers (e.g., styrene)
- ORGANIC_MATERIAL and INORGANIC_MATERIAL: non-polymer components of material formulations
- MATERIAL_AMOUNT: quantitative descriptions of materials used

Annotation was conducted in three iterative rounds using the Prodigy annotation platform. Each round involved refining annotation guidelines and re-annotating previously processed documents. To expedite annotation, pre-annotation using entity dictionaries was employed where available.

Table 7.7 Evaluation results of multiple BERT-based encoders on test sets from publicly available materials science NER datasets.

BERT-based encoder	CHEMDNER (%) [17]			Inorganic synthesis recipes (%) [25]			Inorganic abstracts (%)			ChemRxnExtractor (%)		
	P	R	F_1	P	R	F_1	P	R	F_1	P	R	F_1
MaterialsBERT	70.1	68.2	69.2	69.1	68.3	68.6	85.3	86.7	86	73.5	69.5	71.4
PubMedBERT	71.5	69	70.2	69.9	65.3	67.6	84	86.2	85	68.1	59.5	63.6
MatBERT	71.7	66.9	69.2	68.6	67.7	68.2	85.6	86.7	86.2	67.4	58	62.4
BioBERT	70.6	65.7	68	64.4	63.7	64	85.6	87.1	86.4	74.8	65.4	69.8
ChemBERT	72.5	66.4	69.4	66.8	64.3	65.6	83.2	86.4	84.8	65	64	64.4
BERT-base	71.2	65.7	68.4	62.4	60.3	61.4	81	81.9	81.4	57.7	54.9	56.2

Source: [24] / Springer Nature / CC BY 4.0.

To quantify annotation consistency, the authors computed inter-annotator agreement using Fleiss's Kappa (0.885) and pairwise Cohen's Kappa (ranging from 0.864 to 0.906), indicating high levels of agreement and annotation reliability.

Unlike conventional BIO tagging schemes, the authors employed a simplified IO labeling strategy, justified by prior evidence indicating negligible performance gains for BERT-based models when using BIO versus IO tagging.

The PolymerAbstracts corpus captures a diverse and representative sample of polymer literature and is well aligned with the property extraction goals of the study. The inclusion of both specific (POLYMER) and generalized (POLYMER_CLASS) categories facilitates subsequent normalization steps and supports downstream analyses such as trend detection across polymer families (see following text).

Importantly, this dataset not only supports the immediate training of the NER model but also serves as a reusable benchmark for future NLP models in polymer informatics. It is publicly available via the authors' GitHub repository, contributing to the broader community's efforts in reproducible and scalable materials data extraction.

4. **Model architecture and property record extraction pipeline**
Having established the annotated corpus and domain-specific language model, the next critical component of Shetty et al.'s pipeline is the end-to-end infrastructure for extracting structured material property records from unstructured text. This involves several stages: text preprocessing, tokenization, entity recognition, name normalization, and heuristic-based relation extraction. The overall system architecture is depicted in Figure 7.8.

The source corpus consists of ~2.4 million HTML-formatted abstracts obtained from major publishers. Preprocessing entailed stripping HTML tags, normalizing superscripts and subscripts into LaTeX-style syntax, and standardizing

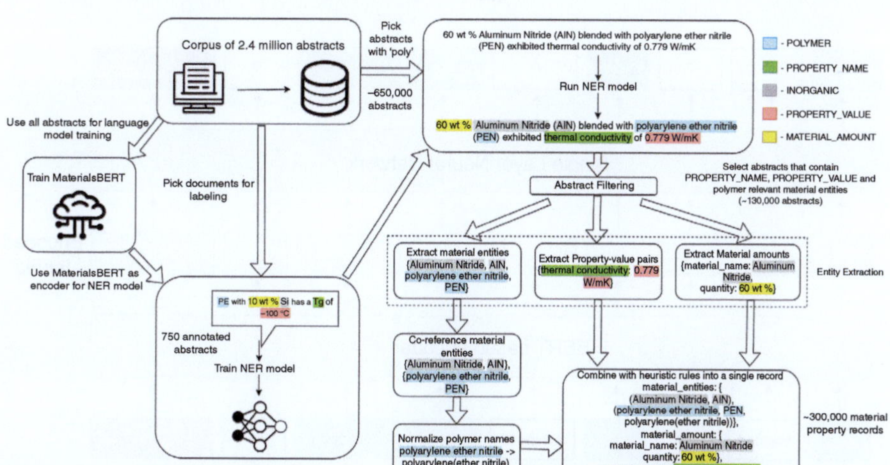

Figure 7.8 Overview of the pipeline for extracting material property records from scientific abstracts. *Source*: [24] / Springer Nature / CC BY 4.0.

Unicode characters to remove redundant representations of common symbols. This cleaning step ensures consistency in numerical formats, particularly for values with scientific notation.

Tokenization follows the WordPiece algorithm used in BERT and its derivatives. This subword-based strategy enables representation of rare and domain-specific terminology while maintaining a compact vocabulary. For example, the polymer name "resultant PLLA" may be tokenized into ["result," "##ant," "P," "##LL," "##A"], where the first subword token is used as the anchor for NER labeling.

The NER model utilizes a BERT-based encoder (specifically MaterialsBERT) followed by a single-layer linear classifier with a softmax activation to predict the entity class for each token. The training objective is to minimize cross-entropy loss between predicted and true labels for tokens in the annotated PolymerAbstracts dataset. Dropout regularization ($p = 0.2$) is applied to mitigate overfitting.

The model architecture is illustrated in Figure 7.9, where contextual embeddings produced by the encoder serve as input to the classifier. During inference, only complete entity matches are counted as true positives, thus enforcing stricter evaluation criteria and ensuring high-quality predictions.

One of the most intricate challenges in entity recognition is resolving co-references—cases where a material is referred to by multiple names or abbreviations within a document. The authors employ abbreviation detection from ChemDataExtractor [26] and fuzzy string matching (using Levenshtein distance ≤ 1) to unify references to the same material within each abstract. For example, "polyvinyl chloride" and "PVC" would be correctly linked.

Additionally, normalization of polymer names is crucial due to widespread nomenclature variability. A dictionary-based normalization system was employed to convert alias variants to a canonical name. The normalization process followed the methodology outlined in earlier work by the same group,

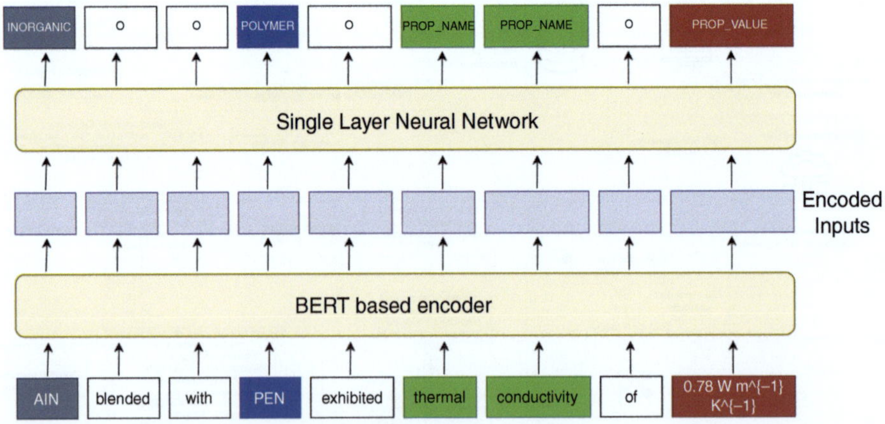

Figure 7.9 A BERT-based encoder transforms each input token into a contextual embedding, which is passed through a single-layer neural network to predict its entity type. *Source*: [24] / Springer Nature / CC BY 4.0.

allowing polymers such as "poly(ethylene)" and "poly-ethylene" to be harmonized as "polyethylene."

The final step involves constructing structured material property records by associating materials (e.g., polymers, monomers) with property name–value pairs. Due to the lack of annotated relation data, the authors rely on heuristic rules: If a material and property pair occurs in the same sentence, they are linked. If no such co-occurrence exists, all materials in the abstract are associated with the property. Units and numerical values are parsed using regular expressions, and standardized units are assigned for each property type. For instance, electrical conductivity values reported in different units (e.g., S/cm, S/m) are normalized to a common unit.

This pipeline, while heuristic-based, is remarkably efficient: It processed ~130,000 polymer-relevant abstracts and extracted ~300,000 structured material property records within 60 hours on a single 16-GB GPU.

5. **Data scale, quality, and insights from extracted records**

A major outcome of the pipeline is the construction of a large-scale structured dataset of polymer property records. The authors benchmark the scope, diversity, and potential scientific utility of this dataset through a series of quantitative analyses.

From the original 2.4 million-paper corpus, the system identified ~650,000 abstracts as polymer-relevant (via the presence of keywords like "poly"), of which 130,000 contained both material and property information. From these, ~300,000 material property records were extracted. This scale is notable when compared with PoLyInfo, a widely used curated polymer database, which at the time of writing contained ~492,000 property records. While PoLyInfo is the product of years of manual curation, the NLP-based pipeline produced a comparable volume of data in just 60 hours—demonstrating the feasibility of automated, high-throughput data extraction [27].

This work summarizes the number of data points extracted for key properties such as molecular weight, glass transition temperature (T_g), electrical conductivity, tensile strength, and elongation at break. Notably:

- Molecular weight: 9,053 records for 2,623 unique neat polymers.
- T_g: 6,155 records for 1,732 unique neat polymers.
- Tensile strength: 4,382 records for 926 unique neat polymers.
- Elongation at break: 1,499 records for 413 unique neat polymers.

These numbers are significant in comparison to prior datasets used for training property prediction models. For example, Doan Tran et al. used 672 data points for tensile strength prediction, while Shetty et al.'s pipeline generated 926 data points.

Figure 7.10 of the paper illustrates the average values of four properties (T_g, tensile strength, electrical conductivity, and molecular weight) across five major polymer classes: polyurethane, polyimide, polyamide, polyester, and polysiloxane.

Key insights include: Polyimides and polyamides exhibit higher T_g values, consistent with their rigid backbone structures. Polyimides have both high

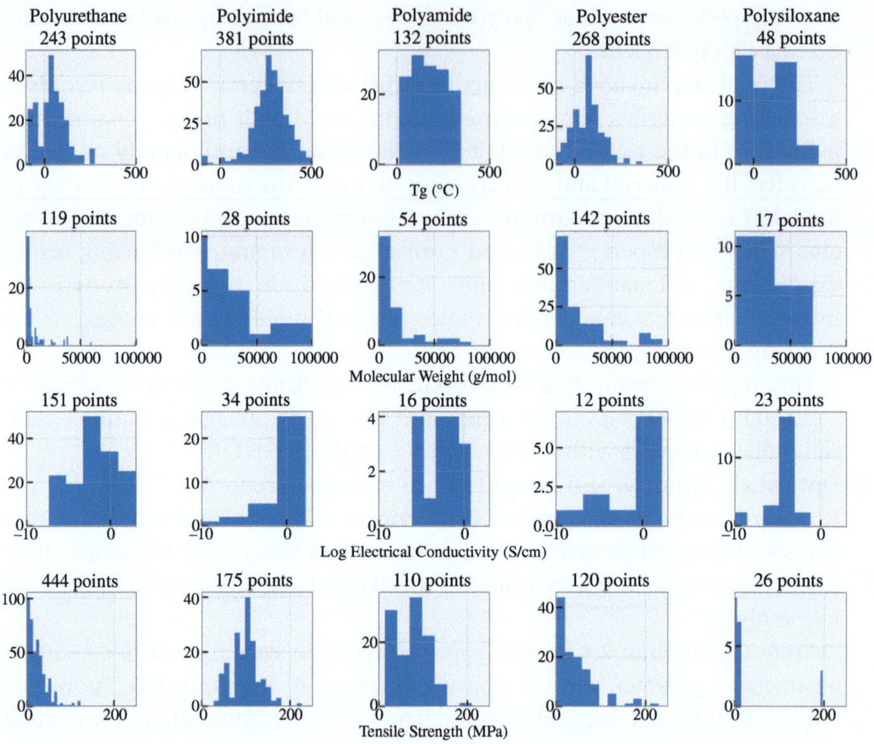

Figure 7.10 The most frequently mentioned polymer classes alongside the most commonly reported properties in the analyzed corpus. *Source*: [24] / Springer Nature / CC BY 4.0.

tensile strength and low electrical conductivity, validating their use in insulation applications. Molecular weight distributions resemble a power-law decay, reflecting synthetic limitations on long-chain polymers.

Another noteworthy visualization is Figure 7.11, which plots tensile strength versus elongation at break for polymer films, confirming the well-known "strength–ductility trade-off" in materials science: materials with high tensile strength often exhibit low ductility, and vice versa. These findings illustrate the capacity of NLP-extracted data to replicate known physical phenomena and support materials informatics research without manual annotation.

6. **Application-specific knowledge discovery and temporal analysis**
Beyond global property distributions, Shetty et al. demonstrate that NLP-extracted data can uncover domain-specific trends and validate theoretical expectations in three key application areas: polymer solar cells, fuel cells, and supercapacitors.

For polymer solar cells, the authors extracted performance metrics such as power conversion efficiency (PCE), short-circuit current (J_{sc}), open-circuit voltage (V_{oc}), and fill factor (FF). Figure 7.12 compares trends in NLP-extracted data (left) with manually curated data (right) from a prior study.

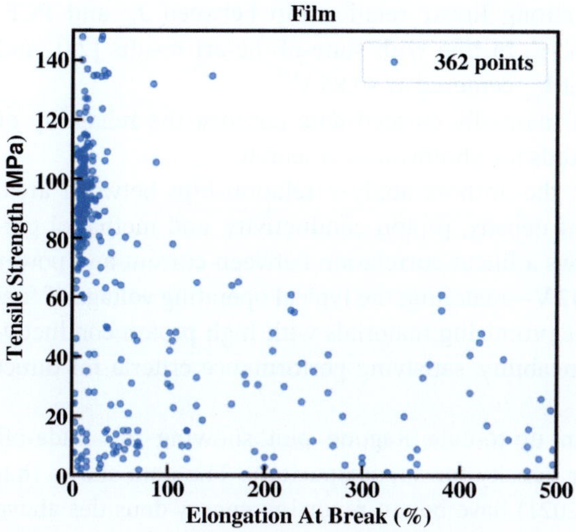

Figure 7.11 Materials exhibiting high elongation at break typically show lower tensile strength, while those with high tensile strength tend to have reduced elongation at break. *Source*: [24] / Springer Nature / CC BY 4.0.

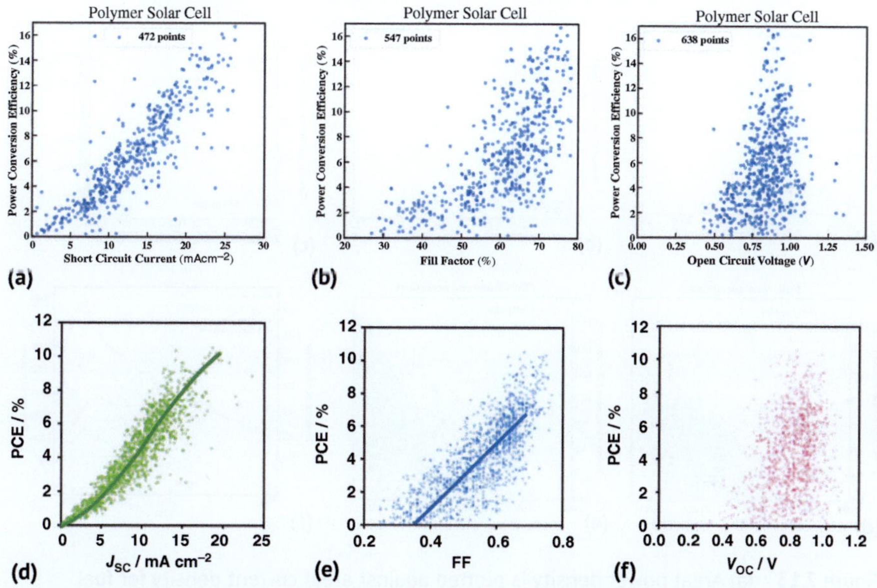

Figure 7.12 (a) PCE is plotted against short-circuit current based on NLP-extracted data; (b) PCE is plotted against FF using automatically extracted values; (c) PCE is shown versus open-circuit voltage from the extracted dataset; (d) PCE versus short-circuit current based on manually curated literature data; (e) manually extracted values illustrate the relationship between PCE and FF; and (f) open-circuit voltage is plotted against PCE using manually collected data. *Source*: [24] / Springer Nature / CC BY 4.0.

Observations include a strong linear relationship between J_{sc} and PCE, PCE values peaking at ~16.7%, in line with state-of-the-art results [28], and Gaussian-like distribution of V_{oc} centered at ~0.85 V.

These consistencies with manually curated data confirm the reliability of NLP-extracted property records for photovoltaic research.

In the case of fuel cells, the authors analyze relationships between areal power density, areal current density, proton conductivity, and methanol permeability. Figure 7.13a shows a linear correlation between current and power density, with a slope of ~0.42 V—matching the typical operating voltage of fuel cells. Figure 7.13b identifies promising materials with high proton conductivity and low methanol permeability, satisfying performance criteria for direct methanol fuel cells.

Figure 7.13c provides an up-to-date Ragone plot showing the trade-off between energy and power density for supercapacitors. The data shows that newer publications (2020–2021) have begun reporting energy densities above 10 Wh/kg, surpassing historical expectations and aligning with recent experimental breakthroughs.

Figure 7.13d–f present a temporal analysis of extracted data. They show: a shift from fullerene to non-fullerene acceptors in polymer solar cells, with

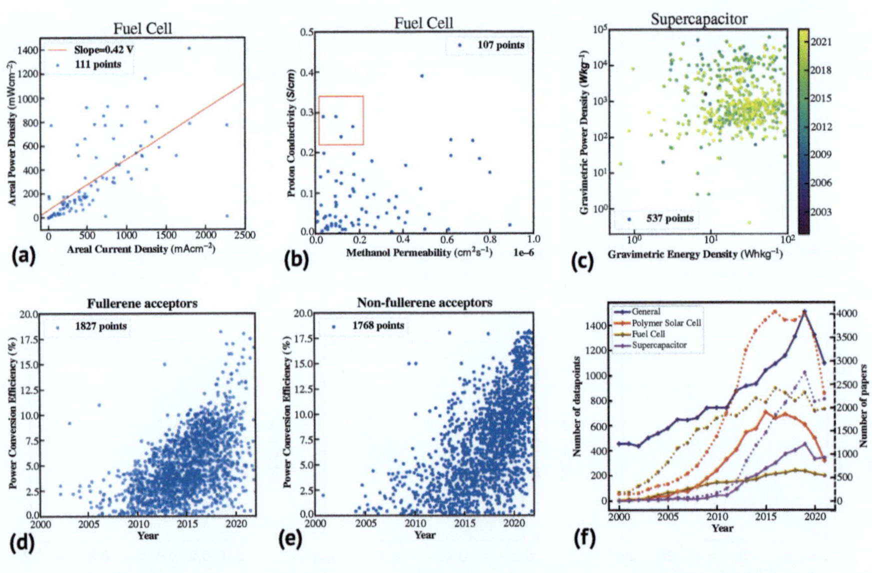

Figure 7.13 (a) Areal power density is plotted against areal current density for fuel cells, with a best-fit line slope of 0.42 V representing the typical operating voltage; (b) proton conductivity versus methanol permeability for fuel cells, where the red box indicates the optimal performance region; (c) a Ragone plot for supercapacitors displays the relationship between energy density and power density based on recent data; (d) PCE trends over time are shown for fullerene-based acceptors in polymer solar cells; (e) PCE over time is plotted for polymer solar cells with non-fullerene acceptors; and (f) temporal trends in the number of extracted data points are shown, with dashed lines indicating the corresponding publication volume for each application. *Source*: [24] / Springer Nature / CC BY 4.0.

higher PCE values for the latter; a decline in both the number of publications and property records during 2020–2021, attributed to COVID-19 disruptions (Figure 7.13f). These application-specific and time-dependent analyses exemplify the utility of automated extraction pipelines for both retrospective studies and prospective materials design.

The case study presented in this chapter illustrates a comprehensive and methodologically rigorous pathway for realizing automated extraction of chemical and property information from scientific literature, with a specific focus on the polymer domain. Central to this approach is the integration of domain-adapted language modeling, custom-curated ontologies, and scalable pipeline design—each component strategically aligned to address the distinct linguistic and structural challenges posed by polymer-related texts.

References

1 Raccuglia P, Elbert KC, Adler PDF, Falk C, Wenny MB, Mollo A, et al. Machine-learning-assisted materials discovery using failed experiments. Nature. 2016;533(7601):73–6.

2 Zhang H, Fu H, Zhu S, Yong W, Xie J. Machine learning assisted composition effective design for precipitation strengthened copper alloys. Acta Materialia. 2021;215:117118.

3 Xue D, Balachandran PV, Hogden J, Theiler J, Xue D, Lookman T. Accelerated search for materials with targeted properties by adaptive design. Nature Communications. 2016;7:1–9.

4 Wen C, Zhang Y, Wang C, Xue D, Bai Y, Antonov S, et al. Machine learning assisted design of high entropy alloys with desired property. Acta Materialia. 2019;170:109–17.

5 Segler MHS, Preuss M, Waller MP. Planning chemical syntheses with deep neural networks and symbolic AI. Nature. 2018;555(7698):604–10.

6 Wen C, Wang C, Zhang Y, Antonov S, Xue D, Lookman T, et al. Modeling solid solution strengthening in high entropy alloys using machine learning. Acta Materialia. 2021;212:116917.

7 Zhang Y, Wen C, Wang C, Antonov S, Xue D, Bai Y, et al. Phase prediction in high entropy alloys with a rational selection of materials descriptors and machine learning models. Acta Materialia. 2020;185:528–39.

8 Jiang X, Jia B, Zhang G, Zhang C, Wang X, Zhang R, et al. A strategy combining machine learning and multiscale calculation to predict tensile strength for pearlitic steel wires with industrial data. Scripta Materialia. 2020;186:272–7.

9 Kononova O, He T, Huo H, Trewartha A, Olivetti EA, Ceder G. Opportunities and challenges of text mining in materials research. iScience. 2021;24(3):102155.

10 Court CJ, Cole JM. Auto-generated materials database of Curie and Néel temperatures via semi-supervised relationship extraction. Scientific Data. 2018;5(1):1–12.

11 Swain MC, Cole JM. ChemDataExtractor: A toolkit for automated extraction of chemical information from the scientific literature. Journal of Chemical Information and Modeling. 2016;56(10):1894–904.

12 Kim E, Huang K, Jegelka S, Olivetti E. Virtual screening of inorganic materials synthesis parameters with deep learning. npj Computational Materials. 2017;3(1):53.

13 Kim E, Jensen Z, van Grootel A, Huang K, Staib M, Mysore S, et al. Inorganic materials synthesis planning with literature-trained neural networks. Journal of Chemical Information and Modeling. 2020;60(3):1194–201.

14 Krallinger M, Rabal O, Lourenço A, Oyarzabal J, Valencia A. Information retrieval and text mining technologies for chemistry. Chemical Reviews. 2017;117(12):7673–761.

15 Lu Y, Ji D, Yao X, Wei X, Liang X. CHEMDNER system with mixed conditional random fields and multi-scale word clustering. Journal of Cheminformatics. 2015;7:1–5.

16 Turian G, Ratinov L-A, Bengio Y. Word representations: A simple and general method for semi-supervised learning. In: Hajič J, Carberry S, Clark S, Nivre J, editors. Proceedings of the 48th Annual Meeting of the Association for Computational Linguistics [Internet]. Association for Computational Linguistics; 2015. pp. 384–94. Available from: https://aclanthology.org/P10-1040/

17 Krallinger M, Rabal O, Leitner F, Vazquez M, Salgado D, Lu Z, et al. The CHEMDNER corpus of chemicals and drugs and its annotation principles. Journal of Cheminformatics. 2015;7:1–17.

18 Hettne KM, Stierum RH, Schuemie MJ, Hendriksen PJ, Schijvenaars BJ, van Mulligen EM, et al. A dictionary to identify small molecules and drugs in free text. Bioinformatics. 2009;25(22):2983–91.

19 Krallinger M, Leitner F, Rabal O, Vazquez M, Oyarzabal J, Valencia A. CHEMDNER: The drugs and chemical names extraction challenge. Journal of Cheminformatics. 2015; 7(Suppl 1):S1.

20 Hawizy L, Jessop DM, Adams N, Murray-Rust P. ChemicalTagger: A tool for semantic text-mining in chemistry. Journal of Cheminformatics. 2011;3:1–13.

21 Schwartz AS, Hearst MA. A simple algorithm for identifying abbreviation definitions in biomedical text. In: Altman RB, Keith Dunker A, Hunter L, Jung TA, Klein TE, editors. Biocomputing 2003: Proceedings of the Pacific Symposium. Singapore: World Scientific; 2002. pp. 451–62.

22 Tetko IV, Lowe DM, Williams AJ. The development of models to predict melting and pyrolysis point data associated with several hundred thousand compounds mined from PATENTS. Journal of Cheminformatics. 2016;8:1–18.

23 Gu Y, Tinn R, Cheng H, Lucas M, Usuyama N, Liu X, et al. Domain-specific language model pretraining for biomedical natural language processing. ACM Transactions on Computing for Healthcare (HEALTH). 2021;3(1):1–23.

24 Shetty P, Rajan AC, Kuenneth C, Gupta S, Panchumarti LP, Holm L, et al. A general-purpose material property data extraction pipeline from large polymer corpora using natural language processing. npj Computational Materials. 2023;9(1):52.

25 Kononova O, Huo H, He T, Rong Z, Botari T, Sun W, et al. Text-mined dataset of inorganic materials synthesis recipes. Scientific Data. 2019;6(1):203.

26 Mavračić J, Court CJ, Isazawa T, Elliott SR, Cole JM. ChemDataExtractor 2.0: Autopopulated ontologies for materials science. Journal of Chemical Information and Modeling. 2021;61(9):4280–9.

27 Otsuka S, Kuwajima I, Hosoya J, Xu Y, Yamazaki M. PoLyInfo: Polymer database for polymeric materials design. In: 2011 International Conference on Emerging Intelligent Data and Web Technologies [Internet]. IEEE; 2011. pp. 22–9.

28 Zhang M, Ma X, Zhang H, Zhu L, Xu L, Zhang F, et al. Metallated terpolymer donors with strongly absorbing iridium complex enables polymer solar cells with 16.71% efficiency. Chemical Engineering Journal. 2022;430:132832.

8

Case Studies of Alloy Information Extraction

8.1 Introduction

This chapter introduces the development and application of automated text mining pipelines for extracting structured data from scientific literature on alloys. With the rapid growth of publications in materials science, traditional manual data curation has become increasingly inefficient. To address this challenge, natural language processing (NLP) and machine learning (ML) techniques are employed to systematically capture chemical compositions, processing routes, and material properties from thousands of articles. The chapter highlights rule-based and semi-supervised approaches for named entity recognition (NER), relation extraction, and table parsing, which enable high-accuracy construction of databases without requiring large annotated corpora. By integrating these methods, the resulting datasets not only support fundamental understanding of composition–property relationships but also facilitate predictive modeling for alloy design.

8.2 Automated Pipeline for Superalloy Data by Text Mining

AI/ML is transforming materials research by changing the paradigm from a "trial and error" to a data-driven methodology, thereby accelerating the discovery of new materials [1–10]. Well-characterized data remains a prerequisite for the success of AI/ML. Currently, there are two main sources of scientific data: (1) experimental and calculated results from a researcher's own laboratory and (2) data collected from papers published by other researchers. The scientific literature contains a vast amount of peer-reviewed, and largely high-quality, reliable data. Yet, manual data extraction with expert knowledge is time-consuming and labor-intensive for the tens of thousands of articles communicated using free-flowing natural language [11]. With ever-increasing number of new publications, maintaining and updating a database manually becomes increasingly difficult for the individual researcher.

Therefore, developing methods for automatically extracting data rapidly and accurately has increasingly become a necessity.

Recently, several pipelines have been introduced that use NLP techniques to automatically extract organic and inorganic chemical substance data from chemistry and materials science articles [12–16]. The NER and relation extraction tasks are widely considered the critical components of such literature-based data extraction. Kim et al. employed neural-network and parse-based approaches to identify and extract synthesis parameters, reporting an F_1-score of 81% across more than 640,000 journal articles [17]. ChemDataExtractor was developed to identify chemical named entities and extract compound relations from a very large corpus (hundreds of thousands of articles) using a dictionary-driven ML approach combined with multiple grammar rules [12]. Court and Cole applied "ChemDataExtractor" together with a modified Snowball algorithm to extract Curie and Néel temperatures for magnetic materials, achieving an estimated overall Precision of 73% from a corpus of 68,078 articles [16]. Because ferroelectrics and antiferroelectrics also use the terms "Curie" and "Néel" for their transition temperatures, the resulting database—although focused on magnetic materials—also contains entries for nonmagnetic materials where those terms are used differently.

Superalloys are extensively used in turbine blades and vanes for advanced aeroengines and industrial gas turbines. Designing alloys requires knowledge of their properties—particularly transition temperatures on multicomponent phase diagrams—alongside chemical composition and synthesis conditions. With roughly 20,000 articles on superalloys, applying supervised deep-learning (DL) approaches to NER or relation extraction is challenging because these methods need large, hand-labeled training datasets. Even semi-supervised approaches such as Snowball require a specified number of labeled seed samples to initiate learning, which makes it difficult to obtain both high Precision and high Recall simultaneously [18]. Hence, to accelerate data-driven superalloy design [19–22], extraction and assimilation of existing data from the literature is crucial.

Wang et al. [23] proposed an automatic NLP pipeline that captures both chemical composition and property data for superalloys into a single dataset, enabling global analysis based on superalloys using data extracted from 14,425 journal articles [23]. Specifically, they developed a rule-based NER approach and a heuristic, distance-based text algorithm for extracting multiple relations—both of which require no labeled samples and are therefore suitable for small corpora. They also implemented a general table-parsing and relation-extraction algorithm to handle tabular data. The alloy NER achieved an F_1-score of 92.07%, substantially outperforming a BiLSTM-CRF model (55.54%) and the ChemDataExtractor tool (24.86%). For text relation extraction of the γ' solvus temperature, their method attained an F_1-score of 79.37%, compared with 33.21% and 43.28% for the Snowball and modified Snowball semi-supervised algorithms, respectively. Because the distance-based algorithm does not require labeled seeds to extract multiple relations, it performs better on small corpora. In contrast, Snowball suffers from low Recall when the seed tuple set is too small, since its sentence clustering becomes ineffective at covering all sentence variants, limiting the number of extracted tuple relations. The table-parsing and relation-extraction tool achieved an F_1-score of 95.23%. Overall,

Wang et al. [23] automatically extracted a dataset of 2,531 instances covering chemical compositions and physical properties, e.g., γ' solvus temperature, density, solidus temperature, and liquidus temperature, from 14,425 articles obtained from Elsevier and other publishers. They interrogated this database to reveal trends that align with established superalloy behavior. The database, however, does not include synthesis or processing conditions nor experimental details such as measurement uncertainties, all of which are important for superalloy development. To assess the predictive value of the extracted data, they trained a data-driven ML model to predict γ' solvus temperatures for 15 superalloys that were not present in the extracted dataset but were reported later (in 2020 and 2021); the predictions achieved a relative error of 2.27%. These ML results therefore demonstrate the pipeline's potential and indicate that the text-mined database can be a valuable resource for superalloy development.

All source code from this study is available at https://github.com/MGEdata/SuperalloyDigger, and a web-based toolkit with usage examples and adaptation guides is hosted at http://SuperalloyDigger.mgedata.cn. Although developed for superalloys, the pipeline and toolkit offer a practical, effective approach for extracting data from the literature to accelerate data-driven materials design more broadly.

1. **Extraction strategy**
   The automated text mining pipeline for superalloys comprises several stages—document download, preprocessing, table parsing, text classification, NER, table and text relation extraction, and interdependency resolution—as schematically illustrated in Figure 8.1. Beginning from a corpus of scientific articles scraped in XML (eXtensible Markup Language), HTML, or plain-text formats, Wang et al. [23] preprocessed the archived raw files to assemble complete document records and remove irrelevant content.

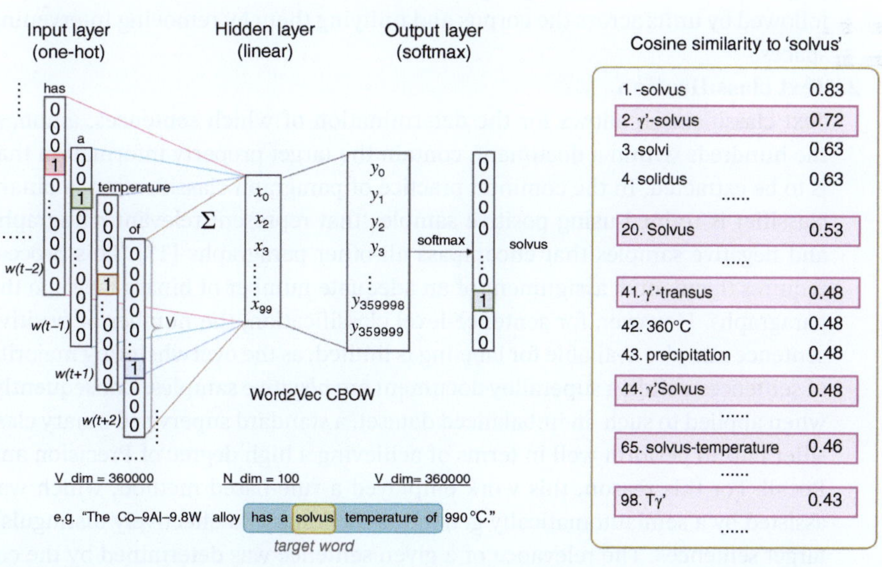

Figure 8.1 Schematic workflow of the automated text mining pipeline. *Source*: [23] / Springer Nature / CC BY 4.0.

The scientific articles used in this work were taken largely from Elsevier publisher. Combined with the CrossRef search application programming interface (API) [24] and the Web of Science search engine, Wang et al. [23] compiled a list of DOIs (digital object identifiers) for superalloy articles. A total of 14,425 journal articles corresponding to those DOIs were then programmatically archived in plain-text, XML, and HTML formats through Elsevier's Scopus and ScienceDirect APIs (https://dev.elsevier.com/) and the extended scraping utilities of ChemDataExtractor.

The first preprocessing step for HTML and XML files was to isolate relevant document domains, extract raw text, and merge any fragmented segments to produce complete document records. Semantic paragraph markups from HTML/XML were parsed and concatenated into plain-text documents represented as ordered lists of paragraphs. For tables, individual cells were treated as separate text domains and stored in nested lists that preserve the original table structure for downstream table parsing. Specifically for Elsevier content, the full-text API provides access to XML and plain-text formats for individual articles; in this study, XML files were used primarily for table extraction and plain-text files for body text extraction. The next step removed extraneous metadata—DOI, article ID, title, journal, publication details, and numerous URLs, because such metadata hinder NER and relation extraction. The pipeline therefore programmatically filtered out metadata by keyword and retained only the abstract and body of each full text. Additionally, inconsistent notations for values and units, such as "1039 **°C**" and "1039□**°C**" (here, "□" represents the space between 1,039 and **°C**), were identified and standardized to reduce ambiguity in downstream task. The latter notation with a space was split into "1039" and "**°C**" by the Natural Language Toolkit (NLTK) during word tokenization [25]. Wang et al. [23] addressed this by using regular expressions to locate all numeric values followed by units across the corpus and unifying them by removing intervening spaces.

2. **Text classification**

   Text classification allows for the determination of which sentences, amongst the hundreds within a document, contain the target property information that is to be extracted. In the common practice of paragraph classification, a binary classifier is trained using positive samples that represent relevant paragraphs and negative samples that encompass all other paragraphs [17]. This process requires the manual assignment of an adequate number of binary labels to the paragraphs. However, for sentence-level classification, the number of positive sentence samples available for labeling is limited, as the overwhelming majority of sentences within a superalloy document are negative samples. Consequently, when applied to such an imbalanced dataset, a standard supervised binary classifier fails to perform well in terms of achieving a high degree of Precision and Recall. For this reason, this work employed a rule-based method, which was assisted by a semiautomatically generated dictionary, to effectively distinguish target sentences. The relevance of a given sentence was determined by the co-occurrence of a superalloy name, a target property specifier, and its corresponding value and unit.

To provide an example, the γ' solvus temperature is considered the primary property of interest in this study. This particular property specifier can be expressed in various textual forms, all of which need to be captured during the sentence analysis. For instance, "γ' solvus temperature" can also be written as "γ'-solvus temperature." In order to generate a comprehensive synonym dictionary, which consisted of approximately 14,000 unlabeled full-text articles, Word2Vec continuous bag of words (CBOW) is utilized in the code genism (https://radimrehurek.com/gensim/) [26]. This technique leverages information about the co-occurrence of words by assigning high-dimensional vectors, known as embeddings, to each word in the text corpus, thereby preserving their syntactic and semantic relationships. Assuming there are $V = 360000$ unique words in the vocabulary of the whole superalloy corpus, Word2Vec CBOW loops through all words in the training text and uses its one-hot encoding as an input for a neural network with window size 10. The weights of the hidden layer are given by a $(V \times N)$-dimensional matrix, where N is the dimensionality of the embedding space (set to 100 in this case), into which the words are embedded.

Following the training process, the top-100 words most similar to the target property term "solvus" were screened by calculating the cosine similarity based on the newly obtained Word2Vec embedding model. As shown in Figure 8.2, the output of the Word2Vec CBOW model is a list of the top-100 similar words, which are ranked in descending order of their syntactic and semantic relationship to "solvus" within the corpus. A synonym dictionary for the target term "solvus" was then formed by manually selecting the most likely synonyms from these 100 words using expert knowledge, as highlighted by the words in pink in Figure 8.2. This dictionary subsequently allows for the determination that a sentence is a target sentence if a word from the synonym list and a property

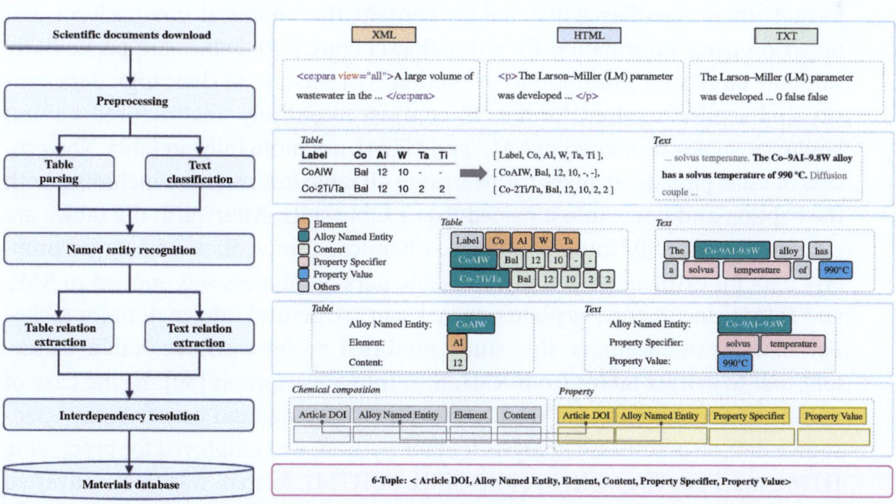

Figure 8.2 Word2Vec CBOW model and top-100 words with cosine similarity in descending order of syntactic and semantic relationships. *Source:* [23] / Springer Nature / CC BY 4.0.

Table 8.1 Recognition rule for property value and unit.

Examples for property value and unit	Rule for writing (regular expression)
1,050 °C, 850–950 °C, > 950 °C	^\W{0,1}[7-9][0-9]{2}(\.[0-9]{1,2})?\S*°C$ ^\W{0,1}1[0-9]{3}\S*°C$
1,050 K, 850–1,180 K, > 1,620 K	^\W{0,1}[7-9][0-9]{2}(\.[0-9]{1,2})?\S*K$ ^\W{0,1}1[0-9]{3}\S*K$

value with a specific unit appear concurrently within it. The corresponding property value can be either a single number or a range, and its unit may be expressed as °C or K (Kelvin), which can be recognized by regular expression, as illustrated in Table 8.1.

For the evaluation of sentence classification performance, the metrics of Precision, Recall, and F_1-score based on the confusion matrix were used as the metrics for sentence classification. Precision evaluates the proportion of correctly classified instances among those classified as positive [27]. In contrast, Recall quantifies the number of correct positive predictions made out of all actual positive cases [28]. The F_1-score, calculated using Eq. 8.1, represents the harmonic mean of Precision and Recall, and it is a metric frequently used when learning from imbalanced datasets because it weights both measures equally [29]. When evaluated on a randomly sampled test set of 30 articles, comprising approximately 3,000 sentences, the proposed sentence classification method exhibited a Precision of 88.46%, a Recall of 97.87%, and an F_1-score of 92.93%,

$$F_1\text{-score} = \frac{2 \times \text{Precision} \times \text{Recall}}{\text{Precision} + \text{Recall}} \tag{8.1}$$

Table parsing converts the complete table caption and body into a structured format, then classifies which tables contain the chemical composition and target property information for extraction (Figure 8.3). Tables are particularly valuable targets for information extraction (IE) because of their high data density and semi-structured format. In contrast to entirely unstructured natural language, tables formatted in XML and HTML are more interpretable. The process of table parsing involves transforming all table information, including both the caption and body, into a nested list of table cells. Afterward, the tables are classified to identify those containing chemical composition and target property data for extraction. Initially, the table parsing process was applied to XML and HTML documents, as plain text lacks any structural information for tables. For Elsevier publications, this study modified an open-source "table_extractor" tool to extract tables from XML files into a list format [30]. In the case of publications containing HTML files, pandas, a popular and user-friendly open-source data structure and analysis tool for Python, was employed for processing HTML markup. The table, in either XML or HTML format, was then converted into a cell-list format organized by rows and table captions. Next, Wang et al. [23] carried out table classification to filter out tables that contained chemical compositions and target properties. Similar to text classification, the keyword

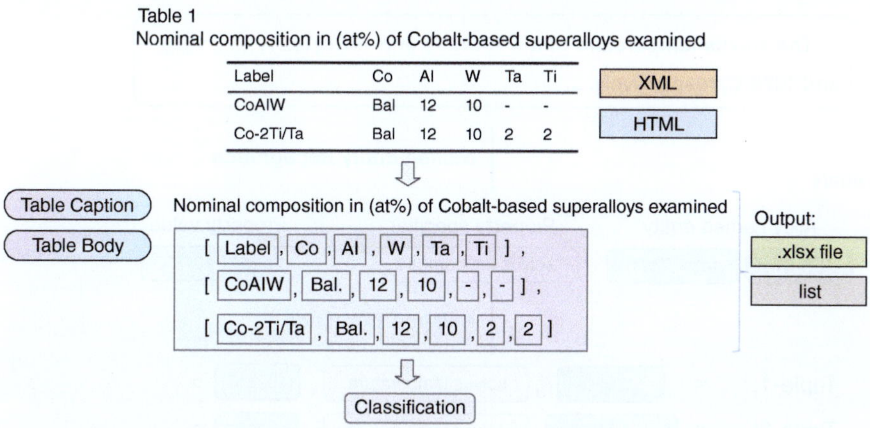

Figure 8.3 Schematic diagram of table parsing process. *Source*: [23] / Springer Nature / CC BY 4.0.

"composition" and property-specific terms, such as "γ' solvus temperature," were searched for within the table captions. The workflow for this process is depicted in Figure 8.3. Ultimately, 9,158 tables were successfully extracted from approximately 14,000 articles, with a manual inspection of 4,593 table cells from 20 articles showing an F_1-score of 98.8%. Table classification identified 5,327 tables related to composition and 114 tables related to solvus temperature.

NER methods are employed to identify the alloy named entity, property specifiers, and property values within the English-language text and tables. Following this, relation extraction is performed to identify specific tuple relations linking element content and properties. The final step, interdependency resolution, links the fragmented chemical composition and property data for a given material. The extracted tuple entities, which include the article DOI, alloy named entity, chemical element, content, property specifier, and property value, are then automatically compiled into a highly structured format, forming the basis of a materials database.

3. **Named entity recognition**

The extraction of chemical composition and property data from superalloy literature can be summarized as a six-tuple relation extraction, where the six-tuple consists of the article DOI, alloy named entity, chemical element, content, property specifier, and property value. The alloy named entity is typically described in terms of elemental composition (e.g., Co–9Al–9.8W or 8Al1W2Mo), superalloy designations (e.g., ERBOCo-0 or U720Li), or pronouns (e.g., this alloy). Chemical elements can be identified based on the periodic table, while the content provides the value and unit for each corresponding element. The property specifier refers to the name of the target property, such as the γ' solvus temperature or density, and the property value indicates the value and unit for each property. NER technology is crucial for identifying alloy named entities in English-language text and tables, which is essential for subsequent relation extraction [31]. This study, based on an automatically archived

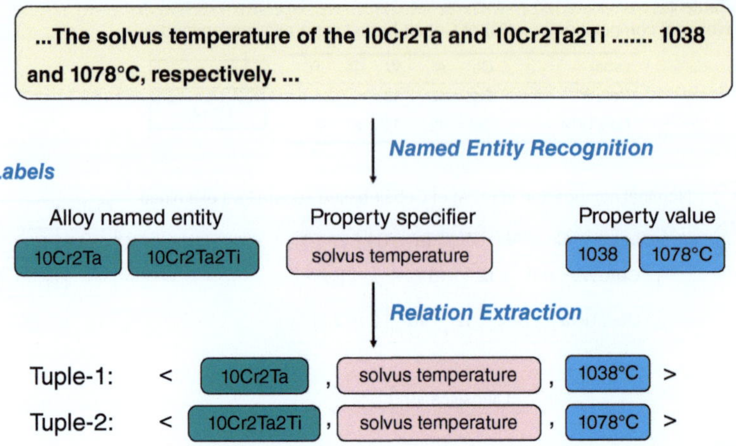

Figure 8.4 NER technology to provide entity sequence for subsequent relation extraction. *Source*: [23] / Springer Nature / CC BY 4.0.

corpus of approximately 14,000 superalloy documents, explored a rule-based NER method following text and table classification. The γ' solvus temperature was chosen as an illustrative example of property-specifier recognition in superalloys (Figure 8.4), which provides entity sequences for the subsequent relation extraction.

Multiple specialized grammar rules are tailored to recognize specific types of superalloy information. Table 8.2 outlines nine common writing patterns for superalloys. These include three patterns specific to nickel-based superalloys, such as IN738LC, Udimet 720Li, and Hastelloy C276, as well as six general patterns that apply to all superalloys, such as ERBOCo-0, alloy 718, CoWAlloy2, and 8Al1W2Mo. Each pattern corresponds to specific grammar rules of natural language, and nine distinct regular expressions were designed in Python to define these patterns. For instance, the elements and corresponding percentages in "8Al1W2Mo" and "2Nb2Re" can be recognized by the regular expression "^[0-9]+.?[0-9]{0,2}[A-JL-Z]." The designation "FGH98," consisting of capital letters and numbers, is recognized by the pattern "^[A-Z]\S+[0-9]$." These rules were applied to 545 sentences (approximately 19,000 words) from 283 articles, and the resulting Precision, Recall, and F_1-scores are provided in Table 8.3. The recognition procedure for property values during text classification is illustrated by the rule in Table 8.1, and its Precision, Recall, and F_1-score on 845 sentences are also shown in Table 8.3.

This work also used the BiLSTM-CRF model for NER tasks [32]. Usually, a BiLSTM network with a CRF (conditional random field) layer, namely the BiLSTM-CRF model, can be used for NER tasks [32]. Figure 8.5 illustrates the neural architecture of the BiLSTM-CRF model. BiLSTM is a bidirectional recurrent neural network (RNN) that utilizes LSTM cells to address long-term dependencies in text data, capturing more semantic contextual information [33]. The input to BiLSTM consists of a layer of word embeddings, which

Table 8.2 In-domain word patterns, examples, and rules for writing alloy named entities and property value and unit.

Category	Pattern	Examples	Rules for writing (regular expression)
Alloy named entities	1	9W–0.08B, Co–9Al–7.5W	[0-9]{0,2}(\.[0-9]{1,2})?[A-Z][a-z]?\- [0-9]{0,2}(\.[0-9]{1,2})?[A-Z][a-z]?
	2	SLM-1170 °C, TMS-138A, ERBOCo-0	^[A-Z]+[a-z]*\-[0-9]\w*
	3	Alloy 718, 282 alloy, 9.5W alloy	\s([0-9A-Z]+\w*)s+\S*alloys?|\ s+\S*alloys?\s([0-9A-Z]+\w*)\s
	4	CoWAlloy2, RRHT3, IN718	^[A-Z]\S+[0-9]$
	5	8Al1W2Mo, 8W, 718Plus	^[0-9]+\.?[0-9]{0,2}[A-JL-Z]
	6	Hayness 188, IN 738LC, Incoloy 800HT	[A-Z]+[a-z]*\s+[A-Z]*[0-9]{2,}[A-Z]*
	7	IN738LC, U720Li	^[A-Z]+[0-9]+[A-z]+
	8	Udimet 720Li, Incoloy 25-6Mo	[A-Z]+[a-z]*\s+\d+\-?\d*[A-Za-z]+
	9	Hastelloy C276, Inconel X-750	[A-Z]+[a-z]*\s+[A-Z]+\-?[0-9]+
Property value entities (with unit)	1	1,050 °C, 850–950 °C, >950 °C	^\W{0,1}[7-9][0-9]{2} (\.[0-9]{1,2})?\S*°C$ ^\W{0,1}1[0-9]{3}\S*°C$
	2	1,050 K, 850–1,180 K, >1,620 K	^\W{0,1}[7-9][0-9]{2} (\.[0-9]{1,2})?\S*K$ ^\W{0,1}1[0-9]{3}\S*K$

Table 8.3 Precision, Recall, and F_1-score of the NER.

Category		Precision	Recall	F_1-score	Test set
Alloy named entity	This work	90.58%	93.60%	92.07%	545 Sentences
	BiLSTM-CRF	51.99%	36.53%	42.91%	545 Sentences
	ChemDataExtractor	36.52%	18.84%	24.86%	545 Sentences
Property value (with unit)		85.71%	99.25%	91.98%	845 Sentences

transforms each word into a dense, fixed-length vector. The output of BiLSTM gives the corresponding probabilities for all possible labels for each word in a sequence, which are then passed into the CRF layer. The CRF layer accounts for label correlations within neighboring words and jointly decodes the most probable sequence of labels for the input sentence [34].

To train such a BiLSTM-CRF model, phrase-level labels were applied using the "BIO" (Begin–Inside–Outside) sequence labeling method on 545 sentences

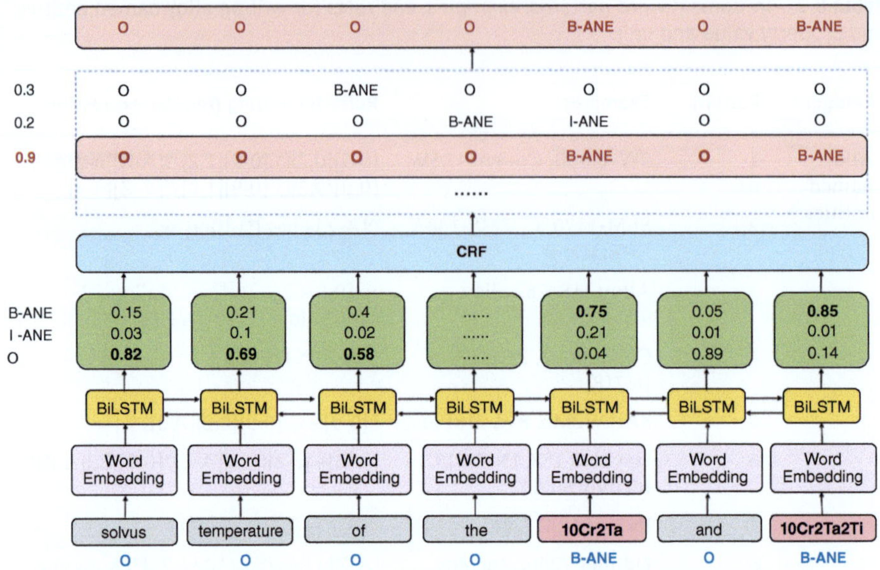

Figure 8.5 BiLSTM-CRF model for entity recognition of alloy names. The sequences of sentences and their "BIO" tags split by word are input into the BiLSTM model, followed by a CRF model to predict the target token sequences with the highest probability. *Source*: [23] / Springer Nature / CC BY 4.0.

from 283 articles by humans [35]. "B" is used for the beginning of a named entity, "I" for the inside part of a named entity, and "O" for unrelated words. For example, the alloy named entity "Inconel 718" would be labeled as "B-ANE I-ANE." The BiLSTM-CRF model was trained using fivefold cross-validation and achieved categorical Precision of 70.50%, Recall of 45.81%, and an F_1-score of 55.54% on a test set comprising about 109 sentences (approximately 4,000 words).

Additionally, the NER tool in "ChemDataExtractor" was also used to perform superalloy NER. Compared with the BiLSTM-CRF model and "ChemDataExtractor," the proposed rule-based method performs better for alloy named entity (Table 8.3). The BiLSTM-CRF model, due to its large parameter space, suffered from underfitting when trained on the small labeled corpus. ChemDataExtractor, which uses rule-based and dictionary-based methods for recognizing chemical substances, did not perform as well for superalloys due to differences in the rules and dictionaries used.

4. **Text relation extraction**

Relation extraction identifies and resolves ambiguities in semantic relationships between two entities in unstructured text data [36]. For property extraction from superalloy articles, the relation is represented as a quaternary tuple: ⟨article DOI, alloy named entity, property specifier, property value⟩. The article DOI can be archived during the retrieval and preprocessing stages. One of the most challenging aspects of superalloy property extraction is handling multiple-relation extraction from a single sentence [37]. It is common for several alloy named entities (≥1) to be mentioned alongside their corresponding property values for a specific property within one sentence. Alternatively, a single alloy named entity may be reported with multiple properties (≥1) and their respective values (≥1). This complexity leads to several obstacles in relation extraction, particularly when working with a limited superalloy corpus. Supervised relation extraction algorithms typically require large labeled datasets, usually exceeding approximately 70,000 samples [38]. Even semi-supervised methods demand a certain number of labeled samples as seeds to initiate the learning process. To address this issue, this study proposes a distance-based algorithm that does not require labeled samples to handle multiple-relation extraction; the workflow of the relation extraction is shown in Figure 8.6a. In feature-based relationship extraction methods, the number of words and word sequences between entities can act as the main syntactic features [39]. Therefore, the number of entities and distance between entities provide a basis for evaluating the relationship dependence.

After NER, target sentences are organized by the form of the entity sequences with position index for alloy named entity, property, and value. The shortest distance matching algorithm is applied where (1) the number of alloy named entities is not equal to that of property values ($n \neq k$ as flow 1 in Figure 8.6a); and (2) number of property specifiers is not equal to that of property values ($p \neq k$ as flow 2 in Figure 8.6a), $n = k, p > 1$ as flow 3, and $p = k,\ n > 1$ as flow 5. The shortest distance matching algorithm adopts a greedy strategy, and the pseudocode is given in Figure 8.7. Taking the conditional branch $n \leq k$ in Figure 8.7 as an example, for each alloy named entity, N_i, the distance between each property value entity K_i and N_i is calculated from Eq. 8.2 to find the closest property value entity K_m to the current N_i.

$$\text{Distance}(x_1, x_2) = |\text{pi}(x_1) - \text{pi}(x_2)| \tag{8.2}$$

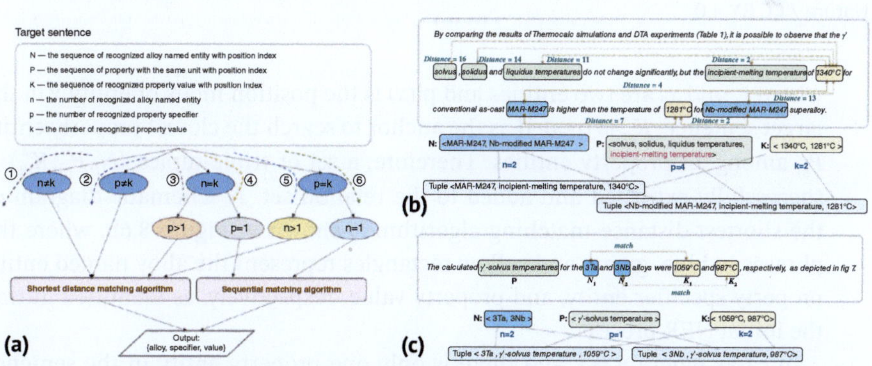

Figure 8.6 (a) Algorithm flowchart of relation extraction by shortest distance matching and sequential matching. The number of words and word sequences between entities can act as the main syntactic features in feature-based relationship extraction methods. (b) Schematic of shortest distance matching algorithm. (c) Schematic of sequential matching algorithm. The phrases in blue, green, and yellow rectangles are recognized as alloy named entity, property specifier, and property value by NER, respectively.

```
Shortest distance matching algorithm
Input:
    N — the sequence of recognized alloy named entity with position index
    P — the sequence of property with the same unit with position index
    K — the sequence of recognized property value with position index
    n — the number of recognized alloy named entity
    p — the number of recognized property specifier
    k — the number of recognized property value
Algorithm:
    Tuple_property =<>
    if n ≤ k:
        for each N_i in N:
            for each K_i in K:
                K_m = get_min_route(N_i → K_i) // P_m and K_m has the shortest distance
            for each P_i in P:
                P_m = get_min_route(K_m → P_i) // N_i and P_m has the shortest distance
            triple_tuples add <N_i, P_m, K_m>
    if n > k:
        if p=1:
            for each K_i in K:
                for each N_i in N:
                    N_m = get_min_route(K_i → N_i) // K_i and N_m has the shortest distance
                triple_tuples add <N_m, P, K_i>
        else:
            for each K_i in K:
                for each P_i in P:
                    P_m = get_min_route(K_i → P_i) // K_i and P_m has the shortest distance
                for each N_i in N:
                    N_m = get_min_route(P_m → N_i) // P_m and N_m has the shortest distance
                triple_tuples add <N_m, P_m, K_i>
Output: Tuple_property list
```

Figure 8.7 Pseudocode of shortest distance matching algorithm. *Source*: [23] / Springer Nature / CC BY 4.0.

where x_1 and x_2 are two entities and pi(x) is the position index of entity x in the target sentence. K_m is treated as the anchor to search the closest property entity P_m among all property entities. Therefore, a set of triple-tuples $\langle N_i, P_m, K_m \rangle$ is successfully extracted and added to the relation set. A schematic diagram of the shortest distance matching algorithm is shown in Figure 8.6b, where the phrases in blue, green, and yellow rectangles represent the alloy named entity, property specifier entity, and property value, respectively, as identified during the initial NER process.

In cases where $n = k$ and there is only one property entity in the sentence ($p = 1$ as flow 4), the sequential matching algorithm is performed to match the alloy named entity and property value for the specified property in order; the same is true for $p = k, n = 1$, the same sequential matching process is employed. Figure 8.8 shows the pseudocode of the sequential matching algorithm, and a schematic diagram of it is shown in Figure 8.6c.

Sequential matching algorithm

```
Input:
          N — the sequence of recognized alloy named entity with position index
          P — the sequence of property with the same unit with position index
          K — the sequence of recognized property value with position index
          n — the number of recognized alloy named entity
          p — the number of recognized property specifier
          k — the number of recognized property value

Algorithm:
          Tuple_property = <>
          if p = 1 and n = k:
                for i in range (0, n):
                     triple_tuples add <N_i, P, K_i >
          if n = 1 and p = k:
                for i in range (0, p):
                     triple_tuples add <N, P_i, K_i >

Output:   Tuple_property list
```

Figure 8.8 Pseudocode of sequential matching algorithm. *Source*: [23] / Springer Nature / CC BY 4.0.

This relation extraction algorithm was applied to 458 target sentences classified from approximately 14,000 articles. In total, 680 γ' solvus temperature instances were automatically extracted. After manually inspecting 329 randomly selected sentences, the Precision, Recall, and F_1-score for the relation extraction algorithm regarding γ' solvus temperature were 75.86%, 83.22%, and 79.37%, respectively.

This work also used the original "Snowball" algorithm [18] and modified "Snowball" algorithm [16] with seed sizes of 50 and 100 to extract property–tuple relations. The Snowball system is a semi-supervised algorithm designed to generate patterns and extract tuples from text documents, particularly in cases of limited labeled samples [18]. Snowball introduces a strategy for evaluating the quality of pattern and tuple extracted based on DIPRE [40]. Modified Snowball algorithm [16] is capable of handling quaternary relations and performs clustering based on the ordering and number of entities, achieving higher precision with fewer seeds than the original Snowball algorithm. In this study, Wang et al. [23] manually labeled 329 sentences containing γ' solvus temperature information and extracted 467 tuples in both binary and quaternary forms to evaluate the Snowball and modified Snowball algorithms. The binary tuples consist of the superalloy named entity, property value, and their contextual information between different entity categories, while the quaternary tuples include the superalloy named entity, property specifier, property value, property unit, and their contextual information between different entity categories. The initial number of seeds significantly impacts the algorithm's performance, so 50 and 100 manually labeled tuples were used as seeds to train the Snowball and modified Snowball systems, respectively. The trained Snowball and modified Snowball systems were then applied to extract relations from the remaining corpus, and the Precision, Recall, and F_1-score of both algorithms were calculated through manual inspection.

The parameters used for manual tuning to evaluate the Snowball and modified Snowball algorithms on the test set are provided in Tables 8.4 and 8.5, while Table 8.6 presents the Precision, Recall, and F_1-score of both algorithms for different seed sizes.

Table 8.4 Parameters used for evaluating Snowball on the test set.

Parameter	Value	Descriptor
τ_{sim}	0.5	Minimum degree of match
τ_t	0.5	Minimum tuple confidence
τ_{sup}	2	Minimum pattern support
I_{max}	4	Number of iterations of Snowball
W_{left}	0.2	Weight for the left context
W_{middle}	0.6	Weight for the middle context
W_{right}	0.2	Weight for the right context

Table 8.5 Parameters used for evaluating modified Snowball on the test set.

Parameter	Value	Descriptor
τ_{sim}	0.1	Minimum similarity between sentences
τ_t	0.1	Minimum confidence of a relationship
Prefix_weight	0.1	Weight of the prefix
Middle_weight	0.8	Weight of the middle
Suffix_weight	0.1	Weight of the suffix
Prefix_length	2	Number of tokens in phrase prefix
Suffix_length	2	Number of tokens in phrase suffix
Learning_rate	0.5	How fast new confidences update

Table 8.6 Precision, Recall, and F_1-score of the Snowball system and modified Snowball system.

Category		Precision	Recall	F_1-score	Validated on
Snowball	Seeds = 50	78.13%	21.46%	33.67%	329 Sentences
	Seeds = 100	78.87%	24.03%	36.84%	329 Sentences
Modified Snowball	Seeds = 50	60.71%	18.78%	28.69%	329 Sentences
	Seeds = 100	66.67%	32.04%	43.28%	329 Sentences

The method presented higher Recall and F_1-score than the "Snowball" algorithm, as shown in Table 8.6. The Recall for "Snowball" was lower than that of this method because the seed tuples used to initiate the learning process were too few. As a result, each cluster of sentence forms contained a limited number of training tuples, making the sentence clustering process in "Snowball" ineffective. This inefficiency prevented the algorithm from covering all sentence forms, and consequently, only a small number of tuple relations were extracted, resulting in very low Recall. In contrast, the distance-based algorithm, which does not require labeled samples to handle multiple-relation extraction, performed better in such small-corpus conditions.

5. **Table relation extraction**

Table parsing converts the complete table data, including the table caption and body, into a structured format, represented as a nested list of table cells. It then classifies which tables contain chemical composition and target property information to be extracted. As a result of the table parsing process, 5,327 composition tables and 114 tables containing solvus temperature data were obtained. Table relation extraction identifies specific tuple relations between the content of elements and their corresponding properties. For instance, during composition extraction, the table direction ("by row" or "by column") is first determined by estimating the row or column locations of the chemical elements within the table body. Next, the table caption is checked for the presence of any alloy named entities. Figure 8.9 illustrates the schematic diagram corresponding to this process. In the case of a "by row" approach, if multiple alloy named entities are detected or none are specified in the table caption, the alloy named entities and elements are identified based on the row and column indices of each table cell, referring to the alloy_named_entity$_{body}$ and element_heading sequences, respectively. However, if only one alloy named entity is found in the table caption, the elements are identified based on the column index of each table cell from the element_heading sequence. The final output of the table relation extraction process is a set of composition tuples. The pseudocode for

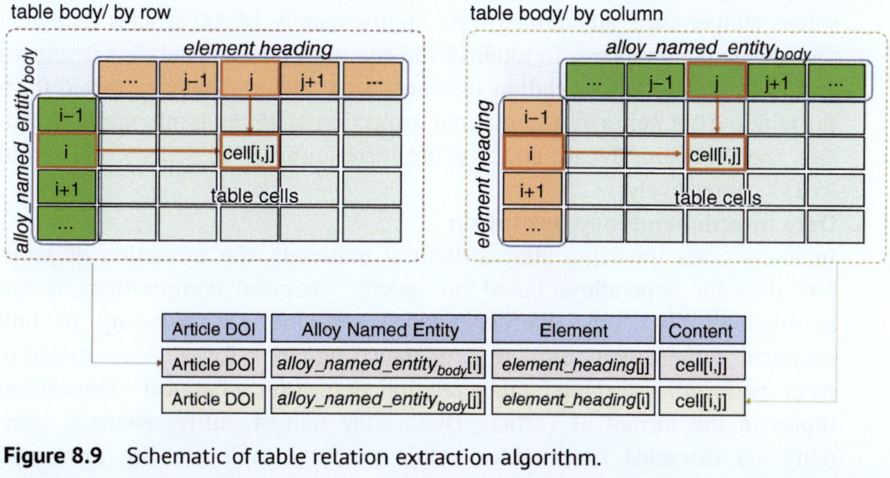

Figure 8.9 Schematic of table relation extraction algorithm.

```
Table relation extraction algorithm
Input:
    files — the table files in .xlsx format
Algorithm:
    read table_caption and table_body from files
    define Tuple_composition = <article_DOI, alloy_named_entity, element, content>
    // acquire alloy named entity in table caption
    alloy_named_entity_caption = alloy_named_entity_recognize (table_caption)
    // determine the pattern of the table
    // direction defines the table is 'by row' or 'by column'; index_target is the row or column location of element
    direction, index_target = judge_table_pattern (table_body)
    if alloy_named_entity_caption != 1:
        alloy_named_entity_body= alloy_named_entity_recognize_by_column (table_body)
        for each table cell with row index i and column index j:
            if direction == 'by row':
                Tuple_composition add <article_DOI,alloy_named_entity_body[i], element_heading[j], cell[i,j]>
            else:
                Tuple_composition add <article_DOI,alloy_named_entity_body[j], element_heading[i], cell[i,j]>
    else:
        for each table cell with row index i and column index j:
            if direction == 'by row':
                Tuple_composition add <article_DOI,alloy_named_entity_caption, element_heading[j], cell[i,j]>
            else:
                Tuple_composition add <article_DOI,alloy_named_entity_caption, element_heading[i], cell[i,j]>
Output: Tuple_composition list
```

Figure 8.10 Pseudocode of table relation extraction algorithm. *Source*: [23] / Springer Nature / CC BY 4.0.

the table relation extraction algorithm can be found in Figure 8.10. The property extraction process follows the same steps as composition extraction, with the only difference being that the objects identified by NER are drawn from the table headings instead.

The above relation extraction algorithm was applied to a total of 5,441 tables, comprising 5,327 tables on chemical composition and 114 tables on property γ' solvus temperature, all sourced from approximately 14,000 articles following the table parsing process. In total, 12,703 composition–tuple relation instances and 579 property–tuple relation instances were automatically extracted from the tables. After performing a manual inspection of 45 randomly sampled articles, the Precision, Recall, and F_1-score were found to be 90.89%, 100%, and 95.23%, respectively.

6. **Data interdependency resolution**

   In many cases involving alloy structural materials, the extraction of property data for superalloys based on specific chemical compositions is not permissible. Therefore, interdependency resolution is necessary to link chemical composition and property data fragments for a given material. After performing text and table relation extraction, chemical composition tuples in the format of ⟨article_DOI, alloy_named_entity, element, content⟩ are extracted from tables, while property tuples in the format of

⟨article_DOI, alloy_named_entity, property_specifier, property_value⟩ are derived from both text and tables. Among the approximately 14,000 articles, Wang et al. [23] automatically extracted a total of 12,703 chemical composition instances and 1,259 γ' solvus temperature instances, with 680 instances coming from text and 579 from tables. To integrate these data fragments, chemical composition and property instances associated with the same alloy named entity and article DOI are merged into a complete record, forming a six-tuple: ⟨article DOI, alloy named entity, element, content, property_specifier, property_value⟩. This integration is accomplished using a divide-and-conquer strategy, as illustrated in Figure 8.11. The detailed algorithm is depicted in Figure 8.12. For $\text{Tuple}_{\text{composition}}$ and $\text{Tuple}_{\text{property}}$ from text and table relation extraction, all the tuples are divided into three sets ($\text{DOIs}_{\text{intersection}}$, $\text{DOIs}_{\text{composition}}-\text{DOIs}_{\text{intersection}}$, and $\text{DOIs}_{\text{property}}-\text{DOIs}_{\text{intersection}}$) according to whether they possess the same article DOIs. For the tuples in the set $\text{DOIs}_{\text{composition}}-\text{DOIs}_{\text{intersection}}$, no corresponding property information is found. For the tuples in the set $\text{DOIs}_{\text{property}}-\text{DOIs}_{\text{intersection}}$, no chemical composition information is extracted. For the tuples in $\text{DOIs}_{\text{intersection}}$, chemical composition and property information are both extracted from one article at the same time. These tuples are further divided into three sets based on whether they share the same alloy named entity. Finally, tuples in set $\text{alloy\_named\_entity}_{\text{intersection}}$ are joined into the complete record with both chemical composition and property for one alloy named entity of each article DOI, while tuples in $\text{alloy\_named\_entity}_{\text{composition}}-\text{alloy\_named\_entity}_{\text{intersection}}$ or $\text{alloy\_named\_entity}_{\text{property}}-\text{alloy\_named\_entity}_{\text{intersection}}$ can only obtain chemical composition or property information for one alloy named entity. In total, for γ' solvus temperature property, 743 complete records were generated from 12,703 chemical composition instances and 1,259 property instances across the approximately 14,000 articles.

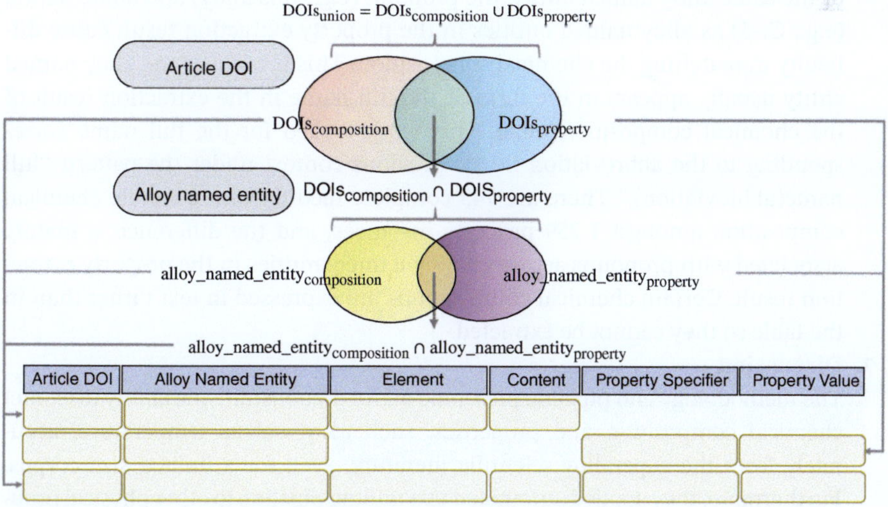

Figure 8.11 Schematic of interdependency resolution algorithm by divide-and-conquer strategy.

```
Interdependency resolution algorithm
Input:
    Tuple_composition and Tuple_property from text and table relation extraction
Algorithm:
    define Dict_all = <article_DOI, alloy_named_entity, element, content, property_specifier, property_value>
    // acquire all DOIs in Tuple_composition and Tuple_property, get the intersection and union between them
    DOIs_composition = gather_doi(Tuple_composition)
    DOIs_property = gather_doi(Tuple_property)
    DOIs_union = DOIs_composition ∪ DOIs_property
    DOIs_intersection = DOIs_composition ∩ DOIs_property
    For article_doi in DOIs_union:
        If article_doi in (DOIs_composition − DOIs_intersection):
            Dict_all[article_DOI] append <alloy_named_entity_composition, element, content>
        If article_doi in (DOIs_property − DOIs_intersection):
            Dict_all[article_DOI] append <alloy_named_entity_property, property_specifier, property_value>
    // acquire corresponding alloy named entities in Tuple_composition through the DOIs in DOIs_intersection
    alloy_named_entity_composition = name_from_tuple(Tuple_composition, DOIs_intersection)
    // acquire corresponding alloy named entities in Tuple_property through the DOIs in DOIs_intersection
    alloy_named_entity_property = name_from_tuple(Tuple_property, DOIs_intersection)
    alloy_named_entity_union = alloy_named_entity_composition ∪ alloy_named_entity_property
    alloy_named_entity_intersection = alloy_named_entity_composition ∩ alloy_named_entity_property
    For alloy_named_entity in alloy_named_entity_union:
        If alloy_named_entity in alloy_named_entity_intersection:
            Dict_all[article_DOI] append <alloy_named_entity, element, content, property_specifier, property_value>
        If alloy_named_entity in (alloy_named_entity_composition − alloy_named_entity_intersection):
            Dict_all[article_DOI] append <alloy_named_entity, element, content>
        If alloy_named_entity in (alloy_named_entity_property − alloy_named_entity_intersection):
            Dict_all[article_DOI] append <alloy_named_entity, property_specifier, property_value>
Output: Dict_all .json or .xlsx file
```

Figure 8.12 Pseudocode of interdependency resolution algorithm. *Source*: [23] / Springer Nature / CC BY 4.0.

It is worth mentioning that during the course of concatenating on the basis of the same alloy named entity, the pronoun (e.g., this alloy) and abbreviation (e.g., Cr-5) as alloy named entities in the property extraction result cause difficulty in matching the chemical composition. This is because the alloy named entity usually appears in the form of the full name in the extraction result of the chemical composition table. This work looked for the full name corresponding to the abbreviation in the previous context under the pattern "full name(abbreviation)." There are 743 complete records with concrete chemical composition amongst 1,259 property instances, and the difference is mainly associated with pronouns acting as alloy named entities in the property extraction result. Certain chemical compositions are expressed in text rather than in the table so they cannot be extracted.

7. **Discussion**

The methodology and pipeline presented above demonstrate the ability to extract chemical composition and properties, such as γ' solvus temperature, accurately from the superalloy scientific literature, even for a limited size corpus. Furthermore, they applied automated text mining pipeline to other physical properties of superalloys, including density, solidus temperature, and liquidus temperature, by regenerating the synonym dictionary of the property specifier based on the pretrained word embedding model and adjusting the writing rules for the value and unit. In total, this work obtained 5,136 property instances of γ' solvus temperature (1,259), density (2,296), solidus temperature (793), liquidus temperature (788), and 12,703 chemical compositions from 14,425 superalloy articles. The average Precision, Recall, and F_1-score for γ' solvus temperature, density, solidus temperature, and liquidus temperature are 83.67%, 93.08%, and 88.13%, respectively. Among these, 2,531 instances successfully matched with their chemical compositions, and the Precision, Recall, and F_1-score were validated on 30 randomly sampled articles for chemical composition and each property.

For a relatively small corpus such as superalloys, this work has presented a rule-based NER method and an effective distance-based heuristic multiple-relation extraction algorithm for the pipeline to overcome the drawback of limited training corpus labels. This work achieved an F_1-score of 92.07% for alloy named entity, and an average F_1-score of 77.92% for relation extraction for γ' solvus temperature, density, solidus temperature, and liquidus temperature. The pipeline does not require any labeled corpus to achieve high Precision and Recall, avoiding the underfitting problem of supervised and semi-supervised learning with low Recall caused by insufficient labeled corpus. Moreover, the common table processing tool with table-parsing and relation-extraction algorithm performed well with an F_1-score of 95.23%. Therefore, the methodology presented here is expected to perform well for subject-oriented IE, even for small corpus, as lack of adequate labeled data often creates problems in the use of supervised or semi-supervised learning methods.

From the perspective of superalloy development, Figure 8.13a shows the γ' solvus temperature trends of cobalt-based and nickel-based superalloys arranged by years. The reported highest γ' solvus temperature of a nickel-based superalloy is 1,308 °C in the year 2012 by Pang et al. [41], while that for the highest cobalt-based superalloy is 1,269 °C in the year 2017 by Lass [42], which was referenced by Li et al. in the year 2019 [43]. An Ashby chart showing the superalloys as a function of γ' solvus temperature and density is plotted in Figure 8.13b. The superalloys in blue circles with high Ni content and Ta content have a higher γ' solvus temperature than other superalloys, whereas superalloys in orange and pink circles without Ni content exhibit relatively low γ' solvus temperatures. This is consistent with the reported behavior [44, 45]. Figure 8.13c shows that the addition of W in the ternary superalloy Co–9Al–xW promotes an increase in the γ' solvus temperature. This is because W tends to accumulate in the γ'-phase and occupy B-sites of the A_3B ordered phase. Thus, for Co–9Al–10W, there is a distribution of values ranging from 980 to 1,060 °C in three articles. In Figure 8.13d, for the Co–Ni–Al–Mo-based superalloy, the γ' solvus temperature and density are significantly increased by the addition of Ta compared with Nb, which is consistent with the results reported by Lass [45]. After adding Ti, the value of γ' solvus temperature is further increased because Ti is a forming element of γ'-phase, and its promotion effect on the γ' solvus temperature is greater than that of Co. Therefore, the extracted data support the known behavior of superalloys.

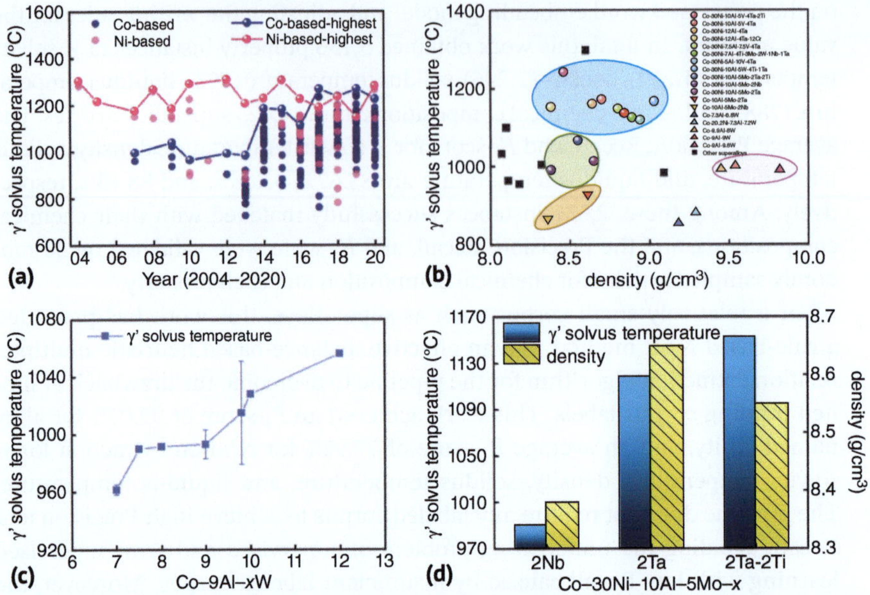

Figure 8.13 Materials insights from the extracted data. (a) Extracted dataset of γ' solvus temperature published from year 2004 to 2020. (b) Ashby chart for γ' solvus temperature and density data. (c) Effect of element W on γ' solvus temperature of Co–9Al–xW alloy. (d) Effect of different elements (Nb, Ta, and Ti) on γ' solvus temperature and density of Co–30Ni–10Al–5Mo–x alloy.

As latent knowledge regarding future discoveries can lie in past publications, this work next examined the values of the extracted data to provide actionable insights for materials discovery and constructed a data-driven model from the extracted 743 records with chemical composition and γ' solvus temperature. Of the 743 records, 259 cobalt-based and 73 nickel-based compounds were retained after data screening and cleaning by removing duplicates and errors.

After the automatic data extraction by the pipeline, 743 instances with both chemical composition and γ' solvus temperature were obtained. For some superalloys, the extracted γ' solvus temperatures for the same superalloy present differences. On the one hand, this work does not incorporate synthesis or processing conditions and other experimental aspects, including measurement uncertainties; on the other hand, for the same superalloy, some temperatures are calculated but some are by experiment, and some are reported in a range. Figure 8.14a presents the distribution of γ' solvus temperature for the Co–9Al–9W, Mar-M247, U720Li, IN738LC, CMSX-4, and CMSX-10 superalloys with γ' solvus temperatures of 993 ± 9, $1{,}206 \pm 28$, $1{,}160 \pm 20$, $1{,}168 \pm 31$, $1{,}286 \pm 30$, and $1{,}343 \pm 9$ **°C**, respectively, with means and standard deviations.

In order to use these data for further analysis, some data preprocessing steps were manually performed as follows:

1. When different γ' solvus temperatures for the same superalloy were extracted from both the tables and text, the data from the tables were prioritized and the text-based data were excluded.

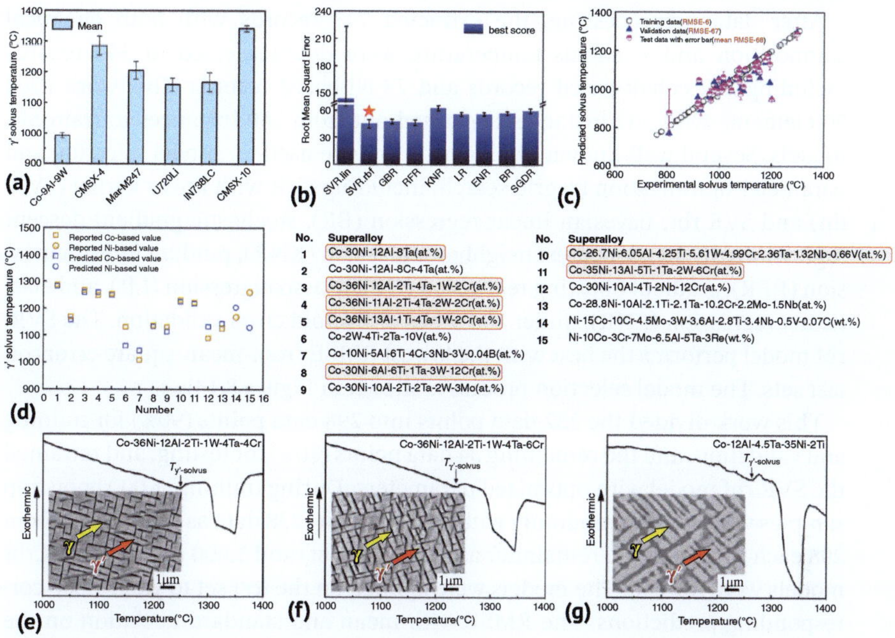

Figure 8.14 Results from a machine-learned model of γ' solvus temperature, and predictions for unexplored Co-based superalloys with high γ' solvus temperature. (a) The γ' solvus temperatures distribution for certain superalloys. (b) RMSE for model selection by fivefold cross-validation. (c) The predicted values compared to values extracted from the pipeline for the SVR.rbf model showing the behaviors of the training and test datasets. (d) The measured and predicted γ' solvus temperatures of 15 superalloys recently reported in 2020 and 2021 that were not part of the database. (e) The microstructure and DSC heating curve for alloy Co–36Ni–12Al–2Ti–1W–4Ta–4Cr. (f) The microstructure and DSC heating curve for alloy Co–36Ni–12Al–2Ti–1W–4Ta–6Cr. (g) The microstructure and DSC heating curve for alloy Co–12Al–4.5Ta–35Ni–2Ti. *Source*: [23] / Springer Nature / CC BY 4.0.

2. When simultaneously extracting the experimental and calculated γ' solvus temperatures for a specific superalloy, the experimental property value was retained, and the calculated data were excluded.
3. When multiple different γ' solvus temperatures for a specific superalloy were obtained from different articles simultaneously, the value with the highest frequency of occurrence was retained, and the other values were excluded.
4. When the γ' solvus temperature or composition value for a specific superalloy was presented as a range (e.g., 1,140–1,150 °C), the mean value (1,145 °C) of this range is retained.
5. The units of composition and γ' solvus temperature were standardized to atomic percentages and degrees Celsius, respectively.

The entire composition space consists of Co, Al, W, Ni, Ti, Cr, Ta, B, Mo, Re, Nb, Si, V, Fe, Hf, Ru, Ir, Cu, Pt, and C, and this work built a prediction model via support vector regression (SVR) with a radial basis function kernel (SVR.rbf) for γ' solvus temperature of superalloys.

After data preprocessing, the extracted 743 records with both chemical composition and γ' solvus temperature were down-selected to 340 records including 262 cobalt-based records and 78 Ni-based records. This work used 20 elements with 332 instances separated out from 340 instances to train ML models. Several well-known ML algorithms were used for model selection and parameter optimization by grid search, including SVR with linear kernel (SVR.lin) and SVR.rbf, Bayesian linear regression (BR), stochastic gradient descent regression (SGDR), *k*-nearest neighbor regression (KNR), random forest regression (RFR), gradient boosting regression (GBR), lasso regression (LR), and elastic net regression (ENR), under 100 times of fivefold cross-validation. The SVR.rbf model performs the best with the lowest RMSE (root-mean-square error) on test sets. The model selection process is shown in Figure 8.14b.

This work divided the 332 data points into 298 data points (90%) for training and validation, and the remaining 34 data points (10%) for testing, and retrained the SVR.rbf model with optimized parameters. During training, 1,000 bootstrap samples were chosen randomly with replacement (238 data as training set from 298 each time, and the remainder as validation set) and 1,000 different SVR.rbf models were trained. The models were applied on the test set to yield 1,000 corresponding predictions. The RMSE with mean and standard deviation on the test set is shown in Figure 8.14c (for the training set, the uncertainties are from the 1,000 bootstrap samples).

The reported 15 superalloys from 12 different subsequently published articles in the years 2020 and 2021 are shown in Table 8.7. The mean relative error between the reported and predicted γ' solvus temperatures by the SVR.rbf model is 2.27%.

Figure 8.14d shows the reported and predicted γ' solvus temperatures of the 15 superalloys reported, and the relative error of the temperatures in orange box is less than 1%. The microstructures and DSC heating curves for Co-36Ni-12Al-2Ti-1W-4Ta-4Cr, Co-36Ni12Al-2Ti-1W-4Ta-6Cr and Co-12Al-4.5Ta-35Ni-2Ti are shown in Figure 8.14e–g, respectively.

In summary, this work has proposed an automated data extraction pipeline for superalloys to generate a structural database by NLP, including scientific document download, preprocessing, table parsing and text classification, NER, relation extraction of text and table, respectively, and interdependency resolution automatically. The extracted entities with a total of 2,531 instances covering the physical properties of γ' solvus temperature, density, solidus temperature, and liquidus temperature, were compiled into a highly structured material database containing article DOI, alloy named entity, chemical element, content, property specifier, and property value. For subject-oriented text mining task of a small corpus like superalloys, a practical rule-based NER method and an effective heuristic multiple-relation extraction algorithm were proposed for the pipeline to overcome the obstacle of limited training corpus labels, and achieved an F_1-score of 92.07% for alloy named entity, and an average F_1-score of 77.92% for relation extraction of γ' solvus temperature, density, solidus temperature, and liquidus temperature. This work also introduced a general-purpose table processing tool that integrates table-parsing and relation-extraction algorithms,

Table 8.7 The reported superalloys from subsequently published articles.

Index	Superalloy	Reported γ' solvus temperature (°C)	Predicted γ' solvus temperature (°C)	Relative error (%)	Source
1	Co–30Ni–12Al–8Ta (at.%)	1,286	1,281.96	0.31	[46]
2	Co–30Ni–12Al–8Cr–4Ta (at.%)	1,169	1,156.80	1.04	[46]
3	Co–36Ni–12Al–2Ti–4Ta–1W–2Cr (at.%)	1,266.5	1,258.56	0.62	[21]
4	Co–36Ni–11Al–2Ti–4Ta–2W–2Cr (at.%)	1,247.6	1,258.41	0.86	[21]
5	Co–36Ni–13Al–1Ti–4Ta–1W–2Cr (at.%)	1,252.8	1,249.15	0.29	[21]
6	Co–2W–4Ti–2Ta–10V (at.%)	1,131	1,060.60	6.22	[47]
7	Co–5Al–10Ni–6Ti–4Cr–3Nb–3V–0.04B (at.%)	1,008	1,042.97	3.46	[48]
8	Co–6Al–3W–30Ni–6Ti–12Cr–1Ta (at.%)	1,135	1,132.49	0.22	[49]
9	Co–10Al–2W–30Ni–2Ti–2Ta–3Mo (at.%)	1,130	1,113.67	1.44	[50]
10	Co–6.05Al–5.61W–26.7Ni–4.25Ti–4.99Cr–2.36Ta–1.32Nb–0.66V (at.%)	1,230	1,224.49	0.44	[51]
11	Co–13Al–2W–35Ni–5Ti–6Cr–1Ta (at.%)	1,220	1,217.40	0.21	[52]
12	Co–10Al–30Ni–4Ti–12Cr–2Nb (at.%)	1,089	1,131.92	3.94	[53]
13	Co–28.8Ni–10Al–2.1Ti–2.1Ta–10.2Cr–2.2Mo–1.5Nb (at.%)	1,125	1,143	1.60	[54]
14	Ni–15Co–10Cr–4.5Mo–3W–3.6Al–2.8Ti–3.4Nb–0.5V–0.07C (wt.%)	1,166.8	1,203.3	3.12	[55]
15	Ni–10Co–3Cr–7Mo–6.5Al–5Ta–3Re (wt.%)	1,259.4	1,129	10.35	[56]
	Mean relative error			2.27	—

achieving strong performance with an F_1-score of 95.23%. The resulting database was then used to develop a data-driven model for predicting the γ' solvus temperature of superalloys published in 2020 and 2021—compounds not included in the original corpus. The model demonstrated excellent predictive accuracy, with a relative error of just 2.27%, underscoring the potential of leveraging historical scientific literature to generate actionable insights for materials discovery through text mining. The complete pipeline is publicly available at: https://github.com/MGEdata/SuperalloyDigger. A web-based toolkit is also available at: http://SuperalloyDigger.mgedata.cn, for online use. Automated text mining methods and tools to extract literature data for superalloys (and other alloy materials) have not previously been reported. The pipeline and toolkit are not customized for superalloys, so they present a general method for text extraction on alloy materials.

8.3 Alloy Synthesis and Processing by Semi-supervised Text Mining

The discovery of materials with targeted properties requires a seamless, integrated approach combining experiments, theory, and computations. The paradigm of Composition–Synthesis/Processing–Structure–Property–Performance [57, 58] often serves as a guide to exploration toward this task. The challenge in discovery is that the large materials space consists of innumerable combinations of components and structures, which are strongly determined by the potentially synthesized route [57]. Over the last few years, ML has guided the search for new materials using a data-driven approach [2, 59–63]. For example, materials synthesis is beginning to see dramatic improvements in efficiencies due to the integration of ML capabilities and robotic control of synthetic planning and automated experiments for flow reactors, photovoltaic films, organic synthesis, and perovskites by mobile robotics [64–69]. However, the design of alloys with desired properties requires dealing with not only increasing chemical and structural complexity, but also a myriad of processing routes. The materials space is just too vast for today's synthesis capabilities. Hence, this work successfully uses data science to identify optimal synthesis-design routes and produces a desired alloy from a specified set of starting materials [70, 71].

Early approaches toward materials data extraction from scientific articles have essentially been manual in nature [7]. The dramatic development of text mining and NLP techniques has made it possible to convert scientific text into ML-oriented data collections [16, 17, 72]. Recently, NLP pipelines for automatic data extraction from journal articles of chemical composition and properties of organic and inorganic chemical compounds, as well as superalloys and aluminum alloys, have been introduced [12–16, 23]. Alloy synthesis and processing information is usually described in the form of continuous events, and the actions are sequentially dependent. There are various types of actions, flexible expressions, and different conditions and parameters, and meanwhile, continuous synthesis and processing events are often mixed with a large number of discussions on experimental phenomena and intermediate products, which bring great challenges to actions and parameters extraction. Nevertheless, mature DL provides powerful capabilities to analyze unstructured data and identify features automatically. The well-documented libraries make use of DL more accessible. Kim et al. labeled 20 articles (~5,200 words) on oxide materials and trained a neural network to recognize synthesis parameters with an F_1-score of 81% [17]. Kononova and coworkers manually annotated the operation entities from 834 solid-state synthesis paragraphs of 750 papers and trained a BiLSTM network with a CRF layer (BiLSTM-CRF model) with an F_1-score of 90% [73]. Huo et al. designed a qualitative topic extraction method related to experimental protocols rather than recognizing detailed processing parameters. They clustered the sentences into topics and then trained a classification model to predict the latent topics of unseen experimental sentences with an F_1-score above 90% [74]. Despite numerous advantages, a DL model uses a few thousand to millions of parameters. To train a DL-based NLP NER and IE model requires many high-quality annotations. For alloys with structures and properties strongly determined by synthesis and processing routes, the limited amount of corpus and lack of high-quality annotations severely hinder accurate and efficient extraction.

As a core material for the most advanced aeroengines and industrial gas turbines, the synthesis and processing of superalloys impacts their design with desired microstructure and properties [19–21, 75]. An earlier study introduced an NLP pipeline designed to extract both chemical compositions and property data from 14,425 superalloy-related articles published prior to 2020 [23]. To address the challenge of limited labeled data, a rule-based NER approach combined with a distance-based heuristic IE algorithm was proposed, achieving high Precision and Recall without requiring large annotated corpora. Under such conditions, rule-based methods demonstrated favorable efficiency, particularly given the relatively low diversity of entity categories, which could be effectively handled through domain expertise. However, when extracting synthesis and processing information, the associated entity categories and their interrelations tend to exhibit greater complexity and variability in expression. In such scenarios, rule-based approaches often become inflexible and labor-intensive to maintain. Moreover, while supervised DL methods offer improved adaptability, they typically necessitate the creation of extensive labeled datasets, as well as costly reannotation when the target domain shifts. To balance scalability with annotation efficiency, recent efforts have adopted semi-supervised learning strategies, which leverage a small amount of labeled data alongside large volumes of unlabeled text to enhance model performance with reduced manual intervention.

This work introduces a semi-supervised text mining method to extract the parameters corresponding to the sequence of actions of synthesis and processing from a corpus. This makes it possible with less domain-specific experience and corpus annotation to achieve relatively high IE performance for superalloy synthesis and processing, i.e., this work extract details of the synthesis process. A semi-supervised recommendation algorithm for token-level actions and a multilevel bootstrapping algorithm for chunk-level actions are developed for a small corpus with few annotations so that a small number of seeds are required to initiate the learning process. The F_1-score of action entity recognition reaches 89.28%, much higher than the 74.95% achieved via the BiLSTM-CRF model. In total, 9,853 superalloy synthesis and processing actions with chemical compositions are automatically extracted from a corpus of 16,604 superalloy articles from Elsevier and other publishers.

To evaluate the accuracy and diversity of the extracted results, this work visualized the data from multiple perspectives to distill scientific insights and analyzed superalloy synthesis processes to determine which are of wide current interest, and show how temperatures for solution and aging treatments are correlated. This work also determined the transition probabilities from one action to another in a given synthesis process. A superalloy synthesis factor combining solution temperature (S_c), aging temperature (A_c), and aging time (A_t), in form of $(A_t \times S_c^{0.5})^{0.5} + A_c$, is inferred by symbolic regression (SR), illustrating a positive correlation with γ'-phase coarsening. This synthesis factor derived from text mining significantly improves the performance of the data-driven γ' size prediction model on the superalloys reported subsequently in 2023. Thus, semi-supervised text mining enables us to complement data-driven approaches for understanding the relationships between synthesis and structures.

1. **Extraction strategy**
   The core stages of automated text mining pipeline for superalloys synthesis and processing involve action dictionary generation, NER, and dependency parsing, in addition to several necessary NLP stages such as article retrieval and preprocessing, paragraph classification, table parsing, and interdependency resolution. The schematic overview of the synthesis extraction is shown in Figure 8.15. For scientific article retrieval and preprocessing, the raw archived corpus was parsed and organized in paragraphs. After paragraph classification, the paragraphs related to the concrete synthesis procedures were automatically selected. Action dictionary generation can generate token-level and chunk-level synthesis actions semiautomatically. NER methods are designed to recognize the action entities based on the generated dictionary. Dependency parsing establishes specific tuple relationships for actions and parameters in terms of latent semantics, and interdependency resolution resolves the linkage between chemical composition of mentioned samples and their actions in the specific synthesis process. Finally, the extracted superalloy synthesis information containing the article DOI, sample composition, and

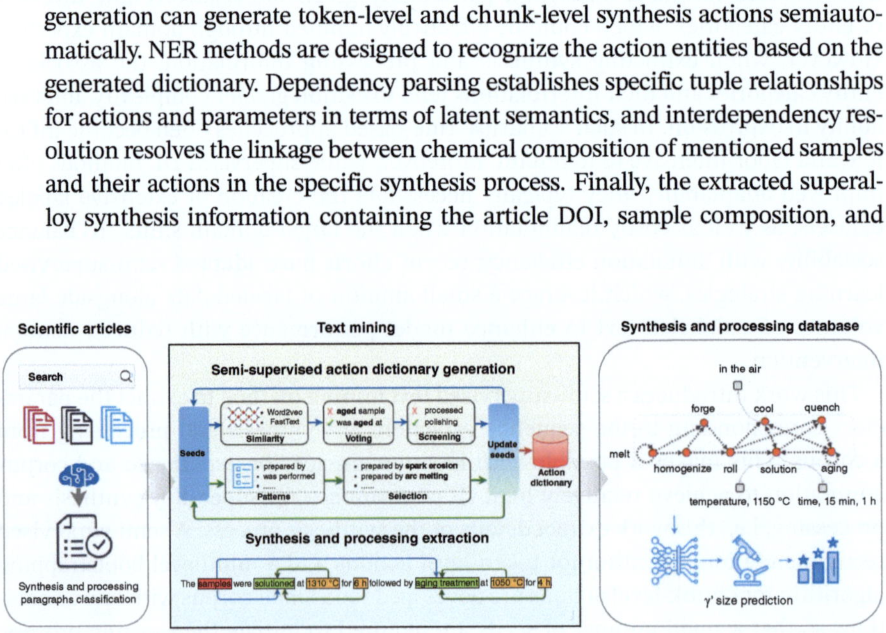

Figure 8.15 The schematic overview of synthesis extraction. *Source*: [76] / Springer Nature / CC BY 4.0.

synthesis action sequence with parameters is automatically compiled into structured (Comma-separated Values [CSV]) and semi-structured (JavaScript Object Notation [JSON]) formats to form a materials database for reuse.

2. **Semi-supervised action dictionary generation**
It is usual for target entities to be recognized by DL models trained on large corpora with hundreds of thousands of tokens. This requires accurate labels for each category of entity and appropriate annotation strategy for different types of corpuses [73, 77]. There are thousands of synthesis and processing superalloy actions discussed in the superalloy literature, although the number of articles is only about 16,000. The synthesis and processing actions in superalloy corpora are described in token-level and chunk-level entities depending on the phrase length, in contrast to chemical synthesis where mainly token-level action entities are involved. Moreover, according to the position of action in the superalloy process routine, the description for the same action is also different, e.g., aging treatment may exist in diverse forms such as primary aging, secondary aging, etc. This introduces further challenges in the manual fine labeling of entities. This work therefore proposes a semi-supervised method to allow for a rapid startup by generating a complete process action dictionary based on the literature corpus for further action entity recognition, which only needs to provide a small number of seed actions based on experience to go with the few rounds of manual checks. Table 8.8 gives some examples for token-level and chunk-level action entities with their POS (part-of-speech) and syntactic templates. Token-level action entities mainly involve nouns and verbs, whereas chunk-level entities belong to noun and verb phrases (NPs and VPs).

Table 8.8 Examples for the parsing grammar of token-level and chunk-level action entities.

Category	Entity POS	Syntactic templates	Examples
Token-level	VB	VB[DNP]\sTO <action>	... and allowed to **soak** for ...
	VBD/VBN/JJ	VB[DNP][\sRB]* <action>	... were mechanically **ground** ...
	VBG	VB[DN]\sIN <action>	... followed by **aging** at ...
Chunk-level	NP/VP	NP\sVB[GZ] <action>	... sample was **solution treated** ...
	NP/VP	VB[DGN]\sIN <action>	... prepared by **vacuum induction melting** ...
	NP	VB[DGN]\sTO <action>	... subject to a **solution heat treatment** ...
	NP/VP	IN\sNN\sIN <action>	... by method of **directional solidification** ...
	NP	<action> VBD\sVBN	... **solution heat treatment** was performed on ...

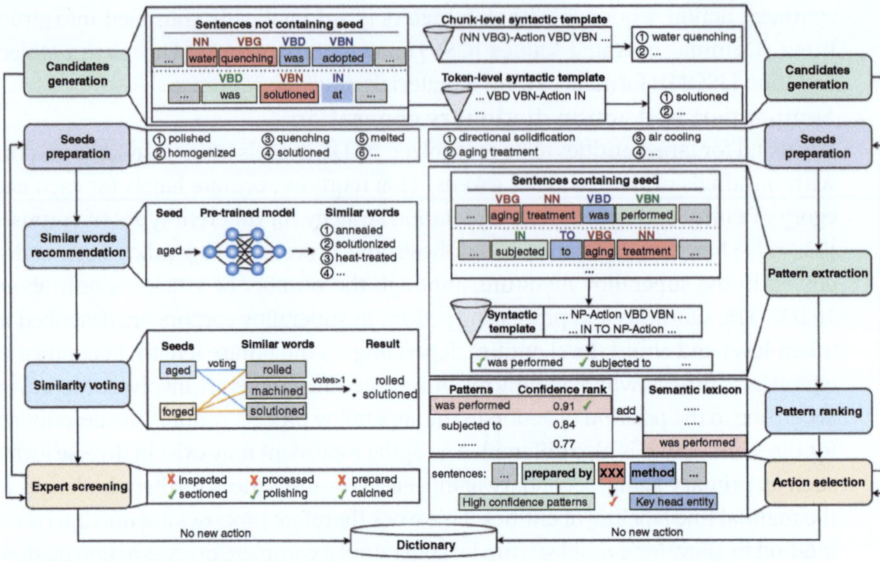

Figure 8.16 Schematic workflow of the semi-supervised action dictionary generation method. *Source*: [90] / Springer Nature / CC BY 4.0.

The proposed semi-supervised action dictionary generation method is shown in Figure 8.16. The token-level entity recommendation algorithm (left) involves four stages of candidate generation, seed preparation, similar words recommendation, similarity voting, and expert screening. The multilevel bootstrapping algorithm for chunk-level action (right) comprises the stages of candidate generation, seed preparation, pattern extraction, pattern ranking, and action selection.

3. **Token-level action recommendation algorithm**

Candidate entities, starting seeds, and two pretrained embedding models are required before execution of the token-level action recommendation algorithm. The candidate entities can provide the largest collection of potential actions according to the POS of the token entity and the syntactic template of its pre-context and post-context from all positive paragraphs, as illustrated in Table 8.8. As for starting seeds, the more seeds provided initially, the fewer iterations of the algorithm, but the overall generated dictionary size remains the same. Here, only 20 seeds are provided by experts to start the algorithm. The token-level starting seeds are further used to obtain similar words based on the word embedding models Word2Vec and FastText. These were initially pretrained on approximately 16,000 unlabeled full-text superalloy articles and used for calculating the cosine similarity between the candidate entity and starting seed. Word2Vec helps find the most similar syntactic and semantic words, and FastText constructs word embeddings from character-level *n*-gram representations to find words with greater morphological similarity [26, 78–80]. The details of word embedding models can be found in Word embedding Model part. Following the recommendation of similar entities, a voting process is performed to obtain action entities with high confidence. In this process, a seed is considered to have voted for the entity if the similarity by Word2Vec between the candidate entity and the seed is higher

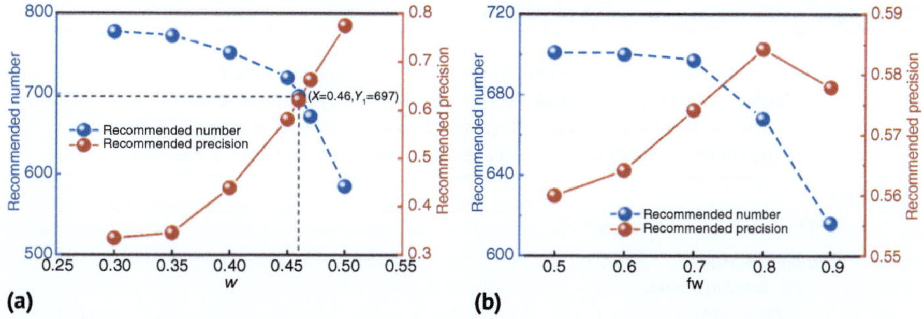

Figure 8.17 Parameter optimization for (a) *w* and (b) fw in semi-supervised token-level action recommendation algorithm. *Source*: [76] / Springer Nature/ CC BY 4.0.

than the threshold *w*, or if the similarity by FastText between the candidate entity and seed is higher than fw. The thresholds *w* and fw need to be optimized during the application of the algorithm. This affects the recommended precision and number of newly generated actions in each iteration. Thus, these two parameters are adjusted to trade-off a balance. The parameter optimization for *w* and fw in the semi-supervised token-level action entity recommendation algorithm can be found in Figure 8.17. As *w* and fw increase, the size of the generated dictionary gets smaller, and the recommended precision gets higher. When *w* is equal to 0.46, there are 697 recommended tokens with little manual intervention. When fw is equal to 0.7, the recommended number and precision are both high.

Entities that receive more than two votes and belong to a candidate are selected and passed to the expert for manual screening. The entities screened out by the expert are added back to the seeds and participate in the next iteration. If there is no new action entity generated and passed to the expert, the iteration will end. The pseudocode of the token-level actions generated algorithm is shown in Figure 8.18. The final generated dictionary for token-level actions contains 717 action entities (including 20 starting seeds and recommended actions) and is used for the subsequent action entity recognition.

4. **The multilevel bootstrapping algorithm for chunk-level actions**
The original multilevel bootstrapping algorithm was published by Ellen to generate a dictionary for NPs, such as the names of people, companies, or locations [81, 82]. This work first extends this multilevel bootstrapping algorithm to generate suitable NPs and VPs for synthesis and processing action entities. Moreover, the original algorithm assigns scores to entities according to the type of pattern, not taking the extraction frequency of each entity into consideration, so it is hard to distinguish different entities obtained from the same pattern. It improves this multilevel bootstrapping algorithm to generate both the collections of chunk-level action entities and patterns simultaneously and optimize the entity confidence score method. The modified multilevel bootstrapping algorithm for chunk-level actions also starts with candidate generation and seed preparation. The candidate chunk-level entities, which are generated by the POS of chunk-level actions and the syntactic template of its pre-context and post-context in Table 8.8, can provide the largest collection of potential actions in the action selection stage.

Token-level entity recommendation algorithm

```
Input:
        seeds — The verbs initially provided by expert for the algorithm to start
        candidates — The verbs that have the context match the syntactic template.
        wv_t — The similarity threshold for voting based on embedding similarity of Word2vec
        ft_t — The similarity threshold for voting based on embedding similarity of FastText

Algorithm:
        tuple_recoms= <>; tuple_Sims_all= <>

        Iteration starts:
          For seed in seeds:
            sims_seed = <>
            top_n = get_similar_word(seed) // Get 300 most similar words by Word2vec
            For word in top_n:
              wv_sim = cosine_similarity(seed,word) // Compute similarity by Word2vec
              ft_sim = cosine_similarity(seed,word) // Compute similarity by FastText
              If wv_sim > wv_t,or ft_sim > ft_t:
                tuple_Sims_all add word
          For ent in tuple_Sims_all:
            If tuple_Sims_all.count(ent) > 2 and ent in candidates:
              tuple_recoms add ent
          post_processed_tuple_recoms = screen_by_expert(tuple_recoms)
          If Post_processed_tuple_recoms:
            seeds add post_processed_tuple_recoms
          else:
            Iteration ends

Output:   seeds list
```

Figure 8.18 Pseudocode of token-level entity recommendation algorithm. *Source*: [76] / Springer Nature / CC BY 4.0.

The starting seeds are provided by experts and contain both NPs and VPs, here 20 seeds picked from five articles were manually provided.

A pattern is the word sequence in front of or behind the seed entity, and it is the most important contextual feature for chunk-level action generation. Pattern extraction is designed to find all patterns that have a co-occurrence constraint with seeds and are also consistent with the syntactic template. The corresponding sentence that contains the seed is divided into three parts: "before," the seed, and "after." The "before" is a sequence of tokens located in front of the seed entity and the "after" is a sequence of tokens behind the seed. The parameter of window_size is used to control the "before" and "after" size and depends on the length of the syntactic templates. For example, the window_size of syntactic template IN\sNN\sIN is 3 and that of VB[DNP]\sTO is 2. If the POS of a "before" or "after" matches any customized syntactic template, the corresponding "before" or "after" sequence will be added to the set of patterns. The patterns are obtained when all the sentences that contain seeds are traversed during an iteration.

Pattern ranking recommends the best pattern amongst all patterns after pattern extraction. During each iteration, the confidence associated with the pattern is calculated with Eqs. 8.3 and 8.4 [81, 82], where Pattern$_i$ stands for the ith pattern, F_i is the number of unique seeds hit by Pattern$_i$, N_i is the total number of unique chunk-level entities that Pattern$_i$ can extract, R_i represents the precision (probability) of the pattern to extract relevant information, and $\text{Score}(\text{Pattern}_i)$ balances the reliability (R_i) and frequency (F_i),

$$R_i = F_i/N_i \tag{8.3}$$

$$\text{Score}(\text{Pattern}_i) = R_i \times \log_2 F_i \tag{8.4}$$

The best pattern with the highest confidence score will then be added to the semantic lexicon. Table 8.9 shows the best patterns in the first six iterations of mutual bootstrapping.

After pattern ranking, the patterns in the semantic lexicon are then used to select optimal chunk-level actions. From a probability perspective, an entity that is hit by more patterns will be more likely to be an action entity. Therefore, another confidence score, $\text{Score}(\text{Entity}_i)$, is defined to reflect the frequentist likelihood that a candidate chunk-level entity is extracted by patterns in the semantic lexicon, shown in Eq. 8.5,

$$\text{Score}(\text{Entity}_i) = \sum_{k=1}^{N_i} (1 + \text{Score}(\text{Pattern}_k) + k \times \text{Count}_{\text{seed}}) \tag{8.5}$$

Table 8.9 The best extraction patterns in the first six iterations.

Iteration	Best pattern	Confidence score	Extracted entities
1	1. Terminated by 2. Followed by 3. Of superalloy during 4. Prepared via	1. 2.456 2. 2.036 3. 1.0 4. 0.925	"Furnace cooling," "air-cooling," "water quench," "water cooling," "ultrasonic cleaning," "air cool," "water-quenching," "oil quenching," "mechanical polishing," "final polishing"
2	Following by	0.773	"Rapid cooling," "water quenched," "cloth polishing," "dimple grinding," "heat treatment," "hot isostatic pressing," "two-step aging treatment," "twin-jet electropolishing," "vibratory polishing," "standard heat treatment," "ion milling," "vacuum arc remelting," "air quench," "electro-polishing"
3	Consolidated via	2.0	"Air quenching," "20 min cooling," "annealing process," "electrolytic polishing," "electrolytic etching," "solution treatment," "two-step ageing," "diffusion heat treatment," "long-term thermal exposure," "gas fan quenching," "controlled cooling," "electro-etching," "fine polishing," "natural cooling," "immediate water quenching," "secondary aging," "electrochemical polishing," "vibration polishing," "diffusion treatment," "two-stage aging process," "twin-jet electrochemical polishing," "gas fan cooling," "slow cooling," "forced air cooling," "long term thermal exposure"

(Continued)

Table 8.9 (Continued)

Iteration	Best pattern	Confidence score	Extracted entities
4	From bars after	1.584	"Hot isostatically pressing," "diamond polishing," "chemical etching," "rapid quenching," "oil cooling," "ion beam thinning," "drop-casting," "vacuum heat treatment," "furnace cooled," "ice water quenching," "double aging," "isothermal hold," "annealing treatment," "cold water quenching," "chemical etch," "furnace cool," "two stage ageing," "cold-water quenching," "hot rolling," "sequential polishing"
5	Treated via Treatment contained Purified via Treatment involves Re-prepared by	1. 1.0 2. 1.0 3. 1.0 4. 1.0 5. 1.0	"Two stage ageing process," "thermomechanical treatments," "rapid gas fan quench," "surface grinding," "electrolytic double jet thinning," "two bar argon cooling," "standard polishing," "vacuum annealing," "argon gas quenching," "ebsd scanning," "second aging," "two step aging heat treatments," "twin-jet electro-polishing," "controlled cool," "isothermal forging," "gas cooling," "electro-slag re-melted," "electrochemical etching," "ageing treatment," "final air cooling," "argon quenching," "continuous cooling," "air cooled," "spray quenching," "immediate water quench," "room temperature brine quench," "jet electropolishing," "rapid air cooling," "fan cooling," "hot-air drying," "double ageing," "final polish," "fine colloidal silica polishing," "oil quench," "hot-isostatic pressing," "subsequent forging," "precipitate aging heat treatment," "hot working," "furnace-cooling," "rapid water cooling," "10 min cooling"
6	Subjected to	0.792	"Twin-jet polishing," "solid solution treatment," "aging heat treatment," "homogenization treatment," "homogenization heat treatment," "long-term aging," "primary aging," "optical microscopy," "light grinding," "argon ion milling," "thermal treatment," "aluminizing treatment," "cold rolling," "further heating," "two-stage aging," "tensile tests," "several heat treatments"

where N_i is the number of patterns in the semantic lexicon that can successfully extract the entity. For each pattern, $\text{Count}_{\text{seed}}$ is the number of seeds that can match the pattern. The k is the weight of $\text{Count}_{\text{seed}}$ and affects the number and precision of recommended entities in each iteration. This also requires to be optimized to get a better token dictionary with higher accuracy and a larger size. The $\text{Score}(\text{Pattern}_k)$ can be calculated using Eq. 8.4. The entity with the highest score is then likely to be selected.

Whether the entity with the highest score can eventually be added back to the seeds depends on the following constraints: a minimum confidence threshold, T_c, for action selection, with the entity satisfying the criteria score $\geq T_c$, and the new entity is lemmatized and added back to the seeds for next iteration. If the highest Score(Entity$_i$) in an iteration falls below T_c, the iteration ends. The parameter optimization for T_c and k in the multilevel bootstrapping algorithm can be found in Figure 8.19. When $T_c = 2$ and $k = 1$, the recommended number and precision of chunk-level actions are both high. The pseudocode of the multilevel bootstrapping algorithm is shown in Figure 8.20.

The semi-supervised recommendation algorithm and multilevel bootstrapping algorithm, used for generating token-level and chunk-level action entities, were applied to a total of 14,487 target paragraphs, which were classified from approximately 16,000 articles. As a result, 697 new token-level action entities (in addition to the 20 initial seed entities) and 1,199 chunk-level action entities were compiled into the action dictionary.

This action dictionary can then be utilized for NER of synthesis and processing actions within the superalloy corpus. Figure 8.21 illustrates the action NER workflow, which involves POS tagging, POS screening, and relaxed matching. The input sentence is first parsed using POS tagging to identify all verbs, NPs, VPs, and their respective contexts. The tagging results are then passed into the POS screening process to identify entities that align with contextual POS rules and share a similar context to the actions listed in the dictionary, as demonstrated during the semi-supervised action dictionary generation. Since the same action can be expressed in various ways (e.g., "aged," "aging treatment," and "two-step aging"), a relaxed matching strategy proves particularly effective in recognizing more latent entities during NER. If an entity, after POS screening, matches a chunk-level action or contains token-level actions from the generated dictionary, it will be recognized as an action entity.

In addition to action entities, synthesis parameters with units are also crucial for synthesis and processing. Regular expressions were employed to recognize these entities, as previously reported [23]. This study also compared the action entity extraction method with the BiLSTM-CRF model. The action NER was validated on 348 sentences (approximately 8,800 words), which were randomly

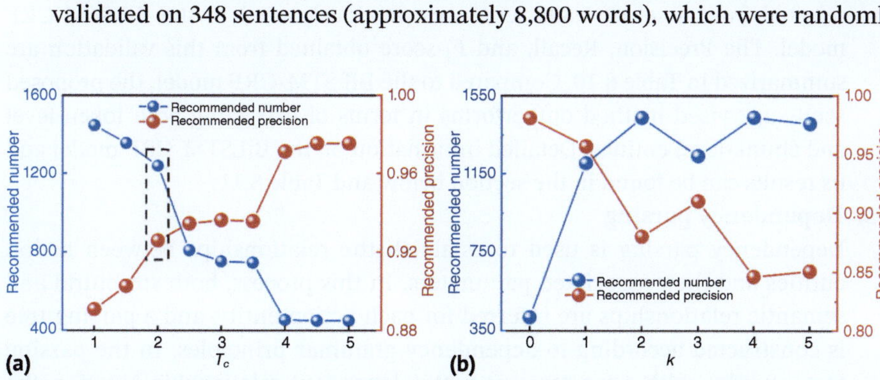

Figure 8.19 Parameter optimization for (a) T_c and (b) k in the multilevel bootstrapping algorithm. *Source*: [76] / Springer Nature / CC BY 4.0.

```
Multi-level bootstrapping algorithm
Input:
        seeds — the starting chunk-level entities provided by expert
        sentences — all superalloy corpus belongs to synthesis paragraphs
        action_dict — the dictionary of actions entities
        syntactic_template — syntactic grammar used to match the context of chunk-level entity
        pos_rules — rules of part-of-speech for actions
        window_size — the size of context around entity
        τ_c— confidence threshold of chunk-level entity
        key_heads — confidence threshold of chunk-level entity
Algorithm:
        dict_ep= {}; all_extraction_patterns = []
        sents_pos = pos_tag(sentences) // Obtain the POS of each word in the sentence
        candidate_entities = get_candidate_entities(sentences, sents_pos, pos_rules)
        For each iteration:
            For seed in seeds:
                For sentence in sentences:
                    pos_bef, pos_after, contexts_bef, contexts_after= obtain_context_of_seed
                                                    (seed, sentences, sents_pos, widow_size)
                    If pos_bef matches any template in syntactic_template:
                        dict_ep [seed] add contexts_bef
                    elif pos_aft matches any template in syntactic_template:
                        dict_ep [seed] add contexts_aft
            best_ep = get_best_patterns(dict_ep) ) // Obtain patterns with highest confidence score
            all_extraction_patterns add best_ep
            candidate_entities = get_best_entities (candidate_entities, all_extraction_patterns, τ_c)
            methods = actions_selection(seeds, action_dict, key_heads, candidate_entities)
            If not methods:
                Iteration ends
            else:
                seeds add methods
Output: seeds list
```

Figure 8.20 Pseudocode of multilevel bootstrapping algorithm. *Source*: [76] / Springer Nature / CC BY 4.0.

selected from a set of 1,308 sentences previously used in the BiLSTM-CRF model. The Precision, Recall, and F_1-score obtained from this validation are summarized in Table 8.10. Compared to the BiLSTM-CRF model, the proposed semi-supervised method outperforms in terms of extracting both token-level and chunk-level entities. Detailed information on the BiLSTM-CRF model and its results can be found in the section below and Table 8.11.

5. **Dependency parsing**

   Dependency parsing is used to establish the relationships between action entities and their associated parameters. In this process, both structural and semantic relationships are inferred for each action entity, and a parsing tree is constructed according to dependency grammar principles. In the parsing tree, an edge with a tag represents the dependent relationship between the starting entity and its subordinate entity. The tag on the directed edge indicates the syntactic role in the dependency relation. As shown in Figure 8.22,

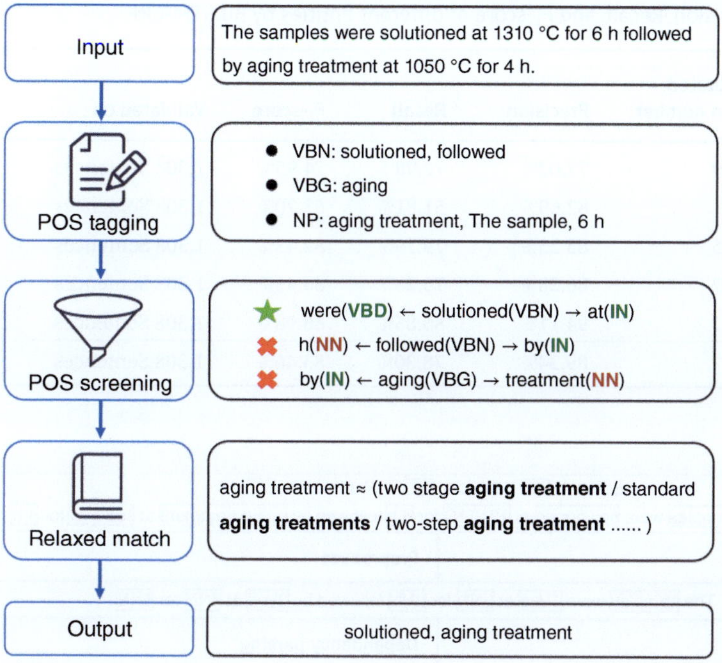

Figure 8.21 NER for action entities by POS tagging, POS screening, and relaxed match.

Table 8.10 Precision, Recall, and F_1-score of the action NER by this work and BiLSTM-CRF.

Category		Precision	Recall	F_1-score	Validated on
Action	This work	90.58%	88.03%	89.28%	348 Sentences
	BiLSTM-CRF	77.02%	72.98%	74.95%	1,308 sentences (fivefold cross-validation)
Parameter (with unit)	This work	98.49%	94.91%	96.67%	348 Sentences
	BiLSTM-CRF	94.11%	80.85%	86.98%	1,308 sentences (fivefold cross-validation)

after performing NER, the sentence is parsed into subject, action, and parameter entities. Based on the entity category and its POS, the original action and parameter entities are replaced with more standardized formats to ensure accurate capture of structural and semantic relations. The replacement follows these rules: for entities with a POS of VBN or VBD, the entity is replaced by "id" + "Ved" (e.g., "1Ved"), where "id" refers to the order in which the entity appears in the sentence. Entities with a POS of VBG are replaced by "id" + "Ving" (e.g., "2Ving"), and for NP entities, the format is "id" + "NP" (e.g., "1NP").

Table 8.11 Precision, Recall, and F_1-score of different entities by BiLSTM-CRF.

Category	Annotated token number	Precision	Recall	F_1-score	Validated on
ACT	2,459	77.02%	72.98%	74.95%	1,308 Sentences
MAT	533	82.69%	51.81%	63.70%	1,308 Sentences
DSC	1,365	85.25%	79.79%	82.43%	1,308 Sentences
PRO	1,277	86.28%	75.28%	80.41%	1,308 Sentences
PV	4,367	94.11%	80.85%	86.98%	1,308 Sentences
In total		89.34%	78.30%	83.46%	1,308 Sentences

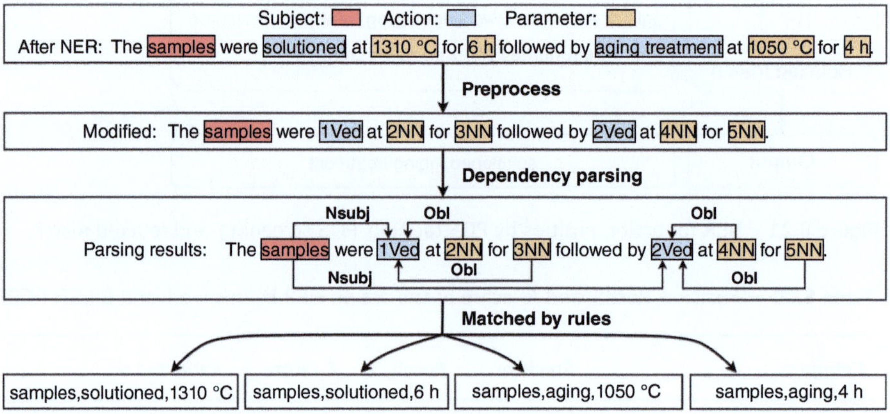

Figure 8.22 Schematic overview of dependency parsing process.

After preprocessing, the sentence is parsed to obtain a dependency parse tree and three-tuple relations among entities using the Stanford CoreNLP package [83]. Among the various types of relations, Nsubj is defined as the relationship between a subject and a verb, which represents the relation from the sample to the action when the verb corresponds to an action entity. Obl is defined as the relationship between an object and a verb, representing the relation from a parameter to an action when the verb is part of the action entity and the object corresponds to the parameter entity. This matching rule is used to interpret the results of the dependency parsing and yield three-tuples with the target entities.

In total, this study has captured 55,206 actions from 16,604 superalloy articles. Of these, 13,211 actions are associated with specific synthesis parameters. The Precision, Recall, and F_1-score for action extraction are presented in Table 8.12, based on manual validation of 30 randomly sampled articles. For each paragraph, the captured action tuples with parameters can be linked into an action sequence, such as ⟨arc melted → homogenized → rolled → solutioned (1,250 °C, 5 h) → air cool → aged (1,000 °C, 3 h) → air cool⟩.

Table 8.12 Precision, Recall, and F_1-score of token-/chunk-level actions and parameters.

Category	Extracted instance number from all superalloy articles	Precision (%)	Recall (%)	F_1-score (%)	Validated on
Token-level action	25,243	92.85	90.28	91.55	30 Articles (348 sentences)
Chunk-level action	29,963	88.23	85.71	86.95	30 Articles (348 sentences)
Action + parameter	13,211	81.31	84.09	82.68	30 Articles (348 sentences)
Action in total	55,206	90.58	88.03	89.28	30 Articles (348 sentences)

The concrete chemical composition information for each sample was obtained from tables by SuperalloyDigger pipeline [23]. From the ~16,000 articles, it automatically extracted a total of 20,368 chemical composition instances. This work performed interdependency resolution to map the composition of samples with the synthesis and processing routine. In total, 9,853 complete records combining composition data with corresponding synthesis actions and parameters were constructed by integrating 20,368 composition instances and 13,211 action–parameter instances. The proposed semi-supervised extraction strategy, together with tree-based dependency parsing, effectively addresses the challenges posed by limited labeled corpora, enabling the accurate capture of synthesis and processing information for superalloys. The extracted results are subsequently evaluated, and the structured data are visualized to facilitate scientific interpretation and reveal underlying trends.

The coverage and diversity of the extracted results are first analyzed through visualization, with a focus on temporal distribution and the categorization of synthesis and processing actions. The heatmap (Figure 8.23) illustrates the frequencies of various actions, such as quenching, aging, cutting, solution treatment, and cooling, across the years 2004–2021. For each category (the row in Figure 8.23), there are multiple subcategories of synthesis actions. For instance, in the "casting" category, subcategories include "investment casting," "ingot casting," and "single crystal casting." The semi-supervised text mining approach demonstrates a strong ability to capture the diverse linguistic expressions associated with synthesis and processing actions. Analysis of the temporal trends reveals a significant increase in research activity related to superalloy technologies since 2013, with growing attention on processes such as quenching, aging, cutting, solution treatment, cooling, and polishing. These heat treatment processes are critical in influencing phase morphology and the resulting mechanical properties of superalloys. The selection of optimal treatment parameters continues to be an area of active research, as they have a direct impact on

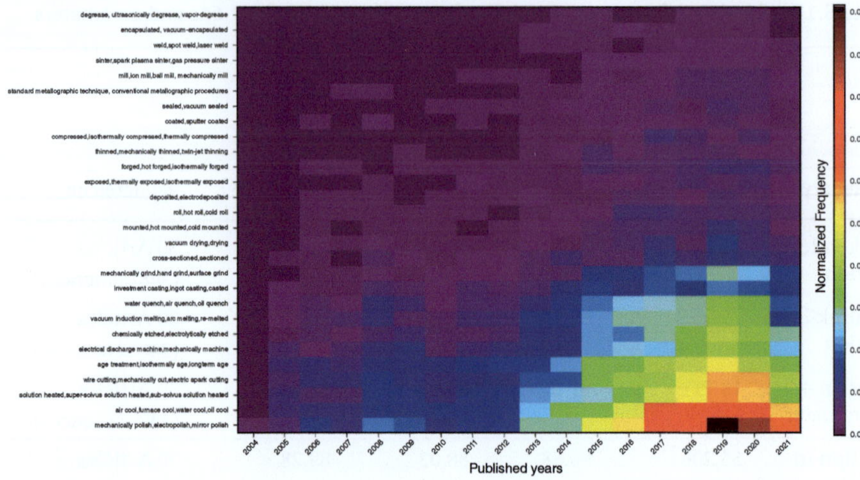

Figure 8.23 A heatmap of the frequencies of actions reported from 2004 to 2021.

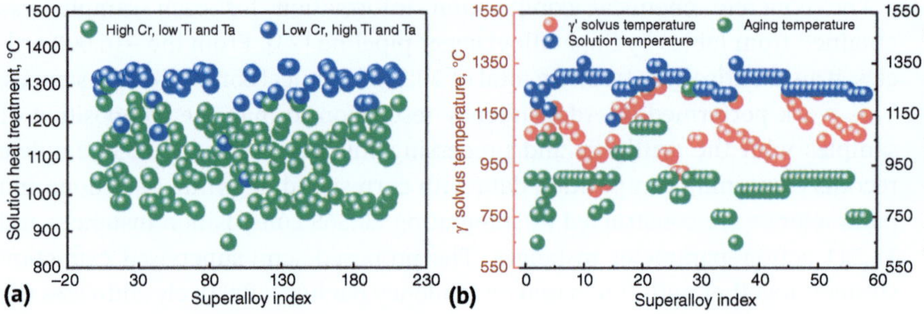

Figure 8.24 Solution and aging treatment temperatures for various superalloys. (a) The solution temperatures dependent on the compositions of Cr, Ti, and Ta. (b) The γ' solvus, aging treatment, and solution treatment temperatures for each alloy.

microstructural features—such as size and distribution—and key performance metrics like strength, hardness, and ductility.

Temperature plays a crucial role in ensuring the proper melting of precipitated phases and the precipitation of strengthening phases during synthesis. Figure 8.24 depicts temperatures during solution and aging treatments for various superalloys. In Figure 8.24a, the extracted dataset, which includes both solution temperature and composition data, is categorized into two classes based on the relative contents of Ta, Ti, and Cr. The "High Cr, Low Ti and Ta" class refers to superalloys with Cr content higher than 6%, and the combined Ta and Ti content not lower than 3%. Conversely, the "Low Cr, High Ti and Ta" class includes superalloys with Cr content lower than 6%, and the combined Ta and Ti content higher than 3%. Superalloys with low Cr and high Ti and Ta tend to have higher solution temperatures compared to those with high Cr and low Ti and Ta. This is because the γ' solvus temperature increases significantly with the addition of Ta and Ti, but decreases with Cr. As a

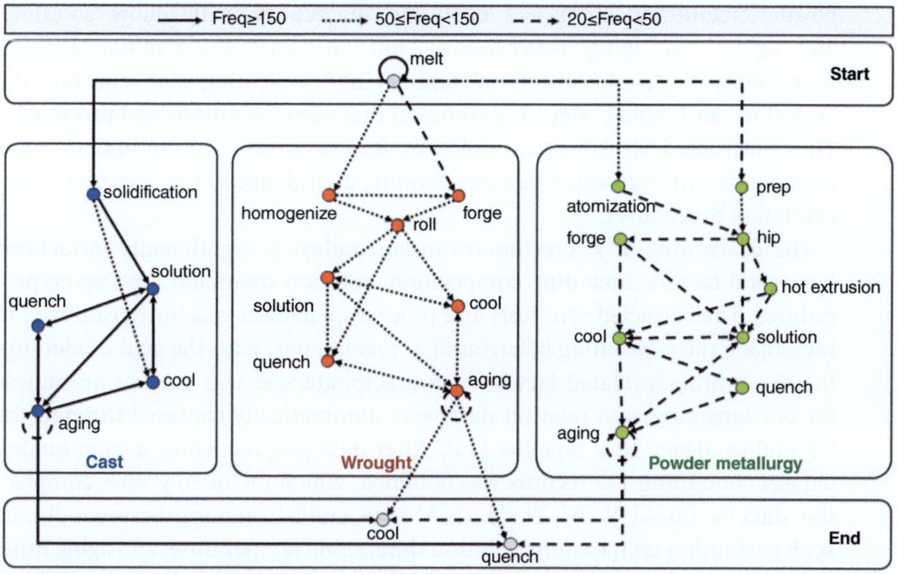

Figure 8.25 The flowchart of transition probabilities from one action to another for cast, wrought, and power metallurgy superalloys.

result, alloys with high Ti and Ta require higher solution temperatures (above 1,250 **°C**), consistent with the results reported by Chen et al. [46]. Coupled with the previously extracted γ' solvus temperature dataset [23], Figure 8.24b shows the relationship between γ' solvus temperature, aging treatment, and solution treatment temperatures. As expected, the γ' solvus temperature for each alloy lies between its solution and aging temperatures, consistent with well-established rules [84].

In materials synthesis, experimental steps typically follow a specific sequence defined by the underlying synthesis methodology. To capture this procedural structure, the extracted actions were linked in chronological order, forming an action sequence for each article. Transition probabilities between successive actions within these sequences were then computed, enabling the construction of a Markov chain representation that illustrates common pathways and transitions among experimental steps, as shown in Figure 8.25. In the flowchart, the directed graph consists of nodes and directed edges, where each node represents an experimental action and each edge represents a transition from one action to another. A solid black line indicates that the transfer from one action to another occurs with a frequency of ≥ 150, while two dashed lines represent transfer frequencies of 50 ≤ frequency < 150 and 20 ≤ frequency < 50. Higher transfer frequencies indicate a greater likelihood of inferring the subsequent action from the previous action in the synthesis process.

Figure 8.25 presents three types of synthesis processes: casting, wrought, and powder metallurgy. The extracted data also reveal sequences with adjacent relations, such as in casting: "solidification" → "solution" → "aging" → "quenching" → "solution" → "aging," etc., which align with expert intuition. For powder metallurgy, actions like "cool" and "quench" typically follow "solution" and "aging," but "aging" never follows "hot extrusion." Bidirectional edges are found between "cool," "solution," and "aging," indicating that repeating the "solution" and "aging" steps is a common practice in synthesis and processing. The constructed Markov chain in Figure 8.25 captures the experimental steps across different synthesis processes, lending confidence to the accuracy of the extraction procedures.

The coarsening of γ' precipitates in superalloys is significantly influenced by several factors, including composition, solution treatment, and aging procedures. The extracted synthesis and processing dataset was further utilized to investigate the coarsening behavior of γ' precipitates, with the goal of identifying key synthesis-related factors. The precipitate size was used as an indicator of coarsening, and relevant data were automatically captured through the SuperalloyDigger NLP pipeline [23]. After data preprocessing, a high-quality dataset containing 137 records was obtained, which included γ' size, composition data for Co, Al, W, Ni, Ti, Cr, Ta, Mo, Re, and Nb, and synthesis conditions such as solution temperature, solution time, aging temperature, and aging time. SR was then adopted to capture the explicitly expressed synthesis factor for γ' size by genetic programming SR (GPSR) implemented in the gplearn code [85]. The complexity and mean absolute error (MAE) were used as metrics to evaluate the generated formulas. Here, complexity refers to the number of arithmetic operators, including addition, subtraction, division, multiplication, and square root. A total of 60,000 candidate formulas were generated and sorted using MAE under the same complexities, as shown in Figure 8.26a. There are nine mathematical formulas (marked A–I in Figure 8.26a) located at the Pareto front with low complexity and MAE. The specific formulas are shown in Table 8.13. While the accuracy of these formulas is limited due to the white-box modeling approach, the frequent appearance of terms such as A_t, S_c, A_c, and the term $(A_t \times S_c^{0.5})^{0.5}$ suggests that these factors may significantly influence γ' size. To further investigate their relationship with γ' size, scatter plots were created using certain terms coupled with these three synthesis parameters. In particular, the terms $(A_t \times S_c^{0.5})^{0.5} + A_c$, $(A_t \times S_c^{0.5})^{0.5} + A_c{}^{0.5}$, and $(A_c \times A_t)^{0.5} + S_c^{0.5}$ were considered in order to study their relationship with γ' size through a scatter plot. In particular, $(A_t \times S_c^{0.5})^{0.5} + A_c$ shows a positive correlation, i.e., exponential growth, with γ' size as shown in Figure 8.26b.

The study of precipitate evolution is crucial for materials design. Classical physical models are used to predict the ripening behavior of particles. The LSW (Lifshitz–Slyozov–Wagner) theory assumes a very dilute environment, with no interactions between particles, to predict the ripening behavior of γ' precipitates [86]. Ardell extended this framework by incorporating the influence of finite precipitate volume fraction into diffusion-controlled coarsening kinetics, resulting in the modified LSW (MLSW) theory. According to classical kinetic theory, γ' size coarsening without coalescence is predicted to obey Eq. 8.6 [87–89]:

$$\langle r_t \rangle^3 - \langle r_0 \rangle^3 = \frac{8D_0 \exp\left(\frac{-Q}{\mathrm{RT}}\right) \sigma C_i^{\gamma}(1 - C_i^{\gamma}) V_m}{9\mathrm{RT}\left(C_i^{\gamma'} - C_i^{\gamma}\right)^2}(t - t_0) \tag{8.6}$$

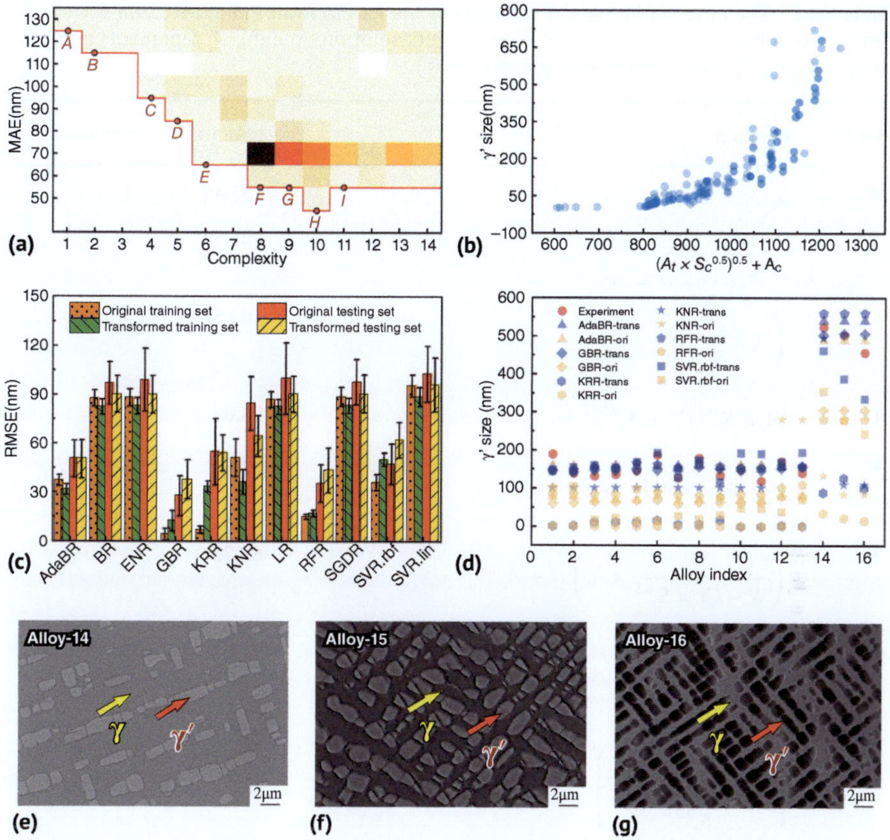

Figure 8.26 The generated superalloy synthesis factor by SR which greatly improves γ' size prediction performance. (a) Pareto front of MAE versus complexity among 60,000 mathematical formulas shown via density plot. (b) Scatter plot of γ' size versus $\left(A_t \times S_c^{0.5}\right)^{0.5} + A_c$. (c) RMSEs for model selection under original and transformed feature spaces by fivefold cross-validation. (d) The measured and predicted γ' sizes of 13 superalloys newly reported in 2023 and three superalloys which were synthesized. (e) The microstructure for alloy Co–29.6Ni–10.8Al–2Ti–2.5W–1.6Ta–1Mo–3.5Cr. (f) The microstructure for alloy Co–30Ni–10.4Al–1.5Ti–1.6W–3Ta–1Mo–4.9Cr. (g) The microstructure for alloy Co–29.9Ni–10.4Al–1.9Ti–1W–3.3Ta–1.1Mo–5.2Cr–0.8Re.

As shown in Eq. 8.6, T is the aging temperature (A_c) and t is the aging time (A_t), but the equation is established under ideal conditions, assuming that all phases are dissolved at the solution temperature. Additionally, an elevated S_c reduces the residual dendritic segregation of refractory elements (such as Re and W) and suppresses the formation of deleterious topologically close-packed (TCP) phases in Ni-based superalloys. Unsuitable S_c promotes the formation of TCP, which reduces the concentration of certain solid solution strengthening elements such as Cr, Mo, W, and Re in the γ-phase. Lower concentrations of these elements result in a reduced coarsening rate of the γ'-phase. Therefore, S_c impacts the coarsening of γ'-phase.

To investigate this factor, this work constructed an ML-based γ' size prediction model with $(A_t \times S_c^{0.5})^{0.5} + A_c$ and compositions (transformed feature space).

Table 8.13 The nine mathematical formulas at the Pareto front in Figure 8.26a. S_c represents solution temperature (°C), S_t represents solution time (h), A_c represents aging temperature (°C), and A_t represents aging time (h).

Point	Formula	MAE (nm)	Complexity
A	$S_c^{0.5}$	135.46	1
B	$\frac{S_c - A_t}{\text{Al}}$	121.54	2
C	$\left(A_t \times \frac{S_c}{S_c^{0.5}}\right)^{0.5}$	104.58	4
D	$(A_t \times S_c^{0.5})^{0.5} + A_c^{0.5}$	98.09	5
E	$\left(\frac{A_c}{\text{W} - \text{Ni}} \times A_t\right)^{0.5} + S_c^{0.5}$	72.40	6
F	$\left(\frac{S_c}{\text{W} - \text{Cr} \times S_c^{0.5}} \times A_t\right)^{0.5} + A_c^{0.5}$	63.76	8
G	$\left(\frac{A_c}{\text{W} - \text{Cr} \times S_c^{0.5}} \times A_t\right)^{0.5} + A_c^{0.5} + \text{Ta}$	67.07	9
H	$\left(\frac{A_c}{\text{W} - \frac{\text{Ni} \times \text{Al}}{S_c^{0.25}}} \times A_t\right)^{0.5} + A_c^{0.5}$	58.59	10
I	$\left(\frac{A_c}{\text{W} - \left(\text{Cr} + \frac{\text{Ni} \times \text{Al}}{\text{Ti}}\right)} \times A_t\right)^{0.5} + (A_c^{0.5} - 0.555\text{Nb})$	62.97	11

The comparison also involved using the original feature space, including solution temperature, solution time, aging temperature, aging time, and composition. The ML models were trained and evaluated using cross-validation, as shown in Figure 8.26c. In general, models utilizing the transformed feature space showed smaller RMSEs compared to those using the original feature space. These ML models were then applied to predict the γ' size for 13 newly reported superalloys from articles published in 2023, as well as three synthesized superalloys that were not included in the dataset extracted by the pipeline (Figure 8.26e–g and Table 8.14). The average of RMSEs between the reported/experiments and predicted γ' sizes amongst all models with transformed features is 83, much lower than 143.63 using original features (Figure 8.26d). This substantial improvement in model performance suggests that $(A_t \times S_c^{0.5})^{0.5} + A_c$ is a critical synthesis factor for determining the γ' size in superalloys.

In recent years, large-scale language pretraining models, such as GPT (Generative Pretraining Transformer), have significantly impacted the field

Table 8.14 The reported and experimental data including composition, solution temperature S_c, solution time S_t, aging temperature A_c, and aging time A_t. The γ' size was represented by γ' equivalent radius ($\langle r \rangle$, unit: nm) in the corresponding literature. The γ' sizes of alloys 6–16 were extracted from figures using GetData Graph Digitizer.

Index	Superalloy (at.%)	S_c(°C)	S_t(h)	A_c(°C)	A_t(h)	Reported γ' size (nm)	Predicted γ' size (nm)	Relative error (%)	Source
1	Co–33.8Ni–8.7Al–1Ti–2.1W–0.9Ta–2.1Mo–0.3Nb–13.5Cr	1,200	24	850	1,000	189	147.39	28.22	[90]
2	Co–30Ni–8.2Al–2.7Ti–2W–1.3Ta–0.9Nb–14.2Cr	1,200	24	850	1,000	139	142.33	2.34	[90]
3	Co–29.6Ni–10.8Al–2Ti–2.5W–1.6Ta–1Mo–3.5Cr	1,225	24	1,100	168	522.8	502.7	3.99	Exp
4	Co–30Ni–10.4Al–1.5Ti–1.6W–3Ta–1Mo–4.9Cr	1,225	24	1,100	168	501.2	506.12	0.97	Exp
5	Co–29.9Ni–10.4Al–1.9Ti–1W–3.3Ta–1.1Mo–5.2Cr–0.8Re	1,225	24	1,100	168	456.3	506.12	9.84	Exp
6	Co–30Ni–8Al–3Ti–3W–1Ta–12Cr	1,240	50	850	1,000	131.6	147.39	10.71	[90]
7	Co–40Ni–8Al–3Ti–3W–1Ta–12Cr	1,240	50	850	1,000	135.7	147.39	7.93	[90]
8	Co–20Ni–8Al–3Ti–3W–1Ta–12Cr	1,240	50	850	1,000	145	147.62	1.78	[90]
9	Co–30Ni–10Al–3Ti–3W–1Ta–12Cr	1,240	50	850	1,000	187.2	147.64	26.79	[90]
10	Co–30Ni–8Al–3Ti–3W–1Ta–16Cr	1,240	50	850	1,000	127.7	147.39	13.36	[90]
11	Co–30Ni–8Al–6Ti–3W–1Ta–12Cr	1,240	50	850	1,000	178.5	156.55	14.01	[90]
12	Co–30Ni–8Al–6Ti–3W–3Ta–12Cr	1,240	50	850	1,000	133.4	151.49	11.94	[90]
13	Co–30Ni–8Al–3Ti–3W–1Ta–2Nb–12Cr	1,240	50	850	1,000	149.1	146.62	1.68	[90]
14	Co–30Ni–8Al–1Ti–3W–1Ta–2Nb–12Cr	1,240	50	850	1,000	118.5	146.62	19.18	[90]
15	Co–30Ni–8Al–3Ti–3W–1Ta–3Mo	1,240	50	850	1,000	170.2	152.29	11.75	[90]
16	Co–30Ni–8Al–3Ti–1Ta–3Mo	1,240	50	850	1,000	140.3	156.05	10.09	[90]
Mean relative error								10.91	—

Source: Adapted from [91].

of NLP [92–94]. These models are trained on vast amounts of unannotated text, enabling them to be fine-tuned for specific NLP tasks. Essentially, they create a "well-read" black box that can interpret language at a high level and perform a wide range of tasks within that language. ChatGPT, a specific implementation of the GPT model, was trained using Reinforcement Learning from Human Feedback (RLHF) and demonstrates impressive abilities in conversational interaction [95]. It can handle dialog, answer follow-up questions, admit mistakes, and even reject inappropriate requests. However, despite these advancements, ChatGPT and similar models have limitations. The scale of these models requires substantial computational resources and access to large, well-organized corpora for training, which may limit their accessibility. Furthermore, these models are highly sensitive to input phrasing—slight rephrasing can lead to different outputs. In the context of materials science, GPT models also struggle to solve the complex correlations between composition, synthesis, and properties, or to summarize extracted databases and automatically build models for mining physical factors related to target properties. The absence of complete and structured data exacerbates this issue. AI models like ChatGPT primarily learn from vast amounts of text data, but they do not inherently possess the capability to extract structured data. While they can provide general information and summarize existing knowledge, extracting specific details and organizing them into a structured database for quantitative prediction models is a more complex task that ChatGPT cannot yet achieve in materials science.

This work introduces a semi-supervised text mining method that operates with a small corpus and low cost to extract action sequences and their parameters related to synthesis and processing conditions. This method automatically generates a machine-learnable dataset containing synthesis actions and parameters, chemical compositions, and γ'-phase size. The dataset has then been used to capture an explicitly expressed synthesis factor for predicting γ'-phase coarsening. The synthesis factor derived from text mining significantly enhances the performance of the data-driven γ' size prediction model. This strategy is simple to apply to specific problems, enabling the extraction of synthesis actions and parameters from scratch, without the need for fine-tuning or pretraining large corpora.

In conclusion, this work demonstrates how knowledge of materials synthesis and processing, as reported in the literature, can be extracted through text mining. The code for this semi-supervised text mining pipeline is available at: https://github.com/MGEdata/Action_extractor. A web-based toolkit is also available at: http://superalloydigger.mgedata.cn/#/spre_extractor, for online use. This open-source code and toolkit can be generalized to other alloys. As the volume of scientific literature continues to grow, it is inevitable that NLP will become a promising tool for extracting and learning from both published and unpublished works, providing a machine-readable and AI-usable format.

6. **Article retrieval and preprocessing**

The scientific articles on superalloys used in this study were published prior to 2022, and their full texts were automatically retrieved in XML format using Elsevier's Scopus and ScienceDirect APIs (https://dev.elsevier.com/), along with the extended scraping package of ChemDataExtractor [12]. Corpus preprocessing and table parsing were carried out using the SuperalloyDigger pipeline, as described in previous work [23]. In total, the pipeline processed 16,604 articles, comprising over 0.4 million paragraphs and 6,644 composition tables.

7. **Paragraph classification**

To identify paragraphs containing information on alloy synthesis, a total of 1,885 paragraphs from 80 journal articles were manually annotated with binary labels. Positive samples indicate that the paragraphs contain synthesis and processing information, while negative samples correspond to paragraphs that are not related. The labeled paragraphs were then split in a 9:1 ratio, with 90% of the corpus used for training and validation, and the remaining 10% reserved for testing. A binary logistic regression classifier was subsequently trained using the scikit-learn package [96], as shown in Figure 8.27a.

Three different feature extraction methods were compared for paragraph classification: bag of words, TF–IDF (term frequency–inverse document frequency), and BERT (Bidirectional Encoder Representations from Transformers). The BERT model was pretrained on 16,604 superalloy articles. Each paragraph in the articles was represented by a feature vector derived from bag of words, TF–IDF, and BERT, concatenated with a simple binary heuristic vector to distinguish whether the section title contained keywords like "Experiments" or "Methods."

The accuracy and F_1-score by different feature extraction methods are shown in Figure 8.27b. The highest overall F_1-score of 96.35% was obtained using TF–IDF. Figure 8.27c illustrates the learning curves of the logistic regression model.

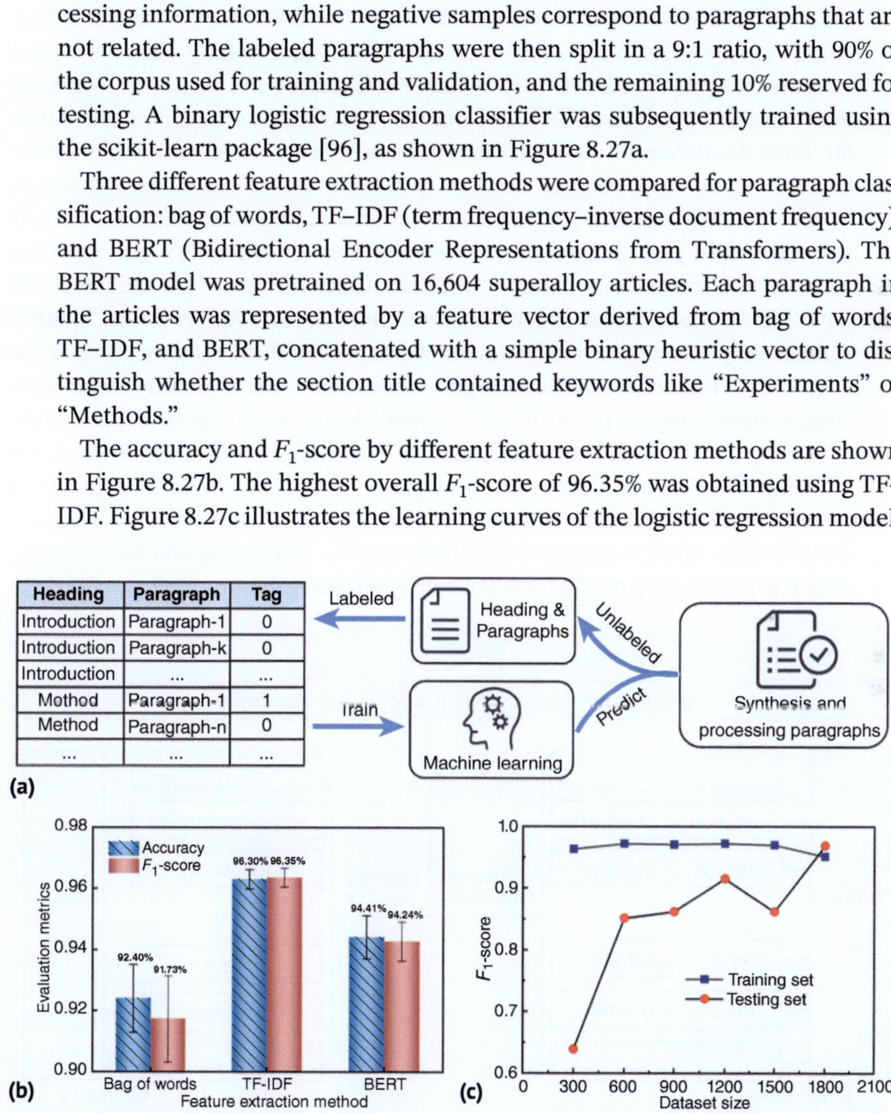

Figure 8.27 The paragraph classification process and results. (a) The pipeline of paragraph classification. (b) Precision and F_1-score of logistic regression classifier by different feature extraction methods. (c) Learning curves of model on training and testing sets. *Source*: [76] / Springer Nature / CC BY 4.0.

The F_1-scores reached approximately 95% when the training dataset size was 1,800. The logistic regression model was trained on 1,885 labeled paragraphs using TF–IDF as the feature representation, and was subsequently used to predict all paragraphs in the entire corpus. After removing paragraphs with fewer than 20 words and those predicted as negative, approximately 14,487 positive paragraphs related to synthesis and processing were identified.

8. **Word embedding Model**
Two word embedding models were pretrained on approximately 16,000 unlabeled full-text superalloy articles using Word2Vec CBOW and FastText [80, 97]. The Word2Vec model for superalloys had already been built and validated in previous work [23]. Word2Vec captures the semantic similarity between word pairs based on their context. If two words are semantically similar, their Word2Vec similarity score will be high, such as for words like "solutioned," "aged," and "forged." For FastText, each word is represented as a sum of its character *n*-grams [79, 80]. FastText embeddings can capture subword structure, multiple word senses, and uncertainty. For example, words with the same root, such as "arc-melted," "induction-melted," and "pre-melted," will have high cosine similarity with one another [97]. Word2Vec helps identify syntactically and semantically similar words, while FastText constructs word embeddings from character-level *n*-gram representations to identify words with similar morphological properties [26, 78–80]. As shown in Figure 8.28, FastText uses a bag of *n*-grams as additional features to capture partial information about the local word order, with words sharing the same prefix or suffix considered to have similar meanings. Therefore, combining the Word2Vec and FastText models helps identify entities with similar syntax, semantics, and morphology, which is particularly useful for identifying synthesis and processing actions.

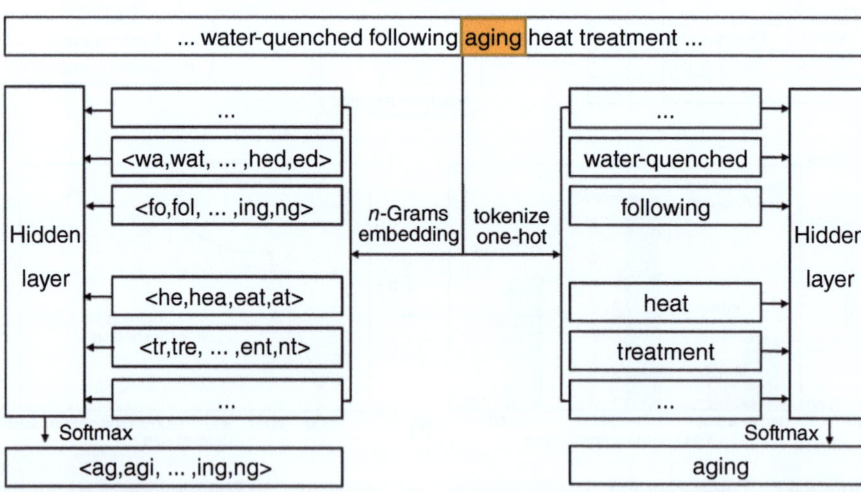

Figure 8.28 Model architecture of Word2Vec CBOW model and FastText CBOW model for a sentence in superalloy corpus. *Source:* [76] / Springer Nature / CC BY 4.0.

9. **BiLSTM-CRF model**
   The BiLSTM-CRF model was also employed for NER tasks. LSTM (long short-term memory) is a variant of RNN that excels at capturing both forward and backward contexts. In this model, the traditional softmax layer is replaced by a CRF layer to capture the interdependencies between labels. To train the BiLSTM-CRF model, five entity labels were designed: action (ACT), superalloy name (MAT), sample descriptor (DSC), material property (PRO), and property value (PV).

   A total of 1,308 sentences from 84 articles were randomly sampled from synthesis paragraphs to ensure that a diverse range of synthesis and processing types was included. All annotations were carried out by a single materials scientist. During annotation, the "BIO" sequence labeling method was applied to mark the entities in the sentences.

   The annotated sentences were split into training (80%), validation (10%), and testing (10%) sets. The validation set was used for hyperparameter optimization. The final model achieved a Precision of 89.34%, Recall of 78.30%, and F_1-score of 83.46% on the testing set, which included approximately 131 sentences (around 4,500 words). The Precision, Recall, and F_1-scores for each entity category are shown in Table 8.11. For action entity, the Precision, Recall, and F_1-score are only 77.02%, 72.98%, and 74.95%.
10. **Evaluation metrics**
    Precision, Recall, and F_1-score were used as the evaluation metrics for paragraph classification, NER, and dependency parsing. The corresponding results are shown in Eqs. 8.7, 8.8, 8.1 [27–29],

$$\text{Precision} = \frac{\text{TP}}{\text{TP} + \text{FP}} \tag{8.7}$$

$$\text{Recall} = \frac{\text{TP}}{\text{TP} + \text{FN}} \tag{8.8}$$

11. **Interdependency resolution**
    To merge the extracted composition and action data fragments, specific rules were developed based on the typical writing conventions of superalloy experts. Initially, the composition table was extracted to identify all alloy names and their compositions within the article using the method outlined in previous work [23]. Next, different strategies were applied based on the characteristics of the composition table and the synthesis paragraph, as described below:

    If no alloy named entity is recognized in the sentence containing the action entity, but an alloy named entity is identified in the preceding sentence within the same paragraph, the alloy named entity will be associated with the action entity.

    If no alloy name is recognized in the sentence preceding the action and no alloy name from the composition table appears anywhere in the paragraph, it is assumed that every alloy listed in the composition table is associated with the action entity.
12. **Symbolic regression**
    To capture the explicitly expressed synthesis factor for γ' size, SR analysis was conducted using the Python library, gplearn. Several hyperparameters needed optimization, including pc (probability of crossover), ps (probability of mutation), ph (population size), pp (probability of parent selection), and

Table 8.15 The hyperparameters employed in gplearn for GPSR.

Parameter	Value
Population size	5,000
Generations	20
Stopping criteria	0.1
pc	0.5, 0.6 (step = 0.025)
ps	0.01, 0.1 (step = 0.01)
ph	ps
pp	0.1, 0.2
Function set	add, sub, mul, div, sqrt
Parsimony coefficient	0.0005, 0.0015 (step = 0.0005)
Tournament size	20
Metric	MAE
Constant range	(−1,1)

Source: Adapted from [85].

the parsimony coefficient. A grid search approach was employed to evaluate performance across a space containing 5 pc values, 10 ps values, 10 ph values, 2 pp values, and 3 parsimony coefficients. Detailed hyperparameter settings are provided in Table 8.15. In total, 3,000 hyperparameter combinations were tested during the SR process, and for each combination, the populations evolved over 20 generations, resulting in a total of 60,000 candidate formulas.

13. **Prediction model for γ' size**
The entire dataset was randomly split, with 70% allocated for model training and validation, and the remaining 30% reserved for model testing. Several regression models were employed, including SVR.lin and SVR.rbf, GBR, BR, KNR, adaptive boosting regression (AdaBR), kernel ridge regression (KRR), RFR, SGDR, ENR, and LR. For parameter optimization, fivefold cross-validation was performed 100 times on the 70% training dataset. Once the parameters were optimized, all models were retrained and used to predict the γ' sizes for new alloys.
14. **Superalloy synthesis and characterization**
The alloys were synthesized from raw metals with a purity exceeding 99.95%. A 40 g ingot was prepared using vacuum arc melting, with the metal being melted eight times. The ingots were then sealed in a quartz tube under an argon atmosphere. Solution heat treatment at 1,225 °C for 12 hours was applied to all experimental alloys followed by air cooling to eliminate the composition segregation. The solutioned samples were further cut and aged at 1,100 °C for 168 hours followed by water cooling. All samples were prepared using the standard metallographic method and chemically etched for a few seconds in a solution of $HCl:H_2O:HNO_3$ in a 1:1:1 ratio. A Zeiss GeminiSEM 300 field-emission scanning electron microscope (SEM) in backscattered electron imaging mode was used to observe the γ/γ' microstructure, and an

energy-dispersive X-ray spectroscopy detector was employed to determine the alloy composition. The γ' sizes of alloys are measured by a computer vision framework by Liu et al. [75].

References

1 Rickman JM, Lookman T, Kalinin SV. Materials informatics: From the atomic-level to the continuum. Acta Materialia. 2019;168:473–510.

2 Lookman T, Balachandran PV, Xue D, Yuan R. Active learning in materials science with emphasis on adaptive sampling using uncertainties for targeted design. npj Computational Materials. 2019;5(1):21.

3 Xue D, Xue D, Yuan R, Zhou Y, Balachandran PV, Ding X, et al. An informatics approach to transformation temperatures of NiTi-based shape memory alloys. Acta Materialia. 2017;125:532–41.

4 Xue D, Balachandran PV, Hogden J, Theiler J, Xue D, Lookman T. Accelerated search for materials with targeted properties by adaptive design. Nature Communications. 2016;7:1–9.

5 Wen C, Zhang Y, Wang C, Xue D, Bai Y, Antonov S, et al. Machine learning assisted design of high entropy alloys with desired property. Acta Materialia. 2019;170:109–17.

6 Segler MHS, Preuss M, Waller MP. Planning chemical syntheses with deep neural networks and symbolic AI. Nature. 2018;555(7698):604–10.

7 Raccuglia P, Elbert KC, Adler PDF, Falk C, Wenny MB, Mollo A, et al. Machine-learning-assisted materials discovery using failed experiments. Nature. 2016;533(7601):73–6.

8 Granda JM, Donina L, Dragone V, Long DL, Cronin L. Controlling an organic synthesis robot with machine learning to search for new reactivity. Nature. 2018;559(7714):377–81.

9 Gesmundo NJ, Sauvagnat B, Curran PJ, Richards MP, Andrews CL, Dandliker PJ, et al. Nanoscale synthesis and affinity ranking. Nature. 2018;557(7704):228–32.

10 Butler KT, Davies DW, Cartwright H, Isayev O, Walsh A. Machine learning for molecular and materials science. Nature. 2018;559(7715):547–55.

11 Tshitoyan V, Dagdelen J, Weston L, Dunn A, Rong Z, Kononova O, et al. Unsupervised word embeddings capture latent knowledge from materials science literature. Nature. 2019;571(7763):95–8.

12 Swain MC, Cole JM. ChemDataExtractor: A toolkit for automated extraction of chemical information from the scientific literature. Journal of Chemical Information and Modeling. 2016;56(10):1894–904.

13 Krallinger M, Rabal O, Lourenço A, Oyarzabal J, Valencia A. Information retrieval and text mining technologies for chemistry. Chemical Reviews. 2017;117(12):7673–761.

14 Kim E, Jensen Z, van Grootel A, Huang K, Staib M, Mysore S, et al. Inorganic materials synthesis planning with literature-trained neural networks. Journal of Chemical Information and Modeling. 2020;60(3):1194–201.

15 Kim E, Huang K, Jegelka S, Olivetti E. Virtual screening of inorganic materials synthesis parameters with deep learning. npj Computational Materials. 2017;3(1):53.

16 Court CJ, Cole JM. Auto-generated materials database of Curie and Néel temperatures via semi-supervised relationship extraction. Scientific Data. 2018;5:180111.

17 Kim E, Huang K, Tomala A, Matthews S, Strubell E, Saunders A, et al. Machine-learned and codified synthesis parameters of oxide materials. Scientific Data. 2017;4:170127.

18 Agichtein E, Gravano L. Snowball: Extracting relations from large plain-text collections. In: Fifth ACM Conference on Digital Libraries [Internet]. Association for Computing Machinery; 2000. pp. 85–94. Available from: https://doi.org/10.1145/336597.336644

19 Ruan J, Xu W, Yang T, Yu J, Yang S, Luan J, et al. Accelerated design of novel W-free high-strength Co-base superalloys with extremely wide γ/γ' region by machine learning and CALPHAD methods. Acta Materialia. 2020;186:425–33.

20 Liu Y, Wu J, Wang Z, Lu XG, Avdeev M, Shi S, et al. Predicting creep rupture life of Ni-based single crystal superalloys using divide-and-conquer approach based machine learning. Acta Materialia. 2020;195:454–67.

21 Liu P, Huang H, Antonov S, Wen C, Xue D, Chen H, et al. Machine learning assisted design of γ'-strengthened Co-base superalloys with multi-performance optimization. npj Computational Materials. 2020;6(1):1–9.

22 Jiang X, Yin HQ, Zhang C, Zhang RJ, Zhang KQ, Deng ZH, et al. An materials informatics approach to Ni-based single crystal superalloys lattice misfit prediction. Computational Materials Science. 2018;143:295–300.

23 Wang W, Jiang X, Tian S, Liu P, Dang D, Su Y, et al. Automated pipeline for superalloy data by text mining. npj Computational Materials Science. 2022;8(1):9.

24 Lammey R. CrossRef's text and data mining services. Learned Publishing. 2014;27(4):245–50.

25 Bird S, Klein E, Loper E. Natural Language Processing with Python: Analyzing Text with the Natural Language Toolkit. Sebastopol, CA: O'Reilly Media, Inc.; 2009.

26 Mikolov T, Sutskever I, Chen K, Corrado GS, Dean J. Distributed representations of words and phrases and their compositionality. In: Burges CJ, Bottou L, Welling M, Ghahramani Z, Weinberger KQ, editors. Advances in Neural Information Processing Systems. New York: Curran Associates, Inc.; 2013. pp. 3111–9.

27 Sniegula A, Poniszewska-Mararida A, Chomatek L. Study of named entity recognition methods in biomedical field. Procedia Computer Science. 2019;160(2018):260–5.

28 Goutte C, Gaussier E. A probabilistic interpretation of precision, recall and F-score, with implication for evaluation. In: Losada DE, Fernandez-Luna JM, editors. ECIR 2005: Advances in Information Retrieval Information Retrieval, Lecture Notes in Computer Science, Vol. 3408. Berlin: Springer; 2005. pp. 345–59.

29 Japkowicz N. Why question machine learning evaluation methods. In: AAAI Workshop on Evaluation Methods for Machine Learning [Internet]. Association for the Advancement of Artificial Intelligence; 2006. pp. 6–11.

30 Jensen Z, Kim E, Kwon S, Gani TZH, Román-Leshkov Y, Moliner M, et al. A machine learning approach to zeolite synthesis enabled by automatic literature data extraction. ACS Central Science. 2019;5(5):892–9.

31 Nadeau D, Sekine S. A survey of named entity recognition and classification. Lingvisticae Investigationes. 2007;30(1):3–26.

32 Huang Z, Xu W, Yu K. Bidirectional LSTM-CRF models for sequence tagging. arXiv preprint arXiv:1508.01991; 2015.

33 Gers FA, Schmidhuber J, Cummins F. Learning to forget: Continual prediction with LSTM. Neural Computation. 2000;12(10):2451–71.

34 Lafferty J, McCallum A, Pereira FCN. Conditional random fields: Probabilistic models for segmenting and labeling sequence data. In: Eighteenth International Conference on Machine Learning [Internet]. Morgan Kaufmann Publishers Inc.; 2001. pp. 282–9.

35 Reimers N, Gurevych I. Optimal hyperparameters for deep LSTM-networks for sequence labeling tasks. arXiv preprint arXiv:1707.06799; 2017.

36 Zhou G, Su J, Zhang J, Zhang M. Exploring various knowledge in relation extraction. In: 43rd Annual Meeting of the Association for Computational Linguistics (ACL'05) [Internet]. Association for Computational Linguistics; 2005. pp. 427–34.

37 Sorokin D, Gurevych I. Context-aware representations for knowledge base relation extraction. In: 2017 Conference on Empirical Methods in Natural Language Processing [Internet]. Association for Computational Linguistics; 2017. pp. 1784–9.

38 Takanobu R, Zhang T, Liu J, Huang M. A hierarchical framework for relation extraction with reinforcement learning. In: Thirty-Third AAAI Conference on Artificial Intelligence and Thirty-First Innovative Applications of Artificial Intelligence Conference and Ninth AAAI Symposium on Educational Advances in Artificial Intelligence [Internet]. AAAI Press; 2019. pp. 7072–9.

39 Bach N, Badaskar S. A review of relation extraction. Literature Review for Language and Statistics II. 2007;2:1–15.

40 Brin S. Extracting patterns and relations from the world wide web. In: Atzeni P, Mendelzon A, Mecca G, editors. WebDB 1998: The World Wide Web and Databases, Lecture Notes in Computer Science, Vol. 1590. Berlin: Springer; 1998. pp. 172–83.

41 Pang HT, Zhang L, Hobbs RA, Stone HJ, Rae CMF. Solution heat treatment optimization of fourth-generation single-crystal nickel-base superalloys. Metallurgical and Materials Transactions A. 2012;43(9):3264–82.

42 Lass EA. Application of computational thermodynamics to the design of a Co-Ni-based γ'-strengthened superalloy. Metallurgical and Materials Transactions A. 2017;48(5):2443–59.

43 Li W, Li L, Antonov S, Feng Q. Effective design of a Co-Ni-Al-W-Ta-Ti alloy with high γ′ solvus temperature and microstructural stability using combined CALPHAD and experimental approaches. Materials & Design. 2019;180:107912.

44 Ooshima M, Tanaka K, Okamoto NL, Kishida K, Inui H. Effects of quaternary alloying elements on the γ′ solvus temperature of Co–Al–W based alloys with fcc/L12 two-phase microstructures. Journal of Alloys and Compounds. 2010;508(1):71–8.

45 Lass EA, Sauza DJ, Dunand DC, Seidman DN. Multicomponent γ′ -strengthened Co-based superalloys with increased solvus temperatures and reduced mass densities. Acta Materialia. 2018;147:284–95.

46 Chen Y, Wang C, Ruan J, Yang S, Omori T, Kainuma R, et al. Development of low-density γ/ γ′ Co–Al–Ta-based superalloys with high solvus temperature. Acta Materialia. 2020;188:652–64.

47 Liu X, Chen Z, Chen Y, Yang S, Pan Y, Lu Y, et al. Multicomponent Co-Ti-based superalloy with high solvus temperature and low lattice misfit. Materials Letters. 2021;284:128910.

48 Tirado FLR, Taylor SV, Dunand DC. Low-density, W-free Co–Nb–V–Al-based superalloys with γ/ γ′ microstructure. Materials Science and Engineering: A. 2020;796:139977.

49 Zhuang X, Li L, Feng Q. Effects of Al, Cr, and Ti on the oxidation behaviors of multi-component γ/γ′ CoNi-based superalloys. In: Tin S, et al. , editors. Superalloys 2020, The Minerals, Metals & Materials Series. Cham: Springer; 2020. pp. 870–9.

50 Baler N, Pandey P, Palanisamy D, Makineni SK, Phanikumar G, Chattopadhyay K. On the effect of W addition on microstructural evolution and γ′ precipitate coarsening in a Co–30Ni–10Al–5Mo–2Ta–2Ti alloy. Materialia. 2020;10:100632.

51 Yu J, Wang C, Chen Y, Wang C, Liu X. Accelerated design of L12-strengthened Co-base superalloys based on machine learning of experimental data. Materials & Design. 2020;195:108996.

52 Wen M, Sun Y, Yu J, Yang Y, Zhou Y, Sun X. Effects of Al content on microstructures and high-temperature tensile properties of two newly designed CoNi-base superalloys. Journal of Alloys and Compounds. 2020;835:155337.

53 Baler N, Pandey P, Chattopadhyay K, Phanikumar G. Influence of thermomechanical processing parameters on microstructural evolution of a gamma-prime strengthened cobalt based superalloy during high temperature deformation. Materials Science and Engineering: A. 2020;791:139498.

54 Cao BX, Kong HJ, Ding ZY, Wu SW, Luan JH, Jiao ZB, et al. A novel L12-strengthened multicomponent Co-rich high-entropy alloy with both high γ′-solvus temperature and superior high-temperature strength. Scripta Materialia. 2021;199:113826.

55 Lv S, Chen J, He X, Jia C, Wei K, Qu X. Investigation on sub-solvus recrystallization mechanisms in an advanced γ-γ' nickel-based superalloy GH_4151. Materials. 2020;13(20):4553.

56 Ai C, Xi L, Wang B, Sun Y, Zhang J, Su H, et al. Investigation on solution heat treatment response and γ′ solvus temperature of a Mo-rich second generation Ni based single crystal superalloy. Intermetallics. 2020;125:106896.

57 Isayev O. Text mining facilitates materials discovery. Nature. 2019;571(7763):42–3.

58 Choudhary K, DeCost B, Chen C, Jain A, Tavazza F, Cohn R, et al. Recent advances and applications of deep learning methods in materials science. npj Computational Materials. 2022;8(1):59.

59 Agrawal A, Choudhary A. Perspective: Materials informatics and big data: Realization of the "fourth paradigm" of science in materials science. APL Materials. 2016;4(5):053208.

60 Curtarolo S, Hart GLW, Nardelli MB, Mingo N, Sanvito S, Levy O. The high-throughput highway to computational materials design. Nature Materials. 2013;12(3):191–201.

61 Ramprasad R, Batra R, Pilania G, Mannodi-Kanakkithodi A, Kim C. Machine learning in materials informatics: Recent applications and prospects. npj Computational Materials. 2017;3(1):54.

62 Zakutayev A, Wunder N, Schwarting M, Perkins JD, White R, Munch K, et al. An open experimental database for exploring inorganic materials. Scientific Data. 2018;5:1–12.

63 Kirklin S, Saal JE, Meredig B, Thompson A, Doak JW, Aykol M, et al. The Open Quantum Materials Database (OQMD): Assessing the accuracy of DFT formation energies. npj Computational Materials. 2015;1:1–15.

64 Batra R, Song L, Ramprasad R. Emerging materials intelligence ecosystems propelled by machine learning. Nature Reviews Materials. 2021;6(8):655–78.

65 Moosavi SM, Chidambaram A, Talirz L, Haranczyk M, Stylianou KC, Smit B. Capturing chemical intuition in synthesis of metal-organic frameworks. Nature Communications. 2019;10(1):1–7.

66 Schweidtmann AM, Clayton AD, Holmes N, Bradford E, Bourne RA, Lapkin AA. Machine learning meets continuous flow chemistry: Automated optimization towards the Pareto front of multiple objectives. Chemical Engineering Journal. 2018;352:277–82.

67 MacLeod BP, Parlane FGL, Morrissey TD, Häse F, Roch LM, Dettelbach KE, et al. Self-driving laboratory for accelerated discovery of thin-film materials. Science Advances. 2020;6(20):eaaz8867.

68 Dragone V, Sans V, Henson AB, Granda JM, Cronin L. An autonomous organic reaction search engine for chemical reactivity. Nature Communications. 2017;8(1):1–8.

69 Li Z, Najeeb MA, Alves L, Sherman A, Parrilla PC, Pendleton IM, et al. Robot-Accelerated Perovskite Investigation and Discovery (RAPID): 1. Inverse temperature crystallization. Preprint, ChemRxiv; 2019. Available from: https://doi.org/10.26434/chemrxiv.10013090.v1

70 Baldan R, Rocha RLP, Tomasiello RB, Nunes CA, Costa AMS, Barboza MJR, et al. Solutioning and aging of MAR-M247 nickel-based superalloy. Journal of Materials Engineering and Performance. 2013;22(9):2574–9.

71 Ramsperger M, Mújica Roncery L, Lopez-Galilea I, Singer RF, Theisen W, Körner C. Solution heat treatment of the single crystal nickel-base superalloy CMSX-4 fabricated by selective electron beam melting. Advanced Engineering Materials. 2015;17(10):1486–93.

72 Kumar P, Kabra S, Cole JM. Auto-generating databases of yield strength and grain size using ChemDataExtractor. Scientific Data. 2022;9(1):1–11.

73 Weston L, Tshitoyan V, Dagdelen J, Kononova O, Trewartha A, Persson KA, et al. Named entity recognition and normalization applied to large-scale information extraction from the materials science literature. Journal of Chemical Information and Modeling. 2019;59(9):3692–702.

74 Huo H, Rong Z, Kononova O, Sun W, Botari T, He T, et al. Semi-supervised machine-learning classification of materials synthesis procedures. npj Computational Materials. 2019;5(1):1–7.

75 Liu P, Huang H, Jiang X, Zhang Y, Omori T, Lookman T, et al. Evolution analysis of γ' precipitate coarsening in Co-based superalloys using kinetic theory and machine learning. Acta Materialia. 2022;235:118101.

76 Wang W. Alloy synthesis and processing by semi-supervised text mining. npj Computational Materials. 2023;9(1):183.

77 O'Gorman T, Jensen Z, Mysore S, Mahbub R, Huang K, Olivetti E, et al. MS-MENTIONS: Consistently annotating entity mentions in materials science procedural text. In: EMNLP 2021: 2021 Conference on Empirical Methods in Natural Language Processing [Internet]. Association for Computational Linguistics; 2021. pp. 1337–52.

78 Mikolov T, Chen K, Corrado G, Dean J. Efficient estimation of word representations in vector space. In: 1st International Conference on Learning Representations (ICLR 2013): Workshop Track Proceedings [Internet]. ICLR; 2013. pp. 1–12.

79 Athiwaratkun B, Wilson AG, Anandkumar A. Probabilistic FastText for multi-sense word embeddings. In: ACL 2018: 56th Annual Meeting of the Association for Computational Linguistics (Volume 1: Long Papers) [Internet]. Association for Computational Linguistics; 2018. pp. 1–11.

80 Joulin A, Grave E, Bojanowski P, Douze M, Jégou H, Mikolov T. FastText.zip: Compressing text classification models. arXiv preprint arXiv:1612.03651; 2016.

81 Riloff E, Jones R. Learning dictionaries for information extraction by multi-level bootstrapping. In: Sixteenth National Conference on Artificial Intelligence and Eleventh Innovative Applications of Artificial Intelligence Conference [Internet]. American Association for Artificial Intelligence; 1999. pp. 474–9.

82 Riloff E, Phillips W. An introduction to the Sundance and AutosSlog systems. Technical Report UUCS-04-015, School of Computing, University of Utah; 2004. pp. 1–47.

83 Manning C, Surdeanu M, Bauer J, Finkel J, Bethard S, McClosky D. The Stanford CoreNLP Natural Language Processing Toolkit. In: 52nd Annual Meeting of the Association for Computational Linguistics: System Demonstrations [Internet]. Association for Computational Linguistics; 2015. pp. 55–60.

84 Makineni SK, Nithin B, Chattopadhyay K. Synthesis of a new tungsten-free γ- γ' cobalt-based superalloy by tuning alloying additions. Acta Materialia. 2015;85:85–94.

85 Stephens, T. gplearn. https://gplearn.readthedocs.io/en/latest/intro.html

86 Ardell AJ. The effect of volume fraction on particle coarsening: Theoretical considerations. Acta Metallurgica. 1972;20(1):61–71.

87 Lifshitz IM, Slyozov VV. The kinetics of precipitation from supersaturated solid solutions. Journal of Physics and Chemistry of Solids. 1961;19 (1–2):35–50.

88 Wagner C. Theory of the aging of precipitation by dissolution (Ostwald maturation). Bunsen Society for Physical Chemistry. 1961;65:581–91.

89 Calderon HA, Voorhees PW, Murray JL, Kostorz G. Ostwald ripening in concentrated alloys. Acta Metallurgica et Materialia. 1994;42(3):991–1000.

90 Zhuang X, Antonov S, Li W, Lu S, Li L, Feng Q. Alloying effects and effective alloy design of high-Cr CoNi-based superalloys via a high-throughput experiments and machine learning framework. Acta Materialia. 2023;243:118525.

91 Informer Technologies, Inc. GetData Graph Digitizer 2.26: Digitize graphs and plots. Graph Digitizing Software; 2013.

92 OpenAI. GPT-4 technical Report. arXiv preprint arXiv:2308.08774; 2023.

93 Koubaa A. GPT-4 vs. GPT-3.5: A concise showdown. Preprint; 2023. https://doi.org/10.20944/preprints202303.0422.v1

94 Brown TB, Mann B, Ryder N, Subbiah M, Kaplan J, Dhariwal P, et al. Language models are few-shot learners—special version. In: Larochelle H, Ranzato M, Hadsell R, Balcan MF, Lin H, editors. Advances in Neural Information Processing Systems, Vol. 33. New York: Curran Associates, Inc.; 2020. pp. 1877–1901.

95 Shen Y, Heacock L, Elias J, Hentel KD, Reig B, Shih G, et al. ChatGPT and other large language models are double-edged swords. Radiology. 2023;307(2):e230163.

96 Baranwal A, Bagwe BR, Vanitha M. Machine learning in Python: Diabetes prediction using machine learning. In: Velayutham S, editor. Handbook of Research on Applications and Implementation of Machine Learning Techniques. Hershey, PA: IGI Global Scientific Publishing; 2019. pp. 128–54.

97 Thavareesan S, Mahesan S. Sentiment lexicon expansion using Word2vec and fastText for sentiment prediction in Tamil texts. In: MERCon 2020: 6th International Multidisciplinary Moratuwa Engineering Research Conference [Internet]. IEEE Press; 2020. pp. 272–6.

9

Case Studies of Materials Synthesis Information Extraction

9.1 Introduction

This chapter explores the emerging field of automated extraction of materials synthesis information, a crucial step toward bridging computational predictions and experimental realizations in materials science. While large databases such as the Materials Project and open quantum materials database (OQMD) have enabled high-throughput property predictions, synthesis knowledge remains largely locked within unstructured scientific literature. Recent advances in natural language processing (NLP) and machine learning (ML) now make it possible to systematically mine, codify, and structure this hidden knowledge [1–8]. By transforming narrative experimental procedures into machine-readable data, researchers can build scalable synthesis datasets, enable intelligent synthesis planning, and support downstream applications such as knowledge graph construction, robotic experimentation, and closed-loop materials design.

9.2 Machine-learned and Codified Synthesis Parameters of Oxide Materials

As materials informatics continues to advance, a critical frontier lies in the integration of synthesis information into machine-readable and machine-learnable formats. While computational resources such as the Materials Project [9] and OQMD [10] have revolutionized the prediction of materials properties using ab-initio methods, an analogous infrastructure for synthesis data has remained conspicuously absent. This gap presents a major bottleneck in the realization of a truly end-to-end framework for materials discovery—from theoretical prediction to experimental realization.

This section presents an in-depth account of the first three foundational components of the study: (1) the motivation and background for automated synthesis informatics; (2) the design and implementation of the corpus construction and data acquisition pipeline; and (3) the NLP and ML techniques employed to extract synthesis parameters from unstructured text. These components form the methodological and conceptual foundation that enables downstream applications such as synthesis planning, outlier detection, and semantic search.

1. **Gaps in synthesis informatics**
   The materials genome initiative (MGI) has catalyzed a surge in computational materials databases, facilitating rapid screening of candidate materials via high-throughput simulation and modeling. Projects such as the Materials Project [9], OQMD [11], and AFLOWLIB [12] have enabled the prediction of materials properties at scale. However, these resources primarily focus on structure–property relationships derived from first-principles calculations. The transition from promising candidates to synthesized materials, however, requires knowledge of viable synthesis routes—a domain traditionally dominated by tacit, empirical knowledge embedded in the scientific literature.

   Efforts to automate the acquisition of synthesis data have been limited. Early works by Ghadbeigi et al. [13] and Raccuglia et al. [1] employed manual curation or targeted data retrieval, resulting in datasets constrained in scale and scope. While useful, these approaches are neither sustainable nor scalable for general synthesis informatics. In contrast, organic chemistry has benefited from machine-readable reaction databases such as Reaxys [14], which support synthesis prediction and retrosynthesis planning. Materials science lacks a comparable infrastructure.

   Kim et al.'s [15] study addresses this critical gap by autonomously mining synthesis parameters across 30 commonly reported oxide materials. Their aim is to create a community-accessible, machine-readable synthesis dataset that is scalable and extensible. By processing over 640,000 full-text journal articles, this study introduces a paradigm shift in how synthesis data can be aggregated, codified, and deployed for downstream computational workflows.

   This large-scale synthesis extraction is particularly urgent given the diversity and complexity of synthesis routes in inorganic materials. The ability to identify patterns—such as commonly co-used precursors, temperature regimes, or synthesis techniques—requires access to structured data spanning broad chemical spaces. Kim et al.'s [15] work thus positions itself as a cornerstone contribution in closing the materials design loop, where synthesis is informed by both data and computation.
2. **Corpus construction and data acquisition pipeline**
   The corpus construction pipeline begins with the retrieval of over 640,000 full-text journal articles from diverse publishers, made possible through the CrossRef API [16] and publisher-specific routes approved on a case-by-case basis. The articles are downloaded in HTML and PDF formats and then converted to plain text for processing. PDF conversion is handled using the open-source watr-works Scala program, ensuring that the downstream NLP pipeline receives standardized input regardless of the original file format.

   To narrow the focus to synthesis-relevant literature, Kim et al. [15] applied a multistep filtering strategy. Out of the initial corpus, 76,000 articles were identified as relevant to the synthesis of 30 selected oxide systems. This selection was achieved by querying for materials such as $BiFeO_3$, ZnO, or TiO_2, using keyword-based heuristics in combination with structured metadata searches.

A critical innovation in this pipeline lies in the paragraph-level classification system. Since most journal articles contain synthesis information within specific sections—typically "Experimental" or "Methods"—the authors trained a binary logistic regression classifier to distinguish synthesis-related paragraphs from the rest [17]. Each paragraph is represented as a "bag-of-words" vector augmented with simple binary features, such as whether the section title includes terms like "synthesis" or "preparation."

The training data for this classifier included thousands of manually labeled paragraphs from approximately 100 articles. The model achieved an F_1-score of 96%, a metric that balances Precision and Recall, and is particularly suited to imbalanced classification problems where positive (synthesis-related) examples are rare. This high level of performance ensured that only relevant synthesis sections were forwarded to the subsequent NLP stages, significantly improving the quality and efficiency of the downstream text extraction process.

Through this carefully constructed pipeline, the authors were able to reduce noise, focus on synthesis-relevant information, and prepare the input data for the next stage of neural-network-based entity extraction.

3. **NLP and ML techniques for synthesis extraction**

At the heart of the synthesis parameter extraction process is a suite of neural-network-based models augmented with domain-specific heuristics. The approach is grounded in the decoder paradigm: unstructured text is ingested as input, and a structured representation of synthesis parameters is produced as output. This transformation is enabled by a hybrid methodology combining transfer learning, word embeddings, and grammatical parsing.

The core of the model is a hierarchical neural network that assigns categorical labels to individual words based on their surrounding context. Each word is embedded using a vector representation learned through the Word2Vec algorithm [18], which was pretrained on the full corpus of 640,000 synthesis articles. This unsupervised pretraining captures the semantic relationships between domain-specific terms (e.g., "calcine," "anneal," and "autoclave") that are infrequent in general English corpora.

To enhance the model's performance, each word's embedding vector is combined with a set of binary heuristic features derived from rule-based systems and external databases [19–21]. These include known materials (e.g., "Fe_2O_3"), operations (e.g., "sinter"), and measurement units (e.g., "°C," "hours").

This hybrid input is processed by a feed-forward neural network with densely connected layers and a softmax classifier output. The network achieves a categorical accuracy of 86% and an F_1-score of 81% on a human-labeled test set of 1,700 words, significantly outperforming a baseline model trained solely on heuristic rules (baseline F_1: 66%).

The model uses a custom word label schema tailored to materials synthesis, which includes:

1. TARGET: final synthesized material (e.g., "$BiFeO_3$")
2. MATERIAL: nontarget materials (e.g., "$TiCl_4$," "NaOH")
3. OPERATION: synthesis actions (e.g., "sinter," "dissolve")

4. AMOUNT UNIT/CONDITION UNIT: units for quantities and experimental conditions (e.g., "ml," "°C," and "hours")
5. SYNTHESIS APPARATUS: equipment (e.g., "autoclave," "furnace")
6. DESCRIPTOR: morphological terms (e.g., "nanorods," "layered")

This schema enables the model to perform fine-grained annotation of synthesis protocols at the word level. Words are then grouped into higher-level entities through a grammatical parsing step, which leverages SpaCy's dependency parser.

After individual words are labeled, the model resolves them into meaningful entities and relationships. For example, the expression "heated at 500 °C" is parsed to establish a link between the operation ("heated") and the condition ("500 °C") using syntactic dependency trees (Figure 9.1). This relation-building step is critical for producing structured records from free-form text.

The output of this process is a rich set of relational data: operations are associated with materials, conditions, apparatuses, and durations. These records are then aggregated by material system to create structured synthesis profiles. Each record is stored in a machine-readable JSON format, which supports downstream analysis and visualization.

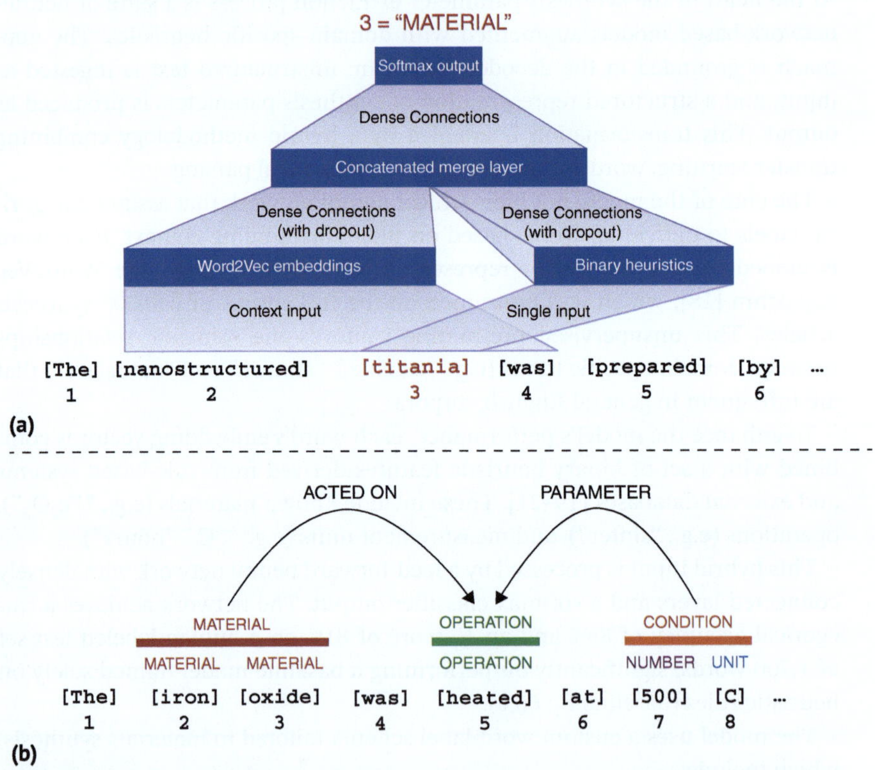

Figure 9.1 Neural-network and parse-based synthesis parameter extraction. (a) A hierarchical neural network assigns word-level labels by combining word embeddings and heuristic features within a five-word contextual window. (b) A grammatical parser resolves word-level labels into phrase-level entities and extracts their relationships based on syntactic dependencies. *Source*: [15] / Springer Nature / CC BY 4.0.

4. **Dataset construction and synthesis parameter codification**

   Following the successful extraction of word-level synthesis entities and their associated relations, the next phase in Kim et al.'s [15] framework involves aggregating and codifying these into a unified dataset organized by material system. Each data record corresponds to one of the 30 oxide materials targeted in the study, with a consistent schema designed for both human interpretability and machine accessibility.

   The authors structured the dataset using a standardized JSON format, where each record is a top-level JSON object with key–value pairs representing different categories of synthesis parameters. The schema includes fields such as:

   1. Name: the material system (e.g., "TiO_2," "ZnO")
   2. Num_papers: the number of source articles used for that material
   3. Associated_operations: a ranked list of the most common synthesis operations (e.g., "calcination," "annealing")
   4. Associated_materials: frequently co-mentioned reactants and solvents
   5. Topics: distribution of document-level topic labels inferred from latent Dirichlet allocation (LDA)
   6. Temperature_kde, Hydrothermal_kde, and Calcine_kde: kernel density estimates (KDEs) representing distributions of synthesis temperatures and times

   This structured approach allows for rapid querying, visualization, and integration with other data sources or tools such as pymatgen or custom experimental planning interfaces.

   One particularly novel aspect of this work is the use of Gaussian KDEs to represent continuous synthesis parameters such as temperature and duration. For example, calcination and hydrothermal conditions are not presented as raw value lists, but as smoothed distributions that highlight central tendencies and common parameter ranges. This approach facilitates immediate visual insight into synthesis norms and anomalies for each material.

   These KDEs serve multiple functions:

   1. Enable estimation of "most likely" synthesis conditions;
   2. support identification of outlier or rare synthesis strategies;
   3. allow quantitative comparison across material systems or synthesis methods.

   Such codified synthesis statistics represent a foundational advancement in automated synthesis informatics, offering a data-driven basis for hypothesis generation in experimental design.

5. **Visualization and technical validation of extracted data**

   To validate the quality of the extracted synthesis data and illustrate its utility, Kim et al. [15] offer a series of compelling visualizations. These not only demonstrate the coherence and richness of the dataset, but also align with known experimental regimes, thereby affirming the reliability of their automated pipeline.

   Figure 9.2 presents a heatmap that cross-references topic distributions with specific material systems. Topics are inferred from article text using LDA, and are labeled with their highest-weighted keywords (e.g., "nanorods," "photocatalysis," and "hydrothermal").

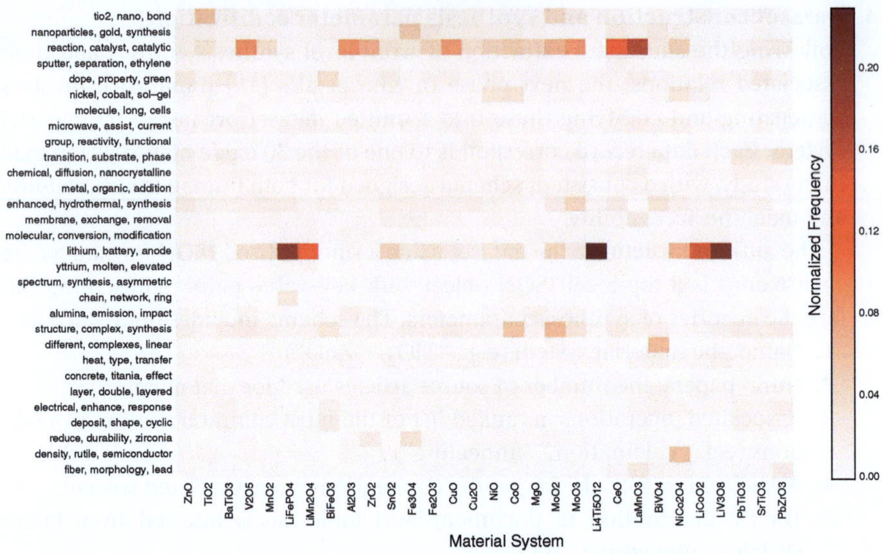

Figure 9.2 Topic and synthesis target distributions within the database. *Source*: [15] / Springer Nature / CC BY 4.0.

Each cell of the heatmap quantifies the normalized frequency of topic occurrence for a given material. For instance, TiO_2 shows strong association with topics related to photocatalysis and nanostructuring, while $BiFeO_3$ aligns with ferroelectricity-related topics.

This visualization confirms that the extracted synthesis sections not only include procedural data but also preserve topical context, enabling cross-disciplinary insights. It also shows that the topic modeling component is effective at distinguishing application domains and synthesis intents.

Figure 9.3 illustrates a heatmap of co-occurring material mentions across the synthesis corpus. For each target oxide, the matrix highlights which other materials are commonly cited in the same synthesis section. The main diagonal (self-references) is suppressed to reveal secondary associations.

For example, the co-mentioning of $Zn(NO_3)_2$ or NaOH with ZnO underscores typical precursor usage patterns. Similarly, the co-occurrence of Bi_2O_3 with $BiFeO_3$ provides data-driven confirmation of synthetic pathways.

This co-mention matrix not only validates chemical intuition but also supports the generation of materials synthesis graphs, where nodes represent materials and edges encode common synthesis relationships—a valuable foundation for knowledge graph construction.

Figure 9.4a,b focus specifically on TiO_2 as a case study, visualizing both temperature and time distributions for hydrothermal and calcination processes.

To validate the learning-based components of the extraction pipeline, Figure 9.5 plots the accuracy and F_1-score of the neural network classifier as a function of training set size. The model shows strong learning behavior, achieving a peak

F_1-score of 81% using only ~5,000 labeled words, which matches benchmarks reported in similar systems such as ChemDataExtractor. This learning curve substantiates the scalability and efficiency of the system, suggesting that modest labeling efforts yield high-quality entity recognition for synthesis-focused literature.

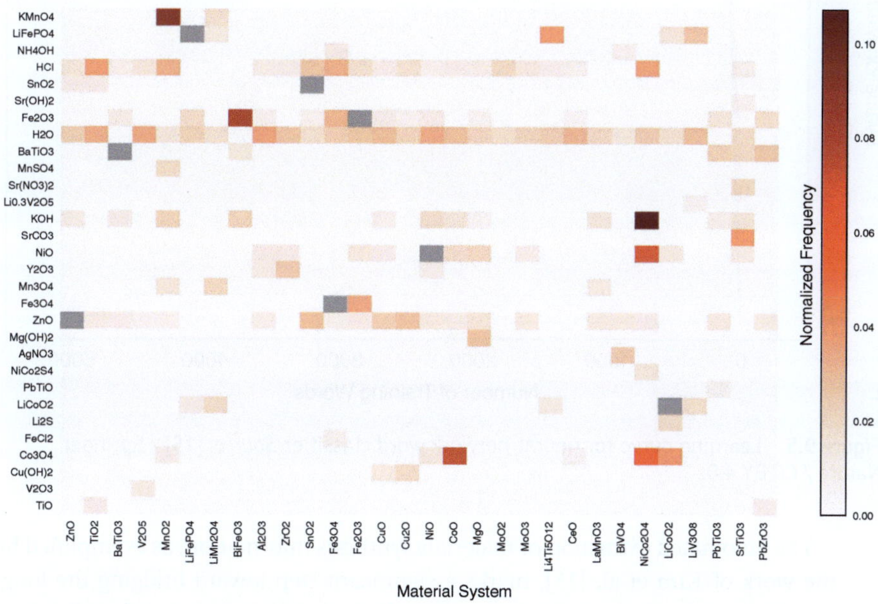

Figure 9.3 Top occurring material mentions per target material system. *Source*: [15] / Springer Nature / CC BY 4.0.

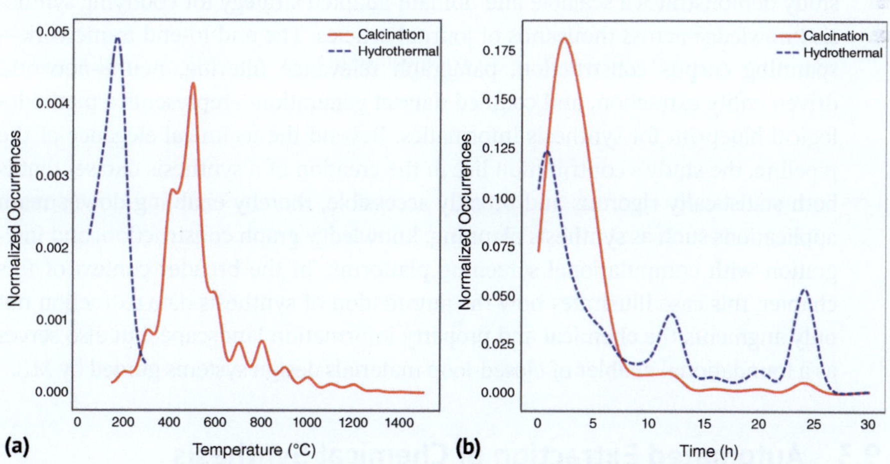

Figure 9.4 Temperature and time distributions for titania. (a) Kernel density estimate of calcination and hydrothermal temperatures for titania. (b) Kernel density estimate of calcination and hydrothermal times for titania. *Source*: [15] / Springer Nature / CC BY 4.0.

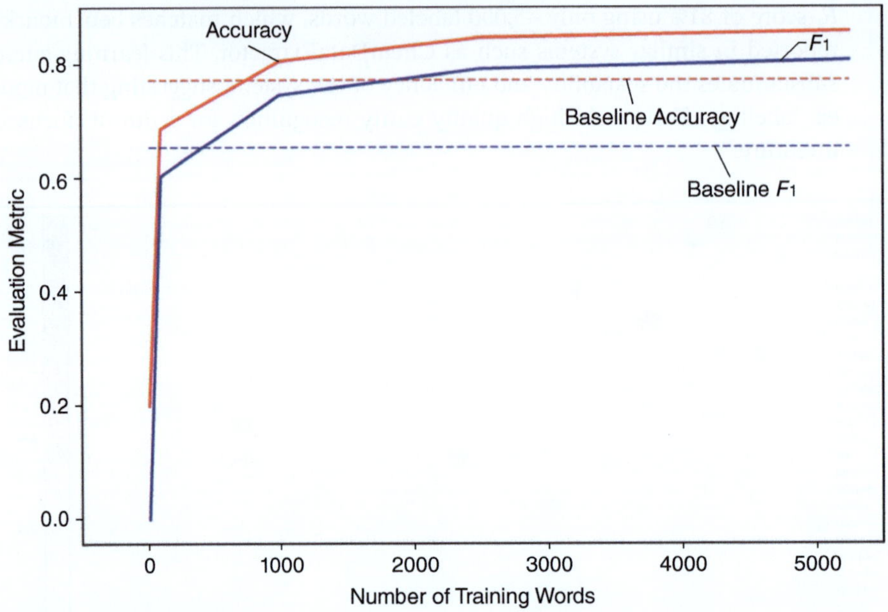

Figure 9.5 Learning curve for neural-network word classifier. *Source*: [15] / Springer Nature / CC BY 4.0.

The automated extraction of materials synthesis information, as exemplified by the work of Kim et al. [15], marks a significant step toward bridging the long-standing gap between computational predictions and experimental realizations in materials science. By applying decoder-based neural architectures to transform unstructured synthesis narratives into structured, machine-interpretable data, this study demonstrates a scalable and domain-adapted strategy for codifying synthesis knowledge across thousands of journal articles. The end-to-end framework—spanning corpus construction, paragraph relevance filtering, neural-network-driven entity extraction, and codified dataset generation—represents a methodological blueprint for synthesis informatics. Beyond the technical elegance of the pipeline, the study's contribution lies in the creation of a synthesis dataset that is both statistically rigorous and broadly accessible, thereby enabling downstream applications such as synthesis planning, knowledge graph construction, and integration with computational screening platforms. In the broader context of this chapter, this case illustrates how the automation of synthesis data extraction not only augments the chemical and property information landscape, but also serves as a foundational enabler of closed-loop materials design systems guided by ML.

9.3 Automated Extraction of Chemical Synthesis Actions From Experimental Procedures

As materials science increasingly adopts data-centric methodologies, the bottleneck in leveraging scientific literature has shifted from access to comprehension. While large-scale materials databases like the Materials Project and OQMD have structured computational data to enable ML, the vast corpus of synthesis knowledge remains largely locked in unstructured, prose-form experimental procedures published in patents and journal articles. These procedures, though rich in domain-specific insights, are expressed in natural language, often ambiguous and heterogeneous, thereby resisting straightforward parsing and automated utilization.

The work of Vaucher et al. [22] represents a significant advance in addressing this challenge. By proposing an end-to-end framework for converting natural language synthesis protocols into structured action sequences, their method serves as a foundational component for downstream applications such as robotic synthesis execution, reaction optimization, and AI-guided materials design. The authors employ a hybrid pipeline that combines rule-based techniques and Transformer-based deep learning to parse and translate experimental synthesis procedures into a sequence of atomic operations. This approach aligns with the broader theme of automated extraction of chemical and property information by addressing the synthesis aspect of the experimental materials life cycle. While many previous studies have focused on extracting chemical entities or physical properties, this work targets the procedural logic underlying chemical synthesis—a key step toward automation and intelligent reasoning over synthetic strategies.

In what follows, the first three thematic components of this study are examined: the motivation for structured extraction, the formalization of synthesis actions, and the hybrid rule-based and data-driven modeling approaches developed by the authors.

Experimental procedures in chemistry are conventionally documented in narrative form, tailored to human interpretation. Such text is characterized by diverse syntactic constructions, implicit references, and contextual assumptions. For instance, "The solution was stirred for 12 hours" may imply several sequential steps (e.g., temperature control, atmospheric conditions) that are not explicitly stated but are assumed by an experienced chemist. As Vaucher et al. observed, this linguistic variability poses a significant obstacle for reproducibility, automated processing, and integration into digital workflows.

The importance of structured synthesis data is amplified in the context of autonomous or semiautonomous laboratory systems. Robotic platforms for organic synthesis, such as those described by Cronin and coworkers [23], depend on precisely defined operation sequences, with no room for interpretive ambiguity. Similarly, ML models for reaction condition optimization and precursor prediction require large, high-quality datasets that include step-by-step procedural details rather than mere summaries of reactants and products.

Vaucher et al. envision a system where a machine can ingest a textual experimental procedure and convert it into a machine-readable sequence of standardized actions. Such a representation would not only enable robotic execution but also foster the development of predictive models for reaction planning, thereby transforming synthesis protocols from passive records to active data assets. This aspiration places their work firmly within the emerging paradigm of automated knowledge extraction, paralleling efforts in materials property prediction, compound discovery, and reaction retrosynthesis [24–27]. Moreover, structured synthesis extraction facilitates the creation of searchable databases of reaction workflows, enabling large-scale statistical analysis of synthesis strategies and offering insights into best practices, process bottlenecks, and novel design heuristics.

A core innovation in this work is the formal definition of a set of synthesis actions that collectively encapsulate the majority of operations encountered in organic synthesis protocols. These actions serve as the target schema for converting free-text procedures into structured, automation-compatible formats. Drawing on extensive empirical analysis of patent literature and synthesis records, the authors design a taxonomy comprising 30 distinct action types, each associated with a predefined set of allowable properties.

The action types range from simple operations such as Add, Stir, and Filter to more nuanced steps like Quench, PhaseSeparation, DrySolution, and MakeSolution. Each action is defined not just by its verb but also by contextual modifiers—such as temperature, duration, atmosphere, concentration, or chemical identity—required to reproduce the step with fidelity. For instance, Stir may carry properties like "for 12 hours at 60 °C under nitrogen," while DrySolution requires a desiccant specification such as "over sodium sulfate" (Table 9.1).

Table 9.1 Action types for information extraction from experimental procedures.

Action name	Description
Add	Add a substance to the reactor
CollectLayer	Select aqueous or organic fraction(s)
Concentrate	Evaporate the solvent (rotavap)
Degas	Purge the reaction mixture with a gas
DrySolid	Dry a solid
DrySolution	Dry an organic solution with a desiccant
Extract	Transfer compound into a different solvent
Filter	Separate solid and liquid phases
MakeSolution	Mix several substances to generate a mixture or solution
Microwave	Heat the reaction mixture in a microwave apparatus
Partition	Add two immiscible solvents for subsequent phase separation
PH	Change the pH of the reaction mixture
PhaseSeparation	Separate the aqueous and organic phases
Purify	Purification (chromatography)
Quench	Stop reaction by adding a substance
Recrystallize	Recrystallize a solid from a solvent or mixture of solvents
Reflux	Reflux the reaction mixture
SetTemperature	Change the temperature of the reaction mixture
Sonicate	Agitate the solution with sound waves
Stir	Stir the reaction mixture for a specified duration
Triturate	Triturate the residue
Wait	Leave the reaction mixture to stand for a specified duration
Wash	Wash (after filtration, or with immiscible solvent)

Action name	Description
Yield	Phony action, indicates the product of a reaction
FollowOtherProcedure	The text refers to a procedure described elsewhere
InvalidAction	Unknown or unsupported action
OtherLanguage	The text is not written in English
NoAction	The text does not correspond to an actual action

Source: [22] / Springer Nature / CC BY 4.0.

Crucially, the taxonomy includes both operational and nonoperational categories. For example, FollowOtherProcedure denotes references to external procedures, while InvalidAction, OtherLanguage, and NoAction provide necessary handles for texts that cannot be parsed or are irrelevant for synthesis reconstruction. This ensures completeness and robustness in real-world parsing tasks where data quality and linguistic noise can vary widely.

A significant challenge in synthesis action extraction is the semantic ambiguity of verbs. For instance, the verb "heat" can signify a SetTemperature action or a time-bound Stir operation, depending on whether it denotes a change in state or a sustained reaction. Similarly, "remove" may map to Concentrate or Filter, depending on context. The authors address this by incorporating syntactic and semantic cues into both their rule-based and learning-based models.

Furthermore, some verbs encapsulate multiple actions. For example, the sentence "The organic layer was dried over sodium sulfate" implies an unspoken CollectLayer action followed by DrySolution. Recognizing and inferring such implicit steps is critical for accurate workflow reconstruction.

The authors introduce a concise, human-readable textual representation of action sequences, where each action instance is specified with its type and all relevant non-empty properties. This representation is bijective, i.e., it can be converted into a structured format and back without loss of information. For example, an extracted action sequence from a sample patent procedure might read:

1. Make solution with compound A, acetic acid, and methanol;
2. add SLN;
3. add acetaldehyde at 0 °C;
4. wait for 30 minutes;
5. add sodium acetoxyborohydride;
6. quench with water.

This linearized yet structured form allows for both machine parsing and human validation, supporting applications ranging from robotic planning to dataset curation.

To operationalize the extraction of synthesis actions from text, the authors implement a hybrid pipeline combining (1) rule-based NLP systems and (2) a Transformer-based deep-learning model for sequence-to-sequence translation.

The initial component of the pipeline uses two rule-based systems to generate preliminary action sequences:

Pistachio-derived model: This model leverages the Pistachio dataset [28], which contains over six million experimental procedures annotated with action labels and properties, using the LeadMine and ChemicalTagger toolkits. These actions, while useful, are limited by their coarse categorization and inability to handle nuanced syntax or implicit steps [29, 30].

Custom NLP rule-based parser: To overcome these limitations, the authors develop a bespoke NLP system tailored to synthesis text. This model identifies verbs and parses syntactic structure to infer action types and arguments. It accounts for complex sentence structures, variations in expression, and compound dependencies, allowing for flexible and precise action labeling. For instance, both "Add A to B" and "To B was added A" are correctly interpreted as Add actions.

Each system outputs a list of action sequences for a given sentence. A heuristic algorithm then combines their outputs to form a high-quality training dataset. Notably, the custom NLP parser supports additional action types such as MakeSolution, which are absent in Pistachio's annotation schema, thereby enriching the expressiveness of the dataset.

To transcend the limitations of rule-based parsing, especially in handling linguistic ambiguity and implicit actions, the authors formulate the task as a machine translation problem. Using the OpenNMT-py framework [31], they train a Transformer-based encoder-decoder model to convert experimental procedure sentences into their corresponding action sequences. This model is first pretrained on a large corpus of 2.76 million sentence–action pairs derived from the combined rule-based systems. This self-supervised phase enables the model to learn domain-specific linguistic patterns and structural regularities in synthesis procedures.

The choice of Transformer architecture, with its multi-head attention and contextual embedding capabilities, allows the model to capture long-range dependencies and syntactic variants. Moreover, restricting the model output to textual action representations (rather than structured graphs) simplifies architecture design and ensures full reversibility.

The combined strategy—pretraining on weakly labeled rule-based data followed by refinement with high-quality manual annotations—offers several advantages:

Data efficiency: pretraining leverages vast unlabeled corpora without requiring extensive manual effort.

Robustness to noise: the model learns to generalize beyond rigid rule sets and handle diverse sentence structures.

Semantic flexibility: the deep-learning model can infer implicit actions and resolve ambiguous verbs based on context.

Domain adaptability: by tuning the model with synthesis-specific annotations, performance is optimized for chemical text rather than general language.

This hybrid approach reflects a broader trend in natural language understanding for scientific applications: combining symbolic and statistical methods to maximize precision and generalizability.

1. **Annotation and data augmentation framework**
While pretraining the model on rule-generated data provides a scalable foundation, high-quality supervised annotations remain indispensable for achieving precision and robustness in downstream applications. Recognizing this, Vaucher et al. designed a meticulous annotation workflow to refine their model with human-curated data, ensuring that it could learn from expert judgment and overcome the limitations of purely rule-based training sets.

The authors developed a custom web-based annotation interface built on the open-source platform Doccano [31]. This interface allows domain experts to view a sentence from an experimental procedure alongside its automatically generated preliminary action sequence, which can then be reviewed, corrected, or rewritten entirely.

Annotators could add new actions not captured by the model, modify existing actions to correct misidentified types or properties, reorder actions to align with procedural chronology, and discard or flag irrelevant text (e.g., NoAction, InvalidAction). To ensure annotation consistency across contributors, a comprehensive set of guidelines was provided, defining each action type and detailing common ambiguities, such as distinguishing between Stir and Wait or properly assigning DrySolution versus DrySolid. All annotations were finally reviewed by a single expert (Vaucher), ensuring high consistency across the 1,764 annotated samples.

To mitigate the risks of overfitting and to expand the diversity of training examples, the team applied a targeted data augmentation protocol. With a 50% probability, annotated samples were augmented by replacing key elements such as: compound names, quantities (e.g., "0.302 mmol"), durations, and temperatures. These substitutions preserved syntactic structure while introducing semantic variation, effectively expanding the model's exposure to plausible synthesis scenarios. The final training dataset after augmentation consisted of 20,000 examples, with duplicates later removed to yield a unique augmented set of 14,168 samples.

2. **Model evaluation and benchmarking**
To quantify the effectiveness of the model and compare various training strategies, Vaucher et al. performed rigorous benchmarking using six metrics and multiple model configurations. These include both rule-based and deep-learning variants.

Model performance was assessed along the following axes (Table 9.2):

1. **Validity (%)**: fraction of syntactically valid output action sequences.
2. **BLEU score**: a standard metric from machine translation that evaluates n-gram overlaps between prediction and reference.
3. **Levenshtein similarity**: normalized edit-distance similarity between predicted and reference action sequences.
4. **100%, 90%, and 75% Accuracies**: fraction of sentences where the predicted action sequence achieved full, near-full, or acceptable partial correctness.

The refined Transformer model, which leveraged both pretraining and human-annotated data, outperformed all other models across every metric. Notably, it achieved a perfect match for 60.8% of test sentences and a 90% or

Table 9.2 Metrics for the extraction of synthesis actions.

Model	Validity (%)	BLEU score (%)	Levenshtein similarity (%)	100% Accuracy (%)	90% Accuracy (%)	75% Accuracy (%)
Combined rule-based model	**100.0**	51.5	60.1	21.9	29.0	42.6
Pretrained translation model	**100.0**	58.6	68.7	24.7	33.2	48.3
Model without pretraining	98.9	64.7	76.4	37.8	47.7	62.8
Refined translation model	99.4	**85.0**	**86.9**	**60.8**	**71.3**	**82.4**

Source: [22] / Springer Nature / CC BY 4.0.

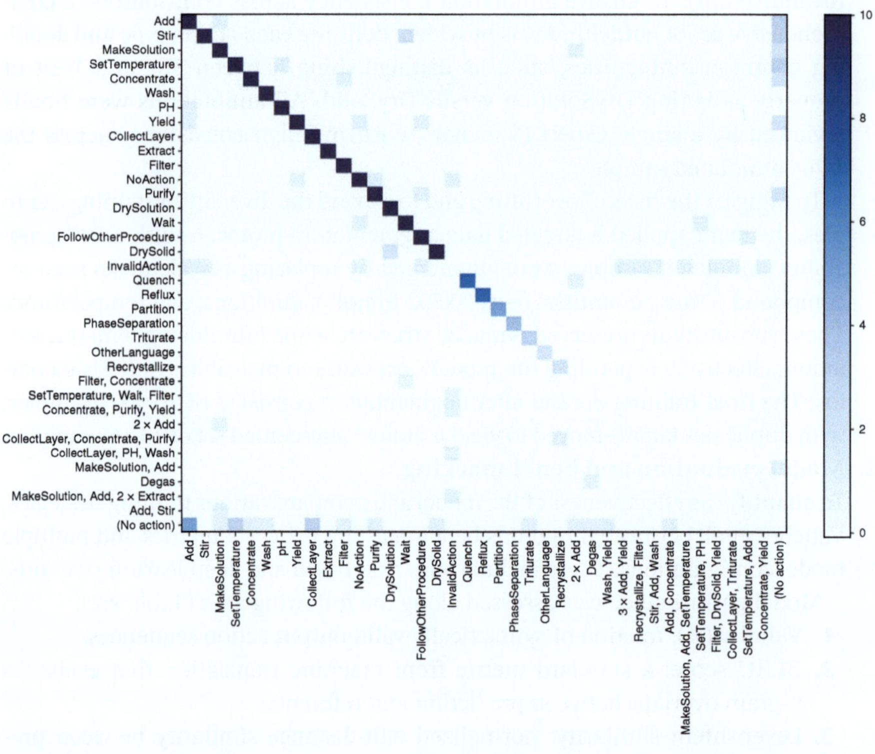

Figure 9.6 Visualization of the correctness of predicted action types. *Source*: [22] / Springer Nature / CC BY 4.0.

better match for over 71%—a substantial leap from the 24.7% baseline achieved by the pretraining-only model.

Figure 9.6 presents a confusion matrix of predicted versus actual action types. Most predictions fell along the diagonal, indicating correct classification, with errors concentrated in similar or interchangeable categories (e.g., Stir vs. Wait,

DrySolution vs. DrySolid). This analysis revealed both the strengths and limits of the model's contextual understanding. Further, Table 9.3 presents qualitative comparisons between annotated ground truth and model predictions.

The work presented in this case study exemplifies a critical advance in the broader pursuit of automated extraction of materials synthesis information. By systematically converting unstructured experimental procedures into structured, machine-readable action sequences, it establishes a scalable paradigm for bridging the gap between human-authored synthesis narratives and computationally actionable formats. This transition is foundational for enabling intelligent synthesis planning, data-driven experimental design, and autonomous laboratory execution in materials science. The integration of rule-based heuristics with Transformer-based sequence modeling offers a robust framework

Table 9.3 Example of extracted action sequences.

After adjusting to pH 1.5 with 10% hydrochloric acid, the ethyl acetate solution was separated, washed with a saturated aqueous sodium chloride and then dried over anhydrous magnesium sulfate.

1. PH with 10% hydrochloric acid to pH 1.5; PhaseSeparation; CollectLayer organic; Wash with saturated aqueous sodium chloride; DrySolution over anhydrous magnesium sulfate.
2. PH with 10% hydrochloric acid to pH 1.5; PhaseSeparation; CollectLayer organic; Wash with saturated aqueous sodium chloride; DrySolution over anhydrous magnesium sulfate.

A solution of solid sodium metal (450 mg, 19.75 mmol) in EtOH at 30 °C was treated with ethyl acetoacetate (103 g, 790 mmol) maintaining temperature.

1. MakeSolution with sodium metal (450 mg, 19.75 mmol) and EtOH; Add SLN; Add ethyl acetoacetate (103 g, 790 mmol) at 30 °C.
2. MakeSolution with **solid** sodium metal (450 mg, 19.75 mmol) and EtOH; Add SLN; Add ethyl acetoacetate (103 g, 790 mmol) at 30 °C.

3-Bromo-2-fluoroaniline (10 g, 52.63 mmol) was dissolved in DCM (100 mL) under nitrogen atmosphere.

1. Add 3-Bromo-2-fluoroaniline (10 g, 52.63 mmol); Add DCM (100 mL) under nitrogen.
2. Add 3-Bromo-2-fluoroaniline (10 g, 52.63 mmol) **under nitrogen**; Add DCM (100 mL) under nitrogen.

Upon complete addition, the reaction mixture was allowed to warm to room temperature and the reaction was stirred for about 12 h.

1. STIR for 12 h at room temperature.
2. **SetTemperature room temperature;** Stir for **about** 12 h.

The residue was crystallized from 60 ml of benzotrifluoride, during this operation, the mixture was briefly boiled with activated carbon and filtered whilst still hot.

1. InvalidAction.
2. Recrystallize **from benzotrifluoride (60 ml); Filter keep precipitate**.

Source: [22] / Springer Nature / CC BY 4.0.

capable of parsing linguistic complexity while capturing procedural semantics. More broadly, the study illustrates how NLP can operationalize latent synthesis knowledge embedded in literature, thereby enriching the materials data ecosystem not only with static properties and compositions, but with dynamic, reproducible workflows. As the field progresses, methodologies such as those explored here will form the basis for standardized synthesis databases, cross-domain knowledge graphs, and fully closed-loop experimental platforms—realizing the full vision of materials informatics as an end-to-end, data-integrated discipline.

References

1 Raccuglia P, Elbert KC, Adler PDF, Falk C, Wenny MB, Mollo A, et al. Machine-learning-assisted materials discovery using failed experiments. Nature. 2016;533(7601):73–6.

2 Zhang H, Fu H, Zhu S, Yong W, Xie J. Machine learning assisted composition effective design for precipitation strengthened copper alloys. Acta Materialia. 2021;215:117118.

3 Xue D, Balachandran PV, Hogden J, Theiler J, Xue D, Lookman T. Accelerated search for materials with targeted properties by adaptive design. Nature Communications. 2016;7:1–9.

4 Wen C, Zhang Y, Wang C, Xue D, Bai Y, Antonov S, et al. Machine learning assisted design of high entropy alloys with desired property. Acta Materialia. 2019;170:109–17.

5 Segler MHS, Preuss M, Waller MP. Planning chemical syntheses with deep neural networks and symbolic AI. Nature. 2018;555(7698):604–10.

6 Wen C, Wang C, Zhang Y, Antonov S, Xue D, Lookman T, et al. Modeling solid solution strengthening in high entropy alloys using machine learning. Acta Materialia. 2021;212:116917.

7 Zhang Y, Wen C, Wang C, Antonov S, Xue D, Bai Y, et al. Phase prediction in high entropy alloys with a rational selection of materials descriptors and machine learning models. Acta Materialia. 2020;185:528–39.

8 Jiang X, Jia B, Zhang G, Zhang C, Wang X, Zhang R, et al. A strategy combining machine learning and multiscale calculation to predict tensile strength for pearlitic steel wires with industrial data. Scripta Materialia. 2020;186:272–7.

9 Jain A, Ong S, Hautier G, Chen W, Richards W, Dacek S, et al. The Materials Project: a materials genome approach to accelerating materials innovation. APL Materials. 2013;1:011002.

10 Saal JE, Kirklin S, Aykol M, Meredig B, Wolverton C. Materials design and discovery with high-throughput density functional theory: The Open Quantum Materials Database (OQMD). JOM. 2013;65:1501–9.

11 Kirklin S, Saal JE, Meredig B, Thompson A, Doak JW, Aykol M, et al. The Open Quantum Materials Database (OQMD): Assessing the accuracy of DFT formation energies. npj Computational Materials. 2015;1:1–15.

12 Curtarolo S, Setyawan W, Wang S, Xue J, Yang K, Taylor RH, et al. AFLOWLIB.ORG: A distributed materials properties repository from high-throughput ab initio calculations. Computational Materials Science. 2012;58:227–35.

13 Ghadbeigi L, Harada JK, Lettiere BR, Sparks TD. Performance and resource considerations of Li-ion battery electrode materials. Energy & Environmental Science. 2015;8(6):1640–50.

14 Goodman J. Computer software review: Reaxys. Journal of Chemical Information and Modeling. 2009;49(12):2897–98.

15 Kim E, Huang K, Tomala A, Matthews S, Strubell E, Saunders A, et al. Machine-learned and codified synthesis parameters of oxide materials. Scientific Data. 2017;4:170127.

16 Lammey R. CrossRef text and data mining services. Science Editing. 2015;2(1):22–7.

17 Pedregosa F, Varoquaux G, Gramfort A, Michel V, Thirion B, Grisel O, et al. Scikit-learn: Machine learning in Python. Journal of Machine Learning Research. 2011;12:2825–30.

18 Mikolov T, Chen K, Corrado G, Dean J. Efficient estimation of word representations in vector space. In: 1st International Conference on Learning Representations (ICLR 2013): Workshop Track Proceedings [Internet]. ICLR; 2013. pp. 1–12.

19 Swain MC, Cole JM. ChemDataExtractor: A toolkit for automated extraction of chemical information from the scientific literature. Journal of Chemical Information and Modeling. 2016;56(10):1894–904.

20 Honnibal M, Johnson M. An improved non-monotonic transition system for dependency parsing. In: 2015 Conference on Empirical Methods in Natural Language Processing [Internet]. Association for Computational Linguistics (ACL); 2015. pp. 1373–8.

21 Kim S, Thiessen PA, Bolton EE, Chen J, Fu G, Gindulyte A, et al. PubChem substance and compound databases. Nucleic Acids Research. 2016; 44(D1):D1202–13.

22 Vaucher AC, Zipoli F, Geluykens J, Nair VH, Schwaller P, Laino T. Automated extraction of chemical synthesis actions from experimental procedures. Nature Communications. 2020;11(1):3601.

23 Steiner S, Wolf J, Glatzel S, Andreou A, Granda JM, Keenan G, et al. Organic synthesis in a modular robotic system driven by a chemical programming language. Science. 2019;363(6423):eaav2211.

24 Wang S, Guo Y, Wang Y, Sun H, Huang J. SMILES-BERT: Large scale unsupervised pre-training for molecular property prediction. In: 10th ACM International Conference on Bioinformatics, Computational Biology and Health Informatics [Internet]. Association for Computing Machinery; 2019. pp. 429–36.

25 Liu B, Ramsundar B, Kawthekar P, Shi J, Gomes J, Nguyen QL, et al. Retrosynthetic reaction prediction using neural sequence-to-sequence models. ACS Central Science. 2017;3(10):1103–13.

26 Thakkar A, Chadimová V, Bjerrum EJ, Engkvist O, Reymond JL. Retrosynthetic accessibility score (RAscore)—rapid machine learned synthesizability classification from AI driven retrosynthetic planning. Chemical Science. 2021;12(9):3339–49.

27 Coley CW, Jin W, Rogers L, Jamison TF, Jaakkola TS, Green WH, et al. A graph-convolutional neural network model for the prediction of chemical reactivity. Chemical Science. 2019;10(2):370–7.

28 Mayfield J, Lowe D, Sayle R. Pistachio [Internet]. Patent; 2018. Available from: https://nextmovesoftware.com/talks/Mayfield_Pistachio_NIHReactions_202105.pdf

29 Hawizy L, Jessop DM, Adams N, Murray-Rust P. ChemicalTagger: A tool for semantic text-mining in chemistry. Journal of Cheminformatics. 2011;3:1–13.

30 Lowe DM, Sayle RA. LeadMine: A grammar and dictionary driven approach to entity recognition. Journal of Cheminformatics. 2015;7:1–9.

31 Klein G, Kim Y, Deng Y, Nguyen V, Senellart J, Rush AM. OpenNMT: Neural machine translation toolkit. arXiv preprint arXiv:1805.11462; 2018.

10

Materials Predictive Modeling with Language-augmented Approaches

10.1 Introduction

The relentless pursuit of novel materials with tailored functionalities remains a cornerstone of technological advancement and sustainable global development. While conventional machine learning (ML) methodologies have significantly accelerated materials research by leveraging structured numerical descriptors, they often inadvertently bypass the vast, nuanced, and context-rich knowledge embedded within the extensive corpus of scientific literature and experimental narratives. This untapped domain holds critical insights—chemical intuition, synthesis pathways, processing nuances, and failure modes—that are frequently best articulated in natural language. To truly unlock the next wave of materials innovation, a paradigm shift is required, moving towards AI systems that can comprehensively understand, integrate, and learn from the full spectrum of materials data. This chapter delves into this transformative frontier. We explore pioneering hybrid AI architectures designed to synergistically fuse the quantitative precision of traditional materials descriptors with the qualitative depth and contextual understanding derived from textual sources.

At the forefront of these advancements are Transformer-based networks, which offer a powerful mechanism for jointly encoding diverse data modalities. This includes structural information, such as SMILES strings for molecules and polymers or graph representations of crystal structures, alongside contextual data meticulously mined and processed from scientific publications. By learning from this holistic data landscape, language-augmented approaches promise not just incremental gains but a significant leap in our ability to predict material properties and guide experimental efforts. Throughout this chapter, we will present compelling case studies that vividly demonstrate these capabilities, showcasing accuracy improvements often in the range of 20–30% over conventional ML techniques when text-derived features are strategically incorporated. We will explore applications in diverse areas, such as enhancing polymer property prediction and optimizing steel design and processing.

10.2 Materials Descriptors and Representations

The efficacy of ML paradigms in predicting material properties or accelerating the discovery of novel compounds is inextricably linked to the fidelity and expressiveness of the material representations employed. These representations, universally termed "descriptors" or "features," fulfill the indispensable role of translating the inherent chemical and physical complexity of materials—often existing in high-dimensional configuration spaces—into a structured, machine-interpretable format, typically a feature vector. The meticulous selection and engineering of these descriptors are therefore of paramount importance, as they must encapsulate the salient attributes governing a material's macroscopic behavior and microscopic interactions. Ideally, such descriptors should be richly informative, computationally tractable, and exhibit robust generalizability across diverse material classes and chemical spaces. This section embarks on a systematic exposition of the various descriptor categories prevalent in contemporary materials informatics.

10.3 Structured Descriptors in Materials Science

10.3.1 Understanding Structured Descriptors

Structured descriptors represent the historical and still fundamental underpinning of materials informatics, having facilitated the application of quantitative modeling and ML in materials science for several decades. These descriptors are formally defined as quantifiable attributes, typically numerical (continuous or discrete) or categorical, that systematically characterize a material's elemental composition, its crystallographic or molecular architecture, or other inherent physical and chemical properties within a well-defined, consistent framework. Their designation as "structured" arises from their natural organization into tabular arrays, where rows might signify distinct material instances and columns correspond to specific, quantified features. This matrix-like format is directly consonant with the input requirements of most conventional ML algorithms, from linear regression to complex neural networks, which are designed to operate on such structured feature vectors.

The significance of structured descriptors in advancing materials science is profound. They furnish a quantitative lexicon for delineating materials, thereby enabling the development of sophisticated mathematical models known as quantitative structure–property relationships. These models can predict material behavior, elucidate the complex interplay between structure and function, and ultimately accelerate the rational design and discovery of custom materials with targeted functionalities. Historically, the evolution of these descriptors has paralleled our deepening understanding of materials at atomic, electronic, and microstructural length scales. Early efforts frequently focused on identifying empirical correlations, such as the Hume-Rothery rules for alloy phase stability, which in hindsight can be viewed as precursors to modern, data-driven, descriptor-based methodologies. The rise of computational materials science, particularly through the use of density functional theory (DFT), marked a turning point. It significantly expanded the scope, dimensionality, and precision of structured descriptors, thereby generating an unprecedented volume of high-quality data for training robust ML models.

It is important to recognize, however, that while essential, structured descriptors inherently capture only certain critical aspects of a material's complex nature. They often reflect an idealized scenario, such as a defect-free single crystal at cryogenic temperatures. As a result, they may overlook important factors related to nonideal synthesis conditions, intricate processing histories, the random influence of defects and interfaces, or the rich qualitative insights and tacit knowledge embedded in the broader scientific literature. This inherent limitation provides strong justification for the language-augmented approaches explored later in this volume, which aim to integrate the precision of structured data with the contextual richness of unstructured textual information. Even so, a solid understanding of structured descriptors remains essential, as they continue to serve as a foundational element in advanced hybrid AI systems.

10.3.2 Element-based Descriptors of Materials Informatics

Compositional descriptors, derived directly from the elemental constitution of a material, represent the most fundamental and universally applicable class of features within the structured descriptor paradigm. Their principal utility and widespread adoption stem from their intrinsic simplicity and ease of derivation, frequently necessitating nothing more than the material's chemical formula—its elemental identity and stoichiometry. These descriptors serve as a first-order approximation for encoding a material's characteristics, providing a foundational input for both theoretical models and data-driven ML algorithms. In materials informatics, they are particularly prevalent in high-throughput computational screening campaigns, initial stages of materials design space exploration, and as baseline features against which more complex descriptor sets are benchmarked. While inherently coarse-grained, as they abstract away structural complexities, their ability to capture the fundamental chemical nature of a material makes them an indispensable starting point in the descriptor hierarchy.

The most elementary compositional descriptor is the stoichiometry itself, often represented as a vector of atomic fractions, x_i, for each constituent element i in the material. These fractional values, normalized to sum to unity, directly quantify the relative abundance of each element. This elemental fraction vector forms the simplest possible feature representation of a material's composition and is directly utilized in various thermodynamic models, such as in the formulation of free energy expressions within the CALPHAD (calculation of phase diagrams) methodology, and serves as a rudimentary input for ML models attempting to correlate composition with physical properties.

Beyond raw stoichiometry, a significantly richer and more informative set of compositional descriptors is engineered by leveraging the intrinsic physicochemical properties of the constituent elements. This approach involves associating each element with a vector of its known attributes and then applying various aggregation schemes to derive a fixed-length feature vector for the multielement material. These elemental attributes can be broadly categorized:

Fundamental identifiers: This category includes the atomic number (Z), which defines the element's unique identity and nuclear charge; the atomic mass (A), crucial for determining macroscopic density and influencing vibrational dynamics (e.g., phonon spectra, isotopic effects in specific applications); and the element's period and group in the periodic table, which provide a coarse-grained classification of its chemical family and electron shell structure. While seemingly basic, these properties are foundational for distinguishing elements within any computational framework.

Atomic size and packing-related parameters: The spatial extent of atoms and ions is captured by various radii, including the atomic radius (r_{atom}), ionic radius (r_{ionic}) (which is sensitive to coordination number and charge state), covalent radius (r_{cov}), and van der Waals radius (r_{vdW}). These descriptors are pivotal in understanding and predicting lattice parameters, solid solubility limits (famously encapsulated in Hume-Rothery rules for alloy formation), the propensity for elements to occupy interstitial versus substitutional sites within a host lattice, and overall packing density. Derived features such as the atomic size mismatch parameter (δ)-often calculated as the standard deviation of atomic radii weighted by composition, e.g., $\delta = \sqrt{\sum x_i(1 - r_i/\overline{r})^2}$ where $\overline{r} = \sum x_i r_i$—are extensively used in alloy design, particularly in the field of high-entropy alloys (HEAs), to predict phase stability (e.g., favoring solid solutions over intermetallic compound formation).

Electronic structure attributes: Properties governing an element's inclination to engage in chemical bonding and exchange electrons are paramount. Electronegativity (χ), a measure of an atom's ability to attract electrons within a chemical bond, is a key descriptor. Various scales exist, including those by Pauling (based on thermochemical data), Mulliken (averaging ionization potential [IP] and electron affinity [EA]), Allred-Rochow (based on effective nuclear charge), and Sanderson (basf^{-1} n electron density), each capturing slightly different facets of this tendency. IPs—the energy required to remove successive electrons (1st IP, 2nd IP, etc.)—and EA—the energy change upon adding an electron provide quantitative measures of an element's redox stability and its likelihood to form cations or anions. These electronic descriptors collectively inform predictions of bond type (ionic, covalent, and metallic character), bond polarity, the magnitude of charge transfer between dissimilar atoms, work function contributions at surfaces, and general chemical reactivity. Similar to atomic size mismatch, an electronegativity mismatch ($\Delta\chi$) parameter, often computed as a weighted standard deviation or range, is a valuable descriptor in alloy design, influencing phase stability and electronic structure.

Valence electron configuration: The number and type of valence electrons (N_{val})—those electrons in the outermost shell available for bonding—are fundamental to chemical interactions. This is often detailed by counts of s, p, d, and f valence electrons. A highly significant and historically successful derived descriptor is the valence electron concentration (VEC), defined as the stoichiometrically weighted average number of valence electrons per atom. VEC has a storied history in physical metallurgy, playing a central role in the Hume-Rothery rules for predicting the phase stability of noble metal alloys and finding connections to the Engel-Brewer theory of alloying. Its utility persists in modern alloy design, for instance, in predicting the propensity for deleterious topologically close-packed (TCP) phases like the σ-phase in superalloys, or for fine-tuning electronic and magnetic properties in transition metal alloys and complex intermetallics.

Thermal properties of elemental phases: Attributes such as the elemental cohesive energy, melting point $\left(T_m^{\text{elem}}\right)$, boiling point $\left(T_b^{\text{elem}}\right)$, and Debye temperature of the element $\left(\Theta_D^{\text{elem}}\right)$ are also frequently incorporated. These elemental properties can serve as proxies for the intrinsic bond strength contributions of different elements within a compound or alloy. They are often used in empirical rules of mixture (e.g., Vegard's law for lattice parameters or simple linear interpolations for melting points, albeit with caution) or as parameters within more sophisticated thermodynamic models. For example, the average melting temperature of constituent elements is sometimes used as a feature in models predicting the melting point or thermal stability of alloys.

To transform these lists of elemental properties—where each element i is characterized by a vector of attributes $P_{i,k}$ (e.g., $P_{i,1}$ = atomic radius of element i, $P_{i,2}$ = electronegativity of element i-into a fixed-length feature vector representative of the overall multielement material, various aggregation strategies are systematically employed. The fundamental goal of these strategies is to distill the collective influence of the constituent elements into a concise yet informative numerical fingerprint, suitable for ingestion by ML algorithms or for use in quantitative heuristic models. The choice of aggregation method is not arbitrary and should ideally reflect underlying physical or chemical intuition about how elemental contributions combine to produce emergent material properties. The most pervasively utilized aggregation methods involve computing statistical moments of the distribution of elemental properties, weighted by their stoichiometric atomic fractions (x_i). The stoichiometrically weighted arithmetic mean (often simply termed the "average"), calculated as $\overline{P}_k = \sum_i x_i P_{i,k}$ for each property k, is a cornerstone. This aggregation captures the mean-field contribution or the expected average value of an elemental property within the material. It is particularly intuitive for properties that might be expected, as a first approximation, to follow a linear rule of mixtures, or for representing the overall "center of mass" of a particular physicochemical attribute within the compositional space.

While the arithmetic mean provides a measure of central tendency, other types of means can be more appropriate depending on the nature of the property being aggregated. The stoichiometrically weighted geometric mean, $\overline{P}_{G,k} = \prod_i (P_{i,k})^{x_i}$ (applicable for positive $P_{i,k}$), is often preferred for properties that are inherently multiplicative or that span many orders of magnitude, as it gives less weight to extreme outliers compared to the arithmetic mean. For instance, it might be considered for averaging quantities like reaction rate constants or certain scaling factors. The stoichiometrically weighted harmonic mean, $\overline{P}_{H,k} = \left(\sum_i x_i P_{i,k}^{-1}\right)^{-1}$, is theoretically more suitable when averaging rates or properties that combine as reciprocals, such as electrical conductivities of components arranged in series, or the effective spring constant of springs in series. These different means (arithmetic, geometric, and harmonic) are specific instances of the more general power mean, and their judicious selection can subtly alter the information captured in the resulting descriptor.

Beyond central tendency, quantifying the dispersion or heterogeneity of elemental properties within the compound is crucial, as this often reflects important physical phenomena such as internal chemical pressure, lattice strain due to atomic size mismatch, or the driving forces for phase separation versus solid solution formation. The stoichiometrically weighted variance $\left(\mathrm{Var}(P_k) = \sum_i x_i \left(P_{i,k} - \overline{P}_k\right)^2\right)$ or its square root, the standard deviation, is a standard measure of this dispersion. A high variance in atomic radii, for example, is a key component of the δ-parameter used to characterize HEAs, signifying substantial local lattice strain. The range of each property (i.e., the difference between the maximum and minimum value of $P_{i,k}$ among the constituent elements) provides a simpler, more direct measure of the spread, though it is sensitive to single extreme outliers. For a more robust measure of dispersion, less susceptible to outliers, the interquartile range (IQR) or the mean absolute deviation (MAD) from the mean can be employed. These dispersion metrics effectively quantify the "chemical diversity" or "physicochemical mismatch" within the material for a given property, often correlating strongly with material stability, miscibility, or the emergence of unique electronic or mechanical behaviors arising from this inherent heterogeneity. Higher-order statistical moments can also be incorporated to capture more nuanced aspects of the elemental property distributions, although their physical interpretation can become less straightforward. The stoichiometrically weighted skewness measures the asymmetry of the distribution. For example, a material composed predominantly of elements with low electronegativity values, but with a small fraction of a highly electronegative element, would exhibit a positively skewed distribution of electronegativities. This asymmetry might indicate specific bonding characteristics or charge localization not captured by mean and variance alone. The stoichiometrically weighted kurtosis quantifies the "tailedness" or peakedness of the distribution, indicating the prevalence of extreme outliers relative to a normal distribution. A highly leptokurtic distribution (high kurtosis) for a property like elemental melting points might suggest that most elements have similar melting points, but one or two have drastically different values, potentially influencing the overall thermal processing window or phase evolution of the material. It is important to note that the vast majority of these aggregation strategies rely on stoichiometric weighting, reflecting the bulk composition. While this is standard, specialized applications might warrant alternative weighting schemes. For instance, if surface properties are of primary interest and surface segregation is known to occur, then weightings based on estimated surface compositions could be employed, though this typically requires additional information beyond the bulk formula and moves toward more complex, spatially resolved featurization.

The careful application of diverse aggregation strategies—from simple averages that capture overall trends, to dispersion measures that quantify chemical heterogeneity, and higher-order moments that describe distributional asymmetries—enables the construction of a rich and multifaceted compositional feature vector.

10.4 Textual and Contextual Features from Literature

The preceding discussion on structured descriptors highlighted their foundational role in materials informatics. However, a vast ocean of knowledge, rich with experimental nuances, synthesis rationales, mechanistic insights, and contextual details, remains encapsulated within the unstructured natural language of scientific literature, patents, and technical documentation. Bridging the gap between this textual domain and quantitative ML necessitates the transformation of natural language into effective numerical representations or features. Once relevant textual passages pertaining to a material's synthesis, characterization, or performance are identified (through processes discussed elsewhere, such as domain-specific information retrieval or targeted corpus construction), the central challenge becomes encoding their semantic and contextual content into a format that ML algorithms can interpret and leverage. This section focuses specifically on the diverse methodologies developed for creating such text representations, ranging from traditional sparse vector models to modern, powerful dense embeddings derived from advanced language models, all crucial for enabling AI to "read" and learn from the collective textual output of the materials science community.

The core objective of text representation in materials science is to generate feature vectors that capture, as effectively as possible, the information embedded in text relevant to a material's properties, its processing history, or observed phenomena. These representations aim to translate the complex, often subtle, meaning conveyed by human language into a structured numerical space where materials described similarly in text are positioned proximally, and where relationships between textual concepts can be algorithmically discerned. The quality of these representations directly dictates the potential of ML models to extract actionable insights, predict outcomes based on textual evidence, or classify documents according to fine-grained materials science criteria.

10.4.1 Sparse Vector Space Models

Historically, the initial forays into representing text numerically relied on sparse vector space models, which, despite their limitations, laid the groundwork for more sophisticated techniques and are still employed for certain applications due to their relative simplicity and interpretability.

The most fundamental of these is the bag-of-words (BoW) model [1]. In this paradigm, a document (or a relevant text snippet) is conceptualized as an unordered collection of its constituent words, disregarding grammatical structure and word sequence. The resulting feature vector for each document resides in a high-dimensional space, where each dimension corresponds to a unique word in the corpus vocabulary. The value in each dimension typically represents the frequency of that word in the document, a raw term frequency (TF). To modulate the importance of terms, TF is often combined with inverse document frequency (IDF). IDF assigns a higher weight to terms that are rare across the entire corpus, thereby emphasizing words that are uniquely characteristic of a specific document. The product, TF-IDF [2], thus highlights terms that are both frequent within a document and discriminative across the collection. In materials science, TF-IDF representations can be effective for tasks like identifying predominant experimental techniques (e.g., "X-ray diffraction" and "scanning electron microscopy") or key material classes (e.g., "perovskite" and "graphene") within research articles or patents. However, the BoW approach suffers from extreme sparsity (most entries in the vector are zero), very high dimensionality (equal to vocabulary size, which can be enormous for scientific corpora), and a critical inability to capture semantic relationships—synonymous terms (e.g., "synthesis" and "fabrication") are treated as entirely distinct, and word order, vital for meaning, is lost.

To address the limitation of word order, *N*-gram models [3] were introduced. Instead of individual words, sequences of *N* consecutive words (e.g., bi-grams like "thin film," tri-grams like "phase transition temperature," or "high specific surface area") are used as the fundamental units or features. This approach allows the model to capture some local contextual information and recognize common multi-word expressions that carry specific meaning within the materials domain. While *N*-grams can improve representational fidelity for domain-specific phraseology, they significantly exacerbate the problems of sparsity and high dimensionality. Moreover, they are still limited in their capacity to understand long-range dependencies or deeper semantic structures within the text.

10.4.2 Dense Vector Representations

The limitations of sparse models catalyzed the development of dense vector representations, commonly known as embeddings. This paradigm shift involves mapping discrete textual units—such as words, sentences, or entire documents—into relatively low-dimensional, continuous (dense) vector spaces. The defining characteristic and principal advantage of embeddings is their ability to capture semantic relationships: words or text segments with similar meanings are designed to be located close to one another in this learned vector space. This not only alleviates the "curse of dimensionality" and sparsity associated with BoW models but also endows ML models with a rudimentary form of semantic understanding.

Pioneering this dense representation were static or non-contextual word embeddings, which generate a single, fixed vector for each word type based on its co-occurrence patterns within large corpora, embodying the distributional hypothesis that words in similar contexts share similar meanings. Foundational models like Word2Vec [4] (with its Skip-gram and CBoW variants learning from local context prediction tasks), GloVe [5] (leveraging global word-word co-occurrence statistics), and FastText [6] (innovatively learning embeddings for character *N*-grams, enabling it to handle out-of-vocabulary words and capture morphological information) have been instrumental. For materials informatics, their true utility is unlocked when trained on domain-specific corpora, leading to specialized embeddings, which adeptly capture nuanced scientific semantics (e.g., associating "calcination" with "high temperature" rather than unrelated concepts). These individual word vectors can then be aggregated, for instance through averaging or TF-IDF weighting, to form rudimentary representations for longer text sequences like phrases or sentences, although this often sacrifices granular word order information.

To more faithfully represent entire sentences and documents, techniques that explicitly model longer sequences have been developed. While simple averaging of constituent word embeddings offers a baseline, more sophisticated approaches directly learn sequence-level representations. Doc2Vec [7] extends the Word2Vec framework to learn fixed-length vectors for variable-length texts by incorporating a unique document identifier during the training process. However, significant advancements have been achieved using Transformer-based architectures; models such as Sentence-BERT (SBERT) [8] and the universal sentence encoder (USE) utilize pretrained Transformers like Bidirectional Encoder Representations from Transformer (BERT) within Siamese or Triplet network structures. These are specifically fine-tuned to output sentence or short document embeddings that reside in a vector space where semantic similarity corresponds to geometric proximity, proving highly effective for tasks like semantic search, clustering of material synthesis protocols, or identifying paraphrased descriptions of material properties within diverse textual sources.

The current state-of-the-art in text representation lies with contextual embeddings from pretrained language models (PLMs), which signify a paradigm shift from static, one-vector-per-word-type embeddings to dynamic, deeply contextualized representations where a word's vector is uniquely determined by its surrounding text. The Transformer architecture, with its powerful self-attention mechanisms capable of weighing the influence of different words in a sequence and capturing long-range dependencies, is the cornerstone of these models. Prominent PLMs like BERT [9] are pre-trained on vast text corpora using self-supervised objectives such as masked language modeling (MLM) and next sentence prediction, enabling them to learn rich linguistic patterns and world knowledge. For materials science, the development and application of domain-specific PLMs like SciBERT [10] (trained on broad scientific texts) or MatSciBERT/MatBERT (further specialized on materials literature) are critical, as these models possess an intrinsic understanding of the specialized vocabulary, syntax, and complex relationships prevalent in materials discourse. The contextualized token embeddings from these PLMs (e.g., the output vector corresponding to a specific material name, or the aggregate representation from a special [CLS] token) can be used directly as features, or the entire PLM can be fine-tuned on specific downstream tasks, such as predicting material properties from their textual descriptions, thereby adapting its learned representations for maximal task-specific performance.

10.4.3 Comparison of Different Methods

Numerical text representations, such as sparse TF-IDF vectors and dense contextual embeddings, are essential for applying ML to materials science literature. These representations allow algorithms to process unstructured text and extract meaningful patterns. For instance, embeddings derived from synthesis descriptions (e.g., chemicals, temperatures, and durations) can be combined with structured features like composition or structure, improving the accuracy of models predicting material properties such as catalytic activity, mechanical strength, or optical performance.

Text representations also enable key applications like document classification, clustering, and semantic search. They support the automatic grouping of research articles by material systems (e.g., perovskite solar cells or battery cathodes) and allow topic modeling to track emerging research areas. Dense embeddings, in particular, enable natural language queries and help identify semantically similar documents, offering advantages over traditional keyword-based methods. Additionally, analyzing relationships between embeddings can reveal new hypotheses by linking materials with similar textual contexts.

Each representation method has its trade-offs. Sparse models like TF-IDF are interpretable and simple but limited in semantic understanding. Dense embeddings from PLMs capture richer contextual meaning and perform better in complex tasks but are often less interpretable and more computationally demanding. Current research aims to develop text representations that are both powerful and practical. When integrated with structured data and other modalities, these representations form a strong foundation for building AI systems that accelerate materials discovery and understanding.

10.5 Atomic Structure and Graph-based Features

The compositional descriptors discussed in Section 10.3.1 provide a fundamental, elemental fingerprint of a material, while the textual features detailed in Section 10.3.2 unlock knowledge from scientific discourse. However, a vast majority of critical material properties—ranging from mechanical strength and electronic conductivity to catalytic activity and optical response—are profoundly dictated by the precise three-dimensional (3D) arrangement of atoms, the nature of their chemical bonds, and the intricacies of their local coordination environments. Simple compositional information or even averaged structural parameters often fail to capture these crucial architectural nuances. Consequently, the development of atomic structure descriptors and graph-based features that explicitly encode this spatial and bonding information into machine-interpretable formats is paramount for building high-fidelity predictive models in materials science. The central challenge lies in converting complex, variable-size sets of atomic coordinates and connectivity patterns into fixed-size, informative feature vectors that inherently respect fundamental physical symmetries and invariances, such as translational, rotational, and permutational invariance of atoms of the same species. Graphs, representing atoms as nodes and interatomic bonds or interactions as edges, emerge as a natural and powerful formalism for this purpose.

10.5.1 From Atomic Coordinates to Graph Representations

The initial step in most graph-based featurization pipelines involves transforming raw structural data—typically atomic coordinates (x,y,z) for each atom and unit cell lattice vectors for crystalline materials—into a graph $G = (V, E)$, where V is the set of nodes representing atoms and E is the set of edges signifying interatomic bonds or interactions. The definition of these graph components is critical for the subsequent feature learning process.

Each atom i in the structure is represented as a node $v_i \in V$, which is then endowed with an initial feature vector, $\mathbf{h}_i$, encoding its intrinsic chemical identity and, potentially, other relevant local properties. Minimally, this vector includes information about the elemental identity, commonly represented as a one-hot encoded vector corresponding to the element's atomic number, or more comprehensively, by a vector compiling various elemental properties such as atomic mass, electronegativity, valence electron configuration, and ionic radius, as detailed in Section 10.3.1. In scenarios where initial quantum mechanical calculations like DFT relaxations have been performed, these node features can be augmented with derived atomic properties, for instance, DFT-calculated partial charges, local magnetic moments, or information gleaned from site-projected density of states.

The edges $e_{ij} \in E$ of the graph symbolize the connections or interactions between pairs of atoms, nodes v_i and v_j. The criteria for establishing these edges are crucial and significantly influence the resulting graph's topology and the information it conveys. A common initial approach to define connectivity involves establishing edges between atoms whose Euclidean distance d_{ij} falls below a predefined cutoff radius, r_c. While straightforward to implement, the selection of an optimal r_c often requires careful consideration of the system's chemistry and density, as a poorly chosen cutoff can either omit important longer-range interactions or introduce spurious connections. To capture interactions at different length scales, some models employ multiple cutoff radii to define distinct neighbor shells. A more robust and adaptive method for defining neighbors involves constructing the Voronoi tessellation of the atomic sites; edges are then typically established between atoms whose Voronoi polyhedra share a common face, a method that naturally identifies nearest neighbors without arbitrary distance parameters. Alternatively, connectivity can be inferred based on chemically intuitive bonding rules, for example, by forming an edge if the interatomic distance is within a specified tolerance of the sum of the covalent radii of the two participating atoms, thereby explicitly attempting to represent chemical bonds. Another strategy involves constructing k-nearest neighbor (k-NN) graphs, where an edge connects an atom to its k closest neighbors, ensuring uniform local graph connectivity, though this may not always align with traditional chemical bonding concepts. Once defined, these edges can also be attributed with feature vectors, $\mathbf{a}_{ij}$. The most prevalent edge feature is the interatomic distance or bond length, d_{ij}. Other potential edge attributes include discretized bond types (e.g., single, double, or aromatic for molecules, or estimates of covalent/ionic character if inferable from electronegativity differences), or bond order indices if available from prior quantum chemical calculations. While angular information is critical for describing 3D structures, it is often captured implicitly through the network of bond distances or more explicitly by advanced graph neural network (GNN) architectures.

A graph representation, if constructed appropriately, inherently handles permutational invariance concerning the arbitrary indexing of atoms, provided the node and edge features are consistently defined. The critical physical requirements of translational and rotational invariance—material properties should not change if the material is simply translated or rotated in space—are typically ensured by utilizing relative geometric information (such as distances and angles) as features, or by the intrinsic design of GNN architectures that learn invariant representations. For periodic crystalline materials, graph construction must correctly account for periodicity by connecting atoms near unit cell boundaries to their periodic images within the defined cutoff radius, effectively creating a representation of an infinite lattice from which a local cluster or supercell graph is often extracted for computational processing.

10.5.2 Classical Graph-derived and Structural Fingerprints

Before the widespread adoption of GNNs, considerable research was dedicated to developing methodologies that compute fixed-size vector representations, or "fingerprints," directly from these atomic graph structures or coordinate sets. These fingerprints could subsequently serve as input to standard ML algorithms like support vector machines or random forests.

Early efforts often drew inspiration from chemical graph theory, primarily applied to organic molecules, adapting various connectivity indices such as the Randić index [11] or the Wiener index [12]. These indices typically quantify aspects of molecular branching, compactness, or overall topology by performing summations over functions of atom degrees (coordination numbers), path lengths within the graph, or interatomic distances. Complementing these were other topological descriptors that provide a more global or averaged structural signature by measuring graph properties like average coordination number, graph density, local clustering coefficients, distributions of shortest path lengths between atoms, or the frequencies of small cyclic motifs within the atomic network.

Attention also turned to matrix-based representations, particularly for molecular systems. For instance, the Coulomb matrix (CM) represents the entire system as a matrix where diagonal elements are related to the nuclear charges of individual atoms (often a polynomial fit), and off-diagonal elements represent the Coulombic repulsion $\left(Z_iZ_j/d_{ij}\right)$ between atom pairs i and j. To achieve permutational invariance with respect to atom indexing and to handle systems with varying numbers of atoms, the sorted eigenvalues of this matrix are commonly used as the feature vector, frequently padded with zeros for smaller systems to ensure a consistent vector length. The Sine Matrix was introduced as a variant designed to be invariant to nuclear permutations without the need for eigenvalue decomposition. These matrix representations demonstrated considerable success in predicting quantum mechanical energies and properties of small- to medium-sized molecules.

Further refinement in structural fingerprinting came from distribution-based methods. The radial distribution function (RDF), or pair correlation function $g(r)$, which statistically describes the probability of finding an atom at a distance r from a reference atom, can be effectively utilized as a structural fingerprint. By discretizing the distance r and using the $g(r)$ values at these discrete points, a feature vector is generated that encapsulates information about coordination shells and average packing densities. Similarly, bond order parameters (BOPs), such as the Steinhardt parameters (Q_l, W_l), offer a sophisticated means to quantify the local orientational symmetry around an atom based on the geometric arrangement of its neighbors, as derived from the graph's connectivity. Averaging these local BOPs over all atoms within a structure yields a global fingerprint that is sensitive to different crystal symmetries (e.g., distinguishing between face-centered cubic (FCC), body-centered cubic (BCC), or hexagonal close-packed (HCP) arrangements) and local motifs like icosahedral order, proving valuable for differentiating polymorphs or distinct condensed phases.

More advanced local environment descriptors were developed to provide highly detailed and invariant representations of an atom's immediate neighborhood. The smooth overlap of atomic positions (SOAP) [13] descriptor has become particularly influential. Conceptually, SOAP involves representing the neighbor density around a central atom using a sum of Gaussian functions centered at each neighbor's position. This density is then expanded in a basis of spherical harmonics and radial basis functions, and the power spectrum of the resulting expansion coefficients forms the rotationally and permutationally invariant SOAP vector. These vectors, computed for each atom, can be averaged to create a global descriptor or used directly within kernel-based ML methods. Another widely used set of local descriptors is the atom-centered symmetry functions (ACSFs), initially developed by Behler and Parrinello for constructing neural network interatomic potentials. ACSFs capture the local atomic environment around an atom through a series of two-body (radial) and three-body (angular) functions, which are carefully designed to be invariant to rotation, translation, and permutation of like atoms. Extending beyond pairwise and three-body interactions, the many-body tensor representation (MBTR) [14] offers a systematic framework for considering k-body correlations. It typically involves constructing distributions of $k = 1$ terms (elemental composition), $k = 2$ terms (inverse distances between atomic pairs), $k = 3$ terms (angles formed by atomic triplets), and potentially higher-order terms such as dihedral angles for $k = 4$, providing a progressively richer description of the material's geometry.

These classical structural descriptors and fingerprints have been instrumental in numerous studies predicting materials properties. However, their design often necessitates considerable domain expertise and chemical intuition, and they may not always capture all relevant structural information optimally or generalize well across diverse material classes or complex structural transformations.

10.5.3 GNNs Learning Features from Atomic Architectures

A paradigm shift in representing and learning from atomic structures occurred with the advent and rapid proliferation of GNNs [15]. GNNs are a powerful class of deep learning models specifically architected to operate directly on graph-structured data. Their suitability for materials representation is manifold: graphs provide a natural and intuitive abstraction for atomic systems; GNNs can learn relevant features directly from the graph structure in an end-to-end manner, thereby obviating the need for extensive manual feature engineering that characterized earlier approaches; they inherently accommodate graphs of varying sizes, corresponding to materials with different numbers of atoms; and their architectures can be explicitly designed to respect fundamental physical invariances.

The central mechanism in most GNNs is an iterative process of message passing. In this process, nodes (atoms) progressively update their feature vectors, or embeddings, by receiving "messages" (information) from their neighboring nodes, aggregating these messages, and then combining this aggregated information with their own current state. Each round of message passing allows a node to incorporate information from progressively larger neighborhoods. After several such rounds, the learned embedding of each node effectively captures detailed information about its local atomic environment extending out to a certain radius, which is related to the number of message-passing layers or iterations in the GNN. This iterative refinement allows the network to learn complex structural patterns and correlations directly from the graph topology and initial node/edge features.

The general message passing framework has given rise to numerous specific GNN architectures, several of which have demonstrated significant utility and impact within materials science. The message passing neural network (MPNN) [16] formulation provides a highly general blueprint, typically defined by distinct message functions that compute information from neighbors, aggregation functions that combine these messages (e.g., sum, mean, and max), and update functions that integrate aggregated messages with a node's previous state to compute its new embedding. Building upon this, graph convolutional networks (GCNs) [17] generalize the concept of convolutional operations from regular grid-like data to the irregular topology of graphs, often defined through spectral graph theory or spatially localized aggregation rules. To enable nodes to selectively focus on more important neighbors during information aggregation, graph attention networks (GATs) [18] incorporate attention mechanisms, which learn to assign different weights or importance scores to messages received from different neighbors, thereby refining the feature learning process.

Beyond these general frameworks, several GNN architectures have been specifically tailored or widely adopted for materials science applications due to their incorporation of physical insights. SchNet [19], for example, employs continuous-filter convolutions where the filter weights are generated by functions of interatomic distances, proving effective for predicting quantum chemical properties of both molecules and, when adapted for periodicity, crystalline materials. The atomistic line GNN (ALIGNN) [20] provides another approach to incorporate angular information by constructing a line graph (where nodes represent bonds and edges signify shared atoms, hence angles) and performing message passing concurrently on both the original atom graph and this line graph. Pioneering work in applying GNNs to crystalline materials includes the crystal graph convolutional neural network (CGCNN) and its subsequent enhancements like MEGNet (materials graph network) [21]. These models typically take elemental features for nodes and interatomic distances for edges as inputs and have demonstrated broad success in predicting a diverse range of properties for inorganic crystal structures. MEGNet further extended this by incorporating global state variables, such as temperature or pressure, directly into the graph learning framework.

To arrive at a prediction for the entire material, which is a graph-level task (e.g., predicting the total formation energy or classifying the material type), the final learned node embeddings $\mathbf{h}_i^L$ (obtained after L layers of message passing) must be aggregated into a single, fixed-size graph-level embedding, $\mathbf{h}_G$. This crucial step, often referred to as the readout phase, can be implemented through simple operations like summing, averaging, or max-pooling the node embeddings across all atoms in the graph. More sophisticated readout functions employ attention mechanisms that learn to weight the contributions of different atoms or local environments based on their relevance to the target property being predicted. This consolidated graph-level vector $\mathbf{h}_G$ then serves as the input to a final feed-forward neural network (or other suitable ML model) for predicting the desired material property. The principal advantages offered by GNNs include their remarkable ability to learn task-relevant structural features directly from raw atomic configurations in an end-to-end manner, their inherent capacity to handle systems of varying sizes, and their consistent demonstration of state-of-the-art predictive performance across numerous material property benchmarks. When applying GNNs to crystalline materials, specific considerations include the robust and efficient handling of periodic boundary conditions during neighbor identification and message passing, as well as developing strategies to incorporate the influence of longer-range interactions that might extend beyond the typical graph cutoff radii, if these are deemed critical for the property being modeled.

The ascendancy of graph-based features, particularly those derived from GNNs, has undeniably revolutionized materials property prediction and accelerated virtual screening. Yet, to fully harness their potential, several inherent limitations of current methodologies must be acknowledged, which in turn illuminate critical directions for future research and development. A primary cluster of challenges revolves around operational efficacy and epistemological transparency. Many contemporary GNN architectures encounter scalability issues when applied to exceedingly large atomic systems or when used to generate extensive dynamic trajectories, imposing constraints due to computational and memory demands. Concurrently, the "black box" nature of deep GNNs often impedes interpretability, making it difficult to extract clear scientific insights or verify that the model has learned physically meaningful structure–property relationships rather than spurious correlations. This is compounded by a strong dependency on large, high-quality, and diverse datasets; GNNs may struggle with sample efficiency, and their ability to generalize to out-of-distribution materials—those with novel chemistries or structural motifs not well-represented during training—remains a significant concern. Further limitations arise in the fidelity of representing complex physical realities. Effectively incorporating crucial long-range interactions (e.g., electrostatics) beyond typical graph cutoffs without incurring prohibitive computational costs is an ongoing challenge.

Similarly, capturing the nuances of material dynamics, finite temperature effects, inherent structural disorder (as in amorphous phases or liquids), and multi-scale phenomena that bridge atomic arrangements to mesoscale characteristics often strains the capabilities of current atom-centric static graph models. While GNNs can be designed for basic geometric invariances, ensuring strict adherence to all relevant physical laws or complex crystallographic symmetries often requires more sophisticated architectural inductive biases than are commonly implemented. The need for continued standardization in graph construction protocols and comprehensive benchmarking also persists to ensure robust and reproducible progress.

Future trends in structural feature learning are vigorously aimed at surmounting these obstacles. We anticipate continued architectural innovations leading to more scalable, computationally efficient, and physically informed GNNs, including those with enhanced capabilities for modeling higher-order interactions and respecting complex symmetries through equivariance. A major thrust is towards improving interpretability and explainability through the development of explainable AI techniques tailored for graph-based materials models, allowing researchers to probe the decision-making processes of these networks. Addressing data limitations, research is intensifying in data-efficient learning strategies, such as transfer learning, meta-learning, few-shot learning, and particularly self-supervised pretraining on vast unlabeled structural databases to create powerful, transferable atomic representations. Moreover, the field is pushing towards GNNs capable of handling multi-scale and multi-physics problems more natively, perhaps through hierarchical architectures or hybrid models that integrate with continuum theories. The advancement of GNN-based generative models for the inverse design of novel materials with targeted properties, while ensuring their physical realism and synthesizability, remains a highly active area. A particularly transformative direction is the pursuit of large-scale, pretrained "foundation models" for atomic structures, which could learn universal structural priors adaptable to a multitude of downstream tasks with minimal specific training. Critically, the ultimate power of structural features will likely be realized through their synergistic integration with other data modalities—compositional, textual, and experimental—allowing for a truly holistic and context-aware understanding of materials, a concept central to the subsequent discussion on multimodal data fusion. Overcoming current limitations and realizing these future prospects will be pivotal for the continued impact of AI in accelerating materials innovation.

10.6 Strategies for Multimodal Data Fusion

The preceding sections have meticulously detailed the distinct categories of material descriptors: structured numerical and categorical features that provide a quantitative basis; textual and contextual features that unlock rich, qualitative knowledge from scientific literature; and atomic structure or graph-based features that encode the intricate architectural arrangements of atoms. While each modality offers a unique and valuable lens through which to view and interpret materials, each also possesses inherent limitations and captures only a partial aspect of a material's multifaceted nature. The overarching goal of multimodal data fusion in materials science is to strategically integrate these diverse information sources, creating synergistic representations that are more comprehensive, robust, and ultimately more predictive than those derived from any single modality in isolation. This approach acknowledges that a holistic understanding of material behavior—and the ability to accurately model structure–process–property relationships—often necessitates the concurrent consideration of information from disparate origins. The central premise is that by combining these complementary perspectives, ML models can uncover deeper insights and achieve superior performance, effectively leveraging the "whole" of available knowledge which is demonstrably greater than the sum of its individual parts. The process of fusing data from multiple modalities can occur at various stages within an ML pipeline, broadly classifiable into early, intermediate, and late fusion strategies. The choice of strategy depends on the nature of the data, the specific task, the complexity of the models employed, and the desired trade-offs between model performance, interpretability, and developmental effort.

10.6.1 Early Fusion Feature-level Integration

Early fusion, also known as feature-level fusion, represents the most direct approach to multimodal integration, where information from different sources is combined at the initial data representation stage. The core concept involves deriving feature vectors independently from each data modality and then concatenating, or otherwise combining, them into a single, augmented feature vector before this unified vector is fed into an ML model. For instance, a material might be represented by a vector formed by appending its TF-IDF representation or sentence embedding (from textual analysis) and its graph-derived structural fingerprint (e.g., an averaged SOAP vector or a GNN-generated graph embedding) to its existing vector of structured compositional and crystallographic descriptors. This unified feature vector then serves as the input for a single predictive model. The principal advantage of this method lies in its conceptual simplicity and ease of implementation, allowing standard ML algorithms to directly explore correlations and interactions across the combined feature space. However, early fusion is not without its challenges. The resulting concatenated feature vectors can become exceptionally high-dimensional and sparse, potentially leading to the "curse of dimensionality" and necessitating robust feature selection or dimensionality reduction techniques. Furthermore, careful normalization and scaling of features originating from different modalities (e.g., numerical structured data versus high-dimensional embedding vectors) are crucial to prevent features with larger magnitudes from unduly dominating the learning process, and the inherent heterogeneity of the combined features might pose difficulties for certain learning algorithms.

10.6.2 Intermediate Fusion Joint Learning and Hybrid Architectures

Intermediate fusion seeks to combine information from different modalities at a deeper level within the model itself, allowing for more complex interactions and shared representation learning. The principle underlying this approach is the development of a unified model architecture capable of processing multiple data streams concurrently, thereby enabling the model to learn shared latent spaces or complex intermodal interactions at its intermediate layers, rather than merely combining precomputed features at the input. Common architectural approaches include neural networks with multiple input branches, each tailored to a specific modality (e.g., a CNN for spectra, a GNN for atomic structures, and a Transformer for text). The outputs of these branches are then merged at one or more points within the network, often through concatenation followed by further layers, or via more sophisticated mechanisms like cross-modal attention. Cross-modal attention is particularly powerful, as it allows different parts of one modality's representation to selectively integrate information from relevant parts of another, learning explicit interdependencies. For example, structured descriptors like temperature can act as global conditioning attributes within GNNs. The primary strength of intermediate fusion is its potential to learn richer, more nuanced interactions and shared representations, often leading to superior predictive performance. However, designing and training these sophisticated hybrid architectures presents greater complexities, requiring careful architectural engineering and often larger datasets to effectively learn the increased number of parameters and intricate intermodal relationships.

10.6.3 Late Fusion Decision-level Integration and Ensemble Methods

Late fusion, also referred to as decision-level fusion or ensemble learning, adopts a strategy where separate ML models are trained for each modality, and their outputs are combined at the prediction stage. The methodology involves training distinct models independently for each data modality or for specific combinations thereof; the predictions generated by these individual models are then combined to produce a final, aggregated prediction. Common techniques for prediction combination include simple averaging, weighted averaging (where weights might reflect the confidence or expected accuracy of each unimodal model), majority voting for classification tasks, or more advanced stacking, where a meta-learner is trained using the outputs of the individual base models as its input features. Late fusion offers several practical benefits: it allows for modular model development, leveraging existing models for each modality. It can be more robust to noise or missing data in one modality and readily permits the combination of heterogeneous model types. The principal limitation is that late fusion may fail to capture subtle, low-level interactions between modalities, as these are not explicitly learned during the individual model training phases, and its effectiveness critically depends on the diversity of base models and the chosen combination strategy.

10.6.4 Multimodal Strategies in Materials Informatics

In the practical application of multimodal data fusion to materials science, several specific strategies leveraging the aforementioned fusion levels have gained considerable traction by combining the unique strengths of different data types. One common and effective approach involves augmenting traditional structured data with textual insights derived from scientific literature. For instance, embedding vectors representing the synthesis methods or experimental conditions detailed in research articles are often concatenated with quantitative compositional and crystallographic features. This allows ML models to implicitly incorporate crucial contextual knowledge—such as subtle processing variations or reported defects—that is typically absent from structured databases, thereby refining predictions for properties known to be sensitive to their processing history or nuanced experimental setup.

Another powerful set of strategies centers on integrating graph representations of atomic structures with various complementary features. GNNs inherently achieve a basic level of fusion by utilizing compositional features (e.g., elemental properties) as initial node attributes. This is often extended by incorporating global structured descriptors, such as experimental temperature or applied pressure, as conditioning inputs or graph-level state variables within GNN architectures like MEGNet, enabling property predictions under diverse environmental conditions. Furthermore, GNN-derived graph embeddings, which encapsulate the learned structural fingerprint of a material, can be subsequently fused with textual features describing its observed performance, known applications, or documented anomalies not directly inferable from its idealized atomic structure alone, often through concatenation or attention-based mechanisms.

The advent of Transformer-based unified multimodal models presents new and highly flexible opportunities for achieving deep, learned integration of diverse data streams. These architectures are being explored for their capacity to process input sequences composed of mixed-modality tokens—for example, a sequence might interleave textual descriptions with discretized numerical properties and simplified structural fingerprints, each appropriately tokenized. The cross-attention mechanisms inherent in Transformers are particularly adept at modeling fine-grained alignments and complex interdependencies between these different informational tokens, potentially allowing for an end-to-end learning of both modality-specific representations and their intricate fusion. While the optimal methods for tokenizing heterogeneous materials data for such models are still evolving, the prospect of a single, powerful architecture learning complex cross-modal relationships is compelling, though often associated with significant computational demands.

Finally, the utilization of materials science knowledge graphs (KGs) offers a structured and semantically rich framework for enhancing multimodal fusion strategies. KGs, which are meticulously constructed by curating and integrating information from both structured databases and knowledge extracted from scientific literature, explicitly represent materials, properties, processes, and their myriad relationships as interconnected entities. These KGs contribute to fusion primarily by providing KG embeddings—learned vector representations of entities and relations—which can serve as an additional, context-rich feature modality. These embeddings, imbued with relational information, can be fused with structured, textual, or graph-structural features to provide a broader understanding. Moreover, KGs can guide the fusion process by highlighting relevant connections between data points from different modalities or by facilitating data augmentation, thereby incorporating curated domain knowledge and complex relational patterns that might be challenging for models to learn implicitly from other data sources alone.

The strategic fusion of diverse data modalities—encompassing structured, textual, and graph-based structural features—holds profound promise for creating exceptionally insightful and powerful predictive models within materials science. However, realizing this transformative potential necessitates navigating a landscape of significant hurdles. Effectively implementing multimodal fusion confronts several inherent challenges, spanning data-centric issues such as managing the pervasive heterogeneity across different data types, ensuring accurate alignment and registration of disparate data sources to specific material instances, and gracefully handling the common problem of missing data in one or more modalities. Concurrently, model-centric difficulties include achieving computational scalability for increasingly complex fusion architectures and ever-larger datasets, enhancing the often-limited interpretability of how these fused models derive their predictions, and methodologically addressing the optimal way to balance the contributions of potentially redundant or variably informative data streams. Despite these obstacles, the impetus for pursuing multimodal integration remains strong, driven by clear and compelling benefits. Fused models consistently demonstrate the potential for enhanced predictive accuracy and improved model robustness when compared to their unimodal counterparts, as they leverage a more complete informational basis. More critically, such integrated approaches facilitate a deeper and more holistic understanding of complex structure–process–property–performance (SPPP) relationships. By enabling ML systems to learn from the synergistic interplay of diverse influencing factors—from atomic configurations and elemental compositions to synthesis narratives and experimental conditions—multimodal fusion can, in turn, lead to the discovery of novel scientific insights and unexpected correlations that might remain hidden when data sources are analyzed in isolation.

The future prospects in this domain are centered on developing more sophisticated, practical, and reliable solutions to overcome current limitations. Key research directions include the creation of advanced, yet interpretable, fusion architectures that can dynamically and adaptively weigh modality importance based on context or task. Self-supervised multimodal pre-training is rapidly emerging as a powerful technique to learn general-purpose material embeddings from vast, unlabeled heterogeneous datasets, fostering better generalization and data efficiency. Furthermore, there is a significant and accelerating drive towards developing inherently multimodal foundation models for materials science, envisioned as large-scale AI systems capable of seamlessly processing, integrating, and reasoning over structured data, scientific text, and atomic structural information within a unified framework. These collective advancements aim to surmount existing data and model complexities, paving the way for AI systems endowed with a truly comprehensive grasp of the material world and thereby significantly accelerating the pace of discovery and innovation in the field.

References

1 Harris ZS. Distributional structure. Word. 1954;10 (2–3):146–62.

2 Sparck Jones K. A statistical interpretation of term specificity and its application in retrieval. Journal of Documentation. 1972;28(1):11–21.

3 Bengio Y, Ducharme R, Vincent P, Jauvin C. A neural probabilistic language model. Journal of Machine Learning Research. 2003; 3(Feb):1137–55.

4 Mikolov T, Chen K, Corrado G, Dean J. Efficient estimation of word representations in vector space [Internet]. arXiv; 2013 [cited 2025 July 14]. Available from: http://arxiv.org/abs/1301.3781

5 Pennington J, Socher R, Manning CD. GloVe: Global vectors for word representation. In: Proceedings of the 2014 Conference on Empirical Methods in Natural Language Processing (EMNLP) [Internet]; 2014 [cited 2025 July 14]. pp. 1532–43. Available from: https://aclanthology.org/D14-1162.pdf

6 Joulin A, Grave E, Bojanowski P, Douze M, Jégou H, Mikolov T. FastText.zip: Compressing text classification models [Internet]. arXiv; 2016 [cited 2025 July 14]. Available from: http://arxiv.org/abs/1612.03651

7 Le Q, Mikolov T. Distributed representations of sentences and documents. In: International Conference on Machine Learning [Internet]. PMLR; 2014 [cited 2025 July 14]. pp. 1188–96. Available from: http://proceedings.mlr.press/v32/le14.html?ref=https://githubhelp.com

8 Reimers N, Gurevych I. Sentence-BERT: Sentence embeddings using Siamese BERT-Networks [Internet]. arXiv; 2019 [cited 2025 July 14]. Available from: http://arxiv.org/abs/1908.10084

9 Devlin J, Chang MW, Lee K, Toutanova K. BERT: Pre-training of deep bidirectional transformers for language understanding. In: Proceedings of the 2019 Conference of the North [Internet]. Minneapolis, MN: Association for Computational Linguistics; 2019 [cited 2024 Mar 20]. pp. 4171–86. Available from: http://aclweb.org/anthology/N19-1423

10 Beltagy I, Lo K, Cohan A. SciBERT: A pretrained language model for scientific text. arXiv preprint arXiv:1903.10676. 2019. Available from: https://arxiv.org/abs/1903.10676

11 Randic M. Characterization of molecular branching. Journal of the American Chemical Society. 1975;97(23):6609–15.

12 Rouvray DH. The rich legacy of half a century of the Wiener index. In: Topology in Chemistry [Internet]. Elsevier; 2002 [cited 2025 July 14]. pp. 16–37. Available from: https://www.sciencedirect.com/science/article/pii/B9781898563761500068

13 Bartók AP, Kondor R, Csányi G. On representing chemical environments. Physical Review B [Internet]. 2013 [cited 2025 July 14];87(18). Available from: https://link.aps.org/doi/10.1103/PhysRevB.87.184115

14 Huo H, Rupp M. Unified representation of molecules and crystals for machine learning. Machine Learning: Science and Technology. 2022;3(4):045017.

15 Micheli A. Neural network for graphs: A contextual constructive approach. IEEE Transactions on Neural Networks. 2009;20(3):498–511.

16 Gilmer J, Schoenholz SS, Riley PF, Vinyals O, Dahl GE. Neural message passing for quantum chemistry. In: International Conference on Machine Learning [Internet]. PMLR; 2017 [cited 2025 July 14]. pp. 1263–72. Available from: https://proceedings.mlr.press/v70/gilmer17a

17 Kipf TN, Welling M. Semi-supervised classification with graph convolutional networks [Internet]. arXiv; 2017 [cited 2025 July 14]. Available from: http://arxiv.org/abs/1609.02907

18 Velickovic P, Cucurull G, Casanova A, Romero A, Lio P, Bengio Y. Graph attention networks. Stat. 2017;1050(20):10–48550.

19 Schütt KT, Sauceda HE, Kindermans PJ, Tkatchenko A, Müller KR. SchNet – A deep learning architecture for molecules and materials. The Journal of Chemical Physics [Internet]. 2018 [cited 2025 July 14];148: 241722. Available from: https://pubs.aip.org/aip/jcp/article/148/24/241722/962591

20 Choudhary K, DeCost B. Atomistic line graph neural network for improved materials property predictions. npj Computational Materials. 2021;7(1):185.

21 Chen C, Ye W, Zuo Y, Zheng C, Ong SP. Graph networks as a universal machine learning framework for molecules and crystals. Chemistry of Materials. 2019;31(9):3564–72.

11

Case Studies of Materials Predictive Modeling

11.1 Introduction

While conventional machine learning (ML) in materials research has excelled using structured numerical descriptors, it often fails to leverage the vast, context-rich knowledge embedded in scientific literature. To address this gap, a transformative approach utilizes hybrid artificial intelligence (AI) architectures, particularly Transformer-based networks, to synergistically fuse quantitative and qualitative data. These language-augmented models are designed to jointly encode structured information, such as Simplified Molecular Input Line Entry System (SMILES) strings or crystal graphs, with nuanced insights mined from textual sources like synthesis pathways and processing details. This chapter primarily provides relevant case studies on utilizing predictive models for materials design.

11.2 Polymer Property Prediction with Language Models

Polymers underpin numerous technological applications in areas such as sustainable materials, advanced electronics, and biomedicine [1, 2]. Their extensive chemical diversity affords the potential to tailor materials to specific functional requirements, yet the combinatorial breadth of polymer structures makes systematic discovery and optimization challenging. Polymer informatics addresses this challenge by applying ML to accelerate property prediction and to enable data-driven inverse design of polymers [3–5].

Christopher Kuenneth et al. [6] delve into the "polyBERT" initiative, a framework that treats polymer SMILES (PSMILES) representations as a formal chemical language. By pre-training a Transformer model on a corpus comprising millions of PSMILES, polyBERT learns the statistical grammar and semantic relationships of polymer representations in an unsupervised manner. The pretrained model maps any PSMILES string to a dense numerical embedding, a "Transformer fingerprint," that encodes learned structural features without manual feature engineering.

These embeddings can be used as inputs to supervised models for predicting a wide range of polymer properties, forming a scalable, end-to-end pipeline for ML-driven polymer informatics. Subsequent sections detail the dataset construction, model architecture, fingerprint analysis, predictive performance, and broader implications of this language-augmented strategy.

11.2.1 Data Preparation

In the polyBERT work, two separate datasets were carefully prepared, one for self-supervised pretraining of the language model and another for supervised training of downstream property prediction models (Figure 11.1).

For language model pretraining, the authors created a very large corpus containing 100 million hypothetical PSMILES strings. As outlined in Figure 11.1c, this workflow started from 13,766 known polymers that had already been synthesized. Using the Breaking Retrosynthetically Interesting Chemical Substructures (BRICS) method implemented in the RDKit toolkit, these polymers were split into 4,424 distinct chemical fragments. These fragments were then systematically combined through both random and enumerative chemical rules to construct the vast library of 100 million hypothetical PSMILES strings. While these strings represent chemically valid polymer structures, the majority had not been previously synthesized, providing a diverse and expansive dataset for polyBERT to learn the fundamental "grammar" of polymer construction beyond the scope of known materials.

For the subsequent supervised learning phase, aimed at training models to predict specific material properties from the polyBERT-generated fingerprints, the authors employed a comprehensive property dataset. This dataset collates 35,517 individual data points covering 29 distinct polymer properties, which include both experimentally measured and computationally derived values. The dataset is broad in scope, containing about 28,061 records for homopolymers, which correspond to

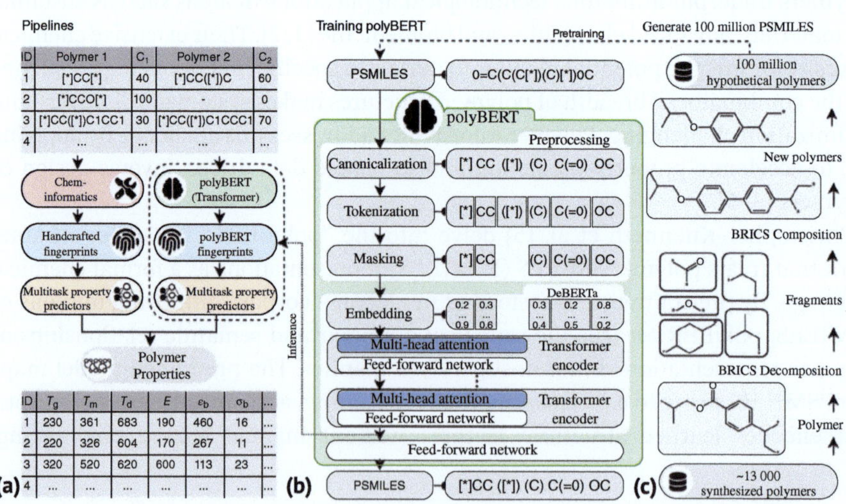

Figure 11.1 Polymer informatics with polyBERT. (a) Pipelines, (b) training polyBERT, (c) generate 100 million PSMILES. *Source*: [6] / Springer Nature / CC BY 4.0.

11,145 unique monomers, and 7,456 records for copolymers, spanning 1,338 distinct copolymer chemistries. Most copolymer entries describe random copolymers with different comonomer ratios. This type of structure is well supported by the selected copolymer fingerprint approach, which sums comonomer fingerprints weighted by their compositions. The property dataset was compiled from earlier studies, published literature, and public repositories, and it served as the ground truth needed to train and carefully assess the property prediction models. The value distributions for the 29 properties are shown as histograms in Figures 11.2–11.7 of the original study.

11.2.2 Model Architecture and Pre-training

The central component of the polyBERT framework is a sophisticated chemical language model, meticulously designed and trained to interpret the structural information encoded in PSMILES strings. This model, based on the high-performing DeBERTa (Decoding-enhanced BERT with disentangled attention) architecture, iteratively processes an extensive corpus of 100 million

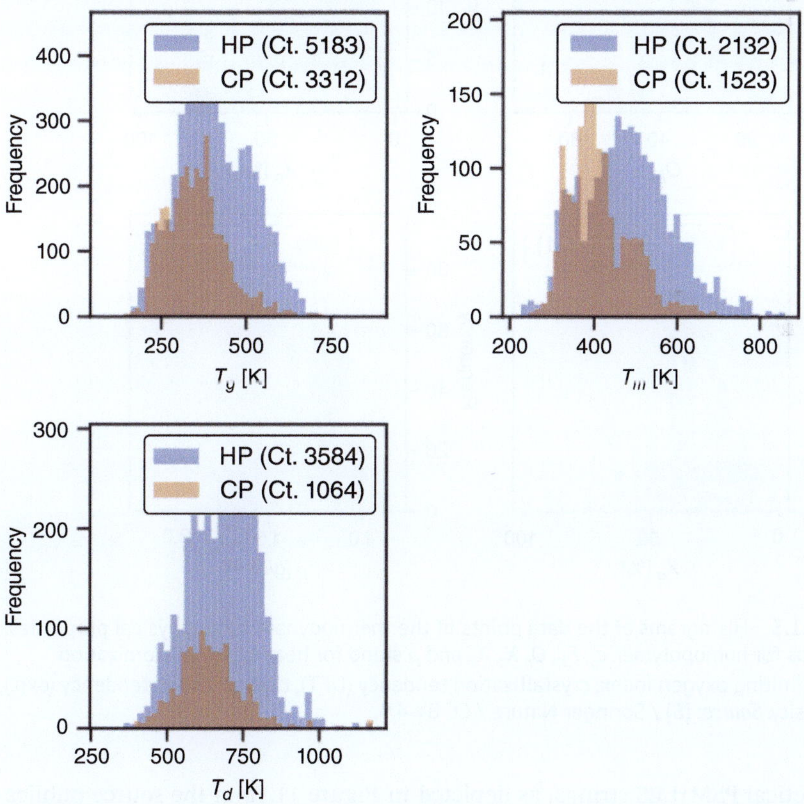

Figure 11.2 Histograms of the data points of the thermal properties. HP and CP stand for homopolymer and copolymer, respectively. T_g, T_m, and T_d stand for glass transition temperature, melting temperature, and degradation temperature. *Source*: [6] / Springer Nature / CC BY 4.0.

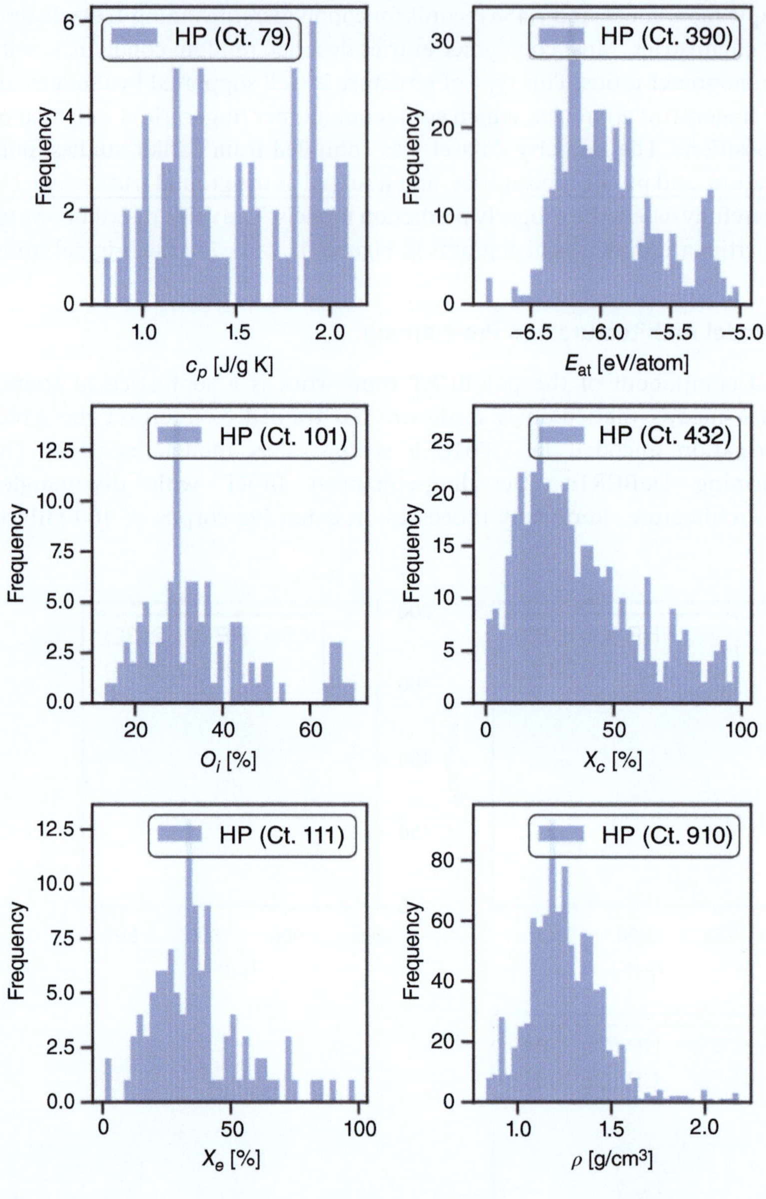

Figure 11.3 Histograms of the data points of the thermodynamic and physical properties. HP stands for homopolymer. c_p, E_{at}, O_i, X_c, X_e, and ρ stand for heat capacity, atomization energy, limiting oxygen index, crystallization tendency (DFT), crystallization tendency (exp.), and density. *Source*: [6] / Springer Nature / CC BY 4.0.

hypothetical PSMILES strings, as depicted in Figure 11.1b of the source publication by Kauwe et al. The selection of DeBERTa was predicated on its demonstrated superiority in various natural language processing benchmarks and internal comparative evaluations against other prominent BERT-like models. The sheer scale of this pretraining dataset places polyBERT among the language models trained

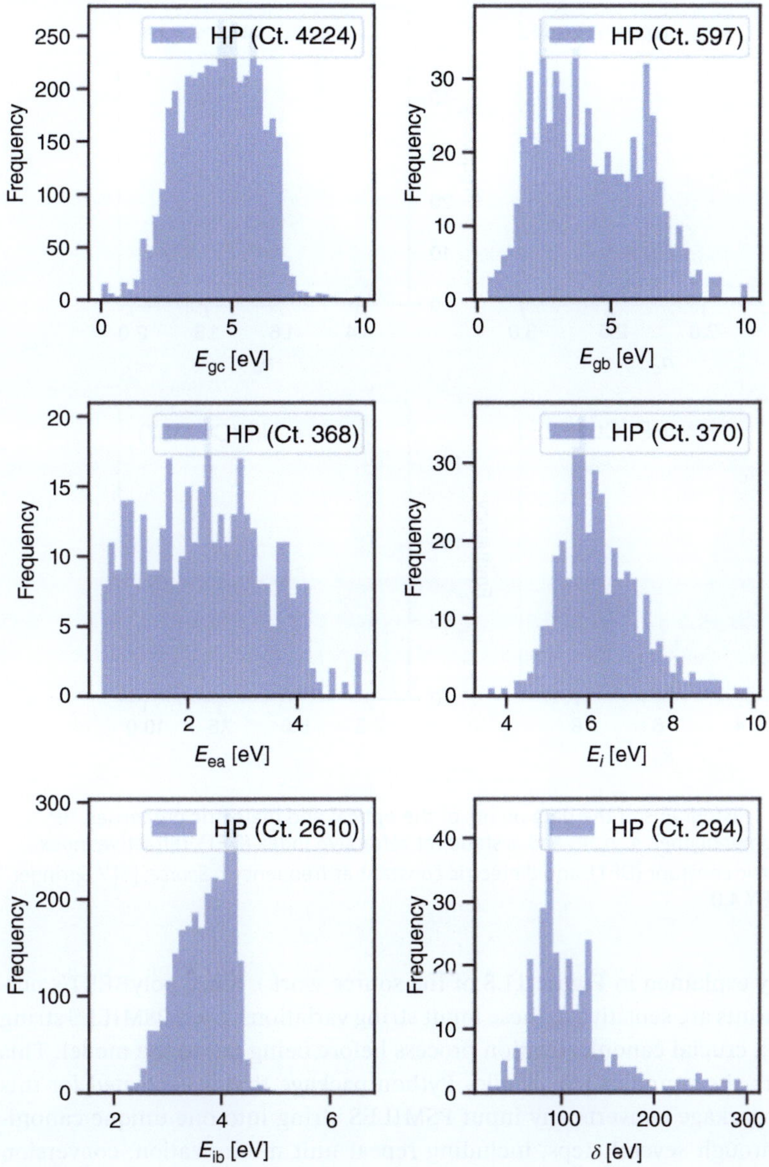

Figure 11.4 Histograms of the data points of the electronic properties. HP stands for homopolymer. E_{gc}, E_{gb}, E_{ea}, E_i, E_{ib}, and δ stand for band gap (chain), band gap (bulk), electron affinity, ionization energy, electronic injection barrier, and cohesive energy density. *Source*: [6] / Springer Nature / CC BY 4.0.

on the largest chemical data corpora, reflecting a trend of leveraging increasingly vast datasets in this domain. A critical prerequisite for the effective functioning of polyBERT is the consistent representation of its input; raw PSMILES strings can exhibit significant variability for the same polymer structure due to differences in starting atom selection, directionality, or the delineation of the repeat unit

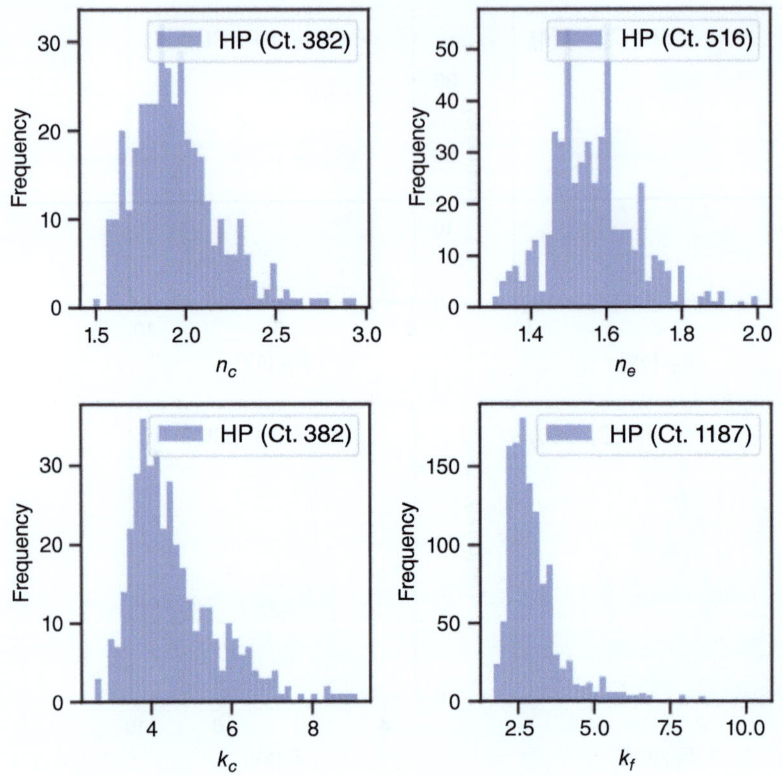

Figure 11.5 Histograms of the data points of the optical and dielectric properties. HP stands for homopolymer. n_c, n_e, k_c, and k_f stand for refractive index (DFT), refractive index (exp.), dielectric constant (DFT), and dielectric constant at frequency f. *Source*: [6] / Springer Nature / CC BY 4.0.

(graphically explained in Figure 11.8 of the source work). Since polyBERT's output fingerprints are sensitive to these input string variations, each PSMILES string undergoes a crucial canonicalization process before being fed to the model. This is done with the canonicalize_psmiles Python package that was created for this work. The package converts any input PSMILES string into one unique canonical form through several steps, including repeat unit minimization, conversion to a periodic representation, and RDKit-based canonicalization. In this way, the language model always receives a consistent structural representation as input.

Once the PSMILES strings are canonicalized, the subsequent stage involves preparing them for the self-supervised learning task through which polyBERT acquires its understanding of polymer chemistry. This begins with tokenization, where the canonical PSMILES strings are segmented into fundamental units using the SentencePiece tokenizer. A carefully constructed vocabulary of 265 distinct tokens is employed, encompassing common PSMILES characters, comprehensive symbols for the 118 elements of the periodic table (both uppercase and lowercase), numerals, and various special characters frequently encountered in SMILES notation (such as [*], (,), and =), thus ensuring full coverage of the PSMILES linguistic domain.

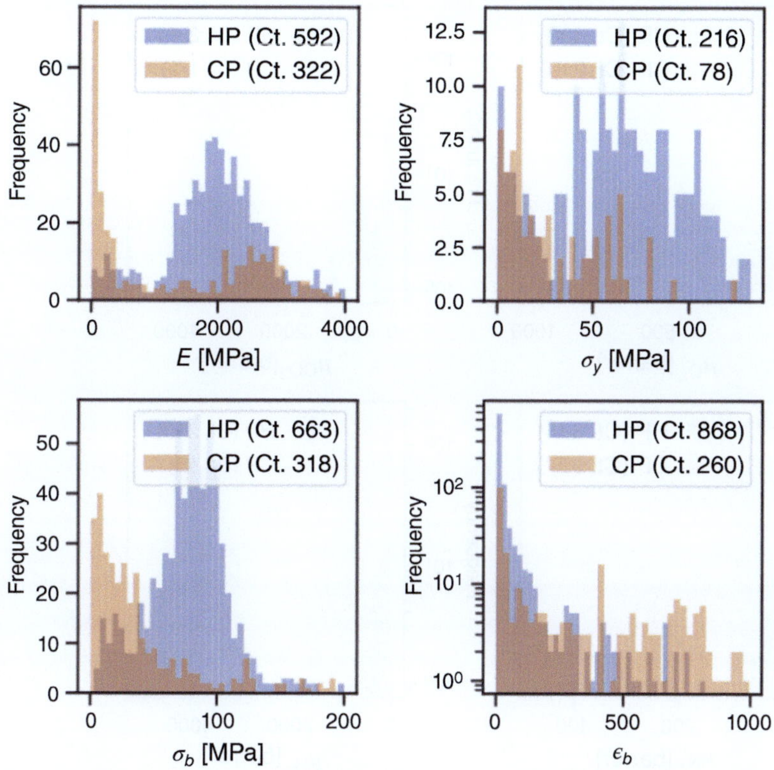

Figure 11.6 Histograms of the data points of the mechanical properties. HP and CP stand for homopolymer and copolymer, respectively. E, σ_y, σ_b, and ε_b stand for Young's modulus, tensile strength at yield, tensile strength at break, and elongation (EL) at break. *Source*: [6] / Springer Nature / CC BY 4.0.

The core of polyBERT's learning process is a self-supervised masked language modeling (MLM) task. In this setup, which resembles a fill-in-the-blanks task, 15% of tokens in each tokenized PSMILES string are randomly swapped with a special [MASK] token. polyBERT is trained to recover the original tokens by using the context from the surrounding unmasked tokens. This objective forces the model to learn detailed relationships, recurring patterns, and implicit rules of polymer construction through updates to the weights in its Transformer encoder layers. For pretraining, 80 million canonical PSMILES strings were used, and another 20 million were held out for validation. Reaching a validation F_1 score above 0.99 shows that polyBERT could reconstruct the masked parts of the polymer sequences with very high accuracy, which reflects a strong learned grasp of the polymer chemical language.

11.2.3 Fingerprints Characterization

The rigorous self-supervised pretraining enables polyBERT to function as an "expert polymer chemical linguist," with its Transformer-based architecture, comprising multi-head self-attention and feed-forward layers, allowing the model

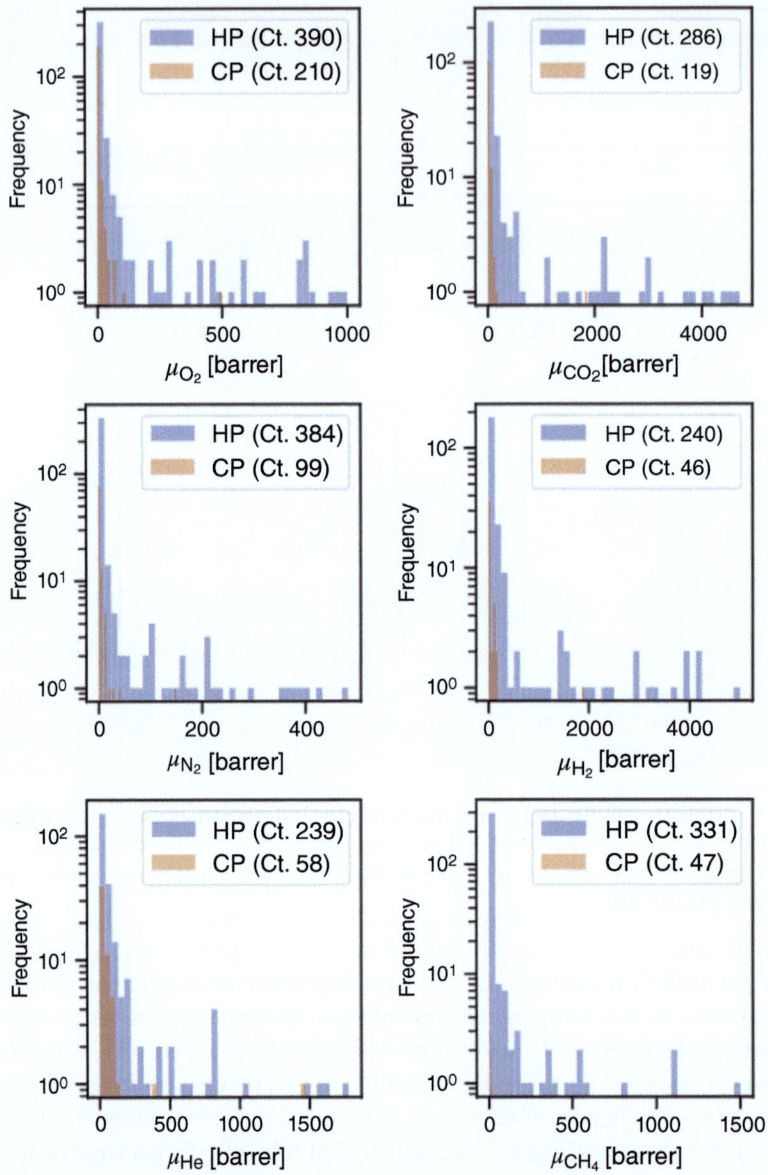

Figure 11.7 Histograms of the data points of the permeability properties. HP and CP stand for homopolymer and copolymer, respectively. μ_{O_2}, μ_{CO_2}, μ_{N_2}, μ_{H_2}, μ_{He}, and μ_{CH_4} stand for O_2 gas permeability, CO_2 gas permeability, N_2 gas permeability, H_2 gas permeability, He gas permeability, and CH_4 gas permeability. *Source*: [6] / Springer Nature / CC BY 4.0.

to capture relationships between tokens representing chemical motifs or atoms within PSMILES strings. Through this mechanism, polyBERT generates context-aware internal representations, from which polymer fingerprints are obtained by averaging token embeddings from the final encoder block (Figure 11.1b), using SentenceTransformers for implementation. These learned polyBERT fingerprints

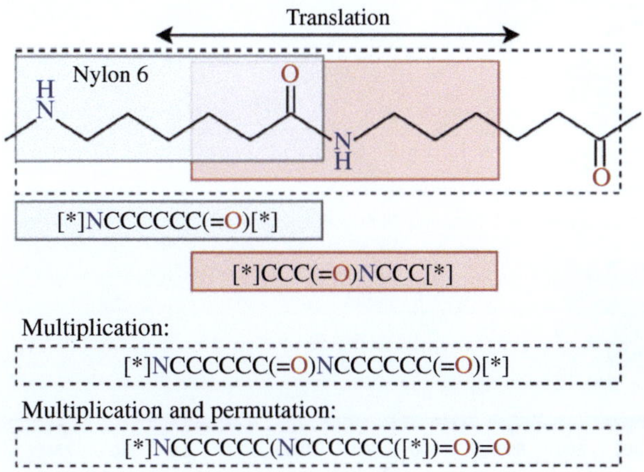

Figure 11.8 Translational, multiplicative, and permutational variances of PSMILES strings. *Source*: [6] / Springer Nature / CC BY 4.0.

were systematically compared with the handcrafted Polymer Genome (PG) fingerprints developed by the Ramprasad Group, which encode polymers across atomic, block, and chain scales but result in sparse, high-dimensional vectors (945 components, 93.9% zeros), unlike the fully dense 600-dimensional polyBERT fingerprints. The authors highlight that such dense and often lower-dimensional representations benefit ML models whose computational costs scale poorly with dimensionality, improving both efficiency and memory usage. To evaluate chemical relevance, the authors employed Uniform Manifold Approximation and Projection (UMAP) visualizations of homopolymers and copolymers (Figure 11.9), showing that polyBERT fingerprints cluster chemically similar polymers, such as poly(but-1-ene) and poly(pent-1-ene), in close proximity, while separating distinct polymers like poly(4-vinylpyridine). This organization parallels the chemically intuitive PG fingerprints and supports the conclusion that distances in polyBERT fingerprint space, such as cosine distances (Figure 11.10), provide meaningful measures of polymer similarity.

More evidence of what polyBERT learned was obtained by inspecting its internal behavior. For several example polymers, the normalized attention maps were added across all attention heads and encoder layers (Figure 11.11a–c), for selected polymers revealed that the model learns to assign high attention scores to chemically significant tokens and their relationships. For instance, high attention was observed for auxiliary tokens marking the beginning ([CLS], _) and end ([SEP]) of PSMILES strings, to connections between these and the first polymer endpoint [*], to next and next-to-next neighbor tokens (highlighting local connectivity), and strong interactions involving the second endpoint [*] token. In poly(4-vinylpyridine), large attention scores were found for the CN (pyridine nitrogen) token extending to its fourth or fifth neighbors, indicating the model recognized the distinct nature and influence of this heteroatom. Additionally, nonnegative matrix factorization of neuron activations within polyBERT's feed-forward network layers (Figure 11.11d–f) indicated that distinct sets of neurons become specialized for processing specific types

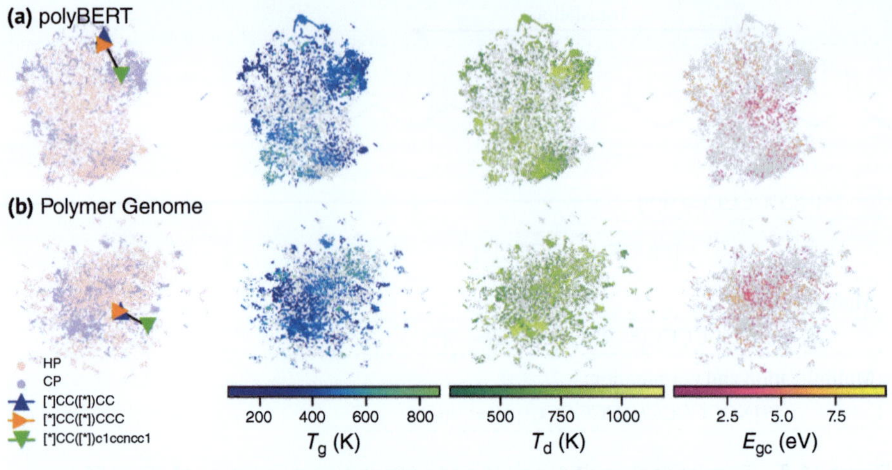

Figure 11.9 Two-dimensional UMAP plot of the fingerprints. (a) Cosine distances calculated using polyBERT. (b) Cosine distances calculated using Polymer Genome. *Source*: [6] / Springer Nature / CC BY 4.0.

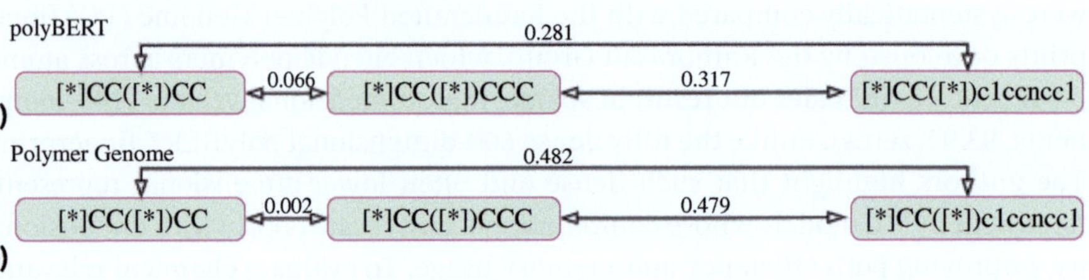

Figure 11.10 Cosine distances between polymer fingerprints. The PSMILES strings [*]CC([*])CC, [*]CC([*])CCC, and [*]CC([*])c1ccncc1 denote poly(but-1-ene), poly(pent-1-ene), and poly(4-vinylpyridine), respectively. *Source*: [6] / Springer Nature / CC BY 4.0.

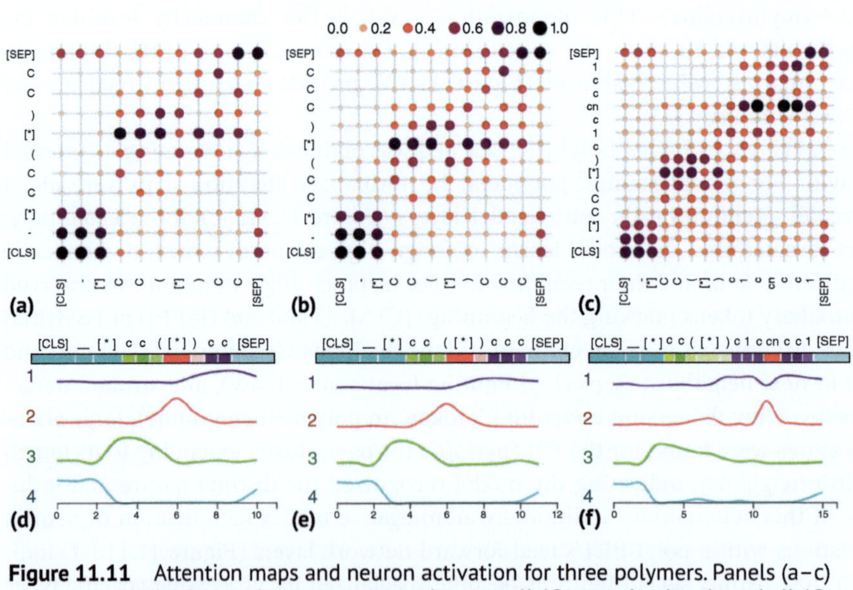

Figure 11.11 Attention maps and neuron activation for three polymers. Panels (a–c) present normalized attention maps, summed across all 12 attention heads and all 12 encoders of polyBERT. Panels (d–f) show the activations of factorized neurons in the feedforward network layers. *Source*: [6] / Springer Nature / CC BY 4.0.

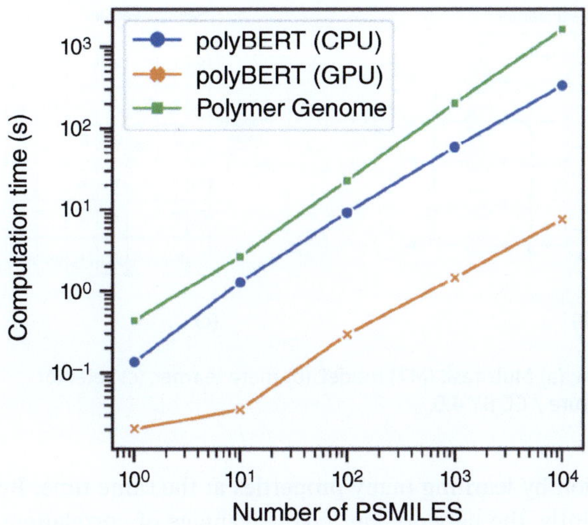

Figure 11.12 Computation times of polymer fingerprints. *Source*: [6] / Springer Nature / CC BY 4.0.

of tokens or chemical motifs (e.g., auxiliary tokens, backbone carbon atoms, side chain atoms, or unique heteroatoms like the CN). Together, these results suggest that the attention layers encode positional and relational information, while the feed-forward layers control how signals are passed and combined, and both contribute to producing detailed polyBERT fingerprints.

In addition to being chemically meaningful and interpretable, polyBERT fingerprints are notable for their high computational efficiency. The study reports in Figure 11.12 that calculating polyBERT fingerprints, particularly with GPU acceleration, is extremely fast. It is more than two orders of magnitude quicker, specifically about 215 times faster, than generating the comparable handcrafted PG fingerprints. This dramatic speed-up, with an estimated total time of 1.06 ms per polymer on a GPU for predicting 29 properties (including fingerprinting and multitask deep neural networks [MT-DNN] inference), is of paramount importance for high-throughput polymer informatics pipelines, particularly those designed to screen vast libraries of candidate polymers or support rapid iterations in generative design loops.

11.2.4 Performance Evaluation and Future Potential

The strongest test of the polyBERT framework is whether it can support accurate property prediction and enable large-scale searches across polymer chemical space. Success on these tasks demonstrates practical value and points to what the approach could enable in future studies.

To evaluate prediction quality in a careful way, polyBERT fingerprints were compared with the widely used handcrafted PG fingerprints using MT DNNs. This model choice matters because MT DNNs have previously achieved leading performance

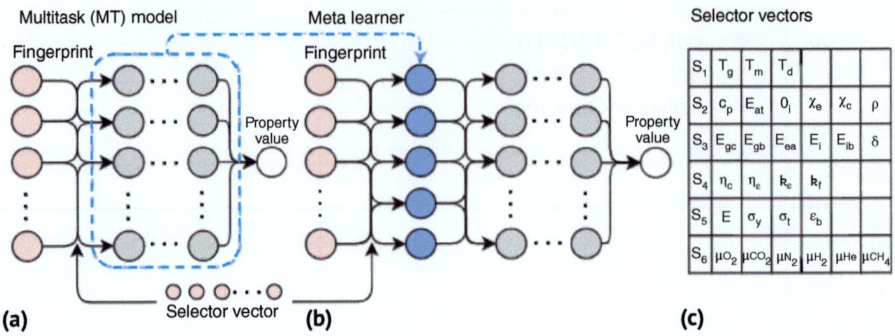

Figure 11.13 Learning workflow. (a) Multitask (MT) model, (b) meta learner, (c) selector vectors. *Source*: [6] / Springer Nature / CC BY 4.0.

for polymer property prediction by learning many properties at the same time. By training on multiple tasks jointly, the network can take advantage of correlations between properties. This shared learning can improve overall accuracy, and it is particularly helpful for properties with limited data as shown in Figure 11.13. The MT-DNN training setup followed common best practices, including fivefold cross-validation to support robustness and estimate generalization, and a consolidating meta-learner that predicts final property values by combining the ensemble of cross-validation model outputs. Neural network hyperparameters were tuned with the Hyperband method to minimize mean squared error (MSE), and dataset splits were stratified using polymer property information.

The comparative performance results, presented in Figure 11.14a, are particularly illuminating. Across the 29 polymer properties investigated, models based on polyBERT fingerprints achieved an overall coefficient of determination (R^2) of 0.80. This was remarkably close to the performance of models using the highly optimized, domain-expert-crafted PG fingerprints, which yielded an overall R^2 of 0.81. The authors noted that this proximity in accuracy was largely consistent across different property categories and individual properties, with polyBERT even slightly outperforming PG fingerprints for a few specific properties (X_c, X_e, and ϵ_b). The standard deviations in R^2 values across cross-validation splits, indicative of the homogeneity of data point distribution in the learning space, were also reported. These findings strongly suggest that the autonomously learned polyBERT representations can effectively match the predictive efficacy of traditional, labor-intensive handcrafted features. The meta-learner models, trained on unseen data and used for all final property predictions (their performance shown in Figure 11.14).

The polyBERT work presents a polymer informatics pipeline that is broadly applicable, very fast, and accurate. It is built around a Transformer model that was pre-trained on the chemical language of PSMILES strings. The combination of high speed and verified accuracy made it possible to predict 29 properties for 100 million hypothetical polymers at high throughput. This produced a large-scale map of the polymer property space shown in Figure 11.14c, which can act as a reference point for later ML studies and help reveal new structure–property relationships. Although the current approach generates fingerprints more than two orders of magnitude faster than handcrafted feature methods while reaching similar accuracy,

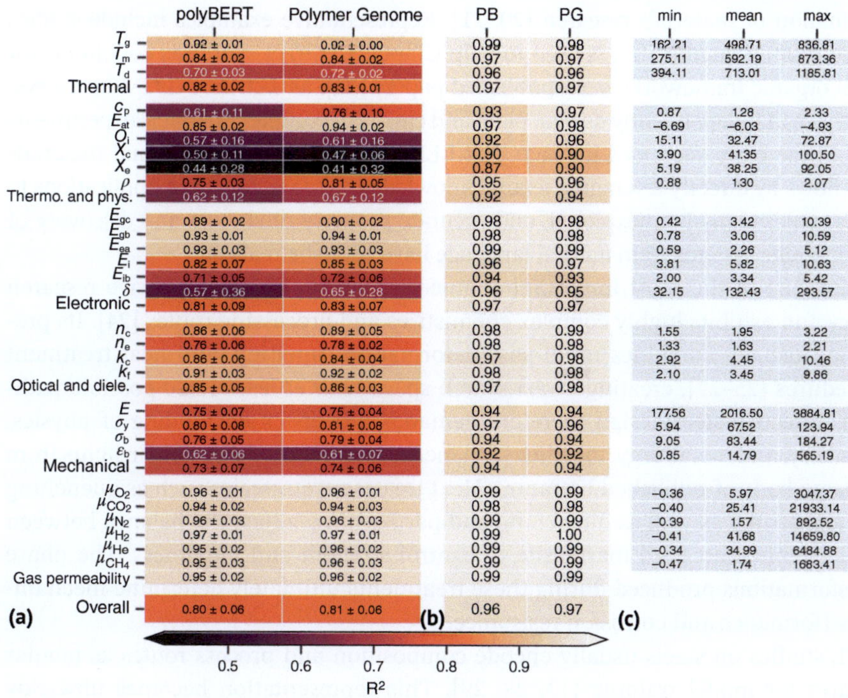

Figure 11.14 Coefficient of determination R^2 performance results for polyBERT (PB) and PG fingerprints. (a) reports the mean R^2 across the five cross-validation validation sets, together with the corresponding standard deviations. (b) shows R^2 values on the meta-learner's test set. Category-averaged R^2 values are listed in the final row of each block, and the overall R^2 values appear in the last block. (c) shows the minimum, mean, and maximum of polyBERT-based property predictions for 100 million hypothetical polymers. *Source*: [6] / Springer Nature / CC BY 4.0.

the authors point to several next steps. These include developing fingerprints that can be decoded for generative inverse design, improving interpretability using more advanced saliency techniques, and expanding the method to small molecules. Overall, this study highlights how modern language modeling can reshape material representations and support much faster discovery cycles in polymer science.

11.3 Steel Design and Process Optimization

In recent years, AI and ML have been used more widely in materials science to accelerate materials discovery and optimization [7–9]. Their performance, however, depends strongly on high-quality structured data and carefully crafted features, which require substantial expert effort. Large language models (LLMs) such as ChatGPT [10], Falcon [11], and BERT [12] have shown strong abilities, supported by massive training data, large neural network capacity, and self-supervised learning. The Transformer architecture with its attention mechanism [13] underpins these models [14], which have been widely adopted in natural language processing [15, 16], code generation [17–19], and automation of materials research [20, 21]. Representative examples include studies where prompt engineering is used to help ChatGPT extract synthesis conditions for metal organic frameworks from published papers [22], as well as a GPT-4-based system that can independently design, plan, and carry out complex chemical experiments [23]. These early successes indicate that LLMs can reshape conventional materials workflows by directly leveraging textual knowledge. To date, however, applications to alloys remain largely unexplored, and accurate property prediction and discovery of new metallic systems from natural language have yet to be realized.

Steel, one of the most important engineering alloys, has an extensive research legacy and exhibits highly complex chemistries and processing routes [24]. Its production usually combines multi-pass deformation, smelting, and heat treatment procedures [25–27], creating a vast search space with at least 1,020 possible pathways. Effective steel design therefore demands a deep understanding of physics, chemistry, and metallurgy, together with methods capable of distilling insights from the large body of published literature. Heat treatment strategies such as quenching and tempering, as well as quenching and partitioning, adjust the balance between soft austenite and hard martensite to control strength and toughness. The phase transformations produced during these treatments ultimately determine mechanical performance and corrosion resistance.

ML studies on steels usually encode composition and process routes as tabular features for model training [13, 28, 29]. This representation becomes unwieldy when proposing new processing sequences, since it produces high-dimensional and sparse feature spaces. A language-based description of diverse process chains is therefore essential. Integrating steel research with LLMs offers a viable way of tackling this challenge.

Shaohan Tian et al. [30] propose an end-to-end pipeline built on a LLM that converts materials text into property data, supporting high-accuracy prediction of mechanical properties and the search for new steels. The central component is a domain-specific LLM designated SteelBERT (Figure 11.15a). Pretrained on a comprehensive corpus of 4.2 million materials science abstracts and 55,000 full-text articles on steel, SteelBERT uses a disentangled attention mechanism to generate accurate knowledge representations from natural language inputs. This allows the full end-to-end pipeline (Figure 11.15b) to generate context-aware vector representations from text that describes composition and complex fabrication processes, so it can capture varied details across processing routes. The model accurately predicted the mechanical properties of 18 new steels reported in 2022 and 2023, reaching coefficient of determination (R^2) values of 78.17% (±3.40%) for yield strength (YS), 82.56% (±1.96%) for ultimate tensile strength (UTS), and 81.44% (±2.98%) for EL. The authors further refined the model by fine-tuning it on a small dataset of 64 austenitic stainless steel (ASS) samples (Figure 11.15c). This optimized model achieved significantly higher R^2 values of 89.85% (±6.17%) for YS, 88.34% (±5.95%) for UTS, and 87.24% (±5.15%) for EL. Using the fine-tuned model, they efficiently explored the large candidate space to identify an improved fabrication route and recommended an additional round of cold rolling followed by tempering. The steel produced with this route showed an excellent property balance, with YS of 960 MPa, UTS of 1,138 MPa, and EL of 32.5%, outperforming previously reported 15Cr ASSs (Figure 11.15d).

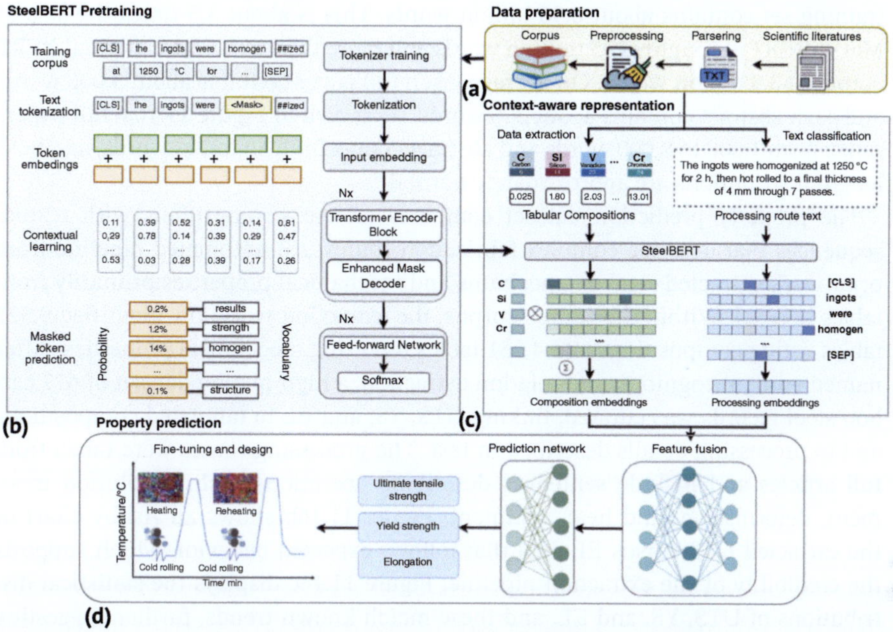

Figure 11.15 Quantitative mechanical properties prediction with SteelBERT. (a) Schematic workflow of data preparation. This workflow encompasses several stages of scientific documents download, corpus preprocessing, table and text information extraction. A corpus of scientific articles is scraped and preprocessed for pretraining SteelBERT. The composition and property information are also extracted for further quantitative mechanical properties prediction. (b) SteelBERT is a linguist specialized in the language of steel materials. After tokenization, the training corpus is passed to the DeBERTa model. Each of the 12 Transformer encoders consists of 12 attention heads. The final dense layer with a softmax activation function identifies the masked tokens. (c) Context-aware representation of steel information. SteelBERT is used to generate embeddings with 768 dimensions for both textual processing routes and chemical compositions. (d) A predictive model for mechanical properties is then constructed using a deep learning network that takes the natural language data as input. It takes embedded composition and text of processing routes as inputs to predict the YS, UTL, and total EL. After fine-tuning on the laboratory dataset, a novel advanced austenitic stainless steel is optimized with outstanding performance. *Source*: [30] with permission of Elsevier.

11.3.1 Corpus Collection

The pretraining corpus consisted of 4.2 million materials science abstracts and 55,000 full-text papers on steel. The collection began with full-text journal articles on steels published up to 2021. To improve robustness, the authors also added abstracts from the wider materials domain up to 2021, including conference papers and patents. Digital object identifiers for roughly 4.2 million abstracts and 55,000 full texts were gathered using the Crossref API, together with searches in Web of Science. The 55,000 full-text papers were obtained as plain text, XML, or HTML through Elsevier APIs and by web-scraping journals from Springer, ASME, and MDPI. For XML and HTML sources, a customized Python parser extracted metadata such as titles, abstracts, authors, keywords, and paragraph-level text. The final SteelBERT training set contains about 0.96 billion words. This is about 3.5 times the size of MatSciBERT at roughly 285 million words and more than 30% of the original BERT corpus at 3.3 billion words. On average, each full paper contains about 3,000 words and each abstract contains about 300 words. As shown in Figure 11.16a, full papers range from 1,000 to 8,000 words and abstracts range from 50 to 500 words, and both length distributions are approximately normal.

The property prediction dataset combines tabular compositions with textual sequences that describe complex fabrication routes. An automated pipeline from prior work extracted steel compositions and mechanical properties primarily from tables [31–33]. Within the source corpus, the extraction workflow identified 9,327 tables listing compositions and 4,251 tables reporting mechanical properties. After named entity recognition and relation extraction, a high-quality dataset of 677 carbon steel records was curated, linking UTS, YS, and EL to tabulated compositions and to processing details described in text. The processing routes were taken from full articles and include sentences describing operations such as solution treatment, deformation, and heat treatment. Figure 11.16b shows an Ashby chart of the extracted UTS versus EL data that follows expected behavior, which supports the credibility of the extraction pipeline. Figure 11.16c displays the statistical distributions of UTS, YS, and EL, and these match known trends, further suggesting accurate extraction. Figure 11.16d presents a heatmap of yearly process frequency

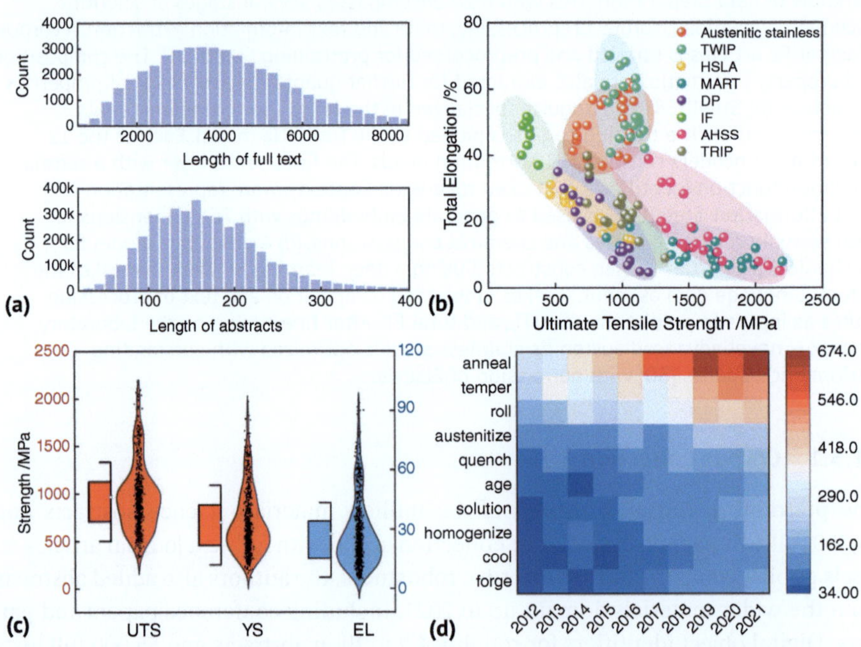

Figure 11.16 Corpus distributions and visualization of extracted data. (a) Histogram of statistical distributions for abstract and full-text lengths. (b) An Ashby chart of the automatically extracted UTS and EL labeled by various steels. (c) Statistical violin plot distributions of the three properties. (d) Heatmap for the frequency of various processing actions. *Source*: [30] with permission of Elsevier.

from 2012 to 2021 for steps such as quenching, aging, forging, solution treatment, and cooling. The heatmap shows rising emphasis on quenching, aging, solution treatment, and cooling, consistent with their roles in phase development and defect reduction in steels.

11.3.2 Pre-training of SteelBERT

SteelBERT was pre-trained on a DeBERTa-based framework [34, 35] using a dataset of about 4.2 million materials science abstracts and 55,000 full-text steel papers, for a total of roughly 0.96 billion words. Compared with encoder-only Transformers such as BERT, ALBERT [36], BART [37], and DeBERTa uses a disentangled attention mechanism that is better at modeling long-range relationships. Instead of describing each token with one vector that mixes content and position information, DeBERTa assigns two vectors to every token. One encodes the token content and the other encodes its relative position, and the relative position embeddings are shared as inputs across all Transformer layers. By separating these parts within self-attention, the model can focus on different information cues more precisely and learn latent representations that better reflect steel specific knowledge.

DeBERTa was therefore selected as the backbone for SteelBERT pretraining. Because the corpus contains special symbols from processing and mathematical expressions that sometimes appear in multiple Unicode forms, characters with little semantic value were removed, and others were mapped to customized tokens that preserve their intended meaning. Although the original DeBERTa uses a large general vocabulary, applying it directly to a specialized materials corpus can introduce tokenization noise. A new tokenizer was thus trained on the steel corpus to construct a domain-specific vocabulary. Despite the smaller training set relative to the original DeBERTa, the vocabulary size was held at 128,100 to retain coverage of nuanced technical terms. SteelBERT has 188 million parameters and uses 12 Transformer encoder layers stacked in sequence, with 12 attention heads in each layer. It is trained with self-supervised MLM, where 15% of tokens are masked. As shown in Figure 11.15b, text is first converted into embeddings and then processed by the Transformer blocks. Two enhanced masked decoders then add absolute position embeddings to further enrich the learned representations. Finally, the hidden state of the CLS token is sent to a prediction head with one hidden layer and a SoftMax activation.

Pretraining took 840 hours on eight NVIDIA A100 GPUs with 40 GB memory each, using a maximum sequence length of 512 tokens and a batch size of 576 sequences. Training continued until the loss no longer improved. The model was optimized with AdamW using $\beta_1 = 0.9$, $\beta_2 = 0.98$, $\varepsilon = 1e^{-6}$, weight decay of $1e^{-2}$, and a linear learning rate decay schedule with a 4.8% warmup ratio and a peak learning rate of $1e^{-4}$. The implementation was built in PyTorch with the Transformers library, and the code is available on the Hugging Face community hub at https://huggingface.co/MGE-LLMs/SteelBERT.

11.3.3 Prediction Model Architecture

This large and wide-ranging corpus, covering steel properties, compositions, and processing routes, was used to develop a DNN for predicting mechanical properties. The overall model structure is presented in Figure 11.17. In this setup, the model takes the composition and the associated processing description for a specific steel as inputs and outputs the target mechanical properties. Inputs are first mapped to embeddings produced by SteelBERT. Each chemical element receives a distinct vector, and the composition vector is formed as a weighted sum of these elemental embeddings according to their weight percentages. Processing routes are encoded from the pooled representation of the special CLS token computed from the text describing each route. Unlike conventional ML that requires manual extraction and alignment of many columns of processing parameters, this context-aware encoding preserves the richness and variability of the original narratives. The text embeddings and composition embeddings are next fed into shared feature layers to learn a common representation, while a separate multilayer perceptron (MLP) is used to extract signals that are specific to each modality. Within this multi-stream design, convolutional neural networks further refine dense information-rich features. Features from all branches are concatenated and transformed by a hybrid MLP to produce an initial prediction. In parallel, a mixed feature matrix that includes the original embeddings propagates both shared and modality-specific information, enriching the final representation. The resulting features are concatenated once more and fed to a terminal multilayer perception to yield the predicted mechanical properties. During training, the model is optimized with AdamW using a linear learning rate decay schedule, and MSE is used as the main loss function.

Optuna [38] was applied to automatically explore hyperparameters over a wide search space with 10^7 candidate configurations. The variables tuned included the number of layers, the number of nodes per layer, the dropout rate, the activation

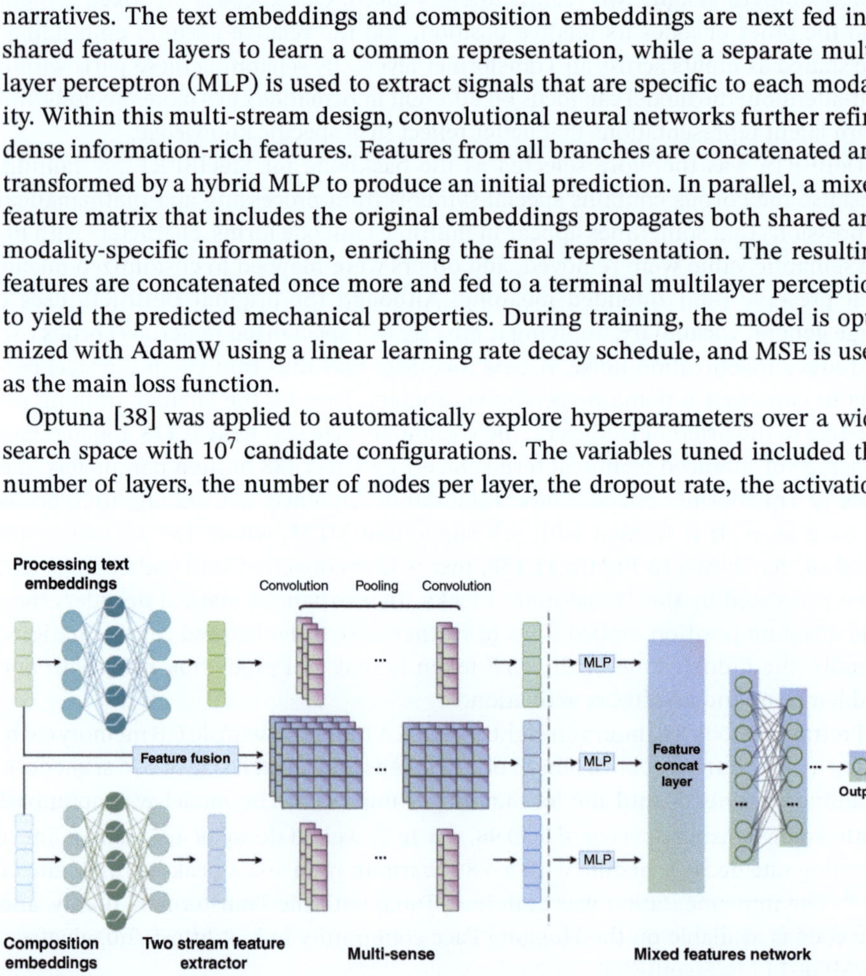

Figure 11.17 Neural network structure of predictive model. *Source*: [30] with permission of Elsevier.

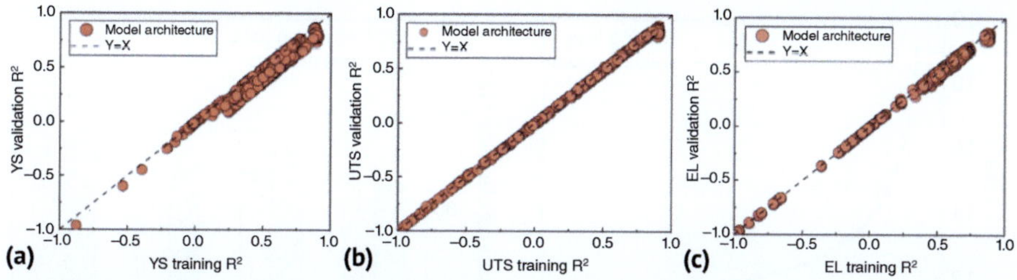

Figure 11.18 Network architecture optimization of the prediction model for YS, UTS, and EL. Each point represents a distinct network architecture variant tailored for mechanical property prediction. The optimization process involved fine-tuning various architectural parameters, including the number of layers, the number of nodes, the dropout rates, the activation functions, the pooling methods, and the number of convolutional neural network channels. The model's performance significantly improved through continuous adjustment and refinement of the network architecture.

function, the pooling method, and the number of CNN channels. Figure 11.18 summarizes this search, where each point represents a distinct model architecture evaluated for predicting YS, UTS, and EL.

11.3.4 SteelBERT Evaluation

In pretraining, SteelBERT was optimized to reconstruct masked word representations by updating weights across its layers. The corpus was divided into 95% for training and 5% for validation, consistent with the split used in the original DeBERTa setting. SteelBERT ran for about 140,000 steps with a minibatch size of 576, while the original BERT training used 1,000,000 steps with a smaller minibatch size of 256. Because the corpus in this work is about 29% of the BERT corpus, a scaled estimate suggests around 130,000 steps. The authors extended training to 140,000 steps to improve results. After 840 hours, the validation loss stabilized at 1.158, as shown in Figure 11.19a.

To evaluate SteelBERT on processing text classification, the authors created a labeled dataset by manually annotating 39,852 sentences drawn from 258 papers. The data were then divided into training, validation, and test splits with a 6:2:2 ratio. This produced a severe class imbalance, with 494 positive examples and 39,358 negative examples, which can make the minority class performance appear worse than it is. To reduce this effect, all positive samples were kept, and the negative class was downsampled at a 1:15 ratio. During each training epoch, 1,581 negative sentences were randomly selected. The classifier consisted of a dropout layer followed by a single MLP layer. As shown in Figure 11.19b, results were benchmarked against several published pretrained models, including MatSciBERT, SteelBERT, BERT, and SciBERT [15]. SteelBERT performed best, reaching 98.06% precision, 94.92% recall, and 96.54% F_1, evaluated on a random set of 52 articles with about 7,940 sentences. To check sensitivity to imbalance handling, negative sampling ratios of 1:5, 1:10, and 1:15 were also explored. In all cases, the model stayed reliable on unseen negative examples, which indicates good robustness.

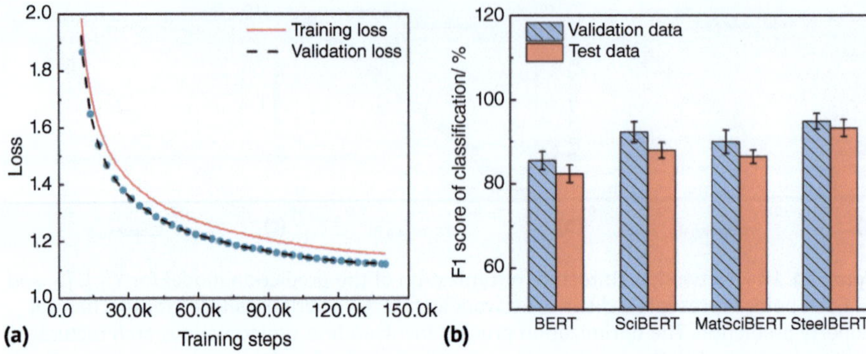

Figure 11.19 SteelBERT evaluation. (a) Visualization of the MLM performance on the validation set relative to the SteelBERT training steps. (b) Processing text classification with different LLMs. *Source*: [30] with permission of Elsevier.

11.3.5 SteelBERT Interpretability

After pretraining on a large collection of materials science papers, SteelBERT captures domain-specific linguistic patterns for steels. As an initial validation, they applied SteelBERT to topic clustering of scientific abstracts. The model generated long text embeddings by converting 87,497 steel-focused abstracts into 768-dimensional vectors. They then applied UMAP [39], to reduce the embedding dimensionality and applied Hierarchical Density-Based Spatial Clustering of Applications with Noise (HDBSCAN), to group the abstracts. UMAP compresses high-dimensional data into a lower-dimensional space while keeping local relationships, and HDBSCAN can separate clusters with varied shapes and sizes, which together support clearer topic clustering.

For embedding each abstract, they used the CLS token representation from SteelBERT to produce a 768 by 87,497 matrix. UMAP reduced these embeddings to two dimensions while retaining key topological relationships. To capture both local structure and more global trends, they set the neighbor count to 700 and the minimum distance to 0.0, which determines how closely points are placed in the low-dimensional projection. For HDBSCAN, they chose a minimum cluster size of 30, used Euclidean distance to measure similarity, and selected clusters with the Excess of Mass (EOM) criterion. Cluster quality was measured using the Davies–Bouldin Index, the Calinski–Harabasz Index, and the Silhouette coefficient. To name the main themes in each cluster, they applied class-based term frequency-inverse document frequency (c-TF-IDF), and selected representative topic words based on their scores.

$$\mathrm{c-TF-IDF}_i = \frac{t_i}{w_i} \times \log \frac{m}{\sum_j^n t_j} \tag{11.1}$$

In this computation, the term frequency for each class is obtained by counting how often a word t appears in class i and then dividing by the total number of words in that class. This normalizes for class length and reduces the influence of words that are broadly frequent within the class. The IDF is then computed by taking the total number of documents across all classes and dividing it by the total count of the word summed over all classes.

The resulting topic clusters are visualized in Figure 11.20a, where distinct topics are represented by different colors, and gray points correspond to data considered noise that could not be assigned to any specific cluster. To form a clear description for each topic, the five highest-scoring words by c-TF-IDF were chosen, as summarized in Figure 11.20b. The resulting clustering produced distinct groups that separate major themes in the steel literature. The clustering quality was also measured and compared with BERT, SciBERT, and MatSciBERT. In these comparisons, SteelBERT performed best overall, which supports its stronger ability to learn steel domain textual knowledge in materials science.

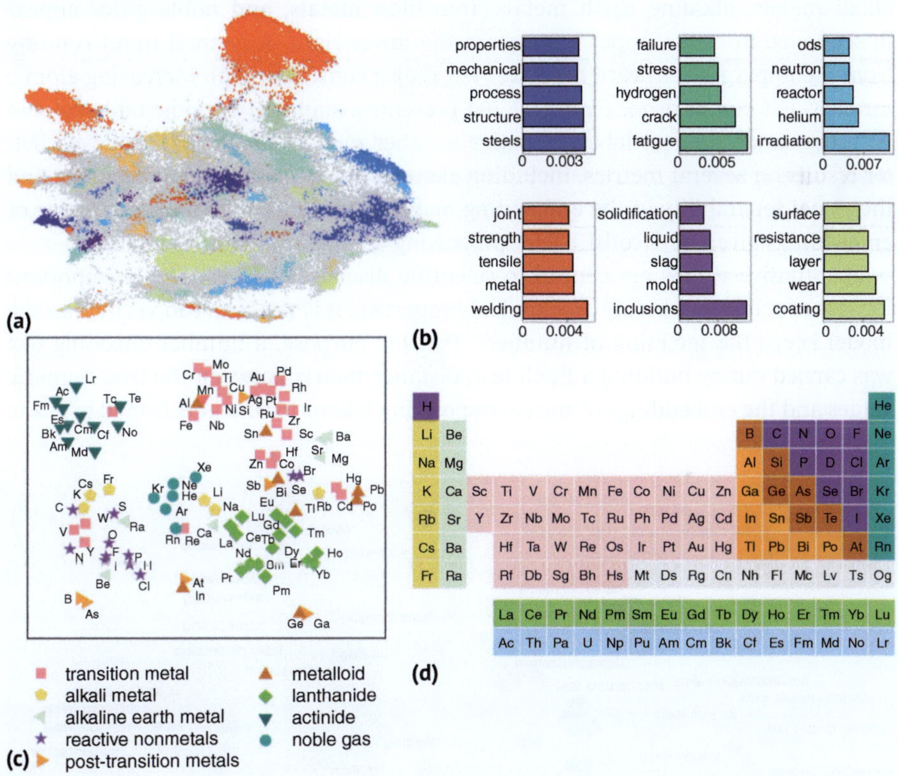

Figure 11.20 SteelBERT interpretability. (a) Visualization of abstract embedding clustering with the SteelBERT model. The collected abstracts related to the steel topic are embedded into a series of 768-length vectors by the pretrained model. The UMAP approach is first used to reduce the high-dimensional data to a two-dimensional feature space. Subsequently, the HDBSCAN method is applied to cluster the resulting vectors based on topic similarity. (b) Topic generation using the c-TF-IDF approach to identify topics within each cluster. (c) A two-dimensional t-SNE projection is utilized to depict word embeddings for 100 chemical elements, which are labeled with its corresponding symbol and grouped by category of transition metal, alkali metal, alkaline earth metal, reactive nonmetals, post-transition metals, metalloid, lanthanide, actinide, and noble gas. Similar elements cluster together, reflecting the periodic table's topology. (d) A periodic table of chemistry using different colors to represent various categories of elements. *Source*: [30] with permission of Elsevier.

To examine whether SteelBERT encodes distinct signatures for individual chemical elements, the authors performed an additional test. SteelBERT generated 768-dimensional embeddings for the symbols of the first 100 elements in the periodic table. They then reduced these embeddings with t-distributed stochastic neighbor embedding (t-SNE), using cosine distance as the similarity metric and a perplexity of 2. The resulting low-dimensional points were plotted, with each point labeled by its element symbol and colored by element group, including transition metal, alkali metal, alkaline earth metal, reactive nonmetal, post-transition metal, metalloid, lanthanide, actinide, and noble gas.

Based on the principle that tokens used in similar contexts tend to learn similar vectors, Figure 11.20c shows that elements from the same chemical group cluster closely together. The broader arrangement also echoes the periodic table, where alkali metals, alkaline earth metals, transition metals, and noble gases appear in sequence from the upper left toward the lower right. A general trend running from the upper right toward the lower left is also consistent with increasing atomic number. For comparison, Figure 11.20d presents a standard periodic table colored with the same category labels. Relative to other models, SteelBERT achieved better results on several metrics, including element group classification accuracy and the visual separability of the embedding map. Figure 11.21 shows the clustering of embeddings given by SteelBERT for processing actions with similar topics.

Quantitative values are central to scientific discourse and especially important for tasks such as predicting mechanical properties; it is necessary to verify that the model keeps the meaning of numbers. For this purpose, a number encoding test was carried out by building a Euclidean distance matrix between the true numeric values and the embeddings of the corresponding tokens for integers from 0 to 2,000.

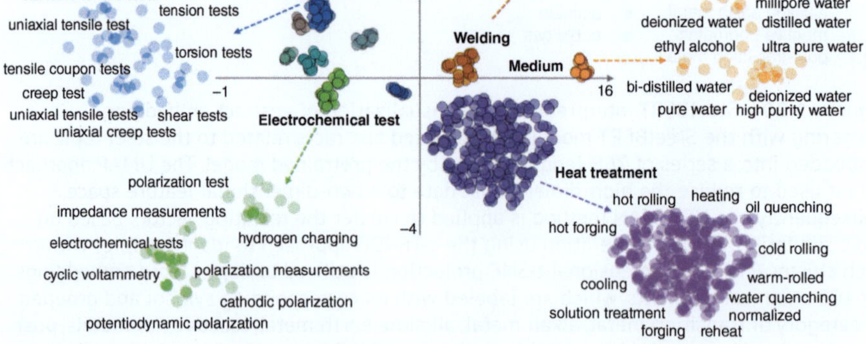

Figure 11.21 Clustering of embeddings for processing actions with similar topics. *Source*: [30] with permission of Elsevier.

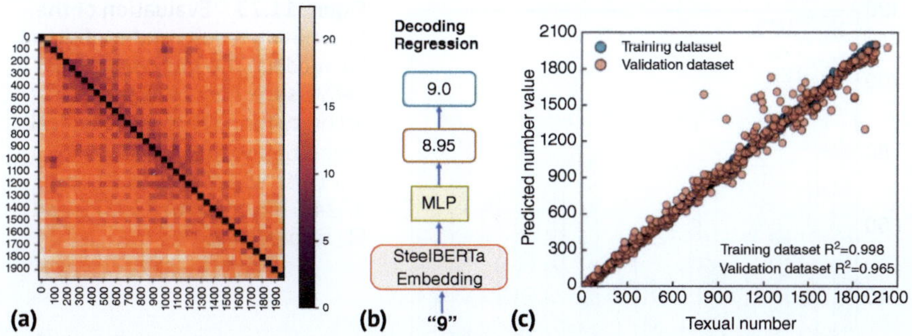

Figure 11.22 Histograms of statistical distributions depicting the length of abstracts and full text. (a) A Euclidean distance matrix heatmap of integer numbers' embeddings with SteelBERT, covering the range from 0 to 2,000. (b–c) A probing model to decode a number from its word embedding across a randomly selected 80% subset of integers within the range from 0 to 2,000, exemplified by the conversion of "100" to 100.0. The resulting plot showcases the model's predictions for all numbers, demonstrating its accurate decoding of numbers within the training and evaluation range, thereby capturing numeracy.

Results from different models are compared in Figure 11.22. For SteelBERT, the matrix exhibits a pronounced dark band along the main diagonal where values match their embeddings that gradually fade away from the diagonal, indicating that the embedding space consistently retains the quantitative relationships among numbers.

As shown in Figure 11.23, the evaluation also included a probing model trained to recover numeric values directly from word embeddings. The probe was trained on a randomly chosen 80% subset of the integers from 0 to 2,000, and it learned mappings such as converting the embedding of the token "100" into the floating-point value 100.0. Its accuracy was then benchmarked against four BERT-based models. Within the training range, SteelBERT achieved a coefficient of determination of 98.35% $\pm$ 0.15% on the training set and 98.11% $\pm$ 0.19% on the validation set, demonstrating reliable decoding and strong numeracy. This outcome is consistent with pretraining on a large scientific corpus and with a large vocabulary that better represents both domain terms and numerical patterns in materials science. Taken together, these results show that SteelBERT handles numeric information with high precision, extending its usefulness beyond purely textual tasks and strengthening its suitability for applications that require joint numerical and textual understanding.

11.3.6 Mechanical Property Prediction

The full predictor was trained end-to-end using 100 random splits of the data, with an 8 to 2 ratio for training and validation. On the training partitions, coefficients of determination R^2 for YS, UTS, and EL were 88.61% $\pm$ 0.99%, 89.14% $\pm$ 0.94%, and 86.29% $\pm$ 1.52%, respectively. On the validation partitions, the corresponding R^2 values were 83.48% $\pm$ 3.62%, 86.39% $\pm$ 2.21%, and 84.85% $\pm$ 1.99%. To further evaluate generalization, an independent test set containing 18 steels drawn from publications in 2022 and 2023 was assembled. Figure 11.24 presents scatter plots for

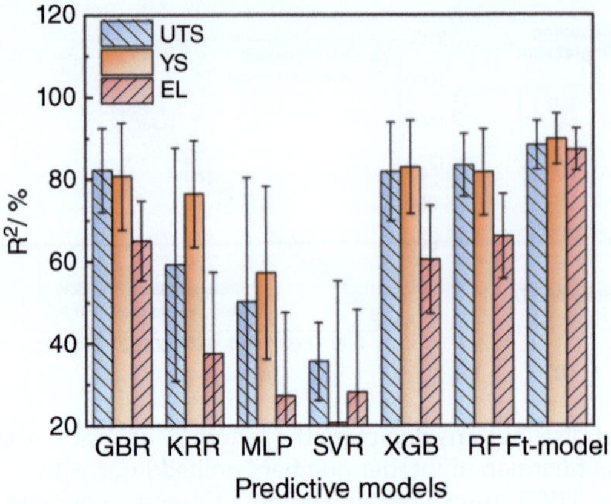

Figure 11.23 Evaluation of the fine-tuning results and traditional ML models. The proposed model delivers the best prediction accuracy for all three mechanical properties, with R^2 of 89.85% (±6.17%), 88.34% (±5.95%), and 87.24% (±5.15%) for YS, UTS, and EL, respectively.

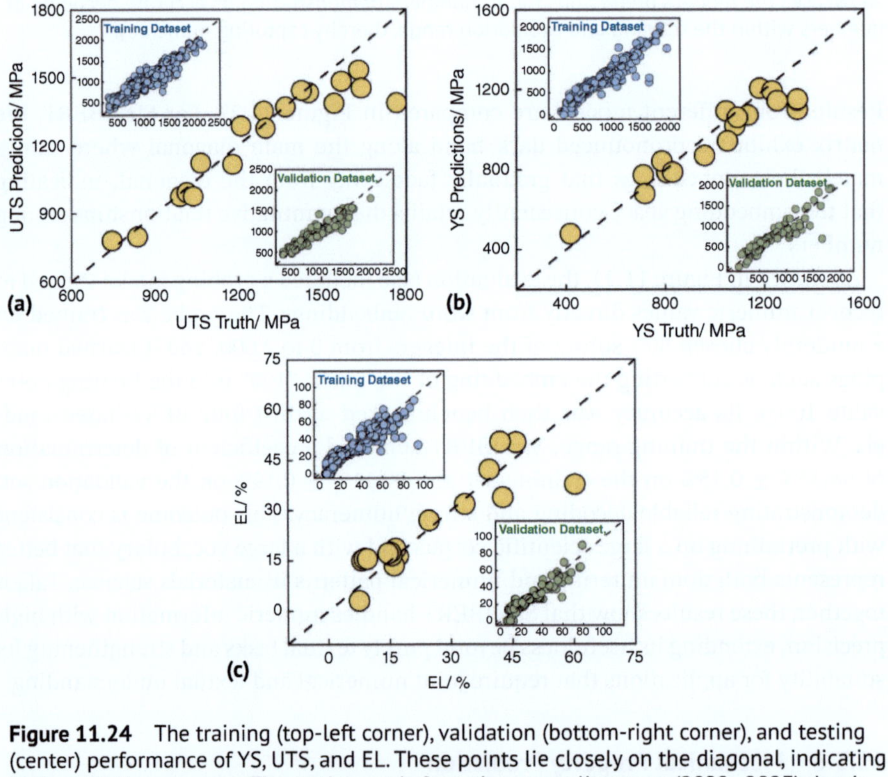

Figure 11.24 The training (top-left corner), validation (bottom-right corner), and testing (center) performance of YS, UTS, and EL. These points lie closely on the diagonal, indicating high prediction accuracy. The testing set is from the recent literature (2022–2023) that is not part of the training and evaluation datasets. *Source*: [30] with permission of Elsevier.

YS, UTS, and EL, with training results shown in the top left, validation results in the bottom right, and test results in the center.

A head-to-head comparison was then conducted by replacing SteelBERT embeddings with those from several open-source LLMs within the same predictive framework. The baselines included BERT, Phi-3.5-Mini-Instruct [40], SciBERT, Mistral-7B-Instruct [41], and LLaMa-3.1 [42], SciBERT, using their text embeddings

Table 11.1 Performance benchmark of different LLM embeddings for predicting steel mechanical properties with MAEs and R^2 scores.

Model	Dataset	YS		UTS		EL	
		R^2/%	MAE/MPa	R^2/%	MAE/MPa	R^2/%	MAE/MPa
BERT	Train	62.65 ± 8.33	150.01 ± 24.44	85.97 ± 2.95	92.47 ± 13.92	76.59 ± 3.94	5.10 ± 0.63
	Validation	43.23 ± 9.63	182.10 ± 24.08	67.60 ± 3.48	138.20 ± 10.74	60.71 ± 2.69	6.89 ± 0.50
	Test	27.00 ± 9.60	173.22 ± 12.45	52.55 ± 5.88	153.64 ± 14.06	57.57 ± 2.14	8.03 ± 0.31
SciBERT	Train	75.14 ± 0.88	119.20 ± 3.90	87.30 ± 1.29	94.21 ± 6.93	78.12 ± 1.55	5.03 ± 0.36
	Validation	58.46 ± 1.28	149.07 ± 4.02	67.30 ± 3.31	150.49 ± 7.43	70.88 ± 2.27	6.59 ± 0.46
	Test	40.66 ± 2.76	156.05 ± 4.27	37.28 ± 7.03	187.59 ± 9.04	55.20 ± 9.76	7.88 ± 0.72
MatSciBERT	Train	65.00 ± 4.00	155.71 ± 9.75	72.88 ± 5.31	148.50 ± 18.25	72.20 ± 9.19	5.61 ± 1.11
	Validation	47.95 ± 8.25	179.75 ± 32.01	51.78 ± 6.86	188.28 ± 18.29	55.98 ± 9.20	6.96 ± 0.87
	Test	21.20 ± 5.90	172.78 ± 6.85	21.40 ± 6.51	189.38 ± 9.64	52.33 ± 10.18	8.73 ± 0.96
SteelBERT	Train	88.61 ± 0.99	67.64 ± 2.41	89.14 ± 0.94	78.53 ± 2.79	86.29 ± 1.52	4.45 ± 0.12
	Validation	83.48 ± 3.62	99.87 ± 6.41	86.39 ± 2.21	97.5 ± 6.65	84.85 ± 1.99	5.14 ± 0.31
	Test	78.17 ± 3.40	92.63 ± 7.77	82.56 ± 1.96	82.29 ± 4.7	81.44 ± 2.98	5.28 ± 0.31
Phi-3.5-mini-Instruct	Train	57.98 ± 6.36	174.38 ± 18.97	69.04 ± 4.74	163.80 ± 12.74	77.88 ± 0.88	5.07 ± 0.25
	Validation	47.55 ± 7.88	198.10 ± 23.44	52.92 ± 3.61	194.94 ± 10.48	73.47 ± 0.24	6.14 ± 0.14
	Test	20.43 ± 7.75	169.11 ± 12.22	23.75 ± 4.71	193.61 ± 6.23	55.47 ± 10.30	8.15 ± 0.60

(Continued)

Table 11.1 (Continued)

Model	Dataset	YS		UTS		EL	
		R^2/%	MAE/MPa	R^2/%	MAE/MPa	R^2/%	MAE/MPa
Mistral-7B-Instruct-v0.2	Train	57.38 ± 1.50	179.22 ± 4.59	63.12 ± 7.07	182.58 ± 20.71	79.85 ± 1.20	4.71 ± 0.34
	Validation	30.35 ± 5.03	212.67 ± 1.77	45.42 ± 4.88	202.36 ± 17.07	66.77 ± 2.70	6.31 ± 0.27
	Test	19.53 ± 7.41	195.40 ± 10.38	19.12 ± 6.31	202.06 ± 11.45	61.55 ± 5.75	8.68 ± 0.84
Llama-3.1-8B	Train	82.90 ± 1.41	88.14 ± 4.30	90.54 ± 1.06	72.53 ± 6.21	76.74 ± 5.66	5.12 ± 0.77
	Validation	58.83 ± 2.84	138.84 ± 5.51	70.19 ± 2.93	129.13 ± 8.26	62.62 ± 5.21	6.64 ± 0.52
	Test	57.23 ± 2.63	136.70 ± 4.65	58.13 ± 2.84	146.34 ± 5.68	64.13 ± 2.73	7.38 ± 0.40
Llama-3.1-8B-Instruct	Train	78.70 ± 3.68	101.99 ± 10.95	83.94 ± 3.34	98.55 ± 12.46	73.34 ± 3.62	5.55 ± 0.59
	Validation	58.71 ± 4.41	151.90 ± 11.55	65.40 ± 1.93	141.89 ± 6.18	61.51 ± 2.61	7.07 ± 0.38
	Test	52.68 ± 4.49	139.04 ± 6.91	57.72 ± 3.35	147.64 ± 4.97	60.66 ± 2.82	7.17 ± 0.23
Llama-3.1-8B-Instruct (ft)	Train	80.75 ± 1.35	101.14 ± 0.98	87.96 ± 0.86	85.44 ± 3.60	76.48 ± 2.84	5.44 ± 0.64
	Validation	62.10 ± 2.20	151.15 ± 6.81	69.08 ± 4.02	131.51 ± 13.34	71.98 ± 2.77	6.04 ± 0.30
	Test	59.30 ± 3.21	124.67 ± 9.84	60.74 ± 1.78	151.79 ± 1.78	67.34 ± 6.36	7.17 ± 0.73
Llama-3.1-70B	Train	77.45 ± 2.23	109.00 ± 5.80	75.58 ± 5.27	121.67 ± 14.10	83.17 ± 1.35	4.34 ± 0.38
	Validation	52.66 ± 2.59	148.35 ± 3.17	59.04 ± 3.80	159.50 ± 8.66	67.11 ± 2.97	6.28 ± 0.35
	Test	48.28 ± 2.31	146.91 ± 4.13	38.19 ± 3.86	178.30 ± 4.88	64.22 ± 2.46	7.16 ± 0.22
Llama-3.1-70B-Instruct	Train	66.80 ± 6.14	149.07 ± 22.41	76.54 ± 4.12	131.78 ± 14.92	71.32 ± 4.84	6.07 ± 0.83
	Validation	49.56 ± 4.56	173.97 ± 13.07	59.54 ± 3.20	172.65 ± 10.70	59.99 ± 5.39	7.12 ± 0.47
	Test	32.41 ± 4.73	159.72 ± 5.83	35.86 ± 5.20	176.06 ± 8.05	58.20 ± 5.41	7.27 ± 0.52

in the same pipeline. Table 11.1 summarizes performance using R^2 and mean absolute error (MAE) for all three properties. SteelBERT consistently outperformed the alternatives, achieving test set R^2 values of 78.17% ± 3.40% for YS, 82.56% ± 1.96% for UTS, and 81.44% ± 2.98% for EL. Notably, the base LLaMA model outperformed its instruction-tuned version, and LLaMA 70B showed weaker results than the smaller 8B model. This pattern implies that very large decoder-only models can excel on general tasks but may perform worse in a specialized domain when broad pretraining reduces the strength of domain-specific signals. The effect may be compounded by embedding dimensionality, since LLaMA 8B produces 4096-dimensional vectors, whereas LLaMA 70B yields 8,192 dimensions, and the downstream network here was designed to balance dataset scale with accuracy and may align better with lower-dimensional embeddings. Additionally, LLaMA 8B Instruct was fine-tuned on the same corpus as SteelBERT using a causal language modeling objective with low-rank adaptation. As reported in Table 11.1, these adapted embeddings provided a modest average gain of about 5% over the base model, likely constrained by the size of the fine-tuning corpus, which can encourage memorization rather than learning general metallurgical principles. If the underlying pretraining does not capture the specialized vocabulary and relationships of steel science, subsequent fine-tuning will struggle to impart these connections. For completeness, the strategy was also benchmarked against a GPT model that was instructed and fine-tuned via its official API on the same 677-record dataset using prompts tailored to the regression task.

SteelBERT performs especially well in property prediction mainly because it is pretrained on a large, domain-focused materials science corpus of 0.96 billion words. This training helps it learn the field vocabulary, semantic links, and context patterns in much greater depth. Its tokenizer also strengthens this advantage. It uses byte-pair encoding (BPE) [43] with a 128k vocabulary, which improves how technical terms and steel-specific expressions are represented. This contrasts with MatSciBERT, which was pretrained on a broader corpus and uses a WordPiece tokenizer with a smaller 31k vocabulary based on the WordPiece algorithm [44]. The wider training scope and the smaller token set can reduce attention to steel-centered knowledge. SteelBERT also benefits from lower-dimensional embeddings, which lower computation cost, speed up convergence, and make optimization easier. In addition, its DeBERTa-based design includes a disentangled attention mechanism and an improved mask decoder. These choices help the model represent token relationships and contextual information with high accuracy. Overall, SteelBERT's strong results come from the combination of domain-specific pretraining, efficient low-dimensional embeddings, and key architectural improvements, making it well-suited for producing embeddings that support accurate property prediction.

11.3.7 Steel Design

A fine-tuning approach was adopted to progressively sharpen predictive performance using compact, task-specific datasets. ASSs are of broad interest for their ductility and their resistance to corrosion and oxidation, although their strength is usually modest. To address this limitation, a literature-trained predictor was used to guide the design of advanced austenitic grades with improved mechanical response.

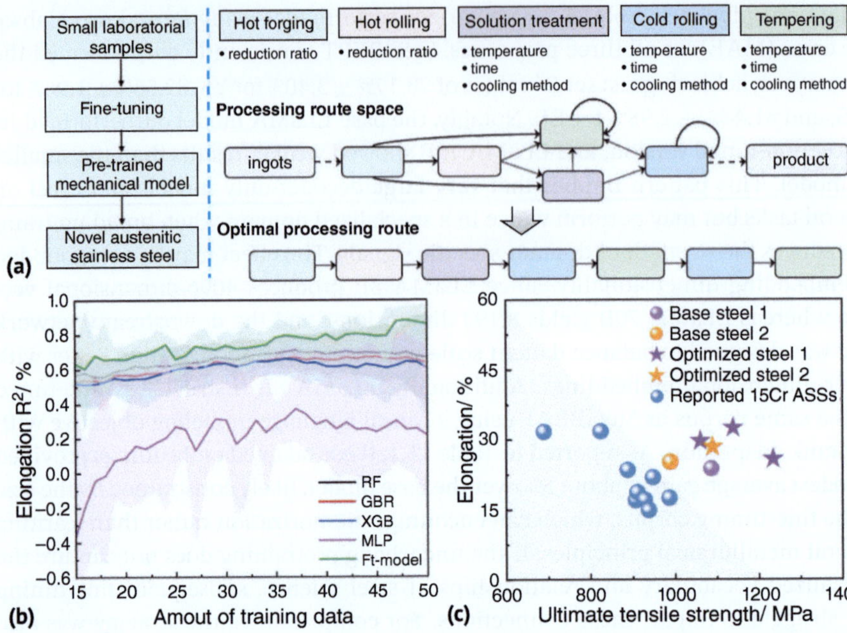

Figure 11.25 Rational design of new steels. (a) Schematic diagram illustrating the fine-tuning process on a small experimental dataset for the exploration of advanced ASSs. (b) Comparison between the fine-tuned model and traditional ML models trained with varying amounts of training data ranging from 15 to 50. (c) Scatter plot of reported 15Cr ASSs, base steels, and optimized steels. *Source*: [30] with permission of Elsevier.

Because the original model was built from a broad corpus, targeted adaptation was required to capture new compositions and processing paths. The model was therefore fine-tuned on an experimental dataset of 64 records for 15Cr ASSs processed by hot forging, solution treatment, cold rolling, and tempering, as shown in Figure 11.25a.

Plates were used as the initial stock. Bullet-shaped ingots weighing 50 kg were hot forged into cuboids measuring 80 × 80 × 630 mm. They were then solution treated at 1,000 and 1,050 °C for four hours, and water quenched to control grain size. Next, cold rolling was applied to achieve thickness reductions between 25 and 80%, followed by tempering at 590 to 780 °C for 5 to 60 minutes. Fine-tuning followed these routes and their accompanying text descriptions by updating only the mixed feature layers in Figure 11.17, while all other network components were kept fixed.

The experimental table was organized into structured inputs and split 8 to 2 into training and test sets. During fine-tuning, the shared specific feature layers were frozen by holding their weights and biases constant, whereas the mixed feature layers were updated to adapt the representation to this dataset. Training ran for 1,000 epochs with AdamW and MSE loss. AdamW used $\beta_1 = 0.9$, $\beta_2 = 0.999$, $\varepsilon = 1e^{-8}$, weight decay $= 1e^{-2}$, and a linear learning rate decay schedule.

Performance was compared with standard ML baselines. These models used features such as elemental weight fractions along with cold rolling reduction, solution treatment settings, and tempering conditions, and their hyperparameters were chosen through grid search. Using and MAE as evaluation measures, they also tested other open-source LLMs under the same two-stage fine-tuning procedure to ensure a fair comparison. The fine-tuned model delivered the best accuracy across all properties, with R^2 of 89.85% ± 6.17% for YS, 88.34% ± 5.95% for UTS, and 87.24% ± 5.15% for EL. As shown in Figure 11.25b for EL, the fine-tuned approach remains stronger than conventional methods even under very limited data regimes with only 15–50 samples.

Two test plates with an identical composition, 15Cr-7.92Ni-0.83Mn-0.45Si-1.99Mo-0.043Nb-1.49Cu, were used as the baseline materials. In Figure 11.25c, they are denoted as base steel 1 in purple and base steel 2 in orange. Both plates underwent the same hot forging and solution treatment, but their starting schedules differed in the single cold rolling reduction step and in the tempering temperature and time. The authors then defined a search space that covered many cold rolling and tempering sequences, parameterized by the number of rolling passes, the reduction ratio, and the treatment temperature. A 20-hour grid search over this space produced four candidate processing routes. The selected optimized routes preserved the first cold rolling reduction of each base steel and added one extra cycle of cold rolling followed by tempering. This produced three variants from base steel 1 and one variant from base steel 2. As shown in Figure 11.25c, these optimized steels achieved strong properties compared with previously reported 15Cr ASSs [45–48].

Figure 11.26a summarizes the tensile results and shows that the optimized steels gained both strength and ductility. Plate specimens measuring 65 × 5 × 1 mm were machined parallel to the rolling direction and tested on a ForceCreate WDML-1 stress corrosion frame at a constant crosshead rate of 0.12 mm/min. The reported properties are the averages of at least three repeats. With one additional cold rolling and tempering cycle, optimized steel 1-1 reached a YS of 960 MPa, an UTS of 1,138 MPa, and an EL of 32.5%. Compared with its base steel, this corresponds to increases of 50 MPa in YS, 52 MPa in UTS, and 8.7% in EL. For the optimized steels, the MAEs between prediction and experiment were 34.34 MPa for YS, 42.16 MPa for UTS, and 1.74% for EL. Figure 11.26b presents the EBSD map of optimized steel 1-1 and shows refined

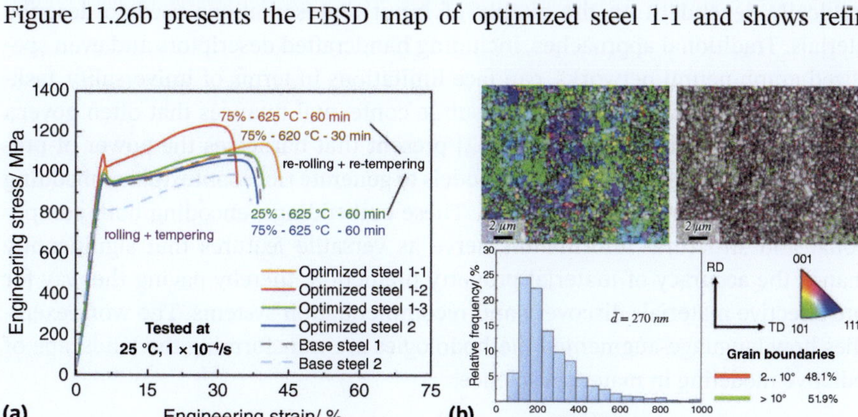

Figure 11.26 Tensile tests and characterization of the steel. (a) Tensile stress-strain plots at 25 °C and 1×10^{-4}/s of ASSs for base and optimized steels. (b) Electron backscatter diffraction map and grain size distribution for optimized steel 1-1 (rolling direction is marked by a vertical arrow). *Source*: [30] with permission of Elsevier.

austenite grains with clear boundaries and an average size of about 270 nm, which likely supports the observed mix of high strength and good ductility.

This study demonstrates accurate prediction of properties for 18 steels first reported in 2022 and 2023. The methodology uses a LLM trained on historical steel literature to encode core metallurgical knowledge, followed by learning from a quantitative dataset to capture composition–process–property relations for new alloy discovery. Validation includes successful prediction for the 18 recent steels and improved accuracy after fine-tuning on 64 laboratory records. Applying the model to process optimization produced a material whose performance surpasses prior reports for 15Cr ASSs.

Steel design remains a complex task that requires an integrated understanding of physics, chemistry, and metallurgy, and despite long study, still demands significant computational and experimental effort. The proposed data-driven pipeline employs textual features encoding to predict mechanical properties more efficiently, grounded in literature and validated experimentally to support targeted development. Relative to traditional ML, it mitigates burdensome steps such as structuring data from intricate processing routes, aligning multistep sequences, and handling high-dimensional sparsity, while delivering stronger regression performance than pure Transformer-decoder language models. Implementation is straightforward, requiring only composition and a text description of processing to obtain predictions, and the approach points to broad potential for natural-language-based features in property prediction beyond steel.

11.4 Transformer-generated Atomic Embeddings to Enhance Prediction Accuracy

The accurate prediction of material properties is a cornerstone of computational materials science, enabling the efficient screening of candidates and the rational design of novel materials [49–51]. While ML has shown great promise, its efficacy is critically dependent on the quality of input representations used to describe materials. Traditional approaches, including handcrafted descriptors and even specialized graph neural networks, can face limitations in terms of universality, task-agnosticism, or the ability to capture subtle contextual nuances that often govern material behavior. Jiaxing Qu et al. [52] present that harnesses the power of pre-trained Transformer-based language models to generate rich, contextual embedding from textual descriptions of materials. These embeddings, encoding both compositional and structural information, serve as versatile features that significantly enhance the accuracy of material property prediction, thereby paving the way for more effective materials discovery and recommendation systems. The work exemplifies how language-augmented methodologies are transforming the landscape of predictive modeling in materials science.

11.4.1 Leveraging Language Models for Material Representation

The central premise of the approach investigated by Jiaxing Qu et al. is that LLMs, pre-trained on extensive corpora of scientific literature, can internalize a deep understanding of materials science concepts, chemical principles, and structural relationships. This learned knowledge can then be distilled into dense vector representations, or embeddings, which serve as powerful, information-rich features for downstream ML tasks, particularly property prediction.

To achieve this, materials are first translated into a textual format that can be processed by these LLMs. This involves creating comprehensive descriptions that encompass both compositional and structural information as shown in Figure 11.27. For instance, compositional information might be represented by the material's chemical formula (e.g., "PbTe"). Crucially, detailed structural information is also converted into natural language using tools like Robocrystallographer, which generates human-readable sentences describing crystal structure, symmetry, and space group. These textual strings, combining both elemental makeup and architectural details, are then fed as input to pretrained language models specifically adapted for the scientific domain, such as MatBERT or MatSciBERT. The output embeddings, typically extracted from one of the model's hidden layers, are dense vectors that provide a contextualized numerical representation of the input material, capturing nuanced features learned during the LLM's extensive unsupervised pre-training on materials science text. This method effectively transforms complex material characteristics into a "language" interpretable by advanced AI.

11.4.2 Characterizing Transformer Material Representations

The strength of language-based material embeddings can be seen in how they convert complex material descriptions into compact, low-dimensional vectors that preserve semantic relationships. Qualitative analysis by Jiaxing Qu et al. using UMAP for visualization shows that these embeddings capture meaningful knowledge in materials science. For instance, structure-level embeddings from language models such as MatSciBERT and MatBERT produce distinct, chemically sensible clustering of materials by elemental group, a separation not clearly achieved with traditional structural fingerprints (Figure 11.28a). This demonstrates that the embeddings capture overarching chemical classifications beyond local atomic environments. Further evidence is provided by the proximity of known high-performance thermoelectric (TE) materials to the canonical PbTe in the embedding space, which correctly mirrors established chemical and structural analogies, such as the close relationship between the rock-salt structures of PbTe and SnTe (Figure 11.28b). Together, these

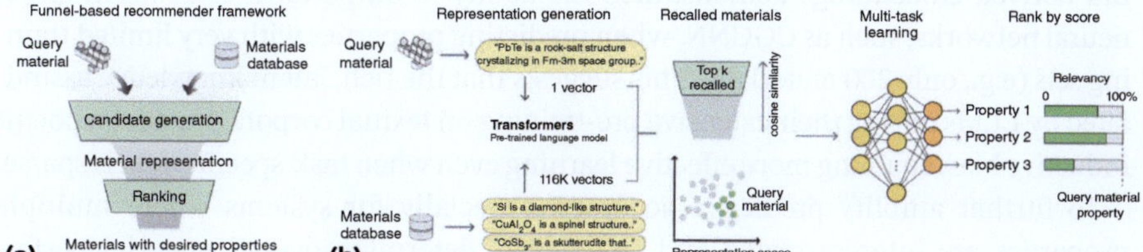

Figure 11.27 Recommender framework for material discovery. (a) Proposed funnel-based recommender framework for recalling and ranking candidate materials by similarity to the query material. (b) Workflow for candidate screening, including language representation construction, candidate recall, and multi-task ranking prediction. *Source*: [52] / Springer Nature / CC BY 4.0.

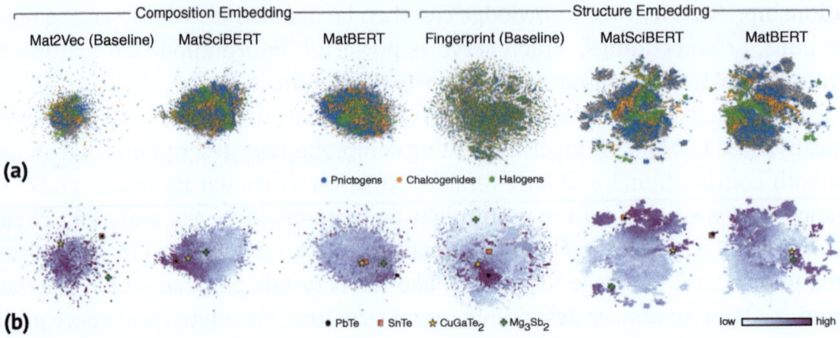

Figure 11.28 UMAP projections of 116k materials using different embedding models. Materials are colored (a) by anionic groups, and (b) by cosine similarity to PbTe under the indicated embedding model. *Source*: [52] / Springer Nature / CC BY 4.0.

visualizations underscore the capacity of language model-derived embeddings to encapsulate essential chemical trends and material interrelationships.

11.4.3 Enhancing Property Prediction Accuracy with Embeddings

The most convincing test of a new material representation is whether it can improve performance on downstream property prediction. In the work by Jiaxing Qu et al., the authors quantified the value of these language-derived embeddings by training MLP regression models to predict six different properties. The evaluation used a dataset of 5,700 materials drawn from the AFLOW database. The results unequivocally demonstrate that these fixed embeddings, when employed as input features, achieve highly competitive prediction accuracy. Notably, structural embeddings extracted from MatBERT yielded robust performance across multiple properties. Similar positive outcomes were reported for seven different material benchmarks from the MatBench dataset (Figure 11.29 of their original paper), reinforcing the general applicability of these language-based features.

A particularly significant finding from their research highlighted the advantage of these pre-trained language embeddings in data-scarce scenarios—a prevalent challenge in materials science where acquiring large labeled datasets can be resource-prohibitive. As illustrated in Figure 11.30 of their work, models trained using these LM-derived embeddings demonstrated the ability to outperform established graph neural networks, such as CGCNN, when predicting properties with very limited training sets (e.g., only 200 materials). This suggests that the rich, latent knowledge assimilated by LLMs during their extensive pre-training on textual corpora provides a potent inductive bias, enabling more effective learning even when task-specific data is sparse.

To further amplify predictive accuracy, especially for systems where multiple properties are intercorrelated and collectively determine overall material performance (a common scenario in designing functional materials like TEs), Jiaxing Qu et al. used a Multi-gate Mixture-of-Experts (MMoE) model. This sophisticated multi-task learning (MTL) architecture, depicted in their Figure 11.31a, utilizes a set of shared "expert" subnetworks along with task-specific "gating" networks.

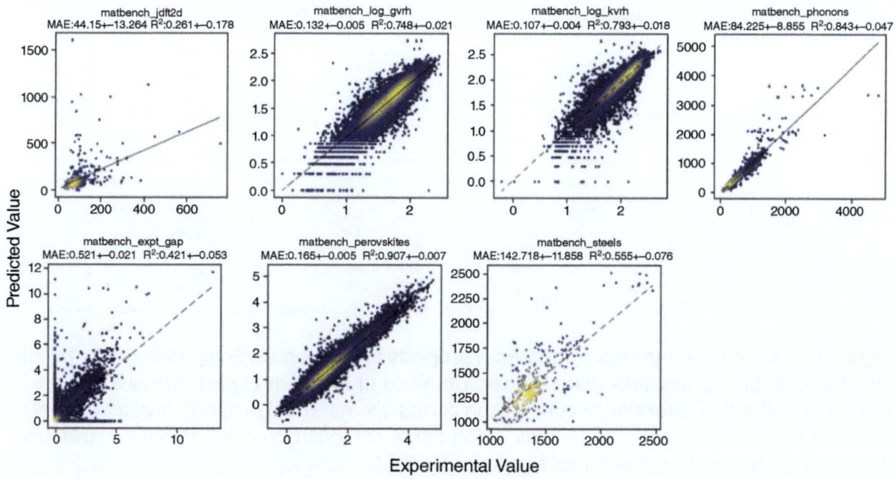

Figure 11.29 Comparison of the model performance of the model with MatBERT language representation against leaderboard models on the MatBench Material Dataset. Dashed lines represent the best linear fits. *Source*: [52] / Springer Nature / CC BY 4.0.

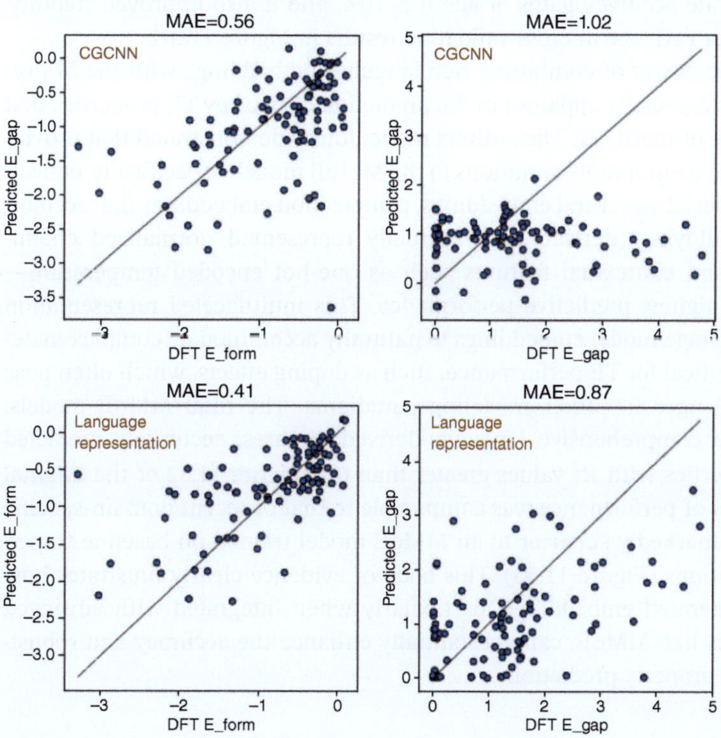

Figure 11.30 Model performance on a small dataset containing 200 training and 100 testing materials for band gap and formation energy predictions. Scatter plots show the comparisons between CGCNN and the current models trained with language representations as input. E_form: formation energy (eV per atom); E_gap: band gap (eV). *Source*: [52] / Springer Nature / CC BY 4.0.

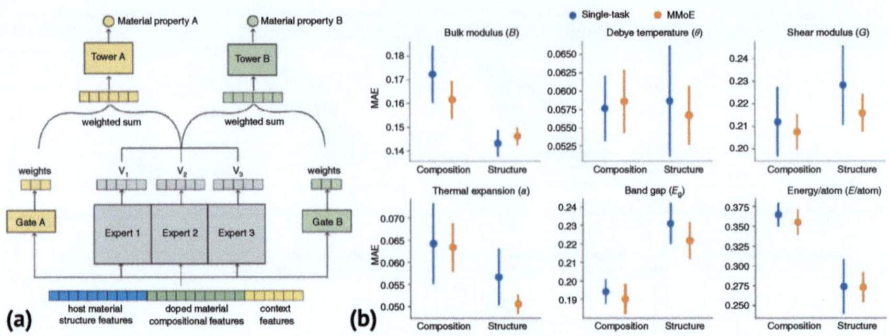

Figure 11.31 MTL framework for material property prediction. (a) Schematic of the MMoE architecture: gating networks learn the contributions of different experts via task-specific tower networks. (b) Performance comparison across six materials property prediction tasks between single-task models and MMoE, using either composition embeddings or structure embeddings. *Source*: [52] / Springer Nature / CC BY 4.0.

This design allows the model to simultaneously predict multiple properties while explicitly learning and leveraging their underlying correlations. Compared with single-task models for the six general material properties, the MMoE method delivered moderate accuracy gains of about 5–10%, and it also improved stability, reflected by lower variance in cross-validation results in Figure 11.31b.

The synergistic power of combining rich language embeddings with the MMoE architecture was especially apparent in the prediction of five key TE properties that govern the figure of merit, zT. The authors meticulously demonstrated that providing comprehensive input representations to the MMoE model—specifically, concatenating host material structural embeddings, composition embeddings that account for doping or alloying (derived from textually represented normalized chemical formulae), and contextual features such as one-hot encoded temperature—resulted in the highest predictive performance. This multifaceted representation enabled the language model embeddings to naturally accommodate complex material variations critical for TE performance, such as doping effects, which often pose significant challenges for other modeling paradigms. The final MMoE models, leveraging these comprehensive language-derived features, accurately predicted all five TE properties with R^2 values greater than 0.7 (Figure 11.32 of the original work). This level of performance was comparable to that of recent domain-specific TE models and markedly superior to an MMoE model trained on baseline fingerprint representations (Figure 11.33). This body of evidence clearly illustrates how Transformer-generated embeddings, particularly when integrated with advanced ML architectures like MMoE, can substantially enhance the accuracy and robustness of material property prediction.

11.4.4 Summary and Future Directions

The robust material representations and heightened property prediction accuracy achieved through Transformer-generated language embeddings, as established in the preceding discussion, enable sophisticated applications that can significantly accelerate materials discovery. The research by Jiaxing Qu et al. compellingly

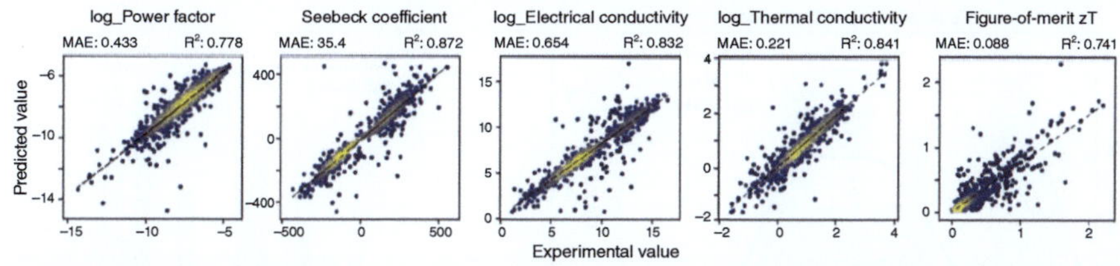

Figure 11.32 Multi-task prediction results of five TE properties from the best performing MMoE model with fivefold cross-validation on the experimental dataset. *Source*: [52] / Springer Nature / CC BY 4.0.

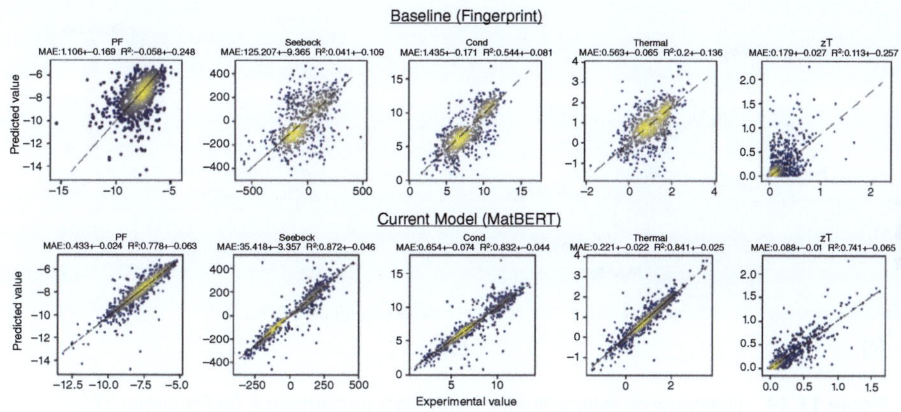

Figure 11.33 Comparisons of baseline representation (structural level, fingerprint) and structural level language representations on prediction of TE properties. *Source*: [52] / Springer Nature / CC BY 4.0.

demonstrates this by deploying these language-augmented features within a systematic materials recommendation and discovery framework. This framework effectively operationalizes the predictive power of language models to navigate the vast materials landscape. Its initial unsupervised candidate recall stage benefits directly from the rich semantic information captured by language model embeddings; the ability to retrieve functionally similar materials, as evidenced by favorable zT distributions and strong recall metrics (Figure 11.34), is a direct consequence of the quality of these features derived from textual descriptions of composition and structure. Subsequently, the candidate ranking stage leverages the accurate multi-property predictions from the MMoE model, which itself is powered by these comprehensive language embeddings. The resulting discovery potential, highlighted by UMAP visualizations showing meaningful clustering of high-performance TE materials and identifying under-explored promising regions, underscores the practical utility of this approach.

The study by Jiaxing Qu et al. demonstrates a language-augmented approach to materials informatics, where diverse material information, including complex structural characteristics, is converted into a linguistic format. This textual representation enables advanced Transformer models to generate highly informative and versatile embeddings. These embeddings not only exhibit competitive or superior performance in property prediction tasks, particularly in data-scarce scenarios, but

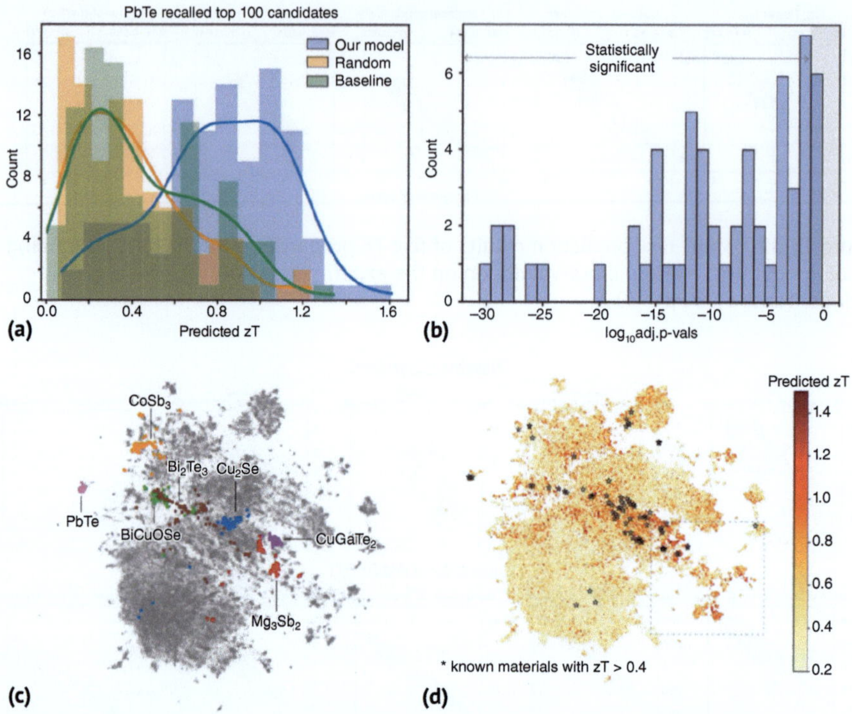

Figure 11.34 Overview of recommender framework performance. (a) Predicted zT distributions of the top 100 candidates recalled for PbTe using MatBERT, the baseline model and random sampling. (b) Adjusted p-values of the top 100 highest-zT candidates and 94 of 100 differ significantly from random. (c) Recall results for seven high-performing thermoelectric materials on the UMAP projection of 116K material representations. (d) UMAP projection overlaid with predicted zT values and known materials with zT > 0.4 are marked by stars. *Source*: [52] / Springer Nature / CC BY 4.0.

also facilitate higher-level discovery workflows by capturing nuanced chemical-structural relationships and representing complex features like doping or crystallographic symmetries. The future of this methodology involves refining LLMs on specialized materials science corpora, incorporating multimodal datasets, and enhancing model interpretability to extract new scientific insights. Ultimately, leveraging these context-aware embeddings within generative frameworks holds significant promise for the de novo design of materials with algorithmically optimized properties, positioning language-augmented modeling as a pivotal tool in the future of computational materials discovery.

11.5 Fine-tuning Foundation Models for Materials Discovery

The discovery of novel materials with enhanced properties is a cornerstone of technological progress, underpinning innovations in sectors from clean energy to healthcare [10, 53, 54]. To navigate the immense and complex materials space, researchers have increasingly turned to computational methods, primarily first-principles simulations for accuracy [3] and ML for speed and data-driven insights. However, ML adoption is hindered by the challenge of creating universal input representations. Human-readable descriptions (e.g., "tetragonal TiO_2") must be converted into machine-readable formats (e.g., CIF files), requiring complex feature engineering that often produces descriptors lacking generality and transferability. These representations, typically derived from idealized structures, also fail to capture the imperfections and variations found in real experimental materials, limiting predictive accuracy in practice. LLMs, such as ChatGPT, present a promising alternative by processing natural language directly, thereby bypassing bespoke feature engineering and using text as a universal input format [55]. Early studies have shown that GPT-based models can perform well in chemical and materials tasks, particularly in data-scarce settings [56, 57]. Yet, the proprietary and closed nature of many leading LLMs restricts accessibility, requiring expensive fine-tuning on external servers and raising concerns about transparency and reproducibility. This underscores the need for powerful, open-source models designed for scientific use. In this context, DARWIN 1.5, developed by Tong Xie et al. [58], represents a key step forward. By fine-tuning a general-purpose LLM through a two-stage strategy, the model is adapted into a flexible and high-performing predictive tool for materials science. DARWIN 1.5 can process natural language inputs directly, enabling it to tackle diverse tasks without task-specific feature engineering, and thus demonstrates a viable path toward universal predictive modeling in the field.

11.5.1 A Two-stage Fine-tuning Strategy for Domain Adaptation

The core innovation of the DARWIN 1.5 framework is a sophisticated two-stage fine-tuning strategy, designed to adapt a general-purpose, open-source foundation model (LLaMA-7B) [59] into a specialized materials science expert as shown in

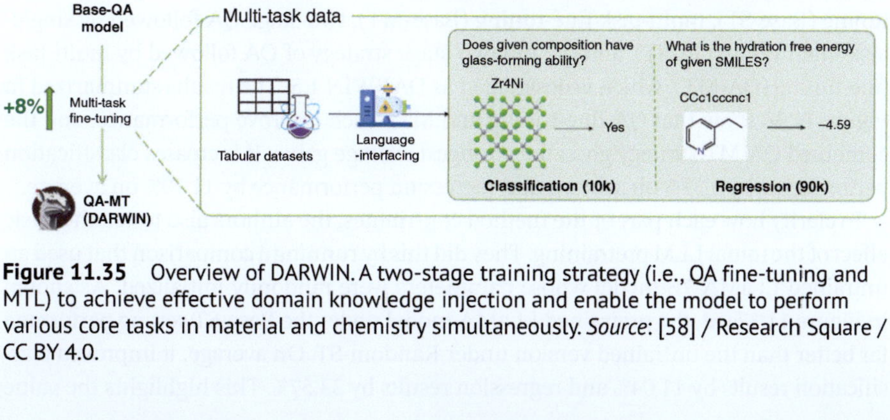

Figure 11.35 Overview of DARWIN. A two-stage training strategy (i.e., QA fine-tuning and MTL) to achieve effective domain knowledge injection and enable the model to perform various core tasks in material and chemistry simultaneously. *Source*: [58] / Research Square / CC BY 4.0.

Figure 11.35. This strategy aims to endow the model with both deep domain knowledge and versatile task-handling capabilities.

The first stage involves question-answering (QA) fine-tuning for knowledge injection. This approach is designed to emulate how human scientists acquire expertise—by reading, interpreting, and reasoning with the vast body of scientific literature. To accomplish this, the SciQAG framework [60] was used to create 332,997 open-ended scientific QA pairs from highly cited materials science papers. Fine-tuning on this dataset allows for the seamless integration of critical "know-how" knowledge, experimental context, and nuanced scientific reasoning directly into the language model. To mitigate catastrophic forgetting of general language capabilities and prevent overfitting to scientific jargon, a comparable number of general QA pairs from the Tulu dataset [61] were mixed into training. This mix helps the model keep strong general language understanding while gaining domain-specific capability.

The second stage employs MTL for generalization and synergy. In contrast to prior efforts that fine-tuned separate LLMs for individual predictive tasks [62], this stage fine-tunes a single model to perform 22 different materials science tasks simultaneously. This approach has demonstrated significant value in materials science for its ability to effectively capture inherent correlations among various material properties [63]. The tasks were drawn from 21 FAIR (Findable, Accessible, Interoperable, and Reusable) datasets [64] and include five classification tasks and 17 regression tasks, spanning different material systems and properties represented by compositions, material names, or SMILES strings. By learning multiple related tasks concurrently, the model is encouraged to develop shared underlying representations and improve its overall generalization capabilities. To enable this, all original datasets were converted into a unified Language-Interfaced Fine-Tuning (LIFT) format [65], using prompt templates that turn structured entries into natural language instructions. For instance, one instruction can ask, "What is the band gap of the given composition?" with the input "$CdCu_2SnS_4$," and the expected output text is "1.37." To improve robustness and reduce hallucinations, 5% artificial counterexamples were also added to the training sets.

11.5.2 An Empirical Evaluation of Fine-tuning Strategies

To systematically assess the effectiveness of their proposed methodology, the study conducted a series of comparative experiments evaluating four distinct fine-tuning strategies applied to the base LLaMA-7B model. These included single-task fine-tuning (Base-ST), multi-task fine-tuning (Base-MT), two-stage QA followed by single-task fine-tuning (QA-ST), and the full two-stage strategy of QA followed by multi-task fine-tuning (QA-MT), which is designated as DARWIN 1.5. The results summarized in Figure 11.36 show that QA fine-tuning and MTL each improve performance, and the combined QA MT strategy gives the strongest average gains. It increases classification performance by 3.38% on average and regression performance by 11.79% on average.

To clarify how each part of the method contributes, the authors also tested the basic effect of the initial LLM pretraining. They did this by running a comparison that used an untrained LLaMA-7B model whose parameters were randomly initialized. As shown in Figure 11.37a, d, the pretrained LLaMA model under the Base-ST setting performed far better than the untrained version under Random-ST. On average, it improved classification results by 11.04% and regression results by 33.57%. This highlights the value

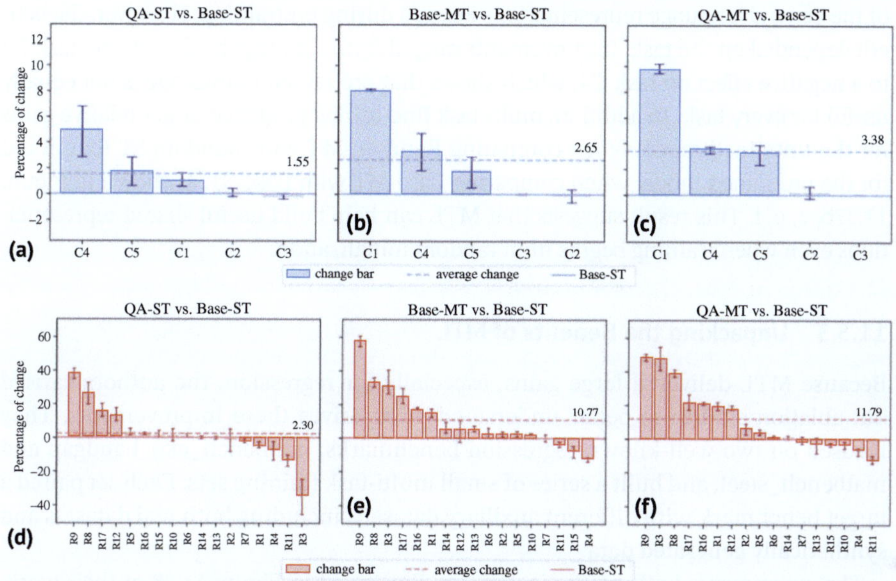

Figure 11.36 Comparison of different fine-tuning strategies on task performance. Base-ST is used as baseline and bars illustrate the difference between models on the specific tasks. The average improvement is indicated by the dashed line. *Source*: [58] / Research Square / CC BY 4.0.

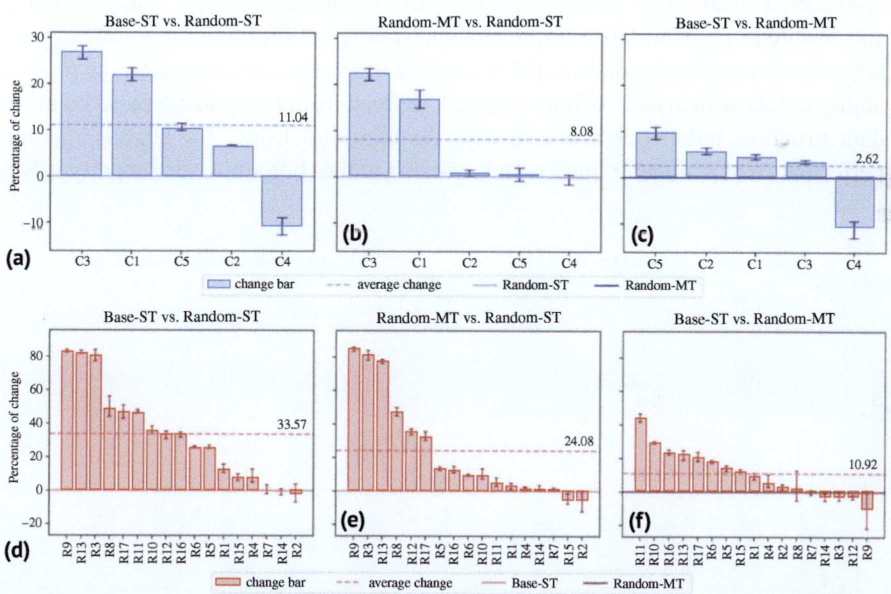

Figure 11.37 Comparative performance of pre-trained and non-pretrained models under single-task and multi-task fine-tuning strategies. (a and d) Single-task fine-tuning results, comparing pretrained (Base-ST) and non-pretrained (Random-ST) models. (b and e) Multi-task fine-tuning results for non-pretrained (Random-MT) versus single-task non-pretrained (Random-ST) models. (c and f) Comparison of pretraining and multi-task fine-tuning effects on overall performance. (a–c) Show results on classification tasks and (d–f) show results on regression tasks. *Source*: [58] / Research Square / CC BY 4.0.

of the general language representations learned during pretraining. However, the benefit depended on the task. Improvements ranged from a large gain of 83.3% on task R9 to a negative effect on task C4, which shows that pretrained knowledge is not equally useful for every task. In addition, multi-task fine-tuning produced larger relative gains for the untrained model when comparing Random-MT with Random-ST than it did for the pretrained model when comparing Base-MT with Base-ST, as shown in Figure 11.37b, c, e, f. This result suggests that MTL can help build useful shared representations even when training begins from random initialization.

11.5.3 Unpacking the Benefits of MTL

Because MTL delivered large gains, especially for regression, the authors carried out ablation studies to better understand what drives these improvements. They focused on two well-known regression benchmarks, matbench_exp_bandgap and matbench_steel, and built a series of small multi-task training sets. Each set paired a target benchmark with different auxiliary datasets, including both real datasets and synthetically generated data.

The experiments with synthetic datasets, illustrated in Figure 11.38 of their work, provided crucial insights. Incorporating auxiliary data with correct material compositions but randomized property values (+Syn 1) led to a significant deterioration in model performance, indicating that the introduction of misleading knowledge is detrimental. Conversely, using auxiliary data with completely fabricated inputs (random alphabetical codes for compositions) and fabricated property values (+Syn 2) still resulted in performance improvement. This result indicates that training on a larger and more varied collection of structured examples can strengthen the model's ability to follow instructions [66], a benefit derived from recognizing patterns in the data structure, independent of authentic domain knowledge. The slightly superior performance of a model trained with fabricated names but authentic property values

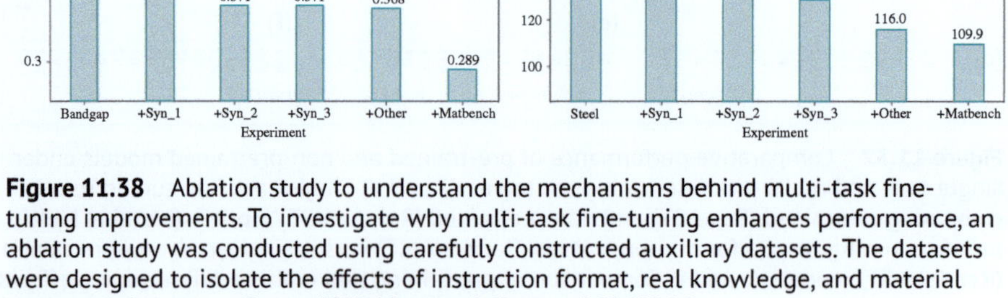

Figure 11.38 Ablation study to understand the mechanisms behind multi-task fine-tuning improvements. To investigate why multi-task fine-tuning enhances performance, an ablation study was conducted using carefully constructed auxiliary datasets. The datasets were designed to isolate the effects of instruction format, real knowledge, and material representation types. *Source*: [58] / Research Square / CC BY 4.0.

(+Syn 3) further indicates that capturing underlying statistical patterns in real property distributions is beneficial.

Using real auxiliary datasets produced much larger improvements than using synthetic alternatives, which points to the importance of training with genuine materials science knowledge. Among the auxiliary datasets, the one incorporating general composition-based representations (+matbench) outperformed the dataset with specialized SMILES-based representations (+other), highlighting the superiority of general representations for effective knowledge transfer. By comparing the performance gains between the best real dataset (+matbench) and the instruction-tuned synthetic dataset (+Syn 2), the study disentangled the components of multi-task fine-tuning benefits. The improvement achieved with +Syn 2 reflects enhanced general problem-solving and instruction-following capabilities, whereas the additional gains from +matbench stem from the successful transfer and integration of domain-specific scientific knowledge. Overall, these ablation results indicate that multi-task fine-tuning helps through two routes: it improves instruction following, and it also brings in real materials knowledge that is difficult to replace with synthetic data.

11.5.4 Benchmarking and Application to Bandgap Prediction

To place the performance of the fully fine-tuned DARWIN 1.5 model, also called QA MT, in context, they benchmarked it against both strong closed-source language models and competitive traditional ML methods. The results in Figure 11.39 show that DARWIN 1.5 is an effective and competitive open-source option. Across most

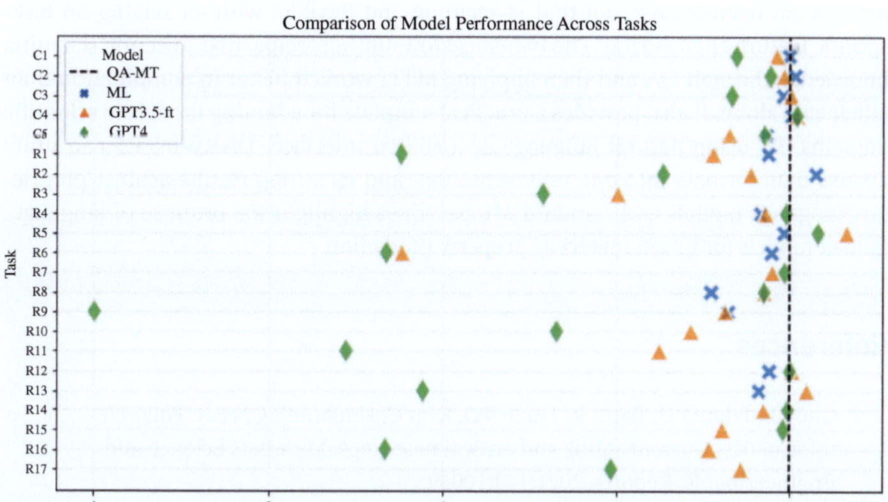

Figure 11.39 Comparison of model performance relative to QA-MT across tasks. Points on the left of the vertical line indicate poorer performance compared to QA-MT. The x-axis represents the performance ratio, where a value of −1 indicates performance 100% worse than QA-MT. For tasks with more than one ML results, the best one is used in this visualization. *Source*: [58] / Research Square / CC BY 4.0.

of the 22 tasks, it performed better than a fine-tuned GPT-3.5 model and a few shots prompted the GPT-4 model. In addition, for tasks where traditional ML baselines were reported, DARWIN 1.5 achieved higher performance in 11 of 14 tasks. This result is notable because it indicates that one unified language model can outperform a collection of separate ML models that often depend on carefully designed, task-specific descriptors.

To show that the final model is useful in a scientifically important setting and can deliver accurate predictions, the study used band gap prediction as a case study. Band gap is a key electronic property that affects whether a material is suitable for transistors [67], photovoltaics [68], and light-emitting diodes (LEDs) [69], so it serves as a strong benchmark for evaluating AI models in materials science. The QA MT model predictions for seven specific compositions that were not seen during training were then compared with several computational approaches and with experimental measurements.

Conventional first-principles approach such as Perdew–Burke–Ernzerhof, or PBE, tend to predict band gaps that are too small, while hybrid functionals such as Heyd–Scuseria–Ernzerhof, or HSE, are usually more accurate but require far more computation. QA MT reached a mean absolute deviation of 0.51 and a root MSE of 0.65. This places its accuracy close to HSE at a much lower computational cost, and it performs clearly better than both PBE and AFLOW, which is a ML model trained using PBE data. Even though QA MT does not match the accuracy of more advanced many-body perturbation methods such as GW, it offers three clear strengths. It can reduce systematic errors because experimental data are included during fine-tuning, it needs only the material composition rather than full structural information, and it can produce predictions almost immediately, which makes it attractive for large-scale screening. This DARWIN 1.5 case study shows that fine-tuning foundation language models can produce an open-source tool that is accurate and flexible, without relying on task-specific feature engineering. The two-stage fine-tuning recipe, first injecting scientific knowledge through QA and then applying MTL, worked better in combination than either step alone. It also provides a practical template for adapting models to scientific domains. By using natural language as a shared interface, DARWIN 1.5 can unify diverse data formats into one representation, and its strong results against proprietary language models and standard ML baselines highlight the promise of language-guided models for broad materials property prediction.

References

1 Chen L, Pilania G, Batra R, Huan TD, Kim C, Kuenneth C, et al. Polymer informatics: Current status and critical next steps. Materials Science and Engineering: R: Reports. 2021;144:100595.

2 Audus DJ, De Pablo JJ. Polymer informatics: Opportunities and challenges. ACS Macro Letters. 2017;6(10):1078–82.

3 Kuenneth C, Rajan AC, Tran H, Chen L, Kim C, Ramprasad R. Polymer informatics with multi-task learning. Patterns [Internet]. 2021 [cited 2025 June 29];2(4):100238. Available from: https://www.cell.com/patterns/fulltext/S2666-3899(21)00058-1?uuid=uuid%3A3ac4e564-753c-4d4e-9ea4-0f9057a9d97a

4 Chen G, Tao L, Li Y. Predicting polymers' glass transition temperature by a chemical language processing model. Polymers. 2021;13(11):1898.

5 Doan Tran H, Kim C, Chen L, Chandrasekaran A, Batra R, Venkatram S, et al. Machine-learning predictions of polymer properties with Polymer Genome. Journal of Applied Physics [Internet]. 2020 [cited 2025 June 29]; 128(17). Available from: https://pubs.aip.org/aip/jap/article/128/17/171104/1062836

6 Kuenneth C, Ramprasad R. polyBERT: A chemical language model to enable fully machine-driven ultrafast polymer informatics. Nature Communications. 2023;14(1):4099.

7 Moon J, Beker W, Siek M, Kim J, Lee HS, Hyeon T, et al. Active learning guides discovery of a champion four-metal perovskite oxide for oxygen evolution electrocatalysis. Nature Materials. 2024;23(1):108–15.

8 Raccuglia P, Elbert KC, Adler PD, Falk C, Wenny MB, Mollo A, et al. Machine-learning-assisted materials discovery using failed experiments. Nature. 2016;533(7601):73–6.

9 Rao Z, Tung PY, Xie R, Wei Y, Zhang H, Ferrari A, et al. Machine learning–enabled high-entropy alloy discovery. Science. 2022;378(6615):78–85.

10 OpenAI, Achiam J, Adler S, Agarwal S, Ahmad L, Akkaya I, et al. GPT-4 Technical Report [Internet]. arXiv; 2023 [cited 2024 Jan 23]. Available from: http://arxiv.org/abs/2303.08774

11 Almazrouei E, Alobeidli H, Alshamsi A, Cappelli A, Cojocaru R, Debbah M, et al. The Falcon series of open language models [Internet]. arXiv; 2023 [cited 2024 Jan 23]. Available from: http://arxiv.org/abs/2311.16867

12 Devlin J, Chang MW, Lee K, Toutanova K. BERT: Pre-training of Deep Bidirectional transformers for language understanding. In: Proceedings of the 2019 Conference of the North [Internet]. Minneapolis, MN: Association for Computational Linguistics; 2019 [cited 2024 Mar 20]. pp. 4171–86. Available from: http://aclweb.org/anthology/N19-1423

13 Wei X, van der Zwaag S, Jia Z, Wang C, Xu W. On the use of transfer modeling to design new steels with excellent rotating bending fatigue resistance even in the case of very small calibration datasets. Acta Materialia. 2022;235:118103.

14 Zhao WX, Zhou K, Li J, Tang T, Wang X, Hou Y, et al. A survey of large language models [Internet]. arXiv; 2023 [cited 2024 Jan 24]. Available from: http://arxiv.org/abs/2303.18223

15 Beltagy I, Lo K, Cohan A. SciBERT: A pretrained language model for scientific text. arXiv preprint arXiv:1903.10676. 2019.

16 Gupta T, Zaki M, Krishnan NA, Mausam. MatSciBERT: A materials domain language model for text mining and information extraction. npj Computational Materials. 2022;8(1):102.

17 Ahmad WU, Chakraborty S, Ray B, Chang KW. Unified pre-training for program understanding and generation [Internet]. arXiv; 2021 [cited 2024 Jan 24]. Available from: http://arxiv.org/abs/2103.06333

18 Chen M, Tworek J, Jun H, Yuan Q, Pinto HP de O, Kaplan J, et al. Evaluating large language models trained on code [Internet]. arXiv; 2021 [cited 2024 Jan 24]. Available from: http://arxiv.org/abs/2107.03374

19 Austin J, Odena A, Nye M, Bosma M, Michalewski H, Dohan D, et al. Program synthesis with large language models [Internet]. arXiv; 2021 [cited 2024 Jan 24]. Available from: http://arxiv.org/abs/2108.07732

20 Merchant A, Batzner S, Schoenholz SS, Aykol M, Cheon G, Cubuk ED. Scaling deep learning for materials discovery. Nature. 2023;624(7990):80–5.

21 Szymanski NJ, Rendy B, Fei Y, Kumar RE, He T, Milsted D, et al. An autonomous laboratory for the accelerated synthesis of novel materials. Nature. 2023;624(7990):86–91.

22 Zheng Z, Zhang O, Borgs C, Chayes JT, Yaghi OM. ChatGPT chemistry assistant for text mining and the prediction of MOF synthesis. Journal of the American Chemical Society. 2023;145(32):18048–62. https://doi.org/10.1021/jacs.3c05819

23 Boiko DA, MacKnight R, Kline B, Gomes G. Autonomous chemical research with large language models. Nature. 2023;624(7992):570–8.

24 Nanda T, Singh V, Singh V, Chakraborty A, Sharma S. Third generation of advanced high-strength steels: Processing routes and properties. Proceedings of the Institution of Mechanical Engineers, Part L: Journal of Materials: Design and Applications. 2019;233(2):209–38.

25 Gao J, Jiang S, Zhang H, Huang Y, Guan D, Xu Y, et al. Facile route to bulk ultrafine-grain steels for high strength and ductility. Nature. 2021;590(7845):262–7.

26 Zhi H, Li J, Li W, Elkot M, Antonov S, Zhang H, et al. Simultaneously enhancing strength-ductility synergy and strain hardenability via Si-alloying in medium-Al FeMnAlC lightweight steels. Acta Materialia. 2023;245:118611.

27 Zhang Y, Wang C, Reddy KM, Li W, Wang X. Study on the deformation mechanism of a high-nitrogen duplex stainless steel with excellent mechanical properties originated from bimodal grain design. Acta Materialia. 2022;226:117670.

28 Zou C, Li J, Wang WY, Zhang Y, Lin D, Yuan R, et al. Integrating data mining and machine learning to discover high-strength ductile titanium alloys. Acta Materialia. 2021;202:211–21.

29 Zhang H, Fu H, He X, Wang C, Jiang L, Chen LQ, et al. Dramatically enhanced combination of ultimate tensile strength and electric conductivity of alloys via machine learning screening. Acta Materialia. 2020;200:803–10.

30 Tian S, Jiang X, Wang W, Jing Z, Zhang C, Zhang C, et al. Steel design based on a large language model. Acta Materialia. 2025;285:120663.

31 Yan R, Jiang X, Wang W, Dang D, Su Y. Materials information extraction via automatically generated corpus. Scientific Data. 2022;9(1):401.

32 Wang W, Jiang X, Tian S, Liu P, Lookman T, Su Y, et al. Alloy synthesis and processing by semi-supervised text mining. npj Computational Materials. 2023;9(1):183.

33 Wang W, Jiang X, Tian S, Liu P, Dang D, Su Y, et al. Automated pipeline for superalloy data by text mining. npj Computational Materials. 2022;8(1):9.

34 He P, Liu X, Gao J, Chen W. DeBERTa: Decoding-enhanced BERT with disentangled attention. arXiv preprint arXiv:2006.03654. 2020.

35 He P, Gao J, Chen W. DeBERTaV3: Improving DeBERTa using ELECTRA-style pre-training with gradient-disentangled embedding sharing. arXiv preprint arXiv:2111.09543. 2021.

36 Lan Z, Chen M, Goodman S, Gimpel K, Sharma P, Soricut R. ALBERT: A lite BERT for self-supervised learning of language representations [Internet]. arXiv; 2020 [cited 2023 Dec 9]. Available from: http://arxiv.org/abs/1909.11942

37 Lewis M, Liu Y, Goyal N, Ghazvininejad M, Mohamed A, Levy O, et al. BART: Denoising sequence-to-sequence pre-training for natural language generation, translation, and comprehension. arXiv preprint arXiv:1910.13461. 2019.

38 Akiba T, Sano S, Yanase T, Ohta T, Koyama M. Optuna: A next-generation hyperparameter optimization framework. In: Proceedings of the 25th ACM SIGKDD International Conference on Knowledge Discovery & Data Mining [Internet]. Anchorage, AK: ACM; 2019 [cited 2024 Jan 10]. pp. 2623–31. Available from: https://dl.acm.org/doi/10.1145/3292500.3330701

39 McInnes L, Healy J, Melville J. UMAP: Uniform manifold approximation and projection for dimension reduction [Internet]. arXiv; 2020 [cited 2023 Dec 9]. Available from: http://arxiv.org/abs/1802.03426

40 Abdin M, Aneja J, Awadalla H, Awadallah A, Awan AA, Bach N, et al. Phi-3 technical report: A highly capable language model locally on your phone [Internet]. arXiv; 2024 [cited 2024 Oct 12]. Available from: http://arxiv.org/abs/2404.14219

41 Jiang AQ, Sablayrolles A, Mensch A, Bamford C, Chaplot DS, Casas D de las, et al. Mistral 7B [Internet]. arXiv; 2023 [cited 2024 Oct 12]. Available from: http://arxiv.org/abs/2310.06825

42 Touvron H, Martin L, Stone K, Albert P, Almahairi A, Babaei Y, et al. Llama 2: Open foundation and fine-tuned chat models [Internet]. arXiv; 2023 [cited 2023 Nov 3]. Available from: http://arxiv.org/abs/2307.09288

43 Sennrich R, Haddow B, Birch A. Neural machine translation of rare words with subword units. arXiv:1508.07909 [Internet]. 2016 June 10 [cited 2019 Dec 22]. Available from: http://arxiv.org/abs/1508.07909

44 Schuster M, Nakajima K. Japanese and Korean voice search. In: 2012 IEEE International Conference on Acoustics, Speech and Signal Processing (ICASSP) [Internet]. IEEE; 2012 [cited 2024 Dec 5]. pp. 5149–52. Available from: https://ieeexplore.ieee.org/abstract/document/6289079/

45 Wan J, Ran Q, Li J, Xu Y, Xiao X, Yu H, et al. A new resource-saving, low chromium and low nickel duplex stainless steel 15Cr–xAl–2Ni–yMn. Materials & Design. 2014;53:43–50.

46 Wan J, Ruan H, Shi S. Excellent combination of strength and ductility in 15Cr-2Ni duplex stainless steel based on ultrafine-grained austenite phase. Materials Science and Engineering: A. 2017;690:96–103.

47 Wan J, Ruan H, Wang J, Shi S. Exploiting the non-equilibrium phase transformation in a 15Cr-2Ni-2Al-11Mn resource-saving duplex stainless steel. Materials & Design. 2017;114:433–40.

48 Zheng Y, Sun H, Yan L, Yang H, Gao K, Pang X, et al. Ferrite effects on the hydrogen embrittlement of 17-4PH stainless steel. Anti-Corrosion Methods and Materials. 2022;69(3):331–8.

49 Choudhary K, DeCost B. Atomistic line graph neural network for improved materials property predictions. npj Computational Materials. 2021;7(1):185.

50 Chen C, Ye W, Zuo Y, Zheng C, Ong SP. Graph networks as a universal machine learning framework for molecules and crystals. Chemistry of Materials. 2019;31(9):3564–72.

51 Yan K, Liu Y, Lin Y, Ji S. Periodic graph transformers for crystal material property prediction. Advances in Neural Information Processing Systems. 2022;35:15066–80.

52 Qu J, Xie YR, Ciesielski KM, Porter CE, Toberer ES, Ertekin E. Leveraging language representation for materials exploration and discovery. npj Computational Materials. 2024;10(1):58.

53 Jablonka KM, Schwaller P, Ortega-Guerrero A, Smit B. Leveraging large language models for predictive chemistry. Nature Machine Intelligence. 2024;6(2):161–9.

54 Wan Y, Liu Y, Ajith A, Grazian C, Hoex B, Zhang W, et al. SciQAG: A Framework for auto-generated science question answering dataset with fine-grained evaluation [Internet]. arXiv; 2024 [cited 2025 July 14]. Available from: http://arxiv.org/abs/2405.09939

55 Wilkinson MD, Dumontier M, Aalbersberg IJJ, Appleton G, Axton M, Baak A, et al. The FAIR Guiding Principles for scientific data management and stewardship. Scientific Data. 2016;3(1):1–9.

56 Chen Y, Qian S, Tang H, Lai X, Liu Z, Han S, et al. LongLoRA: Efficient fine-tuning of long-context large language models [Internet]. arXiv; 2024 [cited 2025 July 14]. Available from: http://arxiv.org/abs/2309.12307

57 Dunn A, Wang Q, Ganose A, Dopp D, Jain A. Benchmarking materials property prediction methods: The Matbench test set and Automatminer reference algorithm. npj Computational Materials. 2020;6(1):138.

58 Xie T, Wan Y, Liu Y, Zeng Y, Wang S, Zhang W, et al. DARWIN 1.5: Large language models as materials science adapted learners [Internet]. arXiv; 2025 [cited 2025 July 14]. Available from: http://arxiv.org/abs/2412.11970

59 Luo R, Sun L, Xia Y, Qin T, Zhang S, Poon H, et al. BioGPT: Generative pre-trained transformer for biomedical text generation and mining. Briefings in Bioinformatics. 2022;23(6):bbac409.

60 Taylor R, Kardas M, Cucurull G, Scialom T, Hartshorn A, Saravia E, et al. Galactica: A large language model for science. arXiv preprint arXiv:2211.09085 [Internet]. 2022 [cited 2025 June 29]; Available from: https://arxiv.org/abs/2211.09085

61 Bran AM, Cox S, Schilter O, Baldassari C, White AD, Schwaller P. ChemCrow: Augmenting large-language models with chemistry tools [Internet]. arXiv; 2023 [cited 2025 June 29]. Available from: http://arxiv.org/abs/2304.05376

62 Rasley J, Rajbhandari S, Ruwase O, He Y. DeepSpeed: System optimizations enable training deep learning models with over 100 billion parameters. In: Proceedings of the 26th ACM SIGKDD International Conference on Knowledge Discovery & Data Mining, Virtual Event [Internet]. ACM; 2020 [cited 2025 July 14]. pp. 3505–6. Available from: https://dl.acm.org/doi/10.1145/3394486.3406703

63 Pei Z, Yin J, Hawk JA, Alman DE, Gao MC. Machine-learning informed prediction of high-entropy solid solution formation: Beyond the Hume-Rothery rules. npj Computational Materials. 2020;6(1):50.

64 Nagasawa S, Al-Naamani E, Saeki A. Computer-aided screening of conjugated polymers for organic solar cell: Classification by random forest. Journal of Physical Chemistry Letters. 2018;9(10):2639–46.

65 Gaulton A, Bellis LJ, Bento AP, Chambers J, Davies M, Hersey A, et al. ChEMBL: A large-scale bioactivity database for drug discovery. Nucleic Acids Research. 2012; 40(D1):D1100–7.

66 Griffiths RR, Greenfield JL, Thawani AR, Jamasb AR, Moss HB, Bourached A, et al. Data-driven discovery of molecular photoswitches with multioutput Gaussian processes. Chemical Science. 2022;13(45):13541–51.

67 Stanev V, Oses C, Kusne AG, Rodriguez E, Paglione J, Curtarolo S, et al. Machine learning modeling of superconducting critical temperature. npj Computational Materials. 2018;4(1):29.

68 Chen K, Zhuang D, Chang JM. SuperCon: Supervised contrastive learning for imbalanced skin lesion classification [Internet]. arXiv; 2022 [cited 2025 July 14]. Available from: http://arxiv.org/abs/2202.05685

69 Gaultois MW, Sparks TD, Borg CKH, Seshadri R, Bonificio WD, Clarke DR. Data-driven review of thermoelectric materials: Performance and resource considerations. Chemistry of Materials. 2013;25(15):2911–20.

12

Retrieval-augmented Generation for Materials Large Language Models

12.1 Introduction

In the dynamic realm of materials research, Conversational AI is emerging as a pivotal tool, reshaping how researchers navigate complex challenges and uncover new knowledge. The integration of large language models (LLMs) into this field marks a significant leap forward, offering unprecedented opportunities to streamline research processes and foster innovation.

LLMs have witnessed exponential growth in capabilities in recent times. Their advanced language processing abilities, such as natural language understanding and generation, enable them to engage in sophisticated interactions. These models are not only capable of answering questions but also of inferring relationships and suggesting novel hypotheses, all of which are invaluable in materials research. Retrieval-augmented generation (RAG) has become an essential component in the utilization of LLMs. By combining the generative power of LLMs with the retrieval of external knowledge, RAG allows for more informed and context-rich outputs. This approach is particularly relevant in materials research, where vast amounts of data from diverse sources, such as research papers, databases of material properties, and experimental records, need to be synthesized.

This chapter explores the design principles of LLM-powered research assistants in materials research. It delves into architectures that incorporate RAG. Through this exploration, the authors aim to provide a comprehensive understanding of how Conversational AI can be harnessed to drive progress in materials research.

12.2 LLM-powered Research Assistants: Design Principles

Materials science, an inherently interdisciplinary field integrating physics, chemistry, biology, and metallurgy, presents a multitude of challenges in experimental and theoretical research. Traditional experimental studies on metallic materials, such as exploring microstructural evolution during casting, forging, and heat treatment, involve complex and labor-intensive processes. Specialized equipment like induction furnaces for synthesis and meticulous sample preparation techniques like electropolishing and chemical etching are essential. Characterization methods, including electron backscatter diffraction (EBSD) and X-ray diffraction (XRD), require precise operation and generate vast amounts of data, which demand in-depth metallurgical knowledge for accurate interpretation. Theoretical research is equally daunting, as incorporating metallurgical theories into computational approaches, whether using density functional theory (DFT) for alloy electronic structure analysis or computational fluid dynamics for simulating molten metal flow, necessitates proficiency in multiple programming languages and computational platforms. Handling uncertainties and making accurate predictions further complicates the research process.

Against this backdrop, LLM-powered research assistants emerge as a promising solution, and their design principles play a crucial role in addressing these challenges in materials science, as shown in Figure 12.1. These research assistants, trained on extensive materials science and metallurgy corpora, are tailored to meet the specific needs of the field. They can rapidly sift through vast amounts of literature, extracting key insights and generating well-informed hypotheses, thus significantly reducing the time-consuming literature review process for researchers.

12.2.1 Design Principles for Literature-related Functions

Semantic-aware Literature Retrieval: The design of LLM-powered research assistants should incorporate advanced semantic understanding techniques. In materials science, terms can have multiple meanings depending on the context. For example, "phase" can refer to a physical phase in thermodynamics or a phase in crystallography. By using deep learning-based semantic models, the assistant can accurately understand the researcher's query intent and retrieve relevant

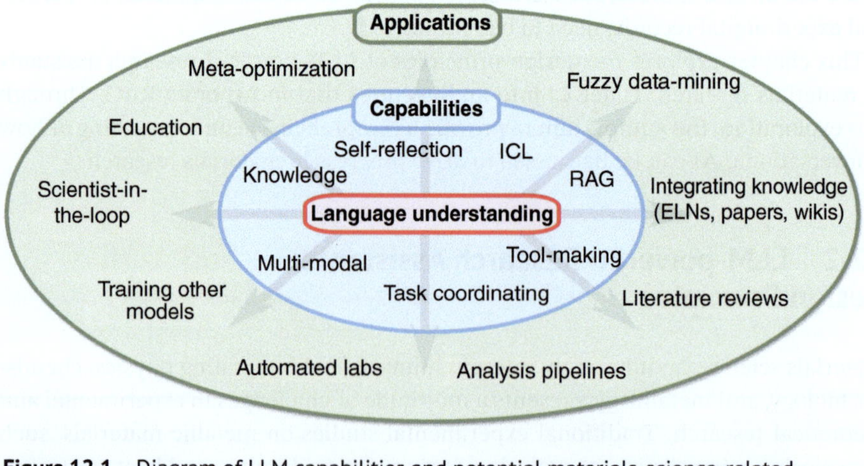

Figure 12.1 Diagram of LLM capabilities and potential materials-science-related applications. These emergent capabilities can be combined with each other and integrated into traditional pipelines (genetic algorithms, databases, etc.) to form the different applications. *Source*: [1] / Royal Society of Chemistry / CC BY 3.0.

literature from a wide range of databases, including academic journals, conference proceedings, and patents.

Knowledge Graph (KG)-enhanced Literature Summarization: To make the vast amount of literature more digestible, the assistant can be designed to construct a KG for materials science. This graph can connect different concepts such as material properties, processing methods, and applications. When summarizing literature, the LLM can draw on the KG to provide structured and comprehensive summaries. For instance, it can show how a particular heat treatment process affects the mechanical properties of a specific metal alloy, and how this relationship is related to other similar materials and processes.

12.2.2 Design Principles for Experimental Design

Multi-objective Optimization for Synthesis Routes: In the experimental design phase, the design of LLM-powered research assistants should focus on leveraging historical research patterns and data to recommend optimal synthesis routes and characterization techniques. The assistant can be equipped with multi-objective optimization algorithms. For example, when suggesting a synthesis route for a new metal-based composite material, it can consider multiple factors simultaneously, such as cost, production time, and the desired material properties. It can compare different synthesis methods like powder metallurgy, chemical vapor deposition, and melt blending and recommend the most suitable one based on the specific requirements of the experiment.

Adaptive Characterization Technique Selection: Different characterization techniques have their own advantages and limitations. The research assistant can be designed to adaptively select the most appropriate characterization methods based on the nature of the material being studied and the research questions. For example, if the research focuses on the surface morphology of a metallic coating, techniques like scanning electron microscopy (SEM) or atomic force microscopy (AFM) might be recommended. If the goal is to analyze the crystal structure at the atomic level, XRD or transmission electron microscopy (TEM) would be more suitable. The assistant can also take into account the availability of equipment and the expertise of the research team when making these recommendations.

12.2.3 Design Principles for Data Analysis

Anomaly Detection and Outlier Handling: During data analysis, these research assistants are designed to quickly process experimental data, identify trends, and cross-reference with theoretical models for more accurate interpretations. An important aspect of data analysis in materials science is dealing with anomalies and outliers. The LLM-powered assistant can use statistical methods and machine learning algorithms to detect unusual data points. For example, in a dataset of material strength measurements, if a value is significantly different from the rest, the assistant can flag it as a potential outlier and suggest methods to verify its validity, such as repeating the measurement or checking for experimental errors.

Data-driven Model Refinement: The research assistant can be designed to continuously refine theoretical models based on experimental data. By comparing the experimental results with the predictions of existing models, the LLM can identify discrepancies and suggest modifications to the models. For instance, if the measured electrical conductivity of a new alloy does not match the predicted value from a conductivity model, the assistant can analyze the data and propose adjustments to the model parameters or even suggest new model structures to better represent the behavior of the material.

12.2.4 Design Principles for Theoretical Research

Hybrid Computational Method Recommendation: For theoretical research, LLM-powered research assistants can assist in algorithm development. The design principles should ensure that they can suggest suitable computational methods for specific problems. In materials science, different computational methods have their own strengths and weaknesses. For example, DFT is good for calculating the electronic structure of materials at the atomic level, while molecular dynamics (MD) is more suitable for simulating the behavior of materials at the mesoscale. The research assistant can recommend a hybrid approach that combines these methods, depending on the research problem. For example, when studying the diffusion of atoms in a metal alloy, it might suggest using DFT to calculate the energy barriers for atom movement and then using MD to simulate the actual diffusion process over time.

Code Generation and Optimization: To help researchers navigate the complexities of programming and platform deployment, the research assistant can be designed to generate code snippets for implementing computational methods. It can also optimize the generated code for better performance. For example, if the researcher wants to perform a computational fluid dynamics simulation of molten metal flow, the assistant can generate Python or Fortran code for setting up the simulation and then optimize the code by suggesting appropriate data structures and algorithms to reduce the computational time.

LLMs themselves possess remarkable capabilities that underpin the effectiveness of these research assistants. They have a strong understanding of natural language structure, enabling them to translate, transpose, generate, and answer questions based on texts. Their ability to perform reasoning and extract patterns from textual and numerical data extends their application beyond language-based tasks, making them competent in programming, managing, and coordinating complex tasks. Compared to traditional models like CNNs, the transformer architecture of LLMs is more amenable to multi-modality, allowing for the integration and processing of text and images, which further broadens the scope of problems that LLM-powered research assistants can address in materials science.

Moreover, the inexhaustible nature of LLMs, unlike human scientists, allows these research assistants to run continuously. This feature is invaluable in automated digital discovery workflows, automated laboratories, and pilot lines. Their flexibility and adaptability, along with the ability to process natural language instructions, retrieve domain knowledge, generate code, and coordinate systems, make them highly attractive tools for materials science researchers.

In conclusion, the design principles of LLM-powered research assistants are crucial for leveraging the unique capabilities of LLMs to meet the specific demands of materials science research. When designed and utilized wisely, these research assistants have the potential to revolutionize materials science research, expedite materials discovery, and conduct large-scale analyses that were previously impractical, thereby opening up new frontiers for scientific exploration and technological innovation in the field.

12.3 Overview of the RAG Framework

LLMs have demonstrated remarkable success, yet they remain constrained by significant limitations—particularly in domain-specific and knowledge-intensive tasks [2]. A key challenge is their tendency to generate "hallucinations" [3], especially when processing queries beyond their training scope or requiring up-to-date information. To address these issues, RAG augments LLMs by retrieving semantically relevant document chunks from external knowledge bases (KBs). By grounding responses in external knowledge, RAG mitigates factual inaccuracies in generated content. This integration has driven widespread adoption, positioning RAG as a critical technology for advancing chatbots and improving the real-world applicability of LLMs.

RAG technology has developed rapidly in recent years (as shown in Figure 12.2), exhibiting distinct stage-specific characteristics in its developmental trajectory during the era of LLMs. In its early stages, the rise of RAG coincided with the advancement of the Transformer architecture, primarily employing pre-trained models (PTMs) to inject external knowledge into language models. Research during this phase focused on optimizing pre-training techniques [5–7]. With the emergence of ChatGPT [8], LLMs demonstrated powerful in-context learning (ICL) capabilities, shifting the focus of RAG research toward providing higher-quality information for LLMs during the inference stage to tackle complex, knowledge-intensive tasks. This marked the beginning of a period of rapid development in related research. Currently, RAG optimization has expanded beyond the inference stage and is increasingly being deeply integrated with LLM fine-tuning techniques.

RAG, as an emerging field, has experienced rapid growth. However, this expansion has not yet been accompanied by a systematic synthesis to clarify its broader trajectory. This section focuses on the integration of RAG into LLMs. Current research tends to prioritize methodological advancements while lacking comprehensive analyses and evaluative reviews of RAG approaches—particularly concerning downstream tasks, datasets, benchmarks, and assessment methods tailored for RAG applications. Overall, the authors endeavor to provide foundational technical concepts and historical developments of RAG, a systematic review of RAG methodologies and their applications in the post-LLM era. This work aims to equip readers and professionals with a detailed, structured understanding of both LLMs and RAG. It seeks to illuminate the evolutionary path of retrieval-augmented technologies,

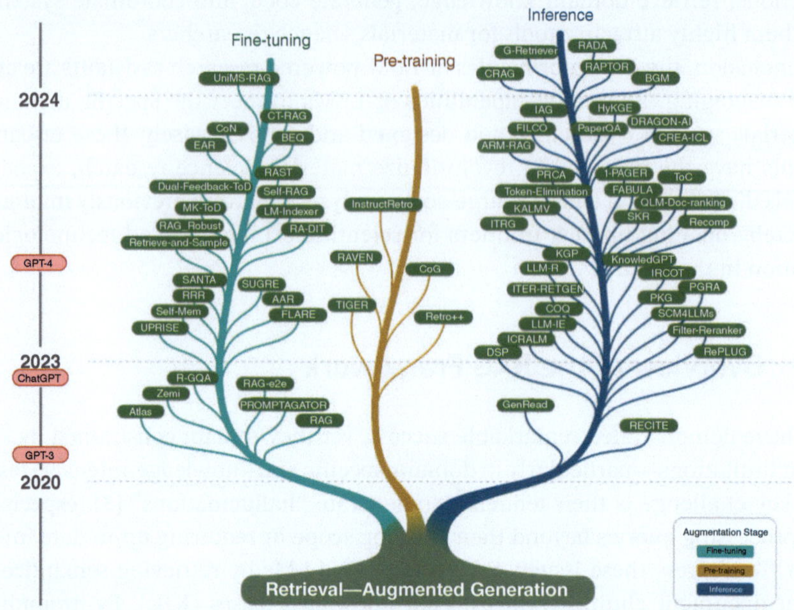

Figure 12.2 Technology tree of RAG research. The stages of involving RAG mainly include pre-training, fine-tuning, and inference. With the emergence of LLMs, research on RAG initially focused on leveraging the powerful ICL abilities of LLMs, primarily concentrating on the inference stage. Subsequent research has delved deeper, gradually integrating more with the fine-tuning of LLMs. Researchers have also been exploring ways to enhance language models in the pre-training stage through retrieval-augmented techniques. *Source*: [4] / arXiv.

evaluate the strengths and weaknesses of various approaches within their respective contexts, and foresee future trends and potential innovations in the field.

Figure 12.3 illustrates a typical RAG application scenario where a user queries ChatGPT about trending news. Since LLMs are constrained by their pretraining data, RAG bridges this knowledge gap by retrieving relevant articles from external databases and integrating them with the original query to generate informed responses. As shown in Figure 12.4, RAG's evolution can be categorized into three progressive stages: Naive RAG (basic implementation), Advanced RAG (enhanced methodologies), and Modular RAG (flexible architectures). While more cost-effective than fine-tuning and outperforming native LLMs, RAG still faces limitations, a key driver behind the development of Advanced and Modular RAG approaches that address Naive RAG's shortcomings through improved retrieval techniques and modular design.

12.3.1 Naive RAG

The Naive RAG paradigm emerged as the foundational methodology in RAG, gaining significant traction following ChatGPT's widespread adoption [8]. This approach established the core "Retrieve-Read" framework that systematically combines three key stages: (1) document indexing, (2) relevant passage retrieval, and (3) context-aware generation.

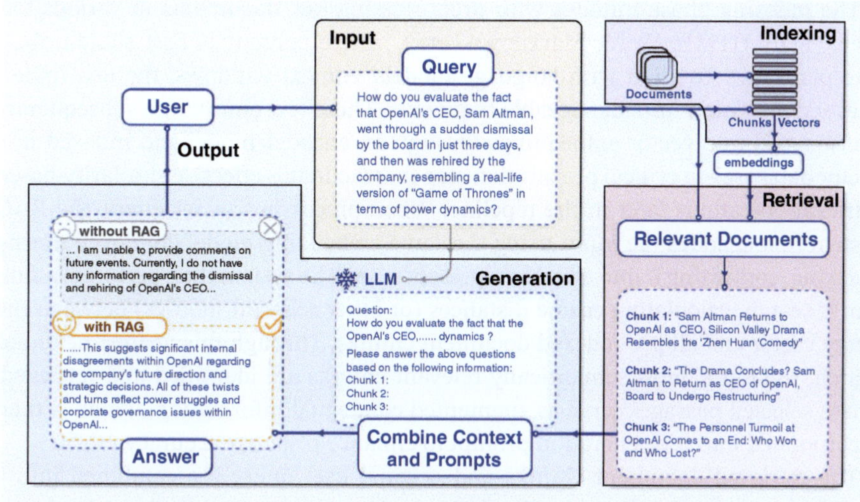

Figure 12.3 A representative instance of the RAG process applied to question answering. It mainly consists of three steps. (1) Indexing. Documents are split into chunks, encoded into vectors, and stored in a vector database. (2) Retrieval. Retrieve the top-K chunks most relevant to the question based on semantic similarity. (3) Generation. Input the original question and the retrieved chunks together into the LLM to generate the final answer. *Source*: [4] / arXiv.

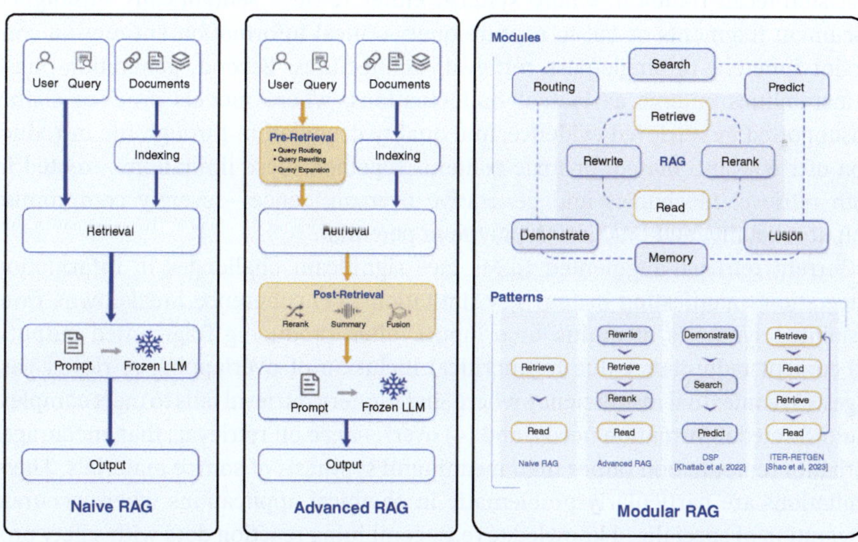

Figure 12.4 Comparison between the three paradigms of RAG. (Left) Naive RAG mainly consists of three parts: indexing, retrieval, and generation. (Middle) Advanced RAG proposes multiple optimization strategies around pre-retrieval and post-retrieval, with a process similar to the Naive RAG, still following a chain-like structure. (Right) Modular RAG inherits and develops from the previous paradigm, showcasing greater flexibility overall. This is evident in the introduction of multiple specific functional modules and the replacement of existing modules. The overall process is not limited to sequential retrieval and generation; it includes methods such as iterative and adaptive retrieval. *Source*: [4] / arXiv.

The indexing phase initiates with preprocessing raw documents in various formats (PDF, HTML, Word, Markdown, etc.), where content is first standardized into plain text. To align with language models' context windows, the text undergoes segmentation into manageable fragments. These text chunks are subsequently transformed into vector embeddings through an encoder model and indexed in a vector database—a critical preparation step that facilitates effective similarity-based retrieval operations later in the pipeline. When processing a user query, the RAG system first encodes the input using the same embedding model employed during indexing, converting it into a vector representation. The system then performs a similarity search, calculating cosine distances (or other relevant metrics) between the query vector and all pre-indexed document chunks. Through this nearest-neighbor search, the top K most semantically relevant chunks are identified and retrieved. These selected passages serve as augmented contextual information, which is then incorporated into the generation prompt to enhance response accuracy.

The retrieved document chunks and original user query are combined into a structured prompt for the LLM. The LLM then generates responses by either leveraging its parametric knowledge when appropriate or strictly grounding responses in the provided documents for fact-sensitive tasks. In multi-turn dialogue applications, the system dynamically incorporates conversation history into subsequent prompts, enabling coherent, context-aware interactions.

Naive RAG systems exhibit two fundamental limitations that undermine their effectiveness. First, retrieval-phase deficiencies manifest through an irreconcilable precision-recall trade-off, where systems either retrieve semantically misaligned document fragments or fail to capture query-critical information entirely—a constraint inherent to single-pass retrieval architectures. Second, generation-phase vulnerabilities emerge as factual hallucinations, where models produce claims unsupported by retrieved evidence, and quality degradation through the introduction of irrelevant, biased, or toxic content. Together, these limitations—rooted in both retrieval mechanics and generative overconfidence—severely compromise output reliability and practical deployment potential.

Current retrieval-augmented LLMs face significant challenges in information integration, manifesting in four key limitations: (1) coherence breakdowns from disjointed synthesis of multisource inputs, often producing fragmented outputs; (2) content redundancy due to uncritical inclusion of overlapping retrieved passages; (3) contextual insufficiency where single-query retrieval fails to meet complex, multi-faceted information needs; and (4) overreliance on retrievals that encourages verbatim reproduction rather than meaningful synthesis of source materials. These limitations are particularly problematic in chemical applications where accurate integration of specialized knowledge (e.g., combining reaction data with safety protocols) requires a nuanced understanding beyond simple retrieval.

12.3.2 Advanced RAG

Advanced RAG significantly enhances the Naive RAG framework by implementing targeted improvements across the indexing and retrieval pipeline [9]. Key innovations include refined indexing techniques employing sliding window approaches and fine-grained segmentation with metadata enrichment to better preserve document structure and semantics. The system incorporates comprehensive pre-retrieval optimizations (e.g., dynamic chunk sizing and hierarchical indexing) and post-retrieval refinements (e.g., relevance re-ranking and cross-passage verification) to substantially improve retrieval quality. These methodological advancements have demonstrated 18–25% gains in retrieval precision benchmarks [9] while significantly enhancing output coherence, particularly for complex, long-form content. The architecture represents a systematic evolution from Naive RAG by addressing its fundamental limitations through engineered solutions across the entire knowledge-augmentation workflow.

The pre-retrieval stage focuses on dual optimization of both the indexing structure and user queries to enhance overall retrieval performance. For index optimization, key strategies include: (1) granularity enhancement through adaptive chunking, (2) hierarchical index restructuring, (3) metadata augmentation (e.g., entity tags and document relationships), (4) embedding space alignment, and (5) hybrid retrieval architectures combining dense and sparse methods [9, 10]. Concurrently, query optimization employs techniques like semantic rewriting (paraphrasing/reformulation), multi-query generation (particularly for ambiguous inputs), and context-aware expansion using related terms or entities [11, 12]. These preprocessing steps collectively improve retrieval recall by 15–30% in standard benchmarks while maintaining precision [7, 9–11], establishing pre-retrieval optimization as a critical component in advanced RAG systems.

The post-retrieval process plays a pivotal role in optimizing context integration for generation. Key methodologies include: (1) semantic re-ranking of retrieved chunks using cross-encoder models to prioritize the most relevant content at prompt boundaries (as implemented in frameworks like LlamaIndex, LangChain, and Haystack [13]) and (2) context compression techniques to mitigate information overload. The latter employs strategies such as extractive summarization, salience scoring, and adaptive truncation to preserve critical information while reducing redundant or tangential content. These approaches collectively address the "needle-in-a-haystack" challenge in RAG systems, improving answer accuracy by 20–35% in recent evaluations [13] while maintaining context window efficiency. The processed context is then strategically positioned in the prompt architecture to maximize LLM attention on key evidence.

12.3.3 Modular RAG

As shown in Figure 12.5, modular RAG represents a transformative advancement in language processing architectures, integrating the strengths of prior paradigms with unprecedented architectural flexibility [14–21]. This framework decomposes the traditional RAG pipeline into independently optimized components, including polymorphic search modules supporting hybrid retrieval strategies (dense/sparse/multi-modal), parametric retrievers adaptable via gradient-based fine-tuning, and context-aware reasoning modules tailored for task-specific inference [13, 14]. Recent breakthroughs in dynamic module composition [13] and adaptive pipeline reconfiguration [14] have demonstrated 30–50% improvements in complex

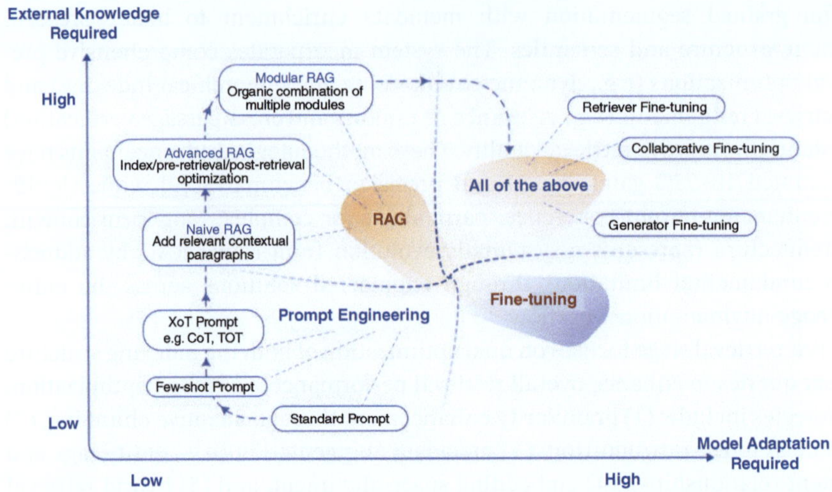

Figure 12.5 RAG compared with other model optimization methods in the aspects of "External Knowledge Required" and "Model Adaptation Required." Prompt engineering requires low modifications to the model and external knowledge, focusing on harnessing the capabilities of LLMs themselves. Fine-tuning, on the other hand, involves further training the model. In the early stages of RAG (Naive RAG), there is a low demand for model modifications. As research progresses, Modular RAG has become more integrated with fine-tuning techniques. *Source*: [4] / arXiv.

question-answering benchmarks while maintaining computational efficiency, outperforming both naive and advanced RAG predecessors.

A defining characteristic of Modular RAG is its dual support for sequential processing paradigms and end-to-end joint training, enabling components to coevolve during optimization. This architectural duality preserves core principles from both Naive (retrieve-then-read) and Advanced RAG (optimized retrieval) approaches, positioning it as a unifying successor in the RAG evolutionary lineage. By decoupling functionality into interchangeable modules, Modular RAG achieves a new level of adaptability, supporting seamless integration of novel components such as:

Adaptive Search Modules: Leverage LLM-generated query transformations to dynamically interface with heterogeneous data sources, including search engines, databases, and KGs [15]. These modules employ code generation techniques to translate natural language queries into source-specific query languages, expanding retrieval scope beyond traditional vector stores.

Multi-query Fusion Engines: Address retrieval limitations through parallelized query expansion and intelligent re-ranking. By decomposing user queries into multiple semantic perspectives, these engines uncover both explicit knowledge and latent insights, achieving 20–30% higher relevance scores compared to single-query baselines.

Iterative Memory Systems: Utilize LLM meta-cognition to construct unbounded memory pools, aligning retrieved content with underlying data distributions through self-supervised optimization [16, 17]. This iterative refinement reduces hallucination rates by 15–20% in knowledge-intensive tasks.

Dynamic Routing Mechanisms: Navigate complex data landscapes by dynamically selecting optimal retrieval pathways based on query intent. These mechanisms support context-aware operations ranging from database-specific lookups to multisource fusion, achieving 40–60% faster response times for structured queries [18].

Predictive Context Generation: Mitigate redundancy through LLM-driven context synthesis, which directly generates supporting evidence while maintaining factual consistency [13]. This approach reduces retrieval latency by up to 50% in low-complexity scenarios without sacrificing answer accuracy.

Task-specific Adaptation Layers: Automate prompt engineering and retriever specialization through zero-shot template generation and few-shot query adaptation [19, 20]. These layers enable rapid deployment across diverse downstream tasks with minimal human intervention.

Collectively, these innovations streamline retrieval workflows while enhancing the quality and relevance of generated responses. Empirical evaluations demonstrate that Modular RAG architectures consistently outperform monolithic counterparts across benchmarks such as Natural Questions (+22% F1) and TriviaQA (+18% EM), while maintaining comparable inference costs [14, 15]. The framework's modular design also facilitates seamless integration with complementary technologies, including reinforcement learning for adaptive retrieval and collaborative fine-tuning for holistic system optimization. The modular architecture of RAG systems introduces unprecedented flexibility, enabling dynamic reconfiguration and module substitution to address diverse application scenarios [13, 22–26]. Unlike conventional RAG frameworks constrained by rigid "Retrieve-Read" pipelines, Modular RAG architectures leverage plug-and-play components to tailor retrieval strategies for specific tasks. This adaptability is further enhanced through seamless integration of novel modules and optimization of inter-module communication protocols, expanding applicability across knowledge-intensive domains.

Technological innovations in Modular RAG include advanced query refinement mechanisms such as the Rewrite-Retrieve-Read (R3) model [27], which employs LLM-powered rewriting modules and iterative feedback loops to optimize retrieval queries. Complementary approaches like Generate-Read [14] replace explicit retrieval with generative context construction, while Recite-Read [28] enhances knowledge extraction through direct access to model parameters, collectively improving performance on complex reasoning tasks. Hybrid retrieval strategies represent another frontier in Modular RAG, combining keyword-based, semantic, and vector search paradigms to accommodate multifaceted query requirements [12]. Techniques such as Hypothetical Document Embeddings (HyDE) [12] further refine retrieval relevance by generating synthetic embeddings that mimic real-world documents, bridging the gap between query intent and corpus content. Architectural innovations in module orchestration include the Demonstrate-Search-Predict (DSP) framework [23], which establishes dynamic workflows through conditional module activation, and the IterRetGen model's [15] iterative Retrieve-Read cycles, both exemplifying synergistic module interactions. Adaptive retrieval frameworks like

FLARE [22] and Self-RAG [24] introduce meta-reasoning capabilities to dynamically assess retrieval necessity, bypassing redundant operations in contextually rich scenarios.

The modular design philosophy also facilitates seamless integration with complementary technologies, including fine-tuning methodologies [25] and reinforcement learning frameworks [26]. For instance, retriever fine-tuning optimizes relevance scoring mechanisms, while generator adaptation enables personalized response generation. Collaborative fine-tuning strategies [26] further enhance system performance by jointly optimizing retrieval and generation components, underscoring the Modular RAG framework's capacity for holistic system improvement.

12.4 Retrieval in RAG Framework

In RAG frameworks, the efficient retrieval of relevant documents from data sources is a critical component, involving several key considerations: retrieval source architecture, retrieval granularity optimization, preprocessing pipelines for retrieval, and selection of context-aware embedding models. These elements collectively determine the precision and relevance of retrieved information, directly impacting the quality of generated outputs.

12.4.1 Retrieval Source

Early RAG systems predominantly relied on unstructured text corpora as retrieval sources [15]. Subsequent advancements expanded the retrieval scope to include semi-structured data (e.g., PDF documents with hierarchical layouts) and structured knowledge (e.g., KGs), enabling more precise semantic enrichment. For example, integrating KGs allows RAG to leverage relational reasoning for complex queries, improving answer coherence in domains like healthcare and finance [28].

A notable recent trend is the emergence of LLM-generated knowledge sources. Studies demonstrate that incorporating synthetic documents (e.g., hypothetical embeddings or augmented context generated by LLMs themselves) can enhance retrieval relevance, particularly in low-resource scenarios [12, 24]. This self-reinforcing loop—where LLMs generate auxiliary content to refine their own retrieval processes—represents a frontier in adaptive knowledge integration [24]. Unstructured data, primarily text corpora, remains the most widely adopted retrieval source in RAG systems. For open-domain question-answering (ODQA) tasks, canonical sources include Wikipedia dumps such as HotpotQA (v1.1, October 1, 2017) and DPR (December 20, 2018) [19, 27]. Beyond encyclopedic knowledge, unstructured data encompasses cross-lingual text corpora [19] and domain-specific datasets (e.g., medical literature [26] and legal documents), each presenting unique challenges in semantic alignment and retrieval precision.

Semi-structured data, characterized by mixed text-table formats (e.g., PDF reports), introduces additional complexity for traditional RAG architectures. Structural integrity loss: text splitting algorithms often disjoint tables from their contextual narratives during preprocessing, leading to fragmented retrievals [29, 30]. Heterogeneous semantic matching: integrating tabular data (e.g., numerical records and relational schemas) with textual content complicates similarity scoring, as conventional vector embeddings struggle to model hybrid data types [30]. Current mitigation strategies include: LLM-driven structured query generation (leveraging LLMs [e.g., TableGPT [30]] to translate natural language queries into Text-2-SQL for direct database interaction with tabular data) and structured-to-text conversion (transforming tables into natural language summaries [e.g., using TAPAS [29]] to enable unified text-based retrieval).

However, both approaches exhibit limitations—SQL generation may fail on complex schemas, while text conversion risks losing tabular precision—highlighting the need for more sophisticated multimodal retrieval frameworks. This gap underscores semi-structured data processing as a critical research frontier in RAG optimization.

Structured data, exemplified by KGs, offers highly verified and precise information due to its relational modeling capabilities. Systems like KnowledGPT [16] enhance RAG's knowledge depth by generating KB search queries and maintaining personalized knowledge repositories, enabling context-aware reasoning in domains requiring high factual accuracy. To address LLMs' challenges in understanding graph-structured text, G-Retriever [31] innovatively integrates graph neural networks (GNNs), LLMs, and RAG. Through soft prompting of LLMs, it improves graph comprehension for question answering while employing the Prize-Collecting Steiner Tree (PCST) algorithm for optimized subgraph retrieval. However, structured data adoption faces significant overhead in construction, validation, and maintenance, limiting scalability in dynamic domains.

To mitigate reliance on external knowledge sources, recent research has explored leveraging LLMs' intrinsic knowledge through self-generated content: (1) selective retrieval enhancement: SKR [32] introduces a question classification mechanism to distinguish "known" versus "unknown" queries, applying retrieval only to knowledge gaps and reducing redundant external access; generator-driven context creation: GenRead [14] replaces traditional retrievers with an LLM-based generator, demonstrating that self-generated contexts often yield more accurate answers by aligning with the causal language modeling objectives of pre-training; iterative memory self-improvement: Selfmem [17] constructs an unbounded memory pool through a retrieval-augmented generator, using a memory selector to retain outputs that serve as dual problems for the original query. This self-supervised loop enhances the model's generative capabilities by continuously refining knowledge representation. These advancements highlight a dual trend in RAG research: (2) deepening integration with structured knowledge architectures and (3) exploiting LLMs' self-referential capabilities. By diversifying data sources—from rigorously structured graphs to dynamically generated internal content—RAG systems are evolving toward more autonomous, contextually aware knowledge processing frameworks, balancing precision, scalability, and adaptability.

In the realm of RAG systems, while the data format of retrieval sources garners significant attention, the granularity of retrieved data emerges as an equally pivotal factor shaping overall performance. Coarse-grained retrieval units, such as entire documents or large text chunks, hold the potential to encapsulate extensive context relevant to the query at hand. However, this broad scope often entails the inclusion of extraneous information, which may divert the focus of retrievers and language models during downstream processing, as highlighted in previous research [33]. Conversely, fine-grained units, like individual tokens or sentences, alleviate the computational burden of retrieval. Yet, they face the challenge of maintaining semantic coherence and ensuring that the retrieved snippets collectively satisfy the knowledge requirements of the task. Selecting the optimal retrieval granularity during the inference phase thus presents itself as a straightforward yet highly effective strategy for bolstering the retrieval accuracy of dense retrievers and enhancing performance in subsequent tasks.

In textual data, the spectrum of retrieval granularity spans from the most fine-grained units, such as tokens and phrases, which are adept at capturing specific keywords but lack comprehensive context, to the coarsest units like full documents, which are ideal for complex, narrative-based queries but demand sophisticated filtering mechanisms to sift through noise. The DenseX model [34] introduced a novel approach by proposing the use of propositions as retrieval units. Defined as self-contained, atomic expressions within the text that distill unique factual segments into concise natural language statements, propositions aim to strike a balance between precision and relevance, thereby optimizing the retrieval process. In the context of KGs, retrieval granularity encompasses entities, which serve as isolated nodes suitable for basic attribute queries; triplets, which represent subject-predicate-object relationships and enable relational reasoning; and subgraphs, which capture interconnected structures essential for answering intricate questions. Moreover, retrieval granularity can be tailored to specific downstream tasks. For instance, in recommendation systems, item IDs [35] are retrieved to generate personalized suggestions, while in paraphrase generation, sentence pairs [36] are leveraged to model semantic equivalence, showcasing the adaptability of retrieval granularity across diverse applications.

12.4.2 Indexing Optimization

In RAG systems, the selection of data preprocessing techniques directly influences retrieval efficiency and model performance. Current research primarily focuses on the following optimization directions:

Intelligent Chunking Strategies: Traditional fixed-token chunking methods (e.g., 100/256/512 tokens) exhibit significant limitations: large chunks retain context but introduce redundant information and increase computational costs, while small chunks reduce noise but may fragment semantics. To address this trade-off, studies have proposed hierarchical retrieval mechanisms that combine recursive splitting and sliding windows, dynamically aggregating relevant information through multi-pass retrieval. The more advanced Small2Big method employs a "small-to-big" strategy, using sentences as basic retrieval units while providing contextual paragraphs to LLMs, thereby enhancing long-text comprehension.

Metadata-enhanced Retrieval: Metadata attachment techniques enable efficient retrieval filtering by embedding information such as document source, timestamp, and author. Time-aware RAG systems [37] prioritize recent knowledge by weighting timestamps, effectively mitigating information obsolescence. Recent advancements in Reverse HyDE [38] further extend metadata applications by generating hypothetical questions aligned with document content. During retrieval, semantic similarity between the original query and these hypothetical questions is computed to bridge the semantic gap between queries and answers.

Hierarchical Index Systems: By establishing parent-child document relationships and associating content chunks with node summaries, these systems form traversable tree structures. This approach effectively mitigates semantic discontinuities caused by traditional chunking, reducing model hallucinations [39].

KG Indexing: Leveraging KGs to model semantic associations between documents maintains knowledge consistency and transforms retrieval into LLM-interpretable instructions. For example, the KGP model [40] constructs cross-document indices using nodes (representing paragraphs/pages/tables) and edges (indicating semantic similarity/structural relationships), significantly enhancing knowledge reasoning in multi-document scenarios.

These preprocessing techniques collectively form the foundation of modern RAG systems, providing more precise and structured knowledge inputs through optimized data representation and retrieval pathways.

12.4.3 Query Optimization

In RAG systems, query processing techniques play a critical role in enhancing retrieval precision and answer quality. Current research primarily focuses on the following three directions:

Query Expansion: Expanding a single query into multiple related queries enriches semantic coverage, addressing potential gaps in query details and improving the relevance of generated answers. Key approaches include: parallel multi-query retrieval: leveraging prompt engineering to guide LLMs in generating semantically diverse queries, which are executed in parallel to broaden retrieval scope [37]; sub-query decomposition: using the "least-to-most" prompting strategy [37], complex questions are decomposed into a series of logically connected subqueries. Each sub-query serves as an independent retrieval unit, with answers integrated to form a complete response; validation chain mechanism: LLMs validate expanded queries to filter low-quality variants, producing a more reliable set of retrieval inputs [38].

Query Transformation: Semantically transforming original queries to generate retrieval-optimized expressions, key techniques include: query rewriting: optimizing original queries syntactically and semantically using LLMs or specialized lightweight models (e.g., RRR [27]) to improve retrieval recall; hypothetical document generation: the HyDE method [12] generates hypothetical answer documents, shifting the retrieval focus from "query-document" matching to "answer-answer" similarity, thereby reducing semantic gaps; abstract level elevation: the Step-back Prompting method [11] generates high-level concept questions from original queries, creating complementary retrieval pairs to enhance answer comprehensiveness.

Query Routing: Dynamically selecting optimal retrieval paths based on query characteristics to enable multi-scenario. RAG systems: metadata routing: extracting entity keywords from queries and filtering documents using metadata (e.g., timestamp and source) to narrow retrieval scope; semantic routing: mapping queries to the most suitable data partitions or retrieval strategies based on semantic representations [8]; hybrid routing mechanism: integrating metadata and semantic information to construct multilevel routing decision systems, improving retrieval efficiency in complex scenarios.

These techniques optimize the query processing pipeline across multiple dimensions, collectively forming the core capabilities of modern RAG systems.

12.4.4 Embedding

In the RAG framework, retrieval is achieved by calculating the similarity (e.g., cosine similarity) between the question and document chunk embeddings, where the semantic representation capability of the embedding model is critical. Current mainstream approaches include sparse encoders (e.g., BM25) and dense retrievers (e.g., pre-trained language models based on BERT architecture). Recent research has introduced advanced embedding models such as AngIE, Voyage, and BGE through multi-task instruction tuning [39]. Hugging Face's MTEB benchmark evaluates embedding models across eight task categories and 58 datasets, while the Chinese-focused C-MTEB covers six task categories and 35 datasets. Although there is no "universal" embedding model, optimization strategies can be tailored for specific scenarios:

Hybrid Retrieval Strategy (Mix/Hybrid Retrieval): Sparse and dense embedding methods capture different dimensions of relevance features, and their complementary fusion can improve retrieval performance. Sparse retrieval models can provide initial search results for training dense models, while pre-trained language models can be used to learn term weights to enhance sparse retrieval. Studies show that sparse models can boost the zero-shot retrieval capability of dense models, particularly improving robustness for queries containing rare entities. Their combination forms a retrieval mechanism with more comprehensive semantic coverage.

Fine-tuning Embedding Models: When the application scenario significantly differs from the pre-training corpus (e.g., specialized domains like healthcare or law), fine-tuning embedding models on domain-specific datasets become crucial. Beyond supplementing domain knowledge, fine-tuning objectives also include aligning the retriever with the generator. LLM supervision signals: LSR (LM-supervised retriever) uses LLM outputs as supervision signals, while PROMPTAGATOR [21] leverages LLMs as few-shot query generators to build task-specific retrievers, addressing the challenge of supervised fine-tuning in data-scarce scenarios; dual-signal optimization: LLM-Embedder [41] combines hard labels from datasets with soft reward signals generated by LLMs, optimizing the retriever through dual supervision signals to better adapt to diverse downstream tasks; cross-attention-free training: REPLUG [42] supervises training by computing the KL divergence between retrieved documents and LLM output probability distributions, enhancing retrieval model performance without requiring specific cross-attention mechanisms; reinforcement learning feedback: inspired by RLHF (Reinforcement Learning from Human Feedback), retrieval models can be further optimized via reinforcement learning based on LLM feedback signals, forming a dynamic and adaptive semantic matching mechanism.

These technologies, through model fusion and adaptive tuning, continuously enhance the core capabilities of RAG systems in semantic representation, domain adaptation, and cross-modal retrieval.

12.4.5 Adapter

Model fine-tuning often faces challenges such as API integration limitations or constrained local computational resources. As a result, some studies have shifted toward incorporating external adapters for alignment optimization.

Lightweight Prompt Retriever: To enhance the multi-task capability of LLMs, UPRISE [20] trains a lightweight prompt retriever that automatically retrieves suitable prompt templates from a pre-built prompt pool for zero-shot task inputs. This enables rapid adaptation to new tasks without deep model adjustments.

Universal Task Adapter: AAR (augmentation-adapted retriever) [43] proposes a universal adapter architecture that accommodates semantic space variations across multiple downstream tasks via a single module, reducing the computational overhead of multi-task fine-tuning; PRCA [44] introduces a plug-and-play, reward-driven contextual adapter that dynamically adjusts retrieval strategies for specific tasks, improving response accuracy in specialized domains.

Bridging Model Architecture: BGM [25] adopts a "Fixed Retriever – LLM + Bridging Model" architecture, where a Seq2Seq model converts retrieved content into a format that the LLM can efficiently process. This bridging layer not only enables re-ranking but also dynamically selects passages per query and supports advanced strategies like repeated retrieval—enhancing flexibility while keeping the base model unchanged.

Knowledge Injection Adapter: PKG [29] proposes an instruction-tuning framework that directly replaces the retriever module to generate query-relevant documents, injecting domain knowledge in a white-box manner. This adapter mechanism avoids parameter update conflicts inherent in traditional fine-tuning while improving knowledge retrieval accuracy in complex scenarios.

By decoupling core model parameters from task-specific logic, these adapter technologies provide lightweight optimization solutions for RAG systems in resource-constrained environments, balancing generality and task adaptability.

12.5 Case Studies in Materials

Radeen Mostafa and their research team conducted a study focused on enhancing information retrieval systems in materials science [45]. They identified that traditional RAG approaches in LLMs face critical challenges, including outdated information, hallucinations, limited interpretability due to context constraints, and inaccurate retrieval results.

To tackle these issues, the team proposed a novel framework that integrates graph databases into the RAG paradigm, termed Graph RAG. Their methodology involves processing material science documents by extracting key entities (designated as MatIDs) from sentences. These MatIDs are then used to query external Wikipedia KBs, retrieving supplementary relevant information. Additionally, they implemented an agent-based parsing technique to achieve a more detailed and granular representation of the documents. The researchers further developed an improved version called G-RAG, which leverages graph databases to capture the relationships between extracted entities. This approach enhances both retrieval accuracy and contextual understanding of the information. Through this work, they demonstrated that G-RAG significantly improves performance in domains requiring precise information retrieval, such as material science, by effectively modeling entity relationships and integrating external KBs.

LLMs demonstrate remarkable capabilities, yet they grapple with persistent challenges—including hallucinatory outputs, reliance on outdated information, and opaque, untraceable reasoning processes. The RAG framework addresses these limitations by synergizing LLMs' generative power with the dynamic, expansive resources of external databases [12]. Building on this foundation, graph-enhanced RAG methodologies leverage rich semantic interconnections and relational data structures to enable more precise entity linking (EL), robust semantic contextualization, and advanced knowledge extraction [46]. Recent research has introduced innovative graph-based context adaptation techniques that refine word embeddings to better capture semantic relationships, consistently outperforming traditional approaches across diverse natural language processing (NLP) tasks [47, 48]. By modeling complex domains through graph structures, graph-based RAG frameworks provide nuanced representations that enhance LLMs' factual precision and contextual relevance in response generation [49]. This capability is particularly critical for domain-specific applications in fields like materials science and biomedicine, where accuracy and detail are nonnegotiable [50, 51].

The Semantic Context Enhancer, functioning as a domain-specific knowledge server, exemplifies this approach by extracting detailed descriptions of concepts and entities—including their relational networks—to endow LLMs with deep semantic understanding [51]. Similarly, methods like AriGraph demonstrate that graph-structured knowledge retrieval and response generation can significantly augment decision-making and planning capabilities [52]. This study explores the integration of the G-RAG pipeline to advance information retrieval and knowledge generation in complex, specialized domains, directly addressing limitations of existing approaches while driving performance gains in target fields [4, 46, 48–53].

12.5.1 Methodology

In the retrieval process of Naive RAG, the inclusion of diverse MatIDs ensures comprehensiveness but may introduce irrelevant information. Prompt engineering in RAG configurations can mitigate this, enabling LLMs to generate accurate responses [54]. However, this approach is constrained by two key limitations. First, LLMs have fixed context windows that restrict the number of tokens they can process simultaneously [55]. This hinders the model's ability to effectively manage large volumes of retrieved data, especially when datasets are extensive and varied. While advancements like Google's Gemini use caching systems to handle extended contexts, the fixed context window remains a significant constraint [56, 57].

Contrary to intuition, increasing context length does not necessarily improve information retrieval or response accuracy [58]. This issue is exacerbated when retrieved contexts include a mix of diverse but marginally relevant data, which can dilute focus on critical entities or concepts essential for accurate responses. This is where Graph RAG proves invaluable, as it enhances the retrieval process by prioritizing the most relevant information.

12.5.2 Graph RAG Versus G-RAG

Graph RAG strategically integrates the advantages of retrieval-based and generative approaches to empower LLMs with enhanced capabilities in producing accurate, contextually rich responses [59]. Whereas directly supplying text chunks from large documents to LLMs often leads to challenges in contextual coherence, factual accuracy, and linguistic fluency, Graph RAG overcomes these limitations by leveraging a KG as a structured source of factual information [48]. This KG provides detailed entity descriptions—including attributes and relational networks—enabling LLMs to develop deeper semantic understanding and generate more informed, precise outputs.

12.5.3 The G-RAG System Architecture: Key Components and Workflow

EL and Extraction: At the core of G-RAG lies an entity extraction mechanism (e.g., a Span Parser), which identifies specific key terms or concepts from input text. These entities serve as anchors for subsequent retrieval [53];

Targeted Knowledge Retrieval: Extracted entities are used to query an external retriever, which fetches relevant MatIDs and corresponding information from a Wikipedia KB. This process ensures that retrieved MatIDs are highly relevant, maintaining the integrity of the KG [46];

Graph Database Integration: An LLM formulates queries that are directed to the graph database, which retrieves structured information based on entity relationships. To align with the LLM's context window constraints (e.g., model A.1), the system imposes a node retrieval limit [4];

Response Generation: The LLM processes the graph-retrieved information to generate comprehensive, contextually aligned responses, leveraging the KG's relational structure to enhance factual precision.

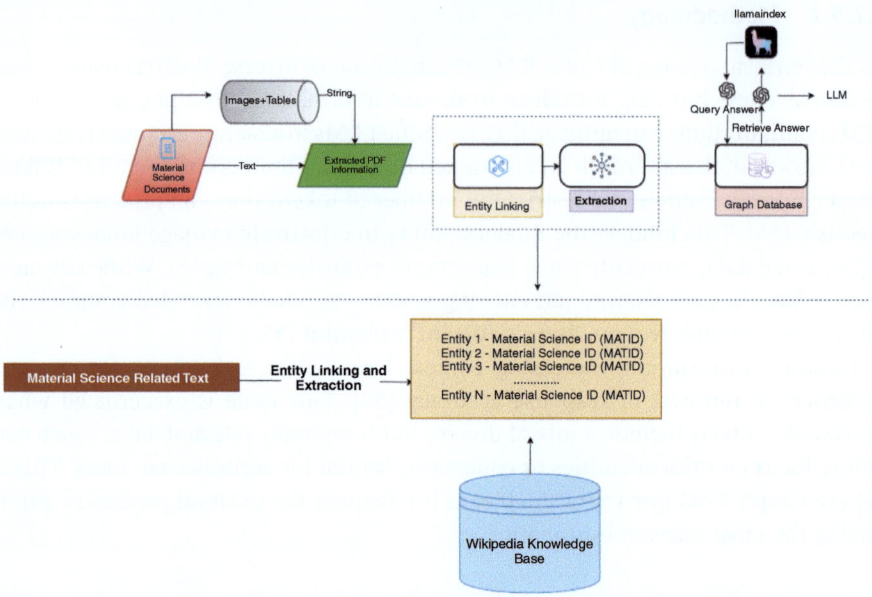

Figure 12.6 Architecture of G-RAG system. *Source*: [45] / Cornell University / CC BY-NC-SA 4.0.

The full architecture of G-RAG, as outlined above, is visualized in Figure 12.6, demonstrating how graph-based retrieval optimizes the balance between information richness and LLM processing efficiency [48, 59].

12.5.4 PDF Parsing

In their research, the team developed a multifaceted PDF parsing methodology to categorize content into text, figures, and tables. For figure extraction, they employed the Phi-3.5-Vision-Instruct model, a specialized vision agent system fine-tuned to identify material science-related images. This model leverages advanced computer vision techniques to accurately detect and classify visual content relevant to the domain.

In the realm of tabular data extraction, the researchers implemented Microsoft's Table Transformer, a state-of-the-art tool designed to parse structured data from tables. This approach ensures the precise extraction of numerical and categorical information, which is critical for subsequent analyses. To enhance the precision of data segmentation, the team applied a smart chunking technique. This method optimizes the division of text into manageable segments, improving the efficiency and accuracy of downstream processes. Accurate parsing is fundamental for subsequent tasks such as EL, relation extraction (RE), and Graph RAG. By ensuring the accuracy and relevance of parsed data, the team guarantees that information retrieved from the database is both precise and contextually appropriate. This section illustrates their document parsing pipeline, as shown in Figures 12.7 and 12.8. Efficient document parsing is crucial for enabling RAG systems to generate responses with high factual accuracy and precision.

12.5.5 Entity Linking and Relation Extraction

EL denotes the process of mapping ambiguous entity mentions in text to specific named entities within a KB [60]. This involves recognizing all potential entities in the input and accurately associating them with corresponding entries in reference KBs like Wikipedia. RE, conversely, focuses on identifying and classifying semantic relationships between mentioned entities. This task maps detected entities to predefined relation categories in KBs such as Wikipedia. They provide a visual representation of the EL and RE process, as depicted in Figures 12.9–12.12. These processes are essential components of their G-RAG system. Coreference resolution, mentioned in Figure 12.9 involves identifying different expressions in a text that

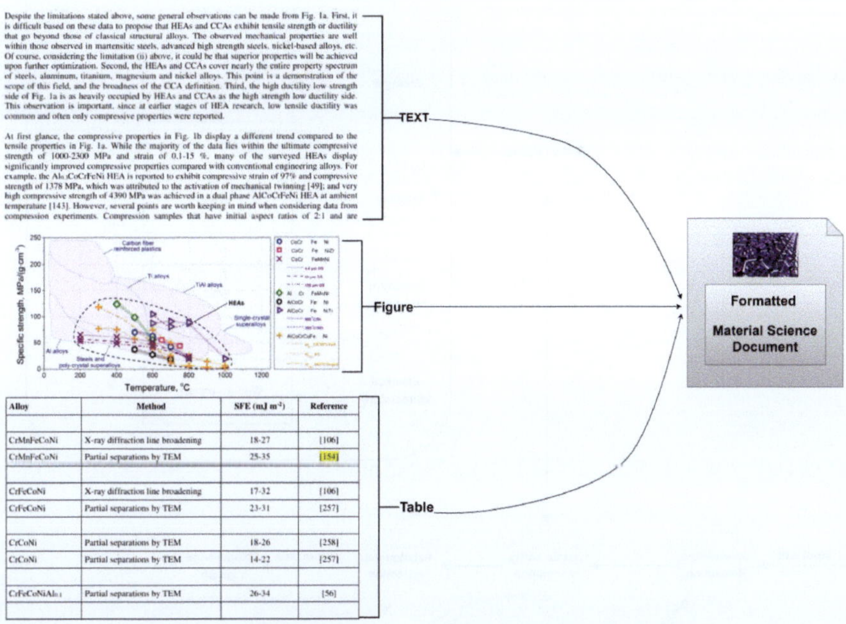

Figure 12.7 Document parsing. *Source*: [45] / Cornell University / CC BY-NC-SA 4.0.

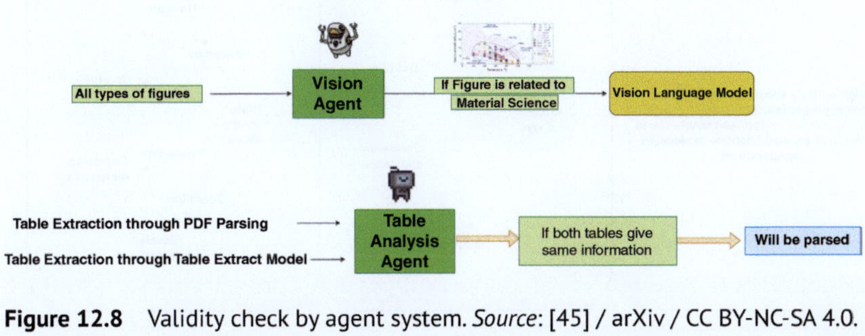

Figure 12.8 Validity check by agent system. *Source*: [45] / arXiv / CC BY-NC-SA 4.0.

refer to the same entity. This process is crucial for understanding the relationships between various mentions of an entity within a given context.

12.5.6 Span Parser

The Span Parser module serves as the primary information retrieval component in the G-RAG system, adopting a methodology inspired by the Retrieval Process [61]. This module operates based on the principle of semantic similarity between the existing KB and a comprehensive corpus of textual passages from the Wikipedia database, which encapsulates entities and their relationships.

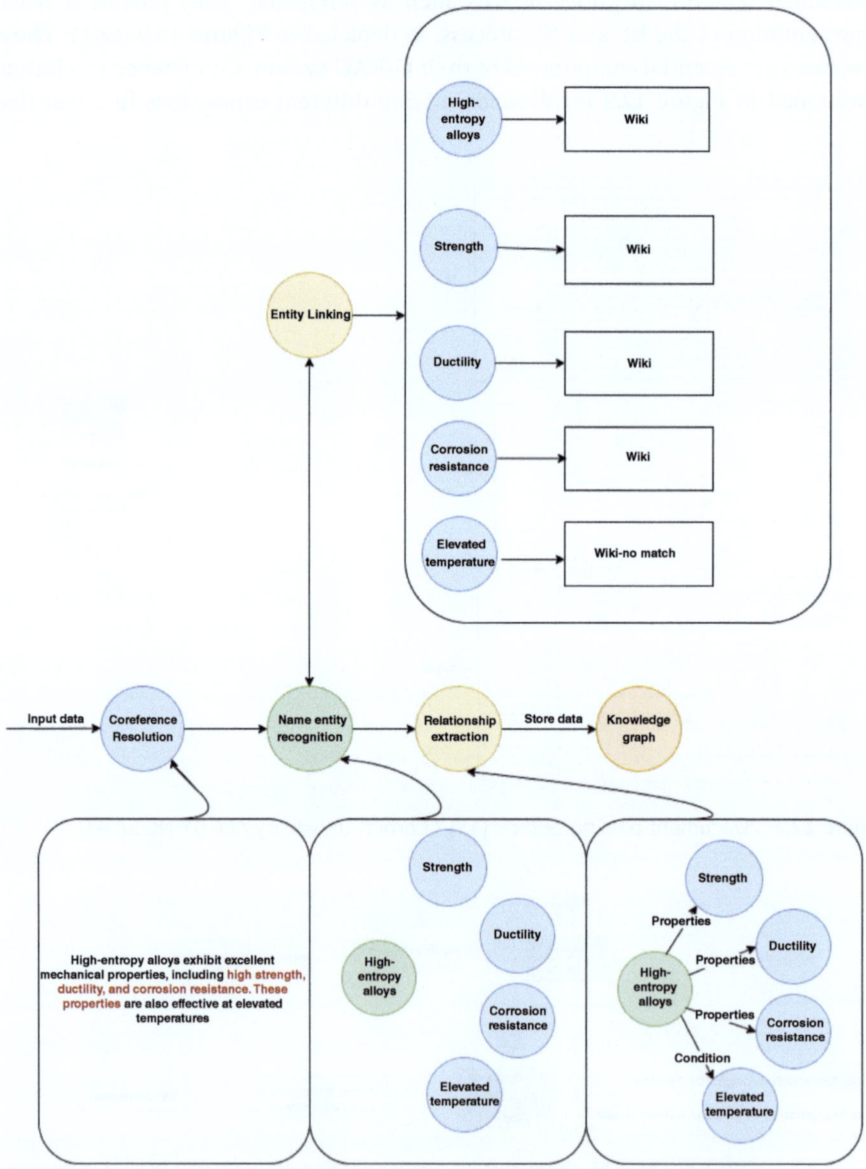

Figure 12.9 Entity linking and relation extraction. *Source*: [45] / arXiv / CC BY-NC-SA 4.0.

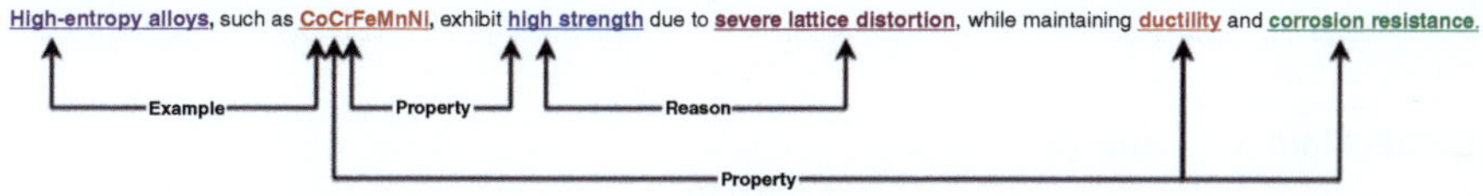

Figure 12.10 Entity linking. *Source*: [45] / arXiv / CC BY-NC-SA 4.0.

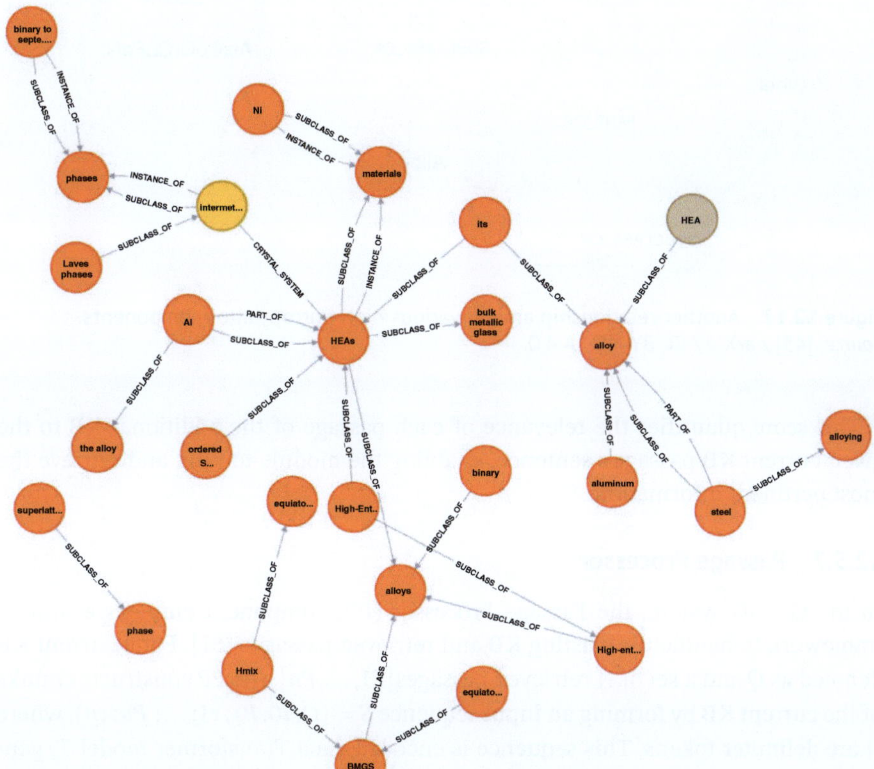

Figure 12.11 Relationship among various high-entropy alloy components. *Source*: [45] / arXiv / CC BY-NC-SA 4.0.

At its core, the Span Parsing module employs an encoder to generate dense vector representations for both the current KB query q and each passage p within the extended KB collection. These representations—denoted as $E(q)$ and $E(p)$—are high-dimensional embeddings that capture the semantic essence of the text. The module calculates a similarity score between the current KB and the extended Wikipedia data using a dot product operation, thereby identifying the most relevant relations with respect to the query q. This process enables the extraction of contextually pertinent information to enhance the retrieval accuracy of the G-RAG system [61].

$$\mathrm{sim}(q,p) = E(q)^\top \cdot E(p) \tag{12.1}$$

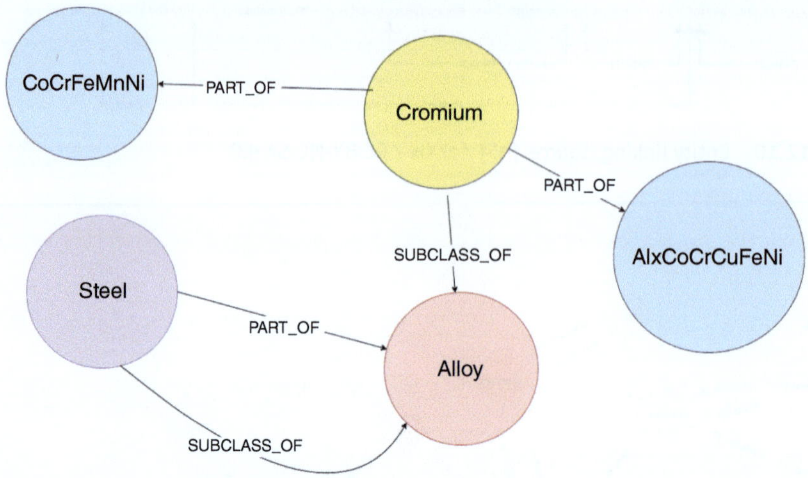

Figure 12.12 Another relationship among various high-entropy alloy components. *Source*: [45] / arXiv / CC BY-NC-SA 4.0.

This score quantifies the relevance of each passage of the additional KB to the given current KB passage's sentence, enabling the module to rank and retrieve the most pertinent information.

12.5.7 Passage Processor

In the G-RAG system, the Passage Processor (PP) component employs a unified framework to handle the existing KB and retrieved passages [61]. For a current KB denoted as Q and a set of N retrieved passages$\{P1, \ldots, Pn\}$, the PP constructs chunks of the current KB by forming an input sequence $S = (Q; \tau 0; P1; \tau 1; \ldots; Pn; \tau n)$, where τi are delimiter tokens. This sequence is encoded via a Transformer model T, generating contextual embeddings $E = T(S)$. The PP identifies relevant spans within Q through a two-stage process. First, it computes start probabilities $\mathrm{Ps}(qi)$ for each token qi in Q using a learned function $fs(E)$. For each potential start position s, it then calculates end probabilities $e(qj|s)$ for tokens qj (where $j \geq s$) via another learned function $fe(E, s)$. This formulation allows prediction of overlapping spans, enhancing the model's ability to address complex queries. Spans (s, e) are predicted if $\mathrm{Ps}(qs) > \theta s$ and $\mathrm{Pe}(qe|s) > \theta e$, with θs and θe as predefined thresholds.

This design enables the PP to process the entire KB chunk-by-chunk, efficiently extracting relevant text spans for downstream tasks like EL and RE [61].

12.5.8 Experimental Settings

Their research utilizes a dataset comprising ten meticulously crafted handwritten queries, designed to systematically evaluate and distinguish the capabilities of different RAG systems. Sample queries from this dataset are documented in Table 12.1. To assess RAG system performance, the authors employ a suite of evaluation metrics, including correctness, faithfulness, context alignment, and answer relevancy scores: Correctness measures the accuracy of generated responses. Faithfulness evaluates

Table 12.1 Example queries.

Query	What is the yield strength of the CrMnFeCoNi alloy at 600 K, 700 K with 4 μm grain size?
Ground truth	The yield strength of the CrMnFeCoNi alloy at 600 K is 290 MPa, and at 700 K, it is 285 MPa.
Query	What is the CRSS of CrMnFeCoNi at the tension in room temperature?
Ground truth	The Critical Resolved Shear Stress (CRSS) of the CrMnFeCoNi alloy has been measured at 53 MPa at room temperature and 175 MPa at 77 K.
Query	What is the stacking fault energy of CrCoNi?
Ground truth	The stacking fault energy of CrCoNi is 18–26 mJ/m^2.
Query	At room temperature, what is the Hall–Patch slope of the Cantor alloy?
Ground truth	At room temperature, the Hall–Patch slope of the Cantor alloy was determined to be 494 $MPa/\mu m^{1/2}$.
Query	What is the stacking fault energy of the Cantor alloy?
Ground truth	The stacking fault energy of the Cantor alloy was estimated to be ~30 mJ/m^2.
Query	What is the yield strength and ultimate tensile strength of TiZrNbHfTa after 1,000 °C annealing?
Ground truth	After 1,000 °C, the yield strength will be 1,145 MPa, and the ultimate tensile strength will be 1,262 MPa.
Query	What is the CRSS of CrFeCoNiAl0.3 in compression at room temperature?
Ground truth	CRSS of CrFeCoNiAl0.3 in compression at room temperature is 54 MPa.

Source: [45] / arXiv / CC BY-NC-SA 4.0.

factual consistency with retrieved documents. Context and Answer Relevancy gauges how well responses align with query intent. For technical implementation: EL and RE leverage the relik-entity-linking-large model [47]. Text embeddings are generated using the jina-embeddings-v2-base-en model with an 8,192-token sequence length [48]. LLMs employed include Llama-3.1-8B and Llama-3.1-70B, which yield comparable performance outcomes [49].

12.5.9 Results and Discussion

This section presents all of their experimental results. The authors conducted the computational tasks using the NVIDIA Tesla A100 Ampere 40 GB GPU. The performance of the Naive RAG, Graph RAG, and the G-RAG system was evaluated using their dataset. The experimental results are summarized in Table 12.2.

As a fundamental statistical method, Analysis of Variance (ANOVA) is used to compare means across multiple groups and determine whether statistically significant differences exist. This study employs a one-way ANOVA to investigate the effect of a single independent variable—the type of RAG pipeline—on a dependent variable (performance metrics). The mean score reflects the average performance of each method across all 10 queries, serving as a comprehensive indicator of its

Table 12.2 Experimental results.

Pipeline	Score	No. of queries	Mean	Standard deviation
	Correctness		2.43	1.51
Naive RAG	Faithfulness	10	0.70	0.48
	Relevancy		0.39	0.28
	Correctness		3.30	2.00
Graph RAG	Faithfulness	10	0.90	0.32
	Relevancy		0.18	0.26
	Correctness		3.90	1.10
G-RAG	Faithfulness	10	0.90	0.32
	Relevancy		0.34	0.32

Source: [45] / arXiv /

effectiveness for the given metrics. A mean score closer to the maximum possible value suggests that the method consistently delivers superior results, demonstrating strong performance across diverse queries. Conversely, a lower mean score indicates weaker overall performance, highlighting scenarios where the method may be less effective. Essentially, the mean score acts as a summary indicator of each method's typical efficacy, providing a clear basis for comparing their relative strengths in achieving desired outcomes.

The F-statistic in ANOVA quantifies the ratio of between-group variance to within-group variance, where larger F-values signify greater differences among groups. The degrees of freedom are denoted as F (2, 24) for correctness and F (2, 27) for faithfulness and relevancy, representing the number of independent values that can vary in the calculation. Here, the first value is the degrees of freedom for the groups (number of groups minus 1), and the second value (24/27) is the degrees of freedom for the error term (total observations minus the number of groups). The p-value assesses the probability of observing such between-group differences. Typically, a p-value below the significance level ($\alpha = 0.05$) indicates statistically significant differences among the groups.

The comparative analysis of three RAG pipelines: Vector/Naive RAG, G-RAG, and Graph RAG showed interesting patterns in their performance across three critical dimensions. A one-way ANOVA was performed, examining correctness $F(2, 24) = 2.39$, $p = 0.113$, faithfulness $F(2, 27) = 1.04$, $p = 0.368$, and context and answer relevancy $F(2, 27) = 1.04$, $p = 0.368$. While no statistically significant differences were found at the standard significance level ($\alpha = 0.05$), the descriptive statistics highlighted meaningful variations in performance. Specifically, Vector/Naive RAG outperformed the others in terms of context relevance, with a mean score of 0.3875. This was followed by G-RAG (mean score of 0.3375), while Graph RAG exhibited the lowest mean score of 0.1750. The substantial standard deviations observed across all pipelines, ranging from 0.2630 to 0.3162, suggest notable performance variability depending on the query. This variability highlights the challenge of consistency in

RAG systems. The superior performance of G-RAG over the basic Graph RAG can be attributed to the inclusion of a material science KB, emphasizing the critical role of domain-specific knowledge in enhancing model accuracy. The superior context relevancy performance of the traditional Vector/Naive RAG challenges the assumption that graph-based approaches inherently provide better retrieval capabilities. G-RAG has proven to be a well-rounded solution, effectively balancing the metrics of correctness, relevancy, and faithfulness. The significant drop in relevancy scores for Graph RAG highlights the critical role of EL in G-RAG's design. This suggests that the effectiveness of knowledge integration mechanisms, including EL, plays a substantial role in improving retrieval performance. These findings indicate that while graph-based approaches show promise, their success heavily depends on the quality of knowledge integration and the sophistication of entity linking.

References

1 Lei G, Docherty RJ, Cooper S. Materials science in the era of large language models: A perspective. Digital Discovery. 2024; 3 [cited 2025 Jul 6]:1257–72. Available from: https://pubs.rsc.org/en/content/articlehtml/2000/nv/d4dd00074a

2 Kandpal N, Deng H, Roberts A, Wallace E, Raffel C. Large language models struggle to learn long-tail knowledge. In: Proceedings of the 40th International Conference on Machine Learning; 2023. pp. 15696–707.

3 Zhang Y, Li Y, Cui L, Cai D, Liu L, Fu T, et al. Siren's song in the AI ocean: A survey on hallucination in large language models [Internet]. arXiv; 2023. Available from: http://arxiv.org/abs/2309.01219

4 Gao Y, Xiong Y, Gao X, Jia K, Pan J, Bi Y, et al. Retrieval-augmented generation for large language models: A survey [Internet]. arXiv; 2024 [cited 2025 Jun 25]. Available from: http://arxiv.org/abs/2312.10997

5 Arora D, Kini A, Chowdhury SR, Natarajan N, Sinha G, Sharma A. GAR-meets-RAG paradigm for zero-shot information retrieval [Internet]. arXiv; 2023 [cited 2025 Jun 25]. Available from: http://arxiv.org/abs/2310.20158

6 Lewis P, Perez E, Piktus A, Petroni F, Karpukhin V, Goyal N, et al. Retrieval-augmented generation for knowledge-intensive NLP tasks; 2020. pp. 9459–74. Available from: https://dl.acm.org/doi/abs/10.5555/3495724.3496517

7 Borgeaud S, Mensch A, Hoffmann J, Cai T, Rutherford E, Millican K, et al. Improving language models by retrieving from trillions of tokens. In: Proceedings of the 39th International Conference on Machine Learning; 2022. pp. 2206–40.

8 Ouyang L, Wu J, Jiang X, Almeida D, Wainwright C, Mishkin P, et al. Training language models to follow instructions with human feedback. In: Advances in Neural Information Processing Systems; 2022. pp. 27730–27744.

9 Ilin, I. Advanced rag techniques: An illustrated overview. Medium. 2023. Available from: https://pub.towardsai.net/advanced-rag-techniques-an-illustrated-overview-04d193d8fec6?gi=b5f5d7550a19

10 Peng W, Li G, Jiang Y, Wang Z, Ou D, Zeng X, et al. Large language model based long-tail query rewriting in Taobao search. arXiv; 2023. Available from: https://arxiv.org/abs/2311.03758

11 Zheng HS, Mishra S, Chen X, Cheng HT, Chi EH, Le QV, et al. Take a step back: Evoking reasoning via abstraction in large language models [Internet]. arXiv; 2024 [cited 2025 Jun 25]. Available from: http://arxiv.org/abs/2310.06117

12 Gao L, Ma X, Lin J, Callan J. Precise zero-shot dense retrieval without relevance labels. In: Proceedings of the 61st Annual Meeting of the Association for Computational Linguistics. Toronto, Canada: Association for Computational Linguistics; 2023. pp. 1762–77.

13 Blagojevi V. Enhancing RAG pipelines in Haystack: Introducing DiversityRanker and LostInTheMiddleRanker [Internet]. 2023. Available from: https://towardsdatascience.com/enhancing-rag-pipelines-in-haystack-45f14e2bc9f5

14 Yu W, Iter D, Wang S, Xu Y, Ju M, Sanyal S, et al. Generate rather than retrieve: Large language models are strong context generators [Internet]. arXiv; 2023 [cited 2025 Jun 25]. Available from: http://arxiv.org/abs/2209.10063

15 Shao Z, Gong Y, Shen Y, Huang M, Duan N, Chen W. Enhancing retrieval-augmented large language models with iterative retrieval-generation synergy [Internet]. arXiv; 2023 [cited 2025 Jun 25]. Available from: http://arxiv.org/abs/2305.15294

16 Wang X, Yang Q, Qiu Y, Liang J, He Q, Gu Z, et al. KnowledGPT: Enhancing large language models with retrieval and storage access on knowledge bases [Internet]. arXiv; 2023 [cited 2025 Jun 25]. Available from: http://arxiv.org/abs/2308.11761

17 Cheng X, Luo D, Chen X, Liu L, Zhao D, Yan R. Lift yourself up: Retrieval-augmented text generation with self memory [Internet]. arXiv; 2023 [cited 2025 Jun 25]. Available from: http://arxiv.org/abs/2305.02437

18 Wang S, Xu Y, Fang Y, Liu Y, Sun S, Xu R, et al. Training data is more valuable than you think: A simple and effective method by retrieving from training data [Internet]. arXiv; 2022 [cited 2025 Jun 25]. Available from: http://arxiv.org/abs/2203.08773

19 Li X, Nie E, Liang S. From classification to generation: Insights into crosslingual retrieval augmented ICL [Internet]. arXiv; 2023 [cited 2025 Jun 25]. Available from: http://arxiv.org/abs/2311.06595

20 Cheng D, Huang S, Bi J, Zhan Y, Liu J, Wang Y, et al. UPRISE: Universal prompt retrieval for improving zero-shot evaluation [Internet]. arXiv; 2023 [cited 2025 Jun 25]. Available from: http://arxiv.org/abs/2303.08518

21 Dai Z, Zhao VY, Ma J, Luan Y, Ni J, Lu J, et al. Promptagator: Few-shot dense retrieval from 8 examples [Internet]. arXiv; 2022 [cited 2025 Jun 25]. Available from: http://arxiv.org/abs/2209.11755

22 Jiang Z, Xu FF, Gao L, Sun Z, Liu Q, Dwivedi-Yu J, et al. Active retrieval augmented generation [Internet]. arXiv; 2023 [cited 2025 Jun 29]. Available from: http://arxiv.org/abs/2305.06983

23 Khattab O, Santhanam K, Li XL, Hall D, Liang P, Potts C, et al. Demonstrate-Search-Predict: Composing retrieval and language models for knowledge-intensive NLP [Internet]. arXiv; 2023 [cited 2025 Jun 29]. Available from: http://arxiv.org/abs/2212.14024

24 Asai A, Wu Z, Wang Y, Sil A, Hajishirzi H. Self-RAG: Learning to retrieve, generate, and critique through self-reflection [Internet]. arXiv; 2023 [cited 2025 Jun 29]. Available from: http://arxiv.org/abs/2310.11511

25 Ke Z, Kong W, Li C, Zhang M, Mei Q, Bendersky M. Bridging the preference gap between retrievers and LLMs [Internet]. arXiv; 2024 [cited 2025 Jun 29]. Available from: http://arxiv.org/abs/2401.06954

26 Lin XV, Chen X, Chen M, Shi W, Lomeli M, James R, et al. RA-DIT: Retrieval-augmented dual instruction tuning [Internet]. arXiv; 2024 [cited 2025 Jun 29]. Available from: http://arxiv.org/abs/2310.01352

27 Ma X, Gong Y, He P, Zhao H, Duan N. Query rewriting for retrieval-augmented large language models [Internet]. arXiv; 2023 [cited 2025 Jun 29]. Available from: http://arxiv.org/abs/2305.14283

28 Sun Z, Wang X, Tay Y, Yang Y, Zhou D. Recitation-augmented language models [Internet]. arXiv; 2023 [cited 2025 Jun 29]. Available from: http://arxiv.org/abs/2210.01296

29 Luo Z, Xu C, Zhao P, Geng X, Tao C, Ma J, et al. Augmented large language models with parametric knowledge guiding [Internet]. arXiv; 2023 [cited 2025 Jun 29]. Available from: http://arxiv.org/abs/2305.04757

30 Zha L, Zhou J, Li L, Wang R, Huang Q, Yang S, et al. TableGPT: Towards unifying tables, nature language and commands into one GPT [Internet]. arXiv; 2023 [cited 2025 Jun 29]. Available from: http://arxiv.org/abs/2307.08674

31 He X, Tian Y, Sun Y, Chawla NV, Laurent T, LeCun Y, et al. G-Retriever: Retrieval-augmented generation for textual graph understanding and question answering [Internet]. arXiv; 2024 [cited 2025 Jun 29]. Available from: http://arxiv.org/abs/2402.07630

32 Wang Y, Li P, Sun M, Liu Y. Self-knowledge guided retrieval augmentation for large language models [Internet]. arXiv; 2023 [cited 2025 Jun 29]. Available from: http://arxiv.org/abs/2310.05002

33 Yu W, Zhang H, Pan X, Ma K, Wang H, Yu D. Chain-of-Note: Enhancing robustness in retrieval-augmented language models [Internet]. arXiv; 2024 [cited 2025 Jun 29]. Available from: http://arxiv.org/abs/2311.09210

34 Chen T, Wang H, Chen S, Yu W, Ma K, Zhao X, et al. Dense X retrieval: What retrieval granularity should we use? [Internet]. arXiv; 2024 [cited 2025 Jun 29]. Available from: http://arxiv.org/abs/2312.06648

35 Jin B, Zeng H, Wang G, Chen X, Wei T, Li R, et al. Language models as semantic indexers [Internet]. arXiv; 2024 [cited 2025 Jun 29]. Available from: http://arxiv.org/abs/2310.07815

36 Wang L, Yang N, Wei F. Learning to retrieve in-context examples for large language models [Internet]. arXiv; 2024 [cited 2025 Jun 29]. Available from: http://arxiv.org/abs/2307.07164

37 Zhou D, Schärli N, Hou L, Wei J, Scales N, Wang X, et al. Least-to-Most prompting enables complex reasoning in large language models [Internet]. arXiv; 2023 [cited 2025 Jun 29]. Available from: http://arxiv.org/abs/2205.10625

38 Dhuliawala S, Komeili M, Xu J, Raileanu R, Li X, Celikyilmaz A, et al. Chain-of-Verification reduces hallucination in large language models [Internet]. arXiv; 2023 [cited 2025 Jun 29]. Available from: http://arxiv.org/abs/2309.11495

39 Li X, Li J. AnglE-optimized text embeddings [Internet]. arXiv; 2024 [cited 2025 Jun 29]. Available from: http://arxiv.org/abs/2309.12871

40 Wang Y, Lipka N, Rossi RA, Siu A, Zhang R, Derr T. Knowledge graph prompting for multi-document question answering [Internet]. arXiv; 2023 [cited 2025 Jun 29]. Available from: http://arxiv.org/abs/2308.11730

41 Gao Y, Sheng T, Xiang Y, Xiong Y, Wang H, Zhang J. Chat-REC: Towards interactive and explainable LLMs-augmented recommender system [Internet]. arXiv; 2023 [cited 2025 Jun 29]. Available from: http://arxiv.org/abs/2303.14524

42 Shi W, Min S, Yasunaga M, Seo M, James R, Lewis M, et al. REPLUG: Retrieval-augmented black-box language models [Internet]. arXiv; 2023 [cited 2025 Jun 30]. Available from: http://arxiv.org/abs/2301.12652

43 Yu Z, Xiong C, Yu S, Liu Z. Augmentation-adapted retriever improves generalization of language models as generic Plug-In [Internet]. arXiv; 2023 [cited 2025 Jun 30]. Available from: http://arxiv.org/abs/2305.17331

44 Yang H, Li Z, Zhang Y, Wang J, Cheng N, Li M, et al. PRCA: Fitting black-box large language models for retrieval question answering via pluggable reward-driven contextual adapter [Internet]. arXiv; 2023 [cited 2025 Jun 29]. Available from: http://arxiv.org/abs/2310.18347

45 Mostafa R, Baig MN, Ehsan MT, Hasan J. G-RAG: Knowledge expansion in material science [Internet]. arXiv; 2024 [cited 2025 Jun 30]. Available from: http://arxiv.org/abs/2411.14592

46 Ma S, Xu C, Jiang X, Li M, Qu H, Yang C, et al. Think-on-Graph 2.0: Deep and faithful large language model reasoning with knowledge-guided retrieval augmented generation [Internet]. ADS; 2024 [cited 2025 Jun 30]. Available from: https://ui.adsabs.harvard.edu/abs/2024arXiv240710805M/abstract

47 Sandhu T. Exploration of word embeddings with graph-based context adaptation for enhanced word vectors – ProQuest [Internet]. University of Windsor; 2024 [cited 2025 Jun 30]. Available from: https://www.proquest.com/docview/3069510526?pq-origsite=gscholar&fromopenview=true&sourcetype=Dissertations%20&%20Theses

48 Edge D, Trinh H, Cheng N, Bradley J, Chao A, Mody A, et al. From local to global: A graph RAG approach to query-focused summarization [Internet]. arXiv; 2025 [cited 2025 Jun 30]. Available from: http://arxiv.org/abs/2404.16130

49 Park C, Lee H, Jeong OR. Leveraging medical knowledge graphs and large language models for enhanced mental disorder information extraction. Future Internet. 2024;16(8):260.

50 Buehler MJ. Generative retrieval-augmented ontologic graph and multiagent strategies for interpretive large language model-based materials design. ACS Engineering Au [Internet]. 2024;4(2) [cited 2025 Jun 30]:241–77. Available from: https://pubs.acs.org/doi/full/10.1021/acsengineeringau.3c00058

51 Peng C, Xia F, Naseriparsa M, Osborne F. Knowledge graphs: Opportunities and challenges. Artificial Intelligence Review [Internet]. 2023;56(11) [cited 2025 Jun 30]:13071–102. Available from: https://link.springer.com/article/10.1007/s10462-023-10465-9

52 Anokhin P, Semenov N, Sorokin A, Evseev D, Kravchenko A, Burtsev M, et al. AriGraph: Learning knowledge graph world models with episodic memory for LLM agents [Internet]. arXiv; 2025 [cited 2025 Jun 30]. Available from: http://arxiv.org/abs/2407.04363

53 Sharma S, Yoon DS, Dernoncourt F, Sultania D, Bagga K, Zhang M, et al. Retrieval augmented generation for domain-specific question answering [Internet]. arXiv; 2024 [cited 2025 Jun 30]. Available from: http://arxiv.org/abs/2404.14760

54 Merth T, Fu Q, Rastegari M, Najibi M. Superposition prompting: Improving and accelerating retrieval-augmented generation [Internet]. arXiv; 2024 [cited 2025 Jun 30]. Available from: http://arxiv.org/abs/2404.06910

55 Chen S, Wong S, Chen L, Tian Y. Extending context window of large language models via positional interpolation [Internet]. arXiv; 2023 [cited 2025 Jun 30]. Available from: http://arxiv.org/abs/2306.15595

56 Hu C, Huang H, Hu J, Xu J, Chen X, Xie T, et al. MemServe: Context caching for disaggregated LLM serving with elastic memory pool [Internet]. arXiv; 2024 [cited 2025 Jun 30]. Available from: http://arxiv.org/abs/2406.17565

57 Wang X, Wang Z, Gao X, Zhang F, Wu Y, Xu Z, et al. Searching for best practices in retrieval-augmented generation [Internet]. arXiv; 2024 [cited 2025 Jun 30]. Available from: http://arxiv.org/abs/2407.01219

58 Liu NF, Lin K, Hewitt J, Paranjape A, Bevilacqua M, Petroni F, et al. Lost in the middle: How language models use long contexts [Internet]. arXiv; 2023 [cited 2025 Jun 30]. Available from: http://arxiv.org/abs/2307.03172

59 Peng B, Zhu Y, Liu Y, Bo X, Shi H, Hong C, et al. Graph retrieval-augmented generation: A survey [Internet]. arXiv; 2024 [cited 2025 Jun 30]. Available from: http://arxiv.org/abs/2408.08921

60 Tedeschi S, Conia S, Cecconi F, Navigli R. Named Entity Recognition for Entity Linking: What Works and What's Next – ACL Anthology. In: Findings of the Association for Computational Linguistics: EMNLP 2021 [Internet]; 2021 [cited 2025 Jun 30]. pp. 2584–96. Available from: https://aclanthology.org/2021.findings-emnlp.220/

61 Orlando R, Cabot PLH, Barba E, Navigli R. ReLiK: Retrieve and LinK, fast and accurate entity linking and relation extraction on an academic budget [Internet]. arXiv; 2025 [cited 2025 Jun 30]. Available from: http://arxiv.org/abs/2408.00103

13

Fine-tuning and Application for Materials Large Language Models

13.1 Introduction

Language serves as a foundational medium for human communication, self-expression, and interaction with machines. The imperative for generalized models arises from the escalating requirement for machines to undertake complex linguistic tasks—encompassing translation, text summarization, information retrieval, and conversational interactions, among others. In recent times, remarkable advancements in language models have been observed, driven primarily by the advent of Transformers [1], enhanced computational capacities, and the accessibility of large-scale training corpora. These breakthroughs have spurred a revolutionary shift, enabling the development of large language models (LLMs) that demonstrate human-like performance across diverse tasks [2, 3]. As state-of-the-art AI systems, LLMs now possess the capability to process and generate coherent textual content [4], while also exhibiting the ability to generalize across multiple linguistic applications [5, 6]. Fine-tuning is a key technology for LLMs, enabling their customization according to the specific requirements of materials research. It enables these models to better understand the domain-specific jargon, concepts, and research paradigms. By adjusting the parameters of pretrained LLMs on materials-related datasets, authors can enhance their performance in tasks like predicting material properties, suggesting synthesis routes, and analyzing experimental results.

13.2 The Evolution, Capabilities, and Adaptation Strategies for LLMs

The evolutionary trajectory of NLP has unfolded from statistical language modeling to neural-based approaches, and further from PLMs to LLMs. Conventional LM historically relied on supervised learning to train task-specific models, whereas PLMs leverage self-supervised learning on massive text corpora [5, 7–9].

This paradigm enables PLMs to acquire generic representations that can be transferred across diverse NLP tasks. Notably, fine-tuning PLMs for downstream applications consistently outperforms traditional LM approaches [10]. The performance gains associated with larger PLMs have driven the transition to LLMs, characterized by exponential increases in model parameters (ranging from tens to hundreds of billions) [11] and training data scales (ranging from gigabytes to terabytes) [11, 12].

This progress has spurred a rapid rise in LLMs in recent literature [6, 11–16]. As illustrated in Figures 13.1 and 13.2, the trend of released LLMs has grown exponentially, with landmark models emerging across different research epochs.

Early explorations in LLMs, including T5 [18] and mT5 [19], primarily relied on transfer learning paradigms. This landscape shifted with GPT-3 [10], which demonstrated that LLMs could exhibit zero-shot generalization to downstream tasks—eliminating the need for task-specific fine-tuning. These models showed the ability to respond to task queries by leveraging prompt-based instructions and exemplars. However, pretrained LLMs often struggle to align with user intent, with zero-shot performance consistently lagging behind few-shot settings. Subsequent research has shown that fine-tuning LLMs on task-instruction datasets [20–23] and aligning them with human preference signals [24, 25] yields dramatic improvements. This approach enhances generalization to unseen tasks, significantly boosting zero-shot capabilities while mitigating misalignment issues.

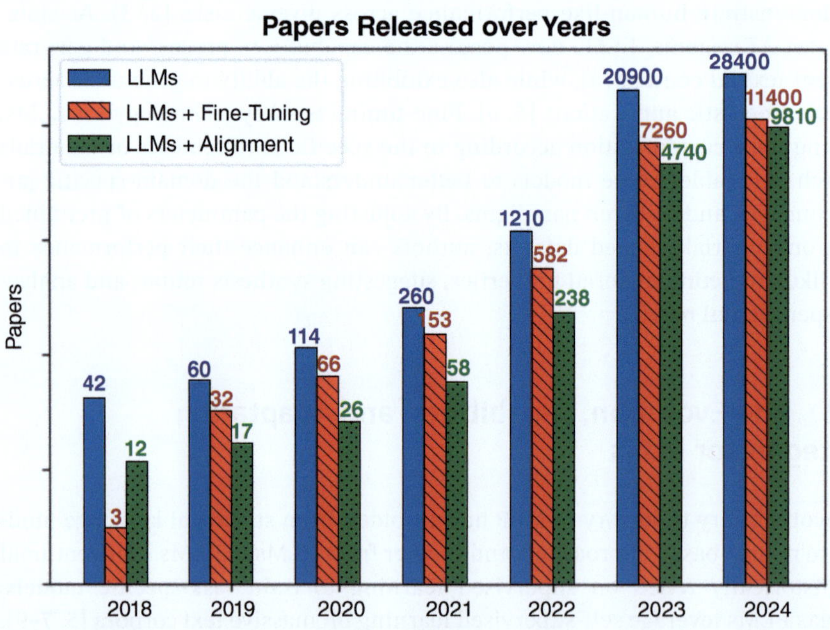

Figure 13.1 The trend of papers released over the years containing the keywords "Large Language Model," "Large Language Model + Fine-tuning," and "Large Language Model + Alignment." *Source*: [17] / Journal of Machine Learning Research / CC BY 4.0.

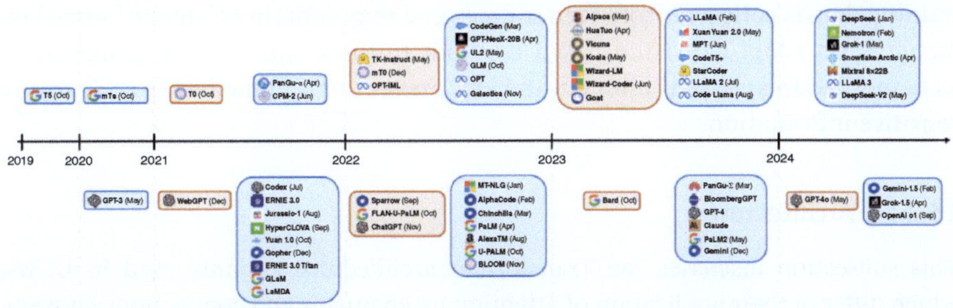

Figure 13.2 Chronological display of LLM releases: blue cards represent "pretrained" models, while orange cards correspond to "instruction-tuned" models. Models on the upper half signify open-source availability, whereas those on the bottom are closed-source. The chart illustrates the increasing trend toward instruction-tuned and open-source models, highlighting the evolving landscape and trends in natural language processing research. *Source*: [17] / Journal of Machine Learning Research / CC BY 4.0.

Beyond their enhanced generalization and domain adaptation capabilities, LLMs have demonstrated emergent abilities—including reasoning, planning, decision-making, in-context learning, and zero-shot answering. Notably, these capabilities emerge naturally from their enormous scale, even though pretrained LLMs are not explicitly trained to acquire such attributes [26–28]. This emergent functionality has driven the widespread adoption of LLMs across diverse fields, including multi-modal systems, robotics, tool manipulation, question answering, and autonomous agent development. Advancements in these domains have been achieved through two primary approaches: task-specific training [29–35] and optimized prompting strategies [36].

While LLMs exhibit human-like capabilities across diverse tasks, these advancements are accompanied by significant trade-offs: slow training and inference speeds, intensive hardware demands, and elevated operational costs. Such constraints have both limited their broad adoption and spurred innovation in architectural design [16, 36–38] and training methodologies [39–44]. To address these challenges, researchers have focused on optimizing LLM efficiency through approaches such as parameter-efficient tuning [41, 44, 45], network pruning [45, 46], model quantization [47, 48], knowledge distillation, and context length interpolation [49–52].

13.2.1 Data Preprocessing

The quality of training data is essential for optimal results, and effective filtering approaches include classifier-based methods (training classifiers on high-quality data to predict text quality) and heuristics-based methods (using rules related to language, metrics, statistics, and keywords for screening). Data deduplication is a critical preprocessing step for LLM training, as duplicated data can degrade model performance and enhance data memorization, and it can be implemented at the sentence, document, or dataset level. Additionally, since most LLM training data is collected from web sources and may contain private information, many LLMs adopt heuristics-based methods to filter out personal details such as names, addresses, and phone numbers, preventing the models from learning sensitive information.

13.2.2 Architectures

This subsection discusses the Transformer architecture variants used in LLMs, which differ in their application of attention mechanisms and connections between Transformer blocks. The encoder-decoder architecture processes inputs through an encoder that uses self-attention to handle the complete sequence, then passes the intermediate representation to a decoder that generates output step-by-step using cross-attention. In contrast, the causal decoder architecture lacks an encoder and relies solely on a decoder where each predicted token depends only on previous time steps. The prefix decoder, also known as a noncausal decoder, employs bidirectional attention that is not strictly constrained by past information, as illustrated by the noncausal attention mask in Figure 13.3. The mixture-of-experts (MoE) variant features parallel independent experts—feed-forward layers following the attention block—paired with a router to direct tokens to specific experts. As an efficient sparse architecture, MoE achieves performance comparable to dense models while allowing model size expansion without proportional computational cost, as only a subset of experts are activated at any given time [53–55].

13.2.3 Pretraining Objectives

This subsection overviews the pretraining objectives of LLMs, with detailed explanations available in [56]. Full language modeling follows an autoregressive paradigm, where the model predicts subsequent tokens based on preceding context, as illustrated in Figure 13.4. Prefix language modeling adopts a noncausal training approach: a random prefix is selected, and the loss is calculated using only the remaining target tokens, as shown in the same figure. Masked language modeling involves randomly masking tokens or token spans, requiring the model

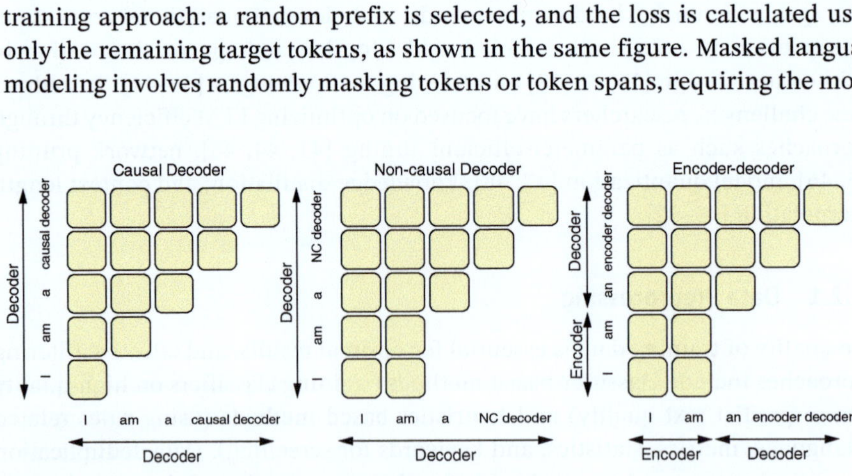

Figure 13.3 An example of attention patterns in language models. *Source*: [56] / Proceedings of Machine Learning Research.

to predict masked elements by leveraging both past and future contexts, with an example provided in Figure 13.4. Unified language modeling [57] integrates causal, noncausal, and masked training objectives: unlike traditional masked modeling, its attention mechanism is unidirectional (left-to-right or right-to-left) rather than bidirectional when processing masked contexts. As illustrated in Figure 13.5, the diagram exemplifies the distinct training phases and inference processes in LLMs. In this chapter, the term "alignment tuning" specifically denotes the process of aligning models with human preferences, though it is noteworthy that the literature occasionally employs the term "alignment" for varying contextual purposes.

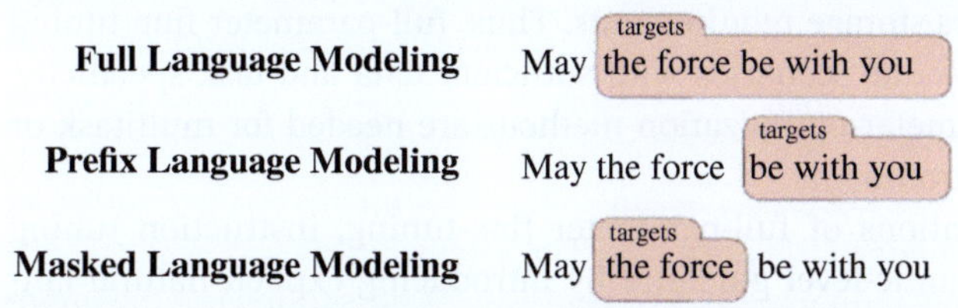

Figure 13.4 An example of language model training objectives. *Source*: [56] / Proceedings of Machine Learning Research.

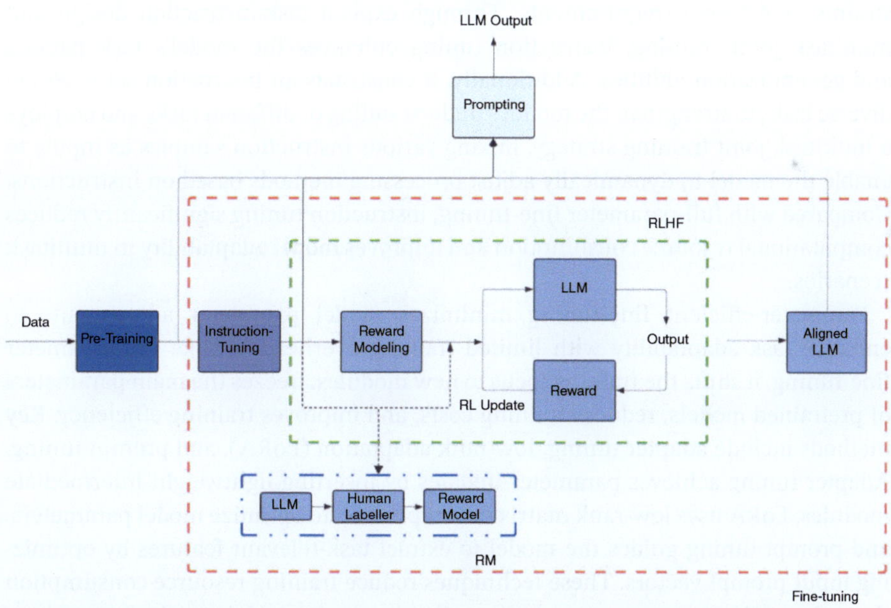

Figure 13.5 A basic flow diagram depicting various stages of LLMs from pretraining to prompting/utilization. Prompting LLMs to generate responses is possible at different training stages like pretraining, instruction tuning, or alignment tuning. "RL" stands for reinforcement learning, "RM" represents reward modeling, and "RLHF" represents reinforcement learning with human feedback. *Source*: [17] / Journal of Machine Learning Research / CC BY 4.0.

13.2.4 Supervised Fine-tuning

Full-parameter fine-tuning is a fundamental adaptation method in supervised learning, where all parameters of a pretrained model are optimized to fit various downstream tasks. This approach utilizes task-specific "input–output pairs" and optimizes weight matrices through backpropagation guided by cross-entropy loss, enabling the model to better adapt to target tasks. However, as model scales continue to expand, the computational cost of full-parameter iterative updates has increased significantly, limiting its practical application (as it primarily suits scenarios with abundant computing resources). Over-optimization may cause the model to forget general knowledge acquired during pretraining, leading to poor performance in cross-domain tasks. Storing complete parameter copies for each independent task also significantly increases storage requirements. Thus, full-parameter fine-tuning is more suitable for high-value scenarios with sufficient data and task specificity, while more efficient parameter optimization methods are needed for multitask or resource-constrained environments.

To overcome the limitations of full-parameter fine-tuning, instruction tuning emerges. It provides semantic-level guidance by introducing explicit natural language instructions, optimizing the model's task adaptation capabilities. Unlike traditional "input–output pair" supervised learning, instruction tuning uses "instruction-response" task examples, requiring the model to understand instruction semantics and parse multilevel information such as task types, domain constraints, and format requirements. Through explicit task-instruction design and multitask joint training, instruction tuning enhances the model's task parsing and generalization abilities. Additionally, it constructs an instruction set covering diverse tasks to strengthen the model's understanding of different tasks and employs a multitask joint training strategy, mixing various instruction samples as inputs to enable the model to dynamically adjust processing methods based on instructions. Compared with full-parameter fine-tuning, instruction tuning significantly reduces computational resource consumption and improves model adaptability in multitask scenarios.

Parameter-efficient fine-tuning minimizes model parameter adjustments to enhance task adaptability with limited training overhead. Unlike full-parameter fine-tuning, it shifts the training focus to new modules, freezes the main parameters of pretrained models, reduces training costs, and improves training efficiency. Key methods include adapter tuning, low-rank adaptation (LoRA), and prompt tuning. Adapter tuning achieves parameter updates by inserting lightweight intermediate modules, LoRA uses low-rank matrix decomposition to optimize model parameters, and prompt tuning guides the model to extract task-relevant features by optimizing input prompt vectors. These techniques reduce training resource consumption through precise parameter regulation, allowing basic models to flexibly address various downstream tasks by stacking specific adaptation modules or prompt vectors, which significantly enhances the feasibility of model deployment in resource-constrained scenarios and multitask expansion capabilities.

13.2.5 LLM Evaluation

As the field of LLMs undergoes rapid advancement, the development of robust evaluation frameworks has become critical to accurately measure the capabilities and limitations of these complex AI systems. Evaluation serves multiple key functions: benchmarking model performance across diverse tasks, identifying areas for improvement, and ensuring that advancements in LLM technology align with ethical and practical standards. There is a consensus in academic and professional circles of LLM evaluation that a multifaceted approach is essential for comprehensively assessing the strengths and limitations of these advanced AI systems.

The evaluation of the Qwen2.5 base language model in a Qwen blog highlights the importance of using multiple benchmarks to comprehensively assess model performance across various domains [58]. Platforms like Hugging Face provide a suite of datasets for this purpose including Instruction-Following Eval (IFEval), BigBench Hard (BBH), MATH [59], General-Purpose Question Answering (GPQA) [60], and Multimodal Unsupervised Scene Representation (MuSR) [61]. These datasets cover a broad spectrum of tasks from language modeling to mathematical problem-solving, ensuring a comprehensive evaluation of model capabilities. Models like Qwen2.5 are evaluated using a diverse range of datasets, encompassing general tasks such as MMLU and HellaSwag, specialized tasks in mathematics and science (e.g., GPQA and MATH), coding tasks (including HumanEval and MBPP), and multilingual capabilities via datasets like Multi-Exam and Multi-Translation.

A comprehensive evaluation of language model performance typically requires a combination of multiple benchmarks. These benchmarks should reflect real-world scenarios and cover different domains and language complexities. Evaluations include various tests to measure a model's ability to handle extended conversations and execute diverse tasks. By leveraging these diverse datasets and evaluation methods, researchers can effectively benchmark language models, guide their development toward practical applications, and ensure they meet ethical and practical standards. This subsection explores benchmarking in two parts: general tasks and specialized tasks. General tasks aim to assess the broad capabilities of LLMs across a wide range of disciplines and skills. These benchmarks are crucial for evaluating an LLM's foundational knowledge and general reasoning abilities. They help determine a model's capacity to understand and generate text in various contexts, ensuring it has a solid grasp of basic language fundamentals. Datasets such as MMLU, ARC, and HellaSwag are commonly used for general evaluations. Specialized tasks focus on evaluating LLMs in specific domains such as mathematics, programming, and natural language understanding. These benchmarks are designed to assess a model's proficiency in specialized fields, providing deeper insights into its expert-level skills. Specialized tasks are crucial for identifying domain-specific strengths and weaknesses, ensuring that models can effectively apply their knowledge to real-world scenarios (Table 13.1).

Table 13.1 This table provides an exhaustive compilation of datasets pertinent to the evaluation of LLMs. These datasets span a diverse array of tasks, from general to domain-specific, aiming to holistically assess the performance of LLMs across various scenarios. The table delineates the publication timeline, the number of examples, and the access points (URLs) for each dataset, facilitating researchers in procuring and utilizing these resources.

Collection	Task	Publication time	Examples	URL
MMLU	General	2021	15,908	https://huggingface.co/datasets/cais/mmlu
ARC	General	2018	7,787	https://huggingface.co/datasets/allenai/ai2_arc
HellaSwag	General	2019	59,950	https://huggingface.co/datasets/Rowan/hellaswag
GLUE	Natural language understanding	2018	1,485,043	https://huggingface.co/datasets/nyu-mml/glue
SuperGLUE	Natural language understanding	2019	196,309	https://huggingface.co/datasets/aps/super_glue
GSM8K	Science and mathematics	2021	17,584	https://huggingface.co/datasets/openai/gsm8k
Theoremqa	Science and mathematics	2023	800	https://huggingface.co/datasets/TIGER-Lab/TheoremQA
Humaneval	Code	2021	164	https://huggingface.co/datasets/openai/openai_humaneval
MBPP	Code	2021	1,401	https://huggingface.co/datasets/google-research-datasets/mbpp
AGIEval	Exam	2023	8,062	https://github.com/ruixiangcui/AGIEval
GAOKAO-Bench	Exam	2023	2,811	https://github.com/OpenMLab/GAOKAO-Bench
TruthfulQA	Other	2021	1,634	https://huggingface.co/datasets/truthfulqa/truthful_qa
BBH	Other	2022	6,511	https://huggingface.co/datasets/lukaemon/bbh

Source: [62] / with permission of Springer Nature.

13.3 Case Studies for Materials LLM Fine-tuning

Jablonka et al. demonstrated that the small datasets prevalent in chemistry have driven the development of sophisticated machine learning approaches requiring chemical knowledge integration, which demand specialized expertise to develop [63]. To address this, they proposed adapting GPT-3—an LLM trained on massive Internet text—by fine-tuning it to answer chemical questions in natural language with correct responses, enabling its application to diverse tasks in chemistry and materials science. Comparisons with dedicated machine learning models across applications (ranging from molecular/materials properties to chemical reaction yields) showed that fine-tuned GPT-3 performs comparably to or even outperforms conventional techniques, particularly under low-data conditions. Inverse design is achievable by simply inverting questions. The method's ease of use and high performance for small datasets could revolutionize the fundamental approach to machine learning in chemical and materials sciences, making pretrained LLM querying a routine way to bootstrap projects by leveraging collective knowledge in foundation models or establish baselines for predictive tasks. To demonstrate the broad applicability of models, they conducted benchmarking across diverse datasets and scientific domains, spanning materials property prediction, synthesis protocols, and materials design (as outlined in Figure 13.6).

For all main text results, authors utilized the smallest GPT-3 Ada variant accessible via the OpenAI API. Across all case studies, fine-tuning employed consistent parameters: eight epochs and a learning rate multiplier of 0.02. Unless specified otherwise, error bands denote the standard error of the mean.

To assess GPT-3's data efficiency against baselines, authors fitted learning curves to power laws. Using these curves, authors identified the data point where the top-performing baseline matched the GPT-3 approach's accuracy at the first learning curve point (evaluated via Cohen's kappa [κ] metric, surpassing random performance). For SMILES validation, authors used the "is valid" method from the GuacaMol package [57], which leverages RDKit parsing to determine validity.

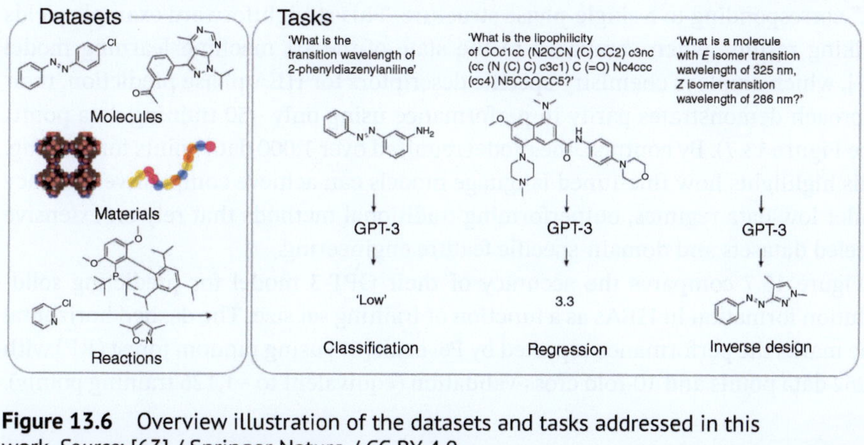

Figure 13.6 Overview illustration of the datasets and tasks addressed in this work. *Source*: [63] / Springer Nature / CC BY 4.0.

Figure 13.6 illustrates the datasets and tasks in this work. They benchmarked GPT-3 on chemical datasets covering molecules, materials, and reactions, investigating tasks like classification (e.g., predicting "high/low" classes), regression, and inverse design (molecule prediction). The metal–organic framework (MOF) rendering uses iRASPA [64].

13.3.1 Language-interfaced Fine-tuning for Classification and Regression

Before delving into the diverse applications in detail, let us first explore the practical implementation of fine-tuning the GPT-3 model, using a simple yet highly complex example as an illustration. High-entropy alloys (HEAs), as a novel category of structural metals, have attracted significant research interest recently [65]. Intriguingly, the number of possible metal combinations in HEAs is virtually infinite. From a practical perspective, determining whether a given combination of metals will form a solid solution or multiple phases is of great importance.

Therefore, the question authors aim to address is: "what is the phase of composition of the HEA?" The model is expected to provide text completion from the set of possible answers {single phase, multiphase}. Fine-tuning of GPT-3 for such a task follows a general process as described in previous studies [66]. First, a dataset is prepared, which contains pairs of HEA compositions (as prompts) and their corresponding phases (as completions). This dataset serves as the basis for training the model to adapt to this specific chemical problem. Then, during the fine-tuning process, the pretrained GPT-3 model adjusts its parameters according to the patterns and relationships in this dataset, enabling it to better answer questions related to the phase determination of HEAs.

In extended data Table 13.2, authors present the question–answer dataset utilized for fine-tuning the GPT-3 model. These entries consist of experimentally validated phase determinations for HEAs, where each question specifies an alloy composition and the answer designates its phase state (single phase or multiphase). Leveraging the OpenAI API, model tuning typically concludes within minutes, yielding a specialized model—for instance, inputting "Sm0.75Y0.25" prompts the text completion "1," corresponding to a single-phase structure. This straightforward example yields striking results: when compared to the state-of-the-art machine learning model [67], which employs chemistry-specific descriptors for HEA phase prediction, their approach demonstrates parity in performance using only ~50 training data points (see Figure 13.7). By contrast, the model required over 1,000 data points for training. This highlights how fine-tuned language models can achieve competitive accuracy under low-data regimes, outperforming traditional methods that rely on extensive labeled datasets and domain-specific feature engineering.

Figure 13.7 compares the accuracy of their GPT-3 model for predicting solid-solution formation in HEAs as a function of training set size. The dashed horizontal line marks the performance reported by Pei et al. [67] using random forest (RF) with 1,252 data points and 10-fold cross-validation (equivalent to ~1,126 training points).

Table 13.2 Data efficiency comparison of best-performing GPT-3-based approaches with best-performing baselines.

Group	Benchmark	Publication year	Best non-DL	Best DL baseline
Molecules	Photoswitch transition wavelength	2022	1.1 (n)	1.2 (t)
	Free energy of solvation	2014	3.1 (g)	1.3 (t)
	Solubility	2004	1.0 (x)	0.002 (m)
	Lipophilicity	2012	3.43 (g)	0.97 (t)
	HOMO–LUMO gap	2022	4.3 (x)	0.62 (t)
	OPV PCE	2018	0.95 (n)	0.76 (t)
Materials	Surfactant free energy of adsorption	2021	1.4 (xj)	0.37 (t)
	CO_2 Henry coefficients	2020	0.40 (x)	12 (t)
	CH_4 Henry coefficients	2020	0.52 (xmo)	0.60 (t)
	Heat capacity	2022	0.22 (mo)	0.76 (c)
	HEA phase	2020	24 (prf)	9.0 (c)
	Bulk metallic glass formation ability	2006	0.98 (a)	0.62 (mod)
	Metallic behavior	2018	0.52 (a)	0.46 (mod)
Reactions	C–N cross-coupling	2018	2.9 (drfp)	
	C–C cross-coupling	2022	0.98 (n)	

Source: [63] / Springer Nature / CC BY 4.0.

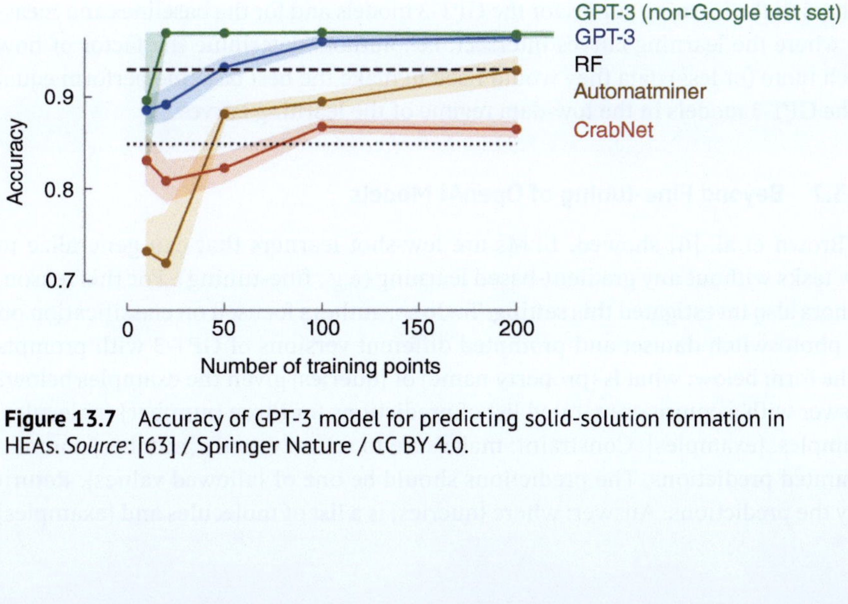

Figure 13.7 Accuracy of GPT-3 model for predicting solid-solution formation in HEAs. *Source*: [63] / Springer Nature / CC BY 4.0.

The dotted line represents a simple rule-based baseline ("if present in composition, classify as single-phase; else multiphase"). The yellow line shows results from Automatminer [68], which employs automated machine learning (via TPOT [69]) to select optimal featurization and modeling approaches using composition data. Authors also evaluated CrabNet [70] (red line), a composition-based neural network with default settings. Their GPT-3 model (blue line, error bands indicate s.e.m.) achieves comparable accuracy to Lederer et al.'s RF model with only ~50 training points. The green line shows performance on a separate test set containing only compositions without exact matches in Google search results, confirming robust generalization.

These findings prompted us to investigate whether analogous results could be replicated for other properties. Thus, authors examined a diverse range of molecular, material, and chemical reaction properties—specifically targeting applications where conventional machine learning methods have been established and widely adopted as field benchmarks. Table 13.2 contrasts the performance of a fine-tuned GPT-3 model against baselines by analyzing learning curves and identifying intersection points—quantifying the data ratio required for baselines to match GPT-3's low-data performance. The analysis covered diverse properties, molecules: HOMO–LUMO gaps, water solubility, and organic photovoltaic performance; materials: alloy characteristics, MOFs, and polymer properties; and reactions: two key organic cross-coupling reactions. Findings show that in low-data regimes, GPT-3 typically matches or outperforms conventional machine learning models with fewer data points. In high-data scenarios, baselines often catch up, likely because GPT-3's pretrained inductive biases (e.g., chemical correlations) become less critical when datasets are large enough to capture inherent patterns directly.

For best comparison, authors also split baselines into (pretrained) deep-learning (DL)-based baselines (including MolCLR68, ModNet69, CrabNet62, and TabPFN70) and baselines not using (pretrained) DL approaches (*n*-gram, Gaussian process regression [GPR], XGBoost, RFs, and automated machine learning optimized for materials science) on hand-tuned feature sets. For the analysis in Table 13.2, authors fit the learning curves for the GPT-3 models and for the baselines and measure where the learning curves intersect, i.e., authors determine the factor of how much more (or less) data they would need to make the best baseline perform equal to the GPT-3 models in the low-data regime of the learning curves.

13.3.2 Beyond Fine-tuning of OpenAI Models

As Brown et al. [6] showed, LLMs are few-shot learners that can generalize to new tasks without any gradient-based learning (e.g., fine-tuning). For this reason, authors also investigated this setting. To do so, authors focused on classification on the photoswitch dataset and prompted different versions of GPT-3 with prompts of the form below: what is {property name} of {queries} given the examples below? Answer with a comma-separated list of predictions for those {number} molecules. Examples: {examples}. Constraint: make sure to return exactly {number} comma-separated predictions. The predictions should be one of {allowed values}. Return only the predictions. Answer: where {queries} is a list of molecules and {examples}

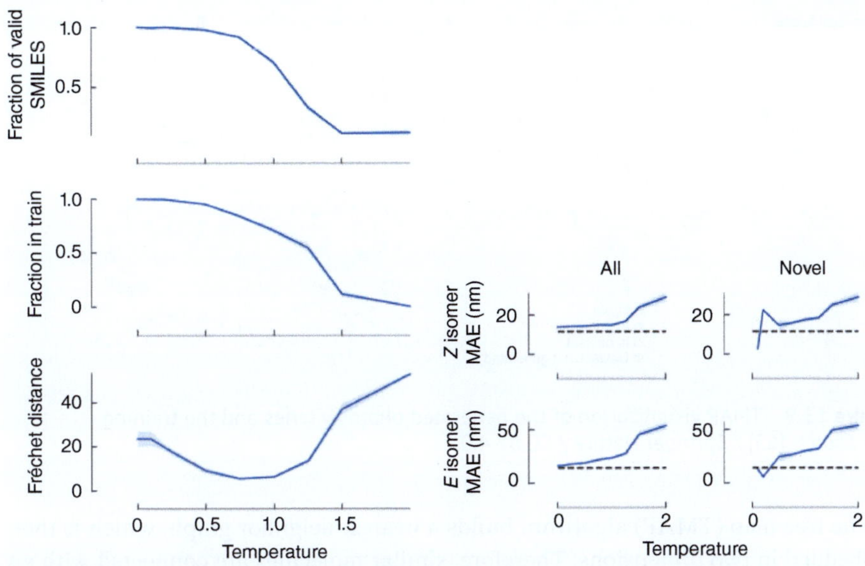

Figure 13.8 Photoswitch inverse design metrics as a function of temperature. *Source*: [63] / Springer Nature / CC BY 4.0.

is a list of examples of molecules with associated properties. As evident from Figure 13.8, with this approach, authors find good results, competitive or even outperforming the GPR baseline. Interestingly, authors also observe a stronger dependence on the representation of the molecules when they use this prompt-based approach without fine-tuning. However, a fundamental limitation of this approach is that the context length of the model limits the number of examples one can provide. For this reason, the learning curves also add at different points (with larger models having larger context sizes and different representations needing a different number of tokens). In addition to the OpenAI models, authors also investigated Anthropic's Claude-v1 model. These encouraging results motivate further work using other prompting strategies, including soft prompting, such as instruction prompt tuning [10].

The fraction of valid SMILES indicates the fraction of generated SMILES that can successfully be parsed using RDKit (note that it does not plateau at 0, but approximately 0.1) [71]. Authors then determine the fraction of those that are already part of the training set and find that at low temperature GPT-3 tends to restate molecules from the training set. To quantitatively capture the similarity of the distribution of the generated molecules to the ones from the training set, authors compute the Fréchet ChemNet distance [72], which quantifies both diversity and distribution match [18] and goes through a minimum at intermediate temperatures. For quantifying how well the generated molecules match the desired transition wavelengths, authors use the GPR models reported by [19] to predict the transition wavelengths. The dashed horizontal lines indicate those models' mean absolute error (MAE). Across all temperatures, authors found high average synthesizability (synthetic accessibility [SA], score smaller than 3). Error bands indicate standard error of the mean.

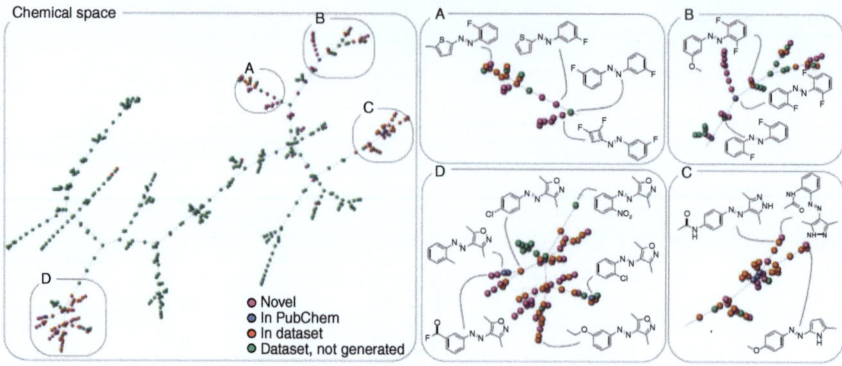

Figure 13.9 TMAP visualization of the generated photoswitches and the training set. *Source*: [63] / Springer Nature / CC BY 4.0.

The tree map (TMAP) algorithm builds a nearest-neighbor graph, which is then embedded in two dimensions. Therefore, similar molecules are connected with an edge. Authors color the points depending on whether they are part of the original dataset but not generated (green) or part of the dataset and generated by their model (orange). Their models can also generate molecules that have not been part of the photoswitch dataset (note that the model was only trained on 92 molecules from this database). In some cases, those molecules have been reported before and are part of the PubChem database (blue) or are not part of the PubChem database (pink). From Figure 13.9, it can be seen that the generated molecules are sometimes substitutions for molecules in the dataset. In other cases, newly generated molecules introduce a completely new scaffold. For this visualization, authors used the TMAP [73] algorithm on photoswitch molecules described using MinHash fingerprint with 2,048 permutations [74].

13.3.3 Regression

Regression tasks pose a greater challenge than classification, as they require predicting continuous properties like the Henry coefficient for gas adsorption in porous materials. Since the pretrained language model cannot directly predict real numbers (without architectural/training changes), authors leverage a practical workaround: using Henry coefficients rounded to a feasible accuracy for training, assuming GPT-3 can interpolate these values. Alternatively, this could be framed as a classification task with fine-grained bins. While tuning GPT-3 for regression demands more data than classification, it still approaches state-of-the-art performance. However, the key advantage here—beyond training simplicity—diminishes compared to low-data scenarios, as traditional methods can compete when sufficient labeled data is available.

Authors use data reported by Moosavi et al. [75]. Authors also use the model reported by Moosavi et al. [75], for which they extensively optimized the hyperparameters. Notably, this model can achieve good data efficiency because it has been trained on local environments. Additionally, this model is a bootstrapped ensemble, which authors replicated for their baseline. To ensure a fair comparison, authors only trained on MOFs and not covalent–organic frameworks (COFs) and zeolites. Additionally, authors only considered those MOFs for which they could determine an MOFid [76]. Note that, similar to Moosavi et al. [75], authors did not utilize advanced splitting strategies other than stratification on the gravimetric heat capacity. Additionally, they investigated a composition-only approach (reduced formula of the primitive cell). As a baseline for this, authors considered CrabNet [76].

13.3.4 Stretching the Limits

The results on photoswitches illustrate the potential of LLMs for chemistry. To obtain more insight into whether authors can trust these GPT-3 predictions, they carried out some experiments where they tried to stretch the limits.

Authors have established that GPT-3 delivers robust results regardless of molecular representation (IUPAC names, SMILES, or SELFIES). But can it interpret an abstract molecular representation authors devised? A prior study used a coarse-grained machine learning approach to design dispersants—linear copolymers with four monomer types and chain lengths of 16–48 units, spanning a 58-million-member design space [77]. The key objective was identifying dispersants with optimal binding free energy by determining ideal polymer length and monomer sequences. Notably, GPT-3 lacked preexisting knowledge of these coarse-grained polymers' properties or representations. Yet, when queried with questions like "What is the adsorption free energy of coarse-grained dispersant AAAABBBBDDDDAAAACCCC?" or tasked with inverse design ("Give me a structure... with a free energy of 17"), the model outperformed the approach in [77] for adsorption free energy prediction. It also succeeded in inverse design, generating monomer sequences that matched target compositions and achieved the desired free energy with a mean percentage error of ~22%.

For photoswitches, GPT-3 can generate novel molecules distinct from the training set. To explore its generative limits, authors focused on HOMO–LUMO gap prediction—a property accurately calculable via semiempirical quantum mechanics (GFN2-xTB [75]) and documented for 665,000 molecules in the QMugs dataset [76]. This allows rigorous validation of generated molecules' electronic properties. Authors demonstrate that training GPT-3 on just 500 samples enables reasonable estimation of HOMO–LUMO gaps in the QMugs dataset. Inverse design training allows generating novel molecules (not in the training or QMugs datasets) by asking for specific gap values (e.g., 3.5 eV).

Extrapolation tests show GPT-3 can generate molecules with HOMO–LUMO gaps > 4 eV, despite no training data exceeding 3.5 eV. Authors queried the model over 1,000 times with Gaussian-sampled gaps around 4 eV, and quantum calculations confirmed a significant fraction of generated structures met the target. Similar results emerged for photoswitches: fine-tuned GPT-3 generated molecules absorbing at lower energies, extending beyond a training library limited to < 350 nm wavelengths. Inspired by these results, authors conducted an inverse design experiment to create molecules with HOMO–LUMO gaps > 5 eV—properties far beyond the training data [78]. The QMugs database shows an average bandgap of ~2.58 eV (Figure 13.10), with only a few molecules exceeding 5 eV. Hence, this is a challenging inverse design problem, as only a few materials in the database have the desired properties. Here, their experiment is the quantum calculation, and authors typically assume that they can evaluate hundreds to thousands of materials in a reasonable time. From a machine learning point of view, a set of thousands of materials is in a very-low-data regime. However, from an experimental point of view, this is a large but sometimes doable effort.

Authors initiated the process by training with hundreds of randomly selected molecules from the QMugs dataset, whose bandgap distribution mirrored that of the database (blue curve in Figure 13.10). Authors then queried the model ~1,000 times for molecules with HOMO–LUMO gaps sampled from a normal distribution (mean = 4 eV, SD = 0.2 eV). Evaluation of the generated molecules (green curve) showed a distribution shift toward higher gaps. Through four iterative rounds of retraining with new data and querying for increasingly higher gaps, authors successfully achieved the target of designing molecules with HOMO–LUMO gaps > 5 eV, as illustrated in Figure 13.10.

This study demonstrates that a GPT-3-based machine learning system excels in addressing diverse chemical questions—including those incompatible with conventional linear representations like SMILES. Compared to traditional machine learning, it offers distinct advantages: its unified approach relies on natural-language question formatting for both model training and deployment across applications. This sets a new benchmark for future ML research, where novel models must at least surpass this straightforward yet effective framework.

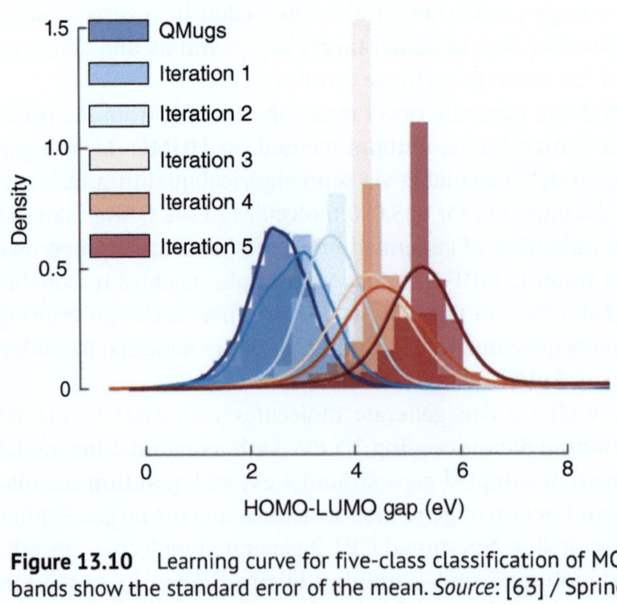

Figure 13.10 Learning curve for five-class classification of MOF heat capacities. Error bands show the standard error of the mean. *Source*: [63] / Springer Nature / CC BY 4.0.

13.4 LLM Evaluation Example

Mirza et al.'s [79] study can serve as a typical case to illustrate the evaluation paradigm and capability boundaries of LLMs in the field of chemistry. Aiming to address the gap in applying LLMs to chemical knowledge reasoning, this research developed an automated evaluation framework, ChemBench, to systematically compare the performance of LLMs and human chemists, providing critical empirical evidence for understanding the "capabilities" and "limitations" of AI in professional domains.

To systematically assess the capabilities of LLMs in chemical sciences and identify areas for improvement, robust evaluation frameworks are essential for tracking progress and mitigating potential risks. Currently, LLM development relies on standardized benchmarks like BigBench [80] and the LM Eval Harness [81], which offer limited coverage of chemical tasks—BigBench includes only two chemistry-related tasks among 204, while LM Eval Harness lacks chemistry-specific evaluations entirely. In the absence of widely adopted benchmarks, researchers developing chemical language models [63, 82–84] often resort to language-interfaced [85] tabular datasets from sources such as MoleculeNet [86, 87], Therapeutic Data Commons [88], safety databases [89], or MatBench [68]. However, these evaluations typically focus on narrow tasks like predicting molecular properties (e.g., solubility, toxicity, melting temperature, or reactivity) or reaction outcomes, providing only a partial perspective on the models' broader chemical understanding.

Here, authors introduce ChemBench, a comprehensive benchmarking framework (Figure 13.11) designed to assess the capabilities and limitations of state-of-the-art models in the chemical sciences. Their benchmark comprises 2,788 question–answer pairs, sourced from a combination of 1,039 manually curated and 1,749 semiautomatically generated items, covering a broad spectrum of topics taught in undergraduate and graduate chemistry curricula. ChemBench evaluates key competencies, including reasoning, knowledge retention, and intuitive understanding, and is applicable to any text-generating system—including those augmented with external tools. To provide meaningful context for model performance, authors conducted a survey of 19 chemistry experts, who evaluated a subset of the benchmark under realistic conditions (e.g., permitting tool use such as web searches).

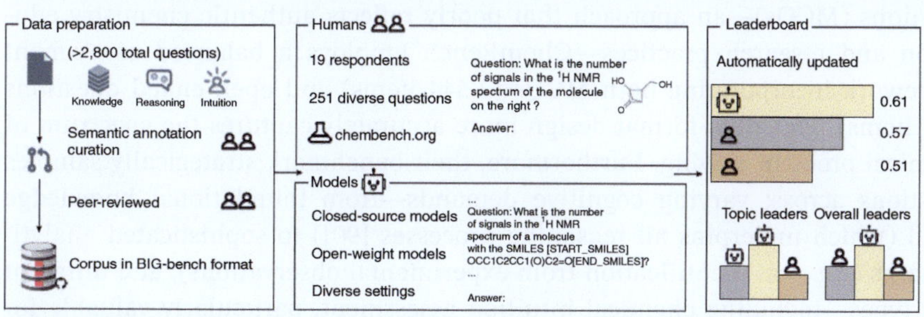

Figure 13.11 Overview of the ChemBench framework. *Source*: [79] / Springer Nature / CC BY 4.0.

This allows for direct comparison between human chemists of varying specializations and AI systems, offering insights into how well current models perform relative to domain experts.

Different components of the ChemBench framework: the framework's foundation is the benchmark corpus comprising thousands of questions and answers that authors manually or semiautomatically compiled from various sources in a format based on the one introduced in the BigBench benchmark (see extended data Figure 13.11). Questions are classified on the basis of topics, required skills (reasoning, calculation, knowledge, and intuition), and difficulty levels. Authors then used this corpus to evaluate the performance of various models and tool-augmented systems using a custom framework. To provide a baseline, authors built a web application that they used to survey experts in chemistry. The results of the evaluations are then compiled in publicly accessible leaderboards, which authors propose as a foundation for evaluating future models.

13.4.1 Benchmark Corpus

Authors constructed their benchmark corpus by aggregating questions from diverse sources, including original manually crafted items, university examinations, and semiautomatically generated questions derived from curated chemical databases, implementing rigorous quality control through multistage review where each question was verified by at least two independent scientists alongside the original curator, supplemented by automated validation checks. The resulting dataset spans a comprehensive thematic scope from foundational chemistry (e.g., general chemistry) to specialized subfields (e.g., inorganic, analytical, and technical chemistry), as illustrated in Figure 13.12. To enable granular performance analysis, authors annotated questions along two key dimensions: required skills (categorized as knowledge recall, reasoning, calculation, intuition, or multimodal combinations) and difficulty level (assessed by human annotators to facilitate nuanced model evaluation across proficiency tiers), ensuring ChemBench not only assesses breadth of chemical understanding but also discerns how models handle varying cognitive demands. Research design: establishing a "Golden Standard" for chemical evaluation.

Unlike conventional benchmarks that predominantly rely on multiple-choice questions (MCQs)—an approach that poorly reflects authentic chemistry education and research practices—ChemBench employs a balanced assessment framework incorporating both MCQs (2,544 items) and open-ended questions (244 items). This dual-format design more accurately captures the spectrum of chemical problem-solving. Furthermore, their benchmark strategically samples questions across: varying cognitive demands—from foundational knowledge recall (which underpins all reasoning processes [90]) to sophisticated analytical tasks (e.g., ion identification from experimental observations); and different skill types—including chemical intuition assessment, particularly valuable for applications requiring human-aligned reasoning like hypothesis generation or

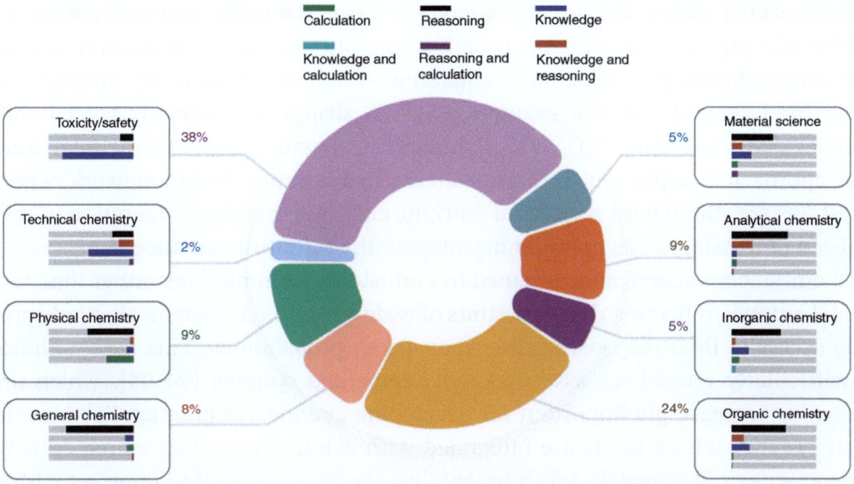

Figure 13.12 Distribution of topics and required skills. The distribution of questions across various chemistry topics, along with the primary skills required to address them. The topics were manually classified, showing a varied representation across different aspects of chemistry. Each topic is associated with a combination of three key skills: calculation, reasoning, and knowledge, as indicated by the colored bars. ChemBench samples encompass diverse topics and diverse skills, setting a high bar for LLMs to demonstrate human-competitive performance across a wide range of chemistry tasks. *Source*: [79] / Springer Nature / CC BY 4.0.

optimization tasks [91]. This multidimensional approach ensures comprehensive evaluation of both technical knowledge and higher-order chemical thinking.

To address practical constraints in large-scale evaluations, authors developed ChemBench-Mini, a carefully curated subset of 236 questions representing the diversity of their full benchmark. This streamlined version responds to the significant cost barriers in comprehensive assessments, as demonstrated by Liang et al. [92] who reported API call expenses exceeding $10,000 for single evaluations on established benchmarks like HELM. While no subset can fully capture the complete corpus, ChemBench-Mini was designed through rigorous methodological selection to maintain balanced topic coverage across chemical subdisciplines, proportional representation of different skill requirements, and inclusion of both question formats. All questions in this condensed benchmark were human-validated through responses from their volunteer chemistry experts, ensuring maintained quality despite the reduced scale.

13.4.2 Benchmark Suite Design

Scientific text often contains specialized elements (e.g., molecules, equations) that require distinct processing compared to natural language. While models like Galactica [93] have introduced specialized encodings for such content, current benchmarking suites fail to accommodate these scientific representations. To bridge this gap, ChemBench incorporates semantic encoding for domain-specific elements—chemicals, units, and equations—enabling models to differentiate them from general text. For example, SMILES strings are explicitly tagged with [START_SMILES] and [END_SMILES] markers, signaling their unique treatment. This approach is implemented through an extensible annotation framework, where questions are stored with structured markup, ensuring seamless integration of specialized processing while maintaining adaptability for future enhancements.

ChemBench is specifically designed to evaluate text completions rather than raw model outputs, reflecting the constraints of widely used LLM systems that only provide access to final responses rather than token probabilities. This design choice is particularly crucial for assessing tool-augmented systems [94–96], which are becoming increasingly important for developing chemical copilot applications. In such systems, where LLMs are integrated with external tools like search APIs or code executors, the model's token probabilities become difficult to interpret within the broader system context. Since text completions represent the actual outputs that would be deployed in real-world applications [97], they serve as the most relevant and practical basis for evaluation in their benchmark.

13.4.3 Overall System Performance

Authors conducted a comprehensive evaluation of leading LLMs on the ChemBench corpus, including both standalone and tool-augmented systems, to assess current capabilities in chemical sciences. As shown in Figure 13.13, their analysis reveals several key findings: the top-performing LLM (o1-preview) achieved nearly twice the accuracy of their best human expert; most models surpassed average human performance; and the open-source Llama-3.1-405B-Instruct demonstrated competitive performance with leading proprietary models. These human benchmarks were established through expert evaluations conducted via their custom web platform (https://www.chembench.org), which recorded the full range of human performances (worst, best, and average). As shown in Figure 13.14, performance across distinct topic areas of ChemBench-Mini, visualized via radar plots for both models and human evaluators, further contextualizes these aggregate results. The results suggest that modern LLMs, including emerging open-source options, can outperform human experts on standardized chemical knowledge assessments, while highlighting the rapid progress in chemical AI capabilities.

Their evaluation reveals two key limitations in current chemical LLMs: knowledge retention and scaling effects. First, models struggle with knowledge-intensive questions, suggesting they have not memorized critical chemical facts. This deficiency persists even when using retrieval-augmented generation systems like PaperQA2, likely because essential chemical knowledge often resides in specialized databases (e.g., PubChem, Gestis) rather than general literature—a pattern mirrored by human experts who similarly relied on these resources. These findings highlight the need for either training on more specialized chemical datasets or better integration with domain-specific databases. Second, authors observe a clear correlation between model size and performance, consistent with trends in other domains, indicating that continued scaling could yield further improvements in chemical reasoning capabilities.

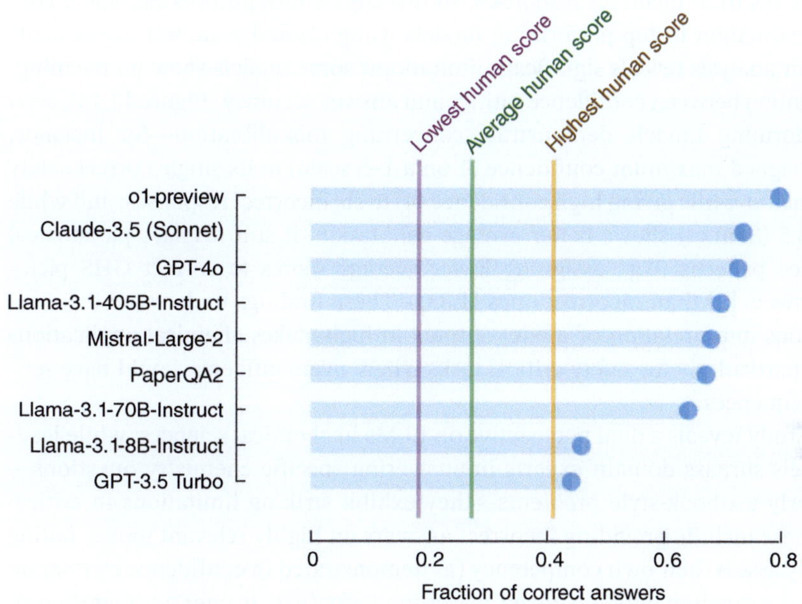

Figure 13.13 Performance of models and humans on ChemBench-Mini. *Source*: [79] / Springer Nature / CC BY 4.0.

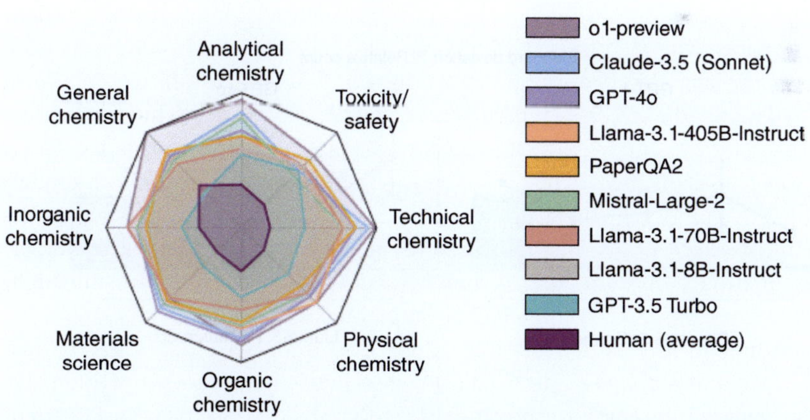

Figure 13.14 Performance of the models and humans on the different topics on ChemBench-Mini. The radar plot shows the performance of the models and humans on the different topics of ChemBench-Mini. Performance is measured as the fraction of questions that were answered correctly by the models. The best score for every dimension is 1 (all questions answered correctly) and the worst is 0 (no question answered correctly). A larger colored area indicates a better performance. This figure shows the performance on ChemBench-Mini. *Source*: [79] / Springer Nature / CC BY 4.0.

13.4.4 Confidence Estimation

A critical question for practical applications is whether LLMs can reliably assess their own competency in answering chemical questions—a capability that could mitigate risks from incorrect responses. To investigate this, authors evaluated confidence estimation in top-performing models using ordinal-scale self-assessments [97]. Their analysis reveals significant limitations: some models show no meaningful correlation between confidence ratings and answer accuracy (Figure 13.15); even high-performing models demonstrate concerning miscalibration—for instance, GPT-4 assigned maximum confidence (1 on a 1–5 scale) to its single correct safety profile answer while giving high confidence (4) to six incorrect responses; and while Claude-3.5 (Sonnet) shows better average calibration, it still displays paradoxical confidence patterns (e.g., assigning lower average scores to correct GHS pictogram answers [2] than incorrect ones [1.83]). These findings underscore the need for cautious interpretation of model outputs in high-stakes chemical applications [98, 99], particularly for safety-critical tasks where overconfidence could have serious consequences.

Their study reveals a dual perspective on LLMs in chemical sciences: while leading models surpass domain experts in answering specific chemistry questions—particularly textbook-style problems—they exhibit striking limitations in critical areas. These include providing incorrect answers on highly relevant topics, failing to reliably assess their own competency (as demonstrated in confidence estimation tests), and struggling with advanced reasoning tasks (e.g., isomer enumeration or NMR peak prediction) [98, 99]. The models' superior performance over average humans on memorization-based questions challenges traditional chemistry education paradigms, suggesting a need to shift focus from rote learning to critical

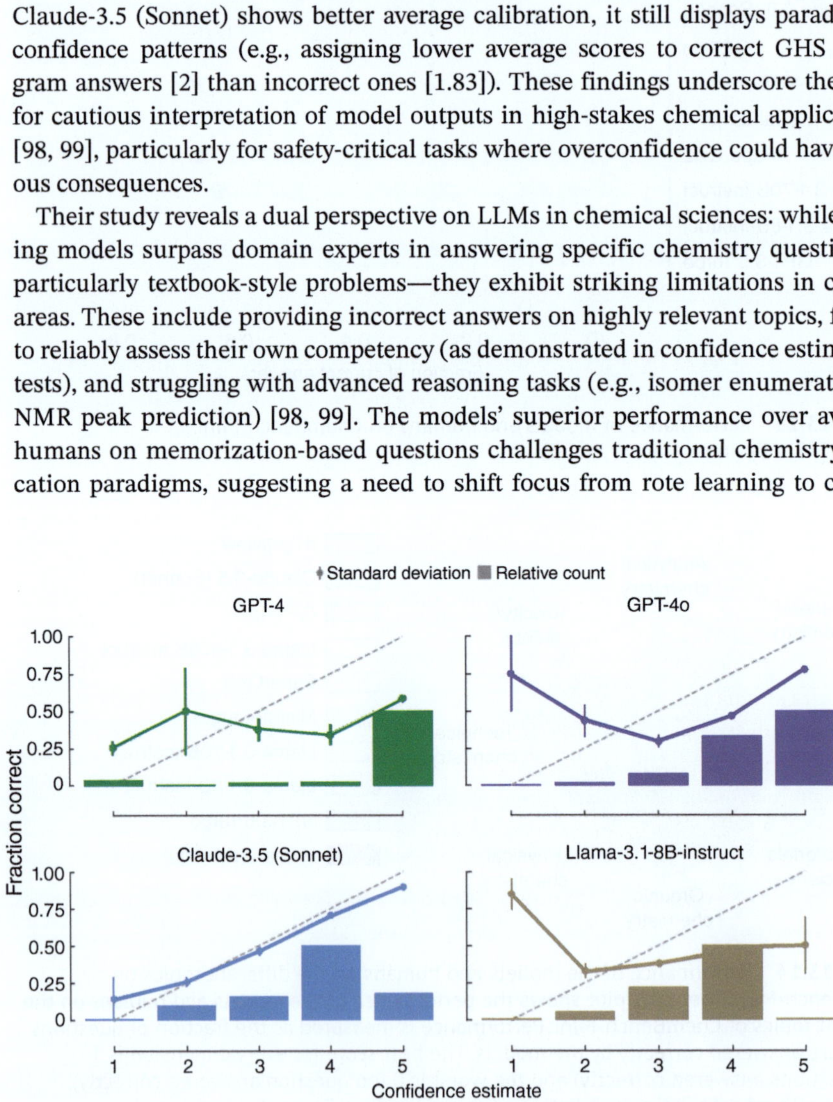

Figure 13.15 Reliability and distribution of confidence estimates. *Source*: [79] / Springer Nature / CC BY 4.0.

thinking skills. Furthermore, their analysis reveals significant performance variability across chemical subfields and question types, highlighting the limitations of current evaluation frameworks that primarily assess narrow property prediction tasks rather than scientific reasoning or real-world applicability. These findings underscore the necessity for more nuanced benchmarks that better reflect chemical reasoning demands, and improved human–model interaction frameworks to compensate for models' poor self-assessment capabilities—particularly crucial for safety-sensitive applications where overconfidence could have serious consequences [99].

Mirza et al.'s [100] research, through the "mirror" of ChemBench, clearly reflects LLMs' advantages and limitations in chemistry: they are efficient "knowledge transporters" but not yet qualified "scientific collaborators." Breakthroughs in professional AI depend not only on algorithmic progress but also on deep understanding of disciplinary essence—from knowledge representation to systematic adaptation of expert thinking patterns. This provides a concrete analytical dimension for discussing "how AI reshapes scientific discovery paradigms" and empirical foundations for constructing interdisciplinary research methodologies.

References

1 Chernyavskiy A, Ilvovsky D, Nakov P. Transformers: "The end of history" for natural language processing? In: Oliver N, Pérez-Cruz F, Kramer S, Read J, Lozano JA, editors. ECML PKDD 2021: Machine Learning and Knowledge Discovery in Databases, Lecture Notes in Computer Science, Vol. 12977. Cham: Springer; 2021. pp. 677–93. Available from: https://doi.org/10.1007/978-3-030-86523-8_41

2 Wang A, Pruksachatkun Y, Nangia N, Singh A, Michael J, Hill F, et al. SuperGLUE: A stickier benchmark for general-purpose language understanding systems. arXiv preprint arXiv:1905.00537; 2020. Available from: http://arxiv.org/abs/1905.00537

3 Adiwardana D, Luong MT, So DR, Hall J, Fiedel N, Thoppilan R, et al. Towards a human-like open-domain chatbot. arXiv preprint arXiv:2001.09977v3; 2020. Available from: https://arxiv.org/abs/2001.09977v3

4 Agüera y Arcas, B. Do large language models understand us? Daedalus. 2022;151(2):183–97. Available from: https://direct.mit.edu/daed/article/151/2/183/110604/Do-Large-Language-Models-Understand-Us

5 Radford A, Wu J, Child R, Luan D, Amodei D, Sutskever I. Language models are unsupervised multitask learners. OpenAI Blog. 2019;1(8):9.

6 Brown TB, Mann B, Ryder N, Subbiah M, Kaplan J, Dhariwal P, et al. Language models are few-shot learners. arXiv preprint arXiv:2005.14165v4; 2020. Available from: https://arxiv.org/abs/2005.14165v4

7 Peters ME, Neumann M, Iyyer M, Gardner M, Clark C, Lee K, et al. Deep contextualized word representations. arXiv preprint arXiv:1802.05365v2; 2018. Available from: https://arxiv.org/abs/1802.05365v2

8 Lewis M, Liu Y, Goyal N, Ghazvininejad M, Mohamed A, Levy O, et al. BART: Denoising sequence-to-sequence pre-training for natural language generation, translation, and comprehension. arXiv preprint arXiv:1910.13461; 2019. Available from: http://arxiv.org/abs/1910.13461

9 Devlin J, Chang MW, Lee K, Toutanova K. BERT: Pre-training of deep bidirectional transformers for language understanding. arXiv preprint arXiv:1810.04805; 2018.

10 Nigam A, Friederich P, Krenn M, Aspuru-Guzik A. Augmenting genetic algorithms with deep neural networks for exploring the chemical space. arXiv preprint arXiv:1909.11655; 2020. Available from: http://arxiv.org/abs/1909.11655

11 Raffel C, Shazeer N, Roberts A, Lee K, Narang S, Matena M, et al. Exploring the limits of transfer learning with a unified text-to-text transformer. The Journal of Machine Learning Research. 2020;21(1):5485–551. Available from: https://dl.acm.org/doi/10.5555/3455716.3455856

12 Xue L, Constant N, Roberts A, Kale M, Al-Rfou R, Siddhant A, et al. mT5: A massively multilingual pre-trained text-to-text transformer. arXiv preprint arXiv:2010.11934; 2021. Available from: https://arxiv.org/abs/2010.11934

13 Zhang Z, Gu Y, Han X, Chen S, Xiao C, Sun Z, et al. CPM-2: Large-scale cost-effective pre-trained language models. arXiv preprint arXiv:2106.10715; 2021. Available from: http://arxiv.org/abs/2106.10715

14 Le Scao T, Fan A, Akiki C, Pavlick E, Ilić S, Hesslow D, et al. BLOOM: A 176B-parameter open-access multilingual language model. arXiv preprint arXiv:2211.05100; 2023. Available from: http://arxiv.org/abs/2211.05100

15 Zhang S, Roller S, Goyal N, Artetxe M, Chen M, Chen S, et al. OPT: Open pre-trained transformer language models. arXiv preprint arXiv:2205.01068; 2022. Available from: http://arxiv.org/abs/2205.01068

16 Chowdhery A, Narang S, Devlin J, Bosma M, Mishra G, Roberts A, et al. PaLM: Scaling language modeling with pathways. arXiv preprint arXiv:2204.02311; 2022. Available from: http://arxiv.org/abs/2204.02311

17 Chowdhery A, Narang S, Devlin J, Bosma M, Mishra G, Roberts A, et al. PaLM: Scaling language modeling with pathways. The Journal of Machine Learning Research. 2023;24(1):11324–436. Available from: https://dl.acm.org/doi/10.5555/3648699.3648939

18 Brown N, Fiscato M, Segler MH, Vaucher AC. GuacaMol: Benchmarking models for de novo molecular design. Journal of Chemical Information and Modeling. 2019;59(3):1096–108. Available from: https://pubs.acs.org/doi/10.1021/acs.jcim.8b00839

19 Griffiths RR, Greenfield JL, Thawani AR, Jamasb AR, Moss HB, Bourached A, et al. Data-driven discovery of molecular photoswitches with multioutput Gaussian processes. Chemical Science. 2022;13(45):13541–51. Available from: https://pubs.rsc.org/en/content/articlehtml/2022/sc/d2sc04306h

20 Chung HW, Hou L, Longpre S, Zoph B, Tay Y, Fedus W, et al. Scaling instruction-finetuned language models. arXiv preprint arXiv:2210.11416; 2022. Available from: http://arxiv.org/abs/2210.11416

21 Sanh V, Webson A, Raffel C, Bach SH, Sutawika L, Alyafeai Z, et al. Multitask prompted training enables zero-shot task generalization. arXiv preprint arXiv:2110.08207; 2022. Available from: http://arxiv.org/abs/2110.08207

22 Wang Y, Mishra S, Alipoormolabashi P, Kordi Y, Mirzaei A, Arunkumar A, et al. Super-NaturalInstructions: Generalization via declarative instructions on 1600+ NLP tasks. arXiv preprint arXiv:2204.07705; 2022. Available from: http://arxiv.org/abs/2204.07705

23 Wang Y, Kordi Y, Mishra S, Liu A, Smith NA, Khashabi D, et al. Self-instruct: Aligning language models with self-generated instructions. arXiv preprint arXiv:2212.10560; 2023. Available from: http://arxiv.org/abs/2212.10560

24 Ouyang L, Wu J, Jiang X, Almeida D, Wainwright CL, Mishkin P, et al. Training language models to follow instructions with human feedback. arXiv preprint arXiv:2203.02155; 2022. Available from: http://arxiv.org/abs/2203.02155

25 Touvron H, Martin L, Stone K, Albert P, Almahairi A, Babaei Y, et al. Llama 2: Open foundation and fine-tuned chat models. arXiv preprint arXiv:2307.09288; 2023. Available from: http://arxiv.org/abs/2307.09288

26 Webb T, Holyoak KJ, Lu H. Emergent analogical reasoning in large language models. arXiv preprint arXiv:2212.09196; 2023. Available from: http://arxiv.org/abs/2212.09196

27 Wei J, Tay Y, Bommasani R, Raffel C, Zoph B, Borgeaud S, et al. Emergent abilities of large language models. arXiv preprint arXiv:2206.07682; 2022. Available from: http://arxiv.org/abs/2206.07682

28 Boiko DA, MacKnight R, Gomes G. Emergent autonomous scientific research capabilities of large language models. arXiv preprint arXiv:2304.05332; 2023. Available from: http://arxiv.org/abs/2304.05332

29 Parisi A, Zhao Y, Fiedel N. TALM: Tool augmented language models. arXiv preprint arXiv:2205.12255; 2022. Available from: http://arxiv.org/abs/2205.12255

30 Driess D, Xia F, Sajjadi MSM, Lynch C, Chowdhery A, Ichter B, et al. PaLM-E: An embodied multimodal language model. arXiv preprint arXiv:2303.03378; 2023. Available from: http://arxiv.org/abs/2303.03378

31 Izacard G, Lewis P, Lomeli M, Hosseini L, Petroni F, Schick T, et al. Atlas: Few-shot learning with retrieval augmented language models. arXiv preprint arXiv:2208.03299; 2022. Available from: http://arxiv.org/abs/2208.03299

32 Zhang B, Soh H. Large language models as zero-shot human models for human-robot interaction. In: 2023 IEEE/RSJ International Conference on Intelligent Robots and Systems (IROS) [Internet]. IEEE Press; 2023. pp. 7961–8. Available from: http://arxiv.org/abs/2303.03548

33 Ye Q, Xu H, Xu G, Ye J, Yan M, Zhou Y, et al. mPLUG-Owl: Modularization empowers large language models with multimodality. arXiv preprint arXiv:2304.14178; 2024. Available from: http://arxiv.org/abs/2304.14178

34 Wang W, Chen Z, Chen X, Wu J, Zhu X, Zeng G, et al. VisionLLM: Large language model is also an open-ended decoder for vision-centric tasks. arXiv preprint arXiv:2305.11175; 2023. Available from: http://arxiv.org/abs/2305.11175

35 Yang R, Song L, Li Y, Zhao S, Ge Y, Li X, et al. GPT4Tools: Teaching large language model to use tools via self-instruction. arXiv preprint arXiv:2305.18752; 2023. Available from: http://arxiv.org/abs/2305.18752

36 Canisius University. Prompt engineering guide—AI and society initiative at Canisius University [Internet]. 2023. Available from: https://blogs.canisius.edu/generativeai/prompt-engineering-guide/

37 Wang S, Sun Y, Xiang Y, Wu Z, Ding S, Gong W, et al. ERNIE 3.0 Titan: Exploring larger-scale knowledge enhanced pre-training for language understanding and generation. arXiv preprint arXiv:2112.12731. Available from: https://arxiv.org/abs/2112.12731

38 Wang Y, Le H, Gotmare AD, Bui NDQ, Li J, Hoi SCH. CodeT5+: Open code large language models for code understanding and generation. arXiv preprint arXiv:2305.07922; 2023. Available from: http://arxiv.org/abs/2305.07922

39 Rasley J, Rajbhandari S, Ruwase O, He Y. DeepSpeed: System optimizations enable training deep learning models with over 100 billion parameters. In: 26th ACM SIGKDD International Conference on Knowledge Discovery & Data Mining [Internet]. Association for Computing Machinery; 2020. pp. 3505–6. Available from: https://doi.org/10.1145/3394486.3406703

40 Rajbhandari S, Rasley J, Ruwase O, He Y. ZeRO: Memory optimizations toward training trillion parameter models. In: IEEE International Conference for High Performance Computing, Networking, Storage and Analysis [Internet]. IEEE Press; 2020. pp. 1–16. Available from: https://ieeexplore.ieee.org/abstract/document/9355301

41 He J, Zhou C, Ma X, Berg-Kirkpatrick T, Neubig G. Towards a unified view of parameter-efficient transfer learning. arXiv preprint arXiv:2110.04366; 2022. Available from: http://arxiv.org/abs/2110.04366

42 Hu Z, Wang L, Lan Y, Xu W, Lim EP, Bing L, et al. LLM-Adapters: An adapter family for parameter-efficient fine-tuning of large language models. arXiv preprint arXiv:2304.01933; 2023. Available from: http://arxiv.org/abs/2304.01933

43 Lester B, Al-Rfou R, Constant N. The power of scale for parameter-efficient prompt tuning. arXiv preprint arXiv:2104.08691; 2021. Available from: http://arxiv.org/abs/2104.08691

44 Li XL, Liang P. Prefix-tuning: Optimizing continuous prompts for generation. arXiv preprint arXiv:2101.00190; 2021. Available from: http://arxiv.org/abs/2101.00190

45 Ma X, Fang G, Wang X. LLM-Pruner: On the structural pruning of large language models. arXiv preprint arXiv:2305.11627; 2023. Available from: http://arxiv.org/abs/2305.11627

46 Xu R, Luo F, Wang C, Chang B, Huang J, Huang S, et al. From dense to sparse: Contrastive pruning for better pre-trained language model compression. arXiv preprint arXiv:2112.07198; 2021. Available from: http://arxiv.org/abs/2112.07198

47 Xiao G, Lin J, Seznec M, Wu H, Demouth J, Han S. SmoothQuant: Accurate and efficient post-training quantization for large language models. arXiv preprint arXiv:2211.10438v7; 2022. Available from: https://arxiv.org/abs/2211.10438v7

48 Tao C, Hou L, Zhang W, Shang L, Jiang X, Liu Q, et al. Compression of generative pre-trained language models via quantization. arXiv preprint arXiv:2203.10705v2; 2022. Available from: https://arxiv.org/abs/2203.10705v2

49 Pal A, Karkhanis D, Roberts M, Dooley S, Sundararajan A, Naidu S. Giraffe: Adventures in expanding context lengths in LLMs. arXiv preprint arXiv:2308.10882v1; 2023. Available from: https://arxiv.org/abs/2308.10882v1

50 Peng B, Quesnelle J, Fan H, Shippole E. YaRN: Efficient context window extension of large language models. arXiv preprint arXiv:2309.00071v2; 2023. Available from: https://arxiv.org/abs/2309.00071v2

51 Guo M, Ainslie J, Uthus D, Ontanon S, Ni J, Sung YH, et al. LongT5: Efficient text-to-text transformer for long sequences. arXiv preprint arXiv:2112.07916v2; 2021. Available from: https://arxiv.org/abs/2112.07916v2

52 Chen S, Wong S, Chen L, Tian Y. Extending context window of large language models via positional interpolation. arXiv preprint arXiv:2306.15595v2; 2023. Available from: https://arxiv.org/abs/2306.15595v2

53 Fedus W, Zoph B, Shazeer N. Switch transformers: Scaling to trillion parameter models with simple and efficient sparsity. arXiv preprint arXiv:2101.03961v3; 2021. Available from: https://arxiv.org/abs/2101.03961v3

54 Du N, Huang Y, Dai AM, Tong S, Lepikhin D, Xu Y, et al. GLaM: Efficient scaling of language models with Mixture-of-Experts. arXiv preprint arXiv: 2112.06905v2; 2021. Available from: https://arxiv.org/abs/2112.06905v2

55 Ren X, Zhou P, Meng X, Huang X, Wang Y, Wang W, et al. PanGu-Σ: Towards trillion parameter language model with sparse heterogeneous computing. arXiv preprint arXiv:2303.10845v1; 2023. Available from: https://arxiv.org/abs/2303.10845v1

56 Wang T, Roberts A, Hesslow D, Scao TL, Chung HW, Beltagy I, et al. What language model architecture and pretraining objective work best for zero-shot generalization? arXiv preprint arXiv:2204.05832; 2022. Available from: http://arxiv.org/abs/2204.05832

57 Dong L, Yang N, Wang W, Wei F, Liu X, Wang Y, et al. Unified Language Model pre-training for natural language understanding and generation. arXiv preprint arXiv:1905.03197; 2019. Available from: http://arxiv.org/abs/1905.03197

58 QwenTeam. Qwen2.5-LLM: Extending the boundary of LLMs [Internet]. 2024. Available from: https://qwenlm.github.io/blog/qwen2.5-llm/

59 Hendrycks D, Burns C, Kadavath S, Arora A, Basart S, Tang E, et al. Measuring mathematical problem solving with the MATH dataset. arXiv preprint arXiv:2103.03874; 2021. Available from: http://arxiv.org/abs/2103.03874

60 Rein D, Hou BL, Stickland AC, Petty J, Pang RY, Dirani J, et al. GPQA: A graduate-level Google-proof Q&A benchmark. In: First Conference on Language Modeling [Internet]. OpenReview; 2024. pp. 1–31. Available from: https://openreview.net/forum?id=Ti67584b98

61 Sprague Z, Ye X, Bostrom K, Chaudhuri S, Durrett G. MuSR: Testing the limits of chain-of-thought with multistep soft reasoning. arXiv preprint arXiv:2310.16049; 2024. Available from: http://arxiv.org/abs/2310.16049

62 Wang L, Chen S, Jiang L, Pan S, Cai R, Yang S, et al. Parameter-efficient fine-tuning in large language models: a survey of methodologies. Artificial Intelligence Review. 2025;58(8):227. Available from: https://doi.org/10.1007/s10462-025-11236-4

63 Jablonka KM, Schwaller P, Ortega-Guerrero A, Smit B. Leveraging large language models for predictive chemistry. Nature Machine Intelligence. 2024;6(2):161–9. Available from: https://doi.org/10.1038/s42256-023-00788-1

64 Dubbeldam D, Calero S, Vlugt TJ. iRASPA: GPU-accelerated visualization software for materials scientists. Molecular Simulation. 2018;44(8):653–76. Available from: https://doi.org/10.1080/08927022.2018.1426855

65 Que H, Li B, Sun L, Peng Z, Huaning J, Xingguo W et al. Synthesis of two-dimensional transition metal phosphorous chalcogenides and their high-entropy alloys. Nature Synthesis. 2025;4,582–591. https://doi.org/10.1038/s44160-025-00737-3

66 Xie Z, Evangelopoulos X, Omar OH, Troisi A, Cooper AI, Chen L. Fine-tuning GPT-3 for machine learning electronic and functional properties of organic molecules. Chemical Science. 2024;15(2):500–510.

67 Pei Z, Yin J, Hawk JA, Alman DE, Gao MC. Machine-learning informed prediction of high-entropy solid solution formation: Beyond the Hume-Rothery rules. npj Computational Materials. 2020;6(1):50. Available from: https://doi.org/10.1038/s41524-020-0308-7

68 Dunn A, Wang Q, Ganose A, Dopp D, Jain A. Benchmarking materials property prediction methods: The Matbench test set and Automatminer reference algorithm. npj Computational Materials. 2020;6(1):138. Available from: https://doi.org/10.1038/s41524-020-00406-3

69 Le TT, Fu W, Moore JH. Scaling tree-based automated machine learning to biomedical big data with a feature set selector. Bioinformatics. 2020;36(1):250–6. Available from: https://doi.org/10.1093/bioinformatics/btz470

70 Wang AYT, Kauwe SK, Murdock RJ, Sparks TD. Compositionally restricted attention-based network for materials property predictions. npj Computational Materials. 2021;7(1):77. Available from: https://doi.org/10.1038/s41524-021-00545-1

71 GitHub. RDKit: Open-source cheminformatics software [Internet]. 2025. Available from: http://www.rdkit.org/

72 Preuer K, Renz P, Unterthiner T, Hochreiter S, Klambauer G. Fréchet ChemNet distance: A metric for generative models for molecules in drug discovery. Journal of Chemical Information and Modeling. 2018;58(9):1736–41. Available from: https://doi.org/10.1021/acs.jcim.8b00234

73 Probst D, Reymond JL. Visualization of very large high-dimensional data sets as minimum spanning trees. Journal of Cheminformatics. 2020;12(1):12. Available from: https://doi.org/10.1186/s13321-020-0416-x

74 Probst D, Reymond JL. A probabilistic molecular fingerprint for big data settings. Journal of Cheminformatics. 2018;10(1):66. Available from: https://doi.org/10.1186/s13321-018-0321-8

75 Moosavi SM, Novotny BÁ, Ongari D, Moubarak E, Asgari M, Kadioglu Ö et al. A data-science approach to predict the heat capacity of nanoporous materials. Nature Materials. 2022;21:1419–1425.

76 Isert C, Atz K, Jiménez-Luna J, Schneider G. QMugs, quantum mechanical properties of drug-like molecules. Scientific Data. 2022;9(1):273. Available from: https://doi.org/10.1038/s41597-022-01390-7

77 Jablonka KM, Jothiappan GM, Wang S, Smit B, Yoo B. Bias free multiobjective active learning for materials design and discovery. Nature Communications. 2021;12(1):2312. Available from: https://doi.org/10.1038/s41467-021-22437-0

78 Westermayr J, Gilkes J, Barrett R, Maurer RJ. High-throughput property-driven generative design of functional organic molecules. Nature Computational Science. 2023;3(2):139–48. Available from: https://doi.org/10.1038/s43588-022-00391-1

79 Mirza A, Alampara N, Kunchapu S, Ríos-García M, Emoekabu B, Krishnan A, et al. A framework for evaluating the chemical knowledge and reasoning abilities of large language models against the expertise of chemists. Nature Chemistry. 2025;17:1027–34. Available from: https://doi.org/10.1038/s41557-025-01815-x

80 Srivastava A, Rastogi A, Rao A, Shoeb AAM, Abid A, Fisch A, et al. Beyond the imitation game: Quantifying and extrapolating the capabilities of language models. arXiv preprint arXiv:2206.04615; 2023. Available from: http://arxiv.org/abs/2206.04615

81 Sutawika L, Gao L, Schoelkopf H, Biderman S, Tow J, Abbasi B, et al. EleutherAI/lm-evaluation-harness: Major refactor [Internet]. Zenodo; 2023. Available from: https://zenodo.org/records/10256836

82 Ahmad W, Simon E, Chithrananda S, Grand G, Ramsundar B. ChemBERTa-2: Towards chemical foundation models. arXiv preprint arXiv:2209.01712; 2022. Available from: http://arxiv.org/abs/2209.01712

83 Cai X, Lai H, Wang X, Wang L, Liu W, Wang Y, et al. Comprehensive evaluation of molecule property prediction with ChatGPT. Methods. 2024;222:133–41.

84 Frey NC, Soklaski R, Axelrod S, Samsi S, Gomez-Bombarelli R, Coley CW, et al. Neural scaling of deep chemical models. Nature Machine Intelligence. 2023;5(11):1297–1305. Available from: https://doi.org/10.1038/s42256-023-00740-3

85 Dinh T, Zeng Y, Zhang R, Lin Z, Gira M, Rajput S, et al. LIFT: Language-interfaced fine-tuning for non-language machine learning tasks. In: Koyejo S, Mohamed S, Agarwal A, Belgrave D, Cho K, Oh A, editors. Advances in Neural Information Processing Systems, Vol. 36. New York: Curran Associates, Inc.; 2022. pp. 11763–84.

86 Wu Z, Ramsundar B, Feinberg EN, Gomes J, Geniesse C, Pappu AS. MoleculeNet: a benchmark for molecular machine learning. Chemical Science. 2018;9(2): 513–30. Available from: https://doi.org/10.1039/C7SC02664A

87 Huang Y, Zhang R, He X, Zhi X, Wang H, Li X, et al. ChemEval: A comprehensive multi-level chemical evaluation for large language models. arXiv preprint arXiv:2409.13989; 2024. Available from: http://arxiv.org/abs/2409.13989

88 Huang K, Fu T, Gao W, Zhao Y, Roohani Y, Leskovec J, et al. Therapeutics Data Commons: Machine learning datasets and tasks for drug discovery and development. arXiv preprint arXiv:2102.09548; 2021. Available from: http://arxiv.org/abs/2102.09548

89 Zhao H, Tang X, Yang Z, Han X, Feng X, Fan Y, et al. ChemSafetyBench: Benchmarking LLM safety on chemistry domain. arXiv preprint arXiv:2411.16736; 2024. Available from: http://arxiv.org/abs/2411.16736

90 Hu X, Chen J, Li X, Guo Y, Wen L, Yu PS, et al. Towards understanding factual knowledge of large language models [Internet]. ICLR; 2023. pp. 1–36. Available from: https://openreview.net/forum?id=9OevMUdods

91 Zhang J, Lehman J, Stanley K, Clune J. OMNI: Open-endedness via models of human notions of interestingness. arXiv preprint arXiv:2306.01711; 2024. Available from: http://arxiv.org/abs/2306.01711

92 Liang P, Bommasani R, Lee T, Tsipras D, Soylu D, Yasunaga M, et al. Holistic evaluation of language models. arXiv preprint arXiv:2211.09110; 2023. Available from: http://arxiv.org/abs/2211.09110

93 Taylor R, Kardas M, Cucurull G, Scialom T, Hartshorn A, Saravia E, et al. Galactica: A large language model for science. arXiv preprint arXiv:2211.09085; 2022.

94 Schick T, Dwivedi-Yu J, Dessì R, Raileanu R, Lomeli M, Hambro E, et al. Toolformer: language models can teach themselves to use tools. In: 37th International Conference on Neural Information Processing Systems [Internet]. Curran Associates, Inc.; 2023. pp. 68539–51. Available from: https://proceedings.neurips.cc/paper_files/paper/2023/hash/d842425e4bf79ba039352da0f658a906-Abstract-Conference.html

95 Karpas E, Abend O, Belinkov Y, Lenz B, Lieber O, Ratner N, et al. MRKL systems: A modular, neuro-symbolic architecture that combines large language models, external knowledge sources and discrete reasoning. arXiv preprint arXiv:2205.00445; 2022. Available from: http://arxiv.org/abs/2205.00445

96 Yao S, Zhao J, Yu D, Du N, Shafran I, Narasimhan K, et al. ReAct: Synergizing reasoning and acting in language models. arXiv preprint arXiv:2210.03629; 2023. Available from: http://arxiv.org/abs/2210.03629

97 Xiong M, Hu Z, Lu X, Li Y, Fu J, He J, et al. Can LLMs express their uncertainty? An empirical evaluation of confidence elicitation in LLMs. arXiv preprint arXiv:2306.13063; 2024. Available from: http://arxiv.org/abs/2306.13063

98 Miret S, Krishnan NMA. Are LLMs ready for real-world materials discovery? arXiv preprint arXiv:2402.05200; 2024. Available from: http://arxiv.org/abs/2402.05200

99 Li B, Qi P, Liu B, Di S, Liu J, Pei J, et al. Trustworthy AI: From principles to practices. ACM Computing Surveys. 2023;55(9):177:1–46.

100 Mirza A, Alampara N, Kunchapu S, Ríos-García M, Schilling-Wilhelmi M, Aneesh A et al. chem-bench version v0.2.0. Zenodo; 2024. https://doi.org/10.5281/zenodo.14010212

14

Materials Agents for Autonomous Research

14.1 Introduction

As a strategic and basic industry, new materials are an important support for modern industrial system and new industrialization, and an important industrial direction to accelerate the development of new quality productive forces and solely promote high-quality development [1–4]. However, for a long time, the research and development of new materials have relied on theoretical exploration and experience accumulation, which needs to undergo repeated experiments, resulting in long research and development cycles and high costs. "AI + materials" empowers through data-driven professional large models, greatly shortens the research and development cycle of new materials, reduces research and development costs, and significantly improves research and development efficiency [1–6]. The effective computing power of technology in dealing with high-dimensional and cross-scale complex system problems can also realize the prediction, simulation, and analysis of materials composition, structure, and performance, and empower research and development, manufacturing, and application, which is of great significance for scientific and technological innovation and industrial development in the field of materials [7–11].

Principles governing autonomy of agents, search of hypothesis spaces, knowledge, sequential improvement, and statistical models in scientific discovery systems had already been articulated more than two decades ago. In Figure 14.1, the development trend of autonomous agents based on LLMs is presented. For materials, autonomous discovery is being fueled by two concurrent developments: the need for new technological materials and the progress in AI [12]. Recent advances in rapid generation of materials data have created a new frontier where computational and experimental materials research is intersecting with AI [7, 13–16]. This new wave of AI in materials has a theme of acceleration, enabled either by bypassing tools or methods via surrogate models, or by identification of new materials via adaptive schemes that combine models with decision-making approaches.

To accelerate research efforts, traditional human-centric experimentation has a significant opportunity to undergo a paradigm shift. Autonomous materials research

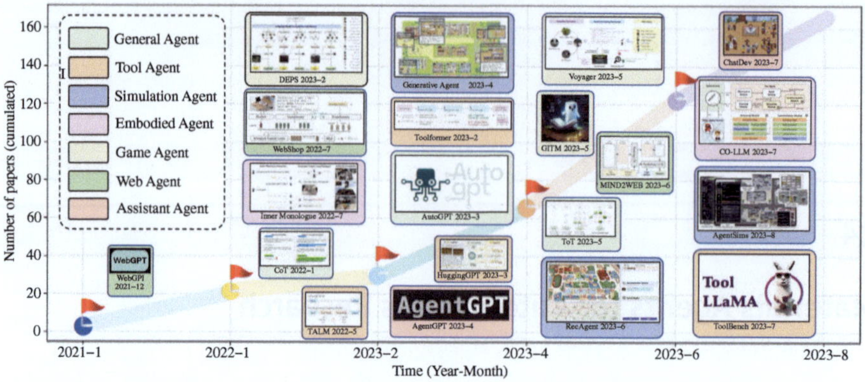

Figure 14.1 The growth trend in the field of LLM-based autonomous agents. *Source*: [17] / Springer Nature / CC BY 4.0.

systems enable scientists to fail smarter, learn faster, and spend fewer resources in their research. Nowadays, successful integration of LLMs into laboratory workflows has demonstrated their power for NLP, autonomous task execution, and collaborative problem-solving [16, 18–21]. In the world of materials, this can mean designing, developing, and deploying automated experimental platforms that perform synthesis and characterization, with the goal of using AI and ML to optimize a specific material property, either through a more efficient "exploratory" method, or through capabilities to explore new design space that was previously not considered. This chapter overviews the materials agents for autonomous research, covering their system architectures—including LLM-based robotic lab integration and heterogeneous system collaboration mechanisms. We detail core technical modules, including natural language interfaces for equipment control, Transformer models for experimental planning, and reinforcement learning for iterative optimization.

14.2 Architecture of Autonomous Materials Research Agents

The main framework of the autonomous materials research agent system needs to break the traditional linear process, build a dynamic closed-loop framework of perception, decision, and execution, and realize data-driven intelligent iteration. As shown in Figure 14.2, the framework is oriented to materials science problems, and integrates artificial intelligence, robotics, and automation equipment to form a research system with autonomous evolution ability. A large model-based AI Agent system can be divided into four components: LLM, planning, memory, and tool use. AI Agent may become the beginning of a new era, and its infrastructure can be simply divided as: Agent = LLM + planning skills + memory + tool usage, where LLM acts as the brain of the Agent, providing reasoning and planning in this system [17, 22].

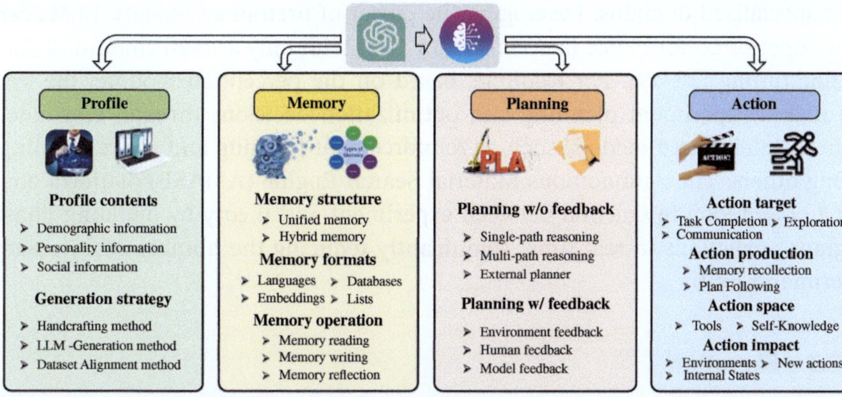

Figure 14.2 The main framework of the autonomous materials research autonomous system. *Source*: [17] / Springer Nature / CC BY 4.0.

14.2.1 Perception

The core purpose of the perception module is to extend the perceptual space of the subject from the text-only domain to textual, auditory, and visual modalities [23, 24]. Through various sensors or interfaces, information from the external environment is collected, which can be text, image, sound, video, and other forms. The collected raw data is also preliminarily processed to improve data quality and facilitate subsequent analysis and understanding. Nowadays, LLM-based agents have the basic ability to communicate with humans through text input and output. Meanwhile, instruction-tuned LLMs can exhibit superior zero-shot instruction understanding and generalization, eliminating the need for task-specific fine-tuning. On this basis, LLM-based agents are also widely used in materials science. For example, autonomous systems utilize high-throughput sensors and online characterization devices, such as X-ray diffractometers and spectrometers, to collect experimental data in real time, build a comprehensive perception of the material state, and significantly accelerate the speed of material discovery [25].

14.2.2 Decision-making

Similar to the human brain, it is mainly composed of an LLM, which not only stores knowledge and memory, but also undertakes functions such as information processing and decision-making, and can present the process of reasoning and planning, which can deal with unknown tasks well. After receiving the information processed by the perception module, the brain module first turns to storage, retrieving in knowledge from memory. These results help the Agent to plan, reason, and make decisions. In addition, the module is capable of memorizing the observations, thoughts, and actions of the Agent in the form of summaries, vectors, or other data structures. At the same time, it can also update knowledge such as common sense and domain knowledge for future use [17]. In addition to this, LLM-based agents can also exploit their inherent generalization and transfer capabilities to delve into other specialized domains. Leveraging the power of pretrained models, LLMs can show superior performance in downstream tasks with only a small amount of data for fine-tuning [26–29]. For example, based on the perception module, the system makes experiment planning and optimization decisions through embedded artificial intelligence models, such as reinforcement learning and active learning, among others. The Autonomous Material Search Engine (AMASE) platform enables a closed-loop interaction between experiment and theory by updating phase diagram predictions in real time, significantly reducing the number of necessary experiments [30].

14.2.3 Execution

After perceiving the environment, the human brain will integrate, analyze, reason, and make decisions on the perceived information. Subsequently, they use the nervous system to control their bodies to make adaptive or creative actions. When an Agent has a brain-like structure with knowledge, memory, reasoning, planning, and generalization capabilities, and multimodal perception, it is also expected to have a variety of human-like actions to cope with its surroundings. During the construction of the Agent, the execution module receives the sequence of actions sent by the brain module and performs actions that interact with the environment.

The rise and development of large Transformer-based language generation models have endowed LLM-based agents with inherent language generation capabilities. The quality of the text they generate is outstanding in various aspects such as fluency, relevance, diversity, and controllability. As a result, LLM-based agents can be exceptionally powerful language generators. When faced with complex tasks, humans use tools to simplify the task-solving process and improve efficiency, thus saving time and resources. Similarly, if agents also learn to use and exploit tools, it is possible to complete complex tasks more efficiently and with higher quality.

Specialization tools allow LLMs to enhance their expertise, adapt their domain knowledge, and better fit the needs of specific domains. LLM-based agents not only require the use of tools, but are also well-suited for tool integration. Using the rich world knowledge accumulated through the pretraining process and chain-of-thought (CoT) prompts [31, 32], LLM has shown remarkable reasoning and decision-making capabilities in complex interactive environments, which helps the Agent to decompose and process the user-specified tasks in an appropriate manner. When agents are combined with tools, the barrier to tool use can be lowered, so as to fully unleash the creative potential of human users. For example, the decision results are quickly implemented through an automated experimental platform, such as a robot synthesis system, to form new experimental data, which are fed back to the perception module to complete the closed loop. This efficient execution mechanism greatly shortens the experiment cycle and improves the research efficiency [30].

In other words, an AI Agent is equipped to plan, make decisions, and call up tools. Its role is defined through prompt engineering so that users can customize the Agent's behavior to meet specific requirements. In-context learning allows an AI Agent to accumulate experience and evolve so that its actions become increasingly consistent, logical, and effective over time. Techniques such as CoT and tree-of-thought approaches allow to decompose complex tasks into smaller, more manageable subtasks. By interacting with external tools, the plans devised by AI Agents are translated into actionable steps and executed effectively. Compared with traditional experimental systems, the autonomous materials research agent system has significant advantages in experimental efficiency, data processing ability, and decision-making intelligence. Traditional systems rely on manual operation, long experiment cycle, and data processing lag, and it is difficult to achieve real-time optimization. Through the closed-loop architecture, the autonomous system realizes the automation and intelligence of the experimental process, which significantly improves the efficiency and quality of material research.

14.2.4 Integration Framework of LLMs and Robotic Laboratories

In the field of materials science research, the integration of LLMs and robotics laboratories is reshaping the traditional research and development paradigm. This subsection will deeply analyze the integration mechanism of the two, explore how to achieve seamless connection of data, knowledge, and physical experiments through technological collaboration, and build an efficient and intelligent materials research system.

14.2.4.1 The Role of LLMs in Agent

Throughout the entire workflow of the AI Agent, the Agent receives environmental information, writes it into short-term memory, then processes and analyzes it, generates an action plan, and finally the execution component executes the plan and makes adjustments and optimizations based on environmental feedback. LLMs with their outstanding NLP capabilities and knowledge reasoning mechanisms, have become highly potential intelligent tools in the field of materials science research [22, 33]. With the Generative Pretrained Transformer (GPT) series models [34], as well as open-source LLMs, such as Meta AI [35, 36], LLaMA [27, 29, 37], Falcon [38], and others, being typical representatives, their core functions are deeply integrated into the entire life cycle of materials research and development, as shown in Figure 14.3 [39, 40].

In the literature analysis stage, although materials data can be systematically generated through computational and physical experiments, a substantial amount of historical data is trapped in published literature. An efficient data is continually released in scientific journal articles, but this data frequently exists in unstructured natural language text formats, posing challenges for immediate utilization by modern informatics that rely on the availability of structured datasets. LLMs can quickly sort out a vast number of academic papers and patents, identify key research achievements, technical bottlenecks, and emerging trends [41–43]. For example, Gupta et al. [44] presented an approach to employing LLM- and NER-based pipelines, specifically designed to automate the extraction of property data of polymers from the full-text contents of journal articles as shown in Figure 14.4. Leveraging a corpus of 2.4 million full-text articles, their method successfully identified and processed around 681,000 polymer-related articles, resulting in the extraction of over one million records corresponding to 24 properties of over 106,000 unique polymers. The potential of LLMs lies in their ability to produce not just structured outputs such as property data extraction, but also assist with property predictions, material design, and synthesis recipe recommendations.

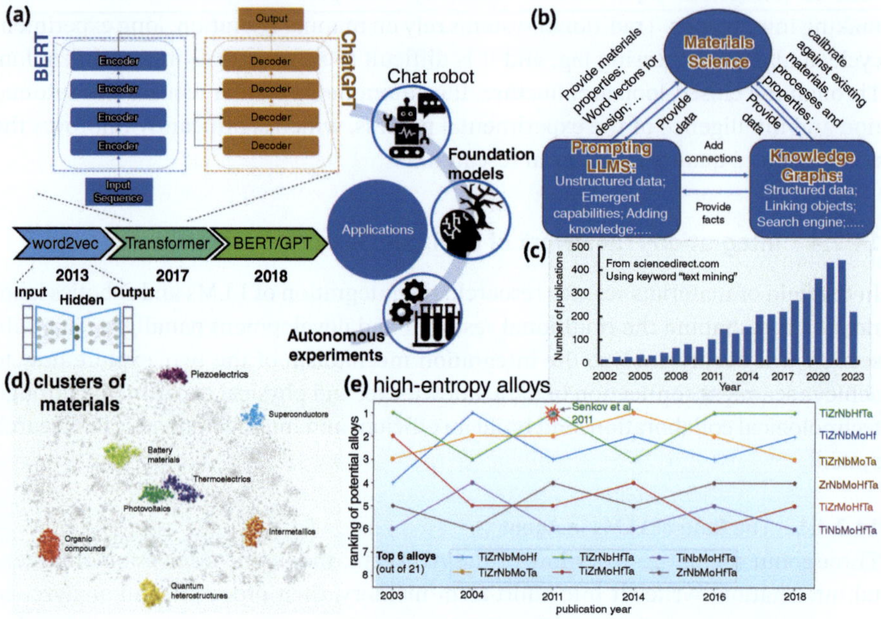

Figure 14.3 LLMs are deeply integrated into the entire cycle of materials research and development. (a) The structures of encoder and decoder. (b) The overview of LLM in materials research. (c) The analysis of word frequency "text mining." (d) The clusters of materials. (e) The analysis of ranking of potential of alloys. *Source*: [45] / with permission of Elsevier.

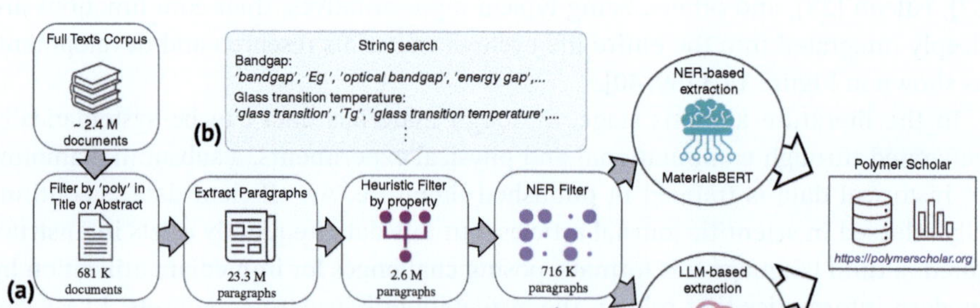

Figure 14.4 Overall workflow to extract polymer property data by LLMs. (a) The source of the corpus content. (b) Property-specific paragraphs are selected by a heuristic filter based on string matching and dictionary lookup. (c) Extractable named entities. *Source*: [44] / with permission of Springer Nature.

In terms of hypothesis generation and experimental scheme design, scientific hypotheses, which contain predictions, explanations, and insights to be verified, are the key early steps of scientific discovery. Based on the Bit-flip Attack, Stanford University in the United States once proposed the Bit-flip method: that is, first identify the hypothesis (Bit) commonly contained in the existing methods in the research field, and then propose a new idea different from this hypothesis (Flip) [46]. LLMs based on the pretrained knowledge system and user input requirements, realize the automatic generation of innovative research hypotheses through the semantic understanding framework and logical reasoning algorithms. O'Neill et al. [47] centered on the idea of combining LLM with the aforementioned Bit-flip method to generate new scientific hypotheses as shown in Figure 14.5. Based on a large number of conference papers in the field of computer science, the authors used OpenAI's Chatgpt-o1 model to extract more than 5,000 sets of Bit, Flip, Spark, and chain-of-reasoning information, forming the HypoGen dataset. Among them, the definitions of Bit and Flip information are as described above. The Spark information contains a summary of the core insights of a paper, and the chain-of-reasoning information contains a description of the thinking process from Bit to Flip generated based on the full text of a paper. Finally, the LLaMA model fine-tuned by the HypoGen dataset can generate novel insights and descriptions of the thinking process during reasoning based on the input Bit information.

The performance of a language model largely depends on the quality of the dataset used for training. Therefore, the creation of datasets is a key factor in promoting progress in various deep-learning fields such as computer vision, graph learning, and NLP [48, 49]. Especially in the field of materials science, the high variability of text in specific domains and the development of multimodal language models require the development of datasets that combine other modalities such as graphs, tables, and images. In this way, scientific information can be represented in multiple

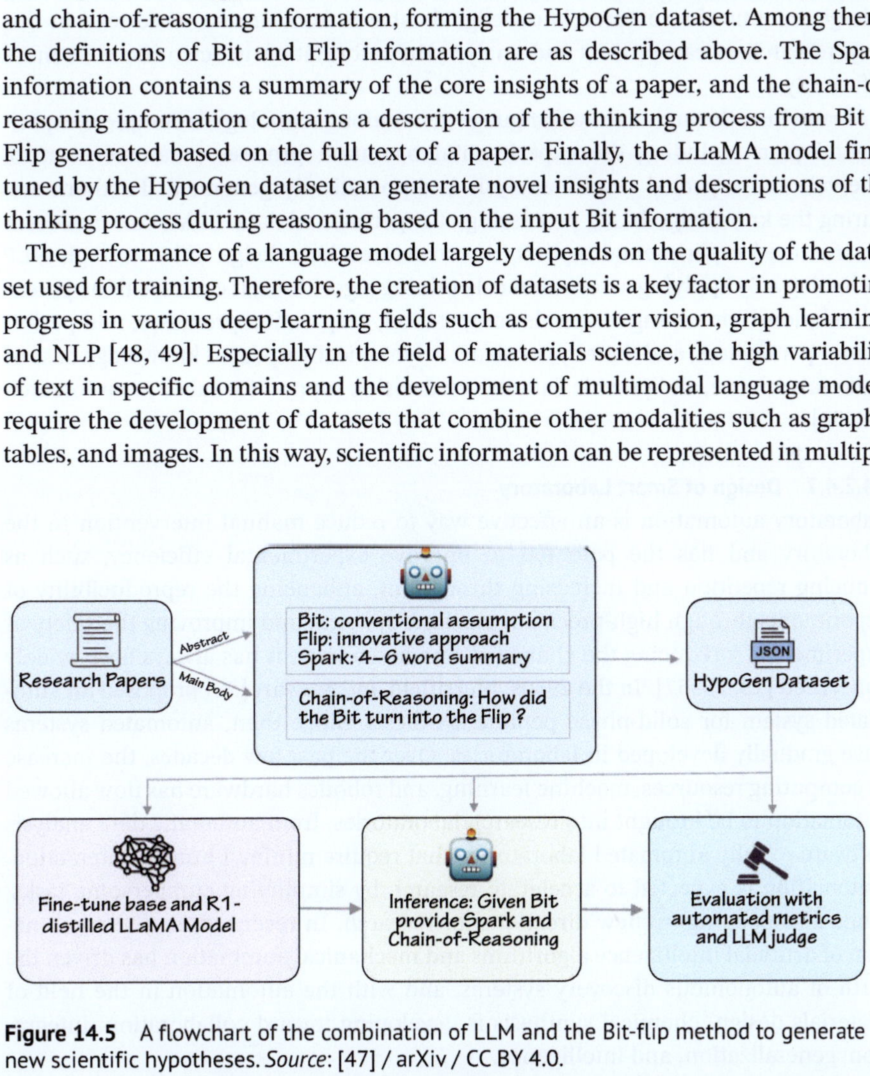

Figure 14.5 A flowchart of the combination of LLM and the Bit-flip method to generate new scientific hypotheses. *Source*: [47] / arXiv / CC BY 4.0.

patterns, thereby establishing a more powerful language model [50–52]. Based on the professional knowledge base and user requirements, LLMs are utilized to conduct semantic parsing of existing literature, patents, experimental reports, etc., and construct a database including components, structure, performance, and process. For instance, in 2024, Prof. Su and his team [7] from the University of Science and Technology Beijing proposed a prediction framework named SteelBert based on an LLM. This model was pretrained on a comprehensive corpus containing 4.2 million abstractions related to materials science and 55,000 full texts in the steel field, and is capable of accurately expressing textual knowledge of steel materials. Besides, Dagdelen et al. [53] presented a simple approach to joint named entity recognition and relation extraction and demonstrated how pretrained LLMs (GPT-3, Llama-2) can be fine-tuned to extract useful records of complex scientific knowledge. This method represents a simple, accessible, and highly flexible approach to obtaining a large amount of structured knowledge databases from unstructured text. In this way, researchers can process and analyze the information in scientific texts more efficiently.

Meanwhile, through natural language generation technology, a complete experimental plan including the experimental operation process, parameter control interval, and expected results analysis is automatically generated. Furthermore, during the knowledge-based processing of experimental data, LLMs can transform unstructured experimental reports into structured knowledge units through NLP technology. By applying association rule mining algorithms, it can reveal the potential relationships among materials composition, preparation processes, and performance parameters, and thereby construct a dynamically updated knowledge base of materials science, and provide data-driven decision support for subsequent research [44, 54].

14.2.4.2 Design of Smart Laboratory

Laboratory automation is an effective way to reduce manual intervention in the laboratory and has the potential to improve experimental efficiency, such as reducing repetition and increasing throughput, enhancing the reproducibility of experiments through high-precision machine systems, and improving the safety of experiments by reducing the chance of human contact. It has always been widely concerned [13, 55–57]. In the 1960s, Merrifield and Stewart [58] proposed an automated system for solid-phase peptide synthesis. Since then, automated systems have gradually developed in laboratories. Over the past few decades, the increase in computing resources, machine learning, and robotics hardware has now allowed automation to be brought into research laboratories: from automatic data analysis software to fully automated laboratories that require minimal human interaction. Automation is expected to accelerate research by simplifying cumbersome tasks, while also opening up new directions for research. In recent years, the combination of artificial intelligence algorithms and mechanical automation has driven the birth of autonomous discovery systems, and with the automation in the field of materials design, chemical synthesis is developing toward collaboration, integration, generalization, and intelligence.

In 2022, Shang and coworkers [18] from the University of Science and Technology of China established an all-round artificial intelligence chemistry laboratory, which can (1) propose scientific hypotheses and generate experimental plans using existing knowledge, and (2) carry out complete experimental processes (synthesis, characterization, and performance testing) for various chemical tasks; (3) a prediction model is established by using theoretical calculations and experimental data feedback to conduct all-round chemical research intelligently. Its overall architecture is shown in Figure 14.6. This artificial intelligence chemistry laboratory consists of three modules, including the machine reading module that acquires existing chemical knowledge by automatically reading a large number of publications in chemical literature, the mobile robot module that generates experimental data by performing various chemical experiments, and the computational brain module that generates physical/theory-based prediction models by conducting theoretical calculations.

Along this line, to build an intelligent AI laboratory, it is first necessary to establish a reliable and controllable physical platform. This platform is typically composed of automated synthesis equipment, sensor networks, actuators, and mobile

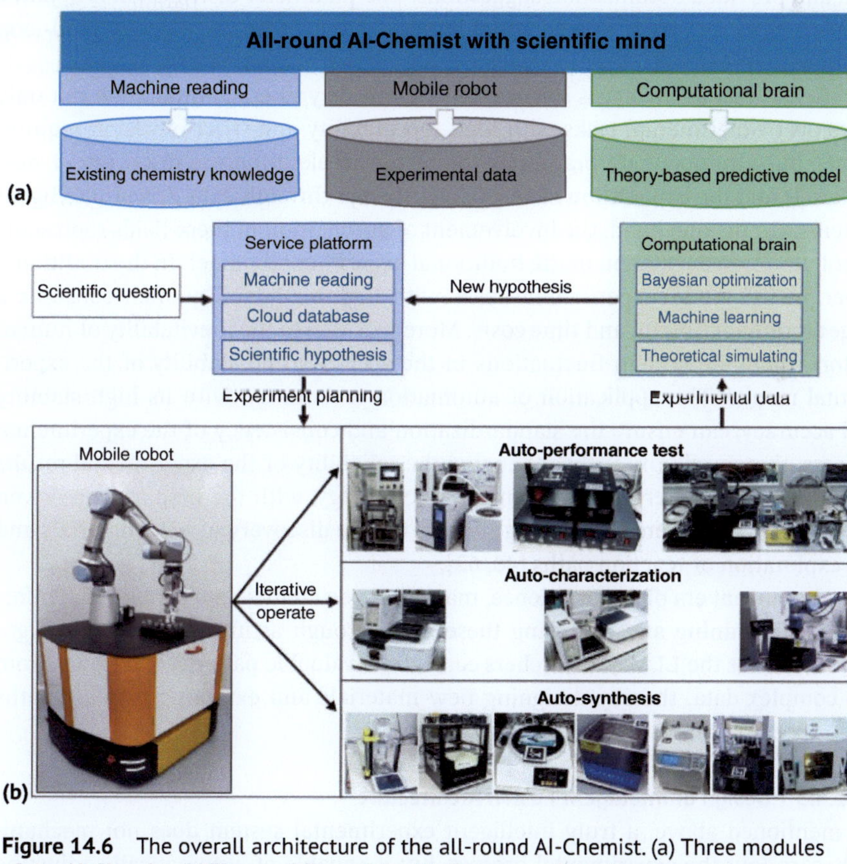

Figure 14.6 The overall architecture of the all-round AI-Chemist. (a) Three modules of the AI-Chemist. (b) The workflow of the AI-Chemist and the functions of each module. *Source*: [18] / Oxford University Press / CC BY 4.0.

platforms [40, 59]. These hardware configurations build an experimental execution platform with high throughput, high consistency, and high repeatability, laying a solid foundation for subsequent data collection and intelligent analysis. In addition, real-time sensing and data collection are the key bridges for AI empowerment. In the intelligent experimental system, by integrating multiple sensors and analysis modules, the robot experimental system can achieve real-time perception and quantitative characterization of the experimental state [60]. Through such closely integrated sensing systems, the robot laboratory not only achieves full-time perception of the experimental environment, but also provides a solid data foundation for the next step of intelligent decision-making and autonomous optimization. Finally, a truly intelligent experimental system is not merely about mechanically executing the experimental process; rather, it should be capable of automatically adjusting strategies and continuously optimizing results based on experimental feedback. The AI system recommends synthetic paths through literature mining or historical databases (such as the Retrosynthesis tool), and recommends initial reaction conditions based on previous success rates [61, 62]. During the experiment, the AI continuously predicted and analyzed the reaction bottlenecks based on real-time data and provided optimization suggestions. The parameter distribution is updated after multiple rounds of experimental iterations to achieve intelligent closed-loop optimization [63].

In fields such as materials science and chemical synthesis, automation not only takes over experimental tasks with high repeatability and strict precision requirements, but also opens up brand-new paths for accelerating the discovery of new materials and the exploration of reaction pathways through data-driven intelligent systems. On the one hand, the involvement of automation in these fields represents a profound transformation of the traditional experimental model. In the traditional experimental mode, experimental tasks with high repeatability often consume a large amount of human and time costs. Moreover, due to the inevitability of human factors, there are certain fluctuations in the accuracy and stability of the experimental results. The application of automation technology, with its high stability and accuracy, can ensure the standardization and consistency of the experimental process, thereby significantly enhancing the reliability of the experimental results [59, 60]. On the other hand, automation technology, with the help of data-driven intelligent systems, provides strong support for the discovery of new materials and the exploration of reaction paths [40, 62].

In the current era of AI for Science, massive experimental data contain rich information. By mining and analyzing these data through artificial intelligence algorithms such as the LLMs, researchers can extract valuable patterns and trends from the complex data, thereby designing new materials and exploring reaction paths more specifically [63, 64].

14.2.4.3 Design of Intelligent Fusion Architecture

As mentioned above, a truly intelligent experimental system does not mechanically carry out the experimental process, but is capable of automatically adjusting strategies based on experimental feedback and continuously optimizing the results.

With the advent of the era of large models, large models can be divided into four fields: basic large models, domain large models, agent platforms and applications, and AI applications and integration [34]. At present, for engineering fields such as materials and chemistry, the scenarios they face are all related to actual production and preparation. Researchers have always been focusing on the field of agents and believe that agents are the last mile for large models to be truly implemented [65]. However, how are these agent frameworks designed and constructed, and how do agents achieve deep collaboration among the knowledge reasoning ability of LLMs, the physical execution ability of robot laboratories, and the feedback optimization ability of experimental data [19, 40, 59]?

In 2022, Yao et al. [66] from Princeton University presented a general paradigm to combine reasoning and acting with language models for solving diverse language reasoning and decision-making tasks. Currently, the ReAct framework [66] is the mainstream decision-making model for Agents. Traditional ReAct can be regarded as a collection of few-shot prompts, thoughts, actions, and observations. It is a commonly used prompt structure for calling tools, reasoning, and planning, reasoning first and then executing. It performs specific actions based on the environment and provides the thinking process (Thought), as shown in Figure 14.7.

As shown in Figure 14.7, LLM serves as the core implementation of the entire ReAct, and Tools provide the functions or interfaces that need to be executed. ReAct can be simply understood as a complete set of processes of thinking–acting–observing [66]. The following takes an example as a starting point to better understand the underlying logic of the Agent. Taking the question "What is steel?" as an example, LLM would decompose the task hierarchically. It would first identify the need to determine the fundamental nature of steel as a material, followed by an inquiry into its composition, properties, and applications. Through this process of reasoning, the LLM would decide to first invoke a tool to retrieve the definition of steel. This action would yield the information that steel is an alloy primarily composed of iron with a controlled amount of carbon. Recognizing that the task remains incomplete, the LLM would proceed to call upon another tool to gather data on the properties and industrial applications of steel. The result of this step would provide insights such as steel's high mechanical strength, good ductility, and its widespread

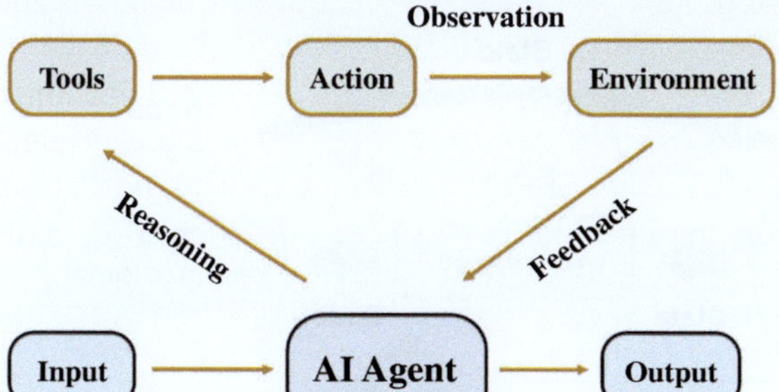

Figure 14.7 The core flowchart of ReAct combined with LLMs.

use in construction, manufacturing, and transportation sectors. Ultimately, by synthesizing the outcomes of these tool-based queries, the LLM would generate a comprehensive response: steel is an iron-based alloy containing carbon and other elements, known for its favorable mechanical properties and extensively utilized in engineering and industrial applications.

On 13 June 2023, OpenAI introduced the Function Calling capability for the first time in the GPT-4 Turbo model, taking a crucial step for LLMS to break through the boundaries of plain-text interaction [67, 68]. Function Calling aims to provide LLMs with a powerful and flexible way to interact with external code or external services. For instance, ChatGPT model can invoke external functions or services through function calls to provide more contextual information for its own text generation capabilities. The model itself does not execute the function. Instead, it generates parameters including the function name and the parameters required to execute the function. The specific execution is still implemented by the OpenAI application programming interface (API). After the emergence of function calls, there have been more practical advancements and developments in the LLMs field. OpenAI has summarized the widely recognized Agent architecture as: Agent = LLM + Memory + Planning skills + Tool usage as a typical representative of AI Agents. With the emergence of the LLMs + Function Calling approach, the Agent research and development framework has gradually added the Function Calling feature. With the advancement of these technologies, multi-agent research and development frameworks are constantly emerging. The following is an introduction to three typical multi-agent frameworks of LangGraph [69, 70], AutoGen [67], and Swarm [71].

14.2.4.3.1 LangGraph

LangGraph is an open-source framework developed by the LangChain team, specifically designed for building stateful and Multi-agent Systems (MASs). As shown in Figure 14.8, it implements complex dynamic workflows through graph structures and is particularly adept at integrating with LLMs. For graph structures, nodes represent execution steps, which can be LLMs calling, utility functions, or custom logic. Edges are used to define the flow rules between nodes and support conditional judgments, such as selecting branches based on LLMs output. The three major elements of a graph structure are: nodes, edges, and states [69, 70]. Nodes represent execution

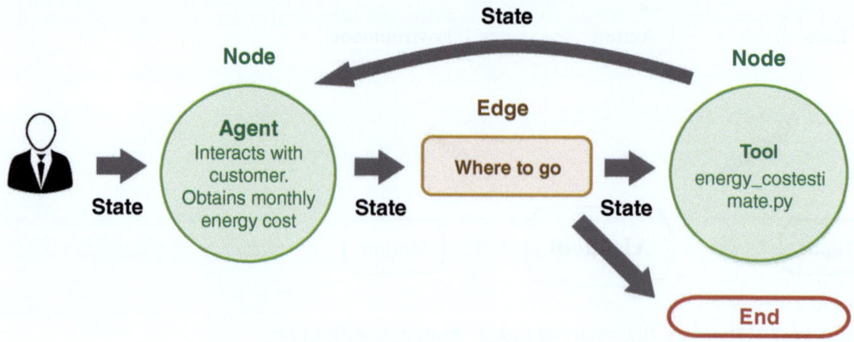

Figure 14.8 The flowchart of LangGraph.

steps and can be divided into: (1) Agent node: for example, calling LLMs to handle problems and knowledge retrieval, among others, to encapsulate the capabilities of independent agents; (2) Tool node: calls specific tools such as search APIs and database queries, among others, to perform specific computing or processing tasks; and (3) END node: an identifier indicating the end of the process. The edge is used to define the flow rules between nodes and supports conditional judgments, such as selecting branches based on the LLMs output to determine which node to jump to next. The status runs through the entire process, recording data or interaction status, such as the messages input by users, thereby driving the flow between nodes. The graph thinking of LangGraph decomposes complex tasks into orchestrable nodes and achieves efficient collaboration through dynamic routing and shared states. The codes of the LangGraph are available at: https://langchain-ai.github.io/langgraph.

14.2.4.3.2 AutoGen

AutoGen [67], as an open-source programming framework of Microsoft Research, provides complete tools and best practices for building scalable, flexible, and debuggable agent collaboration applications. As shown in Figure 14.9, the framework of AutoGen is divided into three core layers: (1) Core layer, which provides basic components and runtime environment, including basic abstractions such as message communication, message-driven Agents, Tools, and logs, among others, as well as operational support for distributed Agent systems; (2) AgentChat layer, which is a high-level API on top of Core. It includes various pre-built agent types and message types; and (3) Extensions layer, as a third-party extension component, which is used to improve the ecosystem, e.g., third-party LLM components, code executors, Tools, and more preset agents, among others. People can also develop extensions by themselves and submit them to the community. Besides, it is equipped with developer tools such as AutoGen Studio and Bench. The codes of the AutoGen are available at: https://github.com/microsoft/autogen [67].

14.2.4.3.3 Swarm

Swarm is a lightweight framework for multi-agent orchestration launched by OpenAI. Like a conductor, it coordinates multiple AI Agents to collaboratively complete complex tasks. The core concepts of the Swarm framework are Agent and Handoff. The Agent is the basic unit in the Swarm framework. It encapsulates a set of instructions and tools and allows for the option to transfer the conversation

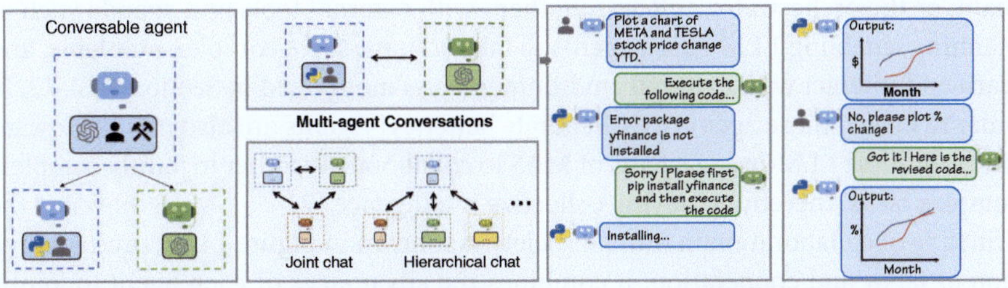

Figure 14.9 The flowchart of AutoGen. *Source*: [67] / OpenReview / CC BY 4.0.

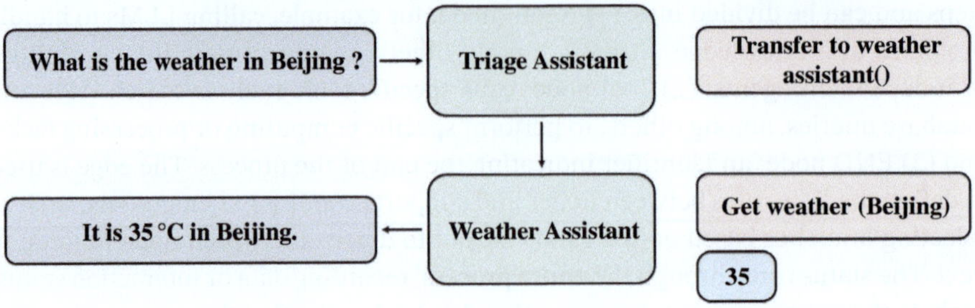

Figure 14.10 The core function of Swarm: the interaction and Handover among multiple agents.

to another Agent, as shown in Figure 14.10. An Agent can be regarded as an independent entity responsible for performing specific tasks [71]. They can accomplish these tasks in the following ways: (1) obtaining information: an Agent can call functions and APIs to obtain the required information; (2) data processing: Agents can process the obtained information, such as analyzing, transforming, and storing it; and (3) communication with other Agents: Agents can communicate with others by sending messages and calling functions to collaboratively complete tasks. Handover refers to the process of transferring the execution control rights between Agents. When an Agent finishes its task, it can transfer control to another Agent to continue performing the task. The Handover mechanism enables Agents to work collaboratively and jointly complete complex tasks. The codes of the Swarm are available at: https://github.com/openai/swarm.

The emergence of Swarm marks a new era in the development of MASs. It simplifies the complex collaboration process, allowing developers to focus on creative problem-solving rather than being troubled by cumbersome frameworks. Although Swarm is still in the experimental stage at present, it has already demonstrated great potential. It is not merely a framework, but rather a new way of thinking about AI collaboration.

14.2.5 Coordination Mechanisms of Heterogeneous Systems

With the development of LLMs, as the intelligent coordination hub, they can integrate multisource heterogeneous data in the domain, and collaborate across scale machines and multiple roles to form AI Agent. Agentic AI leverages LLMs as the brain, or the orchestrator, integrating them with external tools and agenda such as planning, enabling LLM-based agents to take actions, solve complex problems, and learn and interact with external environments. As mentioned in Section 14.2.4.3, in order to enable these agents to collaborate better, AI Agents are also moving toward using multiple LLM-based agents of MAS to collaborate together to jointly complete complex tasks, thereby achieving collective intelligence [72–75]. MAS draws on the principle of collaboration in human society. As shown in Figure 14.11, through division of labor and cooperation, it combines the advantages of each agent to jointly complete the task. Therefore, multi-agent coordination is the key mechanism that underlies system integration and a force multiplier of MAS. In particular, Shen and

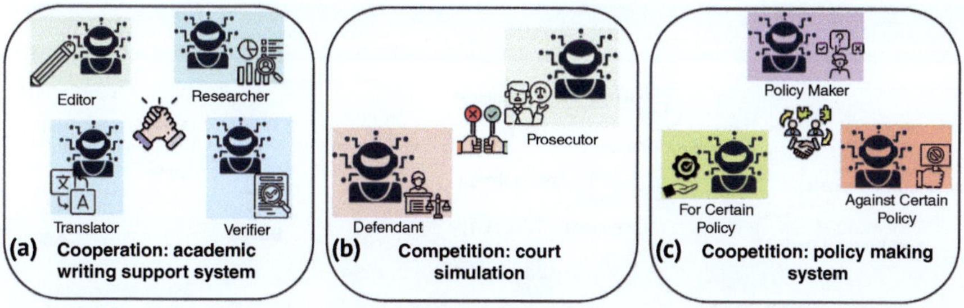

Figure 14.11 Examples of collaboration types: (a) cooperation type, (b) competition type, and (c) coopetition type. *Source*: [74] / arXiv.

coworkers [72] from the University of Technology Sydney gave the definition of multi-agent coordination as agents interacting and making decisions for the overall system-level performance, including resolving their conflicted interests. This subsection will delve into the multi-agent collaboration mechanism based on LLMs, especially exploring the types and structures of collaboration.

14.2.5.1 Collaboration Types

14.2.5.1.1 Cooperation

Cooperation is the backbone of MAS, enabling agents to work together effectively toward shared goals [74]. Cooperation in LLM-based MAS occurs when agents align their individual objectives (o_i) with a shared collective goal (Θ_{collab}), working together to achieve a mutually beneficial outcome: $\Theta_{\text{collab}} = \bigcup_{i=1}^{n} o_i$ [74]. Based on these findings, Liu and coworkers [24] from Tsinghua University introduced a framework, denoted as AgentVerse, capable of leveraging agents' group cooperation to manage more intricate scenarios. This framework can dynamically adjust its composition according to the current state, aiming to facilitate optimal decision-making and execution. In their article, for a given goal $g \in \vartheta$, a specific agent Mr is prompted as a "recruiter," similar to a human resources manager, and Mr dynamically generates a set of expert descriptions based on g. Different agents receive these different expert descriptions and then form an expert group $M = \text{Mr}(g)$ on the given goal g. Thus, it enables AgentVerse to hire the most suitable team based on the current state, so as to make better decisions in future rounds. Similarly, Hong et al. [76] proposed a pioneering multi-agent framework MetaGPT based on standardized operating procedures (SOPs) and combined with professional knowledge as shown in Figure 14.12. MetaGPT integrates a system of multiple GPT models, each of which is assigned specific roles and tasks. It aims to form a collaborative software entity by assigning different roles to GPTs to complete complex tasks. Unlike traditional single models, MetaGPT assigns different roles to multiple GPT models, enabling each model to focus on the tasks it excels at. This design not only enhances the ability to handle complex tasks, but also strengthens the flexibility and adaptability of the system. In addition, MetaGPT is highly scalable. As the complexity of tasks increases, the system's capabilities can be expanded by adding more GPT models. This scalability enables MetaGPT to adapt to constantly changing requirements and environments.

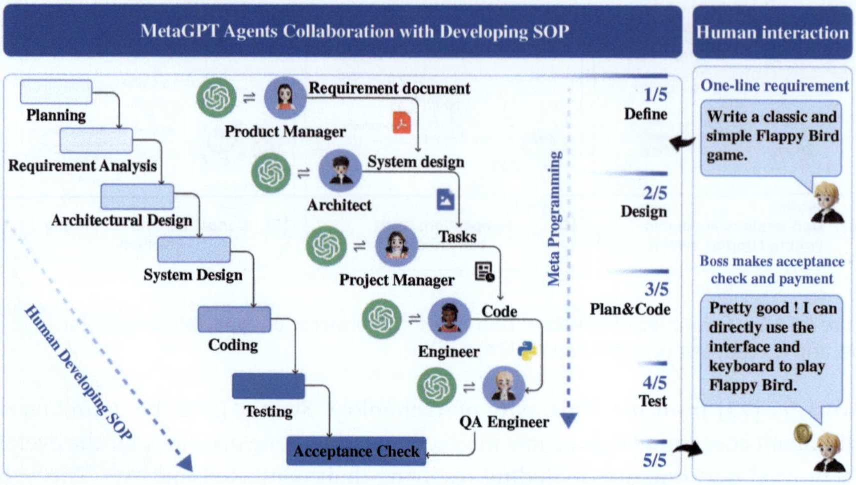

Figure 14.12 The core components overview of MetaGPT. *Source*: [76] / LYRASIS.

14.2.5.1.2 Competition

Competition in multi-agent coordination refers to the process of managing and resolving situations where multiple agents attempt to access shared resources or perform actions in a way that leads to conflicts or collisions [14]. When such competition occurs, agents will give priority to their goals. There may be oppositions and conflicts among these targets, and effective conflict resolution ensures smooth and coordinated behavior while meeting the overall goals of the system. Therefore: $\Theta_{\text{collab}} = \left\{ o_i | o_i \neq o_j, \vartheta_i \neq j \right\}$ [74]. In LLM-based MAS, competition encourages agents to develop more advanced reasoning capabilities and more creative problem-solving abilities, and enhances the adaptability of the system by testing the limits of each agent's capabilities. However, excessive competition may lead to conflicts and wastage of resources. Therefore, it is necessary to rationally design the competition mechanism, balance the relationship between competition and collaboration, and ensure the maximization of the overall interests of the system [72–75]. For example, Chen et al. [14] from Tsinghua University proposed the LLMARENA multi-agent framework to construct various dynamic multi-agent game environments. Combined with the TrueSkill scoring system and reward mechanism, this scoring system is based on Bayesian theory, which not only records the winning and losing results but also assesses the relative skill levels of the agents. Defeating highly skilled opponents can earn one higher scores; otherwise, points will be deducted. Meanwhile, the quality of a single action of the agent is measured through the concept of "reward" in reinforcement learning. The application of reinforcement learning in agents is described in detail in Section 14.3.3. LLMARENA systematically reveals the ability distribution and shortcomings of LLMs in multi-agent competition through multidimensional game environments, dynamic rules, and quantitative evaluations, providing a standardized assessment benchmark for the multi-agent interaction capabilities of LLMs. In other words, competitive approaches offer several advantages, such as promoting robustness, strategic adaptability, and complex problem-solving capabilities in MAS. However, competition also brings challenges, including potential conflicts that require mechanisms to ensure competition remains constructive and aligned with the overall system's goals. Effective coordination among agents is of great importance, particularly for competitive–cooperative types.

14.2.5.1.3 Coopetition

The core challenge of multi-agent collaboration systems is how to achieve a balance between collaboration and competition among agents, and how to enable agents to adapt and learn according to different tasks and roles. In MASs, the collaboration and competition among agents are the core interaction modes that drive the evolution of the system. The two are both opposing and unified, jointly shaping the performance, stability, and adaptability of the system [72–75]. For instance, the mixture-of-experts (MoE) framework [77] is also applicable to the coopetition type as shown in Figure 14.13. It was proposed by Jacobs et al. in 1991 [78]. This model broke the traditional single network method. Its core idea is to combine multiple specialized networks (experts) together. Each expert is proficient in handling specific subtasks within a broader range of problems and guides the input data to the most appropriate expert through a gated network. In 2020, the Transformer structure demonstrated outstanding performance and excellent scalability in the field of language models [79]. Since then, the Google GShard project has integrated MoE into the Transformer. Through in-depth design and experiments, it has significantly improved the effect and performance, laying the foundation of the MoE architecture in the field of large models [77, 80].

Therefore, in the field of materials science, it is particularly important to transform the multi-objective optimization algorithms of existing applications from the level of machine learning into the optimization of multi-agents and establish a unified framework between individual agents and the global goals of the system. Meanwhile, it is difficult for the existing black-box models, such as reinforcement learning, to explain the logic of the cooperation–competition strategy, and it is necessary to develop explainable AI to enhance the trust among agents.

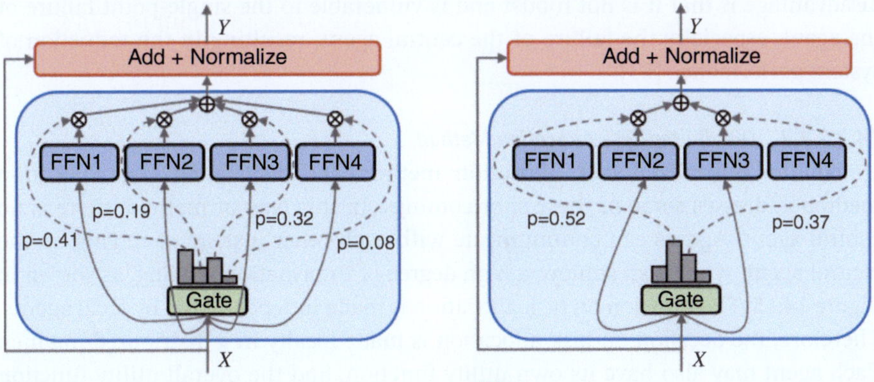

Figure 14.13 The frameworks of (a) dense and (b) sparse MoEs. *Source*: [77] / IEEE.

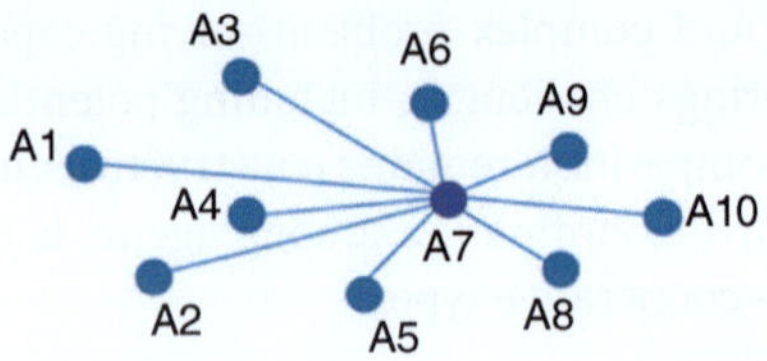

Figure 14.14 A centralized system, with agent A7 being the central coordinator. *Source*: [81] / IEEE.

14.2.5.2 Communication Structures

With the rapid development of intelligent technology, MASs are widely used in all walks of life. Especially when solving distributed task problems, due to the complexity of the system, it is difficult for a single agent to complete it independently. The existence of multiple agents improves the efficiency of task execution and also enhances the robustness of the system. In the process of using MASs to complete multiple tasks, the problem of task allocation occurs: what tasks are assigned to which agents, how the agents communicate with each other, and how to define the behavior of each agent to achieve the best performance of the system [72, 81]. The emergence of task allocation technology has solved the above problems. At present, the communication structure of multi-agent collaboration can be divided into three major categories: (1) centralized task allocation methods, (2) distributed task allocation method, and (3) learning-based task allocation method.

14.2.5.2.1 Centralized Task Allocation Methods

In the method of centralized task allocation, there is a central coordinator agent that centrally processes the task allocation information of all agents and makes decisions, as shown in Figure 14.14. This agent manages the negotiations of other agents and decides on the tasks assigned to them. In these cases, most of the time, the global utility function will be considered [82].

This method has high computational complexity and high requirements for the reliability of the system. It is only applicable to systems with a small number of agents and is not suitable for dynamic environments. Since the tasks are centrally allocated by the central controller, conflicts in the task allocation process are avoided, and the optimal solution to the task allocation problem can be found. The disadvantage is that it is not robust and is vulnerable to the single-point failure of the agent, especially the failure of the central agent, resulting in the reduction of system performance [83].

14.2.5.2.2 Distributed Task Allocation Method

Unlike the centralized task allocation method, the distributed task allocation method addresses some of these shortcomings. In this type of method, there is no central agent. Agents can communicate with each other instead of relying on the central agent, which can achieve a high degree of information sharing, as shown in Figure 14.15. The decision on task allocation is made independently by local agents. Therefore, the decision on task allocation is made locally in a distributed manner. Each agent may also have its own utility function, and the overall utility function may be approximate [81, 82].

The advantages of the distributed method are that it has certain robustness. The failure of a single agent has a relatively small impact on the overall performance

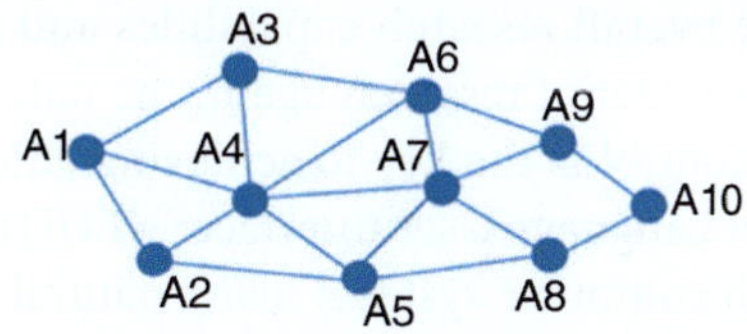

Figure 14.15 A decentralized system. *Source*: [82] / Association for Computing Machinery.

of the system. Moreover, agents communicate with each other and have good scalability. It is very suitable for large-scale systems. At the same time, there are also certain deficiencies. Only approximate solutions to the task allocation problem can be obtained, and local task allocation will lead to task allocation conflict [84]. Therefore, how to effectively solve the conflicts among various agents is still the direction that needs to be improved at present.

14.2.5.2.3 Learning-based Task Allocation Method

Learning-based task allocation algorithm is a type of algorithm that solves task allocation problems through machine learning methods. This type of algorithm usually combines learning models, enabling task allocation to self-optimize based on historical experience or environmental changes, and adapt to dynamic and complex task allocation scenarios [85]. The learning-based task allocation algorithm optimizes the task allocation strategy in a data-driven manner. Compared with traditional heuristic algorithms and optimization algorithms, it has stronger adaptive ability and global optimization ability. It also provides powerful solution tools for task allocation problems through methods such as reinforcement learning [75, 86, 87], deep learning [87, 88], and supervised learning [89], which are suitable for complex, dynamic, and large-scale optimization scenarios. Although these methods pose challenges in terms of data requirements, computing resources, and model generalization capabilities, with the advancement of technology, their application prospects in many fields are very broad.

At present, due to the great potential shown by reinforcement learning in task allocation algorithms, agents achieve the maximization of long-term cumulative returns through continuous interaction with the environment. Reinforcement learning has advantages such as no need for pre-precise modeling and processing of high-dimensional nonlinear data, and has achieved great success in solving various dynamic decision-making problems. The ideas such as reinforcement learning are applied to the learning, optimization, and control of MASs, providing effective methods and technical means for solving the complex decision-making problems of MASs [75, 86, 87]. The specific application of reinforcement learning in agents is introduced in detail in Section 14.3.3.

14.3 Core Techniques in Agents

14.3.1 Natural Language Interaction Interfaces for Equipment Control

The AI-driven research paradigm, supported by large models, big data, and high computing power, is deeply embedded in the entire research process. It will drive significant changes in the governance model of research organizations and greatly enhance the overall research capabilities and efficiency [90, 91]. Therefore, in the autonomous material research agent, the natural language interaction interface of equipment control is the key to achieving efficient human–machine collaboration [59]. Natural Language User Interface (NLUI) is a user interface that allows users to interact with computer systems using natural language [92]. This interface makes users feel as if they are having a conversation with another person rather than communicating with a machine. The design purpose of NLUI is to improve the naturalness and convenience of human–computer interaction [93]. It is an important research direction in the field of artificial intelligence, combining various technologies such as NLP, speech recognition, and machine learning [94]. For instance, researchers issue instructions through natural language, such as "Raise the temperature of the reaction kettle to 80 °C and maintain it for 30 minutes." The interface needs to convert such unstructured text into precise instructions executable by the device. This process relies on NLP technology. Key parameters in the instructions (such as "reaction kettle," "80 °C," and "30 minutes") are located through named entity recognition, and then a semantic parsing model is used to map them to the device control protocol.

In other words, directly controlling experimental equipment through natural language not only significantly lowers the operational threshold, but also holds the promise of achieving knowledge-driven generation and dynamic adjustment of experimental processes [95]. However, mapping human language "seamlessly" into complex, precise, and executable experimental operations is no easy task and involves breakthroughs in multiple technical bottlenecks [96]. From the perspective of system implementation, the natural language control interface faces three core challenges: the first is intention recognition, i.e., how to accurately understand the scientific research intentions expressed by users, especially to extract structured task goals in the highly specialized context of materials science [97]. Second, there is instruction mapping, which converts the high-level semantic goals into specific operation instructions that the underlying devices can execute, including the sequence of actions, parameter configurations, and device invocations [98, 99]. Finally, there is safety control to ensure that the automatically generated and executed instructions do not cause experimental risks, equipment damage, or violations of laboratory regulations [40].

This subsection will focus on the above three core issues, systematically analyze the application practice and research progress of current natural language control technology in the automation of material and chemical experiments, and explore its engineering implementation path and challenges in combination with typical system examples.

14.3.1.1 Intent Recognition

Intent recognition is an important task in NLP. It aims to determine the intention or purpose expressed in the user's input statement. Simply put, intent recognition is to semantically understand the user's words in order to better answer the user's questions or provide relevant services. In NLP, intent recognition is usually regarded as a classification problem, i.e., classifying the input statement into predefined intent categories to identify its intention. These categories can be various tasks, queries, requests, etc., such as search, purchase, consultation, and command, among others [100–102].

14.3.1.1.1 Semantic Understanding Based on Pretrained LLMs

The semantic understanding system based on LLM mainly relies on the synergy of the NLP technology and prompt engineering [103–105]. Its core logic can be decomposed into the following levels. The first is the input parsing layer. The user input is decomposed into subword units, such as Byte-pair Encoding of GPT-4, through the Tokenizer of the LLM to generate the token sequence. Meanwhile, the Attention mechanism is utilized to capture the semantic associations between words. For example, long-distance dependencies are modeled through Multi-head Attention. The second is the intent induction layer, which maps the input to the intent label space based on the pretrained semantic vector space, such as the cross-modal feature space of CLIP. Furthermore, to ensure the structuring and standardization of the intention output, this layer uses Prompt templates to guide the LLM. Through the preset format framework, such as "User intention classification: {Synthesis/analysis/optimization}, Target substance properties: {Electrical conductivity/thermal stability}, Molecular type: {Organic small molecules/polymers}," which prompts the model to output structured intentions according to the specified dimensions and labels, thereby providing a clear instruction basis for subsequent tool calls or task executions. Finally, the induced intentions are transformed into tool interface parameters through Function Calling protocols, such as OpenAI Function Calling. Bran et al. [106] from the University of Rochester have developed ChemCrow, an LLM-based chemistry agent designed to perform various tasks such as organic synthesis, drug discovery, and materials design. ChemCrow integrates 13 expert-designed chemical tools, as shown in Figure 14.16, and operates by prompt LLM (GPT-4) to provide specific instructions on tasks and required formats. And using various software packages and software related to chemistry, a set of tools was created, and then these tools and user prompts were provided to the LLM. LLM then determines its path, tool selection, and input through an automatic and iterative CoT process before reaching the final answer.

14.3.1.1.2 Combined With the Intent Classifier

In the semantic understanding system based on LLM, some architectures will introduce lightweight intent classifier modules as prescreening or auxiliary mechanisms to improve the accuracy and efficiency of intent induction [107]. By defining the classification description, the problem classifier can infer the matching classification using LLM based on user input and output the classification result, providing more accurate information to downstream nodes [108]. In traditional NLP, the classification process relies on handcrafted features and machine learning algorithms, which often have difficulty capturing the complexity and nuances of language, especially for large and unstructured datasets. For instance, researchers use the pretrained BERT model to solve the binary text classification problem. In text classification, the BERT model is utilized to encode the entire sentence, and then the most likely

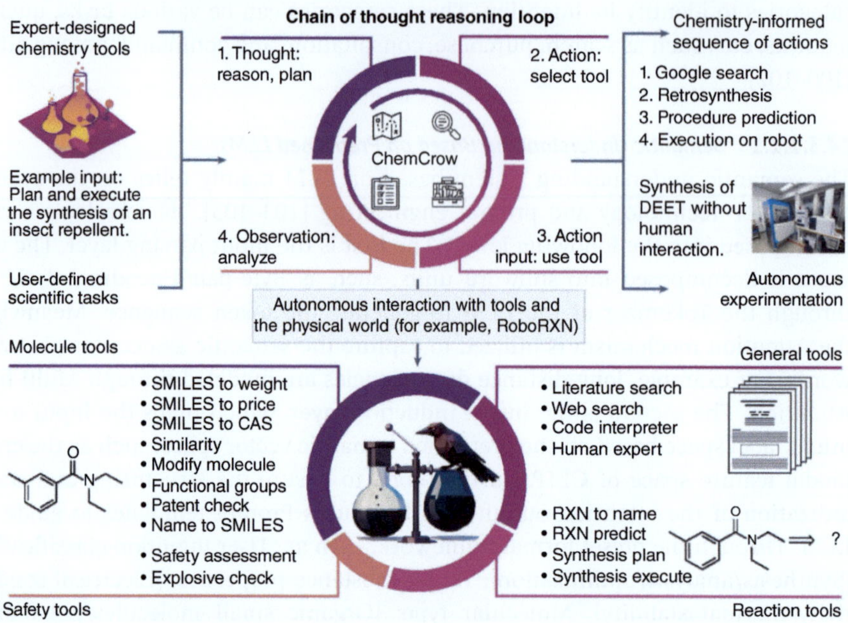

Figure 14.16 An overview of the task-solving process for ChemCrow: (a) the task-solving process and (b) toolsets implemented in ChemCrow. *Source*: [106] / Springer Nature / CC BY 4.0.

intent category is predicted through the classification layer, thereby classifying the text into predefined categories or labels [109]. Jia et al. [110] proposed a text sentiment classification method BERT-CBLBGA based on BERT. The BERT-CBLBGA framework uses the BERT, TextCNN (convolutional neural network), BiLSTM (bidirectional long short-term memory networks), and BiGRU (bidirectional gated recurrent unit) models to process the text. It first uses BERT to generate word embedding of the text, then extracts the local and global sentiment features of the text through TextCNN, BiLSTM, and BiGRU, and uses the self-attention mechanism to fuse the outputs of these models to enhance the ability to extract text sentiment classification features. Finally, a classifier is used to classify the extracted features, as shown in Figure 14.17.

14.3.1.1.3 Context Awareness and Task Memory

Multi-round conversational agents driven by LLM represent a significant leap in conversational artificial intelligence, capable of conducting complex and context-aware interactions across multiple rounds, which is crucial for applications such as customer service and personal assistants [111]. In the multi-round interaction scenarios of agents, context awareness and task memory are the core technologies for achieving natural and coherent dialogs, which maintain interaction continuity by managing dialog history and modeling user intentions. LLMs, such as ChatGPT and Qwen, are crucial for multi-round conversations and they can achieve coherent responses during long interactions. LLM is trained on a massive text corpus and is capable of predicting subsequent content based on previous dialog rounds,

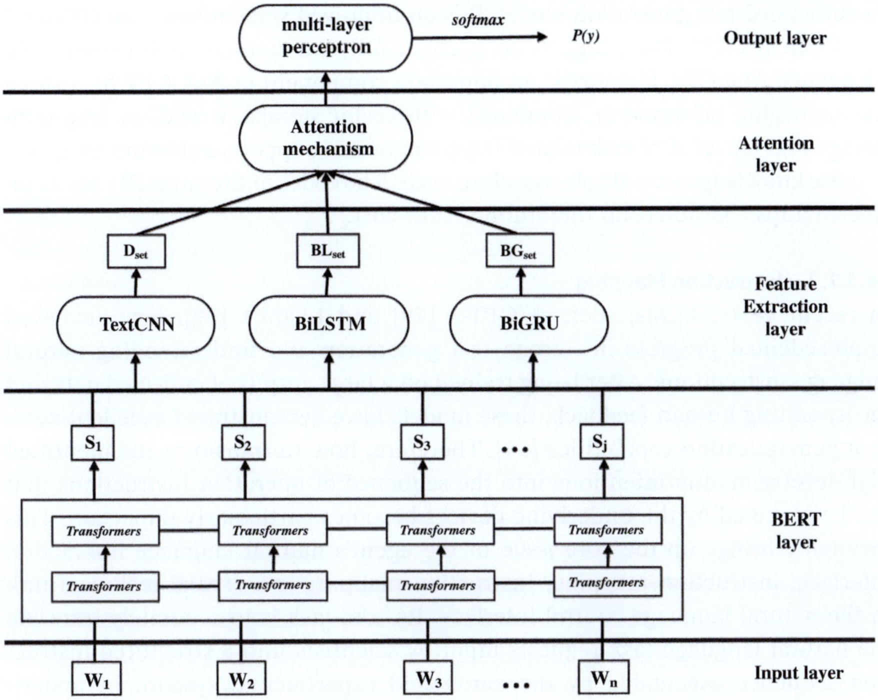

Figure 14.17 The framework of BERT-CBLBGA. *Source*: [110] / with permission of Elsevier.

handling complex language structures, and maintaining the context of the dialog. This is crucial for generating appropriate responses throughout the extended interaction process. Agents in multi-round systems process user input based on LLMs, understand user intentions, and manage the dialog process. They utilize previous dialog rounds and external resources, such as databases and API, to provide the latest information.

In the process of implementing context awareness and task memory, typical tools such as LangChain Memory and AutoGPT Memory play an important role. In 2022, American Computer Scientist Harrison Chase proposed LangChain, an open-source framework for developing applications based on LLMs. Its design goal is to simplify the process of integrating LLMs with external data sources, tools, and computing capabilities, thereby building more powerful intelligent applications. The Memory component of LangChain is a flexible dialog memory framework that can store various forms of information. Whether it is historical dialog content, real-time communication information, or external file materials, they can all be remembered and referenced when needed. It can also serve as an efficient retrieval system. It can be based not only on computer memory, but also on vector databases, and find the most relevant information to the input according to the semantic relevance. Through such a design, the Memory component not only improves the quality and efficiency of the conversation, but also makes the AI more accurate in understanding user needs. The codes of the LangChain are available at: https://github.com/langchain-ai/langchain. Besides, in 2023, Toran Bruce Richards released AutoGPT, an automated text generation model. It is an improved version based on GPT and uses OpenAI's GPT-4 language model to create fully autonomous and customizable AI Agents. AutoGPT Memory is an important component in AutoGPT for storing and managing information. Combined with vector storage, it realizes long-term storage and retrieval of task-related information and supports autonomous agents to reuse knowledge in multiple rounds of tasks. The codes of the AutoGPT are available at: https://github.com/Torantulino/Auto-GPT.

14.3.1.2 Instruction Mapping

In recent years, LLMs, such as GPT-3 [79] and LLaMA [37], have achieved unprecedented progress in human text generation and understanding natural language instructions. After being trained on a large corpus of collected data and incorporating human feedback, these models have demonstrated excellent zero-shot generalization capabilities [34]. Therefore, how to transform the identified high-level semantic intentions into the sequence of operation instructions that can be executed by the underlying devices becomes particularly important. This inevitably brings up the core issue of the agent's natural language interaction interface: instruction mapping. Instruction mapping is a crucial technical link in the natural language control interface. Its core task is to accurately translate the natural language task requests input by scientists into a structured instruction sequence executable by the automated experimental system. In materials science experiments, this mapping process is particularly complex because experimental actions are not only highly specialized but also often involve precise coordinated operations of multiple heterogeneous devices. This process usually requires traversing multiple semantic levels: from statement-level task identification, to function-level instruction generation, and then to device-level command invocation. For instance, the user inputs "Please stir at room temperature for 10 minutes, then heat to 80 °C and continue the reaction for 2 hours." So, this mapping involves multiple steps such as semantic extraction, task decomposition, and action normalization, ensuring that high-level semantic intentions can be reliably translated into physical operations.

14.3.1.2.1 The Transformation from Natural Language to Experimental Logic

Natural language expressions are usually rich in ambiguity and context dependence [112]; for example, "Use an appropriate acid to pickle the metal sample, remove the surface oxide layer, and then clean it with a common solvent." This expression must be broken down by the system into precise operational steps, including (1) action recognition, such as "acid washing," "removal of the oxide layer," and "cleaning"; (2) parameter extraction, such as the pickling time and the temperature; and (3) sequential dependencies, such as make sure that "acid washing" is completed before "cleaning with other solutions." To achieve this semantic transformation, the system usually employs the following techniques: (1) establish matching rules from natural language to control commands for common instructions; (2) use the sequence labeling model to identify the relationship among actions, objects, and parameters in the statement. Sequence labeling models [113, 114] are a fundamental and crucial model architecture in NLP, used to label each word or subword in the text with a specific label. For instance, pretrained language models such as BERT [109], RoBERTa [115], and T5 (Text-to-Text Transfer Transformer) [116] extract context representations and support complex semantics. This type of model plays the role of extracting key information such as operations, objects, and parameters from the instruction language when controlling experimental equipment in natural language, and is one of the core technologies in the instruction mapping process; and (3) convert the task expression into intermediate-layer representations such as JSON/YAML for subsequent invocation.

14.3.1.2.2 Standardization and Structuring of Operation Instructions

Experimental equipment usually does not understand natural language but relies on specific programming commands or API interfaces. Therefore, instruction mapping must generate a standard structure that can be recognized by the lower-level execution engine. The system needs to construct a set of action semantic mapping tables covering the experimental process to achieve a one-to-one or many-to-one correspondence between natural language instructions and equipment operation instructions [117, 118].

14.3.1.2.3 Process Decomposition and Step Sequencing of Complex Tasks

In materials science experiments, the experimental process is usually composed of multiple steps and is accompanied by strict logical dependencies, such as preconditions and parallel control [13, 19–21]. Therefore, the instruction mapping module must have the ability to model the logic of task execution. This includes the division of experimental steps, such as sample addition–reaction–analysis, the sorting of the sequence and dependencies among various operations, as well as the inheritance and passing of parameters, to ensure that semantics such as "use the solution obtained from the previous reaction" can be accurately understood and executed. In addition, complex tasks also require the system to have basic control flow capabilities, such as supporting conditional judgments like "terminate the experiment if the reaction temperature exceeds the threshold." To achieve this goal, modern automation platforms usually introduce task graphs, state machines, or scripting languages, such as LabScript and YamL-based workflows, to express and schedule the complete experimental process logic, thereby ensuring that natural language instructions can be accurately and safely mapped into executable experimental operation sequences [119, 120].

In other words, as the central link of the natural language experimental interface, the instruction mapping directly determines whether the system can "understand" and "correctly execute" the experimental intentions of researchers. An efficient instruction mapping system needs to possess semantic parsing accuracy, action standardization capabilities, and complex process management capabilities, ensuring the correctness of instructions while achieving seamless control of heterogeneous devices. This technology is driving the evolution of materials science experiments from "code-driven" to "language-driven," laying the foundation for the realization of true autonomous scientific research.

14.3.1.3 Safety Control

In the natural language-driven experimental automation system, safety control is not only a necessary guarantee to prevent physical layer accidents, but also a fundamental module to ensure that the system operation complies with regulations, experimental ethics, and equipment protection norms [121, 122]. Since the experimental instructions input by researchers through natural language may have ambiguity, incompleteness, or even potential dangers, the system must have the ability to identify, assess, and intervene in safety risks, so as to achieve truly controllable and reliable autonomous experimental operations.

The experimental control system driven by natural language must strike a balance between high degree of freedom and high security. For this purpose, multiple cutting-edge platforms have constructed systematic security guarantee strategies.

In terms of identifying hazardous substances and reactions, ChemCrow [106] integrates secure databases and tools such as PubChem and RDKit during the synthesis planning stage, which can automatically detect potential dangerous reactions or prohibited substances. If high-risk molecular structures are identified, it will refuse to generate execution instructions. Specifically, the ControlledChemicalCheck tool compares the Chemical Abstracts Service (CAS) number or Simplified Molecular Input Line Entry System (SMILES) representation of the input molecule with multiple identified lists of chemical weapons and precursors. If the input molecule is not on these lists, it calculates the maximum similarity between it and the molecules in the database. A warning is issued when the similarity is greater than 0.35. Once the molecules are in these lists, ChemCrow will immediately stop executing the instructions. In addition, ChemCrow queries the PubChem database using a molecular identifier such as generic name, IUPAC name, or CAS number to determine whether the Globally Harmonized System (GHS) rating of a molecule is "Explosive." If a molecule is detected to potentially explode, a warning or error message will be issued to prevent the execution of relevant instructions.

In terms of the verification of action feasibility, the CLAIRify method proposed by Yoshikawa et al. from the University of Toronto combines automatic iterative prompts with program verification to ensure that the generated experimental plans are effective and comply with environmental constraints [5]. With the help of environmental modeling and physical simulation functions, it verifies in advance whether the experimental path is accessible and whether there is a risk of mechanical collision or liquid overflow. Furthermore, many systems also introduce the Verifier module, which performs static analysis or symbolic execution on the code during the LLM output stage to ensure logical correctness and device compatibility. In terms of ensuring human–machine collaboration, systems such as ChemCrow [106] and Coscientist [16] require users to make final confirmations before critical operations and clarify the boundaries of responsibility. For high-risk experiments, permission control and hierarchical authorization mechanisms are also introduced. For example, the activation of specific high-temperature modules requires the approval of the administrator. It ensures that the application of AI in high-risk fields such as chemistry is both innovative and compliant with scientific ethics and legal norms by clarifying risk awareness, defining responsibility boundaries, and strengthening ethical review. Finally, in terms of the experimental log and tracking mechanism, such systems comprehensively record the user's natural language input, generated code, device execution process, and experimental results, forming an auditable operation chain.

In other words, the safety control mechanism is a necessary prerequisite for the implementation of the natural language experimental interface system in real materials science experiments. It is not only related to the protection of personnel and equipment, but also the foundation for achieving the credibility, stability, and sustainability of the automated system. In the future, as AI systems undertake more autonomous decision-making functions in laboratories, safety control modules will further evolve toward intelligence. By integrating with knowledge graphs, reinforcement learning, and safety simulation technologies, they will achieve an evolution from passive interception to active prevention.

14.3.2 Design of Transformer Models for Experimental Planning

In the field of materials science research, the integration of LLMs and robotics laboratories is reshaping the traditional research and development paradigm. This subsection will deeply analyze the integration mechanism of the two, explore how to achieve seamless connection of data, knowledge, and physical experiments through technological collaboration, and build an efficient and intelligent materials research system.

14.3.2.1 Research Background and Task Definition

Deep learning is a collection of algorithms in machine learning, and its emergence has brought revolutionary improvements to many traditional machine learning and pattern recognition tasks [123]. Deep learning simulates the human brain to learn the structure of input data, output data, and the potential relationships between them. It can perform simulation calculations without relying on complex or even nondeductible mathematical formulas. This technology is widely applied in various scenarios such as materials, chemistry, and biology [124–126]. With the development of technology, Vaswani et al. proposed the Transformer network model in 2017 [127]. Its innovative self-attention mechanism enables the model to process input sequences in parallel, which allows the model to better capture long-distance dependencies in the sequences and greatly accelerates the training process. Transformer was originally developed for NLP tasks, modeling complex relationships in sequences using self-attention mechanisms, thereby effectively capturing long-term dependencies and interactions. Subsequently, the Transformer has been widely applied in fields such as materials and chemistry. It can not only accelerate basic scientific research works such as literature retrieval, data processing, translation, and reading, but also optimize material experimental plans, predict material properties, and even provide intelligent recommendations and innovative design plans.

In the autonomous materials research system, experimental planning is a key central link connecting scientific goals with physical experiments. Its task is to automatically formulate reasonable, executable, and scientifically valuable experimental plans based on research goals, prior knowledge of materials, and experimental history [128].

This task not only requires the system to have scientific reasoning ability, but also needs to consider the feasibility of experimental operations and resource constraints. It is a typical multi-objective and cross-modal decision-making problem. Traditional experimental planning mostly relies on expert experience or rule-based heuristic search methods, and is often inadequate when facing large-scale material spaces and complex experimental process flows. In recent years, large-scale language models based on Transformer, such as the GPT series, BERT, and T5, have shown great potential in the modeling of scientific research tasks due to their outstanding language understanding and generation capabilities [129]. For its powerful context modeling ability and self-attention mechanism, the Transformer model can not only handle complex experimental texts, but also extract causal relationships and logical structures in highly unstructured data environments, bringing a new paradigm change to experimental planning. The core tasks of experimental planning include goal parsing, path planning, action sequencing, and parameter suggestion [106, 130]. Meanwhile, to ensure that the generated scheme has physical feasibility and engineering controllability, the planning system also needs to embed materials science knowledge, experimental constraints, and process feedback mechanisms [40]. Therefore, introducing the Transformer model into the experimental planning system is not only an extended application of NLP technology, but also an important breakthrough point for constructing the intelligent scientist agent system [131].

14.3.2.2 Model Input and Output

For the experimental planning task, the input and output design of the Transformer model needs to simultaneously satisfy the naturalness of language expression and the rigor of experimental logic. The core lies in how to effectively encode unstructured scientific research objectives, material information, and operational context, and generate structured or semi-structured scheme outputs that conform to experimental logic.

14.3.2.2.1 Input Representation

In the experimental planning task of materials and chemicals research, the effective operation of the Transformer model depends on the precise modeling of the input semantics. Typical inputs can usually be summarized into three types of core semantic fragments. First is the identification of research objectives and experimental intent. It is necessary to have the LLM analyze and determine the user's purpose or intention from the dialog content input by the user, such as text, voice, and other forms. Second is the experimental context, including information on existing material types, equipment configurations, historical experimental records, and procedural details. Finally, the model may incorporate external knowledge base, such as known formulations, reaction conditions, or performance data extracted from literature, patents, or databases. These components are jointly represented through natural language prompts, structured templates, or embedded context, enabling the model to construct a semantically rich and logically coherent input representation that supports the generation and optimization of high-quality experimental workflows.

In the input design of the Transformer model, to achieve the transformation from the original text to the structured representation and simultaneously achieve a dynamic balance between the naturalness of language expression and the rigor of experimental logic, the following steps are often adopted. Tokenization is the most fundamental step in the tasks of NLP, which processes the text content into the smallest basic unit, namely token, for subsequent processing. For example, subword segmentation techniques such as Byte-pair Encoding (GPT) [34], WordPiece (BERT) [109], and SentencePiece (T5) [116] are adopted to balance the lexical coverage and semantic granularity. They convert text data into a form that machines can understand, just like breaking down a complex dish into simple ingredients. In addition, position encoding is achieved through sine/cosine functions or learnable parameter (BERT) encoding sequence order to solve the disorder problem of the self-attention mechanism and ensure that the natural logic of word order in the sentence is correctly modeled. Besides, for some cross-modal tasks, such as text-image generation, the naturalness of image semantics, such as color and object spatial relationships, is often retained through the fusion of visual feature embedding (such as the CLIP model) and text embedding [132]. Meanwhile, the consistency of multimodal data in the vector space is ensured through standardized projection layers.

Here are some specific research cases. Weininger [133] used SMILES input specification of drugs as the text data input model. Compared with using traditional drug descriptors, using the SMILES sequence of molecular structure as input can describe the molecular structure more accurately and avoid the problem of information loss in the encoding of drug descriptors at the same time. The word embedding technology can map each molecular fragment in the molecular structure into a real number vector. Through the combination and transformation of these vectors, the Transformer model can better learn the features and patterns in the molecular structure, and thereby predict the drug synergy more accurately. On this basis, Schwaller et al. [134] proposed a Molecular Transformer model based on the Transformer structure, which predicted drug chemical reactions through the Bayesian deep-learning training method and achieved high accuracy. Wang et al. [135] introduced the BERT semi-supervised model SMILES-BERT based on the Transformer and fine-tuned the pretrained model to be applied to different molecular property prediction tasks. Tetko et al. [136] proposed a method using an enhanced NLP Transformer model to determine the feasibility of chemical reactions and improve the efficiency of synthetic planning by evaluating the predicted results of synthetic steps.

This information can be used to construct input sequences through natural language concatenation, structured template filling, or embedding prompt engineering. Natural language splicing can flexibly express the experimental goals and conditions, and is close to the experimental plans written by humans. Structured templates ensure the consistency and interpretability of the input format, enabling the model to better understand the predefined grammar rules and task requirements. Prompt engineering, on the other hand, guides the model to generate coherent, relevant, and experiment-compliant outputs by ingeniously embedding contextual information, constraints, or demonstration cases in the input. The combination of multiple methods ensures that the input representation is both semantically rich and closely consistent with the subsequent experimental planning tasks, improving the accuracy and practicability of the model generation experimental process.

14.3.2.2.2 Output Representation

Output representation is a key link in realizing the connection of Transformer models to real experimental systems. Its core task is to accurately convert the vector representations within the model into experimental planning texts or structured instructions that are both naturally readable, logically rigorous, and operable. In the experimental planning of materials science, this process needs to seek an accurate balance between the naturalness of language and the logic of experiments to ensure that the generated results not only conform to human scientific research writing and interaction habits but can also be directly parsed and executed by the robot experimental platform [118, 137, 138].

In terms of language naturalness, the output design of the Transformer model usually adopts multiple decoding strategies to optimize the generation quality. It is well known that the generation process of LLM is autoregressive. That is to say, after completing one inference, the LLM needs to determine what the next token is, and then combine the current sequence with the next token for the next inference. In the process of determining what the next token is, numerous strategies are involved, such as greedy search, auxiliary generation, and sampling, among others. The relevant generation modes are mainly divided into two categories, namely non-beam methods and beam methods. Among the non-beam methods, there are the methods that we are most familiar with, such as greedy search and sampling. While the beam method, e.g., beam search [139–141], is a commonly used search algorithm and is often used in tasks such as text generation, text summarization, and machine translation in NLP [142]. Its core idea is to maintain a candidate set (beam) during the search process, and the elements in this set are sorted according to certain sorting rules, usually by their current scores. At each moment when the sequence is generated, a probability distribution of the word list size is produced. Therefore, the one with the highest probability needs to be selected as the final solution, as shown in Figure 14.18. This method retains multiple candidate paths, such as beam width = 3–5, effectively avoiding greedy decoding falling into local optimum, reducing problems such as repetition and ambiguity, and achieving a good balance between diversity and rationality in the generated results.

Meanwhile, in order to further control the creativity and determinism generated by the Transformer, temperature scaling [141, 143–145] is a key parameter. In 2014, Sutskever et al. [143] first applied Softmax Temperature to control and generate output distribution in the Seq2Seq framework in their paper, which was the basis of subsequent temperature regulation techniques. In 2017, Guo et al. [146] proposed Temperature Scaling in Modern Neural Networks as a post-processing method to calibrate the softmax probability distribution of the model to better reflect the prediction confidence level, which is widely used for the optimization of the output confidence level of classification models. In 2019, Radford et al. [34] from OpenAI widely used temperature scaling to control the determinism and diversity of sampling output in the GPT series models. The results showed that at low temperature scaling ($T < 1$), the LLM distribution was more acute, the model was more deterministic, the output was more conservative, and the repeatability was high. It is suitable for tasks that require high precision and determinism, such as summarization and code generation. Besides, at high temperature scaling ($T > 1$), LLMs have a

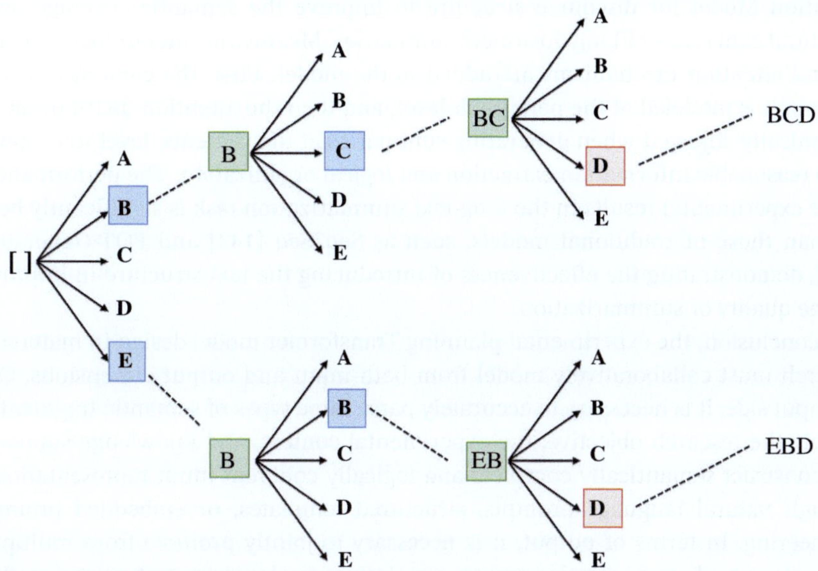

Figure 14.18 Example of beam search.

flatter distribution, generate more random and diverse contents, and are suitable for creative generation tasks, such as writing stories, generating poems, and other creative text generation. In 2020, Holtzman et al. [141] conducted an in-depth analysis of the impact of temperature parameters on the diversity and quality of text generation. The paper pointed out that models are prone to generating repetitive and mechanical text at low temperatures, which is a core literature for understanding the trade-off between temperature regulation and natural language generation.

So, Temperature Scaling is often used to control the creativity level of the text generated by LLM [144, 145]. By adjusting this parameter, the probability distribution of LLM can be influenced, making the text more concentrated or diverse. Therefore, for the actual application in the experimental planning task, when generating relevant contents such as explanation of experimental purpose or literature background, low Temperature Scaling ($T < 1$) needs to be used to ensure accuracy and consistency. When generating hypothetical schemes or alternative paths, it is necessary to increase the Temperature Scaling ($T > 1$) to enhance creativity and divergence.

Furthermore, enhancing semantic coherence is another key path to ensure output quality. The introduction of Discourse-aware Attention Model [147] and latent variable modeling methods, such as VAE-Transformer [148], can effectively handle language challenges such as pronoun referrals in material naming, causal sequences of experimental steps, and context switching. For example, when generating a multistep experimental plan, the model can correctly identify the pointing relationship of phrases such as "this solution" and "the above conditions," maintaining the coherence of the process. Similar technologies have been widely applied in the task of generating scientific abstracts. Cohan et al. [142] proposed a Discourse-aware

Attention Model for discourse structure to improve the semantic coverage and structural coherence of long document summaries. Meanwhile, hierarchical encoders and attention mechanisms are added to the model. First, the content of each paragraph is modeled at the paragraph level, and then the attention distribution is dynamically adjusted when generating summaries at the sentence level to achieve more reasonable information extraction and logical organization. The performance of the experimental results in the long-text summarization task is significantly better than those of traditional models, such as Seq2Seq [143] and POT-Generator [149], demonstrating the effectiveness of introducing the text structure in improving the quality of summarization.

In conclusion, the experimental planning Transformer model design in materials research must collaboratively model from both input and output dimensions. On the input side, it is necessary to accurately parse three types of semantic fragments, namely the research objective, the experimental context, and knowledge support, and construct semantically complete and logically coherent input representations through natural language prompts, structured templates, or embedded prompt engineering. In terms of output, it is necessary to jointly promote from multiple dimensions such as generation strategy optimization, language coherence modeling, logical traceability, multidimensional evaluation system, and controlled generation design, ultimately achieving that the generated content is both natural and smooth, as well as scientifically rigorous, meeting the dual requirements of human readability and system executability [5].

14.3.2.3 Model Structure Design and Improvement Strategies

In the experimental planning tasks of materials science, the structural design of the Transformer model not only determines the ability of semantic understanding and process generation, but also directly affects its interpretability, executability, and scalability in complex scientific research tasks. To this end, the current research has carried out in-depth optimization at multiple levels such as infrastructure selection, functional enhancement modules, and training data design, with the expectation of constructing a high-performance planning agent system adapted to the experimental environment.

1. **From the general framework to structure-aware modeling**
   The task attributes of experimental planning determine that its processing of input semantics is not limited to linear language structures, but often involves complex material formulations, equipment configurations, and multistep experimental logics. Based on this, the model structure design can be divided into two types of paths. The first is about LLMs based on the standard Transformer architecture, such as BART (Bidirectional Autoregressive Transformer) [150], T5, and FLAN-T5 [116], among others. The encoder-decoder architecture with Transformer as the core is the basic scheme for processing experimental texts. Its core advantage lies in capturing long-distance dependencies through the self-attention mechanism, and it is suitable for sequence-to-sequence tasks such as experimental step generation and parameter description. This type of model is represented by BART, T5, and FLAN-T5, and acquires powerful semantic modeling capabilities through large-scale text pretraining.

- **BART:** as shown in Figure 14.19, BART integrates the respective characteristics of BERT's bidirectional encoder and GPT's left-to-right decoder, and is based on the standard Seq2seq Transformer model. This makes it more suitable than BERT for the scenario of text generation. BART captures complex semantic associations in the experimental context through a bidirectional encoder, and combines an autoregressive decoder to generate a logically coherent sequence of operations [150]. Nishimura et al. [151] proposed the structured relevant work generation task in their paper, aiming to automatically generate relevant work parts that include the organization of research topics and novelty statements. This method utilizes LLMs (such as BART) to generate structured paragraphs and introduces new datasets and automatic evaluation metrics to assess the structure and novelty of the generated text. Therefore, the bidirectional encoder of BART is helpful for understanding complex contexts. Combined with the autoregressive decoder, it can generate a well-structured operation sequence and is suitable for formulating primary experimental plans.
- **T5 and FLAN-T5:** T5 was released by Google in 2020 and for the first time systematically unified all NLP tasks into a "text-to-text" format [116]. That is to say, whether it is translation, classification, summarization, or question answering, both the input and output are processed as strings [116]. FLAN-T5 is an enhanced version of T5 and was proposed by Google in 2022. The core improvement is the large-scale, multitask instruction fine-tuning, which makes the model better at following the natural language instructions provided by humans [152]. T5 and FLAN-T5 support the task of optimizing experimental parameters through a unified text-to-text modeling framework.

The limitations of the model of this type of general framework lie in its insufficient processing ability for unstructured data, such as charts and molecular structures, and its weak cross-modal integration ability of experimental logic. For example, in the experimental planning involving three-dimensional crystal structures, the standard Transformer has difficulty in directly associating text descriptions with spatial topological features, resulting in problems such as the omission of key spatial operation details in the generated experimental steps.

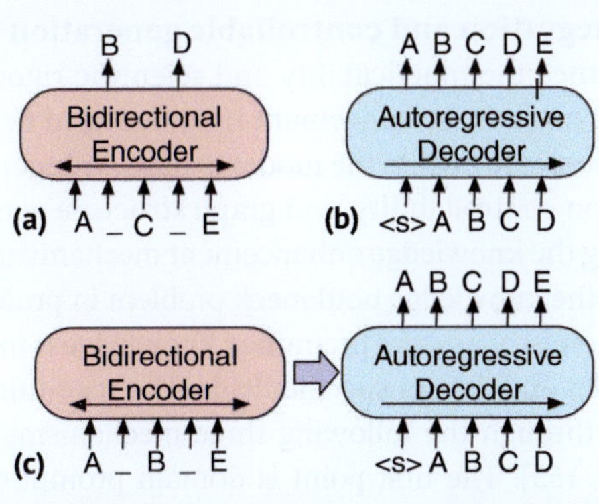

Figure 14.19 The encoder-decoder architectures of (a) BERT, (b) GPT, and (c) BART. *Source*: [150] / arXiv.

However, the materials composition, reaction pathways, and parameter relationships in experimental data often present graph structure characteristics, such as molecular bond diagrams and device connection diagrams. It is difficult for traditional sequence models to efficiently capture such structured semantics. Therefore, the structure-aware model that integrates graph neural networks (GNNs) has become a research hotspot. This type of architecture enhances the modeling ability of hierarchical and correlated data by introducing Inductive Bias. One of them is about the integration of Transformer and graph attention mechanism. Due to the lack of an effective mechanism for encoding graph structure information in the Transformer, it lags behind mainstream GNN variants in graph-level prediction tasks. Ying et al. [153] proposed the Graphormer model, a graph representation learning model based on the standard Transformer, which effectively captures graph structure information through three structural coding methods: centrality coding, spatial coding, and edge coding. Experiments show that Graphormer significantly outperforms mainstream GNN models in tasks such as the Open Graph Benchmark Large-scale Challenge (OGB-LSC) quantum chemistry dataset, MolHIV, and MolPCBA. Moreover, it is theoretically proved that its expression ability can cover classical GNN variants such as Graph Isomorphism Networks (GINs) and Graph Convolutional Networks (GCNs). It provides a new paradigm for the application of Transformer in the field of graphs. In addition, a hybrid Transformer–Graph framework proposed by Madani et al. [152] breaks through the limitations of traditional GNNs in materials property prediction by explicitly modeling four-body interactions and transfer learning, especially performing well in scenarios with scarce data. This framework provides an efficient tool for accelerating the discovery of new materials, such as high-strength alloys and stable energy storage materials, and opens up a new path for the application of GNNs in many-body physics.

In other words, the development from general frameworks to structure-aware modeling reflects the upgrade of multimodal reasoning models from text generation to semantic understanding. Such structure-aware models demonstrate higher accuracy and generalization ability when predicting the performance of new materials or planning complex reaction paths.

2. **Knowledge integration and controllable generation**

In order to enhance the practicability and scientific rigor of the experimental planning model, multiple enhancement modules need to be integrated outside the main architecture to enable the model to have stronger knowledge guidance ability, instruction controllability, and graph structure generation ability.

First, regarding the knowledge enhancement mechanism: Transformer models often encounter the knowledge bottleneck problem in professional fields such as materials and chemistry, i.e., the pretrained knowledge is insufficient to cover the experimental rules and data in specific fields. The scientific rigor of the model is often enhanced through the following three mechanisms by injecting external knowledge [154, 155]. The first point is domain prompt engineering. Through manually designed or automatically generated domain prompts, it guides the model to focus on the semantic space of the experimental target [13, 20]. For instance, in the design and preparation of the stainless steel process route, if one inputs the prompt template, "Based on the {material type} system, design an {experimental type} plan with {performance indicators} ≥ {threshold}, and the constraint condition: {prohibited components}," the model can generate targeted process flows and steps. The second point is the embedding of the knowledge graph. For example, the entity relationships in the materials database (such as Materials Project), such as the relationship between elements, atomic radii, and bond energies, are encoded as low-dimensional vectors and injected into the model through the cross-modal fusion layer [156, 157]. The third point is to introduce Cross-attention over Database. Establish a cross-attention mechanism between the fragments of the literature or materials database and the hidden layer of the model, enabling the model to dynamically retrieve relevant experimental knowledge, such as reaction conditions and existing formulas, during the generation process.

Second, through the constraints of control instructions: the actual experimental steps often need to meet the requirements of grammatical legitimacy and physical feasibility. The following control strategies gradually enhance the output of the agent system reliability from soft constraints to hard rules. Prefix-tuning and Soft Prompting fine-tune the model's behavior direction in a particular task by applying lightweight controls to the output space. Constrained Decoding introduces hard constraint rules during the decoding process, such as naming rules and parameter ranges, to prevent the generation of invalid or dangerous experimental instructions. Template-aware Decoding combines experimental templates such as YAML and LabScript to guide the model to output structured scripted instruction sequences [106].

Finally, it is about the generation of the experimental planning process, achieving the construction from linear sequences to structured processes. The real experimental process usually contains complex control structures, such as conditional jumps (if or else), parallel step, and intermediate result dependency. This makes it difficult for traditional linear sequence generation methods to accurately express the actual logic of the experiment. See et al. [158] combined pointer networks and generation networks. Through the pointer mechanism, they directly copied keywords from the original text, and at the same time utilized the generation network to generate fluent sentences, balancing the issues of accuracy and fluency of the summary. Meanwhile, the coverage mechanism was introduced. By tracking the attention distribution of the generated content, it inhibits the repeated generation of the same content, alleviates the redundancy problem in the summary, and avoids the generation of repetitive fragments. This pointer-generator network combines the generative language model with the pointer mechanism, and is capable of flexibly switching between the original text and the context, reflecting the structure of the process more accurately. In addition, the relation tagging method [159] automatically adds semantic tags such as dependency, sequence, and condition between experimental operations, enabling the model to automatically recognize and construct flowcharts, thereby enhancing the execution ability of the experimental automation system.

3. Database construction: multisource fusion and automatic expansion

The performance of the Transformer model is highly dependent on high-quality training data. In the field of materials science, labeled data is scarce. Therefore, multiple strategies need to be adopted to jointly construct the training corpus. To break through this bottleneck, a three-dimensional data construction strategy of multisource acquisition–intelligent annotation–synthetic expansion needs to be adopted to form a high-quality training corpus system covering multiple modalities and multiple tasks [160].

The first point is to construct a professional literature dataset, and heterogeneous data sources need to be constructed into standardized corpora. Professional literature is the core carrier of experimental knowledge, but the problem of fragmented data formats is significant, and structured datasets need to be constructed through cross-modal parsing and entity alignment [161]. At present, Mishra et al. [162] pretrained the LLaMa model using a large materials science corpus containing over 30 billion tokens, and then fine-tuned the instructions and tasks to develop the LLaMat model. This model performs outstandingly in NLP and structured information extraction related to materials science. It can effectively extract information from the texts and tables of scientific literature and has achieved a major breakthrough in the construction of databases in the field of materials science. Besides, Liu et al. [163] proposed a pipeline to construct high-quality datasets for text mining in materials science. This method aims to provide an effective solution for the high-quality acquisition of materials science text mining datasets in a pipeline-based manner. This pipeline consists of four steps: traceable automatic literature acquisition, downstream task-driven literature preprocessing, label definition and data annotation, and text data enhancement. This method reduces the high overhead in the processes of obtaining, processing, annotation, and expansion of large-scale text data, thereby achieving the acquisition of high-quality datasets for text mining in materials science. Nowadays, University of Science and Technology Beijing and other departments have launched a national-level materials data platform named the National Materials Big Data Center; this data platform aims to promote the innovative development of the new materials industry and is committed to becoming a core resource hub in the field of new materials.

The second point is to achieve automatic data annotation. Through methods such as rule engines and template systems, experimental steps are extracted from unstructured literature and matched to standard operation types [160]. Wang et al. [164] from the University of Chinese Academy of Sciences proposed a BERT-based model for extracting knowledge from unstructured domain texts, aiming to solve the problem of automatically extracting specific information from specialized texts in specific domains.

In addition, using LLMs such as GPT to generate a large amount of synthetic texts with experimental rationality is also one of the methods for constructing training corpora. In the field of materials science, the experimental steps in traditional material data acquisition are cumbersome. The limitations of long cycles and high costs often make real experimental data relatively scarce [1, 3].

In the task of planning experiments in materials science, generating a large amount of synthetic texts with experimental rationality with the help of LLMs such as GPT has become an important means to expand the training dataset and improve the generalization ability of the model [165, 166]. By carefully designing prompt engineering, the model can generate experimental process descriptions that conform to scientific logic, such as materials preparation, reaction conditions, and detection methods. Besides, prompt engineering combines template fill-in-the-blank and domain knowledge base control to generate the structure, ensuring clear semantics and ease of parsing. Meanwhile, the rule engine and knowledge base are utilized to conduct logical verification of material tests and security screening on the generated texts to enhance data quality. This type of synthetic data not only enriches the coverage of experimental types and enhances the model's processing ability for rare experimental structures, but also supports cross-modal training of natural language and control instructions, improving the execution feasibility and semantic consistency of the experimental instruction generation model. Related works such as ChemCrow [106], SCALMS [15], and Schwaller et al. [167], among others, have all demonstrated the potential and practical application value of large models in generating material experimental processes.

In conclusion, the structural design of the Transformer model for the materials experiment planning task requires systematic planning from three aspects: the selection of the underlying architecture, the functional enhancement mechanism, and data support. Through the reasonable combination of the standard architecture and the structure-aware model, the enhanced design of the knowledge and control module, as well as the supporting strategy of multisource data construction, the model can not only achieve smooth and natural generation of experimental language, but also ensure the executability, rationality, and scientificity of the results. These improvements have laid a solid foundation for building material research agents that truly possess the ability of autonomous planning and decision-making.

14.3.3 Reinforcement Learning Strategies for Iterative Optimization

In Chapter 3 of this book, we talked about RL of LLMs. RL is a trial-and-error learning method. The learning system learns the optimal behavioral strategy by interacting with the environment, obtaining environmental feedback signals, and adaptively adjusting based on the feedback signals, in order to maximize its long-term cumulative rewards. Figure 14.20 shows the basic framework of reinforcement learning. The RL methods mentioned above basically take the decision-making problem of a single agent as the main research object. However, in many real-life scenarios, there are often multiple agents simultaneously within a system. Each agent needs to not only interact with the environment but also cooperate or compete with other agents. Meanwhile, the change of the environmental state is no longer determined by a single agent, but jointly determined by the behaviors of all agents [75, 168]. The core technology of multi-agent reinforcement learning (MARL) is an extension

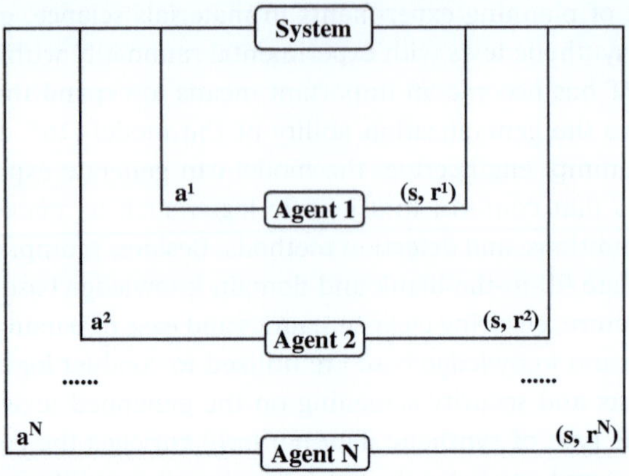

Figure 14.20 The basic framework of reinforcement learning in multi-agent scenarios.

of RL, focusing on scenarios where multiple agents learn and make decisions in a shared environment. Therefore, in autonomous materials research, experimental design and parameter tuning often involve complex high-dimensional spaces, uncertain feedback, and high trial-and-error costs. RL offers a data-driven strategy optimization paradigm that can gradually learn the optimal strategy through multiple rounds of trials, and is used for automatic exploration of experimental conditions, formula adjustment, and performance improvement. Compared with static optimization algorithms, RL can dynamically adapt to experimental feedback and make decision optimizations under incomplete information [169, 170].

14.3.3.1 MARL in Different Tasks

In cooperative tasks, MARL focuses on the mutual coordination of behaviors among agents. The concept of multi-agent collaboration was first proposed by Claus and Boutilier in [171]. In this article, Claus and Boutilier proposed the famous Coordination Game to elaborate on the importance of multi-agent collaboration, and its general meaning is: although agents have exactly the same interests, since agents make decisions independently, when there are multiple equilibrium solutions in the game, they may not reach an agreement with each other, thus preventing the learning algorithm from learning the equilibrium strategy. Therefore, each agent in the game must consider the behaviors of other agents and keep in step with each other. Based on this idea, Claus and Boutilier proposed the Joint Action Learners (JAL) algorithm and introduced the mechanism of policy modeling in it to track the policy changes of other agents. Similarly, the Optimal Adaptive Learning (OAL) algorithm proposed by Wang and Sandholm is the first and currently the only algorithm with complete convergence guarantee [172]. This algorithm is based on a brand-new equilibrium selection mechanism, Biased Adaptive Play, which can converge to the optimal equilibrium strategy of cooperative Markov games with a 100% probability.

In competitive tasks, the main purpose of MARL is how to enable agents to gain as much benefit as possible in the competition with rivals. In 1947, the Scientific Giant von Neumann proposed the Minimax Principle for solving two-person competitive problems in his work named The Theory of Game and Economic Behavior [173], which states that the benefits obtained by an agent are maximized by itself while being minimized by its opponent. It is precisely based on this principle that Littman proposed the reinforcement learning algorithm Minimax-Q in Markov games [174]. At each state, the Minimax-Q algorithm models the competition process between the agent and the opponent as a general game and solves the minimum and maximum strategies of the game for the agent to guide the choice of its actions. After this, Littman proposed the Friend-or-Foe (FoF) algorithm for solving multi-person competitive problems [175]. The Minimax Principle is the best strategy selection method when the opponent's strategy is assumed to be optimal. However, in many cases, the opponents of the agent are not so intelligent, and they do not always minimize the agent's interests. For such suboptimal opponents, the principle of minimum and maximum is not the optimal solution. In blog theory, solving the optimal response strategy and the regret minimization method are common approaches for suboptimal opponents.

These methods have also been introduced into MARL. For instance, the Nonstationary Converging Policy (NSCP) algorithm proposed by Weinberg and Rosenschein maintains the opponent's policy model for the agent [176]. At each state, this algorithm calculates the corresponding optimal response policy based on the opponent's policy model. This is used for action selection and value function update.

In hybrid tasks, cooperation and competition among agents coexist, thus posing greater challenges to MARL. The primary issue among them is how to define the learning objectives of MARL in hybrid learning tasks. Because the interests among agents are intertwined, it is impossible to maximize the interests of each agent simultaneously. Inspired by the Minimax-Q algorithm, Hu and Wellman introduced the core concept Nash Equilibrium in game theory into MARL and proposed the NashQ learning algorithm (NashQ) [177, 178]. Nash Equilibrium is the optimal solution of the game when all agents are in an absolutely rational state. In Nash Equilibrium, any unilateral strategy change will cause a loss of utility value for the agent. The proposal of NashQ has opened up new ideas for MARL. After that, the concepts of other types of equilibrium solutions have also been successively introduced into MARL, forming the family of equal-based MARL. For instance, Greenwald and Hall proposed the correlated Q-learning algorithm (CEQ) based on the concept of Correlated Equilibrium [179]. Compared with Nash Equilibrium, Correlated Equilibrium allows for correlation in the action choices of agents. Furthermore, based on the concept of Stackelberg Equilibrium, Kononen [180] proposed the asymmetric Q-learning algorithm (Asymmetric Q), which solved the multi-agent learning problem in the case of information asymmetry.

MARL often involves multiple agents learning and making decisions in a dynamic and interactive environment. It is confronted with challenges such as instability, dimensionality disaster, and collaboration and competition among agents. However, with the continuous advancement of algorithms, MARL has shown great potential and prospects in multiple complex application fields.

14.3.3.2 Online Learning and Security Exploration Mechanism

In autonomous materials research, although RL has great potential to explore experimental space and optimize material properties, its practical application faces a key challenge: exploration may bring some dangers. That is, the essence of RL lies in trial-and-error learning. The model may propose high-risk or even infeasible experimental operations without fully understanding the experimental boundaries and safety constraints, such as heating up to dangerous temperatures, using flammable or restricted chemicals, and overloading equipment. This may not only lead to the failure of the experiment, but also pose a threat to the safety of the experimental equipment and personnel [13, 18–20]. To effectively deal with this problem, some key mechanisms need to be introduced in the design of the agent systems, such as: Safe Reinforcement Learning (Safe RL) [181], Proximal Policy Optimization (PPO) [182], and human–machine collaboration and human intervention mechanisms, among others. The introduction of these mechanisms not only ensures the availability of RL in the real material experiment system, but also enhances the robustness and credibility of the closed-loop optimization system, providing a safe and reliable intelligent core for self-driven laboratories' autonomous materials research.

14.3.3.2.1 Safe Reinforcement Learning

In RL, an agent learns a strategy through interaction with the environment to maximize the long-term cumulative return. However, traditional reinforcement learning algorithms often do not consider the security of agent behaviors when optimizing returns, resulting in potentially unsafe behaviors during the training or deployment process. Safe RL was proposed precisely under this background [181]. It aims to optimize the rewards while ensuring that the agent's behavior complies with certain security constraints. Its core objective is to maximize task rewards while ensuring that the strategy meets specific security constraints during the training and deployment processes. This type of method is particularly suitable for high-risk, high-cost, and irreversible actual environments, such as material experiments, autonomous driving, medical robots, or energy systems, among others.

First of all, in Safe RL, safety constraints are the criteria that determine whether the agent's behavior is safe. Typically, safety constraints can be reflected in the need to clearly define the physical, chemical, legal, or ethical boundaries of the system, represented by formulas, rules, or probabilistic models. These constraints can be hard constraints, i.e., constraints that must be strictly adhered to, or soft constraints, i.e., they can be violated within a certain range but at a cost. In terms of materials and chemical experiments, physical and chemical safety rules, such as material thermal stability thresholds, pressure limits, and toxicity judgment rules, are embedded in the RL policy network. Hard constraints can be adopted or the reward values for rule-violating behaviors can be reduced through penalty functions, thereby guiding the RL to avoid dangerous operations [19, 20].

Second, in traditional reinforcement learning, the optimization objective is usually to maximize the expected return. However, in secure reinforcement learning, we need risk-sensitive optimization objectives. Specifically, Risk can be measured in various ways, such as Variance, Value at Risk (VaR), Conditional Value at Risk (CVaR), and other measurement methods [181]. As shown in Eq. 14.1, by introducing risk measurement, security reinforcement learning can prevent agents from making high-risk decisions in certain uncertain situations:

$$\left[\text{maximize } E\left[\sum_{t=0}^{\infty} \gamma^t R^t\right] \text{subject to Risk}(G_t) \leq \delta\right] \tag{14.1}$$

Finally, in order to guide the agent to learn safe behaviors, Reward Shaping is a common technique [183, 184]. The specific approach is to modify the reward function to a combination that considers both returns and security simultaneously as shown in Eq. 14.2:

$$R'(s,a,s') = R(s,a,s') - \lambda \text{Cost}(s,a,s') \tag{14.2}$$

Among them, $R(s,a,s')$ is the original reward function, $\text{Cost}(s,a,s')$ is the safety cost when the agent enters the state s' after taking action (a) in the state s, and λ is the cost weight. In this way, when the agent adopts unsafe behavior, it will be punished, thereby prompting the agent to learn safer strategies.

14.3.3.2.2 Proximal Policy Optimization

The PPO algorithm is a reinforcement learning algorithm proposed by OpenAI in 2017 [182]. It is regarded as the state-of-the-art method in the field of reinforcement learning at present and is also one of the most widely applicable algorithms. The core idea is to limit the variation range between the old and new strategies when updating the strategies, thereby avoiding the problem of unstable performance caused by drastic changes in the strategies. PPO achieves a balance between the stability and convergence of the training process by introducing the clipped objective function to control the step size of policy update while maintaining the optimization efficiency.

For the current strategy $\pi_\theta(a_t|s_t)$ and the old strategy $\pi_{\theta_{\text{old}}}(a_t|s_t)$, calculate the action value advantage function A_t, and define the objective function as in Eq. 14.3:

$$L^{\text{PPO}}(\theta) = E_t\left[\min\left(\frac{\pi_\theta(a_t|s_t)}{\pi_{\theta_{\text{old}}}(a_t|s_t)}\right)A_t,\ \text{clip}\left(\frac{\pi_\theta(a_t|s_t)}{\pi_{\theta_{\text{old}}}(a_t|s_t)}, 1-\epsilon, 1+\epsilon\right)A_t\right] \tag{14.3}$$

Among them, the clip function limits the probability ratio of the old and new strategies within the interval of $[1-\epsilon, 1+\epsilon]$ to avoid performance degradation caused by excessive strategy update amplitude. In addition, PPO allows for multiple iterative optimizations of the same batch of sampling data (typically 3–10 rounds), improving sample efficiency by reusing historical data. This is particularly important in scenarios where sample acquisition costs are high, such as robot control.

PPO improves sample efficiency while ensuring the stability of policy updates by pruning the objective function and multi-round data utilization. It has become one of the mainstream algorithms in the current practical applications of reinforcement learning, especially suitable for actual scenarios that require a balance between exploration and security and have limited computing resources.

14.3.3.2.3 Human–Machine Collaboration and Human Intervention Mechanisms

For exploration areas with high risks or high uncertainties, the system should introduce a manual confirmation process, retain the final decision-making power of experts over key experimental procedures, and ensure that RL does not operate independently in a black-box state [13, 18–20].

In RL, traditional policy updates are usually based on the reward signals generated by environmental interactions. However, this process often faces problems such as low sample efficiency, risks in exploration, and sparse rewards in real complex tasks. Therefore, introducing Human Feedback has become an important way to improve the quality and security of strategy learning. By collecting human preference, instruction, or correction signals in real time, the agent can dynamically adjust its behavior during the policy learning process. Meanwhile, based on observing the evolution of the agent's behavior, humans gradually transform from operation executors to policy supervisors or task designers. Thus, a closed-loop iterative mechanism of coevolution between humans and machines is formed as shown in Figure 14.21.

The key collaboration models are as follows: the first is supervision and control. Humans set high-level goals, while machines independently perform tasks and update their status regularly. The second one is shared control. Humans and machines share task control rights in real time and allocate decision-making power according to task requirements. The third one is the fully collaborative system. Humans and machines dynamically adjust their roles based on real-time feedback.

For example, based on the Reinforcement Learning from Human Feedback (RLHF) architecture proposed by OpenAI, namely the three-stage process of generating content through manual evaluation, training the reward model, and using it for strategy optimization, the alignment training of the LLM InstructGPT has been achieved [183]. DeepMind's InstructRL framework further proposes to combine natural language instructions with human preferences to achieve policy fine-tuning, demonstrating the guiding ability of Human Feedback on the direction and diversity of policies [185]. Furthermore, there are branches such as Safe RL and Cooperative Reinforcement Learning, which emphasize the realization of feedback-driven dynamic role adjustment and risk control in human–machine shared decision-making.

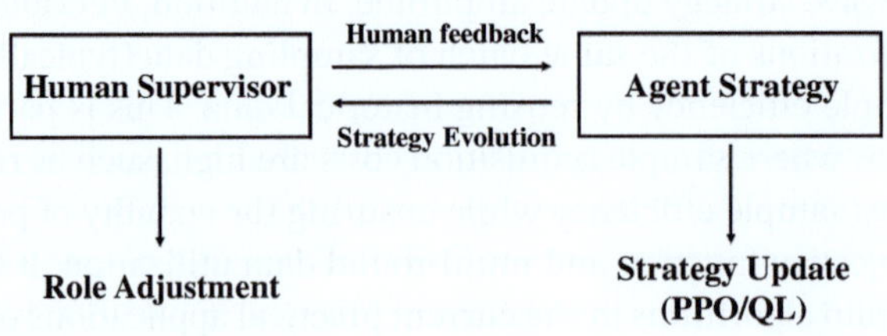

Figure 14.21 Schematic diagram of the closed-loop iterative mechanism of human–machine coevolution.

The model presents a closed-loop system of feedback, response, and adjustment. Humans and machines jointly act as part of the learning system, constantly updating each other's role boundaries and behavioral strategies, thereby achieving efficient, safe, and controllable reinforcement learning.

References

1 Wang WY, Zhang S, Li G, Lu J, Ren Y, Wang X, et al. Artificial intelligence enabled smart design and manufacturing of advanced materials: The endless Frontier in AI^+ era. Materials Genome Engineering Advances. 2024;2(3):e56.

2 Hart GLW, Mueller T, Toher C, Curtarolo S. Machine learning for alloys. Nature Reviews Materials. 2021;6(8):730–55.

3 Ramprasad R, Batra R, Pilania G, Mannodi-Kanakkithodi A, Kim C. Machine learning in materials informatics: recent applications and prospects. npj Computational Materials. 2017;3(1):54.

4 Xie J. Prospects of materials genome engineering frontiers. Materials Genome Engineering Advances. 2023;1(2):e17.

5 Yoshikawa N, Skreta M, Darvish K, Arellano-Rubach S, Ji Z, Bjørn Kristensen L, et al. Large language models for chemistry robotics. Autonomous Robots. 2023;47(8):1057–86.

6 Raccuglia P, Elbert KC, Adler PDF, Falk C, Wenny MB, Mollo A, et al. Machine-learning-assisted materials discovery using failed experiments. Nature. 2016;533(7601):73–6.

7 Tian S, Jiang X, Wang W, Jing Z, Zhang C, Zhang C, et al. Steel design based on a large language model. Acta Materialia. 2025;285:120663.

8 Li S, Dong Z, Jin J, Pan H, Hu Z, Hou R, et al. Optimal design of high-performance rare-earth-free wrought magnesium alloys using machine learning. Materials Genome Engineering Advances. 2024;2(2):e45.

9 Jiang X, Wang Y, Jia B, Qu X, Qin M. Using machine learning to predict oxygen evolution activity for transition metal hydroxide electrocatalysts. ACS Applied Materials Interfaces. 2022;14(36):41141–8.

10 Xue D, Balachandran PV, Hogden J, Theiler J, Xue D, Lookman T. Accelerated search for materials with targeted properties by adaptive design. Nature Communications. 2016;7(1):11241.

11 Wen C, Wang C, Zhang Y, Antonov S, Xue D, Lookman T, et al. Modeling solid solution strengthening in high entropy alloys using machine learning. Acta Materialia. 2021;212:116917.

12 Ghafarollahi A, Buehler MJ. SciAgents: Automating scientific discovery through multi-agent intelligent graph reasoning. arXiv preprint arXiv:2409.05556; 2024. Available from: https://arxiv.org/abs/2409.05556

13 Ghafarollahi A, Buehler MJ. Automating alloy design and discovery with physics-aware multimodal multiagent AI. Proceedings of the National Academy of Sciences of the United States of America. 2025;122(4):e2414074122.

14 Chen J, Hu X, Liu S, Huang S, Tu WW, He Z, et al. LLMArena: Assessing capabilities of large language models in dynamic multi-agent environments. In: 62nd Annual Meeting of the Association for Computational Linguistics (Volume 1: Long Papers) [Internet]. Association for Computational Linguistics; 2024. pp. 13055–77. Available from: https://aclanthology.org/2024.acl-long.705

15 Jia S, Zhang C, Fung V. LLMatDesign: Autonomous materials discovery with large language models. arXiv preprint arXiv:2406.13163; 2024. Available from: https://arxiv.org/abs/2406.13163

16 Gottweis J, Weng WH, Daryin A, Tu T, Palepu A, Sirkovic P, et al. Towards an AI co-scientist. arXiv preprint arXiv:2502.18864; 2025. Available from: https://arxiv.org/abs/2502.18864

17 Wang L, Ma C, Feng X, Zhang Z, Yang H, Zhang J, et al. A survey on large language model based autonomous agents. Frontiers of Computer Science. 2024;18(6):186345.

18 Zhu Q, Zhang F, Huang Y, Xiao H, Zhao L, Zhang X, et al. An all-round AI-Chemist with a scientific mind. National Science Review. 2022;9(10):nwac190.

19 Boiko DA, MacKnight R, Kline B, Gomes G. Autonomous chemical research with large language models. Nature. 2023;624(7992):570–8.

20 Song T, Luo M, Zhang X, Chen L, Huang Y, Cao J, et al. A multiagent-driven robotic AI chemist enabling autonomous chemical research on demand. Journal of the American Chemical Society. 2025;147(15):12534–45.

21 Kang Y, Kim J. ChatMOF: An artificial intelligence system for predicting and generating metal-organic frameworks using large language models. Nature Communications. 2024;15(1):4705.

22 Russell SJ, Norvig P. Artificial Intelligence: A Modern Approach, 3rd edition, Global edition, Prentice Hall Series in Artificial Intelligence. Upper Saddle River, NJ: Pearson Education, Inc.; 2016.

23 Qian C, Liu W, Liu H, Chen N, Dang Y, Li J, et al. ChatDev: Communicative agents for software development. arXiv preprint arXiv:2307.07924; 2023. Available from: https://arxiv.org/abs/2307.07924

24 Chen W, Su Y, Zuo J, Yang C, Yuan C, Chan CM, et al. AgentVerse: Facilitating multi-agent collaboration and exploring emergent behaviors. arXiv preprint arXiv:2308.10848; 2023. Available from: https://arxiv.org/abs/2308.10848

25 Lu JM, Wang HF, Guo QH, Wang JW, Li TT, Chen KX, et al. Roboticized AI-assisted microfluidic photocatalytic synthesis and screening up to 10,000 reactions per day. Nature Communications. 2024;15(1):8826.

26 Hu EJ, Shen Y, Wallis P, Allen-Zhu Z, Li Y, Wang S, et al. LoRA: Low-rank adaptation of large language models. arXiv preprint arXiv:2106.09685; 2021. Available from: https://arxiv.org/abs/2106.09685

27 Wu C, Gan Y, Ge Y, Lu Z, Wang J, Feng Y, et al. LLaMA Pro: Progressive LLaMA with block expansion. arXiv preprint arXiv:2401.02415; 2024. Available from: https://arxiv.org/abs/2401.02415

28 He J, Zhou C, Ma X, Berg-Kirkpatrick T, Neubig G. Towards a unified view of parameter-efficient transfer learning. arXiv preprint arXiv:2110.04366; 2021. Available from: https://arxiv.org/abs/2110.04366

29 Touvron H, Martin L, Stone K, Albert P, Almahairi A, Babaei Y, et al. Llama 2: Open foundation and fine-tuned chat models. arXiv preprint arXiv:2307.09288; 2023. Available from: https://arxiv.org/abs/2307.09288

30 Liang H, Wang C, Yu H, Kirsch D, Pant R, McDannald A, et al. Real-time experiment-theory closed-loop interaction for autonomous materials science. arXiv preprint arXiv:2410.17430; 2024. Available from: https://arxiv.org/abs/2410.17430

31 Wei J, Wang X, Schuurmans D, Bosma M, Ichter B, Xia F, et al. Chain-of-Thought prompting elicits reasoning in large language models. arXiv preprint arXiv:2201.11903; 2022. Available from: https://arxiv.org/abs/2201.11903

32 Kojima T, Gu SS, Reid M, Matsuo Y, Iwasawa Y. Large language models are zero-shot reasoners. arXiv preprint arXiv:2205.11916; 2022. Available from: https://arxiv.org/abs/2205.11916

33 Park JS, O'Brien JC, Cai CJ, Morris MR, Liang P, Bernstein MS. Generative agents: Interactive Simulacra of human behavior. arXiv preprint arXiv:2304.03442; 2023. Available from: https://arxiv.org/abs/2304.03442

34 Achiam J, Adler S, Agarwal S, Ahmad L, Akkaya I, Aleman FL, et al. GPT-4 technical report. arXiv preprint arXiv:2303.08774; 2023. Available from: https://arxiv.org/abs/2303.08774

35 Barrault L, Chung YA, Meglioli MC, Dale D, Dong N, Duppenthaler M, et al. Seamless: Multilingual expressive and streaming speech translation. arXiv preprint arXiv:2312.05187; 2023. Available from: https://arxiv.org/abs/2312.05187

36 Grauman K, Westbury A, Torresani L, Kitani K, Malik J, Afouras T, et al. Ego-Exo4D: Understanding skilled human activity from first- and third-person perspectives. arXiv preprint arXiv:2311.18259; 2023. Available from: https://arxiv.org/abs/2311.18259

37 Touvron H, Lavril T, Izacard G, Martinet X, Lachaux MA, Lacroix T, et al. LLaMA: Open and efficient foundation language models. arXiv preprint arXiv:2302.13971; 2023. Available from: http://arxiv.org/abs/2302.13971

38 Grcić M, Gadetsky A, Brbić M. Fine-grained classes and how to find them. arXiv preprint arXiv:2406.11070; 2024. Available from: https://arxiv.org/abs/2406.11070

39 Zhao C, Habule M, Zhang W. Large language models (LLMs) as research subjects: Status, opportunities and challenges. New Ideas in Psychology. 2025;79:101167.

40 MacLeod BP, Parlane FGL, Morrissey TD, Häse F, Roch LM, Dettelbach KE, et al. Self-driving laboratory for accelerated discovery of thin-film materials. Science Advances. 2020;6(20):eaaz8867.

41 Shetty P, Rajan AC, Kuenneth C, Gupta S, Panchumarti LP, Holm L, et al. A general-purpose material property data extraction pipeline from large polymer corpora using natural language processing. npj Computational Materials. 2023;9(1):52.

42 Gu Y, Tinn R, Cheng H, Lucas M, Usuyama N, Liu X, et al. Domain-specific language model pretraining for biomedical natural language processing. ACM Transactions on Computing for Healthcare. 2022;3(1):1–23.

43 Trewartha A, Walker N, Huo H, Lee S, Cruse K, Dagdelen J, et al. Quantifying the advantage of domain-specific pre-training on named entity recognition tasks in materials science. Patterns. 2022;3(4):100488.

44 Gupta S, Mahmood A, Shetty P, Adeboye A, Ramprasad R. Data extraction from polymer literature using large language models. Communications Materials. 2024;5(1):269.

45 Pei Z, Yin J, Zhang J. Language models for materials discovery and sustainability: Progress, challenges, and opportunities. Progress in Materials Science. 2025;154:101495.

46 Rakin AS, He Z, Fan D. Bit-flip attack: Crushing neural network with progressive bit search. In: 2019 IEEE/CVF International Conference on Computer Vision (ICCV) [Internet]. IEEE Press; 2019. pp. 1211–20. Available from: https://ieeexplore.ieee.org/document/9009040/

47 O'Neill C, Ghosal T, Răileanu R, Walmsley M, Bui T, Schawinski K, et al. Sparks of science: Hypothesis generation using structured paper data. arXiv preprint arXiv:2504.12976; 2025. Available from: https://arxiv.org/abs/2504.12976

48 Tran HD, Kim C, Chen L, Chandrasekaran A, Batra R, Venkatram S, et al. Machine-learning predictions of polymer properties with Polymer Genome. Journal of Applied Physics. 2020;128(17):171104.

49 Wu C, Chen L, Deshmukh A, Kamal D, Li Z, Shetty P, et al. Dielectric polymers tolerant to electric field and temperature extremes: Integration of phenomenology, informatics, and experimental validation. ACS Applied Materials & Interfaces. 2021;13(45):53416–24.

50 Kononova O, He T, Huo H, Trewartha A, Olivetti EA, Ceder G. Opportunities and challenges of text mining in materials research. iScience. 2021;24(3):102155.

51 Foppiano L, Castro PB, Ortiz Suarez P, Terashima K, Takano Y, Ishii M. Automatic extraction of materials and properties from superconductors scientific literature. Science and Technology of Advanced Materials: Methods. 2023;3(1):2153633.

52 Cheung JJ, Zhuang Y, Li Y, Shetty P, Zhao W, Grampurohit S, et al. PolyIE: A dataset of information extraction from polymer material scientific literature. arXiv preprint arXiv:2311.07715; 2023. Available from: http://arxiv.org/abs/2311.07715

53 Dagdelen J, Dunn A, Lee S, Walker N, Rosen AS, Ceder G, et al. Structured information extraction from scientific text with large language models. Nature Communications. 2024;15(1):1418.

54 Batra R, Song L, Ramprasad R. Emerging materials intelligence ecosystems propelled by machine learning. Nature Reviews Materials. 2020;6(8):655–78.

55 Zhang Q, Hu Y, Yan J, Zhang H, Xie X, Zhu J, et al. Large-language-model-based AI agent for organic semiconductor device research. Advanced Materials. 2024;36(32):2405163.

56 Ding Q, Miret S, Liu B. MatExpert: Decomposing materials discovery by mimicking human experts. arXiv preprint arXiv:2410.21317; 2024. Available from: http://arxiv.org/abs/2410.21317

57 Zhang H, Song Y, Hou Z, Miret S, Liu B. HoneyComb: A flexible LLM-based agent system for materials science. arXiv preprint arXiv:2409.00135; 2024. Available from: http://arxiv.org/abs/2409.00135

58 Pedersen SL, Jensen KJ. Instruments for automated peptide synthesis. In: Jensen KJ, Tofteng Shelton P, Pedersen SL, editors. Peptide Synthesis and Applications, Methods in Molecular Biology, Vol. 1047. Totowa, NJ: Humana Press; 2013. pp. 215–24. Available from: https://link.springer.com/10.1007/978-1-62703-544-6_15

59 Häse F, Roch LM, Aspuru-Guzik A. Next-generation experimentation with self-driving laboratories. Trends in Chemistry. 2019;1(3):282–91.

60 Burger B, Maffettone PM, Gusev VV, Aitchison CM, Bai Y, Wang X, et al. A mobile robotic chemist. Nature. 2020;583(7815):237–41.

61 Segler MHS, Waller MP. Neural-symbolic machine learning for retrosynthesis and reaction prediction. Chemistry—A European Journal. 2017;23(25):5966–71.

62 Granda JM, Donina L, Dragone V, Long DL, Cronin L. Controlling an organic synthesis robot with machine learning to search for new reactivity. Nature. 2018;559(7714):377–81.

63 Coley CW, Thomas DA, Lummiss JAM, Jaworski JN, Breen CP, Schultz V, et al. A robotic platform for flow synthesis of organic compounds informed by AI planning. Science. 2019;365(6453):eaax1566.

64 Venkatasubramanian V, Mann V. Artificial intelligence in reaction prediction and chemical synthesis. Current Opinion in Chemical Engineering. 2022;36:100749.

65 Xuan VD, Vo H, Murphy D, Nguyen HD. AgentSGEN: Multi-agent LLM in the loop for semantic collaboration and GENeration of synthetic data. arXiv preprint arXiv:2505.13466; 2025. Available from: https://arxiv.org/abs/2505.13466

66 Yao S, Zhao J, Yu D, Du N, Shafran I, Narasimhan K, et al. ReAct: Synergizing reasoning and acting in language models. arXiv preprint arXiv:2210.03629; 2022. Available from: https://arxiv.org/abs/2210.03629

67 Wu Q, Bansal G, Zhang J, Wu Y, Li B, Zhu E, et al. AutoGen: Enabling next-gen LLM applications via multi-agent conversation. arXiv preprint arXiv:2308.08155; 2023. Available from: http://arxiv.org/abs/2308.08155

68 Gim I, Lee S-s, Zhong L. Asynchronous LLM function calling. arXiv preprint arXiv:2412.07017; 2024. Available from: https://arxiv.org/abs/2412.07017

69 Duan Z, Wang J. Exploration of LLM multi-agent application implementation based on LangGraph + CrewAI. arXiv preprint arXiv:2411.18241; 2024. Available from: https://arxiv.org/abs/2411.18241

70 Wang J, Duan Z. Agent AI with LangGraph: A modular framework for enhancing machine translation using large language models. arXiv preprint arXiv:2412.03801; 2024. Available from: https://arxiv.org/abs/2412.03801

71 Feng S, Wang Z, Goyal P, Wang Y, Shi W, Xia H, et al. Heterogeneous swarms: Jointly optimizing model roles and weights for multi-LLM systems. arXiv preprint arXiv:2502.04510; 2025. Available from: https://arxiv.org/abs/2502.04510

72 Sun L, Yang Y, Duan Q, Shi Y, Lyu C, Chang YC, et al. Multi-agent coordination across diverse applications: A survey. arXiv preprint arXiv:2502.14743; 2025. Available from: https://arxiv.org/abs/2502.14743

73 Dorri A, Kanhere SS, Jurdak R. Multi-agent systems: A survey. IEEE Access. 2018;6:28573–93.

74 Tran K-T, Dao D, Nguyen M-D, Pham Q-V, O'Sullivan B, Nguyen HD. Multi-agent collaboration mechanisms: A survey of LLMs. arXiv preprint arXiv:2501.06322; 2025. Available from: https://arxiv.org/abs/2501.06322

75 Li Y-C, Liu Z-J, Hong Y-T, Wang J-C, Wang J-R, Li Y, et al. Multi-agent reinforcement learning based game: A survey. Acta Automatica Sinica. 2025;51(3):540–58.

76 Hong S, Zhuge M, Chen J, Zheng X, Cheng Y, Zhang C, et al. MetaGPT: Meta programming for a multi-agent collaborative framework. arXiv preprint arXiv:2308.00352; 2023. Available from: https://arxiv.org/abs/2308.00352

77 Cai W, Jiang J, Wang F, Tang J, Kim S, Huang J. A survey on mixture of experts in large language models. arXiv preprint arXiv:2407.06204; 2024. Available from: https://arxiv.org/abs/2407.06204

78 Jacobs RA, Jordan MI, Nowlan SJ, Hinton GE. Adaptive mixtures of local experts. Neural Computation. 1991;3(1):79–87.

79 Brown TB, Mann B, Ryder N, Subbiah M, Kaplan J, Dhariwal P, et al. Language models are few-shot learners. arXiv preprint arXiv:2005.14165; 2020. Available from: https://arxiv.org/abs/2005.14165

80 Liu J, Tang P, Wang W, Ren Y, Hou X, Heng PA, et al. A survey on inference optimization techniques for mixture of experts models. arXiv preprint arXiv:2412.14219; 2024. Available from: https://arxiv.org/abs/2412.14219

81 Skaltsis GM, Shin HS, Tsourdos A. A survey of task allocation techniques in MAS. In: 2021 International Conference on Unmanned Aircraft Systems (ICUAS) [Internet]. IEEE Press; 2021. pp. 488–97. Available from: https://ieeexplore.ieee.org/document/9476736/

82 Zhang K, Collins EG, Shi D. Centralized and distributed task allocation in multi-robot teams via a stochastic clustering auction. ACM Transactions on Autonomous and Adaptive Systems. 2012;7(2):1–22.

83 Whitbrook A, Meng Q, Chung PWH. Reliable, distributed scheduling and rescheduling for time-critical, multiagent systems. IEEE Transactions on Automation Science and Engineering. 2018;15(2):732–47.

84 Wu H, Li H, Xiao R, Liu J. Modeling and simulation of dynamic ant colony's labor division for task allocation of UAV swarm. Physica A: Statistical Mechanics and Its Applications. 2018;491:127–41.

85 Gupta O, Raskar R. Distributed learning of deep neural network over multiple agents. Journal of Network and Computer Applications. 2018;116:1–8.

86 Gupta A, Chaurasiya VK. Reinforcement learning based energy management in wireless body area network: A survey. In: 2019 IEEE Conference on Information and Communication Technology [Internet]. IEEE Press; 2019. pp. 1–6. Available from: https://ieeexplore.ieee.org/document/9066260/

87 Zhu C, Dastani M, Wang S. A survey of multi-agent deep reinforcement learning with communication. Autonomous Agents and Multi-Agent Systems. 2024;38(1):4.

88 Hausknecht M, Stone P. Deep recurrent Q-learning for partially observable MDPs. arXiv preprint arXiv:1507.06527; 2015. Available from: https://arxiv.org/abs/1507.06527

89 Iqbal S, Costales R, Sha F. ALMA: Hierarchical learning for composite multi-agent tasks. In: Koyejo S, Mohamed S, Agarwal A, Belgrave D, Cho K, Oh A, editors. Advances in Neural Information Processing Systems, Vol. 36. New York: Curran Associates, Inc.; 2022. pp. 7155–66. Available from: https://proceedings.neurips.cc/paper_files/paper/2022/file/2f27964513a28d034530bfdd117ea31d-Paper-Conference.pdf

90 Jumper J, Evans R, Pritzel A, Green T, Figurnov M, Ronneberger O, et al. Highly accurate protein structure prediction with AlphaFold. Nature. 2021;596(7873):583–9.

91 Zunger A. Inverse design in search of materials with target functionalities. Nature Reviews Chemistry. 2018;2(4):0121.

92 Hirschberg J, Manning CD. Advances in natural language processing. Science. 2015;349(6245):261–6.

93 Jurafsky D, Martin JH. Speech and Language Processing: An Introduction to Natural Language Processing, Computational Linguistics, and Speech Recognition, Prentice Hall Series in Artificial Intelligence. Upper Saddle River, NJ: Prentice Hall; 2000.

94 Zhang S, Dong L, Li X, Zhang S, Sun X, Wang S, et al. Instruction tuning for large language models: A survey. arXiv preprint arXiv:2308.10792; 2023. Available from: https://arxiv.org/abs/2308.10792

95 Coley CW, Eyke NS, Jensen KF. Autonomous discovery in the chemical sciences part I: Progress. Angewandte Chemie International Edition. 2020;59(51):22858–93.

96 Dinh T, Zeng Y, Zhang R, Lin Z, Gira M, Rajput S, et al. LIFT: Language-interfaced fine-tuning for non-language machine learning tasks. arXiv preprint arXiv:2206.06565; 2022. Available from: https://arxiv.org/abs/2206.06565

97 Liang J, Huang W, Xia F, Xu P, Hausman K, Ichter B, et al. Code as policies: Language model programs for embodied control. arXiv preprint arXiv:2209.07753; 2022. Available from: https://arxiv.org/abs/2209.07753

98 Kim E, Jensen Z, van Grootel A, Huang K, Staib M, Mysore S, et al. Inorganic materials synthesis planning with literature-trained neural networks. Journal of Chemical Information and Modeling. 2020;60(3):1194–201.

99 Prein T, Pan E, Jehkul J, Weinmann S, Olivetti EA, Rupp JLM. Language models enable data-augmented synthesis planning for inorganic materials. arXiv preprint arXiv:2506.12557; 2025. Available from: https://arxiv.org/abs/2506.12557

100 Rodrigues A, Hegazy M, Naeem A. From intent discovery to recognition with topic modeling and synthetic data. arXiv preprint arXiv:2505.11176; 2025. Available from: https://arxiv.org/abs/2505.11176

101 Liu Y, Liu G, Wang J, Zhang R, Niyato D, Sun G, et al. LAMeTA: Intent-aware agentic network optimization via a large AI model-empowered two-stage approach. arXiv preprint arXiv:2505.12247; 2025. Available from: https://arxiv.org/abs/2505.12247

102 Qian Y, Huang J, Wang Y, Yu S, Zhou KZ, Mao J, et al. WHEN TO ACT, WHEN TO WAIT: Modeling structural trajectories for intent triggerability in task-oriented dialogue. arXiv preprint arXiv:2506.01881; 2025. Available from: https://arxiv.org/abs/2506.01881

103 Vatsal S, Dubey H. A survey of prompt engineering methods in large language models for different NLP tasks. arXiv preprint arXiv:2407.12994; 2024. Available from: https://arxiv.org/abs/2407.12994

104 Sahoo P, Singh AK, Saha S, Jain V, Mondal S, Chadha A. A systematic survey of prompt engineering in large language models: Techniques and applications. arXiv preprint arXiv:2402.07927; 2024. Available from: https://arxiv.org/abs/2402.07927

105 Li H, Leung J, Shen Z. Towards goal-oriented prompt engineering for large language models: A survey. arXiv preprint arXiv:2401.14043; 2024. Available from: https://arxiv.org/abs/2401.14043

106 Bran AM, Cox S, Schilter O, Baldassari C, White AD, Schwaller P. Augmenting large language models with chemistry tools. Nature Machine Intelligence. 2024;6(5):525–35.

107 Yang Y, Sun H, Li J, Liu R, Li Y, Liu Y, et al. MindLLM: Pre-training lightweight large language model from scratch, evaluations and domain applications. arXiv preprint arXiv:2310.15777; 2023. Available from: https://arxiv.org/abs/2310.15777

108 Liu C, Zhang H, Zhao K, Ju X, Yang L. LLMEmbed: Rethinking lightweight LLM's genuine function in text classification. In: 62nd Annual Meeting of the Association for Computational Linguistics (Volume 1: Long Papers) [Internet]. Association for Computational Linguistics; 2024. pp. 7994–8004. Available from: https://aclanthology.org/2024.acl-long.433

109 Devlin J, Chang MW, Lee K, Toutanova K. BERT: Pre-training of deep bidirectional transformers for language understanding. In: NAACL-HLT 2019 Conference [Internet]. Association for Computational Linguistics; 2019. pp. 4171–86. Available from: http://aclweb.org/anthology/N19-1423

110 Jia K, Meng F, Liang J, Gong P. Text sentiment analysis based on BERT-CBLBGA. Computers and Electrical Engineering. 2023;112:109019.

111 Guan S, Xiong H, Wang J, Bian J, Zhu B, Lou JG. Evaluating LLM-based agents for multi-turn conversations: A survey. arXiv preprint arXiv:2503.22458; 2025. Available from: https://arxiv.org/abs/2503.22458

112 Cheng J, Reddy S, Saraswat V, Lapata M. Learning structured natural language representations for semantic parsing. In: 55th Annual Meeting of the Association for Computational Linguistics (Volume 1: Long Papers) [Internet]. Association for Computational Linguistics; 2017. pp. 44–55. Available from: http://aclweb.org/anthology/P17-1005

113 Lample G, Ballesteros M, Subramanian S, Kawakami K, Dyer C. Neural architectures for named entity recognition. In: 2016 Conference of the North American Chapter of the Association for Computational Linguistics: Human Language Technologies [Internet]. Association for Computational Linguistics; 2016. pp. 260–70. Available from: http://aclweb.org/anthology/N16-1030

114 Huang Z, Xu W, Yu K. Bidirectional LSTM-CRF models for sequence tagging. arXiv preprint arXiv:1508.01991; 2015. Available from: https://arxiv.org/abs/1508.01991

115 Liu Y, Ott M, Goyal N, Du J, Joshi M, Chen D, et al. RoBERTa: A robustly optimized BERT pretraining approach. arXiv preprint arXiv:1907.11692; 2019. Available from: https://arxiv.org/abs/1907.11692

116 Raffel C, Shazeer N, Roberts A, Lee K, Narang S, Matena M, et al. Exploring the limits of transfer learning with a unified text-to-text transformer. arXiv preprint arXiv:1910.10683; 2019. Available from: https://arxiv.org/abs/1910.10683

117 Tenorth M, Beetz M. KnowRob: A knowledge processing infrastructure for cognition-enabled robots. The International Journal of Robotics Research. 2013;32(5):566–90.

118 Misra D, Langford J, Artzi Y. Mapping instructions and visual observations to actions with reinforcement learning. In: 2017 Conference on Empirical Methods in Natural Language Processing [Internet]. Association for Computational Linguistics; 2017. pp. 1004–15. Available from: http://aclweb.org/anthology/D17-1106

119 Coleman T, Casanova H, Ferreira da Silva R. Automated generation of scientific workflow generators with WfChef. Future Generation Computer Systems. 2023;147:16–29.

120 Starkey PT, Billington CJ, Johnstone SP, Jasperse M, Helmerson K, Turner LD, et al. A scripted control system for autonomous hardware-timed experiments. Review of Scientific Instruments. 2013;84(8):085111.

121 Zeller M. Safety assurance of autonomous systems using machine learning: an industrial case study and lessons learnt. INCOSE International Symposium. 2023;33(1):320–33.

122 Abeywickrama DB, Fisher M, Wheeler F, Dennis L. Autonomy and safety assurance in the early development of robotics and autonomous systems. arXiv preprint arXiv:2501.18448; 2025. Available from: https://arxiv.org/abs/2501.18448

123 Fukushima K, Miyake S. Neocognitron: A new algorithm for pattern recognition tolerant of deformations and shifts in position. Pattern Recognition. 1982;15(6):455–69.

124 Bengio Y. Learning deep architectures for AI. Foundations and Trends in Machine Learning. 2009;2(1):1–127.

125 LeCun Y, Bengio Y, Hinton G. Deep learning. Nature. 2015;521(7553):436–44.

126 Goodfellow I, Bengio Y, Courville A. Deep learning: The MIT Press, 2016, 800 pp, ISBN: 0262035618. Genetic Programming and Evolvable Machines. 2018;19(1–2):305–7.

127 Vaswani A, Shazeer N, Parmar N, Uszkoreit J, Jones L, Gomez AN, et al. Attention is all you need. arXiv preprint arXiv:1706.03762; 2017. Available from: https://arxiv.org/abs/1706.03762

128 King RD, Rowland J, Oliver SG, Young M, Aubrey W, Byrne E, et al. The Automation of Science. Science. 2009;324(5923):85–9.

129 Li J, Wang Q, Mao Z, Guo J, Yang Y, Zhang Y. Improving Chinese spelling check by character pronunciation prediction: The effects of adaptivity and granularity. arXiv preprint arXiv:2210.10996; 2022. Available from: http://arxiv.org/abs/2210.10996

130 Moás PM, Lopes CT. Automatic quality assessment of Wikipedia articles—A systematic literature review. ACM Computing Surveys. 2024;56(4):1–37.

131 Yao Z, Lum Y, Johnston A, Mejia-Mendoza LM, Zhou X, Wen Y, et al. Machine learning for a sustainable energy future. Nature Reviews Materials. 2022;8(3):202–15.

132 Radford A, Kim JW, Hallacy C, Ramesh A, Goh G, Agarwal S, et al. Learning transferable visual models from natural language supervision. arXiv preprint arXiv:2103.00020; 2021. Available from: https://arxiv.org/abs/2103.00020

133 Weininger D. SMILES, a chemical language and information system. 1. Introduction to methodology and encoding rules. Journal of Chemical Information and Computer Sciences. 1988;28(1):31–6.

134 Schwaller P, Laino T, Gaudin T, Bolgar P, Hunter CA, Bekas C, et al. Molecular transformer: A model for uncertainty-calibrated chemical reaction prediction. ACS Central Science. 2019;5(9):1572–83.

135 Wang S, Guo Y, Wang Y, Sun H, Huang J. SMILES-BERT: Large scale unsupervised pre-training for molecular property prediction. In: 10th ACM International Conference on Bioinformatics, Computational Biology and Health Informatics [Internet]. Association for Computing Machinery; 2019. pp. 429–36. Available from: https://dl.acm.org/doi/10.1145/3307339.3342186

136 Tetko IV, Karpov P, Van Deursen R, Godin G. State-of-the-art augmented NLP transformer models for direct and single-step retrosynthesis. Nature Communications. 2020;11(1):5575.

137 Roch LM, Häse F, Kreisbeck C, Tamayo-Mendoza T, Yunker LPE, Hein JE, et al. ChemOS: An orchestration software to democratize autonomous discovery. PLoS ONE. 2020;15(4):e0229862.

138 Choi J, Lee B. Accelerating materials language processing with large language models. Communications Materials. 2024;5(1):13.

139 Graves A. Sequence transduction with recurrent neural networks. arXiv preprint arXiv:1211.3711; 2012. Available from: http://arxiv.org/abs/1211.3711

140 Wu Y, Schuster M, Chen Z, Le QV, Norouzi M, Macherey W, et al. Google's neural machine translation system: Bridging the gap between human and machine translation. arXiv preprint arXiv:1609.08144; 2016. Available from: https://arxiv.org/abs/1609.08144

141 Holtzman A, Buys J, Du L, Forbes M, Choi Y. The curious case of neural text degeneration. arXiv preprint arXiv:1904.09751; 2020. Available from: http://arxiv.org/abs/1904.09751

142 Cohan A, Dernoncourt F, Kim DS, Bui T, Kim S, Chang W, et al. A discourse-aware attention model for abstractive summarization of long documents. In: 2018 Conference of the North American Chapter of the Association for Computational Linguistics: Human Language Technologies, Volume 2 (Short Papers) [Internet]. Association for Computational Linguistics; 2018. pp. 615–21. Available from: http://aclweb.org/anthology/N18-2097

143 Sutskever I, Vinyals O, Le QV. Sequence to sequence learning with neural networks. arXiv preprint arXiv:1409.3215; 2014. Available from: https://arxiv.org/abs/1409.3215

144 Peeperkorn M, Kouwenhoven T, Brown D, Jordanous A. Is temperature the creativity parameter of large language models? arXiv preprint arXiv:2405.00492; 2024. Available from: https://arxiv.org/abs/2405.00492

145 Xie J, Chen AS, Lee Y, Mitchell E, Finn C. Calibrating language models with adaptive temperature scaling. arXiv preprint arXiv:2409.19817; 2024. Available from: https://arxiv.org/abs/2409.19817

146 Guo C, Pleiss G, Sun Y, Weinberger KQ. On calibration of modern neural networks. In: 34th International Conference on Machine Learning, Vol. 70 [Internet]. JMLR.org; 2017. pp. 1321–30.

147 Bowman SR, Vilnis L, Vinyals O, Dai A, Jozefowicz R, Bengio S. Generating sentences from a continuous space. In: 20th SIGNLL Conference on Computational Natural Language Learning [Internet]. Association for

Computational Linguistics; 2016. pp. 10–21. Available from: http://aclweb.org/anthology/K16-1002

148 Zhao T, Zhao R, Eskenazi M. Learning discourse-level diversity for neural dialog models using conditional variational autoencoders. In: 55th Annual Meeting of the Association for Computational Linguistics (Volume 1: Long Papers) [Internet]. Association for Computational Linguistics; 2017. pp. 654–64. Available from: http://aclweb.org/anthology/P17-1061

149 Chen W, Ma X, Wang X, Cohen WW. Program of thoughts prompting: disentangling computation from reasoning for numerical reasoning tasks. arXiv preprint arXiv:2211.12588; 2022. Available from: https://arxiv.org/abs/2211.12588

150 Lewis M, Liu Y, Goyal N, Ghazvininejad M, Mohamed A, Levy O, et al. BART: Denoising sequence-to-sequence pre-training for natural language generation, translation, and comprehension. In: 58th Annual Meeting of the Association for Computational Linguistics [Internet]. Association for Computational Linguistics; 2020. pp. 7871–80. Available from: https://www.aclweb.org/anthology/2020.acl-main.703

151 Nishimura K, Saito K, Hirasawa T, Ushiku Y. Toward structured related work generation with novelty statements. In: Fourth Workshop on Scholarly Document Processing (SDP 2024). Association for Computational Linguistics; 2024. pp. 30–57.

152 Madani M, Lacivita V, Shin Y, Tarakanova A. Accelerating materials property prediction via a hybrid Transformer Graph framework that leverages four body interactions. npj Computational Materials. 2025;11(1):15.

153 Ying C, Cai T, Luo S, Zheng S, Ke G, He D, et al. Do transformers really perform bad for graph representation? arXiv preprint arXiv:2106.05234; 2021. Available from: https://arxiv.org/abs/2106.05234

154 Wang G, Hu J, Zhou J, Liu S, Li Q, Sun Z. Knowledge-guided large language model for material science. Review of Materials Research. 2025;1(2):100007.

155 Li H, Zhang R, Min Y, Ma D, Zhao D, Zeng J. A knowledge-guided pre-training framework for improving molecular representation learning. Nature Communications. 2023;14(1):7568.

156 Bai X, He S, Li Y, Xie Y, Zhang X, Du W, et al. Construction of a knowledge graph for framework material enabled by large language models and its application. npj Computational Materials. 2025;11(1):51.

157 Da L, Shah PM, Liou KR, Zhang J, Wei H. GE-Chat: A graph enhanced RAG framework for evidential response generation of LLMs. arXiv preprint arXiv:2505.10143; 2025. Available from: https://arxiv.org/abs/2505.10143

158 See A, Liu PJ, Manning CD. Get to the point: Summarization with pointer-generator networks. In: 55th Annual Meeting of the Association for Computational Linguistics (Volume 1: Long Papers) [Internet]. Association for Computational Linguistics; 2017. pp. 1073–83. Available from: http://aclweb.org/anthology/P17-1099

159 Zhou W, Huang K, Ma T, Huang J. Document-level relation extraction with adaptive thresholding and localized context pooling. arXiv preprint arXiv:2010.11304; 2020. Available from: https://arxiv.org/abs/2010.11304

160 Yan R, Jiang X, Wang W, Dang D, Su Y. Materials information extraction via automatically generated corpus. Scientific Data. 2022;9(1):401.

161 Mooney RJ, Bunescu R. Mining knowledge from text using information extraction. SIGKDD Explorations Newsletter. 2005;7(1):3–10.

162 Mishra V, Singh S, Ahlawat D, Zaki M, Bihani V, Grover HS, et al. Foundational large language models for materials research. arXiv preprint arXiv:2412.09560; 2024. Available from: https://arxiv.org/abs/2412.09560

163 Liu Y, Liu DH, Ge XY, Yang ZW, Ma SC, Zou ZY, et al. A high-quality dataset construction method for text mining in materials science. Acta Physica Sinica. 2023;72(7):070701.

164 Wang Z, Li Y, Zhu Z. BERT-based knowledge extraction method of unstructured domain text. arXiv preprint arXiv:2103.00728; 2021. Available from: https://arxiv.org/abs/2103.00728

165 Nadas M, Diosan L, Tomescu A. Synthetic data generation using large language models: Advances in text and code. arXiv preprint arXiv:2503.14023; 2025. Available from: https://arxiv.org/abs/2503.14023

166 Xie T, Wan Y, Huang W, Yin Z, Liu Y, Wang S, et al. DARWIN series: Domain specific large language models for natural science. arXiv preprint arXiv:2308.13565; 2023. Available from: https://arxiv.org/abs/2308.13565

167 Ali RSAE, Meng J, Khan MEI, Jiang X. Machine learning advancements in organic synthesis: A focused exploration of artificial intelligence applications in chemistry. Artificial Intelligence Chemistry. 2024;2(1):100049.

168 Gronauer S, Diepold K. Multi-agent deep reinforcement learning: A survey. Artificial Intelligence Review. 2022;55(2):895–943.

169 Hernandez-Leal P, Kartal B, Taylor ME. A survey and critique of multiagent deep reinforcement learning. Autonomous Agents and Multi-Agent Systems. 2019;33(6):750–97.

170 Nguyen TT, Nguyen ND, Nahavandi S. Deep reinforcement learning for multiagent systems: A review of challenges, solutions, and applications. IEEE Transactions on Cybernetics. 2020;50(9):3826–39.

171 Claus C, Boutilier C. The dynamics of reinforcement learning in cooperative multiagent systems. In: Fifteenth National/Tenth Conference on Artificial Intelligence/Innovative Applications of Artificial Intelligence. American Association for Artificial Intelligence; 1998. pp. 746–52.

172 Wang X, Sandholm T. Reinforcement learning to play an optimal Nash equilibrium in team Markov games. In: 16th International Conference on Neural Information Processing Systems. The MIT Press; 2002. pp. 1603–10.

173 Leonard RJ. Theory of games and economic behavior. History of Political Economy. 2006;38(1):189–90.

174 Littman ML. Markov games as a framework for multi-agent reinforcement learning. In: Eleventh International Conference on Machine Learning 1994 [Internet]. Elsevier; 1994. pp. 157–63. Available from: https://linkinghub.elsevier.com/retrieve/pii/B9781558603356500271

175 Littman ML. Friend-or-Foe Q-learning in General-Sum games. In: Eighteenth International Conference on Machine Learning. Morgan Kaufmann Publishers, Inc.; 2001. pp. 322–8.

176 Weinberg M, Rosenschein JS. Best-response multiagent learning in non-stationary environments. In: Third International Joint Conference on Autonomous Agents and Multiagent Systems (AAMAS 2004) [Internet]. IEEE Press; 2004. pp. 506–13.

177 Hu J, Wellman MP. Nash Q-learning for general-sum stochastic games. Journal of Machine Learning Research. 2003;4:1039–69.

178 Casgrain P, Ning B, Jaimungal S. Deep Q-learning for Nash equilibria: Nash-DQN. arXiv preprint arXiv:1904.10554; 2019. Available from: https://arxiv.org/abs/1904.10554

179 Greenwald A, Hall K. Correlated-Q learning. In: Twentieth International Conference on Machine Learning. AAAI Press; 2003. pp. 242–9.

180 Kononen V. Asymmetric multiagent reinforcement learning. In: IEEE/WIC International Conference on Intelligent Agent Technology (IAT 2003) [Internet]. IEEE Computer Society; 2003. pp. 336–42. Available from: http://ieeexplore.ieee.org/document/1241094/

181 Hao J, Yang T, Tang H, Bai C, Liu J, Meng Z, et al. Exploration in deep reinforcement learning: From single-agent to multiagent domain. IEEE Transactions on Neural Networks and Learning Systems. 2024;35(7):8762–82.

182 Schulman J, Wolski F, Dhariwal P, Radford A, Klimov O. Proximal policy optimization algorithms. arXiv preprint arXiv:1707.06347; 2017. Available from: https://arxiv.org/abs/1707.06347

183 Fu J, Zhao X, Yao C, Wang H, Han Q, Xiao Y. Reward shaping to mitigate reward hacking in RLHF. arXiv preprint arXiv:2502.18770; 2025. Available from: https://arxiv.org/abs/2502.18770

184 Ng AY, Harada D, Russell SJ. Policy invariance under reward transformations: Theory and application to reward shaping. In: Sixteenth International Conference on Machine Learning. Morgan Kaufmann Publishers, Inc.; 1999. pp. 278–87.

185 Hu H, Sadigh D. Language instructed reinforcement learning for human-AI coordination. arXiv preprint arXiv:2304.07297; 2023. Available from: https://arxiv.org/abs/2304.07297

15

Case Studies of Materials Agents

15.1 Introduction

In this chapter, we introduce in-depth case studies of autonomous discovery, which are analyzed in combination with a detailed performance index system, comparing autonomous agents with traditional manually led research in terms of efficiency, accuracy, and scalability, highlighting the transformative potential of this emerging paradigm in the field of materials science.

15.2 AtomAgents: Alloy Design and Discovery Through Multi-agents

In the field of metal alloy design, multi-scale problems rely on expert experience and require the integration of a series of processes such as knowledge retrieval, computational simulation, and experimental verification. However, such traditional methods rely on expert experience and have a slow process. Most of the existing machine learning models are only targeted at specific goals, such as the prediction of a single material property. Furthermore, there are deficiencies such as insufficient integration of cross-disciplinary knowledge and weak adaptability, making it difficult to cope with unforeseen design challenges. Besides, the properties of metallic materials are often determined by crystal defects. The research on them requires atomic-scale simulation and the correlation with macroscopic properties, but it faces challenges such as high computational costs and fragmented multimodal data. Existing studies lack intelligent systems that can automatically integrate multimodal knowledge and dynamically optimize design strategies. Based on the above issues, Ghafarollahi and Buehler from the Massachusetts Institute of Technology proposed AtomAgents to explore how they can achieve full-process automation in alloy design and discovery through multi-agent collaboration [1]. By combining LLMs with knowledge retrieval, multimodal data integration, and physics-based simulations, AtomAgents effectively perform alloy design. Computational experiments then reveal the proficiency in material property calculations and defect analysis, enabling the autonomous design of alloys with superior properties compared to pure metals.

15.2.1 System Architecture of AtomAgents

In the AtomAgents platform, the entire system is composed of three core modules: brain, perception, and action. AtomAgents adopts a three-tier architecture of physical perception, multi-agent collaboration, and multimodal fusion, with LLM as the decision-making core and integrates cross-domain toolchains to automate alloy design, as depicted in Figure 15.1. First comes the brain module, mainly composed of the core LLM, which is responsible for planning tasks, logical reasoning, and decision-making. The second is the perception module, which collects and processes multimodal data from different sources, such as text, images, simulation results, etc. Finally, there is the action module, which performs specific tasks such as code writing, simulation, and image analysis based on the instructions from the brain module. Figure 15.1 shows the overall framework of the multimodal multi-agent system. As can be seen in Figure 15.1, the brain module is at the core position and is interconnected with the perception and movement modules, forming a dynamic cyclic feedback system. This architecture enables the system to continuously iterate, learn, and optimize the alloy design scheme.

Figure 15.2 shows the detailed structural schematic diagram of the AtomAgents system. AtomAgents is composed of multiple agents, and the multi-agent collaboration process is automatically designed through a closed loop of perception, reasoning, action, and feedback. The core decision-making layer agents of the AtomAgents system consist of Users, Engineers, Scientists, and Group Managers, and conduct logical reasoning, task planning, and decision-making based on GPT-4. For the tool execution layer, it includes planning tools (Admin/Planner/Critic agents), computing tools (LAMMPS atomic simulation), knowledge retrieval tools (literature data extraction), encoding tools (Python script generation), and image analysis tools (multimodal data parsing). Each tool is dynamically invoked through the AutoGen framework. Especially in the computing tool module, molecular dynamics simulation is carried out with LAMMPS as the core and combined with density functional theory (DFT) and machine learning potential, such as Moment Tensor Potential (MTP), high-precision

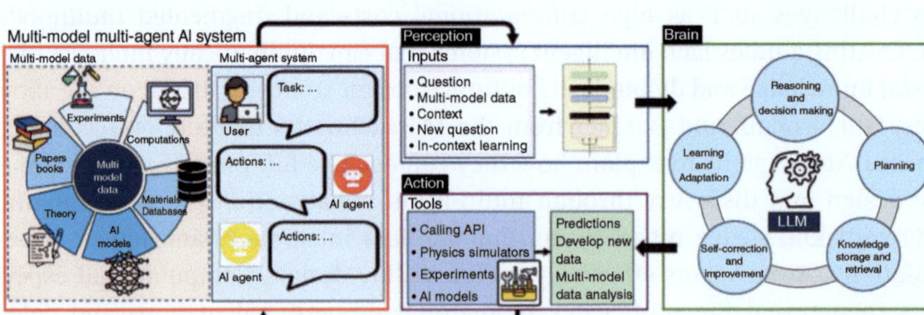

Figure 15.1 The overall framework of the multimodal multi-agent system. *Source*: [1] / National Academy of Sciences.

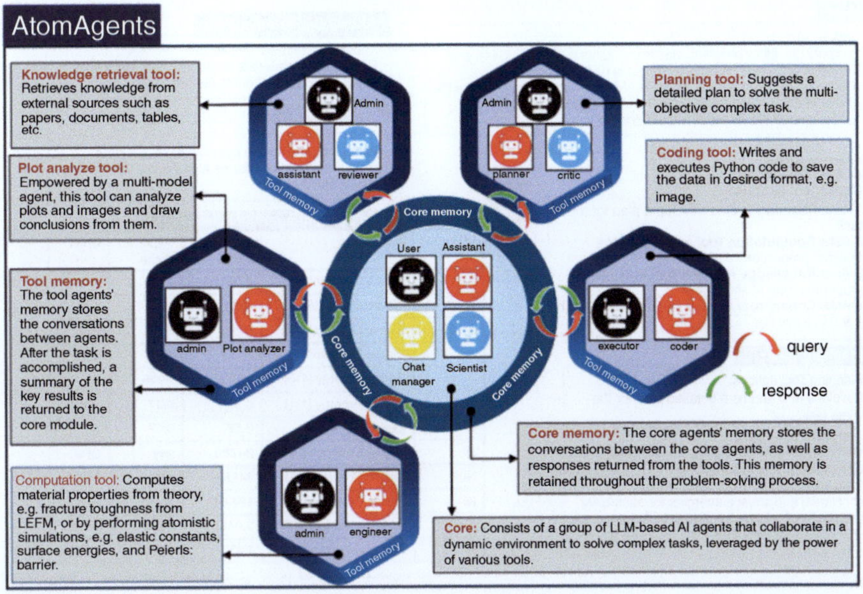

Figure 15.2 The overall framework of AtomAgent. *Source*: [1] / National Academy of Sciences.

calculation of defect behavior at the atomic scale is achieved. In summary, the entire problem-solving process is managed by an artificial intelligence (AI) agent, including plan definition, simulation execution, knowledge retrieval, and result storage.

15.2.2 Experimental Design and Verification

15.2.2.1 Experiment I: Material Properties Calculation and Knowledge Retrieval

As shown in Figure 15.2, to verify the accuracy and automation ability of AtomAgents in the calculation of complex material properties, the authors designed AtomAgents to calculate the lattice constants, elastic constants (C11, C12, and C44), and the surface energies of the crystal planes [110] and [111] of aluminum (Al) and nickel (Ni). And compare it with the data of Mishin's paper to demonstrate the integration ability of the multi-agent system for cross-modal data, including simulated numerical values and literature knowledge.

The specific process is as follows. Firstly, after receiving the task from the user, the planning tool breaks down the task into three stages: simulation calculation, literature retrieval, and data storage. Planner formulates the steps, and Critic verifies the integrity of the scheme. Then, the core agent begins to execute the plan, using the potential energy of Al99.eam.alloy and Ni99.eam.alloy, calculating the material properties through LAMMPS, and the knowledge retrieval tool automatically extracts the reference values in Mishin's paper. Finally, the encoding tool generates CSV files between the simulation results and the literature data to achieve structured storage of numerical comparisons.

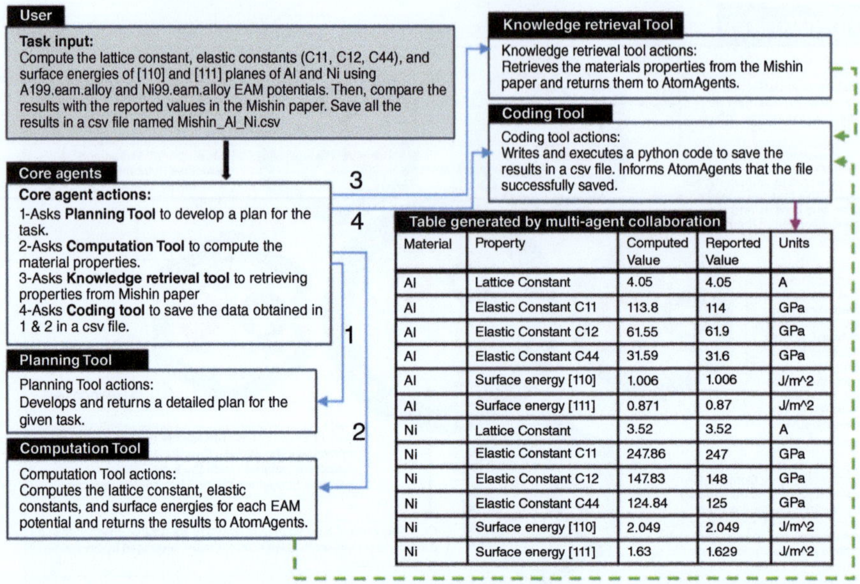

Material	Property	Computed Value	Reported Value	Units
Al	Lattice Constant	4.05	4.05	A
Al	Elastic Constant C11	113.8	114	GPa
Al	Elastic Constant C12	61.55	61.9	GPa
Al	Elastic Constant C44	31.59	31.6	GPa
Al	Surface energy [110]	1.006	1.006	J/m^2
Al	Surface energy [111]	0.871	0.87	J/m^2
Ni	Lattice Constant	3.52	3.52	A
Ni	Elastic Constant C11	247.86	247	GPa
Ni	Elastic Constant C12	147.83	148	GPa
Ni	Elastic Constant C44	124.84	125	GPa
Ni	Surface energy [110]	2.049	2.049	J/m^2
Ni	Surface energy [111]	1.63	1.629	J/m^2

Figure 15.3 Overview of the multi-agent collaboration to solve the complex task posed in Experiment I. *Source*: [1] / National Academy of Sciences.

15.2.2.2 Experiment II: Analyzing Screw Dislocation Core Structure

As shown in Figure 15.3, to verify the multimodal analysis ability of atomic agents for defect structures, the authors demonstrated the agent system's ability to parse atomic simulation images and correlate with physical mechanisms by comparing the dislocation core structures of tungsten with body-centered cubic structures under two Embedded Atom Method potential energies.

After receiving the task from the user, the core agents call the planning tool to create a plan for the task as shown in Figure 15.4. Firstly, the planning tool formulates the task decomposition plan, including dislocation generation, differential displacement (DD) map calculation, and image analysis. Then, the calculation tool performs the simulation through LAMMPS to generate the DD map. Finally, the analysis tool utilizes the multimodal agent equipped with GPT-4o to parse the DD map through context learning, by providing it with an image including polarized and unpolarized cores with corresponding labels as shown in Figure 15.5.

AtomAgents, through the collaboration of multimodal agents, has achieved an automated process from atomic simulation to defect structure analysis, providing an intelligent analysis tool for the study of alloy plasticity mechanisms and demonstrating the crucial role of context learning in enhancing the material science understanding of AI models.

15.2.2.3 Experiment III: Fracture Toughness in Alloy Systems

As shown in Figure 15.6, to explore the influence of niobium (Nb) concentration on the fracture toughness of NbMo alloy, the authors tested the integration ability of atomic agents in cross-scale mechanical problems (from atomic simulation to macroscopic performance prediction). Meanwhile, test the applicability of machine learning MTP in complex alloy systems.

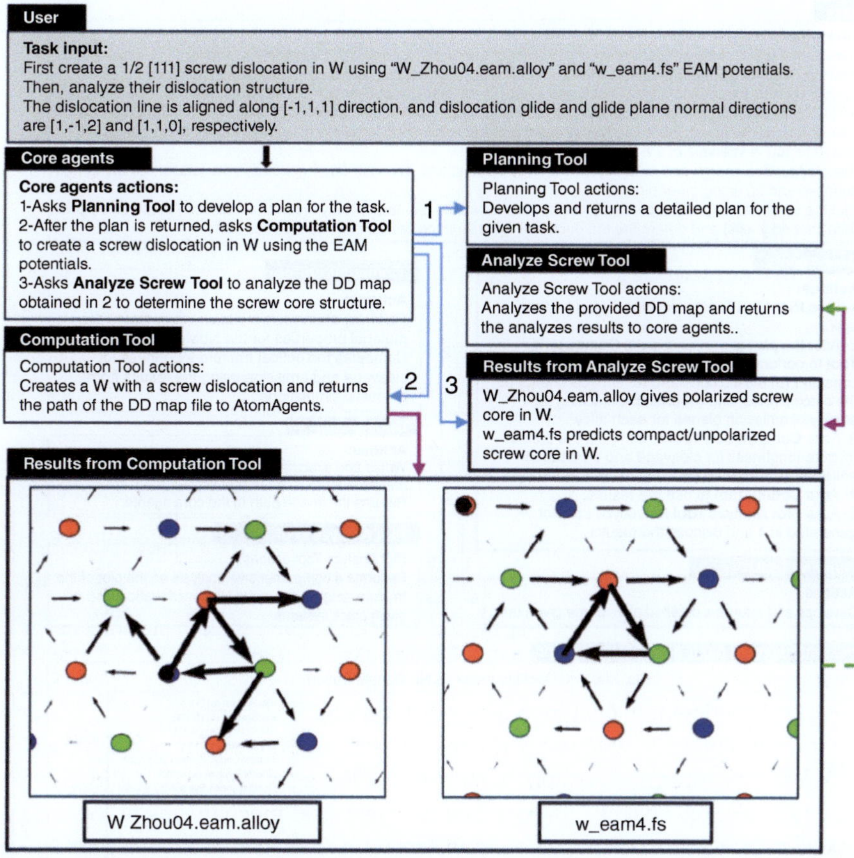

Figure 15.4 Overview of the multi-agent collaboration to solve the complex task posed in Experiment II. *Source*: [1] / National Academy of Sciences.

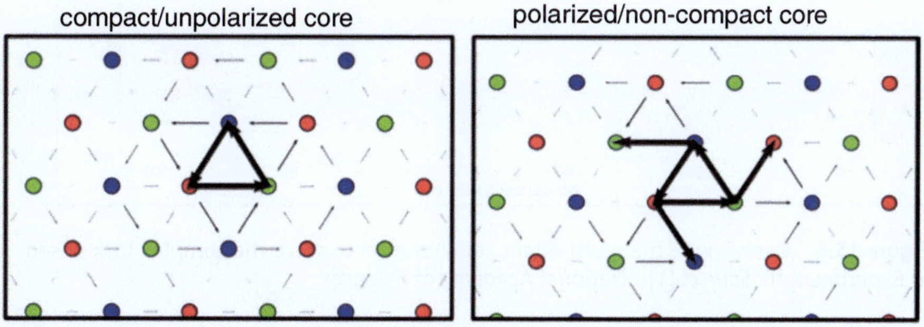

Figure 15.5 The image displays a dislocation displacement map template. *Source*: [1] / National Academy of Sciences.

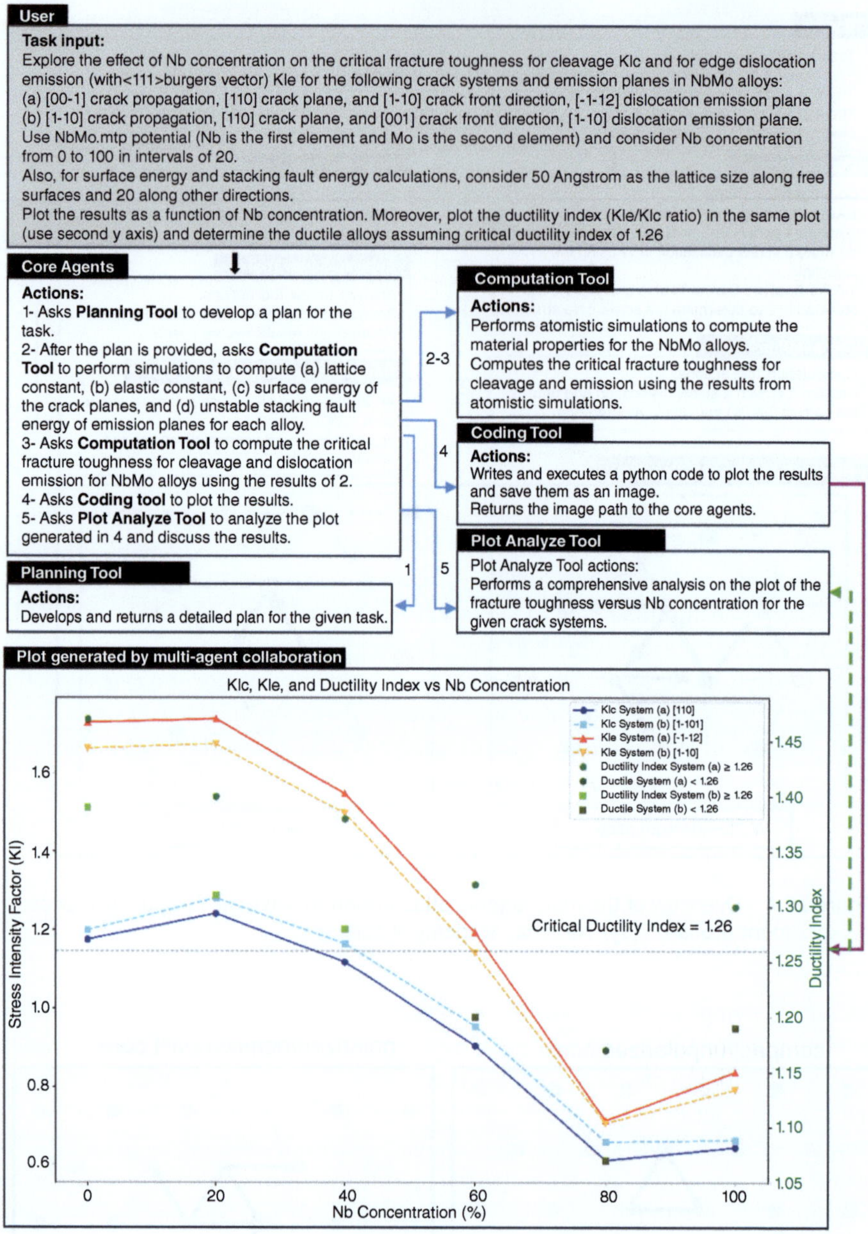

Figure 15.6 Overview of the multi-agent collaboration to solve the complex task posed in Experiment III. *Source*: [1] / National Academy of Sciences.

After receiving the task from the user, the core agents call the planning tool to create a plan for the task. Firstly, the planning tool formulated the task chain: calculate the surface energy and elastic constant of the alloy, predict the fracture toughness based on linear elastic fracture mechanics, and finally conduct the plotting analysis.

Then, the calculation tool uses the MTP potential energy to perform atomic simulations, sampling five configurations for each combined gold to balance the calculation cost and accuracy. Finally, the coding tool generates the relationship graph between fracture toughness and concentration, and the analysis tool parses the chart trend through GPT-4o.

AtomAgents has successfully achieved the full-process automation of theoretical models, atomic simulations, data visualization, and result interpretation, verifying the feasibility of multi-scale mechanical analysis. This framework can not only quantitatively predict the variation of alloy properties with composition, but also reveal the microscopic mechanism through image analysis, providing a cross-scale toolchain for the rational design of high-performance structural materials.

15.2.2.4 Experiment IV: Material Properties Calculation and Knowledge Retrieval

As shown in Figure 15.7, to verify the scientific hypotheses generated by the LLMs and to construct the correlation between the Peierls barrier and the standard deviation of energy changes through atomic simulation, and to explore new metrics for alloy design, the authors posed the following questions to the AtomAgents:

> *It is hypothesized that there is positive correlation between the Peierls barrier and the potential energy change standard deviation due to dislocation movement. Can you test this hypothesis using atomistic simulations in NbMo binary system for Nb concentrations between 10 and 90 in intervals of 10.*
>
> *Use "NbMo.mtp" potential (Nb is the first element and Mo is the second) with 40 cpus, 20 replicas, and 20 samples for each composition. The dislocation line is aligned along [-1, 1, 1] with size 5, and dislocation glide and glide plane normal directions are [1, -1, 2] and [1, 1, 0] with sizes 6 and 10, respectively.*

After receiving the task from the user, the core agents call the planning tool to create a plan for the task. Then the core agents start executing the plan by using computation tool to compute the Peierls barrier and the SD of the potential energy changes for a set of alloys. Subsequently, Coding tool is used to plot the results which are then analyzed by the Plot analyze tool via a multimodal agent. In the second task, the results of the first task are plotted via the coding agent to show the variation of the Peierls barrier against Nb concentration. The final analysis from the multi-agent model finalizes the task [1].

AtomAgents has successfully achieved the full-process automation of LLM hypothesis generation, atomic simulation verification, and data visualization analysis, providing a new theoretical basis for alloy design. This framework not only reduces the computational cost of traditional trial-and-error simulation, but also demonstrates the potential of AI to drive innovative hypotheses in materials science, opening up new paths for the rational design of high-entropy alloys and fault-resistant materials.

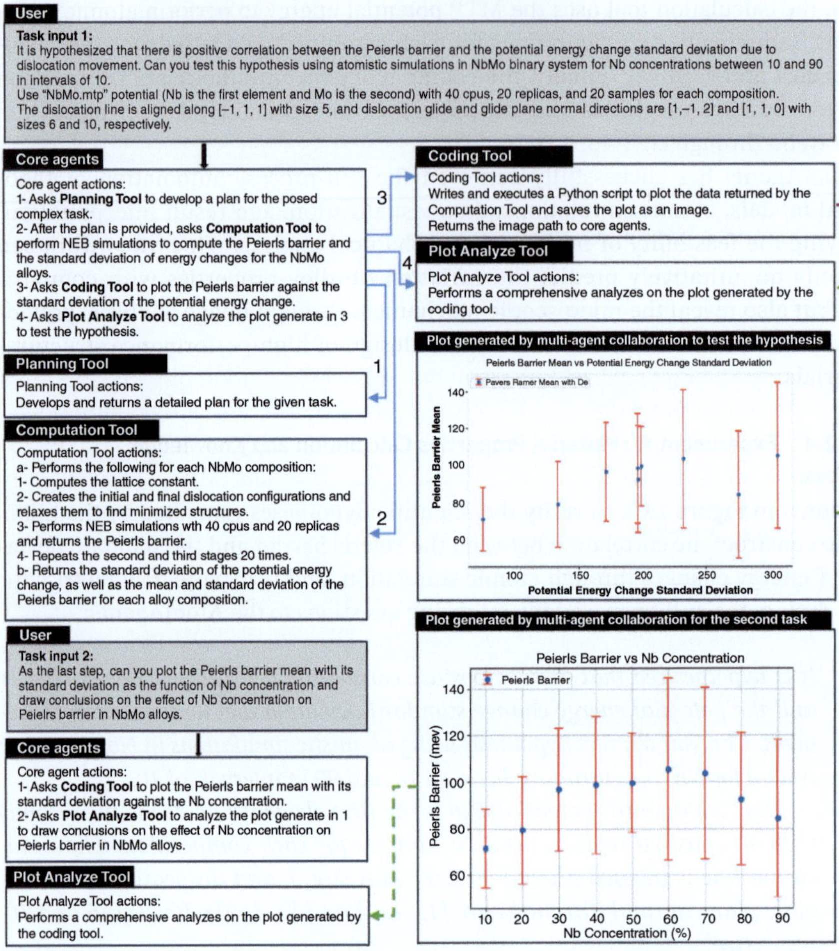

Figure 15.7 Overview of the multi-agent collaboration to solve the complex task posed in Experiment IV and then a follow-up task. *Source*: [1] / National Academy of Sciences.

15.3 Coscientist: Autonomous Chemical Research with LLMs

LLMs, especially the Transformer-based models, have developed rapidly in recent years and have been successfully applied in multiple fields such as natural language, biology, chemistry research, and code generation. OpenAI has achieved a major breakthrough in this field through the extreme expansion of models. Its GPT-4, released on March 14, 2023, demonstrates outstanding problem-solving capabilities and can handle various challenges, such as chemical-related issues. Meanwhile, significant progress has been made in the automation of chemical research, covering the autonomous discovery and optimization of organic reactions, the development of automated flow systems and mobile platforms, etc. The combination of laboratory automation technology and powerful LLMs has opened up the way for the development of systems capable of independently designing and conducting scientific experiments. Against this background, on December 20, 2023, Boiko et al. published an article "Autonomous chemical research with LLMs" in *Nature*. An AI system named Coscientist is introduced [2].

15.3.1 System Architecture and Module Functions

Coscientist [2] is an AI system driven by GPT-4, which can independently design, plan, and execute complex experiments, and combine LLMs with tools such as the Internet and document search, code execution, and experiment automation. Its unique architecture and the collaborative operation among various modules enable it to achieve the independent design, planning, and execution of complex experiments. As shown in Figure 15.8, the system mainly consists of five major modules: Planner, Web searcher, Code execution, Docs searcher, and Automation. The functions of each module are clear and they cooperate with each other.

1. **Planner Module:** The Planner module, as the intelligent hub of the entire system, is built based on GPT-4. It is responsible for receiving and parsing user input. By calling the four instructions of Google, Python, Documentation, and Experiment [2], it coordinates other modules to plan and advance the entire experimental process. When planning a chemical synthesis experiment, Planner will, based on the synthesis goals input by the user, invoke the corresponding instructions, integrate the information provided by each module, and formulate a complete experimental plan.

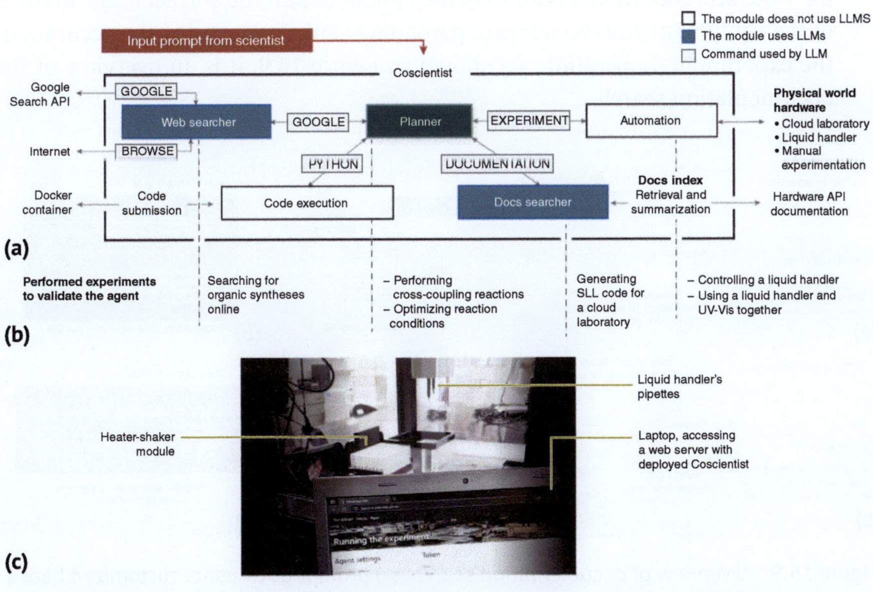

Figure 15.8 The Coscientist system's architecture: (a) the multiple component modules of Coscientist; (b) types of experiments performed to demonstrate the capabilities when using individual modules or their combinations; (c) image of the experimental setup with a liquid handler. *Source*: [2] / Springer Nature / CC BY 4.0.

2. **Web Searcher Module:** The Web searcher module is driven by GPT-4 and is mainly responsible for retrieving information about experiments on the Internet using Google commands. When planning the synthetic route of compounds, Web searcher can search for various relevant chemical reaction information, experimental conditions, research literature, etc., providing rich knowledge sources for experimental design. It is also a large model by itself, with powerful information retrieval and processing capabilities, and performs well in synthetic planning tasks, such as achieving excellent results in the synthetic experiments of acetaminophen, aspirin, nitroaniline, and phenolphthalein.
3. **Code Execution Module:** The Code execution module is controlled by Python instructions. It is an isolated Docker container that provides an independent Python execution environment for computationally related tasks. During the experiment, when tasks such as data processing and reaction condition calculation are required, the Code execution module can run the corresponding Python code to ensure the smooth progress of the calculation tasks and avoid potential risks to the user's machine at the same time.
4. **Docs Searcher Module:** The Docs searcher module is controlled by the Documentation directive and is used for text retrieval and document understanding. It can locate the technical documents of the experimental equipment, such as the programming manual of the manipulator and the operation guide of the experimental instrument, extract the key information through text mining, and provide the experimental parameters and operation details for the Planner module. When conducting experiments with specific experimental equipment, the Docs searcher module can help the system obtain the correct usage method of the equipment and the relevant parameter settings, ensuring the accuracy of the experimental operation. As shown in Figure 15.9, it is an overview of the documentation search.

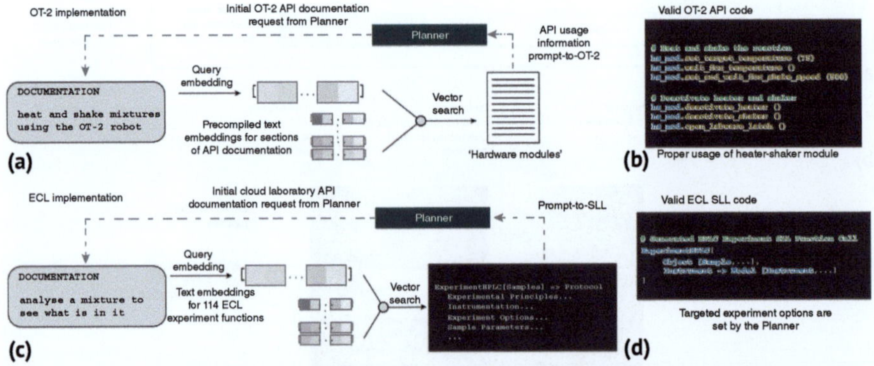

Figure 15.9 Overview of documentation search: (a) prompt documents customized based on specific tasks; (b) example for heating the vibrating screen module using OT-2; (c) example of function conversion and language conversion; (d) example of valid ECL SLL code for performing HPLC. *Source*: [19] / Springer Nature / CC BY 4.0.

5. **Automation Module:** The Automation module is responsible for automatically connecting the actual experimental equipment through API, converting the experimental plan formulated by Planner into equipment control code and distributing it for execution, thereby completing the experimental operation. In the "cloud laboratory" scenario, the Automation module can remotely control the pipetting robot to precisely transfer reagents, mix reactants, etc., according to the preset experimental plan, achieving the automation operation of the experiment.

The various modules of the Coscientist system work closely together, forming a complete closed loop of experimental execution from information retrieval, experimental planning, code calculation to equipment control, enabling it to efficiently complete various complex scientific experimental tasks.

15.3.2 Experimental Verification and Performance

In this article, the author demonstrates the versatility and performance of Coscientist in six tasks:

15.3.2.1 Planning the Chemical Synthesis of Known Compounds Using Publicly Available Data

As shown in Figure 15.10, to verify the performance of the web search module, the author designed a test set consisting of seven compounds, allowing different models

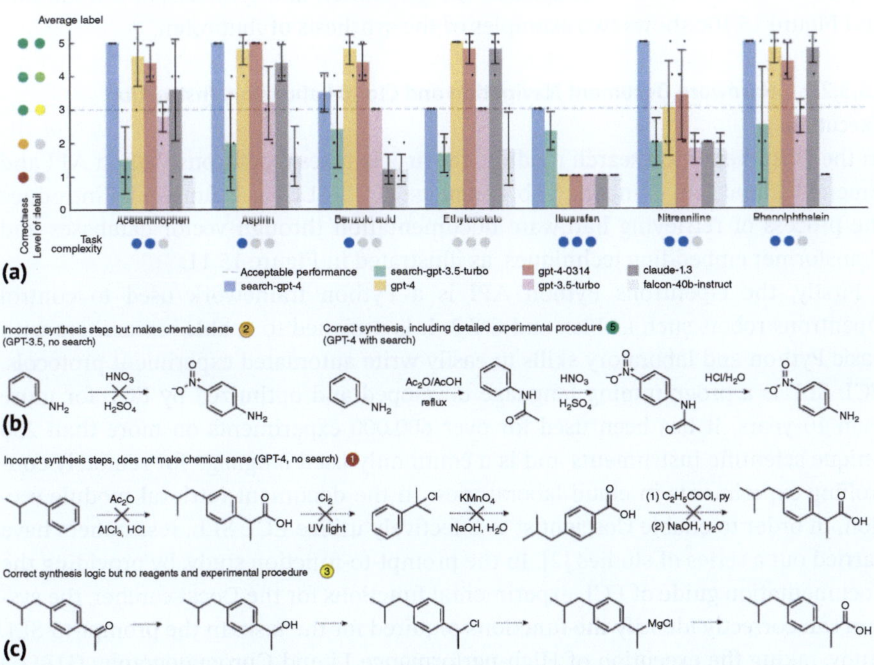

Figure 15.10 Coscientist's capabilities in chemical synthesis planning tasks. (a) Comparison of various LLMs on compound synthesis benchmarks. (b) Examples of generated syntheses of nitroaniline. (c) Example of generated syntheses of ibuprofen. *Source*: [2] / Springer Nature / CC BY 4.0.

Table 15.1 Scoring criteria for the compound synthesis steps provided by different models.

Score	Specific description
1	It is for incorrect responses or failure to follow instructions.
2	It is for extremely vague or unfeasible descriptions.
3	It is for a correct chemistry description that does not include step-by-step procedure.
4	It is for a detailed and chemically accurate description but without reagent quantities.
5	It is for a very detailed and chemically accurate procedure description.

Source: [2] / Springer Nature / CC BY 4.0.

to provide the compound synthesis steps and score them. The specific scoring criteria are summarized in Table 15.1. For the selection of LLMs, the author chose the best open-source models at that time based on the OpenLLM ranking list, including OpenAI's GPT-3.5 and GPT-4, Anthropic's Claude 1.328 and eagle-40b-instruct2.

As shown in the results of Figure 15.10a, the GPT-4-driven Web Searcher performed well in the synthetic planning, achieving the highest score in the synthetic experiments of acetaminophen, aspirin, nitroaniline, and phenolphthalein, and was the only model that reached an acceptable score in the synthesis of ibuprofen. Figure 15.10b shows two examples of the generation and synthesis of nitroaniline and Figure 15.10c shows two examples of the synthesis of ibuprofen.

15.3.2.2 Hardware Document Navigation and Cloud Laboratory Instruction Execution

In the Documentation search module, the article takes Opentrons Python API and Emerald Cloud Lab Symbolic Lab Language (ECL SLL) as examples to introduce the process of retrieving hardware documentation through vector databases and Transformer embedding techniques, as illustrated in Figure 15.11.

Firstly, the Opentrons Python API is a Python framework used to control Opentrons robots such as Flex and OT-2. It is designed to enable researchers with basic Python and laboratory skills to easily write automated experiment protocols. ECL SLL is a programming language developed and optimized by ECL for more than 10 years. It has been used for over 600,000 experiments on more than 230 unique scientific instruments and is a commonly used language for remotely controlling experiments in cloud laboratories. In the document retrieval module section, in order to enable Coscientist to effectively utilize ECL SLL, researchers have carried out a series of studies [2]. In the prompt-to-function study, by providing the documentation guide of ECL experimental functions for the Docs searcher, the system can correctly identify the functions required for the task. In the prompt-to-SLL study, taking the execution of High-performance Liquid Chromatography (HPLC) experiments as an example, the system will select appropriate functions. After the document is processed by the GPT-4 model, Planner receives the usage information and generates the experimental code in SLL format, which is successfully

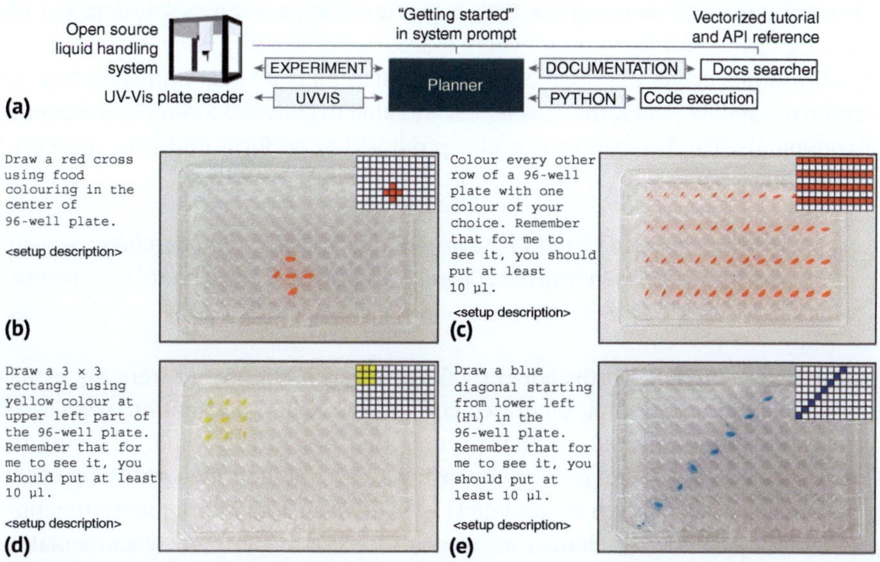

Figure 15.11 Robotic liquid handler control capabilities and integration with analytical tools. (a) Coscientist's configuration. (b) Drawing a red cross. (c) Coloring every other row. (d) Drawing a yellow rectangle. (e) Drawing a blue diagonal. *Source*: [2] / Springer Nature / CC BY 4.0.

executed in ECL. However, some problems also emerged during the experiment. For instance, air bubbles were injected along with the analyte solution during the experiment, which also indicates the significance of developing automated quality control technology for cloud laboratories.

Further exploration beyond the initial HPLC experiment revealed that the Coscientist could identify the correct ECL Experiment function to perform the requested operation. The prompt:

> *"Write a command for Emerald Cloud Lab to collect an absorbance spectrum of sample Object [Sample, "id: aXRlGnRLm4nO"]. Set the Instrument option to Model [Instrument, PlateReader, "Lunatic"] and leave all other options at their default values."*

It was correctly interpreted as a request to run an absorbance spectroscopy experiment. A syntactically valid command, including the specified sample and instrument model, was produced. When provided with similar prompts that did not stipulate those other options to be left at their default values, the Coscientist frequently included other options with mixed results.

Similarly, the prompt:

> *"Generate an Emerald Cloud Lab command to make a stock solution consisting of 10 g of Model [Sample, "Sodium Chloride"] and 10 g of Model [Sample, "EDTA, Reagent grade"] and 1 L of Model [Sample, "Milli-Q water"]. Mix for 17 minutes at 30 Celsius."*

It was interpreted as using the ECL function ExperimentStockSolution, but the generated command had minor syntax inaccuracies. Notably, the Coscientist produced multiple sequential ECL experiment commands for this multistep request. Despite the syntax issues, the Coscientist was able to generate a complete sequence of commands. The Coscientist was also prompted to perform multiple, sequential sample transfers:

> *"Write an Emerald Cloud Lab command to transfer 100 microliters of water into a 2 mL tube, then transfer 200 microliters of absolute ethanol into the same tube."*

In this case, the Coscientist correctly filled in assumptions that were not explicitly specified in the prompt, identified correct source and destination wells based on those assumptions, and produced syntax that was largely valid. Allowing the Coscientist to iterate led to more complex behavior as it identified recurring operations and generated its own streamlined code with a custom Mathematica function.

In some cases, the Coscientist attempted to access experiment documentation via the Google command opposed to the Documentation command. Importantly, performance was substantially better when querying supplied plaintext versions (Documentation command) compared to the web search strategy (Google command).

15.3.2.3 Precisely Control Liquid Handling Instruments with Low-level Instructions

In the experimental verification section of the Controlling laboratory hardware, the author selected Opentrons OT-2 as the open-source liquid processor by providing the Planner with the page documentation for its Getting Started. Verify the ability of the system to precisely control liquid handling instruments by having it execute natural language instruction tasks such as "*Color every other row with the selected color.*"

As shown in Figure 15.11b–e, the research team first conducted a simple specific board layout experiment, generating accurate experimental protocols through natural language prompts such as "*Color every other row with the selected color,*" and the robot's execution results were highly consistent with the prompt.

Meanwhile, to evaluate the ability of the system to integrate multiple modules simultaneously, a toy task was studied and designed: three color solutions of red, yellow, and blue were placed, respectively, in three wells of a 96-well plate, and the system was required to determine the colors and their positions without prior information. The Coscientist first prepares a small sample from the original solution, then calls the ultraviolet-visible measurement module to obtain the spectral data of each well (stored as a NumPy array), and generates Python code to analyze the maximum absorption wavelength. Combined with the guiding prompt of "*How different colors absorb light,*" ultimately, the positions of each color were accurately inferred, demonstrating the collaborative capabilities of multiple hardware tools and data analysis capabilities.

15.3.2.4 Solve Complex Scientific Tasks That Require the Simultaneous Use of Multiple Hardware Modules and the Integration of Various Data Sources

To assess Coscientist's ability to solve complex scientific tasks involving the integration of multiple modules and data sources, researchers evaluated Coscientist's capacity to plan catalytic cross-coupling experiments (Figure 15.12) using Internet data, perform necessary calculations, and ultimately write code for liquid processors. To increase complexity, Coscientist is required to use the OT-2 heater-shaker module released after the deadline of GPT-4 training data collection. The available commands and operations provided to Coscientist are shown in Figure 15.12a.

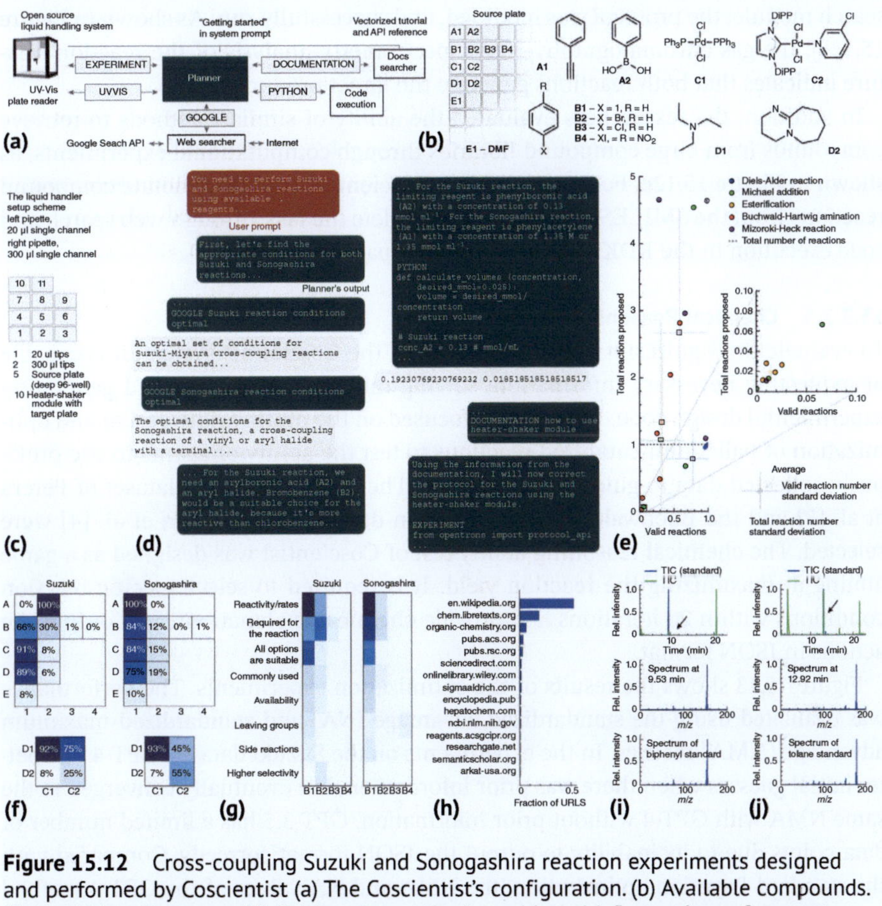

Figure 15.12 Cross-coupling Suzuki and Sonogashira reaction experiments designed and performed by Coscientist (a) The Coscientist's configuration. (b) Available compounds. (c) Liquid handler setup. (d) Solving the synthesis problem. (e) Comparison of reagent selection performance with a large dataset. (f) Comparison of reagent choices across multiple runs. (g) Overview of justifications made. (h) Frequency of visited URLs. (i) Mass spectra at 9.53 min. (j) Mass spectra at 12.92 minutes. *Source*: [2] / Springer Nature / CC BY 4.0.

Firstly, as shown in Figure 15.12d, Coscientist searches the Internet for the information, stoichiometry, and conditions of the required reaction, and selects the correct coupling partners for the corresponding reaction. When designing and conducting experiments, Coscientist's strategy varied with the number of runs, as shown in Figure 15.12f, and no chemical errors occurred in the system. Furthermore, the model can provide reasonable explanations for specific choices as shown in Figure 15.12g, demonstrating its ability to apply concepts such as reactivity and selectivity. Although the Web searcher visited multiple websites as shown in Figure 15.12h, most of the data came from Wikipedia, the journals of the American Chemical Society and the Royal Society of Chemistry in the UK. Subsequently, Coscientist calculated the required volumes of all reactants and wrote the Python protocol for the OT-2 robot. Subsequently, the OT-2 document was consulted through the Docs search module, the protocol was modified, and successfully run. As shown in Figure 15.12i,j, the gas chromatography–mass spectrometry analysis of the reaction mixture indicates that both reactions generate the target products.

In addition, the researchers evaluated the ability of similar methods to retrieve compounds from large compound libraries through computational experiments, as shown in Figure 15.12e. For each reaction, Coscientist needs to generate compound reactions from the SMILES database and complete the task through web search and code execution in the RDKit cheminformatics package.

15.3.2.5 Chemical Reasoning Capabilities

To evaluate the significant reasoning ability of the Coscientist system, it is capable of requesting necessary information, solving multistep problems, and generating experimental design code. Researchers focused on the multivariate design and optimization of palladium-catalyzed reactions to test the ability of agents to use previously collected data to guide future actions. The Suzuki reaction dataset of Perera et al. [3] and the Buchwald–Hartwig reaction dataset of Ahneman et al. [4] were selected. The chemical reasoning ability test of Coscientist was designed as a game aiming at maximizing the reaction yield. It is required to select specific reaction conditions within 20 iterations and provide chemical explanations, and output the actions in JSON format.

Figure 15.13 shows the results of the optimization experiments. The performance was evaluated using the standardized advantage (NA) and standardized maximum advantage (NMA) metrics. In the experiments on the Suzuki dataset, GPT-4 had better initial guesses when there was prior information, but eventually converged to the same NMA with GPT-4 without prior information. GPT-3.5 has a limited number of data points due to its inability to output the JSON format correctly. Compared with the standard Bayesian optimization, the NMA and NA values of the GPT-4 method are higher, and the NA value of the Bayesian optimization remains almost unchanged, which may be due to the exploration/utilization of equilibrium differences. In the Buchwald–Hartwig dataset experiment, GPT-4 performed similarly using compound names or SMILES strings, and the reactivity of compounds could be inferred through SMILES strings. However, in some cases, this model demonstrates the ability to infer the reactivity of these compounds merely by providing their SMILES strings.

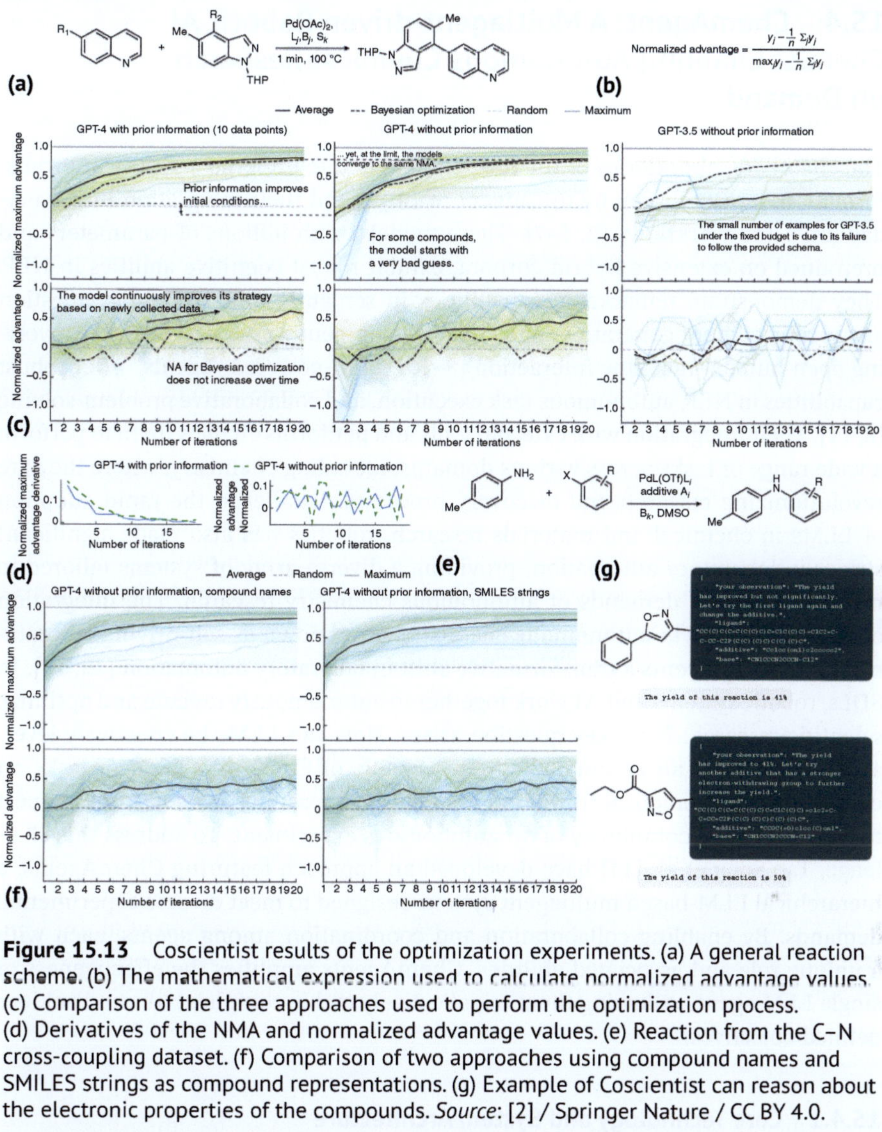

Figure 15.13 Coscientist results of the optimization experiments. (a) A general reaction scheme. (b) The mathematical expression used to calculate normalized advantage values. (c) Comparison of the three approaches used to perform the optimization process. (d) Derivatives of the NMA and normalized advantage values. (e) Reaction from the C–N cross-coupling dataset. (f) Comparison of two approaches using compound names and SMILES strings as compound representations. (g) Example of Coscientist can reason about the electronic properties of the compounds. *Source*: [2] / Springer Nature / CC BY 4.0.

Compared with autonomous agent systems such as ChemCrow and Auto-GPT, Coscientist demonstrates the multi-functionality, efficiency, and interpretability of LLMs in scientific research. It realizes autonomous experimental design through tool integration and reduces reliance on human resources. Coscientist is not merely a chemical tool; it is a crucial step for LLMs to move from "language understanding" to "intervention in the physical world," marking a fundamental transformation in the role of AI in scientific discovery. In the future, how the autonomous agent system can integrate a larger reaction database, optimize the prompt strategy, and expand to more materials and chemical fields are the issues that need to be considered at present.

15.4 ChemAgent: A Multiagent-driven Robotic AI Chemist Enabling Autonomous Chemical Research on Demand

The chemistry laboratories of the future are poised to undergo a significant transformation, largely driven by the advent of LLMs and their integration into everyday tasks and workflows [2, 5–7]. These models, with billions of parameters and pretrained on extensive hybrid corpora, exhibit robust cognitive abilities in NLP. They demonstrate remarkable capabilities in semantic understanding, question answering, content generation, and code development across various tasks involving open human–machine interaction [8–10]. LLM-powered "agents" offer robust capabilities in NLP, autonomous task execution, and collaborative problem-solving [8, 11]. Their integration with external tools and platforms enables them to perform a wide range of tasks across various domains, including chemistry, where they are revolutionizing research and discovery processes. Parallel to the rapid adoption of LLMs in chemical and materials research, robotics has also made significant strides in laboratory automation, providing a diverse array of systems tailored to meet the complex demands of autonomous chemistry research. The integration of robotics and AI in autonomous chemistry, also known as self-driving laboratories (SDLs), represents a transformative shift in laboratory automation [12, 13]. In SDLs, robotic systems and AI work together to autonomously execute and optimize scientific research. For a key question arises: How can LLMs be effectively leveraged to construct an automated system capable of executing chemical tasks on demand? It is impractical to rely on a single LLM prompt or a one-time request to handle the full complexity of an autonomous experiment. To address this challenge, Tao Song et al. [14] have developed an approach featuring ChemAgents, a hierarchical LLM-based multiagent system designed to meet diverse experimental demands. By enabling collaboration and coordination among agents, each with a unique role and associated abilities, ChemAgents amplifies the effectiveness of single LLM agents, allowing them to collectively tackle complex tasks and achieve defined objectives.

15.4.1 Core Technology and System Architecture

15.4.1.1 The ChemAgents Architecture

The author uses the open-source Llama-3.1-70B as the basis of LLM and implements the construction of the agent in Huawei's open-source AI framework MindSpore [14]. As shown in Figure 15.14, the ChemAgents architecture includes a central coordinating entity, namely the Task Manager agent, which interacts with human researchers and manages four specific roles of agents: Literature Reader, Experiment Designer, Computation Performer, and Robot Operator. The Task Manager has its own preset system prompt, in which its roles, tasks, specific functions, and interface descriptions of all entities it can use (including low-level agents and their tools),

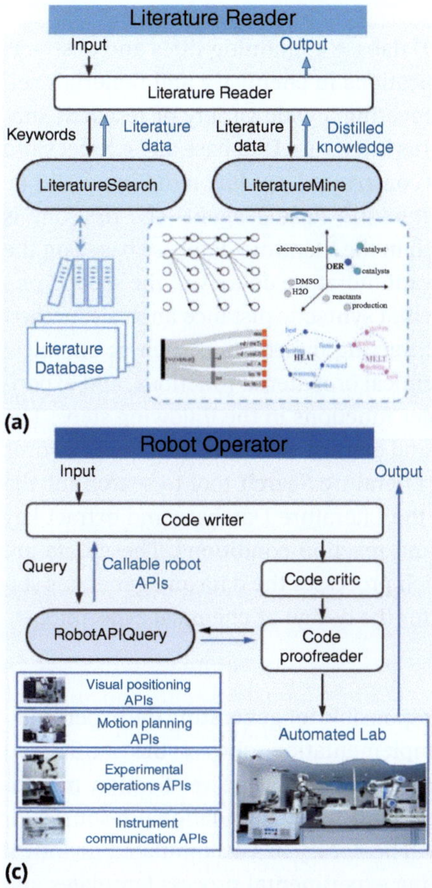

Figure 15.14 Schematic illustration of the architectures of the Literature Reader (a), Experiment Designer (b), Robot Operator (c), and Computation Performer (d) agents. *Source*: [14] / with permission of American Chemical Society.

as well as the message formats for workflows and communications are defined. By leveraging the capabilities of LLM and LangGraph, the Task Manager interacts with human researchers in natural language. It explains the requirement description, plans and decomposes tasks, and guides and manages the agents of specific roles to perform the specified operations using the corresponding basic resources, promoting the automated experimental workflow of the requirement description. Each role-specific agent has its own LLM instances and built-in tools, operating on one of four foundational resources of the system.

The Task Manager as the central coordinating unit interacts with humans and decomposes tasks, and invokes the other four agents to work collaboratively. Here, we will introduce in detail the agents of the four major roles.

15.4.1.2 Literature Reader

The authors constructed a local Literature Database containing titles and abstracts of approximately 1.2 million scientific publications in chemistry and materials science subjects, downloaded from Crossref, covering a wide variety of research subjects, topics, and fields. In order to analyze this Literature Database and extract valid knowledge, the Literature Reader agent was constructed, as shown in Figure 15.14a. There are two built-in tools used by the Literature Reader agent. The first one is Literature Search, which is designed to search in the literature database based on the keywords input by the user and return relevant literature data. And the second one is Literature Mine, which uses the unsupervised syntactic distance analysis method to mine literature data and is capable of extracting information such as chemical substances, physicochemical properties, chemical or material functions, and experimental conditions [15]. The Literature Reader functions in the following steps: The Task Manager receives the user's questions and extracts and transmits the keywords to the Literature Reader. Then, it uses the Literature Search tool to search for the titles and abstracts of relevant literature in the Literature Database and extract key information such as chemical substances and reaction conditions. These data are then passed to the Literature Mine tool, which processes the data and generates statistics, which can serve as prior knowledge for the design of chemical experiments.

15.4.1.3 Experiment Designer

The Experiment Designer is the core agent responsible for generating the experimental process in the ChemAgents system. Its implementation relies on the multi-agent sub-architecture, protocol library integration, and expert rule verification mechanism as shown in Figure 15.14b. On the one hand, a key foundational resource for this robotic AI chemist is a Protocol Library. The library has accumulated historical experimental data of the laboratory, including experimental process templates and experimental station configuration protocols, covering scenarios such as synthesis, characterization, and performance testing. These protocols are stored in the XML format and describe experimental steps in natural language. On the other hand, it is about the collaboration between Protocol critic and Protocol writer, verifying and optimizing the experimental process generated by Protocol writer to ensure that it complies with laboratory operation norms and safety rules. Based on predefined expert rules, such as the sequence of reagent addition and safety operation restrictions of the instrument, it analyzes the rationality of the process through LLM, corrects potential errors, such as the risk of using volatile solvents at high temperatures, and formats and outputs the final experimental steps. Protocol critic verifies, corrects, and formats the process based on predefined expert rules to ensure its compliance with laboratory operation norms. The final output experimental process can be passed to the Robot Operator for conversion into robot execution code, achieving the automatic generation and optimization from task description to experimental steps.

15.4.1.4 Robot Operator

The Robot Operator is the role agent in the ChemAgents system that is responsible for converting the experimental process into the executable code of the robot. As a multi-agent system integrating three LLM-driven sub-agents, namely Code writer, Code critic, and Code proofreader, Figure 15.14c illustrates the architecture of the Robot Operator agent. The Robot Operator functions as follows: The Code writer receives the natural language Experiment process generated by the Experiment Designer, acquires the available API interfaces of the robots and experimental stations in the automated laboratory through the RobotAPIQuery tool, and parses and maps the experimental steps into the corresponding Python code framework. Code critic conducts logical verification of the code based on predefined robot code expert rules and corrects issues such as hardware conflicts and parameter out-of-bounds. Code proofreader calls the RobotAPIQuery tool to check the syntax, spelling, and format of the code to ensure compliance with the Python code specification and API interface requirements. This intelligent agent relies on a hardware environment that includes two robots (mobile robots and desktop mechanical arms) and 20 automated experimental stations. It realizes the control of the robots and experimental stations through standardized API interfaces. The interfaces encapsulate the details of the underlying hardware, enhance code reusability and system scalability, and finally send the verified code to the back-end control system to execute the experiments.

15.4.1.5 Computation Performer

The Computation Performer is a role agent responsible for computational modeling and data-driven optimization in the ChemAgents system, and it is constructed based on Llama-3.1-70B. It integrates ModelSearch, ModelFuse tools, Bayesian optimizer and deep learning computing platform, and integrates key components and technical processes such as model retrieval, fusion training, optimization algorithms, and computing platform. Figure 15.14d illustrates the architecture of the Computation Performer agent. The Computation Performer operates as follows: first, the Computational task manager receives input information, including experimental task keywords, a detailed task description, and (if applicable) robotic experimental results. Then, it invokes the ModelSearch tool to obtain a suitable pretrained model. Based on the specific task, it determines the subsequent actions to be taken and executes them sequentially to generate machine learning codes. This process may involve the use of the ModelFuse tool and the Bayesian optimizer, depending on the task specification. Finally, prediction results are obtained by executing the generated code on a deep learning computing platform.

15.4.2 Experimental Tasks and Performance Verification

15.4.2.1 Make and Measure Tasks

This type of task is a kind of basic experimental task that the ChemAgents system can perform, aiming to complete compound synthesis and specified characterization operations according to user instructions, and verify the system's precise execution ability of standardized experimental procedures. The following is the experimental verification of this type of tasks.

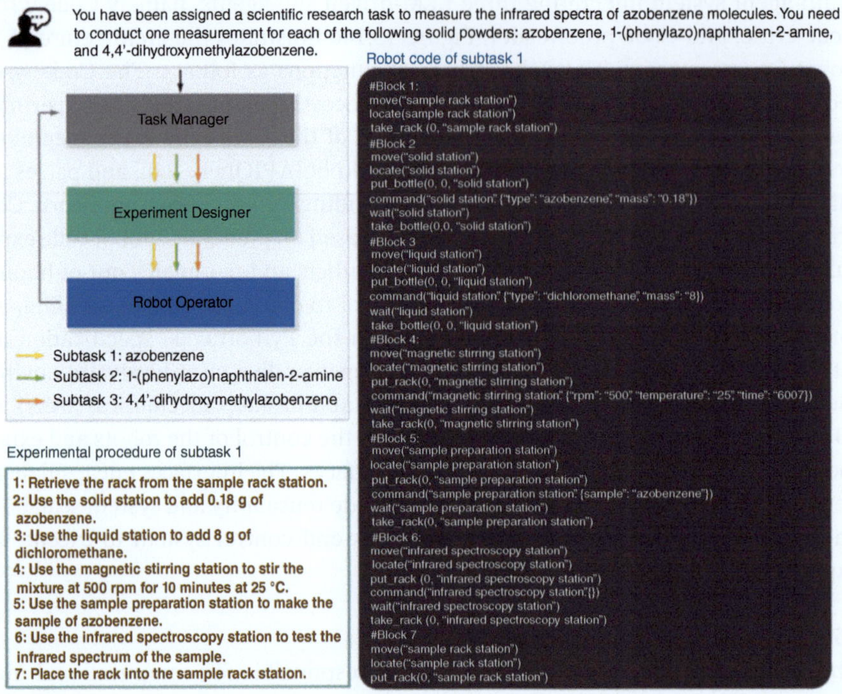

Figure 15.15 The workflow of ChemAgents and intermediate results on Task 1. *Source*: [14] / with permission of American Chemical Society.

As shown in Figure 15.15, Task 1 takes the infrared spectrum characterization of azobenzene molecules as an example. Task Manager receives the input prompt:

> *"You have been assigned a scientific research task to measure the infrared spectra of azobenzene molecules. You need to conduct one measurement for each of the following solid powders: azobenzene. 1-(phenylazo) naphthalen-2-amine and 4.4'-dihydroxymethylazobenzene."* [14]

And it drives Experiment Designer and Robot Operator to complete Task 1. Task 1 is divided into three subtasks, each corresponding to a different type of molecule (indicated by different colors). Taking subtask 1 (azobenzene) as an example, the experimental procedure generated by Experiment Designer and the code generated by Robot Operator are shown. The sample preparation step (step 5) includes adding 80 μL of sample solution onto a CaF_2 substrate and waiting for natural drying.

As shown in Figure 15.16, Task 2 involved synthesizing researcher-specified metal oxides and then characterizing them using powder X-ray diffraction (PXRD) [14].

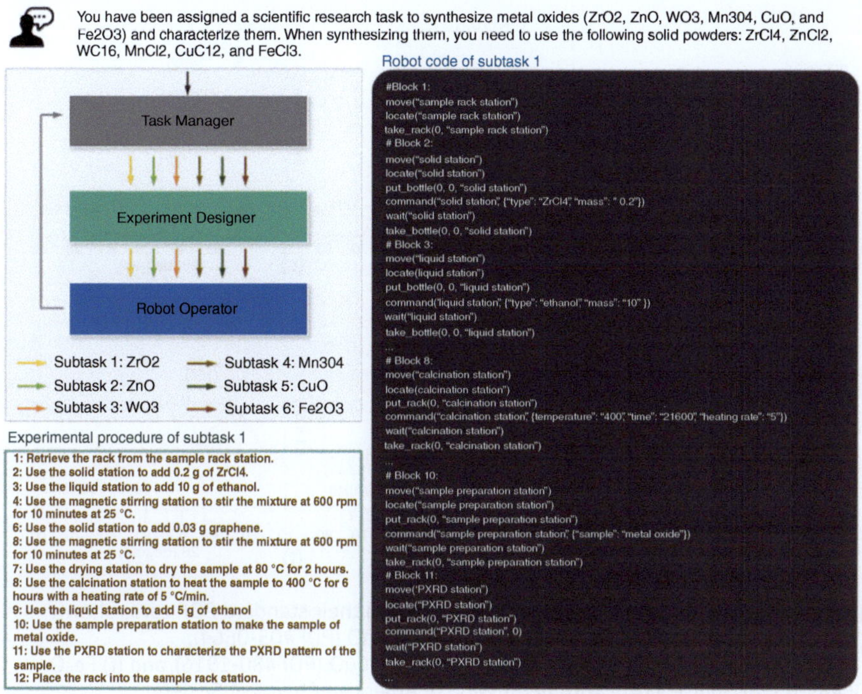

Figure 15.16 The workflow of ChemAgents and intermediate results on Task 2. *Source*: [14] / with permission of American Chemical Society.

The Task Manager was instructed to synthesize metal oxides and characterize by PXRD using the following prompt:

> *"You have been assigned a scientific research task to synthesize metal oxides (ZrO_2, ZnO, WO_3, Mn_30_4, CuO, and Fe_20_3) and characterize them. When synthesizing them, you need to use the following solid powders: $ZrCl_4$, $ZnCl_2$, WCl_6, $MnCl_2$, $CuCl_2$, and $FeCl_3$."* [14]

Task Manager receives the input prompt and drives Experiment Designer and Robot Operator to complete Task 2. Task 2 is divided into six subtasks, each corresponding to a different type of metal oxide (indicated by different colors). Taking subtask 1 (ZrO_2) as an example, the experimental procedure generated by Experiment Designer and the code generated by Robot Operator are shown. The sample preparation step (step 10) consists of adding the sample solution dropwise on a quartz substrate and waiting for natural drying. Figure 15.17 shows the PXRD patterns of six metal oxides with their standard PDF cards shown for comparison.

As shown in Figure 15.18, Task 3 involved using lead halide perovskite ($APbX_3$) quantum dots (PQDs) to achieve a wide gamut of high-purity colors through their

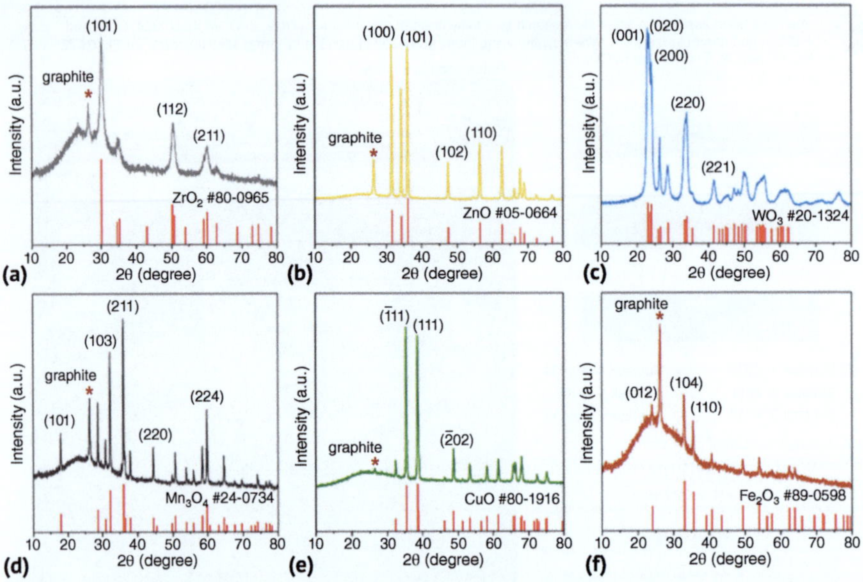

Figure 15.17 PXRD patterns of six metal oxides with their standard PDF cards shown for comparison, namely (a) ZrO_2 (PDF# 80-0965), (b) ZnO (PDF#05-0664), (c) WO_3 (PDF#20-1324), (d) Mn_3O_4 (PDF#24-0734), (e) CuO (PDF#80-1916), and (f) Fe_2O_3 (PDF#89-0598). *Source*: [14] / with permission of American Chemical Society.

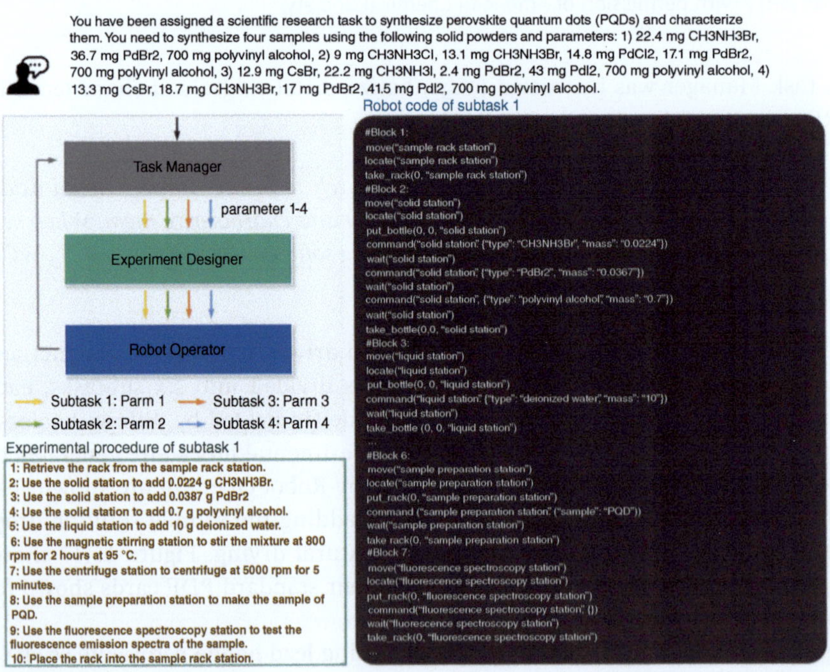

Figure 15.18 The workflow of ChemAgents and intermediate results on Task 3. *Source*: [14] / with permission of American Chemical Society.

tunable emission wavelengths and sharp photoluminescence profiles [14]. The Task Manager was instructed to PQDs and characterized by fluorescence spectroscopy using the following prompt:

> *"You have been assigned a scientific research task to synthesize PQDs and characterize them. You need to synthesize four samples using the following solid powders and parameters: 1) 22.4 mg CH_3NH_3Br, 36.7 mg $PdBr_2$, 700 mg polyvinyl alcohol; 2) 9 mg CH_3NH_3Cl, 13.1 mg CH_3NH_3Br, 14.8 mg $PdCl_2$, 17.1 mg $PdBr_2$, 700 mg polyvinyl alcohol; 3) 12.9 mg CsBr, 22.2 mg CH_3NH_3I, 2.4 mg $PdBr_2$, 43 mg $PdiI_2$, 700 mg polyvinyl alcohol; 4) 13.3 mg CsBr, 18.7 mg CH_3NH_3Br, 17 mg $PdBr_2$, 41.5 mg PdI_2, 700 mg polyvinyl alcohol."* [14]

Task Manager receives the input prompt and drives Experiment Designer and Robot Operator to complete Task 3. Task 3 is divided into four subtasks, each corresponding to a different parameter (indicated by different colors). Taking subtask 1 (Parm1: 22.4 mg CH_3NH_3Br, 36.7 mg $PdBr_2$, 700 mg polyvinyl alcohol) as an example, the experimental procedure generated by Experiment Designer and the code generated by Robot Operator are shown. The sample preparation step (step 8) consists of adding the sample solution dropwise on a quartz substrate and waiting for natural drying.

15.4.2.2 Exploration and Screening Tasks

These tasks necessitate more advanced LLM capabilities compared to the "make & measure" ones. For instance, the Task Manager must have the ability to interpret task descriptions that require a deeper understanding of domain-specific knowledge, such as comprehending and executing instructions to "perform a full factorial experiment" [14]. Similarly, experimental designers must demonstrate flexibility and creativity in adapting to existing protocols to accommodate more complex experimental designs. The following is the experimental verification of this type of tasks.

As shown in Figure 15.19, Task 4 involved the robotic AI chemist performing a full factorial experiment to synthesize graphitic carbon nitrides (g-C_3N_4) and test them for the hydrogen evolution reaction performance [14]. The Task Manager was instructed using the following prompt:

> *"You have been assigned a scientific research task to synthesize graphitic carbon nitrides (g-C_3N_4) and test them for hydrogen evolution reaction performance. You need to conduct different experiments using combinations of three heating temperatures (500, 550, and 600 °C) and three heating durations (3, 4, and 5 hours)."* [14]

Task Manager receives the input prompt and drives Experiment Designer and Robot Operator to complete Task 4. Task 4 is divided into nine subtasks, each corresponding to a different combination of heating temperatures and heating durations (indicated by different colors). Taking subtask 1 (500 °C, three hours) as an example, the experimental procedure generated by Experiment Designer and the code generated by Robot Operator are shown. Pre-prepared H2PtCl6 solutions, containing 3 wt. % Pt, were used.

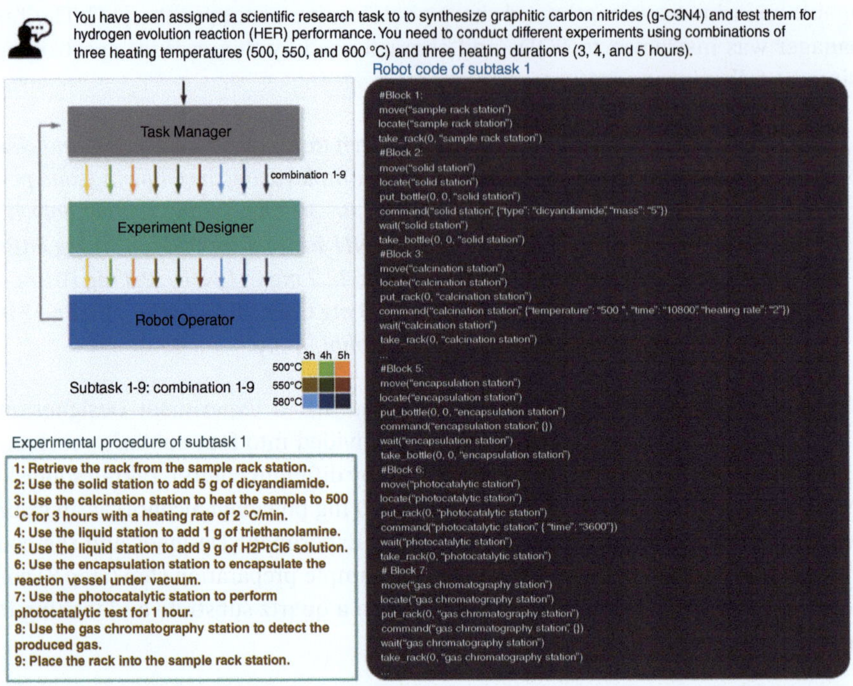

Figure 15.19 The workflow of ChemAgents and intermediate results on Task 4. *Source*: [14] / with permission of American Chemical Society.

As shown in Figure 15.20, Task 5 explored the photocatalytic degradation of organic pollutants in water using bismuth oxyhalides (BiOX, where X = Cl, Br, or I) [14]. The Task Manager was instructed using the following prompt:

> *"You have been assigned a scientific research task to synthesize bismuth oxyhalides (BiOX, where X = Cl, Br, or I) and characterize them, you need to conduct different experiments using combinations of three solid powders (NaX, where X = Cl, Br, or l) and four photocatalytic durations (30, 60, 90, and 120 minutes)."* [14]

Task Manager receives the input prompt and drives Experiment Designer and Robot Operator to complete Task 5. Task 5 is divided into twelve subtasks, each corresponding to a different combination of elements and photocatalytic durations (indicated by different colors). Taking subtask 1 (BiOCl, 30 minutes) as an example, the experimental procedure generated by Experiment Designer and the code generated by Robot Operator are shown.

> *"The tetracycline was dissolved in deionized water in advance to form an aqueous solution with a concentration of 10 μg g^{-1}. Subsequently, the prepared BiOX was mixed with the aqueous solution of TC, and the suspension was placed in the dark under constant stirring for 30 min to reach the adsorption/desorption equilibrium before exposure to light. Photocatalytic degradation of pollutants was executed using a photocatalytic station. The degradation performance of*

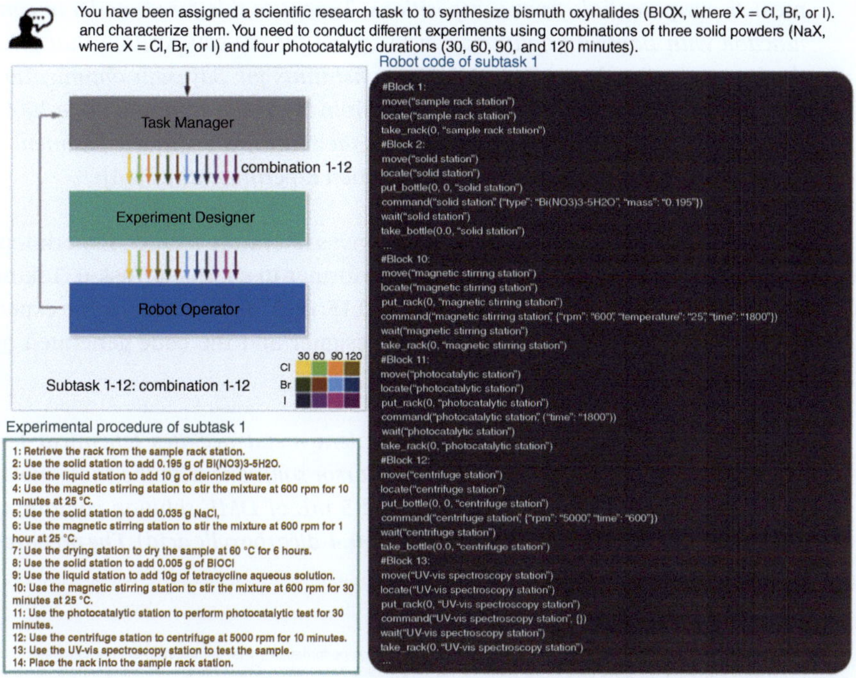

Figure 15.20 The workflow of ChemAgents and intermediate results on Task 5. *Source*: [14] / with permission of American Chemical Society.

> *BiOX on TC was obtained by measuring the change in intensity of the maximum absorption peak at 358 nm using UV-vis spectroscopy and converting the absorbance to concentration according to the Lambert-Beer law."*

15.4.2.3 Discovery and Optimization Tasks

These are the most advanced research tasks that the robotic AI chemist can currently accomplish, requiring the use of all four role-specific agents and their corresponding foundational resources, as illustrated in Figure 15.21 [14]. Take the Task 6 as an example, the authors tasked the robotic AI chemist with discovering metal–organic high-entropy catalysts (MO-HECs) for the oxygen evolution reaction [14]. The Task Manager was instructed using the following prompt:

> *"A scientific research task has been assigned to you to discover high-performance MO-HECs, comprising five metallic elements, for the oxygen evolution reaction. Please use the Literature Reader to identify prevalent metals for the OER in the literature and select the top five recommendations. The automated laboratory is equipped with the necessary precursor solutions and other required chemicals. Please select 100 quinary metal compositions by random sampling, with each metal's proportion ranging from 5% to 35%, and conduct their syntheses and overpotential measurements. After the experimental data were obtained from the 100 automated experiments, use the Computation Performer to optimize the experiments. Use the compositions and measured overpotentials of these 100 experiments, together with a pretrained model, to build a fused model that*

predicts overpotentials from compositions. Then, use this fused model in conjunction with Bayesian optimization to search for the composition that gives the lowest predicted overpotential. The constraints for Bayesian optimization are that the metal proportions must range from 5% to 35% and sum up to 100%. After the Computation Performer provides the results, your final experiment is to synthesize and test this optimal composition experimentally." [14]

Task Manager receives the input prompt and drives Literature Reader, Experiment Designer, Robot Operator, and Computation Performer to complete Task 6. Taking subtask 1 (Co: Ni: Fe: Mn: Cu = 0.32: 0.26: 0.15: 0.15: 0.12) as an example, the experimental procedure generated by Experiment Designer and the code generated by Robot Operator are shown:

"Initially, dissolve the corresponding divalent metal acetates (chemical formula $M(CH_3COO)_2$) in DMF to form precursor solutions with the same molar concentration of 50 mmol/L. Finally, add 5 mL of DMF solution containing 100 mmol/L terephthalic acid (i.e., benzene-1,4-dicarboxylic acid). The prepared

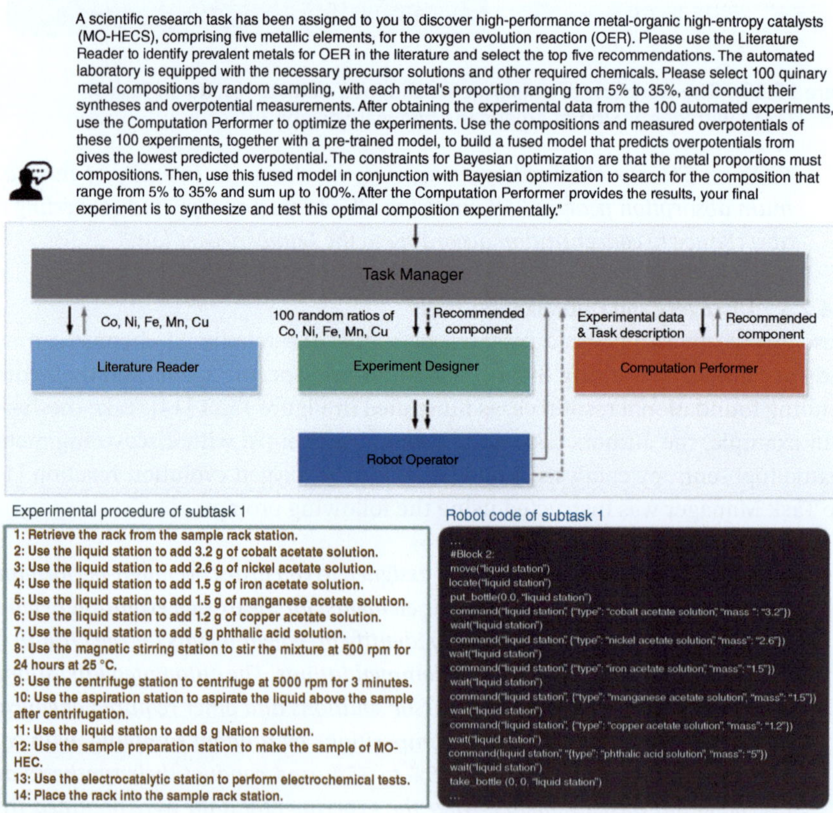

Figure 15.21 The workflow of ChemAgents and intermediate results on Task 6. *Source*: [14] / with permission of American Chemical Society.

catalyst was dispersed in 8 mL of mixed solution (isopropyl alcohol (7.8 mL) and 5 wt% Nafion (0.2 mL)) to prepare the catalyst ink. The sample preparation step (step 12) consists of treatment on the magnetic stirring station for 3 minutes at 500 rpm, the catalyst ink (200 μL) was dropped onto the surface of the carbon paper (CP) with a fixed working area of ca. 1.5 cm^2 and dried naturally. All electrochemical measurements were performed on the electrocatalytic station (CHI760E, CH Instrument) in a standard three-electrode system, with high-entropy metalorganic compound catalyst on CP as the working electrode, graphite rods (φ = 5 mm) as the counter electrode, and Ag/AgCl in saturated KCl solution as the reference electrode. OER tests were performed in 1 M KOH solution with the applied voltage converted to the reversible hydrogen electrode following the equation: $E_{RHE} = E_{Ag/AgCl} + 0.0592 \times pH + 0.197$ V to unify the potential window. All test results were displayed as raw results without any further processing (i.e. no iR correction)."

Figure 15.22 shows the data mining results related to Task 6. Due to the limitation of Pt being a precious metal, the 5 recommended metals for the Literature Reader are Co, Ni, Fe, Mn, and Cu. Then, the robotic AI chemist synthesized 100 MO-HECs with randomly chosen metal compositions and measured their overpotentials, all of which were above 300 mV at a current of 10 Ma/cm^2 [14]. The Computation Performer used this trained fused model in conjunction with its Bayesian optimizer to conduct virtual experiments to identify metal compositions imparting reduced overpotentials to MO-HECs, and the optimal combination of components is shown in Table 15.2.

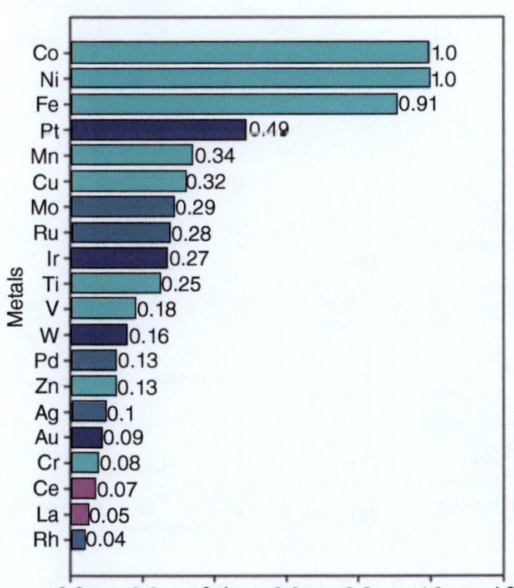

Figure 15.22 Data mining results related to Task 6. *Source*: [14] / with permission of American Chemical Society.

15.4.2.4 Scalability and Adaptability Verification Tasks

To further validate the scalability and adaptability of ChemAgents, the authors implemented Task 7 in a new robotic chemistry laboratory, demonstrating its capability to autonomously conduct organic reactions. Unlike Tasks 1–6, Task 7 is carried out in a new environment, as shown in Figure 15.23, which includes a new robot platform, additional automated workstations, and more powerful experimental capabilities [14].

Table 15.2 Bayesian optimized experimental data for Task 6.

Catalyst no.	Elemental composition (%)					Overpotential @ 10 mA/cm^2 (V vs. RHE)	η_{10}(mV)
	Co	Ni	Fe	Mn	Cu		
1	30.81	34.35	16.44	8.64	9.77	1.4961	266.1

Source: [14] / with permission of American Chemical Society.

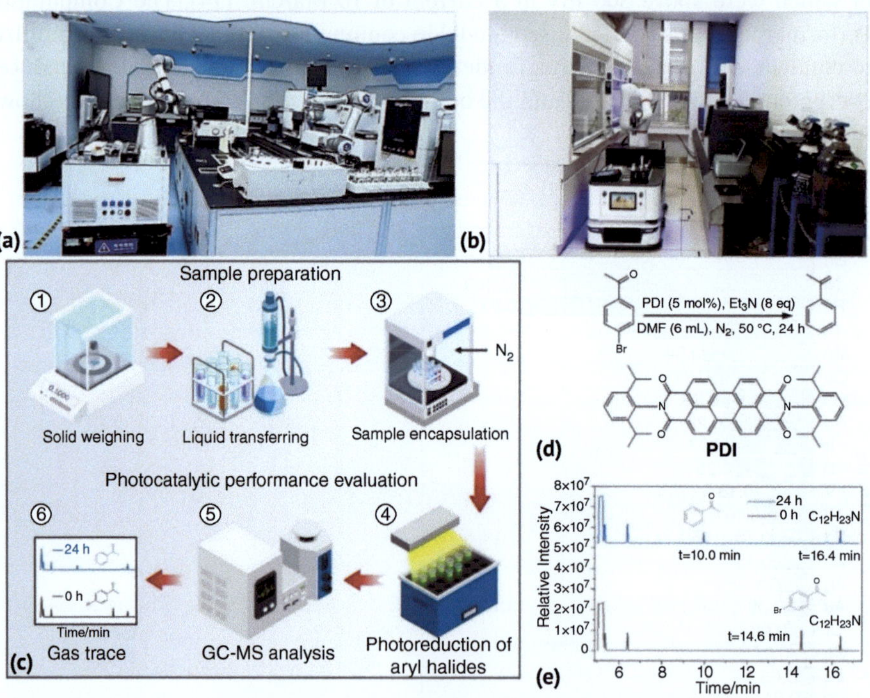

Figure 15.23 (a) The robotic lab in which Tasks 1 to 6 were carried out. (b) The robotic organic chemistry lab in which Task 7 was carried out. (c) The experimental workflow of Task 7. (d) Photoreduction of aryl halides with the PDI catalyst. (e) Gas chromatography of the crude product of dehalogenation of 4-bromoacetophenone with dodecanenitrile as the internal substance at different times: 0 h (black line) and 24 h (blue line). *Source*: [14] / with permission of American Chemical Society.

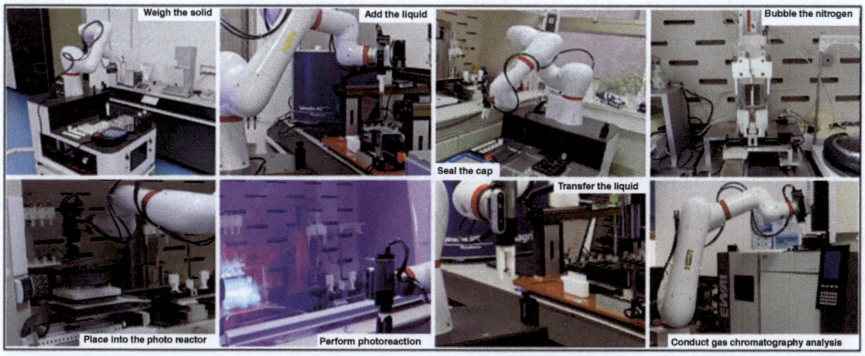

Figure 15.24 Key robotic operations for executing the photocatalytic debromination of aryl bromides and monitoring the reaction products [14].

For this task, as shown in Figure 15.23b, ChemAgents was deployed in the robotic organic chemistry laboratory and adapted to integrate new resources. The Literature Reader agent was instructed to leverage the LLM's capabilities for mining specific chemical information from text data. As shown in Figure 15.23c, the Experiment Designer was updated to accommodate and integrate the lab's experimental instruments, including a solid dispensing station, a liquid transferring station, a sample encapsulation station, a photoreactor station, and a gas chromatography–mass spectrometry (GC–MS) station. Meanwhile, the Robot Operator agent was equipped with an appropriate set of APIs to support the additional functionalities required in this experimental setting. In this task, the robot AI chemist was instructed to carry out the photocatalytic debromination of aryl bromides and to monitor the reaction products. The entire experimental process is shown in Figure 15.23c, and the key operations of the robot are detailed in Figure 15.24.

Therefore, Task 7 demonstrates the deployment capabilities of ChemAgents in the new environment, enabling them to effectively manage laboratory resources and independently perform chemical research tasks, showcasing the scalability and adaptability of ChemAgents.

15.5 ChatMOF: An AI System for Predicting and Generating MOFs using LLMs

In the field of materials science, when applying LLMs to the research of metal–organic frameworks (MOFs), two core challenges are faced: the high complexity of material structures and the scarcity of data in this field [16]. Traditional methods are difficult to effectively handle the material characterization of text input. On the one hand, the porous structure and topological features of MOF materials lack textual characterization methods suitable for LLMs. Moreover, compared with fields such as chemistry and biology, the scale of its dedicated databases is limited and the data forms are diverse, such as geometric properties and synthesis conditions, which makes it difficult for traditional machine learning methods based on structured data to handle them effectively. On the other hand, the reverse design of MOF materials relies on methods such as experimental trial and error or complex calculations, which is often inefficient. Moreover, the knowledge of MOF materials in unstructured texts, such as academic literature, has not been fully exploited and utilized. Yeonghun Kang and his research team proposed the ChatMOF model in 2024, an AI system that uses LLMs (such as GPT-4) to predict and generate MOFs [16]. The development of ChatMOF is precisely aimed at breaking through these bottlenecks, exploring the possibility of LLMs integrating multisource data and toolchains through natural language interaction, and promoting the transformation of material design from experience-driven to AI intelligence-driven. According to the results, ChatMOF can complete tasks such as data retrieval, attribute prediction, and structure generation. The accuracy rates in the search, prediction, and generation tasks reach 96.9%, 95.7%, and 87.5% (GPT-4), respectively, and it can generate materials that meet the needs of users through natural language. It demonstrates the potential and challenges of integrating LLM with databases and machine learning in materials science.

The effectiveness of autonomous LLM agents is predicated on its capability to accurately extract essential details from textual inputs and offer relevant responses, irrespective of the presence of a rigidly structured query [17]. As shown in Figure 15.25, it is the concept and core schematic representation of ChatMOF.

As shown in Figure 15.25a, users can ask queries about the material properties in text form: "*What is the largest cavity diameter of MIBQAR16?*," ChatMOF responds by providing detailed descriptions related to the asked materials. In addition, when the user indicates the need to generate MOF with specific attributes: "Can you generate a structure with an accessible volume fraction with 0.5 and visualize it?" ChatMOF can generate the required material structure accordingly. As depicted in Figure 15.25b, the core design of the ChatMOF system adopts a three-module collaborative framework, achieving an automated closed loop from natural language queries to material design results through the collaborative integration of the Agent, Toolkit, and Evaluator.

15.5.1 Agent Module

The agent module, based on ReAct [18] and MRKL [19] frameworks, acts as an intelligent brain and processes user requirements through a four-step cycle of data parsing, strategy planning, tool invocation, and result reflection. Its specific steps are: Initially, the user's query is established as the objective, followed by systematic planning to determine the steps to meet this objective. Subsequently, ChatMOF decides on the appropriate tool to employ from the available options. After the chosen tool is executed, the observed results serve as the basis for evaluating whether a final answer can be generated. If feasible, the final answer is presented; otherwise, the process cycles back to the thought step to formulate a new strategy [16].

15.5.2 Toolkit Module

The toolkit module, as a functional engine, aims to use various tools to obtain, predictor generate material information, integrating four major types of tools: table-searcher, predictor, generator, and utilities [16].

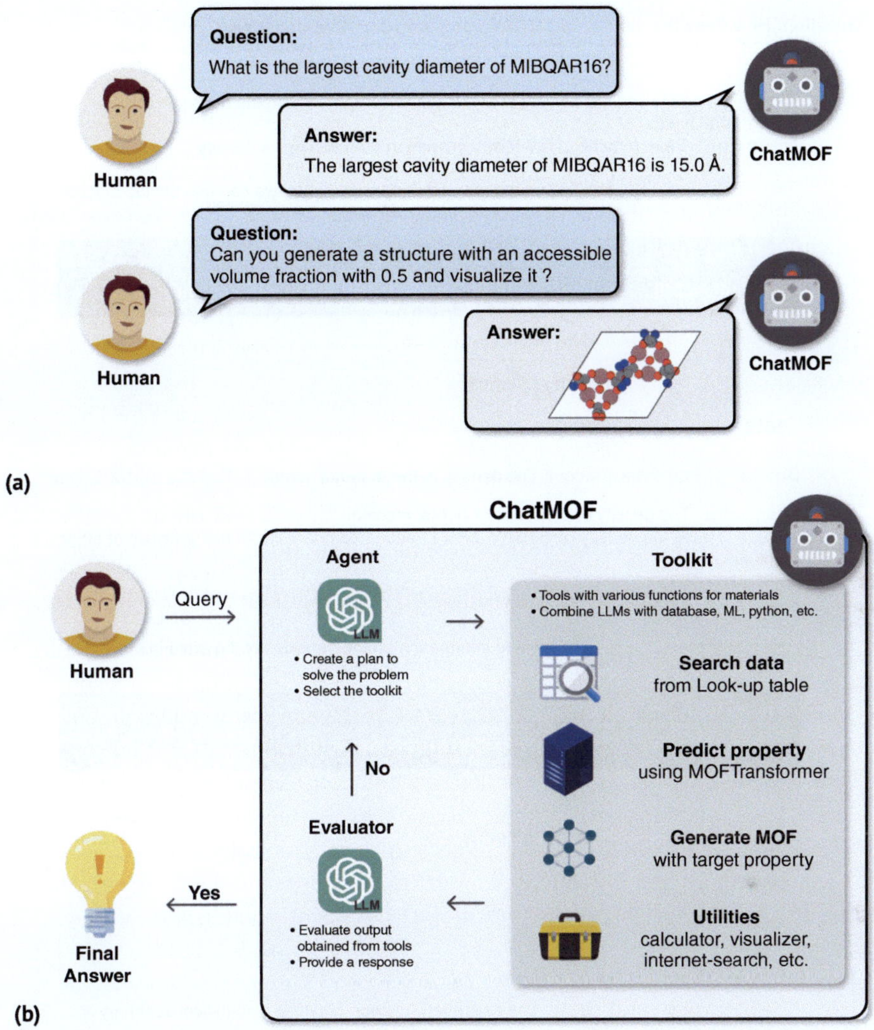

Figure 15.25 The overview of (a) conceptual and (b) schematic representation of ChatMOF. *Source*: [16] / Springer Nature / CC BY 4.0.

First is the table-searcher tool, which involves obtaining the required information from the existing data. These databases are mainly those related to the MOF structure: (1) computational-ready, experimental MOFs (CoREMOF) [20, 21]. The CoREMOF database is an archive of synthetic materials in a subset of CSD MOF [22], covering the calculations of various properties of MOF, including geometric descriptors and (2) quantum MOF (QMOF) database [23]. The QMOF database stores the electrical property data calculated through DFT, such as band gaps, formation energy, the highest occupied molecular orbital, and the lowest occupied molecular orbital. Therefore, if users search for information about a specific MOF contained in these databases, ChatMOF can locate and provide the required information from the pre-tabulated data as shown in Figure 15.26. When responding to

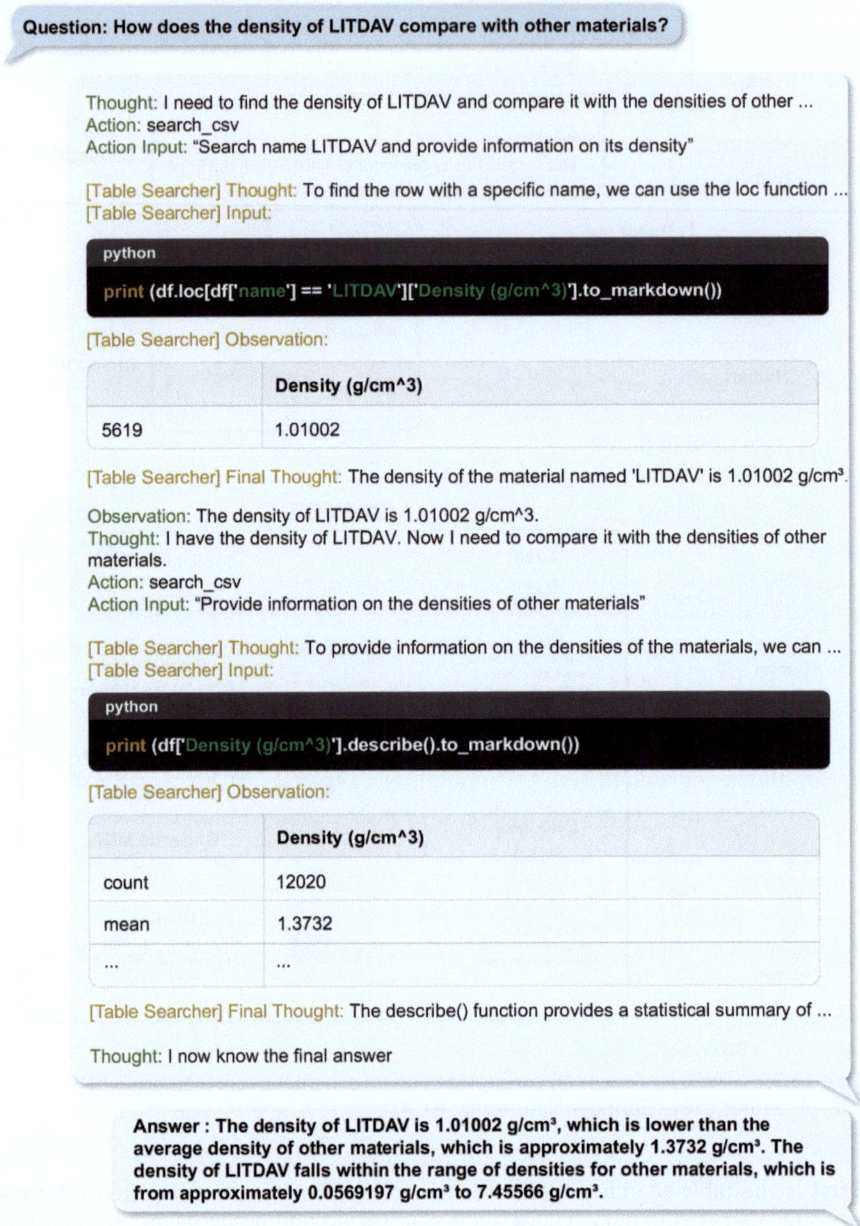

Figure 15.26 Example of a Table Searcher for the question. *Source*: [16] / Springer Nature / CC BY 4.0.

user queries, ChatMOF will automatically determine the best way to extract the necessary data. Subsequently, it creates a Python code to retrieve specific information from the database based on a pre-considered strategy. This code is then executed in the designated executor of ChatMOF. After this operation is completed, ChatMOF automatically determines the subsequent processes required to generate the final answer and then replies to the user as a response to the request.

The second is about the predictor tool. ChatMOF uses the MOFTransformer [24, 25] model developed by this team to make general predictions about the properties of MOF. This model combines local features, such as atoms and chemical bonds, with global features, such as surface area and topological structure, for prediction. After being pre-trained with one million hypothetical MoFs and subsequently fine-tuned for specific properties, MOFTransformer performs exceptionally well in predicting various properties [16]. Furthermore, by retaining the model structure while changing the weights of the model that has been fine-tuned for the required properties, it can conveniently predict the required properties. The key to generating accurate responses is selecting the appropriate fine-tuned model with the MOFTransformer and the material to which it will be applied, based on the query [16]. As shown in Figure 15.27, the predictor sets up a plan to solve the question, an appropriate model, and target material. Based on this, it uses machine learning to predict the value, which is then used to derive the final answer. When the user requests this information, the agent will develop a strategy to handle this query. It chose the most suitable fine-tuning MOFTransformer model named hydrogen diffusivity dilute 77 K. Then, the agent identifies the materials used for prediction and makes the prediction using the selected model. Subsequently, ChatMOF conducts a table search to identify the substance with the highest value determined by the machine learning model.

The third is the generation tool, which is based on the integration of genetic algorithms and LLMs. Figure 15.28 shows how ChatMOF uses genetic algorithms to reverse-design MOF based on user requirements. After receiving the user's questions, the agent will formulate a strategy optimized through the algorithm process based on the genetic algorithm. It can identify the target attributes and determine the loss function that is most suitable for the target. According to the algorithm strategy, ChatMOF will select the parent gene from the existing database based on the predefined loss function through the algorithm. In the adopted genetic algorithm, those parent genes that show high potential based on the required target characteristics will be selected, thereby increasing the probability that the obtained offspring genes exhibit the target attributes more significantly. Then, these offspring are transformed into structure files and their attributes are estimated through machine learning.

This process will be repeated a fixed number of times to generate multiple generations of offspring, and the MOF produced by each generation will get closer and closer to the target. From the created structure, ultimately select the one that best matches the question to respond and reply to the user.

Finally, it is practical tools, which integrate functions such as Internet search, unit conversion, visualization tools, atomic simulation environment libraries, and even simple calculations as shown in Figure 15.25. These functions are mainly achieved by leveraging LangChain, thereby enhancing the overall functionality and practicality of ChatMOF. It has been meticulously designed to perform a wide range of practical functions that go beyond the scope of LLMs. In conclusion, it is not merely a material analysis tool, but a comprehensive system capable of handling a wide range of tasks and operations.

As depicted in Figure 15.29, it shows how ChatMOF can handle complex multistep tasks efficiently. ChatMOF responds to the user's question: "*Provide the CO2*

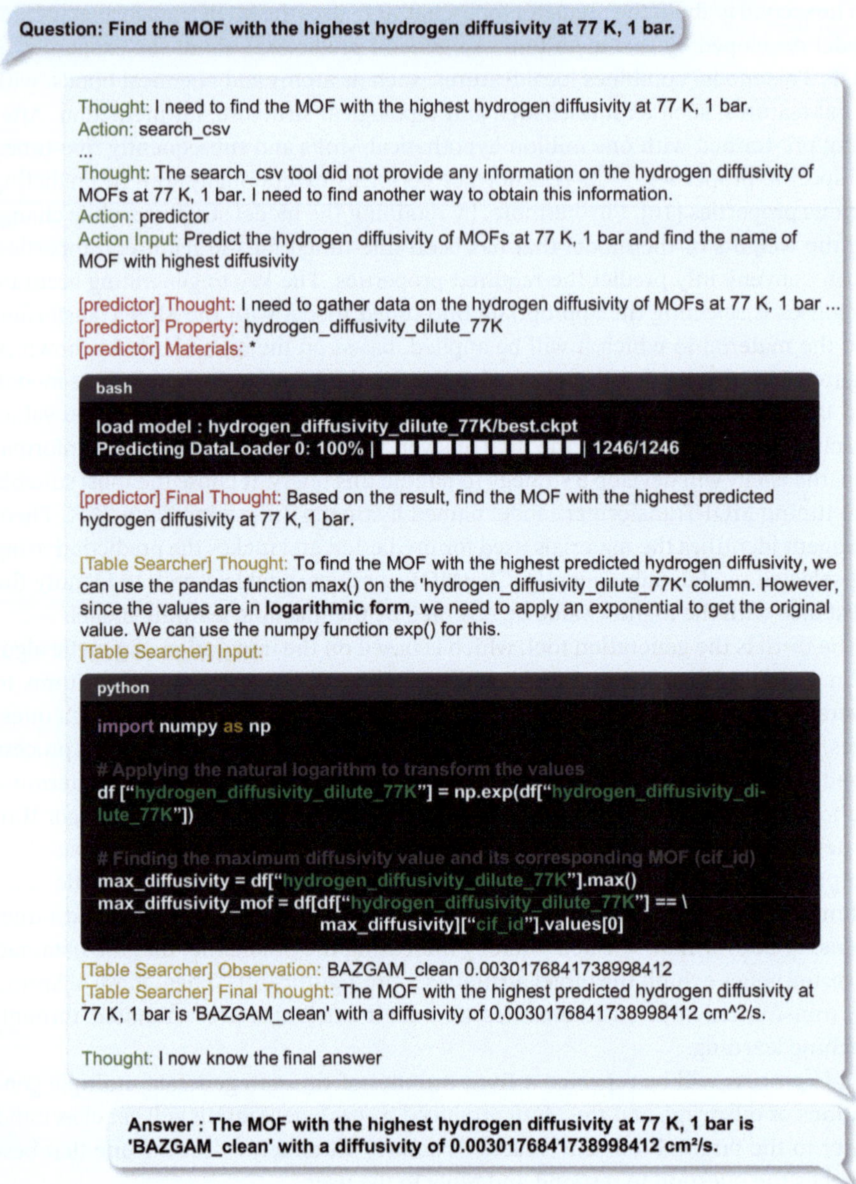

Figure 15.27 Example of a predictor for the question. *Source*: [16] / Springer Nature / CC BY 4.0.

Henry coefficient of XEGKUR at 298 K in mol/cm ³ Pa." Firstly, ChatMOF uses a predictive tool to determine the Henry coefficient of carbon dioxide. Subsequently, it uses a unit conversion tool to convert the unit from mol/KgPa to mol/cm^3Pa. Then, it conducts a table search to obtain the density value of the material. After obtaining the density data, ChatMOF uses the unit conversion tool to convert g/cm^3 to kg/cm^3. Finally, integrate all this information to obtain the final answer.

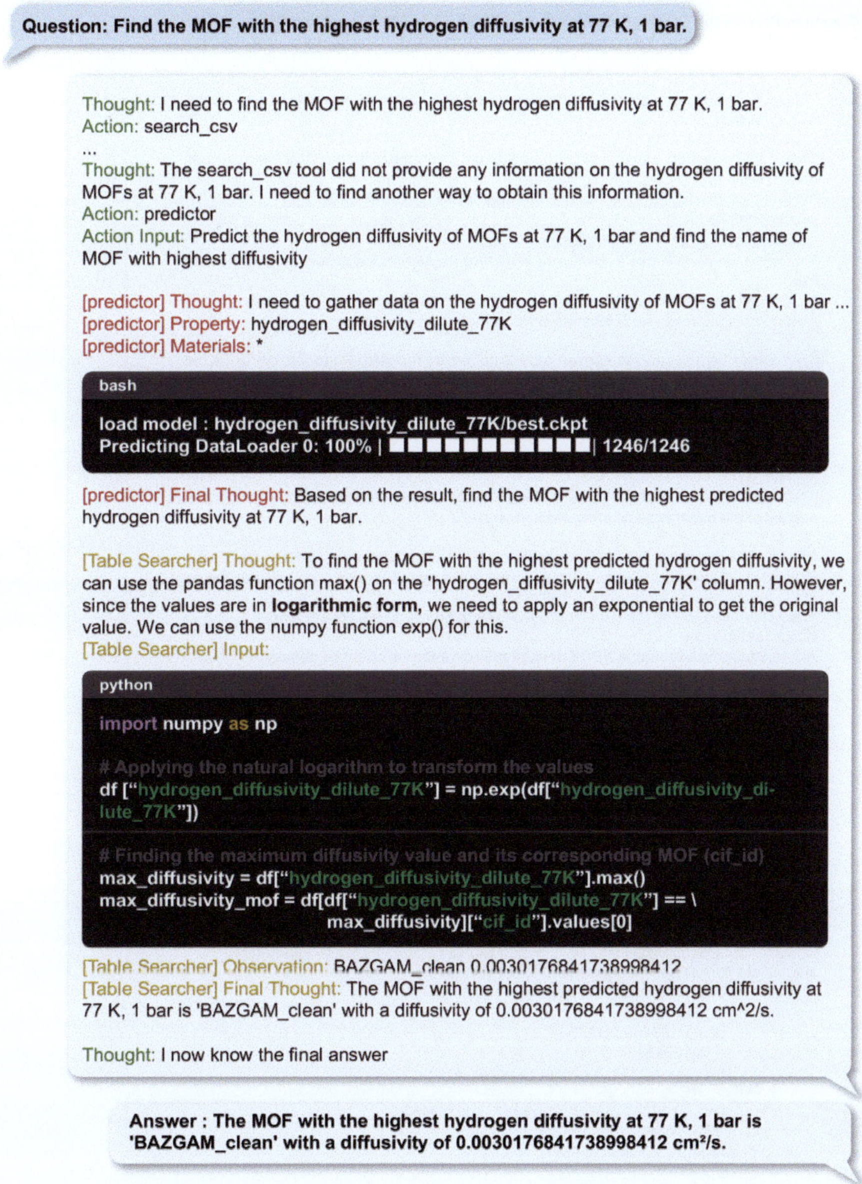

Figure 15.28 Example of a generator for the question. *Source*: [16] / Springer Nature / CC BY 4.0.

As shown in Figure 15.30, it is another example of complex and multistep questions for ChatMOF. These cases highlight that ChatMOF is not only adept at transforming natural language into structured data, but also capable of interpreting and strategically handling complex queries to formulate tailor-made and efficient solutions.

Question: Provide the CO_2 Henry coefficient of XEGKUR at 298 K in mol/cm³Pa.

Thought: I need to find the CO2 Henry's coefficient of XEGKUR at 298K.
Action: search_csv
... *(run search_csv tool)*
Thought: The search_csv tool did not provide any information on the CO2 Henry's coeff...
Action: predictor
Action Input: Predict the CO_2 Henry coefficient for XEGKUR at 298K
... *(run predictor tool)*
Observation: 0.026577507595890823
Thought: I have the Henry's coefficient in mol/Kg·Pa, but I need to convert it to mol/cm^3-Pa.
Action: unit_converter
Action Input: Convert -3.62769 mol/Kg·Pa to mol/cm3Pa

[Unit converter] Thought: To convert the H2 uptake from mol/kg·Pa to mol/cm^3·Pa, we need to convert the mass from kilograms to cubic centimeters. This requires the knowledge of the density of the metal-organic framework.
[Unit converter] Unit: The unit to be converted from is mol/kg·Pa and the desired target unit is mol/cm^3·Pa.
[Unit converter] Equation: H2 uptake (mol/cm^3·Pa) = H2 uptake (mol/kg·Pa) * Density (kg/cm^3)
[Unit converter] Information: The additional information required for this conversion is the density of the metal-organic framework in kg/cm^3.

Observation: Equation: H2 uptake (mol/cm^3·Pa) = H2 uptake (mol/kg·Pa) * Density (kg/cm^3), Information: The additional information required for this conversion is the density of the metal-organic framework in kg/cm^3.
Thought: I need to find the density of XEGKUR to complete the conversion.
Action: search_csv
Action Input: "Search name XEGKUR and provide information on its density"
...
Observation: The density of XEGKUR is 1.03463 g/cm^3.
Thought: I need to convert the density from g/cm^3 to kg/cm^3.
Action: unit_converter
Action Input: "Convert 1.03463 g/cm^3 to kg/cm^3"

[Unit converter] Thought: To convert the density from g/cm^3 to kg/cm^3, we need to convert the mass unit from grams to kilograms. This can be done by using the conversion factor of 1 kg = 1000 g.
[Unit converter] Unit: The unit to be converted from is g/cm^3 and the desired target unit is kg/cm^3.
[Unit converter] Equation: Density (kg/cm^3) = Density (g/cm^3) / 1000
[Unit converter] Information: No additional information is required for this conversion.

Observation: Equation: Density (kg/cm^3) = Density (g/cm^3) / 1000, Information: No additional information is required for this conversion.
Thought: I can now calculate the density in kg/cm^3.
Action: Python_REPL
Action Input: print (1.03463 / 1000)
Observation: 0.00103463
Thought: I can now calculate the Henry's coefficient in mol/cm^3Pa.
Action: Python_REPL
Action Input: print (0.026577507595890823 * 0.00103463)
Observation: 2.749788668393652e-05
Thought: I now know the final answer

Answer : The CO2 Henry's coefficient of XEGKUR at 298 K is 2.75e-05 mol/cm³Pa.

Figure 15.29 Example of a complex and multistep question. *Source*: [16] / Springer Nature / CC BY 4.0.

Figure 15.30 Example of a complex and multistep question "What is the relationship between H_2 uptake and accessible volume in MOFs?" *Source*: [16] / Springer Nature / CC BY 4.0.

15.5.3 Evaluation Module

The evaluator module acts as a quality gatekeeper and needs to conduct logical verification and result integration for the tool outputs of search tasks, prediction tasks, and generation tasks. The author used GPT-4 to create questions about ChatMOF, generated various sentences about the given attributes of MOF, aiming to minimize human participation, improve the fairness of measurement, and generate a wide variety of questions. As shown below is the question generation prompt for search and prediction tasks.

> *"I have a model that generates the structures with specific properties, and you need to generate a question to match. The question should be as diverse as possible, and should be something a human would ask the model. The property should be one of [bandgap, hydrogen uptake (100 bar), hydrogen diffusivity, O_2 diffusivity, N_2 diffusivity, O_2 uptake, N_2 uptake, thermal stability, CO_2 henry coefficient, surface area, largest cavity diameter, pore limiting diameter, pore volume, volume fraction]. You add temperature 298 K and 1 bar condition for each question with property listed in ['diffusivity', 'uptake', and 'henry coefficient'].*
>
> *Example:*
> - *What are the top 10 materials with the highest bandgap?*
> - *What is the bandgap of [mat]?*
> - *Can you show the materials with a hydrogen uptake greater than [value] at 298 K, 1 bar?*
>
> *Generate questions.*
> *Begin!"*

After creating the prompt, select the CoreMOF refcode and value for the property in the [mat] and [value] fields. For 298 K and 1 bar, select the appropriate temperature and pressure for the property. Among the properties, those that can be obtained from the lookup table are classified as search tasks, and the rest are classified as prediction tasks.

As shown below is the question generation prompt for generation tasks.

> *"I have a model that generates the structures with specific properties, and you need to generate a question to match. The question should be as diverse as possible, and should be something a human would ask the model. The property should be one of [bandgap, hydrogen uptake (100 bar), hydrogen diffusivity, O_2 diffusivity, N_2 diffusivity, O_2 uptake, N_2 uptake, thermal stability, CO_2 henry coefficient, surface area, largest cavity diameter, pore limiting diameter, pore volume, volume fraction]. You add temperature 298 K and 1 bar condition for each question with property listed in ['diffusivity', 'uptake', and 'henry coefficient'].*

Example:
- *Can you generate the structures with highest PLD?*
- *Can you generate the structures with band-gap near 0.4?*
- *Can you generate the structures with smallest bandgap?*

Generate 30 questions.
Begin!"

In this study, accuracy measurement ensures that the reasoning process of AI from formulating a problem-solving plan to selecting a toolkit and then interpreting the results is not affected by logical differences. Meanwhile, the accuracy analysis of ChatMOF uses three labels: True, False (token limit exceeded), and False (logic error). True label indicates that the process of ChatMOF correctly matches the logic required to generate accurate answers. When the token count in a LLM exceeds the maximum allowable value of 4,000, thereby hindering further progress, the False (token limit exceeded) label is used. Finally, the False (logic error) is used to specify the situation where the logical error of ChatMOF leads to incorrect responses or exceptions. Figure 15.31 shows the accuracy measurements for the three tasks using ChatMOF with GPT-4. The accuracy of 100 sample questions for the search and prediction tasks and 10 sample questions for the generation tasks was measured. For the search and prediction tasks, their accuracy rates were as high as 96.9% and 95.7%, respectively. For the generation task, the accuracy rate is 87.5%.

From the perspective of ChatMOF, it demonstrates the huge potential of generative AI in the field of materials science, especially in the prediction and generation of MOFs. This system is not only capable of handling simple information retrieval tasks, but also can generate new material structures that meet specific requirements. In the future, with the increase in model capacity and the structural improvement of data sharing on online platforms, the performance of ChatMOF is expected to be further enhanced, laying a solid foundation for significant progress in MOF research. Currently, the online experience site for the ChatMOF model is: https://chatmof-online.streamlit.app/.

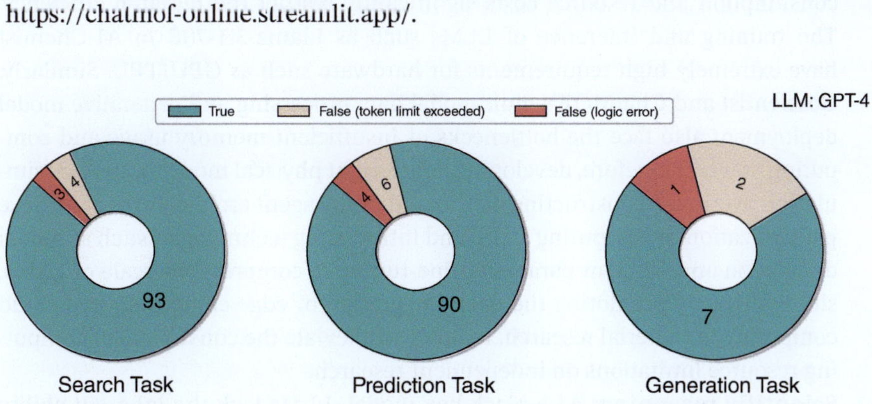

Figure 15.31 Depiction of accuracies for three tasks using GPT-4 mode. *Source:* [16] / Springer Nature / CC BY 4.0.

15.6 Summary and Prospect

NLP and LLMs have significantly enhanced the ability to analyze, extract, and interpret valuable information from extensive materials science literature. The agents derived from LLMs can understand the intentions of users and independently design, plan, and utilize tools to perform complex tasks. These characteristics are particularly beneficial for materials science research, which is an interdisciplinary field characterized by numerous complex and time-consuming activities. At present, autonomous materials research agents have demonstrated great potential in the field of materials, showcasing their advantages in automated experimental design and complex task execution through a series of intelligent processes, including data collection, experimental design, characterization, preparation, and material intelligent platforms. However, Leveraging Agents in materials science presents considerable challenges due to the domain's complexity and specificity. Key issues include computing resource limitations, scientific reasoning, and safety risk.

1. **Computing resource limitation:** The training, fine-tuning, and deployment of LLMs require a large amount of computing resources, which poses a challenge to small research teams and institutions. In terms of computing resource limitations, existing autonomous material research agents are confronted with multiple challenges and optimization spaces. Firstly, the training, fine-tuning, and deployment of LLMs require a large amount of computing resources. For most research teams, a trade-off often needs to be made between model accuracy and task applicability, which poses a challenge to small research teams and institutions. When AtomAgents rely on tools such as LAMMPS for atomic simulation, DFT calculations, and large-scale molecular dynamics simulations consume a significant amount of computing power. Especially when dealing with complex systems such as high-entropy alloys, the computing time consumption and resource costs significantly restrict the research efficiency. The training and inference of LLMs such as Llama-3.1-70B in AI-Chemist have extremely high requirements for hardware such as GPU/TPU. Similarly, Coscientist and ChatMOF's multimodal data processing and generative model deployment also face the bottlenecks of insufficient memory usage and computing power. Therefore, developing lightweight physical models to reduce simulation overhead, constructing distributed multi-agent architectures to achieve parallelization of computing tasks, and introducing technologies such as model distillation and efficient parameter fine-tuning to compress the scale of LLMs, simultaneously promoting the deep integration of edge computing and cloud computing in material research, in order to alleviate the constraints of computing resource limitations on independent research.
2. **Scientific reasoning:** As a black-box model, LLMs lack the inherent ability to evaluate the reliability of generated or retrieved data. In terms of Scientific reasoning, the future development of autonomous materials research agents will revolve around breakthroughs in multidimensional capabilities. From the existing literature, the training corpora of LLMs are mainly composed

of general text data, which often lack domain-specific terms and precise scientific knowledge. LLMs may generate inaccurate, misleading, or fictional information, including incorrect values and unrealistic materials or processes. Therefore, how to enhance the causal reasoning ability of LLMs for physical mechanisms becomes particularly important. Furthermore, with the launch of DeepSeek-R1, it has achieved state-of-the-art "reasoning" capabilities by leveraging large-scale reinforcement learning based on proximal policy optimization. Therefore, the interpretability of scientific reasoning remains a key challenge. It is necessary to construct a transparent reasoning chain to make the agent reasoning process meet the rigor requirements of materials science research. Establishing a credibility assessment mechanism for LLMs in the field of materials science and developing a comprehensive credibility assessment framework are necessary measures to mitigate this risk. In the future, through reinforcement learning, the integration of symbolic reasoning and neural networks, the construction of cross-domain knowledge graphs, and the closed-loop optimization of dynamic hypotheses and verifications, we will promote the evolution of agents from "data processors" to "scientific discoverers," achieving a leap in reasoning from phenomenon observation to mechanism understanding in material research.

3. **Safety risk:** Agents based on LLMs introduce additional security risks when interacting with the physical world. Users may deliberately use the LLMs intelligent system to produce dangerous chemicals, or accidentally generate toxic by-products or unstable compounds during legal scientific tasks. In terms of Safety risk, the future development of autonomous material research agents requires the construction of a protection system from three aspects: technical design, experimental operation, and ethical norms. If the physical simulation relied upon by AtomAgents in alloy design lacks a toxicity prediction module, environmentally harmful materials may be produced. It is necessary to embed a material safety assessment model in the agent architecture and screen high-risk components in combination with the literature knowledge graph. In the automated laboratory of AI-Chemist, if the robot operation is not integrated with the chemical safety database, it may lead to the leakage of corrosive reagents or the loss of control of reactions. Through the linkage of sensor data and LLM inference, the compliance of the operation is verified before the experiment is executed. In the future, a full-chain safety closed loop covering design, simulation, and experimentation needs to be established. Technically, lightweight safety assessment algorithms should be developed. In terms of process, it is mandatory for agents to output safety risk ratings. Ethically, safety norms for AI material research should be formulated to ensure that agents' decisions comply with international standards. Fundamentally avoid potential safety hazards in material design, experimental operation, and knowledge management.

In summary, the autonomous materials research agent has achieved full-process automation from theoretical design to experimental verification through the deep integration of LLM and multi-agent systems, significantly accelerating the discovery of new materials. In the future, breakthroughs need to be made in complex material modeling, interdisciplinary integration, intelligent experiments, etc., to promote the integrated development of materials science towards prediction, design, and manufacturing, and provide innovative solutions for fields such as materials, chemistry, and physics.

References

1 Ghafarollahi A, Buehler MJ. Automating alloy design and discovery with physics-aware multimodal multiagent AI. Proceedings of the National Academy of Sciences of the United States of America. 2025;122(4):e2414074122.

2 Boiko DA, MacKnight R, Kline B, Gomes G. Autonomous chemical research with large language models. Nature. 2023;624(7992):570–8.

3 Perera D, Tucker JW, Brahmbhatt S, Helal CJ, Chong A, Farrell W, et al. A platform for automated nanomole-scale reaction screening and micromole-scale synthesis in flow. Science. 2018;359(6374):429–34.

4 Ahneman DT, Estrada JG, Lin S, Dreher SD, Doyle AG. Predicting reaction performance in C–N cross-coupling using machine learning. Science. 2018;360(6385):186–90.

5 Bran AM, Cox S, Schilter O, Baldassari C, White AD, Schwaller P. Augmenting large language models with chemistry tools. Nature Machine Intelligence. 2024;6(5):525–35.

6 Zheng Z, Zhang O, Borgs C, Chayes JT, Yaghi OM. ChatGPT chemistry assistant for text mining and the prediction of MOF synthesis. Journal of the American Chemical Society. 2023;145(32):18048–62.

7 Zheng Z, Alawadhi AH, Chheda S, Neumann SE, Rampal N, Liu S, et al. Shaping the water-harvesting behavior of metal–organic frameworks aided by fine-tuned GPT models. Journal of the American Chemical Society. 2023;145(51):28284–95.

8 Zhao WX, Zhou K, Li J, Tang T, Wang X, Hou Y, et al. A survey of large language models [Internet]. arXiv; 2023 [cited 2025 Jun 25]. Available from: https://arxiv.org/abs/2303.18223

9 OpenAI, Achiam J, Adler S, Agarwal S, Ahmad L, Akkaya I, et al. GPT-4 Technical Report [Internet]. arXiv; 2023 [cited 2025 Jun 25]. Available from: https://arxiv.org/abs/2303.08774

10 Kaddour J, Harris J, Mozes M, Bradley H, Raileanu R, McHardy R. Challenges and applications of large language models [Internet]. arXiv; 2023 [cited 2025 Jun 25]. Available from: https://arxiv.org/abs/2307.10169

11 Song Y, Xiong W, Zhu D, Wu W, Qian H, Song M, et al. RestGPT: Connecting large language models with real-world RESTful APIs [Internet]. arXiv; 2023 [cited 2025 Jun 25]. Available from: https://arxiv.org/abs/2306.06624

12 MacLeod BP, Parlane FGL, Morrissey TD, Häse F, Roch LM, Dettelbach KE, et al. Self-driving laboratory for accelerated discovery of thin-film materials. Science Advances. 2020;6(20):eaaz8867.

13 Abolhasani M, Kumacheva E. The rise of self-driving labs in chemical and materials sciences. Nature Synthesis. 2023;2(6):483–92.

14 Song T, Luo M, Zhang X, Chen L, Huang Y, Cao J, et al. A multiagent-driven robotic AI chemist enabling autonomous chemical research on demand. Journal of the American Chemical Society. 2025;147(15):12534–45.

15 Zhang B, Xiao H, Ye G, Song Z, Han T, Sharman E, et al. Label-free data mining of scientific literature by unsupervised syntactic distance analysis. Journal of Physical Chemistry Letters. 2024;15(1):212–9.

16 Kang Y, Kim J. ChatMOF: An artificial intelligence system for predicting and generating metal-organic frameworks using large language models. Nature Communications. 2024;15(1):4705.

17 Jablonka KM, Ai Q, Al-Feghali A, Badhwar S, Bocarsly JD, Bran AM, et al. 14 examples of how LLMs can transform materials science and chemistry: A reflection on a large language model hackathon. Digital Discovery. 2023;2(5):1233–50.

18 Yao S, Zhao J, Yu D, Du N, Shafran I, Narasimhan K, et al. ReAct: Synergizing reasoning and acting in language models [Internet]. arXiv; 2022 [cited 2025 Jun 21]. Available from: https://arxiv.org/abs/2210.03629

19 Karpas E, Abend O, Belinkov Y, Lenz B, Lieber O, Ratner N, et al. MRKL systems: A modular, neuro-symbolic architecture that combines large language models, external knowledge sources and discrete reasoning [Internet]. arXiv; 2022 [cited 2025 Jun 25]. Available from: https://arxiv.org/abs/2205.00445

20 Chung YG, Haldoupis E, Bucior BJ, Haranczyk M, Lee S, Zhang H, et al. Advances, updates, and analytics for the computation-ready, experimental metal–organic framework database: CoRE MOF 2019. Journal of Chemical & Engineering Data. 2019;64(12):5985–98.

21 Chung YG, Camp J, Haranczyk M, Sikora BJ, Bury W, Krungleviciute V, et al. Computation-ready, experimental metal–organic frameworks: A tool to enable high-throughput screening of nanoporous crystals. Chemistry of Materials. 2014;26(21):6185–92.

22 Moghadam PZ, Li A, Wiggin SB, Tao A, Maloney AGP, Wood PA, et al. Development of a Cambridge structural database subset: A collection of metal–organic frameworks for past, present, and future. Chemistry of Materials. 2017;29(7):2618–25.

23 Rosen AS, Iyer SM, Ray D, Yao Z, Aspuru-Guzik A, Gagliardi L, et al. Machine learning the quantum-chemical properties of metal–organic frameworks for accelerated materials discovery. Matter. 2021;4(5):1578–97.

24 Kang Y, Park H, Smit B, Kim J. A multi-modal pre-training transformer for universal transfer learning in metal–organic frameworks. Nature Machine Intelligence. 2023;5(3):309–18.

25 Park H, Kang Y, Kim J. Enhancing structure–property relationships in porous materials through transfer learning and cross-material few-shot learning. ACS Applied Materials & Interfaces. 2023;15(48):56375–85.

16
Challenges and Future Developments

16.1 Challenges

Natural language processing (NLP) and large language models (LLMs) offer powerful new tools for analyzing and extracting insights from the vast materials science literature. However, applying LLMs effectively in this field faces significant challenges due to its complexity. Key hurdles include improving the models' numerical understanding, quantitative prediction, structural interpretation, and overall scientific reasoning.

16.1.1 Numerical Understanding

Interpreting numerical data embedded in textual narratives is a prerequisite for resolving quantitative problems such as the prediction of material properties. Elevating the numerical proficiency of these models constitutes a complex endeavor that demands a coordinated approach involving dataset construction and architectural design as well as strategic task planning and training optimization. The first priority is creating high-quality, materials-specific numerical datasets to provide accurate training data. After this, fine-tuning models on domain-focused numerical tasks will be essential to ensure the precise interpretation and use of this quantitative information.

16.1.2 Quantitative Prediction

Deriving robust quantitative correlations linking material composition and processing history to final properties based exclusively on textual inputs remains a formidable hurdle for contemporary models. Current research investigates various solutions, such as developing integrated frameworks that directly combine materials-focused language encoders with property-prediction networks. Beyond architectural modifications, the predictive accuracy can be significantly augmented through the deployment of advanced AI agents. These autonomous entities facilitate integration with established computational resources, including finite element modeling software and thermodynamic simulators, to bolster predictive capabilities.

16.1.3 Efficiency and Resource Optimization

The prohibitive computational expenditures associated with training massive models necessitate the development of efficient parameter architectures that maintain high performance standards. Knowledge distillation represents a viable methodology for equilibrating model size and efficacy by training a compact student model to emulate a larger teacher architecture and thereby preserving essential reasoning faculties. This approach effectively compresses information from complex systems into smaller, efficient structures without major performance loss. Optimizing these models involves balancing size, dataset variety, and domain focus. Future research should explore hybrid methods that combine efficient designs with specialized tuning and advanced training techniques to create scalable, practical solutions for materials science.

16.1.4 Scientific Reasoning

General text data dominates the training corpora of most language models, which results in a deficit of the specialized terminology and precise scientific knowledge requisite for materials science. This deficiency often leads to the generation of inaccurate or hallucinated content, including erroneous numerical values and descriptions of nonphysical materials. Retrieval-augmented generation serves as a potent mitigation strategy by permitting models to access verified and high-quality data sources prior to formulating a response. Enhancing intrinsic scientific reasoning via advanced training protocols is equally critical. The DeepSeek-R1 model has recently gained attention for applying advanced reasoning capabilities through large-scale reinforcement learning. Its multistage training pipeline utilizes reasoning-oriented reinforcement learning and rejection sampling to achieve performance levels comparable to proprietary leaders on complex tasks. Furthermore, the DeepSeek project demonstrated the efficacy of distillation by utilizing its large model to train smaller and highly efficient variants that excel in reasoning tasks. Consequently, the integration of reinforcement learning into materials science applications holds the promise of significantly elevating model reasoning and establishing more reliable AI-assisted research workflows.

Despite these challenges, NLP and LLMs have already made significant progress in materials design. They help speed up discovery by analyzing large datasets to uncover hidden patterns and generate new ideas that can guide experiments. Promising results have been realized in the prediction of material properties and the identification of optimal compositions as well as the suggestion of effective processing conditions, especially within alloy development and polymer engineering. Ongoing advancements are poised to further elevate these capabilities by refining numerical reasoning and quantitative prediction alongside structural interpretation. Future progress will likely require smoothly integrating language models with computational and experimental tools. This would generate real-time insights to support quick, well-informed decisions. Such evolution is expected to simplify the design process and drive new discoveries, significantly reducing the time and cost of finding new materials.

16.2 Future Developments

The accelerated evolution of artificial intelligence is fundamentally transforming the materials science landscape by unlocking novel capacities in informatics and computational modeling alongside autonomous experimentation and cross-disciplinary synergy. To fully exploit this trajectory, it is imperative to construct a resilient infrastructure for materials data circulation that guarantees frictionless integration and rigorous standardization. The continued maturation of computational power and algorithmic sophistication, including deep learning and reinforcement learning as well as generative models, serves to expedite material discovery and augment predictive fidelity. Federated learning offers a strategic avenue for training LLMs by facilitating secure collaboration while maintaining data privacy. At the same time, intelligent automated experimental systems are improving high-throughput synthesis and characterization. These smart, closed-loop setups are increasing the efficiency of research and development. Innovations such as digital twins and multiscale modeling are dismantling historical spatiotemporal constraints to permit real-time simulation and exacting control over manufacturing processes. Ultimately, the endurance of this innovation and its transition into industrial utility demand the cultivation of a hybrid workforce possessing dual competency in artificial intelligence and materials engineering. The profound embedding of AI will not only hasten the identification of next-generation materials but also act as the catalyst for sustainable and intelligent manufacturing.

16.2.1 Materials Data Circulation Infrastructure

The future progress of AI in this field depends on gathering large amounts of high-quality, multi-scale data, including datasets designed for LLMs and full records of algorithm development. Applying the FAIR principles—ensuring data is Findable, Accessible, Interoperable, and Reusable—is increasingly important for encouraging collaboration between academia and industry. Creating centralized national big data platforms could transform how industries manage and use materials information in the global economy. This would improve data access for researchers and manufacturers, speeding up product development and supporting better, more sustainable material choices. Future work should focus on building integrated cloud platforms that automate the entire data process—from collection and storage to analysis and application—to establish a comprehensive, multilayered system. It is also essential to develop tools for data identification, citation, evaluation, and secure sharing to overcome technical barriers and build trust in shared resources. Together, these steps will change how research is done and accelerate the industrial use of new materials.

Furthermore, the adoption of standardized data sharing protocols is requisite for ensuring compliance with international regulatory frameworks and enhancing cross-border interoperability to guarantee seamless information flow between distinct jurisdictions. This requires the adoption of common data standards, uniform formats, and clear tracking of data origins. The current lack of globally accepted guidelines for collecting, storing, and sharing materials data significantly hinders progress within the field. There is an urgent requirement for standardized data schemas and consistent file architectures as well as internationally sanctioned APIs for federated databases and robust protocols for communication between software platforms and machinery. The rapid pace of technological adoption renders the task of maintaining currency with field developments increasingly arduous. While checklists have been proposed to guide the assessment of projects containing substantial machine learning components, determining whether specific AI technologies possess sufficient maturity for standardization remains difficult. Consequently, achieving a data standardization framework that is simultaneously informative and exhaustive as well as adaptable persists as a formidable and unresolved challenge in materials science.

16.2.2 Advances in AI Power, Algorithms, Models, and Tools

The relentless progression of artificial intelligence technologies underpinned by augmented computational capacities and sophisticated algorithmic architectures is precipitating a paradigm shift within materials science. These technological strides empower researchers to address historically intractable challenges and substantially expedite the timeline for materials discovery and design. This integration is substantiated by state-of-the-art computational infrastructure, where specialized hardware, including graphics processing units and high-performance computing clusters alongside nascent quantum platforms facilitates the massive data processing requisite for data-intensive modeling and rapid simulation.

AI has moved far beyond classical methods, now using deep learning, reinforcement learning, and generative models to predict properties, optimize synthesis, and design new materials. Tools like GANs and VAEs are key for creating materials with specific traits, such as high strength. Platforms like the Materials Project and libraries like TensorFlow have made AI more accessible to researchers. Cloud systems enable scalable, on-demand computing, speeding up material optimization. Future advances are expected through new design approaches—such as reflection, planning, and multi-agent collaboration—which will help automate complex workflows and drive autonomous discovery.

16.2.3 Federated Training of Materials Science LLMs

The implementation of federated training for LLMs in materials science denotes a critical methodological evolution. This decentralized strategy facilitates privacy-compliant collaboration across disparate data silos to expedite discovery and refine preparation workflows while simultaneously enhancing material performance. Future efforts will likely focus on combining high-throughput simulations with AI-driven optimization and multiscale modeling to design materials with precise properties. Integrating data from experiments, sensors, and large-scale simulations will speed up the discovery of new materials and improve the efficiency and repeatability of how they are made. Federated approaches will also help tailor materials for different industries—such as aerospace, automotive, energy, and healthcare—driving functional innovation. These efficient methods support global sustainability goals by using resources wisely and reducing waste.

The main advantage of federated training is its ability to enable collaboration among multiple parties while protecting confidential data, allowing models to be improved without sharing sensitive raw information. This distributed framework promotes scalable model evolution and international cooperation to yield more robust and adaptable predictive systems. However, significant hurdles persist, including the complexity of harmonizing heterogeneous data sources and the exorbitant computational demands associated with training massive architectures as well as the necessity to maintain accuracy as datasets evolve. Despite these challenges, the ongoing advancement of federated learning holds great potential to transform industrial practices, accelerate breakthroughs in material design, and improve sustainability in global manufacturing.

16.2.4 Intelligent Experimental Techniques

The trajectory of experimental materials science is increasingly defined by the convergence of advanced robotics and artificial intelligence alongside cloud-centric data architectures, which collectively facilitate the creation of intelligent robotic systems. These next-generation platforms will have improved sensitivity and multidimensional motion awareness, along with multi-objective recognition and self-guided decision-making. By integrating smart robotics with high-throughput preparation and testing, researchers can create physics-aware systems that are adaptive, intelligent, and efficient. These systems use real-time data from multiple sensors and multiscale models to continuously track experiments, giving immediate feedback to refine procedures dynamically. By applying deep learning to collected data and multi-objective optimization, an intelligent platform can independently adjust each step of the process, forming a self-optimizing workflow for materials research and development.

As these intelligent technologies mature, autonomous experimental systems are poised to adopt modular architectures that allow for the flexible assembly of components tailored to specific investigative scenarios. This modularity enables the full process automation of complex tasks across diverse environments, including solid and liquid as well as gas phase systems, by accommodating multiple stages and adjustable parameters. The integration of visual assistance technologies further augments robotic operational efficiency and minimizes unproductive iterations to significantly elevate overall experimental throughput. Furthermore, the incorporation of metaverse concepts and digital twin technologies permits the visualization of experimental data within virtual and augmented reality environments, which allows researchers to immerse themselves in the process and guide experiments with unprecedented directness.

An intelligent cloud platform acts as the central nervous system for scheduling and control, where data and optimization algorithms alongside decision logic are orchestrated. This connected model improves how experimental data is stored, processed, and shared, while also offering strong support for remote monitoring and coordinated resource management. The adoption of blockchain technology can further fortify data security and traceability to ensure transparency and tamper resistance throughout the experimental lifecycle, while next-generation artificial intelligence methods continue to drive the self-learning capabilities and autonomous optimization of the system.

16.2.5 Multiscale Modeling and Digital Twin Technology

Future materials design frameworks must move beyond traditional limits by enabling methods that work across scales and self-managing software for truly integrated modeling. Machine-learning-boosted molecular dynamics can already simulate millions of atoms with near-quantum accuracy, allowing detailed study of structure-property links from micro to nano and atomic scales. On-demand, scalable computing will support distributed, efficient automated calculations. Merging data-driven techniques with computational materials engineering will overcome current multiscale modeling bottlenecks and allow designs to be optimized before physical production.

A systematic collection of core materials data is essential for this progress. This data should include multidimensional processing variables, diverse environmental conditions, and in-service performance. Integrating computational engineering with machine learning will help build a dynamic digital knowledge base that captures the complex links and changes between composition, structure, processing, and performance. Advances in database technology—capable of handling multi-source, varied data in real time—will allow immediate simulation and visualization of preparation processes. By better connecting physical systems to information models and using real-time interaction, digital twins can be developed for rapid online optimization and precise manufacturing control. Together, these steps will move materials science and engineering toward full digitalization and intelligent automation.

16.2.6 Building an Interdisciplinary Workforce

The widespread use of AI in materials science requires a workforce skilled in both traditional fields like metallurgy or chemistry and modern algorithmic technologies. Professionals need to combine expertise—materials scientists should learn machine learning, data analysis, and simulation tools, while AI practitioners must understand material properties, production methods, and testing.

To meet this need, educational programs must evolve to offer training that blends materials science with data science and computing. Partnerships among universities, research centers, and industry can help develop a workforce ready to handle the challenges of integrating AI. Continued support for this interdisciplinary approach will ensure future experts can fully use AI to speed up the creation of new materials. Building this skilled, collaborative talent pool is essential for advancing the next generation of materials science and engineering.

Index

Note: Page numbers in *italics* and **bold** refers to figures and tables respectively.

f

g

h

i

k

l

m

n

Metal Powder–Based Additive Manufacturing

Kun Zhou and Changjun Han

WILEY VCH

Authors

Prof. Kun Zhou
Nanyang Technological University
50 Nanyang Avenue
Singapore - 639798

Prof. Changjun Han
South China University of Technology
381 Wushan Road
Tianhe District
Guangzhou
China - 510641

Library of Congress Card No.: applied for

British Library Cataloguing-in-Publication Data
A catalogue record for this book is available from the British Library.

Bibliographic information published by the Deutsche Nationalbibliothek
The Deutsche Nationalbibliothek lists this publication in the Deutsche Nationalbibliografie; detailed bibliographic data are available on the Internet at <http://dnb.d-nb.de>.

Print ISBN: 978-3-527-34631-8
ePDF ISBN: 978-3-527-82222-5
ePub ISBN: 978-3-527-82223-2
oBook ISBN: 978-3-527-82224-9

Cover Image © Christian Bay/Getty Images
Cover Design Wiley

Typesetting Straive, Chennai, India
Printing and Binding CPI Group (UK) Ltd, Croydon, CR0 4YY

C9783527346318_191124

Contents

Biography

Kun Zhou is a Professor of Mechanical Engineering in the School of Mechanical and Aerospace Engineering at Nanyang Technological University, Singapore. His research interests focus on mechanics of materials and additive manufacturing. He has founded the journal *Smart Manufacturing* and serves as its Editor-in-Chief. He is also a co-founder of *Journal of Micromechanics and Molecular Physics* and serves as its Co-Editor-in-Chief. He was elected Fellow of Institution of Mechanical Engineers, Royal Aeronautical Society, Royal Society of Chemistry, Institute of Physics, and Institute of Materials, Minerals & Mining.

Changjun Han is an Associate Professor in the School of Mechanical and Automotive Engineering at South China University of Technology, China. He achieved his Ph.D. from Huazhong University of Science and Technology in 2018 and subsequently worked as a postdoctoral fellow in the School of Mechanical and Aerospace Engineering at Nanyang Technological University, Singapore for over two years. He has rich experience in additive manufacturing of metallic materials. He serves as an Associate Editor of Smart Manufacturing.

Preface

Additive manufacturing (AM), commonly known as three-dimensional (3D) printing, has emerged as a revolutionary technology capable of fabricating parts with complex geometries in a layer-by-layer manner, thereby granting unprecedented design freedom. It has evolved from a tool primarily employed for rapid prototyping into a well-recognized technology for producing functional parts with highly desirable properties. This technology provides a cost-effective option for the low-volume production of highly specialized parts for a wide variety of applications in the aerospace, art, biomedical, automotive, electronics, manufacturing, energy, building and construction, marine, "oil and gas industries".

Powder-based AM is a major subset of the AM family, in which parts are using powders as feedstock materials. Within this category, powder bed fusion (PBF), directed energy deposition (DED), and binder jetting are presently the predominant techniques. In particular, laser powder bed fusion (LPBF), electron beam melting (EBM), laser-based directed energy deposition (L-DED), and metal binder jetting (MBJ) are commonly adopted powder-based AM processes for printing metal parts.

The production of defect-free and structurally sound parts through metal powder-based AM requires an in-depth understanding of the feedstock materials, the printing processes, and the underlying mechanisms of defect formation in the printed parts. Notably, metal powder-based AM is multidisciplinary in nature and entails key concepts from different subjects, including materials science, engineering, powder metallurgy, physics, and chemistry. Therefore, understanding the fundamentals of metal powder-based AM may constitute a daunting challenge for novice researchers.

This book introduces the fundamentals of powder preparation processes and metal powder-based AM techniques while emphasizing their significance in various industrial applications. For more experienced AM practitioners, the recent progress of metal powder-based AM techniques is reviewed along with noteworthy case studies, which feature the wide applicability of metal powder-based AM techniques.

Chapter 1 presents a brief introduction to metal powder-based AM. The history and fundamentals of AM are first highlighted, followed by seven categories of AM techniques with their representative processes and commercial companies. Next, different types of post-processing treatment techniques for metal AM parts are illustrated. Powder properties and characterization methods are also discussed comprehensively. Finally, the challenges and future trends of metal powder–based AM are outlined. Researchers who are unfamiliar with metal powder–based AM techniques are encouraged to peruse this chapter to attain a solid understanding of these principles.

Chapter 2 presents the fundamentals of the preparation processes of metal powders for AM, specifically, atomization, mechanical mixing, the reduction process, and powder modification. A comprehensive review of the relationships between the preparation processes and powder properties is provided. For more experienced researchers, this chapter serves as a concise handbook in which the key concepts underlying AM powder metallurgy are succinctly explained.

Chapters 3–6 delve into various metal powder–based AM techniques, namely, LPBF, EBM, L-DED, and MBJ. Chapter 3 focuses on the LPBF technique, with regard to its history, fundamentals, printing process, metallurgical defects, powder materials, equipment, and the microstructures and mechanical properties of the representative materials investigated for the printed parts. In particular, the characteristics of popular LPBF-printed mechanical metamaterials, which typically correspond to a high Young's modulus, high shear and bulk moduli, and zero or negative Poisson's ratio, are comprehensively evaluated.

In Chapter 4, the EBM technique is introduced with an overview of its history, fundamentals, processing characteristics, powder materials, equipment, and the microstructures and mechanical properties of the most extensively studied materials for the printed parts. The preheating and melting processes during EBM printing are also described, and their influence on the surface quality of the printed parts is discussed.

Chapter 5 expounds on the L-DED technique in terms of its history, fundamentals, deposition process, metallurgical defects, powder materials, equipment, and the microstructures and mechanical properties of the representative materials for the printed parts. Additionally, the unique capability of L-DED to fabricate multi-materials and functionally graded materials is highlighted.

Chapter 6 is devoted to the MBJ technique, with respect to its history, fundamentals, printing process, powder and binder materials, equipment, and the microstructures and mechanical properties of the representative materials for the printed parts. The unique ability of MBJ to print parts using refractory metals, magnesium and its alloys, and magnetic alloys, which are difficult to print using the abovementioned three techniques, is also examined.

Chapter 7 reports on the latest advances in metal powder–based AM products in the aerospace, biomedical, automotive, molding and tooling, energy, marine, oil and gas, and jewelry industries. This chapter focuses on providing insights of both academic and industrial relevance to readers and keeping them abreast of recent applications involving metal AM.

The completion of this book could not have been possible without the excellent teamwork displayed by the members of our research group, with whom our discussions have always been fruitful. We greatly appreciate the efforts of Liming You, Chengcheng Wang, Haiyang Fan, Yujia Tian, Jiazhao Huang, Boyuan Li, Zhuohong

Zeng, Devesh Kripalani, Priyanka Vivegananthan, Asker Jarlöv, Feng Liu, Bin Liu, Liming Tan, Yung Zhen Lek, Kiran Raj, Pengfei Tan, Ming Gao, Ran An, Terence Zhi Xiang Hong, Yang Qi, Weiming Ji, Wei Fan, Meixin Zhou, Chao Cai, and Wei Zhu, who contributed valuable feedback during the preparation of this book. We are also deeply grateful to our collaborators, who have rendered us tremendous support in conducting research pertaining to metal powder–based AM.

Singapore, December 2022

Kun Zhou
Changjun Han

1 Introduction

This chapter presents a brief introduction to metal powder–based additive manufacturing (AM). The history and fundamentals of AM are first highlighted. Seven categories of AM techniques with their representative processes and commercial companies are then presented, followed by an overview of the types of metal powder–based AM techniques. Next, the types of post-processing treatments for metal AM parts are illustrated. Powder properties and characterization methods are also discussed comprehensively. Finally, the challenges and future trends of metal powder–based AM are outlined.

1.1 History and Fundamentals of AM

Additive manufacturing (AM), commonly known as three-dimensional (3D) printing, has emerged as a revolutionary technology capable of fabricating parts with complex geometries in a layer-by-layer manner, thereby granting extensive freedom of design (Martin et al. 2017; Kelly et al. 2019; Zhang et al. 2019). The term "additive manufacturing" describes "the process of joining materials to make parts from 3D model data, usually layer upon layer, as opposed to subtractive manufacturing and formative manufacturing methodologies", according to the ISO/ASTM 52900 standard (ISO and ASTM International 2015). ISO and ASTM are the abbreviations for the International Standards Organization and the American Society for Testing and Materials, respectively. The most commonly used term before AM was "rapid prototyping", which is currently considered outdated because of the increasingly extensive applications of the process.

The advent of AM marks an important milestone in the history of manufacturing. In 1987, Stereolithography Apparatus-1, the first commercial AM system, was launched by 3D Systems in the United States. The system enabled the printing of 3D parts from a computer-aided design (CAD) model for the first time. Other AM techniques were introduced for commercial purposes soon thereafter, which include fused deposition modeling, selective deposition lamination, selective laser sintering, and laminated object manufacturing. Machines operating based on the inkjet printing process began appearing on the market in 1996. In 1998, Optomec delivered the first laser engineered net shaping metal powder system capable of producing

Metal Powder–Based Additive Manufacturing, First Edition. Kun Zhou and Changjun Han.

near-fully dense metal parts. Meanwhile, in 1999, a selective laser melting system was initiated by Fockele & Schwarze of Germany.

The expiry of old AM patents, such as the fused deposition modeling technique (which expired in 2009), has led to a surge in the number of low-cost personal AM machines on the consumer market, some of which can even be assembled by hand. Moreover, the rise of online platforms, such as open-source communities where AM design files are shared freely, further improves the accessibility of 3D printing technology. In 2014, the number of AM patents that expired has reached a peak, which has led to a greater variety of low-cost machines flooding the consumer market.

Over the years, AM has evolved from a tool primarily used for visualization (i.e. rapid prototyping) into a well-recognized technology for producing functional parts with desirable properties. Before printing, a 3D model is constructed via CAD and mathematically sliced into ultrathin printed layers along the build direction. Subsequently, the layers are printed according to their predefined shapes, with consecutive layers bonding to each other.

AM also provides a cost-effective option for low-volume customized production, which differs from conventional mold jetting methods. Parts comprising multiple components can be redesigned as single units and fabricated efficiently without assembly (Chua and Leong 2017). The distinct ability of this process to manufacture complex shapes and structures has already rendered it invaluable for producing prototypes or parts in industries including aerospace, military, biomedical, automotive, electronics, energy, molding, building and construction, marine and offshore, education, art, robotics, environment, and social culture.

AM is also perceived as an environmentally sustainable manufacturing technology, as it can potentially reduce up to 525.5 Mt of total carbon dioxide emissions by 2025 compared to conventional processes (Gebler et al. 2014; Ribeiro et al. 2020). Furthermore, in terms of product development, AM can reduce up to 70% of development costs and time-to-market by up to 90% compared to conventional processes (Gibson et al. 2014). With such remarkable benefits, the global market for AM is projected to reach nearly US$23 billion by 2023, with a compound annual growth rate of 22% (Tan et al. 2020).

1.2 AM Techniques

AM technologies have been classified into seven broad categories according to the ISO/ASTM 52900 standard, namely, vat photopolymerization, material jetting, material extrusion, powder bed fusion (PBF), directed energy deposition (DED), binder jetting, and sheet lamination techniques (ISO and ASTM International 2015), as shown in Figure 1.1. Table 1.1 provides an overview of the representative processes and commercial companies of these AM categories.

The vat photopolymerization technique utilizes radiation (e.g. ultraviolet [UV] and visible light) to selectively polymerize liquid photosensitive resins in a vat to form high-resolution solid 3D structures. It includes various processes such as stereolithography (Hull 1986), digital light processing (Kuang et al. 2019), continuous liquid interfase production (Tumbleston et al. 2015), two-photon photopolymerization

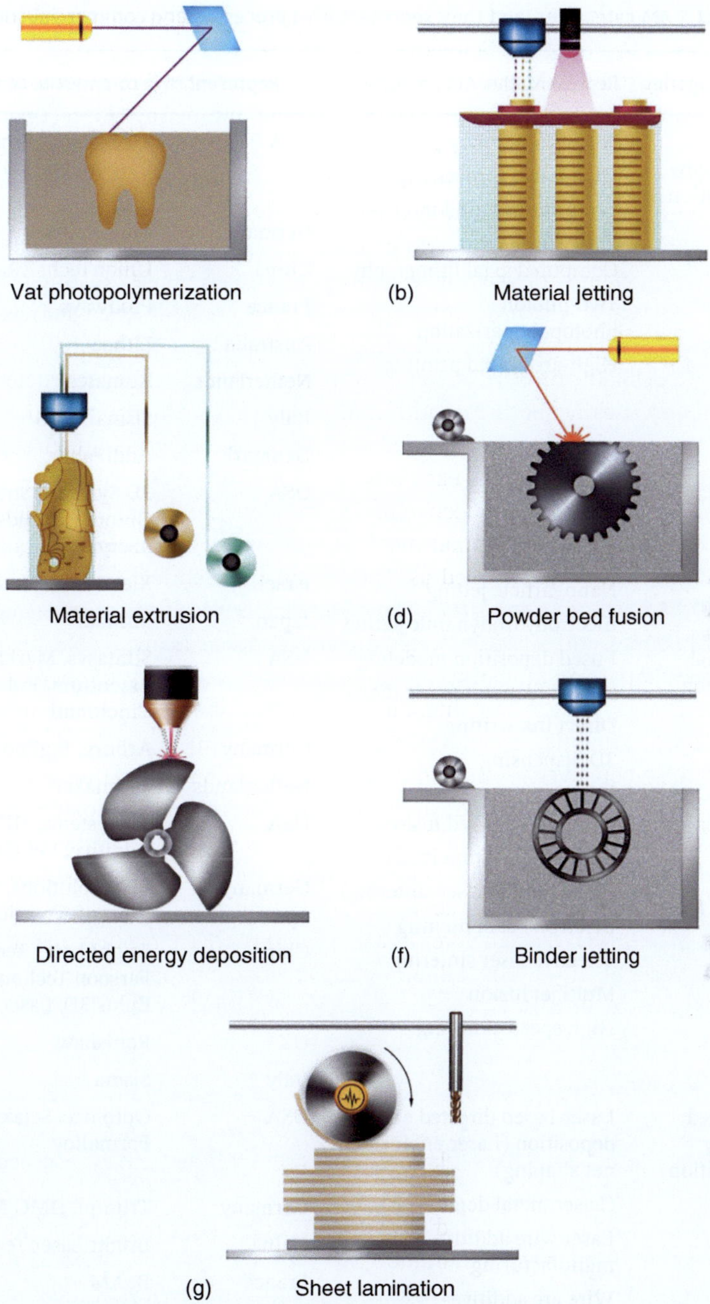

Figure 1.1 Schematics of the seven broad categories of AM technologies according to the ISO/ASTM 52900 standard. Source: ISO and ASTM International (2015)/ISO.

Table 1.1 AM categories, and their representative processes and commercial companies.

AM categories	Representative AM processes	Representative commercial companies	
Vat photopoly-merization	Stereolithography Digital light processing Continuous liquid interfase production Computed axial lithography Two-photon photopolymerization High-area rapid printing	USA	3D Systems, Stratasys, Carbon, Formlabs
		Germany	Envisiontec
		China	UnionTech, ZRapid
		France	Prodways
		Australia	Lithoz
		Netherlands	Admatec
		Italy	Sisma, DWS
		Denmark	AddiFab
Material jetting	PolyJet MultiJet Aerosol jet NanoParticle jetting Electrohydrodynamic jetting	USA	3D Systems, Stratasys, Optomec, Solidscape, nScrypt
		Israel	Xjet, Nano Dimension
		Japan	Mimaki
Material extrusion	Fused deposition modeling (Fused filament fabrication) Direct ink writing 3D dispensing (3D plotting)	USA	Stratasys, Markforged, Essentium, Robocasting, Cincinnati
		Germany	Arburg, BigRep
		Netherlands	Ultimaker
Powder bed fusion	Laser powder bed fusion (Selective laser melting) (Direct metal laser sintering) Electron beam melting Selective laser sintering Multi jet fusion High speed sintering	USA	3D Systems, HP Inc., GE Additive, Velo3D
		Germany	SLM Solutions, EOS, Trumpf, Voxeljet
		China	Bright Laser Technologies, Farsoon Technologies, Eplus 3D, Laseradd
		UK	Renishaw
		Italy	Sisma
Directed energy deposition	Laser-based directed energy deposition (Laser engineered net shaping) (Laser metal deposition) Laser wire additive manufacturing Wire arc additive manufacturing Electron beam freeform fabrication (Electron beam additive manufacturing)	USA	Optomec, Sciaky, Addere, Formalloy
		Germany	Trumpf, DMG Mori
		China	Bright Laser Technologies
		France	BeAM
		Australia	AML3D
		Netherlands	MX3D
		Japan	Sodick
		South Korea	InssTek
		Spain	Addilan

Table 1.1 (Continued)

AM categories	Representative AM processes	Representative commercial companies	
Binder jetting	Binder jetting	USA	3D Systems, HP Inc., Exone, Desktop Metal, Digital Metal
		Germany	Voxeljet
		China	Long Yuan
		UK	Raplas
Sheet lamination	Laminated object manufacturing Ultrasonic additive manufacturing Selective deposition lamination Composite-based additive manufacturing	USA	Fabrisonic, Evolve Additive, Impossible Objects

(Saha et al. 2019), high-area rapid printing (Walker et al. 2019), and computed axial lithography (Kelly et al. 2019). Representative companies for vat photopolymerization include 3D Systems (USA), Carbon (USA), Envisiontec (Germany), Lithoz (Australia), and UnionTech (China).

Figure 1.1a illustrates the top-down stereolithography process, a typical vat photopolymerization technique. Stereolithography is a common vat photopolymerization technique that utilizes an UV laser source positioned above or underneath a resin vat to selectively cure the exposed layer.

The material jetting technique is analogous to traditional two-dimensional (2D) inkjet printing, in which liquid materials (e.g. photosensitive resins, thermoplastics, wax, and reactive materials) are deposited from inkjet printheads onto a build platform through either a drop-on-demand or a continuous approach and subsequently solidified through photopolymerization, cooling, etc. This technique is employed in commercialized systems such as PolyJet, MultiJet, Aerosol Jet, and NanoParticle Jet, as well as non-commercialized systems based on electrohydrodynamic jetting.

Figure 1.1b shows a schematic of the material jetting technique. For example, in the PolyJet system, multiple depositions of photosensitive build materials and support materials (to form the support structures within the printed parts) in the form of droplets are made by nozzles in the printhead before undergoing UV curing. During post-processing, the support structures are removed through mechanical (cutting with a water jet system) and/or chemical (dissolving with a solvent) procedures. Representative companies of material jetting include Stratasys (USA), 3D Systems (USA), Optomec (USA), Xjet (Israel), and Nano Dimension (Israel).

The material extrusion technique typically involves a continuous extrusion of polymer filaments, viscous inks, or even polymer pellets through a heated nozzle onto a build platform, on which the extruded molten material subsequently solidifies. Fused deposition modeling (also known as fused filament fabrication) and direct ink writing are the key representative processes in this category. The former utilizes thermoplastic filaments and pellets as feedstock materials, while the latter employs viscous inks such as pastes and concentrated polymer solutions (Truby and Lewis 2016). Recent studies have also established the significant potential of fused deposition modeling in printing metal parts. For example, in one such study, printable filaments were prepared through the extrusion of a polyolefin-based binder mixed with 59 vol.% Ti–6Al–4 V powder and then fabricated into parts using the Renkforce 1000 printer (Zhang et al. 2020). In addition, solvent debinding, thermal debinding, and sintering were also conducted to densify the printed parts. Representative companies of material extrusion include Stratasys (USA), Markforged (USA), Essentium (USA), Robocasting (USA), and Ultimaker (Netherlands).

Figure 1.1c illustrates the fused deposition modeling process, a representative material extrusion technique. The build material and the support material, in the form of thermoplastic filaments, are heated to the molten state in the nozzles, extruded, and solidified on the build platform. The support structures are removed through mechanical and/or chemical procedures during post-processing.

The PBF technique utilizes a heat source (e.g. electron beam or laser beam) to coalesce metal, polymer, or ceramic powder particles in a powder bed to build 3D objects, and it can theoretically process any powder–based materials on the condition that the corresponding powder particles can be fused or melted through heating. The technique can be categorized into laser powder bed fusion (LPBF) (also widely known as selective laser melting), electron beam melting (EBM), selective laser sintering, multi jet fusion, and high speed sintering processes. LPBF and EBM are mainly employed to print pure metals, alloys, and metal matrix composites (Han et al. 2020a, 2020b), while the selective laser sintering, multi jet fusion, and high speed sintering processes are typically implemented to treat polymers and their composites. Representative commercial companies of PBF include SLM Solutions (Germany), EOS (Germany), Trumpf (Germany), 3D Systems (USA), GE Additive (USA), Renishaw (UK), Bright Laser Technologies (China), and Farsoon Technologies (China).

Figure 1.1d presents a typical PBF process, LPBF, which utilizes a laser beam to selectively melt metal powder particles in the powder bed. The molten metal particles coalesce and solidify to form the printed part. While unmelted powder particles can serve as support, support structures are generally required to be printed to ensure the good printability of overhanging structures of a part.

The DED technique utilizes a laser beam, an electron beam, or an electric arc to melt metal powders or wires upon their deposition along the printing paths (Bermingham et al. 2015; Li et al. 2021). The technique includes processes such as laser-based directed energy deposition (L-DED) (also widely known as laser engineered net shaping and laser metal deposition), laser wire additive manufacturing (LWAM), wire arc additive manufacturing (WAAM), and electron beam freeform fabrication (EBF^3, also known as electron beam AM). The L-DED process is commonly employed to print parts using powders, while the other four processes employ wires for printing. Examples of representative companies for DED include Optomec (USA), Sciaky (USA), DMG Mori (Germany), Trumpf (Germany), BeAM (France), and Bright Laser Technologies (China).

Figure 1.1e illustrates a typical DED process, L-DED, in which metal powder particles are delivered through channels in a nozzle and melted by a laser beam upon deposition onto a freeform substrate. The molten metal particles subsequently solidify to form the printed part.

The binder jetting technique utilizes one or more inkjet printheads to deposit droplets of a liquid polymer binder onto ceramic or metal powder particles in a powder bed and selectively glue them together to build 3D objects (Mostafaei et al. 2017). The droplets can be deposited through a single-pass or multi-pass printing strategy. Post-processing procedures, such as curing, de-powdering, infiltration, and sintering, are often required for green parts printed through this technique. Examples of commercialized companies specializing in binder jetting include Exone (USA), Hewlett–Packard (USA), Desktop Metal (USA), Digital Metal (USA), 3D Systems (USA), Voxeljet (Germany), and Long Yuan (China).

Figure 1.1f demonstrates a metal binder jetting (MBJ) process, in which a layer of metal powder particles is spread by a roller across the build platform, and a liquid binder is then selectively deposited to bond specific regions of the powder layer.

The sheet lamination technique can be employed to fabricate 3D objects by stacking and laminating thin sheets of materials (e.g. paper, metal sheets, ceramic tapes, woven fiber composite sheets, and thermoplastic foils) through different bonding (e.g. adhesive bonding, thermal bonding, and ultrasonic welding) and cutting (e.g. computer numerical control [CNC] milling, laser cutting, and water jet cutting) strategies. According to the ISO/ASTM standard, representative processes of the sheet lamination technique include laminated object manufacturing and ultrasonic AM. Laminated object manufacturing was the first commercialized sheet lamination process and was initially implemented to bond kraft paper. Subsequently, the applicability of this process for bonding plastic/metal tapes and foils has been investigated. Meanwhile, the ultrasonic AM process is commonly utilized for printing parts from metal foils and sheets (Dehoff and Babu 2010). In addition to the two aforementioned processes, a novel sheet lamination process, friction stir AM, is currently in development. In this process, metal plates are utilized as feedstock materials. A stirring pin is inserted into a newly added metal plate at high rotation rates, and the resultant friction between the stirring pin and the plate produces heat that softens and bonds the plate to the previous plate. Representative companies for sheet lamination include Fabrisonic (USA), Evolve Additive (USA), and Impossible Objects (USA).

Figure 1.1g presents a diagram of the sheet lamination process, ultrasonic AM, in which each layer of a metal foil is laid and bonded to the previous layer through ultrasonic welding. This process involves the actuation of a cylindrical sonotrode with an ultrasonic transducer to induce a scrubbing motion of the sonotrode. At the same time, a downward force is applied to the metal foils. The sonotrode rolls along the length of the foils due to the downward force, and ultrasonic vibrations are applied along the width of the foils through the sonotrode. Friction can be produced between the metal foils through the interaction of the downward force and ultrasonic vibrations. Bonding is achieved through disruption of the oxide layers between the metal foils utilizing friction, thereby promoting nascent metal-to-metal contact. The metal layer is subsequently cut through CNC milling to obtain the desired geometry, and the cycle is repeated until the metal part is completed.

1.3 Metal Powder–Based AM

Powder–based AM is an important subset of the AM family for manufacturing parts using powders as feedstock materials, and it employs the PBF, DED, and binder jetting techniques. In particular, LPBF, EBM, L-DED, and MBJ are commonly adopted powder–based AM processes for printing metal parts. LPBF, EBM, and L-DED share similar features, such as the usage of high-energy heat sources, localized melting, and microstructural evolution upon solidification. Applying metal powder–based AM in producing structurally sound, defect-free, and reliable parts requires an in-depth understanding of existing printing techniques, the physical and chemical processes involved during printing, feedstock materials, process control methods, and underlying mechanisms of common defects and their prevention.

Metal powder–based AM excels in the following aspects: (i) its ability to recycle feedstock materials, (ii) its relatively high manufacturing accuracy as compared to wire- and sheet-printing, and (iii) the capacity of certain powder–based AM processes (MBJ and EBM) to print parts without requiring support structures because of the support provided by the surrounding unfused or partially melted powder particles. Notably, metal powder–based AM plays an indispensable role in various domains, including the aerospace, biomedical, automotive, molding and tooling, energy, jewelry, marine, oil and gas, and repair and remanufacturing industries.

1.4 Post-Processing

It is challenging to employ metal powder–based AM to directly fabricate parts with properties and surface characteristics that satisfy application requirements. Most metal powder–based AM techniques require post-processing treatment to obtain the desired properties in the printed parts (ISO and ASTM International 2015). Post-processing is crucial for addressing the main issues of AM parts, such as high surface roughness, high porosity, dimensional deviations with respect to the models, and substandard mechanical properties for industrial applications.

In certain metal powder–based AM processes (PBF and DED), a metal substrate where parts can be printed onto is often required. Therefore, the post-processing process of wire cutting, typically using electric discharge machining (EDM) which generates a pulse discharge between a tool electrode and a target object (comprising printed parts and the substrate) for cutting, is required to separate the printed parts from the substrate. This process is also commonly employed to assign desired geometries to printed parts for various purposes (e.g. tensile, fatigue, or fracture toughness testing).

Table 1.2 Post-processing for metal powder–based AM.

Categories	Representative techniques
Surface quality improvement	Manual grinding
	Machining
	Sandblasting
	Shot peening
	Mechanical polishing
	Chemical polishing
	Chemical etching
	Laser shock peening
	Laser polishing
Residual stress relief and defect reduction	Stress relief annealing
	Hot isostatic pressing
Aesthetic improvement	Spray painting
	Electroplating

Table 1.2 exhibits representative post-processing techniques for metal powder–based AM. In accordance with their respective objectives, these techniques can be classified into the categories of surface quality improvement, residual stress relief and defect reduction, and aesthetic improvement.

1.4.1 Surface Quality Improvement

Printed metal components typically undergo post-processing treatment to improve their surface quality (i.e. reduction in surface roughness), which includes manual grinding, machining, sandblasting, shot peening, mechanical and chemical polishing, chemical etching, laser shock peening, and laser polishing.

Manual grinding typically involves abrasive sandpapers with different grit sizes. However, such treatment is only applicable for prototypes or small batches of parts due to its low repeatability and high dependence on the skill of the operator. In addition, any support structures attached to the parts (e.g. overhangs) should first be removed before the grinding process.

Machining utilizes power-driven machines and cutting tools to reduce the surface roughness of printed parts. In particular, CNC is a commonly adopted precision machining process characterized by stable machining quality and high flexibility, machining accuracy, and productivity. Notably, CNC allows for the customization of control programs in each working task to establish control over various tools (e.g. lathes, milling machines, and grinders) in reducing the surface roughness of printed parts.

Sandblasting is the process of removing rust, oxides, and oil contaminants from a surface with high-speed sand (e.g. copper ore sand, emery sand, quartz sand, or iron sand) propelled by compressed air. Sandblasting operations can be categorized into dry-type sandblasting and liquid blasting processes. The dry-type sandblasting involves pure abrasive propellants capable of removing large amounts of surface material without contamination. In contrast, liquid blasting utilizes a mixture of abrasives and liquid, which removes only small amounts of material but introduces contamination to the parts.

Shot peening is a cold working process that bombards the surface of a material with a stream of small shots (i.e. spherical particles of metal or ceramic) with controlled intensity and coverage. It can increase surface hardness and extend the service life of a part by creating an induced compressive stress layer to enhance its fatigue resistance. Additionally, the surface roughness of printed parts can be reduced and their surface grains can be refined. The shot peening process is commonly performed using air blast systems or centrifugal blast wheels.

Polishing is the process of creating a smooth and specular surface through mechanical or chemical methods. Mechanical polishing methods include magnetically driven abrasive polishing, hydrodynamic cavitation abrasive finishing, and ultrasonic cavitation abrasive finishing.

Magnetically driven abrasive polishing utilizes a slurry comprising magnetic and abrasive materials in a viscous liquid to polish a surface. Hydrodynamic cavitation abrasive finishing is a novel surface modification process employing hydrodynamic cavitation along with abrasives to remove surface irregularities and decrease the surface roughness of a workpiece. In ultrasonic cavitation abrasive finishing, the application of ultrasound and micro-abrasives produces a synergistic effect, in which the former induces the cavitation effect to remove partially melted powder particles on a surface, while the latter serves as bubble nucleation sites to increase the overall cavitation intensity, which further contributes to gradual surface erosion through high-velocity abrasive collisions.

In chemical polishing, a ground sample is immersed in a polishing agent or swabbed with a chemical solution until a clean surface is obtained. Electropolishing is a representative electrochemical polishing process for producing smooth surfaces, and it is accomplished by creating an electrochemical cell in which the printed part is charged anodically. A varying current density is established across the material surface, and it is higher at the peaks and lower at the valleys of the surface topography. The relatively high current density at the protruding points on the surface causes these sites to rapidly dissolve, which levels the surface. However, the effectiveness of electropolishing is limited by the accessibility of counter electrodes in printed parts containing tight spaces and exhibiting complex geometries.

Chemical etching is a post-processing treatment in which chemical reactions occur at the interface between a printed part and a chemical solution, leading to changes in the former's surface roughness. This approach is particularly applicable for treating printed parts with open porous structures.

Laser-based treatment methods utilize laser sources to remove materials and improve the surface accuracy of AM metal parts. Laser-based treatment includes laser shock peening and laser polishing.

In laser shock peening, a pulsating laser beam is directed onto the surface of a metal part, generating shock waves induced by the ablation of a sacrificial layer on the surface. These shock waves travel throughout the surface layer of the part to cause surface grain refinement and plastic deformation and induce compressive residual stress, which improves the resistance of a material toward fatigue crack initiation and propagation. The underlying mechanism of residual stress generation is similar to that of conventional shot peening, and plastic compression is achieved by the passage of shock waves.

Laser polishing, also known as laser remelting, can enhance the surface accuracy of AM parts while avoiding ablation. In this process, a high-power laser source irradiates the material surface with low-frequency pulses at high scanning speeds to induce local surface melting on the order of nanometers to micrometers. Laser remelting is an eco-friendly process that can improve surface accuracy and reduce surface porosity without incurring any loss of surface materials.

1.4.2 Residual Stress Relief and Defect Reduction

Parts printed by metal powder–based AM processes such as PBF and DED are often subjected to residual stress, which must be relieved through heat treatment before the parts are suitable for industrial usage. The most commonly adopted heat treatment is stress relief annealing, which is a special annealing process that minimizes the residual stress within printed metal parts. This treatment is conducted by heating the parts to a specific temperature below their recrystallization temperature followed by air-cooling. Low-temperature stress relief annealing has a low impact on the microstructure and mechanical properties of a material (Wang et al. 2016), while high-temperature stress relief annealing may refine grains, produce a low dislocation density, and alter the mechanical properties of a material (Xiong et al. 2017). Therefore, low-temperature stress relief annealing is preferable to high-temperature stress relief annealing because it is desirable not to change the microstructure of printed parts during the annealing process.

Hot isostatic pressing (HIP) heat treatment, a thermomechanical treatment process involving the simultaneous implementation of a high temperature (up to 2000 °C) and high isostatic pressure (up to 200 MPa) in a specially constructed vessel with gas as the pressure-transmitting medium, is employed to reduce the porosity of printed parts and improve their densification. Argon gas is the most common pressure-transmitting medium in HIP. Under a high temperature and pressure, the internal pores within a printed part tend to collapse, thereby leading to its densification.

The principal factor distinguishing HIP from other heat treatment techniques is its use of an inert gas as a pressure-transmitting medium to produce uniform microstructural changes on the part surface. The isostatic pressure in HIP arises from the gas atoms colliding with the surface of the part, during which each gas atom is akin to the hammer in a forge. These atomic "hammers" reliably and consistently reach the entire part surface, which corresponds to a uniform pressure.

1.4.3 Aesthetic Improvement

Aesthetics is an indispensable aspect of printed parts. Therefore, improving the aesthetic quality of printed products can further increase their value. Generally, the "appearance, anti-corrosion, anti-aging, and anti-slip" properties of parts should be considered. Commonly employed methods for achieving aesthetic improvement include spray painting and electroplating.

Spray painting is a painting technique in which paint particles are atomized and sprayed onto a surface. It includes various methods such as air gun spraying, electrostatic spray painting, airless gun spraying, automated linear spraying, and automated flatline spraying. Spray painting is often utilized to improve the appearance of a printed part by applying a smooth and flat coating of the desired color(s) on its surface.

Electroplating is a process that applies a metal coating on a part by external electric fields. In a salt solution containing the metal, the cations of the metal are reduced to atoms through the electrode reaction, and the atoms are subsequently deposited on the surface of the part (that acts as the cathode) to form the coating. Aside from enhancing the appearance of printed parts, the electroplating process can improve their resistance to oxidation, wear, and corrosion, as well as their electrical conductivity. Metals such as chromium, zinc, copper, and nickel are applicable for electroplating.

1.5 Powder Properties and Characterization Methods

Powders form the basis of powder-based AM, and their quality determines the printability and performance of the final parts. While the Technical Committee 119 of ISO has developed and published numerous standards for powder metallurgy since 1967, no standard defined specifically for powder-based AM exists, and thus previously defined metallurgy standards on powder characterization methods are adopted.

Powder sampling must be conducted before any powder characterization can be performed. Sampling methods include scoop sampling, conical pouring and quartering, and chute splitting. In scoop sampling, a scoop of powder is obtained for sampling. Conical pouring and quartering involve pouring the powder onto a flat horizontal surface and dividing the heap into four samples by a cross-shaped cutter. In chute splitting, a chute splitter is employed for sample division. After sampling, a powder can be characterized in terms of its particle morphology, particle size, particle size distribution, density, flowability, chemical composition, and microstructure.

1.5.1 Particle Morphology

The morphology of powder particles can be determined via a standard glossary developed by the British Standards Institute (British Standards 2955 Glossary of Terms Relating to Powders). Particle morphology, which includes shapes and surface features, is dependent on the preparation processes during powder production. The shape of a particle can be described using the following terms: spherical (globular-shaped), acicular (needle-shaped), angular (sharp-edged or roughly polyhedral-shaped), crystalline (a geometric shape freely developed in liquid), dendritic (branched crystalline-shaped), fibrous (regularly or irregularly thread-like),

Table 1.3 Summary of particle size indices and particle shape indices and their descriptions.

Particle size index	Description
Martin's diameter	Length of the line that bisects the area of the particle image (all particles are measured in the same direction).
Feret's diameter	Maximum length of a particle measured in a fixed orientation.
Projected area diameter	Diameter of a circle with the same area as the 2D image of a particle.
Longest diameter	Maximum diameter of a particle.
Perimeter diameter	Diameter of a circle having the same circumference as the perimeter of a particle.
Maximum horizontal intercept	Length of the longest line that can be drawn through a particle in a fixed direction.
Particle shape index	**Description**
Elongation factor	Aspect ratio that is the ratio of the side lengths of an enveloping rectangle that has the minimum area around the cross-section of a particle.
Bulkiness factor	Ratio of the area of a projected particle to the area of the enveloping rectangle.
Surface factor	Sphericity that is used to compare the surface of a particle and the surface of a sphere of equivalent volume.

lamellar (plate-like), granular (equidimensional but irregularly shaped), irregular (lacking symmetry), or modular (round but irregularly shaped).

Table 1.3 summarizes the particle size indices and particle shape indices that can be used to quantitatively describe the morphology of a particle (Allen 1997). Figure 1.2 presents representative surface features and defects of powder particles, which include "splat caps", pores, elongated shapes, breakage, agglomeration, irregular shapes, and small satellites (Mostafaei et al. 2021).

The morphology of powder particles can be characterized by optical microscopy and scanning electron microscopy (SEM). Optical microscopy allows for the counting and measurement of particles at maximum magnification values ranging from 500 to 1500, depending on the model of the apparatus used. Meanwhile, a typical SEM is applicable for imaging surface features with a magnification of up to 3×10^6 and at a resolution of tens of nanometers, thereby enabling the evaluation of grain sizes and second phases on unetched surfaces. Furthermore, when equipped with a backscatter electron detector, an SEM device can facilitate the observation of microstructures on unetched surfaces.

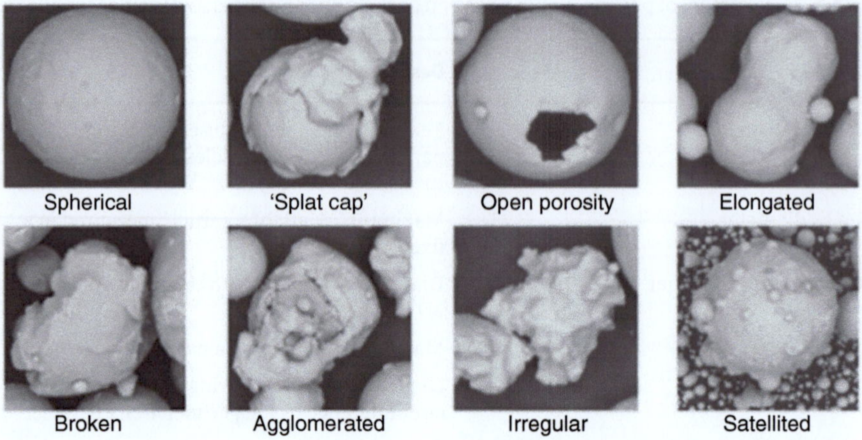

Figure 1.2 Micrographs of powder particles exhibiting various surface features and defects. Source: Mostafaei et al. (2021)/reproduced with permission from Elsevier.

1.5.2 Particle Size Distribution

The particle size distribution of an AM powder is one of its most important characteristics. A volume distribution with respect to particle size can be plotted, as shown in Figure 1.3. The area under the distribution curve to the left of the vertical line $x = D_i$ corresponds to the percentage of the total powder particles that are of sizes smaller than or equal to a specific size gauge D_i. For example, $D_i = 20\,\mu m$ indicates that $i\%$ of all particles are smaller than or equal to 20 μm. The size gauges D_{10}, D_{50}, and D_{90} are the most commonly used indicators. The span, given by $(D_{90} - D_{10})/D_{50}$, is sometimes selected to represent the width of a Gaussian particle size distribution. Additionally, the mean, median, and mode of a particle size distribution, which

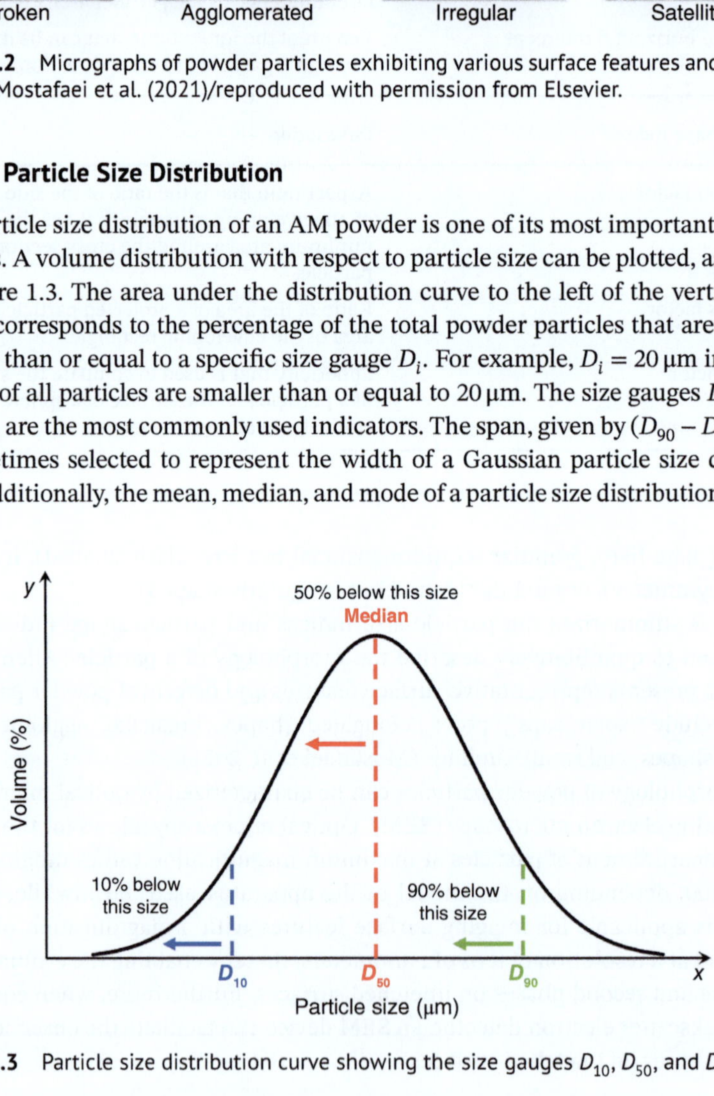

Figure 1.3 Particle size distribution curve showing the size gauges D_{10}, D_{50}, and D_{90}.

correspond to the average particle size, D_{50}, and the peak of the distribution curve, respectively, can be determined.

Particle size distribution is typically measured by the laser diffraction method. This method serves as a convenient and rapid analysis for a broad range of particle sizes. A typical laser diffraction measuring instrument contains a laser source, a particle dispersion module, a particle delivery module, and a detector. When the laser beam is blocked by a particle, part of the light scatters. An angle between the propagation direction of the original and scattered light is formed. A larger particle size results in a smaller angle. Upon being illuminated by parallel laser beams, particles of the same size in a sample deflect light at an identical angle. The scattered light is directed onto a sensor located on its focal plane, forming a series of concentric rings. The intensity of each ring, which corresponds to a specific angle of scattering, is measured to evaluate the volume size distribution of the constituent particles of the sample.

The particle size distribution of a sample can also be quantitatively measured from images captured by optical microscopy or SEM. The images of the constituent particles are usually post-processed before their equivalent diameters are determined, and the entire procedure may be time-consuming. An SEM machine developed by ASPEX Corporation is equipped with an automated feature analysis module, which is capable of quantifying the sizes of thousands of particles within several hours.

Sieve analysis, which involves the separation of particles according to their sizes, is also commonly employed to obtain the particle size distribution of a sample owing to its simplicity and low cost. A typical sieving unit comprises a series of sieves stacked on top of a shaker exhibiting rotary and tapping motions, with each sieve incorporating a phosphor bronze or stainless steel wire mesh cloth woven in a square mesh pattern. The sieves are stacked in the order of decreasing mesh sizes, with the sieve possessing the largest mesh size on top. Standard sieve sizes are specified in ISO standards 565 and 3310/1, ASTM standard E 11, and CIS standard GOST 3584.

1.5.3 Density

The packing density of a powder bed is defined as the ratio of the volume of its constituent powder particles to its total volume. It is a key parameter for assessing the packing efficiency of a powder bed in AM. A powder bed with a large packing density minimizes the porosity of the printed parts, thereby improving their mechanical properties. The packing density of a powder bed is influenced by a variety of factors such as the characteristics of the powder particles (size, size distribution, and shape), bulk properties (e.g. Young's modulus and hardness), and powder spreading parameters (powder layer thickness and spreading velocity). For bimodal powder mixtures containing both small and large particles, the maximum packing density can be achieved with specific ratios of the two types of powder particles, as shown in Figure 1.4.

The apparent density of a powder, with units of g/cm^3, is defined as the mass per unit volume of its loose powder particles. The Hall funnel method is the most prevalent method for measuring the apparent density of a powder, and the powder is

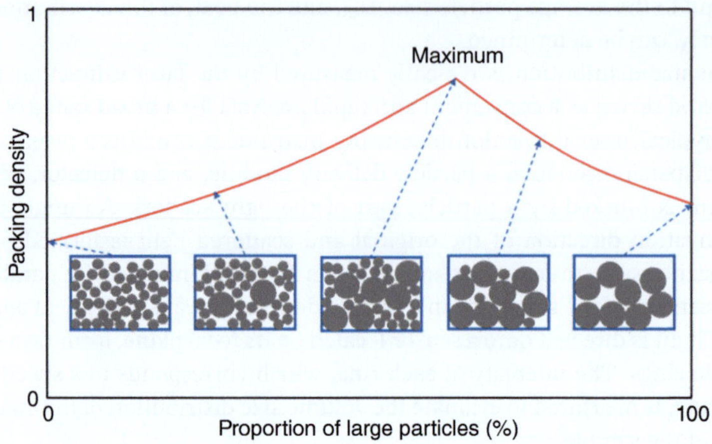

Figure 1.4 Variation of the packing density with the proportion of large powder particles for bimodal powder mixtures. Source: Adapted from German (1992).

poured through a funnel to fill a 25 cm^3 container. The apparent density is obtained by dividing the mass of the contents in the container by its volume. Generally, the apparent density of a powder decreases with its particle size and increases with its particle surface roughness. A powder with a wide size distribution possesses a relatively high apparent density since the space between its coarse powder particles is filled by smaller powder particles.

The tap density of a powder is defined as its density when its container is tapped or vibrated under specified conditions. The tap density (unit: g/cm^3) of a powder is obtained by dividing the mass of the tapped powder in a container by its volume. ISO 3953 established a method for measuring the tap density of a powder, in which a 100 ml (with increments of 0.2 ml) glass graduated cylinder is typically utilized. The tap density of a powder is dependent on the size distribution, shape, and surface roughness of its particles and is always higher than the free-flow apparent density.

The skeletal density, which is often measured through pycnometric methods, is the true density of a powder in its solid state. In pycnometry, the ratio of the mass of a powder to its volume is determined based on the volumetric displacement of a fluid medium. Gas pycnometry, in which helium or nitrogen is employed as the fluid medium, is commonly practiced since the tiny gas atoms can occupy the defects within a solid powder. Meanwhile, liquid pycnometry provides an alternative approach in which highly penetrative liquids (ethanol, oils, butanol, acetone, etc.) are selected as the fluid medium. The number of defective powder particles exhibiting cracks, satellite particles, and pores can be estimated based on the measured skeletal density of a powder.

1.5.4 Flowability

The flowability of a powder is often construed as its ability to flow in a desired manner. In powder-based AM systems, flowability determines the feeding and spreading behaviors of a powder. It is influenced by the particle size distribution, density, particle morphology, and moisture content of the powder. For example, fine powder particles (smaller than 10 μm) typically flow poorly and may even become stuck, while a powder that contains both fine and large particles exhibits good flowability. Due to its relative ease of measurement and correlation to powder flow behaviors, the Hausner ratio of an AM powder, expressed as its tap density divided by its apparent density (Hausner 1967), is commonly used to quantify its flowability. Since both the apparent and tap densities are related to interparticle friction, the Hausner ratio also represents interparticle friction, which directly influences flowability. Generally, a Hausner ratio smaller than 1.25 corresponds to a free-flowing powder.

The flowability of a powder can be assessed based on its flow time, which is defined as the duration for 50 g of the powder to flow through a funnel with a 2.5 mm diameter opening under the influence of gravity (Hall flowmeter). The flow time of a powder is expressed in units of g/s and can be readily compared with those of different powders for a quick assessment of its flowability. Another common flowability parameter is the angle of repose α, which is defined as the angle between a horizontal surface and the slope of a powder heap (Figure 1.5). To measure this parameter, a powder is allowed to fall freely, typically through a fixed funnel, and deposited onto a horizontal surface. Similarly to the Hausner ratio, the angle of repose is related to interparticle friction and cohesion, and a large angle of repose indicates poor powder flowability.

Spherical powder particles usually exhibit better flowability than their irregularly shaped counterparts. A high moisture content can increase powder stickiness and reduce powder flowability. Therefore, it is recommended to dry the powder before printing. Although the flowability of a powder is mostly an intrinsic property, it is also dependent on the interactions between the powder and the environment of AM systems. The interactions between a powder and various AM systems thus contribute to different powder flow behaviors. Therefore, extensive testing with respect to different AM equipment is necessary to evaluate powder flowability.

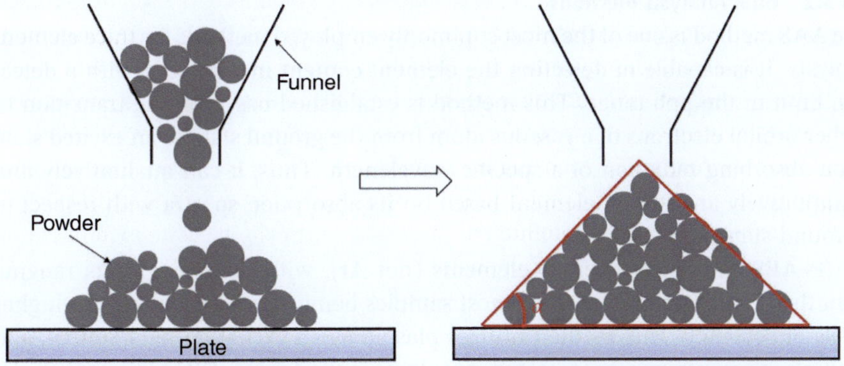

Figure 1.5 Schematic of a fixed funnel method for measuring the angle of repose of a powder.

1.5.5 Chemical Composition

The chemical composition analysis of a powder involves both surface and bulk analyses. Surface analysis methods include energy dispersive spectroscopy (EDS), electron probe microanalysis (EPMA), auger emission spectrometry (AES), X-ray photoelectron spectrometry (XPS), and secondary ion mass spectrometry (SIMS). Meanwhile, bulk analysis methods cover atomic absorption spectrometry (AAS), inductively coupled plasma atomic emission spectroscopy (ICP–AES), X-ray fluorescence (XRF), infrared (IR) spectroscopy, and inert gas fusion. These two categories of analytical methods are further elaborated in the following subsections.

1.5.5.1 Surface Analysis Methods

When coupled with SEM, EDS can be utilized to identify chemical compositions on the microscale and evaluate the chemical composition on a particle surface. Since a SEM/EDS system requires conductive samples, this method is more suitable for evaluating the particle size and chemical composition of metal powders than polymer powders. Meanwhile, EPMA provides qualitative and quantitative analyses for elements with atomic numbers greater than or equal to 11 (Na) with detection limits of approximately 1 μm. However, the detection sensitivity of EPMA is relatively poor for light elements with atomic numbers ranging from 5 (B) to 10 (Ne).

AES can facilitate the compositional analysis of almost every element (except H and He) on a powder surface with a sensitivity of 0.1–1.0 at.%. Meanwhile, XPS, a surface-sensitive quantitative spectroscopic method based on the photoelectric effect, can be conducted in tandem with ion-etching to identify the elements constituting the surface of a material and their chemical states, as well as the overall electronic structure and density of electronic states within the material. Finally, a variety of information pertaining to the surface, subsurface, or bulk composition of a powder can be acquired through SIMS, with a detection limit in the parts per billion (ppb, in units of ng/ml^3) to parts per million (ppm, in units of μg/ml) range for elements.

1.5.5.2 Bulk Analysis Methods

The AAS method is one of the most commonly employed methods for trace element analysis. It is capable of detecting the element content in a powder with a detection limit in the ppb range. This method is established based on the transition of higher orbital electrons of a gaseous atom from the ground state to an excited state upon absorbing radiation of a specific wavelength. Thus, it can qualitatively and quantitatively analyze an element based on its absorption spectra with respect to a ground-state atomic vapor.

ICP–AES can detect over 70 elements (not Ar), with detection limits ranging from the ppm to ppb range, with most samples being aqueous solutions at higher levels of accuracy. This method utilizes plasma as an excitation light source, and multiple elements can be detected simultaneously. In an ICP–AES system, the sample is first atomized and subsequently enters a plasma channel in the form of an aerosol. Finally, the aerosol sample is fully vaporized, ionized, and excited in a high-temperature inert atmosphere, thereby emitting radiation corresponding to the characteristic spectra of its constituent elements, from which the element content can be determined.

XRF is an efficient and nondestructive method that employs high-energy X-rays or gamma-rays to bombard a sample and induce the emission of secondary X-rays characterized by its constituent elements. These secondary X-rays are captured by a detector installed within an XRF system, which converts such signals into useful information regarding the element content of the sample. This method has a resolution limit of approximately 100 ppm.

IR spectroscopy is usually performed to identify functional groups in organic molecules, and the samples can be in the form of solids, liquids, or thin films. In the sample, molecular vibrations, in the form of stretching and bending of covalent bonds, occur at specific frequency values, which correspond to troughs in the IR spectra. Specific chemical bonds and functional groups in the sample are identified based on these troughs. IR spectroscopy has also been employed to characterize surface oxides on metal specimens (Jasinski and Iob 1988). Oxidation of a powder, particularly metal powder, is detrimental to powder-based AM processes. When a metal powder melts during the printing process, oxide layers form on the powder surfaces due to oxidation and hinder the consolidation of the printed parts, thereby resulting in the so-called "balling effect". Hardened oxide films also impede proper surface wetting and lead to poor adherence of the printed layers, which may induce porosity in the printed parts. Another type of powder contamination exists in the form of hydroxide films, which typically form on powder surfaces due to their ability to absorb moisture under high humidity. In contrast to solidified oxide layers that are often hard and brittle, hydroxide films are generally highly viscous. Therefore, the hydroxide films increase the tendency of particles to agglomerate and inhibit particle flow within the powder bed. Finally, irradiation of water layers adsorbed onto powder surfaces facilitates the dissociation of hydrogen atoms from the water molecules, and entrapped gas pores are formed upon the solidification of the melt pool, causing melt pool sputtering.

Inert gas fusion is a quantitative method for measuring the contents of H, N, and O elements in metal samples (Holt and Goodspeed 1963). Samples are first weighed and placed in a graphite crucible, in which they are heated to a molten state. Meanwhile, H_2, N_2, and O_2 molecules released from the samples are separated to evaluate the weight percentage of each element.

1.5.6 Microstructure

X-ray diffraction (XRD) is an X-ray spectrographic method primarily for identifying crystal structures and properties, including the number of phases in a metal powder sample, with a sensitivity of 3–5 vol.% for each phase component. During XRD analysis, a sample is bombarded with X-rays at different incident angles θ, and the intensity of the diffracted X-rays is measured. Constructive interference of the diffracted X-rays results in the formation of distinguishable intensity peaks. The condition for constructive interference to occur is described by Bragg's law, $n\lambda = 2d\sin\theta$, where n is the diffraction order (an integer), λ is the wavelength of the incident radiation, and d is the spacing between successive crystal planes (Skoog et al. 2006). By incorporating the knowledge of reflection angles for specific phases and indexing the positions of the peaks, the crystal structure of a sample can be determined.

Focused ion beam (FIB) microscopy is a technique for imaging and milling material surfaces, in which a beam of cations is utilized to form a raster image of a surface or to remove layers from it. Typically, a dual FIB/SEM approach is adopted for powder characterization, in which the FIB setup is mainly employed for milling purposes. Such an approach can also be undertaken to directly visualize the internal grain structure and porosity of powder particles.

Transmission electron microscopy (TEM) is a highly effective analytical tool for imaging a phase in powder form to obtain specific information (size and shape of the phase particles, crystallographic orientation, etc.). In TEM analysis, the specimens must be processed to reach a size of less than 100 nm to be electron transparent. Various specimen preparation methods are available, which include bulk preparation by thinning and dimpling and final polishing by an ion polisher or via electropolishing. A more advanced method for preparing TEM samples is to conduct FIB microscopy in conjunction with SEM to retrieve a sample from a bulk specimen, which is subsequently attached to a copper or molybdenum grid that can be inserted into the TEM facility. The obtained TEM diffraction patterns can be used to determine whether a material is a single crystal (sharp diffraction patterns), polycrystalline (diffraction rings), or amorphous (hazy rings).

Electron backscatter diffraction (EBSD) is a scanning electron microscope-based microstructural-crystallographic characterization method commonly employed when studying crystalline or polycrystalline materials. This method can provide details on the structure, crystal orientation, phase, and strain in powder materials. During an EBSD measurement, a flat and polished crystalline specimen is placed at a largely tilted angle (about 70° relative to the normal incidence of the electron beam) in an SEM chamber. A camera, equipped with a phosphor screen and integrated with a digital frame grabber, is inserted into the chamber and approaches the surface of the inclined specimen. The sample is then bombarded by electrons, which may backscatter and exit at various Bragg's angles, thereby undergoing diffraction and forming Kikuchi bands corresponding to different diffracting crystal planes of the lattice. As a result, the microstructural maps of the sample, which spatially describe its crystal orientation, can be obtained and utilized to examine the micro-texture of the sample. Such maps provide abundant information about a sample, such as its grain orientation, grain boundary, and grain size distribution. Additionally, various statistical tools can also be used to evaluate the average misorientation, grain size, and crystallographic texture of the sample.

Thermal analysis methods are mainly employed for tracking changes in the chemical properties of a sample over a range of temperatures, and they include thermogravimetric analysis (TGA), differential thermal analysis (DTA), and differential scanning calorimetry (DSC). These methods also reveal information regarding the exothermic and endothermic events that occur within the sample. In TGA, the mass of a sample is recorded as a function of temperature, which provides information on its content according to the loss of volatile elements and the potential formation of oxides. In addition, TGA allows for *ex situ* analysis of the response of a powder subjected to increasing temperatures during PBF processes. Meanwhile, DTA is a qualitative method in which a sample and an inert reference material are simultaneously heated, and the temperature difference between them is recorded as a function of temperature or time, which can be used to identify potential phase transitions. Generally, DTA analysis is also complemented by TGA and DSC. Finally, DSC is a quantitative method that records the difference in heat capacity between the sample and a reference material as a function of temperature.

1.6 Challenges and Future Trends of Metal Powder–Based AM

Present challenges and future trends of metal powder–based AM are centered on developing printing materials and processes, *in situ* monitoring, numerical simulations, and standardization.

The main challenges surrounding the development of printing materials are present in the following aspects: (i) tuning the powder composition to achieve the desired properties of the final parts (e.g. multifunctional, strong, ductile, and lightweight); (ii) controlling the stresses and distortion within the printed parts; (iii) achieving a comprehensive understanding of metallurgical principles; (iv) tuning microstructures and properties in different regions of a printed part by manipulating process parameters and ensuring reproducibility; (v) reducing anisotropy in the mechanical properties of the printed parts; (vi) preventing the formation of metallurgical defects; and (vii) establishing a database of printable materials. In addition, the lack of specialized metal powder preparation processes hinders the standardization of AM powder properties (see Chapter 2 for further details).

Large-scale manufacturing, hybrid manufacturing, and multiple-energy-source-aided manufacturing are projected to be popular future research trends concerning the development of printing processes. For example, large-scale PBF manufacturing utilizes multiple laser sources and optical systems to fabricate parts extensively. Hybrid manufacturing refers to the integration of a powder–based AM technique (PBF or DED) with a subtractive manufacturing system (e.g. milling or grinding), offering a significant improvement of the printed parts in terms of their dimensional accuracy and surface quality. Multiple-energy-source-aided manufacturing incorporates external energy fields (e.g. electromagnetic fields, ultrasonic fields, and electric fields) into a powder–based AM system. Such energy fields induce strong vibrations within the melt pools during the printing process, which can effectively refine the microstructures of the printed parts and reduce metallurgical defects within them.

The challenges of quality control during printing processes primarily lie in the areas of *in situ* monitoring, process sensing, and adaptive control, which are critical in regulating such processes to achieve high productivity while avoiding common defects in the printed parts. For example, a key aspect of process control is the availability of well-tested real-time models that serve as media for relating the process variables with the desired properties. Notably, adequately standardized phenomenological modeling can reveal the most important factors that affect the quality of metal parts, such as temperature and velocity fields, cooling rates, and solidification parameters. Therefore, it is a powerful tool for assessing the effects of process variables on part quality prior to production. Furthermore, the development and testing of phenomenological models provide a solid foundation for the eventual construction of digital twins of physical objects in metal powder–based AM. With the incorporation of genetic algorithms and other global search algorithms, the utility of such digital twins can be greatly expanded to tailor part attributes, optimize production variables, reduce defects, and improve part quality. However, refining such virtual representations requires considerable time, effort, and resources.

Contemporary numerical simulations of AM are primarily conducted to provide insights on powder-spreading behaviors on the powder bed and heat transfer phenomena, residual stress development, distortion, microstructure evolution, and porosity formation in AM parts. However, there appears to be a lack of agreement regarding the characteristics of an acceptable model since the efficacy of each model is assessed according to its computational cost and/or the accuracy of its corresponding results. Such criteria are difficult to implement due to the different resources available in each research project and the variations between model outputs.

AM modeling research is directed at understanding the underlying mechanisms governing different phenomena, developing a computational model for describing specific processes, and establishing a process–structure–property relationship. One should note that adopting a feedback-based approach is necessary, in which earlier simulation results are utilized in subsequent simulations until a model has successfully satisfied specific requirements (low distortion, desired microstructure, high cycle fatigue, etc.).

Material–structure–performance integrated AM could be explored to achieve high-performance and multifunctional AM parts (Gu et al. 2021). Such technology relies on developing more digitized materials and structures, which could be accomplished with the Materials Genome Initiative, standardization of formats, and a systematic printability database. High mechanical performance and multifunctionality can be simultaneously achieved for an AM metal part through the innovative design of materials and structures.

The standardization of metal powder–based AM is impeded by critical challenges spanning many aspects, which include cost, design, software, materials, traceability, machine constraints, printing process monitoring, mechanical properties, repeatability, scalability, validation, standards, quality, inspection, post-processing, and tolerances. A primary challenge is the unsatisfactory development of qualification and certification process for various AM applications because of high costs, long lead times, and inadequate standards, rules, and regulations. Other challenges arise due to the lack of *in situ* measurements for real-time closed-loop process control systems, insufficient data, and poor process reliability and repeatability due to limited research on heat source–powder interactions, heat transfer, vaporization, defect formation, quantitative metallography, etc. Moreover, the large-scale production of AM parts using commercially available AM equipment is highly difficult.

While the market expansion of AM technologies requires increased standardization and improved quality control of AM products, a more comprehensive understanding of the printing processes and practical innovations must first be achieved.

Despite being a revolutionary method for facilitating customized part fabrication and niche applications, metal powder–based AM requires further research and development to achieve the mass production of parts at a reduced cost.

1.7 Summary

This chapter briefly discusses the development of AM technology, popular metal powder–based AM techniques, post-processing methods, representative metal powder characteristics and their characterization methods, and challenges and future trends of metal powder–based AM.

AM technology is classified into seven broad categories, i.e. vat photopolymerization, material jetting, material extrusion, PBF, DED, binder jetting, and sheet lamination techniques. Among them, PBF, DED, and binder jetting are the predominant powder–based AM techniques. In particular, LPBF, EBM, L-DED, and MBJ are the most commonly adopted powder–based AM processes for printing metal parts.

Most metal powder–based AM products require post-processing to obtain the desired properties. Surface quality improvement methods include manual grinding, machining, sandblasting, shot peening, mechanical and chemical polishing, chemical etching, laser shock peening, and laser polishing. In addition, heat treatment methods, such as stress relief annealing and HIP, can be employed to minimize residual stress within the printed parts and reduce their porosity. Finally, aesthetic improvement approaches include spray painting and electroplating.

Powders are the principal component of powder–based AM, and they are characterized in terms of their morphology, size, size distribution, density, flowability, chemical composition, and microstructure. The quality of a powder is determined by these characteristics, which determine the printability and performance of the final parts.

Present challenges and future trends of metal powder–based AM are proposed in terms of developing printing materials and processes, *in situ* monitoring, numerical simulations, and standardization. Furthermore, as a revolutionary method for facilitating customized part fabrication and certain niche applications, metal powder–based AM technologies require further development to mass-produce parts at a reduced cost.

References

Allen, T. (ed.) (1997). *Particle Size Measurements*, 5e. Berlin: Springer.

Bermingham, M.J., Kent, D., Zhan, H. et al. (2015). Controlling the microstructure and properties of wire arc additive manufactured Ti–6Al–4V with trace boron additions. *Acta Materialia* 91: 289–303.

Chua, C.K. and Leong, K.F. (2017). *3D Printing and Additive Manufacturing: Principles and Applications*, 5e. Singapore: World Scientific.

Dehoff, R.R. and Babu, S.S. (2010). Characterization of interfacial microstructures in 3003 aluminum alloy blocks fabricated by ultrasonic additive manufacturing. *Acta Materialia* 58 (13): 4305–4315.

Gebler, M., Uiterkamp, A.J.S., and Visser, C. (2014). A global sustainability perspective on 3D printing technologies. *Energy Policy* 74: 158–167.

German, R.M. (1992). Prediction of sintered density for bimodal powder mixtures. *Metallurgical Transactions A* 23 (5): 1455–1465.

Gibson, I., Rosen, D., and Stucker, B. (2014). *Additive Manufacturing Technologies: 3D Printing, Rapid Prototyping, and Direct Digital Manufacturing*, 2e. Berlin: Springer.

Gu, D., Shi, X., Poprawe, R. et al. (2021). Material-structure-performance integrated laser-metal additive manufacturing. *Science* 372 (6545): eabg1487.

Han, C., Babicheva, R., Chua, J.D.Q. et al. (2020a). Microstructure and mechanical properties of (TiB+TiC)/Ti composites fabricated in situ via selective laser melting of Ti and B_4C powders. *Additive Manufacturing* 36: 101466.

Han, C., Fang, Q., Shi, Y. et al. (2020b). Recent advances on high-entropy alloys for 3D printing. *Advanced Materials* 32 (26): 1903855.

Hausner, H.H. (1967). Friction conditions in a mass of metal powder. *International Journal of Powder Metallurgy* 13 (4): 7–13.

Holt, B.D. and Goodspeed, H.T. (1963). Determination of nitrogen, oxygen, and hydrogen in metals by inert gas fusion. A manometric method. *Analytical Chemistry* 35 (10): 1510–1513.

Hull, C.W. (1986). Apparatus for production of three-dimensional objects by stereolithography. US Patent 4,575,330, filed 08 Augst 1984, issued 11 March 1986.

ISO and ASTM International (2015). ISO/ASTM 52900: Additive manufacturing – General principles – Terminology.

Jasinski, R. and Iob, A. (1988). FTIR measurements of iron oxides on low alloy steel. *Journal of the Electrochemical Society* 135 (3): 551.

Kelly, B.E., Bhattacharya, I., Heidari, H. et al. (2019). Volumetric additive manufacturing via tomographic reconstruction. *Science* 363 (6431): 1075–1079.

Kuang, X., Wu, J., Chen, K. et al. (2019). Grayscale digital light processing 3D printing for highly functionally graded materials. *Science Advances* 5 (5): eaav5790.

Li, B., Zheng, H., Han, C., and Zhou, K. (2021). Nanotwins-containing microstructure and superior mechanical strength of a Cu-9Al-5Fe-5Ni alloy additively manufactured by laser metal deposition. *Additive Manufacturing* 39: 101825.

Martin, J.H., Yahata, B.D., Hundley, J.M. et al. (2017). 3D printing of high-strength aluminium alloys. *Nature* 549 (7672): 365–369.

Mostafaei, A., Kimes, K.A., Stevens, E.L. et al. (2017). Microstructural evolution and magnetic properties of binder jet additive manufactured Ni-Mn-Ga magnetic shape memory alloy foam. *Acta Materialia* 131: 482–490.

Mostafaei, A., Elliott, A.M., Barnes, J.E. et al. (2021). Binder jet 3D printing – process parameters, materials, properties, modeling, and challenges. *Progress in Materials Science* 119: 100707.

Ribeiro, I., Matos, F., Jacinto, C. et al. (2020). Framework for life cycle sustainability assessment of additive manufacturing. *Sustainability* 12 (3): 929.

Saha, S.K., Wang, D., Nguyen, V.H. et al. (2019). Scalable submicrometer additive manufacturing. *Science* 366 (6461): 105–109.

Skoog, D.A., Holler, F.J., and Crouch, S.R. (2006). *Principles of Instrumental Analysis*, 6e. Boston: Cengage Learning.

Tan, L.J., Zhu, W., and Zhou, K. (2020). Recent progress on polymer materials for additive manufacturing. *Advanced Functional Materials* 30 (43): 2003062.

Truby, R.L. and Lewis, J.A. (2016). Printing soft matter in three dimensions. *Nature* 540 (7633): 371–378.

Tumbleston, J.R., Shirvanyants, D., Ermoshkin, N. et al. (2015). Continuous liquid interface production of 3D objects. *Science* 347 (6228): 1349–1352.

Walker, D.A., Hedrick, J.L., and Mirkin, C.A. (2019). Rapid, large-volume, thermally controlled 3D printing using a mobile liquid interface. *Science* 366 (6463): 360–364.

Wang, X., Zhang, W., Liu, Z. et al. (2016). Improvement on room-temperature ductility of 6.5 wt% Si steel by stress-relief annealing treatments after warm rolling. *Materials Characterization* 122: 206–214.

Xiong, X., Hu, S., Dang, N., and Hu, K. (2017). Effect of stress-relief annealing on microstructure, texture and hysteresis curve of mechanically cut non-oriented Fe-Si steel. *Materials Characterization* 132: 239–247.

Zhang, D., Qiu, D., Gibson, M.A. et al. (2019). Additive manufacturing of ultrafine-grained high-strength titanium alloys. *Nature* 576 (7785): 91–95.

Zhang, Y., Bai, S., Riede, M. et al. (2020). A comprehensive study on fused filament fabrication of Ti-6Al-4V structures. *Additive Manufacturing* 34: 101256.

2

Metal Powder Preparation Processes

The geometrical, physical, and chemical characteristics of a metal powder play vital roles in its printability and the quality of additively manufactured products. For example, the shape, size, and size distribution of powder particles influence their flowability and packing density during the powder spreading process. In addition, the laser absorptivity and thermal conductivity of powder particles affect their melting and solidification process during printing. Other important factors include the powder surface roughness, oxidation, impurities, and moisture. All these factors heavily depend on how the metal powder is prepared.

As illustrated in Figure 2.1, commercial metal powders are mainly prepared through atomization, mechanical mixing, reduction process, and powder modification. Some metal powders can be produced via multiple processes. A preparation method is commonly chosen by considering its cost and the physical and chemical properties of the materials involved.

Figure 2.2 shows the morphology of Ti–6Al–4V powder particles prepared through the different processes (Sun et al. 2016, 2017a; Dawes et al. 2015). Some of these processes produce powder particles with spherical or near-spherical shapes. In contrast, others produce irregularly shaped powder particles that require further processing to improve their degree of sphericity, such as plasma spheroidization (see Section 2.4.1 for further details). This chapter focuses on the fundamentals and characteristics of the preparation processes of metal powders for additive manufacturing (AM) purposes. Furthermore, it provides a comprehensive understanding of the relationships between the preparation processes and the powder properties.

2.1 Atomization

Atomization is the most widely adopted process for the mass production of pre-alloyed powders for AM. It typically includes gas atomization, water atomization, plasma atomization, and the plasma rotating electrode process. The fundamental principle of atomization is the utilization of a high-pressure gas jet or water stream, or rotating forces to disperse a stream of molten metal into droplets, which solidify to form spherical powder particles. Overall, this process involves melting, atomization, and solidification.

Metal Powder–Based Additive Manufacturing, First Edition. Kun Zhou and Changjun Han.

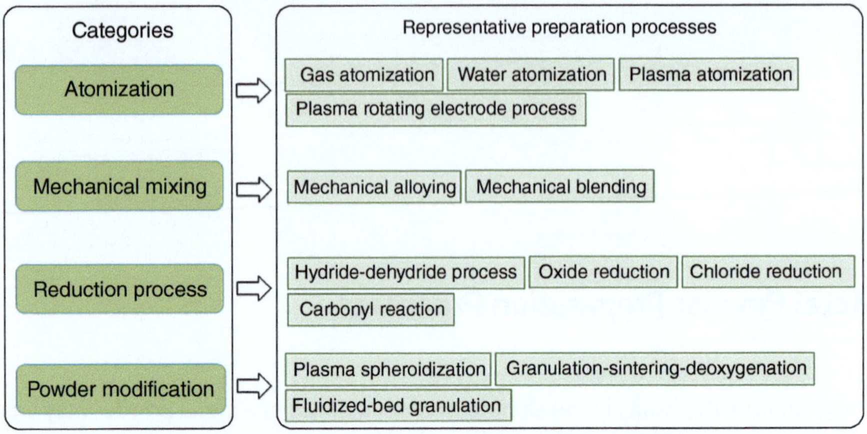

Figure 2.1 Overview of the preparation processes for metal powders.

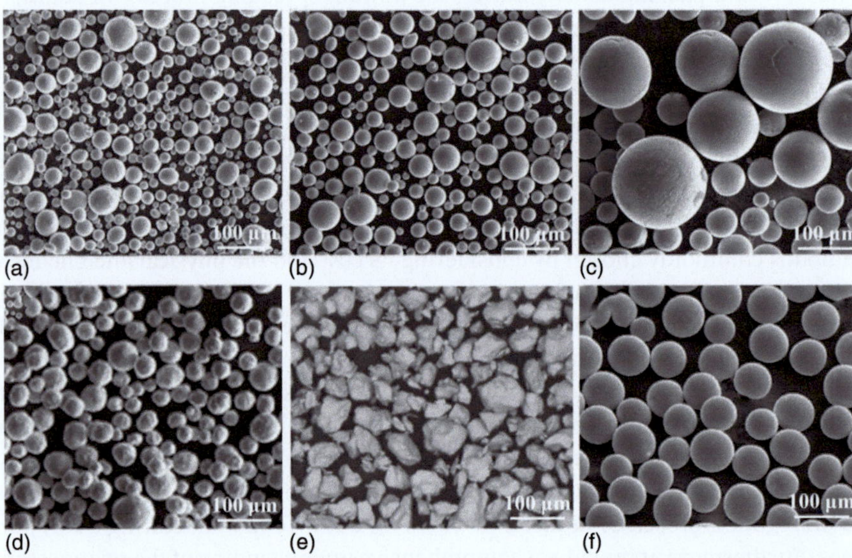

Figure 2.2 SEM images showing the morphology of Ti–6Al–4V powder particles produced through different processes. (a–c) Gas atomization, plasma atomization, and the plasma rotating electrode process, respectively. Source: Sun et al. (2017a)/reproduced with permission from Springer Nature. (d) Oxide reduction. Source: Sun et al. (2016)/reproduced with permission from Elsevier. (e) Hydride–dehydride process. Source: Dawes et al. (2015)/reproduced with permission from Johnson Matthey Plc. (f) Plasma spheroidization. Source: Sun et al. (2017a)/reproduced with permission from Springer Nature.

2.1.1 Gas Atomization

Gas atomization is the most extensively applied process for the production of fine metal powders for AM, e.g. ferrous alloys, titanium alloys, nickel alloys, aluminum alloys, copper alloys, cobalt alloys, magnesium alloys, tin alloys, titanium–aluminum alloys, high-entropy alloys (HEAs), metallic glasses, and

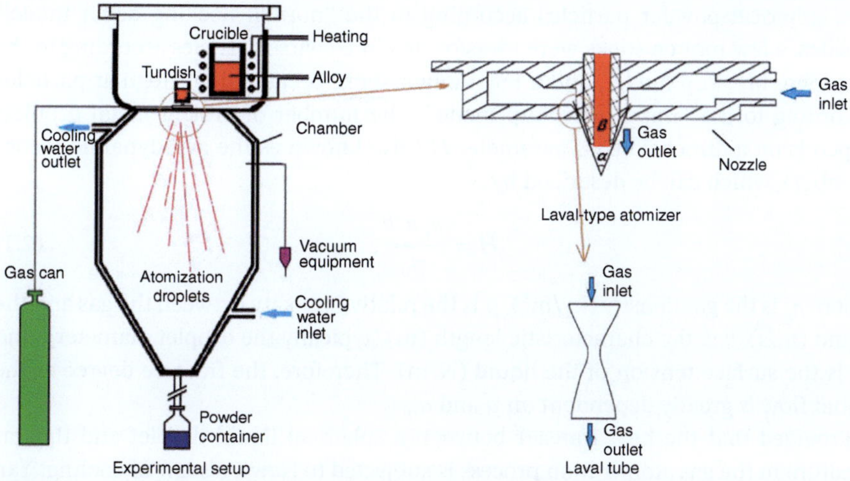

Figure 2.3 Schematic diagram of gas atomization. The parameter α is the angle of the nozzle airflow, and β is the cone angle of the metal tube. Source: Reproduced from Zheng et al. (2018)/with permission from Taylor & Francis.

pure metals. The underlying mechanism involves converting part of the kinetic energy of a high-velocity gas jet ejected through a nozzle to the surface energy of metal droplets disintegrated from molten metal streams. The metal droplets then spheroidize and solidify to form powder particles.

A schematic of the experimental setup for gas atomization is illustrated in Figure 2.3 (Zheng et al. 2018). A typical equipment setup for gas atomization consists of a gas supply system, a melting system, a nozzle, an atomization chamber, a powder collection container, and a gas extraction system (Ünal 2006). First, alloy or pure metal ingots are melted in a crucible, and the molten material is transferred to a tundish from which it is dispensed as a molten stream. Subsequently, the stream is disintegrated into metal droplets by a high-velocity gas using an atomizer. At this stage, the impact between the gas and the stream creates a shockwave that breaks down the molten stream into metal droplets. Finally, the metal droplets solidify to form powder particles when they fly onto the chamber wall. During this process, the chamber is filled with inert gas (e.g. argon) to avoid oxidation of the metal droplets. Meanwhile, cooling water is supplied through the chamber walls to maintain the temperature of the chamber. As a result, gas atomization produces high-purity powders with a high degree of sphericity and low oxygen content.

The complicated breakup process of the metal droplets exerts a significant influence on the shape and size of the solidified powder particles. A work by Wei et al. (2017) presented the breakup process of TA15 titanium alloy droplets during electrode-induction-melting gas atomization. The TA15 droplets first formed oval molten pieces due to the complex mechanical effect of the high-velocity gas. Then, thin molten flakes were decomposed into anomalous molten bands due to the surface tension of the liquid. After further liquid contraction, regular molten sticks were produced. Finally, most regular molten sticks were decomposed into fine spherical powder particles according to the “normal necking down mode”. Besides, a few molten sticks were transformed into coarse particles according to the “interfere breakup mode”, and a few molten sticks became fine irregular particles according to the “impact breakup mode”. The number of broken metal droplets depends on a dimensionless parameter H (also known as the aerodynamic Weber number), which can be described by

$$H = \frac{\rho_g \mu^2 b}{\sigma_0}, \tag{2.1}$$

where ρ_g is the gas density (kg/m^3), μ is the relative velocity between the gas and the liquid (m/s), b is the characteristic length (m) (typically the droplet diameter), and σ_0 is the surface tension of the liquid (N/m). Therefore, the fracture degree of the liquid flow is greatly dependent on μ and σ_0.

Provided that the heat transfer between a spherical liquid droplet and the gas medium in the gas atomization process is subjected to Newton's law of cooling, Yan et al. (2011) showed that the cooling rate of a spherical droplet can be calculated by

$$\frac{\mathrm{d}T}{\mathrm{d}t} = \frac{6h(T_1 - T_c)}{D\rho_1 C}, \tag{2.2}$$

where T is the temperature (K), t is time (s), h is the coefficient of heat transfer, T_1 is the initial temperature of the liquid droplet (K), T_c is the temperature of the cooling medium (K), D is the powder particle size (μm), ρ_1 is the density of the liquid (kg/m^3), and C is the specific heat of the liquid (J/kg/K). For a specific material, the size of the powder particle obtained is inversely proportional to the cooling rate. Thus, the particle size distribution of a powder can be tailored through the change in the cooling rate by controlling the atomization process parameters.

The atomization process parameters (such as type of gas, gas pressure, gas flow rate, and gas-to-metal ratio) should be optimized to obtain a metal powder with a high degree of sphericity, a homogeneous distribution of elements, and specific particle sizes as shown in Table 2.1. The gas flow rate generally depends on the gas pressure, temperature, and nozzle cross-sectional area. The gas pressure and mass flow rate are determined by the gas delivery system and hardware. For a particular gas nozzle design, the average particle size of a powder is regulated by the gas pressure and the melt flow rate through the nozzle diameter and suction pressure. A decrease in the melt flow rate may speed up the production of fine powder particles.

Gas atomization has been utilized to produce powders of multi-component alloys, e.g. HEAs and metallic glasses. Figure 2.4a presents an inverse pole map of gas-atomized CoCrFeMnNi HEA powder particles with a mean grain size of 3.6 μm based on electron backscatter diffraction (EBSD) analysis (Wang et al. 2019). Dendrite structures rich in Fe, Cr, and Co were observed by energy dispersive spectroscopy (EDS) mapping (Figure 2.4b).

The powder particle size affects the phase and morphology of HEA powders significantly. It was found that gas-atomized $Al_{0.6}$CoCrFeNi powder particles with a minimum size of over 75 μm possessed duplex structures consisting of face-centered cubic (FCC) and body-centered cubic (BCC) phases. In contrast, smaller powder particles only formed the BCC phase (Zhou et al. 2018a). The rapid cooling in the case

Table 2.1 Summary of processing parameters for gas atomization.

	Parameters				
Powder	**Gas pressure (MPa)**	**Gas flow rate**	**Gas-to-metal ratio**	**Gas type**	**Average particle size (μm)**
Sn alloy[a)]	0.2–1.0	27–97 m^3/h	0.14–0.56	Nitrogen	—
Al alloy[b)]	1.4	1.4 Nm^3/min	—	Nitrogen	—
316L[c)]	2.76–6.89	2.75–13.1 kg/min	0.36–1.67	Nitrogen	32.5–34.8
Fe–Ni[d)]	2	1.4 Nm^3/min	—	Nitrogen	48
Al-Si[e)]	0.2–0.45	28.62 g/s	0.54–074	Nitrogen	60–96
Al-Si10-Mg[f)]	2–4	10–54 kg/min	5.32–23.97	Nitrogen	25.67–31.75
Al–Mn[g)]	0.4–1.4	—	—	Argon	44–59
TA15[h)]	2–8	—	—	Argon	45
AlCoCrFeNi[i)]	2.2	—	—	Nitrogen	27
$((Fe_{0.6}Co_{0.4})_{0.75}B_{0.2}Si_{0.05})_{96}Nb_4$[j)]	1.6	513–780 kg/h	1.6–9.6	Argon	35
$Mg_{63}Zn_{32}Ca_5$[k)]	0.4	—	—	—	—

a) (Lagutkin et al. 2004);
b) (Hong et al. 1999);
c) (Anderson and Terpstra 2002);
d) (Feng and Qiu 2012);
e) (Goudar et al. 2017);
f) (Gao et al. 2019);
g) (Allimant et al. 2009);
h) (Wei et al. 2017);
i) (Yang et al. 2017);
j) (Ciftci et al. 2018);
k) (Zhao et al. 2014).

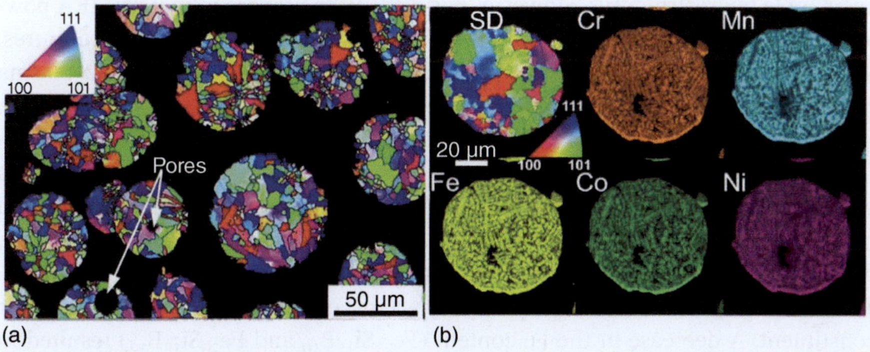

Figure 2.4 EBSD and EDS analysis of CoCrFeMnNi HEA powder particles. (a) The polycrystalline structures of the particles and (b) the corresponding elemental distribution of a particle. Source: Wang et al. (2019)/reproduced with permission from Elsevier, CC BY-NC-ND 4.0.

Table 2.2 Characteristics of Fe–Si–B metallic glass alloy powders produced via gas atomization.

Powder	Particle size (μm)			Apparent density (%)	Tap density (%)	True density (g/cm^3)	Phase
	D10	*D50*	*D90*				
$Fe_{80}Si_{12}B_8$	4.1	17.5	75.5	38.9	64.5	7.26	αFe(Si) + Fe_2B
$Fe_{75}Si_{15}B_{10}$	4.2	17.2	70.9	40.4	69.7	7.04	αFe(Si) + Fe_2B + Fe_3Si
$Fe_{70}Si_{18}B_{12}$	4.1	14.0	48.2	37.1	64.3	6.84	αFe(Si) + Fe_3Si
$Fe_{60}Si_{24}B_{16}$	3.7	12.8	44.6	38.4	63.9	6.63	$Fe_{4.9}Si_2B$
$Fe_{50}Si_{30}B_{20}$	2.6	10.1	40.2	37.1	64.6	6.33	FeSi + FeB

Source: Adapted from Alvarez et al. (2018).

of smaller powder particles suppressed the eutectic solidification and improved the solubility of Al in CoCrFeNi to generate the BCC structure. Additionally, a reduction in particle size enhances the powder's surface quality. For example, CoCrFeNiMn powder particles with a maximum size below 53 μm were observed to possess a higher degree of sphericity and a smoother surface than those with a minimum size exceeding 100 μm (Park et al. 2018).

The constituent elements and their ratios in an alloy significantly influence the phase composition of HEA powders prepared by gas atomization. For alloys in the AlCoCrFeNi category, an increase in the Al content tends to result in a BCC-dominant structure. The introduction of Si into AlCoCrFeNiCu could accelerate the structural transition from the FCC phase to the BCC phase during gas atomization (Yang et al. 2017). Alloys belonging to the CoCrFeNi category exhibit FCC structures even with the addition of Mn, Mo, and C elements. A high cooling rate (10^4–10^5 K/s) is favorable for preventing compositional segregation and restricting the atomic ordering required to generate intermetallic compounds. As a result, gas-atomized HEA powders generally exhibit uniform element distributions and simple microstructures. However, it is difficult to prepare refractory HEA powders using gas atomization because of their extremely high melting temperatures. Spherical refractory HEA powder particles can be prepared through the plasma rotating electrode process as well as milling and thermal plasma spheroidization (for details on the process, refer to Sections 2.1.4 and 2.4.1).

Table 2.2 summarizes the characteristics of $Fe_x(Si_{15}B_{10})_{1-x}$ metallic glass powders prepared by gas atomization (Alvarez et al. 2018). The low apparent density indicated high interparticle friction because of the fine powder particle size. On the contrary, the high tap density showed an efficient vibration packing of the powder particles. The dominant phase in the $Fe_{80}Si_{12}B_8$ alloy was α-Fe(Si), in which Fe_2B was a minor constituent. A decrease in the Fe content ($Fe_{75}Si_{15}B_{10}$ and $Fe_{70}Si_{18}B_{12}$) resulted in ordered Fe_3Si compounds.

Various gas atomization techniques have been developed by modifying the gas atomizers, such as vacuum-induction-melting, free-fall, close-coupled, electrode-induction-melting, and plasma-melting-induction-guiding gas atomization (Sun

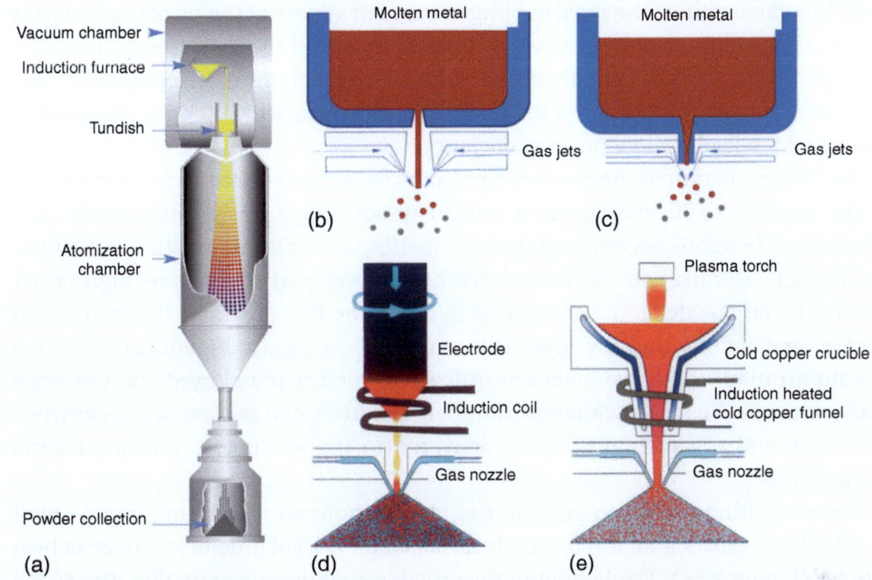

Figure 2.5 Schematic illustrations of different gas atomization techniques. (a) Vacuum-induction-melting. Source: Adapted from Gerling et al. (2004). (b, c) Free-fall and close-coupled processes, respectively. Source: Reproduced from Sun et al. (2017a)/with permission from Springer Nature. (d, e) Electrode-induction-melting and plasma-melting-induction-guiding, respectively. Source: Reproduced from Gerling et al. (2004)/with permission from John Wiley & sons.

et al. 2017a; Gerling et al. 2004). Vacuum-induction-melting gas atomization combined with inert gas atomization is the universal process for preparing nonreactive metal powders with low oxygen content. In this process, a vacuum-induction-melting unit is integrated with an inert gas atomization unit (Figure 2.5a). The feedstock materials are melted through electromagnetic induction, in which electricity is applied to the crucible in a vacuum or an inert gas atmosphere. The molten metal is poured into a tundish when the required melt uniformity and chemical composition have been obtained. Afterward, a fine metal stream is channeled from the tundish orifice to the atomization nozzle, where it is subsequently subjected to a high-pressure gas jet and atomized.

In free-fall gas atomization, a stream of molten metal falls freely under the influence of gravity from the crucible, and it interacts with a gas field generated by the impact of gas jets (Figure 2.5b). At the point of impact, the molten stream breaks into metal droplets that solidify to produce spherical powder particles of various sizes.

Close-coupled gas atomization has been developed to produce metal powders with refined microstructures, extended solid solubility, and amorphous phases. The gas flow in the close-coupled atomizer immediately impacts the exiting molten metal stream (Figure 2.5c). The distance between the gas exit and the melt stream within the close-coupled atomizer is smaller than that in the free-fall atomizer. Therefore, the maximum kinetic energy of the gas can be transferred to a smaller volume of the metal stream. In addition, the close spatial coupling between the gas and the melt flow can facilitate the rapid cooling of the melt stream at the tip of the atomizer nozzle. Controlling the interactions between the gas and the melt stream using the free-fall design is more convenient than doing so through the close-coupled design. However, it is difficult to control the powder particle size distribution in the free-fall design, which leads to low process efficiency.

The contamination of the produced powders by ceramic particles from the lining of the tundish or furnace poses a critical disadvantage in the abovementioned processes. Therefore, electrode-induction-melting and plasma-melting-induction-guiding gas atomization methods have been developed to prepare high-purity reactive metal powders. The distinction between the two processes lies in the heat source used for melting the feedstock materials. In electrode-induction-melting gas atomization, inductive coils are utilized to melt a pre-alloyed rod electrode that comprises the feedstock metal material. The front-end section of the electrode is melted, and the molten metal is delivered to the gas nozzle for atomization (Figure 2.5d).

Plasma-melting-induction-guiding gas atomization uses a plasma gun to melt metals. The feedstock material is melted in a water-cooled crucible by plasma heat sources (Figure 2.5e). The bottom of the crucible contains a system that directs the molten metal to the gas nozzle while providing additional heating. Therefore, a well-defined melt stream is achieved through this technique.

Figure 2.6 exhibits the morphology of different metal powders prepared by the various gas atomization processes (Singh et al. 2001; Ouyang et al. 2007; Löber et al. 2014; Zhang et al. 2009). Notably, the powders produced by the close-coupled, electrode-induction-melting, and plasma-melting-induction-guiding processes possess a higher degree of sphericity than the powders produced by free-fall gas atomization. However, satellite particles are attached to the surfaces of larger droplets produced by these processes, which reduces powder flowability. The presence of satellite particles can be attributed to the inevitable merging of fine powder with coarser powder. As a result, fine powder particles solidify more rapidly than their coarser counterparts during the atomization processes, and they undergo greater acceleration in the high-velocity gas flow, consequently impacting or welding themselves onto larger molten droplets.

Generally, during gas atomization, the powders produced from alloys with a low density (e.g. aluminum alloys) may have a higher amount of satellite particles than alloys with a high density (e.g. steels). When the droplets of low-density alloys are ejected from the nozzle, their travel trajectories may be more random due to larger acceleration driven by the high-velocity gas flow, which creates a higher chance for finer powder particles to attach onto larger ones to become satellite particles.

Inert gases with higher thermal conductivity (e.g. helium) may be beneficial in facilitating the solidification of fine particles before they impact larger particles to form satellite particles. Nevertheless, argon gas is used instead of helium gas due to the practical cost consideration. Another potential approach for minimizing the formation of satellite particles is to eject the molten metal droplets from the nozzle at supersonic speeds to expedite solidification.

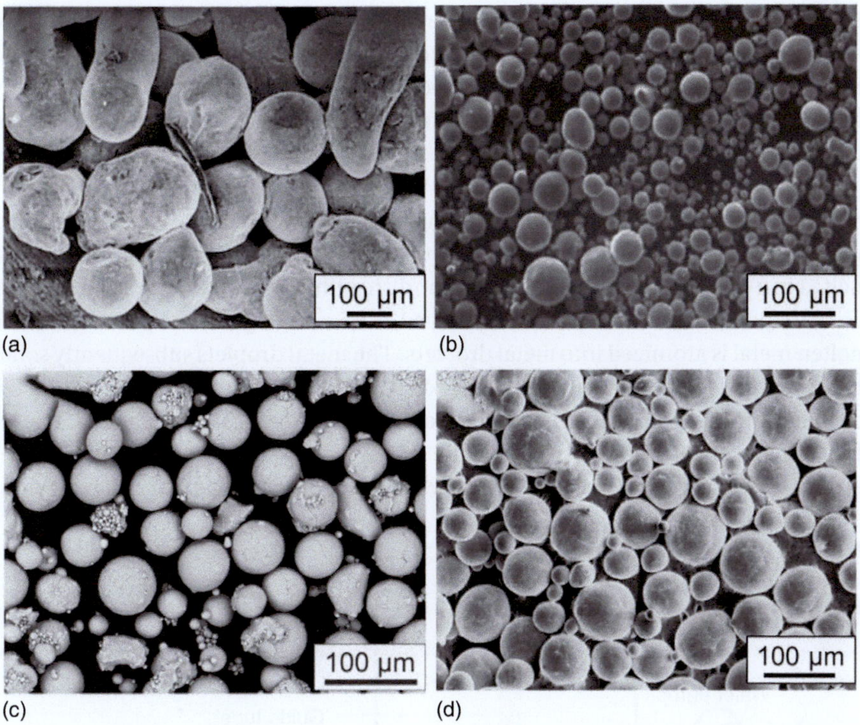

Figure 2.6 Powder morphology produced by different gas atomization processes. (a) Free-fall gas atomization of aluminum powder. Source: Singh et al. (2001)/reproduced with permission from Taylor & Francis. (b) Close-coupled gas atomization of copper powder. Source: Ouyang et al. (2007)/reproduced with permission from Elsevier. (c) Electrode-induction-melting gas atomization of TiAl powder. Source: Löber et al. (2014)/reproduced with permission from Elsevier. (d) Plasma-melting-induction-guiding gas atomization of TiAl powder. Source: Zhang et al. (2009)/reproduced with permission from Elsevier.

2.1.2 Water Atomization

Water atomization is usually cheaper than other atomization methods due to the lower cost of the atomization medium (water), the lower energy used for pressurization compared to gases, and the ability to achieve very high productivity. Water atomization mainly produces ferrous metal powders, such as stainless steels, tool steels, and soft magnetic powders. However, it is also sometimes used to produce nonferrous metal powders, such as copper, nickel, and their alloys, as well as precious metals.

Low powder purity and irregular powder particle shapes are the main limitations of water atomization, particularly for water-reactive metals (e.g. aluminum and magnesium alloys), pure metals, and oxidation-sensitive metals. Nevertheless, at a high cooling rate, the surface oxides of water-atomized powder could be reduced to an average thickness comparable to that of gas-atomized powder (Neikov et al. 2004). In addition, water atomization is generally not used for pure metals and alloys with a melting point below 500 °C, as the ultrarapid solidification process does not provide enough time for the metal droplets to become spherical after they are ejected from the nozzle (Eisen et al. 1998). Consequently, these metal particles have highly irregular shapes, which are undesirable for AM. Commercially, zinc is considered to be the metal with the lowest melting point that can be prepared viably by water atomization.

Water atomization is similar to gas atomization in terms of the underlying mechanisms, as schematized in Figure 2.7 (Pasupathy et al. 2016). After a metal feedstock is melted in a crucible, it is transferred to a tundish with a nozzle at the bottom through which it flows downward. Water is sprayed from a high-pressure pump until the molten metal is atomized into metal droplets. The metal droplets subsequently solidify and form powder particles that are then sieved and separated. Nozzles adopted for water atomization are either multiple discrete nozzles or annular slit nozzles that are concentric with reference to the metal stream. Generally, water atomizer designs conform to a "V" configuration with a minimum of two nozzles positioned

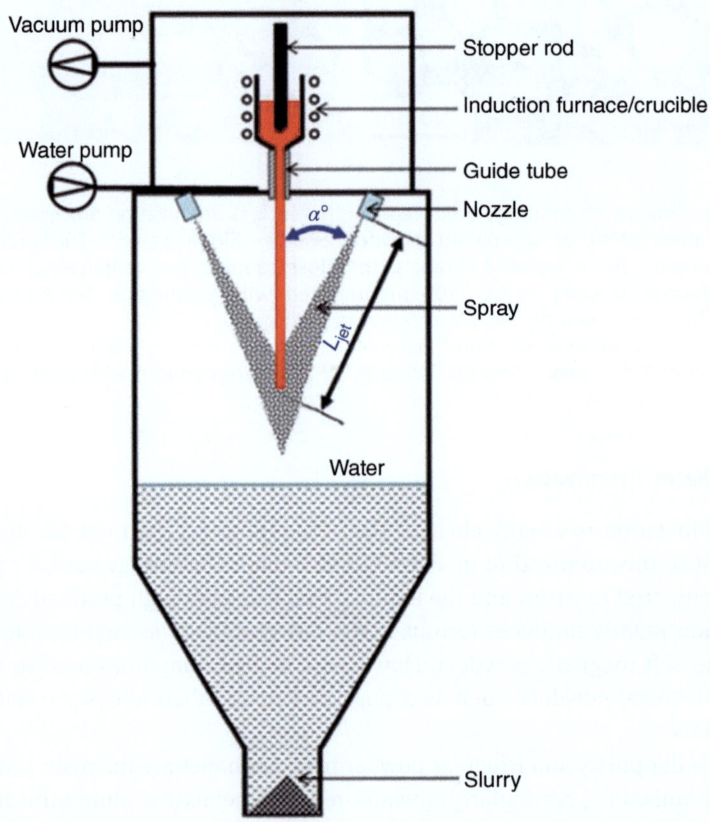

Figure 2.7 Schematic diagram of water atomization. The parameter α is the apex angle between the axis of the water jet nozzle and the molten metal stream, and L_{jet} is the length of the water jet. Source: Reproduced from Pasupathy et al. (2016)/with permission from Taylor & Francis.

symmetrically with respect to the axis of the metal stream. The free-fall configuration is adopted in common set-ups in which the molten metal exits from the bottom of the tundish and falls under gravity before colliding with the water jets.

The critical variables in water atomization include the thermophysical properties of the metal and water, geometric parameters, and process parameters. For example, thermophysical properties, such as the density and melting point of the metal, as well as the surface tension, viscosity, and degree of superheat of the molten metal, can be considered. Geometric parameters can involve the nozzle diameter, jet geometry, and the apex angle α between the axis of the water jet nozzle and the molten metal stream. Process parameters include the water pressure, water flow rate, melt flow rate, and metal stream length. The water pressure is believed to be the most effective parameter in determining the powder particle size.

Water atomization is characterized by the high density of the atomization medium (water), high cooling rate, and steam generation in the water/melt stream contact area. The high density of water results in an increase in linear momentum and kinetic energy. The steam film surrounding a metal droplet reduces the amount of heat that the droplet loses to the environment; once the film is broken, the cooling rate becomes substantially higher. The metal droplets are broken up by overheated compressed steam. The breakdown force of the steam film on a metal droplet depends on the physical conditions in the contact area, namely, the temperature and density of the steam, the water pressure, and the water flow rate.

Water atomization is primarily applied to generate powder with ferrous compositions. However, it is also employed to prepare powders from various nonferrous alloys, such as those of copper, nickel, and zinc. For example, Figure 2.8 shows a Fe–Si–B–C–P magnetic powder produced by water atomization (Liu et al. 2011). The powder exhibited both dendritic and spherical shapes because of different cooling conditions. The small Fe–Si–B–C–P droplets experienced a rapid cooling rate and solidified into irregularly shaped powder particles. In contrast, the large droplets had sufficient time to become spherical because of surface tension.

Compared to gas atomization, water atomization usually produces powder with irregular particle shapes, thereby resulting in the reduction of both the packing density and flowability of the powder. As shown in Figure 2.9a,b, gas-atomized 17-4 PH powder particles exhibited a more spherical shape and narrower particle size distribution than water-atomized 17-4 PH powder particles, which corresponded to a higher packing density and densification (Hausnerova et al. 2017). Figure 2.9c presents the torque values of both powders. A stable torque value indicates a homogeneous mixing and/or uniform dispersion of a powder within a binder that is often needed for torque measurement. The increase in torque value was marginal at lower solid loading values (up to 30 vol%) but significant at higher solid loading values. Such an increase resulted from more friction between the powder particles. The torque of the gas-atomized powder was higher than that of the water-atomized powder for small particles (sizes of 3–8 μm), while an inverse relationship could be observed for larger particles (sizes of 11–20 μm). The inverse relationship could be attributed to the irregular morphology of the water-atomized powder particles generating more friction than the gas-atomized powder particles.

Figure 2.8 SEM images showing the morphology of Fe–Si–B–C–P magnetic powder produced by water atomization. (a) Powder particles with different shapes, (b) dendritic powder particles, (c) a spherical powder particle, and (d) the internal microstructure of the powder. Source: Liu et al. (2011)/reproduced with permission from Elsevier.

Both powders displayed pseudo-plastic behaviors with a decrease in viscosity as the shear rate increased. The temperature sensitivity of the powders could be determined from the relationship between their activation energy and the solid loading at various shear conditions. Both the gas- and water-atomized powders exhibited similar activation energies at different shear rates (Figure 2.9d).

Inconel 625 powders produced by the two atomization processes were characterized by X-ray microtomography to determine their particle size distribution, morphology, and internal porosity at high resolution (Mostafaei et al. 2018). Cross-section images of the water- and gas-atomized powder particles were obtained from X-ray microtomography scanning, which revealed their morphology and internal porosity as shown in Figure 2.10a,b. These pores were generated during the solidification stage of the atomization process. As a result, the water-atomized powder particles exhibited higher porosity and circularity indices than their gas-atomized counterparts (Figure 2.10c,d).

2.1.3 Plasma Atomization

Plasma atomization, a powder production technique jointly developed by Pyro-Genesis Inc. and Hydro-Quebec, utilizes multiple plasma arcs to atomize

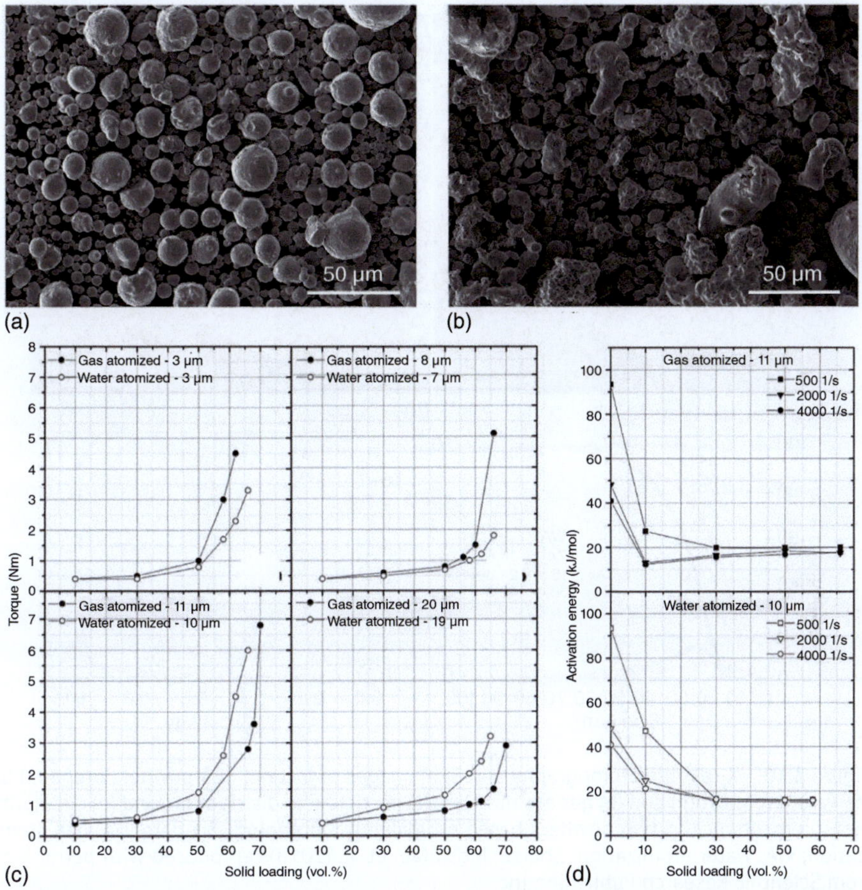

Figure 2.9 Morphology of 17-4 PH powders prepared by (a) gas atomization and (b) water atomization; (c) torque values of gas- and water-atomized powders; (d) activation energy of typical gas- and water-atomized powders produced at different shear rates. Source: Hausnerova et al. (2017)/reproduced with permission from Elsevier.

high-purity spherical powders, such as titanium, molybdenum, copper, and nickel alloy powders. Unlike high-pressure gas and water atomization, metal wires are selected as the feedstock material for plasma atomization. A schematic of the plasma atomization process is displayed in Figure 2.11 (Yolton and Froes 2015). A metal wire is fed through a hot zone created by the plasma torches, where it is melted and atomized by the plasma arc. The resultant droplets are then solidified to form spherical powder particles, with cooling rates ranging from 10^2 to 10^3 K/s.

During this process, a large amount of electrical energy is supplied to the plasma torches and converted to thermal energy, thereby generating a powerful jet of hot ionized inert gas. The high kinetic energy of the gas jet subsequently facilitates the atomization of the metal droplets. A unique feature of the plasma atomization process compared to gas atomization is that the produced powder particles are highly spherical and have almost no satellite particles attached to them (Figure 2.2b). This

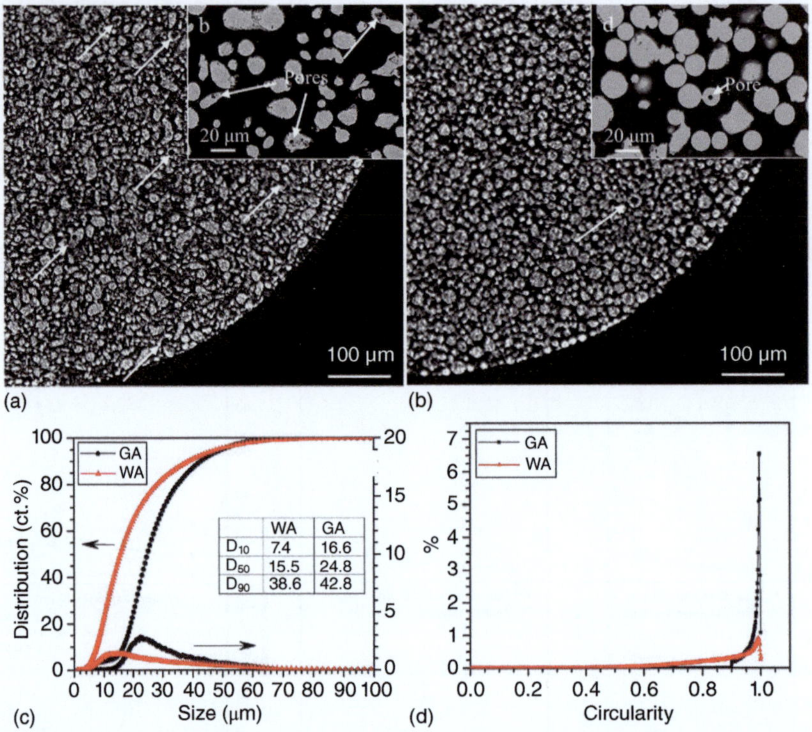

	WA	GA
D_{10}	7.4	16.6
D_{50}	15.5	24.8
D_{90}	38.6	42.8

Figure 2.10 X-ray microtomography and SEM images of Inconel 625 powders prepared by (a) water atomization and (b) gas atomization; (c) particle size distribution, and (d) circularity measurements of the two powders based on Malvern Morphologi 3D data. GA, gas atomization; WA, water atomization. Source: Mostafaei et al. (2018)/reproduced with permission from Scientific Research Publishing Inc.

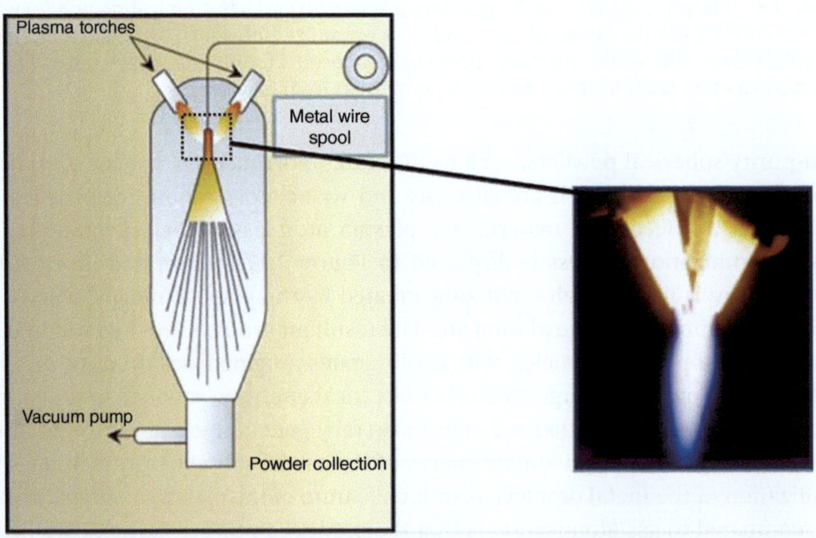

Figure 2.11 Schematic illustration of plasma atomization. Source: Yolton and Froes (2015)/reproduced with permission from Elsevier.

powder formation results from the extended hot zone generated by the plasma jets, which enables the droplets to remain in the molten state for a sufficiently long time so that powder particles with a high degree of sphericity can be obtained because of surface tension.

The particle morphology and particle size distribution of plasma-atomized powders can be influenced by the nozzle design, the velocity of the particles leaving the hot zone, and the distance across which the metal droplet solidifies. Compared to gas/water atomization, the plasma atomization process endows powders with superb characteristics for powder-based AM, such as increased flowability, higher density, and reduced porosity. The increased flowability of powders can be attributed to their relatively spherical morphology. Powders prepared by plasma atomization are also highly pure, as they have not been contaminated by any containers used to store the molten feedstock materials. Note that the feedstock material for plasma atomization can only be in the form of wires.

2.1.4 Plasma Rotating Electrode Process

The plasma rotating electrode process is a centrifugal atomization method that utilizes a rotating feedstock bar. It is a modified version of the rotating electrode process introduced in the early 1970s, in which the electric arc is substituted by plasma as the heat source. During centrifugal atomization, the molten metal flows onto the rotating disk/atomizer rotor and is sheared into a thin film at high speeds. Under the centrifugal force, the film is then flung from the edge of the disk/rotor in the form of droplets, ligaments, or sheets, depending on the rotating speed, the flow rate of the molten metal, and the atomizer rotor geometry. Gases with high thermal conductivity, such as helium, can be utilized to achieve the desired cooling rates. In addition to the rotating speed and plasma torch, another one of the key limitations of this technique is the high cost incurred, which is attributed to the enormous quantities of helium required.

A schematic of the plasma rotating electrode process is presented in Figure 2.12 (Chen et al. 2018; Sun et al. 2017a). Helium plasma is utilized to melt one end of a rapidly rotating electrode bar. The molten droplets are spun off from the surface of the rotating electrode by centrifugal forces and subsequently solidify to form flying spherical powder particles in a helium environment. Melting and atomization occur in a stainless-steel chamber at a positive gas pressure. The rotating electrodes are pre-alloyed bars that are 64 mm in diameter and are rotated at a rate of 20 000–30 000 rpm. The cooling rates are typically lower than 100 K/s, which are dependent on the gas atmosphere and the molten droplet size. The merits of the plasma rotating electrode process lie in its ability to produce highly pure and spherical powder particles.

Three spray models, namely, the direct drop formation, ligament disintegration, and film disintegration models, are illustrated in Figure 2.13 (Liu et al. 2018). A statistical equation of the actual spray model is given by

$$H = \frac{\mu^{0.17} Q \rho^{0.71} \omega^{0.6}}{\gamma^{0.88} D^{0.68}}, \tag{2.3}$$

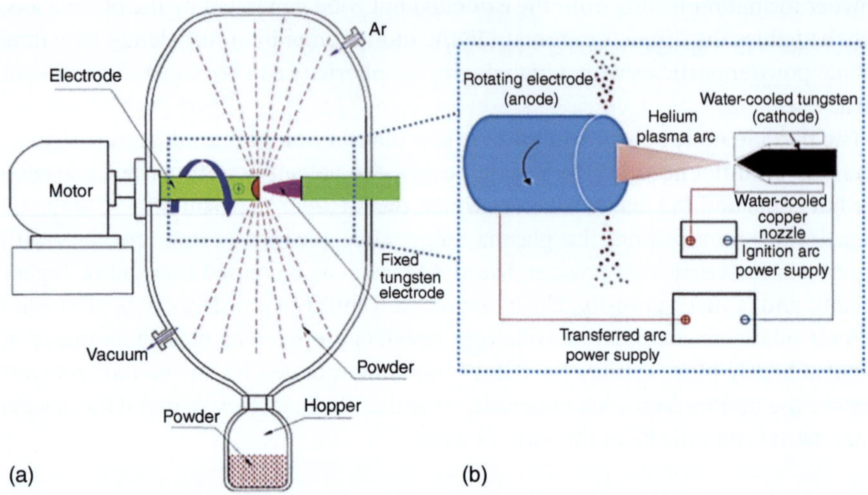

Figure 2.12 Schematic illustration of the plasma rotating electrode process. (a) Overview of the process. Source: Reproduced from Chen et al. (2018)/with permission of Elsevier. (b) Details of the process. Source: Reproduced from Sun et al. (2017a)/with permission of Springer Nature.

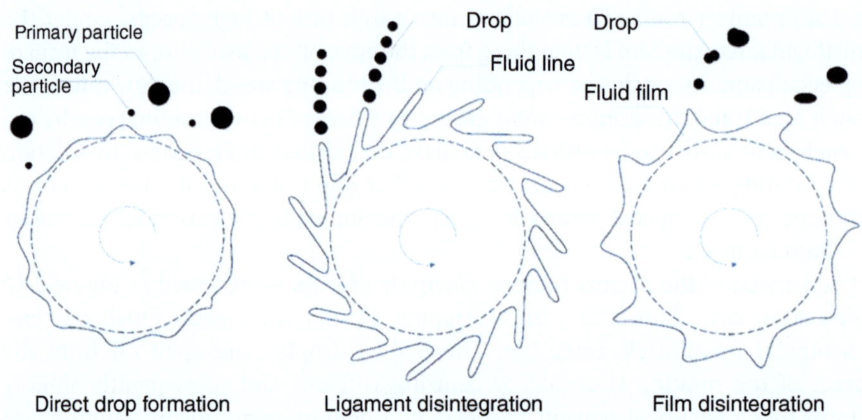

Figure 2.13 Representative centrifugal spray models of direct drop formation, ligament disintegration, and film disintegration. Source: Reproduced from Liu et al. (2018)/with permission from Elsevier.

where H is a dimensionless parameter, μ is the viscosity of the molten metal (Pa·s), Q is the melting rate of the metal (m^3/s), ρ is the density of the molten metal (kg/m^3), ω is the rotation speed of the rotor (rad/s), γ is the surface tension of the molten metal (N/m), and D is the diameter of the rotating bar (m). If H is smaller than 0.07, the main spray model is the direct drop formation model; if H is between 0.07 and 1.33, the main spray model is the ligament disintegration model; if H is greater than 1.33, the main spray model is the film disintegration model.

The plasma rotating electrode process is commonly applied to produce Fe-based, Ti-based, and Ni-based powders. Ti–6Al–4V powder particles produced by this method are spherical and possess good flow and packing features. The particle size distribution of a powder is determined by the material, electrode diameter, and rotation speed of the rotor. The average particle size D_{50} (μm) produced through the plasma rotating electrode process can be described by

$$D_{50} = \frac{K}{S\sqrt{D_e}}, \tag{2.4}$$

where K is a constant dependent on the surface tension and density of the material, S is the rotation speed (rad/s), and D_e is the electrode diameter (m). The production of fine powders can be boosted by increasing S and D_e. However, larger D_e and higher S values correspond to a higher accuracy requirement for the electrode dimensions to minimize out-of-balance forces. In addition, an increase in the melting rate leads to a remarkable rise in the yield of fine powders (e.g. with a size range below 45 μm).

Powders with different particle size distributions exhibit different levels of surface oxidation. Spherical pre-alloyed Nb–Si powders with particle size distributions of 45–75, 75–180, 180–250, and 250–380 μm were produced through the plasma rotating electrode process (Guo et al. 2017). An oxygen-enriched layer with a thickness of 5.14 nm was formed on the surface of the pre-alloyed powders. The oxygen content was discovered to decrease exponentially with the etching depth. Meanwhile, the external surface of the Nb–Si powder was Nb-depleted and (Ti, Si)-enriched, which could be attributed to the low thermodynamic stability of Nb_2O_5.

Figure 2.14 displays the morphology and particle size of Ti–6Al–4V powders produced by gas atomization, the plasma rotating electrode process, and plasma atomization (Chen et al. 2018). Compared to the Ti–6Al–4V powder particles prepared by gas atomization and plasma atomization, the powder prepared by the plasma rotating electrode process possessed a low argon content and porosity. Additionally, the powder prepared by gas atomization yielded more satellite particles than those produced by the plasma rotating electrode process and plasma atomization (for further details, refer to Section 2.1.1).

The particle size plays a vital role in determining the morphology and porosity of the atomized powder (Figure 2.14g,h). Powders produced by the plasma rotating electrode process possess fewer gas pores than gas-atomized and plasma-atomized powders because the molten droplets in the former process are formed by centrifugal forces instead of high-pressure gases. Moreover, large powder particles usually possess a higher gas content than small particles. This phenomenon can be attributed to the following two factors: first, during gas and plasma atomization processes, large droplets disintegrate into several small droplets that are concomitant with the breaking up of gas bubbles. Thus, no gas pores are formed. As a corollary, small powder particles possess fewer gas pores than large ones.

Second, during the plasma rotating electrode process, the surface tension of the original droplets from the electrode is lower than that of other droplets in flight. Consequently, the large original droplets can easily capture the surrounding argon gas. In comparison, the small droplets in flight undergo a high cooling rate, which

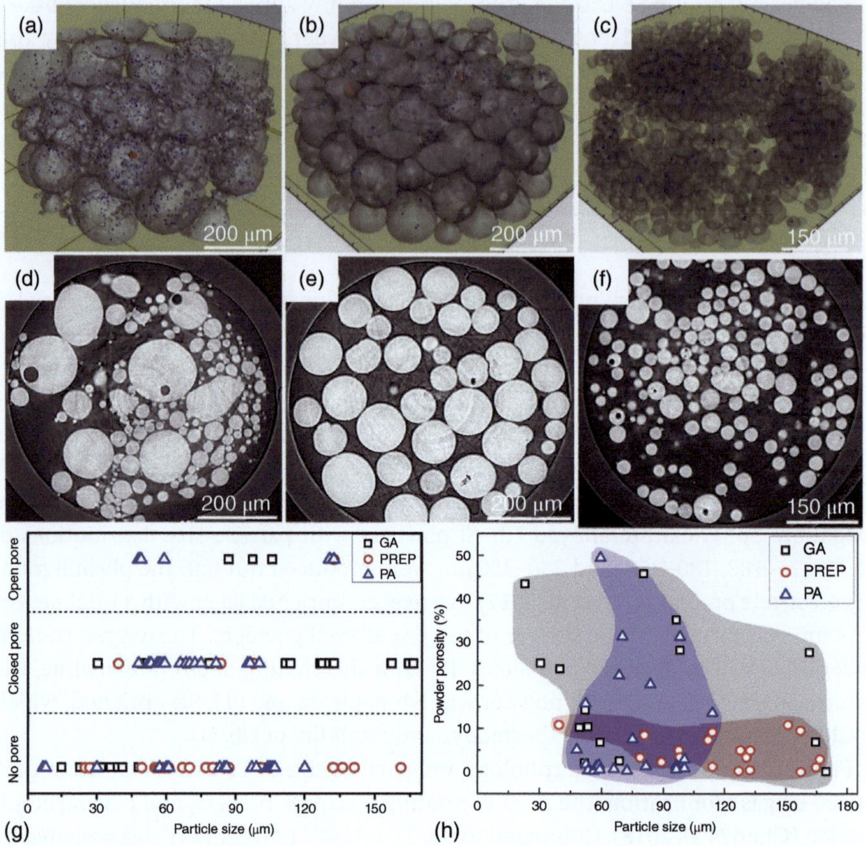

Figure 2.14 Morphology of Ti–6Al–4V powders produced by different atomization processes. (a–c) Computed tomography (CT) analysis showing the 3D reconstructed models of the powder from gas atomization, the plasma rotating electrode process, and plasma atomization, respectively, and (d–f) SEM images showing the cross-sectional morphology of the powders from the three processes, respectively. The particle size of the powders: (g) relationship between the particle size and pore characteristics; (h) statistic of porosity with respect to the particle size. GA, gas atomization; PA, plasma atomization; PREP, plasma rotating electrode process. Source: Chen et al. (2018)/reproduced with permission from Elsevier.

results in a higher surface tension and a lower capacity to entrap gases. Table 2.3 compares the properties of different powders prepared by the plasma rotating electrode process and gas atomization.

2.2 Mechanical Mixing

Mechanical mixing is another widespread method to produce metal powders for AM. Based on the principles of the solid-state mixing process, raw materials are converted into powder by external mechanical forces. Mechanical mixing processes for metal powder production can be divided into two categories, namely,

Table 2.3 Comparison of the plasma rotating electrode process and gas atomization for different powders.

Powder	Characteristics	Plasma rotating electrode process	Gas atomization
Ti–6Al–4V[a)]	Powder volume (mm^3)	2.329	3.877
	Pores volume (mm^3)	0.000399	0.002117
	Porosity (%)	0.017	0.055
	Average diameter (μm)	72	94
Ti–6Al–4V[b)]	Flowability (s/50g)	29.6	33.5
	Apparent density (g/cm^3)	2.59	2.38
	Average particle size (μm)	105	70
	Argon gas content (μg/g)	0.16	0.77
	Porosity (%)	0.08	0.20
Ni718[c)]	Powder flow rate (kg/h)	2.21	2.02
	Average dendrite spacing (μm)	5.0	3.8
	Laves phase (wt%)	15.82	12.46
	γ matrix (wt%)	2.51	5.82
TiAl[d)]	Dendritic microstructure (%)	52	0
	Rosette microstructure (%)	28	100
	Featureless microstructure (%)	20	0

a) (Ahsan et al. 2011);
b) (Chen et al. 2018);
c) (Zhong et al. 2016);
d) (Fuchs and Hayden 1992).

mechanical alloying and mechanical blending (Suryanarayana 2001). The main difference between them is that material transfer involves mechanical alloying but not mechanical blending. Table 2.4 summarizes the typical metal composite powders prepared through mechanical alloying and mechanical blending. Table 2.5 lists the advantages and disadvantages of the two processes.

Mechanical alloying is a high-energy milling process for producing fine metal composite powders with tailored microstructures (Arami and Simchi 2007; Krasnowski and Kulik 2008; Suryanarayana 2001). As schematized in Figure 2.15 (Han et al. 2020b), the process begins with mixing different powders in a specific ratio. Subsequently, the powder mixture is loaded into a grinding bowl containing the grinding medium (generally steel balls). The main part of this method involves fracturing and cold welding a mixture of metallic or nonmetallic powders in a highly activated ball milling process. When the powder particles are trapped between the colliding grinding balls, the materials experience large plastic strain under extreme hydrostatic compression, which results in their fracture. As the process continues, fracturing and welding repeatedly occur, which results in the successive refinement of the powder particles.

Table 2.4 Comparison of mechanical alloying and mechanical blending for the powder preparation of metal matrix composites in powder-based AM.

Process	Composite powder	Particle size
Mechanical alloying with high energy	SiC/Fe[a)]	Fe: 20 μm; SiC: 27 μm
	TiC/316L[b)]	316L: 45 μm; TiC: 50 nm/1 μm
	Al_2O_3/Al[c)]	Al: 17 μm; Al_2O_3: 10 nm
	TiC/Al–Si10–Mg[d)]	Al–Si10–Mg: 30 μm; TiC: 50 nm
	TiC/Ti[e)]	Ti: 23 μm; TiC: 50 nm
	TiB/Ti[f)]	Ti: 49 μm; TiB_2: 3.5–6 μm
	TiC/TiAl[g)]	Ti: 30 μm; Al: 16 μm; Graphite: 30 μm
	TiB_2/TiAl[h)]	TiAl: 28 μm; TiB_2: 3–5 μm
	MoTiAl/Al_2O_3/CNT[i)]	MoTiAl: 1–45 μm; Al_2O_3: 140 nm; CNT: 2-15 nm
Mechanical blending with low energy	HA/316L[j)]	316L: 28 μm; HA: <100 nm
	WC/Fe[k)]	Fe: 5–30 μm; WC: 5–15 μm
	WC/Ni718[l)]	Ni718: 15–45 μm; WC: 25–45 μm
	TiB_2/Ni625[m)]	Ni625: 11–45 μm; TiB_2: 5–12 μm
	SiC/Al12Si[n)]	Al12Si: 20–63 μm; SiC: 11–45 μm
	TiB_2/Al–Si10–Mg[o)]	Al–Si10–Mg: 19.3–74.8 μm; TiB_2: 2–5 μm/58 nm
	TiB_2/Ti–6Al–4V[p)]	Ti–6Al–4V: 20–70 μm; TiB_2: 2.71 μm
	B_4C/Ti[q)]	Ti: 16–59 μm; B_4C: 500 nm
	HA/Ti[r)]	Ti: 30.8 μm; HA: 50 nm

a) Song et al. (2014);
b) AlMangour and Grzesiak (2016);
c) Han et al. (2016);
d) Gu et al. (2015);
e) Gu et al. (2012);
f) Attar et al. (2014);
g) Gu et al. (2009);
h) Li et al. (2017);
i) Zhou et al. (2018b);
j) Wei et al. (2015);
k) Gu et al. (2018);
l) Rong and Gu (2016);
m) Zhang et al. (2016);
n) Li et al. (2016);
o) Aversa et al. (2017);
p) Cai et al. (2019);
q) Han et al. (2020);
r) Han et al. (2018).

Table 2.5 Summary of the advantages and disadvantages of mechanical alloying and mechanical blending.

Process	Advantages	Disadvantages
Mechanical alloying with high energy	• Fast and simplified • Decent dispersion of reinforcements	• Particle deformation • Uncertainty in tailoring the powder particle size • Poor powder flowability • Introduction of undesirable phases and contamination
Mechanical blending with low energy	• Fast and simplified • Nearly retains the shape, particle size, and distribution of the powder • Decent powder flowability	• Poor dispersion for nano-sized powder particles • Limited to the preparation of alloy powders

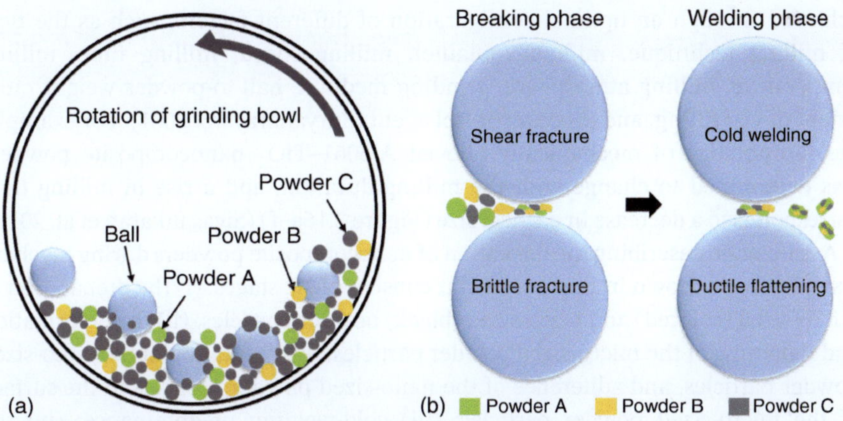

Figure 2.15 Schematic of mechanical alloying. (a) Process, and (b) mechanism. Source: Reproduced from Han et al. (2020)/with permission from John Wiley & sons.

Shaker milling, planetary ball milling, and attritor milling are typical high-energy milling techniques for mechanical alloying (Neikov et al. 2009). Shaker milling is the most commonly utilized technique for laboratory experiments and alloy screening, in which 0.2–20 g of powder is milled at a time. A typical setup for shaker milling comprises a vial containing grinding balls and powders. The vial is fixed by a clamp that swings back and forth several thousand times per minute. The materials used for the vials are stainless steel, hardened steel, zirconia, and alumina. The standard machine model is the SPEX mill, which is manufactured by SPEX CertiPrep.

Planetary ball milling can process a few hundred grams of powder at a time. In contrast to shaker milling, the vials are installed on a rotating support disk and rotated about their axes. The centrifugal forces produced by the vials and the supporting disk are opposite to one another because of their opposite rotating directions. As a result, friction and impact effects are induced, and they act on the powder and grinding balls in the vials. The materials used for the vials and grinding balls include chrome steel, chromium–nickel steel, agate, silicon nitride, sintered corundum, zirconia, tungsten carbide, and plastic polyamide. The energy transferred during planetary ball milling is lower than that during SPEX milling due to the lower impact frequency in the former technique.

An attritor mill consists of a vertical drum with a series of impellers. By setting the impellers at right angles to one another, they activate the ball charge, thereby reducing the powder particle size. Such size reduction is derived from the impact between (i) the grinding balls, (ii) the grinding balls and the container wall, and (iii) the grinding balls, the agitator shaft, and the impellers. As a result, large quantities of powder (about 0.5–40 kg) can be processed in one milling cycle. In addition, the speed of the grinding media in attritor mills is lower than in SPEX mills, and thus the energy transferred during attritor milling is relatively low.

In mechanical alloying, the required microstructure of the powder can be achieved through an optimal configuration of different factors, such as the type of milling technique, milling container, milling speed, milling time, milling temperature, milling atmosphere, grinding medium, ball-to-powder weight ratio, extent of vial filling, and process control agent (Suryanarayana 2001). For example, the morphology of mechanically alloyed Al6061–TiO_2 nanocomposite powders was determined to change with the milling duration, and a rise in milling time corresponds to a decrease in particle size (Figure 2.16a–f) (Sivasankaran et al. 2010).

A schematic describing the formation of nanocomposite powders during mechanical alloying is shown in Figure 2.16g. It consists of six stages: (i) the preparation of micro-sized (colored) and nano-sized (black) powder particles; (ii) the deformation and flattening of the micro-sized powder particles, fragmentation of the nano-sized powder particles, and adherence of the nano-sized powder particles to the surface of the micro-sized powder particles; (iii) cold-welding predominance; (iv) the formation of equiaxed powder particles; (v) random welding orientation and the formation of necks; and (vi) attaining equilibrium among breaking-up, cold welding, and fracturing.

At the initial stage of milling, the apparent density of powders decreases continuously and reaches a minimum value after 10 hours. Then, it begins to increase with the milling time until a plateau is reached. This change in the apparent density is attributed to the cold welding of powder particles during the initial milling stage and the fracturing of powder particles during the later stages. The apparent density first reduces and then increases because of the fragmentation and morphological variations of the powder particles, which are both induced by cold welding.

Mechanical alloying applies to a wide range of pure metals and alloys, particularly refractory metal materials. It is also useful for developing new materials, such as metal matrix composites, HEAs, metallic glasses, and intermetallic compounds. For example, the milling time is crucial in determining the phase composition, grain size, and lattice strain of HEA powder particles. Increasing the milling time weakens

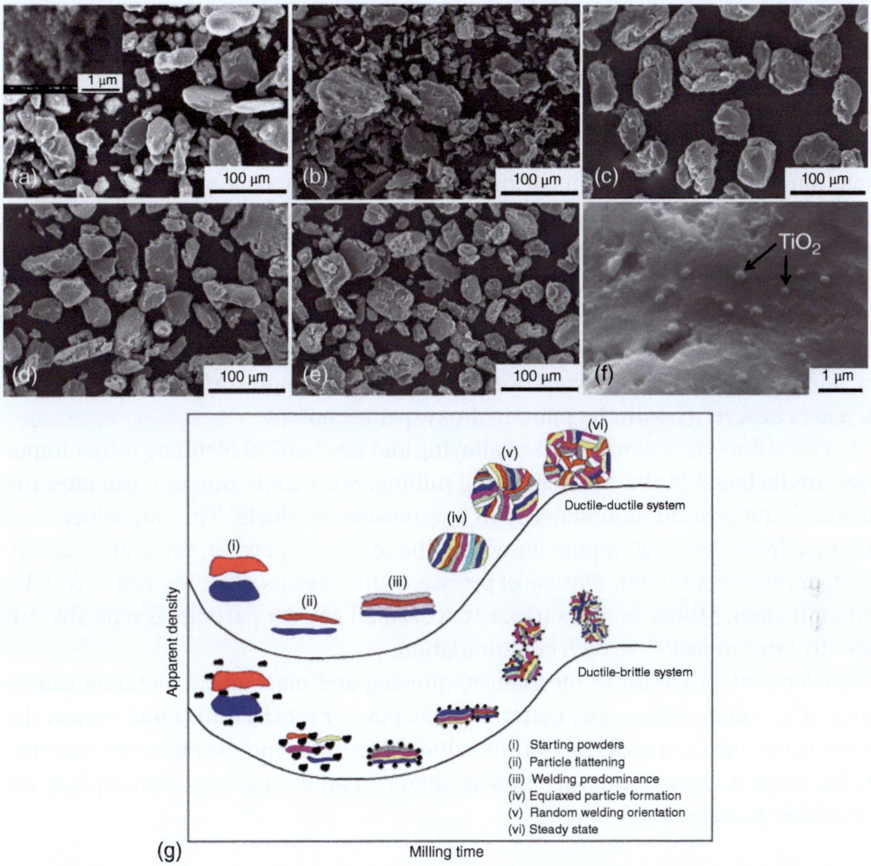

Figure 2.16 SEM images displaying the morphology of Al6061–TiO_2 composite powders with different milling durations. (a) 1 hour, (b) 5 hours, (c) 10 hours, (d) 20 hours, and (e) 40 hours, in which (f) is the enlarged image of (e). (g) Schematic illustration of the relationship between the milling time, morphology, and apparent density of different types of powders. Source: Sivasankaran et al. (2010)/reproduced with permission from Elsevier.

the diffraction peaks of $Al_{0.4}CoFeNiSi_{0.4}$ powder (Zhang et al. 2018), thereby posing difficulty in identifying the phase composition of HEA powders. A prolonged milling time also reduces the grain size and increases the lattice strain within the powder particles because of the increased grain boundary fraction and lattice deformation. Additionally, the repetition of fracturing and welding during the process accelerates the formation of solid-solution phases. Consequently, elements with low melting points or brittle crystal structures are dissolved preferentially.

In addition, due to the intensive impaction of the powder particles, mechanical alloying allows powders of nano sizes and multiple phases (e.g. metastable crystalline phases, quasi-crystalline phases, and supersaturated solid phases) to be achieved. However, the powder particles usually exhibit irregular shapes, which are detrimental to their flowability for AM. This problem can be resolved by employing the plasma spheroidization technique to endow the powder particles with a spherical shape for better printability (see Section 2.4.1 for further details).

Mechanical blending requires lower energy and a shorter time than mechanical alloying for powder preparation. Additionally, material transfer hardly occurs when blending powders with different shapes. Therefore, mechanical blending becomes increasingly preferable for preparing metal matrix composite powders. However, due to their high aspect ratio, dispersing nanoparticles via mechanical blending is problematic because the nanoparticles can agglomerate under strong van der Waals forces. For example, Figure 2.17 shows the powder mixture of titanium and hydroxyapatite prepared by mechanical blending, whereby titanium powder particles are covered with hydroxyapatite nanoparticles (Han et al. 2017). The agglomeration of the nano-hydroxyapatite powder occurs because of its high surface activity, which increases in severity with the nano-hydroxyapatite content.

A critical drawback of mechanical alloying and mechanical blending is that impurities are included in the powder during milling. Such contamination can alter the chemical composition and stability of the powder products. The impurities may originate from chemical impurities within the feedstock powder, the milling equipment, or the environment. The use of process control agents may also lead to powder contamination. Moreover, the surface area of small powder particles is typically sufficiently large to facilitate such contamination.

Powder contamination in mechanical alloying and mechanical blending can be reduced by using: (i) the same material as the powder for the mill container and the grinding medium, (ii) materials for the grinding medium and the mill container that are harder and stronger than the powder, and (iii) an inert gaseous atmosphere for the milling process.

2.3 Reduction Process

Reducing metal oxides and metallic salts is one of the most common methods of producing metal powders. Using ores and industrial metallurgical waste as raw materials is an economically viable strategy. The advantages of the reduction process include a simple operation procedure, easy control of technological parameters, high production efficiency, and low cost, thereby rendering it suitable for the industrial production of metal powders. Below, four typical reduction processes are introduced, namely, the hydride–dehydride process, oxide reduction, chloride reduction, and carbonyl reactions.

2.3.1 Hydride–Dehydride Process

Unlike the atomization processes, the hydride–dehydride process includes crushing, milling, and screening to process larger blocks of metal feedstock into finer powder particles without melting them. Such a technique is particularly suitable for producing powders of materials that are brittle when exposed to hydrogen gas, such as titanium, zirconium, vanadium, and tantalum. The hydride–dehydride process is

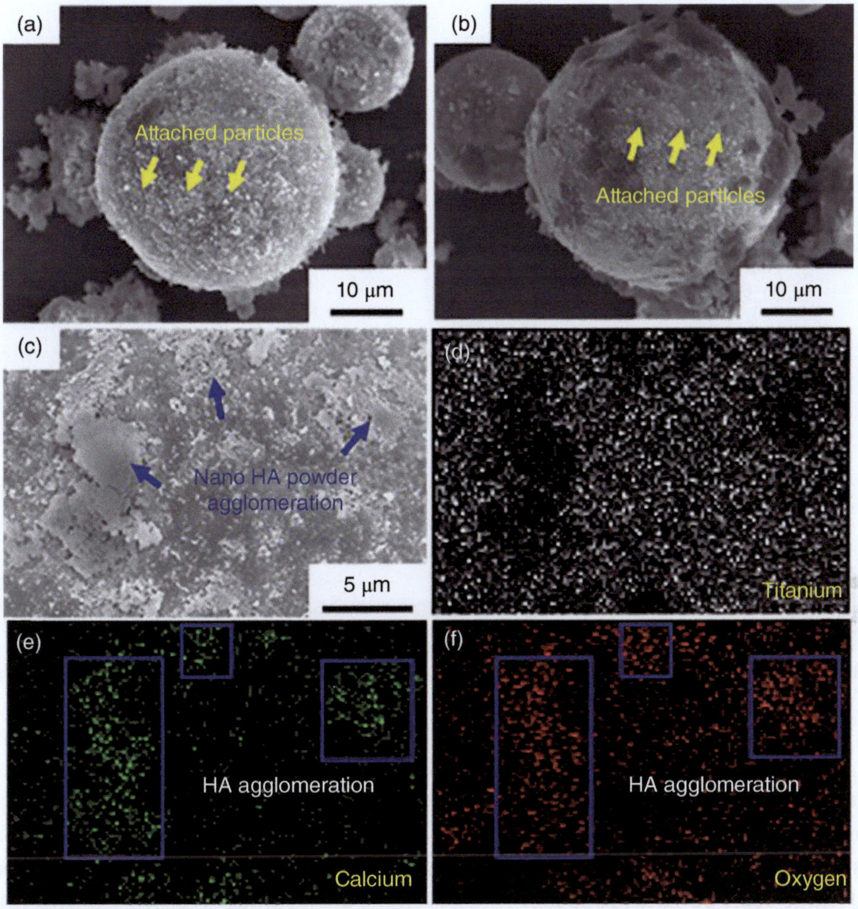

Figure 2.17 SEM micrographs of mechanically blended powders. (a) Titanium–2 wt% hydroxyapatite, (b) titanium–5 wt% hydroxyapatite, and (c) the surface of a titanium–5 wt% hydroxyapatite powder particle. (d–f) EDS mapping of the powder particle showing Ti, Ca, and O, respectively. Source: Han et al. (2017)/reproduced with permission from Elsevier.

most adopted for synthesizing powders of titanium and its alloys, such as commercially pure titanium and Ti–6Al–4V. A schematic of the hydride–dehydride process for obtaining titanium or Ti–6Al–4V powder is illustrated in Figure 2.18 (Agripa and Botef 2019). Titanium or Ti–6Al–4V hydrides are first produced in a hydride unit filled with hydrogen gas through heating at a high temperature of 300–700 °C. The hydrogenated titanium blocks are crushed, milled, and sieved to produce the powder. Subsequently, the powder is returned to the hydride unit for the dehydride process, where it is reheated within a high vacuum environment at about 350 °C to remove any excess hydrogen (McCracken et al. 2011). After the hydride–dehydride process, the titanium or Ti–6Al–4V powder becomes very brittle and can be ground into fine powder particles. Mechanical crushing in ball mills with screening procedures enables the manufacturing of powder with specific particle sizes.

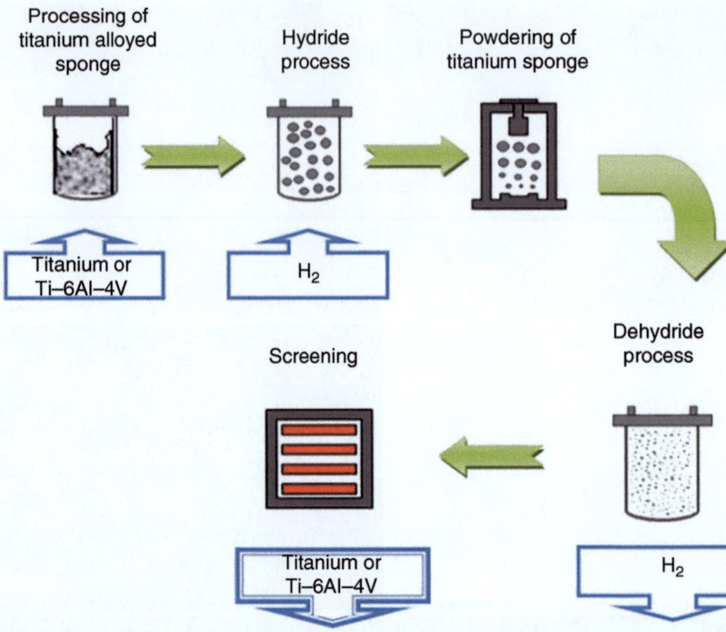

Figure 2.18 Schematic of the hydride–dehydride process in producing titanium or Ti–6Al–4V powder. Source: Agripa and Botef (2019)/IntechOpen/CC BY-3.0.

The hydride–dehydride process is a feasible approach for tailoring the particle size and chemical composition of powders for different requirements and applications. In addition, compared to the atomization processes, the hydride–dehydride process is a simple procedure with low cost and energy consumption, and it can be utilized for large-scale powder production. However, the disadvantages associated with this process include the underlying challenges of hydrogen purification and the high reactivity of titanium toward the residual hydrogen, which results in fragile final products. Table 2.6 compares titanium powders prepared by the hydride–dehydride and plasma rotating electrode processes (Lou et al. 2015). The powder particles of the former approach exhibit a smaller size, lower density, higher oxygen content, and more irregular shapes than the latter.

Table 2.6 Comparison of titanium powders prepared by the hydride–dehydride and plasma rotating electrode processes.

Process	Oxygen content (wt%)	D_{10} (μm)	D_{50} (μm)	D_{90} (μm)	Apparent density (g/cm^3)	Tap density (g/cm^3)
Hydride–dehydride	0.33	20.84	47.69	91.03	1.42	2.08
Plasma rotating electrode process	0.09	107.24	153.48	218.67	2.74	2.95

Source: Reproduced from Lou et al. (2015)/with the permission of Springer Nature.

The oxidation of hydride–dehydrided powders may occur due to the raw materials and the hydride–dehydride process. The oxygen content in powder particles primarily depends on the specific surface area of the powder. Once a powder is exposed to air, the large specific surface area instantly causes the powder to oxidize, and thus a high oxygen content is obtained. For hydride–dehydrided titanium powder, a Ti–O passivation layer could be induced in the form of TiO_2, Ti_2O_3, and TiO, where the oxygen content in the oxide layer could reach up to 23.1 wt% (Zhang et al. 2021). When the Ti–O layer was produced on the powder surface, the oxygen content could not decrease unless the passivation layer was deoxidized by certain reductants (e.g. calcium vapor).

Titanium powder produced by the hydride–dehydride process is usually modified to obtain a cost-effective powder with good flowability. For example, hydride–dehydrided titanium powder was modified by the ball milling process (Xu et al. 2019). The powder was milled for two to six hours using a planetary ball milling machine, and the production efficiency of the powder was found to be 65–70%. As shown in Figure 2.19a, the powder particles exhibited irregular shapes and sharp edges before milling, but they became near-spherical with sharp edges after two hours of milling (Figure 2.19b). When the milling time was increased to four hours, the powder particles were near-spherical with no apparent angularity (Figure 2.19c). However, the powder particles became flatter after six hours (Figure 2.19d). The average particle size of the modified powder was found to be about 30 μm after ball milling and sieving. As the milling time was extended, the apparent density and tap density of the powder increased from 1.26 to 1.64 g/cm^3 and from 1.88 to 2.10 g/cm^3, respectively.

The hydride–dehydride process also allows for the preparation of metal matrix composite powders. For example, a Ti/TiC composite powder could be prepared by directly introducing TiH_2 and TiC powders to pure titanium powder during the dehydride process (Peillon et al. 2015). The TiH_2 powder played a vital role in controlling the powder particle size. Moreover, the hydride–dehydride process is the most common approach for preparing powders containing Ti, Zr, and Hf elements. Zr–Ti–Nb foams with different levels of Nb content were fabricated from powders produced by the hydride–dehydride process. Such foams exhibited excellent biocompatibility with subcutaneous and bone tissues (Maya et al. 2012).

Apart from the standard hydride–dehydride process, a hydride–milling–dehydride system has been developed to produce zirconium powder (Figure 2.20a) (Kelley and McDeavitt 2013). The design of this process was based on the requirement to dehydride zirconium hydride at a temperature of 1000 °C. Figure 2.20b,c show the morphologies of the powder particles under different milling times. It was demonstrated that an increase in the milling time reduced the average powder particle size and improved the uniformity of the powder particle size.

2.3.2 Oxide Reduction

The oxide reduction process can be conducted to prepare pure metal powders, such as iron, cobalt, copper, and tungsten, through the reduction reactions of metal oxide ores by gaseous hydrogen, carbon monoxide, or dissociated ammonia. It is a simple

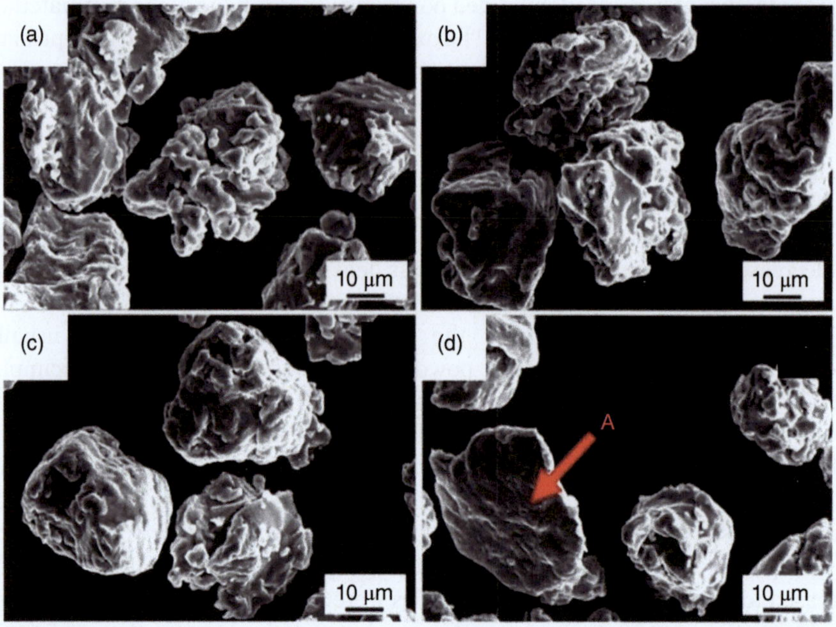

Figure 2.19 SEM images displaying the morphology of hydride–dehydrided titanium powders produced after various milling durations. (a) 0 hour, (b) 2 hours, (c) 4 hours, and (d) 6 hours. The symbol "A" indicates a relatively flat surface. Source: Xu et al. (2019)/reproduced with permission from Elsevier.

process that produces powders with a well-controlled particle size distribution, good formability, and high sinterability by treating the oxide ores and tailoring the reduction atmosphere and milling time.

The reduction reactions for metal oxides can be described by the following equations:

$$MO + H_2 \rightarrow M + H_2O, \tag{2.5}$$

$$MO + CO \rightarrow M + CO_2, \tag{2.6}$$

$$MO + C \rightarrow M + CO, \tag{2.7}$$

where M denotes the metal element. Equation (2.6) describes an indirect carbothermic reduction at low temperatures, and Equation (2.7) represents a direct carbothermic reduction at high temperatures.

The electrolysis process consists of key components such as an electrolyte, a power supply, and electrodes. Metal elements are deposited on the cathode of an electrolytic cell in the form of powder through the electrolysis of the electrolyte, which is an aqueous salt solution. The composition and viscosity of the electrolyte, current density, and temperature will affect the deposition process. The merits of the process are the high purity of the powder produced, good control over the powder particle size, and the ability to produce ultrafine powder. However, the high power consumption

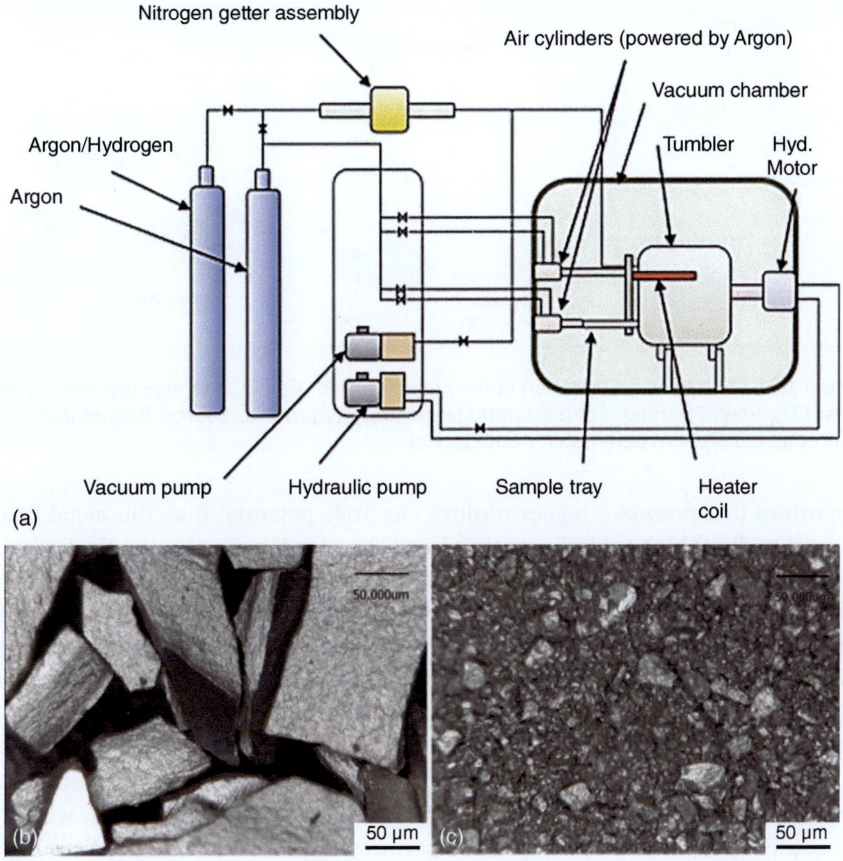

Figure 2.20 (a) Schematic of a hydride–milling–dehydride system. Resultant morphology of zirconium powder after milling for (b) 45 minutes and (c) 24 hours. Source: Kelley and McDeavitt (2013)/reproduced with permission from Elsevier.

of this process corresponds to high cost, and thus it is not widely employed in powder production.

A novel electrolytic technique, named the Fray–Farthing–Chan Cambridge process, has been developed to produce low-cost titanium powder with high purity (Fray et al. 2009). This process utilizes molten salt electrolysis to extract metals from their oxides. Figure 2.21 shows a graphical representation of the process. The titanium oxide acts as a cathode, and graphite acts as an anode. The two electrodes are immersed in a molten salt bath at a temperature of 800–1000 °C. Then, oxide ions depart from the titanium oxide cathode and are carried across the molten salt to the anode, where they react to form carbon dioxide and carbon monoxide. Once the carbon dioxide and carbon monoxide bubbles have dispersed, pure titanium is retained. This novel extraction technique can be applied to produce spherical titanium powder for AM.

Oxides, chlorides, and zirconates are commonly employed in the electrolysis process. The main requirements for electrolytes include: (i) the exclusion of metal

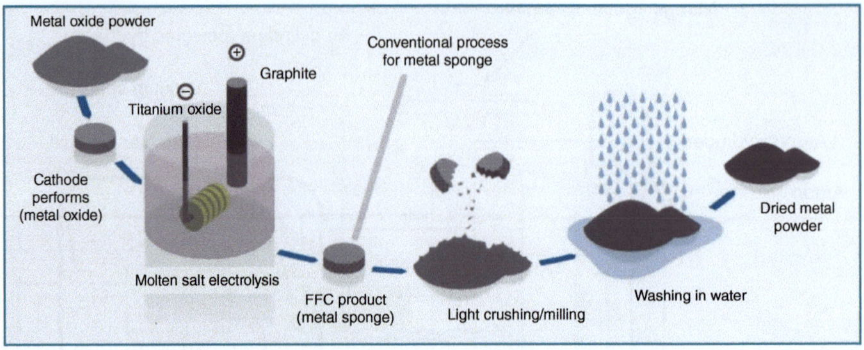

Figure 2.21 Schematic illustration of the Fray–Farthing–Chan Cambridge electrolysis process. FFC, Fray–Farthing–Chan Cambridge electrolysis process. Source: Reproduced from Fray et al. (2009)/with permission from Elsevier.

impurities that possess a higher positive electrode potential than the metal being electrolyzed; (ii) high solubility of the electrolyte with the metal being electrolyzed; (iii) low viscosity and high fluidity of the electrolyte at the temperature of the electrolysis process, which is advantageous for the discharge of the anodic gas and the uniformity of the electrolyte composition; (iv) a low melting point of the electrolyte to reduce the required electrolytic temperature; (v) high conductivity of the molten electrolyte; and (vi) high chemical stability of the electrolyte in both solid and liquid states.

The electrolysis process has been applied to produce copper, chromium, and manganese powders. In such cases, a dense and brittle deposit is produced, which is subsequently crushed to form powder particles. The electrolysis process is capable of yielding high-purity powders with controllable sizes. However, it is limited to producing pure metal powders, as the production of alloyed powders may require additional measures, such as washing, drying, reducing, annealing, and crushing of the powders.

2.3.3 Chloride Reduction

The chloride reduction process is employed to produce metal powders by thermally decomposing metal chlorides. The TiRO^{TM} process, a relatively new method developed by CSIRO for producing pure titanium powder, is introduced herein (Doblin et al. 2012). This cost-effective process produces pure titanium in a continuous fluidized bed in which $TiCl_4$ reacts with magnesium powder. The process flow diagram can be divided into two stages. The first stage involves the chemical reactions between $TiCl_4$ and magnesium powder particles in a fluidized bed to produce small titanium metal powder particles, which are homogeneously dispersed with spheroidal $MgCl_2$ powder particles. Overall, $TiCl_4$ reacts with magnesium and is thermally reduced to a solid $MgCl_2$/Ti composite in the fluidized bed reactor. The fluidized bed reactor employs argon as a sweeping gas to feed liquid $TiCl_4$ and magnesium powder to the bed.

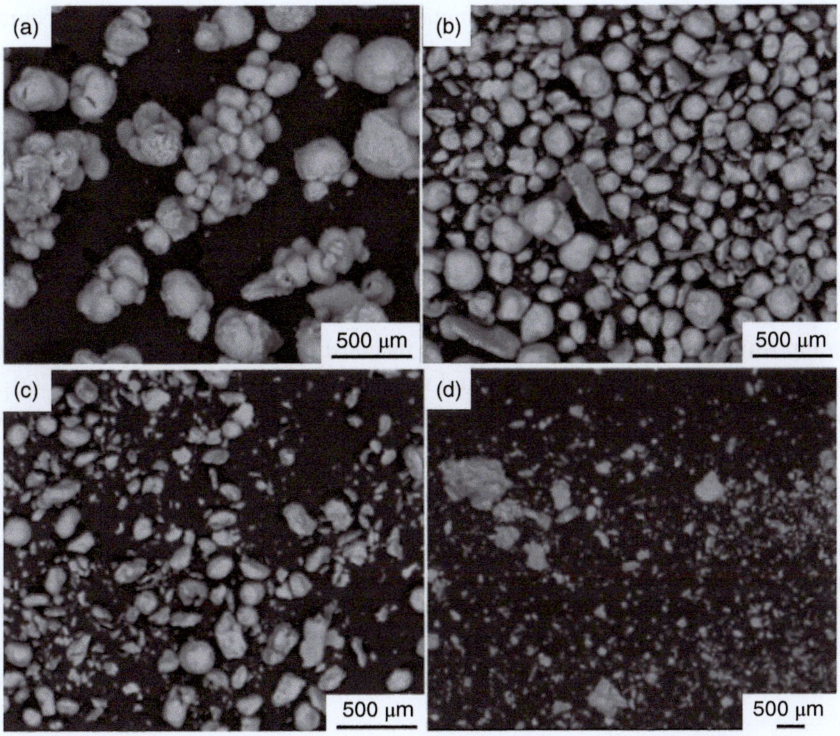

Figure 2.22 SEM images showing the morphology of titanium powder produced by the TiRO^{TM} process after different milling times. (a) 0 minutes, (b) 0.5 minutes, (c) 2 minutes, and (d) 10 minutes. Source: Doblin et al. (2013)/reproduced with permission from Trans Tech Publications Ltd.

The second stage includes a vacuum distillation process in which high-purity titanium powder is separated from the $MgCl_2/Ti$ composite. The vacuum-distilled powder is then grounded and screened, and titanium powder particles up to a maximum size of 125 mm are selected. The process has low energy consumption, high powder yield, and low labor requirements. As a result, the TiRO^{TM} process can produce titanium powder that meets commercially pure titanium grade 3 standards.

However, the as-prepared titanium powder particles after vacuum distillation are still too large to be used for powder bed fusion AM. Fortunately, a high-shear milling process can reduce the powder particle size (Doblin et al. 2013). After milling for 0.5 minutes, the powder particles were more densely packed, and some powder particles were observed to deform into platelets. In addition, with an increase in milling time, the number of smaller powder particles increased while more powder particles were flattened (Figure 2.22). However, a long milling time may be disadvantageous because the agglomeration of powder particles may occur.

2.3.4 Carbonyl Reactions

The carbonyl reaction is extensively utilized for chemical deposition. It involves the high-pressure and high-temperature treatment of metals by carbon monoxide, which produces metal carbonyls. The metal carbonyls then undergo thermal decomposition to produce the metal powders and recycle the carbon monoxide. The process can be described by the following reaction:

$$M_xB_z + yCO \rightarrow zB + M_x(CO)_y \rightarrow xM + yCO, \quad (2.8)$$

in which M and B represent the metal and ballast (e.g. oxygen, salt residue, and impurities), respectively; x, y, and z refer to the coefficients that depend on the metal species; and $M_x(CO)_y$ is the carbonyl of the metal M.

The carbonyl reaction is a popular chemical process for producing nickel and iron powders. It can produce micro- and nano-sized powders because the thermal decomposition of a metal carbonyl refines its constituent metal. In such a process, nickel reacts with carbon monoxide and forms gaseous nickel carbonyl ($Ni(CO)_4$), which can be decomposed back into refined nickel metal at a moderate temperature with the recovery of the carbon monoxide. Fine nickel powder can then be prepared through thermal shock decomposition. Meanwhile, iron carbonyl from iron powder/scraps can be thermally decomposed to prepare spherical iron powder with an average particle size of 10 μm (Antony and Reddy 2003), which is the ideal material for fabricating chemical power supply plates and can also be employed as catalysts. However, additional efforts are still required to determine the applicability of carbonyl-derived powders for metal powder–based AM.

2.4 Powder Modification

In addition to the commonly used powder preparation processes introduced in Sections 2.1–2.3, there are preparation processes aimed at modifying metal powders to have a high degree of sphericity, such as plasma spheroidization, granulation–sintering–deoxygenation, and fluidized-bed granulation. In particular, the plasma spheroidization process has been increasingly applied to produce spherical metal powders for powder-based AM.

2.4.1 Plasma Spheroidization

Plasma spheroidization is one of the new advanced techniques developed for producing metal powders for AM. This technique allows powder particles to be reshaped into the ideal spherical shape, and the resultant powder particles usually exhibit comparable sphericity to gas-atomized powder particles. Therefore, it is extremely suitable for treating nonspherical and irregular powders with low contamination under high temperatures (3000–10 000 K) and rapid quenching rates (about 10^6 K/s). Plasma spheroidization can effectively produce high-purity powders with high degrees of sphericity and small particle sizes. In addition to the ability to alter the particle shape, plasma spheroidization also reduces the internal porosity of a powder, enhances its flowability and purity, and increases its apparent density. Most importantly, this process can be conducted during the post-treatment of recycled AM powders.

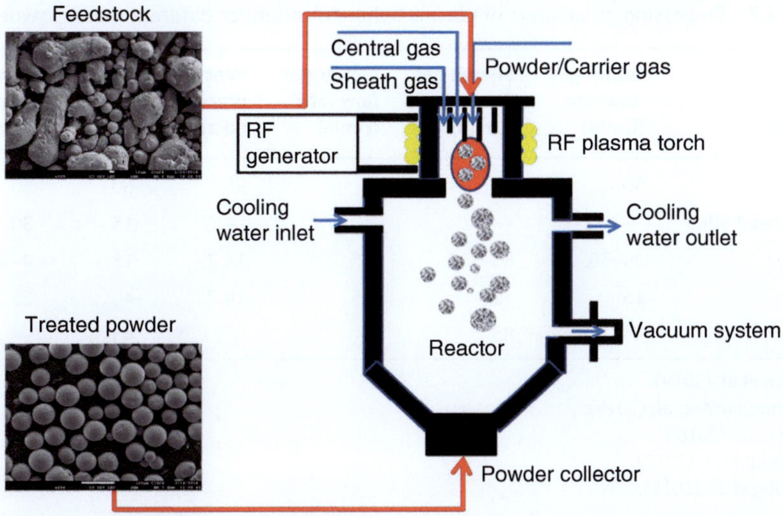

Figure 2.23 Schematic illustration of induction plasma spheroidization. RF stands for radio frequency. Source: Wang et al. (2017)/reproduced with permission from Elsevier.

Nevertheless, preparing fine spherical alloy powder with low oxygen content in a cost-effective manner still poses a great challenge for plasma spheroidization. The level of impurity in the spheroidized powder mainly depends on the feedstock. Another potential issue is the rapid evaporation of alloying elements with low melting points at plasma temperatures. Generally, the plasma spheroidization process is not utilized in large-scale powder production due to its low productivity and high cost.

As schematically illustrated in Figure 2.23, the plasma device primarily consists of plasma generating, powder feeding, and powder collecting systems, reactor and cooling chambers, and a cyclone separator (Wang et al. 2017). An induction plasma torch is formed via an electromagnetic coupling of the input electrical energy and the discharge gas medium. An oscillating magnetic field is formed by an alternating current of radio frequency range passing through a suitable coil. This magnetic field redirects an ionized gas load flowing within the discharge cavity, which provides ohmic heating to maintain the plasma temperature. Based on this principle, gases (such as argon) are purged and ionized to generate a high-temperature plasma torch. While passing through the plasma torch, the irregularly shaped powder particles are melted to form metal droplets. These droplets finally form spherical powder particles through rapid solidification.

The plasma spheroidization process includes four stages. The first stage is the pretreatment of raw powder. A vacuum drying cabinet is utilized to reduce moisture within the raw powder. The dried powder is then sieved to optimize the particle size distribution. The second stage is the generation of plasma after the system is purified via air evacuation and argon gas purging. The third stage is the spheroidization of the pretreated raw powder, which is fed into the plasma reactor chamber. When the powder passes through the high-temperature zone, it is completely melted and forms

Table 2.7 Processing parameters of plasma spheroidization for different metal powders.

Powder	Center gas flow rate (l/min)	Sheath gas flow rate (l/min)	Carrier gas flow rate (l/min)	Reactor pressure (kPa)	Plasma power (kW)	Powder feed rate (g/min)
Ta[a)]	30	85	4	95	60	50
Nb-based alloy[b)]	—	—	—	—	15	30
W[c)]	20–30	85–115	3.5	14.7	15	4–25
W[d)]	15	55	5.5	14.7	40	25
TiAl[e)]	—	85	—	—	45	—

a) Qin et al. (2019);
b) Goncharov et al. (2019);
c) Zi et al. (2018);
d) Wang et al. (2017);
e) Tong et al. (2015).

spherical droplets. The dispersing gas converts these droplets into smaller droplets, which are then cooled and solidified to form regular spherical powder particles. Finally, the spheroidized powder is gathered by the powder collector.

Plasma spheroidization applies to many materials, such as titanium-based, aluminum-based, and niobium-based alloys, pure titanium, and pure tungsten. Feedstock materials used in plasma spheroidization can be powders produced by the hydride–dehydride process, the electrolysis process, or any other process that produces irregularly shaped powder particles. Table 2.7 lists the processing parameters of plasma spheroidization for different metal powders.

As one of the most critical processing parameters, the powder feed rate influences the heat absorption and the duration for which the powder particles remain in the flame area. The latter effect is crucial because the powder particles cannot be melted outside the high-temperature area. The influence of the powder feed rate on the morphology and size distribution of tungsten powder particles produced by plasma spheroidization was investigated (Figure 2.24a) (Zi et al. 2018). Once the feed rate was lowered to 15–20 g/min, a small number of spherical powder particles were produced (Figure 2.24b,c). Most tungsten powder particles were spherical when the feed rate was below 15 g/min (Figure 2.24d–f). When the feed rate was below 10 g/min, powder particle aggregation was limited, and the spheroidization efficiency of the tungsten powder exceeded 95%. Moreover, the particle size distribution curve of the tungsten powder spheroidized at feed rates below 10 g/min was narrower than the distribution curves of raw powders and the spheroidized powders at higher feed rates. As a result, some aggregated powder particles could not maintain their original morphology under surface tension and ultimately split into smaller spherical powder particles. A decrease in the feed rate increased the spheroidization efficiency of the tungsten powder particles but reduced their particle size.

Figure 2.25 presents the property analysis of raw and spheroidized tungsten powders (Wang et al. 2017). After spheroidization, both the basic flow energy (BFE) and

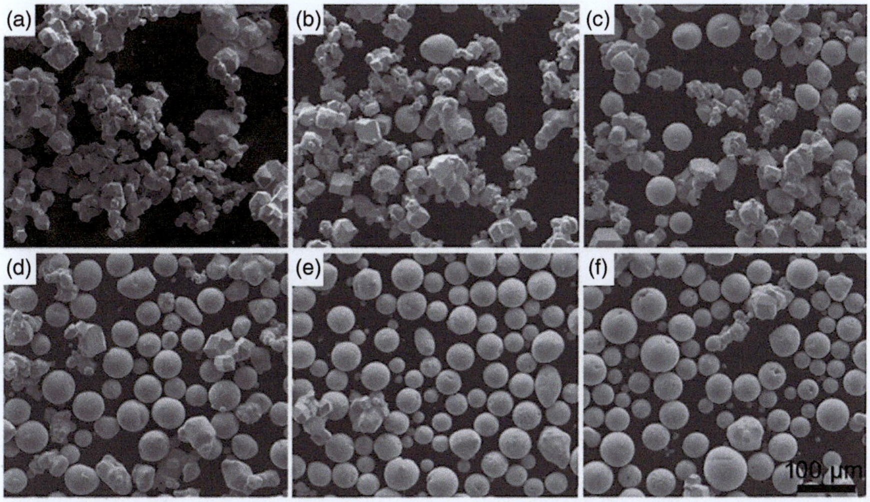

Figure 2.24 SEM images showing the morphology of tungsten powder before and after plasma spheroidization. (a) Raw powder, and (b–f) spheroidized powder produced at feed rates of 25, 15, 10, 6, and 4 g/min, respectively. Source: Zi et al. (2018)/reproduced with permission from Taylor & Francis.

the specific energy (SE) of the powder were significantly reduced (Figure 2.25a). Larger BFE and SE implied strong cohesion of the powder. The aerated flow energy (AE) of the spheroidized tungsten powder was lower than that of the raw powder, as shown in Figure 2.25b. A low AE indicated low cohesive strength among the powder particles and resulted in higher flow performance. The compressibility of these two powders is shown in Figure 2.25c. The compressibility percentage of the spherical powder was larger than that of the raw powder. The delivery of powder with high compressibility is relatively unaffected by non-cohesive powder particles with a low viscosity. A remarkable distinction in permeability could be observed between the raw and spheroidized powders (Figure 2.25d). A lower pressure drop resulted in higher powder permeability. Larger gaps were presented among the spherical powder particles than among the raw powder particles, which provided a way for air to escape from the powder container.

Recently, spherical NbMoTaW refractory HEA powders were prepared by spray drying and plasma spheroidization (Liu et al. 2021). They exhibited a homogeneous particle size distribution in the range of 10–60 μm and a BCC phase. In addition, the powders achieved an ultrahigh microhardness of over 10 GPa, which is 50% higher than their bulk counterparts. This new process for preparing spherical refractory HEA powders holds great potential for applications in metal powder-based AM.

2.4.2 Granulation–Sintering–Deoxygenation

Recently, a novel approach, granulation–sintering–deoxygenation, has been developed to prepare low-cost spherical titanium powder for AM. As its name suggests,

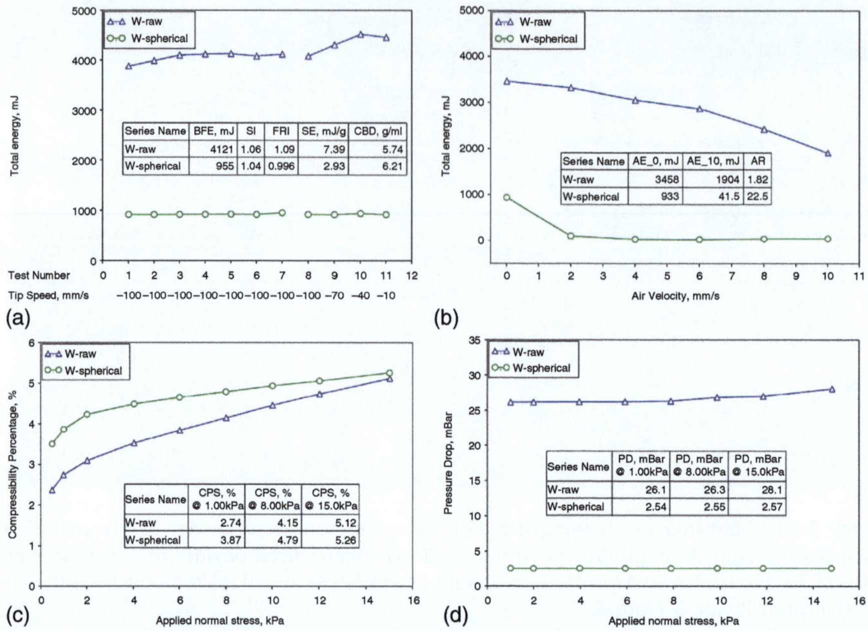

Figure 2.25 Property analysis of raw and spheroidized tungsten powders. (a) Basic flow, (b) aeration, (c) compressibility, and (d) permeability. Source: Reproduced from Wang et al. (2017)/with permission from Elsevier.

the granulation–sintering–deoxygenation process combines the underlying mechanisms of a few popular low-cost processes to prepare powders, such as granulation, sintering, and deoxygenation.

In granulation–sintering–deoxygenation, low-cost scrap titanium hydride powder blended with other pure metal powders can be selected as the raw material. During the granulation process, titanium hydrides containing other alloying elements are milled into fine powder particles and subsequently granulated via spray-drying to form spherical granules with the required sizes. Then, the spherical granules are sintered to produce high-density titanium alloy particles. In the deoxygenation process, the pre-densified spherical titanium alloy powder is deoxygenated to meet the industry standard of oxygen concentration. Overall, the granulation–sintering–deoxygenation process consumes less energy than atomization methods.

Figure 2.26 shows a schematic illustration of a typical granulation–sintering–deoxygenation process for producing Ti-6Al-4V powder (Sun et al. 2016). In the study, Ti-6Al-4V hydride powder with a maximum particle size of 74 μm, produced from Ti-6Al-4V scrap by cleaning, hydrogenation, milling, and sizing, was used as the raw material. The powder was milled in a solvent to reduce its particle size down to 10 μm (Figure 2.27a). Afterward, the slurry was delivered to a spray dryer filled with argon gas to produce spherical granules comprising fine Ti-6Al-4V hydride powder particles, which were then sieved for post-processing.

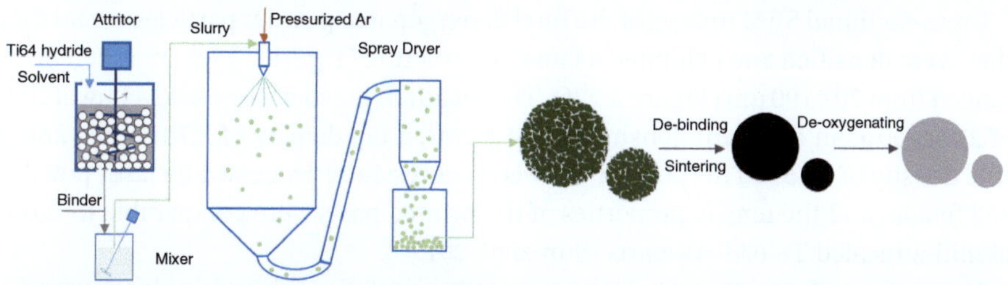

Figure 2.26 Schematic illustration of the granulation–sintering–deoxygenation process for the production of spherical Ti-6Al-4V powder. Source: Reproduced from Sun et al. (2016)/with permission from Elsevier.

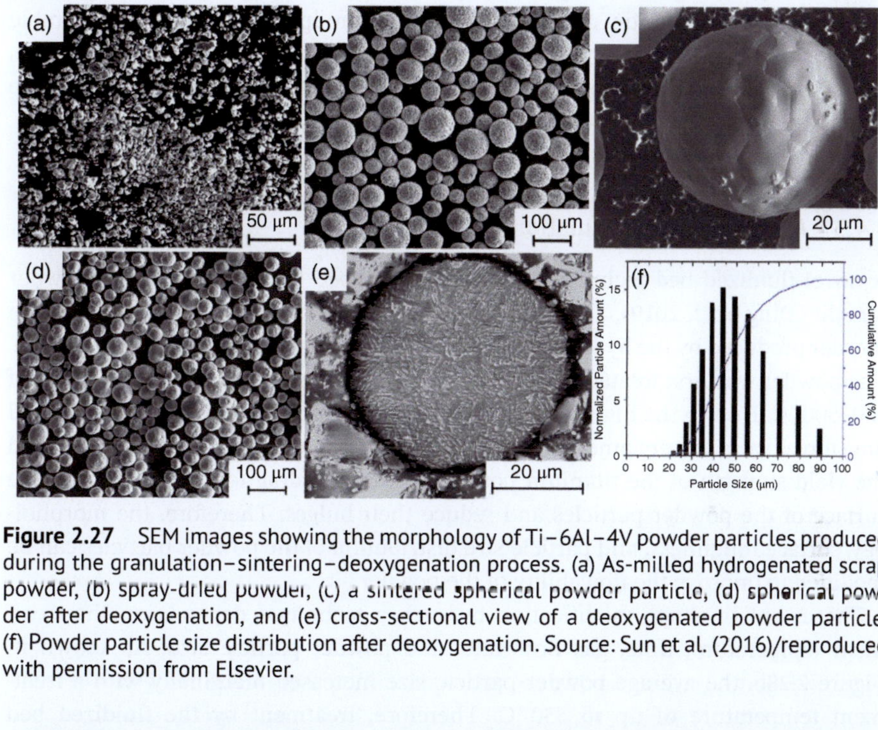

Figure 2.27 SEM images showing the morphology of Ti-6Al-4V powder particles produced during the granulation–sintering–deoxygenation process. (a) As-milled hydrogenated scrap powder, (b) spray-dried powder, (c) a sintered spherical powder particle, (d) spherical powder after deoxygenation, and (e) cross-sectional view of a deoxygenated powder particle. (f) Powder particle size distribution after deoxygenation. Source: Sun et al. (2016)/reproduced with permission from Elsevier.

The spherical granules were thermally de-bounded at a temperature between 200 and 400 °C and then sintered in a furnace filled with argon gas. A calcium oxide separator was utilized in the debinding and sintering processes to prevent sintering amongst the granules. After sintering, the calcium oxide separator was leached by dilute hydrochloric acid (HCl) and washed with water. The sintered Ti-6Al-4V powder was then deoxygenated using calcium metal at a temperature of 750 °C for 12 hours in a tube furnace filled with flowing argon gas. The spray-dried and sintered granules, as well as the deoxygenated powder particles, were found to exhibit a high degree of sphericity (Figure 2.27b–d). The spray drying parameters can be optimized to increase the yield of powder particles with specific sizes to fulfill certain requirements in various applications.

Cross-sectional SEM images of the final deoxygenated powder particles reveal that they were densified and exhibited a lamellar structure (Figure 2.27e), and their sizes ranged from 20 to 90 μm (Figure 2.27f). The resultant powder possessed a flowability of 27.5 s/50 g, an apparent density of 2.34 g/cm^3, a tap density of 2.78 g/cm^3, and a true density of 99.5%. This powder has been successfully processed by laser powder bed fusion, and the tensile properties of the printed parts were comparable to those of mill-annealed Ti–6Al–4V parts (Sun et al. 2017b).

The surfaces of powder particles prepared by granulation–sintering–deoxygenation are not as smooth as those prepared by gas atomization, plasma atomization, and the plasma rotating electrode process. However, it is a common practice to reuse spherical titanium powder after printing, and thus the surface of its particles will become rougher after several printing cycles. The surface quality of powders produced via the granulation–sintering–deoxygenation approach can be comparable to that of recycled Ti–6Al–4V powder produced by atomization, which indicates that the powder prepared by granulation–sintering–deoxygenation can fulfill the surface quality requirements for metal powder–based AM.

2.4.3 Fluidized-bed Granulation

A novel fluidized bed technique has been developed to modify titanium powder for AM (Ding et al. 2019). This study used a fluidized bed reactor to treat titanium powder produced by the hydride–dehydride process, as schematized in Figure 2.28a. The powder was first treated in a fluidized bed reactor at different temperatures of 400–500 °C. During the high-temperature fluidizing process, collision, friction, and drag forces were present among the powder particles. The high temperature reduced the yield strength of the titanium powder so that the drag forces could grind the surface of the powder particles and reduce their bulges. Therefore, the morphology, surface roughness, and particle size distribution of the powder particles can be modified to improve the flowability of the powder.

The extent of shape modification by such a technique depends on the treatment temperature, time, gas flow rate, and powder particle size. As shown in Figure 2.28b, the average powder particle size increased marginally with a treatment temperature of up to 550 °C. Therefore, treatment by the fluidized bed technique influenced the powder particle size marginally. The flowability of the powder after the treatment was measured to be 35.2 s/50 g, which was similar to that of the most commonly utilized AM powders (Figure 2.28c). The modified surface morphology of the resultant powder particles (Figure 2.28a) reduced the frequency of multiple-point interactions, which was attributed to their improved smoothness and decreasing number of small powder particles. Such an effect decreased the surface friction among the powder particles, thereby leading to a smoother powder flow. However, increasing the treatment temperature to 600 °C adversely affected the powder flowability (Figure 2.28c), which was a result of minor sintering occurring among individual powder particles at treatment temperatures exceeding 500 °C.

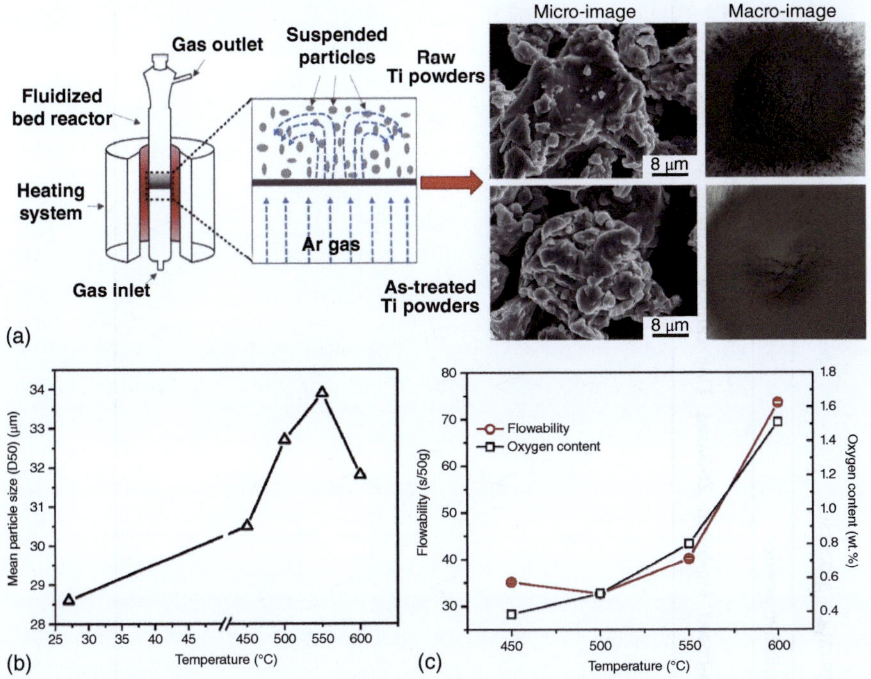

Figure 2.28 (a) Schematic illustration of the fluidized-bed treating process for titanium powder produced by the hydride–dehydride process and the powder morphology before and after treatment, (b) average powder particle size as a function of temperature, and (c) powder flowability and oxygen content as functions of temperature. Source: Ding et al. (2019)/ reproduced with permission from Elsevier.

2.5 Summary

This chapter presents the fundamentals and characteristics of the preparation processes of metal powders for AM. Determining the desired properties of an AM powder is a critical step in selecting a suitable powder preparation process to enable the printing of high-quality parts for industrial applications. Representative AM metal powders developed by the various preparation processes are summarized in Table 2.8. Currently, atomization processes are the most popular class of methods for producing a variety of metal powders. Most metal powders used in powder-based AM are prepared by gas atomization. However, additional research is required to understand the powder properties important for ensuring the repeatable manufacturing of metal parts.

The preparation processes used by representative commercial companies for powder production are presented in Table 2.9. Many industrial companies have focused on developing atomization methodologies and ensuring process stability to produce metal powders for AM purposes. Most companies are devoted to the development of metal powders through gas atomization.

Table 2.8 Summary of representative AM metal powders developed by various preparation processes.

Process	Representative powder													
	Fe-based	**Ti-based**	**Ni-based**	**Al-based**	**Cu-based**	**Nb-based**	**Co-based**	**Mg-based**	**Zr-Based**	**TiAl**	**Cemented carbide**	**Refractory metal**	**HEA**	**Metallic glass**
Gas atomization	√	√	√	√	√		√	√		√	√	√	√	√
Water atomization	√		√	√	√		√						√	√
Plasma atomization		√			√		√							
Plasma rotating electrode process	√	√	√			√	√		√	√			√	
Mechanical mixing	√	√	√	√	√	√	√	√		√	√	√	√	
Hydride–dehydride		√							√			√		
Oxide reduction	√		√		√	√	√	√	√					
Chloride reduction	√	√												
Electrolysis process	√	√	√		√	√			√					
Carbonyl reaction	√		√				√							
Plasma spherodization		√	√	√		√	√			√	√	√		

Table 2.9 The preparation processes used by representative commercial companies for powder production.

Process	Representative commercial companies
Gas atomization	ATI Powder Metals (USA), Crucible Materials Corp (USA), Iowa Powder Atomization Technologies (USA), Puris LLC (USA), Ametek specialty metal products (USA), Carpenter Technology Corp (USA), ALD Vacuum Technologies GmbH (Germany), TLS Technik (Germany), Affinity International (China), Osaka Titanium Technologies (Japan), H.C. Starck GmbH (Germany), Höganäs Sweden AB (Sweden), Sandvik Materials Technology (Sweden), LPW Technology Ltd (UK)
Water atomization	H.C.Starck GmbH (Germany), Höganäs Sweden AB (Sweden), LPW Technology Ltd (UK)
Plasma atomization	AP&C (Canada), LPW Technology Ltd (UK), ARCAM AB (Sweden)
Plasma rotating electrode process	Timet (USA), LPW Technology Ltd (UK), Phelly Materials (China), Baoji Orchid Ti (China)
Hydride–dehydride	Affinity International (China), Teledyne Wah Chang (USA), Global Titanium (USA), Ametek specialty metal products (USA), Osaka Titanium Technologies (Japan)
Chloride reduction	CSIRO (Australia)
Plasma spherodization	Tekna Plasma Systems (Canada), Global Advanced Metals (USA)

Metal powder preparation techniques have existed for decades. Before the advent of metal AM techniques, high costs and a lack of industrial demand are the main reasons behind the unremarkable development of new powder preparation processes. The wide application of AM techniques in various industries significantly stimulates the supply and demand market for metal powders, which provides great opportunities for developing novel powder preparation processes in the future.

References

Agripa, H. and Botef, I. (2019). Modern production methods for titanium alloys: a review. In: *Titanium Alloys-Novel Aspects of Their Manufacturing and Processing* (ed. M. Motyka). London: IntechOpen.

Ahsan, M.N., Pinkerton, A.J., Moat, R.J., and Shackleton, J. (2011). A comparative study of laser direct metal deposition characteristics using gas and plasma-atomized Ti–6Al–4V powders. *Materials Science and Engineering A* 528 (25, 26): 7648–7657.

Allimant, A., Planche, M.P., Bailly, Y. et al. (2009). Progress in gas atomization of liquid metals by means of a De Laval nozzle. *Powder Technology* 190 (1, 2): 79–83.

AlMangour, B. and Grzesiak, D. (2016). Selective laser melting of TiC reinforced 316L stainless steel matrix nanocomposites: influence of starting TiC particle size and volume content. *Materials and Design* 104: 141–151.

Alvarez, K.L., Martín, J.M., Ipatov, M., and Gonzalez, J. (2018). Soft magnetic amorphous alloys (Fe-rich) obtained by gas atomisation technique. *Journal of Alloys and Compounds* 735: 2646–2652.

Anderson, I.E. and Terpstra, R.L. (2002). Progress toward gas atomization processing with increased uniformity and control. *Materials Science and Engineering A* 326 (1): 101–109.

Antony, L.V. and Reddy, R.G. (2003). Processes for production of high-purity metal powders. *JOM Journal of the Minerals Metals and Materials Society* 55 (3): 14–18.

Arami, H. and Simchi, A. (2007). Reactive milling synthesis of nanocrystalline Al–Cu/Al_2O_3 nanocomposite. *Materials Science and Engineering A* 464 (1, 2): 225–232.

Attar, H., Bönisch, M., Calin, M. et al. (2014). Selective laser melting of *in situ* titanium–titanium boride composites: processing, microstructure and mechanical properties. *Acta Materialia* 76: 13–22.

Aversa, A., Marchese, G., Lorusso, M. et al. (2017). Microstructural and mechanical characterization of aluminum matrix composites produced by laser powder bed fusion. *Advanced Engineering Materials* 19 (11): 1700180.

Cai, C., Radoslaw, C., Zhang, J. et al. (2019). *In-situ* preparation and formation of TiB/Ti–6Al–4V nanocomposite via laser additive manufacturing: microstructure evolution and tribological behavior. *Powder Technology* 342: 73–84.

Chen, G., Zhao, S., Tan, P. et al. (2018). A comparative study of Ti–6Al–4V powders for additive manufacturing by gas atomization, plasma rotating electrode process and plasma atomization. *Powder Technology* 333: 38–46.

Ciftci, N., Ellendt, N., Barreto, E.S. et al. (2018). Increasing the amorphous yield of {(Fe0.6Co0.4) 0.75 B0.2Si0.05} 96Nb4 powders by hot gas atomization. *Advanced Powder Technology* 29 (2): 380–385.

Dawes, J., Bowerman, R., and Trepleton, R. (2015). Introduction to the additive manufacturing powder metallurgy supply chain. *Johnson Matthey Technology Review* 59 (3): 243–256.

Ding, W., Chen, G., Qin, M. et al. (2019). Low-cost Ti powders for additive manufacturing treated by fluidized bed. *Powder Technology* 350: 117–122.

Doblin, C., Chryss, A., and Monch, A. (2012). Titanium powder from the TiRO™ process. *Key Engineering Materials* 520: 95–100.

Doblin, C., Freeman, D., and Richards, M. (2013). The TiRO™ process for the continuous direct production of titanium powder. *Key Engineering Materials* 551: 37–43.

Eisen, W.B., Ferguson, B.L., German, R.M. et al. (1998). *Powder Metal Technologies and Applications*. United States: ASM Handbook.

Feng, Y. and Qiu, T. (2012). Preparation, characterization and microwave absorbing properties of FeNi alloy prepared by gas atomization method. *Journal of Alloys and Compounds* 513: 455–459.

Fray, D., Farthing, T.W., and Chen, G.Z. (2009). Metalysis leads charge for change in titanium production. *Metal Powder Report* 64 (9): 6–11.

Fuchs, G. and Hayden, S. (1992). Microstructural evaluation of as-solidified and heat-treated γ-TiAl based powders. *Materials Science and Engineering A* 152 (1, 2): 277–282.

Gao, C., Xiao, Z., Zou, H. et al. (2019). Characterization of spherical AlSi10Mg powder produced by double-nozzle gas atomization using different parameters. *Transactions of Nonferrous Metals Society of China* 29 (2): 374–384.

Gerling, R., Clemens, H., and Schimansky, F.P. (2004). Powder metallurgical processing of intermetallic gamma titanium aluminides. *Advanced Engineering Materials* 6 (1, 2): 23–38.

Goncharov, I.S., Razumov, N.G., Silin, A.O. et al. (2019). Synthesis of Nb-based powder alloy by mechanical alloying and plasma spheroidization processes for additive manufacturing. *Materials Letters* 245: 188–191.

Goudar, D.M., Srivastava, V.C., and Rudrakshi, G.B. (2017). Effect of atomization parameters on size and morphology of Al-17Si alloy powder produced by free fall atomizer. *Engineering Journal* 21 (1): 155–168.

Gu, D., Wang, Z., Shen, Y. et al. (2009). *In-situ* TiC particle reinforced Ti–Al matrix composites: powder preparation by mechanical alloying and selective laser melting behavior. *Applied Surface Science* 255 (22): 9230–9240.

Gu, D., Meng, G., Li, C. et al. (2012). Selective laser melting of TiC/Ti bulk nanocomposites: influence of nanoscale reinforcement. *Scripta Materialia* 67 (2): 185–188.

Gu, D., Wang, H., Dai, D. et al. (2015). Rapid fabrication of Al-based bulk-form nanocomposites with novel reinforcement and enhanced performance by selective laser melting. *Scripta Materialia* 96: 25–28.

Gu, D., Ma, J., Chen, H. et al. (2018). Laser additive manufactured WC reinforced Fe-based composites with gradient reinforcement/matrix interface and enhanced performance. *Composite Structures* 192: 387–396.

Guo, Y., Jia, L., Kong, B. et al. (2017). Microstructure and surface oxides of rapidly solidified Nb-Si based alloy powders. *Materials and Design* 120: 109–116.

Han, Q., Setchi, R., and Evans, S.L. (2016). Synthesis and characterisation of advanced ball-milled Al Al_2O_3 nanocomposites for selective laser melting. *Powder Technology* 297: 183–192.

Han, C., Wang, Q., Song, B. et al. (2017). Microstructure and property evolutions of titanium/nano-hydroxyapatite composites in-situ prepared by selective laser melting. *Journal of the Mechanical Behavior of Biomedical Materials* 71: 85–94.

Han, C., Li, Y., Wang, Q. et al. (2018). Titanium/hydroxyapatite (Ti/HA) gradient materials with quasi-continuous ratios fabricated by SLM: material interface and fracture toughness. *Materials and Design* 141: 256–266.

Han, C., Babicheva, R., Chua, J.D.Q. et al. (2020a). Microstructure and mechanical properties of (TiB + TiC)/Ti composites fabricated in situ via selective laser melting of Ti and B4C powders. *Additive Manufacturing* 36: 101466.

Han, C., Fang, Q., Shi, Y. et al. (2020b). Recent advances on high-entropy alloys for 3D printing. *Advanced Materials* 32: 1903855.

Hausnerova, B., Mukund, B.N., and Sanetrnik, D. (2017). Rheological properties of gas and water atomized 17-4PH stainless steel MIM feedstocks: effect of powder shape and size. *Powder Technology* 312: 152–158.

Hong, S.J., Kim, T.S., Kim, H.S. et al. (1999). Microstructural behavior of rapidly solidified and extruded Al-14wt% Ni-14wt% Mm (Mm, misch metal) alloy powders. *Materials Science and Engineering A* 271 (1, 2): 469–476.

Kelley, R. and McDeavitt, S. (2013). Design of an integrated system to recycle Zircaloy cladding using a hydride–milling–dehydride process. *Nuclear Engineering and Design* 263: 47–58.

Krasnowski, M. and Kulik, T. (2008). Nanocomposites obtained by mechanical alloying in Fe–Al–Ti–C system. *Journal of Alloys and Compounds* 448 (1, 2): 227–233.

Lagutkin, S., Achelis, L., Sheikhaliev, S. et al. (2004). Atomization process for metal powder. *Materials Science and Engineering A* 383 (1): 1–6.

Li, X., Kong, C., Becker, T., and Sercombe, T. (2016). Investigation of interfacial reaction products and stress distribution in selective laser melted Al12Si/SiC composite using confocal Raman microscopy. *Advanced Engineering Materials* 18 (8): 1337–1341.

Li, W., Yang, Y., Liu, J. et al. (2017). Enhanced nanohardness and new insights into texture evolution and phase transformation of TiAl/TiB2 *in-situ* metal matrix composites prepared via selective laser melting. *Acta Materialia* 136: 90–104.

Liu, Y., Niu, S., Li, F. et al. (2011). Preparation of amorphous Fe-based magnetic powder by water atomization. *Powder Technology* 213 (1–3): 36–40.

Liu, Y., Liang, S., Han, Z. et al. (2018). A novel model of calculating particle sizes in plasma rotating electrode process for superalloys. *Powder Technology* 336: 406–414.

Liu, B., Duan, H., Li, L. et al. (2021). Microstructure and mechanical properties of ultra-hard spherical refractory high-entropy alloy powders fabricated by plasma spheroidization. *Powder Technology* 382: 550–555.

Löber, L., Schimansky, F.P., Kühn, U. et al. (2014). Selective laser melting of a beta-solidifying TNM-B1 titanium aluminide alloy. *Journal of Materials Processing Technology* 214 (9): 1852–1860.

Lou, J., Gabbitas, B., Zhang, D., and Yang, F. (2015). Effects of initial powder compact thickness, lubrication, and particle morphology on the cold compaction behavior of Ti powder. *Metallurgical and Materials Transactions A* 46 (8): 3646–3655.

Maya, A.A., Grana, D.R., Hazarabedian, A. et al. (2012). Zr–Ti–Nb porous alloys for biomedical application. *Materials Science and Engineering C* 32 (2): 321–329.

McCracken, C.G., Barbis, D.P., and Deeter, R.C. (2011). Key characteristics of hydride–dehydride titanium powder. *Powder Metallurgy* 54 (3): 180–183.

Mostafaei, A., Hilla, C., Stevens, E.L. et al. (2018). Comparison of characterization methods for differently atomized nickel-based alloy 625 powders. *Powder Technology* 333: 180–192.

Neikov, O.D., Vasilieva, G.I., Sameljuk, A.V., and Krajnikov, A.V. (2004). Water atomised aluminium alloy powders. *Materials Science and Engineering A* 383 (1): 7–13.

Neikov, O., Naboychenko, S.S., and Dowson, G. (ed.) (2009). *Handbook of Non-Ferrous Metal Powders: Technologies and Applications*. Amsterdam: Elsevier.

Ouyang, H., Chen, X., and Huang, B. (2007). Influence of melt superheat on breakup process of close-coupled gas atomization. *Transactions of Nonferrous Metals Society of China* 17 (5): 967–973.

Park, T.G., Lee, S.H., Lee, B. et al. (2018). Fabrication and characterization of CoCrFeNiMn high entropy alloy powder processed by gas atomization. *Archives of Metallurgy and Materials* 63 (2): 1055–1059.

Pasupathy, M., Martín, J.M., Rivas, A. et al. (2016). Effect of the solidification time on the median particle size of powders produced by water atomisation. *Powder Metallurgy* 59 (2): 128–141.

Peillon, N., Fruhauf, J.B., Gourdet, S. et al. (2015). Effect of TiH_2 in the preparation of MMC Ti based with TiC reinforcement. *Journal of Alloys and Compounds* 619: 157–164.

Qin, Q., Yang, F., Shi, T. et al. (2019). Spheroidization of tantalum powder by radio frequency inductively coupled plasma processing. *Advanced Powder Technology* 30 (8): 1709–1714.

Rong, T. and Gu, D. (2016). Formation of novel graded interface and its function on mechanical properties of WC_{1-x} reinforced Inconel 718 composites processed by selective laser melting. *Journal of Alloys and Compounds* 680: 333–342.

Singh, D., Koria, S.C., and Dube, R.K. (2001). Study of free fall gas atomisation of liquid metals to produce powder. *Powder Metallurgy* 44 (2): 177–184.

Sivasankaran, S., Sivaprasad, K., Narayanasamy, R., and Iyer, V.K. (2010). An investigation on flowability and compressibility of AA 6061_{100-x}-x wt.% TiO_2 micro and nanocomposite powder prepared by blending and mechanical alloying. *Powder Technology* 201 (1): 70–82.

Song, B., Dong, S., and Coddet, C. (2014). Rapid *in situ* fabrication of Fe/SiC bulk nanocomposites by selective laser melting directly from a mixed powder of microsized Fe and SiC. *Scripta Materialia* 75: 90–93.

Sun, P., Fang, Z., Xia, Y. et al. (2016). A novel method for production of spherical Ti–6Al–4V powder for additive manufacturing. *Powder Technology* 301: 331–335.

Sun, P., Fang, Z., Zhang, Y., and Xia, Y. (2017a). Review of the methods for production of spherical Ti and Ti alloy powder. *JOM Journal of the Minerals Metals and Materials Society* 69 (10): 1853–1860.

Sun, P., Fang, Z., Zhang, Y., and Xia, Y. (2017b). Microstructure and mechanical properties of Ti–6Al–4V fabricated by selective laser melting of powder produced by granulation-sintering-deoxygenation method. *JOM Journal of the Minerals Metals and Materials Society* 69 (12): 2731–2737.

Suryanarayana, C. (2001). Mechanical alloying and milling. *Progress in Materials Science* 46 (1, 2): 1–184.

Tong, J., Lu, X., Liu, C. et al. (2015). Compact process for the preparation of microfine spherical high-niobium-containing TiAl alloy powders. *JOM Journal of the Minerals Metals and Materials Society* 67 (3): 573–579.

Ünal, R. (2006). The influence of the pressure formation at the tip of the melt delivery tube on tin powder size and gas/melt ratio in gas atomization method. *Journal of Materials Processing Technology* 180 (1–3): 291–295.

Wang, L., Wu, J., and Zhang, D. (2017). Properties evolution of additive manufacture used tungsten powders prepared by radio frequency induction plasma. *International Journal of Refractory Metals and Hard Materials* 67: 90–97.

Wang, P., Huang, P., Ng, F.L. et al. (2019). Additively manufactured CoCrFeNiMn high-entropy alloy via pre-alloyed powder. *Materials and Design* 168: 107576.

Wei, Q., Li, S., Han, C. et al. (2015). Selective laser melting of stainless-steel/ nano-hydroxyapatite composites for medical applications: Microstructure, element distribution, crack and mechanical properties. *Journal of Materials Processing Technology* 222: 444–453.

Wei, M., Chen, S., Liang, J., and Liu, C. (2017). Effect of atomization pressure on the breakup of TA15 titanium alloy powder prepared by EIGA method for laser 3D printing. *Vacuum* 143: 185–194.

Xu, W., Xiao, S., Lu, X. et al. (2019). Fabrication of commercial pure Ti by selective laser melting using hydride-dehydride titanium powders treated by ball milling. *Journal of Materials Science and Technology* 35 (2): 322–327.

Yan, M., Wang, J.Q., Schaffer, G.B., and Qian, M. (2011). Solidification of nitrogen-atomized $Al_{86}Ni_6Y_{4.5}Co_2La_{1.5}$ metallic glass. *Journal of Materials Research* 26 (7): 944–950.

Yang, C.C., Chau, J.L.H., Weng, C.J. et al. (2017). Preparation of high-entropy AlCoCrCuFeNiSi alloy powders by gas atomization process. *Materials Chemistry and Physics* 202: 151–158.

Yolton, C.F. and Froes, F.H.S. (2015). Conventional titanium powder production. *Titanium Powder Metallurgy* 21–32.

Zhang, H., He, X., Qu, X., and Zhao, L. (2009). Microstructure and mechanical properties of high Nb containing TiAl alloy parts fabricated by metal injection molding. *Materials Science and Engineering A* 526 (1, 2): 31–37.

Zhang, B., Bi, G., Nai, S. et al. (2016). Microhardness and microstructure evolution of TiB_2 reinforced Inconel 625/TiB_2 composite produced by selective laser melting. *Optics & Laser Technology* 80: 186–195.

Zhang, B., Duan, Y., Cui, Y. et al. (2018). Improving electromagnetic properties of $FeCoNiSi_{0.4}Al_{0.4}$ high entropy alloy powders via their tunable aspect ratio and elemental uniformity. *Materials and Design* 149: 173–183.

Zhang, C., Yang, F., Chen, C., and Guo, Z. (2021). The microstructure and mechanical properties of extra low interstitials (ELI) Ti–6Al–4V alloys manufactured from hydride–dehydride (HDH) powder. *Journal of Alloys and Compounds* 873: 159757.

Zhao, Y., Si, J., Song, J. et al. (2014). Synthesis of Mg–Zn–Ca metallic glasses by gas-atomization and their excellent capability in degrading azo dyes. *Materials Science and Engineering B* 181: 46–55.

Zheng, M., Zhang, S., Hu, Q. et al. (2018). Microstructural characterisation of CuAgZr powder particles produced by argon gas atomisation. *Powder Metallurgy* 61 (3): 231–240.

Zhong, C., Chen, J., Linnenbrink, S. et al. (2016). A comparative study of Inconel 718 formed by high deposition rate laser metal deposition with GA powder and PREP powder. *Materials and Design* 107: 386–392.

Zhou, S., Zhang, P., Xue, Y. et al. (2018a). Microstructure evolution of $Al_{0.6}CoCrFeNi$ high entropy alloy powder prepared by high pressure gas atomization. *Transactions of Nonferrous Metals Society of China* 28 (5): 939–945.

Zhou, W., Sun, X., Kikuchi, K. et al. (2018b). Carbon nanotubes as a unique agent to fabricate nanoceramic/metal composite powders for additive manufacturing. *Materials and Design* 137: 276–285.

Zi, X., Chen, C., Wang, X. et al. (2018). Spheroidisation of tungsten powder by radio frequency plasma for selective laser melting. *Materials Science and Technology* 34 (6): 735–742.

3

Laser Powder Bed Fusion

In recent years, laser powder bed fusion (LPBF) has gained the most attention from academia and industry among metal powder–based additive manufacturing (AM) techniques. The small laser spot size and powder particle size in LPBF enable the high-precision printing of metal parts. In addition, the ultrahigh cooling rates result in refined microstructures, contributing to the excellent mechanical performance of the printed parts. This chapter discusses the history, fundamentals, process characteristics, powder materials, and equipment of LPBF, as well as microstructure and mechanical properties of LPBF-printed parts, including mechanical metamaterials.

3.1 History

LPBF utilizes high-energy laser beams to melt metal powders. The patent for LPBF was first filed to the German Patent and Trade Mark Office in 1997 by F & S Stereolithographietechnik GmbH with Fraunhofer ILT and later published in 1998 (Meiners et al. 1998). The ASTM International F42 standards committee initially classified selective laser melting (SLM) under the category of laser sintering, but it was later rectified as the two processes are fundamentally different. Both SLM and laser sintering are now grouped under powder bed fusion (PBF). Another name for LPBF is direct metal laser sintering (DMLS), as dictated by the EOS company. The first test systems of DMLS were developed by EOS and installed in 1994. The first commercial EOSINT M 250 system was established in 1995, enabling complex and massive parts to be built with high accuracy and good surface quality.

LPBF rapidly became a commercial success after the advent of the EOSINT M 250 system, and the development of the materials and process advanced rapidly because of the EOS–ERD cooperation. In 1997, an improved version of the system was introduced, which allowed the part to be built with a 50 μm layer thickness, thus significantly enhancing the surface quality by reducing the stairstep effect. A significant milestone in part quality improvement was achieved in 2001 with the introduction of DirectSteel 20, a steel powder that can be printed with a layer thickness of only 20 μm. The same layer thickness was also achieved in 2002 for the bronze-based DirectMetal 20 powder.

Metal Powder–Based Additive Manufacturing, First Edition. Kun Zhou and Changjun Han.

In 2004, a novel system employing a solid-state fiber laser, EOSINT M 270, was first introduced to print stable and high-quality metal parts. Then, boosted by the increasing development of PBF, other companies started to develop their commercial LPBF equipment. For example, Concept Laser presented a machine named M3 Linear at the end of 2001 and coined the term "LaserCusing"; SLM Solutions introduced a more popular designation named SLM for their machines, which is commonly known among contemporary researchers and engineers; Trumpf introduced their commercial system named Trumaform LF 250 in 2003, which used a disc laser and included two separate build chambers with a preheating temperature of up to 500 °C.

3.2 Fundamentals

The LPBF process consists of several steps, from computer-aided design (CAD) model preparation to removing fabricated parts from the build platform. Before the CAD data are uploaded to the LPBF machine to produce parts, the stereolithography (STL) files must be processed to provide support structures for overhanging features and generate slice data for the laser scanning of individual layers. The printing process starts with spreading a thin layer of metal powder on a substrate plate by a recoater (roller or blade). After the powder is applied, one or multiple laser sources with high energy densities are employed to melt the selected areas. Once the laser scanning is completed, the build platform is lowered, and the next layer of powder is deposited on top. The laser then scans a new layer, and the process is repeated until the parts are completely printed.

The build chamber is protected with an inert gas flow (e.g. argon or nitrogen) during the printing process to prevent oxidation. In addition, the homogeneous gas that flows across the build area helps to remove the condensate (e.g. spatters and condensed metal vapor) produced during the powder melting process and plays a vital role in determining the quality and performance of the printed parts (Ferrar et al. 2012).

Once the process is completed, loose powder is removed from the chamber, and the parts can be separated from the substrate plate manually or by electrical discharge machining. The entire process is automated, except for the CAD data preparation process and the removal of the printed parts from the build platform. Schematic illustrations of the LPBF machine and produced melt pools are presented in Figure 3.1.

The distinct advantages of LPBF are presented as follows:

- The process involves a fine laser beam with a high beam quality, which carries sufficient energy to melt metals and ensures high printing accuracy. Its high manufacturing precision makes it suitable for directly printing complex structures with delicate features and smooth surfaces.
- The melt pools possess ultrahigh cooling rates, which vary with process parameters and materials in the range of 10^3–10^8 K/s, resulting in ultrafine microstructures and superior mechanical properties of the printed parts.

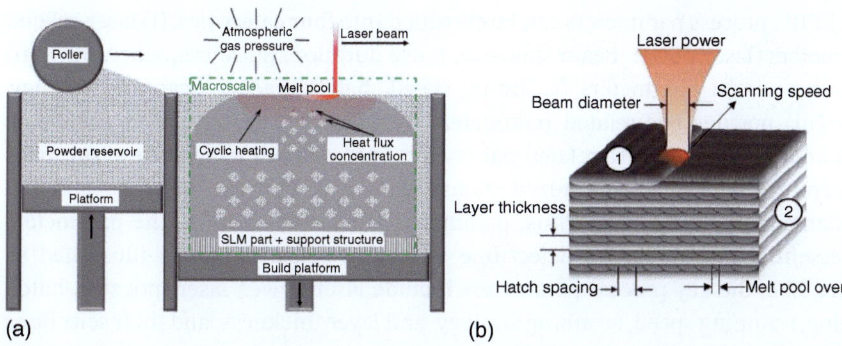

Figure 3.1 (a) Schematic illustration of an LPBF machine. Source: Reproduced from Meier et al. (2017)/with permission of Begell house Inc. (b) Schematic illustration of melt pools. Note that the scanning strategy depicted in this figure is bidirectional scanning with 90° rotation between each powder layer, and numbers 1 and 2 represent the horizontal and vertical build planes, respectively. Source: Reprinted with permission from a work by Vilaro et al. (2008).

- The process can manufacture parts with high relative densities (above 99%) from various metallic materials, such as titanium and its alloys, iron and steel alloys, nickel alloys, cobalt alloys, aluminum alloys, and copper and its alloys, to achieve high structural integrity.
- The cost and lead time of the production cycle can be reduced, especially for high value-added but low-production volume parts.

Meanwhile, the limitations of LPBF are:

- High residual stress is produced in parts resulting from the large temperature gradients. Therefore, post-heat treatment is often required to relieve the residual stress.
- Adding support structures to parts with overhanging features is necessary, and post-processing is required to remove these support structures manually.
- The size of printed parts is restricted due to the limited volume of the build chamber.
- The types of available powder materials are limited. Printing high-reflectivity powder materials, such as pure copper and pure aluminum, is challenging through infrared laser.
- The process is characterized by low production rates due to the powder recoating process.

Some limitations can be alleviated by developing industrial-oriented LPBF facilities with multiple laser beams and an extended build chamber. For example, SLM Solutions Group AG recently launched the industrial-scale NXG XII 600 equipped with envelope dimensions of 600 mm × 600 mm × 600 mm and 12 laser beams of 1000 W each, achieving a production rate 20 times faster than a standard single-laser LPBF machine. Moreover, ultrafine powder particles (e.g. <10 μm) employed in LPBF may introduce safety hazards such as particle inhalation and explosions.

In LPBF, process parameters can be classified into four categories: (i) laser-related parameters (laser power, beam spot size, pulse duration, pulse frequency, etc.), (ii) scanning-related parameters (scanning speed, hatch spacing, scanning strategy, etc.), (iii) powder bed-related parameters (powder bed density, layer thickness, etc.), and (iv) temperature-related parameters (powder bed temperature, substrate temperature, etc.). The optimized parameters vary with the machine and the powder material employed. Thus, parametric studies to optimize the parameters are essential for producing defect-free and fully dense parts. As illustrated in Figure 3.1b, the key process parameters include laser power, laser spot size, hatch spacing, scanning speed, scanning strategy, and layer thickness, and their selections are interdependent.

The laser absorptivity of different powder materials depends on the laser wavelength, which is further determined by the type of laser selected, as illustrated in Figure 3.2. Three types of lasers are commonly used in LPBF equipment: CO_2 laser, fiber laser, and diode-pumped neodymium-doped yttrium aluminum garnet (Nd:YAG) laser (Figure 3.2a). The CO_2 laser has relatively low power and absorptivity for metals and is generally selected for polymer-based LPBF. The fiber and Nd:YAG lasers perform similarly in metal printing and are widely utilized for commercialized metal-based LPBF equipment. Nevertheless, the applicability of the above two types of lasers is limited to certain metal materials. For example, iron and steel exhibit high laser absorptivity with respect to the fiber laser, which is a quality prevalent in modern LPBF equipment. In contrast, aluminum, copper, silver, and gold exhibit low laser absorptivity or high optical reflectivity at the laser wavelength of 1.07 μm. Fortunately, the laser absorptivity of certain reflective materials (such as copper) can be significantly enhanced using a green laser source, as shown in Figure 3.2b. Nevertheless, the green laser is still pending commercialization for LPBF equipment.

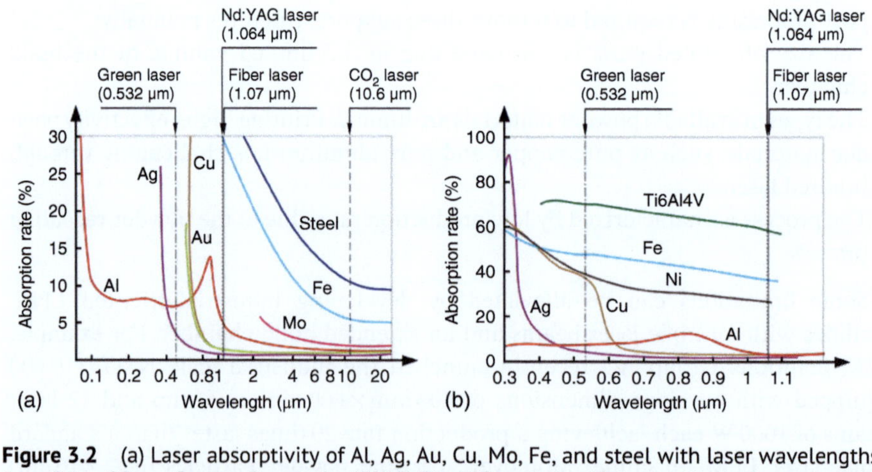

Figure 3.2 (a) Laser absorptivity of Al, Ag, Au, Cu, Mo, Fe, and steel with laser wavelengths from 0.1 to 20 μm. Source: Bergström (2018)/David Bergström. Note that the horizontal axis is logarithmic. (b) Laser absorptivity of Ag, Cu, Al, Ni, Fe, and Ti–6Al–4V with laser wavelengths from 0.3 to 1.1 μm. Source: Adapted from Alsaddah et al. (2021).

The hatch spacing refers to the overlap between adjacent melt tracks and determines the extent of metallurgical bonding between the tracks. Small hatch spacing often leads to excessive overlap and thus increases the fabrication time. On the other hand, large hatch spacing imposes limitations on the maximum layer thickness and increases the propensity of gap formation due to the lack of intra-layer bonding.

The layer thickness should be sufficiently large for a part to be manufactured cumulatively. However, the excessive thickness can introduce metallurgical defects, such as balling and lack of fusion, thus degrading the structural integrity and properties of the printed part.

The laser power controls the amount of energy delivered to the irradiated material. Highly reflective materials (such as pure copper) require high power to be melted completely; otherwise, the resultant partial melting will cause the melt pool space to be filled incompletely upon solidification, thus introducing defects. In addition to regulating the melting efficiency, the laser power governs the size and continuity of a melt track, thereby affecting the structural integrity of the parts.

The scanning speed controls melting and solidification rates. The low speed extends the duration of irradiated spot maintained in a molten state, thereby slowing down the solidification process and affecting the microstructure and size of the melt track.

The powder material, laser spot size, and layer thickness are usually fixed for any given build, leaving the laser power, scanning speed, and hatch spacing to be optimized. However, the adjustable range of the hatch spacing is narrow in terms of practical applications, in which a balance must be achieved between the build rate and part quality. Hence, the combination of laser power and scanning speed is crucial for determining the process outcomes during LPBF printing, as illustrated in Figure 3.3.

The scanning strategy defines the spatial movement pattern of the laser beam. Various scanning strategies have been developed in pursuit of printing components with near-full density and controllable crystallographic texture. Four basic patterns, namely, unidirectional, bidirectional, stripe/paintbrush scanning, and island/chessboard/checkboard scanning, are commonly employed in LPBF for every single layer (Figure 3.4).

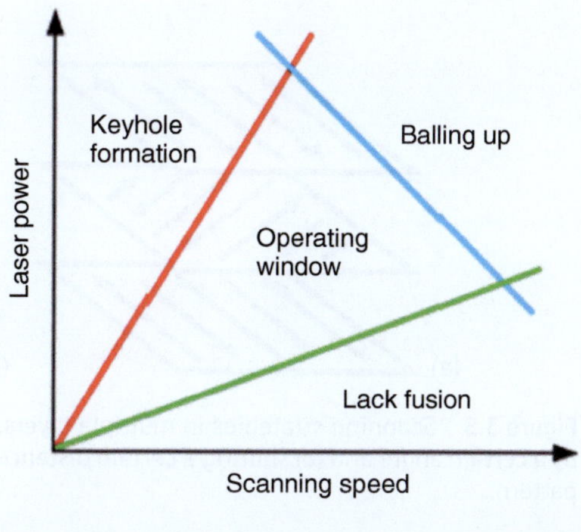

Figure 3.3 Effects of laser power and scanning speed on the process outcomes during LPBF printing.

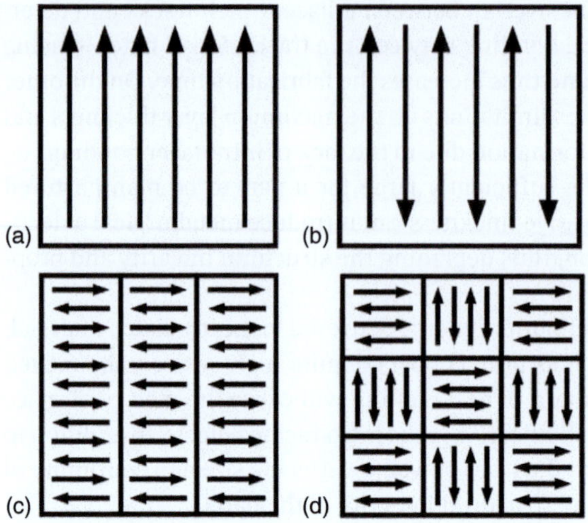

Figure 3.4 Four main basic scanning patterns in a single layer. (a) Unidirectional scanning, (b) bidirectional scanning, (c) stripe/paintbrush scanning, and (d) island scanning.

Unidirectional scanning is the most straightforward pattern in which the scanning process is performed in the same direction, which generally leads to the least densification but the strongest texture of a melt track. Bidirectional scanning is conducted by scanning in opposite directions for two adjacent lines, thus exhibiting a zigzag pattern. Stripe scanning is carried out by scanning bidirectionally within each parallel stripe. Island scanning adopts a checkerboard pattern of alternating bidirectional fills, which reduces temperature gradients in the scan plane by uniformly distributing the process heat.

Various scanning strategies can be employed for adjacent layers, including a rotation of the scan plane by a certain angle and a shift of the starting point by a certain distance between the layers (Figure 3.5). The choice of the scanning strategy also significantly affects the residual stress produced in printed parts. For example, the rotation of the scan plane on successive layers at an angle of 67° is commonly implemented, which can reduce warping, the number of defects, and the anisotropy of the printed parts. In addition, a remelting strategy (i.e. scanning a layer with the same path multiple times) can improve the surface quality of the printed parts.

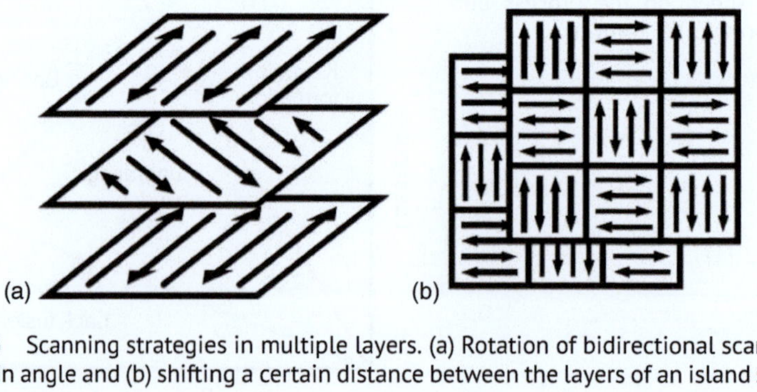

Figure 3.5 Scanning strategies in multiple layers. (a) Rotation of bidirectional scan vectors by a certain angle and (b) shifting a certain distance between the layers of an island scanning pattern.

The laser power, spot size, scanning speed, and powder bed temperature determine the energy input needed to melt the powder and form a desirable part. A lower laser power requires lower scanning speeds to ensure proper particle fusion. The hatch spacing should be carefully selected to provide sufficient melt pool overlap between adjacent tracks of fused material to ensure robust mechanical properties. The optimized hatch spacing is often set at $0.7\,\omega_0$, where ω_0 is the diameter of the laser beam at its focus point (Cloots et al. 2016). The LPBF processing window changes for different metals because of variations in surface tension and viscosity of the melt and the melting point, chemical composition, laser absorption, and thermal conductivity of metals.

One of the metrics commonly employed for optimizing process parameters to obtain near-full dense parts is energy density, which is calculated based on a linear, areal, or volumetric approach. The volumetric approach is widely encountered in scientific literature.

The linear energy density E_l (J/mm) can be described by Gu et al. (2012b)

$$E_l = \frac{P}{v}, \tag{3.1}$$

in which P is the beam power (W), and v is the scanning speed (mm/s).

The surface energy density E_s (J/mm^2) that is applied to the surface of the powder bed can be described by Das et al. (2010)

$$E_s = \frac{P}{vd}, \tag{3.2}$$

in which d is the laser beam diameter on the substrate (mm).

The volumetric energy density E_v (J/mm^3) can be described by Kruth et al. (2005) and Thijs et al. (2010)

$$E_v = \frac{P}{vht}, \tag{3.3}$$

in which h is the hatch spacing (mm), and t is the layer thickness (mm).

A variation of the volumetric energy density E_v (J/mm^3) can be described by Liu et al. (2015)

$$E_v = \frac{4P}{\pi v d^2}. \tag{3.4}$$

The optimization of process parameters is based on the energy density. However, these formulae do not consider other factors such as the thermal conductivity of the powder material, the time interval between printing consecutive layers, the scanning patterns and hatch spacing, the gas flow rate and direction within the process chamber, the scan offset at the corners and boundaries, and the number of times that a layer is scanned if remelting is applied.

3.3 Printing Process

LPBF involves heating and melting the metal powders with laser beams and rapidly solidifying the molten material to form the desired parts with complex thermal histories. Several critical phenomena are known to occur during the melting process,

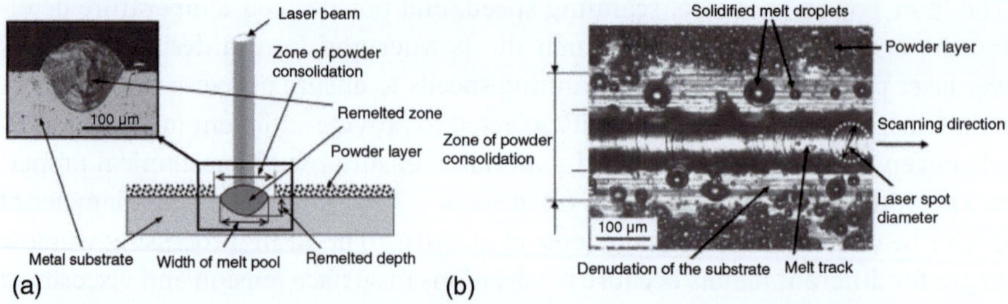

Figure 3.6 A melt pool produced by LPBF from the 316L stainless steel powder on a steel substrate. (a) Typical cross-sectional view and (b) top view. The laser power, scanning speed, and layer thickness were fixed at 50 W, 0.10 m/s, and 40 µm, respectively. Source: Yadroitsev et al. (2010)/reproduced with permission from Elsevier.

such as the absorption of the laser irradiation by the powder materials, heat transfer, melt flow, evaporation of materials, balling, spattering, phase transformation, and chemical reactions.

3.3.1 Melt Pool

Laser–matter interaction in LPBF is a controlled heating process in which the laser serves as a controllable heat source. When a laser beam impinges on a metal powder surface, the optical properties of the powder material determine the reflection, absorption, and transmission of light. The fraction of the laser beam absorbed by the surface of the metal powder is transformed into heat through electron–phonon interactions, and such heat raises the temperature of the surface to achieve the required molten state of the metal. Eventually, a tiny melt pool is produced on the surface of the powder bed (Figure 3.6) (Yadroitsev et al. 2010).

At the initial stage of LPBF, when the powder is still in its solid state, thermal conductivity is the main material property that influences the outcome. During laser processing, only a fraction of the radiation is absorbed by the particles at the surface of the loose powder layer. The rest of the radiation penetrates the voids between the powder particles of the layer and interacts with the underlying particles. Common heat transfer mechanisms (conduction, convection, and radiation) govern heat distribution within the powder layer. The intensity of the laser radiation decreases with its penetration depth into the powder layer. Metals with high thermal conductivity and reflectivity, such as aluminum and copper alloys, are generally harder to process than iron and steel alloys, titanium and its alloys, and nickel-based alloys. Therefore, higher laser intensities and shorter dwell times, which can be estimated by dividing the laser spot size by the laser scanning speed, are generally adopted to minimize conduction losses and obtain a sufficiently high temperature to create a melt pool.

There are three melting modes in LPBF that depend on the processing conditions, i.e. the conduction mode (Figure 3.7a), keyhole mode (Figure 3.7b), and transition mode (Figure 3.8). The threshold value for determining the mode depends on material properties and laser characteristics. As the power density increases above the

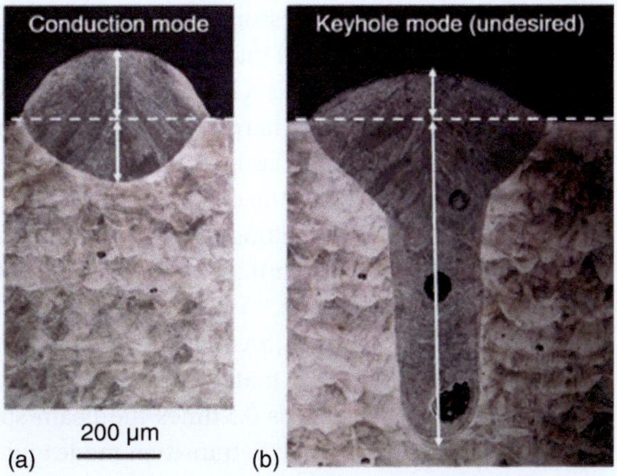

Figure 3.7 Representative melt pool morphology of LPBF-printed 316L stainless steel showing different melting modes. (a) Conduction mode and (b) keyhole mode. Source: Metelkova et al. (2018)/reproduced with permission from Elsevier.

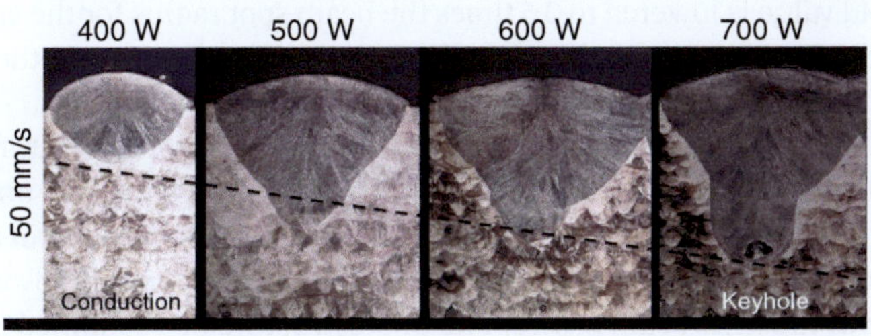

Figure 3.8 Representative melt pool morphology displaying the transition from the conduction mode to the keyhole mode of LPBF-printed 316L stainless steel with increasing laser power. Source: Metelkova et al. (2018)/reproduced with permission from Elsevier.

threshold value, the melting mode changes from the conduction mode to the transition mode to the keyhole mode.

Heat conduction is the dominant heat transfer mechanism for conduction mode melting. When the power density is lower than a particular threshold value, heat conduction is the dominant heat transfer mechanism, particularly for determining the depth of the melt pool created. The cross-section of melt pools created in the conduction mode is generally semicircular, but the depth of the melt pool is smaller than or equal to its half-width (Figure 3.7a).

Note that the vaporization of metals can be ignored in the conduction mode in conventional laser welding. However, it has been claimed to play a significant role in the conduction mode for LPBF (Patel and Vlasea 2020), which can be attributed to the reduced beam size in LPBF corresponding to a higher energy density and, therefore, the possibility of vaporization.

Heat convection is the dominant heat transfer mechanism for keyhole mode melting. When the energy density exceeds a threshold value, a deep vapor cavity forms within the molten metal due to intense localized heating and vaporization of the material on the surface of the molten metal. The depth of a melt pool is controlled by the recoil momentum pressure (also known as recoil pressure) generated by the vaporization of the melt pool material. The top region of a keyhole mode melt pool resembles an hourglass and is wide due to the radially outward Marangoni flow. In contrast, its bottom region is narrow and resembles the shape of a keyhole. The melt pool depth in keyhole mode laser melting is typically greater than the half-width of the melt pool defined at the top of a given melt pool (Figure 3.7b).

The transition mode of melting in LPBF lies between the conduction and keyhole modes (Figure 3.8), in which the dominance of conduction or convection depends on the processing conditions.

The threshold for these three modes depends on the laser absorptivity and thermal conductivity of the material. For materials with average absorptivity and conductivity, the threshold is 0.5 times the beam spot radius for the vaporization depth from conduction mode to transition mode and 0.8 times the beam spot radius for the vaporization depth from transition mode to keyhole mode. However, for high-reflectivity and high-thermal conductivity materials such as aluminum alloys, the threshold value is lowered to 0.5 times the beam spot radius for the vaporization depth from the transition mode to the keyhole mode. In contrast, the transition mode melting can occur with surface vaporization alone.

The melt pool is characterized by melt pool temperature, temperature gradient, lifetime of melt, melt pool dimensions, melt viscosity, melt flow, melt pool stability, and balling and spattering due to melt pool instability. The formation of a melt pool and its characteristics are fundamentally determined by the energy absorbed by the powder bed as the laser beam passes. The microstructure of a solidified melt pool is divided into three regions: (i) the melt pool core, where the cells are finer and exhibit a more equiaxed structure, (ii) the melt pool boundary, where the cells are coarser and mostly elongated, and (iii) the heat-affected zone.

The melt pool temperature and temperature gradient are primarily determined by the energy density. For example, the maximum temperature increases with the linear energy density but slightly decreases with the laser scanning speed (Dai and Gu 2014). The most comprehensive understanding of the temperature profiles has been achieved through computational modeling. As shown in Figure 3.9, the highest temperature gradients appeared within the melt pool soon after the laser scanning commenced (Khairallah et al. 2016). For a laser power of 200 W and a scanning speed of 1.5 m/s, the surface melt pool shape settled into a quasi-steady state of 225 μs after scanning. Another investigation illustrated that temperatures, thermal gradients, and cooling rates in the range of 1000–4000 K, 5–20 K/μm, and 1×10^6–4×10^7 K/s, respectively, were measured for the melt pools of Ti–6Al–4V through a high-speed imaging thermography method with a resolution of 20 μm (Hooper 2018). They were observed to vary with the process parameters, part geometry, and location in the scan path. The temperature gradient in the melt pool is steeper in materials with low thermal conductivity and increases linearly with the applied laser power. The lifetime of the melt pool refers to the duration between the powder particles in the local region melting and solidifying. The melt pool lifetime can be extended by increasing the laser power or decreasing the scanning speed.

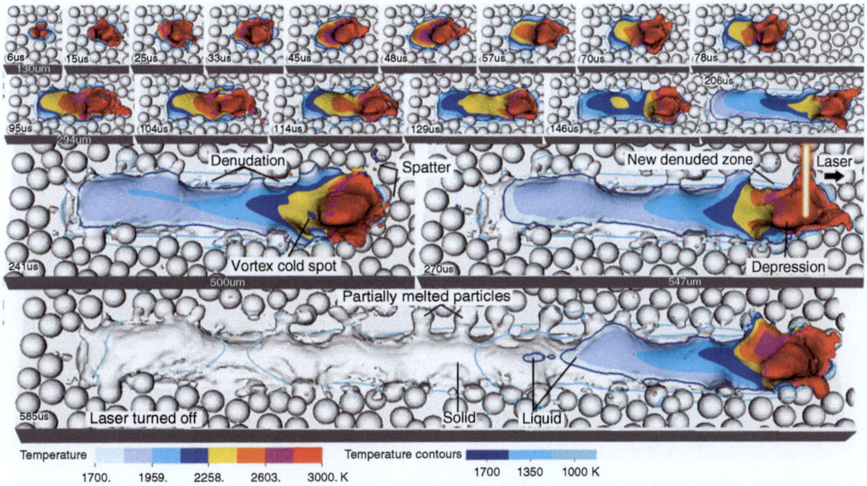

Figure 3.9 Time snapshots presenting the surface temperature evolution in a melt pool of 316L stainless steel. The scanning speed and power of the laser were 1.5 m/s and 200 W, respectively. Source: Khairallah et al. (2016)/reproduced with permission from Elsevier.

The dimensions of the melt pool include its length, width, and depth, which are functions of the absorbed energy density. The influence of linear energy density E_l on the melt pool length and width is more significant than that on the melt pool depth.

Melt flow can be applied to describe the dynamics of melt pools. Mass transfer in melt pools occurs because of the thermocapillary flow, and large temperature gradients produce surface tension gradients within melt pools. The sign of the surface tension gradient $\mathrm{d}\gamma/\mathrm{d}T$ determines the direction of the melt flow, where γ is the surface tension and T is the temperature. As shown in Figure 3.10, if $\mathrm{d}\gamma/\mathrm{d}T < 0$, the flow extends radially outward, and a desirable shallow and broad melt pool is produced; if $\mathrm{d}\gamma/\mathrm{d}T > 0$, the flow is directed radially inward, and a narrow/deep melt pool is formed.

Melt pool stability is responsible for the quality of LPBF-fabricated parts. The range of stable zones in melt pools increases with the laser power, and it becomes narrower for materials with higher thermal conductivity (Yadroitsev et al. 2010). The instability of melt pools results in an irregular and/or discontinuous track, which is derived from hydrodynamic instability and capillary (Rayleigh) instability. The hydrodynamic instability of a melt pool is driven by the Marangoni effect. The Marangoni instability of the flow within a melt pool can be estimated from the dimensionless Marangoni number M_a:

$$M_a = \frac{\mathrm{d}\gamma}{\mathrm{d}T}\frac{\mathrm{d}T}{\mathrm{d}r}\frac{L}{2\eta\delta}, \tag{3.5}$$

where $\mathrm{d}T/\mathrm{d}r$ is the temperature gradient, L is the characteristic length of the melt pool, η is the viscosity, and δ is the thermal diffusivity (Lu et al. 2004). The capillary instability implies the break-up of a liquid cylinder with a high aspect ratio to reduce the surface energy. It occurs when the total surface area of the melt pool is larger than that of a sphere of the same volume and when the viscosity of the liquid is too low.

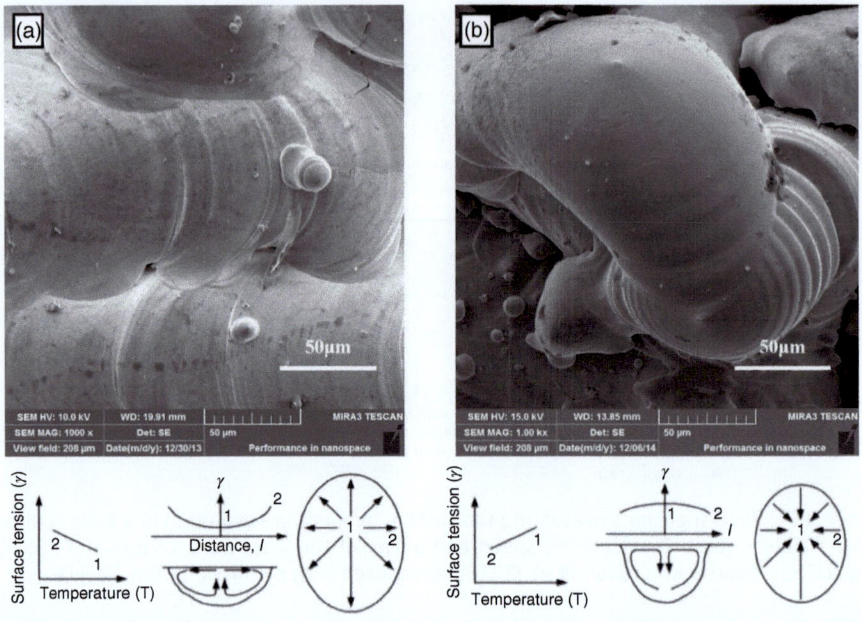

Figure 3.10 Effect of melt flow on the melt track formation of tungsten printed by LPBF. (a) $d\gamma/dT < 0$ and (b) $d\gamma/dT > 0$. Source: Zhou et al. (2015a)/reproduced with permission from Elsevier.

Therefore, it is vital to control the ratio of the length l and the width d of the melt to meet $l/d < 2.1$ (Kruth et al. 2004). The break-up time t is equal to

$$t = \{0.3433\sqrt{\gamma/[\rho(W/2)^3]}\}^{-1}, \tag{3.6}$$

where ρ is the liquid density, and W is the diameter of the melt pool. The expression reveals that alloying elements with only a change in γ do not significantly alter the dynamics of the capillary instability because such a change only affects t on the order of one-tenth of a millisecond. However, a higher viscosity increases the break-up time and results in larger droplets with larger spacings between the droplets. The melt pool instability easily results in a balling phenomenon.

3.3.2 Balling

Balling is characterized by the melt track shrinking and breaking up into a row of liquid spheres to reduce the surface energy due to surface tension when the molten material does not wet the underlying substrate.

Ellipsoidal and spherical balls can be observed in the balling phenomenon of LPBF. The ellipsoidal balls are large with equivalent diameters of ~500 μm, which are detrimental to the quality of LPBF parts. In contrast, the spherical balls are much smaller, with diameters of ~10 μm, and they have little impact on the printed parts (Li et al. 2012). The large-sized ellipsoidal balls can be produced due to a limited amount of molten metal formed and a low undercooling degree of melt

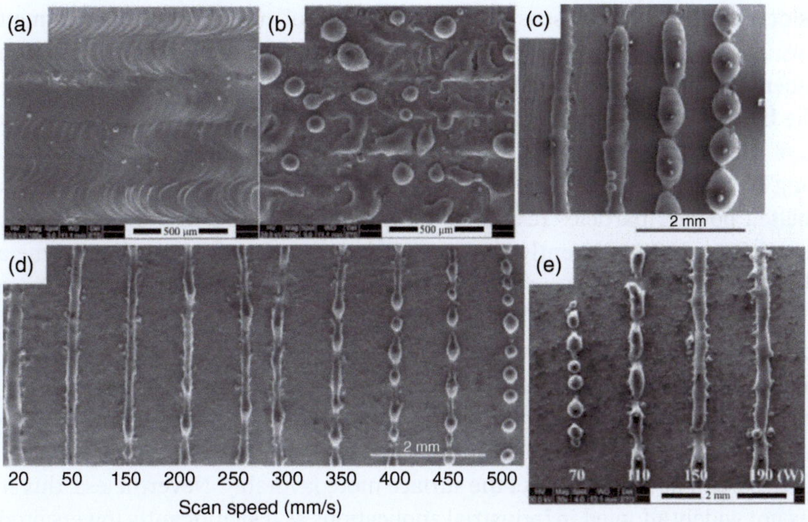

Figure 3.11 Balling phenomenon in LPBF resulting from different mechanisms due to changes in various process parameters. (a) 0.1% and (b) 10% oxygen content, (c) layer thickness (increasing from left to right), (d) scanning speed, and (e) laser power. Source: Li et al. (2012)/reproduced with permission from Springer Nature.

at a low laser power; the small-sized spherical balls are formed by laser-induced splashes due to excessively high energy input (Gu and Shen 2009).

The balling phenomenon severely degrades the material properties and printed part geometry. It severely impedes the uniform spreading of fresh powder onto the previously solidified layer and can potentially induce porosity and even delamination resulting from poor interlayer bonding combined with residual stress. It might even jeopardize the powder-spreading process if the size of the balls is sufficiently large to interfere with the movement of the recoater.

The formation and morphology of balling are influenced by parameters including oxygen content, layer thickness, scanning speed, and laser power during SLM printing, as summarized by Li et al. (2012) (Figure 3.11).

The balling phenomenon can be alleviated by decreasing the oxygen content in the atmosphere (Figure 3.11a,b), as oxidation of the melt pool can reduce the wettability of the liquid at the wetting surface. A large layer thickness is unfavorable for wetting melt pools (Figure 3.11c), as the laser energy absorbed in a unit volume of powder and the contact between powder particles and the substrate are both insufficient. Under a given laser power, increasing the scanning speed results in forming a narrower melt track, which eventually breaks into discontinuous balls (Figure 3.11d). For a given scanning speed, a lower laser power results in a smaller amount of liquid forming, while a higher laser power facilitates the wetting and spreading of melt pools (Figure 3.11e).

In contrast, excessive energy density resulting from a combination of high laser power and low scanning speed may cause self-balling. This combination results in excessive liquid formation and a long melt pool lifetime. Consequently, a considerably lower melt viscosity, a higher degree of superheat of the low melting phase, and an enhanced Marangoni effect are achieved, thereby producing a large number of individual small balls.

The balling phenomenon often occurs during the solidification of the first melt pool of the first layer on a cold substrate, which is also known as "first-line scan balling". The high and steep thermal gradient between the molten material and unmelted powder particles results in a significant increase in the surface tension of the melt pool. Consequently, the melt pool tends to break into spherical metallic agglomerates to minimize surface energy. First-line scan balling can be alleviated by increasing the powder bed temperature. The presence of a low-temperature gradient in this instance leads to lower surface tension of the melt, which restricts the capillary instability and reduces the tendency of first-line scan balling.

In addition to optimizing the process parameters, the balling phenomenon can also be alleviated to a certain extent by remelting the surface, through which the metal balls can be melted to wet the surface more favorably. Nevertheless, this strategy is not widely adopted in industrial applications as it significantly lowers productivity due to the prolonged scanning time.

3.3.3 Spattering

Spattering is a common phenomenon caused by the interaction between the laser and powder bed during the LPBF process, which can lead to the high-speed escaping of powder particles and metal droplets from melt pools. The heat in melt pools is difficult to transmit to the surrounding materials, which results in the temperature of the exposed powder particles exceeding the melting temperature. A further increase in temperature induces the vaporization of the material. During this stage, the rapidly moving vaporized materials expand and generate recoil pressure on the melt pools. While a low recoil pressure facilitates the flattening of the melt pools, a high recoil pressure leads to the removal of molten material by melt expulsion and creates a metallic jet. The ejected metallic jets are disintegrated by the metallic vapor and broken down into micro-droplets when passing through the laser irradiation field to form spatters.

At the starting point of the scanning path, the laser began to strike the powder bed. However, the melt pool was not produced yet, and the vaporization and glare (i.e. the light intensity) were insignificant, as shown in Figure 3.12 (Liu et al. 2015). When the laser beam moved to the middle of the scanning path, the melt pool reached a dynamic equilibrium state, and the glare became steady. When the laser beam moved to the end of the scanning path, it slowed down, and the powder particles absorbed more energy, which dramatically enhanced the vaporization and glare.

The intensity of spattering increases with an increase in the energy density. At a low energy density, the spattering behavior is extremely insignificant as the temperature in the melt pools does not exceed the vaporization point, and almost no metallic vapor is generated. With an increase in the energy density, the extent of the metallic jet, droplets, and sideways spattering increases (the unmelted powder particles around the melt pools are dispersed due to the impact of the metallic vapor, which is

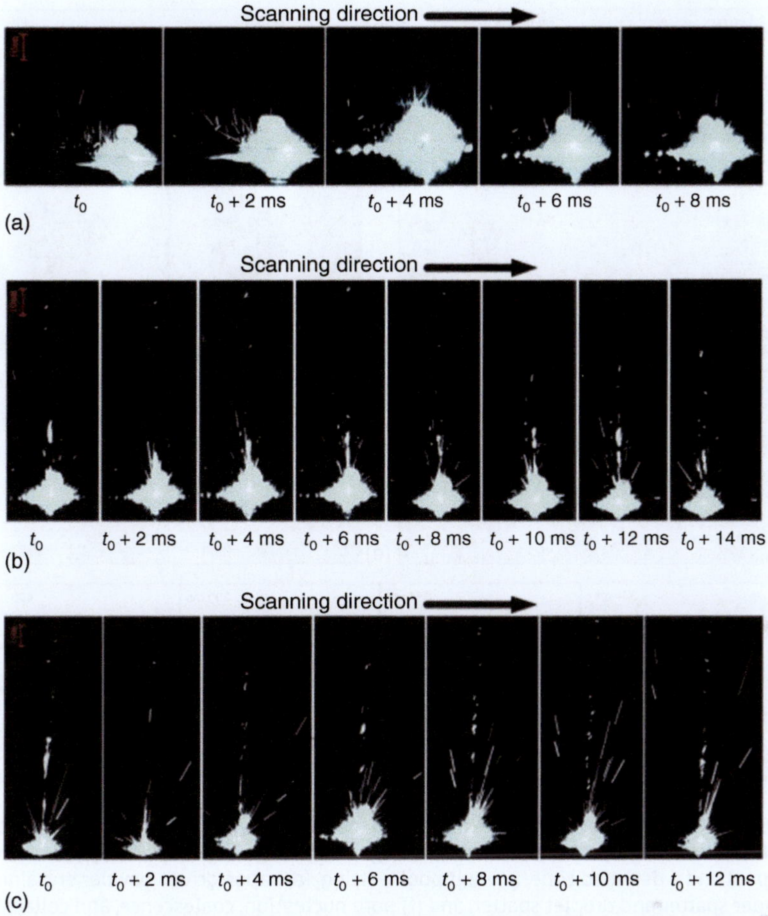

Figure 3.12 Spattering behaviors during the LPBF printing of 316L under different energy inputs. (a) 0.26×10^6 W/cm^3, (b) 0.52×10^6 W/cm^3, and (c) 0.78×10^6 W/cm^3. Source: Liu et al. (2015)/reproduced with permission from Elsevier.

known as sideways spattering). At an extremely high energy density, the intensity of spattering oscillates periodically, which can be attributed to the produced metallic vapor absorbing a portion of the incident laser energy and forming a cloud consisting of small metal particles. This cloud arises from the condensation of hot metallic vapor, preventing most of the laser radiation from reaching the metal powder bed effectively.

Liquid droplets and unmelted powder particles around melt pools are expulsed by the recoil pressure generated on melt pools resulting from the melt vaporization. The spatters resulting from melt expulsion are spherical, and they mix with the metal powder and spread onto the previously solidified layer, resulting in the balling phenomenon. The spattering caused by the expulsion of unmelted particles is also known as satellites. When such satellites fall on the powder bed, their excessively large sizes prevent them from fully melting by the laser beam, thereby becoming

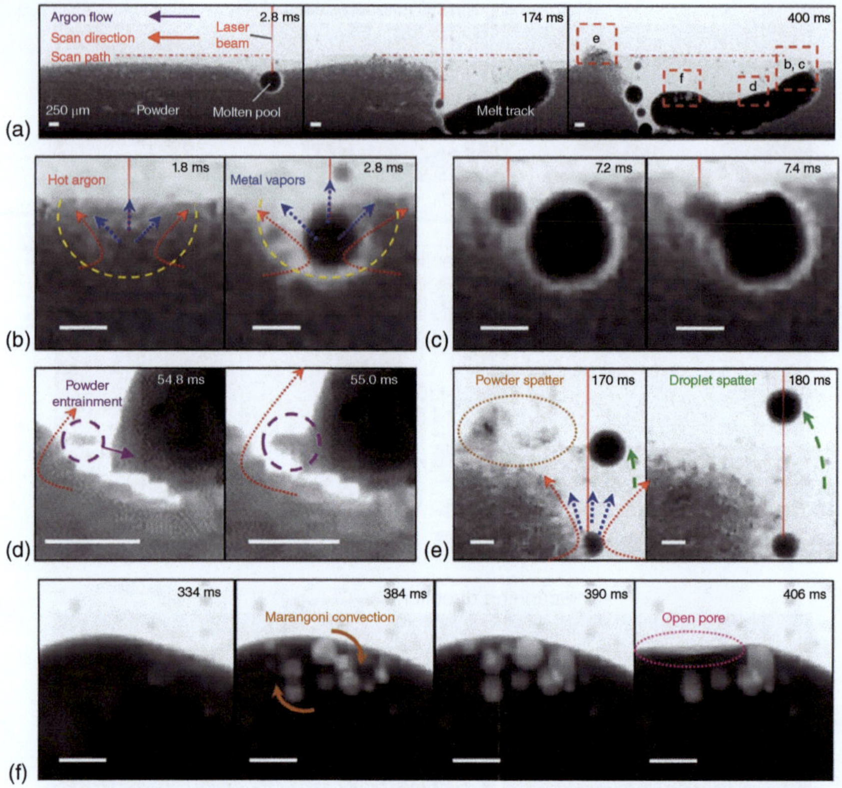

Figure 3.13 Time-series radiographs acquired during the LPBF printing of an Invar 36 single-layer melt track. (a) Melt track morphology at three key stages of LPBF, (b) formation of a melt pool and a denuded zone, (c) melt pool wetting, (d) vapor-driven powder entrainment, (e) powder spatter and droplet spatter, and (f) pore nucleation, coalescence, and collapse. All scale bars are 250 μm. Source: Leung et al. (2018)/reproduced with permission from Springer Nature, CC BY 4.0.

inclusions that form irregular pores in the subsequent layer. Additionally, oxides of a few micrometers in size on the spatter surface resulted from oxidation of the most volatile alloying elements enriched on it (Simonelli et al. 2015).

A comprehensive evolution of a single melt track can be elucidated by an *in situ* and operando synchrotron X-ray imaging method, as shown in Figure 3.13 (Leung et al. 2018). Invar 36 powder appears light gray, while the melt pool and melt track are dark gray because their effective density is almost twice that of the powder. Hence, they attenuate more X-rays. In Figure 3.13a, the laser consolidated powder particles into a liquid melt pool (t = 2.8 ms). The melt pool subsequently evolved into a melt track that extended toward the bottom of the powder bed. As the melt track cooled, it bent upward and formed pores in the last solidified region of the melt track (t = 400 ms).

Multiple steps are involved when a melt pool transforms into a melt track. During the initial stages of LPBF, a melt pool with a diameter of 100 μm formed below the powder bed surface and rapidly grew into a sphere with a diameter of 250 μm. Powder spattering removed a significant amount of powder particles ahead of the laser beam, forming a denuded zone in front of the melt pool, corresponding to fewer powder particles available for the subsequent powder consolidation. As a result, the laser beam moved ahead and formed a new melt pool further along the scan path. The growth rate of this newly formed melt pool was higher than the scanning speed, thus resulting in the laser beam heating the melt pool while lowering its surface tension. Consequently, the newly formed melt pool coalesced with the first melt bead via wetting, revealing a key track growth mechanism.

Serious spattering often induces metallurgical defects (e.g. porosity and cracks) in printed parts. Effective methods, such as optimizing laser process parameters and adding nanoparticles to powder, have been demonstrated to alleviate the spattering during the LPBF process.

3.4 Metallurgical Defects

Metallurgical defects in LPBF, such as pores, cracks, and warping, are commonly observed in the final printed metal parts. These defects reduce the part density and act as stress risers, which adversely affect the properties and performance of the parts. The classification, formation mechanisms, and suppression strategies of these defects are comprehensively introduced in this section.

3.4.1 Porosity

Parts printed by LPBF typically contain more pores than those manufactured by conventional methods (Sames et al. 2016), which severely limits their applications. The buoyant force cannot effectively eliminate pores in melt pools (Weingarten et al. 2015), despite being a commonly employed mechanism that eliminates pores from a liquid. The high drag force induced by the strong melt flow in the LPBF process traps pores within the melt pools. Based on pore formation mechanisms, porosity in LPBF-printed parts can be classified into gas pores, keyhole pores, lack-of-fusion pores, lack-of-material-induced pores, and balling-induced pores.

Gas pores are produced by entrapped gases and are spherical. Firstly, if a metal powder has a low packing density, the gases between the powder particles may dissolve into the melt pools. The dissolved gases cannot escape from the surface of the melt pools because of the high cooling rate during solidification, and thus they form spherical pores. Secondly, gas bubbles can be induced at a high laser energy input due to the vaporization of volatile elements, which can be far beneath the surface at the bottom of the melt pools. These gas bubbles cannot escape from the surface, and thus they form regular spherical pores during the rapid solidification of the LPBF process. Thirdly, gas pores can be inherited from gas-atomized powder (Figure 3.14a). During powder preparation, inert gases, such as argon and helium, are inevitably introduced into the powder materials. As a result, gas pores are

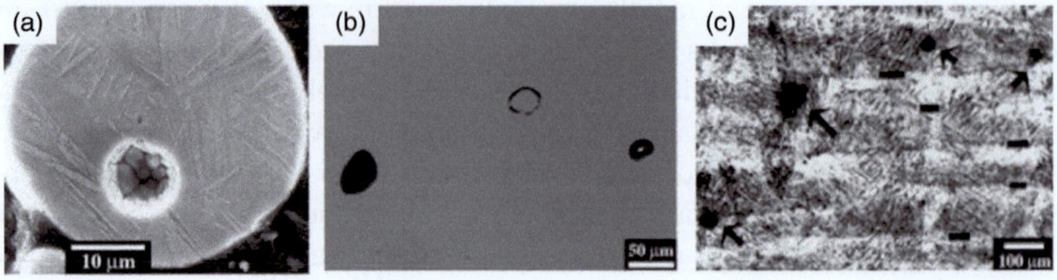

Figure 3.14 Typical morphology of gas pores in: (a) a Ti–6Al–4V powder particle, and (b, c) LPBF-printed Ti–6Al–4V samples. Source: Qiu et al. (2013)/reproduced with permission from Elsevier.

randomly distributed in an LPBF part, and their typical morphology is shown in Figure 3.14b,c.

Keyhole pores are generally produced at the start/end points of a steep and deep melt pool after its solidification when the laser energy density is excessive (Cunningham et al. 2019). As shown in Figure 3.15, pores form at laser turning points because of the emergence and collapse of a deep keyhole depression, which are caused by the deceleration and acceleration of the galvanometer-based scanning mirrors (a common mechanical component for controlling the scanning of the laser beam) during the turn. As the laser accelerates away from the turning point, the keyhole depression

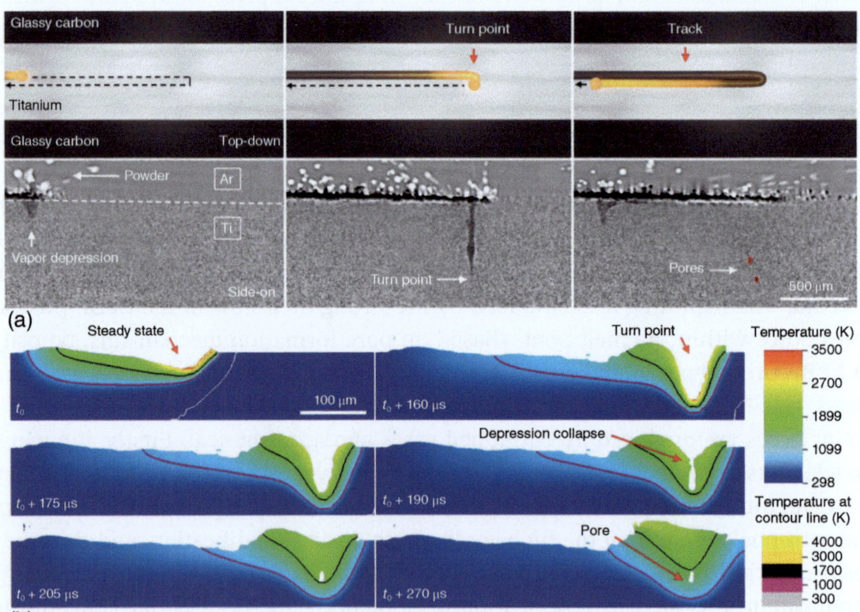

Figure 3.15 Formation of keyhole pores harnessed by *in situ* X-ray imaging and multi-physics modeling. (a) Description of a laser turning point in Ti–6Al–4V performed at a laser power of 200 W and a scanning speed of 1000 mm/s; (b) cross-sections of turning point dynamics from multi-physics simulation. Source: Martin et al. (2019)/reproduced with permission from Springer Nature, CC BY 4.0.

collapses and molten metal fills the void, thereby trapping the argon gas and ultimately forming a pore as the material solidifies (Martin et al. 2019).

More specifically, as the laser beam approaches the end of the track and decelerates, the vapor depression depth increases, which is a direct result of the increase in laser dwell time as the scanning speed decreases for the execution of the turn. More energy is absorbed by the track locally, causing the depth to increase. A more pronounced increase in depth occurs as the turn is completed, which is attributed to the accumulated residual heat being higher than at the entrance of the turning point. Immediately after reaching the maximum depth of the melt pool, the vapor depression depth rapidly decreases as the laser beam accelerates away from the turning point. The significant increase in vapor depression depth due to overheating followed by the rapid collapse gives rise to keyhole formation, as the liquid metal cannot fill the deep vapor depression before rapidly solidifying due to the extreme cooling rate ($\sim 10^6$ K/s). This behavior is universal and occurs regardless of the material at turning points during LPBF.

The mitigation of keyhole pore formation must satisfy the following mechanical and physical requirements: (i) the vapor depression must not transit into the keyhole regime at the turning points, (ii) the laser power must be controlled with no rapid oscillations to maintain a relatively stable energy output, and (iii) the laser power must not rapidly increase when accelerating out of the turning point into the preheated region.

Lack-of-fusion pores with irregular shapes are produced because of the lack of energy input, as shown in Figure 3.16. The metal powder is not fully melted and is thus unable to sufficiently bond to the adjacent solidified layer. The low laser energy input may also lead to the small width of melt pools, which results in an insufficient overlap between the melt tracks. The inadequate overlap induces the presence of unmelted powder particles between the tracks. In the spreading process of a new layer, it becomes difficult to fully remelt these powder particles. Therefore, lack-of-fusion holes are generally formed between the layers of LPBF-printed parts.

The lack of powder materials within a printing layer, which may be derived from uneven powder spreading on solidified printing layers or the powder bed, can induce irregular pores. For example, in a local area where defects such as warping and swelling deformation were generated, the surface of a solidified layer became rough (Zhou et al. 2015b). The rough surface directly contributed to poor powder spreading for the subsequent few layers and poor flow of the molten metal, hence forming interlayer irregular pores that led to significant multilayer defects in the continuous printing process.

Moreover, the oxidation of materials with a high affinity toward oxygen, such as Al–Si10–Mg, can lead to the localized delamination of layers and the formation of irregular pores. An oxide layer is usually formed on the surface of a part due to residual oxygen in the LPBF process. As the wettability of the surface decreases, the molten metal flow is blocked, leading to poor bonding between the layers and introducing fusion defects (Zhang et al. 2017).

Balling-induced pores are derived from the shrinkage of melt pools and spattering, which are caused by incomplete melt flow and excess energy inputs, respectively.

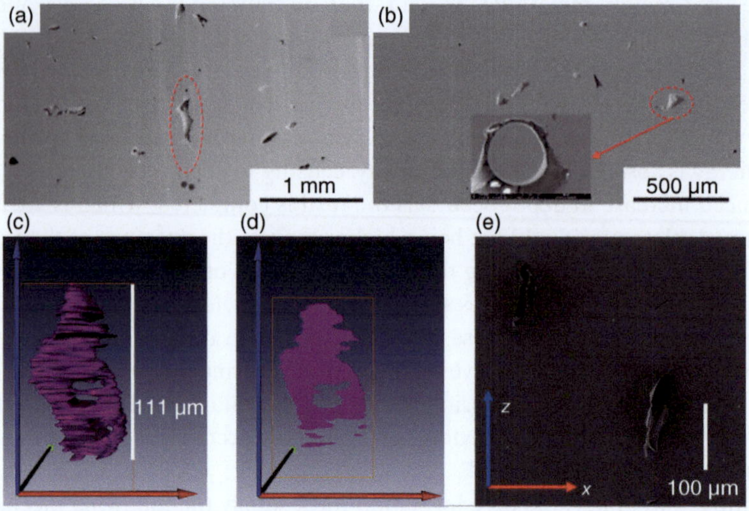

Figure 3.16 Secondary electron images for lack-of-fusion pores in LPBF-printed Al–Si–Mg alloys with volumetric energy densities of (a) 28.66 J/mm^3 and (b) 31.75 J/mm^3. Source: Yang et al. (2018a)/reproduced with permission from Elsevier. Morphology of lack-of-fusion pores in LPBF-printed Co–Cr–Mo alloys. (c) 3D reconstructed top view in the *x–z* plane, (d) cross-section image, and (e) SEM image of the top view. Source: Zhou et al. (2015b)/ reproduced with permission from Elsevier.

Particles from excessively large spattering may not be completely melted, thereby creating irregular pores in the following layers. In most studies, pores can be suppressed by optimizing the process parameters of LPBF. The porosity increased with an increase in the scanning speed. The pores produced from the combination of higher laser power and larger scanning speed were smaller and more circular in shape, which was attributed to the balling effect and high thermal stress cracking; pores produced under lower energy density were filled with unmelted powder particles due to insufficient melting.

The pore motion during printing needs to be understood. One representative research unveiled that the micropore motion is governed by the competition between the temperature gradient-induced thermocapillary force and the melt flow-induced drag force (Hojjatzadeh et al. 2019). As presented in Figure 3.17a, in the region near the laser beam, the micropores moved toward the depression zone and escaped from the melt pool. This region is called the laser interaction domain. Further away from the laser interaction domain, the micro-pores were circulated within the melt pool. This region is called the circulation domain. Between these two regions, a region called the transition domain existed. In this region, the micropores moved randomly, i.e. occasionally moving toward the surface of the melt pool and escaping, while at other times circulating.

In the laser interaction domain, the temperature gradient exhibited an average value of 6.5×10^7 K/m with a direction approximately normal to the melt pool boundary. Such a high temperature gradient resulted in a thermocapillary force at least three times higher than the melt flow-induced drag force. The thermocapillary

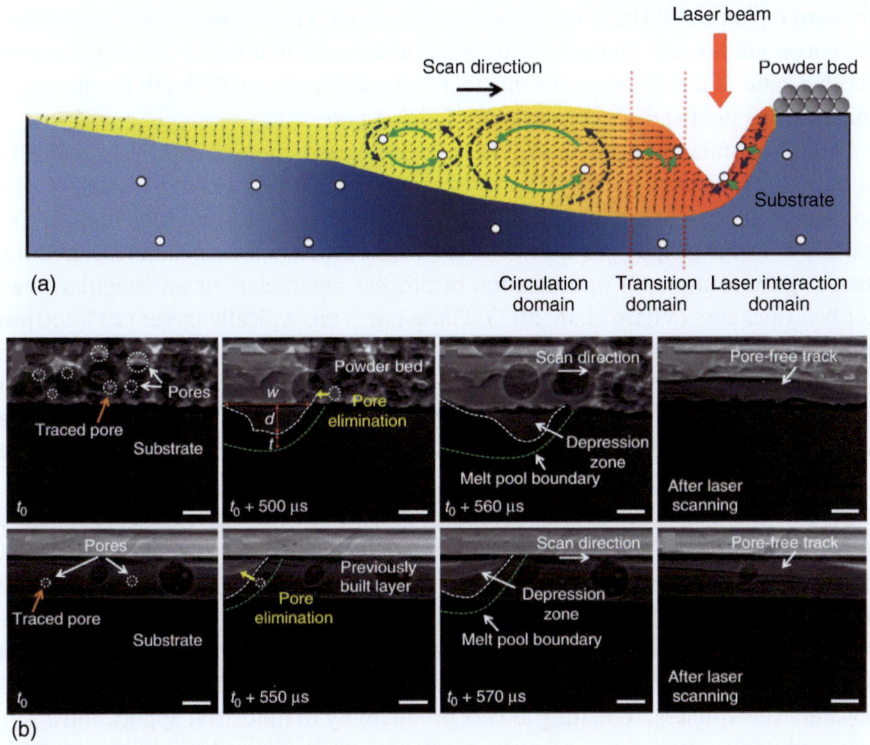

Figure 3.17 (a) Schematic illustration of the dynamics of pore motion and mechanisms of pore elimination during LPBF, and (b) dynamic X-ray images showing the elimination of pores through the application of thermocapillary forces. All scale bars are 50 μm. Source: Hojjatzadeh et al. (2019)/reproduced with permission from Springer Nature, CC BY 4.0.

force drove the melt around the micropore to flow from the hot region to the cold region, resulting in the micropore movement in the opposite direction. Thus, the thermocapillary force is the dominant force in the laser interaction domain, which drives the micropores to move in the direction of the temperature gradient, while the drag force arising from the melt flow controls micropore motion in the circulation domain.

The pore elimination mechanism driven by the thermocapillary force can serve as an effective approach to eliminate micropores during the LPBF process. It has been validated to eliminate micropores in the feedstock powder and the previously printed layer to achieve a pore-free track, as shown in Figure 3.17b. The proper laser process parameters are determined based on the following two general guidelines. First, the temperature gradient in the laser interaction domain is sufficiently high to overcome the melt flow-induced drag force. The temperature gradient around the laser interaction domain can be estimated by the difference between the boiling temperature T_b and melting temperature T_m of the material over the thickness t of the liquid layer around the depression zone, i.e. $(T_b - T_m)/t$. For a given material, a smaller t indicates a higher temperature gradient. Second, the area of the high-temperature gradient region should be reasonably large to provide a high possibility of encountering pores. This area is a large laser interaction domain, defined by the width *w* over depth *d* ratio of the depression zone, *w*/*d*. An excessively small depth *d* indicates a shallow melt pool and may cause lack-of-fusion pores.

Careful optimization and control of the printing parameters are indispensable to minimizing the porosity of LPBF-printed parts. In general, a relative density of 99% and above can be reached by LPBF for various materials. It was reported that the formation of linear-shaped pores in Ti–6Al–4V was attributed to insufficient melting, resulting from improper optimization of process parameters or an irregular powder bed thickness (Vilaro et al. 2011). These pores are typically larger (100–150 μm in length) than spherical pores. Irregular or clustered pores may lead to stress concentration. Therefore, the pores are expected to significantly degrade mechanical properties compared to spherical pores (Maskery et al. 2016). Similar results have been obtained through synchrotron radiation micro-tomography for LPBF-printed 316L stainless steel (Carlton et al. 2016).

On the other hand, a remelting approach, in which a layer is melted a second time before depositing the next powder layer, was validated to reduce residual porosity in LPBF-printed 316L from 0.77% to 0.036% (Yasa and Kruth 2011). In the study, the irregular-shaped pores between the melt pools were shown to close once the material was remelted. Moreover, remelting led to further grain refinement due to the relatively high heat conduction in the consolidated material compared to the initial powder. Nevertheless, remelting lowers productivity in industrial applications.

In addition to process parameter optimization, post-processing can also be done to minimize porosity. A representative study indicated that most of the residual pores in LPBF-printed Ti–6Al–4V can be effectively eliminated by applying hot isotropic pressing (HIP) (Qiu et al. 2013). Nevertheless, HIP cannot suppress open pores and excessively large closed pores. Furthermore, the geometrical accuracy of the LPBF-printed parts may also be negatively affected by the usage of HIP (see Chapter 1 for further details on HIP).

3.4.2 Cracks and Warpage

Macro- and micro-cracks often exist in LPBF-printed parts. Macro-cracks may be related to cold cracks resulting from the brittleness of the material and the delamination of adjacent printing layers within parts due to residual stresses when the parts rapidly cool down to room temperature. On the other hand, micro-cracks may be induced during solidification due to thermal stresses and the formation or dissolution of precipitates along the grain boundaries. The crack density depends mainly on the laser power because the temperature profile depends more on the laser power than the scanning speed; hence, steeper temperature gradients produced by high laser energy densities increase the magnitude of thermal stress, thereby promoting solidification cracking.

Residual stresses are internal stresses within a solid material. Such stresses are present even after all external loading forces have been eliminated, resulting from the equilibrium obtained after the plastic deformation of the material (Withers and

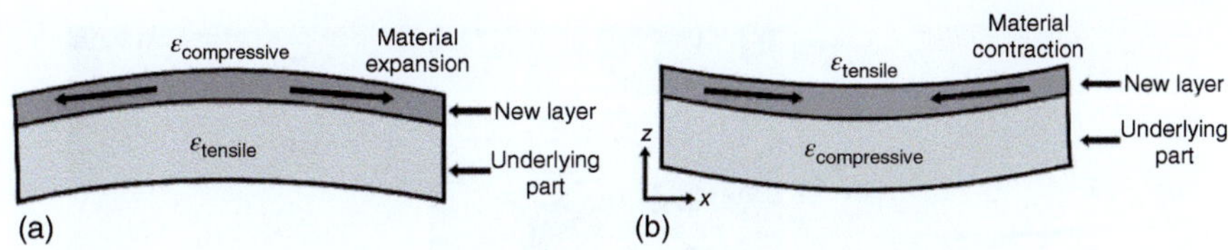

Figure 3.18 Basic mechanisms of stress and plastic deformation development during LPBF. (a) During heating and thermal expansion of the new layer and (b) during cooling and thermal contraction of the new layer. Source: Adapted from Mercelis and Kruth (2006).

Bhadeshia 2001). Residual stresses are usually classified by the length scale on which they operate. Type I residual stresses are macroscopic stresses that act on the geometry of parts and may cause global distortion. Type I stresses directly impact the fatigue properties of the produced materials, resulting in distortion during or after production. Therefore, they are extensively researched. Type II residual stresses are microstresses acting on an individual grain, often known as intergranular stresses. These stresses form because of local microstructural effects, such as grain-to-grain differences in slip behavior. Type III residual stresses are at the atomic scale and are considered misfit stresses due to vacancies, substitutional atoms, etc.

Large and anisotropic Type I stresses commonly exist in LPBF parts because of the large thermal gradients present during processing. Figure 3.18 presents a simplified case of a part produced through LPBF, during which entire layers are melted instantaneously (Mercelis and Kruth 2006). The layers printed up to this point, referred to as the "underlying part" in Figure 3.18, are cooled uniformly with a temperature gradient in the build direction. When a new layer is added and heated above the temperature of the underlying part, the new layer first expands uniformly. However, this expansion is restricted by the much cooler underlying region, resulting in compressive stresses in the new layer and tensile stresses in the underlying region (Figure 3.18a). When the heat source is removed, the new layer cools down rapidly, and the underlying region cannot accommodate such a high rate of contraction, resulting in tensile stresses in the new layer and compressive stresses in the previous layer (Figure 3.18b).

More specifically, during the solidification of the liquid melt, the cooler underlying material restricts the contraction of the melt and pulls on the solidifying region (opposite to the contraction direction); meanwhile, the rapidly solidifying melt tends to pull the underlying material inward during solidification. Since the material has sufficiently melted into the underlying part and assuming that strong metallurgical bonding occurs, strain must be accommodated between these regions and form residual stresses. As described, the stresses in the new layer are primarily tensile, and the resulting residual stress imposed on the underlying material is predominantly compressive. During solidification, the plastic strain that developed can be thought of as hydrostatic contraction, directed inward from every point of the liquid melt. It is noted that nonuniform melt pool geometries and solidification rates can be attributed to the inherent temperature gradient and melt pool motion altering the magnitude of stresses developed in each direction.

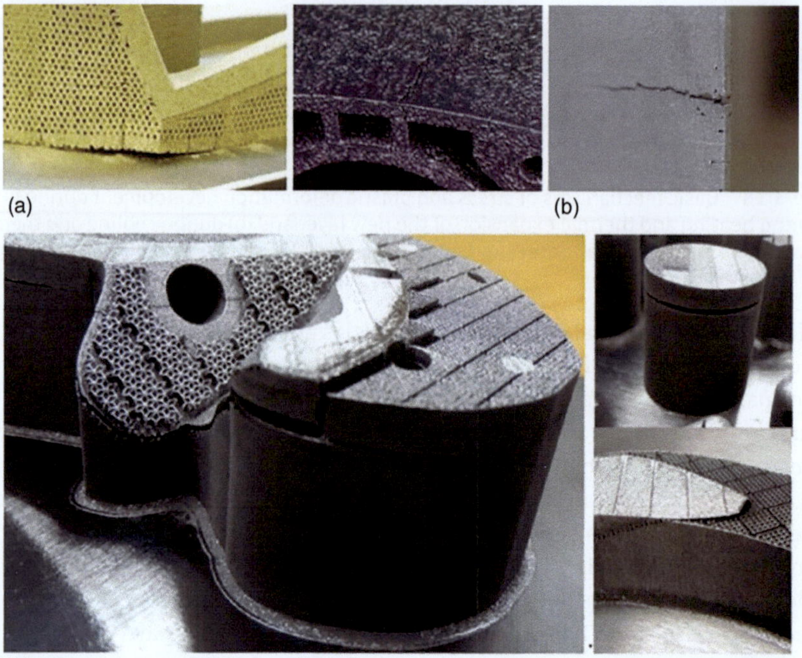

(a) (b) (c)

Figure 3.19 Cracks induced by residual stresses in LPBF-printed metal parts. (a) Tool steel 1.2709. Source: Zaeh and Branner (2010)/reproduced with permission from Springer Nature. (b) Ti–6Al–4V. Source: Parry et al. (2016)/reproduced with permission from Elsevier. (c) Ti–6Al–4V. Source: Yadroitsev and Yadroitsava (2015)/reproduced with permission from Taylor & Francis.

LPBF-printed parts are constantly under tension, with maximum stress located at the surface of the parts. The final residual stress on the parts increases with the number of layers built. The residual stress distribution along the build direction for a part removed from its substrate consists of a zone of tensile stresses just below the upper surface, followed by a large zone subjected to compressive stress, and subsequently a tensile stress zone at the bottom. The magnitude of the stresses depends on the part height and the stiffness and height of a substrate. The typical cracks resulting from residual stresses are shown in Figure 3.19.

The process parameters of LPBF have a critical influence on the magnitude and distributions of residual stress in the parts. Among these parameters, the scanning strategy effect has been investigated extensively as the scan vector length plays a significant role in determining the residual stress. For example, an island scanning strategy with a short scanning length can be advantageous for obtaining low residual stress (Li et al. 2018c). In addition, a suitable rotation angle of the laser scanning direction for the adjacent powder layer can also reduce residual stress (Cheng et al. 2016).

The underlying mechanisms of the scanning strategies to reduce residual stresses are explained as follows. The largest planar stress component generated is parallel to the scan vector, and it increases with the scan vector length. Through a coupled thermo-mechanical simulation, the longitudinal stress (parallel to the scan vector) increases with the scan vector length, and it is the main contributor to residual stress due to the presence of a larger thermal gradient parallel to the scan vector (Parry et al. 2016). Additionally, residual stress and plastic strain are reduced at the end of scan vectors when using the alternate scanning strategy because of the reduced temperature gradients at the end of each vector. Based on these findings, the designed scanning strategies should avoid long scan vector lengths and instead emphasize on orientating the direction of scan vectors uniformly to produce an isotropic stress field in the component.

The effects of laser power, scanning speed, and layer thickness on the residual stress of LPBF-printed Inconel 718 thin-walled parts have been investigated (Chen et al. 2019a). In the study, the residual stress in an LPBF-printed Inconel part was observed to increase with an increase in laser power from 250 to 450 W, and its maximum von Mises stress increased by 22%. The residual stress decreased with an increase in scanning speed from 500 to 1000 mm/s, and the maximum von Mises stress decreased by 14%. The residual stress decreased with an increase in layer thickness from 20 to 60 μm, and the maximum von Mises stress was reduced by 33%. As shown in Figure 3.20, the nature of the residual stress changed from compressive to tensile along the deposition direction, and its magnitude increased with the deposition height. The maximum stress occurred at both ends of the interface between the part and substrate, while the second-largest stress occurred near the top center of the part.

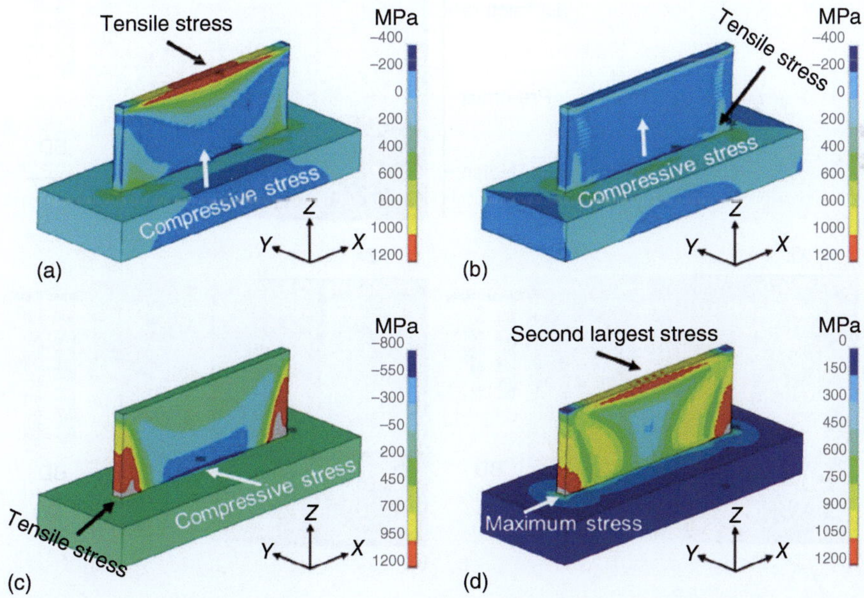

Figure 3.20 Stress distribution in a thin-walled part. (a) X-component stress, (b) Y-component stress, (c) Z-component stress, and (d) von Mises stress. Source: Reproduced from Chen et al. (2019a)/with permission from Emerald Publishing Limited.

The residual stress of LPBF-printed parts can be measured through destructive testing methods, including microhardness testing and the contour method, and nondestructive methods such as X-ray diffraction (XRD) and neutron diffraction.

For alloys without significant precipitates and dominated by a single matrix phase, the shape of their microhardness indents can be utilized to quantify the residual stress. However, the microhardness only reveals information about the stresses near the tested surfaces.

Both XRD and neutron diffraction can be conducted to measure bulk stress variation. However, the two technologies are relatively expensive. They require specialized equipment, and the penetration depth of the radiation beams of both processes is limited, especially for bulky parts with constituent elements of high atomic numbers.

The contour method measures the deflection of the printed surfaces to obtain their residual stresses and can provide results comparable to those obtained by neutron diffraction. It was utilized to measure the residual stress of LPBF-printed Ti–6Al–4V along three orientations, as shown in Figure 3.21 (Vrancken et al. 2014b). The stress field in the *XY* sample exhibited tension at the center, which was balanced by compressive stresses at the left and right edges. In the *XZ* and *ZX* samples, there were compressive stresses at the center and tensile stresses near the top and bottom edges.

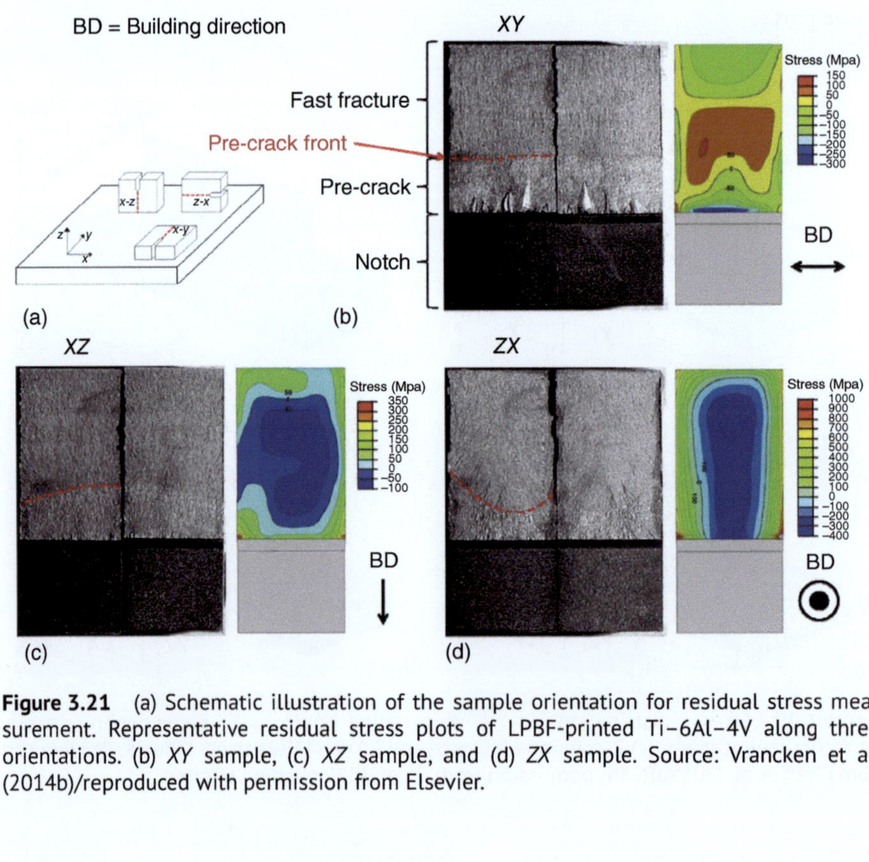

Figure 3.21 (a) Schematic illustration of the sample orientation for residual stress measurement. Representative residual stress plots of LPBF-printed Ti–6Al–4V along three orientations. (b) *XY* sample, (c) *XZ* sample, and (d) *ZX* sample. Source: Vrancken et al. (2014b)/reproduced with permission from Elsevier.

The residual stress distributions and magnitudes provided clear insight into the behavior of the material. The pre-crack fronts of the *XY* and *XZ* samples were both relatively straight. The residual stress of those samples was lower than their yield stress, as shown in Figure 3.21a,b. Moreover, at the notch where the pre-crack was initiated, the maximum and minimum stress differed by ~300 MPa. As the pre-crack grew, the residual stress distribution was readjusted, and the difference decreased. For the *ZX* sample, the maximum and minimum stress at the notched edge differed by >1 GPa. This result indicates that one part of the crack front may be experiencing tensile stress sufficiently high to cause crack growth, and the other parts experienced almost no stress or even compressive stress. The uneven stress distribution resulted in the crack propagating faster along the sides and led to the curved shape of the pre-crack front, as shown in Figure 3.21c.

Large residual stresses bring about great challenges to the design and manufacturing of LPBF parts. The residual stress-induced warpage is especially severe for overhanging regions printed without any support structures beneath them. Support structures are useful for overhanging features because they bear the weight of the feature and compensate for any distortion along the horizontal direction. Moreover, a support structure facilitates heat dissipation as the loose powder possesses poor thermal conductivity. It also provides additional material stock to the base plate for more precise part removal. Although support structures decrease residual stress values, stress-relieving heat treatment is still required (Mishurova et al. 2018).

Despite the advantages of support structures, it is difficult to remove them from internal structural features, such as cooling channels, since these structural features generally lack accessibility. Moreover, the removal of supports may incur additional post-processing and material costs and may cause potential damage to the parts.

For certain crack-prone materials, the presence of residual stress can accelerate the formation of cracks. It is widely accepted that the enrichment of low-melting-point alloy components and precipitates along grain boundaries can produce intergranular cracks with lengths of a few microns up to 100 μm and more. For example, the hot cracking mechanism aggravated by residual stress in an LPBF-printed nickel-based superalloy IN738LC was revealed (Cloots et al. 2016). During cooling, the emerging grains were oriented preferentially along the build direction, which is the direction in which rapid crystallization occurs the most. Because of the rapid solidification, the solidus temperature was lowered until a pronounced critical temperature range (CTR) was characterized, whose boundaries were described by the liquidus and solidus temperatures (Figure 3.22). However, the complete homogenization of the liquid alloy within the CTR was impossible.

Consequently, low-melting-point films with a high concentration of Zr, a product of constitutional undercooling, covered the emerging grains. This led to brittle grain boundaries, resulting in the grains being unable to transfer the residual tensile stress or accommodate the shrinkage caused by the cooling melt. Depending on the liquid film dimension and the strain at the grain boundary, a separation between two adjacent grains may occur, which is perceived as a solidification crack (scenario I in Figure 3.22). In addition, residual stress allows the preexisting solidification crack

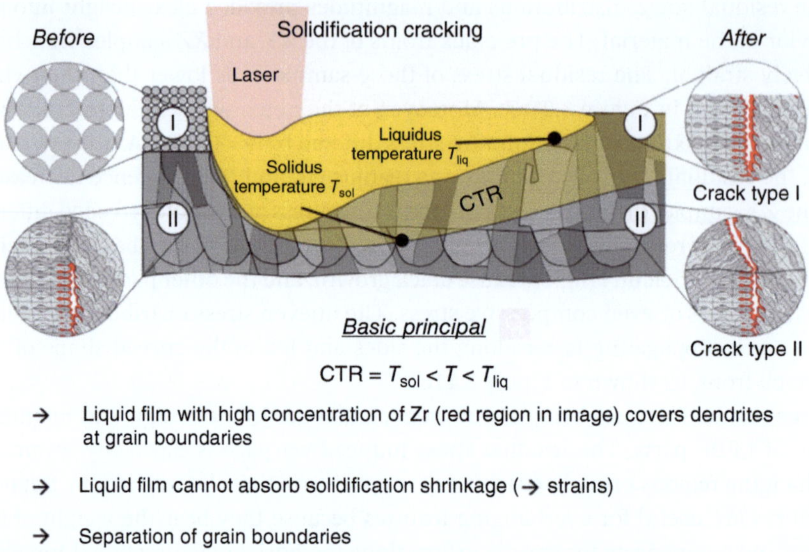

Figure 3.22 Schematic illustration of the solidification crack initiation in an LPBF-printed IN738LC alloy due to residual stress. Source: Reproduced from Cloots et al. (2016)/with permission from Elsevier.

to serve as a nucleus for continuous crack growth in the subsequent layers, as illustrated in scenario II.

In general, residual stress is undesirable since it results in shape distortion of the printed parts. Residual stress and warpage of LPBF-printed parts can be reduced by (i) applying stress-relief heat treatment, (ii) rescanning with lower laser energy densities to effectively stop crack propagation, (iii) preheating substrates to lower the temperature gradient, and (iv) shortening melt tracks with the island/checkerboard scanning strategy and scanning with a rotating scan pattern. However, retaining a reasonable residual stress level in LPBF-printed parts may enhance their hardness (Gu and Meiners 2010), demonstrating that residual stress is not always detrimental to part quality and may be beneficial depending on the applications.

3.5 Powder Materials

Powder quality is one of the critical factors determining the quality of LPBF parts. Poor powder quality can produce defects in the final part, including pores, cracks, inclusions, residual stress, and suboptimal surface roughness, which compromises the throughput. Powder characteristics are governed by the morphology, particle size, surface chemistry, moisture, packing density, and rheological and thermal properties, which affect feedstock behavior and subsequent printing processes. In general, smooth flowability and high powder packing density are desirable to ensure successful material deposition and part densification during the LPBF process.

Powder particles produced from different atomization techniques exhibit various particle morphologies and sizes, as discussed in Chapter 2. Highly spherical particles are often desired in the LPBF process because of their high powder packing density and rheological performance. Hence, qualified powder typically consists of primarily spherical particles with few irregularly shaped or angular grains. Given the same particle size distribution, spherical particles pack more efficiently than irregular particles, resulting in higher packing densities.

In addition to powder morphology, particle size distribution is commonly utilized to quantify powder particle sizes in terms of their volume composition. The particle size distribution fluctuates over various stages of the LPBF process, such as powder storage, spreading, and recycling, thus inducing variations in the feedstock behavior. Powder coarsening can result from pre-sintering of the particles near the melting zone, at which droplets of molten metal ejected from melt pools may adhere to unfused particles, forming larger and less spherical powder particles. The increase in powder aggregation with a higher number of build cycles disrupts the flow and packing performance of the recycled powder for subsequent printing operations. Therefore, before using the used feedstock, necessary sieving procedures are performed during powder recycling operations to minimize size deviations from the original particle distribution array.

The study of granulometry variation has gained traction in the LPBF process and other powder-based AM techniques as changes in powder size can influence material processability throughout the printing process. A typical powder particle size distribution of powder used for LPBF is 10–45 μm, which aims to achieve high part density and surface quality under available laser energy densities (Spierings et al. 2011). However, a slight deviation from the above range of 20–63 μm is also acceptable in some cases.

Fine particles smaller than 10 μm are not preferred as they can contribute to poor powder flowability. As the percentage of fine powder particles increases (corresponding to finer average particle size), the Carney flow time begins to increase until no flow occurs, which can be attributed to increased friction between particles and agglomeration due to static charge. Moreover, fine powder particles are an inhalation hazard. Therefore, the volume percentage of powder particles smaller than 10 μm should be limited in real applications.

Powder contamination is an underlying issue with LPBF, especially during the processing of highly reactive feedstocks, such as magnesium, titanium, and aluminum alloys. Contaminant levels of just a few parts per million can significantly affect the final part quality. The prolonged exposure of reactive feedstock to external environments, interstitial gas intrusions, and the proximity to heat irradiation during part forming may trigger oxidation reactions (Starr et al. 2012). In addition, the drop in water vapor pressure at elevated build chamber temperatures can trigger the formation of hydroxide layers to produce oxides upon crystallization.

Although the LPBF process is often performed under a controlled inert atmosphere in which O_2 concentrations are much lower than 0.15%, oxygen pickup and inclusion are unavoidable, especially in chemically reactive powder materials (Strondl et al. 2015). The oxygen content of recycled Ti–6Al–4V powder was observed to rise above 50% of its original composition, which resulted in a 14% decrease in overall part toughness. Additionally, the reduction in ductility was also accompanied by an increase in part porosity from 0.17% to 0.36% following powder re-usage, which was caused by decreased powder flowability due to changes in the particle shape.

The addition of inhibitors to curb oxidation issues, such as phosphorus and carbon, has been reported to provide deoxidizing effects, which helped reduce the surface tension of melt pools, thereby preventing balling in Fe-based alloys (Kruth et al. 2004). Deoxidizing treatment methods can also be integrated into powder recycling solutions to retain the recycled feedstock quality.

If moisture is adsorbed onto the surface of a powder, its flow behavior may be adversely affected. For alloys susceptible to oxygen and hydrogen pickup, such as titanium and aluminum, the decomposition of hydrogen and oxygen from the water vapor during melting may increase these elements in the resultant solution. Irradiation contact with adsorbed water layers also facilitates the dissociation of hydrogen atoms from water molecules during laser–powder interaction, which can produce entrapped gas pores upon melt pool solidification, thereby resulting in melt pool spattering. Moisture adsorption on the powder surface can contribute to slow-flowing powder due to polarity effects. Depending on the situation, it may be helpful to heat a powder at low temperatures or under a vacuum to improve its flowability. The drying step can aid in the removal of residual moisture from used powder.

The powder materials for LPBF can be classified into pure metals, pre-alloys, and multicomponent mixed metals/alloys (Gu et al. 2012a). Pre-alloyed powders are usually produced through atomization processes, and multicomponent mixed powders can be obtained through mechanical alloying or blending. LPBF can be used to print a broad spectrum of metal powder materials, including Fe-based alloys, Ti-based alloys, Al-based alloys, Ni-based alloys, Co-based alloys, Cu-based alloys, high-entropy alloys (HEAs), metallic glass alloys, and their composites, which are commonly employed in various industrial applications. A summary of these materials amenable to LPBF is provided in Table 3.1, although it may not be exhaustive due to active ongoing research in the LPBF community.

The powder particle shape, size, and size distribution strongly influence laser absorption, powder spreading, powder bed density, and powder bed thermal conductivity. Finer powder particles provide a greater specific surface area and absorb laser energy more efficiently than coarser particles. Powder bed densities typically range between 50% and 60% for most commercially available powders. Generally, a higher powder packing density corresponds to a higher powder bed thermal conductivity and better mechanical properties of parts. The typical effects of powder granulometry on 316L powder performance for LPBF and the final part performance are presented in Table 3.2.

During the powder spreading process, three dominant kinds of deposition mechanisms have been identified: the wall effect, percolation effect, and cohesion effect, which compete with each other during the spreading process and cumulatively affect the packing density of the powder layer, as shown in Figure 3.23 (Chen et al. 2019b).

Table 3.1 Summary of metal powder materials amenable to LPBF.

Materials	Commercial equivalents
Titanium	Commercially pure Ti, Ti–6Al–4V, Ti–6Al–4V ELI, Ti–6Al–7Nb, Ti–24Nb–4Zr–8Sn, Ti–13Zr–Nb, Ti–13Nb–13Zr, Ti–Nb, Ti–Ta, Ti–Nb–Ta, Ti–5.5Al–3.4Sn–3Zr, TiAl
Iron	Pure Fe, 316L, 304L, 314S, M2, H13, 17-4 PH, H20, 15-5 PH, Inox904L, AISI Maraging 300, X110CrMoVAl 8-2, GP1, Fe–Ni, Fe–Ni–Cu–P, Fe–Ni–Cr, Fe–Al, Fe–Cr–Al, ultrahigh carbon steel, EH36
Aluminum	Pure Al, Al–Si10–Mg, Al–Si–7Mg, AlSi12, AlSi20, AlSi50, Al–20Si–5Fe–3Cu, Al–Zn–Mg–Cu, Al–Cu–Mg, Al6061, Al7075, Al2024, Al2195, Al2219, Al2618, Al4047, ScalmalloyRP
Nickel	Inconel 718, Inconel 625, Chromel, Hastelloy X, IN738LC, Nimonic 263, MAR-M 247, NiTi
Cobalt	Co–Cr–Mo, Co–Cr–W, Stellite 21
Copper	Pure Cu, Cu–9Al, Cu–10Sn, Cu–Ni, K220, C18400, NiAl bronze, Brass
Tantalum	Pure Ta
Molybdenum	Pure Mo
Magnesium	Pure Mg, AZ31, AZ91D, ZK60, Mg–9Al, WE43, Mg–Mn
Tungsten	Pure W, 80W–20Fe, 90W–7Ni–3Fe, W–6Ta
Zirconium	Pure Zr
Zinc	Pure Zn
Gold	Pure Au, 3N (18k), 24 Karat gold
Silver	Pure Ag, Sterling silver
High-entropy alloys	CoCrFeMnNi, AlCoCrFeNi, CoCrFeNi, CoCrFeNiTi, CoCrFeNiC, AlCrFeNiV, AlCrCuFeNi, NiCrWFeTi, AlCoFeNiSmTiVZr, MoNbTaW
Metallic glass alloys	Fe-based, Zr-based, Ti-based, Ti/Zr-based

The wall effect can reduce the packing density of the powder layer and is categorized into static and dynamic wall effects. The static wall effect emerges when a stress dip exists at the bottom of the moving powder pile during the powder spreading process, forcing the powder layer to be uniformly deposited regardless of the powder pile size variation along the rake and spreading directions. In addition to the static wall effect of the rake and baseplate, the force-arches act as dynamic walls. The static and dynamic wall effects lead to more voids in the powder layer, thereby causing the packing density of the thin powder layer to be lower than those of powder heaps packed in large containers. This phenomenon becomes more prominent with a decrease in layer thickness or an increase in particle size. The percolation effect exists in bimodal powder particles, which leads to particle segregation within the powder layer, thus reducing the packing density. Therefore, although using bimodal

Table 3.2 Effects of powder granulometry on the performance of LPBF-printed 316L parts. Note that grade herein refers to the classification of the powder size distribution.

Performance	Key findings
Part density	• A negatively skewed powder containing fine particles led to higher part densities than a finer Gaussian grade at a lower layer thickness[a)] • The use of powders with fine granulometry can provide better densification of parts than coarse grades[a)] • Finer particles tend to vaporize at low speed and high energy density, resulting in keyhole porosities[b)] • The more efficient packing behavior of multimodal powders contributed to high part densification compared to Gaussian powders[c)]
Mechanical properties	• The negatively skewed grades exhibited high ductility but low tensile strength as compared to powders with a Gaussian size distribution[d)] • The presence of fine particles induced rapid melting and a prolonged melt pool life, producing coarse grains that reduced the strength of the printed part[d)] • Powder with a narrower size distribution exhibited higher mechanical strength but smaller elongation of the printed part as compared to the powder with a broader size distribution[e)]
Surface quality	• Powders with fine granulometry generated better surface quality over coarser grades; more widely distributed powder (0–45 μm) exhibited better side surface roughness than narrower grades (10–45 μm)[f)] • A positively skewed powder with a higher composition of fine particles generated a smoother melt pool surface than a negatively skewed powder of a similar particle size range[g)]

a) (Spierings and Levy 2009);
b) (Liu et al. 2011);
c) (Olakanmi et al. 2012);
d) (Spierings et al. 2011);
e) (Liu et al. 2011);
f) (Spierings et al. 2011);
g) (Lee and Zhang 2015).

powders improves the packing density for large volumes of powder particles, it is not necessarily applicable to the thin powder layer in LPBF. The cohesion effect caused by the van der Waals force can be estimated by the bonding number *K*. This effect leads to particle agglomeration, reduces the packing density, and becomes more prominent with a decrease in particle size.

3.6 Equipment

A typical LPBF equipment setup comprises an optical system, an elevator system, a recoating system, an air circulation system, and software and control systems. The optical system is the core module in LPBF equipment.

Lasers play a vital role in the LPBF technique and its equipment development. Of the many laser sources discovered over the years, carbon dioxide and Nd:YAG lasers

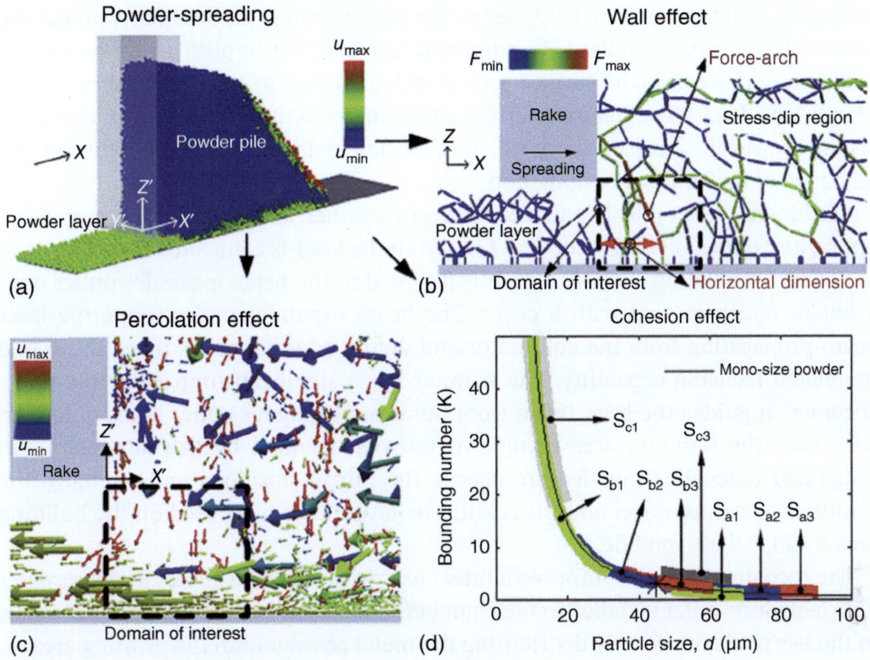

Figure 3.23 Fundamental mechanisms of powder spreading in LPBF revealed through the discrete element method. (a) Velocity field u of the powder pile during the powder spreading process. (b, c) Close-up views of the region near the bottom of the powder pile illustrating the dynamic wall effect and percolation effect. The lines in (b) refer to the magnitude of the contact forces F between neighboring powder particles. (d) Illustration of the cohesion effect, where $S_{a,b,c}$ denote different powder samples with different size distributions. Source: Reproduced from Chen et al. (2019b)/with permission from Elsevier.

have dominated metal AM from the late 1980s to the early 2000s. Afterward, fiber, disk, and diode lasers have become the lasers of choice for simplifying and improving the process (Dutta Majumdar and Manna 2011). A fiber, diode, or disk laser with a wavelength of ~1 μm is more efficient than a CO_2 laser with a wavelength of ~10 μm because of the greater laser absorptivity of the metal at shorter wavelengths.

Laser output power, spatial distribution, time dependence (pulsed or continuous), beam diameter, and beam divergence affect the rate at which energy can be delivered to the material and the spatial distribution of energy. Wavelength is a key factor in determining the suitability of a laser for a particular application. A correct wavelength must be chosen at which the selected material can absorb the energy, or else there will be no transfer of energy into the material. Consequently, no heating or state variation will occur (Figure 3.2). The angle of incidence of the beam to the material and its polarization characteristics determine how much of the incident laser light is reflected from its surface. The spatial distribution of the laser beam on the surface of a material affects the maximum irradiance obtained and the resultant temperature distribution. A Gaussian laser beam with the maximum intensity at its center is usually employed in LPBF. Therefore, the temperature of a produced melt pool surface is higher at the center than at its boundary, producing a radial temperature gradient. This temperature gradient implies a surface tension gradient, which drives the surface profile and mass flow in a melt pool through the Marangoni flow. Most commercialized LPBF manufacturers use continuous-wave lasers. However, a pulsed laser can form a flat melt pool, thereby reducing the tendency of the balling phenomenon.

Additionally, the optical system consists of a scanner, a collimator, beam expander optics, and optical fiber. The optical fiber from the laser is connected to a collimator (a device for aligning beams), which is fastened to the beam expander optics with a holder and protected with a cover. The beam expander optics shape the laser beam propagating from the collimator and define how it reaches the scanner and its related focusing capability. The scanner has a significant impact on the part's accuracy. It guides the laser beam from the beam expander optics along a defined path over the building area using two mirrors operated by galvanometers. An integrated autocalibration feature checks the mirror positions and regulates the positions to reduce offset and gain drift. The laser beam is focused on the building area by an F-theta module.

The recoating system comprises a linear axis with a stepper motor and a recoater with a blade or roller installed in the chamber. The stepper motor moves the recoater in the x–y plane, uniformly distributing the metal powder onto the printing area.

In general, both soft and hard recoaters are commonly employed. A soft recoater possesses blades made from silicone, rubber, or soft carbon fiber, which can deform slightly from a collision due to their flexibility. This type of recoater is ideal for printing different parts on the same build platform; even if one of the parts is deformed or is built inaccurately, the other parts will continue to be produced, and the printing process does not have to be stopped. A hard recoater has blades made of high-speed steel or ceramic, which exert pressure on the powder and do not offer much tolerance for part deformation compared to the soft recoater. When the hard recoater collides with a part, the printing process stops promptly to prevent solidified segments of the part from being dragged across the powder bed. If one of the parts is deformed or is not built successfully, it may be preferable to stop the printing process to prevent a collision. Achieving a uniform powder thickness in each layer is essential to control the quality of the part built on the powder bed. A sufficient amount of powder for building each layer is supplied by dropping a controlled powder volume through a recoater. A powder delivery system should maximize the powder's flowability and minimize the formation of a particle cloud and the shear forces over the previous layer of the build.

The elevator system consists of building, dispensing, and collecting subsystems, of which each subsystem contains a platform mounted in a duct with a linear guide. During the building process, the building system lowers the building platform by the thickness of a single layer after each exposure. Once the building process is completed, the operator can move the building platform with the completed part to the home position. The dispenser platform conveys the metal powder required for the building process from the dispenser duct to the process chamber. In the collecting system, excess metal powder falls onto the collector platform in the collector duct. The operator can elevate the collector platform and push the excess metal powder into the dispenser duct to be reused.

The air circulation system ensures that the process chamber is filled with an inert gas to prevent the metal powder from oxidizing. In the inert gas cylinder, injection nozzles produce a continuous flow of inert gas that removes metal vapor, liquid droplets from spattering, and dust from the inert gas atmosphere. The inert gas is extracted via an extraction nozzle or groove and fed to the recirculating filter unit for purification. The gas flow keeps the protective glass window free of the process by-products detrimental to process reliability.

The software and control systems use a large number of sensors to monitor the temperatures of the chamber and build platform, the status of the laser and scanning system, cooling system, electrical system, powder bed, melt pool, level of oxygen, gas flow, humidity, etc., thereby ensuring the regular operation of the aforementioned systems.

A variety of mature commercial LPBF equipment manufacturers have emerged all over the world, such as EOS (Germany), SLM Solutions (Germany), GE Additive (USA), 3D Systems (USA), Velo3D (USA), Renishaw (UK), Bright Laser Technologies (China), and Farsoon Technologies (China). A non-exhaustive list of the equipment developed by representative LPBF manufacturers and their available materials is summarized in Tables 3.3 and 3.4.

The development of LPBF equipment aims to increase productivity, improve process stability and part quality, reduce cost and time, print at a large scale, and implement part multifunctionality. For example, SLM Solutions has developed the NXG XII 600 machine equipped with twelve 1000 W lasers for high-volume production. It is up to 20 times faster than a standard single-laser system and 5 times faster than a quad-laser machine, with a building rate of 1000 cm^3/h and an unrivaled output of 10 000 kg of parts produced per year. Additionally, the equipment enables the fabrication of short-range overhangs without supports down to an angle of 5° and long-range overhangs of several centimeters down to 10°. Similarly, the Sapphire® System developed by Velo3D can achieve angles as small as 10°, enabling support-free geometries and significantly less post-processing.

LPBF equipment can incorporate preheating modules for substrates, typically up to a temperature of 200 °C. Note that the TruPrint 5000 (Trumpf) and FormUp™ 350 (AddUp) systems can preheat up to 500 °C, and that the AconityMIDI (Aconity3D) even possesses high-temperature preheating of up to 1200 °C. According to Table 3.3, parts with a length greater than 500 mm can be printed via certain LPBF equipment developed for large-scale production, such as SLM®500 and SLM®800 (SLM Solutions), X LINE 2000R (GE Additive), BLT-S510/S600/S800 and BLT-C600/C1000 (Bright Laser Technologies), EP-M650 (E-Plus-3D), and DiMetal-500 (Laseradd). Hybrid additive and subtractive manufacturing equipment (OPM250L and OPM350L) is predominantly developed by Sodick.

3.7 Typical Materials Used in LPBF

3.7.1 Titanium and Its Alloys

Titanium and its alloys are widely employed in the chemical, aerospace, and biomedical industries due to their high strength-to-weight ratio, excellent biocompatibility

Table 3.3 Summary of typical commercialized equipment for LPBF.

Company	Equipment	Building volume (mm^3)	Laser power (W)	Scanning speed (m/s)	Focus diameter (μm)
EOS (Germany)	M 100	Ø 100 × 95	200	7	40
	M 290	250 × 250 × 325	400		100
	M 300-4	300 × 300 × 400	400 (4 lasers)		100
	M 400	400 × 400 × 400	1000		90
	M 400-4	400 × 400 × 400	400 (4 lasers)		100
SLM Solutions (Germany)	SLM®125	125 × 125 × 125	400	10	70–100
	SLM®280	280 × 280 × 365	400/700/ 700 + 1000		80–115
	SLM®500	500 × 280 × 365	400/700		80–115
	SLM®800	500 × 280 × 800	400/700		80–115
	NXG XII 600	600 × 600 × 600	1000 (12 lasers)	—	160
Trumpf (Germany)	TruPrint 1000	Ø 100 × 100	200 (2 lasers)	—	30–55
	TruPrint 3000	Ø 300 × 400	500	—	100–500
	TruPrint 5000	Ø 300 × 400	500 (3 lasers)	—	100–500
DMG Mori (Germany)	LASERTEC 12 SLM	125 × 125 × 200	200/400	—	35
	LASERTEC 30 SLM 2nd Gen	300 × 300 × 300	600/1000	—	70–200
Aconity3D (Germany)	AconityONE	Ø 400 × 400	400 (4 lasers)/1000	4	80–500
	AconityMIDI	Ø 170 × 250	400 (2 lasers)/1000		80–500
GE Additive (USA)	Mlab cusing/Mlab cusing R	50 × 50 × 80; 70 × 70 × 80; 90 × 90 × 80	100	7	50
	Mlab cusing 200R	100 × 100 × 100	200		75
	M2 cusing/M2 cusing Multilaser	250 × 250 × 350	400 (2 lasers)		50
	X LINE 2000R	800 × 400 × 500	1000 (2 lasers)		100–500

Table 3.3 (Continued)

Company	Equipment	Building volume (mm^3)	Laser power (W)	Scanning speed (m/s)	Focus diameter (μm)
3D Systems (USA)	DMP Flex 100	100 × 100 × 80	100	—	—
	ProX® DMP 200	140 × 140 × 100	300	—	—
	ProX® DMP 300	250 × 250 × 300	500	—	
	DMP Flex 350	275 × 275 × 380	500	—	—
	DMP Factory 350	275 × 275 × 420	500		—
Velo3D (USA)	Sapphire® System	Ø 315 × 400	1000 (2 lasers)	—	—
Renishaw (UK)	AM 400	250 × 250 × 300	400	—	70
	RenAM 500M	250 × 250 × 350	500	—	—
	RenAM 500S/500Q	250 × 250 × 350	500 (4 lasers)	10	80
Bright Laser Technologies (China)	BLT-S210	105 × 105 × 200	500	7	—
	BLT-S310/S320	250 × 250 × 400	500 (2 lasers)		
	BLT-S400	400 × 250 × 400	500 (2 lasers)		
	BLT-S450	400 × 400 × 500	500 (4 lasers)		—
	BLT-S510	500 × 500 × 1000	500 (4 lasers)		
	BLT-S600	600 × 600 × 600	500 (4 lasers)		—
	BLT-S800	800 × 800 × 600	500 (6 lasers)		
	BLT-A160/A160D	160 × 160 × 100	200 (2 lasers)		—
	BLT-A300/A320	250 × 250 × 300	500 (2 lasers)		—
	BLT-C600	600 × 600 × 1000	1000/2000/4000	—	
	BLT-C1000	1500 × 1000 × 1000	2000/4000/6000	—	—
Farsoon Technologies (China)	FS121M	120 × 120 × 100	200	15.2	40–100
	FS271M	275 × 275 × 340	500	15.2	70–200
	FS301M	305 × 305 × 400	500 (2 lasers)	15.2	75–200
	FS421M	425 × 425 × 420	500 (2 lasers)	10	70–200

Table 3.3 (Continued)

Company	Equipment	Building volume (mm^3)	Laser power (W)	Scanning speed (m/s)	Focus diameter (μm)
E-Plus-3D (China)	EP-M150T	Ø 150 × 80	200	8	40–60
	EP-M150Pro	Ø 150 × 240	200/500 (2 lasers)		60
	EP-M150	Ø 150 × 120	200/500 (2 lasers)		40–60
	EP-M250	250 × 250 × 300	200/500		—
	EP-M250Pro	262 × 262 × 350	500 (2 lasers)		70
	EP-M260	266 × 266 × 390	500/1000 (2 lasers)		70–100
	EP-M300	305 × 305 × 450	500/1000 (2 lasers)		90–130
	EP-M450	455 × 455 × 500	500 (2 lasers)		80–120
	EP-M650	655 × 655 × 800	500 (4 lasers)		80–120
Laseradd (China)	DiMetal-50	50 × 50 × 50	70	7	20
	DiMetal-300	250 × 250 × 300	500		—
	DiMetal-500	500 × 250 × 300	500 (2 lasers)		—
Sisma (Italy)	MYSINT 100	Ø 100 × 100	200	—	55
	MYSINT 100 RM Dual laser	Ø 100 × 100	200 (2 lasers)	—	55
	MYSINT 300	Ø 300 × 400	500	—	100–500
AddUp (France)	FormUp™ 350	350 × 350 × 350	500 (2 lasers)	10	—
Matsuura (Japan)	LUMEX Avance-25	256 × 256 × 300	500/1000	—	—
	LUMEX Avance-60	600 × 600 × 500	1000	—	—
Aurora Labs (Australia)	S-Titanium Pro	200 × 200 × 250	300	—	—
	RMP1	Ø 450 × 400	—	—	—
Sodick (Japan)	OPM250L	250 × 250 × 250	500	—	—
	OPM350L	350 × 350 × 350	500	—	—

Table 3.4 Summary of available powders for the commercialized LPBF equipment.

Company	Available powders
EOS (Germany)	• M 100: CoCr, 316L, Ti–6Al–4V, W • M 290: Al–Si10–Mg, CoCr, Maraging steel, IN625, IN718, 316L, 17-4 PH, Ti–6Al–4V, Ti–6Al–4V ELI, CP Ti • M 300-4, M 400: IN718, Maraging steel, Ti–6Al–4V, Al–Si10–Mg • M 400-4: Al–Si10–Mg, Maraging steel, IN718, 316L, Ti–6Al–4V, CP Ti, Ti–6Al–4V ELI
SLM Solutions (Germany)	Al–Si10–Mg, AlSi7Mg0.6, AlSi9Cu3, Ti–6Al–4V ELI, TA15, CP Ti, IN625, IN 718, IN939, Ni-Cr-Fe, 316L, 15-5 PH, 17-4 PH, H13, Invar 36, Co–Cr–Mo, CuSn10, CuNi2SiCr
Trumpf (Germany)	AlSi9Cu3, Al–Si10–Mg, AlSi12, 316L, 630, CoCr, bronze, IN625, IN718, Ti2, Ti–6Al–4V, tool steel
DMG Mori (Germany)	Ti–6Al–4V, Al–Si10–Mg0.5, Co–Cr–Mo, 316L, 17-4 PH, IN625, IN718, IN738, tool steel, CM 24 LC
Aconity3D (Germany)	316L, 17-4 PH, maraging steel, 15-5 PH, H13, Hastelloy X, IN625, IN718, IN939, AlSi9Cu3, AlSi12, AlSi7Mg, Al–Si10–Mg, Ti–6Al–4V, Ti6Al7Nb, CP Ti, CoCr
GE Additive (USA)	• Mlab cusing/Mlab cusing R: 316L, Al–Si10–Mg, Ti–6Al–4V ELI, CP Ti, bronze, 17-4 PH, CoCr, 18 karat 3N, 18 karat 4N, 18 karat 5N, Pt alloy, Ag alloy • Mlab cusing 200R: 316L, Al–Si10–Mg, Ti–6Al–4V ELI, CP Ti, bronze, 17-4 PH, IN718, maraging steel, CoCr • M2 cusing/M2 cusing Multilaser: 316L, Al–Si10–Mg, F357, Ti–6Al–4V ELI, CP Ti, hot-work steel, bronze, 17-4 PH, IN718, IN625, CoCr • X LINE 2000R: Al–Si10–Mg, Ti–6Al–4V ELI, IN718
3D Systems (USA)	• DMP Flex 100: CoCr, 17-4 PH, 316L • ProX® DMP 200: CoCr, 17-4 PH, maraging steel, AlSi12, 316L, IN625 • ProX® DMP 300: CoCr, 17-4 PH, maraging steel, AlSi12 • DMP Flex 350, DMP Factory 350: CoCr, 17-4 PH, maraging steel, AlSi12, AlSi7Mg0.6, Al–Si10–Mg, CP Ti, Ti alloys, IN625, IN718, 316L

Table 3.4 (Continued)

Company	Available powders
Velo3D (USA)	IN718, Ti–6Al–4V
Renishaw (UK)	Ti–6Al–4V ELI, Al–Si10–Mg, CoCr, 316L, IN625, IN718, maraging steel
Bright Laser Technologies (China)	• BLT-S210: Ti alloy, Al alloy, high-temperature alloy, CoCr, stainless steel, high-strength steel, tool steel, Cu alloy, W alloy, Mg alloy, Ta • BLT-S310/S320, BLT-S400, BLT-A300/A320: Ti alloy, Al alloy, high-temperature alloy, stainless steel, high-strength steel, tool steel, Cu alloy • BLT-S450, BLT-S4510, BLT-S600, BLT-S800, BLT-C600: Ti alloy, Al alloy, high-temperature alloy, stainless steel, high-strength steel, tool steel • BLT-A160/A160D: CoCr, Ti alloy • BLT-C1000: Ti alloy, high-temperature alloy, stainless steel, high-strength steel
Farsoon Technologies (China)	• FS121M: 316L, Co–Cr–Mo–W, Co–Cr–Mo, 17-4PH, CuSn10 • FS271M: 316L, 17-4PH, 15-5PH, 420, Co–Cr–Mo–W, Co–Cr–Mo, Ti–6Al–4V, TA15, Al–Si10–Mg, 18Ni300, CuSn10, GH3536, IN718, IN625 • FS301M: Al–Si10–Mg, Ti–6Al–4V, 316L • FS421M: Al–Si10–Mg, Ti–6Al–4V, 316L, IN718, GH3536, TA15
E-Plus-3D (China)	• EP-M150T, EP-M150Pro, EP-M150, EP-M250: Ti alloy, CoCr • EP-M250Pro, EP-M260, EP-M300, EP-M450: Ti alloy, Al alloy, Ni alloy, maraging steel, stainless steel, CoCr, Cu alloy • EP-M650: Ti alloy, Al alloy, Ni alloy, die steel, stainless steel, Co–Cr–Mo
Laseradd (China)	316L, CoCr, Ti–6Al–4V, high-temperature alloys, precious alloys, die steel

Table 3.4 (Continued)

Company	Available powders
Sisma (Italy)	• MYSINT 100, MYSINT 100 RM Dual laser: Gold, silver, platinum, titanium alloys, bronze, steel alloys, nickel alloys, CoCr • MYSINT 300: Titanium alloys, aluminum alloys, steel alloys, nickel alloys, CoCr
AddUp (France)	Ti–6Al–4V, Al–Si10–Mg, 316L, 18Ni300, IN625, IN718
Matsuura (Japan)	316L, 630, Ti–6Al–4V, maraging steel, CoCr, IN718, Al–Si10–Mg
Aurora Labs (Australia)	316L, 304, 309, bronze, CP Ti, Ti–6Al–4V, IN718, iron, NiSiB
Sodick (Japan)	Maraging steel, 420, 630, 316L, IN718

and corrosion resistance, and low density (Banerjee and Williams 2013). In addition, the great variety of alloy composition and related microstructure, and the allotrope of Ti render Ti-based alloys one of the most interesting materials for academia and industry.

When the other process parameters were kept constant, the microstructure of LPBF parts fabricated from commercially pure (CP) Ti depended on the laser scanning speed (Gu et al. 2012b). The phase constitutions and microstructural characteristics of pure Ti parts underwent successive changes by increasing the laser scanning speeds: relatively coarsened lath-shaped α (100 mm/s) → refined acicular-shaped martensitic α' (200 mm/s) → further refined zigzag-structured martensitic α' (300–400 mm/s), as shown in Figure 3.24. In another study, the microstructures exhibited a phase transformation from the β phase to relatively coarse plate-like α grains when the laser power and scanning speed were less than 100 W and 100 mm/s, respectively, which was attributed to the occurrence of energy thermalization (Attar et al. 2014). Upon increasing the laser scanning speed beyond 100 mm/s, thermal and kinetic undercooling increased, leading to the formation of refined α' from β.

Ti–6Al–4V primarily comprises a hexagonal close-packed α phase and a body-centered cubic (BCC) β phase at room temperature. During the LPBF process, the α phase transforms into the β phase when heated above the β-transus temperature, while the β phase transforms to the primary α phase at a low cooling rate or to a martensitic α' phase at a high cooling rate. The LPBF-printed Ti–6Al–4V alloy generally exhibited a fine-grained acicular α' martensitic structure, as shown in Figure 3.25 (Thijs et al. 2010). The elongated prior β grains larger than 100 μm indicated epitaxial growth in the build direction. An increased heat input led to a larger width of the melt track and coarser prior β grains. As the printing progressed, the Ti_3Al phase (α_2 phase) precipitates appeared during consecutive reheating to

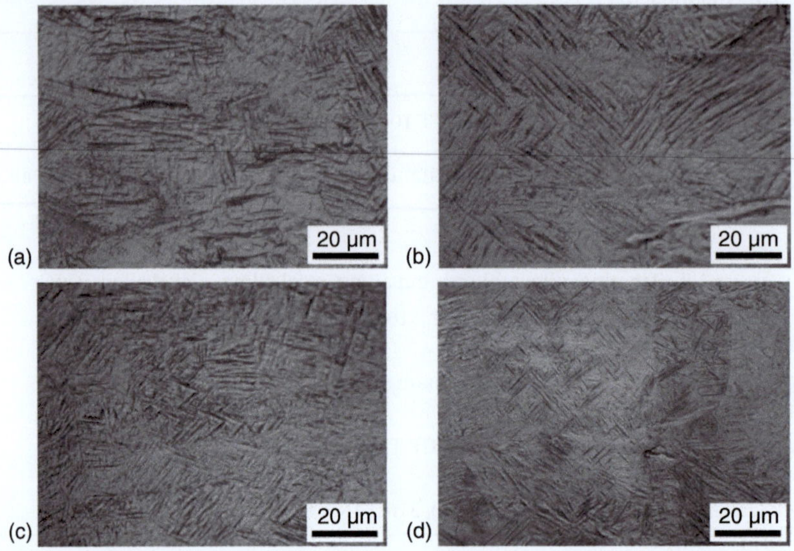

Figure 3.24 Microstructures of pure Ti dependent on LPBF process parameters. (a) Coarse lath-shaped α phase observed at a scanning speed of 100 mm/s, (b) refined acicular α′ martensite at 200 mm/s, and (c, d) further refinement of the α′ martensite at 300 and 400 mm/s, respectively. Source: Gu et al. (2012b)/reproduced with permission from Elsevier.

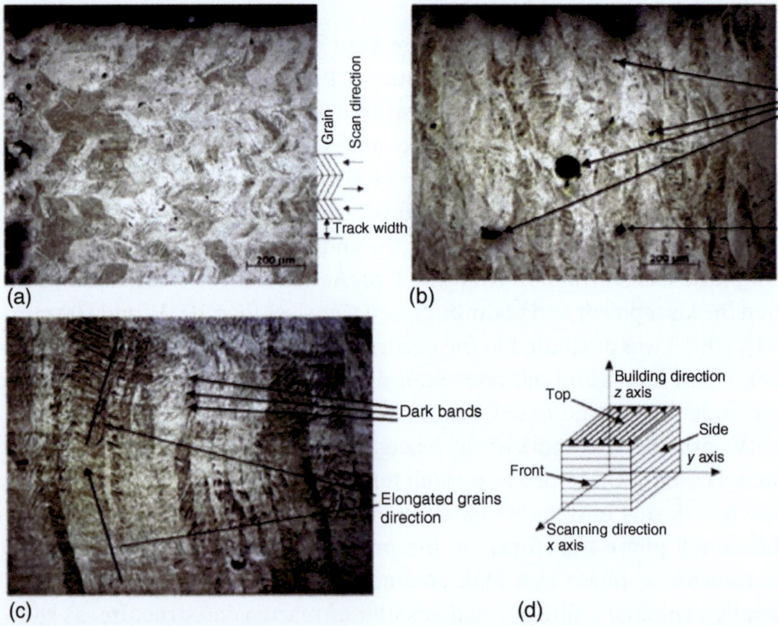

Figure 3.25 Microstructure of a Ti–6Al–4V alloy printed by LPBF. (a) Fine-grained herringbone structure corresponding to the scan direction, and (b) front and (c) side views of elongated prior β grains. (d) Schematic illustration of the sample orientation. Source: Thijs et al. (2010)/reproduced with permission from Elsevier.

500–600 °C. The high temperature gradients and short interaction time led to a small number of such precipitates. The lamellar spacing of the α phase decreased with the increase in cooling rate. Meanwhile, the α phase experienced *in situ* growth, and its amount depended on the local thermal cycles. Significant variation of the α phase lamellar spacing was observed in a single layer (Martina et al. 2012), which is a function of the peak temperature below the β transus. The thickness of the α phase at the grain boundary delineating the prior β grains depends on the cooling rate.

The Ti–6Al–4V sample fabricated under fast cooling conditions consists of acicular α′ martensite that usually displays high yield strength but has limited ductility. Heat treatment is often required to transform the martensitic α′ phase to α + β dual phases to achieve a compromise between strength and ductility. The width of the α lamella plate mainly depends on the heat treatment temperature. The prior β grains transformed to lamellar α + β, α-Widmanstätten colonies, and α′ martensite when subjected to furnace cooling, air cooling, and water quenching, respectively (Vrancken et al. 2012). The lamellar α + β and α-Widmanstätten colonies resulted in high ductility but low strength, while the α′ martensite imparted high strength but low ductility. A previous study proposed a stress-relief treatment at 650 °C to form a very fine microstructure consisting of both α and α′ phases (Wycisk et al. 2015). The thickness of an individual α-lamella was measured to be below 1 μm and slightly coarsened to ~4 μm under HIP, as shown in Figure 3.26. Additionally, it was discovered that HIP conducted at 920 °C for four hours and a pressure of 103 MPa could

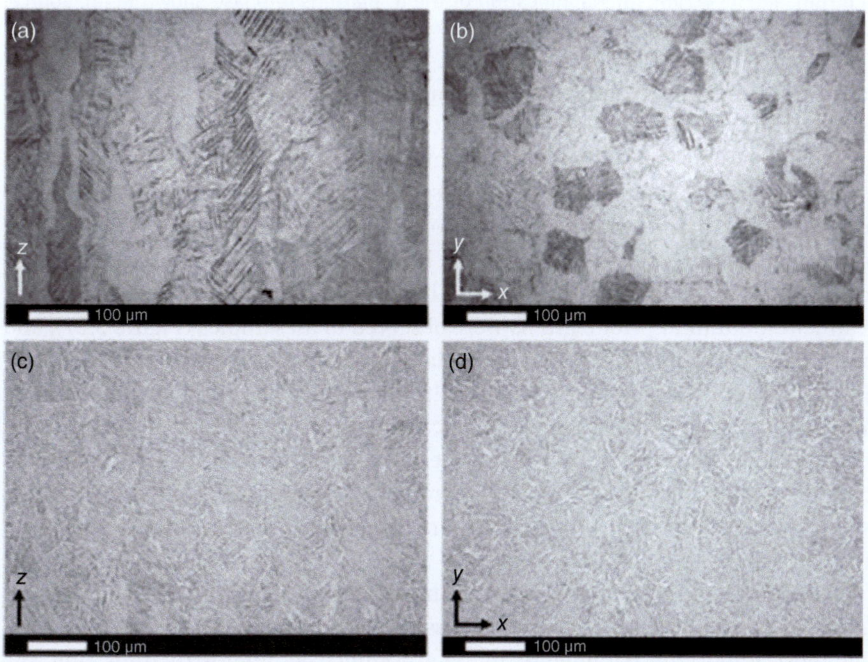

Figure 3.26 Microstructures of a Ti–6Al–4V alloy. (a, b) Stress-relieved condition after heat treatment at 650 °C for 3 h in a vacuum and (c, d) HIP condition. Source: Wycisk et al. (2015)/reproduced with permission from Frontiers Media S.A.

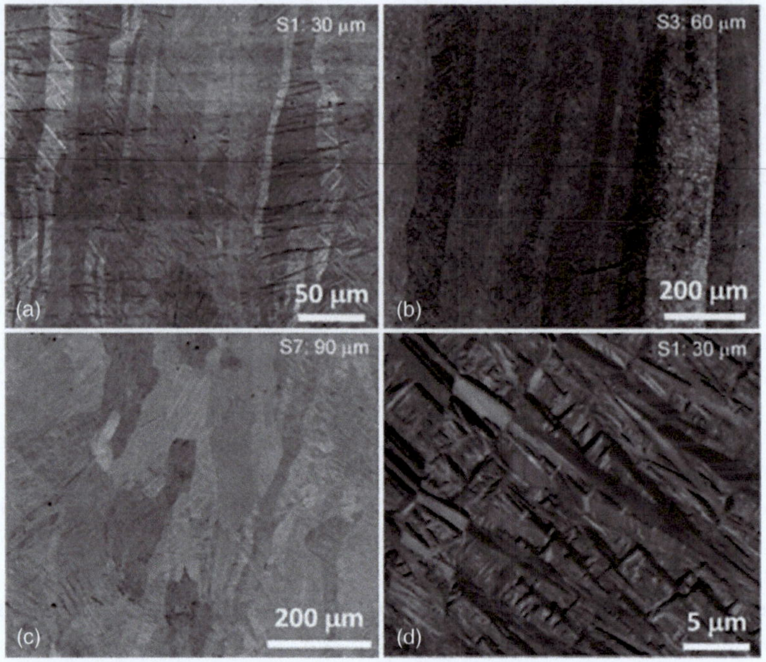

Figure 3.27 Effect of layer thickness on the microstructure of a Ti–6Al–4V alloy. (a) 30 μm, (b) 60 μm, and (c) 90 μm. (d) Magnification of the α′ martensite phase shown in (a). Source: Xu et al. (2015)/reproduced with permission from Elsevier.

transform the as-printed martensitic α′ microstructure into α and β phases with a simultaneous reduction in porosity (Qiu et al. 2013).

The type of deposition (single track or multilayer) and the LPBF process parameters, such as layer thickness, laser focal offset distance, and energy density, critically influence the microstructure of Ti–6Al–4V parts (Xu et al. 2015). Optimizing process parameters could decompose the brittle martensitic α′ phase into a more ductile α + β microstructure, thus potentially avoiding the need for subsequent heat treatment. Columnar prior β grains were observed in all samples (Figure 3.27), and the α′ martensitic phase was present at a layer thickness of 30 μm. However, the α′ phase was found to decompose into an ultrafine lamellar α + β structure for layer thicknesses of 60 and 90 μm. This specific microstructure resulted from an *in situ* heat treatment at ~400 °C induced by the buildup of consecutive layers. The top eight layers consisted of the α′ phase regardless of the process parameters selected, as there were insufficient successive layers to reach the required temperature for decomposition.

Several investigations have focused on the direct relations between the mechanical properties of LPBF parts and the processing parameters or their microstructures. The most important parameters in determining the microstructures and the resulting properties are factors that influence the thermal history and cooling rates. Metallurgical defects, build location, and scanning strategies can directly affect the mechanical properties. The highest strength can be achieved if the process conditions result in a very fine martensitic α′ microstructure, which can be induced by the extremely high cooling rates of the LPBF process. According to the Hall–Petch relationship, grain refinement contributes to the increase in yield strength. The distorted hexagonal lattice structure of the α′ martensite is more robust than that of the lamellar α due to its fine lath width, although it does not necessarily reduce the ductility. LPBF-printed Ti–6Al–4V has increased tensile strength compared to its cast or wrought counterparts. Conventionally processed Ti–6Al–4V with α + β phases exhibit a considerably lower elongation than pure Ti because of the blocking of the twinning deformation modes. With careful adjustment of the LPBF parameters, such as the layer thickness and volumetric energy density, the temperature cycle during the fabrication of parts may be controlled to allow for the decomposition of the α′ martensite phase, leading to a favorable high strength and high ductility combination.

In situ fabrication via LPBF allows pure metals or powder particle reinforcements to be added to the titanium matrix to develop novel titanium alloys or titanium matrix composites, such as Ti/hydroxyapatite composites (Han et al. 2017a, 2018a), (TiB + TiC)/Ti composites (Han et al. 2020a), Mo_2C/Ti–6Al–4V composites (Cai et al. 2021b), Ti–Nb alloys (Zhao et al. 2020), and Ti–Ta alloys (Zhao et al. 2019, 2021). Notably, titanium matrix composites reinforced with TiB and TiC were fabricated *in situ* via LPBF (Han et al. 2020a). Varying contents of B_4C from 0 to 5 wt% were added to pure titanium to prepare blended powders for LPBF. The microstructure of the composite indicated that both the dendritic and cellular structures were arranged by coalescent clusters composed of TiB whiskers and TiC particles. In contrast, the clusters were formed through a self-joining phenomenon of the TiB whiskers that were mechanically locked with the TiC particles (Figure 3.28). The composites with the addition of 1 wt% B_4C exhibited the highest ultimate tensile strength of 946 MPa, yield strength of 762 MPa, and elastic modulus of 128 GPa, which were 62.4%, 49.2%, and 15.3% higher than those of the Ti matrix, respectively. The remarkable enhancement in mechanical strength was attributed to the synergistic effect of dispersion strengthening and grain refinement strengthening.

The microstructures can be tailored by the content of the additives, which has a profound effect on the mechanical properties. For example, the addition of 10 wt% Mo to Ti–6Al–4V completely suppressed the β to α′ martensite transformation during rapid solidification by reducing the β-transus temperature from 995 to 900 °C. Thus, a combination of high strength and good ductility was achieved through the microstructure of a fully β-Ti matrix with dispersed Mo particles (Vrancken et al. 2014a).

Titanium–tantalum (Ti–Ta) alloys are endowed with the excellent mechanical properties of Ti and the osseointegration ability of Ta, and they have emerged as an attractive candidate for bone substitute materials. The role of Ta on the phase transformation, microstructure evolution, mechanical properties, and corrosion resistance of Ti–Ta alloys *in situ* manufactured via LPBF has been studied (Zhao et al. 2019). The results indicated that as the Ta content increased, martensite transformation was gradually suppressed in the printed Ti–Ta alloys due to the β stabilization effect of Ta. Furthermore, with increasing Ta content,

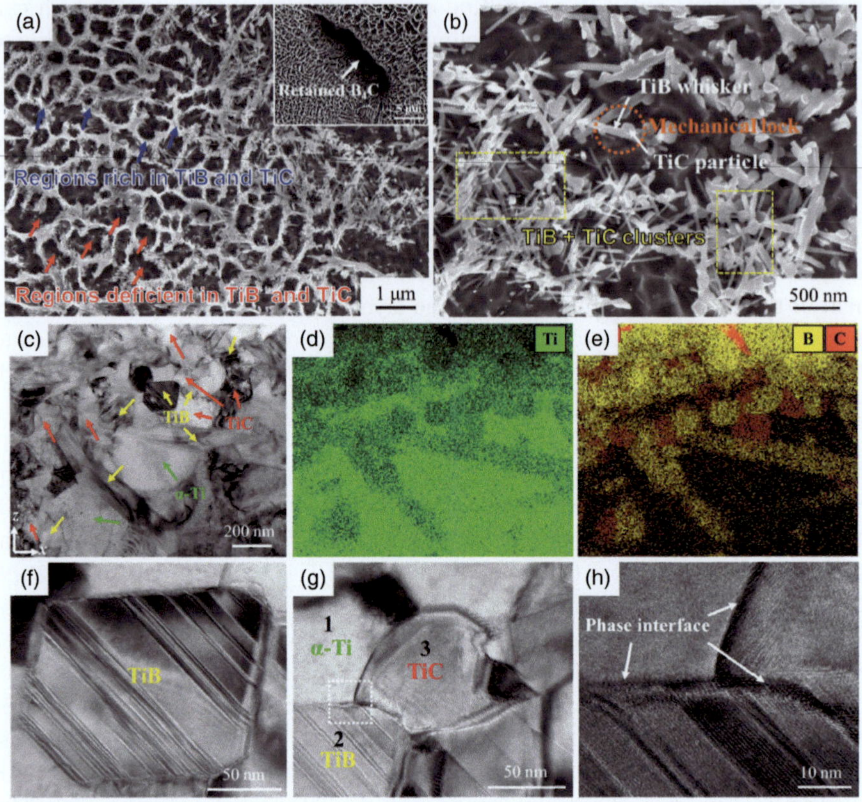

Figure 3.28 SEM images of the cellular microstructure of (TiB + TiC)/Ti composites. (a) Regions rich and deficient in the TiB whiskers and TiC particles, and the insert showing a retained B_4C particle and (b) agglomeration of TiB whiskers and TiC particles. Phase identification in the printed composites by TEM-EDS analysis: (c) TEM bright-field image; energy dispersive spectroscopy (EDS) mapping for the bright-field image showing element distributions of (d) Ti and (e) B and C; TEM images of (f) the TiB phase, (g) the coexistence of TiC, TiB, and α-Ti phases, and (h) an enlarged view of the area indicated by the white dotted rectangle in (g) displaying clear phase interfaces. Source: Han et al. (2020a)/reproduced with permission from Elsevier.

the microstructure evolved from lath α grains to acicular α′ + primary cellular β grains, which led to improvements in the yield strength from 582 to 880 MPa, the tensile strength from 672 to 1186 MPa, and the microhardness from 257 to 353 HV. Moreover, the corrosion resistance of printed Ti–Ta alloys was also improved by introducing a few pits to the surface due to the rising amount of Ta_2O_5 identified through X-ray photoelectron spectroscopy (XPS).

In addition, a Ti–Ta gyroid scaffold with 25 wt% Ta was printed through LPBF. It achieved excellent elastic admissible strain (EAS), bioactivity, and *in situ* bone regeneration capability (Zhao et al. 2021), as shown in Figure 3.29. The printed scaffold with 90% porosity exhibited a good combination of low elastic modulus (1.8 GPa) and high compressive yield strength (55.5 MPa), resulting in a superb EAS (3.03%) suitable for the reconstruction of cancellous bone. The mechanisms of

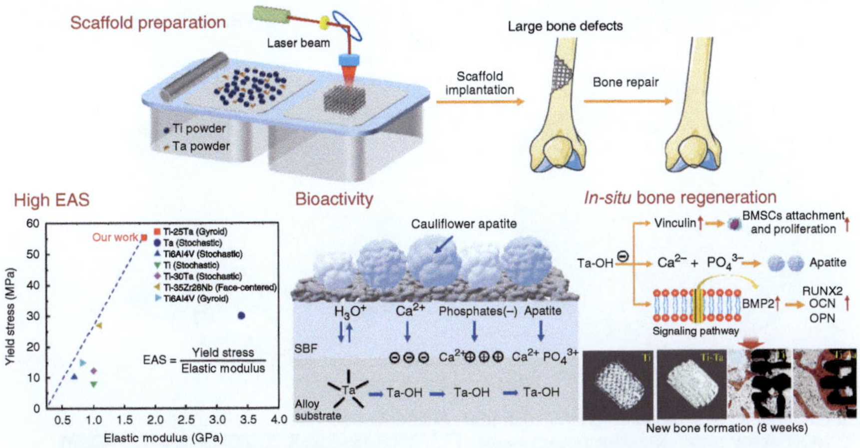

Figure 3.29 Schematic illustration of the titanium–tantalum gyroid scaffold with superb EAS, bioactivity, and *in situ* bone regeneration capability. Source: (Zhao et al. 2021)/reproduced with permission from Elsevier.

the high EAS were ascribed to the formation of β(Ti, Ta) solid solution, ultrafine β grains accompanied by nanocrystalline α′ grains, and the existence of dislocations and stacking faults. In addition, bone-like apatite was spontaneously induced onto the surface of the printed Ti–Ta alloy due to the generation of a self-passivating Ta_2O_5 film, indicating a good biomineralization ability. Compared to pure Ti, the printed Ti–Ta alloy exhibited an enhanced expression of vinculin, earlier cell extension, increased nuclei density, superior cell proliferation, and the up-regulated expression of osteogenesis genes. Studies on animals further validated that the printed Ti–Ta scaffold was capable of reinforcing bone integration and accelerating bone regeneration. These findings provided a promising strategy for treating bone defects through implementing LPBF-processed metallic scaffolds.

3.7.2 Aluminum Alloys

Al alloys offer a good compromise between strength and density. Al alloys are relatively inexpensive and are therefore employed in applications where high mechanical performance and being lightweight are simultaneously required. Cast Al alloys, such as Al–Si10–Mg and AlSi12, are promising candidates for LPBF. The representative microstructures of Al–Si alloys depend on the composition and solidification condition (Dinda et al. 2013). For instance, the microstructure of LPBF-printed Al–Si10–Mg parts is typically cellular with occasional side branches; the cellular primary Al was decorated with fibrous Si particles, and Mg_2Si was precipitated out (Thijs et al. 2013).

An example of the representative microstructure of LPBF-printed Al–Si10–Mg with overlapping melt pools in the vertical and horizontal directions is shown in Figure 3.30 (Aboulkhair et al. 2016). Along the build direction, α-Al solidifies with a columnar morphology with continuous segregations of inter-dendritic Si at the

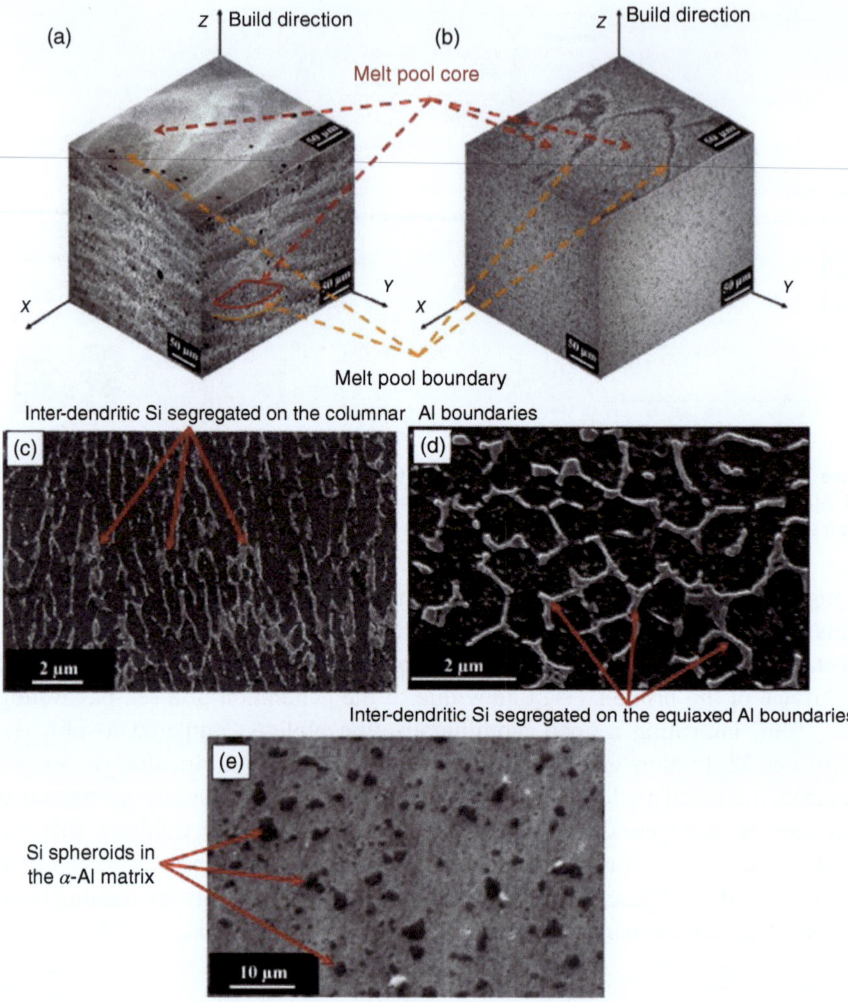

Figure 3.30 Isometric views of the microstructure of an Al–Si10–Mg alloy. (a) As-built and (b) after heat treatment, (c) elongated α-Al in the *x*–*z* plane, (d) equiaxed α-Al grains in the *x*–*y* plane, and (e) Si spheroids in the α-Al matrix after T6 heat treatment (solid solution + aging heat treatment). Source: Aboulkhair et al. (2016)/reproduced with permission from Elsevier.

boundaries (Prashanth et al. 2014). The solidification front rejected Si and deposited it into the liquid, thereby increasing its Si content during solidification. According to the Al–Si phase diagram, the solubility of Si in Al decreases as the temperature drops. High cooling rates in LPBF increased the solubility of Si in the Al matrix to ~7 wt% instead of the expected 1.6 wt% (Li et al. 2015a). Therefore, the formation of a cellular structure is promoted when α-Al solidifies first, leaving the residual Si to segregate at the cellular boundaries. A similar observation was made for the extended solubility of Cu and Mg in Al in AA-2024 (Zhang et al. 2016).

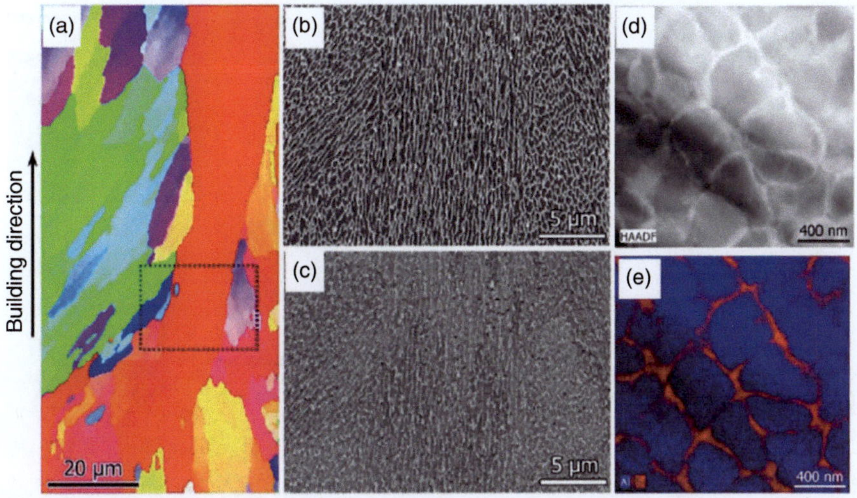

Figure 3.31 (a) Electron backscatter diffraction (EBSD) image of the grains of Al–Si10–Mg produced by LPBF with columnar cells growing parallel to the build direction. The microstructure of the dashed region in (a) obtained using (b) a secondary electron detector and (c) a backscatter electron detector. (d) A scanning transmission electron microscopy (STEM) image of the cells in the LPBF-printed Al–Si10–Mg and (e) the corresponding Al–Si EDX map. Source: Wu et al. (2016)/reproduced with permission from Elsevier.

The columnar morphology was formed as a result of the tendency of the solidifying material to perpetuate in the thermal gradient direction. The columnar cells were hundreds of microns long and 20 μm wide, as shown in Figure 3.31 (Wu et al. 2016). Scanning electron microscope (SEM) images of higher magnifications showed sub-cells in the order of 500 nm.

The inhomogeneous microstructure resulted from the liquid oscillations and temperature profile within a single melt pool. The temperature at the center of the melt pools was higher than that of other regions. Within this temperature range, an inhomogeneous microstructure with Al- and Si-rich phases was formed. The rapid cooling helped retain the Al-rich supersaturated matrix and the Si-rich nano-sized particles. The fine microstructure, alongside the abovementioned solidification mechanism, led to the formation of Al–Si10–Mg parts with a uniform dispersion of the alloying elements.

This type of microstructure validates the sequence suggested by the calculated phase diagram for Al–Si10–Mg using Calphad (Takata et al. 2018). Moreover, the microstructure was dependent on the size of the printed samples. Smaller samples in the 0.1–0.3 mm range contained Si particles inside the columnar Al grains, indicating that Si precipitation occurred during LPBF. In smaller samples, the melt pools were surrounded by regions with unmelted powder particles. Those regions exhibited lower thermal conductivity than those with a solidified material surrounding the melt pools. The lower thermal conductivity reduced the heat flow efficiency, imposing a relatively low rate of solidification, thereby prolonging the duration allowed for the precipitation of Si in the columnar Al grains at elevated temperatures.

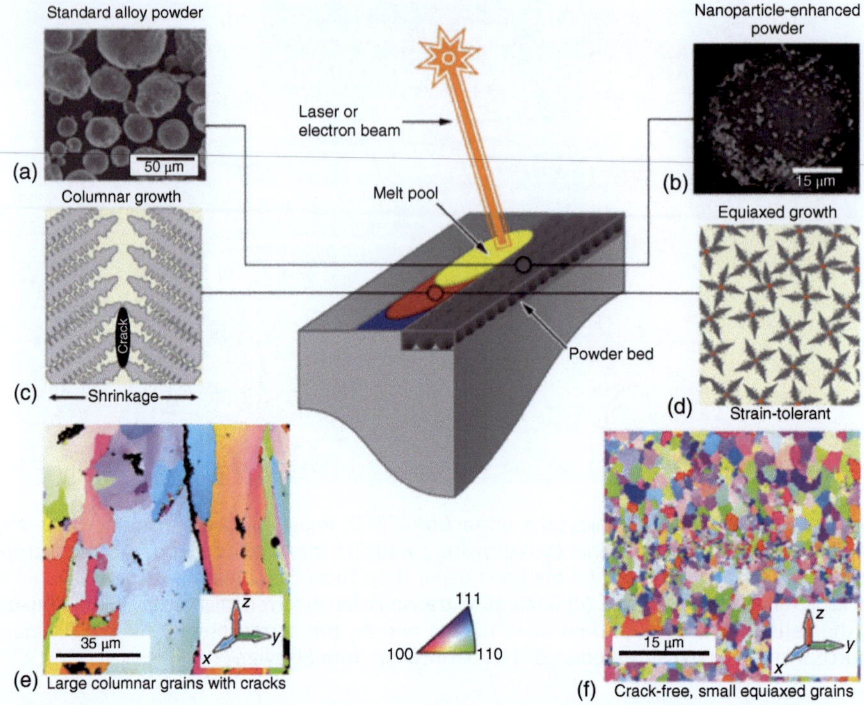

Figure 3.32 Lattice-matched nanoparticles were added to high-strength Al alloys to replace the elongated grains with equiaxed grains and avoid cracking. (a) Standard alloy powder and (b) powder after adding the nanoparticles. (c) Columnar growth and (d) equiaxed growth after the addition. (e) Columnar grains with cracks were replaced with (f) an equiaxed crack-free microstructure. Source: Martin et al. (2017)/reproduced with permission from Springer Nature.

A distinction between the microstructures was generated from the keyhole and conduction modes of melting (Qi et al. 2017). The keyhole mode contained finer grains at the melt pool boundary than the conduction mode because of the higher aspect ratio of the melt pool in the former mode.

The elongated or columnar microstructure in LPBF-printed Al alloys possesses anisotropic mechanical properties. The columnar grain growth promotes the formation of cracks during solidification (Figure 3.32). The addition of lattice-matched nanoparticles to the powder before printing can control the solidification and promote the formation of fine equiaxed grains to achieve high-strength Al alloys (Martin et al. 2017).

High thermal gradients and extreme cooling rates foster the epitaxial growth of columnar grains during LPBF processing, leading to strong crystallographic texture in most Al alloys. Such a texture results in mechanical anisotropy and crack susceptibility. The crystallographic texture in LPBF-printed Al alloys is derived from the directional solidification within the melt pools. The melt pool shape, the heat flow direction at the liquid–solid interface, and the solidification rate are crucially affected by process parameters and the thermophysical properties of the materials (Gu et al. 2012a).

During the early stages of solidification, the average grain growth direction is the same as the solidification front direction, typically perpendicular to the melt pool boundaries (i.e. the direction of the largest thermal gradient). However, thermal gradients are predominantly formed opposite the build direction, depending on the melt pool width-to-depth ratio, as constitutional undercooling is hindered in most Al alloys due to their high thermal conductivity and propensity toward rapid solidification (Martin et al. 2017). These conditions create morphological grain texture, with the longitudinal cross-section of LPBF parts exhibiting elongated grain structures originating from the melt pool boundaries. These elongated grains are either aligned to the build direction or slanted toward the melt pool center. Such morphological grain texture is accompanied by crystallographic texture in metals and is characterized by "easy-growth" directions.

Numerous studies have validated the coupling relationship between the morphology and crystallographic textures of grains formed at the melt pool boundaries. For example, elongated grains consisting of sub-cells of identical orientation were found in an Al–Si10–Mg alloy (Wu et al. 2016). This predominant grain orientation gave rise to a <100> fiber texture component along the scan direction (Thijs et al. 2013), while an AlSi12 alloy also exhibited a similar texture (Suryawanshi et al. 2016). Additionally, the elongated grain structures of the Al–Si10–Mg and Al–Mg–Cu alloys were demonstrated to possess a dominant <100> texture along the build direction (Takata et al. 2017). Such differences were attributed to variations in the geometry of melt pools and the thermal gradients.

Equiaxed grains possess no predominant crystallographic texture, which is advantageous for minimizing mechanical anisotropy. The addition of suitable heterogeneous elements increases the density of nucleation sites in melt pools and facilitates the columnar-to-equiaxed transition of the grain structure. This approach is effective in promoting refined texture-free melt pool grain structures in several Al alloys, including Al–Mg–Zr, AA-2xxx series, Al6061, and Al7075.

Additionally, the intensity of crystallographic textures depends on the combination of different individual melt pools and tracks. During the partial remelting of the neighboring tracks, any refined equiaxed structures formed in the latter stage of solidification can be reduced at the expense of elongated textured grains. The hatch spacing, layer thickness, and scanning strategy selected may influence texture intensity (Thijs et al. 2013; Qi et al. 2017). Inter- and intra-layer rotation scanning strategies have been proven effective in weakening the texture. Melt pools formed by the moving laser source are elongated in shape, and rotating the scan direction reduces the amount of remelted material compared to uni/bi-directional scanning strategies. This strategy can promote the retention of the randomly orientated equiaxed grain structures formed in solidified melt pools (Figure 3.33).

To address the challenges of crack suppression and microstructure control in wrought aluminum alloys, a novel Ti-modified Al–2.25Cu–1.8Mg alloy was developed through thermodynamic calculations of the cracking susceptibility factor,

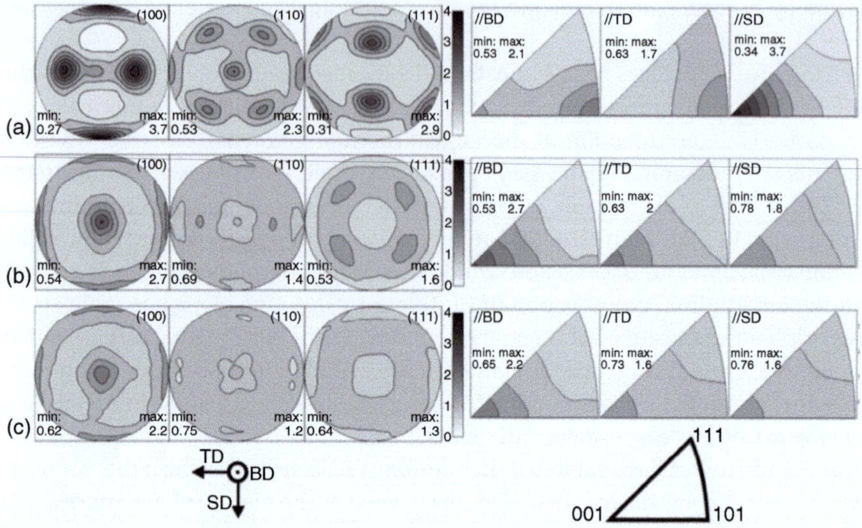

Figure 3.33 Contour pole figures and inverse pole figures of LPBF-printed Al–Si10–Mg samples were produced using (a) unidirectional, (b) bidirectional, and (c) chessboard scan strategies. Source: Reproduced from Thijs et al. (2013)/with permission from Elsevier.

growth-restriction factor, and phase diagram based on the Scheil solidification model, as shown in Figure 3.34a–c (Zhang et al. 2021). Before modification, the Al–Cu–Mg alloy exhibited a textured coarse microstructure with hot tearing cracks distributed along its columnar grains (Figure 3.34d). The addition of 1.5 wt% Ti effectively promoted the columnar-to-equiaxed grain transition and grain refinement due to the formation of Al_3Ti nanoparticles (Figure 3.34e). The LPBF-processed Ti-modified Al–Cu–Mg alloy was crack-free and possessed superior mechanical properties, such as a high ultimate tensile strength of 426.4 MPa, a yield strength of 293.2 MPa, and an elongation to failure of 9.1%. In contrast, the ultimate tensile strength of the LPBF-fabricated Al–Cu–Mg alloy without Ti modification is only 173.2 MPa, with an elongation to failure of less than 2% due to the high level of hot cracking (Figure 3.34f).

3.7.3 Nickel Alloys

Nickel-based alloys possess excellent corrosion resistance and mechanical properties at elevated and low temperatures, making them critical superalloys for industrial applications. Additionally, these alloys exhibit outstanding wear resistance, magnetic properties, and high electrical resistance.

Among the refractory nickel-based alloys, Inconel alloys are resistant to corrosion by chlorine, fluorine, and solutions containing ions of these elements at elevated temperatures. In addition, Inconel alloys have displayed better resistance to progressive oxidation at 1100 °C and in an oxidizing-sulfide atmosphere at 850 °C as they are not vulnerable to vapor, ammonia, or sulfur-containing gas. Due to these properties,

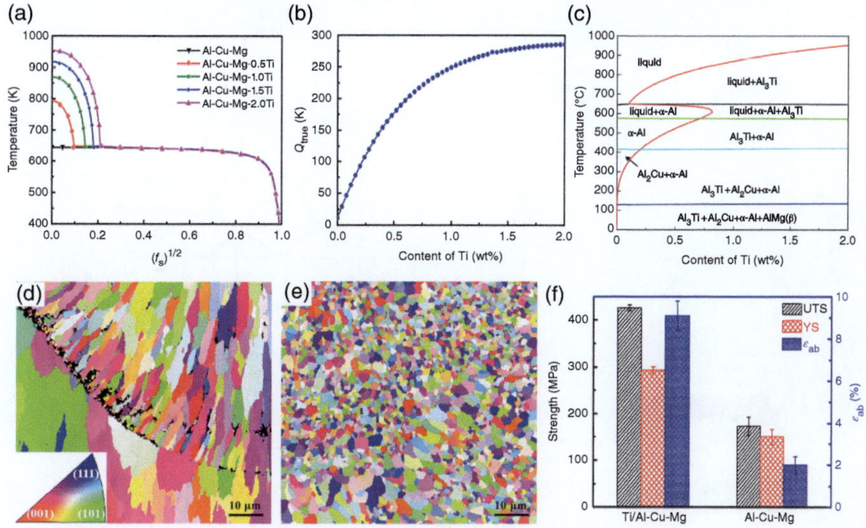

Figure 3.34 (a) Relationship between temperature and the fraction solid f_s, (b) variation in the growth restriction factor Q_{true} with Ti content, and (c) the phase diagram of Al–Cu–Mg–xTi alloys; IPF maps of (d) Al–Cu–Mg and (e) Ti/Al–Cu–Mg samples, and (f) their mechanical properties. Source: Zhang et al. (2021)/reproduced with permission from Elsevier.

Inconel alloys are selected to manufacture steam and gas turbines and parts for the aerospace and nuclear industries.

The formation of precipitates and intermetallic compounds in nickel-based alloys is critical to their mechanical properties. The austenite γ and Laves phases are dominant in the LPBF-built Inconel 718, with almost no other phases, while subsequent heat treatment can promote the precipitation of the γ′ and γ″ phases (Amato et al. 2012; Li et al. 2018b). Figure 3.35a presents a magnified 3D transmission electron microscope (TEM) image of the alloy, which exhibits a variety of dense precipitates within the columnar [100] arrays composed of apparent array boundaries. The horizontal surface contains oblate ellipsoidal or spheroidal particles as large as 250 nm immersed in a dense field of precipitates <10 nm in diameter inside or at the boundaries of equiaxed arrays. Well-defined columnar arrays of elongated precipitates coincident with (100) are observed in the vertical reference plane parallel to the build direction.

Figure 3.35b provides a schematic composite displaying these features as they relate to the scan geometry and melt pool-related layer development, as well as the columnar arrays formed by the directional, epitaxial-like solidification pertaining to the high temperature gradients during the LPBF process. The stacks of disc-like or oblate spheroidal γ″ precipitates appear to be created in successively formed melt pools due to their short interaction time and highly conductive heat transfer, which is possibly facilitated by the argon or nitrogen gas environments or their corresponding thermal conductivities.

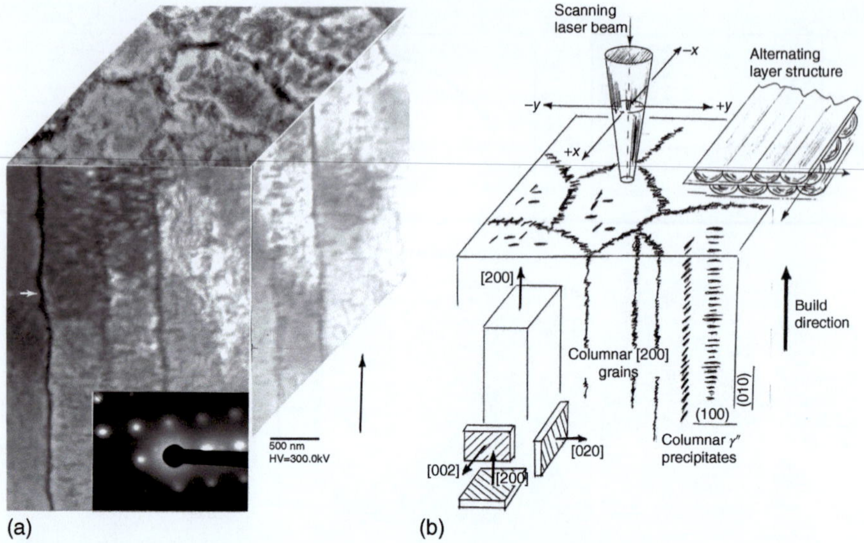

Figure 3.35 (a) 3D TEM image composite view for an as-fabricated cylinder along the *z* axis with HIP treatment. Inset: selected area electron diffraction (SAED) pattern corresponding to a γ [0 0 1] zone in the vertical reference plane. (b) A schematic illustrating the directional solidification-induced columnar γ″ precipitates and the melt pool development in LPBF. Source: Amato et al. (2012)/reproduced with permission from Elsevier.

Comparatively, Inconel 625, a solid solution-strengthened superalloy, is primarily strengthened by Mo and Nb elements. The as-built microstructure of an LPBF-printed Inconel 625 superalloy was austenite with high solution distortions but no carbide precipitation (Li et al. 2015b). After heat treatment, a large number of NbC and MoC carbide precipitates were produced from the matrix. When the samples were treated at 700 °C, the Nb element precipitated and the γ″ (Ni_3Nb) phase was formed. Therefore, the lattice constants decreased accordingly. When the temperature rose to 850 °C, the γ″ phase transformed into a new stable structure of orthorhombic δ (Ni_3Nb). The inverse pole figure (IPF) of the *y*–*z* and *x*–*y* planes showed that the individual grains with different crystal orientations were represented by different colors (Figure 3.36). Some of the larger elongated grains were several hundred micrometers long and grew across a few layers. A strong texture could be identified for the LPBF-printed sample.

The nucleation of γ′-Ni_3(Al, Ti) phase precipitates and the unexpected δ-Ni_3Nb precipitates were observed because of the significant microsegregation of niobium caused by the rapid heating and cooling cycles (Idell et al. 2016). The nucleation of δ precipitates was restricted when the local content of niobium was less than 6.8% molar fraction. By increasing the laser energy, the coarsening effect led to a larger size of the γ′ precipitates. As the sample was built up, the previous layer experienced an aging treatment, which further promoted the formation of the γ′ phase.

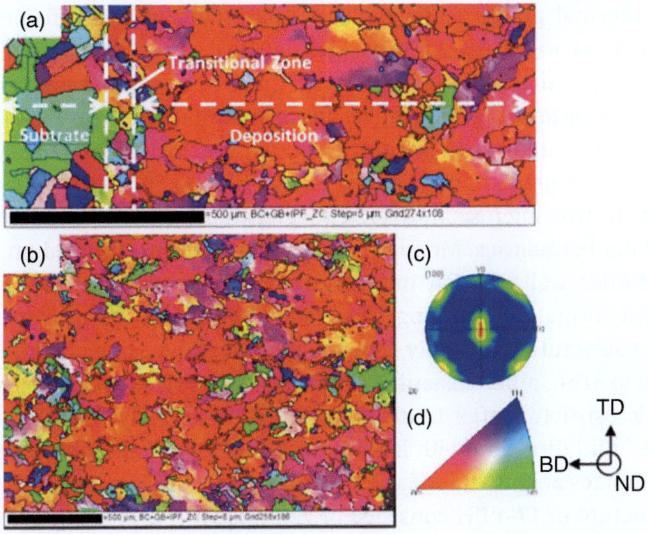

Figure 3.36 Microstructure morphology of an LPBF-printed Inconel 625 superalloy. Inverse pole figure (IPF) colored map of (a) the *y*–*z* plane and (b) the *x*–*y* plane. (c) (100) pole figure, and (d) the index map of the IPF, including the reference coordinates. Source: Li et al. (2015b)/reproduced with permission from Elsevier.

3.7.4 Iron Alloys

Iron-based alloys are the most common engineering materials in industries. Steels available for LPBF are predominantly common austenitic stainless steels (316L and 304L), maraging steel (18Ni-300), precipitation-hardening stainless steels (17-4 PH, AISI: 630 and 15-5 PH), martensitic cutlery steels (AISI 420), tool steels (H13), etc. Besides steels, iron-based alloys and intermetallics, such as Fe–Ni, Fe–Ni–Cu–P, Fe–Ni–Cr, Fe–Al, and Fe–Cr–Al alloys, have also been processed through LPBF.

Most research publications on the LPBF processing of iron-based alloys are based on 316L stainless steel. Abe et al. conducted preliminary tests on this material in 2001, but they were unsuccessful due to balling effects (Abe et al. 2001). The first paper that reported the successful printing of 316L stainless steel through LPBF was published in 2005 (Jandin et al. 2005). In this study, a ytterbium fiber laser with a wavelength of 1065 nm was used to process the 316L stainless steel powder. The experiment indicated that low laser power and high scanning speed led to incomplete melting of the powder material and resulted in components with high porosity. However, 316L stainless steel could achieve 99.9% relative density with LPBF (Tolosa et al. 2010).

Figure 3.37 displays a typical hierarchical microstructure obtained in LPBF-printed 316L (Wang et al. 2018). The 316L alloy was highly heterogeneous, comprising millimeter-sized grains to nanometer-sized precipitations. Compared to conventional pure metals and alloys fabricated by casting, LPBF-printed parts display a gradient transition of crystal orientations within each grain (the grain boundary is defined by a misorientation >10°), as shown in Figure 3.37b. Because of the high thermal gradient and rapid cooling, there was insufficient time for the atomic lattices to be regularly packed, leading to the orientation transition gradient. Moreover, the low- and high-angle grain boundaries were highly dense and irregular (Figure 3.37g,h). Unlike alloys manufactured through casting or molding, the fabricated parts exhibit a cellular microstructure because of the unique temperature gradient and cooling rate conditions in LPBF. Under scanning transmission electron microscopy (STEM), segregations of Mo and Cr were detected near the cellular boundaries. Moreover, nano-inclusions precipitated preferentially along the cellular walls, which induced dislocation pinning and promoted the formation of deformation twinning. The elemental segregation near the cell walls of the cellular structure was highly dependent on the processing conditions.

In addition to 316L, other austenitic stainless steels typically possess a completely austenitic microstructure after LPBF. For example, 304L alloys exhibited elongated grains in the build direction with a 100% austenitic phase and no precipitation of chromium carbides at grain boundaries (Abd-Elghany and Bourell 2012). However, the microstructure of 17-4 PH consisted of 72% austenite and 28% martensite with highly twinned martensite plates and untwinned regions near the martensitic grains with a high density of stacking faults and dislocations (Facchini et al. 2010). The retained austenite was located between the martensite plates, which mainly resulted from residual thermal stresses in the material formed under rapid cooling during LPBF, stabilizing the metastable austenitic phase (LeBrun et al. 2015).

In AISI 420, an austenitic phase may exist due to austenite reversion (Krakhmalev et al. 2015). The top layers revealed a mixed microstructure of fresh martensite and 21% of retained austenite. In contrast, the center of the specimens appeared to be composed of tempered martensite with ~57% austenite. Austenite reversion was also reported to occur during aging precipitation hardening of 17-4 PH, during which the diffusion of Ni and Cu in conjunction with the formation of precipitates led to a similar effect (LeBrun et al. 2015). However, with heat treatment (>550 °C for four hours), the austenite generated in LPBF was observed to partially transform to martensite since the relief of residual stresses might enable the austenite to martensite transformation during post-treatment cooling. The LPBF-printed maraging steel (18Ni-300) also exhibited austenite reversion with Ni-rich reverted austenite shells around the retained austenite regions during aging, as shown in Figure 3.38 (Jägle et al. 2014).

Austenitic stainless steels include γ-austenite and δ-ferrite when processed by LPBF but are generally fully austenitic when manufactured conventionally. In austenitic stainless steels printed by LPBF, no clear trend can be identified regarding yield strength and ultimate tensile strength as linear or volumetric heat input functions. 17-4 PH and 15-5 PH are two widely studied precipitation-hardening stainless steel materials for the LPBF printing process. When processed in a nitrogen atmosphere, the as-printed parts consist of a mixture of austenite (50–75 vol%) and martensite (25–50 vol%), in contrast to the mostly martensite phase (92 vol%) formed in the parts when processed in an argon atmosphere (Rafi et al. 2014; Murr et al. 2012). Heat treatment results in the desired precipitation of Cu-rich particles within a martensitic matrix. The LPBF-printed parts generally exhibit lower yield

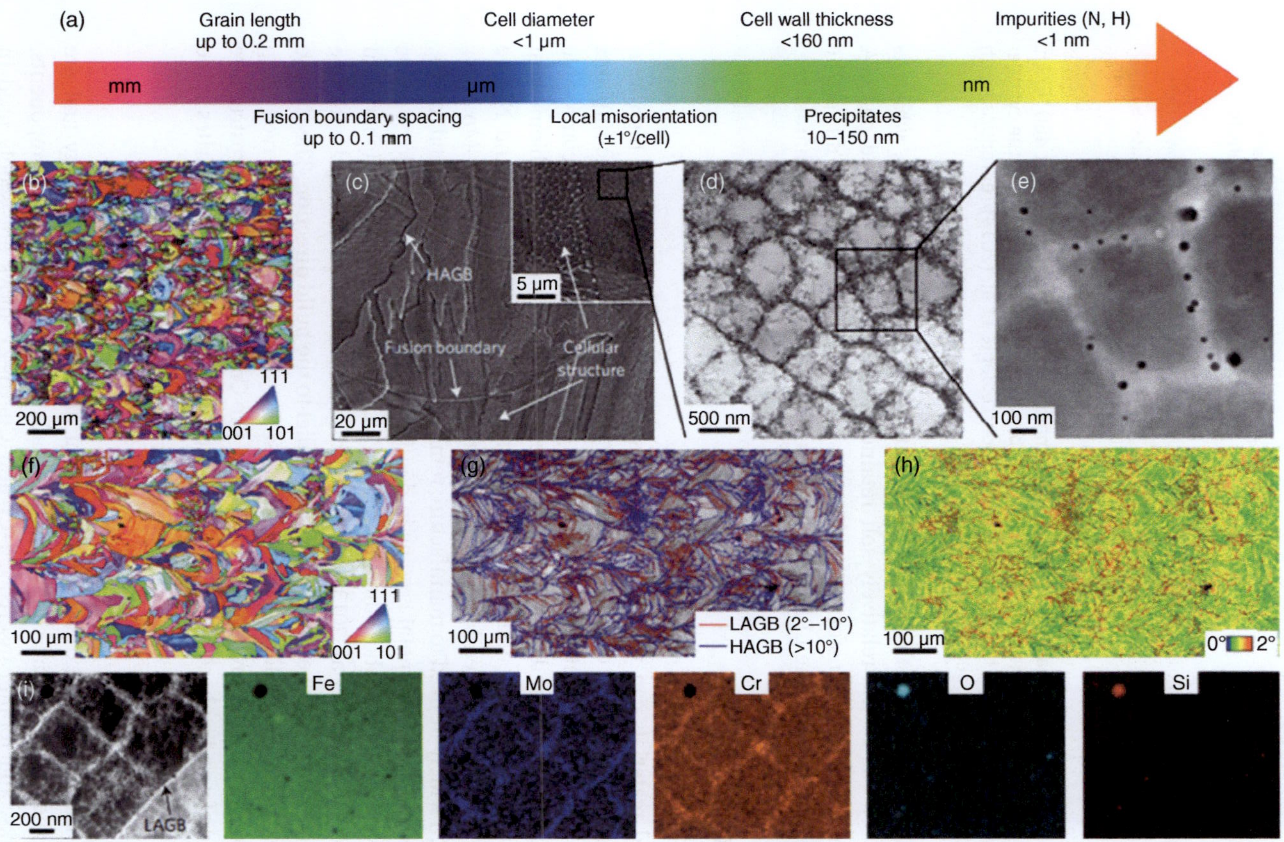

Figure 3.37 A typical microstructure of an LPBF-printed 316L stainless steel. (a) Schematic of microstructures at various length scales, (b) cross-sectional EBSD IPF map of the printed sample showing the grain orientations, (c) cross-sectional SEM image revealing fusion boundaries, high-angle grain boundaries, and solidification cellular structures, (d) TEM image of solidification cells, (e) high-angle annular dark-field (HAADF) image of the solidification cells shown in (d), (f) EBSD IPF map at a higher magnification, (g) EBSD image quality map with grain boundaries, (h) map of the kernel average misorientation, and (i) HAADF image presenting the segregation of Mo and Cr from the solidified cellular walls and low-angle grain boundaries. Source: Wang et al. (2018)/reproduced with permission from Springer Nature.

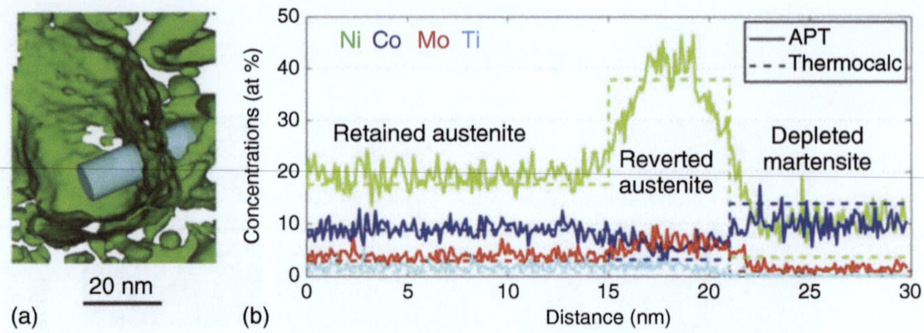

Figure 3.38 Re-austenitization of LPBF-printed maraging steel during aging at 480 °C/5 h. (a) Location of atom probe tomography (APT) and (b) concentration profile along the cylinder shown in (a) for the Ni, Co, Mo, and Ti elements. Source: Jägle et al. (2014)/reproduced with permission from Cambridge University Press.

strength and hardness than their heat-treated wrought counterparts, which is most likely due to the relatively soft retained austenite in the printed parts. A strain-induced phase transformation from austenite to martensite occurs during the plastic deformation of the LPBF-printed parts, resulting in a high strain hardening.

3.7.5 Others

3.7.5.1 Cobalt Alloys

Co-based materials can withstand extreme service conditions such as high temperatures and corrosive environments. Their applicability in orthopedic implants has also been demonstrated. These materials are among the hardest biocompatible alloys with high wear and corrosion resistance, and they are employed in biomedical applications, such as dental, joint, and fracture fixation. Co–Cr–Mo and Ni-free Co–Cr–W alloys produced by LPBF are commonly investigated. Co–Cr–Mo alloys have been widely utilized to manufacture removable partial dentures, metal frames, and porcelain-fused-to-metal crowns in dentistry. For Co–29Cr–6Mo alloys, dense builds were obtained when the input energy of the laser scan was higher than 400 J/mm^3 (Takaichi et al. 2013). A unique microstructure was obtained with fine cellular dendrites in the elongated grains parallel to the build direction. The γ phase was dominant in the build, and a preferential <001> orientation was observed along the build direction at low energy input. For Co–Cr–W alloys, the microstructure consisted of γ and ε phase in a square-like pattern with fine cellular dendrites at the borders (Lu et al. 2015).

3.7.5.2 Copper Alloys

Copper and its alloys exhibit high electrical and thermal conductivity and are some of the most essential conducting materials utilized in electrical components, heat sinks, and tooling inserts. Cu–Cr and Cu–Cr–Zr alloys are widely used in aerospace and nuclear industries; however, binary Cu–Cr alloys are known to suffer from dynamic embrittlement due to the segregation of sulfur elements at grain boundaries. Minor additions of Zr and Ti can improve the hot ductility of

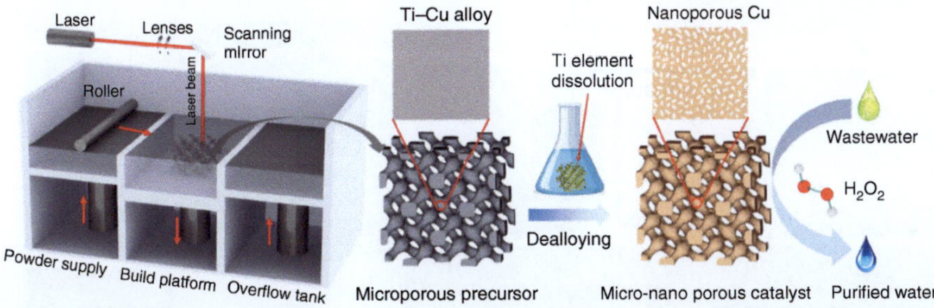

Figure 3.39 Schematic of the preparation procedure of a micro- and nanoporous Cu catalyst via LPBF and dealloying and its application as a heterogeneous Fenton-like reagent for wastewater remediation. Source: Reproduced from Cai et al. (2021a)/with permission from American Chemical Society.

Cu–Cr alloys by forming zirconium and titanium sulfides. The microstructure of LPBF-printed Cu–Cr–Zr–Ti alloys consisted of grains elongated along the build direction with sizes ranging from 30 to 250 μm (Popovich et al. 2016).

A representative study combined LPBF with chemical dealloying to develop a hierarchical micro- and nanoporous catalyst for functional applications (Cai et al. 2021a). A diamond cellular structure with micropores, which was a typical triply periodic minimal surface (TPMS) cellular structure, was printed as alloying precursors via LPBF from mechanically mixed Ti and Cu powders. Subsequently, chemical dealloying was utilized to leech Ti from the printed Ti–Cu structure precursors to create a nanoporous Cu-encapsulating catalyst. Finally, the catalyst was utilized as a heterogeneous photo-Fenton-like reagent to degrade dye pollutants in wastewater (Figure 3.39).

The dealloyed catalyst still retained the microscale porous framework of the as-printed structure precursors, while its surface color changed from silver to a shade of copper after dealloying (Figure 3.40a). The color variation stemmed from the corrosion dissolution of Ti and the self-assembly diffusion of Cu on the surface of the structure precursors upon dealloying. The dealloyed catalyst was covered with 3D open-nanopore structures several micrometers thick, which comprised continuous ligaments and pores with sizes in the range of 10–90 nm (Figure 3.40b,c). The hierarchically interconnected micro- and nanoporous structures contributed to the catalyst having a remarkable degradation efficiency and attractive stability. This work manifests the applicability of the developed catalyst to wastewater purification in practical applications. Furthermore, it inspires incorporating AM and chemical synthesis to design high-performance metal catalysts with tunable hierarchical micro- and nanopores for functional applications.

3.7.5.3 Magnesium Alloys

Magnesium-based materials are primarily employed to develop lightweight structures because of their low density. Hence, they are in high demand in the automotive, aerospace, and biomedical industries. Notably, the complete melting of a single magnesium track using a miniature LPBF system in an inert argon gas atmosphere was

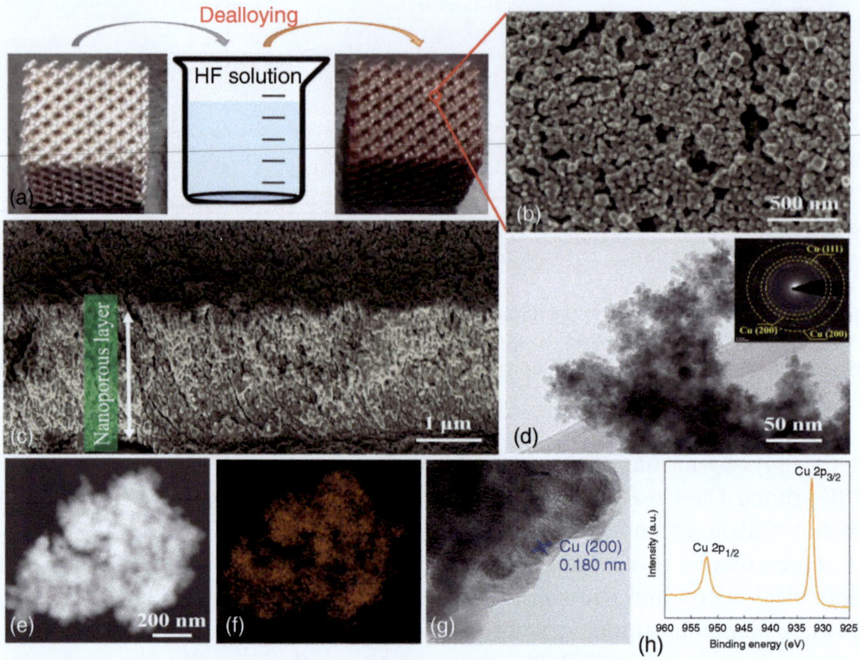

Figure 3.40 Characterization of the catalyst. (a) Photographs of the structure before and after dealloying, (b) plan-view morphology and (c) cross-sectional morphology of the nanoporous copper layer, (d) TEM image of the nanoporous copper layer and corresponding selected area electron diffraction; EDS mapping profile of (e) the original image and (f) Cu element distribution; (g) high-resolution TEM image of the boundary of a cluster of the nanoporous copper layer, and (h) XPS spectra of the Cu 2p orbitals of the nanoporous copper layer. Source: Reproduced from Cai et al. (2021a)/with permission from American Chemical Society.

achieved (Ng et al. 2010). Additionally, the interactions between the laser source and Mg powder tracks under different laser powers, scanning speeds, and irradiation modes (i.e. continuous wave and pulsed mode) have been investigated (Ng et al. 2011). Additionally, several investigations have focused on developing processing windows for Mg and its alloys, such as Mg–9%Al, AZ91D, ZK60, and WE43 (Savalani and Pizarro 2016; Zhang et al. 2012; Wei et al. 2014). The microstructures of Mg alloys consist of equiaxed α-Mg grains with sizes of 1–20 μm, which are often accompanied by the formation of partially or fully divorced eutectic phases (the separation of eutectic phases) homogenously distributed along the grain boundaries.

The difficulties faced in the LPBF processing of Mg alloys mainly consist of two aspects. On the one hand, ductile oxide layers form rapidly due to the high affinity of Mg for oxygen during LPBF. As a result, the Mg oxide film can reduce the bonding strength between the powders, which results in various metallurgical defects. On the other hand, Mg exhibits a narrow temperature range (440 °C) between its boiling point and melting point and is thus easily evaporated under a high-power laser input. Such vaporization can increase the surface roughness and defects (such as pores and cracks) of printed parts. Moreover, Mg alloy powders are highly flammable, and extra caution must be taken when processing them through LPBF.

3.7.5.4 High-Entropy Alloys

HEAs are emerging as a novel frontier in the metal materials community and, hence, have gained extensive research interests in academia and industry. LPBF has been introduced to print HEA parts because of its capability to produce complex components with outstanding mechanical properties (Han et al. 2020b). Efforts have been expended on the LPBF printing of HEAs, such as AlCoCrFeNi, AlCoCuFeNi, AlCrCuFeNi, CoCrFeNi, CoCrFeMnNi, CoCrFeNiC, CoCrFeNiTiMo, MoNbTaW, and NiCrWFeTi. Combinations of different element types and proportions can form different solid solution phases. The high cooling rates in LPBF lead to unique and refined microstructures for the printed HEA parts. For example, the laser energy density influences the microstructural evolution of LPBF-printed CoCrFeMnNi and AlCoCrFeNi HEA parts (Li et al. 2018a; Niu et al. 2019). An increase in the energy density from 37 to 185 J/mm^3 did not change the face-centered cubic (FCC) phase of CoCrFeMnNi but decreased its lattice parameter through the vaporization of Mn. The high cooling rates in LPBF led to large degrees of undercooling in melt pools, thus resulting in sub-micro cellular grains in the CoCrFeMnNi HEA parts. Presently studied HEAs exhibit a wide range of mechanical properties with a yield strength of 402–773 MPa, an ultimate tensile strength of 480–1178 MPa, an elongation of 8–34%, and a microhardness of 212–829 HV.

3.8 Mechanical Metamaterials for LPBF

Metamaterials are introduced in this section, as most of them are fabricated by LPBF. In contrast, only a few are fabricated by electron beam melting to study metamaterials by metal AM.

Porous structures, also called cellular structures, are developed through the biomimicry of natural cellular materials, including wood, cork, and human bone, for their high strength-to-weight ratios (Banhart and Seeliger 2008). Due to its high printing resolution, LPBF has been extensively used to fabricate porous structures with customized shapes and controllable pore geometries. Porous structures can be classified into two categories, i.e. foam and lattice structures. Foam structures can be prepared by injecting gas or mixing a foaming agent into a molten metal or polymer at a low cost. However, irregular pores are often obtained during the preparation of foam structures, often resulting in an inconsistent mechanical response that must be rectified via conservative designs (i.e. with a high safety factor). Lattice structures refer to two-dimensional (2D) or 3D geometries constructed by rationally arraying unit cells. A unit cell is the minimal architecture required to construct a lattice structure. In early studies, lattice structures were often referred to as cellular solids or structures due to the formation of interconnected networks by solid struts or plates (Gibson and Ashby 1997). These artificial lattice structures are subsequently included in the field of metamaterials.

3.8.1 Fundamentals of Mechanical Metamaterials

The initial concept of metamaterials originated from a work that employed an artificially twisted structure to polarize an electromagnetic wave (Bose 1898). Metamaterials possess extraordinary physical properties, such as electromagnetic/acoustic cloaking, a zero/negative Poisson's ratio, and a negative refractive index. Therefore, metamaterials have been progressively applied in optics, acoustics, mechanics, etc. They can be categorized into mechanical metamaterials, electromagnetic metamaterials, acoustic metamaterials, and thermal metamaterials (Figure 3.41) (Fan et al. 2021).

Mechanical metamaterials exhibit unusual mechanical properties, which are determined by their geometries instead of their material composition. Therefore, they are the most extensively fabricated structures by LPBF among all the metamaterial categories. They can be further classified into three categories based on their fundamental elastic constants, i.e. Young's modulus, shear/bulk modulus, and Poisson's ratio (Yu et al. 2018), as shown in Figure 3.42.

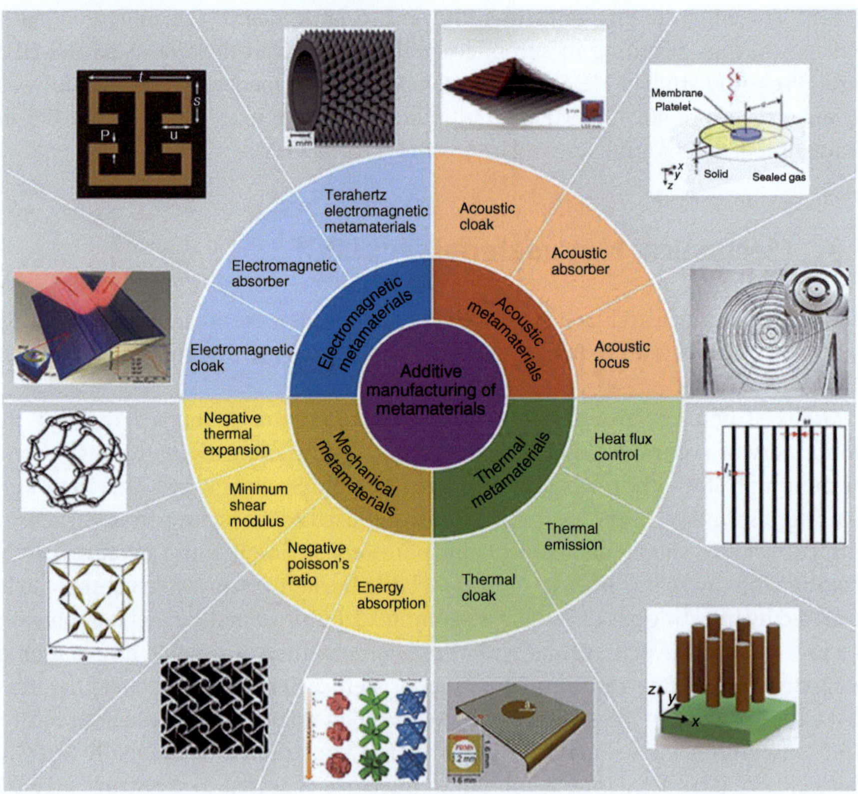

Figure 3.41 Classification of metamaterials according to their functionalities. Source: Fan et al. (2021)/reproduced with permission from Elsevier.

3.8.2 Mechanical Metamaterials with High Young's Modulus

Mechanical metamaterials with a high Young's modulus include high-strength-to-weight lattices and their derivatives, such as origami-inspired metamaterials and cellular origami structures.

High-strength-to-weight lattices consist of strut-based (open-cell) and shell-based (usually in closed-cell) structures. They are the earliest known mechanical metamaterials inspired by the crystalline structures of metals.

Origami-inspired structures include three main types of designs: the Miura-ori pattern, the nonperiodic Ron Resch pattern, and the square twist pattern. In addition, the Kirigami pattern is considered a special type of origami metamaterial design.

Cellular origami metamaterials result from the 3D manipulation of the Miura-ori pattern. The study on the behavior of pattern transformation metamaterials focuses

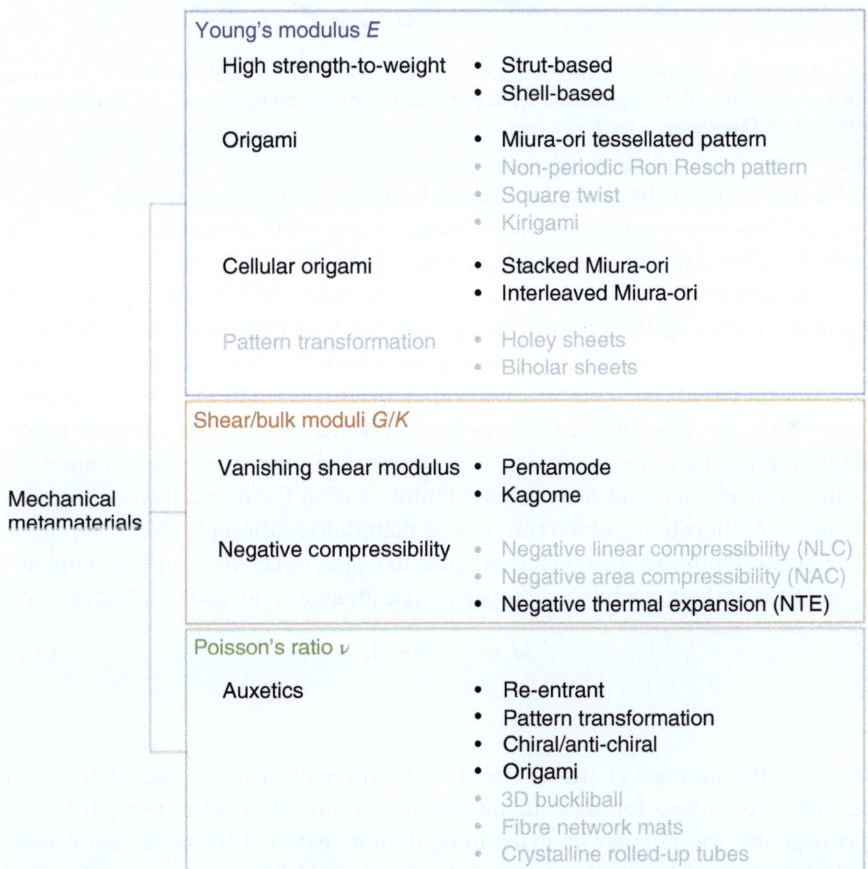

Figure 3.42 Classification of mechanical metamaterials. LPBF-printed mechanical metamaterials are in black.

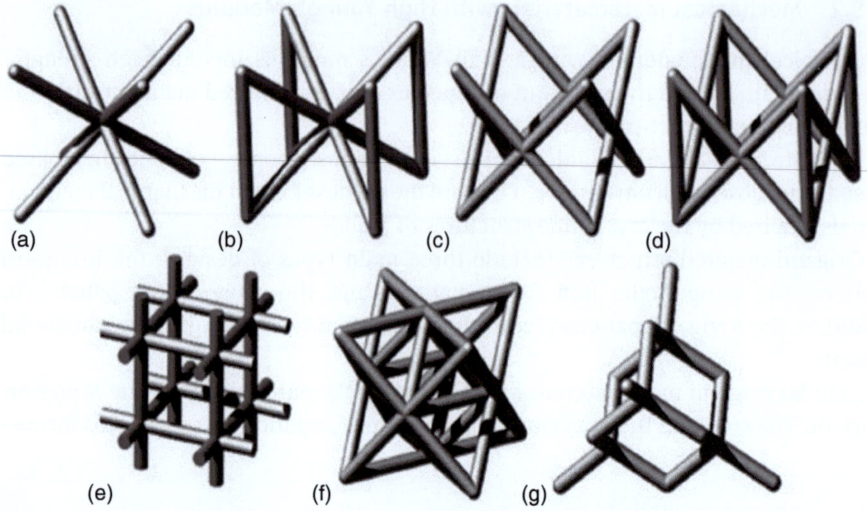

Figure 3.43 Strut-based lattice structures. (a) BCC, (b) BCCZ, (c) FCC, (d) FCCZ, (e) cubic, (f) octet-truss, and (g) diamond. Source: Reproduced from Maconachie et al. (2019)/with permission from Elsevier.

on the tunability of the stiffness. Pattern transformation metamaterials, such as holey and bipolar sheets, can induce changes in material properties based on the threshold deformation in their mechanical designs.

The fundamental strut-based lattice structures consist of BCC and FCC unit cell topologies, including their variations (Figure 3.43a–d) (Maconachie et al. 2019). These unit cells are named after analogous crystalline structures, while other types of strut-based topologies include cubic, octet-truss, and diamond structures (Figure 3.43e–g). The simplicity of these strut-based designs has attracted much attention previously. Recent studies of strut-based topologies have sought to maximize the efficiency of material distribution within the unit cell space through topological optimization, which can only be materialized through AM.

The Maxwell number M is widely adopted to characterize the strut-based topologies in terms of their mechanical properties (Deshpande et al. 2001). It is given by

$$M = s - 2n + 3, \tag{3.7}$$

$$M = s - 3n + 6, \tag{3.8}$$

where s is the number of struts, and n is the number of nodes. Equations (3.7) and (3.8) are applied for truss topologies in 2D and 3D spaces, respectively. If M is negative, the number of struts to equilibrate external forces is insufficient, resulting in bending stress due to moments developed in the struts, which leads to bending-dominated behavior. Meanwhile, suppose M is greater than or equal to zero. In that case, external loads are equilibrated by axial tension and compression in the struts, indicating that no bending occurs at the nodes and rendering such structures stretch-dominated (Leary et al. 2018). Due to these phenomena, stretch-dominated

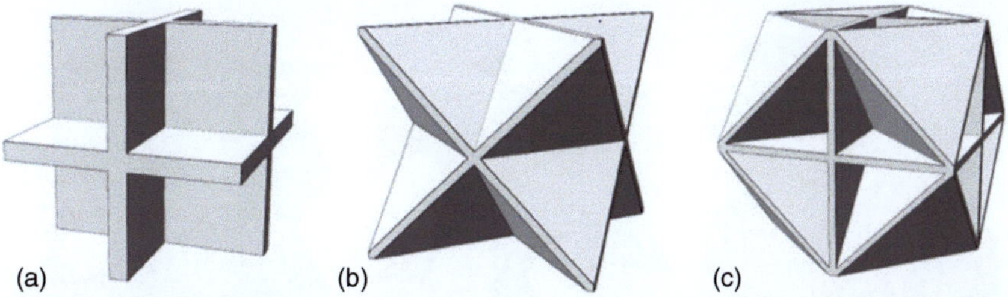

Figure 3.44 Representative topologies of the unit cells for shell lattice structures. (a) Cubic, (b) octet, and (c) Cubic + octet structures. Source: Reproduced from Berger et al. (2017)/with permission from Springer Nature.

structures are stiff and strong but lightweight, while bending-dominated structures are compliant.

LPBF has been employed to fabricate strut-based lattice structures, such as BCC, FCC, and cubic structures. The effect of cell topologies (BCC, FCC, and cubic structures) on the compressive properties of LPBF-printed structures for bone implant applications was investigated through numerical and experimental methods (Han et al. 2017b). The numerical results indicated that different unit cell topologies resulted in distinct stress distributions on the bearing struts of scaffolds, while the unit cell size directly determined the stress value. The LPBF-printed BCC structure was coated with silk fibroin/gentamicin through the electrophoretic technique to prevent postoperative infections of LPBF-printed bone substitutes (Han et al. 2017c). With the development of design optimization tools, the structural and mechanical properties of a topologically optimized Ti–6Al–4V cellular interbody fusion cage were investigated to maintain spine segmental integrity (Lin et al. 2007).

The lightweight design enables a decrease in density and an increase in the mechanical strength of LPBF-printed strut-based structures. The low-velocity impact responses of LPBF-printed Ti–6Al–4V and 316L sandwich structures have been investigated through localized drop weight testing (Mines et al. 2013). The impact resistance of Ti–6Al–4V BCC lattice was comparable to that of the conventionally manufactured honeycomb sandwich panels. However, the printing quality, such as the surface finish, of the 316L BCC sandwich structure was superior to that of the Ti–6Al–4V structure, although the former exhibits a lower specific strength.

Shell lattice structures are another type of cellular structure with unit cells composed of plates instead of struts (Berger et al. 2017). This type of lattice structure is fundamentally a closed-cell lattice formed by thin walls. Representative unit cells for shell lattice structures are shown in Figure 3.44, including cubic, octet, and cubic + octet structures. For the same density, shell lattice structures have demonstrated superior elastic properties relative to strut-based structures (Tancogne-Dejean et al. 2018). Nevertheless, unlike open-cell strut-based structures, LPBF processing of closed-cell shell lattice structures encounters difficulty with powder removal.

Recently, TPMS-based structures have gained much popularity and have been extensively fabricated through LPBF. They are characterized by locally minimized

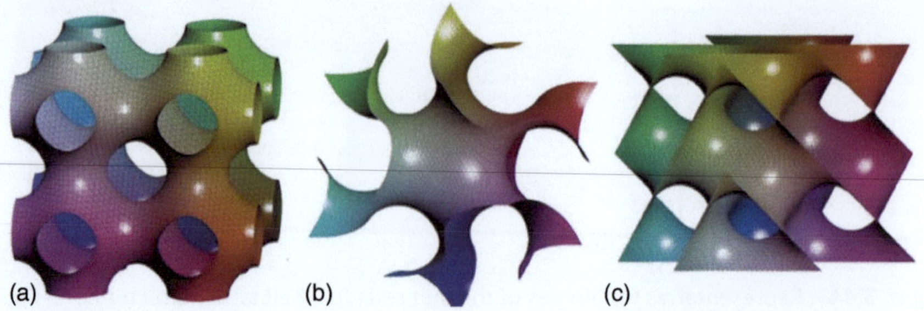

Figure 3.45 Schematic drawings and skeletal graphs of TPMSs. (a) Primitive, (b) gyroid, and (c) diamond surfaces. Source: Reproduced from Han and Che (2018)/with permission from John Wiley & sons.

surface area for a given boundary such that the mean curvature at each point on the surface is zero. Fundamental types of TPMS are primitive, gyroid, and diamond surfaces (Figure 3.45) (Han and Che 2018). A primitive TPMS possesses two intertwined congruent labyrinths, each with the shape of an inflated tubular version of the simple cubic lattice. A gyroid TPMS separates the space into two oppositely congruent labyrinths of passages that run through the surface in a gyrating pattern. It is the only known embedded TPMS with triple junctions and no lines of reflectional symmetry. Finally, a diamond TPMS has two intertwined congruent labyrinths, each with the shape of an inflated tubular version of the diamond bond structure.

The surface of the three aforementioned TPMS models possesses a mean curvature of zero, which can be extended indefinitely in the three periodic directions to provide a concise description of many physical structures. A parametric TPMS is characterized by the Weierstrass formula to generate its coordinates, which is defined as

$$\begin{cases} x = \mathrm{Re} \int_{\omega 0}^{\omega 1} \mathrm{e}^{i\theta}(1-\omega^2)R(\omega)\mathrm{d}\omega \\ y = \mathrm{Re} \int_{\omega 0}^{\omega 1} \mathrm{e}^{i\theta}(1+\omega^2)R(\omega)\mathrm{d}\omega \\ z = \mathrm{Re} \int_{\omega 0}^{\omega 1} \mathrm{e}^{i\theta}(2\omega)R(\omega)\mathrm{d}\omega \end{cases}, \tag{3.9}$$

where θ is an angle known as the Bonnet, ω represents a complex variable, and $R(\omega)$ represents a function that varies with different surfaces. For primitive, gyroid, and diamond surfaces, $R(\omega)$ can be expressed as

$$R(\omega) = \frac{1}{\sqrt{1 - 14\omega^4 + \omega^8}}. \tag{3.10}$$

Alternatively, an approximate TPMS can be formulated by an implicit function, and each point on its surface has a constant value. The periodic surface φ in the approximate TPMS is expressed as

$$\varphi(r) = \sum_{k=1}^{K} A_k \cos\left[\frac{2\pi(h_k \cdot r)}{\lambda_k} + p_k\right] = C, \tag{3.11}$$

where r represents the position vectors in Euclidean space, A_k represents the amplitude factor, h_k is the kth grid vector in reciprocal space, which is a common vector used in the concept of the reciprocal lattice, λ_k represents the periodic wavelength, P_k represents the phase offset, and C is a constant. General TPMS formulae represent the topologies of primitive, gyroid, and diamond TPMS, in which

$$\varphi\,(r) = \cos(X) + \cos(Y) + \cos(Z) = 0, \tag{3.12}$$

$$\varphi(r) = \sin(X)\cos(Y) + \sin(Z)\cos(X) + \sin(Y)\cos(Z) = 0, \tag{3.13}$$

$$\varphi(r) = \cos(X)\cos(Y)\cos(Z) - \sin(X)\sin(Y)\sin(Z) = 0, \tag{3.14}$$

where $X = 2\pi x$, $Y = 2\pi y$, and $Z = 2\pi z$.

The TPMS can be converted to a solid structure from a minimal surface to a periodic lattice based on two methods. In the first method, a TPMS solid lattice can be constructed by considering one of the subdomains divided by the minimal surface to be solid. In contrast, the other subdomain is considered void. Each of these subdomains can be considered a lattice of which volume fraction can be calculated by dividing the volume of the solid domain over the total volume. The other method uses a sheet network enclosed by two minimal surfaces evaluated at two different constants. Various parameters, such as periodicity and relative density, can be altered to tune the mechanical performance of the TPMS structures.

TPMS lattice structures possess advantages over strut-based topologies in terms of manufacturability and applicability in bone fixation. In particular, the continuous inclination angle of cell walls in TPMS structures can provide self-supporting features, which are suitable for the LPBF printing process. The three main types of TPMS and an I-graph-Wrapped Package graph (I-WP) lattice structures printed with Ti–6Al–4V have been studied to investigate their static and fatigue performance for bone-substituting applications (Bobbert et al. 2017). The diamond TPMS with a porosity of 60% and I-WP TPMS with a porosity of 44% can achieve a fatigue endurance limit of up to three times the previously developed AM bone substitutes. Meanwhile, the Ti–6Al–4V primitive pore geometry samples with three levels of porosity were fabricated by LPBF to investigate their mechanical properties (Soro et al. 2019). The mechanical properties of the best sample can achieve a Young's modulus of 22.3 GPa, yield strength of 160 MPa, and pore size of 596 μm.

The gyroid type of TPMS has also been reported to achieve a specific energy absorption almost three times greater than BCC structures for a given porosity level. Therefore, a gyroid TPMS structure was designed with varying volume fractions from 5% to 15% and printed through LPBF using Ti–6Al–4V powder (Yang et al. 2018b). The experimental uniaxial compression results confirmed that the center of the inclined struts was highly susceptible to fracture due to the stress concentrations and existing tensile stress of the upper surface resulting from bending deformation. Additionally, the fatigue properties of the LPBF-printed gyroid TPMS cellular structure were studied to evaluate its long-term performance in a dynamic bio-skeletal environment (Yang et al. 2020). The high-cycle fatigue results indicated that the fatigue ratio of the gyroid TPMS cellular structure reached 0.35 for the as-built samples and 0.45 for the sandblasted samples. These fatigue ratios were relatively

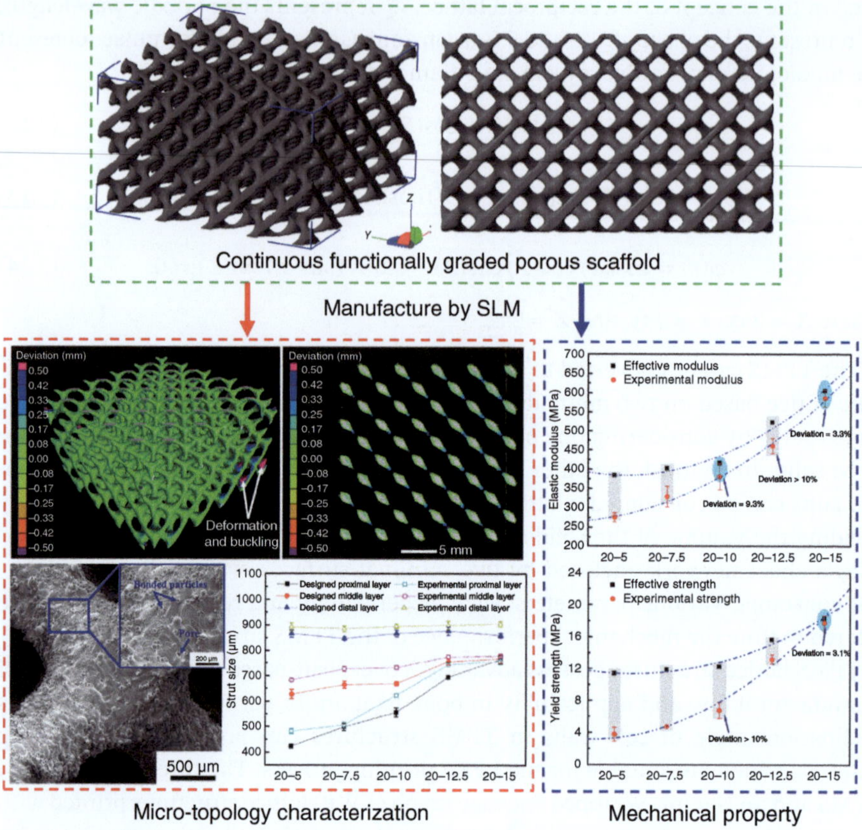

Figure 3.46 Continuous functionally graded porous titanium scaffolds manufactured by LPBF for bone implants. Source: Han et al. (2018b)/reproduced with permission from Elsevier.

high compared to most other bending-dominated lattice structures because of the smooth surface transitions between the struts.

The diamond TPMS has been studied to investigate the mechanical properties of the LPBF-printed parts. Functionally graded porous scaffolds (FGPSs) designed with the diamond unit cell over a wide range of grade volume fractions were printed by LPBF, as shown in Figure 3.46 (Han et al. 2018b). The experimental results indicated that the elastic modulus and yield strength of the LPBF-printed titanium FGPSs could be tailored to a range of 0.28–0.59 GPa and 3.79–17.75 MPa, respectively. Additionally, the effect of unit cell size on the printability, strut dimensions, stress and strain distributions, mechanical properties, and energy absorption capability of the diamond FGPSs was investigated (Yang et al. 2020). It was found from experiments and simulations that an increase in unit cell size reduces the Young's modulus and yield strength, which was attributed to the increase in the strut diameters and the geometric effect caused by the size of the unit cell.

The origami structure was inspired by the Japanese art of paper folding for decoration, and it is now a novel design strategy for ultralight and customizable

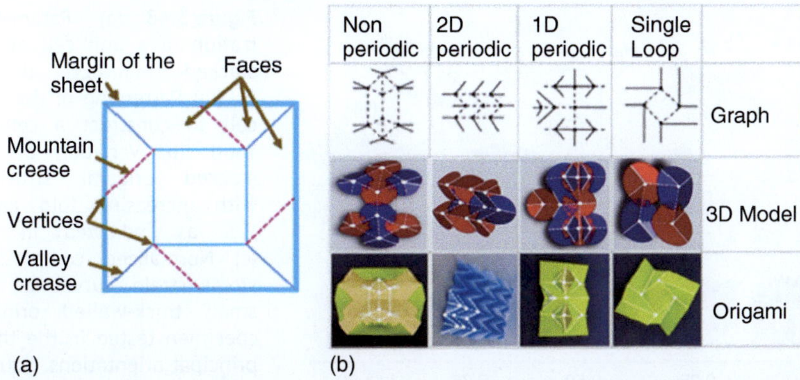

Figure 3.47 Fundamentals of origami-inspired metamaterials. (a) Schematic of a pinwheel crease pattern illustrating the various origami concepts. Source: Reproduced from Peraza-Hernandez et al. (2014)/with permission of IOP Publishing Ltd. (b) Classification of action origami (i.e. origami models that exhibit motion in their final assembled state) in terms of the spherical mechanisms employed in nonperiodic, 2D periodic, 1D periodic, and single loop networks. Source: Bowen et al. (2013)/reproduced with permission from American Society of Mechanical Engineers.

mechanical metamaterials. Fundamentally, an origami structure is constructed by folding and assembling a planar material to form a three-dimensional structure. The final geometry depends on the number, order, and orientation of the folds that can be characterized by the creases and vertices (Figure 3.47).

Three classic origami patterns exist, namely, Miura-ori, nonperiodic Ron Resch, and square twist origami. Mathematically, the geometry of a simple periodically folded Miura-ori is a herringbone pattern consisting of identical unit cells of convex mountain creases and concave valley creases with four coordinated ridges. Vertices are formed when the four creases intersect, and four adjacent vertices bind congruent parallelograms arranged with inversion symmetry. The underlying mechanics of the 2D deformation of a Miura plate can be explained by the one-dimensional beam theory because the effective bending stiffness of its unit cell is singular in nature. Miura-based mechanical metamaterials exhibit negative in-plane Poisson's ratios and positive out-of-plane Poisson's ratios under deformation. The design parameters and mechanical properties of an LPBF-printed Miura-ori structure are presented in Figure 3.48 (Harris and McShane 2020). Vertical walls with variations in the angles in each layer form different design parameters for the design optimization of Miura-ori mechanical metamaterials.

3.8.3 Mechanical Metamaterials with High Shear and Bulk Moduli

Mechanical metamaterials with high shear/bulk moduli constitute two categories, i.e. vanishing shear modulus and negative-compressibility metamaterials. Vanishing shear modulus metamaterials include pentamode and Kagome structures. They are obtained when the shear modulus G approaches zero and the bulk modulus K is much greater than the shear modulus. The unique elastic modulus of these metamaterials represents an ideal type of flowing structure that is incompressible. The

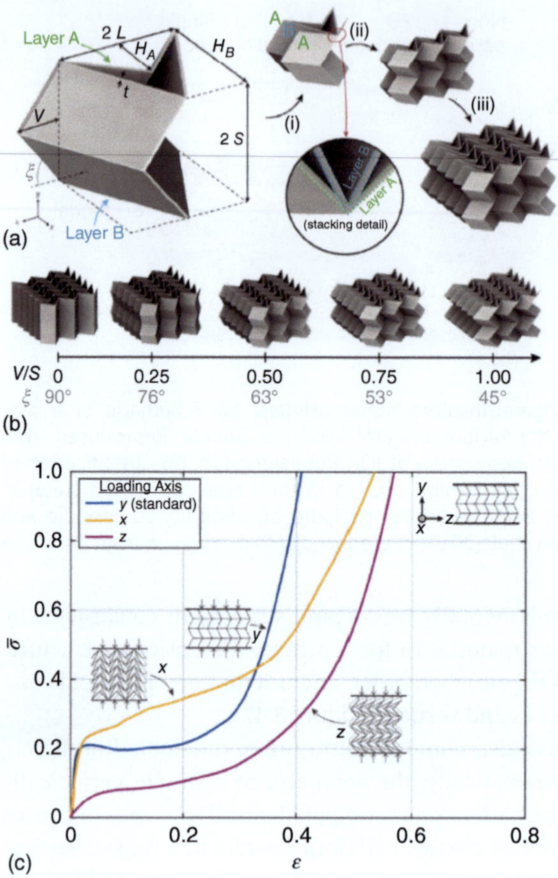

Figure 3.48 (a) Parameterization of a unit cell of the stacked Miura-ori design. (i)–(iii) Patterning of the unit cell to construct a cellular solid. (b) Variation of the stacked origami structure with increasing fold angle V/S as indicated in (a). (c) Normalized compressive stress–strain curves of the small thick-walled origami specimen tested in the three principal orientations. Source: Repoduced from Harris and McShane (2020)/with permission from Elsevier.

relationship between the parameters G, K, and the Poisson's ratio ν is defined as

$$\nu = \frac{3 - 2(G/K)}{2(G/K) + 6}. \tag{3.15}$$

The value of the Poisson's ratio is 0.5 for the vanishing shear modulus metamaterials because G is much smaller than K.

Pentamode metamaterials are the focus of this section, as they are more extensively studied than Kagome metamaterials in LPBF printing. Pentamode metamaterials exhibit an artificial structure that behaves like a fluid but has the form of a solid. It can support five modes of infinitesimal strain and only one mode of stress. The pentamode metamaterial has a diamond-shaped structure that can decouple mass transport and mechanical properties. For instance, the permeability of a pentamode-based bone scaffold can be adjusted without changing the mechanical properties (Figure 3.49).

As the unit cell of a pentamode metamaterial is formed by a center linkage point with four members arranged in a tetrahedron-like structure, four forces, F_0, F_1, F_2,

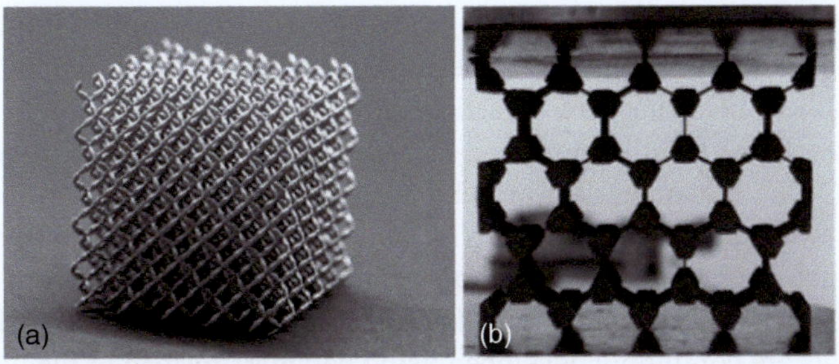

Figure 3.49 LPBF-fabricated mechanical metamaterials with vanishing shear modulus. (a) Pentamode metamaterial. Source: Hedayati et al. (2017)/reproduced with permission from AIP Publishing LLC. (b) Kagome metamaterial. Source: Zhang et al. (2019)/reproduced with permission from Elsevier.

and F_3, are directed toward the members from the linkage point p. The corresponding elastic constants k_0, k_1, k_2, and k_3 can be expressed as:

$$\begin{cases} F_0 = -k_0 p \\ F_1 = k_1(a_1 - p) \\ F_2 = k_2(a_2 - p) \\ F_3 = k_3(a_3 - p) \end{cases}, \tag{3.16}$$

where a_1, a_2, and a_3 are the selecting vectors representing the directional displacement in the unit cell. According to the balance of forces, the equations below can be obtained.

$$\begin{cases} kp = \sum_{i=1}^{3} k_i a_i \\ k = \sum_{i=0}^{3} k_i \end{cases}. \tag{3.17}$$

Following that, the linkage point p lies within the unit cell only if

$$0 < \frac{k_i}{k} < 1 \text{ for } i = 1, 2, 3, \tag{3.18}$$

while the integral of the stress within the unit cell is expressed as:

$$\sigma_l = \sum_{i=1}^{3} k_i a_i \otimes a_i - kp \otimes p. \tag{3.19}$$

The fundamental idea behind pentamode metamaterials is that they only contain normal stress without shear elements in the stress tensors when the ratio of the integral stress and the prescribed stress is constant. Current applications for pentamode metamaterials include acoustic free-space cloaks, which can direct sound waves away instead of colliding with the cloaked object, thereby preventing detection. In addition to the acoustic cloaking application, pentamode metamaterials are suitable for developing ultrahigh strength-to-weight ratio structures.

Negative-compressibility metamaterials possess a range of elastic moduli that satisfy the following conditions: G exceeding zero and K being between $-4G/3$ and 0.

Negative linear compressibility (NLC) and negative area compressibility (NAC) materials are the two categories of negative-compressibility metamaterials, and their mechanical behaviors are investigated, through which the relative volume change of a solid or fluid is measured as a change in pressure. Negative thermal expansion (NTE) metamaterials are characterized by contraction upon heating. Among the few studies investigating NLC and NAC metamaterials through LPBF, the hexagonal and wine-rack-type honeycomb structures have been discussed with analytical and modeling methods (Grima et al. 2011). Nevertheless, NTE metamaterials are commonly developed using polymers instead of metals through LPBF because of their high coefficient of thermal expansion, which provides high functionality and deformability at varying temperatures.

3.8.4 Mechanical Metamaterials with Zero or Negative Poisson's Ratio

Most natural materials possess positive Poisson's ratios. In contrast, materials displaying auxetic behavior are usually counterintuitively described by a zero or negative Poisson's ratio (i.e. the ratio of the transverse to longitudinal strain). Mechanical metamaterials exhibiting such behavior are termed auxetic metamaterials. Under compression and tension along the axial direction, these metamaterials exhibit unconventional behaviors such as contraction and expansion along the transverse direction. The unusual behaviors of these auxetic metamaterials, including improved shear resistance, indentation resistance, fracture resistance, synclastic behavior, variable permeability, and energy absorption, have attracted extensive attention in research. Several representative types of auxetic metamaterials include re-entrant, double-arrowed, chiral, rotating square and triangle (Figure 3.50), missing rib, and origami and Kirigami structures (as mentioned in Section 3.8.2).

In general, the deformation mechanisms of auxetic metamaterials are categorized into three modes: bending-based, rotating-based, and bending-and-rotating hybrid modes.

Bending-based auxetic metamaterials, such as re-entrant structures (Figure 3.50a), consist of six plastic hinges and six connecting members in a unit cell. The auxetic behavior in compression is demonstrated by folding the diagonal ligaments around the plastic hinges that collapse the hourglass shapes into a densified rectangle stacked up by the ligaments (Yang et al. 2015). Similarly, the unfolding of the diagonal ligament transforms the hourglass shape into a hexagonal honeycomb shape, which eventually loses the auxeticity of the re-entrant structure. Similar deformation mechanisms are observed in the 3D re-entrant structures (Figure 3.50b).

The star-shaped re-entrant structure exhibits the same bending mechanism as the hourglass-shaped re-entrant structure (Figure 3.50c) (Grima et al. 2005). However, the hexagonal configuration of the star-shaped re-entrant unit cell provides a uniform deformation pattern at different angles compared to the hourglass-shaped re-entrant structure, which has the advantage of exhibiting anisotropic negative Poisson's ratio. The 2D double-arrowed auxetic structure, as depicted in Figure 3.50d, can achieve a high negative Poisson's ratio along a specific direction. The auxetic behavior of another type of re-entrant structure, the

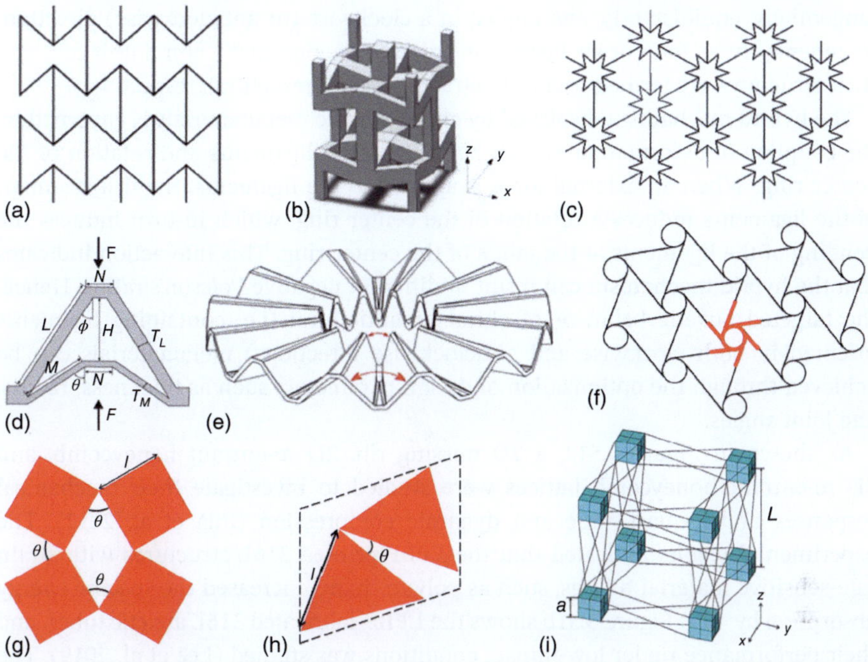

Figure 3.50 Geometrical designs for auxetic metamaterials. (a) 2D re-entrant structure, and (b) 3D re-entrant unit cell. Source: Reproduced from Yang et al. (2015)/with permission of Elsevier. (c) Star-shaped re-entrant structure. Source: Reproduced from Grima et al. (2005)/with permission of Taylor & Francis. (d) Double-arrowed unit cell, and (e) 3D double-arrowed structure. Source: Reproduced from Wang et al. (2016)/with permission of IOP Publishing. (f) Chiral structure. Source: Reproduced from Prall and Lakes (1997)/with permission of Elsevier. (g) Rotating square unit cell. Source: Adapted from Grima and Evans (2000). (h) Rotating triangle unit cell. Source: Reproduced from Grima and Evans (2006)/with permission of Springer Nature. (i) 3D chiral unit cell. Source: Reproduced from Ha et al. (2016)/with permission from IOP Publishing.

triangular re-entrant structure, is governed by its ability to fold or unfold the unit cell through bending deformations (Wang et al. 2016). It can achieve a very high anisotropic negative Poisson's ratio through optimization.

The rotating mechanism of auxetic metamaterials is represented by the rotating square (Figure 3.50g) (Grima and Evans 2000) and triangle unit cells (Figure 3.50h) (Grima and Evans 2006). Most studies have employed material removal methods (e.g. laser cutting) and other 3D printing methods (e.g. fused deposition modeling and selective laser sintering) to fabricate such structures. However, little literature reported on the usage of metal LPBF to produce these metamaterials exists. Hence, this type of auxetic metamaterial will not be detailed in this section.

A hybrid mechanism that combines bending and rotating is demonstrated by the chiral auxetics (Figure 3.50i). The term chiral structure is derived from the spinning nature of the structure, which is not mirror-symmetric but axisymmetric instead (Prall and Lakes 1997). A typical chiral unit cell is constructed by a center ring and ligaments. The ligaments are connected along the center ring's circumference tangentially, equidistantly, and angled in a clockwise (or anticlockwise) direction. In general, three, four, or six ligaments are used to construct a chiral unit cell, and their structures are termed trichiral, tetrachiral, and hexachiral, respectively.

The hybrid mechanism displayed by chiral auxetic metamaterials is governed by the coupling of deformation via the bending of the ligaments and rotation of the center ring. When an external force is applied to the ligaments, the displacement of the ligaments induces a rotation of the center ring, which in turn induces the bending of the ligaments at the joints of the center ring. This interaction indicates that the hybrid mechanism can result in different negative Poisson's ratios. Hence, the targeted auxetic behavior of chiral or anti-chiral (i.e. containing ligaments oriented in both clockwise and anticlockwise directions) metamaterials can be achieved through the optimization of design parameters such as thickness, radius, and joint angles.

As shown in Figure 3.51a, a 2D missing rib, 2D re-entrant honeycomb, and 3D re-entrant honeycomb lattices were studied to investigate their mechanical responses under quasi-static and dynamic compression (Fíla et al. 2017). The experimental results indicated that the LPBF-printed 316L structures with strain rate-sensitive material fillings, such as polyurethane, increased the specific energy absorption by 40%. Figure 3.51b shows the LPBF-fabricated 316L auxetic tubes, and their performance under low-impact conditions was studied (Lee et al. 2019). The results indicated that the auxetic tubes exhibited high specific energy absorption and low deceleration, which was caused by densification, as compared to conventional tubes and honeycomb tubes during crashing (Figure 3.51c).

3.9 Summary

LPBF is one of the most popular powder–based AM processes for printing metal parts. This chapter comprehensively introduces the critical aspects of LPBF, including the history, fundamentals, printing process, metallurgical defects, powder materials, equipment, representative microstructures and mechanical properties of the printed parts, and mechanical metamaterials. Among powder–based AM processes, LPBF utilizes small laser beam diameters and powder with small particle sizes, which enable the fabrication of complex metal parts with the highest part resolution and precise control of the energy input. However, the balling phenomenon and spattering occur during the LPBF process, and metallurgical defects, such as pores, cracks, and warpage, may be generated in the printed parts.

LPBF offers much freedom in the materials domain as it can process titanium and its alloys, aluminum alloys, nickel alloys, iron alloys, cobalt alloys, copper alloys, magnesium alloys, HEAs, etc. Furthermore, most materials can be solely printed by current commercialized LPBF machines developed by many commercial companies worldwide. However, multi-material printing through existing LPBF machines, especially for dissimilar materials, is still challenging for achieving part multifunctionality. In addition, there is a lack of a mature digital powder delivery system for

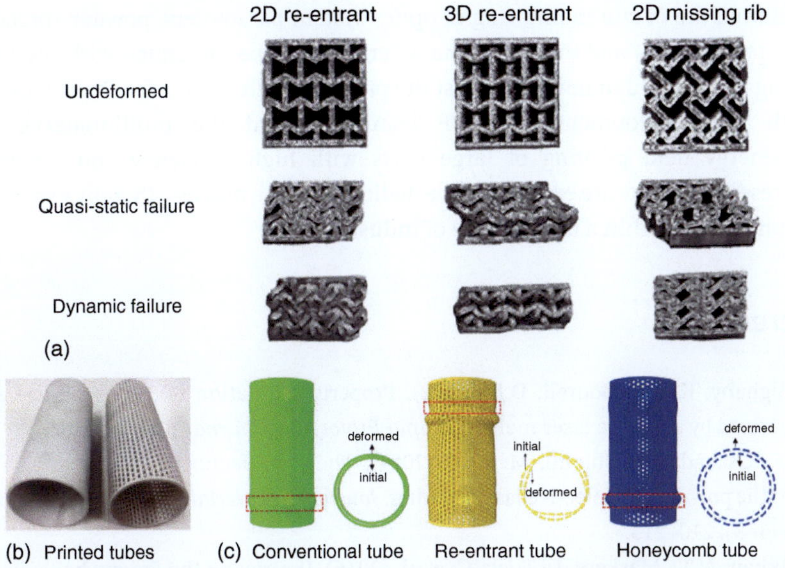

Figure 3.51 (a) Comparison of LPBF-printed auxetic metamaterials with 316L under quasi-static and dynamic loading: 2D re-entrant, 3D re-entrant, and 2D missing rib from left to right. Source: Fíla et al. (2017)/reproduced with permission from John Wiley & Sons. (b) LPBF-printed conventional tube and re-entrant tube, and (c) crash deformation profiles of LPBF-printed tubes: conventional tube, re-entrant tube, and honeycomb tube. Source: Lee et al. (2019)/reproduced with permission from Elsevier.

multi-material LPBF, along with issues such as material cross-contamination and low printing efficiency.

Typical LPBF machines are commonly equipped with a single laser and possess a limited building volume. However, a few leading companies, such as EOS, SLM Solutions, Bright Laser Technologies, and Laseradd, have developed large-scale machines equipped with multiple lasers and optics systems to increase the size of the printed parts and production efficiency. The flow of circulating inert gas within the build chamber and the synergistic incorporation of multiple lasers are critical technologies in large-scale printing. In addition, hybrid LPBF and subtractive manufacturing (LPBF aided by a subtractive system) can be developed to improve the surface quality of the printed parts.

Due to high cooling rates, LPBF-printed parts often exhibit fine microstructure and high mechanical strength compared to their counterparts fabricated by conventional processes. However, the low elongation is attributed to material brittleness and the porosity and microcracks within the materials. From the aspect of material design, the addition of lattice-matched nanoparticles to certain non-weldable powder materials (e.g. Al and Ni alloys) can effectively inhibit cracks and improve ductility by controlling solidification and equiaxed grain growth. Furthermore, additional energy sources, such as magnetic and ultrasonic fields, can be integrated into LPBF, i.e. multi-energy field LPBF, to reduce temperature gradients and residual stresses, minimize pores and cracks, and promote the growth of equiaxed grains.

In addition, the *in situ* monitoring of optics system parameters, powder-spreading on the powder bed, and melt pool characteristics can be integrated with the LPBF printing process to dynamically adjust the parameters through a feedback loop.

With the development of LPBF leaning toward the multi-material and multi-energy field printing of large parts with high efficiency and precision, more research efforts are expected to be dedicated to the process, thereby expanding its applicability within a broad range of industries.

References

Abd-Elghany, K. and Bourell, D.L. (2012). Property evaluation of 304L stainless steel fabricated by selective laser melting. *Rapid Prototyping Journal* 18: 420–428.

Abe, F., Osakada, K., Shiomi, M. et al. (2001). The manufacturing of hard tools from metallic powders by selective laser melting. *Journal of Materials Processing Technology* 111 (1–3): 210–213.

Aboulkhair, N.T., Maskery, I., Tuck, C. et al. (2016). Improving the fatigue behaviour of a selectively laser melted aluminium alloy: influence of heat treatment and surface quality. *Materials & Design* 104: 174–182.

Alsaddah, M., Khan, A., Groom, K., and Mumtaz, K. (2021). Use of 450–808 nm diode lasers for efficient energy absorption during powder bed fusion of Ti6Al4V. *The International Journal of Advanced Manufacturing Technology* 113 (9): 2461–2480.

Amato, K.N., Gaytan, S.M., Murr, L.E. et al. (2012). Microstructures and mechanical behavior of Inconel 718 fabricated by selective laser melting. *Acta Materialia* 60 (5): 2229–2239.

Attar, H., Calin, M., Zhang, L.C. et al. (2014). Manufacture by selective laser melting and mechanical behavior of commercially pure titanium. *Materials Science and Engineering A* 593: 170–177.

Banerjee, D. and Williams, J.C. (2013). Perspectives on titanium science and technology. *Acta Materialia* 61 (3): 844–879.

Banhart, J. and Seeliger, H.W. (2008). Aluminium foam sandwich panels: manufacture, metallurgy and applications. *Advanced Engineering Materials* 10 (9): 793–802.

Berger, J.B., Wadley, H.N.G., and McMeeking, R.M. (2017). Mechanical metamaterials at the theoretical limit of isotropic elastic stiffness. *Nature* 543 (7646): 533–537.

Bergström, D. (2018). The absorption of laser light by rough metal surfaces. Doctoral dissertation, Luleå tekniska universitet.

Bobbert, F.S.L., Lietaert, K., Eftekhari, A.A. et al. (2017). Additively manufactured metallic porous biomaterials based on minimal surfaces: a unique combination of topological, mechanical, and mass transport properties. *Acta Biomaterialia* 53: 572–584.

Bose, J.C. (1898). On the rotation of plane of polarisation of electric wave by a twisted structure. *Proceedings of the Royal Society of London* 63 (389–400): 146–152.

Bowen, L.A., Grames, C.L., Magleby, S.P. et al. (2013). A classification of action origami as systems of spherical mechanisms. *Journal of Mechanical Design* 135 (11): 111008.

Cai, C., Guo, S., Li, B. et al. (2021a). 3D printing and chemical dealloying of a hierarchically micro-and nanoporous catalyst for wastewater purification. *ACS Applied Materials & Interfaces.* 13 (41): 48709–48719.

Cai, C., Qiu, J.C.D., Shian, T.W. et al. (2021b). Laser powder bed fusion of Mo2C/Ti-6Al-4V composites with alternately laminated α′/β phases for enhanced mechanical properties. *Additive Manufacturing* 46: 102134.

Carlton, H.D., Haboub, A., Gallegos, G.F. et al. (2016). Damage evolution and failure mechanisms in additively manufactured stainless steel. *Materials Science and Engineering A* 651: 406–414.

Chen, C., Yin, J., Zhu, H. et al. (2019a). The effect of process parameters on the residual stress of selective laser melted Inconel 718 thin-walled part. *Rapid Prototyping Journal* 25: 1359–1369.

Chen, H., Wei, Q., Zhang, Y. et al. (2019b). Powder-spreading mechanisms in powder-bed-based additive manufacturing: experiments and computational modeling. *Acta Materialia* 179: 158–171.

Cheng, B., Shrestha, S., and Chou, K. (2016). Stress and deformation evaluations of scanning strategy effect in selective laser melting. *Additive Manufacturing* 12: 240–251.

Cloots, M., Uggowitzer, P.J., and Wegener, K. (2016). Investigations on the microstructure and crack formation of IN738LC samples processed by selective laser melting using Gaussian and doughnut profiles. *Materials & Design* 89: 770–784.

Cunningham, R., Zhao, C., Parab, N. et al. (2019). Keyhole threshold and morphology in laser melting revealed by ultrahigh-speed X-ray imaging. *Science* 363 (6429): 849–852.

Dai, D. and Gu, D. (2014). Thermal behavior and densification mechanism during selective laser melting of copper matrix composites: simulation and experiments. *Materials & Design* 55: 482–491.

Das, M., Balla, V.K., Basu, D. et al. (2010). Laser processing of SiC-particle-reinforced coating on titanium. *Scripta Materialia* 63 (4): 438–441.

Deshpande, V.S., Ashby, M.F., and Fleck, N.A. (2001). Foam topology: bending versus stretching dominated architectures. *Acta Materialia* 49 (6): 1035–1040.

Dinda, G.P., Dasgupta, A.K., Bhattacharya, S. et al. (2013). Microstructural characterization of laser-deposited Al 4047 alloy. *Metallurgical and Materials Transactions A* 44 (5): 2233–2242.

Dutta Majumdar, J. and Manna, I. (2011). Laser material processing. *International Materials Reviews* 56 (5, 6): 341–388.

Facchini, L., Vicente, N. Jr., Lonardelli, I. et al. (2010). Metastable austenite in 17–4 precipitation-hardening stainless steel produced by selective laser melting. *Advanced Engineering Materials* 12 (3): 184–188.

Fan, J., Zhang, L., Wei, S. et al. (2021). A review of additive manufacturing of metamaterials and developing trends. *Materials Today* 50: 303–328.

Ferrar, B., Mullen, L., Jones, E. et al. (2012). Gas flow effects on selective laser melting (SLM) manufacturing performance. *Journal of Materials Processing Technology* 212 (2): 355–364.

Fíla, T., Zlámal, P., Jiroušek, O. et al. (2017). Impact testing of polymer-filled auxetics using split Hopkinson pressure bar. *Advanced Engineering Materials* 19 (10): 1700076.

Gibson, L. and Ashby, M. (1997). Introduction. In: *Cellular Solids: Structure and Properties*, Cambridge Solid State Science Series, 1–14. Cambridge: Cambridge University Press.

Grima, J.N. and Evans, K.E. (2000). Auxetic behavior from rotating squares. *Journal of Materials Science Letters* 19 (17): 1563–1565.

Grima, J.N. and Evans, K.E. (2006). Auxetic behavior from rotating triangles. *Journal of Materials Science* 41 (10): 3193–3196.

Grima, J.N., Gatt, R., Alderson, A., and Evans, K.E. (2005). On the potential of connected stars as auxetic systems. *Molecular Simulation* 31 (13): 925–935.

Grima, J.N., Attard, D., Caruana-Gauci, R., and Gatt, R. (2011). Negative linear compressibility of hexagonal honeycombs and related systems. *Scripta Materialia* 65 (7): 565–568.

Gu, D. and Meiners, W. (2010). Microstructure characteristics and formation mechanisms of *in situ* WC cemented carbide based hardmetals prepared by selective laser melting. *Materials Science and Engineering A* 527 (29, 30): 7585–7592.

Gu, D. and Shen, Y. (2009). Balling phenomena in direct laser sintering of stainless steel powder: metallurgical mechanisms and control methods. *Materials & Design* 30 (8): 2903–2910.

Gu, D., Meiners, W., Wissenbach, K., and Poprawe, R. (2012a). Laser additive manufacturing of metallic components: materials, processes and mechanisms. *International Materials Reviews* 57 (3): 133–164.

Gu, D., Hagedorn, Y.C., Meiners, W. et al. (2012b). Densification behavior, microstructure evolution, and wear performance of selective laser melting processed commercially pure titanium. *Acta Materialia* 60 (9): 3849–3860.

Ha, C.S., Plesha, M.E., and Lakes, R.S. (2016). Chiral three-dimensional lattices with tunable Poisson's ratio. *Smart Materials and Structures* 25 (5): 054005.

Han, L. and Che, S. (2018). An overview of materials with triply periodic minimal surfaces and related geometry: from biological structures to self-assembled systems. *Advanced Materials* 30 (17): 1705708.

Han, C., Wang, Q., Song, B. et al. (2017a). Microstructure and property evolutions of titanium/nano-hydroxyapatite composites *in situ* prepared by selective laser melting. *Journal of the Mechanical Behavior of Biomedical Materials* 71: 85–94.

Han, C., Yan, C., Wen, S. et al. (2017b). Effects of the unit cell topology on the compression properties of porous Co-Cr scaffolds fabricated via selective laser melting. *Rapid Prototyping Journal* 23 (1): 16–27.

Han, C., Yao, Y., Cheng, X. et al. (2017c). Electrophoretic deposition of gentamicin-loaded silk fibroin coatings on 3D-printed porous cobalt–chromium–molybdenum bone substitutes to prevent orthopedic implant infections. *Biomacromolecules* 18 (11): 3776–3787.

Han, C., Li, Y., Wang, Q. et al. (2018a). Titanium/hydroxyapatite (Ti/HA) gradient materials with quasi-continuous ratios fabricated by SLM: material interface and fracture toughness. *Materials & Design* 141: 256–266.

Han, C., Li, Y., Wang, Q. et al. (2018b). Continuous functionally graded porous titanium scaffolds manufactured by selective laser melting for bone implants. *Journal of the Mechanical Behavior of Biomedical Materials* 80: 119–127.

Han, C., Babicheva, R., Chua, J.D.Q. et al. (2020a). Microstructure and mechanical properties of (TiB + TiC)/Ti composites fabricated *in situ* via selective laser melting of Ti and B4C powders. *Additive Manufacturing* 36: 101466.

Han, C., Fang, Q., Shi, Y. et al. (2020b). Recent advances on high-entropy alloys for 3D printing. *Advanced Materials* 32 (26): 1903855.

Harris, J.A. and McShane, G.J. (2020). Metallic stacked origami cellular materials: additive manufacturing, properties, and modelling. *International Journal of Solids and Structures* 185: 448–466.

Hedayati, R., Leeflang, A.M., and Zadpoor, A.A. (2017). Additively manufactured metallic pentamode meta-materials. *Applied Physics Letters* 110 (9): 091905.

Hojjatzadeh, S.M.H., Parab, N.D., Yan, W. et al. (2019). Pore elimination mechanisms during 3D printing of metals. *Nature Communications* 10 (1): 1–8.

Hooper, P.A. (2018). Melt pool temperature and cooling rates in laser powder bed fusion. *Additive Manufacturing* 22: 548–559.

Idell, Y., Levine, L.E., Allen, A.J. et al. (2016). Unexpected δ-phase formation in additive-manufactured Ni-based superalloy. *JOM Journal of the Minerals Metals and Materials Society* 68 (3): 950–959.

Jägle, E.A., Choi, P.P., Van Humbeeck, J., and Raabe, D. (2014). Precipitation and austenite reversion behavior of a maraging steel produced by selective laser melting. *Journal of Materials Research* 29 (17): 2072–2079.

Jandin, G., Bertin, J.M., Dembinski, L., and Coddet, C. (2005). Manufacture of stainless steel parts by selective laser melting process. In: *Proceedings of the 2nd International Conference on Advanced Research in Virtual and Rapid Prototyping*, Leiria, Portugal, vol. 28. pp. 431–434.

Khairallah, S.A., Anderson, A.T., Rubenchik, A., and King, W.E. (2016). Laser powder-bed fusion additive manufacturing: physics of complex melt flow and formation mechanisms of pores, spatter, and denudation zones. *Acta Materialia* 108: 36–45.

Krakhmalev, P., Yadroitsava, I., Fredriksson, G., and Yadroitsev, I. (2015). *In situ* heat treatment in selective laser melted martensitic AISI 420 stainless steels. *Materials & Design* 87: 380–385.

Kruth, J.P., Froyen, L., Van Vaerenbergh, J. et al. (2004). Selective laser melting of iron-based powder. *Journal of Materials Processing Technology* 149 (1–3): 616–622.

Kruth, J.P., Mercelis, P., Van Vaerenbergh, J. et al. (2005). Binding mechanisms in selective laser sintering and selective laser melting. *Rapid Prototyping Journal* 11: 26–36.

Leary, M., Mazur, M., Williams, H. et al. (2018). Inconel 625 lattice structures manufactured by selective laser melting (SLM): mechanical properties, deformation and failure modes. *Materials & Design* 157: 179–199.

LeBrun, T., Nakamoto, T., Horikawa, K., and Kobayashi, H. (2015). Effect of retained austenite on subsequent thermal processing and resultant mechanical properties of selective laser melted 17–4 PH stainless steel. *Materials & Design* 81: 44–53.

Lee, Y.S. and Zhang, W. (2015). Mesoscopic simulation of heat transfer and fluid flow in laser powder bed additive manufacturing. In: *Proceedings of the International Solid Freeform Fabrication Symposium*, Austin, TX, USA, pp. 1154–1165.

Lee, W., Jeong, Y., Yoo, J. et al. (2019). Effect of auxetic structures on crash behavior of cylindrical tube. *Composite Structures* 208: 836–846.

Leung, C.L.A., Marussi, S., Atwood, R.C. et al. (2018). *In situ* X-ray imaging of defect and molten pool dynamics in laser additive manufacturing. *Nature Communications* 9 (1): 1–9.

Li, R., Liu, J., Shi, Y. et al. (2012). Balling behavior of stainless steel and nickel powder during selective laser melting process. *The International Journal of Advanced Manufacturing Technology* 59 (9): 1025–1035.

Li, X.P., Wang, X.J., Saunders, M. et al. (2015a). A selective laser melting and solution heat treatment refined Al–12Si alloy with a controllable ultrafine eutectic microstructure and 25% tensile ductility. *Acta Materialia* 95: 74–82.

Li, S., Wei, Q., Shi, Y. et al. (2015b). Microstructure characteristics of Inconel 625 superalloy manufactured by selective laser melting. *Journal of Materials Science and Technology* 31 (9): 946–952.

Li, R., Niu, P., Yuan, T. et al. (2018a). Selective laser melting of an equiatomic CoCrFeMnNi high-entropy alloy: processability, non-equilibrium microstructure and mechanical property. *Journal of Alloys and Compounds* 746: 125–134.

Li, X., Shi, J.J., Wang, C.H. et al. (2018b). Effect of heat treatment on microstructure evolution of Inconel 718 alloy fabricated by selective laser melting. *Journal of Alloys and Compounds* 764: 639–649.

Li, Z., Xu, R., Zhang, Z., and Kucukkoc, I. (2018c). The influence of scan length on fabricating thin-walled components in selective laser melting. *International Journal of Machine Tools and Manufacture* 126: 1–12.

Lin, C.Y., Wirtz, T., LaMarca, F., and Hollister, S.J. (2007). Structural and mechanical evaluations of a topology optimized titanium interbody fusion cage fabricated by selective laser melting process. *Journal of Biomedical Materials Research Part A* 83 (2): 272–279.

Liu, B., Wildman, R., Tuck, C. et al. (2011). *Investigation the Effect of Particle Size Distribution on Processing Parameters Optimisation in Selective Laser Melting Process*, 227–238. Additive Manufacturing Research Group, Loughborough University.

Liu, Y., Yang, Y., Mai, S. et al. (2015). Investigation into spatter behavior during selective laser melting of AISI 316L stainless steel powder. *Materials & Design* 87: 797–806.

Lu, S., Fujii, H., and Nogi, K. (2004). Marangoni convection and weld shape variations in Ar–O_2 and Ar–CO_2 shielded GTA welding. *Materials Science and Engineering A* 380: 290–297.

Lu, Y., Wu, S., Gan, Y. et al. (2015). Investigation on the microstructure, mechanical property and corrosion behavior of the selective laser melted CoCrW alloy for dental application. *Materials Science and Engineering C* 49: 517–525.

Maconachie, T., Leary, M., Lozanovski, B. et al. (2019). SLM lattice structures: properties, performance, applications and challenges. *Materials & Design* 183: 108137.

Martin, J.H., Yahata, B.D., Hundley, J.M. et al. (2017). 3D printing of high-strength aluminium alloys. *Nature* 549 (7672): 365–369.

Martin, A.A., Calta, N.P., Khairallah, S.A. et al. (2019). Dynamics of pore formation during laser powder bed fusion additive manufacturing. *Nature Communications* 10 (1): 1–10.

Martina, F., Mehnen, J., Williams, S.W. et al. (2012). Investigation of the benefits of plasma deposition for the additive layer manufacture of Ti–6Al–4V. *Journal of Materials Processing Technology* 212 (6): 1377–1386.

Maskery, I., Aboulkhair, N.T., Corfield, M.R. et al. (2016). Quantification and characterisation of porosity in selectively laser melted Al–Si10–Mg using X-ray computed tomography. *Materials Characterization* 111: 193–204.

Meier, C., Penny, R.W., Zou, Y. et al. (2017). Thermophysical phenomena in metal additive manufacturing by selective laser melting: fundamentals, modeling, simulation, and experimentation. *Annual Review of Heat Transfer* 20: 241–316.

Meiners, W., Wissenbach, K., and Gasser, A. (1998). Shaped body especially prototype or replacement part production. DE19649865C1, filed 02 December 1996, issued 12 February 1998.

Mercelis, P. and Kruth, J.P. (2006). Residual stresses in selective laser sintering and selective laser melting. *Rapid Prototyping Journal* 12: 254–265.

Metelkova, J., Kinds, Y., Kempen, K. et al. (2018). On the influence of laser defocusing in selective laser melting of 316L. *Additive Manufacturing* 23: 161–169.

Mines, R.A.W., Tsopanos, S., Shen, Y. et al. (2013). Drop weight impact behaviour of sandwich panels with metallic micro lattice cores. *International Journal of Impact Engineering* 60: 120–132.

Mishurova, T., Cabeza, S., Thiede, T. et al. (2018). The influence of the support structure on residual stress and distortion in SLM Inconel 718 parts. *Metallurgical and Materials Transactions A* 49 (7): 3038–3046.

Murr, L.E., Martinez, E., Hernandez, J. et al. (2012). Microstructures and properties of 17-4 PH stainless steel fabricated by selective laser melting. *Journal of Materials Research and Technology* 1 (3): 167–177.

Ng, C.C., Savalani, M.M., Man, H.C., and Gibson, I. (2010). Layer manufacturing of magnesium and its alloy structures for future applications. *Virtual and Physical Prototyping* 5 (1): 13–19.

Ng, C.C., Savalani, M., and Man, H.C. (2011). Fabrication of magnesium using selective laser melting technique. *Rapid Prototyping Journal* 17: 479–490.

Niu, P.D., Li, R.D., Yuan, T.C. et al. (2019). Microstructures and properties of an equimolar AlCoCrFeNi high entropy alloy printed by selective laser melting. *Intermetallics* 104: 24–32.

Olakanmi, E.O., Dalgarno, K.W., and Cochrane, R.F. (2012). Laser sintering of blended Al-Si powders. *Rapid Prototyping Journal* 18: 109–119.

Parry, L., Ashcroft, I.A., and Wildman, R.D. (2016). Understanding the effect of laser scan strategy on residual stress in selective laser melting through thermo-mechanical simulation. *Additive Manufacturing* 12: 1–15.

Patel, S. and Vlasea, M. (2020). Melting modes in laser powder bed fusion. *Materialia* 9: 100591.

Peraza-Hernandez, E.A., Hartl, D.J., Malak, R.J. Jr., and Lagoudas, D.C. (2014). Origami-inspired active structures: a synthesis and review. *Smart Materials and Structures* 23 (9): 094001.

Popovich, A., Sufiiarov, V., Polozov, I. et al. (2016). Microstructure and mechanical properties of additive manufactured copper alloy. *Materials Letters* 179: 38–41.

Prall, D. and Lakes, R.S. (1997). Properties of a chiral honeycomb with a Poisson's ratio of -1. *International Journal of Mechanical Sciences* 39 (3): 305–314.

Prashanth, K.G., Scudino, S., Klauss, H.J. et al. (2014). Microstructure and mechanical properties of Al–12Si produced by selective laser melting: effect of heat treatment. *Materials Science and Engineering A* 590: 153–160.

Qi, T., Zhu, H., Zhang, H. et al. (2017). Selective laser melting of Al7050 powder: melting mode transition and comparison of the characteristics between the keyhole and conduction mode. *Materials & Design* 135: 257–266.

Qiu, C., Adkins, N.J., and Attallah, M.M. (2013). Microstructure and tensile properties of selectively laser-melted and of HIPed laser-melted Ti–6Al–4V. *Materials Science and Engineering A* 578: 230–239.

Rafi, H.K., Pal, D., Patil, N. et al. (2014). Microstructure and mechanical behavior of 17-4 precipitation hardenable steel processed by selective laser melting. *Journal of Materials Engineering and Performance* 23 (12): 4421–4428.

Sames, W.J., List, F.A., Pannala, S. et al. (2016). The metallurgy and processing science of metal additive manufacturing. *International Materials Reviews* 61 (5): 315–360.

Savalani, M.M. and Pizarro, J.M. (2016). Effect of preheat and layer thickness on selective laser melting (SLM) of magnesium. *Rapid Prototyping Journal* 22: 115–122.

Simonelli, M., Tuck, C., Aboulkhair, N.T. et al. (2015). A study on the laser spatter and the oxidation reactions during selective laser melting of 316L stainless steel, Al-Si10-Mg, and Ti–6Al–4V. *Metallurgical and Materials Transactions A* 46 (9): 3842–3851.

Soro, N., Attar, H., Wu, X., and Dargusch, M.S. (2019). Investigation of the structure and mechanical properties of additively manufactured Ti–6Al–4V biomedical scaffolds designed with a Schwartz primitive unit-cell. *Materials Science and Engineering A* 745: 195–202.

Spierings, A.B. and Levy, G. (2009). Comparison of density of stainless steel 316L parts produced with selective laser melting using different powder grades. In: *Proceedings of the Annual International Solid Freeform Fabrication Symposium*. Austin, TX, pp. 342–353.

Spierings, A.B., Herres, N., and Levy, G. (2011). Influence of the particle size distribution on surface quality and mechanical properties in AM steel parts. *Rapid Prototyping Journal* 17: 195–202.

Starr, T.L., Rafi, K., Stucker, B., and Scherzer, C.M. (2012). Controlling phase composition in selective laser melted stainless steels. *Power (W)* 195 (195): 439–446.

Strondl, A., Lyckfeldt, O., Brodin, H., and Ackelid, U. (2015). Characterization and control of powder properties for additive manufacturing. *JOM Journal of the Minerals Metals and Materials Society* 67 (3): 549–554.

Suryawanshi, J., Prashanth, K.G., Scudino, S. et al. (2016). Simultaneous enhancements of strength and toughness in an Al-12Si alloy synthesized using selective laser melting. *Acta Materialia* 115: 285–294.

Takaichi, A., Nakamoto, T., Joko, N. et al. (2013). Microstructures and mechanical properties of Co–29Cr–6Mo alloy fabricated by selective laser melting process for dental applications. *Journal of the Mechanical Behavior of Biomedical Materials* 21: 67–76.

Takata, N., Kodaira, H., Sekizawa, K. et al. (2017). Change in microstructure of selectively laser melted AlSi10Mg alloy with heat treatments. *Materials Science and Engineering A* 704: 218–228.

Takata, N., Kodaira, H., Suzuki, A., and Kobashi, M. (2018). Size dependence of microstructure of AlSi10Mg alloy fabricated by selective laser melting. *Materials Characterization* 143: 18–26.

Tancogne-Dejean, T., Diamantopoulou, M., Gorji, M.B. et al. (2018). 3D plate-lattices: an emerging class of low-density metamaterial exhibiting optimal isotropic stiffness. *Advanced Materials* 30 (45): 1803334.

Thijs, L., Verhaeghe, F., Craeghs, T. et al. (2010). A study of the microstructural evolution during selective laser melting of Ti–6Al–4V. *Acta Materialia* 58 (9): 3303–3312.

Thijs, L., Kempen, K., Kruth, J.P., and Van Humbeeck, J. (2013). Fine-structured aluminium products with controllable texture by selective laser melting of pre-alloyed AlSi10Mg powder. *Acta Materialia* 61 (5): 1809–1819.

Tolosa, I., Garciandía, F., Zubiri, F. et al. (2010). Study of mechanical properties of AISI 316 stainless steel processed by "selective laser melting", following different manufacturing strategies. *The International Journal of Advanced Manufacturing Technology* 51 (5): 639–647.

Vilaro, T., Abed, S., and Knapp, W. (2008). Direct manufacturing of technical parts using selective laser melting: example of automotive application. In: *Proceedings of of 12th European Forum on Rapid Prototyping*, Paris, France (5–6 March 2008).

Vilaro, T., Colin, C., and Bartout, J.D. (2011). As-fabricated and heat-treated microstructures of the Ti–6Al–4V alloy processed by selective laser melting. *Metallurgical and Materials Transactions A* 42 (10): 3190–3199.

Vrancken, B., Thijs, L., Kruth, J.P., and Van Humbeeck, J. (2012). Heat treatment of Ti6Al4V produced by selective laser melting: microstructure and mechanical properties. *Journal of Alloys and Compounds* 541: 177–185.

Vrancken, B., Thijs, L., Kruth, J.P., and Van Humbeeck, J. (2014a). Microstructure and mechanical properties of a novel β titanium metallic composite by selective laser melting. *Acta Materialia* 68: 150–158.

Vrancken, B., Cain, V., Knutsen, R., and Van Humbeeck, J. (2014b). Residual stress via the contour method in compact tension specimens produced via selective laser melting. *Scripta Materialia* 87: 29–32.

Wang, Y., Wang, L., Ma, Z.D., and Wang, T. (2016). Parametric analysis of a cylindrical negative Poisson's ratio structure. *Smart Materials and Structures* 25 (3): 035038.

Wang, Y.M., Voisin, T., McKeown, J.T. et al. (2018). Additively manufactured hierarchical stainless steels with high strength and ductility. *Nature Materials* 17 (1): 63–71.

Wei, K., Gao, M., Wang, Z., and Zeng, X. (2014). Effect of energy input on formability, microstructure and mechanical properties of selective laser melted AZ91D magnesium alloy. *Materials Science and Engineering A* 611: 212–222.

Weingarten, C., Buchbinder, D., Pirch, N. et al. (2015). Formation and reduction of hydrogen porosity during selective laser melting of AlSi10Mg. *Journal of Materials Processing Technology* 221: 112–120.

Withers, P.J. and Bhadeshia, H.K.D.H. (2001). Residual stress. Part 2–nature and origins. *Materials Science and Technology* 17 (4): 366–375.

Wu, J., Wang, X.Q., Wang, W. et al. (2016). Microstructure and strength of selectively laser melted AlSi10Mg. *Acta Materialia* 117: 311–320.

Wycisk, E., Siddique, S., Herzog, D. et al. (2015). Fatigue performance of laser additive manufactured Ti–6Al–4V in very high cycle fatigue regime up to 109 cycles. *Frontiers in Materials* 2: 72.

Xu, W., Brandt, M., Sun, S. et al. (2015). Additive manufacturing of strong and ductile Ti–6Al–4V by selective laser melting via *in situ* martensite decomposition. *Acta Materialia* 85: 74–84.

Yadroitsev, I. and Yadroitsava, I. (2015). Evaluation of residual stress in stainless steel 316L and Ti6Al4V samples produced by selective laser melting. *Virtual and Physical Prototyping* 10 (2): 67–76.

Yadroitsev, I., Gusarov, A., Yadroitsava, I., and Smurov, I. (2010). Single track formation in selective laser melting of metal powders. *Journal of Materials Processing Technology* 210 (12): 1624–1631.

Yang, L., Harrysson, O., West, H., and Cormier, D. (2015). Mechanical properties of 3D re-entrant honeycomb auxetic structures realized via additive manufacturing. *International Journal of Solids and Structures* 69: 475–490.

Yang, K.V., Rometsch, P., Jarvis, T. et al. (2018a). Porosity formation mechanisms and fatigue response in Al-Si-Mg alloys made by selective laser melting. *Materials Science and Engineering A* 712: 166–174.

Yang, L., Yan, C., Han, C. et al. (2018b). Mechanical response of a triply periodic minimal surface cellular structures manufactured by selective laser melting. *International Journal of Mechanical Sciences* 148: 149–157.

Yang, L., Han, C., Wu, H. et al. (2020). Insights into unit cell size effect on mechanical responses and energy absorption capability of titanium graded porous structures manufactured by laser powder bed fusion. *Journal of the Mechanical Behavior of Biomedical Materials* 109: 103843.

Yasa, E. and Kruth, J.P. (2011). Microstructural investigation of selective laser melting 316L stainless steel parts exposed to laser re-melting. *Procedia Engineering* 19: 389–395.

Yu, X., Zhou, J., Liang, H. et al. (2018). Mechanical metamaterials associated with stiffness, rigidity and compressibility: a brief review. *Progress in Materials Science* 94: 114–173.

Zaeh, M.F. and Branner, G. (2010). Investigations on residual stresses and deformations in selective laser melting. *Production Engineering* 4 (1): 35–45.

Zhang, B., Liao, H., and Coddet, C. (2012). Effects of processing parameters on properties of selective laser melting Mg–9% Al powder mixture. *Materials & Design* 34: 753–758.

Zhang, H., Zhu, H., Qi, T. et al. (2016). Selective laser melting of high strength Al–Cu–Mg alloys: processing, microstructure and mechanical properties. *Materials Science and Engineering A* 656: 47–54.

Zhang, B., Li, Y., and Bai, Q. (2017). Defect formation mechanisms in selective laser melting: a review. *Chinese Journal of Mechanical Engineering* 30: 515–527.

Zhang, L., Song, B., Zhao, A. et al. (2019). Study on mechanical properties of honeycomb pentamode structures fabricated by laser additive manufacturing: numerical simulation and experimental verification. *Composite Structures* 226: 111199.

Zhang, J., Gao, J., Song, B. et al. (2021). A novel crack-free Ti-modified Al-Cu-Mg alloy designed for selective laser melting. *Additive Manufacturing* 38: 101829.

Zhao, D., Han, C., Li, Y. et al. (2019). Improvement on mechanical properties and corrosion resistance of titanium-tantalum alloys *in situ* fabricated via selective laser melting. *Journal of Alloys and Compounds* 804: 288–298.

Zhao, D., Han, C., Li, J. et al. (2020). *In situ* fabrication of a titanium-niobium alloy with tailored microstructures, enhanced mechanical properties and biocompatibility by using selective laser melting. *Materials Science and Engineering C* 111: 110784.

Zhao, D., Liang, H., Han, C. et al. (2021). 3D printing of a titanium-tantalum Gyroid scaffold with superb elastic admissible strain, bioactivity and *in-situ* bone regeneration capability. *Additive Manufacturing* 47: 102223.

Zhou, X., Liu, X., Zhang, D. et al. (2015a). Balling phenomena in selective laser melted tungsten. *Journal of Materials Processing Technology* 222: 33–42.

Zhou, X., Wang, D., Liu, X. et al. (2015b). 3D-imaging of selective laser melting defects in a Co–Cr–Mo alloy by synchrotron radiation micro-CT. *Acta Materialia* 98: 1–16.

4

Electron Beam Melting

Electron beam melting (EBM) is a powder bed fusion process that employs an electron beam to selectively melt metal powder layers. This technology provides an alternative for processing materials that are either prone to cracking due to thermal stress or exhibit poor laser absorption during the laser powder bed fusion (LPBF) process. This chapter provides an overview of the history, fundamentals, processing characteristics, powder materials, equipment, microstructures, and properties of EBM-printed parts.

4.1 History

The patent for electron beam melting (EBM) was filed in 1993, which describes the principles of melting electrically conductive powder by an electron beam in a layer-by-layer manner to manufacture complex parts. The pioneering work on its development was conducted at Chalmers University of Technology, Sweden. In 1997, Arcam AB was founded, which further developed and commercialized the fundamental ideas underlying the patent, and the machine models were named EBM. The first commercial model, EBM S12, was released in 2002. Afterward, the company launched Arcam A2 in 2007, which was capable of producing large-scale parts for the aerospace industry.

Meanwhile, an important milestone was achieved in the orthopedic implant industry: Adler Ortho (Italy) released a Conformite Europeenne (CE) – certified Fixa Ti-Por hip implant fabricated through EBM in Europe. After this breakthrough, many CE-certified and Food and Drug Administration (FDA) – approved implants were produced using EBM. From 2009 to 2013, Arcam launched Arcam A1 and Q10 for the mass production of orthopedic implants, Arcam Q20 for the production of aerospace parts, Arcam A2X for processing high-temperature materials, and Arcam A2XX with the largest build chamber for large-scale production. In 2017, Arcam was acquired by GE Additive to further refine the technology for the mass production of metal parts.

Metal Powder–Based Additive Manufacturing, First Edition. Kun Zhou and Changjun Han.

4.2 Fundamentals

EBM schematically resembles laser powder bed fusion (LPBF). Since the energy source is an electron beam, the process is conducted in a vacuum (10^{-4}–10^{-5} mbar) to prevent air from ionizing and deflecting the electron beam. This feature is highly advantageous for fabricating powder materials with a high affinity for oxygen (such as Ti–6Al–4V). Additionally, an elevated temperature (500–1100 °C) is maintained within the chamber to minimize residual stress. Therefore, the in-process distortion, warpage, and even cracking of the parts can be suppressed to a large extent, which is beneficial to printing materials prone to cold cracking, such as γ-TiAl. It can also process materials with high laser reflectivity and thermal conductivity, such as pure copper.

The schematic of an Arcam EBM system and its printing process is shown in Figure 4.1. The substrate plate is supported by a powder bed with a thickness of at least 40 mm for heat insulation to maintain the high building temperature and provide some tolerance for the expansion of the substrate during printing. The EBM process begins with heating the substrate plate to sinter the powder bed surrounding and below the substrate plate to prevent it from being shifted by the moving rake. The initial temperature is determined by the powder characteristics and regulated by the electron beam power. A layer of powder is delivered from the powder hoppers and then spread on the powder bed by the rake. Before melting each layer, the powder is preheated to a partial melting (termed pre-sintering in this chapter) status by rapidly scanning it with a defocused high-energy beam. After the powder layer is selectively melted, the building platform moves downward by one-layer thickness. A new layer of powder is delivered again, and the steps above are repeated until the part is completely printed. Afterward, the printed part is retained in the build chamber to cool it down to a specific temperature (usually set at ~200 °C). Next, helium gas is purged into the chamber to expedite cooling by convection until the part

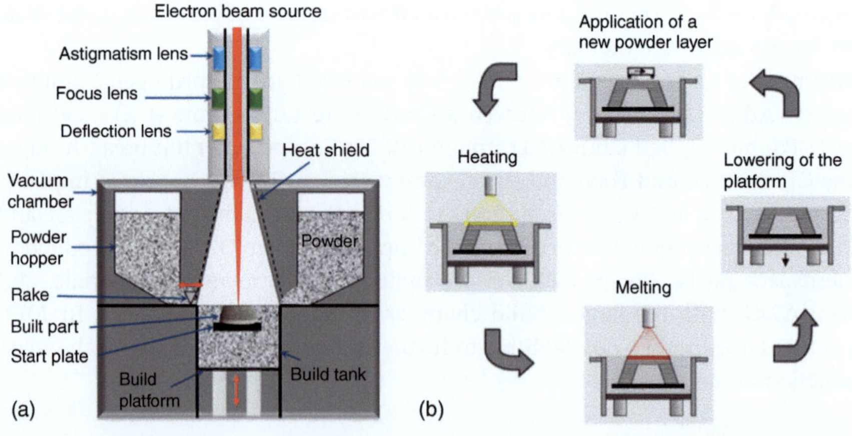

Figure 4.1 (a) Schematic illustration of an Arcam EBM system. Source: Reproduced from Galati and Iuliano (2018)/with permission from Elsevier. (b) Process for printing powder layers. Source: Reproduced from Körner (2016)/with permission from Taylor & Francis.

temperature is lowered below 100 °C, as the cooling rate decreases exponentially in this range if it is left to cool naturally in a high vacuum. After being retrieved from the build chamber, the printed part is covered by a soft agglomerate powder cake within the build tank and can be restored by sandblasting using the same powder.

More than 90% of the partially melted powder particles from the powder cake after printing can be recycled if their chemical composition and physical properties have been largely retained through proper control of the printing and recycling processes. The remaining partially melted powder particles are unrecyclable because they stick to form agglomerates, which can be removed by sieving in the powder recycling process.

The printing process consists of pre-sintering, melting, solidifying, and post-heating. To fabricate high-quality EBM parts, the following parameters must be optimized: beam current, focus offset, scanning speed, line offset, layer thickness, and scanning strategy. The beam current and focus offset determine the spot size, which can be as small as 0.1 mm (Biamino et al. 2011). The speed function index, which affects the instantaneous scanning speed, dynamically controls the translation of the electron beam and the melt pool size. Meanwhile, the line offset parameter regulates the hatch spacing, which is defined as the distance between two parallel lines of scanning.

The scanning pattern defines how the electron beam moves from one point to another. The most adopted scanning pattern is the zigzag pattern, in which the electron beam scans in a snake-shaped manner, and the scanning track is a series of parallel straight lines, which rotate by 90° for each printed layer (Figure 4.2a). The other scanning pattern is the contour pattern, in which the scanning track is a series of concentric lines that begin from the outermost line (Figure 4.2b).

The default scanning strategy combines the zigzag and contour patterns in each printing layer. It typically consists of three contour lines to trace the outer geometry of the part and zigzag lines to fill its inner area (Figure 4.2c).

At the pre-sintering stage, parameters such as the sintering temperature and powder conductivity are preset by the user to define the boundary conditions of the pre-sintering temperature. At the same time, the scanning speed, input current, and scanning strategy should be optimized. At the melting stage, numerous studies have been conducted to explore the optimization of process parameters for various alloys, particularly titanium alloys (Liu et al. 2016b, 2017; Zhao et al. 2016).

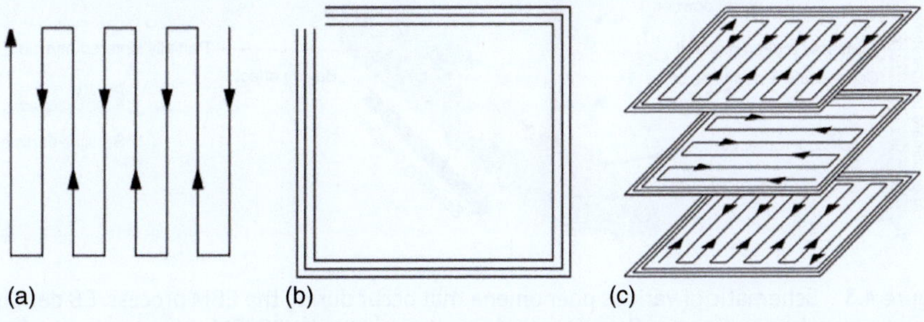

Figure 4.2 Typical scanning patterns used in EBM. (a) Zigzag pattern, (b) contour pattern, and (c) a combination of the two.

These process parameters can influence the size and depth of the melt pool and the thermal distribution, maximum temperature, and cooling rate of the alloys, which determine their densification, crystallographic textures, and mechanical properties. The cooling rates of the melt pool lie in the range of 10^3–10^5 K/s (Al-Bermani et al. 2010; Antonysamy et al. 2013), which are much lower than those of the LPBF process (typically in the range of 10^3–10^8 K/s). Since a fully molten stage is essential for fabricating completely dense parts, an adequate energy density of the input electron beam is required. However, it is difficult to convert the key process parameters, such as the focus offset and speed function, to the ones used in well-established classical energy formulae for LPBF processing. A trade secret of Arcam involves an auto-calculation energy-balancing algorithm embedded in the EBM system, which simultaneously adjusts a few interdependent parameters, such as the scanning speed and beam current, during the printing process. For example, the algorithm increases the scanning speed and beam current concurrently at the turning point of the zigzag pattern to avoid heat accumulation, which helps to improve the quality of the EBM-printed parts.

Each EBM process parameter exerts critical influences on the metallurgical defects (e.g. pores and cracks), microstructural features (e.g. phase and grain size), and surface characteristics (e.g. surface roughness) of the printed parts. Generally, optimizing the EBM process is challenging. At different stages of the layer melting process, many parameters are involved in developing the hatch and contour regions, as shown in Figure 4.3. For example, a contour design was optimized to improve the topographical and microstructural characteristics of the printed parts (Karimi et al. 2020). The scanning speed was established to be one of the most important parameters determining the surface roughness of one-contour parts. The beam current and focus offset influence the porosity of two-contour parts. Low porosity of the printed parts could be obtained by implementing a high beam current and

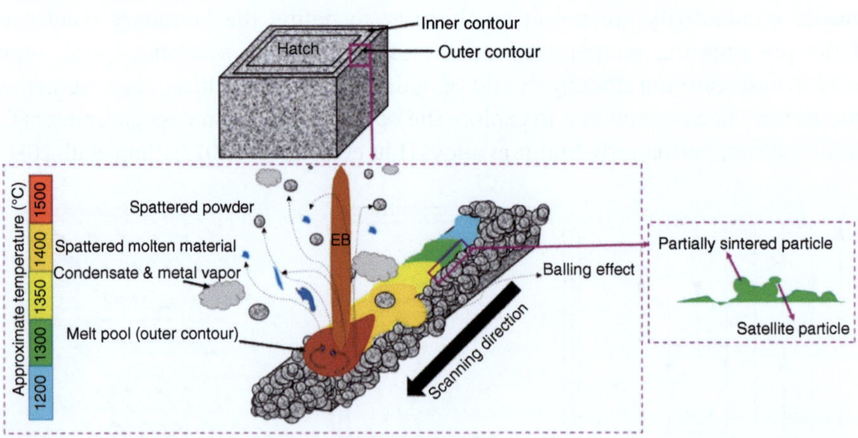

Figure 4.3 Schematic of various phenomena that occur during the EBM process. EB denotes the electron beam. Source: Reproduced from Karimi et al. (2020)/with permission from Elsevier.

a low focus offset, while the scanning speed and beam current could affect their geometrical characteristics.

The advantages of EBM are summarized as follows:

- The electron beam possesses an energy density a few times higher than a laser beam, which corresponds to greater penetration that promotes the uniform melting of highly reflective metals.
- The electromagnetic field generated by an electron beam unit can drive the electron beam to move at high transverse velocities (up to 10^5 mm/s for preheating and 10^4 mm/s for melting) within the printing area, thereby achieving a high build rate.
- The equipment can preheat the powder bed up to 1100 °C to minimize the residual stress within the part, which is a recurring issue in LPBF that results in distortion and cold cracking of the printed part. This feature also enables EBM to possess the capability of *in situ* heat treatment.

Meanwhile, the limitations of EBM are:

- Because of larger powder particles and thicker printing layers, EBM-printed parts typically exhibit higher surface roughness than LPBF-printed components. The printed parts require extensive post-machining to achieve a smooth surface.
- The printing materials currently available for EBM are strictly restricted to non-magnetic metallic materials with high boiling points.
- EBM typically corresponds to lower dimensional accuracy and resolution than LPBF due to electron beams displaying larger spot sizes than laser beams. Thus, the minimum feature size of EBM is usually ~500 μm.
- Loss of alloying elements frequently occurs during printing, especially for light elements with low boiling points (such as Al) in an ultrahigh vacuum environment.
- The equipment required for EBM is even more costly than that for LPBF.
- The selection of equipment manufacturers and commercially available powder materials for EBM is much more limited than that for LPBF, thereby impeding the research on EBM. Because of the multiple stages and numerous process parameters involved in EBM, it takes longer to develop new materials for EBM than for LPBF.

4.3 Preheating and Melting Processes

The EBM printing processes include pre-sintering and melting (i.e. contour melting and hatch melting). Before the melting process, the powder bed is uniformly pre-sintered by using a high-power defocused electron beam to rapidly scan each printing layer at a speed of ~10 m/s (Figure 4.4a). Meanwhile, the beam power is increased to ~3 kW (the recently developed machine model Arcam EBM Spectra H can provide up to 6 kW of power). Maintaining the predefined temperature in the building chamber and slightly pre-sintering the powder particles are critical for increasing their thermal and electrical conductivity. Process instabilities, such as smoking due to the repulsion of charged powder particles, can be prevented by strictly monitoring the pre-sintering process, particularly for powder beds with large proportions of fine powder particles (Lodes et al. 2015). In addition, clusters of sintered powder particles can serve as supports for overhanging structures.

A high temperature stabilizes the powder bed through pre-sintering of the powder particles. The pre-sintering speed is influenced by the surface oxide layer on the powder particles. The high temperatures during pre-sintering play a significant role in determining the microstructure of a part and avoiding distortion and cracking by decreasing its residual stress. Compared to LPBF-printed parts, EBM-printed parts generally exhibit coarse grains due to a low cooling rate and residual stress due to a small thermal gradient.

After pre-sintering, the contours are melted in a quasi-multi-beam mode in which up to 100 points of each contour are quasi-simultaneously melted in a discontinuous manner. Such melting mode is achieved through the electron beam focus rapidly shifting from a point to another (Figure 4.4b), resulting in reduced surface roughness and improved accuracy of the geometry of the printed parts compared to those without contours. The melting procedure is typically performed by hatching (i.e. zigzag pattern). The powder particles are melted by the electron beam at a scanning speed of several meters per second in each printing layer (Figure 4.4c). During melting, the electrons collide with the powder particles, and most of the kinetic energy of the electrons is released in the form of thermal energy that sinters, fuses, and vaporizes the particles.

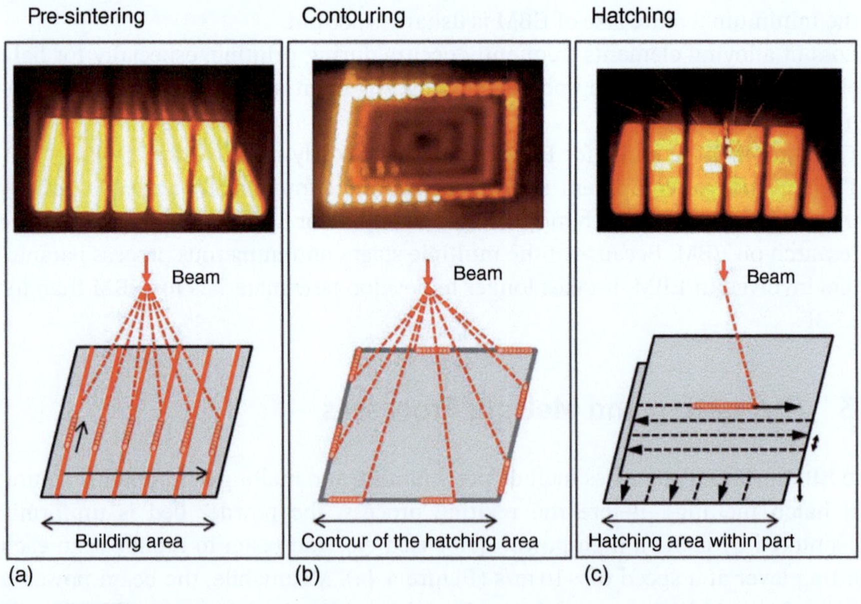

Figure 4.4 Pre-sintering and melting processes in EBM. (a) Pre-sintering by a quasi-multi-beam scan of the entire building zone with a defocused electron beam, (b) quasi-multi-beam contour melting, and (c) hatch melting. Source: Körner (2016)/reproduced with permission from Taylor & Francis.

Nevertheless, "smoking" of powder particles, also known as the pushing phenomenon, occurs in the form of an explosion. It can be caused by (i) the presence of water residue within the powder particles, which is vaporized in a volatile manner when the electrons collide with the powder, (ii) the momentum transferred from the electrons to the pre-sintered powder particles, which disrupts their cohesion, and (iii) the negative electrostatic charge of the powder particles, which leads to large mutually repulsive forces among them. The pushing phenomenon induced by the large momentum imparted by the electrons can be inhibited using larger powder particles. Meanwhile, the slight sintering of the powder particles increases their individual mass, which diminishes the effects of the repulsive forces, thereby reducing the occurrence of smoking.

Sintering of the powder particles occurs as a result of heat transfer during the pre-sintering and melting stages. The amount of heat transferred to the powder particles depends on the process parameters (e.g. the beam power, beam spot size, and scanning speed). When sufficient energy is absorbed by the material, the powder is melted and forms a melt pool. While EBM and LPBF share similar metallurgical processes, the vacuum environment in EBM lowers the boiling point and promotes the vaporization of lightweight elements (Biamino et al. 2011). Therefore, the composition of the raw powder, printed parts, and recycled powder of the two techniques is different.

The amount of energy released by the incident electrons is strongly related to the acceleration voltage, the atomic number of the constituent elements of the powder material, and the angle of incidence of the electron beam (Everhart and Hoff 1971). The released energy melts the powder particles and forms a melt pool that retains its molten state for a few milliseconds. The fluid dynamics parameters involved are primarily capillary forces and back pressure due to vaporization. To print fully dense parts, the depth of the melt pool should be much greater (about 3–4 times more) than the layer thickness to avoid interlayer bonding defects, namely, unmelted powder or non-connected layers.

Composition inhomogeneity results from the selective vaporization of elements with high vapor pressure. The uneven surface tension of the melt due to selective vaporization can cause material displacement and produce a melt surface with varying depths (Guo et al. 2015; Juechter et al. 2014) (Figure 4.5a). In these studies, the degree of composition inhomogeneity in EBM-printed parts depended on the scanning pattern. If the material displacement exceeds the layer thickness, applying a fresh powder layer cannot cover the protrusion of the solidified melt. When the next powder layer is melted, the material displacement will be aggravated if the scanning pattern is not changed (Figure 4.5b), which will eventually terminate the entire process. Elements such as aluminum and zinc exhibit extremely high vapor pressure, which significantly depletes these elements at the melt pool surface under high temperatures. The composition inhomogeneity due to element depletion can be partly mitigated by convection within the melt pool, but element segregation still occurs. As shown in Figure 4.5c, the distribution of aluminum varies across the different layers in an EBM-printed Ti–48Al–2Cr–2Nb alloy.

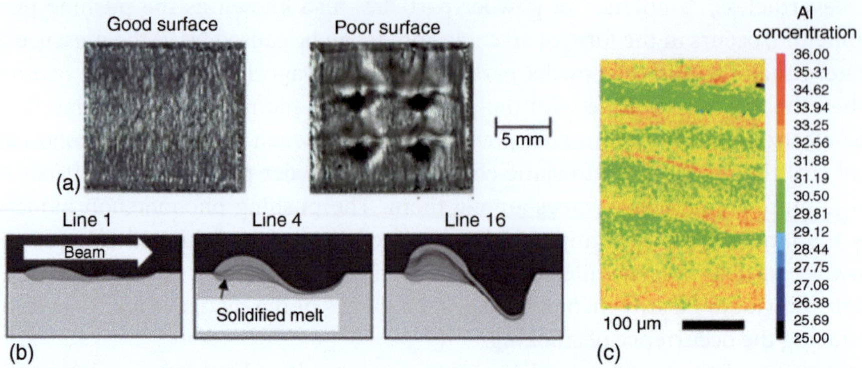

Figure 4.5 (a) Exemplary melt surface and poor melt surface caused by material displacement, (b) simulation results of melt displacement with the electron beam moving from left to right up to 16 times, and (c) element mapping representing the Al distribution in an EBM-printed Ti–48Al–2Cr–2Nb alloy. Source: Reproduced from Körner (2016)/with permission from Taylor & Francis.

A significant distinction between EBM and LPBF lies in the absorption of the beam energy. While incoming photons are absorbed within a depth of only several nanometers of the powder material, electrons can penetrate several micrometers of the powder surface before being absorbed. Consequently, EBM is more suitable for processing powder materials highly reflective to laser (e.g. pure copper and copper alloys).

4.4 Metallurgical Defects

In a controlled vacuum environment with optimized process parameters, parts printed by EBM can achieve a relative density of over 99.8%, which corresponds to a lower porosity level than LPBF products (Frazier 2014). However, since EBM and LPBF involve similar melting processes, metallurgical defects such as pores, delamination, and cracks are still present in EBM-printed parts.

Pores are a major defect in EBM-printed parts. They are formed through the residual gas trapped within the powder, lack of fusion, balling, keyhole formation from element vaporization, and uneven powder spreading. Wang et al. (2019) explored the printability of a CoCrFeMnNi high-entropy alloy (HEA) powder for EBM, and an image of the EBM-printed samples is shown in Figure 4.6a. Many powder particles were observed to be attached to the sides of the cuboid samples (Figure 4.6b), whose surface roughness *Ra* is 41.1 μm. Figure 4.6c presents the morphology of the top surfaces of the samples with different extents of protrusion and numbers of lack-of-fusion pores. The protrusion resulted from turbulent melt pool motion, which was driven by strong forces induced by temperature gradients, Marangoni convection, and evaporation.

To maintain the same fusion condition throughout the different regions of a part, it is necessary to dynamically adjust the line and focus offset, scanning

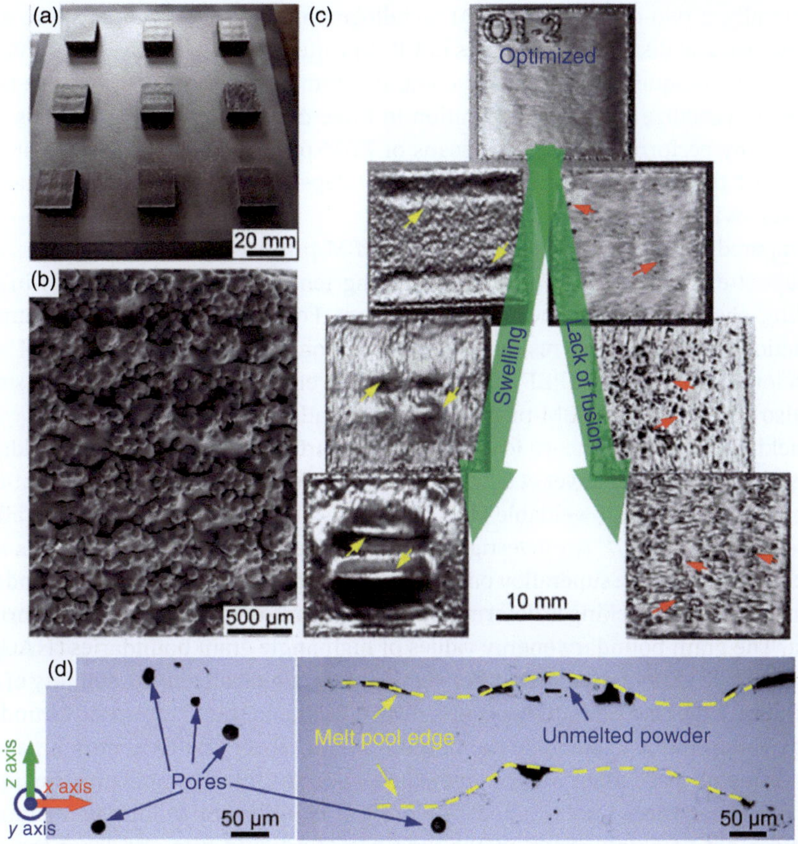

Figure 4.6 (a) EBM-printed HEA samples and (b) SEM image of a rough sample surface along the *z* direction. (c) Top surfaces of the cuboid samples. The yellow and red arrows indicate protrusion and lack-of-fusion pores, respectively. (d) Optical microscope (OM) images displaying the pores. Source: Reproduced from Wang et al. (2019)/with permission from Elsevier.

speed, reference current, and pre-sintering temperature, which is achieved by the auto-calculation algorithm. For example, a combination of large line offsets, high scanning speeds, and low reference currents can lead to shallow melt pools and low building temperatures; a large focus offset can broaden the width of the melt pools. However, the resultant low input energy will be inadequate to form melt pools that are sufficiently deep (Figure 4.6d). In general, there are two types of pores, namely, irregular pores and spherical pores. Once a melt pool cannot penetrate the previously solidified layer, a lack-of-fusion region may occur with the presence of some unmelted powder particles, resulting in irregular pores. During the rapid solidification process, spherical pores are formed from gas bubbles that have previously been trapped within the powder particles. As the EBM process is performed in a vacuum environment, the presence of gas mainly originates from inert gases, such as argon, trapped in the raw powder during the atomization process (see Chapter 1).

Generally, a two-dimensional (2D) metallographic method is employed to analyze the size and distribution of pores in EBM-printed parts. Recently, X-ray characterization techniques, such as micro-computed tomography (micro-CT), have been utilized to visualize the pore distribution in three-dimensional (3D) materials. For example, by performing micro-CT scans of EBM-printed Ti–6Al–4 V parts, it was found that the distribution of defects mainly depends on the scanning strategies (Tammas-Williams et al. 2015).

Compared to LPBF-printed components, EBM-printed components possess low residual stress levels due to the high building temperature (500–1100 °C) in the building chamber during the printing process. For example, using the neutron diffraction technique, the residual stress in Inconel 718 produced by EBM was much lower than that by LPBF (Sochalski-Kolbus et al. 2015). Minor residual stress was also observed in an EBM-printed Ti–6Al–4V alloy (Hrabe et al. 2017).

Cracking is another concern for EBM-printed parts, particularly for non-weldable alloys. In a study by Chauvet et al. (2018), intergranular hot cracking was identified in an EBM-printed non-weldable Ni–Co–Cr–Mo–Al–Ti–B nickel-based superalloy, as shown in Figure 4.7. An investigation of its underlying mechanism reveals that crack formation in the superalloy parts requires the presence of liquid films, and the occurrence of hot cracking is determined by the degree of grain boundary misorientation. The grain boundary energy values of high-angle grain boundaries (HAGBs) and low-angle grain boundaries differ significantly, which affects the stability of the liquid film. Cracking preferably occurs in HAGBs due to their large grain boundary energy values that can stabilize the liquid film. Moreover, the local enrichment of trace elements, particularly boron, plays an important role in stabilizing the liquid film at temperatures lower than its theoretical equilibrium solidus temperature. Although hot cracking occurs in the presence of a liquid film and large thermal stress, it has been demonstrated that the thermal stress within an EBM-printed non-weldable nickel-based airfoil can be redistributed by optimizing the scanning strategy (Lee et al. 2020). Other scanning strategies, such as random raster scanning and spot melting, can mitigate cracking or distortion within EBM-printed parts.

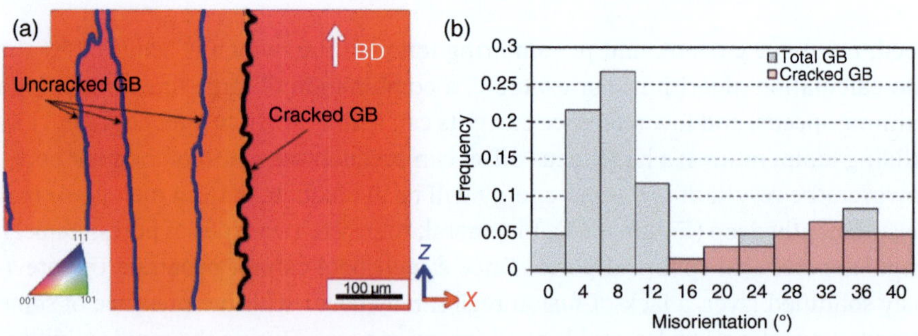

Figure 4.7 (a) Electron backscatter diffraction map indicating cracks propagating along the high-angle grain boundaries (>15° in black). Low-angle grain boundaries are delineated in blue (5° < angle < 15°). (b) Histogram displaying the distribution of the grain boundaries (gray). The distribution of the cracked grain boundaries is delineated in red. Source: Reproduced from Chauvet et al. (2018)/with permission from Elsevier.

The surface roughness and near-surface defects are primary features in the contour zones of printed layers. Generally, the high surface roughness of EBM-printed parts, particularly those with inclined geometrical features, leads to high stress concentration and subsequent crack initiation. The surface defects of the printed parts include partially melted particles located near the surface and some irregularly shaped melted particles generated from the powder bed being sputtered by electrons, which occurs when the electron beam possesses a high energy density. Because of such undesirable surface features, most EBM-printed parts require post-processing to improve their surface quality.

4.5 Powder Materials

Before optimizing the process parameters for EBM, the powder should be characterized to determine its suitability. An ideal powder should exhibit good flowability (<25 s/50 g, similar to Arcam's supplied powder), high apparent density (over 50% of the density of the bulk material), no internal pores from the production process, and no fine particles (<10 μm). Generally, pre-alloyed powders produced through gas atomization or the plasma rotating electrode process are used in EBM. The powder particles are spherical, and the standard particle sizes are in the range of 45–100 μm. The possibility of using powders with smaller particle sizes (25–45 μm) in EBM was investigated (Karlsson et al. 2013), and no significant distinction between the properties of the parts printed using standard powders and powders with smaller particle sizes were observed. However, the latter possesses the potential to improve the surface quality of the printed parts. It was suggested that the material of the substrate should be identical or similar to that of the powder because it prevents contamination and ensures proper bonding.

Powders can be recycled and reused in the EBM process. The effect of the number of powder recycling cycles on the properties of Ti-6Al-4V powder was examined (Tang et al. 2015a), which revealed that the particle size distribution became narrower with an increase in the number of recycling cycles. Additionally, the powder particles were rougher and less spherical. However, the flowability of the powder was found to increase after each cycle. Although the main characteristics of the powder underwent minor changes after each cycle, no powder spreading/smoking was observed after 21 cycles. Nevertheless, recycling a powder increases its oxygen content, which improves the yield strength and ultimate tensile strength of the printed parts.

Commercialized materials for Arcam EBM include Ti-6Al-4V, Ti-6Al-4V ELI, commercially pure titanium, F75 Co–Cr–Mo, Inconel 718, pure titanium, pure copper, highly alloyed tool steel, and TiAl. Based on the currently limited selection of materials available for EBM compared to LPBF, extensive studies have been conducted to explore and develop other materials for EBM, such as Fe-based, Ni-based, and Ti-based alloys, HEAs, and their composites. A summary of these materials is presented in Table 4.1.

Table 4.1 Summary of commercialized and experimental powder materials for EBM.

Material group	Materials	Commercial materials from Arcam
Ti-based	Pure Ti, Ti–6Al–4V, Ti–6Al–4V ELI, Ti–24Nb–4Zr–8Sn, TA15, Ti–Nb, Ti–Mo, Ti–Nb–Ta–Zr	Ti–6Al–4V, Ti–6Al–4V ELI
Fe-based	Pure Fe, 316L, H13, Cr–Mn–Ni TRIP steel, Fe–18Cr–2W–0.5Ti–0.3Y2O3	Highly alloyed tool steel
Al-based	Al–Si10–Mg	—
Ni-based	IN718, IN625, Haynes 282, CMSX-4, K417, Rene 142, Rene-N5, 713ELC	IN718
Co–Cr-based	Co–Cr–Mo, Co–Cr–Ni–W	Co–26Cr–6Mo–0.2C
TiAl	Ti–47Al–2Cr–2Nb, Ti–48Al–2Cr–2Nb, Ti–45Al–7Nb–0.3W	Ti–48Al–2Cr–2Nb
Pure metals	Ti, Cu, Nb, Fe, W	Ti, Cu
HEAs	CoCrFeMnNi, AlCoCrFeNi, CoCrFeNiTiMo, AlCrMoNbTa	—
Others	NiTi, Cu–25Cr composites, 35 wt.% NiBSi + 64 wt.% WCp composites	—

4.6 Equipment

An EBM system generally consists of an electron beam unit, a vacuum chamber, a powder feeder system, and a computerized control system (Figure 4.1). The electron beam module comprises an upper column for electron generation and a lower column that contains magnetic lenses for beam formation and deflection. In the upper column, a hot tungsten filament or a lanthanum hexaboride cathode emits electrons under a voltage of ~60 kV, and the beam current is in the range of 1–50 mA. In contrast to an electron beam generated by a tungsten filament, a beam generated by a lanthanum hexaboride cathode has high quality. However, the cathode is more susceptible to "smoking". Therefore, it is only used for titanium processing. The lower column of the electron beam module includes astigmatism lenses, focus lenses, and deflection lenses, which control the shape, size, and position of the beam on the substrate, respectively.

The vacuum chamber consists of a build container, a powder feeder, and a raking system. The platform of the build container can move along the z axis, and it includes a processing platform for preheating and printing. Two hoppers for powder storage are available, and the powder is fed to the chamber plate by gravity. The raking system controls the degree of spreading of the powder, which forms layers with a thickness of 0.05–0.2 mm during printing. The process is conducted in a vacuum environment (10^{-4}–10^{-5} mbar) to prevent air from interfering with the electron beam. In addition, helium gas is typically purged to the chamber with a low pressure of 10^{-3} mbar to neutralize the negative charges on the powder particles, thereby preventing electrostatic charging and smoking, which may result in process termination.

Currently, the supply of commercialized EBM equipment is dominated by Arcam, which was acquired by GE Additive in 2017. Commercialized EBM machines

Table 4.2 Summary of the specifications of EBM machine models produced by Arcam.

Equipment	Build volume (mm³)	Beam power (kW)	Process temperature (°C)	Machine size ($D \times W \times H$) (mm³)	Materials
Arcam EBM Q10plus	200 × 200 × 180	3	—	1066 × 2060 × 2608	Ti–6Al–4V, CoCr, pure Ti, pure Cu
Arcam EBM Q20plus	Ø 350 × 380	3	—	1300 × 2400 × 2945	Ti–6Al–4V
Arcam EBM A2X	200 × 200 × 380	3	600–1100	900 × 1850 × 2200	Ti–6Al–4V, Ni 718, TiAl
Arcam EBM Spectra H	Ø 250 × 430	6	600–1100	1328 × 2344 × 2858	Ti–6Al–4V, Ni 718, highly alloyed tool steel
Arcam EBM Spectra L	Ø 350 × 430	4.5	700	1328 × 2344 × 2858	Ti–6Al–4V, pure Cu

include Arcam EBM Spectra L, Arcam EBM Spectra H, Arcam EBM Q10plus, Arcam EBM Q20plus, and Arcam EBM A2X, and their specifications are summarized in Table 4.2. The acquisition of Arcam by GE has enhanced the core competency of EBM in orthopedic and aerospace applications, with over 100 000 clinical evaluations and FDA-approved hip implants being produced as of 2018 and over 250 TiAl turbine blades being installed in each Boeing 777X aircraft engine (GE9X) in 2019. In addition, during 2018–2019, EBM systems Spectra H and Spectra L were released, which featured increased power, larger build volume, improved productivity, a closed powder-handling system, and an automatic beam calibration function. In 2019, GE Additive opened its 15 000 m² Arcam EBM Center of Excellence in Gothenburg, Sweden, to boost its production capacity and R&D efforts to meet the ever-increasing demands for EBM-printed components.

Other representative commercial companies that are developing/have developed EBM systems are Sailong (S200 Production EBM 3D Printer and Y150 Biomedical EBM 3D Printer) and QBEAM (QbeamLab, QbeamMed, and QbeamAero) of China, Freemelt (Freemelt ONE) of Sweden, and JEOL (JAM-5200EBM) of Japan. Some research groups have also attempted to develop their respective EBM systems. For instance, the *iwb* application center (Germany) developed an EBM system with an electron beam power of 10 kW by modifying an existing electron beam welding system.

4.7 Microstructures and Mechanical Properties

In general, EBM is performed at high building temperatures (e.g. 300 °C for pure copper, 600 °C for Ti, and 1100 °C for TiAl), which significantly affect the microstructures and mechanical properties of the printed parts. Additionally, the mechanical properties of the parts are height-dependent as a result of the *in situ* heat treatment conditions during the printing process. After the final solidification of the first printed layer, the building temperature measured by the thermocouple placed underneath the substrate plate fluctuates significantly. A fixed building temperature can only be achieved after 10–20 layers are printed. The region near the melting surface approaches the solidification temperature of the printed material several times during the printing process, and the material may undergo several rounds of phase transformation.

With an increase in the number of printed layers, their solidification is influenced by complex thermal cycles. The preheating process in EBM results in a consistently high temperature throughout the printed parts, which eliminates residual stress within them and endows them with a good combination of strength, plasticity, and performance homogeneity. Because of their refined microstructures, EBM-printed parts exhibit mechanical properties comparable to or superior to their wrought counterparts. However, their strength is generally inferior to that of their counterparts printed by laser powder bed fusion and laser-based directed energy deposition.

4.7.1 Titanium and Its Alloys

Presently, titanium and its alloys, which are considered difficult-to-machine materials in conventional manufacturing, are mostly utilized for EBM. The processibility of these materials with EBM has been extensively verified for ($\alpha + \beta$)-type and β-type titanium alloys.

Ti–6Al–4V is an ($\alpha + \beta$)-type titanium alloy comprising 6 wt.% Al (an α-stabilizing element), 4 wt.% V (a β-stabilizing element), and other interstitial elements such as O. A representative printed Ti–6Al–4V sample, as shown in Figure 4.8, consists

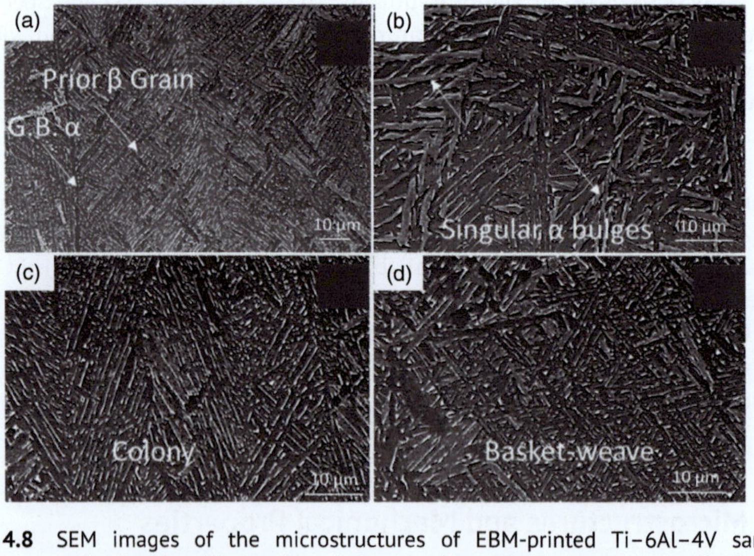

Figure 4.8 SEM images of the microstructures of EBM-printed Ti–6Al–4V samples. (a) Columnar prior β grains, (b) singular α bulges, and ($\alpha + \beta$) microstructures with (c) colony and (d) basket-weave morphology. G.B. denotes a grain boundary. Source: Tan et al. (2015)/ reproduced with permission from Elsevier.

of (i) columnar prior β grains which are delineated by α grain boundaries and (ii) a transformed α/β microstructure with the Widmanstätten pattern and lamellar colony, as well as singular α bulges, within the prior β grains. The formation of the α grain boundary along the grain boundaries of the prior β grains indicates the β-to-α phase transformation of the sample. Consequently, the microstructure formed in an EBM-printed Ti–6Al–4V sample differs from that of an LPBF-printed Ti–6Al–4V sample, in which the martensitic α' phase is commonly present as a result of transformation from the β phase.

In a study by Guo et al., α′ martensite was only observed near the surface of an EBM-printed Ti–6Al–4V sample, implying that the prior β phase first transforms into the martensitic α′ phase before decomposing into (α + β) phases (Guo et al. 2015). The thickness of the α-lamellae was found to be dependent on the energy input, building temperature, and cooling conditions. In another study by Scharowsky et al., the thickness of the α-lamellae of a Ti–6Al–4V sample was observed to increase with the energy input and building temperature. However, the yield stress decreased as a result (Scharowsky et al. 2015).

Generally, the cooling rate of an EBM-printed product decreases with the building height, thereby resulting in an anisotropic microstructure along its height. For example, in a related study, an equiaxed-to-columnar transition of the prior-β grains in an EBM-printed Ti–6Al–4V sample was observed to occur as the number of build layers increased. Such a transition was accompanied by an increased thickness of the retained rod-like β phase (Figure 4.9) (Tan et al. 2015). It was suggested that the relatively high heat conductivity of the substrate had led to a high degree of super-cooling, resulting in the equiaxed-to-columnar transition of the grain morphology. The mechanical properties of the printed parts vary within them due to the height dependence of local thermal conditions.

The mechanical properties of EBM-printed Ti–6Al–4V alloys are summarized in Table 4.3. With optimized process parameters, EBM can produce parts possessing strength and elongation comparable to those of the wrought products (Murr et al. 2009). A decrease in the thickness of the α lamellae corresponds to an increase in the mechanical strength of the printed parts according to the Hall–Petch relationship. EBM-printed Ti–6Al–4V parts with smaller grain sizes and thinner α lamellae can exhibit enhanced tensile strength. Post-heat treatment (e.g. hot isostatic pressing [HIP]) of the printed parts above the β transus temperature resulted in the desired acicular or Widmanstätten morphology. As the size of small EBM-printed Ti–6Al–4V parts increases, their strength decreases, and their ductility increases, which has been attributed to the increase in the thickness of the α lamellae. EBM-printed Ti–6Al–4V parts were found to exhibit slightly lower yield strength along their height than across the building plane, which may be explained by their anisotropic microstructures and the metallurgical bonding defects between the layers (de Formanoir et al. 2016).

The presence of nonspherical lack-of-fusion pores, which contain unmelted or partially melted particles, is detrimental to the mechanical properties of the printed parts with respect to the x–y plane due to the sharp edges of the pores acting as stress risers when the applied tension is perpendicular to the z direction. Consequently, the

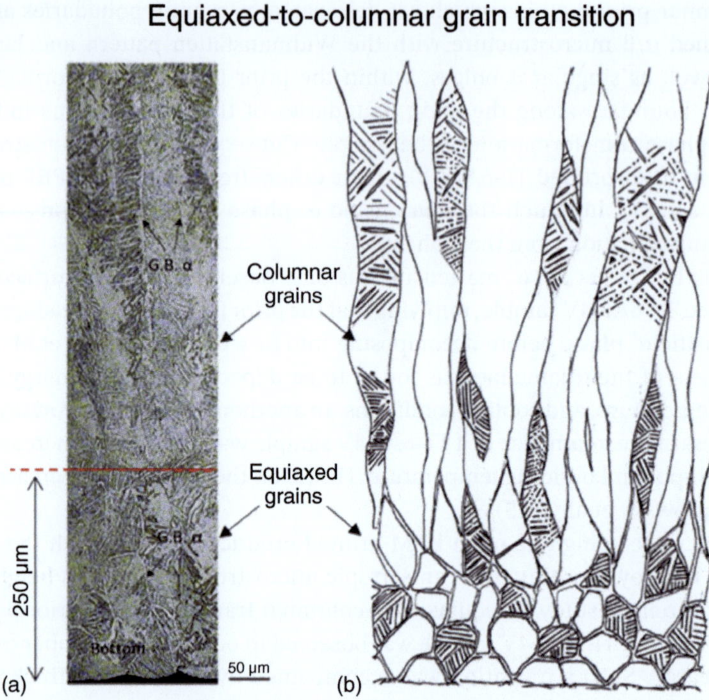

Figure 4.9 (a) OM image and (b) schematic representing the equiaxed-to-columnar transition of prior β grains in an EBM-printed Ti–6Al–4V alloy. Source: Reproduced from Tan et al., (2015)/with permission of Elsevier.

Table 4.3 Summary of the mechanical properties of EBM-printed Ti–6Al–4V alloys.

Ultimate tensile strength (MPa)	Yield strength (MPa)	Elongation (%)	Young's modulus (GPa)	Hardness (GPa)
950–990[a)]	910–940	14–16	120	—
910[b)]	830	—	118	3.21
1100–1400[c)]	—	12–25	—	4.3
1150–1200[d)]	1100–1150	16–25	—	3.6–3.9
994–1029[e)]	883–938	11.6–13.6	—	—
930[f)]	865	12–17	—	—
945–965[g)]	823–852	13.2–16.3	—	3.19–3.27
990–1180[h)]	900–1100	18–23	—	—

a) Parthasarathy et al. (2010)
b) Facchini et al. (2009)
c) Gaytan et al. (2009)
d) Murr et al. (2009)
e) Al-Bermani et al. (2010)
f) Scharowsky et al. (2015)
g) Tan et al. (2015)
h) Zhao et al. (2016).

printed samples subjected to tension perpendicular to the z direction exhibited 28% lower elongation than those subjected to tension along the z direction.

Another titanium alloy extensively investigated for EBM is the β-type Ti–24Nb–4Zr–8Sn (Ti2448) alloy, which was developed for biomedical applications because of its low Young's modulus and high tensile strength (Hao et al. 2007). Such excellent properties were attributed to the α-phase precipitates located at the β grain boundaries due to high-temperature preheating, as shown in Figure 4.10 (Liu et al. 2016a). The microstructure of EBM-printed parts can also facilitate fatigue crack deflection in their columnar grain regions (Liu et al. 2017). In this study, Ti2448 porous structures with a porosity of 70% were discovered to exhibit excellent mechanical properties, with a strength-to-modulus ratio over twice that of Ti–6Al–4V porous structures with the same porosity (Liu et al. 2016b). Additionally, the Ti2448 porous structures displayed high normalized fatigue strength due to their super-elastic properties and large plastic regions at their fatigue crack tips.

There are generally four types of TiAl microstructures, namely, the near-γ, duplex lamellar, near-fully lamellar, and fully lamellar structures. Owing to their low density, high specific yield strength, high specific stiffness, and favorable creep properties at elevated temperatures of 600–900 °C, γ-TiAl alloys are highly adopted in various industrial applications. For example, commercialized γ-TiAl alloys have been employed to manufacture the low-pressure turbine blades of GEnx engines that power the Boeing 787 and 747–8 aircrafts.

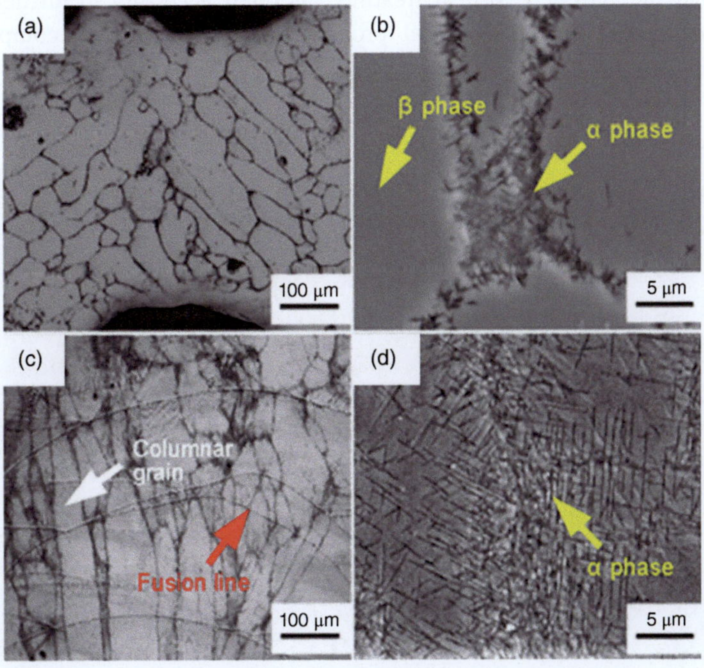

Figure 4.10 Microstructure of an EBM-printed Ti2448 alloy. (a) OM and (b) SEM images taken perpendicular to the z direction. (c) OM and (d) SEM images taken along the z direction. Source: Liu et al. (2016a)/reproduced with permission from Elsevier.

TiAl is favorably processed by EBM because the building temperature can reach up to 1100 °C. Therefore, crack formation is eliminated within the printed parts, which commonly occurs during the LPBF process because the ductility of TiAl is insufficient to counteract the high thermal stress during the rapid solidification process (Tang et al. 2015b). An EBM-printed γ-TiAl alloy was found to exhibit equiaxed small grains (~2 μm) with lamellar γ/α2-Ti_3Al colonies displaying an average spacing of 0.6 μm and a microhardness of 4.1 GPa (Murr et al. 2010). However, Al evaporates significantly during the printing process because of its low melting point. An EBM-printed Ti–48Al–2Cr–2Nb alloy was reported by Biamino et al. to undergo 1 wt% Al vaporization (Biamino et al. 2011). The as-built alloy possessed an extremely fine microstructure (Figure 4.11a), but full recrystallization in near-equiaxed grains with sizes smaller than 20 μm was obtained after HIP, as shown in Figure 4.11b,c. Subsequent heat treatment at 1320 °C for two hours resulted in a duplex microstructure containing ~40 vol.% of a lamellar phase with an average grain size of ~100 μm surrounded by fine equiaxed grains (Figure 4.11d). Consequently, the yield strength, ultimate tensile strength, and elongation of the treated alloy at room temperature were measured to be ~350 MPa, ~470 MPa, and 1%, respectively. At high temperatures (500–800 °C), its elongation was increased to beyond 5% without significantly sacrificing its strength.

The as-built γ-TiAl alloys exhibit a higher fatigue threshold and strength than their counterparts produced through conventional processes, such as casting and powder metallurgy. Post-heat treatment is often necessary to further refine the microstructure and mechanical properties of the as-printed components.

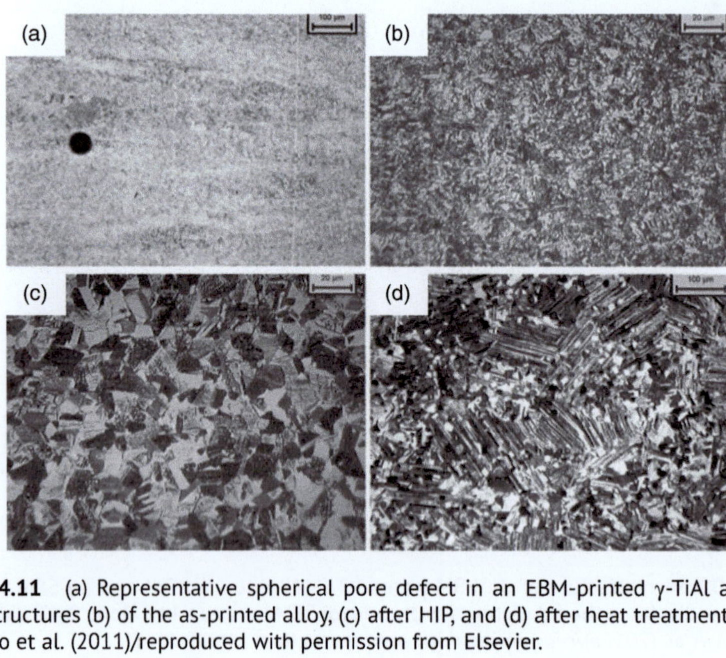

Figure 4.11 (a) Representative spherical pore defect in an EBM-printed γ-TiAl alloy and microstructures (b) of the as-printed alloy, (c) after HIP, and (d) after heat treatment. Source: Biamino et al. (2011)/reproduced with permission from Elsevier.

Nevertheless, TiAl alloys are still inferior to nickel-based alloys in terms of fatigue crack growth resistance, ductility, and fracture toughness.

4.7.2 Nickel Alloys

Nickel-based alloys can be categorized into weldable and non-weldable types. Non-weldable alloys are difficult to process by EBM and LPBF as most contain a substantial proportion of the γ' phase, which strengthens the alloy but renders it highly prone to cracking. Among weldable nickel-based alloys, Inconel 625 and Inconel 718 have been extensively explored for EBM processing. Inconel 625 is widely utilized in columnar-grained gas turbines because of its high Cr content, strength, processability, and corrosion resistance.

Figure 4.12 shows the representative microstructures of EBM-printed Inconel 625 and Inconel 718 parts (Murr et al. 2011; Körner 2016). Figure 4.12a shows equiaxed grains in the horizontal plane of the printed Inconel 625 alloy (Murr et al. 2011). Precipitates (indicated as black) could be observed on both the boundaries and interiors of the grains. The black precipitates along the grain boundaries were considered to be of the γ' phase. Comparatively, columnar grains could be found in the vertical plane of the printed alloy (Figure 4.12b). The EBM-printed Inconel 718 alloy often consisted of a face-centered-cubic (FCC) γ-matrix strengthened by several phases

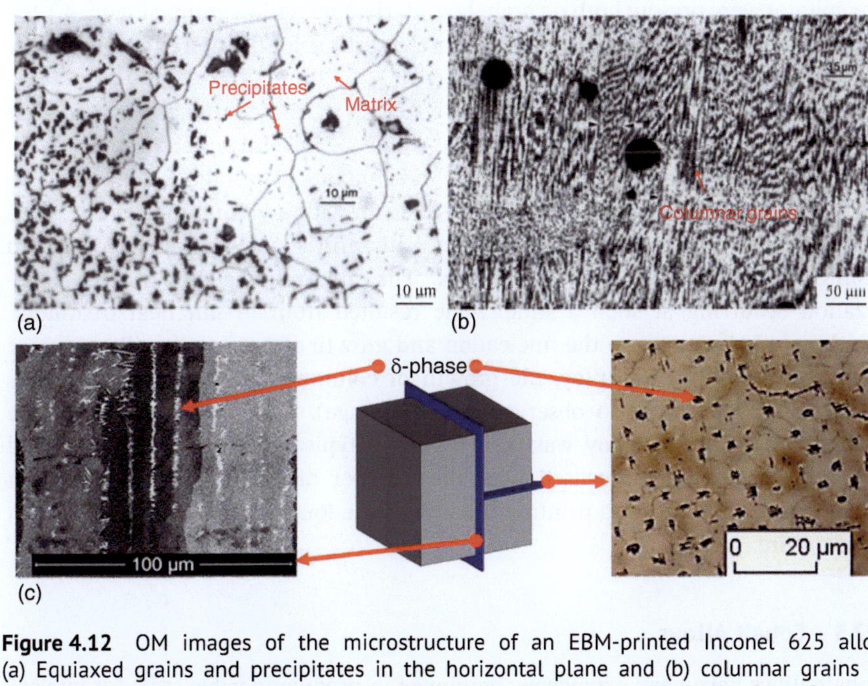

Figure 4.12 OM images of the microstructure of an EBM-printed Inconel 625 alloy. (a) Equiaxed grains and precipitates in the horizontal plane and (b) columnar grains in the vertical plane. Source: Murr et al. (2011)/reproduced with permission from Springer Nature. (c) δ-phase precipitation in the interdendritic zones with columnar architecture in an EBM-printed Inconel 718 alloy. Source: Korner (2016)/reproduced with permission from Taylor & Francis.

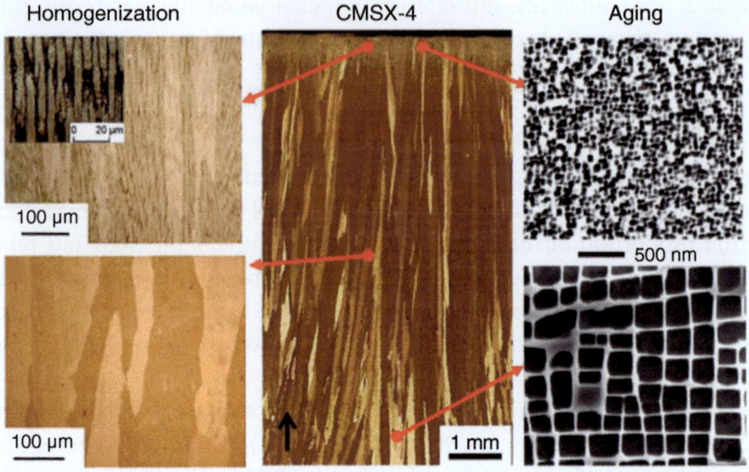

Figure 4.13 Microstructure of a nickel-based alloy CMSX-4 fabricated by EBM with *in situ* heat treatment. Source: Korner (2016)/reproduced with permission from Taylor & Francis.

containing carbides and intermetallic compounds (e.g. FCC γ'-Ni_3 (Al, Ti, Nb) and ordered tetragonal γ'-Ni_3Nb) (Strondl et al. 2008). Note that the columnar γ grains in the matrix have their <001> orientation parallel to the build direction. Lines of the δ precipitates were present both on grain boundaries and within grains (Figure 4.12c). Such precipitates originated from cellular or dendritic solidification and were mainly formed in highly segregated zones.

Certain non-weldable nickel-based alloys, such as the columnar-grained alloy Rene 142 and the single-crystalline alloy CMSX-4, have also been processed by EBM. In another study by Murr et al., γ' precipitates in printed Rene 142 alloys were found to exhibit an average size of ~275 nm with a volume content of ~59% (2013). Figure 4.13 indicates that the dendritic arm spacing of an EBM-printed CMSX-4 alloy was only a few microns. The micro-segregation and homogenization occurring at such a small scale resulted from *in situ* heat treatment, which could also promote the nucleation and growth of precipitates (Ramsperger et al. 2015). Within the alloy, the growth of columnar grains across hundreds of printed layers was also observed (center image). The microstructure of the EBM-printed CMSX-4 alloy was determined to typically include cuboid-shaped γ'-precipitates, which increased with the number of printed layers. Note that aging also occurred during printing because of the long processing time and high temperature.

4.7.3 Cobalt Alloys

Cobalt alloys have been extensively employed in producing valve seats for nuclear power plants, aerospace fuel nozzles and engine vanes, and orthopedic and dental implants because of their excellent mechanical strength at high temperatures, corrosion resistance, wear resistance, and biocompatibility. Figure 4.14 presents the

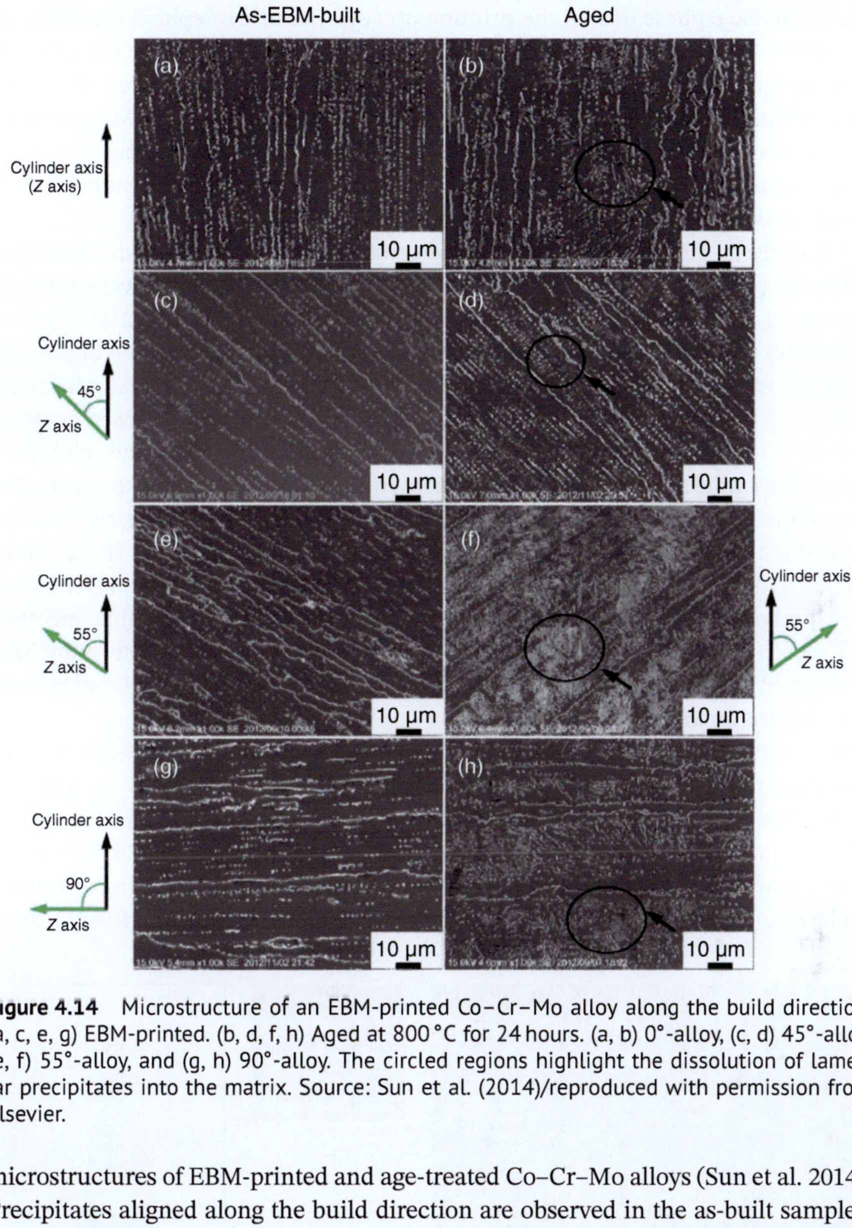

Figure 4.14 Microstructure of an EBM-printed Co–Cr–Mo alloy along the build direction. (a, c, e, g) EBM-printed. (b, d, f, h) Aged at 800 °C for 24 hours. (a, b) 0°-alloy, (c, d) 45°-alloy, (e, f) 55°-alloy, and (g, h) 90°-alloy. The circled regions highlight the dissolution of lamellar precipitates into the matrix. Source: Sun et al. (2014)/reproduced with permission from Elsevier.

microstructures of EBM-printed and age-treated Co–Cr–Mo alloys (Sun et al. 2014). Precipitates aligned along the build direction are observed in the as-built samples. The average distance between the arrays of precipitates was measured to be ~3 μm. The succedent aging treatment promoted the dissolution of the lamellar precipitates $M_{23}C_6$ (where M can be Cr, Mo, or Si) into the matrix. It was demonstrated that the precipitates in the lamellar colonies formed after aging were M_2N. The lamellar structures of the M_2N precipitates and the hexagonal close-packed (HCP) ε phase were based on the eutectoid transition ($\gamma \rightarrow \varepsilon + M_2N$).

In Co–Cr–Mo alloys, the γ phase is stable at high temperatures, while the HCP ε phase is stable at low temperatures. Therefore, the γ phase may partially transform into the stable ε phase during the printing process. Because of epitaxial growth of the γ phase with a preferred grain orientation in the <001> direction, the printed parts typically exhibited a pronounced crystallographic texture (Sun et al. 2014). The distribution of the $M_{23}C_6$ precipitates within EBM-printed Co–Cr–Mo products resulted from segregation induced by cellular solidification. After aging at 800 °C for 24 hours, the γ phase was transformed into the ε phase, and the grains tended to be equiaxed.

In a study by Tan et al. (2018), an EBM-printed Co–Cr–Mo alloy exhibited fine columnar γ-Co grains with cellular dendrites. Carbide precipitates were formed because of the segregation of carbon into the interdendritic zones and grain boundaries of the alloy. The columnar grains along the building direction in the x–z plane are represented in Figure 4.15a, each of which comprises cellular dendrites with secondary arms. The carbide precipitates at the grain boundaries generated successive thin films that were 100–200 nm thick, thereby separating the columnar grains. The carbide precipitates within the interdendritic zones tended to constitute long chains/clusters along the primary dendrites. At low quantities below 5 vol.%, the carbide precipitates at the grain boundaries and interdendritic zones were difficult to detect via traditional X-ray analysis. Figure 4.15b,c show a representative atom probe tomography (APT) tip sample from the carbide thin film and the APT-reconstructed volume of the four main elements, respectively. The study has determined that Mo-enriched M_xC_y(I) was an M_6C-type carbide, and Cr-enriched M_xC_y(II) was an $M_{23}C_6$-type carbide.

In general, the mechanical properties of the printed parts exhibit strong anisotropy due to epitaxial solidification. In a study by Xiang et al. (2019), EBM-printed Co–26Cr–6Mo–0.2C samples subjected to a strain rate of 1.67×10^{-4}/s along the horizontal direction succumbed to premature fracture at 334 MPa with an elongation of only 0.23%. In contrast, samples tested along the building direction displayed an ultimate tensile strength of 957 MPa with an elongation of 10.5%. Sun

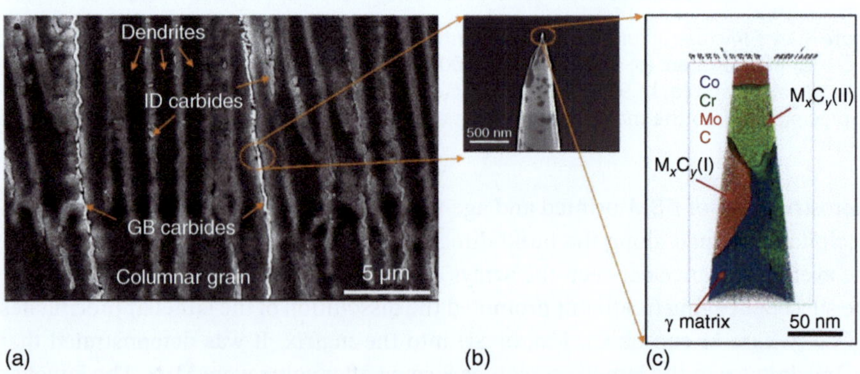

Figure 4.15 (a) SEM image of columnar grains with dendrites in an EBM-printed Co–Cr–Mo alloy. ID carbides: interdendritic carbides. (b) SEM image of an atom probe tomography (APT) tip sample from the carbide thin film in (a). (c) An APT-reconstructed volume featuring two types of carbides and the γ-matrix. Source: Tan et al. (2018)/reproduced with permission from Elsevier.

et al. (2014) demonstrated that at a temperature of 700 °C, a Co–26Cr–6Mo–0.2C sample printed with a deviation of 55° from the *z* axis possessed the highest ultimate tensile strength of 806 MPa at a strain rate of 1.5×10^{-4}/s.

4.7.4 Iron Alloys

316L stainless steel is a representative iron-based alloy investigated for EBM. Its printability was comprehensively studied by Wang et al. (2018), and the microstructure of EBM-printed 316L stainless steel products was found to exhibit heterogeneity along the building direction, with cellular dendrites and melt pool boundaries present at the top area (Figure 4.16). Few defects were observed at the layer boundaries, which demonstrated that the electron beam was sufficiently strong to fuse neighboring layers (Figure 4.16a). The grains were columnar-shaped and parallel to the building direction because the largest thermal gradient was along the same direction during the printing process. The spacing between adjacent melt pool boundaries was ~100 μm (Figure 4.16b), and typical intragranular cellular sub-grains with an average size of ~1 μm were obtained (Figure 4.16c).

Figure 4.16d shows the side view of the melt pool and columnar grain boundaries near the top region of the EBM-printed 316L alloy. The columnar grains grew epitaxially across layer boundaries, i.e. epitaxial growth and columnar-shape formation, and their width values were 25–50 μm. They also exhibited a pronounced <001>

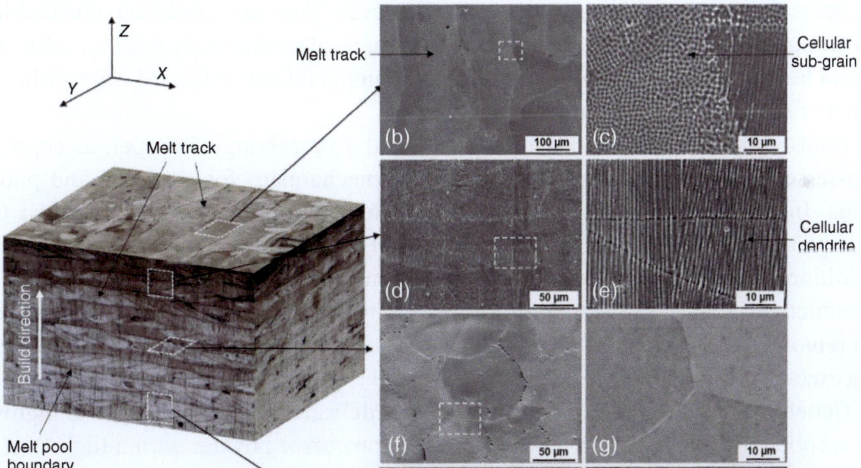

Figure 4.16 Microstructure of an EBM-printed 316L alloy. (a) 3D micrograph under OM; SEM images of (b, c) the *x*–*y* plane at the top region, (d, e) the *x*–*z* plane at the top region, (f, g) the *x*–*y* plane at the middle region, and (h, i) the *x*–*z* plane at the bottom region. Source: Wang et al. (2018)/reproduced with permission from Elsevier.

texture, which implies that the alloy possessed anisotropic mechanical properties. Moreover, intragranular dendrites that grew in the same direction as the columnar grains were observed (Figure 4.16e). The average width of the dendrites was ~1 μm, and cellular sub-grains were identified from their cross-sectional view, which was barely visible below the top region (Figure 4.16f–i). The yield strength, ultimate tensile strength, and elongation of the printed 316L alloy were measured to be in the range of 316–396 MPa, 437–652 MPa, and 9.6–35.2%, respectively. While its strength was higher than its conventional cast and wrought counterparts, it displayed lower elongation. Additionally, the tensile properties of the printed parts along the building direction were superior to those perpendicular to the building direction.

Microstructural analysis of an H13 tool steel printed by EBM indicated that the printed samples lacked macropores and exhibited strong interlayer bonding (Cormier et al. 2004). The hardness of the samples was measured to be 48–50 HRC. The cooling efficiency and dimensional accuracy of an EBM-printed H13 insert with complicated cooling geometries were determined to be higher than those of conventionally cooled core inserts (Rännar et al. 2007).

4.7.5 Others

4.7.5.1 Copper and Its Alloys

Due to their excellent electrical and thermal conductivity, copper and its alloys are promising candidates for applications in electrical and thermal management systems (e.g. heat exchanger devices). However, they are currently unsuitable printing materials for LPBF due to their high thermal conductivity, which results in rapid heat dissipation. Additionally, copper is highly reflective toward lasers, which further hinders the LPBF printing process.

EBM is preferred to LPBF with respect to the processing of copper, as copper possesses different absorption and reflection mechanisms for electrons and photons. In EBM, the amount of energy deposited within the printing material is adequate to consolidate even pure copper (thermal conductivity of 400 W/mK). In addition, copper powder can be rapidly pre-sintered because of its high electrical conductivity. However, the high sintering activity of copper may result in difficulties in removing the powder if it is processed under high pre-sintering temperatures for an excessively long time.

Generally, handling and storing copper is a delicate process as copper is highly susceptible to oxidation. An atomized copper precursor powder with a high Cu_2O concentration was processed by EBM (Ramirez et al. 2011), and copper oxide precipitates were found to contribute to the high hardness and strength of the printed parts. EBM-printed pure Cu parts typically consist of equiaxed precipitate–dislocation cell arrays (1–3 μm) along the building direction, while they comprise elongated or columnar-like arrays with a length of 12–60 μm and corresponding spatial dimensions of 1–3 μm along the transverse directions. In a study that explored directional solidification in EBM-printed parts, a precipitation–dislocation interaction was observed with an increase in their hardness from a base-plate value of 57–88 HV, which is representative of EBM-printed copper products (Guschlbauer et al. 2018).

A process window for dense and crack-free pure copper EBM-printed parts at a building temperature of 530 °C was identified. The EBM-printed copper parts were measured to exhibit a yield strength of 78 MPa, an ultimate tensile strength of 177 MPa, and an elongation to fracture of 59.3%. Cracks in the tensile samples were found to depend on the location of the parts on the substrate and often occur when printed under high-line energy densities.

4.7.5.2 High-Entropy Alloys

EBM has also been adopted to print HEA parts, and the selected HEAs include AlCoCrFeNi, CoCrFeNiTi, and CoCrFeMnNi. EBM-printed AlCoCrFeNi HEA parts have showed the same phase constitutions but distinctive microstructures in the top and bottom zones owing to various temperature distributions caused by preheating and heat transfer (Shiratori et al. 2016). Additionally, dual-phase (B2 + body-centered cubic [BCC]) and FCC precipitates were found within the printed parts. The (Al, Ni)-rich and (Cr, Fe)-rich phases correspond to the B2 and BCC phases, respectively. Preheating at 950 °C led to the precipitation of the FCC phase at the boundaries of the B2 and BCC phases. The decreased temperature gradient within the printed parts also facilitated the BCC-to-FCC phase transformation. Based on a review by Han et al., EBM-printed HEA products could achieve a yield strength of 205–769 MPa, an ultimate tensile strength of 497–1074 MPa, an elongation of 1.2–63%, and a microhardness of 157–500 HV (Han et al. 2020).

Fabricating HEA parts via the *in situ* alloying of at least four premixed pure metal powders through EBM and LPBF remains challenging. In particular, the rapid cooling of melt pools inhibits the convection and diffusion of elements within the metal powders, thereby resulting in the inhomogeneity of element distribution within the printed layers and the anisotropic microstructure of the printed parts.

4.8 Summary

In essence, EBM features the application of a high-energy electron beam to fabricate geometrically complex metal parts in a vacuum environment at an elevated building temperature. This chapter comprehensively covers the historical development, fundamentals, processing, metallurgical defects, powder characteristics, and commercial equipment for EBM, as well as microstructures and mechanical properties of the most extensively studied materials for EBM-printed parts.

A vacuum environment is crucial for processing materials with a high affinity toward oxygen (such as titanium and its alloys). Meanwhile, an elevated building temperature can minimize residual stress and cracking within the printed parts, which is ideal for processing heat-resistant materials prone to cracking (e.g. TiAl). Hence, EBM is a preferred approach for fabricating biomedical implants and aerospace components subjected to stringent requirements. Moreover, the exceptional electron absorptivity of EBM powders enables the processing of materials exhibiting high optical reflectivity (e.g. copper and its alloys).

Compared to LPBF, EBM has disadvantages including low resolution and rough surface of the printed parts, high equipment cost, the requirement for large quantities of printing powder, and significant loss of certain alloying elements. Moreover, EBM-printed parts tend to possess common metallurgical defects such as pores and cracks, although such defects are less pronounced than those in their counterparts printed by LPBF. A major impediment to the widespread adoption of EBM is its small material library, which is partially attributed to the limited number of EBM machine models available worldwide. Existing commercial EBM materials are almost exclusively provided by Arcam, such as Ti–6Al–4V, pure Ti, Co–Cr–Mo, Inconel 718, pure copper, highly alloyed tool steel, and TiAl.

Since generating an electron beam is inherently simpler and less costly than generating a laser beam, the building speed and volume in EBM can be scaled up substantially in the future for ultrafast manufacturing. Future research on EBM should also focus on better process control to obtain higher beam power and quality. With greater precision regarding beam control and the optimization of beam scanning algorithms to prevent "smoking", finer powders and thinner printing layers can be implemented to print parts with higher resolution and a smoother surface finish.

References

Al-Bermani, S.S., Blackmore, M.L., Zhang, W., and Todd, I. (2010). The origin of microstructural diversity, texture, and mechanical properties in electron beam melted Ti-6Al-4V. *Metallurgical and Materials Transactions A* 41 (13): 3422–3434.

Antonysamy, A.A., Meyer, J., and Prangnell, P.B. (2013). Effect of build geometry on the β-grain structure and texture in additive manufacture of Ti6Al4V by selective electron beam melting. *Materials Characterization* 84: 153–168.

Biamino, S., Penna, A., Ackelid, U. et al. (2011). Electron beam melting of Ti-48Al-2Cr-2Nb alloy: microstructure and mechanical properties investigation. *Intermetallics* 19 (6): 776–781.

Chauvet, E., Kontis, P., Jägle, E.A. et al. (2018). Hot cracking mechanism affecting a non-weldable Ni-based superalloy produced by selective electron beam melting. *Acta Materialia* 142: 82–94.

Cormier, D., Harrysson, O., and West, H. (2004). Characterization of H13 steel produced via electron beam melting. *Rapid Prototyping Journal* 10: 35–41.

Everhart, T.E. and Hoff, P.H. (1971). Determination of kilovolt electron energy dissipation vs penetration distance in solid materials. *Journal of Applied Physics* 42 (13): 5837–5846.

Facchini, L., Magalini, E., Robotti, P., and Molinari, A. (2009). Microstructure and mechanical properties of Ti-6Al-4V produced by electron beam melting of pre-alloyed powders. *Rapid Prototyping Journal* 15: 171–178.

de Formanoir, C., Michotte, S., Rigo, O. et al. (2016). Electron beam melted Ti–6Al–4V: microstructure, texture and mechanical behavior of the as-built and heat-treated material. *Materials Science and Engineering A* 652: 105–119.

Frazier, W.E. (2014). Metal additive manufacturing: a review. *Journal of Materials Engineering and Performance* 23 (6): 1917–1928.

Galati, M. and Iuliano, L. (2018). A literature review of powder-based electron beam melting focusing on numerical simulations. *Additive Manufacturing* 19: 1–20.

Gaytan, S.M., Murr, L.E., Medina, F. et al. (2009). Advanced metal powder based manufacturing of complex components by electron beam melting. *Materials Technology* 24 (3): 180–190.

Guo, C., Ge, W., and Lin, F. (2015). Effects of scanning parameters on material deposition during electron beam selective melting of Ti-6Al-4V powder. *Journal of Materials Processing Technology* 217: 148–157.

Guschlbauer, R., Momeni, S., Osmanlic, F., and Körner, C. (2018). Process development of 99.95% pure copper processed via selective electron beam melting and its mechanical and physical properties. *Materials Characterization* 143: 163–170.

Han, C., Fang, Q., Shi, Y. et al. (2020). Recent advances on high-entropy alloys for 3D printing. *Advanced Materials* 32 (26): 1903855.

Hao, Y.L., Li, S.J., Sun, S.Y. et al. (2007). Elastic deformation behaviour of Ti–24Nb–4Zr–7.9Sn for biomedical applications. *Acta Biomaterialia* 3 (2): 277–286.

Hrabe, N., Gnäupel-Herold, T., and Quinn, T. (2017). Fatigue properties of a titanium alloy (Ti–6Al–4V) fabricated via electron beam melting (EBM): effects of internal defects and residual stress. *International Journal of Fatigue* 94: 202–210.

Juechter, V., Scharowsky, T., Singer, R.F., and Körner, C. (2014). Processing window and evaporation phenomena for Ti–6Al–4V produced by selective electron beam melting. *Acta Materialia* 76: 252–258.

Karimi, P., Schnur, C., Sadeghi, E., and Andersson, J. (2020). Contour design to improve topographical and microstructural characteristics of alloy 718 manufactured by electron beam-powder bed fusion technique. *Additive Manufacturing* 32: 101014.

Karlsson, J., Snis, A., Engqvist, H., and Lausmaa, J. (2013). Characterization and comparison of materials produced by electron beam melting (EBM) of two different Ti–6Al–4V powder fractions. *Journal of Materials Processing Technology* 213 (12): 2109–2118.

Korner, C. (2016). Additive manufacturing of metallic components by selective electron beam melting – a review. *International Materials Reviews* 61 (5): 361–377.

Lee, Y.S., Kirka, M.M., Ferguson, J., and Paquit, V.C. (2020). Correlations of cracking with scan strategy and build geometry in electron beam powder bed additive manufacturing. *Additive Manufacturing* 32: 101031.

Liu, Y., Li, S., Hou, W. et al. (2016a). Electron beam melted beta-type Ti–24Nb–4Zr–8Sn porous structures with high strength-to-modulus ratio. *Journal of Materials Science and Technology* 32 (6): 505–508.

Liu, Y.J., Li, S.J., Wang, H.L. et al. (2016b). Microstructure, defects and mechanical behavior of beta-type titanium porous structures manufactured by electron beam melting and selective laser melting. *Acta Materialia* 113: 56–67.

Liu, Y.J., Wang, H.L., Li, S.J. et al. (2017). Compressive and fatigue behavior of beta-type titanium porous structures fabricated by electron beam melting. *Acta Materialia* 126: 58–66.

Lodes, M.A., Guschlbauer, R., and Koerner, C. (2015). Process development for the manufacturing of 99.94% pure copper via selective electron beam melting. *Materials Letters* 143: 298–301.

Murr, L.E., Esquivel, E.V., Quinones, S.A. et al. (2009). Microstructures and mechanical properties of electron beam-rapid manufactured Ti–6Al–4V biomedical prototypes compared to wrought Ti–6Al–4V. *Materials Characterization* 60 (2): 96–105.

Murr, L. E., Gaytan, S. M., Ceylan, A., Martinez, E., Martinez, J. L., Hernandez, D. H., Machado, B. I., Ramirez, D. A., Medina, F., Collins, S., .& Wicker, R. B. (2010). Characterization of titanium aluminide alloy components fabricated by additive manufacturing using electron beam melting. *Acta Materialia*, 58(5), 1887–1894.

Murr, L.E., Martinez, E., Gaytan, S.M. et al. (2011). Microstructural architecture, microstructures, and mechanical properties for a nickel-base superalloy fabricated by electron beam melting. *Metallurgical and Materials Transactions A* 42 (11): 3491–3508.

Murr, L.E., Martinez, E., Pan, X.M. et al. (2013). Microstructures of Rene 142 nickel-based superalloy fabricated by electron beam melting. *Acta Materialia* 61 (11): 4289–4296.

Parthasarathy, J., Starly, B., Raman, S., and Christensen, A. (2010). Mechanical evaluation of porous titanium (Ti6Al4V) structures with electron beam melting (EBM). *Journal of the Mechanical Behavior of Biomedical Materials* 3 (3): 249–259.

Ramirez, D.A., Murr, L.E., Martinez, E. et al. (2011). Novel precipitate-microstructural architecture developed in the fabrication of solid copper components by additive manufacturing using electron beam melting. *Acta Materialia* 59 (10): 4088–4099.

Ramsperger, M., Mújica Roncery, L., Lopez-Galilea, I. et al. (2015). Solution heat treatment of the single crystal nickel-base superalloy CMSX-4 fabricated by selective electron beam melting. *Advanced Engineering Materials* 17 (10): 1486–1493.

Rännar, L.E., Glad, A., and Gustafson, C.G. (2007). Efficient cooling with tool inserts manufactured by electron beam melting. *Rapid Prototyping Journal* 13: 128–135.

Scharowsky, T., Juechter, V., Singer, R.F., and Körner, C. (2015). Influence of the scanning strategy on the microstructure and mechanical properties in selective electron beam melting of Ti–6Al–4V. *Advanced Engineering Materials* 17 (11): 1573–1578.

Shiratori, H., Fujieda, T., Yamanaka, K. et al. (2016). Relationship between the microstructure and mechanical properties of an equiatomic AlCoCrFeNi high-entropy alloy fabricated by selective electron beam melting. *Materials Science and Engineering A* 656: 39–46.

Sochalski-Kolbus, L.M., Payzant, E.A., Cornwell, P.A. et al. (2015). Comparison of residual stresses in Inconel 718 simple parts made by electron beam melting and direct laser metal sintering. *Metallurgical and Materials Transactions A* 46 (3): 1419–1432.

Strondl, A., Fischer, R., Frommeyer, G., and Schneider, A. (2008). Investigations of MX and γ'/γ'' precipitates in the nickel-based superalloy 718 produced by electron beam melting. *Materials Science and Engineering A* 480 (1–2): 138–147.

Sun, S.H., Koizumi, Y., Kurosu, S. et al. (2014). Build direction dependence of microstructure and high-temperature tensile property of Co–Cr–Mo alloy fabricated by electron beam melting. *Acta Materialia* 64: 154–168.

Tammas-Williams, S., Zhao, H., Léonard, F. et al. (2015). XCT analysis of the influence of melt strategies on defect population in Ti–6Al–4V components manufactured by selective electron beam melting. *Materials Characterization* 102: 47–61.

Tan, X., Kok, Y., Tan, Y.J. et al. (2015). Graded microstructure and mechanical properties of additive manufactured Ti–6Al–4V via electron beam melting. *Acta Materialia* 97: 1–16.

Tan, X.P., Wang, P., Kok, Y. et al. (2018). Carbide precipitation characteristics in additive manufacturing of Co-Cr-Mo alloy via selective electron beam melting. *Scripta Materialia* 143: 117–121.

Tang, H.P., Qian, M., Liu, N. et al. (2015a). Effect of powder reuse times on additive manufacturing of Ti-6Al-4V by selective electron beam melting. *JOM* 67 (3): 555–563.

Tang, H.P., Yang, G.Y., Jia, W.P. et al. (2015b). Additive manufacturing of a high niobium-containing titanium aluminide alloy by selective electron beam melting. *Materials Science and Engineering A* 636: 103–107.

Wang, C., Tan, X., Liu, E., and Tor, S.B. (2018). Process parameter optimization and mechanical properties for additively manufactured stainless steel 316L parts by selective electron beam melting. *Materials & Design* 147: 157–166.

Wang, P., Huang, P., Ng, F.L. et al. (2019). Additively manufactured CoCrFeNiMn high-entropy alloy via pre-alloyed powder. *Materials & Design* 168: 107576.

Xiang, D.D., Wang, P., Tan, X.P. et al. (2019). Anisotropic microstructure and mechanical properties of additively manufactured Co–Cr–Mo alloy using selective electron beam melting for orthopedic implants. *Materials Science and Engineering A* 765: 138270.

Zhao, X., Li, S., Zhang, M. et al. (2016). Comparison of the microstructures and mechanical properties of Ti–6Al–4V fabricated by selective laser melting and electron beam melting. *Materials & Design* 95: 21–31.

5

Laser-Based Directed Energy Deposition

Laser-based directed energy deposition (L-DED) is an additive manufacturing (AM) technique that is fundamentally different from powder bed fusion processes such as laser powder bed fusion (LPBF) and electron beam melting (EBM). L-DED, also commonly known as laser metal deposition (LMD) or laser engineered net shaping (LENS), possesses a unique capability of repairing metal parts. This chapter presents the history, fundamentals, processing characteristics, powder materials, and equipment of L-DED, and it discusses the microstructures and properties of parts printed by L-DED.

5.1 History

L-DED originated from a wire welding technology and was previously known as welding AM (Kratky 1937; Isaac 1942). A patent with early concepts of L-DED was filed in the 1980s, describing a layerwise additive deposition process via laser powder metallurgy (Brown et al. 1982). Subsequently, a method for repairing metal parts through the continuous deposition and melting of powder on their surface was patented (Mehta et al. 1988). In 1988, Hammeke patented a laser spray nozzle, while Buongiorno patented a metal cladding nozzle in 1995 (Hammeke 1988; Buongiorno 1995).

Directed light fabrication was proposed as a form of single-nozzle, powder–based L-DED combining computer-aided design (CAD), computer numerical control (CNC), and laser-based powder deposition for the rapid freeform fabrication of stainless steel parts. In 1995, researchers from Sandia National Laboratories developed a multi-nozzle L-DED system with high powder delivery efficiency, which was commercialized and trademarked as Laser Engineered Net Shaping. In 1997, the Optomec Company also began to promote commercialization of the technology.

Metal Powder–Based Additive Manufacturing, First Edition. Kun Zhou and Changjun Han.

Different L-DED techniques have emerged in the past 20 years with varying degrees of success in terms of commercialization and market/research adoption. A wide range of terminologies has been proposed for L-DED, including LMD, LENS, direct metal deposition (DMD), and laser rapid forming. A new generation of the L-DED technology is currently being developed by several companies under the generic name of laser metal deposition.

5.2 Fundamentals

In L-DED, metal powder and/or wires are directly deposited onto a freeform substrate in a layer-by-layer manner while being melted by a laser beam. A schematic of the L-DED process is shown in Figure 5.1a. The printing head consists of a focused laser source and a powder delivery system in the form of a coaxial, multi-stream, or side nozzle. The laser beam creates a melt pool on a substrate or a pre-deposited layer, which moves along a predefined scanning path. The powder is delivered from powder hoppers through the nozzle by a carrier gas, which is usually argon or helium, and injected into the melt pool, which subsequently cools and solidifies once the laser beam moves away. The created melt pool displays different penetration depths depending on the process parameters (Figure 5.1b). After the deposition of a layer, the print head moves away from the solidified surface by a preset distance before the next layer is deposited.

The aforementioned process is repeated until the entire part is fabricated. Generally, it is desired to use a substrate with a similar chemical composition to the printed material. Due to the comparable thermal properties and suppressed brittle intermetallic phases between the printed materials and the substrate, decent interfacial bonding can be achieved between the first deposited layer and the substrate to avoid delamination.

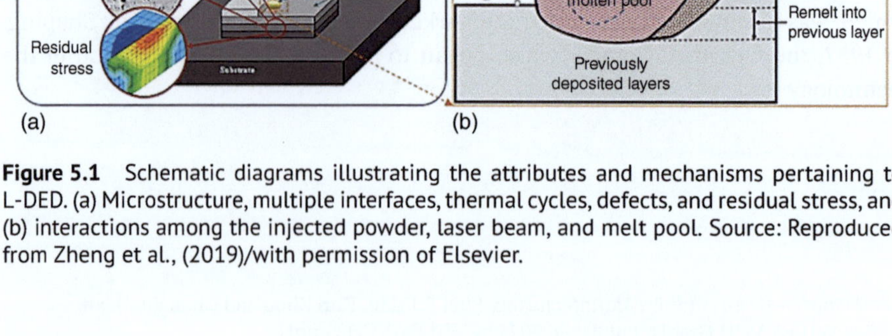

Figure 5.1 Schematic diagrams illustrating the attributes and mechanisms pertaining to L-DED. (a) Microstructure, multiple interfaces, thermal cycles, defects, and residual stress, and (b) interactions among the injected powder, laser beam, and melt pool. Source: Reproduced from Zheng et al., (2019)/with permission of Elsevier.

The advantages of the L-DED process are as follows:

- An L-DED machine can be equipped with multiple powder feeders. Multiple powders can be delivered and printed simultaneously, which enables the fabrication of multi-materials such as functionally graded materials.
- The simultaneous delivery of different powders allows for *in situ* alloying, which can achieve high-throughput screening of the composition of newly developed metal materials.
- The deposition head is installed on a four- or five-axis motion system, which enables printing on a curved surface. L-DED is thus suitable for repairing damaged parts and adding new features to existing parts.
- The L-DED process corresponds to a high building rate, which is favorable for printing relatively large parts.
- L-DED additive and subtractive manufacturing can be simultaneously conducted by sharing the motion control system to improve the dimensional accuracy and surface quality of the printed parts.

The L-DED process presents the following disadvantages:

- Large temperature gradients may induce residual stress within the printed parts, resulting in deformation and cracking.
- Low printing resolution of the machines resulting from large melt pools leads to the printed parts undergoing shape deviation and exhibiting rough surfaces. Thus, post-machining procedures are required to improve the dimensional accuracy and surface quality of the printed parts.
- The design freedom and geometrical complexity pertaining to the parts are limited, particularly for L-DED machines confined to three degrees of freedom.
- The efficiency of powder usage and powder recyclability is low.

The main process parameters of the L-DED process include the laser power, laser spot diameter, scanning speed, powder flow rate, scanning strategy, hatch spacing, and dwell time between each deposited layer. The optimal process parameters depend on both material properties (e.g. laser absorptivity and powder flowability) and machine configurations (e.g. types of nozzles and powder feeders).

The powder particles absorb a fraction of the laser energy during the printing process. The remaining energy is delivered to the substrate to form a melt pool or dissipated through reflection or refraction. The amount of energy absorbed by the powder is determined by the type of laser beam and the powder material, as discussed in Section 3.2.

The amount of energy absorbed by the powder also depends on its flow rate, the size of its particles, and the nozzle geometry. A high powder flow rate or a large particle size results in the melting of more powder particles (Wolff et al. 2020). Single feed nozzles (or side nozzles) installed on one side of the laser beam can result in asymmetric energy absorption and formation of thermal gradients, leading to dimensional inaccuracy and inhomogeneity within the microstructure of the DED-printed parts. Instead, coaxial nozzles are preferred to tackle this issue due to the uniform deposition of powder surrounding the energy source (Svetlizky et al. 2021).

Since mainstream engineering metal materials (e.g. iron-, nickel-, and titanium-based alloys) are highly absorptive with respect to neodymium-doped yttrium aluminum garnet (Nd:YAG), CO_2, and fiber lasers, these three types of lasers are commonly employed in commercial L-DED equipment. Their usage and limitations are discussed in Section 3.2. The laser power determines the amount of energy delivered to the deposited powder and the substrate (see Eq. (5.1) in the subsequent Section 5.3). A high laser power results in a large melt pool and deep penetration into the preceding solidified layers. However, a high laser power can induce spattering of molten droplets and vaporization of alloying elements from the melt pool, which may damage the nozzles and degrade the mechanical properties and surface roughness of the printed parts. In contrast, lower laser power can lead to the formation of defects, such as lack-of-fusion pores and incomplete melting of powders.

The powder flow rate can affect the surface roughness and grain texture of L-DED-printed parts. An increase in the powder flow rate can lead to increased surface roughness, which is attributed to insufficient energy to fully melt the powder particles, resulting in partially melted particles being embedded into the surface of the printed parts (Chen et al. 2021). Additionally, a high powder flow rate can influence the grain texture as it affects the thermal gradient of the melt pool, resulting in equiaxed grains instead of the columnar grain texture commonly observed in AM-printed parts (Zhang et al. 2021). Conversely, a low powder flow rate can lead to insufficient material being deposited onto the preceding layers.

The scanning speed influences the material deposition rate, as well as the melting and solidification rates. A high scanning speed results in a small amount of powder being deposited and a short laser interaction time, and thus a small melt pool is formed (Chen et al. 2021). On the other hand, a continuous increase in the scanning speed can result in hydrodynamic instability of the melt pool, leading to the poor surface quality of the printed parts (Svetlizky et al. 2021).

Scanning strategies for the L-DED process, such as unidirectional and bidirectional scanning with 0° or 90° rotation vectors, spiral patterns, and fractal patterns, are commonly employed in L-DED (Svetlizky et al. 2021). As discussed in Section 3.2, the scanning pattern determines the amount of residual stress within the printed parts and affects their mechanical properties. Additionally, the dwell time between each deposited layer can be tailored to take advantage of the repetitive heating and cooling cycles of L-DED to trigger precipitation hardening, control the microstructure, and reduce residual stress (Amirabdollahian et al. 2021; Guévenoux et al. 2020; Denlinger et al. 2015).

To optimize the printing process, interactions between the various process parameters must be considered because they affect the relative density, surface roughness, microstructure, defects, and mechanical properties of the printed parts. Experimental tools such as the design of experiments (DOEs) can be applied to determine the optimal processing parameters. DOE allows the user to identify the interactions between various process parameters and establish the relationships between these variables and the resultant microstructure, mechanical properties, and surface finish of the printed parts. For instance, the energy density of the laser beam is synergistically dependent on the laser power (100–5000 W), laser spot diameter (up to 4 mm), scanning speed (1–30 mm/s), and powder flow rate (10–100 g/min). Meanwhile, the scanning pattern, hatch spacing, and interlayer cooling time collectively affect the thermal distribution and history of the melt pools. Notably, the cooling rates of L-DED systems typically range between 10^2 and 10^4 K/s (Zheng et al. 2008). The focal plane of the powder flow usually coincides with the substrate, on which the laser beam can be focused or defocused. Scanning configurations, such as raster, offset-out, spiral, and fractal deposition patterns, which dictate the tool path of the printing head, are implemented before the printing process.

5.3 Deposition Process

The deposition process of L-DED comprises energy flow (laser), feedstock flow (metal powder), gas flow (carrier gas and shielding gas), and information flow (numerical control and *in situ* monitoring). Several interconnected physical phenomena occur within a short time span during the process: (i) laser/powder/gas interactions; (ii) initiation, stabilization, and solidification of the melt pool; (iii) thermal transfer activities (i.e. thermal convection in melt pools, thermal conduction from the melt pool to printed layers, substrates, and the atmosphere, and thermal cycling) induced by the layer-wise printing process.

Laser/powder/gas interactions are highly transient and depend on the laser properties (e.g. power, wavelength, and energy profile), powder properties (e.g. energy absorption rate, morphology, and particle size distribution), and atmospheric conditions (e.g. gas type and flow rate). The formation of plasma (i.e. ionized gas), the fourth state of matter, can be induced by the vaporization and ionization of in-flight powder by a high-power laser. The presence of plasma between the powder particles can significantly affect their energy absorption and lower the printing efficiency. Additionally, such plasma can contribute to nonlinear positive pressure effects that restrict the amount of vapor escaping the melt pool, thereby increasing the porosity of the printed parts. Such degradation can be avoided by implementing a proper combination of laser power and powder flow rate to prevent the in-flight powder from overheating. Plasma effects can be further suppressed by employing shielding gases (such as helium and argon) in the L-DED system. The term "specific energy" quantifies energy delivery, which is described in terms of the area energy density E as

$$E = \frac{P}{2r_b V}, \tag{5.1}$$

where P is the laser power, r_b is the laser beam radius, and V is the scanning speed.

The laser/powder interaction (or exposure) time t can be estimated by $t = 2r_b/V$, and t typically varies from 0.2 to 8 ms (Simchi 2006). When in-flight powder ejected from the nozzle is exposed to the laser beam, it undergoes steep heating from the laser beam. Excessive energy input by the laser beam can lead to element loss from the powder in the form of evaporation (Kruth et al. 2004). The in-flight powder particles are primarily cooled by the surrounding carrier gas and shielding gas through thermal convection. Such gases can impede the momentum of the powder particles by applying a drag force on the material jet stream, which decelerates and destabilizes the powder particles during their flight.

The laser beam creates a melt pool on the substrate, and a molten metal stream is injected into the pool, whose dimensions depend on the process parameters. Only a portion of the incident laser energy is absorbed by the powder to initiate and maintain the melt pool, while the rest is lost through reflection. The absorbed energy is dissipated via radiation, convection (shielding/carrier gas), and conduction (substrate and printed layers).

Radiation heat transfer during deposition is governed by the spectral emissivity of the melt pool, which is a function of its superheating temperature and the oxygen concentration within and around it. Therefore, the dimensions and temperature profile of the melt pool are also sensitive to changes in its emissivity (Wang and Felicelli 2006).

In addition, the temperature field and morphology of a melt pool are determined by interfacial forces such as the surface tension and recoil pressure of the metal vapor, which are more prevalent at its surface and beneath its center, respectively. As a result, the highest superheating temperature of a melt pool is obtained at its surface center, which is sufficient to vaporize the molten metal. Consequently, metal vaporization occurs at the surface of the melt pool, resulting in keyhole formation. Once the keyholes collapse, and the melt pool solidifies, ambient gases are trapped within the latter, and pores are formed due to its high solidification rate.

As the temperature of the melt pool rises, the recoil forces within it increase more rapidly than its surface tension. When the vapor recoil force exceeds the liquid–gas surface tension force near the periphery of the melt pool, the liquid within the melt pool is ejected. An increase in bulk temperature near the melt pool can cause it to boil and result in a loss of mass. The tendency for liquid metal ejection to occur as a result of vapor recoil forces increases, and the vapor recoil force F_r can be estimated by

$$F_r = \int_0^{r_m} r\Delta P(r)\mathrm{d}r, \tag{5.2}$$

where r_m is the radial distance at which the melt pool surface reaches its boiling point, and $\Delta P(r)$ is the difference between the local equilibrium vapor pressure and the atmospheric pressure (He et al. 2005). Heat loss due to vaporization of the melt pool can be detrimental to the quality of L-DED products.

During L-DED, substantial variations in the temperature field of the melt pool result in steep density and surface tension gradients along its length and depth, producing free convective and Marangoni-type flows, respectively. Diffusive motion related to turbulence can significantly affect the solute and momentum transport within the melt pool and enhance convective heat transfer. As the melt pool becomes deeper (i.e. penetration depth increases), its vorticity increases, which indicates higher turbulence.

When the surface tension gradient of the melt pool is small, evaporation heat loss becomes non-negligible, and the peak temperature of the melt pool is lowered.

Conversely, a large melt pool surface tension gradient reduces the influence of vaporization heat loss on the peak temperature and shape of the melt pool. In particular, a negative melt pool surface tension/temperature gradient leads to an outward flow toward the periphery of the melt pool, rendering it wide and shallow. In contrast, a positive surface tension/temperature gradient induces a narrow and thick melt pool due to inward flow toward its center. The presence of a mushy zone near the liquid–solid interface at the solidification front also influences the dynamics of the melt pool.

The powder flow rate and angle of attack of the powder particles significantly affect the free surface stability and motion of a melt pool. Meanwhile, small powder particle sizes result in a stable melt pool for L-DED. In contrast, large powder particles can produce a highly agitated and unstable melt pool, corresponding to free surface instability. The agitating motion of an unstable melt pool is attributed to the relatively high momentum inherent in large powder particles, resulting in large displacements within it and oscillations along its free surface.

As the number of deposited layers increases, heat accumulation leads to a higher temperature within the previously deposited layer, and the cooling rate of the melt pool decreases. Thus, more powder particles can be deposited onto the preceding solidified layer, leading to a higher layer thickness. High powder feed rates can also result in the powder particles splashing, which affects the surface quality of the printed parts and damages the deposition head and protective lens.

The dimensions of a melt pool are collectively determined by multiple process parameters, including the laser power, scanning speed, laser spot diameter, and powder flow rate (Figure 5.2). The preheating temperature of the substrate (or previously deposited layers) and interlayer cooling time may also influence the melt pool dimensions (Figure 5.2b) because they affect the energy absorption and temperature distribution of the melt pool by controlling its thermal history. The melt pool width and temperature generally increase with the laser power and decrease with the scanning speed, both of which influence the cooling rate at the liquid–solid interface. As shown in Figure 5.2a, the melt pool dimensions (width and depth) increase almost linearly with the laser power and incident energy density (Zheng et al. 2019).

As the laser power increases, the width of the melt pool increases more rapidly than its depth, indicating that the former is more susceptible to changes in laser power than the latter. If the powder flow rate is preset, the amount of powder delivered and deposited per unit time is almost fixed. For a constant hatch spacing (defined in Chapter 3), the melt pool width and height increase with the laser power, while the layer thickness increases with the powder flow rate and decreases with the scanning speed, as shown in Figure 5.2a,c.

The melt pool width can also be adjusted by altering the laser spot diameter, which determines the area exposed to laser irradiation on a deposited layer. A decrease in the laser spot diameter at a constant laser power increases the energy density of the laser beam, which causes a decrease in the melt pool width and leads to a large contact angle of a single track if excessive powder particles are present. A large contact angle is undesirable as it may induce porosity in the overlapping regions between adjacent tracks.

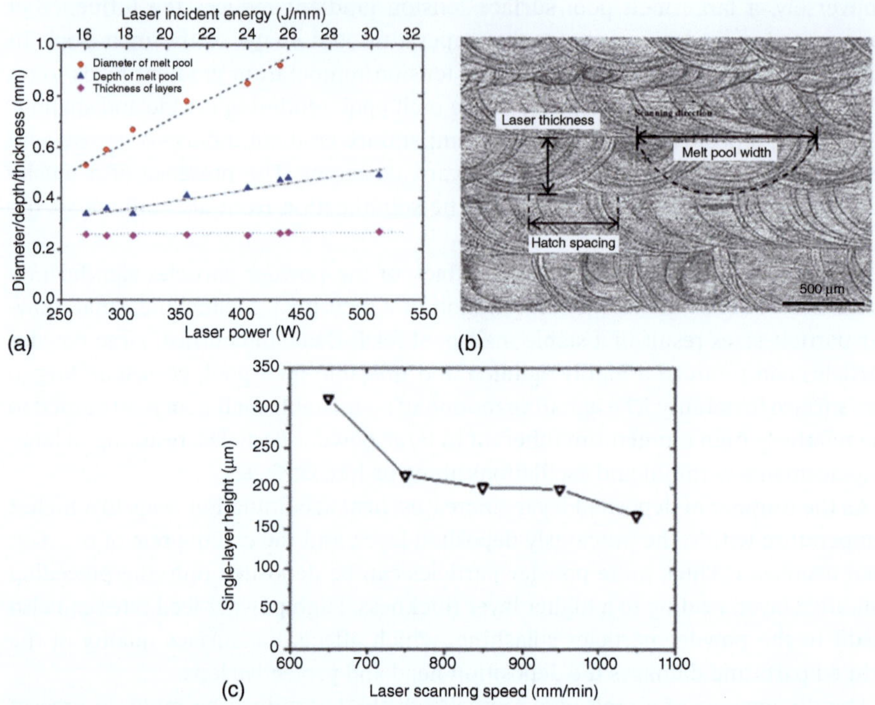

Figure 5.2 (a) Variation in the melt pool size and deposited layer thickness of 316L printed by L-DED with respect to laser power and laser incident energy (laser power/scanning speed). (b) OM image of the typical morphology of a melt pool, which comprises etched patterns. Source: Zheng et al. (2019)/reproduced with permission from Elsevier. (c) Variation in the layer thickness of AISI M4 printed by L-DED with respect to the scanning speed. Source: Reproduced from Shim et al. (2016)/with permission from Elsevier.

A narrow melt pool may collapse before solidifying when the powder is deposited onto an inclined surface. This effect is attributed to a decrease in its cooling rate from excessive heat input. Such collapse may also result from a high powder flow rate and sufficient energy input providing an abundant supply of melt stream, which is injected into the melt pool and pushed outward. Thus, a deposited track with a width greater than the laser spot diameter can be formed.

The free surface of a melt pool governs the cross-sectional profile of a deposited layer. Consequently, the resultant liquid–vapor interface (i.e. melt pool–vapor gas mixture interface) is unstable and in non-equilibrium. Meanwhile, the movement of the laser beam and powder injection results in a slanted melt pool profile (Figure 5.1b). The angle of such inclination is related to the deposition layer thickness and melt pool morphology.

For multi-track deposition, the morphology of a melt pool becomes more asymmetric and slanted because of the overlap of adjacent tracks. For secondary adjacent track deposition, the melt pool is inclined toward the previously deposited track, thereby altering heat transfer and the melt pool wetting behavior. Multi-track deposition is also susceptible to the remelting of adjacent tracks and preceding layers. Moreover, the type of nozzle utilized (coaxial nozzle, multi-stream nozzle, etc.) and the nozzle configuration (lateral injection nozzle, radially symmetric injection nozzle, etc.) affect the morphology of the melt pool by changing its tilt angle.

The solidification of a melt pool is governed by its net heat transfer; it can be classified into three major events: heterogeneous nucleation, mushy-zone heat/mass transfer, and microstructural evolution via intrinsic heat treatment.

The mushy zone contains a two-phase mixture of remnant molten metal and solid particulates; it can be identified as the region between the melt pool and the solidified material, as shown in Figure 5.1b. Natural convection is the primary mode of heat transfer in the mushy zone, as significant density variation exists within it. The mushy zone is a result of melting and solidification occurring between the solidus and liquidus temperatures, as observed by the coexistence of liquid and solid metals.

The heat transfer at the liquid–solid boundary of the melt pool is perhaps the most complex physical process inherent to powder-based L-DED. The increase in mass of blown powder disturbs the transient morphology of the melt pool and contributes to its instability; the trailing edge of the melt pool releases its latent heat of solidification, which further increases the complexity of the process. As a result, the cooling rate is the highest at the liquid–solid interface and decreases closer to the center of the melt pool.

The powder particle momentum can be transferred with minimal momentum loss when impacting the surface of the mushy region, as the powder particles do not rebound or melt, which induces shear stress toward the leading edge of the melt pool. The trailing edge undergoes turbulent flow and a drastic shape change due to the competition between the forward shear stress and upward surface stress induced by Marangoni flow, resulting in disturbance to the temperature distribution during melting. Such disruption of the surface flow and temperature distribution may impede the continuous epitaxial growth of the crystals during rapid solidification.

The initial temperature of the substrate significantly influences the microstructure of the region of a printed part near it because the thermal gradients of the first few deposited layers are determined by the temperature of the substrate and the melt pool. Preheating the substrate reduces residual stress within the printed parts and lowers the risk of cracking and/or thermal distortion. In some cases, partial remelting of the substrate is performed to ensure firm bonding between the substrate and printed part, in addition to the uniformity of the microstructure at their interface.

5.4 Metallurgical Defects

Common metallurgical defects in L-DED parts include porosity, high surface roughness, and cracks. The mechanisms of pore formation in L-DED are similar to those in LPBF, which can be attributed to insufficient melting, entrapped gas, and debris. Lack-of-fusion pores are usually formed in the overlapping region of adjacent tracks when a mismatch between the melt pool size and the hatch spacing exists. The insufficient supply of molten material prevents the gaps between the adjacent tracks from being filled when subsequent layers are printed under low laser powers,

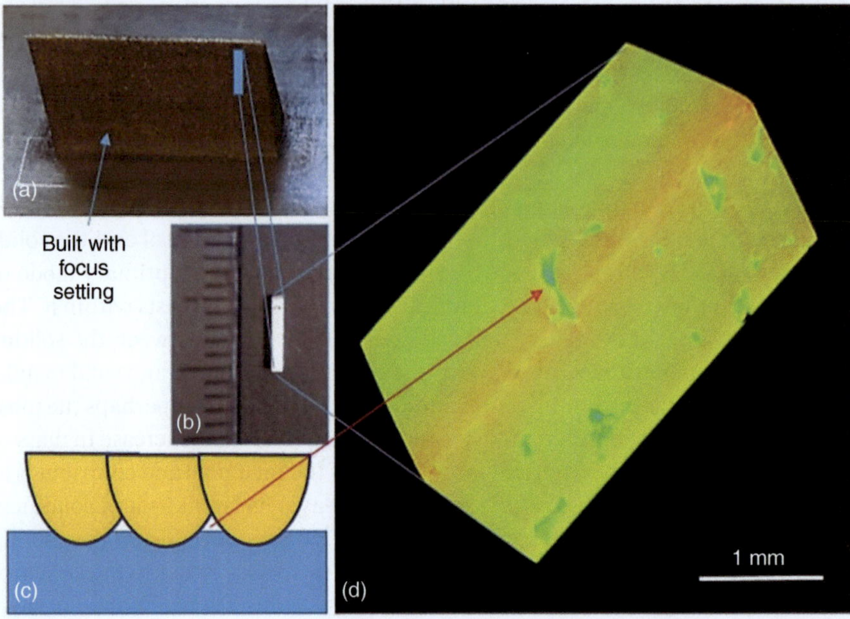

Figure 5.3 Porosity in a 316L sample printed by L-DED measured with X-ray computer tomography (X-CT). (a) Photo of the sample, (b) testing sample removed from the printed sample for X-CT measurement, (c) schematic diagram for the formation of lack-of-fusion pores, and (d) 3D X-CT measured results. Source: Zheng et al. (2019)/reproduced with permission from Elsevier.

high scanning speeds, or high powder feed rates. This phenomenon is particularly prominent during the initial deposition process because this region possesses a higher cooling rate than the other regions. These pores exhibit irregular shapes with coarse inwalls, as presented in Figure 5.3. Lack-of-fusion porosity is rarely observed and can be easily avoided with the proper selection of process parameters.

Gas pores usually exhibit spherical shapes with smooth inwalls. They are randomly distributed within the parts and are either generated through gas entrapment during melting and solidification or by gaseous impurities in the feedstock powder (Wang et al. 2014). The gas porosity in deposits is generally less than 0.5% per unit area.

Porosity arising from unmelted powder particles is most likely present in the initial deposition layers, in which a large amount of heat is expended to initiate the formation of the melt pool and is transferred to the substrate at the beginning of the L-DED process. The effectiveness of increasing the laser power to reduce porosity is limited by the formation of spatters, which reduce the surface quality of the printed part. Meanwhile, higher laser power increases the risk of keyhole pore formation. Porosity can also result from turbulent fluid flow, which occur due to the quick collapse of keyholes and rapid solidification of the melt pool that prevents keyholes from being filled. As a result, keyhole pores primarily appear at the bottom of the melt pool, and this phenomenon is the leading cause of micro-porosity.

In general, porosity can be reduced to a certain extent by optimizing the process parameters, such as implementing a proper combination of the laser power and scanning speed to control the melt pool temperature and cooling rate during L-DED. The laser power, scanning speed, powder flow rate, and hatch spacing are a few of the most important factors in controlling the porosity level.

The poor surface finish of L-DED parts can be attributed to a large melt pool size, spatters attached to the melt track surface, partially melted powder, and the layer-by-layer nature of the fabrication process. Surface roughness values between *Ra* of 4–200 μm have been reported for L-DED-printed Ti–6Al–4V parts, while they vary between *Ra* of 5–10 μm in LPBF-printed Ti–6Al–4V parts (Ahsan et al. 2011). The deposited layers and melt pool boundaries are visible to the naked eye during the deposition process. The surface roughness of L-DED parts can be controlled by process parameters, powder size and morphology, focal distance, laser profile, and laser-powder alignment. For instance, the surface roughness tends to increase with the powder flow rate and the use of powder obtained from the gas atomization process instead of the plasma rotating electrode process.

The degradation of the side surface quality caused by the excessive accumulation of adherent powder particles is shown in Figure 5.4a. The agglomeration of unmelted powder adhering to the side walls of deposited components is related to two types of misalignment: misalignment of the laser beam relative to the focus lens and misalignment of the focused laser beam with respect to the focused powder stream.

Misalignment of the laser beam with the focus lens may cause one side of the component to receive excessive laser power input and vice versa (Figure 5.4d), leading to the accumulation of powder particles on the side surfaces of the printed parts, as indicated by the red arrows in Figure 5.4a.

The misalignment of the laser beam with the powder stream can also result in some of the injected powder particles attaching to the side surface of the builds due to the non-equilibrium kinetic energy and inhomogeneous distribution of injected powder passing through the laser beam region (Figure 5.4e). To prevent this misalignment, the optical system was modified in a study to permit a variable focal distance, which allowed the laser focal plane to be positioned above or below the surface of the workpiece (positive or negative defocusing conditions). The implementation of identical parameters with an adjustable laser focal plane significantly improved the surface quality, as shown in Figure 5.4b.

The unmelted or partially melted powder can also attach to the top surface of a part produced by L-DED. Figure 5.4c displays a contour map detailing the topography of the top surface. The results reveal that the top surface comprised alternating layered ridges originating from the multi-pass deposition process. The top surface also contained unmelted or partially melted powder particles. As the laser spot moved away from the melt pool, the decreasing temperature of the melt pool was insufficient to fully melt the powder particles, causing them to fuse with the cooled-down surface. Some unmelted or partially melted powder particles were delivered to the solidified tracks due to defocused powder delivery or spattering. These powder particles were spatially concentrated along the metal flow trajectories and/or inter-pass boundaries.

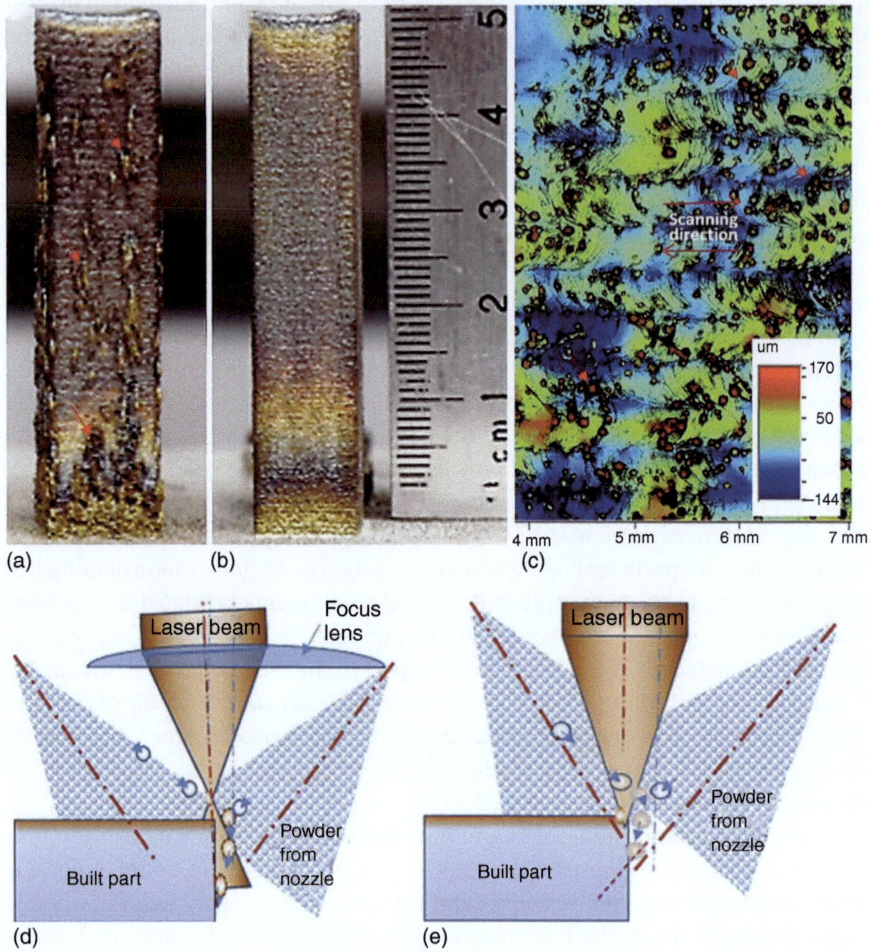

Figure 5.4 Surface quality of 316L samples printed by L-DED. (a) Side surface with adherent powder particles obtained with the laser set to an over-focused condition; (b) side surface without adherent powder particles obtained with an under-focused laser; (c) top layer surface topography illustrated by a contour map with a profilometer, showing unmelted or partially melted powder particles. Schematic diagrams indicating the misalignment of the laser beam with the (d) focus lens and (e) powder feeder. Source: Zheng et al. (2019)/reproduced with permission from Elsevier.

5.5 Powder Materials

The powder properties adopted in L-DED are similar to those in PBF processes, except L-DED employs a wider powder particle size distribution ranging from 50 μm to 150 μm. Gas atomization, water atomization, and plasma rotating electrode processes are typical methods of producing printable powder for L-DED. Therefore, the shape, surface quality, porosity, flowability, purity, and size distribution of the produced powder should be evaluated before usage. L-DED can print a broad spectrum

Table 5.1 Summary of powder materials for L-DED, including commercialized materials and laboratory materials.

Material	Commercial equivalents
Titanium	Commercially pure Ti, Ti–6Al–4V, Ti6242, TC21, TA15
Steel and iron	316L, 304, 254SMO, 2205, 431, 17-4 PH, S53, maraging steels, Hadfield steels, P20, M4, H13, CPM 10V, X42Cr13, Invar steels, G3Si1,4130, 4140, 300M
Aluminum	AlSi5, Al–Si10–Mg, AlSi12, Al–Si–7Mg, AlMgScZr
Nickel	Inconel 718, Inconel 625, Hastelloy, Monel, CMSX-4, René, high-Cr NiCr
Cobalt	Co–Cr–Mo
Copper	CuAl, CuNi, CuAlNi, CuSn, CuSi3, CuCrZr
Tantalum	Pure Ta
Molybdenum	Pure Mo
Magnesium	Magnesium alloys
Tungsten	Pure W

of materials, including iron-, titanium-, aluminum-, nickel-, cobalt-, copper-based alloys, and their composites, which are widely employed in various industrial applications. A summary of printable materials for L-DED is listed in Table 5.1.

L-DED can be equipped with multiple powder feeders, which render it capable of printing multi-materials, hierarchical materials, and functionally graded materials. It can also potentially improve the performance of high-entropy alloy products by increasing the complexity and functionality at user-definable locations. Furthermore, compared to PBF, L-DED is compatible with pre-alloyed high-entropy alloy (HEA) powders and pure elemental powders delivered from multiple powder hoppers simultaneously to print HEA products through *in situ* alloying (Han et al. 2020). This ability eliminates the requirement for a powder preparation process, which incurs a long lead time and high cost. Meanwhile, the flexibility of the composition design of HEAs for materials development is also enhanced.

5.6 Equipment

L-DED simultaneously delivers powder materials and energy, which is accomplished through the controllable integration of powder and energy delivery systems. A typical L-DED machine consists of a laser system, a powder feed system, a motion control system, and auxiliary systems such as monitoring and environmental control systems, as shown in Figure 5.5.

Most L-DED machines employ continuous-wave lasers of infrared wavelength. Some machines allow for an adjustable laser spot diameter, focal length, and laser energy profile that provide versatility during the printing process. The laser beam

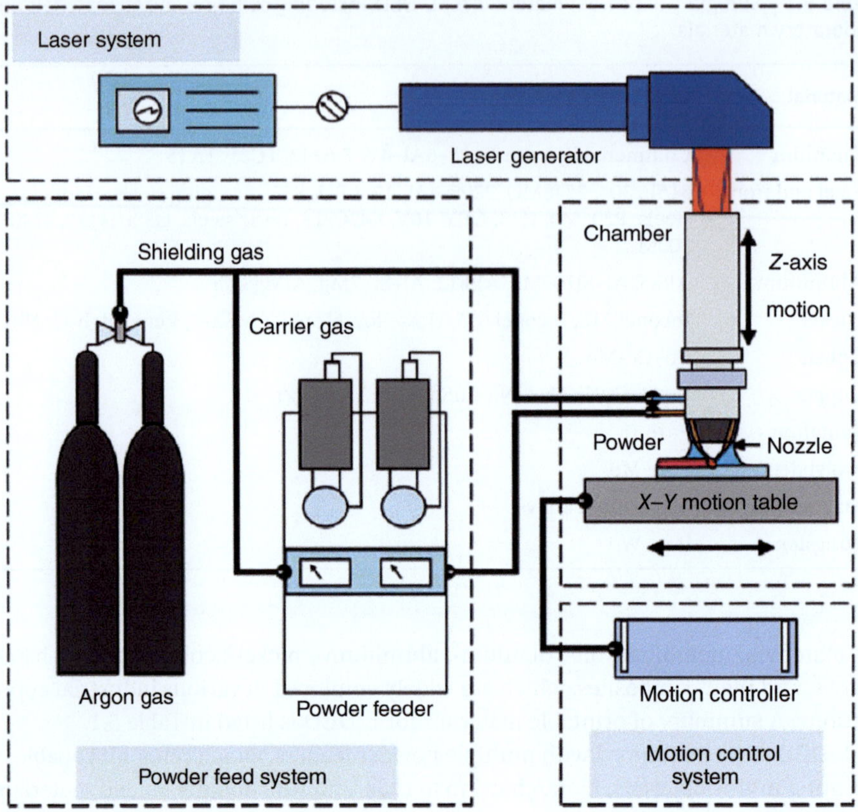

Figure 5.5 System components of an L-DED machine. Source: Kim et al. (2017)/MDPI/CC BY-4.0.

travels through the optical fiber in the printing head and is focused on the deposition region through a planoconvex lens.

For most systems, Nd:YAG and fiber lasers typically exhibit a laser power range of 1–5 kW. However, other types of lasers, such as CO_2 and pulsed-wave lasers, can be employed if sufficient power is available for melting the material. The Nd:YAG laser has been widely applied in many laser-based processing applications, including laser welding, cutting, and coating, since most metals exhibit poor reflectivity toward its incident spectral radiation at a wavelength of ~1067 nm. CO_2-type lasers possess wavelengths of an order of magnitude higher at 10.6 μm and may contribute to a less energy-efficient L-DED process.

In general, the diameter of the laser beam is typically a few millimeters, and the energy intensity of a continuous-wave laser displays a near-Gaussian distribution. Extreme heat fluxes are also present within the melt pool. For example, an L-DED process operating with a laser power of 500 W and a laser spot size of 1 mm produces an average heat flux of over 63.6 kW/cm^2.

Laser attenuation occurs due to the absorption and scattering of laser radiation by the blown powder. The powder stream profile affects both the radial laser power intensity profile and the ratio of the attenuated laser power to the original laser power. For a fixed laser power, increasing the powder flow rate results in a decrease in the mean temperature of the powder stream and a higher degree of laser attenuation for a Gaussian irradiation profile. The effects of laser attenuation can be significant and result in only 75% of the original laser power reaching the melt pool surface (He and Mazumder 2007). To mitigate such an issue, a coherent beam of a wavelength easily absorbed by the powder is desired.

Since the laser optical system and powder feed nozzles are designed to focus on a specific location, a precise stand-off distance must be predefined for optimal deposition. Changes in the focal distance to the substrate caused by misalignment, overbuilding, or in-process distortion can result in variations in the powder feed and laser focus alignment, which may adversely affect the printing process.

The powder delivery performance of the L-DED process can be evaluated by the powder flow rate, which is the mass of powder delivered per unit time. Meanwhile, the powder utilization efficiency can be assessed by the mass ratio of the deposited powder to the delivered powder over a specific time interval. Additionally, the powder utilization efficiency depends on the nozzle geometry, powder delivery angle, laser and powder flow process parameters, and flowability and size distribution of the powder particles. A low deposition efficiency typically corresponds to a low building rate and a significant amount of recycled and/or "scrapped" metallic powder. As a result, the manufacturing cost and complexity associated with the routine cleaning/recycling procedures regarding the L-DED process increase.

As shown in Figure 5.5, the powder delivery system primarily consists of powder dispensing components (i.e. pipes and carrier gases), a printing nozzle (in coaxial, multi-stream, or side nozzle form), and auxiliary components (e.g. mass and gas flow sensors). Metal powder is dispensed from a hopper/container into a connected feeding line, in which it is carried by an inert gas such as argon or helium. The powder stream delivered to work sites may fluctuate at a low frequency, and such irregularities can be detected and rectified by a monitoring system. Notably, the nozzles are specially designed and positioned to ensure that the material stream intersects with the laser beam and is injected into the melt pool.

The angle of attack and spatiotemporal uniformity of the powder jet can be achieved through the proper design of the powder feeding system. The shape of the powder stream, speed of the blown powder, laser intensity distribution, and blown powder temperature distribution are all significantly affected by the feeding angle with respect to the nozzle and the imposed powder feeding rate. However, the deposition uniformity of the powder is nearly independent of the relative laser/substrate speed. The number of coaxially blown powder particles per unit volume, or powder volume concentration, can be modeled by

$$N(r, l) = N_{\text{max}}(l)e^{-2r^2/r_p^2}, \tag{5.3}$$

where r is the distance from the beam centerline, l is the laser/powder working distance, N_{max} is the maximum volume concentration, and r_p is the characteristic powder stream radius (Peyre et al. 2008).

Compared to coaxial nozzle and multi-stream nozzle L-DED systems, maintaining flow uniformity and printing functional multi-materials present a challenge for a side nozzle system. However, for parts with complex geometries, coaxial nozzles and multi-stream nozzles provide relatively high powder deposition efficiency, in which the powder particle stream diameter matches the laser beam diameter.

Since metals exhibit relatively high melting temperatures, the printed parts are highly prone to oxidization during L-DED. Therefore, inert shielding gases such as argon are utilized to protect the deposition area by displacing the oxygen within the printing chamber and powder delivery system to create an inert atmosphere. While oxygen tolerance in L-DED varies with applications and materials, oxygen concentration levels below 10 ppm are typically acceptable in a controlled atmosphere. The chamber is purged at a shielding gas flow rate of 5–20 L/min, with the carrier gas purge rate being governed by the powder feed rate.

Monitoring and control systems are required to ensure a stable laser power output, precise and accurate movement of the printing head and substrate stage, a well-defined material feed rate, and constant chamber pressure and gas concentrations during processing. Monitoring systems capable of detecting scattered or reflected light, such as light waves reflected by turning mirrors or have undergone Rayleigh scattering in the air, exhibit potential for nonintrusive, real-time measurements and assessments of both internal and external laser beam power fluctuations. The laser reflected from the melt pool to the protective glass can be detected by a photodiode, which protects the optical components from being damaged by the reflected laser beams when processing highly reflective materials.

To ensure part quality, a trace oxygen sensor typically monitors the oxygen concentration within the build environment. Electrochemical oxygen sensors can measure concentration on the scale of parts per million, but they require periodic calibration.

In powder delivery systems, real-time monitoring and control of the flow rate can be achieved through continuous weight measurement, camera imaging, and laser diffraction methods. However, due to dull mass flow responses, long delivery routines, and slow sampling rates, a significant delay exists between the changing and stable outputs of the powder flow. Furthermore, controlling the powder flow rate precisely and quickly during the process is challenging.

The selection of process-dependent build attributes requires prior knowledge of the desired part geometry, material composition, density, microstructure, etc. Many characteristics, such as material composition, density, and microstructure, cannot be directly measured. However, certain characteristics of the laser–material interaction and solidified regions of the deposit can be detected to allow for estimations and indirect monitoring of these variables. By coupling imaging systems with a laser deposition system, the melt pool geometry and temperature field can be closely observed, thereby providing the user with control over the properties of the melt pool and the deposition process.

Thermal monitoring of the melt pool can be conducted by implementing infrared pyrometers. The temperature data from such monitoring systems can be used for making on-the-fly adjustments to process parameters or for quality assurance purposes.

Table 5.2 Summary of representative commercialized equipment for L-DED.

Company	Equipment	*XYZ* Travel (mm)	Degree of freedom	Laser power (W)
Optomec (USA)	LENS® 150	150 × 150 × 150	3 axes	400
	LENS® 600	600 × 400 × 400	5 axes	2000
	LENS® 800	800 × 600 × 600	5 axes	3000
	LENS® 1500	900 × 1500 × 900	5 axes	3000
	LENS® 500 HY OA	350 × 325 × 500 500 × 325 × 500	5 axes	2000
	LENS® 860 HY OA	598 × 600 × 610 860 × 600 × 610	5 axes	3000
BeAM (France)	Modulo 250	400 × 250 × 300	5 axes	500
	Modulo 400	600 × 400 × 400	5 axes	2000
	Magic 800	1200 × 800 × 800	5 axes	2000
Trumpf (Germany)	TruLaser Cell 3000	800 × 600 × 400	5 axes	8000
	TruLaser Cell 7040	4000 × 1500/ 2000 × 750/1000	5 axes	6000
FormAlloy (USA)	X & L-Series	200 × 200 × 200; 500 × 500 × 500; 1000 × 1000 × 1000	5 axes	8000
InssTek (South Korea)	MX-Lab	150 × 150 × 150	3 axes	300
	MX-600	450 × 600 × 380	5 axes	1000
	MX-1000	800 × 1000 × 650	5 axes	2000
	MX-Grande	4000 × 1000 × 1000	5 axes	3000
	MPC	300 × 300 × 200	6 axes	500
DMG Mori (Germany)	LASERTEC 65 3D	735 × 650 × 560	5 axes	2500/3000
	LASERTEC 65 3D hybrid	735 × 650 × 560	5 axes	2500/3000
	LASERTEC 125 3D hybrid	1335 × 1250 × 900	5 axes	2500/3000
	LASERTEC 4300 3D hybrid	660 × 400 × 1500	5 axes	2500/3000

Representative commercial L-DED equipment produced by industrial manufacturers, such as Optomec (USA), BeAM (France), Trumpf (Germany), DMG Mori (Germany), FormAlloy (USA), and InssTek (South Korea), and their available powders are summarized in Tables 5.2 and 5.3, respectively.

The Optomec Company provides industry-leading solutions for metal-based AM. For example, Optomec developed LENS® systems to print common engineering materials (stainless steels, tool steels, titanium alloys, cobalt alloys, etc.) and other rarer materials (zirconium, tantalum, tungsten, aluminum, bronze, refractory metals, certain ceramics, etc.). These systems correspond to a layer thickness of

Table 5.3 Summary of available powders for commercialized L-DED equipment.

Company	Available powders
Optomec (USA)	• LENS® 150, LENS® 600, LENS® 800: (Open atmosphere) stainless steel 13-8, 17-4, 304, 316, 410, 420, 15-5PH, AM355, 309, 416, IN625, IN718, IN690, pure copper, bronze, Cu–Ni, GRCOP-84, tool steel H13, S7, A-2, Stellite 6, Stellite 21, Ni-WC, Co-WC (Controlled atmosphere) CP Ti, Ti–6Al–4V, Ti6242, Ti6246. Ti4822, Ti-22Al-23Nb, Alumina, Al 4047, Waspalloy, Hastelloy X, MarM247, Rene 41, Rene 142, pure tungsten, pure molybdenum, pure niobium, TiC, CrC • LENS® 1500: Inconel alloys, stainless steels, titanium alloys • LENS® 500 HY OA: Tool steels, stainless steels, Inconels, Hastelloy, Stellite, tungsten carbide • LENS® 860 HY OA: Inconel alloys, stainless steels, cobalt, tungsten
BeAM (France)	Ti–6Al–4V, 316, Inconel 625, Inconel 718
Trumpf (Germany)	316L, CoCr, bronze, Inconel 625, Inconel 718
FormAlloy (USA)	Inconel 625, Inconel 718, Hastelloy X, Haynes 282, 316L, 15-5 PH, 17-4 PH, maraging steel, H13, Invar 36, Ti–6Al–4V, Co–Cr–Mo, pure copper, C18000, C18150, GRCop-42. GRCop-84
InssTek (South Korea)	• MX-Lab, MX-600, MX-1000, MX-Grande: CP Ti, Ti–6Al–4V, P20 steel, P21 steel, H13, 304, 316, 420, Ni600, Ni625, Ni690, Ni713, Ni718, Hastelloy 22, Hastelloy 276, Cu–Sn, bronze, CoCr, Stellite 21, Stellite 25 • MPC: Ti
DMG Mori (Germany)	• LASERTEC 65 3D, LASERTEC 65 3D hybrid, LASERTEC 125 3D hybrid: 316L, CoCr, Cu–Al, Inconel 625, PH15-5 • LASERTEC 4300 3D hybrid: Ti–6Al–4V

250–750 μm, a minimum wall thickness of 300 μm, a deposition rate of 0.5 kg/hr based on a laser power of 2 kW, and a side surface roughness of 12–25 μm. A hermetically sealed chamber is included, which can be purged with argon to maintain oxygen and moisture levels below 10 ppm. Metal powder feedstock is delivered to the deposition area through Optomec's proprietary powder feed system, which can precisely regulate the mass flow.

The LENS® systems produce near-fully dense materials with mechanical properties comparable to or superior to those of cast materials and, in some cases, forged materials. To overcome the issues of low printing resolution and poor surface quality, L-DED and subtractive manufacturing capabilities are integrated into machines, such as the LENS HY OA series (Optomec) and LASERTEC 3D hybrid series (DMG Mori).

5.7 Microstructure and Mechanical Properties

The microstructural characteristics (e.g. grain morphology and size) of L-DED parts are highly sensitive to their thermal history during the deposition process, which may involve high heating/cooling rates, significant temperature gradients, and complex thermal cycles. Therefore, predicting the microstructures of L-DED parts and their degree of dependence on the process parameters is highly challenging, and overcoming such a challenge is vital for establishing effective microstructural control of L-DED parts with superior mechanical properties.

The solidification mechanisms involved in LPBF are also applicable in L-DED. The temperature gradient G and growth rate R affect the solidification process. They contribute to three major kinds of microstructural morphology within L-DED parts: columnar, columnar + equiaxed, and equiaxed grain morphology. A high temperature gradient promotes the transition of equiaxed grains to columnar grains, and increasing the cooling rate, given by $G \times R$, results in a finer microstructure. For thin-walled parts fabricated via L-DED, the ranges of G and $G \times R$ are approximately 100–200 K/mm and 200–6000 K/s, respectively (Hofmeister et al. 1999). Optimal G and R values depend on the material properties, process parameters, part geometry, printing strategies, environmental/machine conditions, etc. The tensile properties of various metal powder materials printed through L-DED are summarized in Table 5.4.

5.7.1 Titanium and Its Alloys

The microstructural evolution of Ti–6Al–4V parts printed by L-DED strongly depends on the process parameters (e.g. laser power and laser scanning speed), printing strategy, part geometry, and intrinsic properties of the materials involved (e.g. thermal conductivity of the printed material and substrate). The microstructure of such Ti–6Al–4V parts commonly consists of α/α' phases present in a β matrix, namely, primary α, secondary α, plate-like α, colony α, acicular martensite α', and Widmanstätten and basket-weave structures.

In a study, the crystallographic microstructure within a β matrix comprises the acicular martensitic α' phase, a large number of Widmanstätten α-laths structures, and a few β phases between the α-laths (Li et al. 2015). Figure 5.6a,b show needle-like martensitic α' clusters and Widmanstätten α-laths. Within the β grains, the Widmanstätten α-laths structures display basket-weave morphology. Both the needle-like martensitic α-phase clusters and Widmanstätten α-laths exhibit thin and needle-shaped microstructures, which correspond to high strength levels.

Table 5.4 Summary of the tensile properties of various metal materials printed by L-DED.

Material	Yield strength (MPa)	Ultimate tensile strength (MPa)	Elongation (%)
304L[a)]	337	609	48.2
316[b)]	558	639	21
316L[c)]	576	776	33
316L[d)]	405–415	620–660	32–40
17-4 PH[e)]	758	1129	14.1
Fe19Ni5Ti[f)]	—	>1300	>10
AISI 4340 steel[g)]	—	1398.65	1.665
1Cr12Ni2WMoVNb[h)]	—	1227	7.7
Inconel 718[i)]	750	958	17
Inconel 718[j)]	1024	1334	34.1
Inconel 718[k)]	752.17	1074.19	22.46
Inconel 738[l)]	1350	1392	1.13
Inconel 625[m)]	531	840	16
Haynes 282[n)]	633	943	31.6
Ti–6Al–4V[o)]	961	1072	17
Ti–6Al–4V[p)]	945	1041	14.5
Ti–6Al–4V[q)]	1094	1137	4.7
Ti–6Al–2Sn–4Zr–2Mo–0.1Si[r)]	976	1021	8.45
Ti–4Al–1.5Mn[s)]	855	912	10.2
NiTi[t)]	400	480	1.6
TiAl[u)]	421	539	1.7
Al–Si10–Mg[v)]	—	355	8.1

a) (Wang et al. 2016);
b) (Zhang et al. 2014);
c) (Ziętala et al. 2016);
d) (Yadollahi et al. 2015);
e) (Yu et al. 2020);
f) (Kürnsteiner et al. 2020);
g) (Sun et al. 2015);
h) (Wang et al. 2010);
i) (Wang and Shi 2020);
j) (Zhang et al. 2020);
k) (Li et al. 2020);
l) (Ramakrishnan and Dinda 2019a);
m) (Zhang et al. 2016);
n) (Ramakrishnan and Dinda 2019b);
o) (Keist and Palmer 2016);
p) (Carroll et al. 2015);
q) (Todaro et al. 2020);
r) (Wang et al. 2020);
s) (Tian et al. 2014);
t) (Bimber et al. 2016);
u) (Thomas et al. 2017);
v) (Wang et al. 2019).

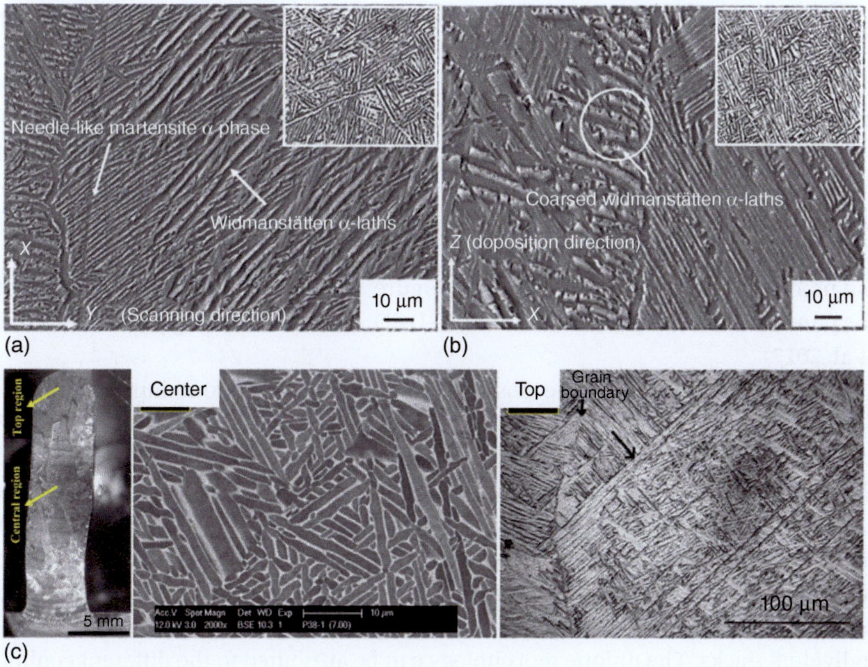

Figure 5.6 Microstructure of Ti–6Al–4V parts fabricated by L-DED. (a) In the *x*–*y* plane and (b) in the *x*–*z* plane. Source: Li et al. (2015)/reproduced with permission from Elsevier. (c) Different regions along the build direction. Source: Baufeld et al. (2011)/reproduced with permission from Elsevier.

Coarse Widmanstätten α-laths can also be observed along the β grain boundaries. The coarse α-laths reduce the strength of the material by weakening the inhibition effect of gliding dislocations. When a high density of α-lamellae precipitates exists along the β grain boundaries, the dislocations glide smoothly, and the yield strength of the material decreases consequently.

The grid structures shown in Figure 5.6c was considered as a martensitic phase (Baufeld et al. 2011). In general, slow cooling leads to the formation of colony structures, while rapid cooling results in the formation of basket-weave structures and induces martensitic transformation. The diversity of microstructures suggests that the microstructural evolution during the L-DED process is controlled by the rate of the final cooling from the β-transus temperature to the ambient temperature. Notably, thermal cycles below the β-transus temperature may have little to no impact on the microstructure. The martensitic α′ structure can subsequently transform into a fine lamellar α + β microstructure through the annealing treatment at a temperature of 700–850 °C.

Post-AM stress-relief treatment is generally recommended for aerospace components printed via L-DED. Such treatment of Ti–6Al–4V parts is typically performed at temperatures below 995 °C in the α + β regions, which does not significantly impact prior β grains but alters the distribution, size, and morphology of the α-phase (Åkerfeldt et al. 2016). Heat treatment at temperatures higher than 995 °C leads to complete dissolution of the α phase and transformation of the morphology of prior β grains from columnar to globular.

A comparative study was conducted to investigate pure titanium parts manufactured by L-DED, LPBF, and casting processes. An analysis of the processing parameters indicated that significantly higher laser power is required in L-DED than in LPBF to obtain a near-full density of the printed parts. The pure titanium parts underwent an allotropic phase transformation at a temperature of approximately 865 °C, at which the $\alpha \rightarrow \beta \rightarrow \alpha$ transformation occurs during the heating and cooling processes (Gey and Humbert 2002). Figure 5.7 displays the microstructures of the pure titanium samples fabricated by L-DED, LPBF, and casting processes (Attar et al. 2017).

The as-cast titanium sample possessed a microstructure containing a mixture of serrated α and fine acicular α phases (Figure 5.7a). In contrast, the LPBF sample exhibited a dominant martensitic α′ phase with lath-type morphology and fine acicular α′ phase in certain regions (Figure 5.7b). In addition, an α matrix embedded with plate-like α phases and Widmanstätten structures was observed in the L-DED samples (Figure 5.7c,d). The melt pool tracks were not visible in the microstructure of pure titanium parts printed by L-DED and LPBF since using pure titanium eliminated compositional segregation or phase variations across the individual tracks. The unique morphology can be attributed to the different cooling

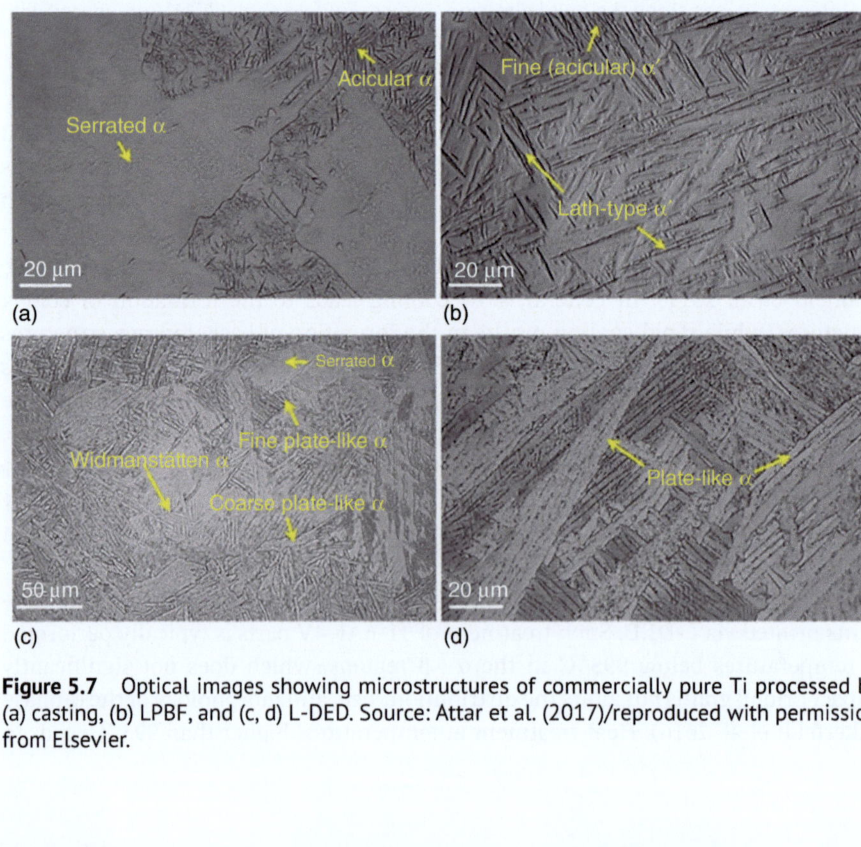

Figure 5.7 Optical images showing microstructures of commercially pure Ti processed by (a) casting, (b) LPBF, and (c, d) L-DED. Source: Attar et al. (2017)/reproduced with permission from Elsevier.

rates required for its formation, i.e. 10^3 K/s for the lath type (Oh et al. 2004) and 10^5–10^6 K/s for the fine acicular type (Yang et al. 2016).

Regarding tensile properties, the Ti–6Al–4V alloy printed by L-DED could achieve a high ultimate tensile strength of 900–1200 MPa. The values of elongation at failure for printed Ti–6Al–4V parts along the transversal direction (*x*–*y* plane) may be 25–33% higher than those along the longitudinal direction (*z* direction) (Keist and Palmer 2016). The lower ductility of the parts along the longitudinal direction was attributed to the β grain boundaries typically aligning perpendicular to the transverse direction, which allows easier crack propagation.

The mechanical properties of a Ti–6Al–4V part printed by L-DED also vary locally along the distance from the substrate. For example, the region close to the top surface of the part can efficiently transfer heat to its ambient environment and undergo fewer thermal shocks from the deposition of subsequent layers. In contrast, the region near the substrate is subjected to more remelting and thermal cycles, and it has a greater heat flux due to conduction from the top layer to the substrate during the printing process.

Figure 5.8 presents the mechanical properties of the specimens tested along the (010), (100), and (001) orientations (Wolff et al. 2016). For both specimens corresponding to the (001) and (010) directions, a clear trend of tensile strength increasing with the distance from the cubic surface can be observed. Meanwhile, the elongation to failure decreases significantly. The specimens tested along the (100) direction exhibited greater strength and elongation.

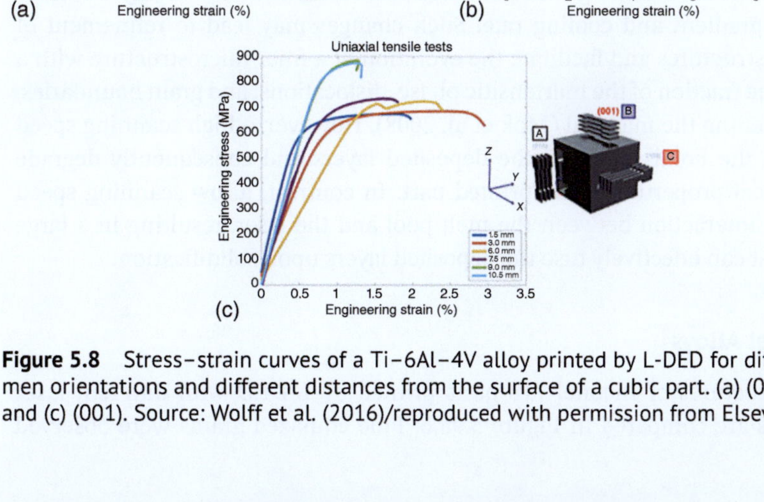

Figure 5.8 Stress–strain curves of a Ti–6Al–4V alloy printed by L-DED for different specimen orientations and different distances from the surface of a cubic part. (a) (010), (b) (100), and (c) (001). Source: Wolff et al. (2016)/reproduced with permission from Elsevier.

The hardness of a Ti–6Al–4V alloy printed by L-DED strongly depends on its thermal history (i.e. temperature, cooling rate, and heat accumulation), which can be manipulated by varying process parameters. The relationship between the microstructure and hardness of an L-DED-printed Ti–6Al–4V part has been investigated (Paydas et al. 2015). Regions with high hardness values of around 350–370 HV were attributed to the martensitic phases formed after rapid cooling, and regions with lower hardness values of 315–350 HV corresponded to the Widmanstatten structure formed at lower cooling rates.

The hardness and microstructure of a prior β grain are functions of the cooling rate of the material below its β-transus temperature instead of the heating duration at temperatures above it (Brandl and Greitemeier 2012). A solution heat treatment at 1200 °C with different heating durations, followed by water quenching from above the β-transus temperature, all resulted in a similar microstructure of basket-weave α and/or martensitic α′, with a microhardness of 333–346 HV. Conversely, a solution heat treatment at 1200 °C for two hours followed by furnace cooling has been determined to substantially decrease the hardness values by introducing the coarsened prior β grains and α-lamellae. On the other hand, heating at 600 °C for four hours followed by furnace cooling was found to increase the average hardness of the printed part due to the precipitation of the Ti_3Al phase (Baufeld et al. 2011).

The melting, solidification, and local thermal history of L-DED parts strongly depend on the laser power and scanning speed. A high laser power usually corresponds to high energy input and low cooling rate, giving rise to the formation of large grain sizes and coarse columnar or lamellar grains. Such large grains decrease the strength and hardness of the printed part. Moreover, high laser power reduces the volume fraction of the martensitic phase as a result of lower cooling rates, thereby improving the ductility and fatigue behavior of the printed part (Zhai et al. 2016).

The powder flow rate directly determines the amount of powder deposited per unit time, which influences the layer thickness and the resultant microstructure of the printed part. A combination of high laser power and low feed rate can effectively suppress the generation of lack-of-fusion pores by ensuring complete melting of the powder.

An increase in the scanning speed or a decrease in the incident energy increases the thermal gradient and cooling rate. Such changes may lead to refinement of the lamellar structures and facilitate the evolution of a finer microstructure with a higher volume fraction of the martensitic phase, dislocations, and grain boundaries, thereby hardening the material (Mok et al. 2008). However, a high scanning speed may weaken the bonds between the deposited layers and consequently degrade the mechanical properties of the printed part. In contrast, a low scanning speed prolongs the interaction between the melt pool and the laser, resulting in a large melt pool that can effectively fuse the deposited layers upon solidification.

5.7.2 Nickel Alloys

The microstructures of Inconel 718 parts printed by L-DED with different laser power values are compared in Figure 5.9a,b. Fine equiaxed grains were observed

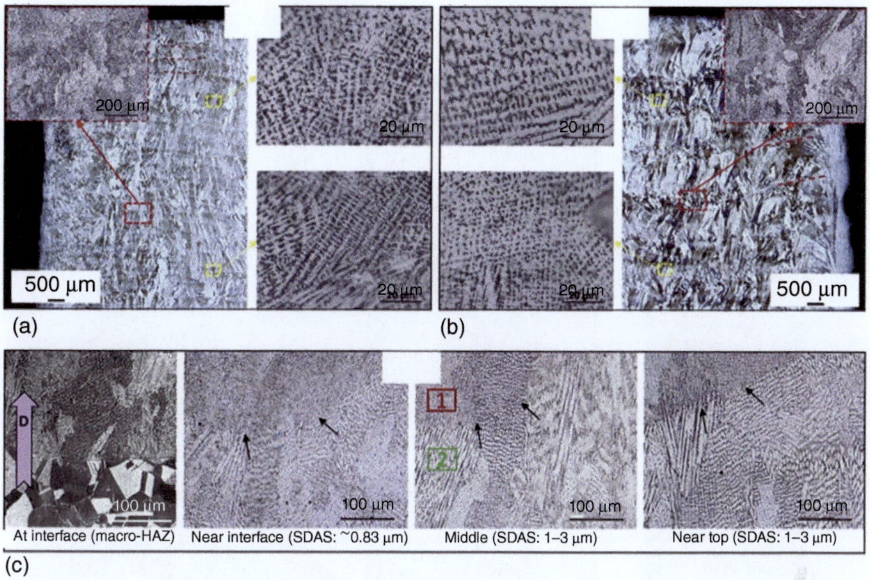

Figure 5.9 Microstructures of Inconel 718 parts fabricated by L-DED under (a) low and (b) high laser power and (c) variation in the microstructure from the bottom to the top of a low-power build. Source: Zhai et al. (2019)/reproduced with permission from Elsevier.

in their layer boundaries/heat-affected zones (HAZs), and columnar grains were distributed within the layers of Inconel 718 samples prepared at both low and high laser power values. In both cases, the columnar grains within the layers were of similar sizes, and a similar dendritic microstructure was observed at higher magnification values. No distinguishable HAZs were found in the top region of the samples fabricated at low laser power, as shown in Figure 5.9c. Meanwhile, finer dendrites were observed in the region close to the substrate with a secondary dendrite arm spacing (SDAS) of ~0.83 µm. Coarser dendrites toward the top of the sample had an SDAS of ~3 µm.

Inconel 718 parts fabricated by L-DED exhibited a banded grain structure, with fine grain zones at the interfaces between their layers containing inclined columnar grains (Figure 5.10a). The fine grain band size of the parts was 150–200 µm at their base, and it decreased to 20 µm at their top. Such a decrease is attributed to the increase in heat accumulation along the build height, which reduces the cooling rate of newly deposited layers and suppresses the formation of fine grains along the build direction.

The fine grain region possessed a more randomized texture than the columnar grains (mainly <001> oriented grains and few frequently <101> oriented grains along the *y* direction). However, the samples exhibited an overall complex randomized texture. Their inverse pole figure (IPF) maps also indicated a banded structure with randomly oriented grains (Figure 5.10b). Such a banded structure was attributed to the layer-wise deposition process, in which remelted regions experienced rapid cooling under a high temperature gradient. Consequently, fine equiaxial

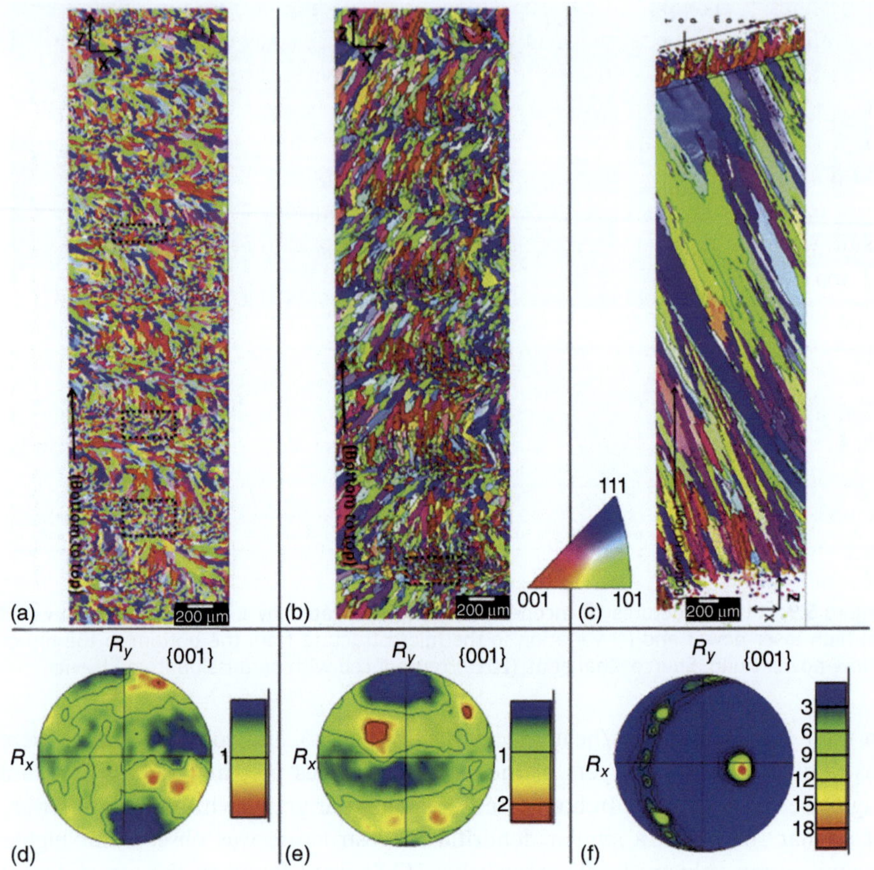

Figure 5.10 Electron backscatter diffraction (EBSD) results displaying the texture of Inconel 718 parts printed by L-DED using different scanning strategies. (a) Unidirectional deposition, (b) bidirectional deposition, (c) bidirectional high-power deposition, and (d–f) their respective pole figures. Source: Parimi et al. (2014)/reproduced with permission from Elsevier.

grains were formed in such regions, while the deposited layers were subjected to relatively slow cooling and formed columnar grains.

Figure 5.10c shows that the build initially began with small columnar grains, which progressively grew larger along the build direction. The appearance of small columnar grains was prominent in the final layers and can be ascribed to the rapid heat transfer between the final layers and the ambient atmosphere. The {001} pole figures of samples indicate a generally weak texture in Figure 5.10d,e. Figure 5.10f indicates that the grains were characteristic of a directionally solidified structure with a preferred orientation. These grains grew epitaxially parallel to the {001} planes, along the direction of the maximum thermal gradient and the favored crystallographic orientation of a face-centered cubic (FCC) structure. The grains tilted toward the resultant heat flux direction (from the sides of the part to its center) as the sides exhibited a higher cooling rate than the center due to heat transfer to the ambient environment via conduction and convection.

The as-deposited Inconel 718 alloy displayed a yield strength of 752 MPa, an ultimate tensile strength of 1047 MPa, and an elongation of 22.46% (Li et al. 2020). The L-DED process induced the precipitation of the γ'' and γ' strengthening phases. The precipitation strengthening and interactions between the precipitates and the dislocations improved the yield strength of the as-deposited L-DED samples. As a result, the yield strength and elongation of the samples were comparable to those of parts fabricated by LPBF.

Inconel 625 parts fabricated by L-DED possessed a unique microstructure in which only columnar grains and primary dendrites were observed at the bottom of each layer (Dinda et al. 2009). The top of the layers displayed a fine dendritic microstructure with secondary arms, as shown in Figure 5.11. It exhibited a slower cooling rate than the bottom of the layers, which allowed for the growth of secondary dendrites. Therefore, a gradual columnar–dendritic microstructural change was observed from the bottom to the top of each layer. The average primary dendrite spacing at the bottom of each layer was approximately 5 μm, and the secondary arm spacing at the top varied between 1.5 and 2.5 μm.

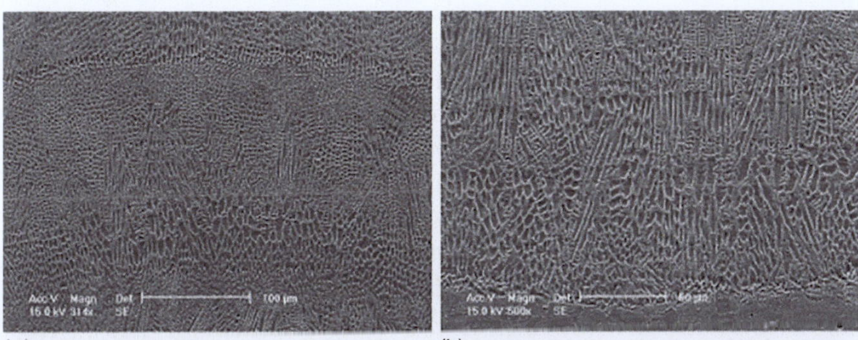

Figure 5.11 Different locations of the microstructure of a single-layer Inconel-625 sample. (a) First layer, (b) second layer, (c) third layer, and (4) fourth layer. Source: Dinda et al. (2009)/ reproduced with permission from Elsevier.

The overall microstructure of the as-printed sample was composed of a nickel-based δ phase along with MC, M_6C, and $M_{23}C_6$ carbides (M represents metal elements). The formation of the Nb-rich MC phase was attributed to the rapid solidification process in L-DED, and these phases were mainly present in the interdendritic regions. While the formation of M_6C and $M_{23}C_6$ was attributed to the exposure of solidified materials at elevated temperatures, the formation of MC phases depended on the formation of carbides during solidification, growth of primary carbides, and high-temperature exposure of previously solidified materials.

Inconel 625 alloys fabricated by L-DED can achieve an ultimate tensile strength of ~920 MPa, a yield strength of ~570 MPa, an elongation of 48%, and a Charpy impact resistance of 104 J (Paul et al. 2007). It was established that the laser scanning pattern does not significantly affect the mechanical properties of Inconel 625 parts. Fracture toughness values ranging from 200 to 255 kJ/m^2 were obtained, and the printed samples could display stable crack growth (Ganesh et al. 2010). In another study, the impact energy of Inconel 625 specimens showed a slight increase after solution treatment (Puppala et al. 2014). The average microhardness of IN625 was measured to be 275 HV at room temperature and 235.7, 214.7, and 202.8 HV at elevated temperatures of 550, 600, and 650 °C, respectively.

5.7.3 Iron Alloys

Austenitic stainless steels, such as 304L and 316L, have been extensively fabricated by L-DED for their excellent combination of strength, ductility, and high corrosion resistance. However, such materials can potentially undergo a deformation-induced phase transformation from an FCC austenitic phase to a body-centered cubic (BCC) martensitic phase under loading. This microstructural transformation results in a material displaying a high strain hardening rate on the macroscale, which endows it with a good synergy of strength and ductility.

In a study by Weng et al., the microstructure of 316L printed by L-DED was determined to be composed of dominant austenite and intergranular skeletal ferrite (Weng et al. 2019). The microstructure of the outer layer of the deposited sample differed from that of its inner region with a transition zone between them, as annotated in Figure 5.12a,b,e,f. The inner region exhibited a dual-phase microstructure comprising dominant austenite γ and inter ferrite δ phases (Figure 5.12c,d,g), while the outer layer contained regular austenite grains (Figure 5.12h). The inhomogeneity in the microstructure is attributed to the Gaussian energy distribution of the laser beam and different cooling conditions in different regions of the sample. It is reasonable to assume that the outer layer was subjected to a lower peak temperature and faster cooling than the inner region when exposed to the laser beam.

The effects of processing parameters on the mechanical properties of 304L stainless steel walls fabricated using L-DED have also been investigated (Wang et al. 2016). Columnar δ-ferrite dendrites were observed to be distributed within the austenite matrix. During the rapid solidification of the austenitic stainless steel, the transformation of δ-ferrite to austenite was seldom fully complete, resulting in

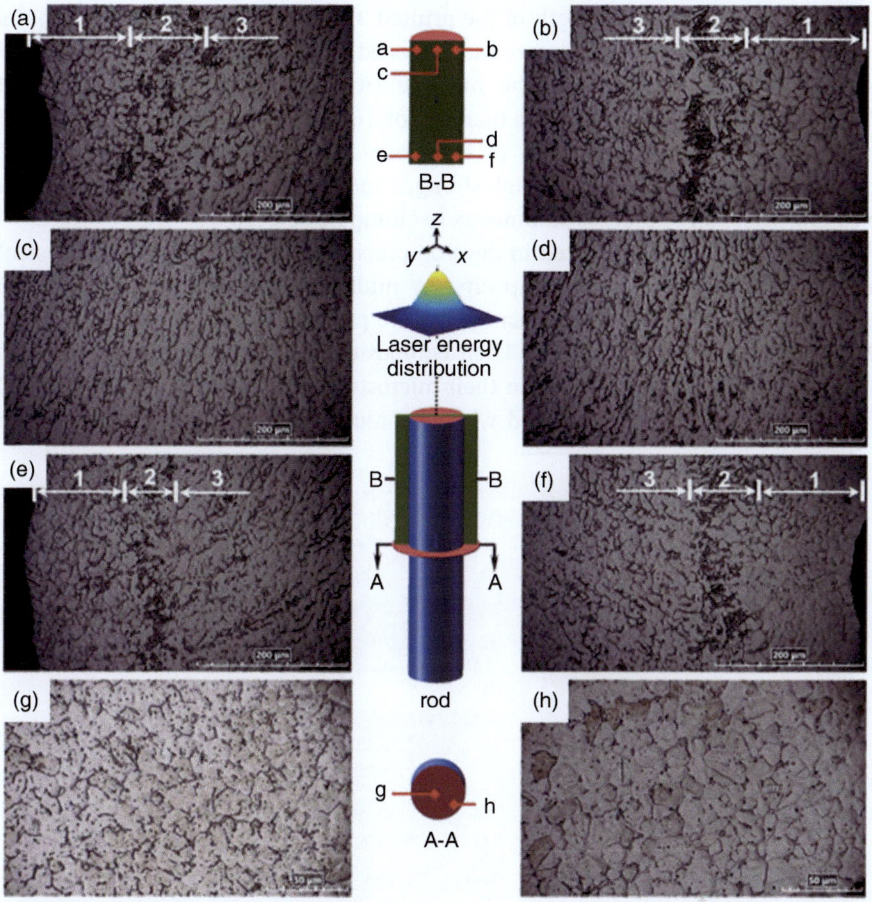

Figure 5.12 Different positions of the microstructure of a 316L part printed by L-DED. (a) Position a, (b) position b, (c) position c, (d) position d, (e) position e, (f) position f, (g) position g, and (h) position h. 1, 2, and 3 indicate the outer layer, transition zone, and internal positions, respectively. The deposited rods were cut into two equal parts, and the upper part was cut longitudinally. Source: Weng et al. (2019)/reproduced with permission from Elsevier.

the formation of residual ferrite dendrites in the austenite matrix upon solidification. At the bottom of the walls, columnar grains that grew along the direction of the predominant heat flow path were formed, which was attributed to the deposition of subsequent printing layers. The grains were more randomly oriented at the top of the walls, resembling an equiaxed texture.

316, 316L, and 304L stainless steel parts produced by L-DED were found to exhibit lower tensile strength when the applied tensile loading direction is parallel to the build direction than when it is normal to the build direction. This physical property was attributed to the different types of microstructures formed.

In one of the studies, a 304L alloy printed by L-DED exhibited an ultimate tensile strength of 734 MPa, a yield strength of 553 MPa, and an elongation of 50%

(Smith et al. 2019). The strength of the printed 304L part was similar to that of its strain-hardened forged counterpart and exceeded the minimum value of a standard annealed wrought 304L part. The high density of dislocations in the printed 304L part led to a yield strength increase of 166–191 MPa, while compositional micro-segregation was predicted to contribute an additional yield strength of 123–135 MPa. A combination of high strength and elongation can be obtained with a careful selection of process parameters to eliminate defects.

L-DED has also been utilized to develop precipitation-hardened maraging steel alloys because of its intrinsic capacity for multi-material printing (Kürnsteiner et al. 2017). Fe–19Ni–xAl graded samples were produced with x ranging from 0 to 25 at%, as shown in Figure 5.13a. Their compositionally graded structure revealed the effect of aluminum addition on their microstructure. Notably, a coarse-grained ferritic microstructure was formed when the aluminum concentration was above

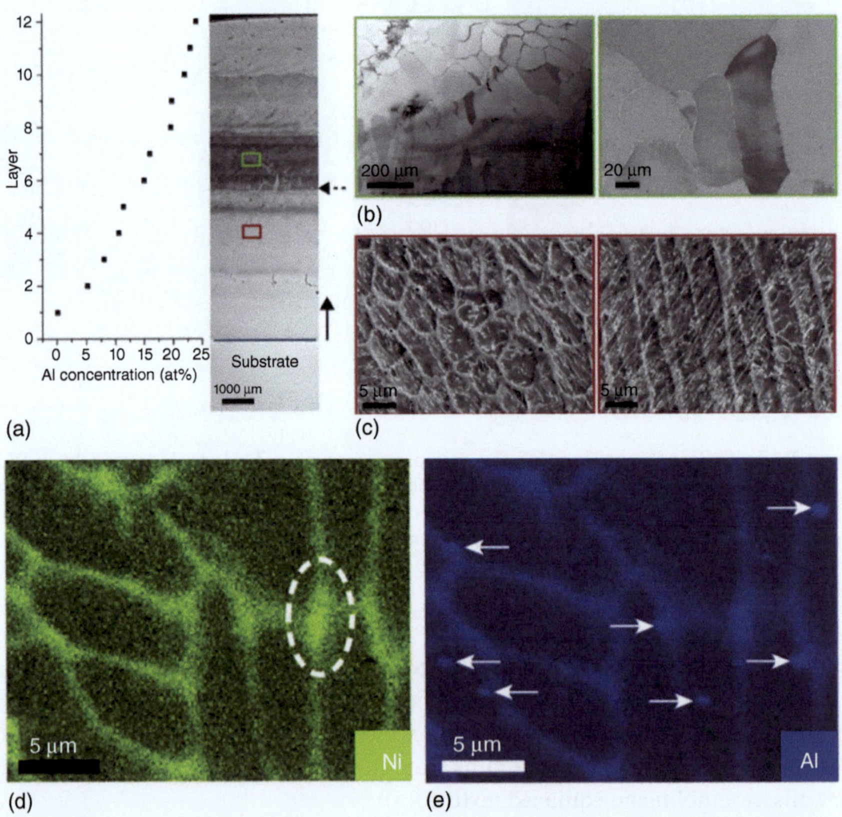

Figure 5.13 (a) Overview of a compositionally graded Fe–19Ni–xAl structure, illustrated by Al concentration plotted against the layer number and the corresponding optical microscope (OM) image, (b) SEM images of coarse-grained ferritic regions with 15 at% Al, and (c) SEM images of fine-grained martensitic regions with 10 at% Al. Energy dispersive spectroscopy (EDS) mappings of the regions with 10 at% Al showing the enrichment of (d) Ni and (e) Al at dendrite boundaries. Source: Kürnsteiner et al. (2017)/reproduced with permission from Elsevier.

15 at% (Figure 5.13b). Meanwhile, a fine-grained martensitic microstructure with fine cellular dendrites was obtained for aluminum concentration below 15 at% (Figure 5.13c).

In the Fe–19Ni–*x*Al graded samples, nickel and aluminum atoms were segregated at the dendrite boundaries during solidification, as shown in Figure 5.13d,e. Additionally, nano-scaled aluminum oxide particles were induced in the martensitic microstructure. Al-rich precipitates began to form once the aluminum content exceeded 5 at%, and the density of precipitation increased with the aluminum concentration. In the compositionally graded samples with an aluminum content ranging from 0 to 25 at%, a martensitic structure was observed up to 15 at% aluminum. A minimum of 5 at% aluminum was required to induce precipitation strengthening.

The layer-wise manufacturing process of L-DED enables the tailoring of the microstructures of the printed samples. A Fe19Ni5Ti bulk sample with hard (dark areas) and soft zones (bright areas) was produced (Figure 5.14a) through idle cooling and cyclic heating (Kürnsteiner et al. 2020), as shown in Figure 5.14b. This compound sample with hard and soft layers (akin to Damascus steel) exhibited enhanced ultimate strength and elongation compared to samples fabricated without interval cooling (Figure 5.14c). The high cooling rate of L-DED allows maraging steels to accommodate a soft martensitic matrix with retained austenite.

The fractions of martensite and austenite in both the hard and soft regions were found to be similar (Figure 5.14d). Note that the martensitic matrix was formed during the internal cooling stage. Subsequent deposition of layers triggered precipitation within the martensite (Figure 5.14e), inducing a high density of precipitates within the martensitic matrix (Figure 5.14g). Meanwhile, titanium atoms were embedded in the austenite and served as a stabilizer (Figure 5.14f). The thermal history of the sample was adjusted through proper pauses at certain deposited layers to control the martensite transformation and achieve *in situ* precipitation strengthening. Such an approach can be extended to fabricating other precipitation-hardened alloys through metal powder–based AM methods.

5.7.4 Others

5.7.4.1 Aluminum Alloys

Research on aluminum alloys printed by L-DED is limited and encounters more challenges than studies based on LPBF. First, aluminum is characterized by a high surface reflectivity and high thermal conductivity, which are unfavorable for material deposition and melt pool formation. Thus, a high laser energy density is required to create and stabilize the melt pool. Second, an excessively high energy density may promote the evaporation of alloy elements with low boiling points, such as zinc and magnesium, which are commonly present in aluminum alloys. The evaporation of such elements can result in material loss, formation of gas pores, and degradation of the mechanical properties of the deposited aluminum alloy. As a result, fluctuations in the alloy composition and hot cracking may occur easily. Third, due to the wide range of solidification temperatures and high thermal expansion coefficients of

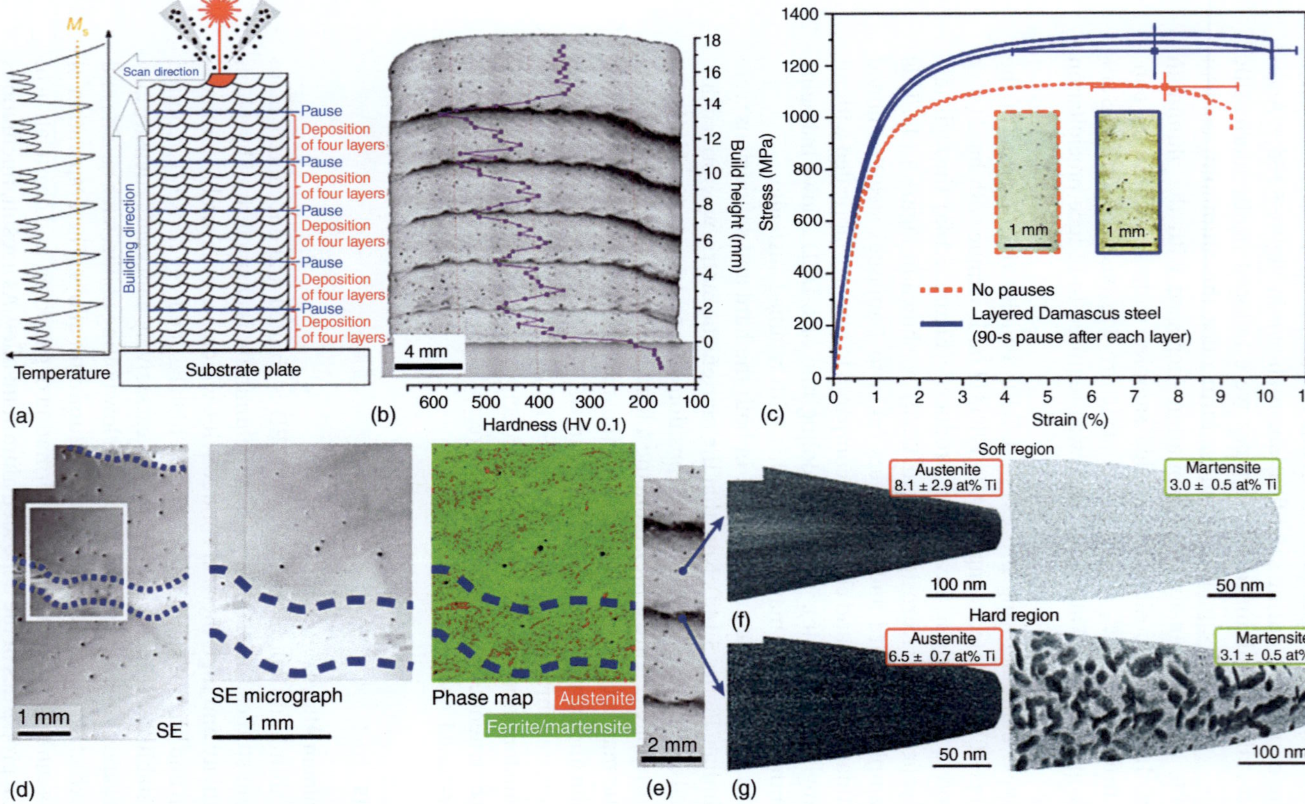

Figure 5.14 (a) Schematic of an L-DED process, including a simple sketch of the temperature profile. The process was paused for the sample to cool after four layers were built. (b) OM image showing the dark zones at the pause positions and their corresponding hardness profiles. (c) Tensile stress–strain responses. (d) SEM image featuring the hard and soft regions and their corresponding phase maps. Atom probe tomography results: (e) OM image indicating the detected positions and Ti atom maps for austenite and martensite in the (f) soft and (g) hard regions. Source: Kürnsteiner et al. (2020)/reproduced with permission from Springer Nature.

aluminum alloys, rapid solidification may be associated with high levels of residual stresses, culminating in cracking and deformation of the as-printed parts. Finally, the low density of aluminum alloys easily makes the powder fly before melting.

A thin-wall Al–Si10–Mg alloy fabricated by L-DED achieved a tensile strength of 360 MPa, an elongation of 12.7%, and it possessed a microstructure consisting of α-aluminum dendrites and eutectic silicon (Wang et al. 2019). In each layer of the Al–Si10–Mg alloy, three representative zones were observed (Figure 5.15a). A fine columnar dendrite zone was created in the upper region of the melt pool due to sufficient heat being dissipated to the atmosphere and previously deposited layers (Figure 5.15b). Meanwhile, a coarse columnar dendrite zone was formed at the bottom of the melt pool due to a low cooling rate, which was attributed to heat supplied by the upper part of the melt pool and heat accumulation in the preceding layers (Figure 5.15c).

The HAZ was induced by the thermal attack from newly deposited layers to previously deposited layers, which acted as an intrinsic heat treatment leading to the precipitation of silicon particles along the aluminum–silicon boundary and their subsequent spheroidization (Figure 5.15d). As a result, a rod-like phase was formed under a high energy input (Figure 5.15e). It comprised a series of small columnar dendrites surrounded by severely altered cellular dendrites (Figure 5.15f).

The densification of an Al–Si10–Mg part fabricated by L-DED is significantly affected by the oxygen concentration of the atmosphere, which is a more prominent factor in L-DED than in LPBF. Its porosity increased to 1.89% under an oxygen level of 900 ppm (Figure 5.15g), and its mechanical properties were modified by changing the SDAS under different heat input values. As the laser power increased from 480 to 910 W, a smaller SDAS was induced under the increased cooling rate, resulting in enhanced strength and elongation of the printed part (Figure 5.15h).

Preheating treatment was applied to the sample printed with a laser power of 1200 W. The cooling rate was reduced due to the suppressed temperature gradient, leading to reduced residual stress and mechanical strength. The microstructure of Al 5083 fabricated by L-DED consisted of both equiaxed and columnar grains, which comprised a small amount of pure hexagonal close packed (HCP)-magnesium phase surrounded by a matrix composed of the FCC-aluminum phase (Svetlizky et al. 2020). The deposited alloy exhibited an ultimate tensile strength of 244 MPa, a yield strength of 117 MPa, and an elongation of 18.3%.

5.7.4.2 Copper and Its Alloys

L-DED has been applied to print pure copper, copper–nickel, GRCop-84, nickel–aluminum–bronze (NAB) alloys, etc. A Cu–9Al–5Fe–5Ni alloy was fabricated through L-DED, which achieved a density of over 99% and a high tensile strength of 769–949 MPa (Li et al. 2021). The microstructure of the printed alloy consisted of a metastable martensite β* phase, a Widmanstätten α phase, and Fe_3Al and NiAl nanoprecipitates (Figure 5.16). In addition, the samples possessed a hierarchical microstructure comprising microscale cellular structures, sub-microscale grains, and nanoscale precipitates and twins.

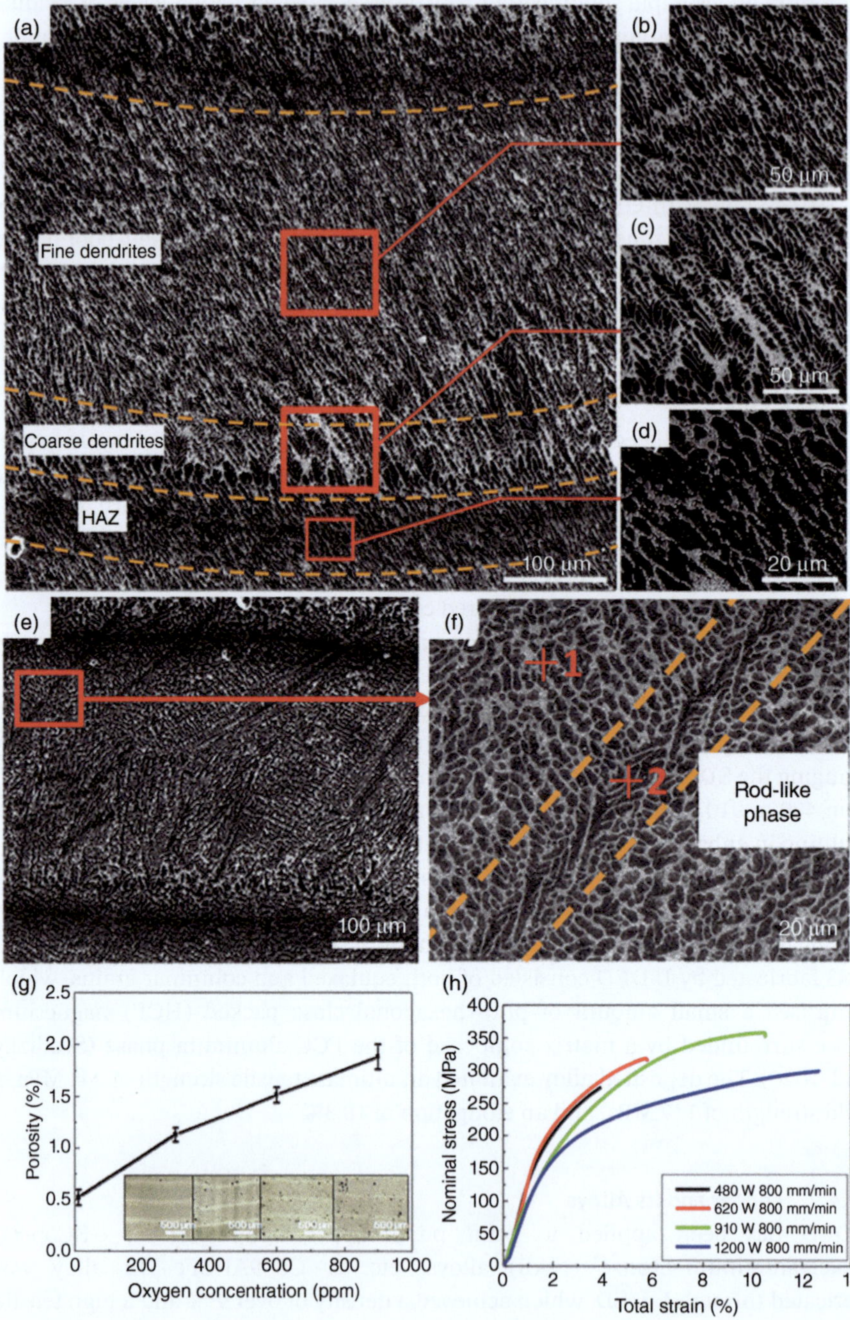

Figure 5.15 (a) SEM image of deposited Al–Si10–Mg with a heat input of 480 W and an enlarged view of its (b) fine columnar dendrite zone, (c) coarse dendrite zone, and (d) HAZ. (e) SEM image of the sample with a heat input of 1200 W and (f) an enlarged view of its rod-like phase. (g) Effect of oxygen concentration on porosity and the corresponding optical micrograph and (h) stress–strain curves with different heat input values. Source: Wang et al. (2019)/reproduced with permission from Taylor & Francis Group.

The cellular structures consisted of strong β* phases and ductile α phases distributed at the boundaries of β* grains (Figure 5.16a,b), which contributed to the high strength and ductility of the printed alloy. In particular, a cellular structure with dimensions of 20–40 μm facilitated Hall–Petch type strengthening through the highly dense α/β* boundaries. The Fe_3Al and NiAl precipitates formed a core/shell structure with a radius of ~50 nm, which was attributed to their similar FCC crystal structures and low lattice mismatch (Figure 5.16c). The high density of hard precipitates impeded dislocation slip, which led to precipitation strengthening.

Aluminum was enriched in the boundaries of adjacent β* plates, thereby reducing the stacking fault energy (SFE) (Figure 5.16e). As a result, nanotwins were formed at the boundaries of β* plates (Figure 5.16d). The twin boundaries comprised partial dislocations, which facilitated dislocation strengthening, and the alloying elements (aluminum, iron, and nickel) in the copper matrix also contributed to solid solution strengthening.

Because of the synergistic effects of the Hall–Petch type strengthening, precipitation strengthening, dislocation strengthening, and solid solution strengthening, the printed alloy possessed a yield strength of 713 MPa, which was 160% higher than its cast counterparts (Figure 5.16f). It also exhibited high strength and good ductility compared to its counterparts prepared by other AM processes.

L-DED is also widely applied in the fabrication of multi-materials such as copper/steel and copper/nickel to harness the corrosion resistance and high thermal conductivity of copper. For example, a copper/steel multi-material was manufactured via L-DED using NAB and 15-5 PH powders, which incorporates the superior corrosion resistance of NAB and the outstanding mechanical performance of 15-5 PH (Figure 5.17a) (Li et al. 2022).

The NAB/15-5 PH interface, which consisted of an interlayer of Fe_xAl dendrites, contained no cracks or lack-of-fusion defects (Figure 5.17b,c). The NAB region comprised equiaxed grains with weak textures, while the 15-5 PH region exhibited columnar dendritic grains growing epitaxially along the build direction (Figure 5.17c). Compared to other copper/steel multi-materials fabricated by AM, the multi-material sample displayed a high ultimate tensile strength of 754.6 MPa in the transverse direction and 854.6 MPa in the longitudinal direction (Figure 5.17e).

In an oceanic environment, a thick passive film was formed on the NAB surface, protecting the underlying material from corrosion. However, Cl-rich seawater can damage the passive Cr_2O_3 film formed in the 15-5 PH region. The average pitting depth of NAB and 15-5 PH was measured to be 0.25 and 1.4 μm, respectively, indicating that NAB has higher pitting resistance than 15-5 PH (Figure 5.17f).

A copper/nickel bimetallic structure was fabricated via L-DED using a GRCop-84 copper alloy and Inconel 718 to enhance the thermophysical performance of aerospace components (Figure 5.18a) (Onuike et al. 2018). A gradual transition between Inconel 718 and GRCop-84 at their diffused interface was achieved using the compositional gradation technique (Figure 5.18b,c). Columnar grain structures were formed at the interfaces with the accumulation of Cr_2Nb precipitates along the grain boundaries and the substrate–deposit interface. The average thermal

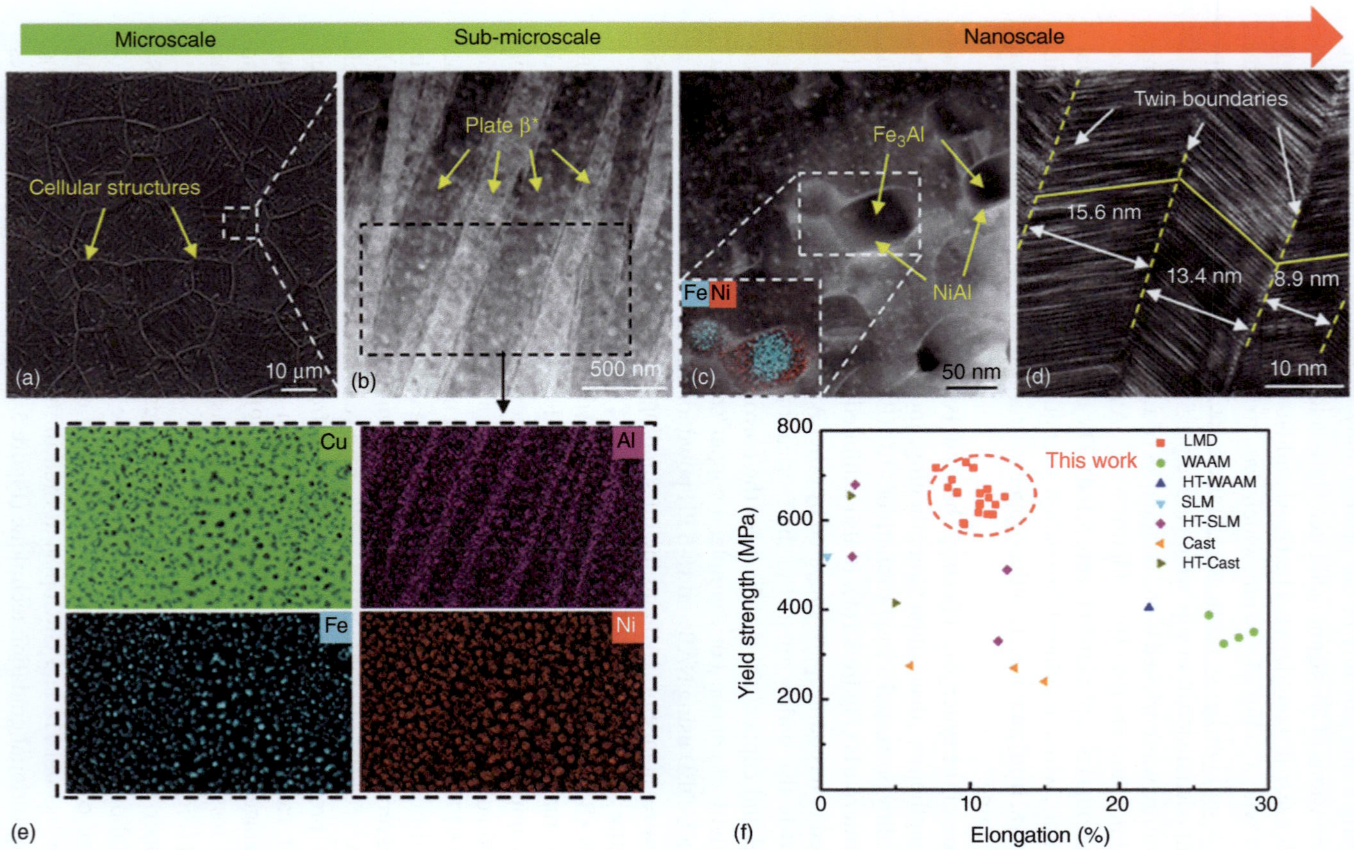

Figure 5.16 Hierarchical microstructure of a printed Cu–9Al–5Fe–5Ni alloy, including (a) cellular structures, (b) the plate martensite β* phase, (c) Fe3Al and NiAl nanoprecipitates, and (d) nanotwins at the boundaries of the β* plates. (e) Elemental distribution in the rectangular region indicated in image (b). (f) Comparison of the mechanical properties of nickel–aluminum–bronze alloys fabricated by L-DED and other processes. Source: Li et al. (2021)/reproduced with permission from Elsevier.

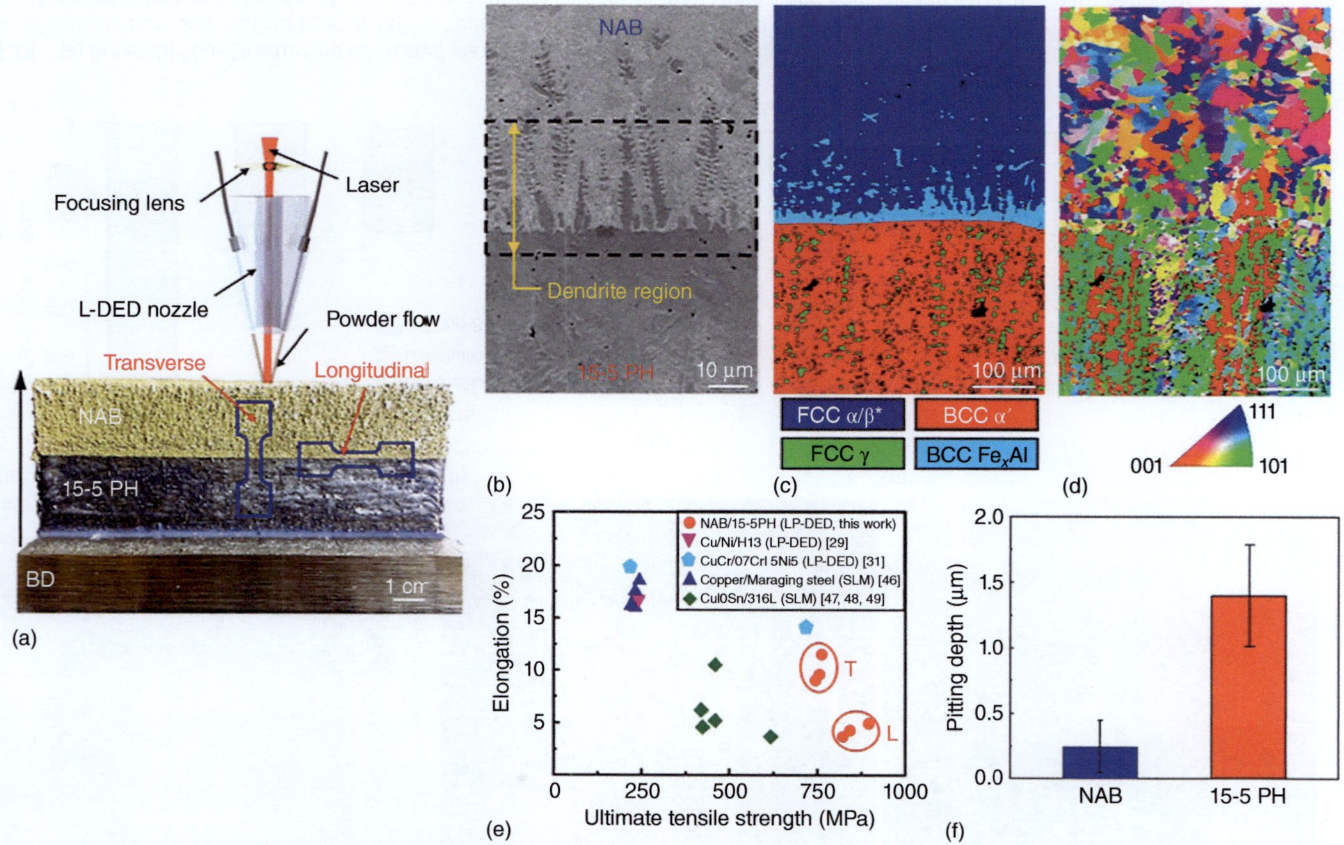

Figure 5.17 Multi-material samples manufactured by L-DED. (a) Schematic of the L-DED process, (b) SEM image, (c) phase distribution, and (d) IPF map of an NAB/15-5 PH multi-material sample. (e) Comparison of the mechanical properties of copper/steel multi-materials fabricated by AM. (f) Comparison of the pitting depths of NAB and 15-5 PH. Source: Li et al. (2022)/reproduced with permission from Elsevier.

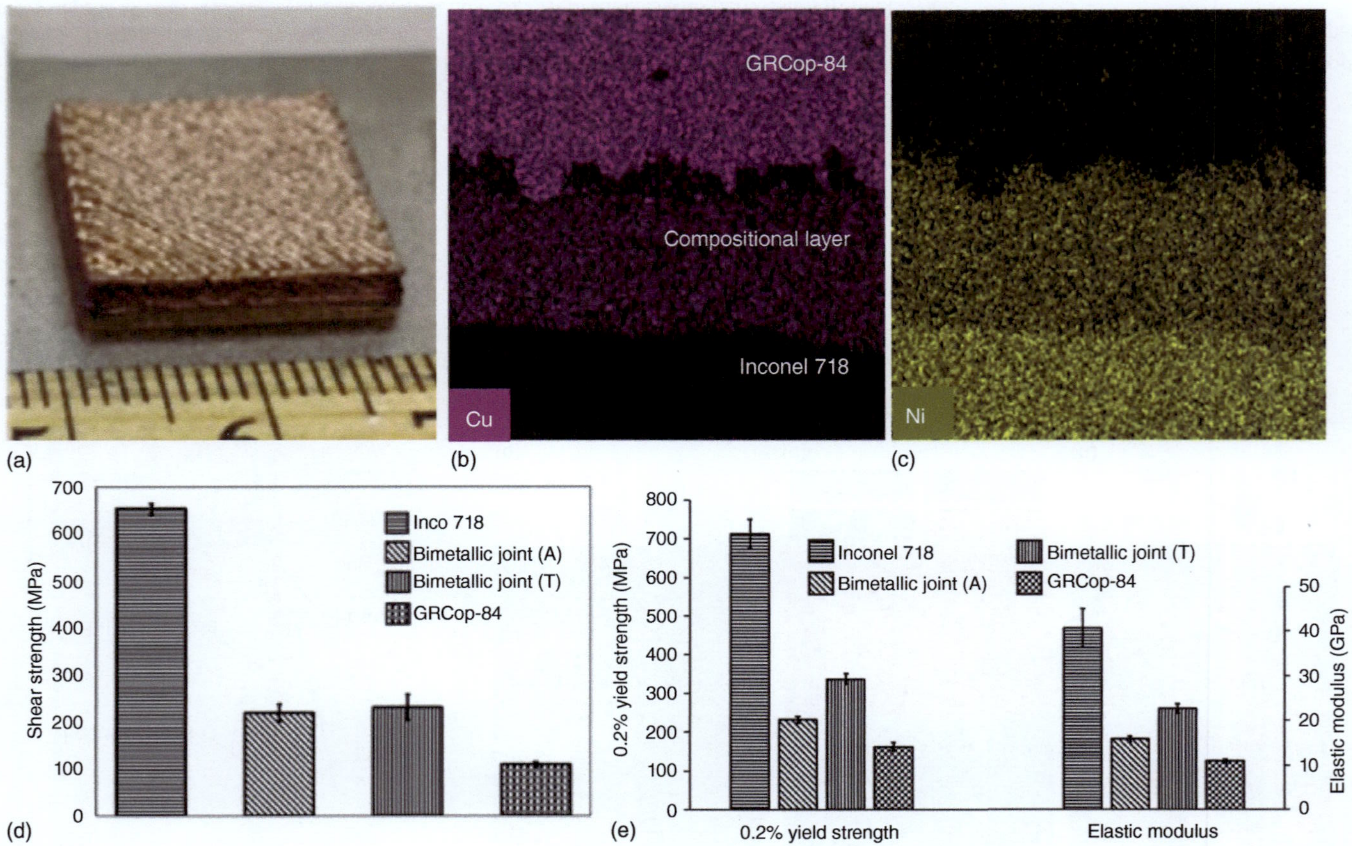

Figure 5.18 (a) Bimetallic Cu–Ni structure fabricated via L-DED; the elemental distribution of (b) Cu and (c) Ni at the Cu–Ni interface indicates the gradual transition of composition. Source: Onuike et al. (2018)/reproduced with permission from ELSEVIER. (d) Shear strength and (e) compression properties of the bimetallic structure before and after one thermal cycle. Source: Reproduced from Onuike and Bandyopadhyay (2019)/with permission from Elsevier.

diffusivity of the bimetallic structure improved by 250% from 3.20 to 11.33 mm^2/s for temperatures ranging from 50 to 300 °C. The thermal conductivity of the bimetallic structures increased by almost 300% compared to that of Inconel 718. A peak hardness value of 2.95 GPa was obtained in the HAZ.

The influence of thermal cycling on the shear and compressive strength of a bimetallic GRCop-84/Inconel 718 joint was evaluated (Onuike and Bandyopadhyay 2019). The shear strength of the bimetallic joint before and after thermal cycling was measured to be 220 and 231 MPa, respectively (Figure 5.18d). After thermal cycling, no crack was observed at the interface of the bimetallic joint, indicating that it did not undergo thermal degradation. On the other hand, the compressive yield strength of the Inconel 718 joint before and after thermal cycling was 232 and 337 MPa, respectively (Figure 5.18e), and such an improvement was attributed to the precipitation of Cr_2Nb particles in the copper matrix and the strengthening effect of gamma phases in Inconel 718.

Pure copper processed by L-DED exhibits a graded microstructure with coarse equiaxed grains of 10–20 μm, Cu_2O macro-precipitates of 200 nm, and Cu_2O nano-precipitates of 2–3 nm due to multiple thermal cycles (Zhong et al. 2018), during which oxygen is solubilized in the melting pool. In the primary thermal cycle, some oxygen precipitates and forms macro-sized Cu_2O, while nano-sized Cu_2O is formed in the later thermal cycles when the remaining oxygen precipitates. The nano-sized precipitates provide better precipitation hardening effects and contribute to the greater hardness of the printed parts.

5.7.4.3 High-Entropy Alloys

L-DED can not only employ pre-alloyed HEA powders but also simultaneously deliver different pure metal powders from multiple hoppers to print HEA products through *in situ* alloying. Elemental powders are delivered, melted, and deposited onto a substrate, forming an alloy product. Homogeneous distribution of the elements can be obtained through optimizing the process parameters and remelting via repetitive scanning, thereby enhancing the flexibility of the composition design of HEAs for materials development. The printing of HEA products through L-DED for powders such as AlCoCrFeNi, AlCrCuFeNi, AlCoFeNiCu, CoCrFeMnNi, and TiZrNbTa has been validated. Such products showed a yield strength of 194–518 MPa, an ultimate tensile strength of 535–660 MPa, an elongation of 19.8–55%, and a microhardness of 195–860 HV (Han et al. 2020).

L-DED exhibits great potential to develop functionally graded HEA parts through the *in situ* alloying of elemental powders. A compositionally graded $Al_xCrCuFeNi_2$ ($0 \leq x \leq 1.5$) sample was fabricated through L-DED to investigate the influence of systematic compositional changes on its microstructural evolution, mechanical properties, and magnetic properties (Figure 5.19a) (Borkar et al. 2016).

With an increase in aluminum content from 0 to 1.5 (mole fraction), the printed $Al_xCrCuFeNi_2$ HEA products exhibited a transition in their crystal structure from overwhelmingly FCC (98%) and minimally BCC (2%) (Figure 5.19b) to predominantly BCC (80%) and substantially FCC (20%) (Figure 5.19c). The volume fraction of the ordered $L1_2$ FCC phase also increased and then decreased.

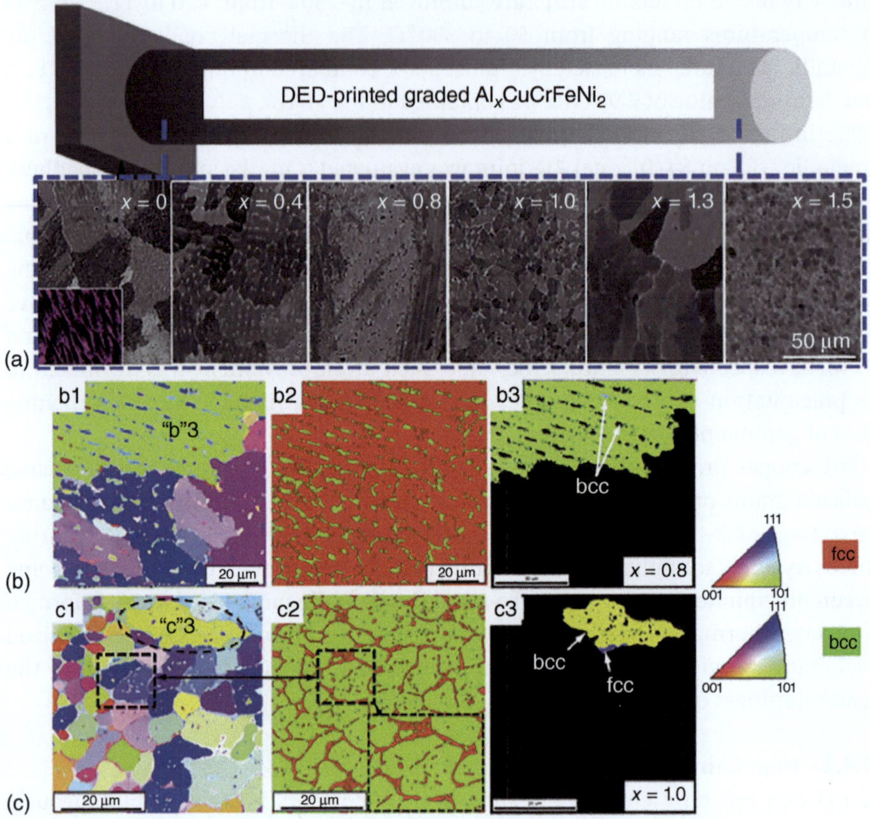

Figure 5.19 Microstructure of compositionally graded $Al_xCuCrFeNi_2$ HEA products printed by L-DED. (a) Morphology and (b) EBSD results for $Al_{0.8}CuCrFeNi_2$: (b1) various orientations and the inverse pole figure map, (b2) phase map demarking FCC and BCC regions, and (b3) the inverse pole figure map of a selected grain in (b1). (c) EBSD results for $AlCuCrFeNi_2$: (c1) various orientations and the inverse pole figure map, (c2) phase map demarking FCC and BCC regions, and (c3) the inverse pole figure map of a selected grain in (c1). Source: Borkar et al. (2016)/reproduced with permission from ELSEVIER.

These phenomena are attributed to the addition of aluminum distorting the FCC lattice, thereby inducing the transition from a close-packed FCC phase to a more loosely packed BCC phase.

5.8 Summary

L-DED delivers feedstock and laser energy simultaneously on a freeform surface in a layer-by-layer manner. Both emerging and established manufacturing corporations have developed many commercial L-DED machines. The modular design of L-DED machines provides flexibility in selecting the printing dimension, laser source, powder feeder, printing chamber, deposition nozzle, etc., to meet practical requirements.

Available materials include various titanium and its alloys, nickel alloys, iron alloys, aluminum alloys, copper alloys, and HEAs.

L-DED has been widely employed to fabricate original parts, add new features to existing parts, and repair damaged parts for aerospace, automotive, biomedical, and maritime applications. Several vital aspects of L-DED, including material development, printing efficiency and accuracy, online monitoring and defect inspection, process mechanism, and standardization and certification, have gained increasing interest from both academia and industries. These aspects inspired the realization of the full potential of L-DED and promoted its market uptake.

Repairing/adding new features to an existing part is a distinctive advantage of L-DED that holds a promising future. Repairing high-value or unique parts through L-DED can significantly reduce costs and lead time. Adding new features such as composite coatings to extend wear resistance or provide biocompatibility enhances the product performance.

Multi-material printing can endow a single product with a combination of multiple desirable properties by depositing different materials at designated locations, which improves its thermal conduction, corrosion resistance, wear resistance, etc. Material development is also an attractive aspect of multi-material printing because its versatile composition control permits the administration of high-throughput experiments and the rapid screening of target materials. Nevertheless, the performance of multi-material delivery devices, the interactions between multi-materials and laser sources, and multi-material printing strategies require further investigation.

Hybrid manufacturing also displays an increasing propensity to be incorporated into L-DED as printing resolutions are typically low due to the large melt pool size. Post-machining modules are expected to be integrated with traditional L-DED machines to manufacture products with the required dimensions directly. Such manufacturing strategies that involve both additive and subtractive manufacturing techniques must be optimized to balance the precision and processing efficiency.

References

Ahsan, M.N., Pinkerton, A.J., Moat, R.J., and Shackleton, J. (2011). A comparative study of laser direct metal deposition characteristics using gas and plasma-atomized Ti–6Al–4V powders. *Materials Science and Engineering A* 528 (25, 26): 7648–7657.

Åkerfeldt, P., Antti, M.L., and Pederson, R. (2016). Influence of microstructure on mechanical properties of laser metal wire-deposited Ti–6Al–4V. *Materials Science and Engineering A* 674: 428–437.

Amirabdollahian, S., Deirmina, F., Harris, L. et al. (2021). Towards controlling intrinsic heat treatment of maraging steel during laser directed energy deposition. *Scripta Materialia* 201: 113973.

Attar, H., Ehtemam-Haghighi, S., Kent, D. et al. (2017). Comparative study of commercially pure titanium produced by laser engineered net shaping, selective laser melting and casting processes. *Materials Science and Engineering A* 705: 385–393.

Baufeld, B., Brandl, E., and Van der Biest, O. (2011). Wire based additive layer manufacturing: comparison of microstructure and mechanical properties of Ti–6Al–4V components fabricated by laser-beam deposition and shaped metal deposition. *Journal of Materials Processing Technology* 211 (6): 1146–1158.

Bimber, B.A., Hamilton, R.F., Keist, J., and Palmer, T.A. (2016). Anisotropic microstructure and superelasticity of additive manufactured NiTi alloy bulk builds using laser directed energy deposition. *Materials Science and Engineering A* 674: 125–134.

Borkar, T., Gwalani, B., Choudhuri, D. et al. (2016). A combinatorial assessment of $Al_xCrCuFeNi_2$ ($0 < x < 1.5$) complex concentrated alloys: microstructure, microhardness, and magnetic properties. *Acta Materialia* 116: 63–76.

Brandl, E. and Greitemeier, D. (2012). Microstructure of additive layer manufactured Ti–6Al–4V after exceptional post heat treatments. *Materials Letters* 81: 84–87.

Brown, C.O., Breinan, E.M., and Kear, B.H. (1982). Method for fabricating articles by sequential layer deposition. US Patent No. 4, 323,756A. Washington, DC: U.S. Patent and Trademark Office, filed 29 October 1979 and issued 06 April 1982.

Buongiorno, A. (1995). Laser/powdered metal cladding nozzle. US Patent No. 5, 477,026A. Washington, DC: U.S. Patent and Trademark Office, filed 27 January 1994 and issued 19 December 1995.

Carroll, B.E., Palmer, T.A., and Beese, A.M. (2015). Anisotropic tensile behavior of Ti–6Al–4V components fabricated with directed energy deposition additive manufacturing. *Acta Materialia* 87: 309–320.

Chen, Y., Clark, S.J., Huang, Y. et al. (2021). *In situ* X-ray quantification of melt pool behaviour during directed energy deposition additive manufacturing of stainless steel. *Materials Letters* 286: 129205.

Denlinger, E.R., Heigel, J.C., Michaleris, P., and Palmer, T.A. (2015). Effect of inter-layer dwell time on distortion and residual stress in additive manufacturing of titanium and nickel alloys. *Journal of Materials Processing Technology* 215: 123–131.

Dinda, G.P., Dasgupta, A.K., and Mazumder, J. (2009). Laser aided direct metal deposition of Inconel 625 superalloy: microstructural evolution and thermal stability. *Materials Science and Engineering A* 509 (1, 2): 98–104.

Ganesh, P., Kaul, R., Paul, C.P. et al. (2010). Fatigue and fracture toughness characteristics of laser rapid manufactured Inconel 625 structures. *Materials Science and Engineering A* 527 (29, 30): 7490–7497.

Gey, N. and Humbert, M. (2002). Characterization of the variant selection occurring during the $\alpha \rightarrow \beta \rightarrow \alpha$ phase transformations of a cold rolled titanium sheet. *Acta Materialia* 50 (2): 277–287.

Guévenoux, C., Hallais, S., Charles, A. et al. (2020). Influence of interlayer dwell time on the microstructure of Inconel 718 laser cladded components. *Optics & Laser Technology* 128: 106218.

Hammeke, A.W. (1988). Laser spray nozzle and method. US Patent No. 4, 724,299A. Washington, DC: U.S. Patent and Trademark Office, filed 15 April 1987 and issued 09 February 1988.

Han, C., Fang, Q., Shi, Y. et al. (2020). Recent advances on high-entropy alloys for 3D printing. *Advanced Materials* 32 (26): 1903855.

He, X. and Mazumder, J. (2007). Transport phenomena during direct metal deposition. *Journal of Applied Physics* 101 (5): 053113.

He, X., Norris, J. T., Fuerschbach, P. W., & DebRoy, T. (2005). Alloying element vaporization and liquid metal expulsion during laser microjoining of stainless steel with short pulse. In: *Trends in Welding Research: Proceedings of the 7th International Conference*, Callaway Gardens Resort, Pine Mountain, Georgia, USA, pp. 109–114.

Hofmeister, W., Wert, M., Smugeresky, J. et al. (1999). Investigating solidification with the laser-engineered net shaping (LENSTM) process. *JOM Journal of the Minerals Metals and Materials Society* 51 (7): 1–6.

Isaac, H. (1942). Method of forming structures wholly of fusion deposited weld metal. US Patent No. 2, 299,747A. Washington, DC: U.S. Patent and Trademark Office, filed 09 April 1941 and issued 27 October 1942.

Keist, J.S. and Palmer, T.A. (2016). Role of geometry on properties of additively manufactured Ti-6Al-4V structures fabricated using laser based directed energy deposition. *Materials and Design* 106: 482–494.

Kim, H., Cong, W., Zhang, H.C., and Liu, Z. (2017). Laser engineered net shaping of nickel-based superalloy Inconel 718 powders onto AISI 4140 alloy steel substrates: interface bond and fracture failure mechanism. *Materials* 10 (4): 341.

Kratky, A. (1937). Production of hard metal alloys. Patent# US, 2076952A, filed 21 August 1935 and issued 13 April 1937.

Kruth, J.P., Froyen, L., Van Vaerenbergh, J. et al. (2004). Selective laser melting of iron-based powder. *Journal of Materials Processing Technology* 149 (1–3): 616–622.

Kürnsteiner, P., Wilms, M.B., Weisheit, A. et al. (2017). Massive nanoprecipitation in an Fe-19Ni-*x*Al maraging steel triggered by the intrinsic heat treatment during laser metal deposition. *Acta Materialia* 129: 52–60.

Kürnsteiner, P., Wilms, M.B., Weisheit, A. et al. (2020). High-strength Damascus steel by additive manufacturing. *Nature* 582 (7813): 515–519.

Li, P.H., Guo, W.G., Huang, W.D. et al. (2015). Thermomechanical response of 3D laser-deposited Ti-6Al-4V alloy over a wide range of strain rates and temperatures. *Materials Science and Engineering A* 647: 34–42.

Li, Z., Chen, J., Sui, S. et al. (2020). The microstructure evolution and tensile properties of Inconel 718 fabricated by high-deposition-rate laser directed energy deposition. *Additive Manufacturing* 31: 100941.

Li, B., Zheng, H., Han, C., and Zhou, K. (2021). Nanotwins-containing microstructure and superior mechanical strength of a Cu–9Al–5Fe–5Ni alloy additively manufactured by laser metal deposition. *Additive Manufacturing* 39: 101825.

Li, B., Han, C., Lim, C.W.J., and Zhou, K. (2022). Interface formation and deformation behaviors of an additively manufactured nickel-aluminum-bronze/15-5 PH multimaterial via laser-powder directed energy deposition. *Materials Science and Engineering A* 826: 142101.

Mehta, P.P., Otten, R.R., and Cooper Jr, E.B. (1988). Method and apparatus for repairing metal in an article. US Patent No. 4, 743,733A. Washington, DC: U.S. Patent and Trademark Office, filed 01 October 1984 and issued 10 May 1988.

Mok, S.H., Bi, G., Folkes, J., and Pashby, I. (2008). Deposition of Ti–6Al–4V using a high power diode laser and wire, Part I: investigation on the process characteristics. *Surface and Coatings Technology* 202 (16): 3933–3939.

Oh, M.S., Lee, J.Y., and Park, J.K. (2004). Continuous cooling β-to-α transformation behaviors of extra-pure and commercially pure Ti. *Metallurgical and Materials Transactions A* 35 (10): 3071–3077.

Onuike, B. and Bandyopadhyay, A. (2019). Bond strength measurement for additively manufactured Inconel 718-GRCop84 copper alloy bimetallic joints. *Additive Manufacturing* 27: 576–585.

Onuike, B., Heer, B., and Bandyopadhyay, A. (2018). Additive manufacturing of Inconel 718—Copper alloy bimetallic structure using laser engineered net shaping (LENS™). *Additive Manufacturing* 21: 133–140.

Parimi, L.L., Ravi, G.A., Clark, D., and Attallah, M.M. (2014). Microstructural and texture development in direct laser fabricated IN718. *Materials Characterization* 89: 102–111.

Paul, C.P., Ganesh, P., Mishra, S.K. et al. (2007). Investigating laser rapid manufacturing for Inconel-625 components. *Optics & Laser Technology* 39 (4): 800–805.

Paydas, H., Mertens, A., Carrus, R. et al. (2015). Laser cladding as repair technology for Ti–6Al–4V alloy: influence of building strategy on microstructure and hardness. *Materials and Design* 85: 497–510.

Peyre, P., Aubry, P., Fabbro, R. et al. (2008). Analytical and numerical modelling of the direct metal deposition laser process. *Journal of Physics D: Applied Physics* 41 (2): 025403.

Puppala, G., Moitra, A., Sathyanarayanan, S. et al. (2014). Evaluation of fracture toughness and impact toughness of laser rapid manufactured Inconel-625 structures and their co-relation. *Materials and Design* 59: 509–515.

Ramakrishnan, A. and Dinda, G.P. (2019a). Direct laser metal deposition of Inconel 738. *Materials Science and Engineering A* 740: 1–13.

Ramakrishnan, A. and Dinda, G.P. (2019b). Microstructure and mechanical properties of direct laser metal deposited Haynes 282 superalloy. *Materials Science and Engineering A* 748: 347–356.

Shim, D.-S., Baek, G.-Y., Seo, J.-S. et al. (2016). Effect of layer thickness setting on deposition characteristics in direct energy deposition (DED) process. *Optical & Laser Technology* 86: 69–78.

Simchi, A. (2006). Direct laser sintering of metal powders: mechanism, kinetics and microstructural features. *Materials Science and Engineering A* 428 (1-2): 148–158.

Smith, T.R., Sugar, J.D., San Marchi, C., and Schoenung, J.M. (2019). Strengthening mechanisms in directed energy deposited austenitic stainless steel. *Acta Materialia* 164: 728–740.

Sun, G., Zhou, R., Lu, J., and Mazumder, J. (2015). Evaluation of defect density, microstructure, residual stress, elastic modulus, hardness and strength of laser-deposited AISI 4340 steel. *Acta Materialia* 84: 172–189.

Svetlizky, D., Zheng, B., Buta, T. et al. (2020). Directed energy deposition of Al 5xxx alloy using Laser Engineered Net Shaping (LENS®). *Materials and Design* 192: 108763.

Svetlizky, D., Das, M., Zheng, B. et al. (2021). Directed energy deposition (DED) additive manufacturing: physical characteristics, defects, challenges and applications. *Materials Today* 49: 271–295.

Thomas, M., Malot, T., and Aubry, P. (2017). Laser metal deposition of the intermetallic TiAl alloy. *Metallurgical and Materials Transactions A* 48 (6): 3143–3158.

Tian, X.J., Zhang, S.Q., and Wang, H.M. (2014). The influences of anneal temperature and cooling rate on microstructure and tensile properties of laser deposited Ti–4Al–1.5 Mn titanium alloy. *Journal of Alloys and Compounds* 608: 95–101.

Todaro, C.J., Easton, M.A., Qiu, D. et al. (2020). Grain structure control during metal 3D printing by high-intensity ultrasound. *Nature Communications* 11 (1): 1–9.

Wang, L. and Felicelli, S. (2006). Analysis of thermal phenomena in LENS™ deposition. *Materials Science and Engineering A* 435: 625–631.

Wang, Y. and Shi, J. (2020). Recrystallization behavior and tensile properties of laser metal deposited Inconel 718 upon *in-situ* ultrasonic impact peening and heat treatment. *Materials Science and Engineering A* 786: 139434.

Wang, Y.D., Tang, H.B., Fang, Y.L., and Wang, H.M. (2010). Effect of heat treatment on microstructure and mechanical properties of laser melting deposited 1Cr12Ni2WMoVNb steel. *Materials Science and Engineering A* 528 (1): 474–479.

Wang, B., Li, J., Sun, J. et al. (2014). Shear localization and its related microstructural evolution in the ultrafine grained titanium processed by multi-axial compression. *Materials Science and Engineering A* 612: 227–235.

Wang, Z., Palmer, T.A., and Beese, A.M. (2016). Effect of processing parameters on microstructure and tensile properties of austenitic stainless steel 304L made by directed energy deposition additive manufacturing. *Acta Materialia* 110: 226–235.

Wang, X., Li, L., Qu, J., and Tao, W. (2019). Microstructure and mechanical properties of laser metal deposited AlSi10Mg alloys. *Materials Science and Technology* 35 (18): 2284–2293.

Wang, F., Lei, L.M., Fu, X. et al. (2020). Effect of heat treatment on microstructures and tensile properties of TA19 alloy fabricated by laser metal deposition. *Materials Science and Engineering A* 782: 139284.

Weng, F., Gao, S., Jiang, J. et al. (2019). A novel strategy to fabricate thin 316L stainless steel rods by continuous directed energy deposition in Z direction. *Additive Manufacturing* 27: 474–481.

Wolff, S., Lee, T., Faierson, E. et al. (2016). Anisotropic properties of directed energy deposition (DED)-processed Ti–6Al–4V. *Journal of Manufacturing Processes* 24: 397–405.

Wolff, S.J., Webster, S., Parab, N.D. et al. (2020). *In-situ* observations of directed energy deposition additive manufacturing using high-speed X-ray imaging. *JOM Journal of the Minerals Metals and Materials Society* 73 (1): 189–200.

Yadollahi, A., Shamsaei, N., Thompson, S.M., and Seely, D.W. (2015). Effects of process time interval and heat treatment on the mechanical and microstructural properties of direct laser deposited 316L stainless steel. *Materials Science and Engineering A* 644: 171–183.

Yang, J., Yu, H., Yin, J. et al. (2016). Formation and control of martensite in Ti–6Al–4V alloy produced by selective laser melting. *Materials and Design* 108: 308–318.

Yu, Z., Zheng, Y., Chen, J. et al. (2020). Effect of laser remelting processing on microstructure and mechanical properties of 17-4 PH stainless steel during laser direct metal deposition. *Journal of Materials Processing Technology* 284: 116738.

Zhai, Y., Lados, D.A., Brown, E.J., and Vigilante, G.N. (2016). Fatigue crack growth behavior and microstructural mechanisms in Ti–6Al–4V manufactured by laser engineered net shaping. *International Journal of Fatigue* 93: 51–63.

Zhai, Y., Lados, D.A., Brown, E.J., and Vigilante, G.N. (2019). Understanding the microstructure and mechanical properties of Ti–6Al–4V and Inconel 718 alloys manufactured by laser engineered net shaping. *Additive Manufacturing* 27: 334–344.

Zhang, K., Wang, S., Liu, W., and Shang, X. (2014). Characterization of stainless steel parts by laser metal deposition shaping. *Materials and Design* 55: 104–119.

Zhang, B., Bi, G., Wang, P. et al. (2016). Microstructure and mechanical properties of Inconel 625/nano-TiB_2 composite fabricated by LAAM. *Materials and Design* 111: 70–79.

Zhang, Y., Yang, L., Lu, W. et al. (2020). Microstructure and elevated temperature mechanical properties of IN718 alloy fabricated by laser metal deposition. *Materials Science and Engineering A* 771: 138580.

Zhang, S., Zhang, Q., Zheng, M. et al. (2021). Grain refinement and improved tensile properties of Ti5Al2Sn2Zr4Mo4Cr titanium alloy fabricated by laser solid forming. *Materials Science and Engineering A* 800: 140388.

Zheng, B., Zhou, Y., Smugeresky, J.E. et al. (2008). Thermal behavior and microstructural evolution during laser deposition with laser-engineered net shaping: part I. Numerical calculations. *Metallurgical and Materials Transactions A* 39 (9): 2228–2236.

Zheng, B., Haley, J.C., Yang, N. et al. (2019). On the evolution of microstructure and defect control in 316L SS components fabricated via directed energy deposition. *Materials Science and Engineering A* 764: 138243.

Zhong, H.Z., Li, C.G., Zhang, X.Y., and Gu, J.F. (2018). The graded microstructures evolving with thermal cycles in pure copper processed by laser metal deposition. *Materials Letters* 230: 215–218.

Ziętala, M., Durejko, T., Polański, M. et al. (2016). The microstructure, mechanical properties and corrosion resistance of 316 L stainless steel fabricated using laser engineered net shaping. *Materials Science and Engineering A* 677: 1–10.

6

Metal Binder Jetting

In metal binder jetting (MBJ), a liquid binder is selectively deposited onto layers of metal powder to bond powder particles and produce green parts. Subsequently, a sintering process is often conducted to densify such green parts. Among metal powder–based additive manufacturing (AM) techniques, MBJ exhibits excellent potential for the high-speed and low-cost fabrication of complex structures. This chapter discusses the history, fundamentals, printing process, raw materials, and equipment of the MBJ technique and the microstructures and mechanical properties of MBJ-printed parts.

6.1 History

Binder jetting, originally termed "three-dimensional printing", was invented at the Massachusetts Institute of Technology. In this method, a glycerin/water binder is jetted onto gypsum-type powder layers through thermal bubble inkjet printheads (Sachs et al. 1992). The technique was first commercialized by Z Corporation, which also introduced full-color printing capabilities to its machine models (Ian Gibson 2015).

In 1996, Extrude Hone licensed the patents from the Massachusetts Institute of Technology to fabricate metal parts, after which the company launched its first binder jetting printer system, ProMetal RTS-300. In 2005, ExOne, a spin-off of Extrude Hone, focused on printing stainless steel parts infiltrated with bronze (Wohlers 2014). One of the popular applications for the MBJ technique is to manufacture sand molds and cores for the casting industry. The interest in MBJ has grown recently, along with increased research and development of large-size part printing and high-throughput production.

6.2 Fundamentals

An illustration of the MBJ process is presented in Figure 6.1 (Mostafaei et al. 2021b). As shown in Figure 6.1a, a layer of powder is first spread on the build bed and compacted by a rotating roller. Next, a liquid binder composed of polymers in a solvent

Metal Powder–Based Additive Manufacturing, First Edition. Kun Zhou and Changjun Han.

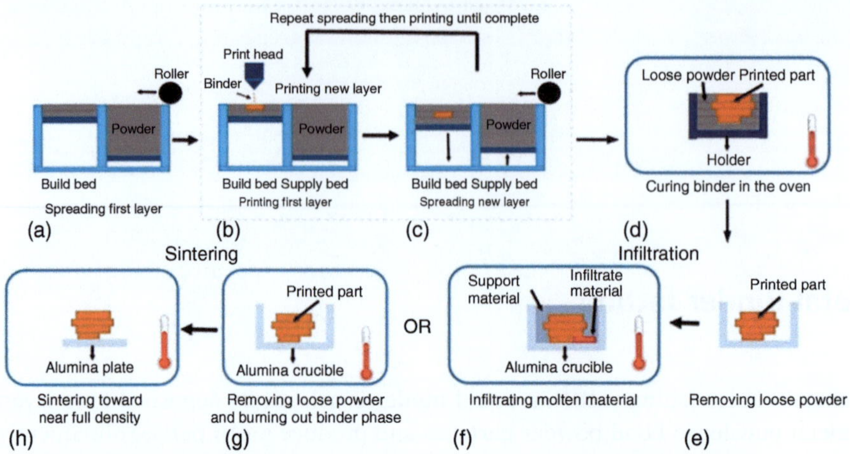

Figure 6.1 Schematic illustration of the MBJ process. (a) Preparing a powder layer on the build bed, (b) printing the binder on the powder layer through a printhead, (c) repeating the powder spreading and printing steps, (d) curing the binder, (e) depowdering the cured green parts, (f) infiltrating, and (g, h) sintering. Source: Mostafaei et al. (2021b)/reproduced with permission from Elsevier.

or aqueous solution is jetted from the printhead onto the powder layer to create a two-dimensional cross-section of the printed part (Figure 6.1b). An electrical heater or ultraviolet-band curing lamp that is installed in the powder spreading system traverses across the powder bed to partially dry or cure the layer at a uniform temperature and prepare it for the spreading of the next layer. Subsequently, the build bed moves downward by the thickness of a single layer (commonly in the range of 15–300 μm), and the powder-spreading and printing steps are repeated until the part is complete (Figure 6.1c).

Once the printing step is completed, a curing step, in which the printed part will be removed from the printer and heated until the binder is adequately dry, will be carried out to infuse the former with "green" strength (Figure 6.1d). Such printed parts have not undergone densification through infiltration or sintering. During a typical curing process, the entire build bed, which contains the loose powder and printed part, is placed in an oven and heated at 180–200 °C.

After curing, the green parts are depowdered in a densification furnace to remove loose powder via manual brushing (Figure 6.1e). The relative density of the green parts ranges from 50% to 60% after the depowdering process. Further densification of the green parts can be achieved through sintering or by infiltrating them with another molten material (Figure 6.1f–h).

MBJ is a less prominent AM technique than powder bed fusion (PBF) and directed energy deposition (DED). The advantages of the MBJ process are:

- MBJ is theoretically compatible with any powder material and boasts the widest range of material selection among all the metal powder–based AM processes. In particular, certain difficult-to-print materials (such as refractory pure metals and alloys) can be effectively printed through MBJ without encountering processing issues present in PBF.

- The MBJ process is conducted at room temperature and in an open atmosphere, thereby eliminating the residual stress, element segregation, and phase transformation, which are all common issues in PBF processes.
- No support structures are required for any part fabricated by MBJ because it is almost free of residual stress. In contrast, support structures are often employed for the overhanging features of PBF-printed parts.
- The build volume of MBJ machines is generally larger than that of other metal powder-bed AM processes.
- MBJ displays excellent high-speed and low-cost manufacturing capabilities and renders the loose powders highly recyclable.

The main disadvantages of MBJ are as follows:

- MBJ is a multistep process that requires post-processing steps (curing and densification). The development of post-processing strategies is still necessary for most MBJ powder materials.
- MBJ-printed green parts often exhibit low relative densities, which may result in significant shrinkage and geometric distortion during their densification.
- MBJ-printed parts exhibit low resolution and high surface roughness.

6.3 Printing Process

The binder plays a critical role in the MBJ process. A liquid binder is jetted onto the powder bed and infiltrates the gaps between the powder particles in each printed layer to create predefined two-dimensional patterns. The viscosity and surface tension of the binder are two essential factors when the binder is jetted through the printhead since the binder should not solidify inside the printhead. Aside from effectively bonding the powder particles, the binder must remain stable at high temperatures of up to several hundred degree Celsius during the curing process. Additionally, the binder should display good wettability with the powder particles and generate minimal residue after pyrolysis. The binder classes have been categorized, namely, acid-based binder, hydration-based binder, metal salt–based binder, inorganic binder, organic binder, phase-changed binder, solvent-based binder, and thermal-cured binder (Utela et al. 2010).

The interactions between the binder and powder particles determine the geometric accuracy, strength, and surface quality of the printed parts before and after post-processing (Bai et al. 2019). A series of infiltration kinetics occur when the binder is jetted onto the powder bed; such infiltration kinetics are mainly influenced by the characteristics of the binder droplets and the powder bed. Characteristics of the binder droplets include their volume, initial velocity, viscosity, and spreading and wetting behaviors, as well as the duration for them to penetrate the powder bed. Meanwhile, the roughness of the powder bed can affect the spreading, wetting, and penetrative behaviors of the binder droplets.

The liquid binder droplets, at the size of a few dozen microns, typically collide with the top surface of the powder particles at a substantial impact velocity (Figure 6.2a), resulting in their spreading across the powder bed and forming liquid networks

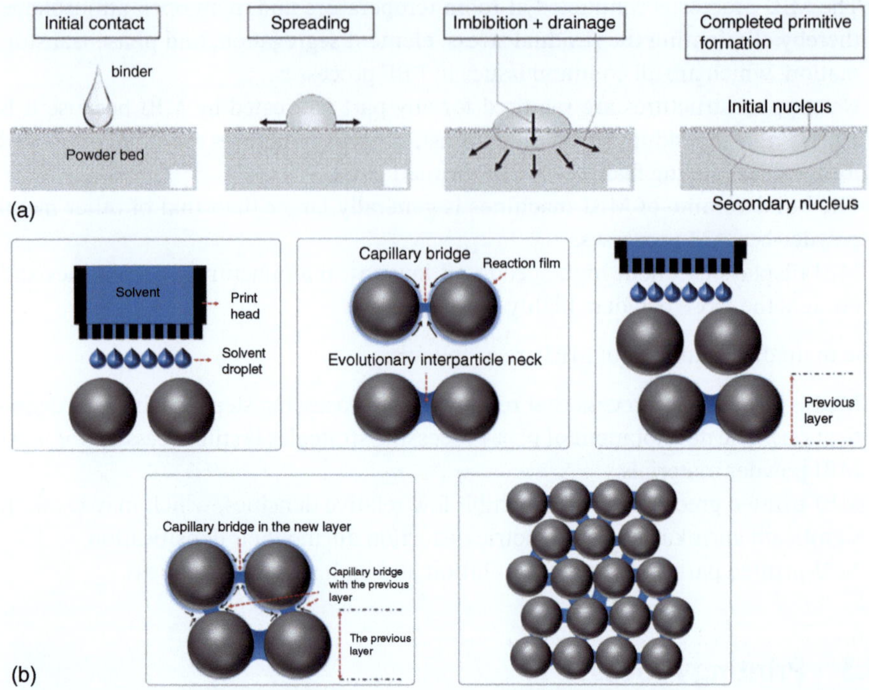

Figure 6.2 Schematic illustration of the binder–powder interactions in MBJ. (a) Interaction between a binder droplet and the powder bed. Source: Reproduced from Bai et al. (2019)/with permission from American Society of Mechanical Engineers. (b) Interaction between binder droplets and powder particles. Source: Salehi et al. (2019a)/reproduced with permission from Elsevier.

that bond adjacent powder particles. Subsequently, the binder droplets penetrate the loose powder particles beneath the top surface under the capillary force effect (Figure 6.2b).

Figure 6.3 presents the different granule formation mechanisms of a binder droplet impacting the powder bed, which include tunneling, spreading, and cratering (Emady et al. 2011). The tunneling mechanism is mainly restricted to fine and cohesive particles. A binder droplet attracts loosely aggregated particles as it penetrates the powder bed and forms spherical granules. Such granules may display protrusions due to the incomplete penetration of the binder droplet.

Spreading and cratering mechanisms are dominant in the case of coarse and free-flowing powder particles. When a slow-moving binder droplet impacts a dense powder bed, the droplet spreads across the surface of powder particles and forms a flat disk (Figure 6.3b). On the contrary, a binder droplet moving at a high velocity can create a crater in the powder bed upon collision before spreading upward along the walls of the crater, gathering powder particles, and finally retracting (Figure 6.3c).

The scanning strategies employed in MBJ directly influence the binder-powder interactions and the resolution of the printed parts. These strategies include vector scanning, raster scanning, and vector tracing with rasterization. In the vector scanning strategy (Figure 6.4a), the binder droplets create a contour and move inward to progressively trace smaller contours until a complete powder layer is printed.

Figure 6.3 Schematic illustration of different granule formation mechanisms. (a) Tunneling, (b) spreading, and (c) cratering. Source: Emady et al. (2011)/reproduced with permission from Elsevier.

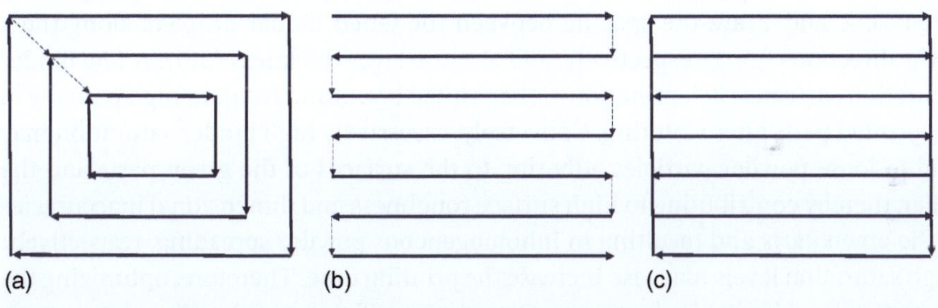

Figure 6.4 Schematic illustration of the main scanning strategies in MBJ. (a) Vector scanning, (b) raster scanning, and (c) vector tracing with rasterization.

This strategy produces the most geometrically accurate parts. However, utilizing only a single nozzle in the printhead increases the printing time (Baker 1997).

The raster scanning strategy (Figure 6.4b) entails the printhead first jetting binder droplets along the x direction. Subsequently, the printhead moves along the y direction and commences to scan a new line along the x direction (Miyanaji et al. 2018). While this strategy may create an undesirable stair-stepping appearance between adjacent lines, the utilization of multiple nozzles results in a higher printing speed. Vector tracing with rasterization uses the same scanning pathway as the vector scanning strategy (Figure 6.4c). The printing speed can be increased by employing all the nozzles in the printhead during the raster scanning process, which also ensures smoother edges on the surfaces of the parts (Baker 1997).

In MBJ, the layer thickness is typically in the range of 15–300 μm, and is determined by the printing resolution and the size of the powder particles. As a rule of thumb, a layer thickness approximately three times the average powder particle size should be implemented to ensure sufficient space for the rearrangement of the powder particles, beyond which an increase in layer thickness may reduce the density of the powder bed. Generally, a highly dense powder bed corresponds to a highly dense green part, which undergoes less shrinkage during the sintering process.

The powder spreading speed, which is comprised of the recoating speed, oscillator speed, roller speed, and roller traverse speed, determines the overall printing efficiency. Specifically, the recoating speed refers to the traveling speed of the recoater, which spreads powder onto the powder bed in the horizontal direction. The oscillator speed is the frequency at which the dispensing mechanism oscillates. The roller speed refers to the rotational speed of the roller, and the roller traverse speed is the speed at which the rotating roller moves across the powder bed. Coordinating these parameters is necessary to ensure the high quality of the printed green parts.

The saturation level of the binder depends on the capacity of the printhead system. The volume of droplets dispensed by printheads can be adjusted, which in turn affects the minimum saturation level and results in droplet overlapping (Miyanaji et al. 2016). The theoretical binder saturation S (%) can be estimated by

$$S = \frac{1000V}{\left(1 - \frac{R}{100}\right)XYZ}, \tag{6.1}$$

where V is the average volume of each binder droplet (pL), R is the powder packing rate (%), X and Y are the spacing between the jetted binder droplets along the x and y directions (μm), respectively, and Z is the layer thickness (μm). A low binder saturation can cause delamination of the printed layers and result in high porosity in the printed parts after sintering. Conversely, excessively high binder saturation may lead to loose powder particles adhering to the surfaces of the green parts and the roller, thereby contributing to high surface roughness and dimensional inaccuracies of the green parts and resulting in inhomogeneous powder spreading, respectively. High saturation levels may also increase the printing time. Therefore, optimizing the saturation level is vital for improving the printing efficiency, enhancing the strength of the green parts, and minimizing the cost.

The level of binder saturation can be influenced by the layer thickness and powder characteristics. For example, if the layer thickness value is smaller than the average size of the powder particles, the amount of binder administered may be excessive. The surplus binder may spill to the sides and produce large parts with a rough surface. On the contrary, a large layer thickness provides sufficient space to accommodate the binder and reduces binder spillage. The porosity of the powder bed and the contact angle between it and the binder jet are the two main factors influencing binder infiltration; an increase in the porosity and contact angle can reduce the infiltration depth and area of contact between the binder and the powder bed, respectively.

After jetting the binder droplets onto the powder bed, the powder bed surface is exposed to the heat emitted from a resistive heater for initial curing. This process is also known as drying. The drying duration is affected by the binder saturation level, binder composition, layer thickness, wettability between the powder and the binder, and powder bed characteristics (e.g. thermal conductivity, surface area, permeability, and packing density). Excessively high drying temperatures or an extended drying duration may lead to the slice cracking of parts during the subsequent powder spreading process, which is attributed to the shear forces produced by the roller and the fragility of the dried layer. During the drying stage, excessive binder is utilized to cleanse the printhead to prevent blockage within its nozzles.

The printed green parts must undergo curing, depowdering, measurement of their green strength, debinding, and infiltration or sintering. A hot isostatic pressing (HIP) step can also be optionally performed.

Curing is conducted to increase the strength of the printed parts. Typically, the powder bed and printed parts are both cured in the same oven to remove the binder within the latter. The curing temperature and duration are critical parameters determined by the binder material, part geometry, volume of the powder bed, etc. The cured green parts should possess sufficient mechanical strength to withstand the subsequent depowdering process, during which loose powder particles are removed manually through brushing, air jetting, vibrating, or vacuuming. In particular, wet depowdering methods (e.g. ultrasonic techniques, microwave-induced boiling, and CO_2 bubble generation in soda water) can be employed when the binder is insoluble within the corresponding fluid.

The green strength is critical to preserving the delicate features and geometries of the printed parts during the depowdering process, during which a green part of ultralow strength may succumb to breakage. However, a green part exhibiting exceptional strength likely contains an excessive amount of binder, which corresponds to a rough surface. Therefore, the strength of such green parts should be evaluated and prioritized during process optimization.

Before the densification process, the binder in the cured green parts must undergo a burnout step, i.e. debinding, because it may impede the sintering and infiltration processes. Typically, the burnout step is performed in the same furnace designated for sintering or infiltration, and the burnout temperature can be determined through differential thermal analysis of the binder. Unfortunately, some oxygen and carbon residue may remain even after the binder decomposes or becomes gaseous (Yan et al. 1993). The oxygen residue may oxidize the surface of the printed parts. In contrast, the carbon residue may form carbides when the burnout procedure is conducted at high temperatures, thereby degrading the mechanical properties of MBJ-printed metal parts sensitive to carbon or oxygen (Miyanaji et al. 2016).

Infiltration is an alternative densification method that utilizes an alloy material to improve the relative density of the printed green parts. The infiltrant material should possess a lower melting temperature than the powder material of the green parts. The dominant mechanism underlying the infiltration process is the capillary action involving the liquid infiltrant and the interspaces of the powder particles. Infiltration can significantly reduce the porosity of the printed green parts and improve their mechanical properties (Do et al. 2015). Note that optimizing the saturation of the infiltrant is critical to achieving high densification of the green parts. If the amount of infiltrant administered exceeds its saturation level, the dimensional accuracy of the printed parts may be compromised. On the other hand, utilizing an insufficient amount of infiltrant results in weak bonding between the powder particles.

Sintering is a basic densification method for processing binder-jetted green parts. The sintering kinetics depend on the chemistry, surface activity, morphology, and size distribution of the powder particles and the sintering atmosphere. Metallic bonds are formed during the sintering process. As a result, the green parts experience shrinkage during sintering, and pores formed during the printing process are filled. Two categories of sintering mechanisms exist, namely, densifying and non-densifying sintering mechanisms.

At lower temperatures, the sintering process is dominated by non-densifying sintering mechanisms, which include diffusion on the surface of the powder particles and vapor transport. Such mechanisms are coupled with particle coarsening and negligible interparticle neck growth. The densifying mechanisms begin to dominate the process as the temperature increases. For metals, the primary densifying mechanisms are grain boundary diffusion and lattice diffusion between the powder particles (Rahaman 2010).

Figure 6.5 shows the three stages of the sintering process (German 2010). The initial stage occurs at a low temperature, and it is dominated by non-densifying sintering mechanisms. No significant densification or powder particle movement occurs at this stage. Notably, the average size of the interparticle necks is less than one third of the average particle size. Due to voids in the powder bed, pores are distributed across the entire part, and its relative density is over 30% lower than its theoretical density.

The intermediate stage occurs when the temperature increases to a certain material-dependent threshold. At this stage, the sintering mechanisms are predominantly grain boundary diffusion and lattice diffusion between the powder particles, which promote interparticle neck growth and the mass movement of powder particles. As the sintering process proceeds, the interparticle necks reach approximately half the average particle size. Additionally, large pores within the part merge to form tubular pores and gradually eliminate the fine pores. In the first two stages, the pores on the surface of the part remain open, allowing gases within it to escape.

At the final stage, the sintering temperature range is similar to the one in the intermediate stage, and the maximum densification effect can be achieved with optimized sintering conditions. Critical aspects of the MBJ sintering process include the heating and cooling rates, the sintering temperature, and the holding time. Note that it is highly challenging to attain the near-full density of the green part even after sintering.

All MBJ-printed green parts are highly porous before sintering. In such parts, the powder particles are closely packed together in the near absence of external forces, and their solid volume fraction is typically below 60%. The low density of the green parts may result in high shrinkage during their densification, and such shrinkage should be considered during the stages of process optimization and part design. Live Sinter, a software solution developed by Desktop Metal, can

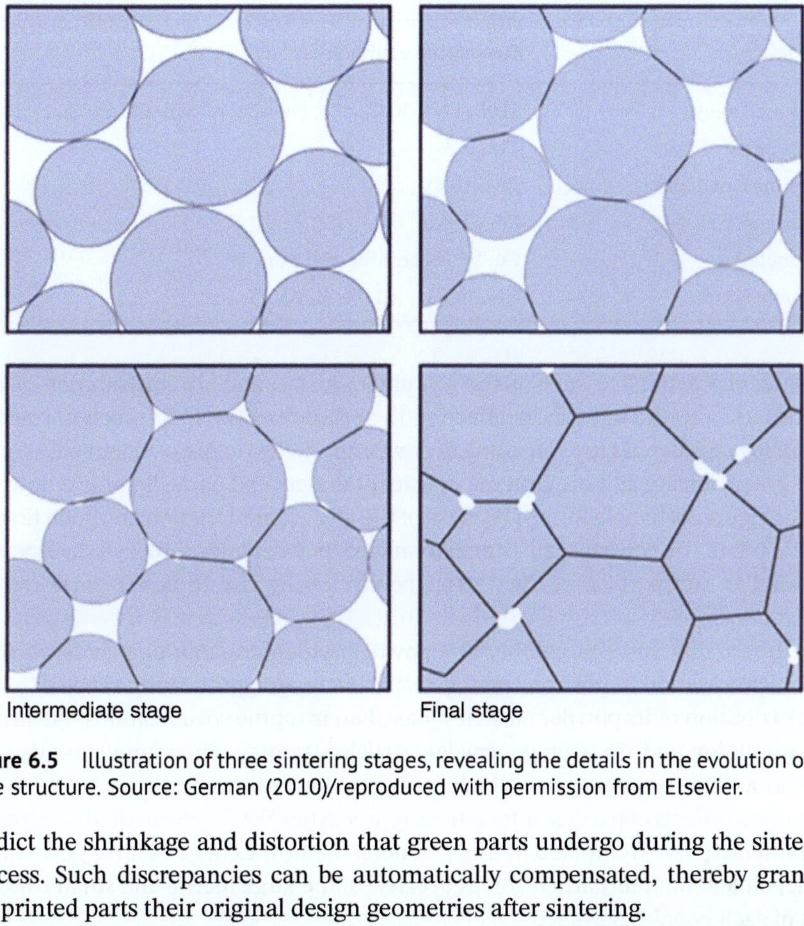

Figure 6.5 Illustration of three sintering stages, revealing the details in the evolution of the pore structure. Source: German (2010)/reproduced with permission from Elsevier.

predict the shrinkage and distortion that green parts undergo during the sintering process. Such discrepancies can be automatically compensated, thereby granting the printed parts their original design geometries after sintering.

6.4 Raw Materials

Different binders are currently being developed and employed for MBJ, including polymeric binders, metal nanoparticle suspensions, metal oxide precursors, metal salts, and metal-organic decomposition (MOD) inks. Ethylene glycol–based binders are the most widely utilized binders because of their excellent solubility in water, high cost effectiveness, long-lasting effects, and good burnout properties. However, selecting an appropriate binder ensures that the printed parts exhibit sufficient mechanical strength. Additionally, the pyrolysis of polymer binders during post-sintering may produce carbon residue, which may degrade the purity and the mechanical, optical, electrical, and thermal properties of the printed parts.

Presently, the types of metal powder materials researched for MBJ applications are fewer than those for PBF and DED. Table 6.1 provides a summary of the most studied powder materials for MBJ. There are also emerging metal materials under development for MBJ, such as magnesium alloys and aluminum alloys.

Table 6.1 Summary of the most investigated powder materials for MBJ.

Materials	Commercial equivalents
Iron-based alloys	316L, 17-4, 304L, 420, Fe–Mn, Fe–Mn–Ca, Fe–Si, H13, D2
Nickel-based alloys	718, 625
Titanium-based alloys	Ti–6Al–4V
Cobalt-based alloys	Co–Cr–Mn
Pure metals	Cu, W, Fe, commercial pure Ti
Magnetic alloys	Ni–Mn–Ga, Nd–Fe–B, Fe–6Si

Powder characteristics such as the morphology, size, and size distribution of the particles are also critical factors affecting the efficacy of the MBJ process. Powder morphology influences the spreading of powder particles on the powder bed, as well as the green density and the sintered density of the printed parts. Spherical powder morphology is preferable in MBJ because of the low internal friction and good flowability it offers. In contrast, an irregular-shaped powder morphology corresponds to a smaller interparticle contact area, thereby giving rise to larger pores in the printed parts.

The flowability and sinterability of a powder, which are dominant properties in the MBJ process and its post-processing, respectively, are significantly affected by the size distribution of its powder particles. Based on machine specifications, the printhead resolution and the average powder particle size may range from sub-microns to 150 μm. Finer powders generally demonstrate lower flowability due to the more pronounced effects of particle interactions (e.g. van der Waals, electrostatic, and capillary interactions). Compared to fine powders, the flowability of coarse powders is less dependent on interparticle forces because of the large inertia and small contact areas of each powder particle.

The powder particle size distinctly impacts the powder bed and green parts. For example, the spreading performance of Inconel 718 powder (<20 μm) was shown to be inferior to that of powders with a larger powder particle size (<53 μm) as a result of more powder agglomeration, creating a large number of pores in the powder bed and eventually reducing the density of the green parts (Turker et al. 2008). However, fine powder particles typically correspond to high powder sinterability, which can be attributed to their higher surface-to-volume ratio and energy state. As a result, fine powder particles are sintered within a shorter time than their larger counterparts, resulting in fast densification.

To tackle the aforementioned issues of employing fine powders, ExOne (USA) has developed a Triple ACT system to improve the dispensing, spreading, and compacting of fine powders (≤10 μm) to print high-quality metal parts displaying high surface quality, dimensional accuracy, and sinterability.

The morphology and size distribution of powder particles can influence the pore size distribution and homogeneity of the powder bed density, which further impacts binder penetration. Powder particles prepared by water atomization or ball milling typically exhibit irregular shapes and may cause heterogeneous powder spreading, which introduces difficulties in attaining complete binder penetration. Small powder particles may also create macropores within the powder bed due to particle agglomeration, thereby increasing the drop penetration time.

6.5 Equipment

The leading companies in the development of commercial binder jetting equipment are ExOne (USA), Voxeljet (Germany), 3D Systems (USA), and Digital Metal (Sweden). Their equipment is capable of printing metals, ceramics, and waxes. Additional MBJ systems have been proposed by Desktop Metal (USA), GE Additive (USA), 3DEO (USA), Hewlett-Packard (HP) (USA), and Long Yuan (China). Among these companies, ExOne boasts the greatest variety of machines that can print using a wide range of metal powders, including stainless steel 316, 420, 17-4 PH, Inconel 625, Inconel 718, pure iron, iron–chrome–aluminum, chromite, cobalt–chrome, and tungsten powders.

GE Additive first announced its venture into the domain of binder jetting in 2017, with the debut of the prototype system H1. The company has since been developing a second generation of the H1 system, which is expected to be commercially available. A summary of standard commercialized equipment for MBJ is listed in Table 6.2.

A significant difference in the equipment of MBJ and PBF is the printhead, which is a core component of the former. Figure 6.6 shows the different types of printheads used to generate liquid binder drops, namely, drop-on-demand (DoD) and continuous-jet (CJ) printheads (Salehi et al. 2018).

Thermal inkjet and piezoelectric printheads are the two types of standard DoD printheads that can produce individual drops on demand. In thermal inkjet printheads, the heater vaporizes a portion of the liquid binder. The subsequent volume expansion of the vaporized binder results in the liquid binder being ejected through the nozzle (Kwon and Kim 2007). Meanwhile, piezoelectric printheads can be utilized to promote ink development because their rheological properties constitute the only requirement for producing a reliable flow of ejected ink.

In CJ printheads, the binder droplets are continuously produced at higher print rates than droplets produced by DoD printheads. In specific areas on the powder bed where the binder does not have to be applied, the inductively chargeable binder droplets can be deflected into a gutter. Compared to DoD printheads, CJ printheads correspond to a slow traverse speed and decent size control over overlapping printed tracks.

HP thermal inkjet printheads, which are installed in HP Metal Jet printers, can facilitate the precise placement of the binder on the powder bed. As shown in Figure 6.7a, each printhead comprises an electrical connector, two independent supply ports for binders, two pressure regulators, and nozzles. There are 5280 nozzles on the top surface of each pressure regulator, and each nozzle is spaced 1/1200 in. from one another. The binder is initially fed into the supply ports and subsequently pumped out by the pressure regulators. Multiple printbars are utilized

Table 6.2 Summary of standard commercialized equipment for MBJ.

Company	Equipment	Build volume (mm)	Throughput (cc/h)	Print resolution	Layer thickness (μm)	Size ($D \times W \times H$) (mm)	Materials and binders
ExOne (USA)	Innovent+	160 × 65 × 65	166	30 μm voxels	30–200	1203 × 1016 × 1434	Materials: 17-4 PH, 304L, 316L, M2 tool steel, Co–Cr, copper, H13 tool steel, Inconel 625, titanium, tungsten. Binder systems: Aqueous, solvent, phenolic
	X1 25PRO	400 × 250 × 250	3600	>30 μm voxels	30–200	2300 × 1800 × 2300	
	M-Flex	400 × 250 × 250	1600	>50 μm voxels	50–200	1675 × 1400 × 1855	
	X1 160Pro	800 × 500 × 400	>10 000	—	30–200	—	
Digital Metal (Sweden)	DM P2500	170 × 150 × 55; 203 × 180 × 69	60; 100	35 μm	—	2900 × 1000 × 1700	316L, 17-4 PH, Ti–6Al–4 V, Ni 625, Ni 247
Desktop Metal (USA)	Shop System™	350 × 222 × 200	700	1600 dpi	40–100	1994 × 762 × 1626	Materials: 17-4 PH
	Production System	490 × 380 × 260	12 000	1200 dpi	—	1900 × 5000 × 1900	Binder systems: Waxes and polymers
HP (USA)	HP Metal Jet	430 × 320 × 200	—	1200 dpi	50–100	—	Materials: 316L. Binder systems: Polymers

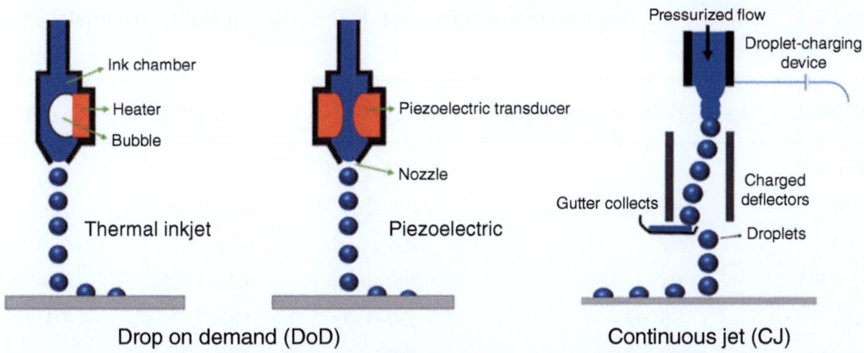

Figure 6.6 Schematic illustration of drop-on-demand and continuous-jet printheads. Source: Salehi et al. (2018)/reproduced with permission from Materials Research Forum.

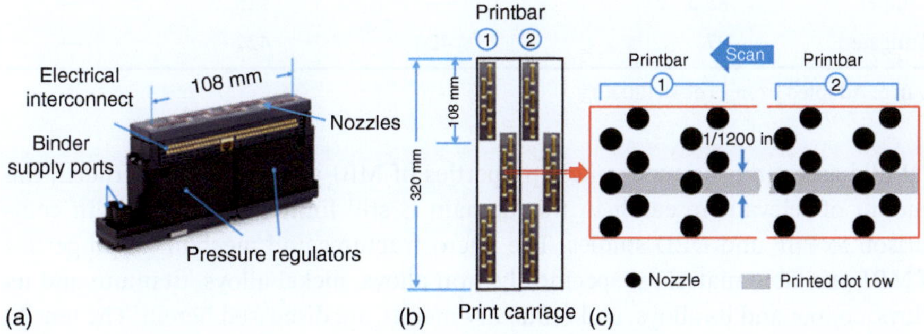

Figure 6.7 HP thermal inkjet printhead in HP Metal Jet printers. (a) Image of the printhead, (b) a print carriage containing two printbars, and (c) schematic illustration of each printbar. Source: Courtesy of HP.

in each printer cartridge to boost productivity. As shown in Figure 6.7b, each printbar contains three printheads arranged in the form of an isosceles triangle and can create a 320 mm-wide print swath. Two printbars (marked as 1 and 2) in the print carriage are aligned to ensure that the nozzles (represented by black dots) can operate along the same line to print a dotted row (indicated by gray lines) with 1/1200 in. between the dots (Figure 6.7c). Consequently, up to four different nozzles can function simultaneously within the same 1200 dpi grid point.

6.6 Microstructure and Mechanical Properties

MBJ is theoretically applicable to printing parts from a wide range of materials because of its inherent aspect of forming parts without material fusion. Compared to laser-based AM (PBF and DED), MBJ can print thermally conductive, hard-to-weld, and refractory materials. Notably, densification of printed parts remains a top priority of the MBJ printing process and the post-treatment of green parts.

Table 6.3 Summary of the tensile properties of MBJ-printed metallic materials after densification.

Material	Relative density (%)	σ_y (MPa)	σ_{UTS} (MPa)	ε (%)
Iron	91.3	30.6	—	—
316L	95–98	180–224	494–582	51–61.9
304L	97	205	579	60
17-4 PH	98	730–1034	900–1317	8.5–12
420	95–99	250–455	682–737	2.3–2.7
Fe–Mn	60.7	190	228	—
Inconel 625	98.9–99.6	327–376	612–644	40.9–47
Ti–6Al–4 V	—	790	890	8
Copper	85.5	—	116.7	—
Tungsten	97	420	427	—

Source: Adopted from Li et al. (2020).

While assessing the mechanical properties of MBJ-printed parts is crucial, the amount of relevant research in this domain is still limited, particularly in comparison to PBF and DED studies. The microstructures and mechanical properties of MBJ-printable materials, specifically, iron alloys, nickel alloys, titanium and its alloys, copper and its alloys, and refractory metals, are discussed herein. The tensile properties of typical MBJ-printable materials are summarized in Table 6.3.

6.6.1 Iron Alloys

Iron alloy powders, such as 304, 316, 347, 420, 430, and 17-4 PH, are the most widely employed powders in MBJ. Since achieving a high packing density of the powder bed in MBJ is challenging, the high porosity in MBJ-printed parts constitutes a common issue. The porosity of the printed parts is associated with several factors, including the selection of the raw powder and the post-sintering process.

The size of the powder particles defines the sinterability of the green parts and their final pore sizes. For example, fine 316 powder with particle sizes of 22–31 μm was found to densify at low sintering temperatures compared to coarser powders with particle sizes of 20–53 and 45–90 μm (Verlee et al. 2012). The resultant densification of the three powders decreased as the average size of their particles increased, and a concomitant increase in large open pores was observed. In addition, irregularly shaped powder particles contributed to a lower powder bed packing density, resulting in the increased porosity of the parts after sintering.

In addition to the size and shape of the powder particles, the post-sintering temperature, duration, and profile can also influence the porosity of the final parts. For example, a higher relative density of ~98% was achieved in 316 parts when the sintering temperature increased from 1255 to 1400 °C, which also corresponded to an increase in their tensile strength from 310 to 520 MPa and their elongation from 21% to 62%.

Achieving high porosity in printed parts may be desirable for applications such as tissue engineering, thereby introducing the necessity of proper porosity control. To this end, Ziaee et al. (2017) employed fine 316L powder to print green parts of varied densities and controlled shrinkage levels via MBJ. Raw 316L powder (smaller than 22 μm) was selected to prepare different powder types, namely, agglomerate powders with particle sizes of 25–45 μm cured at different temperatures and 316L/polyamide mixed powders (average particle size of 55 μm) with different volumetric ratios.

The printed green parts were initially processed at 800 °C to remove the polyamide. Subsequently, they were sintered in an argon/hydrogen gas atmosphere at 1360 °C for 60 minutes. The parts printed from raw 316L powder and an agglomerated powder cured at 185 °C obtained similar spread density values (~56%) and final relative density values (~93%) after sintering (Figure 6.8a). However, a low spread density of 42% was attained for parts printed from an agglomerated powder cured at 100 °C, which exhibited a final density of only 77% after sintering (Figure 6.8b). The polyamide served as a fugitive space holder, and its addition was advantageous for preparing porous structures. Notably, the printed 316L parts with 25% and 33% polyamide displayed a density of 66.8% and 63.3%, respectively, after sintering. High shrinkage of these parts was observed along the build direction, indicating that more pores existed between the powder layers than within them.

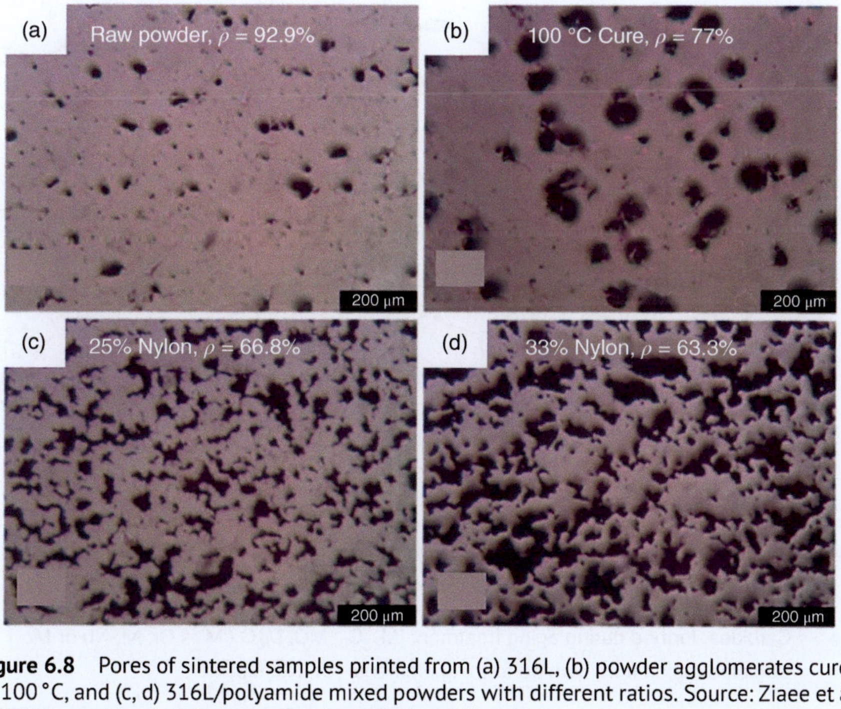

Figure 6.8 Pores of sintered samples printed from (a) 316L, (b) powder agglomerates cured at 100 °C, and (c, d) 316L/polyamide mixed powders with different ratios. Source: Ziaee et al. (2017)/reproduced with permission from Springer Nature.

6.6.2 Nickel Alloys

The weldable nickel-based Inconel 625 and 718 alloys have been widely explored in terms of their compatibility with MBJ and post-sintering properties. For example, Mostafaei et al. (2016) investigated the effects of the sintering temperature on an MBJ-printed Inconel 625 alloy part. At a sintering temperature of 1280 °C, the relative density of the printed part increased from 53% to 99.6%, and its dimensions and volume shrank by ~19% and ~46%, respectively. As the sintering temperature further increased, a liquid phase was formed at the grain boundaries of the printed part, degrading its mechanical properties.

Additionally, grain coarsening occurred in the printed Inconel 625 part when the sintering temperature increased from 1280 °C to 1300 °C, thereby leading to the growth of its grains from 55 to 182 μm. After sintering at 1280 °C, the printed part possessed an ultimate tensile strength of 612 MPa, a yield strength of 327 MPa, and a fracture strain of 40.9%, which are comparable to its cast counterparts. After solution and aging treatment, carbides and Cr_2O_3 precipitates were discovered at the grain boundaries and within the grains, and the printed part possessed a microhardness value of up to ~330 HV, an ultimate tensile strength of ~700 MPa, and an elongation of 30%.

Figure 6.9 shows the schematic of the microstructure evolution of MBJ-fabricated Inconel 625 alloy parts printed from gas- and water-atomized powders (Mostafaei

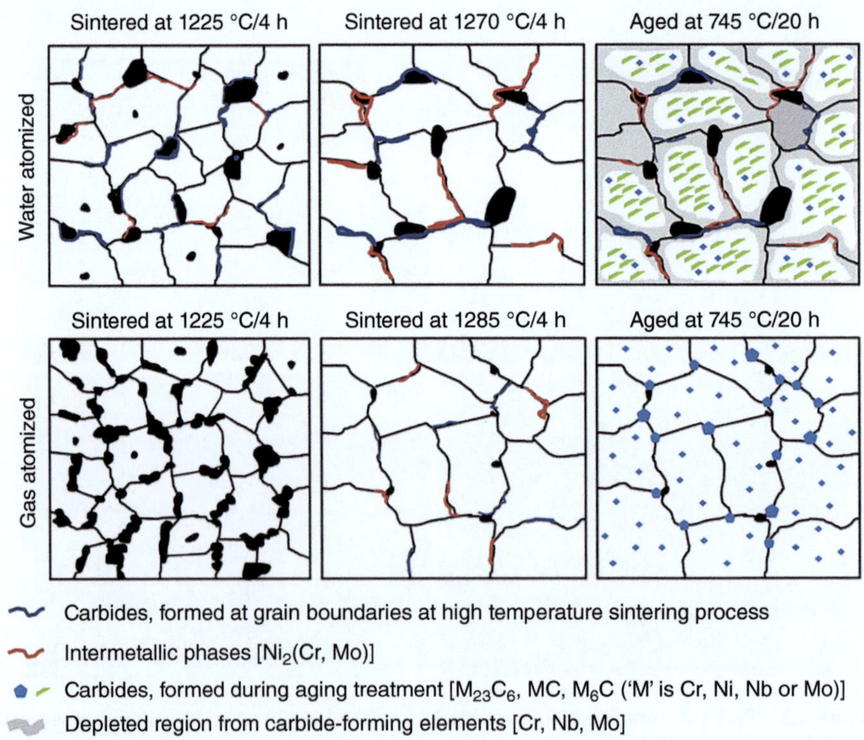

Figure 6.9 Schematic illustration of the microstructure evolution of an MBJ-printed Inconel 625 alloy part using gas- and water-atomized powders. Source: Mostafaei et al. (2017b)/ reproduced with permission from Elsevier.

et al. 2017b). At a low sintering temperature of 1225 °C, the parts printed from the gas-atomized powder displayed a large number of fine pores without any visible precipitation of oxides and carbides. In comparison, the parts printed from the water-atomized powder exhibited fewer large pores at their grain boundaries and smaller pores within their grains, which were accompanied by carbide and oxide precipitates.

A further increase in the sintering temperature promoted grain growth in all the parts and eliminated nearly all pores in the parts printed from the gas-atomized powder. After aging, significant precipitation of carbides and oxides occurred at the grain boundaries of the parts printed from the water-atomized powder, thereby depleting the chromium, niobium, and molybdenum contents around the grain boundaries. In contrast, only a small amount of carbide and oxide particles were detected in the parts printed from the gas-atomized powder.

The effects of the layer thickness and sintering temperature on the properties of MBJ-printed Inconel 718 parts were investigated (Turker et al. 2008). An increase in the layer thickness was found to impede the densification of the printed alloy parts. At a sintering temperature below 1260 °C, the parts displayed high porosity. As the sintering temperature increased to 1280 °C, Nb-rich γ'' precipitates were observed at the grain boundaries and within the grains of the printed alloy parts. The increase in the sintering temperature also resulted in precipitate dissolution along the grains and the consequent formation of residual pits. The relative density and linear shrinkage of the printed alloy parts at a sintering temperature of 1280 °C were 98% and 19.5%, respectively.

The powder characteristics of Inconel 718, such as the sizes of the powder particles and their distribution, also play important roles in determining the mechanical properties of MBJ-printed parts (Nandwana et al. 2017). For example, at the same sintering temperature, smaller powder particles underwent greater linear shrinkage than larger powder particles. Therefore, a powder particle size distribution with a high fraction of small powder particles corresponds to a high powder bed density.

Non-weldable nickel-based superalloys with high amount of ordered Ni_3(Al, Ti)-γ' are desirable for applications in the aerospace and energy domains because of their excellent mechanical performance under high temperatures. However, such non-weldable superalloys are highly susceptible to hot cracking when processed through techniques such as PBF (Chauvet et al. 2018). In the MBJ process, the consolidation of powder particles does not require material melting, thereby allowing residual stress and thermal gradients within the printed parts to be significantly reduced. Therefore, MBJ is superior to PBF in circumventing the issue of hot cracking in non-weldable nickel alloy parts.

For example, the Ni-based RENÉ 108 alloy, also named as CM247LC or Mar M 247, generally possesses a γ' volume fraction of over 60%, which promotes crack formation during PBF. Recently, a green RENÉ 108 part with a relative density of 60–62% was fabricated and further densified, resulting in a relative density of 96% after post-sintering and almost 100% after further HIP treatment (Martin et al. 2021). Consequently, the final part was free of cracks and exhibited thermo-mechanical properties on par with conventional parts manufactured via directional solidification. However, the formation of carbon residue remains an issue for MBJ. A critical debinding operation is necessary to remove such residue, particularly for superalloys containing high amounts of reactive alloying elements such as titanium and tantalum.

6.6.3 Titanium and Its Alloys

MBJ-printed parts composed of titanium and its alloys are mainly investigated for bone implant applications due to their porous structures. A pure titanium implant printed via MBJ was found to possess a stiffness of 2–30 GPa and a compressive strength of 130–180 MPa, which are similar to those of human bones (Xiong et al. 2012). In the study, titanium powder (200 mesh, smaller than 74 μm) and an aqueous polyvinyl alcohol powder binder (20 mesh, smaller than 125 μm) were mixed and employed as the raw material.

In another study, a mixed powder composed of 97 wt% titanium powder with a particle size distribution of 75–90 μm and 3 wt% polyvinyl alcohol powder with particle sizes below 63 μm was prepared (Sheydaeian et al. 2017a). The porosity and shrinkage of the printed parts increased as the layer thickness increased from 80 to 150 μm, while the yield stress only underwent a minor reduction from 175 to 158 MPa.

A combination of MBJ and material extrusion was explored to fabricate porous titanium parts (Sheydaeian et al. 2017b). Sacrificial polymer droplets were jetted periodically onto certain powder layers to create cellular titanium structures with closed cavities. Such structures exhibited an elastic modulus of 2.48–3.55 GPa and a yield strength of 107.65–145.75 MPa, which are comparable to those of cancellous bones. In a similar study, a titanium boride slurry was employed in place of the polymer droplets to manufacture TiB_2/Ti composite parts (Sheydaeian and Toyserkani 2018). The stiffness of the printed composite parts underwent an increase of up to 15.2% when a small fraction (4%) of TiB_2 powder was added to the binder.

The relative density of the MBJ-printed Ti-6Al-4 V part varied from 50% to 95% (Stevens et al. 2018), for which the binder and powder effects could be considered as the main factors leading to such variations (Figure 6.10). The binder droplets disrupted and distorted the powder layer, adversely affecting the quality of the printed parts. Meanwhile, the presence of satellites in the gas-atomized powder reduced

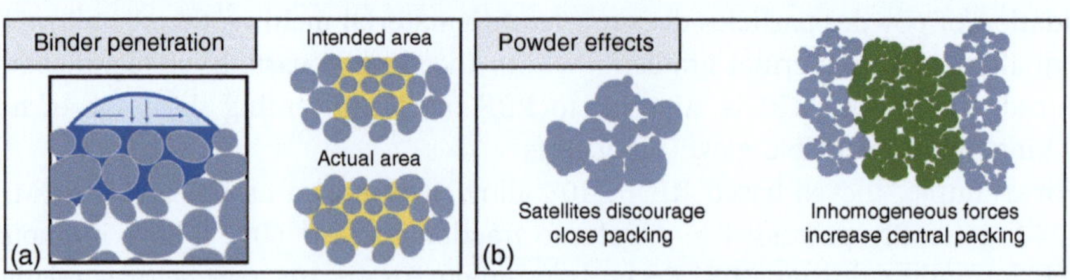

Figure 6.10 Schematic illustration showing (a) the interaction between the binder and the powder bed and (b) powder effects on the part density. Source: Reproduced from Stevens et al. (2018)/with permission from Elsevier.

the packing efficiency. Differences between the absorbed loads on the core areas and the edges could also affect part densification.

6.6.4 Copper and Its Alloys

Copper and its alloys are particularly suitable for manufacturing products via MBJ. Due to their high thermal conductivity and laser reflectivity, it is relatively difficult to process copper and its alloys through laser-based PBF and DED.

The influences of the powder particle size and the ratio of bimodal copper powder mixtures on the density of the final parts and the extent of their shrinkage have been investigated (Bai et al. 2017). In theory, utilizing bimodal or trimodal powder mixtures can increase the density of green parts and reduce their shrinkage during the sintering process, which is achieved by improving the powder bed density and filling the interstitial gaps between coarse particles with fine particles.

In another study, fine copper powder with an average particle size of 15 μm was selected for printing parts via MBJ, which were sintered at 1080 °C (Bai and Williams 2015). The printed parts displayed a mechanical strength of ~117 MPa, which was much lower than that of their injection-molded counterparts (200 MPa).

HIP can be conducted to further densify MBJ-printed copper parts (Kumar et al. 2018), and a maximum increase of 8% in their density can be achieved. For example, different copper parts printed using a 17-μm powder and a bimodal powder (prepared from 5-μm to 30-μm powders) underwent an increase in density from ~84% to ~86% and from ~91% to ~98%, respectively, after HIP. Compared to a fully dense part composed of sintered copper powder with a tensile strength of 220 MPa, the bimodal HIP sample displayed a tensile strength of ~177 MPa and an elongation of ~31%.

In another study, the impact of process-induced porosity on the material properties of copper parts printed through MBJ was investigated (Kumar et al. 2019). The copper parts obtained a relative density of 97.3% after HIP was performed at a temperature of 1075 °C, a pressure of ~207 MPa, and a holding time of two hours. The high relative density of the parts corresponded to a high tensile strength of 176 MPa and a thermal conductivity of ~328 W/mK, which are ~80% and ~85% of those of a wrought copper alloy, respectively.

MBJ-printed copper parts processed separately with a polymeric binder and a MOD ink were compared (Bai and Williams 2018). The MOD ink facilitates the formation of an organometallic compound by introducing ligands to metal salts. Therefore, it could improve the solubility of metal–organic compounds in saturated solvents, thereby increasing the metal content in the binder. Note that the MOD ink is a particle-free ink, and thus nozzle clogging and sedimentation did not hinder the printing process.

Figure 6.11a illustrates the printing, curing, and sintering stages. During the curing stage, nanoparticle precipitates are formed from the MOD ink, and interparticle necks are formed at a sintering temperature of 150–300 °C in the case of copper nanoparticles, thereby enhancing the strength of the green parts. Figure 6.11b shows the cross-sectional images of a printed part processed with MOD ink, revealing large

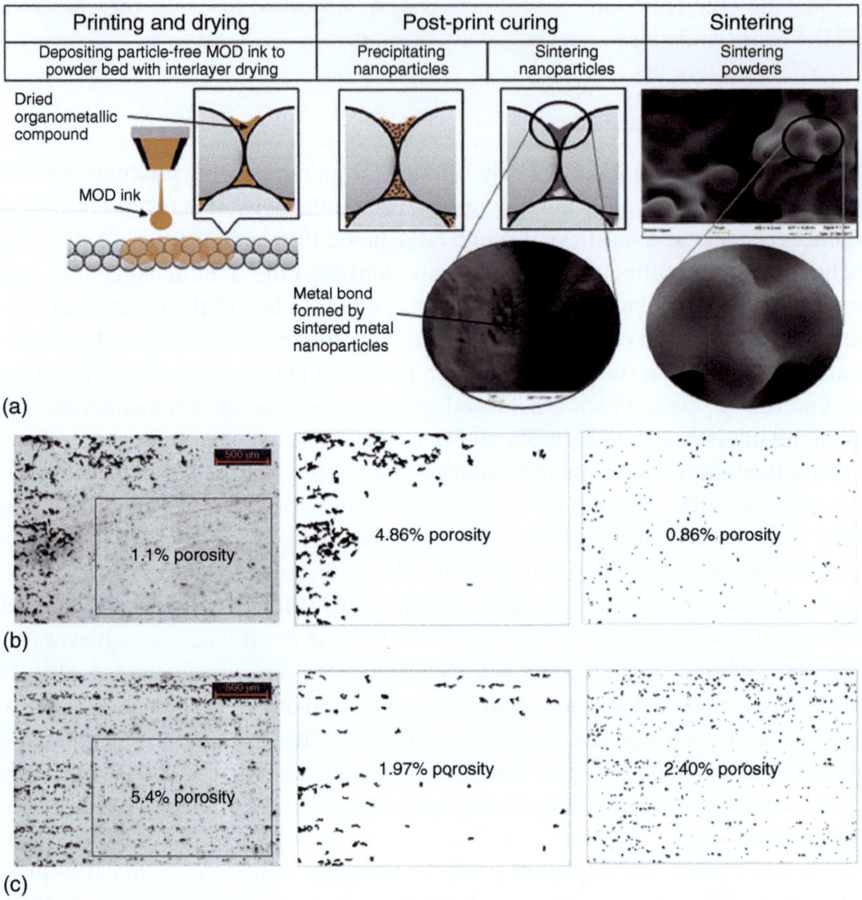

Figure 6.11 (a) Schematic of the printing and curing stages with a MOD ink as the binder and SEM images of the sintered nanoparticles and powder particles. Cross-sectional images displaying the porosity of printed parts processed with (b) a MOD ink and (c) a commercial binder. Source: Bai and Williams (2018)/reproduced with permission from Elsevier.

pores near its surface. Meanwhile, parts printed using a commercial binder possess randomly distributed pores (Figure 6.11c).

6.6.5 Refractory Metals

Refractory pure metals and alloys are often categorized as difficult-to-print materials for laser-based AM. For example, tungsten heavy alloys (WHAs) include refractory tungsten and other alloying elements such as nickel, iron, and copper, which exhibit limited solubility and a wide range of melting temperatures (Iveković et al. 2019). As a result, significant issues regarding elemental vaporization and creation of cracks may arise during the PBF processing of such alloys. The MBJ printing process does not involve metal powder fusion to achieve densification, thereby providing an alternative approach that circumvents the aforementioned issues.

For instance, a green part was printed from a W–Ni–Cu–Fe heavy alloy via MBJ and subsequently dewaxed in a reducing atmosphere at a maximum temperature of 870 °C (Stawovy et al. 2019). After further sintering at 1385 °C, crack-free W–Ni–Cu–Fe parts with a relative density of 99.7% and an average tensile strength of 770 MPa were obtained. The density and mechanical performance of these parts matched the specifications of ASTM B777 for WHAs.

Another refractory material processable by MBJ is tungsten carbide–cobalt (WC–Co). This ceramic–metal composite possesses excellent hardness, fracture toughness, and flexural strength, rendering it particularly effective as a material for machining, cutting, and drilling. A recent study indicates that a combination of binder jetting and post-heating treatment can produce defect-free WC–Co parts with a uniform microstructure and superior mechanical properties (Mostafaei et al. 2021a). Despite the low relative density (21–23%) of the printed green WC–Co parts, a relative density of nearly 100% was achieved through sintering at 1435 °C, followed by HIP under a pressure of 6.1 MPa.

6.6.6 Others

Magnesium alloys, as some of the lightest biodegradable engineering metallic materials, display great potential for fabricating medical implants and lightweight automobile parts. Most research on AM-printed magnesium (Hu et al. 2015) and its alloys, such as WE43 (Li et al. 2018), Mg–Al (Niu et al. 2018), and Mg–Zn (Wei et al. 2019), was conducted using laser-based AM. However, due to their intrinsic properties, such as a strong affinity for oxygen and high vapor pressure, magnesium parts are subjected to oxidization and element evaporation during laser-based AM processes, which may change their chemical composition and cause the formation of gas pores. As a beamless printing technique capable of fabricating parts at near room temperature, MBJ is a suitable candidate for printing parts from magnesium and its alloys. For example, Salehi et al. have printed Mg–Zn–Zr parts via MBJ using capillarity-driven bridging as a novel and efficient tool for assembling powder particles into 3D structures while minimizing the metallurgical complexity (Salehi et al. 2019b). In another work by Su et al. (2021) MBJ was employed to print Mg–Al–Zn (AZ91D) parts, which were subsequently processed through a two-step sintering strategy that drastically enhanced their relative density, mechanical property, and corrosion resistance compared to conventional one-step sintering.

A limited amount of research has been conducted on the MBJ processing of cobalt-based alloys. In a study, an MBJ-printed Co–Cr–Mo alloy part achieved a relative density of ~99.7% and a microhardness of ~322 HV after undergoing HIP (Stoyanov et al. 2016). The effects of binder saturation and post-heat treatment on the relative density and microstructure of Co–Cr–Mo alloy parts were also explored (Sears et al. 2019). The printed parts with a binder saturation level of 95–100% were partially melted at a sintering temperature of 1325 °C. After sintering, the linear shrinkage values along the x, y, and z directions were calculated to be ~19.4%, 18.8%, and 22.3%, respectively. The yield strength, ultimate tensile strength, and elongation of the sintered Co–Cr–Mo parts were measured to be 462 MPa, 730 MPa, and 12%, respectively.

A sintered Co–Cr–W alloy component fabricated through MBJ using powder with an average particle size of ~95 μm exhibited an ultimate tensile strength of ~850 and 885 MPa before and after aging, respectively (Mostafaei et al. 2019). Such a difference can be attributed to variations in the microstructure after the aging treatment, during which the γ phase (face-centered cubic, FCC) was transformed into the ε phase (hexagonal close packing, HCP).

Nickel–manganese-based ferromagnetic shape memory alloys display magnetocaloric and magnetic shape memory effects (Dey et al. 2016). A pre-alloyed angular Ni49.6Mn30.8Ga19.6 ball-milled powder with particle sizes below 63 μm was fabricated by MBJ and sintered at 1020 °C (Mostafaei et al. 2017a). During sintering, interparticle necks were formed, and no manganese was evaporated. Notably, the sintering process could be optimized without incurring any element loss.

A soft magnetic Fe–6Si powder was selected to fabricate geometrically complex, crack-free magnetic parts via MBJ, which obtained a relative density of ~99% after solid-state sintering (Cramer et al. 2019). These parts exhibited reduced core loss at various frequencies, high resistivity, and good magnetic permeability. Notably, their grain size was measured to be 56.3 μm. The printed parts attained an ultimate tensile strength of 434 MPa, an electrical resistivity of 98 μΩ cm, a saturation magnetization of 1.83 T, a low coercivity of 0.4 Oe, and a maximum relative permeability of 10.5. The results indicate that MBJ-printed soft magnetic Fe–6Si parts could be an alternative to silicon steel parts fabricated by chemical vapor deposition.

A neodymium–iron–boron magnetic powder with an average particle size of ~70 μm was synthesized and explored for MBJ purposes (Paranthaman et al. 2016; Li et al. 2017), and Nd2Fe14B phases were identified in both the raw powder and the printed parts. The volume fraction of the printed parts was measured to be ~0.45 with a remanence of ~0.3 T.

High-entropy alloys (HEAs) are potential candidates for printing parts via the MBJ process (Han et al. 2020). However, the amount of research conducted on this topic is still limited. In a study, an AlCoCrFeNi HEA part with a porosity of ~1% was printed through MBJ before sintering (Karlsson et al. 2019), and it was determined to be only stable as a single phase within a narrow temperature range below its melting point. Notably, the as-sintered HEA part displayed a dominant B2 phase (body-centered cubic, BCC), with an additional FCC phase located at its grain boundaries and σ-phase precipitates within its grains. An increase in the annealing temperature further suppressed the expansion of the FCC phase within the HEA part, thereby increasing its yield strength from 1203 to 1461 MPa and its fracture strength from 1996 to 2272 MPa. Additionally, the HEA part obtained an excellent oxidation resistance, with the oxide layers exhibiting a thickness below 5 μm after 20 hours of annealing.

6.7 Summary

MBJ possesses certain advantages over other metal AM methods by eliminating the residual thermal stress in the printed parts and providing a reasonable balance between the productivity and resolution. Additionally, an inert atmosphere is not required for the shaping process. Therefore, compared to PBF processes, extending the build chamber of an MBJ system to print large parts is relatively simple. Furthermore, the applicability of MBJ extends to materials that are difficult to print via PBF, which include non-weldable alloys, pure copper, and refractory metals. However, MBJ merely serves as a complementary technique to PBF and DED instead of their successor, allowing AM to infiltrate traditional metalworking sectors dominated by sintering, such as metal injection molding and powder metallurgy. The existing sintering facilities and parameters can be reutilized for MBJ purposes.

Despite the aforementioned advantages of MBJ, steps must be taken to address its drawbacks, especially with regard to improving the density of the final parts and manipulating their shrinkage during the sintering stage. So far, the amount of research dedicated to the MBJ process is still insufficient, and such inadequacy is manifested in the lack of understanding of its raw materials (powders and binders), building parameters, post-processing (curing and sintering), and numerical modeling.

References

Bai, Y. and Williams, C.B. (2015). An exploration of binder jetting of copper. *Rapid Prototyping Journal* 21: 177–185.

Bai, Y. and Williams, C.B. (2018). Binder jetting additive manufacturing with a particle-free metal ink as a binder precursor. *Materials and Design* 147: 146–156.

Bai, Y., Wagner, G., and Williams, C.B. (2017). Effect of particle size distribution on powder packing and sintering in binder jetting additive manufacturing of metals. *Journal of Manufacturing Science and Engineering* 139 (8): 081019.

Bai, Y., Wall, C., Pham, H. et al. (2019). Characterizing binder–powder interaction in binder jetting additive manufacturing via sessile drop goniometry. *Journal of Manufacturing Science and Engineering* 141 (1): 011005.

Baker, P.R. (1997). Three dimensional printing with fine metal powders. Doctoral dissertation. Massachusetts Institute of Technology.

Chauvet, E., Kontis, P., Jägle, E.A. et al. (2018). Hot cracking mechanism affecting a non-weldable Ni-based superalloy produced by selective electron beam melting. *Acta Materialia* 142: 82–94.

Cramer, C.L., Nandwana, P., Yan, J. et al. (2019). Binder jet additive manufacturing method to fabricate near net shape crack-free highly dense Fe-6.5 wt.% Si soft magnets. *Heliyon* 5 (11): e02804.

Dey, S., Singh, S., Roy, R.K. et al. (2016). Influence of Mn incorporation for Ni on the magnetocaloric properties of rapidly solidified off-stoichiometric NiMnGa ribbons. *Journal of Magnetism and Magnetic Materials* 397: 342–346.

Do, T., Shin, C.S., Stetsko, D. et al. (2015). Improving structural integrity with boron-based additives for 3d printed 420 stainless steel. *Procedia Manufacturing* 1: 263–272.

Emady, H.N., Kayrak-Talay, D., Schwerin, W.C., and Litster, J.D. (2011). Granule formation mechanisms and morphology from single drop impact on powder beds. *Powder Technology* 212 (1): 69–79.

German, R.M. (2010). Thermodynamics of sintering. In: *Sintering of Advanced Materials* (ed. Z.Z. Fang), 3–32. Woodhead Publishing.

Han, C., Fang, Q., Shi, Y. et al. (2020). Recent advances on high-entropy alloys for 3D printing. *Advanced Materials* 32 (26): 1903855.

Hu, D., Wang, Y., Zhang, D. et al. (2015). Experimental investigation on selective laser melting of bulk net-shape pure magnesium. *Materials and Manufacturing Processes* 30 (11): 1298–1304.

Ian Gibson, I.G. (2015). *Additive Manufacturing Technologies 3D Printing, Rapid Prototyping, and Direct Digital Manufacturing*. Springer.

Iveković, A., Montero-Sistiaga, M.L., Vanmeensel, K. et al. (2019). Effect of processing parameters on microstructure and properties of tungsten heavy alloys fabricated by SLM. *International Journal of Refractory Metals and Hard Materials* 82: 23–30.

Karlsson, D., Lindwall, G., Lundbäck, A. et al. (2019). Binder jetting of the AlCoCrFeNi alloy. *Additive Manufacturing* 27: 72–79.

Kumar, A.Y., Bai, Y., Eklund, A., and Williams, C.B. (2018). The effects of hot isostatic pressing on parts fabricated by binder jetting additive manufacturing. *Additive Manufacturing* 24: 115–124.

Kumar, A.Y., Wang, J., Bai, Y. et al. (2019). Impacts of process-induced porosity on material properties of copper made by binder jetting additive manufacturing. *Materials & Design* 182: 108001.

Kwon, K.S. and Kim, W. (2007). A waveform design method for high-speed inkjet printing based on self-sensing measurement. *Sensors and Actuators A: Physical* 140 (1): 75–83.

Li, L., Post, B., Kunc, V. et al. (2017). Additive manufacturing of near-net-shape bonded magnets: prospects and challenges. *Scripta Materialia* 135: 100–104.

Li, Y., Zhou, J., Pavanram, P. et al. (2018). Additively manufactured biodegradable porous magnesium. *Acta Biomaterialia* 67: 378–392.

Li, M., Du, W., Elwany, A. et al. (2020). Metal binder jetting additive manufacturing: a literature review. *Journal of Manufacturing Science and Engineering* 142 (9).

Martin, E., Natarajan, A., Kottilingam, S., and Batmaz, R. (2021). Binder jetting of "hard-to-weld" high gamma prime nickel-based superalloy RENÉ 108. *Additive Manufacturing* 39: 101894.

Miyanaji, H., Zhang, S., Lassell, A. et al. (2016). Process development of porcelain ceramic material with binder jetting process for dental applications. *JOM* 68 (3): 831–841.

Miyanaji, H., Momenzadeh, N., and Yang, L. (2018). Effect of printing speed on quality of printed parts in binder jetting process. *Additive Manufacturing* 20: 1–10.

Mostafaei, A., Stevens, E.L., Hughes, E.T. et al. (2016). Powder bed binder jet printed alloy 625: densification, microstructure and mechanical properties. *Materials and Design* 108: 126–135.

Mostafaei, A., Kimes, K.A., Stevens, E.L. et al. (2017a). Microstructural evolution and magnetic properties of binder jet additive manufactured Ni-Mn-Ga magnetic shape memory alloy foam. *Acta Materialia* 131: 482–490.

Mostafaei, A., Toman, J., Stevens, E.L. et al. (2017b). Microstructural evolution and mechanical properties of differently heat-treated binder jet printed samples from gas-and water-atomized alloy 625 powders. *Acta Materialia* 124: 280–289.

Mostafaei, A., De Vecchis, P.R., Buckenmeyer, M.J. et al. (2019). Microstructural evolution and resulting properties of differently sintered and heat-treated binder-jet 3D-printed Stellite 6. *Materials Science and Engineering C* 102: 276–288.

Mostafaei, A., De Vecchis, P.R., Katerina, A.K. et al. (2021a). Effect of binder saturation and drying time on microstructure and resulting properties of sinter-HIP binder-jet 3D-printed WC-Co composites. *Additive Manufacturing* 46: 102128.

Mostafaei, A., Elliott, A.M., Barnes, J.E. et al. (2021b). Binder jet 3D printing – process parameters, materials, properties, modeling, and challenges. *Progress in Materials Science* 119: 100707.

Nandwana, P., Elliott, A.M., Siddel, D. et al. (2017). Powder bed binder jet 3D printing of Inconel 718: densification, microstructural evolution and challenges. *Current Opinion in Solid State and Materials Science* 21 (4): 207–218.

Niu, X., Shen, H., and Fu, J. (2018). Microstructure and mechanical properties of selective laser melted Mg-9 wt% Al powder mixture. *Materials Letters* 221: 4–7.

Paranthaman, M.P., Shafer, C.S., Elliott, A.M. et al. (2016). Binder jetting: a novel NdFeB bonded magnet fabrication process. *JOM* 68 (7): 1978–1982.

Rahaman, M.N. (2010). Kinetics and mechanisms of densification. In: *Sintering of Advanced Materials* (ed. Z.Z. Fang), 33–64. Woodhead Publishing.

Sachs, E., Cima, M., Williams, P. et al. (1992). Three dimensional printing: rapid tooling and prototypes directly from a CAD model. *Journal of Manufacturing Science and Engineering* 114: 481–488.

Salehi, M., Gupta, M., Maleksaeedi, S., and Sharon, N.M.L. (2018). *Inkjet Based 3D Additive Manufacturing of Metals*. Materials Research Forum LLC.

Salehi, M., Maleksaeedi, S., Nai, S.M.L. et al. (2019a). A paradigm shift towards compositionally zero-sum binderless 3D printing of magnesium alloys via capillary-mediated bridging. *Acta Materialia* 165: 294–306.

Salehi, M., Maleksaeedi, S., Nai, M.L.S., and Gupta, M. (2019b). Towards additive manufacturing of magnesium alloys through integration of binderless 3D printing and rapid microwave sintering. *Additive Manufacturing* 29: 100790.

Sears, J.W., Allen, C., and Holliday, A. (2019). Binder-jet 3D direct metal printing of cobalt chrome moly alloy. AMPM 2019 Conf Phoenix, Arizona.

Sheydaeian, E. and Toyserkani, E. (2018). A new approach for fabrication of titanium-titanium boride periodic composite via additive manufacturing and pressure-less sintering. *Composites Part B: Engineering* 138: 140–148.

Sheydaeian, E., Fishman, Z., Vlasea, M., and Toyserkani, E. (2017a). On the effect of throughout layer thickness variation on properties of additively manufactured cellular titanium structures. *Additive Manufacturing* 18: 40–47.

Sheydaeian, E., Sarikhani, K., Chen, P., and Toyserkani, E. (2017b). Material process development for the fabrication of heterogeneous titanium structures with selective pore morphology by a hybrid additive manufacturing process. *Materials and Design* 135: 142–150.

Stawovy, M.T., Myers, K., and Ohm, S. (2019). Binder jet printing of tungsten heavy alloy. *International Journal of Refractory Metals and Hard Materials* 83: 104981.

Stevens, E., Schloder, S., Bono, E. et al. (2018). Density variation in binder jetting 3D-printed and sintered Ti-6Al-4V. *Additive Manufacturing* 22: 746–752.

Stoyanov, P., Andre, K., Prichard, P. et al. (2016). Microstructural and mechanical characterization of Mo-containing Stellite alloys produced by three dimensional printing. *Procedia Cirp* 45: 167–170.

Su, C., Wang, J., Li, H. et al. (2021). Binder-jetting additive manufacturing of Mg alloy densified by two-step sintering process. *Journal of Manufacturing Processes* 72: 71–79.

Turker, M., Godlinski, D., and Petzoldt, F. (2008). Effect of production parameters on the properties of IN 718 superalloy by three-dimensional printing. *Materials Characterization* 59 (12): 1728–1735.

Utela, B.R., Storti, D., Anderson, R.L., and Ganter, M. (2010). Development process for custom three-dimensional printing (3DP) material systems. *Journal of Manufacturing Science and Engineering* 132 (1): 011008.

Verlee, B., Dormal, T., and Lecomte-Beckers, J. (2012). Density and porosity control of sintered 316L stainless steel parts produced by additive manufacturing. *Powder Metallurgy* 55 (4): 260–267.

Wei, K., Zeng, X., Wang, Z. et al. (2019). Selective laser melting of Mg-Zn binary alloys: effects of Zn content on densification behavior, microstructure, and mechanical property. *Materials Science and Engineering A* 756: 226–236.

Wohlers, T.T. (2014). Wohlers report 2014: 3D printing and additive manufacturing, state of the industry, Annual Worldwide Progress Report. Wohlers Associates Incorporated.

Xiong, Y., Qian, C., and Sun, J. (2012). Fabrication of porous titanium implants by three-dimensional printing and sintering at different temperatures. *Dental Materials Journal* 31 (5): 815–820.

Yan, H., Cannon, W.R., and Shanefield, D.J. (1993). Evolution of carbon during burnout and sintering of tape-cast aluminum nitride. *Journal of the American Ceramic Society* 76 (1): 166–172.

Ziaee, M., Tridas, E.M., and Crane, N.B. (2017). Binder-jet printing of fine stainless steel powder with varied final density. *JOM* 69 (3): 592–596.

7 Applications

The development of metal powder–based additive manufacturing (AM) techniques has progressed rapidly in recent years, and these techniques have been employed in a broad range of industrial applications, indicating their tremendous potential to revolutionize the global manufacturing framework while offering reduced production costs, low energy consumption, and high degrees of customization. This chapter discusses the latest advances in metal powder–based AM products in the aerospace, biomedical, automotive, molding and tooling, energy, jewelry, marine, and oil and gas industries, providing insights of both academic and industrial relevance.

7.1 Aerospace

Research and development sectors within the aerospace industry primarily focus on improving the efficiency of aircraft and reducing air and noise pollution (Behlau 2009). Typical characteristics of aircraft components include complex geometries and large envelope-to-volume ratios, being difficult to machine with high buy-to-fly ratios, small production runs, quick turnaround time, and high performance. Therefore, AM is highly suitable for the fabrication of aircraft components. For example, leading aerospace corporations such as Boeing and Airbus have implemented AM to reduce the production time, construct lighter-weight parts, and lower the operational costs.

Current operations within the aerospace industry include concept modeling, concept prototyping, and low-volume fabrication of complex parts, such as airfoils with embedded cooling channels, engine turbines with intricate internal structures for cooling purposes, and combustion chambers and turbine blades exhibiting thin-walled structures and large envelope-to-volume ratios. Because of these features, the tool path planning algorithm of conventional computer numerical control (CNC) machining is typically complicated and computationally inefficient. Parts manufactured by AM can be quickly shipped and installed in damaged aircraft to resume operation. According to Airbus, the turnaround time for testing or replacing parts can be as short as two weeks.

Metal Powder–Based Additive Manufacturing, First Edition. Kun Zhou and Changjun Han.

Aviation components are typically made of lightweight materials with a high strength-to-weight ratio to improve the fuel efficiency and reduce emissions. Some components are subjected to extreme working conditions, such as ultrahigh/low temperatures or hazardous chemical environments. In particular, engine components are constantly under very high temperatures, and their constituent materials are required to contain flame retardants. Laser-based directed energy deposition (L-DED), powder bed fusion (PBF), and metal binder jetting (MBJ) are popular metal powder–based AM techniques employed in the aerospace industry. These techniques commonly involve materials such as titanium alloys, nickel-based superalloys, aluminum alloys, and special steels.

High-performance aerospace components, such as compressors, blisks, turbines, blades, and airfoils, are usually expensive to manufacture because of the high market value of the materials required (e.g. Ti–6Al–4V and Inconel 718) and the complex manufacturing process involved. Aerospace components may be subjected to corrosion, impact, variable thermal cycles and stresses, and other conditions that could create cracks within them. Fatigue and stress cracks are common initiators of failure in high-performance components that result in them being scrapped. Because of the high cost of materials and labor involved in fabricating such high-value components, it is necessary to repair them whenever possible instead of replacing them.

L-DED is highly suitable for repairing high-value aerospace components. Its basic principle involves utilizing a laser beam to create a metallurgical bond between the damaged surface area of a component and the repair materials (e.g. metal powders). The powder is locally deposited onto the defect area according to its geometry. Compared to conventional repair techniques, L-DED requires a low heat input, leading to the formation of a small heat-affected zone in the surrounding substrate materials, which in turn causes minimal distortion that is ideal for manufacturing thin-walled structures, such as aircraft wing spars and bulkheads.

The company Optomec provides solutions for repairing damaged aircraft engine components through L-DED, which can precisely supply powder to worn-out components and restore their geometry. Integrally bladed rotors, also known as blisks, are high-value components that may each cost US$10 000–100 000 to fabricate. Typically, a single worn-out airfoil results in an entire blisk being scrapped. However, L-DED can be employed to repair such wear. For example, a T700 blisk made of AM355 steel, whose airfoil leading edges had undergone erosion during service, was successfully repaired through L-DED. Figure 7.1a–c illustrates the repair process of the leading edge of a Ti–6Al–4V airfoil, the blisk after deposition, and the blisk after surface finishing. In this case, Optomec selected a cobalt-based wear-resistant material to repair the leading edge.

Fraunhofer ILT has been successfully certified by Rolls-Royce Deutschland for fifteen different repair applications, which include repairing the BR715 high-pressure turbine and BR715 high-pressure compressor front drum. Geometric features on the surface of a BR715 high-pressure turbine case, such as bosses, brackets, and flanges, are repaired locally with just one layer of deposit. Because of the minimized heat input of the L-DED process, distortion can be almost completely avoided. In addition, L-DED can reduce contamination/oxidation by implementing

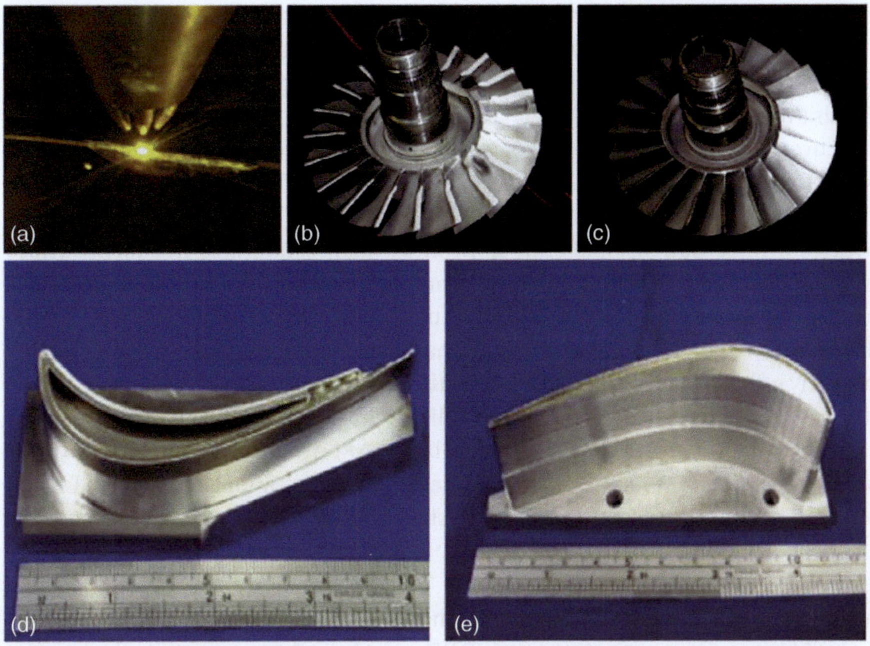

Figure 7.1 Repair of a T700 blisk using the laser engineering net-shaping process. (a) In-process repair of the leading edge of a Ti–6Al–4V airfoil, (b) the blisk after deposition, and (c) the blisk after finishing. Source: Reproduced with permission from Optomec, Inc. (d) Damaged tip and (e) repair of a cast IN738 blade. Source: Xue and Islam (2002)/reproduced with permission from Metal Powder Industries Federation.

proper local gas shielding through a newly developed powder nozzle. Universities and research institutes have also been investigating the usage of L-DED for repair applications in the aerospace industry. Figure 7.1d,e present the tip and seal of a damaged IN738 blade that was repaired through L-DED (Xue and Islam 2002).

RPM Innovations conducted a low-wattage (below 500 W) repair process of a Ti–6Al–4V bearing housing unit within a gas turbine engine using L-DED. The region surrounding the bearing seat was worn beyond standard tolerance limits, and the entire component was initially considered scrap. L-DED was first employed to restore the worn area, followed by machining to meet the tolerance requirements. The housing unit was successfully repaired with no measurable distortion, and its functionality was validated by a test engine. The repair cost was about 50% lower than the price of a new housing unit, and the repair process took only a few days. In comparison, several weeks are required for a new unit to be delivered.

L-DED has also been employed to repair an Inconel 718 gas turbine compressor seal unit, which is a critical high-value component that prevents gas leakage in gas turbine engines (Mudge and Wald 2007). The labyrinth seal in the system underwent excessive wear, which adversely affected the performance of the gas turbine. After being repaired and machined to reach standard tolerance levels, the compressor seal unit was approved for consumer use. The cost of repairing the compressor seal was determined to be 45% of a new unit, indicating the cost-saving capability of L-DED repair processes. Additionally, a turbine airfoil was repaired and remanufactured by L-DED through a novel semi-automated geometric reconstruction algorithm, which also demonstrates the effectiveness of L-DED in repairing defective voids in turbine airfoils (Wilson et al. 2014).

An overhaul and repair method was proposed to restore the complex geometry of expensive aerospace components, which include critical aero-engine components such as thin-curved compressor blades (Yilmaz et al. 2010). This methodology involves a three-dimensional (3D) noncontact digitization and free-form surface reconstruction process using L-DED and milling operations in an automatic hybrid process. It was discovered that the total repair time of the proposed repair methodology is 30% lower than that of conventional repair methods, thereby rendering it a relatively efficient, reliable, and cost-effective approach.

In addition to repair applications, the direct fabrication of aviation components is another essential role of metal powder–based AM in the aerospace industry. The Welding Institute (TWI), an independent research organization, adopted the L-DED process for manufacturing overhanging structures without support mechanisms. A helicopter engine combustion chamber, which is a thin-walled structure, was successfully printed through L-DED (Figure 7.2a). The mechanical testing indicated that the wall density throughout the finished product exceeded 99.5%. Meanwhile, the build time of the part printed by L-DED was about 7.5 hours, while the conventional substrate machining of the same part requires months to complete.

Figure 7.2b presents a turbine housing unit composed of stainless steel fabricated through a LASERTEC 65 3D hybrid manufacturing system developed by DMG Mori (Germany). The dimensions of the turbine housing unit are $180 \times 150\ mm^2$, and the build-up time for deposition and milling were 230 and 76 minutes, respectively. This hybrid manufacturing process produces quality parts by combining the precision of CNC machining and the free-form fabrication capability of the AM process. Figure 7.2c displays a 1/6 scale mixing nozzle of the gas turbine exhaust system of a Bell helicopter, which was manufactured by Optomec through L-DED. Meanwhile, a Ti–6Al–4V airfoil with embedded cooling channels was fabricated by Xue and Islam (2006), as shown in Figure 7.2d.

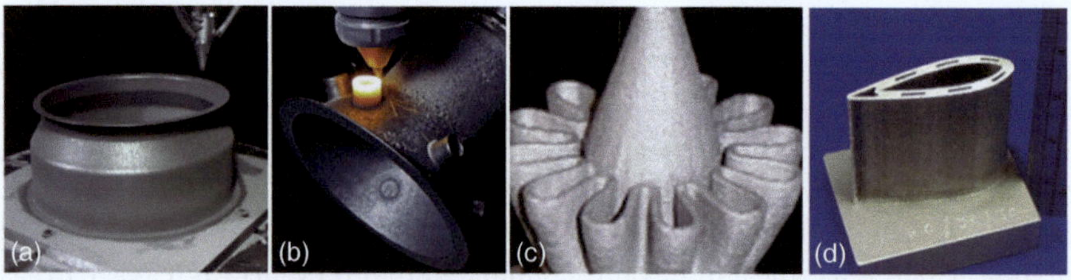

Figure 7.2 (a) IN718 helicopter engine combustion chamber. Source: Reproduced with permission from TWI Ltd. (b) Turbine housing unit. Source: Reproduced with permission from DMG Mori. (c) 1/6 scale mixing nozzle for a gas turbine exhaust system. Source: Reproduced with permission from Optomec, Inc. (d) Ti–6Al–4V airfoil with embedded cooling channels. Source: Xue and Islam (2006)/reproduced with permission from Metal Powder Industries Federation.

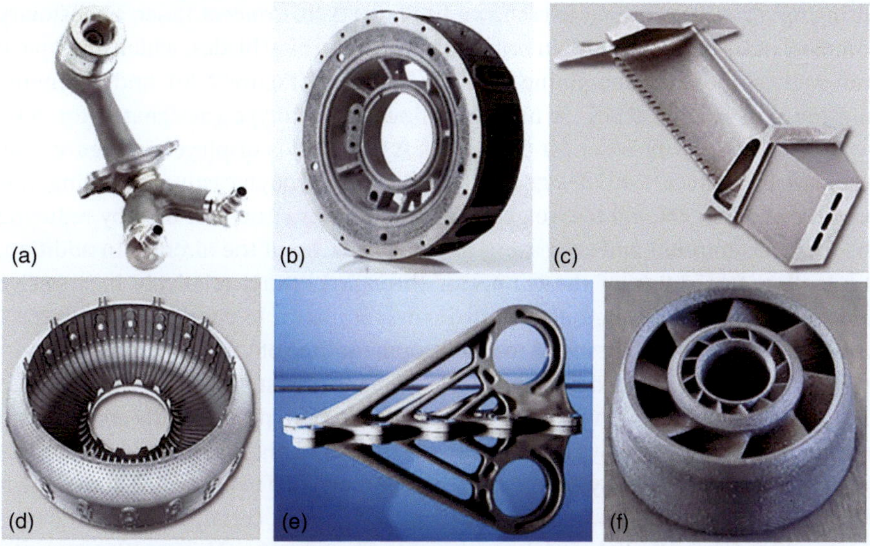

Figure 7.3 (a) GE LEAP engine fuel nozzle fabricated through LPBF. (b) Compressor supporting casing for a gas turbine engine fabricated by EBM. (c) LPBF-printed turbine blade with internal cooling channels, (d) LPBF-printed engine housing unit, and (e) the first LPBF-printed titanium bracket connector for the Airbus A350 XWB. Source: Reproduced with permission from General Electric. (f) Hollow static turbine blade cast with its mold and cores fabricated by MBJ. Source: Reproduced with permission from PROMETAL S.A.R.L.

A wide range of PBF AM materials, including stainless and tool steels, titanium and its alloys, nickel-based alloys, and certain aluminum alloys, are employed in the aerospace industry. For example, General Electric (GE) Aviation manufactures next-generation jet engine components through the laser powder bed fusion (LPBF) process. The newly printed fuel nozzle of a "Leading Edge Aviation Propulsion" (LEAP) engine produced by GE is shown in Figure 7.3a. Compared to conventional fuel nozzle designs, the LPBF-designated nozzle design integrates 18 components with complex passageways, thereby reducing the number of brazes and welds from 25 to 5. As a result, the weight of the LPBF-printed nozzles is 25% less than that of their predecessors. Each LEAP engine includes 19 nozzles, and GE Aviation has set the goal of fabricating 32 000 nozzles per year. This number has been increased to 100 000 by 2020, during which GE has reached full production capability. A total of US$3.5 billion has been invested in new industrial plants to produce LEAP engine nozzles, whose functionality has already been assessed. On a different note, 35% of the components of an advanced turboprop demonstration engine that powers the state-of-the-art Cessna Denali single-engine aircraft are fabricated by AM, which corresponds to a weight reduction of 5% and a specific fuel consumption improvement of 1% (General Electric 2016).

Using titanium alloys, Arcam applies its electron beam melting (EBM) technology to manufacture various functional components, including commercial and military aircraft components, aerospace devices, missiles, and various subsystems (e.g. engines and accessories). An EBM-printed Ti–6Al–4V compressor supporting casing for a gas turbine engine is shown in Figure 7.3b. Concept Laser, a subsidiary company of GE, utilizes LPBF to print Inconel 718 turbine blades, which are typical thin-wall components containing complex channels (Figure 7.3c), and an engine housing unit (Figure 7.3d). A bracket connector prototype (designated for AM) developed by Concept Laser for the Airbus A350 XWB is displayed in Figure 7.3e. Compared to conventional designs that are realized through casting or milling, this prototype boasts 30% lighter weight due to its porous structure, thereby reducing the fuel consumption and increasing the load capacity of the aircraft. In addition, the toolless fabrication of the connector through LPBF is relatively inexpensive and expeditious, requiring only approximately a month to complete. In contrast, conventional designs correspond to a development time of up to six months.

A large build volume and high build rate are characteristics of MBJ, which render the low-volume production of metal-based casting prototypes (typically composed of aluminum and copper alloys, gray and ductile iron, and magnesium and its alloys) a reality. Based on a report by ExOne, MBJ can be employed to fabricate complex gear cases and covers, fuel tanks, transmission housing units, certain components containing draft-free walls, lightweight engine parts, and structural hinges. For example, the mold and cores of a hollow static turbine blade in Airbus A320 were manufactured through MBJ (Figure 7.3f).

The fabrication of spacecraft and satellites has always entailed the most advanced technologies. With the knowledge that AM can satisfy the stringent requirements of the aerospace industry while remaining cost-efficient, an increasing number of companies are venturing into this domain. Figure 7.4a shows a full-scale copper rocket engine component printed through LPBF by the National Aeronautics

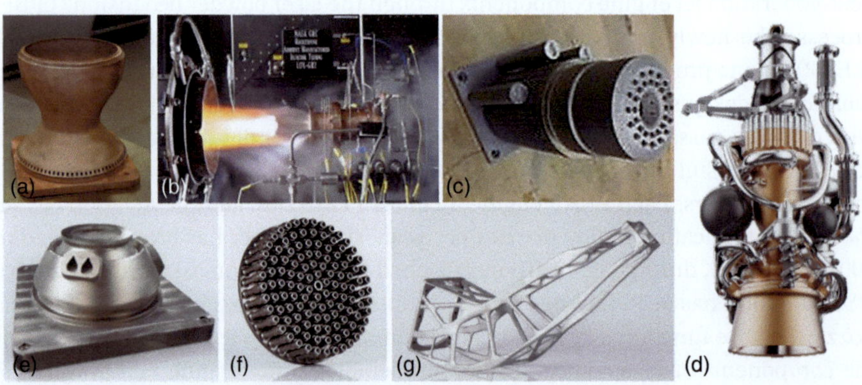

Figure 7.4 Typical LPBF-printed components for the aerospace sector. (a) Copper rocket nozzle, (b) testing a printed rocket nozzle, and (c) rocket injector printed using an Inconel steel powder. Source: Reproduced with permission from NASA. (d) Combustion chamber assembled with printed copper components and conventionally manufactured components. Source: Reproduced with permission from Launcher. (e) Injector head of the all-in-one design of a rocket engine with 122 injection elements using EOS Nickel Alloy IN718 and (f) Ariane 6 baseplate for function integration. Source: Reproduced with permission from ArianeGroup. (g) Al–Si10–Mg satellite bracket. Source: Reproduced with permission from RUAG Group.

and Space Administration (NASA) in 2015, which was designed to operate under extreme temperature and pressure. In 2005, Aerojet Rocketdyne manufactured and conducted the hot-fire test of a rocket engine, whose thrust chamber was composed of a copper alloy and printed through LPBF. The AR1 single-element hot-fire tests were conducted at an excess of 2000 psi, which corresponds to the highest pressure for rocket engine applications. In producing the main injector alone, LPBF offers the potential for a nine-month reduction in part lead time and a 70% reduction in cost.

Figure 7.4b presents a liquid oxygen/gaseous hydrogen rocket injector assembly built through AM and hot-fire tested at the NASA Glenn Research Center. Significant costs may be reduced by decreasing the number of certified parts and processes, such as joining, which are required to manufacture a component. In addition, weight reduction of the product can result in significantly lower fuel consumption when escaping the gravitational force of the Earth. NASA engineers at the Marshall Space Flight Center in Huntsville adopted LPBF to fabricate a rocket injector using an Inconel steel powder (Figure 7.4c). While traditional subscale rocket injectors are composed of four parts and five welds, each of which requires six months to produce and costs over US$10 000, the construction of the LPBF-printed injector was completed in about 40 hours for less than US$5000.

In collaboration with EOS and Additive Manufacturing Customized Machines (AMCM), a US start-up company, Launcher, is searching for more efficient methods of launching small-to-medium-sized payloads into space. Using AMCM's M4K machines, Launcher additively manufactured a rocket engine combustion chamber with an integrated nozzle-and-neck system from a copper alloy. At a height of 86 cm and an outlet nozzle diameter of 41 cm, the product was the largest single-piece combustion chamber for liquid rocket engines in the world (Figure 7.4d).

ArianeGroup is responsible for developing and producing Ariane 6, a new European launch vehicle for placing heavy payloads, such as communication satellites, into geostationary orbit. The internals of a propulsion module are subjected to tremendous forces and extreme conditions, and they are required to attain maximum levels of reliability and precision. The injection head is a core element of a propulsion module, and its role is to feed the fuel mixture to the combustion chamber. The traditional design of the injector head consists of 248 components, which are produced and assembled in different manufacturing steps. Additionally, the conventional production of injector heads requires over 8000 cross holes to be drilled in copper sleeves that are subsequently precisely screwed into 122 injector elements for mixing hydrogen gas. EOS printed a single injector head unit through LPBF, which possessed integrated base and front plates, 122 injection nozzles, and a distribution dome with corresponding feed pipes for hydrogen and oxygen fuels (Figure 7.4e,f). Consequently, a quantum leap in its lead time was achieved, with each iteration reduced from approximately half a year to a few days (a reduction of 131 days) at half the cost.

The Swiss RUAG group aims to develop an extremely lightweight and robust antenna bracket for Sentinel satellites. In collaboration with EOS, such a prototype was printed from Al–Si10–Mg powder through LPBF after topology optimization

(Figure 7.4g). The as-printed product exceeded the minimum rigidity requirement for satellite antenna brackets by over 30%, which renders it capable of transmitting strong radio signals to the Earth even after a turbulent flight, assuming that an ideal antenna position has been attained previously. Such high rigidity can be attributed to the highly uniform stress distribution within the satellite antenna bracket, which was achieved through structural optimization simulations. The use of LPBF significantly reduced the weight of the component from 1600 g to 940 g, corresponding to a decrease of over 40%. The component was certified and approved for use in outer space.

7.2 Biomedical

The metal powder-based AM has been applied in the manufacturing of orthopedic and dental implants, which bridges the gap between biology and engineering through the creation of complex biocompatible and bioactive constructs that take advantage of unique material properties (such as osteoconductivity and osteoinductivity) to promote tissue regeneration and implant acceptance. The application of AM in fabricating biomedical implants benefits from its capability to manufacture highly customizable prosthetic components with complex geometries for effective bone integration. Compared to conventional processes, the overall production of a component through powder-based AM is automated to minimize human error and surgical intervention time. A computed tomography (CT) scan is first performed on the affected zone to reconstruct the replacement tissue and convert it into a model to be printed. A polymer model of the region surrounding an implant is normally fabricated with stereolithography, which allows for necessary adjustments to be made to the implant before it is installed, thereby averting the risk of additional surgical interventions. Such changes also benefit the patients during the postoperative period, as they may recover in a shorter duration.

The orthopedic field is one of the biomedical branches that has received the most attention. A few concerns exist regarding common metallic implants that replace load-bearing joints in the knees and hips. In particular, the release of metallic ions from an implant comprising traditional biomedical materials, such as stainless steel, cobalt–chrome, and titanium alloys, can potentially result in hypertension and various undesirable tissue reactions in the patient. Additionally, it is impossible to avoid the wear-induced damage of a metallic implant during its usage. Moreover, installing dense and highly stiff metallic implants exhibiting isotropic mechanical properties can reduce the amount of load distributed to the host bone. Such changes in load distribution lead to stress shielding, which may result in deterioration of the host bone or premature loosening of the implant. Therefore, optimizing the implant design can mitigate the aforementioned hazards to a patient's health.

Among metal powder-based AM processes, LPBF and EBM have been recognized as the most capable in exerting precise control over internal porous structures and achieving complex geometries with tailored properties, which is a major improvement for orthopedic implants (Han et al. 2020). Furthermore, such porous structures allow for the difference in stiffness between an implant and the host bone to be reduced, which subsequently alleviates stress shielding (Han et al. 2017b, 2018b; Yang et al. 2020). In addition to 316L and cobalt–chrome alloys, titanium and its alloys, such as pure Ti, Ti–6Al–4V, Ti–Nb (Zhao et al. 2020), and Ti–Ta (Zhao et al. 2019, 2021), have been established as the most in-demand metallic biomaterials for printing porous orthopedic structures. The types of unit cells for constructing such porous structures include diamond, hatched, honeycomb-like, cellular, cubic, reticulated mesh, stochastic foam, triangle, hexagonal, rectangular, dodecahedron, rhombic, and triply periodic minimal surface-based structures (Schwarz diamond, Schoen Gyroid, etc.).

The earliest application of LPBF in the orthopedic field occurred in 2005 when an MCP Realizer machine was employed to fabricate a replacement mandible, lumbar vertebra, and tubular femoral bone using 316L stainless steel (Figure 7.5a) (Wehmöller et al. 2005). Figure 7.5b–d present an LPBF-printed Ti–6Al–4V human mandible implant (Bertol et al. 2010), a Co–Cr–Mo talar component for an endoprosthetic ankle device (Liverani et al. 2016), and a customized Ti–6Al–4V ELI frontoparietal-temporal bone implant that precisely restores a patient's cranial defects (Jardini et al. 2014). The endoprosthetic ankle device had undergone a kinematic test on a cadaver leg. The results reveal that the customized articular surfaces of the implant allow it to successfully reproduce natural joint motion, which demonstrates the suitability of LPBF for patient personalization in the scenario of a complete ankle replacement. The fabrication of oral screw-retained implants, hip endoprosthetics, acetabular cups, and other biomedical devices has also validated the applicability of LPBF in the biomedical industry (Vandenbroucke and Kruth 2007; Bertol et al. 2010; Liu et al. 2015).

Figure 7.5e–g show an EBM-printed Co–29Cr–6Mo alloy femoral knee implant made with lattice structures (Murr et al. 2012), acetabular cups printed by Arcam, and a skull plate (Mazzoli et al. 2009), respectively. Using the Arcam EBM technique, Adler Ortho Group launched the CE-certified Fixa Ti-Por acetabular cup in the European market in 2007. Figure 7.5h displays hip stems printed by L-DED for load-bearing purposes (España et al. 2010). Although obtaining the certification for a printed implant from the National Medical Products Administration is a rather long and challenging process, these cases indicate great potential for metal powder–based AM technology to be utilized in manufacturing biomedical implants.

Patients with pelvic sarcomas typically require major bone grafts and implant installation after resectioning the cancerous tissue, but complex implant systems often fail shortly after being installed. For patients who have undergone major tissue removal, customized implants can be extremely beneficial since many large bone defects cannot be easily restored by standard implants. A case report has verified the clinical usage of a printed Ti–6Al–4V pelvic bone implant after removing a giant chondrosarcoma tumor (Chen et al. 2020). The implant possesses a porous structure to mitigate stress shielding (Figure 7.6a,b), and the patient achieved rapid recovery after the implant was installed (Figure 7.6c). No failure or loosening of the implant was observed 12 months after the patient began walking (Figure 7.6d).

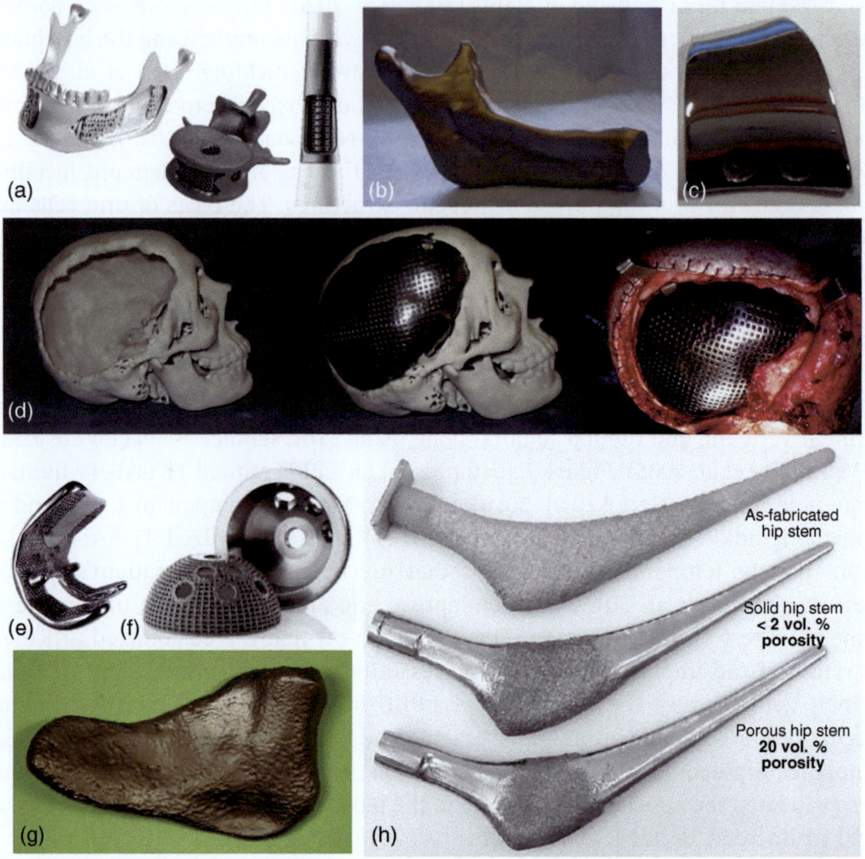

Figure 7.5 Implants fabricated by metal powder–based AM. Applications of LPBF. (a) Replacement mandible, lumbar vertebra, and tubular femoral bone. Source: Wehmöller et al. (2005)/reproduced with permission from Elsevier. (b) Mandible implant. Source: Bertol et al. (2010)/reproduced with permission from Elsevier. (c) Talar component. Source: Liverani et al. (2016)/reproduced with permission from Elsevier. (d) Frontoparietal-temporal bone. Source: Jardini et al. (2014)/reproduced with permission from Elsevier. Use of EBM: (e) Femoral knee implant. Source: Murr et al. (2012)/reproduced with permission from Elsevier. (f) Ti–6Al–4V acetabular cups. Source: Reproduced with permission from General Electric. (g) Skull plate. Source: Mazzoli et al. (2009)/reproduced with permission from Elsevier. (h) Hip stems with designed porosity values fabricated by L-DED. Source: España et al. (2010)/reproduced with permission from Elsevier.

With adherence to metal AM intraoperative guidelines and adequate preoperative planning, high-precision tumor removal can be conducted by most surgeons. Due to the anatomic variations in bone defects among different cancer patients who have undergone tumor removal, a highly personalized implant with soft tissue docks directly benefits their recovery. Printed metallic implants exhibit precise morphology that facilitates the attachment of muscles and ligaments, thereby increasing the stability of the implants and rendering it possible for cancer patients to achieve enhanced recovery after their surgery.

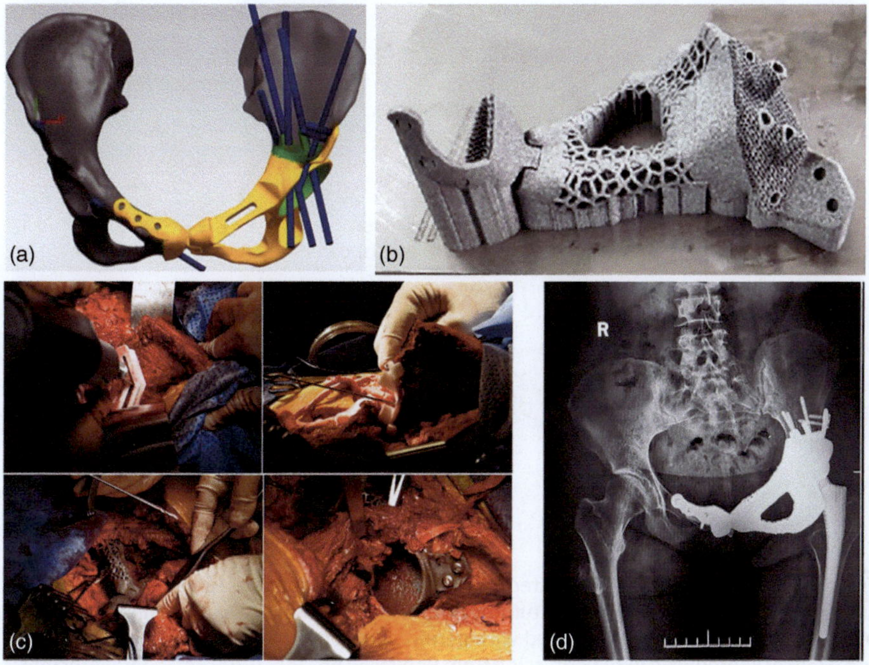

Figure 7.6 (a) Design of an implant model, (b) printed Ti–6Al–4 V implant, (c) surgical procedure of tumor resection and implant installation, and (d) computed tomography (CT) image indicating no failure or loosening of the implant 12 months after the surgery. Source: Chen et al. (2020)/reproduced with permission from SAGE Publications Inc., CC BY-NC 4.0.

Implant dentistry has improved the rehabilitation of edentulous patients, particularly individuals with alveolar bone resorption issues and poor reception toward conventional dentures. A variety of dental implants are available, but limitations on their designs and sizes may sometimes necessitate graft surgeries. The most common biocompatible dental materials are cobalt–chromium-based alloys, such as Co–Cr–Mo, Co–Cr, Co–Cr–Cu, and Co–Cr–W alloys, which have been established to fulfill the mechanical requirements of the ISO 22674:2006 standard with a yield stress of at least 500 MPa (Lu et al. 2015). A core issue that stems from such metallic implants is the release of metallic ions, which may result in diverse hypersensitivity or other undesirable tissue reactions for the patients. A previous *in vitro* study determined that in a simulated saliva environment and cell culture media, fewer Co ions were released from LPBF-printed samples than their cast counterparts (Xin et al. 2012). Co–Cr alloys manufactured via LPBF corresponded to higher cell proliferation and lower biological risks. Several companies, such as EOS, Concept Laser, and Renishaw, have utilized LPBF to produce copings for dental crowns and bridges, as shown in Figure 7.7.

Surface modification and coating are effective approaches for enhancing osseointegration between living tissue and metallic implants. Calcium phosphate ceramic is the most popular coating material for metallic implants, and calcium phosphate

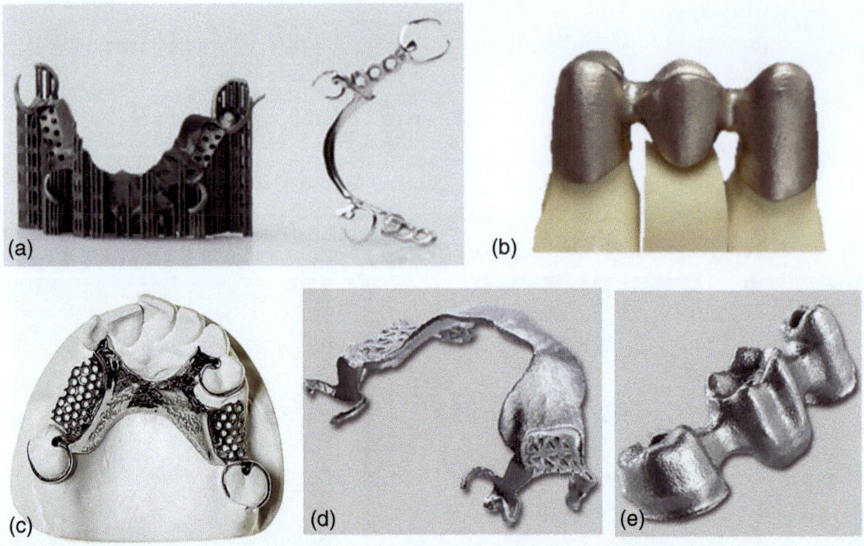

Figure 7.7 Dental implants fabricated by LPBF. (a) Co–Cr removable partial denture with supporting structures after manufacturing and the final product with a polished surface and (b) Dental bridge. Source: Reproduced with permission from EOS. (c) Removable partial denture. Source: Reproduced with permission from Renishaw plc. (d) Ti–6Al–4V dental prosthesis and (e) Co–Cr dental bridge. Source: Reproduced with permission from General Electric.

coatings with graded composition and a controlled porosity can be manufactured through L-DED to boost their biocompatibility and improve osseointegration (Bose et al. 2018). Metallic implants and bioceramic coatings can be simultaneously processed through L-DED. A widely studied bioactive coating material, hydroxyapatite, has been determined to improve the biological inertness of pure titanium (Han et al. 2017a; Han et al. 2018a). Electrophoretic deposition is a liquid-based method with a short processing time. It requires only simple equipment, provides easy control of parameters, and boasts a wide range of possible coating materials. Such materials have been explored to design novel coatings for metallic porous structures (Han et al. 2017c). Specifically, a gentamicin-loaded silk fibroin coating with a thickness of 2.3 μm was successfully prepared and adsorbed onto LPBF-printed porous Co–Cr–Mo bone substitutes through process optimization, which enhanced the initial osteoblastic response of the implants and suppressed bacterial activity.

The development of metal powder–based AM techniques within the biomedical industry is predicted to continue in the foreseeable future due to their ability to align with in-demand medical device components. Most medical devices, such as dental crowns and surgical implants, require miniaturization, and hence AM techniques are highly applicable for fabricating such equipment. Additionally, these techniques can achieve clinical efficacy, for which the as-produced medical components can be specifically tailored to each patient and exhibit excellent properties. Meanwhile, metal powder–based AM companies are also introducing advanced machines for processing metal products rapidly and cost-effectively.

7.3 Automotive

The automotive industry is one of the most competitive business domains, in which a decrease in production time and time to market grants a massive advantage over fellow competitors. Driven by novel design trends and technological evolution, of which esthetics, aerodynamics, safety, and vehicle weight reduction are critical issues, automotive companies are developing new models and innovative measures daily. Metal powder–based AM has been utilized as an important tool in the design and development of small quantities of structural and functional automotive components, such as engine exhausts, driving shafts, gearbox components, and braking systems for low-volume luxury automobiles. Such extensive application of AM can be attributed to its ability to shorten the development cycle and reduce production costs. However, AM is currently only employed for prototyping and the direct manufacturing of non-safety-relevant components in small batches, mainly because the process reliability and consistency of AM are still limited.

Companies and research institutes have successfully applied AM techniques to manufacture functional components for motorsport vehicles. These usually comprise lightweight alloys (e.g. titanium) and possess highly complex structures and low production volumes. For example, CRP Technology of Italy has successfully applied LPBF and EBM to develop various components for motorsports, including Formula One (F1) titanium gearboxes, MotoGP 250R air boxes, motorbike dashboards and supports, camshaft covers for MotoGP engines, reed valves, F1 suspension systems, etc. Compared to their conventionally manufactured counterparts, F1 gearboxes fabricated through AM were approximately 20–25% lighter and 20% more compact while possessing twice the torsion stiffness with minor gear wear and power absorption. For example, the steering knuckle of a race car, which was fabricated by EOS via LPBF, is shown in Figure 7.8a. The weight of the printed steering knuckle was reduced by 35%, and its stiffness was improved by 20% compared to its conventionally manufactured counterpart.

Optomec manufactured a Ti–6Al–4V suspension mounting bracket (Figure 7.8b) and driving shaft spiders for a Red Bull race car through L-DED, achieving a material reduction of over 90% and significantly lowering the production time and cost. Arcam applied EBM to produce Ti–6Al–4V components for race cars, such as gearboxes (Figure 7.8c), suspension parts, and engine parts with lattice structures (Figure 7.8d). Concept Laser has produced steel and aluminum automotive components via LPBF, which include wheel suspension systems, oil pump housing units (Figure 7.8e), engine blocks, exhaust manifolds (Figure 7.8f), and valve blocks. A water pump for motorsports cars was also fabricated through LPBF using Al–Si10–Mg (Figure 7.8g). It exhibited mechanical properties comparable to its conventionally heat-treated counterparts (Vilaro et al. 2008).

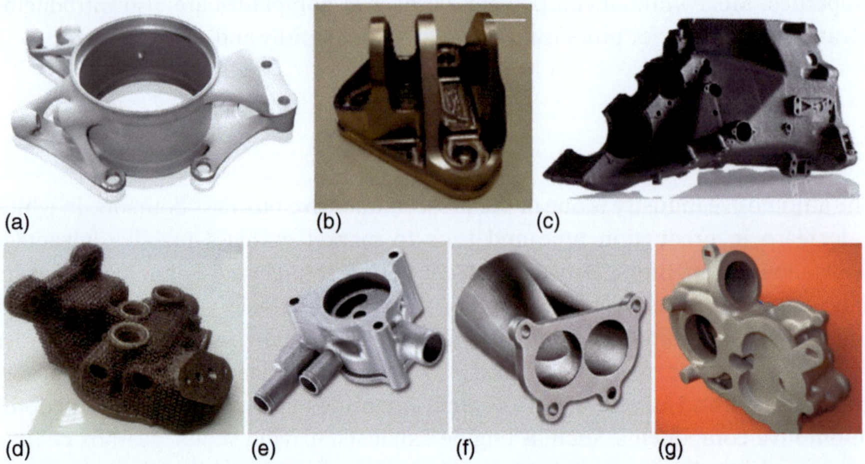

Figure 7.8 (a) Race car steering knuckle produced by LPBF. Source: Reproduced with permission from EOS. (b) Suspension mounting bracket for a Red Bull race car produced through L-DED. Source: Reproduced with permission from Optomec, Inc. (c) Race car gearbox and (d) Ti–6A1–4V engine component with lattice structures fabricated through EBM. (e) Oil pump housing unit and (f) exhaust manifold produced through LPBF. Source: Reproduced with permission from General Electric. (g) Water pump unit for a motorsports car produced through LPBF. Source: Vilaro et al. (2008)/reproduced with permission from European Forum.

Delphi, a top-tier automotive supplier, employs LPBF instead of traditional machining to fabricate diesel pumps. The LPBF process allows the pumps to be manufactured as single units. It bypasses several operations, such as drilling, machining, and chemical deburring, thereby reducing the overall production costs and yielding a final product less prone to leakage. Conventional gravity-casting and machining processes for producing pump housing units typically require a buy-to-fly ratio of 2 : 1. But with LPBF, a significantly lower buy-to-fly ratio of 1.4 : 1 can be achieved for the same units.

Extensive efforts on material development in the automotive industry have been expended. For instance, EOS Maraging Steel MS1 developed by EOS GmbH has a chemical composition similar to that of 1.2709 steel (X3NiCoMoTi 18-9-5), and it provides a viable alternative to the latter in the production of conventional tool inserts. Meanwhile, a TiAl alloy with low density, high specific strength (ratio of elastic modulus to density), and high stiffness (ratio of yield strength to density) designated for EBM has been adopted for fabricating automotive engine components (e.g. engine exhaust valves and pistons) (Murr et al. 2010).

While the rapid prototyping of functional test components remains a highly attractive feature of metal powder–based AM, the production of specialized parts is also actively pursued (e.g. parts involved in vintage automotive restoration). The mass production of automotive components poses a huge challenge for the aforementioned printing processes, i.e. EBM and LPBF. However, small-scale production of automotive metal parts can be easily achieved. In particular, printing a mold from sand or plastic through binder jetting followed by casting is a popular method, and casting large complex parts through such molds can reduce the development time, allowing for multiple design iterations during the prototyping phase. Notably, ExOne produced a batch of five castings from AM molds at €1500 per unit cost. In contrast, conventional patterns, tools, and lost foam casting methods could cost up to €15 000–20 000.

7.4 Molding and Tooling

Metal powder–based AM processes have been employed to manufacture highly geometrically complex molds and tools. Such processes can produce mass-production molds and tools containing 3D conformal cooling channels, which are difficult to fabricate through conventional processes (casting, machining, forging, etc.) and must meet stringent requirements with respect to their properties (hardness, wear resistance, strength, heat conduction, etc.) and service life.

The cooling of parts in molds is the most time-consuming phase of the plastic injection molding process. A decrease in the cooling time boosts the production rate, produces molded parts of higher quality, and reduces wastage. Various conventional techniques, such as implementing bubbler cooling systems and heat pipes and performing complex drilling operations using laminated blocks, have been utilized to maintain a uniform temperature during the cooling process, which is usually tedious and inefficient. Conventional cooling channels are commonly machined in straight lines, rendering them unable to facilitate reliable cooling across a mold cavity due to non-uniform heat conduction. Inhomogeneous cooling results in long cycle durations, warpage, and possible scrapping of the final parts. Conformal cooling molds possess curved cooling channels that align closely with the overall geometry or contours of the mold, thereby maintaining a uniform temperature throughout the structure. While conformal cooling solutions can significantly reduce the total production cost by decreasing the mold cycle time, they typically require sophisticated individual mold designs that constitute a wide variety of unconventional curves, twists, and shapes that must be precisely implemented.

The design and fabrication freedom of metal powder–based AM techniques spurs the application of conformal cooling channels in the molding and tooling industries. Compared to conventional processes, utilizing such techniques can realize complex contour designs on the surface of a mold, particularly for molds with multiple cavities, which are utilized in the areas of injection molding, die casting, blow molding, hot extrusion, and hot stamping (Shinde and Ashtankar 2017). Additionally, AM can be employed to repair or modify existing tools and extend the lifespan or improve the performance of these products.

Presently, LPBF is the most popular metal powder–based AM process for printing molds and tools. Numerous commercial companies have been exploring the applicability of AM in manufacturing geometrically complex molds containing conformal cooling channels (Figure 7.9). For example, B&J Specialty, Inc. utilized a Prox DMP 300 machine developed by 3D Systems to produce a mold containing specialized inner cooling channels optimized through numerical simulations (Figure 7.9b). The printed product corresponded to a reduction in cycle cooling time from 1 minute to 40 seconds, an 86% reduction in temperature variation during cooling, an increased

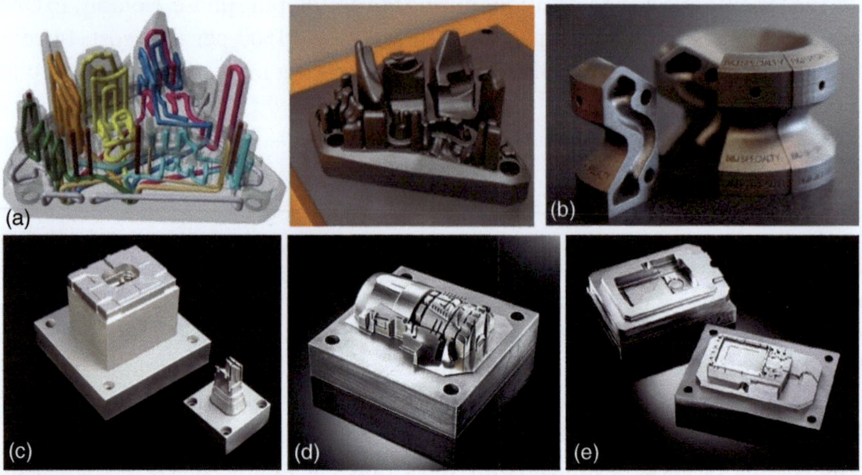

Figure 7.9 Mold products printed through LPBF by commercial companies. (a) Mold with eight complex conformal cooling channels. Source: Reproduced with permission from Renishaw plc. (b) Mold with inner cooling channels optimized through numerical modeling. Source: Reproduced with permission from B&J Specialty, Inc. Molds printed using a hybrid metal 3D printer for (c) waterproof connector, (d) electric driver, and (e) digital camera. Source: Reproduced with permission from Matsuura Machinery Corporation.

mold lifetime, and the delivery of higher-quality parts, thereby decreasing the production time and costs for tool manufacturers and mold operators. Employing a machine-aided LPBF strategy, Matsuura Machinery Corporation printed molds for a waterproof connector, an electric driver, and a digital camera through a hybrid metal 3D printer using Matsuura Maraging II powder (Figure 7.9c–e). The hybrid-manufactured molds have significantly improved surface quality compared to those printed without machining.

The installation of complex conformal cooling channels within mold inserts expedites the molding process and improves the quality of the products. Figure 7.10a presents a finished and polished mold insert with complex cooling channels, which was printed through LPBF by GPI Prototype and Manufacturing Services. The printing process lasted about 39 hours and costs US$3300. The printed mold insert was still in use after 190 000 shots. Oskar Frech GmbH initiated industrial-scale AM over a decade ago and introduced the first Frech Gating System distributors for branch gate die casting in 2007. The main applications of AM in die casting include the fabrication of distributors, prototypes, and inserts for die-casting molds capable of facilitating conformal cooling. With an SLM280 machine, Frech primarily used an aluminum alloy and a tool steel 1.2709 to produce prototypes and inserts for die-casting molds containing complex conformal cooling channels, respectively (Figure 7.10b), which provides a novel and efficient approach for tempering cores, wipers, and even anvils. A cooling time reduction of about 60% was achieved for the printed mold, and the total process cycle time was shortened by 12% consequently. Through LPBF, Additive MFG implemented conformal cooling channels within

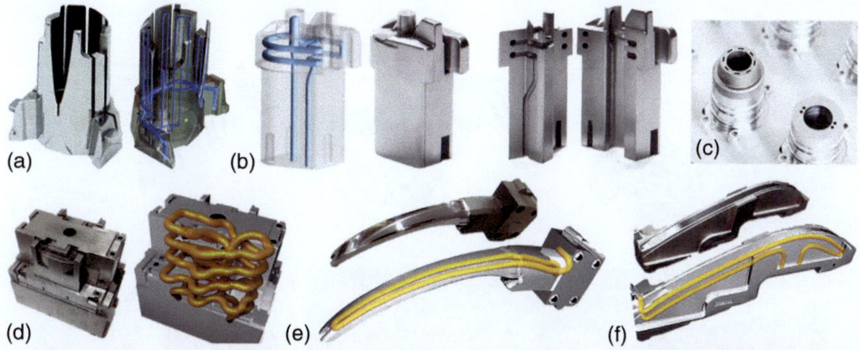

Figure 7.10 Mold inserts with conformal cooling channels printed via LPBF. (a) finished and polished mold. Source: Reproduced with permission from GPI Prototype and Manufacturing Services. (b) Printed mold that could achieve a reduction of about 60% in its cooling time. Source: Reproduced with permission from SLM Solutions. (c) Printed molds with different channels. Source: Reproduced with permission from EOS. Mold inserts for different applications. (d) Contactor bracket, (e) car handle, and (f) car light lens. Source: Reproduced with permission from General Electric.

mold inserts for contactor brackets, car handles, and car light lenses while partnering with Schneider Electric, Volkswagen, and DS Automobiles, respectively. Figure 7.10c–f display these products, which were printed using an EOS M 290 printer. The quality of the injection parts was tremendously improved compared to their conventionally manufactured counterparts.

Conventional tool production processes reach their limitations where the tempering channels can only be drilled in straight lines, which often results in coolants being unable to reach critical hotspots. As a result, the tool may become deformed, and its service life will be reduced. Additionally, partially deficient heat dissipation is responsible for a long cooling time. LPBF powder materials suitable for hot stamping tooling are maraging steels, CL50WS, EOS Stainless Steel CX, and H13 (Chantzis et al. 2020). LBC Laser Bearbeitungs Center GmbH designed and manufactured a tool insert with a significantly optimized tempering performance using an EOS-developed Maraging Steel MS1 powder for the die casting process. During manufacturing, tempering channels were integrated into the tool at defined points, thereby significantly improving its cooling performance and achieving a 20% reduction in its cycle time. A few typical tooling inserts printed through LPBF are shown in Figure 7.11.

Fraunhofer ILT has combined the advantages of conformal cooling channels with the excellent heat conductivity of copper and implemented them within a tool, as shown in Figure 7.11d, thereby reducing its cycle time and minimizing warpage. SLM Solutions also reduced the cycle time of an injection molding tool from 60.5 to 14.7 s by optimizing its conformal cooling insert (Figure 7.11e). Additionally, the cooling time was reduced by about six seconds.

A comparison between AM and conventional manufacturing procedures provides suitable indicators for the future process selection for tool fabrication. For example, in the automotive stamping industry, blank inserts are typically laminated and cut to

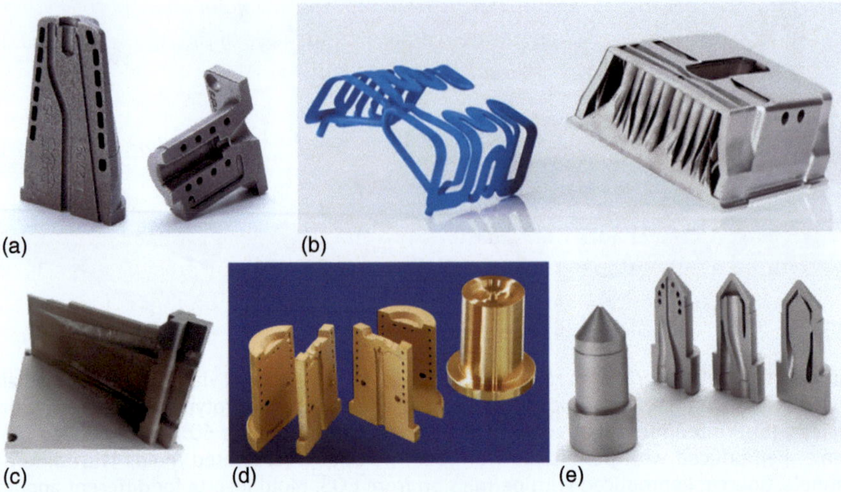

Figure 7.11 Tool inserts printed via LPBF. (a) Printed inserts for die casting, (b) lightweight Audi Hotforming Tool insert with conformal cooling channels, and (c) automotive armrest. Source: Reproduced with permission from EOS. (d) Inserts printed using Hovadur K220. Source: Reproduced with permission from Fraunhofer ILT. (e) Injection molding tool inserts after channel optimization. Source: Reproduced with permission from SLM Solutions.

a near-net shape before undergoing heat treatment to obtain the required properties. Subsequently, the desired surface quality and final tolerance are achieved through proper surface finishing. Meanwhile, tool inserts may be produced through casting processes employing pre-manufactured lost foam models, which may require outsourcing and significantly increase the final tooling lead time. The utilization of LPBF to fabricate cutting tools has been proven to shorten the lead time (Leal et al. 2017). In addition, despite its relatively high processing costs, cost breakdown analyses have indicated that LPBF requires substantially fewer operations than the lost foam process, thereby simplifying the internal logistics.

The cooling performance of an EBM-printed H13 tool insert with conformal cooling channels was found to be superior to that of its traditionally machined counterpart with straight cooling channels (Rännar et al. 2007), which boosts the productivity and the dimensional accuracy of the final products. However, because of a relatively large beam spot size, the EBM process encounters difficulties in printing parts featuring a small radius. Therefore, the shape and dimensions of the cross section of a cooling channel should be considered and optimized before fabricating it via EBM.

In addition to the direct fabrication of molds, tools, and tool inserts, another significant application of AM constitutes the repair and remanufacturing of such instruments. Compared to traditional repair techniques (e.g. tungsten inert gas/plasma and gas tungsten arc welding), L-DED is capable of providing precise, small deposits with good metallurgical properties through a focused and highly concentrated laser beam. For example, a CPM 9 V steel powder was selected to repair the surface of a H13 tool steel plate through L-DED (Kattire et al. 2015), and compressive residual stress was present within the plate, which impeded crack propagation and increased its service life. In addition, a Ni–Co–Mo alloy has been developed as part of a fast, efficient, and cost-effective repair strategy for worn surfaces of maraging steel molds (Grum and Slabe 2004), which increases their durability.

7.5 Energy

Efficient energy conversion, transmission, and storage are the ultimate goals of the energy industry. Therefore, manufacturing methods that can precisely transform materials into functional and efficient energy conversion and storage devices are of paramount importance. AM can reduce the cost and lead time of new products through the rapid development and fabrication of prototypes. In particular, AM can be employed to fabricate compact energy-saving devices with a high surface area, such as heat exchangers. Current applications of metal powder–based AM in the energy sector mostly fall within the domains of thermal energy conversion and production of fuel cells and chemical reactors for energy and fuel processing.

Heat exchangers, which typically exhibit complex geometries that include thin walls and complex fluid passages to achieve efficient heat transfer with a low pressure drop along the flow paths, are commonly fabricated through metal powder–based AM processes. Traditionally manufactured heat exchangers typically comprise brazed pieces of sheet metal with a low yield of parts and limited geometrical complexity. In contrast, AM heat exchangers are lightweight and space-efficient non-assembly systems with increased surface areas. Additionally, they possess welded seams that eliminate the risk of leakage and complex-shaped internal channels for optimized coolant flow and functional integration.

Enhancing the heat transfer performance of heat exchanger products, particularly through miniaturization, is a major research topic in the energy industry. Figure 7.12 displays common commercial heat exchanger products fabricated via AM. Conflux Technology, an Australian company specializing in thermal and fluid dynamic solutions, has introduced a novel heat exchanger designated for AM, namely, the Conflux Core™ heat exchanger (Figure 7.12a). This heat exchanger design has been employed in the automotive and oil and gas industries, as well as in chemical process technologies. The printed heat exchanger is 22% lighter than its conventionally produced counterparts and can dissipate thrice the amount of heat in a given time. Using EOSINT M 280 printers, the Betatype company fabricated heat sink units for automotive LED lights (Figure 7.12b). A total of 384 parts were manufactured in a single batch, and the optimized part design and precise laser scanning paths reduced the production time from 444 hours to only 34 hours. In addition, the unit costs were reduced by 90%. EOS developed a passive heat exchanger with a unique shape resembling a sea anemone that maximizes its surface area (Figure 7.12c).

Velo3D developed a radio flow heat exchanger (Figure 12d) and a diamond heat exchanger (Figure 7.12e) through their unique support-free LPBF technique, which greatly increases the design freedom of complex parts and improves their printability. In collaboration with SLM Solutions, URMA AG successfully employed

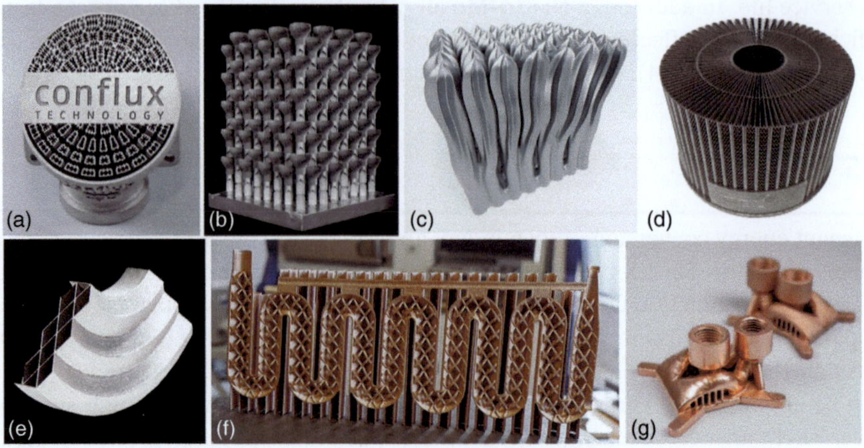

Figure 7.12 Heat exchangers printed through LPBF by commercial companies. (a) Conflux Core heat exchanger, (b) single batch of 384 heat sinks produced by Betatype, and (c) an EOS Corel aluminum heat sink. Source: Reproduced with permission from EOS. (d) Diamond heat exchanger, and (e) radio flow heat exchanger. Source: Reproduced with permission from Velo3D. (f) Copper heat exchanger. Source: Reproduced with permission from URMA AG. (g) LPBF-printed gaming CPU cooler. Source: Reproduced with permission from EOS.

LPBF to fabricate a copper heat exchanger (Figure 7.12f). A complex assembly of parts with different functions was integrated into a single unit that displayed improved heat transfer performance and a low pressure drop. Such freedom in design reveals the potential of AM in optimizing heat exchanger devices with respect to their mechanical integrity, heat transfer performance, and flow behavior.

Imperfect heat dissipation limits the miniaturization of portable computers, electronic devices, and high-power lights. AM can mitigate such a restriction by fabricating thermal management components that dissipate heat efficiently, even in confined regions. In addition, the freedom of design incorporated within AM components may render their thermal management performance superior to that of conventionally manufactured components, often with lower space requirements. For example, EOS and AM Metals GmbH have pushed the limits of AM and developed a state-of-the-art gaming central processing unit (CPU) heat exchanger designated for LPBF (Figure 7.12g). AM Metals manufactured a water-cooled heat exchanger through a flexible EOS M 290 system. The LPBF-printed CPU cooler exhibits a cooling capacity similar to those of other high-end gaming CPU coolers despite being just one-fifth of their size and a quarter of their weight.

Aside from industrial companies, researchers in academia have also made efforts to improve the fabrication of heat exchangers and their applications. For example, a pitot tube incorporated with heat sinks was printed using an Al–Si–Mg powder via LPBF, which achieved a 98% improvement in heat transfer compared to conventional heat sinks (Fasano et al. 2016). Figure 7.13a–c present different designs for printed aluminum heat sinks, namely, Pinfin-Al6061 with a conventional geometry, V-Al6061, which comprises angled ellipses, and Diamond-Al6061, which

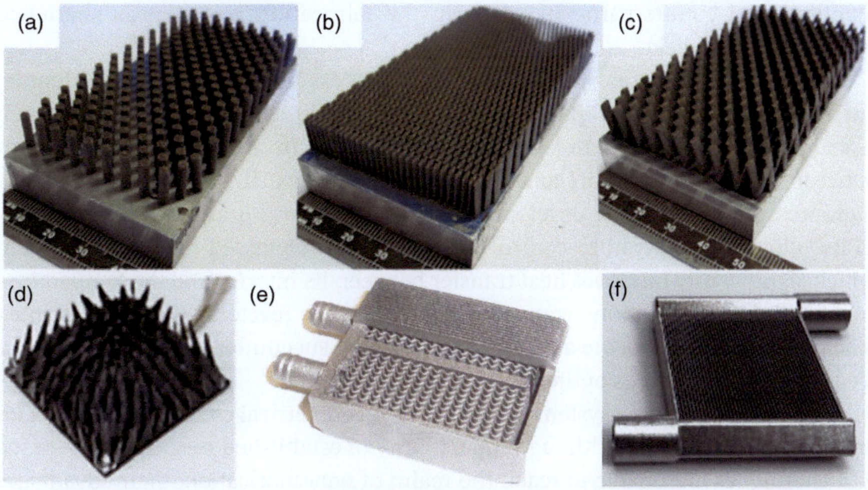

Figure 7.13 Heat sinks fabricated through LPBF. (a) Pinfin-Al6061, (b) Diamond-Al6061, and (c) V-Al6061. Source: Wong et al. (2007)/reproduced with permission from Emerald Publishing Limited. (d) Printed AlSi12 component with a novel design. Source: Dede et al. (2015)/reproduced with permission from American Society of Mechanical Engineers. (e) Compact cooler unit with extremely intricate inner geometry fabricated through LPBF using copper. Source: Neugebauer et al. (2011)/reproduced with permission from Emerald Publishing Limited. (f) Miniaturized air-to-refrigerant heat exchanger prototype fabricated through LPBF. Source: Reproduced with permission from The University of Maryland. Center for Environmental Energy Engineering.

is composed of diamond-shaped pins (Wong et al. 2007). Both the Diamond-Al6061 and V-Al6061 designs cannot be realized through conventional manufacturing methods because of their unique geometrical properties. In particular, Diamond-Al6061 exhibited a tremendous improvement in heat exchange due to its complex fins possessing a large surface area.

The capability of LPBF to fabricate complex 3D heat sink designs was further validated by Wong et al. (2009). They demonstrated that the heat transfer performance of printed heat sinks could be determined mainly by the path of the coolant. Figure 7.13d shows a novel heat sink fabricated through LPBF using AlSi12 powder (Dede et al. 2015). Confined jet cooling tests indicated that the printed novel heat sink design possessed a relatively high coefficient of performance, which can be attributed to a lower jet flow resistance due to its unique blended fin design. Figure 7.13e presents a compact cooler unit fabricated through LPBF with a highly complex inner geometry (Neugebauer et al. 2011). The cooler ($46 \times 59 \times 15\,mm^3$) includes a large number of cooling fins, each with a thickness of 0.3 mm, to facilitate efficient heat exchange. A high surface-to-volume ratio and rapid coolant flow were achieved by modifying the rib-like internal structure of the unit. Intending to develop a new generation of compact, highly efficient heat exchangers, researchers from the University of Maryland designed and produced a miniaturized 1 kW air-to-refrigerant heat exchanger prototype through LPBF (Figure 7.13f), which was 20% lighter and 25% more energy-efficient than its conventionally manufactured counterparts. In addition, the fabrication process was shortened from months to weeks.

High-pressure reactors are employed in process engineering for facilitating chemical reactions under pressure values of below 200 bar. Traditionally, such reactors are built from stainless steel or Hastelloy. In a collaborative endeavor between THALETEC GmbH and the engineering office JUREC, a prototype of a high-pressure reactor was developed and fabricated using an SLM280 machine. LPBF allows a temperature control channel to be integrated within the reactor, which significantly improves heat transfer between its interior and the temperature control fluid. In addition, compared to traditional reactors, the LPBF-printed reactor could accommodate a high pressure before succumbing to damage, and its weight is lower due to its optimized thinner wall design.

In the design of energy systems, chemical and geometrical control on small scales are of great importance. Although the majority of established metal powder–based AM techniques are unable to reach the realm of nanofabrication, their advantages over conventional manufacturing methods and capability to fabricate active materials and functional microstructures have rendered them excellent candidates for applications within fuel-processing chemical reactors. A steel dehydrogenation reactor exhibiting a cubic diamond cellular structure was fabricated through EBM using Ti–6Al–4V powder, as shown in Figure 7.14a (Peters et al. 2015). The optimized cellular structure was determined to enhance the cooling performance of the reactor and facilitate efficient hydrogen gas removal from the surface of the catalysts. Ten reactors were scaled up to form a single hydrogen release unit, and a H_2 generation rate of 9.8 NL/min was achieved. Figure 7.14b shows a novel EBM-printed 3D catalyst support structure employed in catalytic static mixers for hydrogenation reactions (Avril et al. 2017). Catalytic static mixers are derived from static mixer concepts, in which a geometrically optimized mixer is coated with a catalyst and fitted within a tubular flow reactor, as shown in the diagram. Presently, all catalytic static mixers are produced through a combination of EBM and other metal deposition techniques.

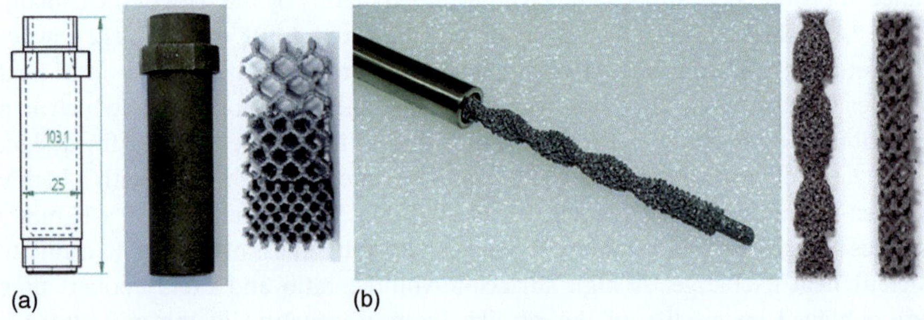

Figure 7.14 Chemical reactor components printed through EBM. (a) Reactor tube with a diamond-shaped porous internal structure. Source: Peters et al. (2015)/reproduced with permission from Royal Society of Chemistry. (b) Support structures of different designs in catalytic static mixers. Source: Avril et al. (2017)/reproduced with permission from Royal Society of Chemistry.

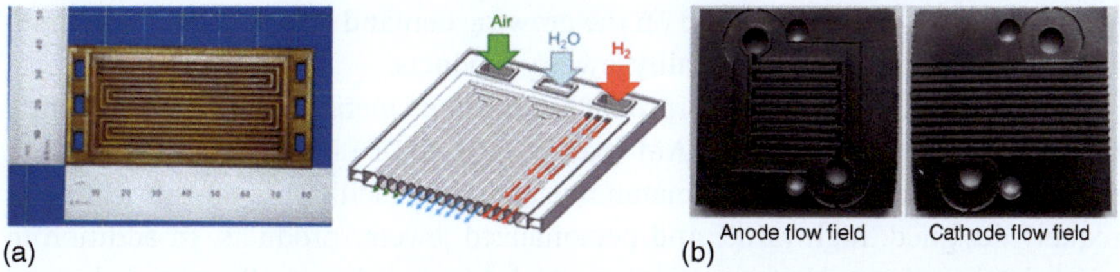

Figure 7.15 Fuel cell components printed through LPBF. (a) Ti–6Al–4V BPP and a sectional view of a BPP solid model with triple serpentine flow fields and straight, parallel internal coolant channels. Source: Gould et al. (2015)/reproduced with permission from The Electrochemical Society. (b) 316L stainless steel BPPs at the anode and cathode of a PEMFC. Source: Dawson et al. (2015)/reproduced with permission from Springer Nature.

Metal powder-based AM can be potentially utilized to manufacture microbial fuel cells as single units. This possibility was explored by using LPBF to manufacture a bio-inspired lattice aluminum alloy anode to improve the functionality of microbial fuel cells, which resulted in a higher power density than other metallic anodes (Calignano et al. 2015). Furthermore, the LPBF-printed anode possessed a 3D macroporous structure with ideal properties for hosting anodophiles, such as low density, large specific surface area, and customizable surface roughness. The cellular lattice structure of the anode resulted in better microbial attachment, enhanced charge transfer between anodophiles and the electrode surface, and facilitated rapid nutrient diffusion without clogging.

Metal bipolar plates (BPPs) are speculated to provide a path to developing a low-cost, compact fuel cell stack (Netwall et al. 2013). As shown in Figure 7.15a, LPBF was employed to print BPPs from a Ti-6Al-4V powder, which were subsequently coated with a low-cost, robust, conductive TiO_2/gold composite to improve their corrosion resistance and minimize their surface resistance (Gould et al. 2015). A single proton-exchange membrane fuel cell (PEMFC) stack comprising two printed BPPs achieved similar fuel efficiency to conventionally manufactured carbon BPPs. One of the largest contributions to the overall cell resistance is the contact resistance originating from the interfaces between the gas diffusion layer and the BPPs. In another study that demonstrates the applicability of LPBF in printing 316L stainless steel BPPs for PEMFCs (Dawson et al. 2015), the interfacial contact resistance and overall fuel efficiency of the PEMFCs containing the printed BPPs were determined to be on par with those of other PEMFCs utilizing conventionally fabricated BPPs, as shown in Figure 7.15b.

7.6 Jewelry

Consumers are influenced by the rapidly advancing digital revolution to expect a continuous and regularly updated selection of innovative products, particularly jewelry. Furthermore, an ever-increasing number of consumers are interested in the unique personalized designs of jewelry pieces. Thus, the future of the high-value jewelry market could well depend on the growing demand for individual and innovative representations of high-quality jewelry products.

While all materials involved in producing precious metal jewelry are inherently expensive, metal powder-based AM (particularly LPBF) introduces a technology shift that can potentially reduce material usage while facilitating the fabrication of uniquely designed, high-value, and personalized jewelry products. In addition to the fabrication of jewelry from stainless steel, titanium-based alloys, nickel-based alloys, aluminum, and copper (Yap and Yeong 2014), LPBF has also been employed to manufacture products composed of precious metals, including 18-karat gold of various colors, silver alloys, and platinum group metals. Notably, rings, bracelets, necklaces, and pendants of complex designs can be printed through LPBF (Yap and Yeong 2014). Cooksongold AM, a division of Cookson Precious Metals Ltd. and Sempsa JP, is a leading company in printing precious metals, having collaborated with EOS to produce gold jewelry products through a specialized LPBF machine. The structures of these products can be modified to accommodate cavities, corresponding to lighter weight and allowing for unique design choices. Figure 7.16 presents an assortment of LPBF-printed gold jewelry pieces from Cookson Precious Metals and Cooksongold.

Flexible joining mechanisms can be incorporated within a design to render it foldable. These mechanisms are composed of multiple small components that interlock at accurately spaced hinge joints, thereby enabling the overall structure to drape and move in a manner akin to that of actual fabric. Such designs are implemented in flexible and customizable bracelets and necklaces.

The visual esthetics of a product is an essential indicator of its value. Thus, the visual appeal of printed fashion products should be on par with that of their traditionally manufactured counterparts. Post-processing of printed jewelry is required since they are typically unattractive otherwise (Figure 7.17a–c) (Klotz et al. 2016). Generally, manual polishing, a time-consuming and costly process, must be performed on printed jewelry instead of mechanical methods to provide it with an intricate and complex surface. Afterward, electroplating or anodizing can be conducted on the jewelry pieces to enhance their esthetics and durability further. Electroplating procedures, such as chrome plating, can be performed on the surface of both polymer and metal components, while anodizing is only applicable to metals. The former can reproduce the appearance and texture of a

Figure 7.16 (a) Cufflinks and (b) a ring are made of 18-karat yellow gold. Source: Reproduced with permission from Cookson Precious Metals Ltd. (c) Sculpture and (d) galleon are both printed in gold. Source: Reproduced with permission from Cookson Precious Metals Ltd.

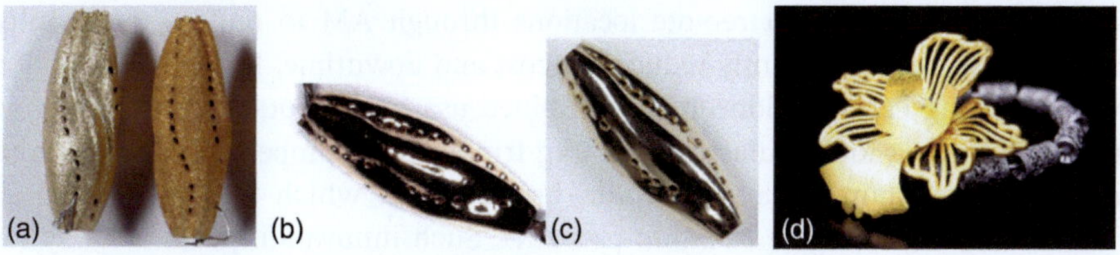

Figure 7.17 Jewelry samples printed by LPBF using 18-karat yellow gold with different polishing steps. (a) As-printed (left) and sand-blasted (right), (b) electro-polished, and (c) hand-polished samples. Source: Klotz et al. (2016)/reproduced with permission from Santa Fe Symposium. (d) "Orchid-Spirit" dual-function bracelet manufactured by LPBF using Ti–6Al–4V and anodized to form gold flower petals. Source: Yap and Yeong (2014)/reproduced with permission from Taylor & Francis Group.

metallic coating and enhance the wear resistance of a component. In contrast, the latter allows the component to be dyed in a variety of colors. Interestingly, different colors of anodized titanium can be obtained simply by varying the thickness of the oxide layer, and no dye is involved. Figure 7.17d shows an anodized "Orchid-Spirit" golden titanium bracelet fabricated through LPBF (Yap and Yeong 2014).

7.7 Marine, Oil, and Gas

The marine, oil, and gas industries have been much slower to embrace metal powder–based AM than the aforementioned industries. The rapid prototyping of tools and complex models, which allows multiple design cycles and quick testing of design concepts, is particularly significant in the oil and gas industries.

Using LPBF has reduced the product development time of a burner for a NovaLT16 gas turbine through rapid prototyping of design concepts and accelerated validation testing, thereby resulting in a cost reduction of over 50%. In addition, compared to investment casting, AM allows for the simplified manufacturing of pumps, turbomachinery, valves, and other vital components, which reduces costs and enhances the mechanical performance of the products. Specifically, GE Oil & Gas Additive Manufacturing Laboratory has installed multiple LPBF machines to fabricate turbomachinery components.

The supply chain management sector in the oil and gas industries may benefit more from the advantages of AM than improvements in part performance and reduction in the production cost (Holmström et al. 2010). With oil and gas assets being deployed in isolated regions for increasingly longer durations, essential components are more likely to break down, become obsolete due to technological changes, undergo changes in quality, or go out of production. Through AM, obsolete components can be reverse-engineered digitally and manufactured on demand, leading to longer asset lifetimes.

As the pursuit of new reserves assigns oil companies to relatively remote locations, the prospect of on-site manufacturing becomes increasingly alluring. Many drilling rig components include multiple parts that must be welded, bolted, or brazed, and fabricating single parts at remote locations through AM to replace their faulty counterparts can significantly reduce the cost and downtime. Rotterdam Additive Manufacturing Lab has pioneered a combined usage of robotic arm wire-arc AM and CNC milling to manufacture a 400 kg triple-bladed propeller prototype from a corrosion-resistant nickel–aluminum–bronze alloy, which was established as the world's first approved AM ship propeller. Such innovation paves the way for the rapid fabrication of replacement parts for ships worldwide without requiring the parts to be ordered, remotely manufactured, and delivered. Subsequently, the printability of the nickel–aluminum–bronze alloy and its multi-material parts through L-DED was investigated for marine and offshore applications (Li et al. 2021, 2022). However, further applications of metal powder–based AM in the industry should still be explored.

Metal powder–based AM techniques are employed by the maritime and marine & offshore sectors for (i) improving the manufacturing process of complex spare parts; (ii) rapid prototyping of parts for installations or retrofitting projects; (iii) replicating the scales of ships, rigs, or equipment models with detailed features; (iv) smoothening workflow and planning; (v) speeding up performance testing; (vi) minimizing errors in new product development; and (vii) reducing storage space and related costs with on-demand local printing.

Researchers from the Singapore Centre for 3D Printing at Nanyang Technological University have utilized EBM in an offshore wind turbine project to fabricate a Pelton turbine system that can harvest renewable hydroelectric energy, which includes a Pelton wheel with an outer diameter of ~210 mm, a penstock, a nozzle, and a regulator (Figure 7.18). Ti–6Al–4V powder was selected to manufacture the turbine wheel because of its excellent corrosion resistance and mechanical strength. Compared to its original counterpart designated for conventional manufacturing, the total part count of the EBM-printed turbine system was reduced from 40 to 3, with a total weight reduction of about 40%. A complete EBM-printed Pelton turbine system containing such non-assembly components can reduce the cost from S$90 000 to S$9000 and the production cycle from six months to two weeks.

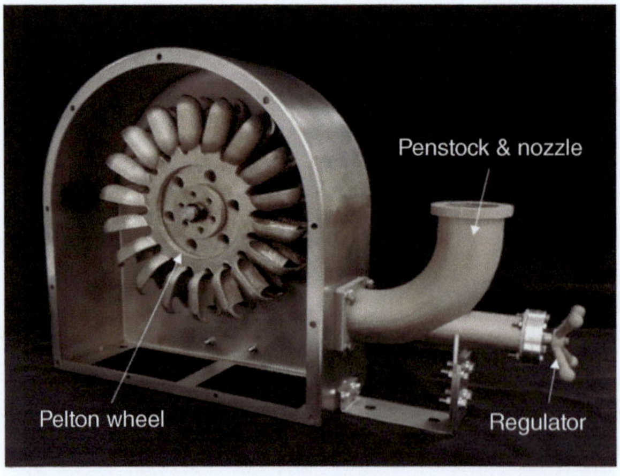

Figure 7.18 EBM-printed Pelton turbine system including a Pelton wheel, penstock, nozzle, and regulator for harvesting renewable hydroelectric energy. Source: Reproduced with permission from Nanyang Technological University.

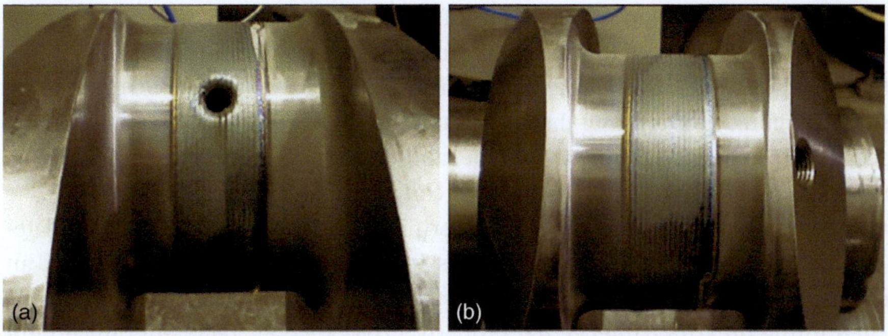

Figure 7.19 Crankshaft repaired by L-DED. (a) Oil-bore region and (b) its bottom side at the lower dead center position. Source: Koehler et al. (2010)/reproduced with permission from Elsevier, CC BY-NC-ND 3.0.

Diesel engines have undergone revolutionary improvements in their design and performance over the last century. The crankshaft is one of the most important components in the engine, which requires regular servicing because it is heavily used. However, repairing the crankshaft via conventional processes is hindered by several factors, including the high cost. Figure 7.19 shows a marine diesel crankshaft repaired using the L-DED process, and good interfacial bonding between the deposited layer and the base metal can be observed. The piston of a four-stroke marine diesel engine is another critical component that can be successfully repaired using L-DED. However, the surface of the piston is prone to damage caused by the high level of forces and temperatures it is subjected to, resulting in the wear of the edges of its grooves. The Kimi Company has not only repaired such pistons by L-DED but also improved the resistance of their grooves against erosion and corrosion.

7.8 Challenges and Risks

Metal powder–based AM has exhibited great potential for applications in the aerospace, biomedical, automotive, molding and tooling, energy, jewelry, and oil and gas industries. It is a cost-effective and time-efficient approach to producing low-volume customized products with complicated geometries and advanced properties. LPBF, L-DED, and EBM are the main processes adopted by the above industries due to their ability to produce dense components with excellent mechanical properties. However, further research and development are still required in terms of design, materials, novel processes and machines, process modeling, and quality control to broaden the range of applications of metal powder–based AM and elevate it to the relatively mainstream technology.

The critical technical challenges of powder–based AM must be overcome through efforts in the research and development of process control, surface finishing, types of heat treatment for the printed components, scalability, processing speed, sustainability, supply chain management and logistics, and qualification and certification.

In particular, the precision and reliability of feedback control systems and predefined metrics must be improved to increase throughput while ensuring consistent quality of the products. Meanwhile, researchers are striving to attain micro- or nano-scale levels of accuracy for refining the surfaces of the printed parts, and the processing speed, energy consumption, water usage, and waste production of AM procedures are expected to be comparable to those of conventional processes for large-volume production. Regarding qualification and certification, the focus should be on material choices, processes, testing, post-processing, and applications. Additionally, an array of business- and market-level challenges exist, such as ethical considerations (e.g. printing firearms), intellectual property/privacy issues, regulatory uncertainties in different countries, limited choices of materials, development of standards, and restrictions of small production runs and scalability.

The risks or uncertainties regarding metal powder–based AM are identified as follows. Intellectual property issues are a primary concern surrounding AM. The ease of access to AM product designs, hardware, and printing materials amplifies the extent of distribution of AM products at the expense of their original creators, particularly through the Internet. Specifically, original digital designs of AM products are difficult to protect through copyright laws, and dealing with the improper usage of such files can be legally challenging, considering that the scope of protection extended to a given work may vary in different countries. Legal issues can also arise when an AM product is manufactured by substandard, incorrectly calibrated, or defective AM equipment or when an original design includes inherent safety hazards. Presently, the development of AM continues to outpace the legal policies pertaining to their usage, thereby introducing uncertainties in terms of user responsibilities and corporate legal obligations. Finally, uncertainties regarding the environmental hazards, toxicity, and chemical degradation associated with the variety of novel print materials introduced for AM still exist, and additional studies are required to determine their long-term effects on humans and the environment.

References

Avril, A., Hornung, C.H., Urban, A. et al. (2017). Continuous flow hydrogenations using novel catalytic static mixers inside a tubular reactor. *Reaction Chemistry & Engineering* 2 (2): 180–188.

Behlau, L. (2009). *Technology Guide: Principles-Applications-Trends*. Springer Science and Business Media.

Bertol, L.S., Júnior, W.K., da Silva, F.P., and Aumund-Kopp, C. (2010). Medical design: direct metal laser sintering of Ti–6Al–4V. *Materials & Design* 31 (8): 3982–3988.

Bose, S., Ke, D., Sahasrabudhe, H., and Bandyopadhyay, A. (2018). Additive manufacturing of biomaterials. *Progress in Materials Science* 93: 45–111.

Calignano, F., Tommasi, T., Manfredi, D., and Chiolerio, A. (2015). Additive manufacturing of a microbial fuel cell – a detailed study. *Scientific Reports* 5 (1): 1–10.

Chantzis, D., Liu, X., Politis, D.J. et al. (2020). Review on additive manufacturing of tooling for hot stamping. *The International Journal of Advanced Manufacturing Technology* 109: 1–21.

Chen, G., Muheremu, A., Yang, L. et al. (2020). Three-dimensional printed implant for reconstruction of pelvic bone after removal of giant chondrosarcoma: a case report. *Journal of International Medical Research* 48 (4): 0300060520917275.

Dawson, R.J., Patel, A.J., Rennie, A.E., and White, S. (2015). An investigation into the use of additive manufacture for the production of metallic bipolar plates for polymer electrolyte fuel cell stacks. *Journal of Applied Electrochemistry* 45 (7): 637–645.

Dede, E.M., Joshi, S.N., and Zhou, F. (2015). Topology optimization, additive layer manufacturing, and experimental testing of an air-cooled heat sink. *Journal of Mechanical Design* 137 (11): https://doi.org/10.1115/1.4030989.

España, F.A., Balla, V.K., Bose, S., and Bandyopadhyay, A. (2010). Design and fabrication of CoCrMo alloy based novel structures for load bearing implants using laser engineered net shaping. *Materials Science and Engineering: C* 30 (1): 50–57.

Fasano, M., Ventola, L., Calignano, F. et al. (2016). Passive heat transfer enhancement by 3D printed pitot tube based heat sink. *International Communications in Heat and Mass Transfer* 74: 36–39.

General Electric (2016). GE tests additive manufactured demonstrator engine for Advanced Turboprop.

Gould, B.D., Rodgers, J.A., Schuette, M. et al. (2015). Performance and limitations of 3D-printed bipolar plates in fuel cells. *ECS Journal of Solid State Science and Technology* 4 (4): P3063.

Grum, J. and Slabe, J.M. (2004). Possibility of introducing laser surfacing into maintenance service of die-casting dies. *Surface and Coatings Technology* 180: 596–602.

Han, C., Fang, Q., Shi, Y. et al. (2020). Recent advances on high-entropy alloys for 3D printing. *Advanced Materials* 32 (26): 1903855.

Han, C., Li, Y., Wang, Q. et al. (2018a). Titanium/hydroxyapatite (Ti/HA) gradient materials with quasi-continuous ratios fabricated by SLM: material interface and fracture toughness. *Materials & Design* 141: 256–266.

Han, C., Li, Y., Wang, Q. et al. (2018b). Continuous functionally graded porous titanium scaffolds manufactured by selective laser melting for bone implants. *Journal of the Mechanical Behavior of Biomedical Materials* 80: 119–127.

Han, C., Wang, Q., Song, B. et al. (2017a). Microstructure and property evolutions of titanium/nano-hydroxyapatite composites in-situ prepared by selective laser melting. *Journal of the Mechanical Behavior of Biomedical Materials* 71: 85–94.

Han, C., Yan, C., Wen, S. et al. (2017b). Effects of the unit cell topology on the compression properties of porous Co-Cr scaffolds fabricated via selective laser melting. *Rapid Prototyping Journal* 23: 16–27.

Han, C., Yao, Y., Cheng, X. et al. (2017c). Electrophoretic deposition of gentamicin-loaded silk fibroin coatings on 3D-printed porous cobalt–chromium–molybdenum bone substitutes to prevent orthopedic implant infections. *Biomacromolecules* 18 (11): 3776–3787.

Holmström, J., Partanen, J., Tuomi, J., and Walter, M. (2010). Rapid manufacturing in the spare parts supply chain: alternative approaches to capacity deployment. *Journal of Manufacturing Technology Management* 21: 687–697.

Jardini, A.L., Larosa, M.A., Maciel Filho, R. et al. (2014). Cranial reconstruction: 3D biomodel and custom-built implant created using additive manufacturing. *Journal of Cranio-Maxillofacial Surgery* 42 (8): 1877–1884.

Kattire, P., Paul, S., Singh, R., and Yan, W. (2015). Experimental characterization of laser cladding of CPM 9V on H13 tool steel for die repair applications. *Journal of Manufacturing Processes* 20: 492–499.

Klotz, U.E., Tiberto, D., and Held, F. (2016). Additive manufacturing of 18-karat yellow-gold alloys. *Santa Fe Symposium on Jewelry Manufacturing Technology*. pp. 255–272.

Koehler, H., Partes, K., Seefeld, T., and Vollertsen, F. (2010). Laser reconditioning of crankshafts: from lab to application. *Physics Procedia* 5: 387–397.

Leal, R., Barreiros, F.M., Alves, L. et al. (2017). Additive manufacturing tooling for the automotive industry. *The International Journal of Advanced Manufacturing Technology* 92 (5): 1671–1676.

Li, B., Han, C., Lim, C.W.J., and Zhou, K. (2022). Interface formation and deformation behaviors of an additively manufactured nickel-aluminum-bronze/15-5 PH multimaterial via laser-powder directed energy deposition. *Materials Science and Engineering A* 829: 142101.

Li, B., Zheng, H., Han, C., and Zhou, K. (2021). Nanotwins-containing microstructure and superior mechanical strength of a Cu–9Al–5Fe–5Ni alloy additively manufactured by laser metal deposition. *Additive Manufacturing* 39: 101825.

Liu, J., Han, G., Pan, S. et al. (2015). Biomineralization stimulated peri-titanium implants prepared by selective laser melting. *Journal of Materiomics* 1 (3): 253–261.

Liverani, E., Fortunato, A., Leardini, A. et al. (2016). Fabrication of Co–Cr–Mo endoprosthetic ankle devices by means of selective laser melting (SLM). *Materials & Design* 106: 60–68.

Lu, Y., Wu, S., Gan, Y. et al. (2015). Investigation on the microstructure, mechanical property and corrosion behavior of the selective laser melted CoCrW alloy for dental application. *Materials Science and Engineering: C* 49: 517–525.

Mazzoli, A., Germani, M., and Raffaeli, R. (2009). Direct fabrication through electron beam melting technology of custom cranial implants designed in a PHANToM-based haptic environment. *Materials & Design* 30 (8): 3186–3192.

Mudge, R.P. and Wald, N.R. (2007). Laser engineered net shaping advances additive manufacturing and repair. *Welding Journal-New York* 86 (1): 44.

Murr, L.E., Gaytan, S.M., Ceylan, A. et al. (2010). Characterization of titanium aluminide alloy components fabricated by additive manufacturing using electron beam melting. *Acta Materialia* 58 (5): 1887–1894.

Murr, L.E., Gaytan, S.M., Ramirez, D.A. et al. (2012). Metal fabrication by additive manufacturing using laser and electron beam melting technologies. *Journal of Materials Science and Technology* 28 (1): 1–14.

Netwall, C.J., Gould, B.D., Rodgers, J.A. et al. (2013). Decreasing contact resistance in proton-exchange membrane fuel cells with metal bipolar plates. *Journal of Power Sources* 227: 137–144.

Neugebauer, R., Müller, B., Gebauer, M., and Töppel, T. (2011). Additive manufacturing boosts efficiency of heat transfer components. *Assembly Automation* 31 (4): 344–347.

Peters, W., Eypasch, M., Frank, T. et al. (2015). Efficient hydrogen release from perhydro-*N*-ethylcarbazole using catalyst-coated metallic structures produced by selective electron beam melting. *Energy and Environmental Science* 8 (2): 641–649.

Rännar, L.E., Glad, A., and Gustafson, C.G. (2007). Efficient cooling with tool inserts manufactured by electron beam melting. *Rapid Prototyping Journal* 13: 128–135.

Shinde, M.S. and Ashtankar, K.M. (2017). Additive manufacturing–assisted conformal cooling channels in mold manufacturing processes. *Advances in Mechanical Engineering* 9 (5): 1687814017699764.

Vandenbroucke, B. and Kruth, J.P. (2007). Selective laser melting of biocompatible metals for rapid manufacturing of medical parts. *Rapid Prototyping Journal* 13: 196–203.

Vilaro, T., Abed, S., and Knapp, W. (2008). Direct manufacturing of technical parts using selective laser melting: example of automotive application. *Proceedings of the 12th European Forum Rapid Prototype*, Paris, France (5–6 March 2008).

Wehmöller, M., Warnke, P.H., Zilian, C., and Eufinger, H. (2005). Implant design and production – a new approach by selective laser melting. In: *International Congress Series*, vol. 1281, 690–695. Elsevier.

Wilson, J.M., Piya, C., Shin, Y.C. et al. (2014). Remanufacturing of turbine blades by laser direct deposition with its energy and environmental impact analysis. *Journal of Cleaner Production* 80: 170–178.

Wong, M., Owen, I., Sutcliffe, C.J., and Puri, A. (2009). Convective heat transfer and pressure losses across novel heat sinks fabricated by selective laser melting. *International Journal of Heat and Mass Transfer* 52 (1–2): 281–288.

Wong, M., Tsopanos, S., Sutcliffe, C.J., and Owen, I. (2007). Selective laser melting of heat transfer devices. *Rapid Prototyping Journal* 13: 291–297.

Xin, X.Z., Xiang, N., Chen, J., and Wei, B. (2012). In vitro biocompatibility of Co–Cr alloy fabricated by selective laser melting or traditional casting techniques. *Materials Letters* 88: 101–103.

Xue, L. and Islam, M.U. (2002). Laser consolidation-a novel one-step manufacturing process from CAD models to net-shape functional components. *International Conference on Metal Powder Deposition for Rapid Manufacturing*, San Antonia, Texas (8–10 April 2002). Providence, RI: MPIF.

Xue, L. and Islam, M.U. (2006). Laser consolidation – a novel one-step manufacturing process for making net-shape functional components. Cost Effective Manufacturing Via Net-Shape Processing, Neuilly-sur-Seine, France. pp. 15-1–15-4.

Yang, L., Han, C., Wu, H. et al. (2020). Insights into unit cell size effect on mechanical responses and energy absorption capability of titanium graded porous structures manufactured by laser powder bed fusion. *Journal of the Mechanical Behavior of Biomedical Materials* 109: 103843.

Yap, Y.L. and Yeong, W.Y. (2014). Additive manufacture of fashion and jewellery products: a mini review: this paper provides an insight into the future of 3D printing industries for fashion and jewellery products. *Virtual and Physical Prototyping* 9 (3): 195–201.

Yilmaz, O., Gindy, N., and Gao, J. (2010). A repair and overhaul methodology for aero-engine components. *Robotics and Computer-Integrated Manufacturing* 26 (2): 190–201.

Zhao, D., Han, C., Li, Y. et al. (2019). Improvement on mechanical properties and corrosion resistance of titanium-tantalum alloys in-situ fabricated via selective laser melting. *Journal of Alloys and Compounds* 804: 288–298.

Zhao, D., Han, C., Li, J. et al. (2020). In situ fabrication of a titanium-niobium alloy with tailored microstructures, enhanced mechanical properties and biocompatibility by using selective laser melting. *Materials Science and Engineering: C* 111: 110784.

Zhao, D., Liang, H., Han, C. et al. (2021). 3D printing of a titanium-tantalum Gyroid scaffold with superb elastic admissible strain, bioactivity and in-situ bone regeneration capability. *Additive Manufacturing* 47: 102223.

Index

Metal Powder–Based Additive Manufacturing, First Edition. Kun Zhou and Changjun Han.